Lecture Notes in Computer Science 623

Edited by G. Goos and J. Hartmanis

W. Kuich (Ed.)

Automata, Languages and Programming

19th International Colloquium
Wien, Austria, July 13-17, 1992
Proceedings

Springer-Verlag
Berlin Heidelberg New York
London Paris Tokyo
Hong Kong Barcelona
Budapest

Series Editors

Gerhard Goos
Universität Karlsruhe
Postfach 69 80
Vincenz-Priessnitz-Straße 1
W-7500 Karlsruhe, FRG

Juris Hartmanis
Department of Computer Science
Cornell University
5149 Upson Hall
Ithaca, NY 14853, USA

Volume Editor

Werner Kuich
Abt. für Theoretische Informatik, Inst. für Algebra und Diskrete Mathematik
Technische Universität Wien, Wiedner Hauptstraße 8-10, A-1040 Wien, Austria

CR Subject Classification (1991): D.1, E.1, F, I.1, G.2, I.2.3, I.3.5

ISBN 3-540-55719-9 Springer-Verlag Berlin Heidelberg New York
ISBN 0-387-55719-9 Springer-Verlag New York Berlin Heidelberg

Printed in Germany

Typesetting: Camera ready by author/editor
Printing and binding: Druckhaus Beltz, Hemsbach/Bergstr.
45/3140-543210 - Printed on acid-free paper

PREFACE

ICALP 92 is the 19th International Colloquium on Automata, Languages, and Programming in a series of meetings sponsored by the European Association for Theoretical Computer Science (EATCS). It is a broadly based conference covering all aspects of Theoretical Computer Science including topics like: computability, automata, formal languages, term rewriting, analysis of algorithms, computational geometry, computational complexity, symbolic and algebraic computation, cryptography, data types and data structures, theory of data bases and knowledge bases, semantics of programming languages, program specification, transformation and verification, foundations of logic programming, theory of logical design and layout, parallel and distributed computation, theory of concurrency, theory of robotics.

ICALP 92 was held at Technische Universität Wien from July 13 to July 17, 1992. The Organizing Committee consisted of G. Baron, P. Kirschenhofer, W. Kuich (Chairman), H. Maurer, H. Prodinger, F. Urbanek.

Previous ICALPs were held in Madrid (1991), Warwick (1990), Stresa (1989), Tampere (1988), Karlsruhe (1987), Rennes (1986), Nafplion (1985), Antwerp (1984), Barcelona (1983), Aarhus (1982), Haifa (1981), Amsterdam (1980), Graz (1979), Udine (1978), Turku (1977), Edinburgh (1976), Saarbrücken (1974), Paris (1972). ICALP 93 will be held in Lund, Sweden, from July 5 to July 9, 1993.

The Programme Committee of ICALP 92 consisted of G. Ausiello, J. Berstel, R. V. Book, B. Buchberger, R. Cori, R. De Nicola, H. Edelsbrunner, S. Even, Ph. Flajolet, D. Harel, M. C. B. Hennessy, N. D. Jones, J. W. Klop, D. C. Kozen, W. Kuich (Chairman), B. Monien, E.-R. Olderog, B. Rovan, A. Salomaa, S. Skyum, A. Tarlecki. It has selected 52 papers from 190 papers submitted in response to the call for papers. These papers came from the following countries: Australia, Austria, Belgium, Canada, China, CSFR, Denmark, Finland, France, FRG, Greece, Hungary, India, Israel, Italy, Japan, Latvia, Lithuania, Netherlands, Poland, Russia, Singapore, Spain, Sweden, Switzerland, Taiwan, UK, Ukraine, USA. Each submitted paper has been evaluated by at least four members of the Programme Committee. The final selection was made during the selection meeting on February 1–2, 1992 in Wien, in which all Programme Committee members except S. Even and D. Harel could participate. Together with six invited presentations all 52 selected papers are contained in this volume. It is a pleasure for the conference chairman to thank the members of the Programme Committee for their evaluation of the papers, and the many referees who assisted in this process. The list of referees given in this volume is as complete as we could achieve, and we apologize for any possible omissions or errors.

The papers in this volume are printed in the order of presentation at ICALP 92 and thus grouped into sessions, most of which are thematic.

Wien, April 1992

Werner Kuich

LIST OF REFEREES

Aalbersberg H.
Aceto L.
Allouche J.-P.
Amtoft T.
Andersen N.
Ardeleanu E.
Arnold A.
Attiya H.
Atzeni P.
Ausiello G.
Autebert J.-M.
Avenhaus J.
Baaz M.
Baeten J.
Baron G.
Bar-On I.
Barringer H.
Bar-Yehuda R.
Bauer
Beauquier D.
Bednarczyk M.
Ben-David S.
Berstel J.
Bertoni A.
Best E.
Białasik M.
Birk Y.
Boasson L.
Bodlaender H.
Boissonnat J.
Bol R. N.
Bolognesi T.
Bonuccelli M. A.
Book R. V.
Borzyszkowski A.
Boucheron St.
Bruyere V.
Buchberger B.
Buntrock G.
Capocelli R.
Cerone A.
Chazelle B.
Choffrut C.
Chor B.
Chretienne P.
Chytil M.
Collins G.
Cori R.
Courcelle B.
Crépeau C.
Crochemore M.
Czaja L.
Dallery Y.
Damgaard I.
D'Amore F.
Dauchet M.
De Boer F.
De Fraysseix H.
Delfinado C.
De'Lignoro U.
De Luca A.
De Nicola R.
Devillers O.
De Vink E.
De Vries F.-J.
Dey T.
Di Battista G.
Diekmann R.
Drabent W.
Dumas Ph.
Duris P.
Dybjer P.
Dybkjaer H.
Edelsbrunner H.
Engelfriet J.
Even S.
Facello M.
Feldmann R.
Feuerstein E.
Flajolet Ph.
Franciosa P. G.
Frandsen G. S.
Fricker C.
Friedman N.
Frougny Ch.
Gallier J.
Gambosi G.
Gastin P.
Gerth R.
Gnesi S.
Goldreich O.
Gomard C.
Goralcik P.
Gorrieri R.
Gouyou-Beauchamps D.
Habel A.
Habib M.
Hagerup T.
Han Chiang-Chih
Hannan A.
Harel D.
Harju T.
Henglein F.
Hennessy M. C. B.
Hertrampf U.
Hilken B.
Hong H.
Honkala J.
Hromkovic
Huizing C.
Hungar H.
Ibarra O.
Inverardi P.
Jacquet Ph.
Jeffrey A.
Jones N. D.
Josko B.
Jourdan M.
Karhumäki J.
Kari J.
Kari L.
Karpinski M.
Katz S.
Kennaway R.
Kenyon C.
Kiehn A.
Kirschenhofer P.
Klarlund N.
Klasing R.
Klop J. W.
Klusener S.
Korach E.
Korec I.
Kortelainen J. M.
Kozen D. C.
Krasucki P.
Krob D.
Kubiak R.
Kuich W.
Kuiper R.
Künneke
Kwiatkowska M. Z.
Larsen K.
Latella D.
Latteux M.
Li A.
Lichtenberger F.
Lin H.
Linna M.
Litman A.
Litovsky I.
Louchard G.
Lüling R.
Maggiolo-Schettini A.
Marchetti-Spaccamela A.
Martelli A.
Masini A.
Mauri G.
Mayoh B. H.
Mazurkiewicz A.
Mehlhorn K.
Meir S.
Melichar B.
Menzel K.
Meyer auf der Heide F.
Middeldorp A.
Mignotte M.
Miltersen P. B.

Miola A.
Mnuk M.
Monien B.
Mücke E.
Mulder H.
Mumbeck W.
Murphy D. V. J.
Mysliwietz
Nanni U.
Naor J.
Nemes I.
Nielsen M.
Nielson F.
Nielson H.
Niemi V.
Niwiński D.
Olderog E.-R.
Orponen P.
Panangaden P.
Pantziou G.
Panzieri F.
Parrow J.
Paterson M. S.
Pawłowski W.
Paz A.
Peine R.
Peleg D.
Pelletier M.
Penaud J. G.
Penczek W.
Perrin D.
Pfalzgraf J.
Pighizzini G.
Pin J. E.
Pinkas B.
Plesnik J.
Pocchiola M.
Ponse A.
Preilowski
Privara I.
Prodinger H.
Protasi M.
Pultr A.
Quaglia P.
Ramme F.
Ramos E.
Régnier M.
Renvall A.
Reutenauer C.
Robert Ph.
Rolletschek H.
Roman J.
Römke Th.
Roscoe A. W.
Rose K.
Rosendahl M.
Rosenstiehl P.
Rössig S.
Roth R. M.
Rovan B.
Rutten J.
Ruzicka P.
Rydeheard D.
Sacibra A.
Sakarovitch J.
Salomaa A.
Salomaa K.
Sangiorgi D.
Santha M.
Schaffer R.
Schenke M.
Schmidt E. M.
Schöning U.
Schreiner W.
Schuster A.
Senizergues G.
Shachnai H.
Shah N.
Shapiro E.
Skyum S.
Slobodova A.
Snyder W.
Sokołowski St.
Sondergaard H.
Spirakis P.
Steffen B.
Steinby M.
Stern J.
Steyaert J. M.
Stifter S.
Stirling C.
Stokkermans K.
Stoughton A.
Sturc J.
Sykora O.
Szałas A.
Szpankowski W.
Talamo M.
Tan T. S.
Tarlecki A.
Teillaud M.
Tel G.
Tirze H.
Tofte M.
Tofts C.
Toyama Y.
Tran N.
Turakainen P.
Tzakalidis A.
Ukkonen E.
Unger W.
Van Draager F.
Van Oostrom V.
Vauquelin B.
Vollmer H.
Vrto I.
Wagner K.
Wagner K. W.
Walker D.
Wanke E.
Waupotitsch R.
Weil P.
Wernisch L.
Wiedermann J.
Williams P.
Winkler F.
Winkowski J.
Winskel G.
Zantema H.
Zielonka W.
Zwiers J.

CONTENTS

Symbolic Computation (Session 9)

Geometric Algorithms (Session 10)

Logic and Models I (Session 11)

Logic and Models II (Session 12)

Time Specification (Session 13)

Concurrency Semantics (Session 14)

Program Development (Session 15)

Process Equivalences (Session 16)

Philosophical Issues in Kolmogorov Complexity

Ming Li*
University of Waterloo

Paul M.B. Vitányi†
CWI and University of Amsterdam

1 Introduction

Five years have passed since we wrote our first survey on Kolmogorov complexity and its applications [10]. This essay is not meant to be an exhaustive survey of the subject, not even of the recent results; that is done thoroughly in our forthcoming book [13] which will appear very soon. Here, we would like to convey to our reader some appealing philosophical ideas by just picking up some pretty shells deposited on the shore by the sea of applications of Kolmogorov complexity. We hope these ideas will be useful or, at least, enjoyable to our reader.

We give preference to ideas and applications that were not (well) covered by our previous articles [10, 11], either due to our ignorance at the time or because the results are new. We also prefer those results that have deeper philosophical or methodological implications. During our narrative, we often venture into strange lands where we are only amateurs or even total strangers. Thus our views might not be completely conventional, but we do hope they are novel and interesting.

Due to space limitation, we refer the reader to [11, 13] for definitions and basic facts of Kolmogorov complexity. For the purpose of reading this article at a conceptual level, it is sufficient to know that Kolmogorov complexity of a finite string x is simply the length of the shortest program, say in FORTRAN[1] encoded in binary bits, which prints x without any input. $C(x)$ is the Kolmogorov complexity of x; $K(x)$ is the prefix Kolmogorov complexity of x where the program for x must be self-delimiting.

2 Should we prefer elementary proofs?

Probabilistic or information–theoretic style proofs have enjoyed major successes in combinatorics and computer science. Our thinking about proofs in computer science parallels the following comments of Kolmogorov [8] about information theory:

> The real substance of the entropy formula [based on probabilistic assumptions about independent random variables] ... holds under incomparably weaker

*Supported by NSERC operating grant OGP-046506. Computer Science Department, University of Waterloo, Waterloo, Ontario, N2L 3G1, Canada. E-mail: mli@math.uwaterloo.ca

†Partially supported by NSERC International Scientific Exchange Award ISE0125663. CWI, Kruislaan 413, 1098 SJ Amsterdam, The Netherlands. E-mail: paulv@cwi.nl

[1] Or in Turing machine codes.

> and purely combinatorial assumptions... Information theory must precede probability theory, and not be based on it. By the very essence of this discipline, the foundations of information theory must have a finite combinatorial character.

From a practical viewpoint the real issue is whether elementary arguments must always be more tedious. We demonstrate through one example that elementary proofs (using Kolmogorov complexity) are not only more intuitive, but also easier. Kolmogorov complexity based arguments, although nonconstructive, are essentially combinatorial in nature without probabilistic assumptions. We use $d(S)$ to denote the number of elements in set S.

A family $\mathcal{D} = \{D_1, D_2, ..., D_j\}$ of subsets of $N = \{1, 2, \ldots, n\}$ is called a *distinguishing family* for N if for any two distinct subsets M and M' of N there exists an i $(1 \leq i \leq j)$ such that $d(D_i \cap M)$ is different from $d(D_i \cap M')$. Let $f(n)$ denote the minimum of $d(\mathcal{D})$ over all distinguishing families for N. The *coin-weighing problem* is to determine $f(n)$. It is known that

$$f(n) \leq (2n/\log n)[1 + O(\log\log n/\log n)]. \tag{1}$$

Equation 1 was independently established by B. Lindström in 1965 and D.G. Cantor and W.H. Mills in 1966. P. Erdös and A. Rényi [5], L. Moser [15], and N. Pippenger [16] have established the following Theorem 1 using various probabilistic and information theory methods (second moment method).

Fix an encoding of the 2^n subsets of N such that each subset is encoded by a binary string of length n. Simplifying notation, we write M as the encoding of M.

Theorem 1 $f(n) \geq (2n/\log n)[1 + O(\log\log n/\log n)]$.

Proof. Choose M such that $C(M|D_2, ..., D_j) \geq n$. Let $d_i = d(D_i)$ and $m_i = d(D_i \cap M)$. By elementary estimates [13], m_i is within range $d_i/2 \pm O(\sqrt{d_i \log d_i})$. Thus, for $1 \leq i \leq j$, m_i can be described using its discrepancy with $d_i/2$, hence

$$C(m_i|D_i) \leq \frac{1}{2}\log d_i + O(\log\log d_i) \leq \frac{1}{2}\log n + O(\log\log n).$$

Since $\mathcal{D}$ is a distinguishing family, given $\mathcal{D}$, the values $m_1, ..., m_j$ determine M:

$$n \leq C(M|D_1, ..., D_j) \leq C(m_1, ..., m_j|D_1, ...D_j) \leq \Sigma_{i=1}^{j}(\frac{1}{2}\log n + O(\log\log n)).$$

This implies the theorem. □

3 The Grue Emerald Paradox

For about two thousand years philosophers have worried about the problem of inductive reasoning. On the one hand, it seems common sense to assume that people learn in the sense that they generalize from observations by learning a 'Law' that governs not only the past observations, but will also apply to the observations in the future. In this sense induction should 'add knowledge'.

Yet how is it possible to acquire knowledge which is not yet present? If we have a system to deduce a general law from observations, then this law is only part of the knowledge contained in this system and the observations. Then, the law does not represent knowledge over and above what was already present, but it represents in fact only a part of that knowledge.

In [7], N. Goodman described the *grue emerald paradox*. Let h be the hypothesis that all emeralds are green. Let k be the hypothesis that all emeralds examined before the the year of 2000 are green and all emeralds examined after 2000 will be blue. Goodman called this color 'grue'. Then both hypotheses are totally confirmed by the experiments so far. How do we develop some tools, or philosophy, to prefer h than k? People have been resorting to farfetched arguments, for example, to prefer time-independent hypothesis (h) than the time-dependent hypothesis (k).

Francis Bacon, in *Sylva Sylvarum* 337, 1627, formulates the power of induction as follows: "The eye of the understanding is like the eye of the sense; for as you may see great objects through small crannies or levels, so you may see great axioms of nature through small and contemptible instances."

Mathematics has come up with an induction principle which has an impeccable derivation, yet allows us to estimate the relative likelihood of different possible hypotheses—which is impossible with the commonly used Pearson-Neyman testing. Consider a discrete sample space Ω. Let $D, H_1, H_2, \ldots$ be a countable set of events (subsets) of Ω. $\mathbf{H} = \{H_1, H_2, \ldots\}$ is called *hypotheses space* . The hypotheses H_i are exhaustive (at least one is true). From the definition of conditional probability, that is, $P(A|B) = P(A \cap B)/P(B)$, it is easy to derive **Bayes' formula** (rewrite $P(A \cap B)$ in two different ways):

$$P(H_i|D) = \frac{P(D|H_i)P(H_i)}{P(D)}. \tag{2}$$

If the hypotheses are mutually exclusive ($H_i \cap H_j = \emptyset$ for all i, j), then

$$P(D) = \sum_i P(D|H_i)P(H_i).$$

Despite the fact that Bayes' rule is just a rewriting of the definition of conditional probability and nothing more, it is its interpretation and applications that are most profound and caused much bitter controversy during the past two centuries. In Equation 2, the H_i's represent the possible alternative hypotheses concerning the phenomenon we wish to discover. The term D represents the empirically or otherwise known data concerning this phenomenon. The term $P(D)$, the probability of data D, may be considered as a normalizing factor so that $\sum_i P(H_i|D) = 1$. The term $P(H_i)$ is called the *a priori* probability or *initial* probability of hypothesis H_i, that is, it is the probability of H_i being true before we see any data. The term $P(H_i|D)$ is called *a posteriori* or *inferred* probability

The most interesting term is the prior probability $P(H_i)$. In the context of machine learning, $P(H_i)$ is often considered as the learner's *initial degree of belief* in hypothesis H_i. In essence Bayes' rule is a *mapping* from *a priori* probability $P(H_i)$ to *a posteriori* probability $P(H_i|D)$ determined by data D. In general, the problem is not so much that in the limit the inferred hypothesis would not concentrate on the true hypothesis, but that the inferred probability gives as much information as possible about the possible hypotheses from only a limited number of data. In fact, the continuous bitter debate between the Bayesian and non-Bayesian opinions centered on the prior probability. The controversy is caused by the fact that Bayesian theory does not say how to initially derive the prior probabilities for the hypotheses. Rather, Bayes' rule only tells how they are to be *updated.* In the real world problems, the prior proabilities may be unknown, uncomputable, or even conceivably non-existent. (What is the prior probability of use of a word in written English? There are many different sources of different social backgrounds living in different ages.) This problem would be solved if we can find a *single* probability distribution to use as the prior distribution in each different case, with approximately the same result as if we had used the real distribution. Surprisingly, this turns out to be possible up to some mild restrictions.

Consider theory formation in science as the process of obtaining a compact description of the past observations. The investigator observes increasingly larger initial segments of an infinite binary sequence X as the outcome of an infinite sequence of experiments on some aspect of nature. To describe the underlying regularity of X, the investigator tries to formulate a theory that governs X, consistent with past experiments. Candidate theories (hypotheses) are identified with computer programs that compute binary sequences starting with the observed initial segment.

First assume the existence of a prior probability distribution μ (actually a measure) over the continuous sample space $\Omega = \{0,1\}^\infty$. Denote by $\mu(x)$ the probability of a sequence starting with x. Given a previously observed data string S, the inference problem is to predict the next symbol in the output sequence, that is, extrapolating the sequence S. In terms of the variables in Equation 2, H_a is the hypothesis that the sequence under consideration continues with a. Data D_S consists of the fact that the the sequence starts with initial segment S. Thus, for $P(H_i)$ and $P(D)$ in Formula 2 we substitute $\mu(H_a)$ and $\mu(D_S)$, respectively, and obtain:

$$\mu(H_a|D_S) = \frac{\mu(D_S|H_a)\mu(H_a)}{\mu(D_S)}.$$

We must have $\mu(D_S|H_a) = 1$ for any a, hence,

$$\mu(H_a|D_S) = \frac{\mu(H_a)}{\mu(D_S)}. \tag{3}$$

Generally, we denote $\mu(H_a|D_S)$ by $\mu(a|S)$. In terms of inductive inference or machine learning, the final probability $\mu(a|S)$ is the probability of the next symbol being a, given the initial sequence S. Obviously we now only need the prior probability μ to evaluate $\mu(a|S)$.

The idea is to approximate the unknown proper prior probability μ. Without too much loss of generality we may as well assume that the measure μ is *enumerable.* That means, there is a Turing machine T which computes a total function $\phi(x,k)$ such that $\phi(x,k+1) \geq \phi(x,k)$ and $\lim_{k\to\infty}\phi(x,k) = \mu(x)$. If μ is recursive then it is also enumerable, but not necessarily the converse. It turns out that the class of all enumerable measures contains a *universal measure*, denoted by $\mathbf{M}$, such that for all μ in this class there exists a constant $c > 0$ such that $\mathbf{M}(x) \geq c\mu(x)$ for all x. We say that $\mathbf{M}$ *dominates* μ. We also call $\mathbf{M}$ the *a priori* probability, since it assigns maximal probability to all hypotheses in absence of any knowledge about them.

Now instead of using Formula 3, we estimate the conditional probability $\mu(y|x)$ that the next segment after x is y by the expression

$$\frac{\mathbf{M}(xy)}{\mathbf{M}(x)}. \tag{4}$$

Now let μ in Formula 3 be an arbitrary computable measure. This case includes all computable sequences. If the length of y is fixed, and the length of x grows to infinity, then it can be shown [18] that

$$\frac{\mathbf{M}(y)/\mathbf{M}(x)}{\mu(y)/\mu(x)} \rightarrow 1,$$

with μ-probability one. In other words, the conditional *a priori* probability is almost always asymptotically equal to the conditional probability. It has also shown by Solomonoff that the convergence is very fast and if we use Formula 4 instead of the real value Formula 3, then our inference is almost as good. We also know that

$$-\log \mathbf{M}(x) = K(x) + O(\log K(x)), \tag{5}$$

That means that $\mathbf{M}$ assigns high probability to simple objects and low probability to complex or random objects. We now come to the punch line: Bayes' rule using the universal prior distribution yields Occam's Razor principle. Namely, if several programs could generate $S0$ then the shortest one is used (for the prior probability), and further if $S0$ has a shorter program than $S1$ then $S0$ is preferred (that is, predict 0 with higher probability than predicting 1 after seeing S). Bayes' rule via the universal prior distribution also gives the so-called indifference principle in case $S0$ and $S1$ have roughly equal length shortest programs which 'explain' $S0$ and $S1$, respectively. The Goodman's grue emerald paradox disappears under this paradigm.

Scientists formulate their theories in two steps: firstly a scientist, based on scientific observations, formulate alternative hypotheses, and secondly a definite hypothesis is selected. The second step is the subject of inference in statistics. Statisticians have developed many different principles to do this, like Occam's Razor principle, the Maximum Likelihood principle, various ways of using Bayes' formula with different prior distributions. No single principle turned out to be satisfiable in all situations. Philosophically speaking, Solomonoff's approach presents an ideal way of solving induction problems. However, due to the non-computability of the universal prior function, such a theory cannot be directly used. Some approximation is needed in the real world applications.

Now we will closely follow Solomonoff's idea, but substitute a 'good' computable approximation to $\mathbf{M}(x)$. This results in Rissanen's **Minimum Description Length** principle [17]. Rissanen not only gives the principle, more importantly he also gives the detailed formulae on how to use this principle. This made it possible to use the MDL principle. The MDL principle can be intuitively stated as follows:

Minimum Description Length Principle. *The best theory to explain a set of data is the one which minimizes the sum of*

- *the length, in bits, of the description of the theory;*
- *the length, in bits, of data when encoded with the help of the theory.*

We now develop this MDL principle from Bayes' rule using the universal distribution $\mathbf{M}(x)$, assuming that P is enumerable. From the Bayes' Formula 2, we must choose the hypothesis H such that $P(H|D)$ is maximized. First we take the negative logarithm on both sides of Equation 2, we get

$$-\log P(H|D) = -\log P(D|H) - \log P(H) + \log P(D)$$

$\log P(D)$ is a constant and hence ignored. Maximizing the $P(H|D)$ over all possible H's is equivalent to *minimizing* $-\log P(H|D)$, or minimizing

$$-\log P(D|H) - \log P(H)$$

Now to get the minimum description length principle, we only need to explain above two terms in the sum properly. According to Solomonoff, when P is enumerable, then we approximate P by $\mathbf{M}$. The prior probability $P(H)$ is set to $\mathbf{M}(H) = 2^{-K(H)\pm O(\log K(H))}$, where $K(H)$ is the prefix-complexity of H. That is, $-\log P(H)$ is about the *length* of a minimum *prefix code*, or program, of hypothesis H.

A similar argument applies to term $-\log P(D|H)$. That is, $2^{-K(D|H)\pm O(\log K(D|H))}$ is a reasonable approximation of $P(D|H)$. The term $-\log P(D|H)$, also known as the *self-information* in information theory and the negative log likelihood in statistics, can now be regarded as the number of bits it takes to redescribe or encode D with an ideal code relative to H. In different applications, the hypothesis H can mean many different things, such as decision trees, finite automata, Boolean formulae, or a polynomial. In general statistical applications, one assumes that H is some model $H(\theta)$ with a set of parameters $\theta = \{\theta_1, \ldots, \theta_k\}$ of precision c, where the number k may vary and influence the descriptional complexity of $H(\theta)$. In such case, we minimize

$$-\log P(D|\theta) - \log P(\theta).$$

Let's consider one example. For each fixed k, $k = 0, \ldots, n-1$, let f_k be the best polynomial of degree k, fitted on points (x_i, y_i) $(1 \le i \le n)$, which minimizes the error

$$error(f_k) = \sum_{i=1}^{n} (f_k(x_i) - y_i)^2.$$

Assume each coefficient takes c bits. So f_k is encoded in ck bits. Let us assume the commonly used Gaussian distribution of the error on y_i's. Thus, given that f_k is the true polynomial,

$$\Pr(y_1, \ldots, y_n | f_k) = \Pi_i \exp(-O((f_k(x_i) - y_i)^2)).$$

The negative logarithm of above is $c' \cdot error(f_k)$ for some computable c'. The MDL principle tells us to choose f_k, $k \in \{0, \ldots, n-1\}$, which minimizes $ck + c' \cdot error(f_k)$.

4 Valiant learning under computable distributions?

Valiant's model [20] provides an excellent framework for studying learnability. Subsequent investigations show many problems intractable (NP-complete) under the original model. Can we adapt the it to obtain a model where more concepts are polynomial time learnable? The philosophy here is that maybe humans just learn a concept under *some restricted class of distributions*, like computable ones (those in our textbooks). Kolmogorov complexity and the Solomonoff-Levin universal distribution allows us to systematically develop a theory of Valiant-style learning under all (semi) computable distributions.

All distributions we have a name for: the uniform distribution, normal distribution, geometric distribution, Poisson distribution, are computable (with computable parameters). Hence the change from distribution-free learning to computable-distribution-free learning is not too restrictive. It turns out that there is a nice mathematical structure in our computable-distribution-free learning case. For example, we can prove completeness results in the sense that there is a single (universal) distribution $\mathbf{m}$ such that if a concept class is learnable under this *single* distribution, they it is learnable under *all* computable distributions. Formally,

Theorem 2 *A concept class C is polynomially learnable under the universal distribution $\mathbf{m}(x)$, iff it is polynomially learnable under each computable distribution P, provided the sample is drawn according to $\mathbf{m}$.*

See [12] for details. In the continuous case, we even have a stronger theorem without needing to sample according to the universal distributions.

Theorem 3 *A concept class C over a continuous sample space is learnable under $\mathbf{M}$ iff it is learnable under each computable measure.*

5 Can we abandon pumping lemmas?

In the current undergraduate formal language courses, it seems that the cumbersome pumping lemmas constitute an important part of the teaching. It may be argued that such lemmas not only obstructs students' ability of viewing the real substance of the proof, but also give them a bad habit (like what 'goto' did to FORTRAN). Further, the usual pumping lemmas do not hold conversely which adds more confusion. Often students need un-aesthetic add-ons like "marked pumping lemma".

It turns out that Kolmogorov complexity is just the right tool to characterize *all* regular languages. It simply makes our intuition of 'finite state'-ness of these languages rigorous and easy to apply.

Theorem 4 (Regular KC-Characterization) *Let $L \subseteq \Sigma^*$, $\chi = \chi_1\chi_2\ldots$ be the* characteristic sequence *of $L_x = \{y|xy \in L\}$. The following statements are equivalent.*
(i) L is regular.
(ii) $\exists c_L, \forall x \in \Sigma^*, \forall n, C(\chi_{1:n}|n) \leq c_L$, c_L depending only on L.
(iii) $\exists c_L, \forall x \in \Sigma^*, \forall n, C(\chi_{1:n}) \leq C(n) + c_L$, c_L depending only on L.
(iv) $\exists c_L, \forall x \in \Sigma^*, \forall n, C(\chi_{1:n}) \leq \log n + c_L$, c_L depending only on L.

PROOF. (i) → (ii) → (iii) → (iv) are simple. To show (iv) → (i), we need,

Claim 5 For each constant c there are only finitely many one-way infinite binary strings ω such that, for all n, $C(\omega_{1:n}) \leq \log n + c$.

D.W. Loveland [14] credits the following result to A. Meyer: For each constant c there are only finitely many $\omega \in \{0,1\}^\infty$ with $C(\omega_{1:n}|n) \leq c$ for all n and each such ω is a recursive real. G.J. Chaitin [3] improved this to $C(\omega_{1:n}) \leq \log n + c$.

By (iv) and Claim 5, there are only finitely many distinct χ's associated with the x's in Σ^*. Define the right-invariant equivalence relation $\sim$ by $x \sim x'$ if $\chi = \chi'$. This relation induces a partition of Σ^* in equivalence classes $[x] = \{y : y \sim x\}$. Thus there are only finitely many $[x]$'s, which implies that L is regular by the Myhill-Nerode theorem: define a finite automaton using one state for each equivalent class, and define transition functions accordingly. □

See [13] for a complete proof, and for CFL's. As a corollary, we have

Lemma 6 (KC-Regularity) *Let L be regular and $L_x = \{y : xy \in L\}$. For each x, if y is the nth string is in L_x, then $C(y) \leq C(n) + c$, for some constant c.*

EXAMPLE 5.1 Consider $\{0^k 1^k : k \geq 1\}$. Set $x = 0^n$ with $C(n) \geq \log n$. Then lexicographically first string in L_x is 1^n, but $C(1^n) = \Omega(\log n)$. Thus KC-Regularity Lemma implies that $\{0^k 1^k : k \geq 1\}$ is not regular.

Comment. Intuitively, when a finite automaton reads 1^n, it has to remember n, but $C(n) \geq \log n$, a *finite* automaton cannot encode these $\log n$ bits.

EXAMPLE 5.2 Prove that $L = \{xx^R w : x, w \in \{0,1\}^* - \{\epsilon\}\}$ is not regular. Set $x = (01)^n$, where $C(n) \geq \log n$. Then, the lexicographically first word in L_x is $y = (10)^n 0$. Hence, $C(y) = \Omega(\log n)$, contradicting the KC-Regularity Lemma.

6 Loschmidt's Paradox

The second law of thermodynamics says that in any thermodynamic process that proceeds from one equilibrium state to another, the entropy of the system + environment either increase or remains unchanged. Fundamentally, this law says that any system, without external influence, goes to maximum disorder.

In 1872, Ludwig Boltzmann (1844-1906), lectured on the foundation of modern statistical thermodynamics. When he mentioned his interpretation of the second law, the physicist Joseph Loschmidt rose to protest. Loschmidt said that the laws governing the motions of all particles are symmetric with respect to time. Thus any system that goes from order to chaos could be made orderly once again by reversing the momentum of each particle. This will not affect the total kinetic energy of the system. In defiance Boltzmann pointed his finger at Loschmidt and said, "You reverse the momenta." [2]. Loschmidt in fact raised the following question: if the system is deterministic, how could its entropy be increased? In fact, Shannon entropy ($H = -\sum_s p_s \log p_s$) or, equivalently, Boltzmann-Gibbs entropy ($S = k \log W$) for a deterministic system cannot increase, because the number of possible states (W) does not increase in a deterministic system by Liouville's Theorem.

With Shannon's information theory, one cannot directly define the entropy of a *single* microscopic state. Sometimes such a definition is actually desired. For instance, in the high temperature superconductivity research, some material like CuO_2 loses magnetic moment below some critical temperature. In such state, the nuclear spins in CuO_2 all line up as in Figure 1.

↑↓↑↓↑↓↑↓↑ .....
↓↑↓↑↓↑↓↑↓ .....
↑↓↑↓↑↓↑↓↑ .....
↓↑↓↑↓↑↓↑↓ .....

Figure 1: Atomic spin in CuO_2 at low temperature

The state represented by Figure 1 is considered to have low entropy, or zero entropy. It is necessary to define entropy for such particular state, and not involve a whole ensemble as information theoretic entropy. It is not convenient to use the Shannon entropy in this situation since we might also like to know the entropy of the state with all arrows up, as in Ferromagnetism, or all arrows down. Or two arrows up and two arrows down alternatively, *etc.*

The above discussion naturally leads to the following new definitions of entropy for physical systems: *algorithmic entropy* and *physical entropy.* As we will demonstrate, the new definitions are not only quantitatively and philosophically correct, but also more direct, intuitive and elementary than the Shannon entropy.

Definition 1 The *algorithmic entropy* of a microscopic state of a system is the Kolmogorov complexity of that state.

The difficulties we had with Shannon entropy disappear in the new definition. Algorithmic entropy of a system can increase, even for a deterministic system, simply because the system evolves over time from regular initial states to more random states. Thus the second law is naturally and rigorously explained with algorithmic entropy. Since most states are random, a random system or even a deterministic system approaches and stays on average in states with maximum entropy. Note that second law was never meant to be true all of the time. The upshot is that the probability of the second law not to hold is very very low.

EXAMPLE 6.1 Regular microscopic states now automatically have well-defined low algorithmic entropies. The state of CuO_2 in Figure 1 naturally has low algorithmic entropy since it can be described by a trivial program of a few bits:

$$\textbf{repeat forever}: \textbf{ print } \uparrow; \textbf{ print } \downarrow .$$

Other easily describable states, such as with all arrows up or alternatively with two arrows up and two arrows down, also naturally have constant algorithmic entropy.

EXAMPLE 6.2 [This example is taken from D. Halliday and R. Resnick, *Fundamentals of Physics*, 3rd extended edition, 1988, page 526] Algorithmic entropy considerations lie at the heart of *adiabatic demagnetization*, an important method that has been used with great success to achieve record low temperatures (near zero Kelvin). In this method, a sample such as a chrome-alum salt (whose atoms may be considered as tiny magnets) is placed in an insulating enclosure at the lowest attainable temperature. A strong magnetic field is applied by an external magnet so

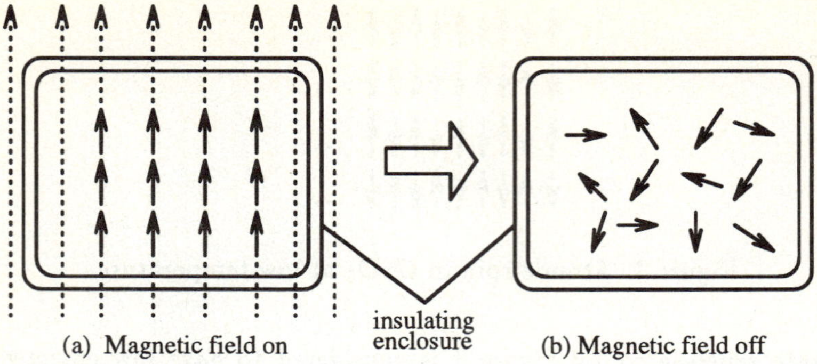

Figure 2: Adiabatic demagnetization to achieve low temperature

that the tiny atomic magnets line up, forming a very ordered state, as Figure 2 (a) shows. Then the magnet is removed so that its field is no longer present. By thermal agitation, the atomic magnets now assume random orientations, as in Figure 2 (b). Kolmogorov complexity, *i.e., algorithmic entropy*, associated with the atomic alignments has clearly increased.

Since the system is adiabatically isolated, no heat can leave or enter. Since the process of removing the magnetic field is almost reversible, there can be no change in entropy in this thermally isolated *reversible* process. Thus the increase of entropy associated with the randomizing of the directions of the atomic magnets must be compensated with the spontaneous lowering of the temperature of the specimen, decreasing the entropy due to thermal agitation by the same amount.

We further justify Definition 1. Given a thermodynamic ensemble $\mathcal{E}$, we prove that Shannon entropy and algorithmic entropy are quantitatively the same.

Theorem 7 *Let $P(s)$ be the probability that the system is in state $s \in \mathcal{E}$. If probabilities $P(s)$ can be recursively generated, then, up to a constant additive factor,*

$$H(P) = \sum_{s \in \mathcal{E}} P(s) K(s), \tag{6}$$

where $H(P) = -\sum_{s \in \mathcal{E}} P(s) \log P(s)$ is the Shannon entropy.

PROOF. By the Noiseless Coding Theorem,

$$H(P) \leq \sum_{s \in \mathcal{E}} P(s) K(s).$$

By a constructive version of the Shannon-Fano code, $K(s) \leq -\log P(s) + O(1)$. It follows that

$$\sum_{s \in \mathcal{E}} P(s) K(s) \leq H(P) + O(1).$$

[13] contains all the complex details. □

Equation 6 shows that algorithmic entropy quantitatively approximates the old Shannon entropy. Thus algorithmic entropy naturally inherits the successes and results of the Shannon entropy. As an example, we prove one such result, the Sackur-Tetrode equation [24].

Ideal gas or Boltzmann gas is a convenient tool to study statistical thermodynamics. Consider a 3-dimensional container with N identical (ideal) gas molecules. Each molecule is considered as an elastic ball with no internal freedom. Imagine we divide the entire container volume into a 3-dimensional array of small cells, each capable of containing one molecule. At any fixed time, a specific cell contains zero or one molecule. To specify N molecules, we use a *phase-space* of $6N$ dimensions. A particular microscopic state corresponds to just one point $P = (p_1, \cdots, p_{3N}, q_1, \cdots, q_{3N})$ in phase-space, where p_i's are coordinates of the positions of N molecules and q_i's specify their momenta. This is called a Hamiltonian system. (In Boltzmann entropy, $S = k \log W$, W is the number of points in our phase-space corresponding to the macroscopic state of entropy S. Liouville's Theorem says any region in the phase-space does not change in volume as the system evolves with time. Thus as a system evolves, W, therefore S, stays constant. S correctly specifies the equilibrium entropy, but it lacks the ability of specifying the dynamic changes of the system, as prescribed by the second law.)

Theorem 8 *(Sackur-Tetrode Equation) Let V be the volume of the container, and ΔV be the volume of each cell. Let Δ_q be the basic unit we record momentum q. Then the entropy of a typical microscopic state of N ideal molecules is given by the following formula:*

$$S \leq N[\log \frac{V}{N\Delta V} + \frac{3}{2} \log \frac{mkT}{(\Delta_q)^2}] + O(1).$$

PROOF. We design a program of above size to specify any given microscopic state, assuming N is given. The positions of molecules can be specified as follows: There are $V/\Delta V$ cells and N molecules. To specify the distances of the molecules, we need $\sum_{i=1}^{N} d_i$ bits, where $\sum d_i = V/\Delta V$. Maximizing this, we conclude that at most $N \log \frac{V}{N\Delta V}$ bits are needed to describe the positions of the N molecules.

The expected value of each component of momentum q of a molecule is $(mkT)^{1/2}$, thus to specify the momentum of such a molecule requires $3 \log \frac{(mkT)^{1/2}}{\Delta_q}$ bits. In total, we need at most $\frac{3}{2} N \log \frac{mkT}{(\Delta_q)^2}$ bits to specify the momenta of N molecules. □

Algorithmic entropy is a definition from the system point of view. It defines the complexity of a system at a particular time. If we look at a system from an observer's angle, W.H. Zurek [24] defined the following 'physical entropy'.

Definition 2 Physical entropy of a system, given observed data d of the system, is $H_d + K(d)$, where $K(d)$ is the Kolmogorov complexity of the observed data d, and H_d is the conditional Shannon entropy or our "ignorance" of the system given d.

Physical entropy reflects the fact that measurements can increase our knowledge about a system. If the system is in a regular state, physical entropy can decrease as we make more and more measurements. In the beginning, we have no knowledge about the state of system, therefore the physical entropy reduces to Shannon entropy and is maximized reflecting our total ignorance. As we make more measurements, if

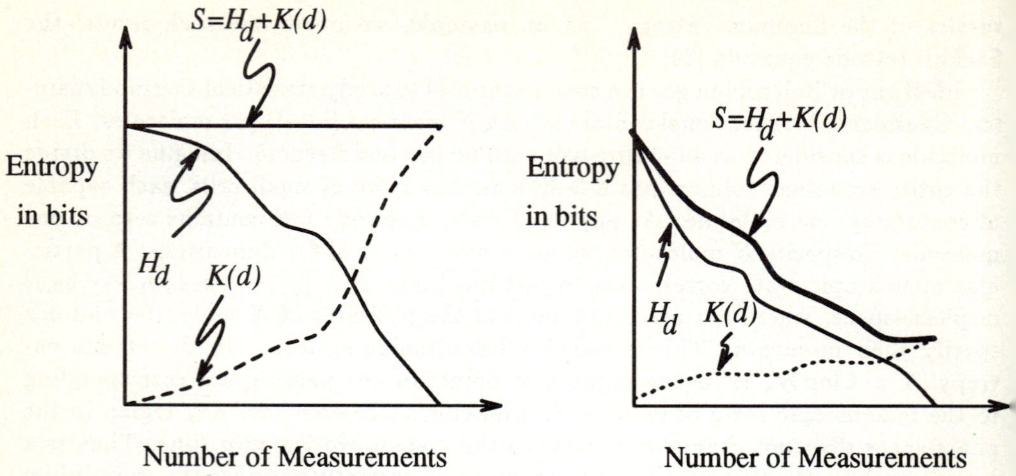

Figure 3: Physical Entropy

the state is not regular, then we cannot achieve compression and the physical entropy remains high. However, if the system is in a regular state, then measurements increases our knowledge about the system and we might be able to compress the data, and the system becomes less unknown and hence physical entropy deminishes. The two pictures in Figure 3 describe these situations.

Notice that the physical entropy in Definition 2 cannot increase if we make no measurements. However, physical entropy can increase if we made sufficient measurements at different time steps as the system evolves and becomes more and more disordered. It is important to distinguish the increase of $K(d)$ with measurements in Figure 3 and the increase of algorithmic entropy with time. In the former case, we have more and more data at hand to describe, therefore $K(d)$ grows until it describes the systems totally. During this process, the system is assumed to be static with the arrow of time frozen. In the latter case, the system evolves to become more and more chaotic with time, according to the second law. Thus if we make sufficient amount of measurements to the system at some later time step, $K(d)$ will be greater, reflecting the fact that the system is now more disordered, or, has greater algorithmic entropy. This increase will stop at the equilibrium of the system. Zurek [25] shows how to use physical entropy to explain Maxwell's demon.

7 Why is our world compressible?

P.C.W. Davies [4] asked the interesting question why the world is compressible. Science, in a sense, may be considered as compression of experimental data. Compression means knowing and comprehending. Why is mathematics compressible to a few axioms? Why is physics (approximately) compressible to a few laws? Why are we humans compressible to a string over alphabet $\{A, C, G, T\}$? Why is the universe compressible, and hence comprehensible, at all?

E. Wigner [22] called this phenomenon "the unreasonable effectiveness" of mathematics in the natural sciences. We are so used to our environment, that above questions probably have never puzzled most of us. If you do not realize the subtleties of these questions, listen to R.W. Hamming [*Coding and Information Theory*, Prentice-Hall, 1987]:

> I have tried, with little success, to get some of my friends to understand my amazement that the abstraction of integers for counting is both possible and useful. Is it not remarkable that 6 sheep plus 7 sheep makes 13 sheep; that 6 stones plus 7 stones make 13 stones? Is it not a miracle that the universe is so constructed that such a simple abstraction as a number is possible? To me this is one of the strongest examples of the unreasonable effectiveness of mathematics. Indeed, I find it both strange and unexplainable.

Is it unreasonable to imagine a total chaotic universe? As we know, a randomly given string (or universe) is most likely to be not algorithmically compressible. How come ours happens to be algorithmically compressible, hence knowable? More remarkably, our universe is not only compressible, it is indeed very *feasibly* compressible, since mankind is good at discovering nature's algorithms. It took only one man and one apple and 13 years to invent Newton mechanics (and calculus); it took only one man and his life time to discover evolution theory; it took only one man and a few years to understand relativity; it took only a few more men and a few more years to formulate quantum mechanics. Yet, we obtain so much from so little: we sent men to the moon with a few laws of Newton; we release gigantic nuclear energy with Einstein's $E = mc^2$; our TV's, radios, X ray, and police radar all depend on just 4 lines of Maxwell equations; and with quantum mechanics and superstring theory, we hope that they explain everything on earth, and in heaven.

Let us look at things from a different perspective. Life, in some sense, also evolves in the direction of minimizing its 'programs', and indeed this turns out to be not only possible but also extremely effective. A DNA molecule can be regarded as a long character string over $\{A, C, G, T\}$. In order to encode proteins, it needs to encode 20 amino acids, plus a 'begin' and an 'end' command to signify the start and the end of an encoding of a protein. Thus to encode each of the 22 different objects, one needs at least three characters which give 64 possible combinations, while two characters only give 16 combinations. Using more than three characters is obviously redundant. Remarkably, indeed Nature uses precisely three characters to encode an amino acid or a 'begin' or an 'end' command, not more and not less. (Note, we do not claim that such sequences have maximum Kolmogorov complexity, in fact they do not.) More remarkably, Nature also dynamically minimizes its programs through Darwinian selection. To demonstrate this point, we describe an experiment performed by S. Spiegelman [19], discussed in D.K. Kondepudi [9]. A Q_β virus replicates using the resources of the cell it infects. Since intra-cellular environment is usually complex and malicious, for successful replication, this virus carries in its RNA the algorithm to synthesize proteins that form a protective coat. Such RNA has about 4500 units. Spiegelman placed such RNA of 4500 units in a friendly environment conducive to replication and let it evolve. Soon, mutations with smaller number of units that could replicate faster arose and replaced the original RNA. This process continued until the RNA was reduced to about 220 units. This is an application of the MDL principle above.

The current laws of physics of this universe only started to hold at approximately

10^{-4}s of the big bang. According to the best cosmological theories, the universe began in an exceedingly simple state. Is this the reason that our current universe is highly compressible? Using a binary string to encode the universe, we can ask the question: what is the Kolmogorov complexity of the world? C.H. Woo [23] argues that it maybe premature to claim that the world is indeed highly (algorithmically) compressible. Even the universe would be maximally random, then our part of the universe could be regular and compressible, since every long string has short compressible substrings. Either way, this will stay unknown, since a random string cannot be proven to be random effectively. Relying on quantum mechanics, R. Penrose in his popular book [*Emperor's New Mind*, Oxford University Press, 1989] has conjectured that the human brain has capabilities superior to that of a Turing machine. Does God really permit us to learn more about Him via proofs imagined by our minds than proofs listed by our hands?

Among other interesting philosophical issues relating Kolmogorov complexity to physics are: Thermodynamics of computing and Maxwell's demon (can one make a perpetual machine of second kind?) [1, 24], chaos theory (can a deterministic system be chaotic?) [6], visual distance, *etc.* The interesting book [25] contains a wealth of related papers. [13] tries to present a complete treatment. [21] contains five surveys on applications of Kolmogorov complexity to structural complexity theory.

References

[1] C.H. Bennett. The thermodynamics of computation—review. *International J. of Theoretical Physics*, 21:905–940, 1982.

[2] R.G. Brewer and E.L. Hahn. Atomic memory. *Scientific American*, 251(6):50–57, 1984.

[3] G.J. Chaitin. Information-theoretic characterizations of recursive infinite strings. *Theoret. Comp. Sci.*, 2:45–48, 1976.

[4] P.C.W. Davies. Why is the physical world so comprehensible? In W.H. Zurek, editor, *Complexity, entropy and the physics of information*, pages 61–70. Addison-Wesley, 1991.

[5] P. Erdös and A. Rényi. On two problems of information theory. *Publ. Hungar. Ac. Sci.*, 8:241–254, 1963.

[6] J. Ford. Chaos: solving the unsolvable, predicting the unpredictable. In M.F. Barnsley and S.G. Demko, editors, *Chaotic dynamics and fractals.* Academic Press, 1986.

[7] N. Goodman. *Fact, Fiction, and Forecast.* Harvard University Press, 1955.

[8] A.N. Kolmogorov. Combinatorial foundations of information theory and the calculus of probabilities. *Russian Mathematical Surveys*, 38(4):29–40, 1983.

[9] D.K. Kondepudi. Non-equilibrium polymers, entropy, and algorithmic information. In W.H. Zurek, editor, *Complexity, entropy and the physics of information*, pages 199–206. Addison-Wesley, 1991.

[10] M. Li and P.M.B. Vitányi. Two decades of applied Kolmogorov complexity: In memoriam A.N. Kolmogorov 1903 - 1987. In *Proc. 3rd IEEE Conference on Structure in Complexity Theory*, pages 80–101, 1988.

[11] M. Li and P.M.B. Vitányi. Kolmogorov complexity and its applications. In J. van Leeuwen, editor, *Handbook of Theoretical Computer Science*, chapter IV, pages 187–254. Elsevier and MIT Press, 1990.

[12] M. Li and P.M.B. Vitányi. A theory of learning simple concepts under simple distributions. *SIAM J. Comput.*, 20(5):915–935, 1991.

[13] M. Li and P.M.B. Vitányi. *An introduction to Kolmogorov complexity and its applications*. Addison-Wesley, 1992. To appear.

[14] D.W. Loveland. A variant of the Kolmogorov concept of complexity. *Information and Control*, 15:510–526, 1969.

[15] L. Moser. The second moment method in combinatorial analysis. In *Combinatorial Structures and their applications*, pages 283–384. Gordon and Breach, New York, 1970.

[16] N. Pippenger. An information-theoretic method in combinatorial theory. *J. Comb. Theory, Ser. A*, 23:99–104, 1977.

[17] J. Rissanen. Modeling by the shortest data description. *Automatica-J.IFAC*, 14:465–471, 1978.

[18] R.J. Solomonoff. Complexity-based induction systems: comparisons and convergence theorems. *IEEE Transactions on Information Theory*, IT-24:422–432, 1978.

[19] S. Spiegelman. An in vitro analysis of a replicating molecule. *American Scientist*, 55(3):221–264, 1967.

[20] L.G. Valiant. A theory of the learnable. *Comm. ACM*, 27:1134–1142, 1984.

[21] O. Watanabe, editor. *Kolmogorov complexity and its relation to computational complexity theory*. Springer-Verlag. EATCS Monographs on Theoretical Computer Science, To appear in 1992.

[22] E. Wigner. The unreasonable effectiveness of mathematics in natural sciences. *Comm. Pure Appl. Math*, 13:1, 1960.

[23] C.H. Woo. Laws and boundary conditions. In W.H. Zurek, editor, *Complexity, entropy and the physics of information*, pages 127–135. Addison-Wesley, 1991.

[24] W.H. Zurek. Algorithmic randomness and physical entropy. *Physical Review, series A*, 40(8):4731–4751, 1989.

[25] W.H. Zurek, editor. *Complexity, entropy and the physics of information*. Addison-Wesley, 1991.

Circuit Complexity and the Expressive Power of Generalized First-Order Formulas

Howard Straubing*

Department of Computer Science, Boston College
Chestnut Hill, Massachusetts, USA 02167

Abstract. The circuit complexity classes AC^0, ACC, and CC (also called pure-ACC) can be characterized as the classes of languages definable in certain extensions of first-order logic. All of the known and conjectured inclusions among these classes have been shown to be equivalent to a single conjecture concerning the form of the formulas required to define the regular languages they contain. (The conjecture states, roughly, that when a formula defines a regular language, predicates representing numerical relations on the positions in a string can be replaced by predicates computed by finite state automata.) Here this conjecture is established in a special case: It is shown that the conjecture holds for the subclasses of AC^0, ACC, and CC defined by restricting all the numerical predicates occurring in the logical formulas to be either unary relations, or the order relation $<$.

1 Introduction

Certain formulas of predicate logic can be interpreted in a natural way in strings over a finite alphabet. First-order variables are interpreted as positions in the string, and the first-order predicate symbols are interpreted as either purely numerical relations on the positions, or as asserting that a certain input letter is found at a given position. Higher-order variables, if there are any, are interpreted as ranging over sets of positions or relations on positions. Thus a sentence (a formula without free variables) defines a language—namely the set of all strings that satisfy the sentence—and a class of sentences defines a class of languages. (See the examples in Section 2, where a precise account of the logical formalism is given.)

This viewpoint has been used in at least two distinct ways: The first is the characterization of computational complexity classes in terms of formal logic. For example, Fagin [F], showed that existential second-order sentences define precisely the languages in NP, and Immerman [I] characterized a large number of complexity classes within P in terms of logic. He showed, among other results, that the nonuniform circuit complexity class AC^0, defined by constant-depth, polynomial-size boolean circuits with unbounded fan-in, consists of exactly those languages defined by first-order sentences with arbitrary numerical predicates.

The second use has been in the classification of finite automata: Büchi [Bu] showed that the regular languages are precisely those defined by second-order sentences with the numerical predicate $x < y$ on first-order variables and with monadic second-order variables. McNaughton [MP] showed that the first-order sentences with the numerical

---

* Supported by NSF Grant CCR-8902369

predicate $x < y$ define exactly the 'star-free' regular languages: These can be characterized as those languages represented by an extended regular expression without the use of the $*$ operator, or alternatively as the class of regular languages whose syntactic monoids contain no nontrivial groups. Straubing, Thérien and Thomas [STT] introduced ***modular quantifiers,*** and showed that using these alone or in combination with the ordinary first-order quantifiers and the numerical predicate $x < y$, one defines classes of regular languages that can also be characterized in terms of their associated syntactic monoids.

These two applications of the logical characterization of language classes turn out to be closely related. Barrington and Thérien [BT] have shown that many circuit complexity classes within NC^1, including NC^1 itself, can be characterized as the behavior of polynomial-length families of programs over finite monoids, where different classes of monoids yield different circuit complexity classes. (NC^1 consists of those languages accepted by polynomial-size log-depth families of bounded fan-in boolean circuits. The subclass CC consists of those languages accepted by polynomial-size constant-depth unbounded fan-in families of circuits in which each gate determines whether the sum of its input bits is divisible by q, for some fixed $q > 0$. The subclass ACC is defined similarly, except here one allows both the $MODq$ gates and unbounded fan-in boolean gates.) These programs can be viewed as generalizations of ordinary finite automata, so that, in a sense, the classification of regular languages according to the structure of the associated syntactic monoid extends to languages within NC^1.

In [BCST] these results on programs, together with McNaughton's theorem on regular languages defined with first-order sentences, are used to give a new proof of Immerman's logical characterization of the circuit complexity class AC^0. In [BCST] and [CDPC], the results of Straubing, Thérien and Thomas on modular quantifiers are used to give analogous characterizations of the classes ACC and CC.

At this point an interesting fact emerges: Let A be a finite alphabet. If $\mathcal{Q}$ is a set of quantifiers and $\mathcal{S}$ a set of numerical predicates, then $\mathcal{Q}[\mathcal{S}]$ denotes the class of languages defined by sentences using quantifiers in $\mathcal{Q}$, numerical predicates in $\mathcal{S}$, and, for each input letter $a \in A$, a monadic predicate Q_a that says that the letter in the given position is a. Let $\mathcal{N}$ be the set of all numerical predicates. The characterizations of AC^0, ACC and CC can then be written

$$AC^0 = \{\exists\}[\mathcal{N}],$$

$$CC = \{\exists_q^r : 0 \leq r < q\}[\mathcal{N}],$$

and

$$ACC = (\{\exists\} \cup \{\exists_q^r : 0 \leq r < q\})[\mathcal{N}],$$

where $\{\exists_q^r : 0 \leq r < q\}$ is the set of modular quantifiers of modulus q. Let $\mathcal{P}$ denote the set of numerical predicates that can be obtained by first-order quantification over predicates of the form $x \equiv 0 \pmod r$ and $x < y$. Let $\mathcal{R}$ denote the class of regular languages over A. The characterization of the regular languages in AC^0 given in [BCST] shows

$$\{\exists\}[\mathcal{N}] \cap \mathcal{R} = \{\exists\}[\mathcal{P}].$$

Remarkably enough, this result, which is equivalent to the circuit lower bounds of Furst, Saxe and Sipser [FSS] and Ajtai [A] was conjectured twenty years ago by McNaughton and Papert [MP], before the connection with circuit complexity was appreciated.

Let T be a set of integers, and let MOD_T be the set of modular quantifiers with moduli in T. It is conjectured that for any $\mathcal{Q} = MOD_T$, or $\mathcal{Q} = \{\exists\} \cup MOD_T$,

$$\mathcal{Q}[\mathcal{N}] \cap \mathcal{R} = \mathcal{Q}[\mathcal{P}].$$

This conjecture is equivalent to most of the conjectured properties of the complexity classes AC^0, ACC and CC. For instance, it implies, among other things, that the AND fucntion of n input bits cannot be computed in CC, and that ACC is properly contained in NC^1. Thus one approach to these open problems of circuit complexity is to study the question of the kind of sentences needed to define regular languages, in the hope of directly establishing the conjecture. (See [CDPC] and [BCST].)

In the present paper the conjecture is shown to hold for formulas in which each numerical predicate is either a monadic predicate or the order relation $<$. More precisely, let $\mathcal{M}$ be the class of all monadic numerical predicates. Then for any $\mathcal{Q} = MOD_T$, or $\mathcal{Q} = \{\exists\} \cup MOD_T$,

$$\mathcal{Q}[\{<\} \cup \mathcal{M}] \cap \mathcal{R} \subseteq \mathcal{Q}[\mathcal{P}].$$

The proof of this theorem relies on semigroup-theoretic constructions and previous work of Barrington and Straubing [BS] on the power of linear-size programs over finite monoids. In Section 2 I give formal definitions of the underlying logical apparatus and a precise statement of the main theorem. Section 3 contains the necessary background concerning finite semigroups, and Section 4 the proof of the main theorem.

2 The Logical Apparatus

Let A be a finite alphabet. Informally, the logical formulas used here are built from two kinds of atomic formulas: For each letter $a \in A$ there is a monadic predicate symbol Q_a, where $Q_a x$ is interpreted to mean 'the letter in position x is a'. The second kind of atomic formula is a *numerical predicate*. This can be of any arity (including 0)—the important property of a numerical predicate θ is that $\theta(x_1, \ldots, x_k)$ depends only on the positions x_i and the length of the string in which the formula is interpreted, and not on the letters that appear in those positions. Larger formulas are built from atomic formulas by applying boolean connectives and quantifiers. There are two kinds of quantifiers: the ordinary existential quantifier $\exists$ and modular quantifiers $\exists^r_q$, where $0 \leq r < q$. $\exists^r_q \Phi(x)$ is interpreted to mean 'there exist exactly r mod q positions x such that $\Phi(x)$'. The universal quantifier $\forall$ can of course be defined in terms of $\exists$.

For example, let $A = \{a, b, c\}$. The set of all strings in A^* that contain an occurrence of a before some occurrence of b is defined by the sentence

$$\exists x \exists y (Q_a x \wedge Q_b y \wedge (x < y)).$$

The set of all strings with an odd number of occurrences of the factor ac is defined by the sentence

$$\exists_2^1 x (Q_a x \wedge \exists y (Q_c y \wedge (y = x + 1))).$$

In these two formulas the numerical predicates are the binary predicates $x < y$ and $y = x + 1$. There are several ways to define the language L consisting of all strings of even length. One can write

$$\exists_2^0 x (x = x).$$

Alternatively, one can introduce a monadic numerical predicate $\theta(x)$ asserting that x is an even-numbered position, and write

$$\forall x (\forall y (y \leq x) \longrightarrow \theta(x)),$$

which says that the last position is even. However, the numerical predicates are allowed to depend on the length of the string in which they are interpreted, so there is a perfectly legal 0-ary numerical predicate α with the interpretation 'the length is even', and thus L is also defined by the sentence

$$\alpha.$$

This framework must be defined more precisely for the purpose of proving theorems about formulas. Let $\{x_1, \ldots, x_r\}$ be a set of variables. A *numerical predicate* with free variables $x_1, \ldots, x_r$ is a subset S of $(2^{\{x_1, \ldots, x_r\}})^*$ such that for all

$$v = T_1 \cdots T_k \in S$$

one has

$$T_i \cap T_j = \emptyset,$$

if $i \neq j$, and

$$\bigcup_{i=1}^{k} T_i = \{x_1, \ldots, x_r\}.$$

A *word structure* with free variables $x_1, \ldots, x_r$ is a string

$$w = (a_1, T_1) \cdots (a_k, T_k) \in (A \times 2^{\{x_1, \ldots, x_r\}})^*$$

such that the string

$$T_1 \cdots T_k$$

satisfies the two conditions above. Formulas are built from numerical predicates and the predicates $Q_a x, a \in A$, by applying boolean operations and quantifiers. A formula with free variables $x_1, \ldots, x_r$ is interpreted in a word structure whose free variables include $x_1, \ldots, x_r$. It remains to define the satisfaction relation $\models$. Let w be a word structure and let N be any numerical predicate whose set of free variables is contained in the set of free variables of w. Then

$$w \models N$$

if and only if the string $\hat{w}$, obtained from w by erasing the letters of A and eliminating the variables that are not free variables of N, belongs to N.

$$w \models Q_a x$$

if and only if w includes a symbol (a, T) where $x \in T$. Boolean operations are interpreted in the usual fashion.

$$w \models \exists x \Phi(x)$$

if and only if x is not a free variable of w, and w includes a symbol (a, T) such that if v is the string that results from w upon replacing (a, T) by $(a, T \cup \{x\})$, then

$$v \models \Phi(x).$$

If the number of symbols with this property is congruent to r modulo q, then

$$w \models \exists_q^r x \Phi(x).$$

Let Ψ be a formula with free variables $x_1, \ldots, x_r$. Then L_Ψ denotes the set of all word structures w with these free variables such that $w \models \Psi$. If Ψ is a *sentence*, that is, a formula without free variables, then L_Ψ is a language in A^*.

The family of languages definable by sentences that use only the quantifier $\exists$ is denoted AC^0; the family of languages definable by sentences that use only modular quantifiers of modulus q is denoted $CC(q)$, and the family of languages definable by sentences that use both modular quantifiers of modulus q and $\exists$ is denoted $ACC(q)$. Observe that $AC^0 = ACC(1)$. Of course, these families were originally defined in terms of constant-depth circuits. The equivalence of the circuit definitions with these logical formulations is proved in [I] for AC^0, in [BCST] for $ACC(q)$, and in [CDPC] for $CC(q)$. Let mAC^0, $mACC(q)$, $mCC(q)$ denote the families obtained by restricting to sentences in which the only numerical predicates are the order relation $<$, and 0-ary or unary (*i.e.*, monadic) predicates, that is, predicates with no more than one free variable. In the notation of Section 1,

$$mAC^0 = \{\exists\}[\{<\} \cup \mathcal{M}],$$

and the other new classes are defined analogously. As in Section 1, let $\mathcal{P}$ denote the class of numerical predicates obtained as first-order formulas in numerical predicates of the form $x \equiv 0 \pmod r$ and $x < y$, and let $\mathcal{R}$ denote the class of regular languages over A. The main result of this paper is:

Theorem 2.1. For all $q > 0$,

$$mACC(q) \cap \mathcal{R} \subseteq (\{\exists\} \cup MOD_{\{q\}})[\mathcal{P}],$$

nd

$$mCC(q) \cap \mathcal{R} \subseteq MOD_{\{q\}}[\mathcal{P}].$$

The corresponding assertion for the classes $ACC(q)$ and $CC(q)$ is known to hold when q is a prime power, but is in general, as noted in Section 1, an important open problem in circuit complexity.

The proof of Theorem 2.1 will be given in Section 4. Some notions concerning finite semigroups, necessary to the proof, will be given in the next section.

3 Background on Finite Semigroups

A *semigroup* is a set together with an associative multiplication. A *monoid* is a semigroup with an identity element; the identity element is denoted 1. A subset of a monoid closed under multiplication forms a subsemigroup of the monoid, and this subsemigroup may well be a group. I denote by $\mathbf{G}_q$ the family of all finite groups that are solvable and whose cardinality divides a power of q; and I denote by $\mathbf{M}_q$ the family of all finite monoids M such that all the groups contained in M belong to $\mathbf{G}_q$. The family $\mathbf{M}_1$ thus consists of all finite monoids containing only trivial groups—such monoids are said to be *aperiodic*.

Programs over finite monoids were defined by Barrington and Thérien [BT] and, implicitly, by Barrington [B]. Here we will only consider *single-scan* programs. A single-scan program π of length n over a finite monoid M consists of a subset X of M, called the set of *accepting values*, and a map

$$\phi : A \times \{1, \ldots, n\} \rightarrow M.$$

On input $a_1 \cdots a_n \in A^n$ the program emits the element

$$\phi(a_1, 1) \cdots \phi(a_n, n)$$

of M and accepts the input if and only if this emitted value is in X. (π can be thought of as a sequence of n *instructions*: each instruction emits an element of M depending on the input letter that is read, and the program accepts or rejects depending on the product of the emitted values.) Thus the program accepts a subset $|\pi|$ of A^n. This apparatus will also be used when the input alphabet is one of the extended alphabets $A \times 2^{\{x_1, \ldots, x_r\}}$ described in Section 2. In this case $|\pi|$ is defined to be the set of word structures accepted by π, rather than the set of all strings over the extended alphabet accepted by π.

Ordinarily, one considers *families of programs,* consisting of a program of length n over a fixed monoid M for each $n \geq 0$. The resulting family thus accepts a language $L \subseteq A^*$. An important special case is that in which there is a single set $X \subseteq M$ of accepting values for all programs in the family, and a single map $\psi : A \rightarrow M$ such that $\phi(a, i) = \psi(a)$ for all i and for all programs in the family. In this case ϕ extends to a homomorphism $\hat{\phi} : A^* \rightarrow M$, and the language accepted by the family is $L = \hat{\phi}^{-1}(X)$. This occurs if and only if L is a regular language. To any regular language L one can associate the smallest finite monoid for which such a homomorphism exists: To do this one defines an equivalence relation $\cong_L$ on A^*, b setting $u \cong_L v$ if and only if for all $x, y \in A^*$,

$$xuy \in L \Leftrightarrow xvy \in L.$$

Regularity of L implies that the index of $\cong_L$ is finite, and it is easy to see tha equivalence is compatible with concatenation of strings. Thus the quotient set

$$M(L) = A^* / \cong_L$$

forms a monoid, called the ***syntactic monoid*** of L, and the map η_L that sends eac string w to its $\cong_L$-class is a homomorphism, called the ***syntactic morphism*** of L Observe that $\eta_L^{-1}(\eta_L(L)) = L$. If $L \subseteq A^+$ (the set of nonempty words over A) i is sometimes more convenient to deal with the ***syntactic semigroup*** $S(L)$ of L; thi is simply the image of A^+ under η_L. (See Eilenberg [E] or Pin [P] for a detaile account of the matters discussed in this paragraph.)

Now let M be a finite monoid and K a finite set. A new monoid $\diamond(M, K)$ is define as follows: The underlying set of $\diamond(M, K)$ is $M \times 2^{M \times K \times M}$, and the multiplicatio is given by

$$(m, X)(m', Y) = (mm', mY \cup Xm'),$$

where

$$mY = \{(mm_1, k, m_2) : (m_1, k, m_2) \in Y\},$$

and

$$Xm' = \{(m_1, k, m_2m') : (m_1, k, m_2) \in X\}.$$

It is easy to check that this multiplication is associative, and that $(1, \emptyset)$ is the identity. Thus $\diamond(M, K)$ is a monoid. $\diamond(M, K)$ and $\diamond_q(M, K)$, defined below, are closely related to the ***Schützenberger product*** and related constructions, and share many of the properties of these products. (see [E], [W]).

Let $q > 0$. Let $\mathbf{Z}_q[S]$ denote the free $\mathbf{Z}_q$-module generated by S. The underlying set of $\diamond_q(M, K)$ is $M \times \mathbf{Z}_q[M \times K \times M]$, and the multiplication is given by

$$(m, X)(m', Y) = (mm', mY + Xm'),$$

where left-multiplication by m and right-multiplication by m' are the unique linear transformations defined by

$$m(m_1, k, m_2) = (mm_1, k, m_2), (m_1, k, m_2)m' = (m_1, k, m_2m').$$

In both $\diamond(M, K)$ and $\diamond_q(M, K)$, projection onto the left coordinate is a homomorphism onto M. Consider the restriction of this homomorphism to a group in $\diamond(M, K)$ or $\diamond_q(M, K)$. It is not difficult to show that in the first case the kernel of this homomorphism is trivial, and that in the second case the kernel is an abelian group of exponent q. It is also easy to show that if M is a group then $\diamond_q(M, K)$ is a group. This gives:

Proposition 3.1. Let $q > 0$. If $M \in \mathbf{M}_q$ then $\diamond(M,K) \in \mathbf{M}_q$ and $\diamond_q(M,K) \in \mathbf{M}_q$. If $M \in \mathbf{G}_q$ then $\diamond_q(M,K) \in \mathbf{G}_q$. ∎

Let

$$r_i = (m_i, (1, k_i, 1)), 1 \leq i \leq n,$$

where this is interpreted either as an element of $\diamond(M,K)$, with the second component a singleton set, or as an element of $\diamond_q(M,K)$, with the second component viewed as a sum with a single term. In the proposition below, the expression $m_i \cdots m_j$ is taken to be 1 if $i > j$.

Proposition 3.2. In $\diamond(M,K)$,

$$r_1 \cdots r_n = (m_1 \cdots m_n, \{(m_1 \cdots m_{i-1}, k_i, m_{i+1} \cdots m_n) : 1 \leq i \leq n\}).$$

In $\diamond_k(M,K)$,

$$r_1 \cdots r_n = (m_1 \cdots m_n, \sum_{1 \leq i \leq n} (m_1 \cdots m_{i-1}, k_i, m_{i+1} \cdots m_n)).$$

Proof. This is straightforward by induction on n. ∎

4 Proof of Theorem 2.1

It will first be shown that every language in $mACC(q)$ or $mCC(q)$ is recognized by a family of single-scan programs over a monoid in $\mathbf{M}_q$ or $\mathbf{G}_q$, respectively. I will then apply the results of Barrington and Straubing [BS] concerning the behavior of linear-size programs over finite monoids.

It should be noted that the class $mCC(1)$ is somewhat anomalous, since application of a modular quantifier of modulus 1 gives a formula that is satisfied by all word structures with the right set of free variables. It follows from this that membership in a language in $mCC(1)$ (or, for that matter, in $CC(1)$) depends only on the length of the string, and thus Proposition 4.1 holds trivially. However the argument given below does not apply in this case.

Proposition 4.1. Let Φ be a formula with free variables $x_1, \ldots, x_r$, in which all numerical predicates are monadic, or $x < y$. If Φ uses only modular quantifiers of modulus q, then L_Φ is accepted by a family of single-scan programs over a monoid in $\mathbf{G}_q$. If Φ uses $\exists$ as well, then L_Φ is accepted by a family of single-scan programs over a monoid in $\mathbf{M}_q$.

Proof. The proof is by induction on the construction of Φ. First, if the assertion is true for Φ_1 and Φ_2, then it is true for $\Phi_1 \wedge \Phi_2$ and $\neg\Phi_1$. In the first case, the program is constructed by taking the cartesian product of the two programs over the direct product of the underlying monoids. In the second case, one keeps the same underlying monoid but changes the set of accepting values from X to $M \backslash X$. The conclusion follows from the fact that $\mathbf{M}_q$ and $\mathbf{G}_q$ are closed under direct products. Note that all the other boolean connectives can be expressed in terms of these two.

I now prove the assertion for quantifier-free formulas. By the preceding remarks concerning boolean operations, it suffices to treat each of the three kinds of formulas

$$Q_a x, \theta(x), x < y,$$

where θ is a numerical predicate. Consider the single-scan program π over a nontrivial monoid M that emits $m \neq 1$ on reading (a, S) if $x \in S$, and emits 1 otherwise. Let the accepting set be $\{m\}$; then $|\pi| = L_{Q_a x_i}$. (This is a slight abuse of notation: More precisely, this construction gives a program π for each length input, and the language accepted by the resulting family is $L_{Q_a x_i}$.) Note that the only requirement is that M be nontrivial, so M can be chosen to be in $\mathbf{M}_q$ for any q or in $\mathbf{G}_q$ if $q > 1$. Let ρ be the single-scan program over a nontrivial monoid M such that the jth instruction emits $m \neq 1$ if and only if $\theta(j)$ and the jth input letter (a, S) satisfies $x \in S$, and emits 1 otherwise. Again let the accepting set be $\{m\}$. Then $|\rho| = L_{\theta(x)}$, and, as before, M can be chosen to be in $\mathbf{M}_q$ for any q or in $\mathbf{G}_q$ if $q > 1$. Finally, if M is a nontrivial monoid then the *wreath product* $M \circ M$ is the monoid whose underlying set is $M^M \times M$, with multiplication given by

$$(f_1, m_1)(f_2, m_2) = (F, m_1 m_2),$$

where, for $m \in M$,

$$F(m) = f_1(m) f_2(m m_1).$$

As is well known (see, for example, [E]) $M \circ M$ is in $\mathbf{M}_q$ (respectively $\mathbf{G}_q$) if M is. Pick $m \in M\backslash\{1\}$. Let $I : M \to M$ be the map that sends every element to 1, and let $J : M \to M$ be the map that sends m to m and all other elements to 1. Consider the single-scan program σ over $M \circ M$ that, upon reading (a, S), emits (I, m) if $x \in S$ and $y \notin S$, $(J, 1)$ if $y \in S$ and $x \notin S$, and $(I, 1)$ otherwise. Let v be a word structure whose free variables include x and y. Upon reading v, σ emits the value (K, m), where $K(1) = m$ if $v \models x < y$, and where $K(1) = 1$ otherwise. Thus if the set of accepting values is $\{(K, m) : K(1) = m\}$, the program accepts $L_{x<y}$. Once again, $M \circ M$ can be chosen to be in $\mathbf{M}_q$ for any $q > 0$, and in $\mathbf{G}_q$ for any $q > 1$.

It remains to study the effect of quantification. Let Φ be a formula with free variables $x, x_1, \ldots, x_r$ for which the assertion holds: Then there is a family $\{\pi_n\}$ of single-scan programs over a monoid M that accepts L_Φ. Furthermore, M is in $\mathbf{M}_q$ or $\mathbf{G}_q$, depending on the quantifiers in Φ. Consider now a single-scan program π'_n over $\diamond(M, K)$, where $K = M$. The jth instruction of π'_n, upon reading (a, S), emits $(m, \{1, m', 1)\})$, where m is the value emitted by the jth instruction of π_n upon reading (a, S), and m' is the value emitted by the jth instruction of π_n upon reading $(a, S \cup \{x\})$. Define the set of accepting values to be all pairs (m, Y) such that Y contains a triple (m_1, m_2, m_3) such that $m_1 m_2 m_3$ is an accepting value for π_n. It follows from Proposition 3.2 that the family π'_n accepts $L_{\exists x \Phi}$. A similar construction gives a family of programs over $\diamond_q(M, K)$ that accepts $L_{\exists_q^r x \Phi}$. (In this case the condition on Y is that the sum of the coefficients of the triples (m_1, m_2, m_3), with $m_1 m_2 m_3$ an accepting value for π_n, be equal to r modulo q.) It follows from Proposition 3.1 that these monoids have the required structure. This proves the Proposition.∎

U_1 denotes the monoid $\{0, 1\}$ with the usual multiplication: $0 \cdot 0 = 0 \cdot 1 = 1 \cdot 0 = 0, 1 \cdot 1 = 1$.

ropostion 4.2. Let $L \subseteq A^*$ be a regular language. If $L \in mACC(q)$ then every roup in $M(L)$ is solvable, and for $t \geq 0$, every group in $\eta_L(A^t)$ has order divid-ıg a power of q. Furthermore, if $L \in mCC(q)$, then $\eta_L(A^+)$ contains no monoid omorphic to U_1.

roof. If $M(L)$ contains a nonsolvable group, then by a result in [BCST], there is ›me $t > 0$ such that $\eta_L(A^t)$ contains a nonsolvable group. Thus it suffices to consider nontrivial group $G \subseteq \eta_L(A^t)$. Suppose $L \in mACC(q)$. It follows from Proposition .1 that L is accepted by a family $\{\pi_n\}$ of single-scan programs over some $M \in \mathbf{M}_q$. et $C = \eta_L^{-1}(G) \cap A^t$. C can be viewed as a finite alphabet, and the restriction of $_L$ to C^* gives a homomorphism ψ from C^* onto G. Let e be the identity of G, and $\in G\backslash\{e\}$. It follows from the definition of the syntactic monoid that there exist $, y \in A^*$ such that if $\eta_L(v) = e$ and $\eta_L(w) = g$, then $xuy \in L$ and $xvy \notin L$, or ice-versa. By suitably modifying the first and last instructions of the program π_n ne obtains a family of programs θ_n that accepts the language $\{u : xuy \in L\}$. If n is multiple of t, then the composition of θ_n with the embedding of C into A^t gives a amily of single-scan programs $\{\rho_m\}$ over M with input alphabet C. It follows from heorem 2.2 of [BS] that if $G \notin \mathbf{G}_q$, then there exist strings $v, w \in C^m$ such that $(v) = e, \psi(w) = g$, but ρ_m emits the same value on v and w. It thus follows that here exist $v', w' \in A^{mt}$ such that $\eta_L(v') = e, \eta_L(w') = g$, and $xv'y, xw'y$ are either oth in L, or both in $A^*\backslash L$, a contradiction. So $G \in \mathbf{G}_q$. It remains to consider what appens when $L \in mCC(q)$. In this case, $M \in \mathbf{G}_q$. Observe that if $\eta_L(A^+)$ contains copy

f U_1, then for some $t > 0$, $\eta_L(A^t)$ contains a copy of U_1. (Take $v, w \in A^+$ mapping nto U_1 and consider $v^{|w|}$ and $w^{|v|}$.) The proof now proceeds exactly as above, naking the same appeal to the results of [BS]. ∎

The theorem now follows from arguments given in [BCST] and [CDPC]. In Theorem ' of [BCST] it is shown that if L is a regular language containing no nonsolvable roups, and such that every group in $\eta_L(A^t)$ has order dividing a power of q, then

$$L \in (\{\exists\} \cup MOD_{\{q\}})[\mathcal{P}].$$

n the proof of Theorem 6.3.1 of [CDPC] it is shown that if, additionally, $\eta_L(A^+)$ ontains no copy of U_1, then

$$L \in MOD_{\{q\}}[\mathcal{P}].$$

5 Additional Remarks

The results here imply that $mACC$ (the union of the $mACC(q)$ over all $q > 0$) is properly contained in NC^1, and that mCC (the union of the $mCC(q)$) does not contain the AND function. In fact, these corollaries follow directly from Proposition 4.1 and the results of [BS]; the additional work done in Proposition 4.2 to characterize all the regular languages in these classes is not needed.

It should be noted that *equality,* rather than just inclusion, holds for the first part of Theorem 2.1, for the simple reason that the predicates $x \equiv 0 \pmod r$ are themselves monadic. Since ordinary first-order quantification over $x \equiv 0$ and $x < y$ is required to form the numerical predicates in $\mathcal{P}$, it is not clear whether equality holds for the second part as well. A more interesting question is whether the quantifier complexity is preserved in passing from the arbitrary monadic sentence to the sentence with numerical predicates in $\mathcal{P}$. There are strong indications that this is true, but I have not verified this.

Of course, the principal open question is whether one can establish these results for numerical predicates of arbitrary arity. Even if this can be done for dyadic numerical predicates the result will be quite significant: The natural uniform versions of the classes $ACC(q)$ and $CC(q)$ are defined using the single numerical predicate $BIT(x,y)$, which is interpreted to mean the xth bit of the binary representation of y is on [BIS]. Thus an extension of Theorem 2.1 to the dyadic case would resolve most of the open questions about the relations among AC^0, CC and ACC in the uniform setting. My suspicion is that the monadic case is very special, and that any extension beyond this will encounter all of the difficulties present in the most general nonuniform setting.

References

[A] M. Ajtai, Σ_1^1 formulae on finite structures, *Annals of Pure and Applied Logic* **24** (1983), 1-48.

[B] D. Barrington, Bounded-width polynomial-size branching programs recognize exactly those languages in NC^1, *J. Comp. Syst. Sci.,* **38** (1989), 150-164.

[Bu] J. Büchi, Weak second-order arithmetic and finite automata, *Z. Math. Math Logik Grundlagen Math.* **6**, (1960), 66-92.

[BCST] D. Barrington, K. Compton, H. Straubing, and D. Thérien, Regular languages in NC^1, to appear in *J. Comp. Syst. Sci.*

[BIS] D. Barrington, N. Immerman and H. Straubing, On uniformity in NC^1, *J. Comp. Syst. Sci.,* **41** (1990), 274-306.

[BS] D. Barrington and H. Straubing, Superlinear lower bounds for bounded-width branching programs, in *Proc. 6th IEEE Structure in Complexity Theory Conference* (1991) 305-314; to appear in *J. Comp. Syst. Sci.*

[BT] D. Barrington and D. Thérien, Finite monoids and the fine structure of NC^1, *JACM* **35**, (1988), 941-952.

[CDPC] H. Straubing, Constant-depth periodic circuits, *International J. Algebra and Computation,* **1** (1991), 49-87.

[E] S. Eilenberg, *Automata, Languages and Machines, vol. B,* Academic Press, New York (1976).

[F] R. Fagin, Generalized first-order spectral and polynomial-time recognizable sets, *SIAM-AMS Proceedings,* vol. 7, American Mathematical Society, Providence (1974).

[FSS] M. Furst, J. Saxe and M. Sipser, Parity, circuits, and the polynomial time hierarchy, *J. Math Systems Theory* **17**, (1984), 13-27.

[I] N. Immerman, Languages that capture complexity classes, *SIAM J. Computing* **16**, (1987), 760-778.

[MP] R. McNaughton and S. Papert, *Counter-free Automata,* MIT Press, Cambridge, Massachusetts (1971).

?] J. E. Pin, *Varieties of Formal Languages*, Plenum, New York (1986).

STT] H. Straubing, D. Thérien and W. Thomas, Regular languages defined with eneralized quantifiers, Proc. 15th ICALP, *Lecture Notes in Computer Science*, **317**, 61-575 (1988).

W] P. Weil, Products of languages with counter, *Theoretical Computer Science*, **76** 1990).

One-Message Statistical Zero-Knowledge Proofs and Space-Bounded Verifier

Alfredo De Santis* **Giuseppe Persiano*†** **Moti Yung‡**

Extended Abstract

Abstract

Traditional mathematical proofs are communicated by a paper or a book as a single message. The reader has to understand the proof and be persuaded by it without any interaction or coordination with the writer of the proof. Statements whose proofs can be written down as a single message and then verified efficiently constitute the class NP.

In modern communication settings, such as distributed and cryptographic scenarios, hiding some or all of the details of a proof while convincing the verifier of the validity of a statement is an essential operation. This is captured by the important concept of *Zero-Knowledge Proofs* which deals with characterizations and techniques of how to convince a polynomial-time machines that a statement is true without giving *any* clue about the proof.

All NP statements have a zero-knowledge proof but, unlike written proofs, these proofs require either interaction between the verifier and the prover [16], or agreement between the verifier and the prover on some initial truly random string independent of the theorem [1, 2] and thus are not "one-message" proofs. In fact, it has been shown [18] that only languages in BPP have "one-message proofs" and thus, if NP had one-message proofs, the seemingly very unlikely conclusion would be that NP$\subseteq$BPP. However, since the traditional one-message proofs minimize coordination and interaction, it is important to characterize when it is still possible to have *meaningful* one-message zero-knowledge proofs for NP.

In this work we show that when a bound on the space of the verifier is known (which is applicable in various computational settings), then all NP statements have *one-message zero-knowledge proofs.* Furthermore, our one-message proofs are *perfect* zero-knowledge, which without the space bound would imply the unlikely event that PH=Σ_2 [3, 11].

---

*Dipartimento Informatica ed Applicazioni, Università di Salerno, 84081 Baronissi (SA), Italy. This work was partially supported by the Italian Ministry of the University and Scientific Research and by CNR Grant # 91.02326.CT12.

†Aiken Computation Laboratory, Harvard University, Cambridge MA, USA. Partially supported by NSF Grant # NSF-CCR-90-07677.

‡IBM Research Division, T.J. Watson Research Center, Yorktown Heights NY, USA.

1 Introduction

Traditional mathematical proofs are communicated by one message in a paper, in a book, or, to some extent, in an extended abstract. The verifying reader can read these full-information proofs, and become convinced without the need to interact or further coordinate actions with the writer of the proof. Theorems whose proofs can be verified in polynomial-time exactly characterize the class NP. In recent years the elegant theory of *Zero-Knowledge Proofs* (introduced by [15] and further developed by [16]) has dealt with the concept of a proof that convinces a polynomial-time verifier of the validity of the statement at hand without releasing any additional useful information about the proof. However, two main drawbacks afflict the proposed zero-knowledge proofs for NP statements.

All current zero-knowledge proofs for NP statements assume that the parties are coordinated: they are able either *to interact* or to actually *share access* to a public resource. These common actions/resources may not always be available practically, while, from a theoretical point of view, it is important to classify when they are indeed necessary. The importance of a non-interactive zero-knowledge proof systems has been first recognized by [2] (where the common shared string was suggested; see also [1]), and since then many papers have dealt with such issues (see [6, 7, 5, 8, 10]) but none of the proof systems proposed can be really considered to give "one-message" non-interactive zero-knowledge proofs *from scratch* as they all rely on previously shared public resources. In fact, as proved by [18] the above additional measures besides the one message sent in such proofs seem necessary unless NP$\subseteq$BPP.

In the original model of [15], zero-knowledge proofs for NP statements [16, 1] are *computational zero-knowledge*. That is, they are shown to be zero-knowledge only under the conjecture that one-way functions exist. This seems to be quite unfortunate as it would be important to have *perfect zero-knowledge* proofs (see [16]); that is proofs whose security does not depend upon any unproven complexity assumption like the existence of one-way functions. Unfortunately, the eventuality that all NP statements have perfect (or even almost perfect) zero-knowledge seems very unlikely. In fact, results of [3] and [11] have shown that a perfect zero-knowledge proof for an NP-complete language would cause the polynomial hierarchy to collapse to its second level, which is widely believed to be unlikely.

Summary of Results. In this work we propose to study the case in which the space available to the verifier is *known*, and otherwise maintain all the properties of the usual model of [15]; more precisely, we assume *space-bounded polynomial-time verifier*, where the space $s = s(n) \geq c \log n$ is known (up to some constant multiplicative factor) to the prover. We successfully address the question of getting rid of the two crucial drawbacks of the usual model: interaction/coordination (which seem needed in the original model), and the need to rely on computational complexity assumptions (which, again, seem needed in the usual model). Here are the main results:

- We construct a *"one-message" proof system* for NP which is *statistical zero-knowledge*. The proof system we design provides proofs for all NP statements. This is *optimal* from a language recognition perspective.

- Our proof system only requires a polynomial (in the size of the problem and the verifier's space) amount of communication as well as polynomial internal verifier computation. This gives to our results a potential applicability to practical situations.
- An important property of our proof systems is that it is sufficient that the (approximately) known space-bound holds *only* for the period of time in which the proof is released. That is, once the verifier has seen the proof, its entire internal state cannot give significant computational advantage to *any* machine (even one with a very large memory and a larger time-bound) which tries to extract information about the assertion. This important property, which we call *Off-Line Zero-Knowledge*, does not hold for the regular zero-knowledge proofs (where a machine that can invert one-way functions can get information about the proof).
- We show that our zero-knowledge proof is also a proof of knowledge, thus making our result applicable to identification and authentication problems.

2 Model and definitions

In this section we give a formal definition for the notion of $s(n)$ known-space zero-knowledge proof system. We start by defining the special types of Turing machine that will model the prover and the verifier.

Definition 1. A *Speaking Turing Machine* is a deterministic Turing machine with five tapes: a two-way read-only *Input* tape, a two-way read-only *Auxiliary* tape, a one-way read-only *Random* tape, a two-way read/write work tape, and a one-way write-only *Communication* tape.
Similarly, a *Listening Turing Machine* is a deterministic Turing machine with four tapes: a two-way read-only *Input* tape, a one-way read-only *Random* tape (modeling coin-flips), a two-way read/write work tape, and a one-way read-only *Communication* tape.
A speaking Turing machine P and a listening Turing machine V constitute a *speaker-listener pair* if they receive the same string as input, P's write-only communication tape is V's read-only communication tape, and both P and V have their respective write or read heads on the same cell of the communication tape.

We stress that the communication in a speaker-listener pair (P, V) models "one-message" systems (in the sequel, we omit one-message, since all our systems have this property), that is from P to V only. In fact, V has no mean of "talking" to P as it can only read and not write the communication tape. The communication tape can only be written and read in a one-way fashion. This property does not limit the generality of the definition as, if the space of V is not limited, V can copy the content of the communication tape onto his work tape and then access it in a two-way fashion. On the other hand, when the work space of V is limited, it is important that either the communication tape be readable in a one-way fashion only, or alternatively, that P does not use the entire tape, but rather rewind, erase, and rewrite from the first cell. (Same holds for the one-way random tapes which actually models coin flips).

efinition 2. Let $s(n) : \mathcal{N} \rightarrow \mathcal{N}$ be any integer function. A speaker-listener pair $P, V)$ constitutes an $s(n)$ *known-space speaker-listener* pair if, whenever a string x given as input to P and V, P receives as an auxiliary input the value $s(|x|)$ and can use at most $s(|x|)$ cells on its work tape.

The computation of a $s(n)$ known-space speaker-listener pair (P, V) on input he string x takes place as follows. P computes and writes the string m_x on the ommunication tape (m_x has length polynomial in the length of the input and he verifier's space). After V reads the message and finishes computing, it enters ither an ACCEPT state or a REJECT state, and then we say that the pair (P, V) ACCEPTS or REJECTS, respectively. We denote by $(\mathrm{P} \rightarrow \mathrm{V})(x)$ the random ariable that associates to ACCEPT (REJECT) the probability (taken over the andom tapes of P and V) that (P, V) accepts (respectively, rejects) the input x.

We are now ready to define the notion of proof systems in the known-space nodel. We follow the definition of proof systems given by [15]. A proof system is a game between two parties, a prover P and a verifier V. The goal of the prover s to convince the verifier of the validity of an assertion given as input (i.e., that a ertain string belongs to a language L). The prover, who possesses a validity proof or is able to compute such a proof, should be able to convince the verifier with very high probability that indeed the true assertion holds. On the other hand, no prover $\tilde{P}$ will be able to use the proof-system in order to convince the verifier, with some non-negligible probability, that a false assertion is valid. In the known-space model the prover and the verifier are modeled by a pair of known-space speaker-listener. Formally:

Definition 3. A $s(n)$ known-space speaker-listener pair (P, V) constitutes a $s(n)$ *known-space proof system* for the language L if

1. COMPLETENESS. For all constants c and for all sufficiently large $x \in L$,

$$\mathbf{Pr}\Big((\mathrm{P} \rightarrow \mathrm{V})(x) = \mathrm{ACCEPT}\Big) > 1 - \frac{1}{|x|^c}.$$

2. SOUNDNESS. For all speaking machines $\tilde{P}$, for all constants c and for all sufficiently large $x \notin L$

$$\mathbf{Pr}\left((\tilde{\mathrm{P}} \rightarrow \mathrm{V})(x) = \mathrm{ACCEPT}\right) < \frac{1}{|x|^c}.$$

In a $s(n)$ *known-space proof system* (P, V), we call P the *prover* and V the *verifier*.

We want next to capture the fact that the proof given by P to V does not leak any information beyond the validation; that is the proof is zero-knowledge. This has been formalized by [15] who introduced the concept of a simulator. A simulator for a proof system (P, V) is an efficient algorithm that, given as input a string $x \in L$ but without interacting with P, outputs a random variable which is *computationally* indistinguishable from what the verifier has seen while interacting with the prover. Proof systems that admit such a simulator are called *computational zero-knowledge*.

A stronger notion is obtained when the simulator is required to output a random variable which is statistically close to the view of the verifier in which case we obtain a statistical zero-knowledge proof system.

In this paper we formalize the concept of statistical zero-knowledge in a direct way using Shannon information-theoretic definition of security. For a $s(n)$ known-space proof system (P, V), we consider an adversary machine $\mathcal{A}$ which tries to compute a certain function about the input. This machine has an extra tape which is either (1) empty or (2) contains information obtained from the verifier. We say that P is zero-knowledge if $\mathcal{A}$ has approximately the same probability of success in computing the function in both cases. We remark that this definition captures the fact that, at any moment of the computation (including at the end of the proof), with very high probability, the information available to the verifier does not give any computational advantage in computing any function about the input assertion; in fact, with overwhelming probability it does not give any information (in the Shannon sense). This captures the *off-line zero-knowledge* property as well, as the zero-knowledge property holds after the interaction is finished. Our definition is similar to the definition of semantic security introduced in the context of secure public-key cryptosystems by Goldwasser and Micali [14]. In the original definition of zero-knowledge [15] the definition is computational and the fact that the state of the verifier cannot give advantage to a polynomial-time machine was technically formalized by showing that its state can also be reconstructed via a simulation of the actual interaction, in which only efficient parties participate.

We refer the reader to the full version of the paper for a formal definition of Zero-Knowledge in our setting.

Notation. If x and y are binary vectors of the same length, then $x \odot y$ denotes the inner product of x and y. If c and m are binary string of the same length, then $c \oplus m$ denotes the bitwise XOR of c and m.

3 The basic $s(n)$ known-space zero-knowledge proof system

In this section we present an $s(n)$ known-space statistical zero-knowledge proof system proof-system for the NP-complete language 3Col4Deg of the 3-colorable graphs with maximum vertex degree four [13] (see also [12] pp. 85–86). We start with an informal account of the protocol. In the next section, we show how to get a $s(n)$ known-space statistical zero-knowledge proof system for all languages in NP, for $s(n) > c \log n$, where c is a constant depending only on the language.

Let G be a 3-colorable graph with maximum vertex degree four over the set of vertices $\{1, 2, \cdots, n\}$. The prover possesses a 3-coloring $C: \{1, 2, \cdots, n\} \rightarrow \{1, 2, 3\}$ of G and the verifier has memory $2s = s(n)$. The prover repeats the following operations n^2 times, each time using independent random decisions.

First, the prover selects a random permutation Π of the colors $\{1, 2, 3\}$. This yields a new coloring $\hat{C}$ for G in the following natural way. If $C(v)$ is the color of v in C, then $\Pi(C(v)) = \hat{C}(v)$ is v's new color in $\hat{C}$ (so far, similar to [16]). Then, the following *encoding scheme* is used. For each vertex v, the prover chooses a random 2-bit string m_v. The string m_v is called the *mask* of v and will be used to hide v's color by xoring m_v. Next, for each vertex v, the prover construct the *information block* IB_v of v. IB_v contains the mask of the $d(v) \leq 4$ vertices adjacent to v along with v's color masked. This encoding scheme has the property that two information blocks IB_u and IB_v relative to two non adjacent vertices v and u give no information about the colors of v and u. On the other hand, if u and v are adjacent, then it is possible to compute their colors. In fact, IB_v contains the mask of u and IB_u contains the mask of v. Once the information blocks have been computed, the prover transfers them to the verifier so that the verifier can read at most two information blocks. This is achieved in the following way. For each information block $IB_v = b_{1v} \cdots b_{r_v v}$ ($r_v \leq 10$), the prover chooses $r+1$ vectors of length s, $a_v, c_{1v}, \cdots, c_{r_v v}$ and computes bits $d_{1v}, \cdots, d_{r_v v}$ such that

$$b_{iv} = (a_v \odot c_{1v}) \oplus d_{1v}, \qquad i = 1, 2, ..., r.$$

Next, the vectors are transmitted in the following way. The prover sends (writes on the communication tape) first a_v for $v = 1, 2, \cdots, n$, then c_{iv} for $v = 1, 2, \cdots, n$ and $i = 1, \cdots, r$, and, finally, then d_{iv} for $v = 1, 2, \cdots, n$ and $i = 1, \cdots, r$.

The verifier, for each of the n^2 iterations, chooses one of two challenges. One is for pairs (x, y) of neighbors and the other is for pairs (u, v) of nodes with a common neighbor (i.e., nodes at distance 2). The first challenge is to check that the graph has been properly colored. This is accomplished by computing IB_x and IB_y from which he reconstructs the two colors and checks for validity and inequality. The second challenge is to verify that a consistent mask is used in the entire neighborhood of a vertex. Given IB_v and IB_u, the verifier can check that the masks corresponding to the common neighbor are identical.

Next we argue this system in detail, and then prove its properties.

Input: A 3-colorable graph $G = (\{1, \ldots, n\}, E)$ with maximum degree four.
Additional Input to P: a 3-coloring $C(\cdot)$ of G and $s(|G|) = 2s$ the amount of space available to V.

P's program

Repeat n^2 times the following Phase

1. Choose a random permutation Π of $\{1, 2, 3\}$ and set $\hat{C}(v) = \Pi(c_v)$ (a color is represented in binary).
   For each vertex v, randomly choose a 2-bit mask m_v.
2. Construct blocks of information.
   For each vertex v, let $v_1, \ldots v_{r_v}$ be v's neighbors.
   Block of information for v, IB_v, is the string $(\hat{C}(v) \oplus m_v) \circ m_{v_1} \circ \ldots \circ m_{v_r}$.
3. Encoding of Information for Memory $2s$.
   For each vertex v
   Let $\text{IB}_v = b_{1v} \ldots b_{r_v v}$.
   Randomly choose $r + 1$ s-bit vectors $a_v, c_{1v}, \ldots, c_{rv}$ and compute bits $d_{1v}, \ldots, d_{rv}$ such that $b_{iv} = (a_v \odot c_{iv}) \oplus d_{iv}$ $\quad i = 1, \ldots, r$.
4. Transfer of Information.
   For each vertex v send $(\#, v, a_v, \#)$.
   For each vertex v send $(\#, v, c_{1v}, c_{2v}, \ldots, c_{rv}, \#)$.
   For each vertex v send $(\#, v, d_{1v}, d_{2v}, \ldots, d_{rv}, \#)$.

V's program

For each phase V randomly chooses between two challenges:

1. Verify coloring is proper.
   Choose an edge (x, y) and store in memory a_x and a_y.
   Compute (on-line) and store the bits
   $$a_x \odot c_{ix}, \qquad i = 1, \ldots, r_x$$
   $$a_y \odot c_{iy}, \qquad i = 1, \ldots, r_y.$$
   Compute (on-line) IB_x and IB_y.
   Compute $\hat{C}(x)$ and $\hat{C}(y)$ and check they are different and in $\{1, 2, 3\}$.
2. Verify coloring is well defined.
   Choose two vertices u, v with common neighbor t, that is $t = v_i = u_j$ for some i, j, and store in memory a_u and a_v.
   Compute (on-line) and store the bits
   $$a_u \odot c_{iu}, \qquad i = 1, \ldots, r_u$$
   $$a_v \odot c_{iv}, \qquad i = 1, \ldots, r_v.$$
   Compute (on-line) IB_u and IB_v.
   Check that $m_{v_i} = m_{u_j}$.

If all checks are successfully passed then ACCEPT else REJECT.

Remark 1: Counting. In order to make the probability that a cheating prover succeeds in proving a false assertion negligible, the prover and verifier have to go through $\Theta(n^2)$ independent iterations of the basic protocol and therefore the verifier needs to have space $s(n) > 2 \log n$ for counting. In the final paper we will prove that a cheating verifier does not obtain any information even if he uses the space devoted for counting to store vectors a_x.

Alternatively, we could count by adding input heads. We provide the verifier with two extra read-only heads on the input tape. Each time one iteration is successfully completed, the first head is moved one step to the right. When it reaches the end of the tape, it goes back to the first position in the tape and the second one is moved one step to the right. When both heads are at the end of the tape, $|G|^2$ iterations have been completed. (In practical applications, of course, this mechanism can be replaced by a built-in counter as part of the verifier hardware).

Remark 2: How to make random decisions. V's only source of randomness is a *one-way read-only* random tape. The random choice of a vertex is usually accomplished by copying on the work tape the next $\lceil \log |V| \rceil$ bits of the random tape but, due to its limited space, the Verifier has to make its choice on-line. That is, to randomly choose a vertex, one input head is moved to the beginning of input tape. Then vertices' names are compared bit by bit with the content of the random tape. At the first mismatch, the next vertex is matched against the next bits on the random tape. If there is no mismatch, then the current vertex is selected. This selection procedure does not give each vertex uniform probability. In fact, for sake of simplicity assume $n = |V|$ to be a power of 2; then, due to the sequential selection process, the i-th vertex has probability $p_i = 1/n - 1/n^2 + 1/n^3 - \ldots (-1)^{i+1} 1/n^i$ of being selected. Thus for large enough n, $1/(2n) \leq p_i \leq 1/n$ for all i (in case n is not a power of 2, then $1/(4n) \leq p_i \leq 2/n$). This means vertices are chosen almost uniformly at random. For challenge 1, the Verifier first chooses a vertex and then chooses one of its neighbor at random, which is enough for our purposes. For hallenge 2, the Verifier chooses one vertex and then two of its neighbors at random. The verifier can also verify that indeed, the graph as a whole is being encoded, by scanning the input against the vertices and edges sent).

3.1 (P, V) is a $s(n)$ zero-knowledge known-space proof system.

We start by proving that the pair (P, V) is a $s(n)$ known-space proof system for the language 3COL4DEG. This proof resemble the one in [16].

Theorem 1. (P, V) is a $s(n)$ known-space proof system for 3COL4DEG.

Proof. It is easy to see that the Completeness holds. Let us now check if the Soundness requirement is met. Suppose that the graph G is not 3-colorable. Then at each iteration at least one of the following two must hold:

- There is at least one edge (x, y) not properly colored (two ends are equal or a non-valid color is used).
- There are at least two vertices u, v with common neighbor t, for which t has different masks in the IB_v and IB_u.

Therefore, in each case P is caught with probability at leat $1/(4n)$ (see remark above) and thus the probability that he is not caught for n^2 iterations is at most

$$\left(1 - \frac{1}{4n}\right)^{n^2} \approx e^{-n/4},$$

which is negligible. ∎

Next we prove the zero-knowledge property.

The proof of the zero-knowledge property is based on the following arguments. If the verifier stores in memory information which is only function of at most two nodes x, y, then he gets no information on the coloring. However, because of the space constraint, no verifier can recover information about more than two vertices.

Let us proceed more formally.

The proof given by the prover P to a verifier that a certain graph G is 3-colorable can be naturally divided into n^2 stages. Within each stage, for each vertex x of G, we call the strings $a_x, c_{1x}, c_{2x}, \cdots, d_{1x}, \cdots$, as the *strings* of x. We say that a verifier V is *well-behaved* if for any t, there exists two vertices x and y of the input graph such that there exists an algorithm that on input the strings of x and y outputs the random variable of the content of the work tape of V at time t; i.e., the content of the work tape of V at time t depends only on the strings of two vertices.

Lemma 1. The prover P is zero-knowledge with respect to a well-behaved verifier.

Proof's sketch. The lemma is proved by observing that there exists an algorithm that, given two vertices x and y of a graph $G \in$3COL4DEG, outputs the strings of x and y with the right probability. In fact, we have the following cases. If x, y are at distance 1 (i.e. adjacent), then the strings of x and y are completely random strings with the only constraint that they define two information blocks which give two different colors for x and y.

If x, y are at distance 2, then their strings are completely random with the only constraint that they define two information blocks that are completely random given that one random mask (relative to the common node) is equal in both information blocks.

If x, y are at distance greater than 2, then the information blocks are independent uniformly distributed random variables. ∎

The above cases, however, may not capture the behavior of a cheating verifier. A cheating verifier might be able to compress the vectors a_v and therefore manage to compute more than two IB's. This will certainly give him some information about the coloring, but we will prove that he has negligible probability of success. This proof relies on the following lower bound on the communication complexity of the inner product function $x \odot y$ (see the work by Chor and Goldreich, and Orlitsky and El Gamal [4, 19]).

Theorem 2. Let A and B have one random s-bit vector each, x and y, respectively. Then, any randomized two-party protocol which correctly computes the inner product $x \odot y$ with probability greater than $1/2+\epsilon$ has communication complexity of at least $\psi(s,\epsilon) = s - 3 - 3\log_2 \epsilon^{-1}$ bits (that must be exchanged).

The above theorem says that essentially s bits have to be exchanged in order to compute the inner product of two s-bit vectors. The following lemma constitutes that core of the proof of the zero-knowledge property. Let us consider the following situation. We want to compute the inner product $x_i \odot y_i$ for $i = 1, 2, 3$. The vectors are given in input in the following way: first x_1, x_2, x_3 and then y_1, y_2, y_3. We ask how much space any algorithm that computes the three inner products with a significant advantage needs to use. The next lemma says that there exists no way of significantly compressing the vectors x_i's so that the three inner products $x_i \odot y_i$ can be computed with probability greater than $1/2 + s^{-d}$. Therefore, any algorithm that correctly computes the three inner products with a significant advantage has to store in memory the vectors x_i and thus uses essentially memory $3s$.

Lemma 2. Let $c < 3$ and $d > 0$ be constants, $\mathcal{A}$ be any compression algorithm that on input a $3s$-bit string outputs a cs-bit string, and $x_1, y_1, x_2, y_2, x_3, y_3$ be randomly chosen vectors of length s. Then, for all sufficiently large s, there exists no algorithm $\mathcal{B}$ that, on input $\mathcal{A}(x_1, x_2, x_3)$ and y_1, y_2, y_3 computes each inner product $x_i \odot y_i$, $i = 1, 2, 3$, with probability at least $1/2 + s^{-d}$.

Proof. Use Theorem 2.

Theorem 3. (P, V) is a $s(n)$ known-space statistical zero-knowledge proof system for 3Col4Deg.

Proof's sketch. Without loss of generality, let us consider the configuration of the machine at time when all the vectors a_v have been read and assume that the machine is in one fixed state. Moreover, we also assume, without loss of generality, that at this point the content of the space is a function only of the bits seen in the current iteration (recall that the information sent by the verifier at each iterations is independent from previous ones). At this point the content of the memory can be seen as the output of an algorithm $\mathcal{A}$ on input vectors $a_{i_1}, \ldots, a_{i_k}$ ($k \leq n$) sent in the current iteration. We show that from this information it is not possible to compute the inner product of more than two vectors a_v with random vectors $c_v(i)$'s and therefore reconstruct more than two IB's. For sake of contradiction, suppose that there is an algorithm $\mathcal{A}$ such that from $\mathcal{A}(a_{i_1}, \ldots, a_{i_k})$ and random vectors $u_{i_1}, \ldots, u_{i_k}$ is possible to compute $a_{i_j} \odot u_{i_j}, a_{i_l} \odot u_{i_l}$, and $a_{i_m} \odot u_{i_m}$ each with probability $\geq 1/2 + s^{-d}$ for some constant $d > 0$ and $1 \leq j, l, m \leq k$. As the Verifier has memory $2s$, it must be the case that $|\mathcal{A}(a_{i_1},\dots,a_{i_k})| \leq 2s$ and therefore we contradict Lemma 2. Thus, with very high probability, in any of the iterations the Verifier can compute no more than two IB's which, as we have already seen, do not give any information (in the Shannon sense) about the 3-coloring. ∎

We note that our proof system remains Zero-Knowledge even if the space is $(3-\epsilon)s$ rather than $2s$, for some $\epsilon > 0$. It is possible to slightly modify the proof system so to maintain the zero-knowledge property even if the space is $(4-\epsilon)s$ for some $\epsilon > 0$.

Open question: An interesting question is whether it is possible to give a larger range of possible space for the verifier, and still maintain the properties of proof-systems and zero-knowledge-ness.

3.2 A $s(n)$ known-space statistical ZKP for all NP

In this section we argue that an NP statement can be proved in statistical zero-knowledge in the $s(n)$ known-space model. As the language 3COL4DEG is NP-complete, all is left to do is to prove that the public reductions from a generic NP statement to an instance of 3COL4DEG can be (probabilistically) checked by a Verifier with limited space.

Informally speaking, this is possible since the reduction operates locally and therefore it is not necessary to entirely store it in memory in order to check its correctness. Instead, the verifier randomly chooses which part of the reduction to check and does the checking in an on-line fashion.

More details on the following theorem will be given in the final version.

4 One-Message Proof of Knowledge

The concept of Zero-Knowledge Proof of Knowledge, implicit in the first works on zero-knowledge [15], was formally introduced by [9, 21]. It is a two person protocol by which one, the Prover, can convince the other, the Verifier, that he *knows* the proof of a certain statement without revealing it. The most direct cryptographic application of such a protocol is identification. In interactive settings, Zero-Knowledge Proof (of Language Membership) are also Proof of Knowledge but in the non-interactive scenario of [2, 1] things are a little more difficult to formalize (see [5]).

In this section we show that the zero-knowledge proofs of membership of the previous sections are also proofs of knowledge. Informally speaking, we say that the Prover gives to the Verifier a zero-knowledge proof of knowledge of the validity of a certain proposition if a polynomial-time machine, for which the space bound does not hold, can extract the proof of the proposition from the Zero-Knowledge Proof that Prover has provided to the Verifier.

Formally, we can consider the following scenario. Let (P,V) be a $s(n)$ known-space speaker-listener pair (P,V) and let $\mathcal{P}$ be any polynomial-time predicate. On input x, P wants to convince that he knows ω for which $\mathcal{P}(x,\omega)$ holds.

Definition 4. A $s(n)$ known-space speaker-listener pair (P,V) constitutes a $s(n)$ *known-space zero-knowledge proof of knowledge* for the language L if (P,V) is a $s(n)$

known-space zero-knowledge proof system for the NP language

$$\{x: \exists \omega_x \text{ s.t. } \mathcal{P}(x, \omega_x) = 1\},$$

and there exists a machine E (for which the space bound needs not to hold) running (interactively with the prover) in time polynomial in the verifier's space (n) and the input size n such that for any machine $\tilde{P}$, any constant c and long enough input x

$$\mathbf{Pr}\left((\tilde{P} \rightarrow V)(x) = \text{ ACCEPT and } (\tilde{P} \leftrightarrow E)(x) \neq \omega_x\right) < |x|^{-c},$$

where the writing "$(\tilde{P} \leftrightarrow E)(x)$" denotes the random variable of E's output after interacting with $\tilde{P}$ on input x.

The next theorem shows that the pair (P, V) of Section 3 is a $s(n)$ known-space statistical zero-knowledge proof for the predicate $\mathcal{P}$ defined as follows

$$\mathcal{P}(G, C) = 1 \iff C \text{ is a 3-coloring of } G$$

This in turn will imply that any NP predicate has a $s(n)$ known-space statistical zero-knowledge proof of knowledge.

Theorem 4. (P, V) is a $s(n)$ known-space zero-knowledge proof of knowledge for all polynomial time predicate.

Proof. All is left to prove is that condition 2 of definition is satisfied. At this aim, we exhibit a polynomial time machine E that on input a graph G and the entire proof provided by $\tilde{P}$ to V outputs a coloring of the graph. Let us consider the machine E that stores in his memory the all the string sent by the Prover, and then computes the coloring. If $\tilde{P}$ convinces V then with a non negligible probability, then all the vertices of the graph are properly colored. Therefore the IB in their totality will provide a coloring for G. ▮

5 Practical Applications

The proof above relies on the fact that the verifier is unable to get the entire message into its memory. There are a few environments in which the difference of bandwidth of entities, may use the idea above in practical scenarios. For example, users which are small smart cards of a known memory size may be the verifiers, or in a network where a message is written into a bounded-size buffer, the nodes may be the verifiers.

One attractive scenario is a network with multiple-access. The prover can send the message as many parallel packets into the media which cannot "accept" all the information. In fact, indeed some communication media resolve the collision of this flooding by choosing an arbitrary packet (either physically, or by selecting a channel among various communication channels). This is, by the way, analogous to "arbitrary CRCW-PRAM". The process is repeated twice for each phase (so only two records can be captured). In this environment we must have $\Omega(n^3)$ phases, since only about $1/n$ of the random checks are meaningful.

In such practical environments, it is important that the proof does not loses its properties if repeated more times. In particular, this also enables the prover and he verifier to carry out the proof without synchronization. As long as the verifier starts listening early enough, the secure proof will be transferred.

Acknowledgements. We would like to thank Oded Goldreich for his comments on an early version of this paper and Ueli Maurer for his suggestions.

References

[1] M. Blum, A. De Santis, S. Micali, and G. Persiano, *Non-Interactive Zero-Knowledge, SIAM Journal on Computing*, vol. 6, December 1991, pp. 1084–1118. (Preliminary version: MIT Research Report MIT/LCS/TM-430, May 1990.)

[2] M. Blum, P. Feldman, and S. Micali, *Non-Interactive Zero-Knowledge Proof Systems and Applications*, Proceedings of the 20th Annual ACM Symposium on Theory of Computing, Chicago, Illinois, 1988.

[3] R. Boppana, J. Hastad, and S. Zachos, *Does co-$\mathcal{NP}$ have Short Interactive Proofs?*, Information Processing Letters, vol. 25, May 1987, pp. 127–132.

[4] B. Chor and O. Goldreich, *Unbiased bits from sources of weak randomness and probabilistic communication complexity*, Siam J. on Computing, April 1988.

[5] A. De Santis and G. Persiano, *Non-Interactive Zero-Knowledge Proofs of Knowledge*, manuscript.

[6] A. De Santis, S. Micali, and G. Persiano, *Non-Interactive Zero-Knowledge Proof-Systems*, in "Advances in Cryptology – CRYPTO 87", vol. 293 of "Lecture Notes in Computer Science", Springer Verlag, pp. 52–72.

[7] A. De Santis, S. Micali, and G. Persiano, *Non-Interactive Zero-Knowledge Proof-Systems with Preprocessing*, in "Advances in Cryptology - CRYPTO 88", vol. 403 of "Lecture Notes in Computer Science", Springer-Verlag, pp. 269–282.

[8] A. De Santis and M. Yung, *Cryptographic Applications of the non-interactive Metaproof and Many-prover Systems,* in "Advances in Cryptology - CRYPTO 90", vol. 537 of "Lecture Notes in Computer Science", Springer-Verlag, pp. 366–377.

[9] U. Feige, A. Fiat, and A. Shamir, *Zero-knowledge Proofs of Identity*, Proceedings of the 19th Annual ACM Symposium on Theory of Computing, New York, 1987, pp. 210–217.

[10] U. Feige, D. Lapidot, and A. Shamir, *Multiple Non-Interactive Zero-Knowledge based on a Single Random String*, Proceedings of the 31st Annual IEEE Symposium on Foundations of Computer Science, St. Louis, MO, Oct. 1990, pp. 308–317.

[11] L. Fortnow, *The Complexity of Perfect Zero-Knowledge*, Proceedings 19th Annual ACM Symposium on Theory of Computing, New York, NY, 1987, pp. 204–209.

[12] M. Garey and D. Johnson, *Computers and Intractability: a Guide to the Theory of $\mathcal{NP}$-Completeness,* W. H. Freeman & Co., New York, 1979.

[13] M. Garey, D. Johnson, and L. Stockmeyer, *Some Simplified NP-complete Graph Problems*, Theor. Comput. Sci., vol. 1, 1976, pp. 237–267.

[14] S. Goldwasser and S. Micali, *Probabilistic Encryption*, J. Computer System Sci., 28 (1984), pp. 270–299.

[15] S. Goldwasser, S. Micali, and C. Rackoff, *The Knowledge Complexity of Interactive Proof-Systems*, SIAM Journal on Computing, vol. 18, n. 1, February 1989.

[16] O. Goldreich, S. Micali, and A. Wigderson, *Proofs that Yield Nothing but their Validity and a Methodology of Cryptographic Design*, Proceedings of 27th Annual Symposium on Foundations of Computer Science, 1986, pp. 174–187.

[17] R. Karp, *Reducibility among combinatorial problems*, in Miller and Thatcher (eds), *Complexity of Computer Computations*, Plenum Press, New York, 1972, pp. 85–103.

[18] Y. Oren, *On the Cunning Power of Cheating Verifiers*, in Proceedings of 28th Annual Symposium on Foundations of Computer, 1987, pp. 462–471.

[19] A. Orlitsky and A. El Gamal, *Average and Ramdomized Communication Complexity*, IEEE Trans. on Info. Theory, vol. 36, pp. 3–16 (1990).

[20] L. Stockmeyer, *Planar 3-colorability is NP-complete*, SIGACT News, vol. 5, 19-25 (1973).

[21] M. Tompa and H. Woll, *Proving Possession of Knowledge and Random Self Reducible Problems*, in Proceedings of 28th Annual Symposium on Foundations of Computer, 1987, pp. 472–482.

Abelian Squares are Avoidable on 4 Letters

Veikko Keränen

Rovaniemi Institute of Technology
Jokiväylä 11, SF-96300 Rovaniemi, Finland

Abstract. *An* abelian square *is a non-empty word of the form* P_1P_2, *where* P_2 *is a permutation of* P_1, *that is, the number of occurrences of each letter is the same in* P_1 *and* P_2. *A word is called* abelian 2-repetition free, *or in short* a-2-free, *if it does not contain any abelian square as a subword. Let* Σ *be the four letter alphabet* $\{a,b,c,d\}$. *We give an example of a uniformly growing endo-morphism* $g: \Sigma^* \to \Sigma^*$; *with* $|g(a)| = 85$; *such that the iteration of* g *yields an a-2-free* ω*-word. In fact, it is proved that the morphism* g *itself is* a-2-free, *that is,* $g(w)$ *is a-2-free for every a-2-free word* w *in* Σ^*. *When proving that this property holds for* g, *one is in practice obliged to use a computer for certain tests. We performed these tests many times by using different computing environments and methods. It is explained how rechecking can be efficiently accomplished.*

The morphism g *is of the form* $g(\sigma(x)) = \sigma(g(x))$ *for all* x *in* Σ, *where* σ *performs a cyclic permutation of letters in* Σ. *As regards morphisms of this form, one may check by using a computer/computers that, for our example morphism* g, *the image word* $g(a)$ *is of minimal length. This checking becomes feasible by the reason that one may restrict the study to only relatively few long prefixes and suffixes of* $g(a)$.

1 Introduction

Let an integer $k \geq 2$ be fixed. A word is called *k-repetition free*, or in short *k-free*, if it does not contain any non-empty subword of the form R^k. Thue [16] showed at the beginning of this century that there exist infinitely long 2-free words over a three letter alphabet . He also showed that 3-free infinite words exist over a binary alphabet. Later, various results concerning avoidability of *k*-repetitions in infinite words have been achieved. In most cases, infinite *k*-free words are constructed by iterating a morphism, which connects closely the theory of repetitions to the theory of L systems; see [14]. Morphisms that preserve the k-freeness of words are sharply characterized in [3,8-11]. A general survey of related results is given in [1]. Fundamental topics are discussed in [12,15].

In [7, p.240] Erdös raised the question whether abelian squares are avoidable in infinitely long words over a four letter alphabet Σ, i.e., whether there exist infinitely many a-2-free words over Σ. Later, Pleasants [13] showed that over five letters there exists an infinite word without any abelian square. On the other hand, it is easily seen that over three letters abelian squares cannot be avoided in words of length ≥ 8. In [6] Entringer et al. showed that every infinite word over a binary alphabet contains arbitrarily long abelian squares. Due to Dekking [4] abelian 4-repetitions [3-repetitions] can be avoided in infinite words over two [three] letters. In recent years the original problem concerning abelian squares over four letters has attracted attention especially in the research of free partially commutative monoids; see for instance [2,5].

The positive solution to the problem posed by Erdös is obtained by showing that the iteration of the endomorphism g_{85}, defined in (8) below, yields the desired result. The subscript 85 refers to the length of $g_{85}(a)$ that, indeed, is unexpectedly high to be minimal. The form of the words $g_{85}(x)$, $x \in \Sigma$, i.e., the cyclic permutation of letters in these words; see (2) or (3); was already used by Pleasants [13, p.269] in connection with five letters.

However, $g_{85}(a)$ differs from the one in [13] as regards the relative number of occurrences of letters: here the number of occurrences in $g_{85}(a)$ is different for every letter. In fact, the Parikh vector of $\alpha = g_{85}(a)$, i.e., $(\#_a(\alpha), \#_b(\alpha), \#_c(\alpha), \#_d(\alpha))$, equals $(19, 21, 27, 18)$, whereas the corresponding vector in [13] is $(7, 2, 2, 2, 2)$. It is an open problem whether there exists an a-2-free endomorphism $g: \Sigma^* \to \Sigma^*$, of another form, such that $g(abcd)$ is shorter than $g_{85}(abcd)$. In Lemma 2 we eliminate one possibility that appeared attractive when searching for such a morphism g - or some other method for constructing an infinite a-2-free word.

2 Preliminaries

An *alphabet* Σ is a finite non-empty set of abstract symbols called *letters*. A *word* over Σ is a finite (unless otherwise indicated) string of letters belonging to Σ, i.e., words [non-empty words] are elements of the free monoid Σ^* [the free semigroup Σ^+] generated by Σ. The *empty word* is denoted by λ.

Let $w = x_0 \cdots x_m$, $x_i \in \Sigma$. The *length* of the word w, denoted by $|w|$, is the number of occurrences of letters in w, i.e., $|w| = m + 1$. The *mirror image* of w, denoted by $\text{mir}(w)$, is the word $x_m \cdots x_0$. Let the alphabet considered be $\Sigma = \{a_0,\ldots,a_n\}$. The number of occurrences of one letter $x \in \Sigma$ in w is denoted by $\#_x(w)$, or simply by $\#_j(w)$ if $x = a_j$. The notation $\psi(w)$ is used to denote the *Parikh vector* of w, i.e., $\psi(w) = (\#_0(w), \ldots, \#_n(w))$.

A word u is called a *subword* of a word w if $w = p\,u\,s$ for some words p and s. The notation $SW(w)$ stands for the set of all subwords of w. If $p = \lambda$ [$s = \lambda$], then u is called a *prefix* [a *suffix*] of w. By $PREF(w)$ [$SUFF(w)$] we mean the set of all prefixes [suffixes] of w. The notation $pref(w)$ [$suff(w)$] denotes an element in $PREF(w)$ [$SUFF(w)$]. In the case $w \neq \lambda$ we write $first(w)$ [$last(w)$] to denote the first [the last] letter of w.

Let $k \geq 2$ be a given integer. A *k-repetition* [an *abelian k-repetition*] is a non-empty word of the form R^k [$P_1 \cdots P_k$, where $\psi(P_\mu) = \psi(P_\nu)$ for all $1 \leq \mu < \nu \leq k$, i.e., P_i:s are permutations of each other]. Instead of [abelian] 2- and 3-repetitions terms [*abelian*] *squares* and [*abelian*] *cubes* are often used. A word or an ω-word (explained below) is called *k-repetition free* [*abelian k-repetition free*, or *k-free in the abelian sense*], or in short *k-free* [*a-k-free*], if it does not contain any k-repetition [abelian k-repetition] as a subword. A word sequence or a set is k-free [a-k-free] if all words in it are k-free [a-k-free].

Subsets of Σ^* are called *languages* over Σ. The *catenation* of languages L and L_1, denoted by LL_1, is the language $\{uv \mid u \in L, v \in L_1\}$. The notation L^i, where L is a language and $i \in \mathbb{N}$, is defined as follows: $L^0 = \{\lambda\}$ and $L^{i+1} = L^iL$. Let $L^* = \bigcup_{i=0}^{\infty} L^i$ and $L^+ = \bigcup_{i=1}^{\infty} L^i$. If a language contains only one word, say w, then we sometimes write w instead of $\{w\}$; especially, $w^* = \{w^i \mid i \geq 0\}$ and $w^+ = \{w^i \mid i \geq 1\}$.

For a language L we write $Q(L)$, where Q is SW, $PREF$ or $SUFF$, to denote the set $\bigcup_{w \in L} Q(w)$. Moreover, for n in $\mathbb{N}$, we write $Q_n(L)$ to denote the words in $Q(L)$ of length n. The notation $pref_n(w)$ [$suff_n(w)$] is used for an element in $PREF_n(w)$ [$SUFF_n(w)$].

A *morphism* h is a mapping between free monoids Σ^* and Σ'^* with $h(uv) = h(u)h(v)$ for every u and v in Σ^*. Especially, $h(\lambda) = \lambda$. A morphism $h: \Sigma^* \to \Sigma'^*$, being compatible with the catenation of words, is uniquely defined if the word $h(a) \in \Sigma'^*$ is (effectively) given for each a in Σ. If $\Sigma' = \Sigma$, we call h an *endomorphism* (and usually write g instead of h). For a morphism h and a language L we define $h(L) =$

$\{h(w) \mid w \in L\}$. A morphism h is termed *uniformly growing* if $|h(a)| = |h(b)| \geq 2$ for every a and b in Σ.

For a given integer $k \geq 2$, a morphism $h: \Sigma^* \to \Sigma^*$ is called *k-free* [*a-k-free*] if $h(w)$ is k-free [a-k-free] for every k-free [a-k-free] w in Σ^*.

As regards the L systems, we specify the following concepts. A *D0L system* is a triple $G = (\Sigma, g, \alpha_0)$, where Σ is an alphabet, $g: \Sigma^* \to \Sigma^*$ is an endomorphism and α_0, called the *axiom*, is a word over Σ. The (*word*) *sequence* $S(G)$ generated by G consists of the words

$$\alpha_0 = g^0(\alpha_0), \quad g^1(\alpha_0), \quad g^2(\alpha_0), \quad g^3(\alpha_0), \quad \ldots \quad ,$$

where $g^i(\alpha_0) = g(g^{i-1}(\alpha_0))$ for $i \geq 1$. The *language* of G is defined by $L(G) = \{g^i(\alpha_0) \mid i \geq 0\}$. Languages [sequences] defined by a D0L system are referred to as *D0L languages* [*D0L sequences*]. D0L systems provide a very convenient way for defining languages and infinite words. Furthermore, if g and α_0 are k-free [a-k-free], then the iteration of g will yield a k-free [a-k-free] D0L sequence.

An *ω-word* is an infinite sequence, from left to right, of letters of an alphabet Σ. Thus an ω-word can be identified with a mapping of $\mathbb{N}_+$ into Σ. One can construct an ω-word, for example, by iterating an endomorphism $g: \Sigma^* \to \Sigma^*$ such that $\lambda \notin g(\Sigma)$ and $g(a) = aw$ for some $a \in \Sigma$, $w \in \Sigma^+$. Such a morphism g is called *prefix preserving* by the reason that $g^i(a)$ is a proper prefix of $g^{i+1}(a)$ whenever $i \geq 0$. An ω-word is obtained as the "limit" of the sequence $g^i(a)$; $i = 0,1,2,\ldots$.

In the sequel Σ denotes the four letter alphabet $\{a, b, c, d\}$. Especially, for all words $w \in \Sigma^*$ below, the Parikh vector of w will be of the form $\psi(w) = (\#_a(w), \#_b(w), \#_c(w), \#_d(w)) = (\#_0(w), \#_1(w), \#_2(w), \#_3(w))$. In many cases it will be convenient to consider Σ as *alphabetically ordered* in the usual way (i.e. $a < b < c < d$) and extend this order to a total order on Σ^+ (i.e. for any u, v in Σ^+, $u < v$ iff v is in $u\Sigma^+$ or $u = pxs$, $v = pys_1$ with $x < y$; $x, y \in \Sigma$; $p, s, s_1 \in \Sigma^*$. This order on Σ^+ is called a *lexicographic order* in [12, p.64]).

Let $\sigma: \Sigma^* \to \Sigma^*$ be the letter-to-letter endomorphism that performs the cyclic permutation

$$\sigma(a) = b, \quad \sigma(b) = c, \quad \sigma(c) = d \quad \text{and} \quad \sigma(d) = a.$$

The inverse of σ, denoted by σ^{-1}, is regarded as a morphism $\sigma^{-1}: \Sigma^* \to \Sigma^*$ given by

$$\sigma^{-1}(a) = d, \quad \sigma^{-1}(b) = a, \quad \sigma^{-1}(c) = b \quad \text{and} \quad \sigma^{-1}(d) = c.$$

Thus, informally, assuming the set $\Sigma = \{a,b,c,d\}$ alphabetically ordered, σ [σ^{-1}] gives the immediate cyclic successors [predecessors] of letters in Σ. We can now use the notation σ^i to mean a morphism also in the case where i is a negative integer. Moreover, for every word w in Σ^* and integers i, j in $\mathbf{Z}$,

(1) $$\sigma^i(w) = \sigma^j(w) \quad \text{whenever} \quad i \equiv j (\text{mod } 4).$$

Concerning morphisms that generate growing D0L sequences (and ω-words) we consider only uniformly growing endomorphisms $g: \Sigma^* \to \Sigma^*$ of the form

(2) $$g(\sigma(x)) = \sigma(g(x)) \quad \text{for all} \quad x \text{ in } \Sigma.$$

Thus, for a given $g(a)$, $g(b) = \sigma(g(a))$, $g(c) = \sigma(g(b))$ and $g(d) = \sigma(g(c))$, that is,

(3) $$g(b) = \sigma(g(a)), \quad g(c) = \sigma^2(g(a)) \text{ and } g(d) = \sigma^3(g(a)).$$

By induction, for a morphism g in (2),

$$g(\sigma^i(x)) = \sigma^i(g(x)) \quad \text{for all} \quad i \text{ in } \mathbf{Z} \text{ and } x \text{ in } \Sigma.$$

From this we see that, whenever $g\colon \Sigma^* \to \Sigma^*$ is of the form (2) and i is in $\mathbf{Z}$, the composite mappings $g\sigma^i$ and $\sigma^i g$ are endomorphisms such that $g\sigma^i = \sigma^i g$. As a summary, recalling (1),

(4) $$g\sigma^i = \sigma^i g = \sigma^j g \quad \text{for all} \quad i, j \text{ in } \mathbf{Z} \text{ with } i \equiv j (\text{mod } 4).$$

Using (4) we see that the composition of morphisms σ^i and g is commutative (and associative). Indeed, it should now be clear that permutation of morphisms in a composition of morphisms σ and g gives back the same morphism. For example,

(5) $$g^{i_0} \sigma^{j_0} g^{i_1} \sigma^{j_1} \cdots g^{i_m} \sigma^{j_n} = g^{i_0+i_1+\cdots+i_m} \sigma^{j_0+j_1+\cdots+j_n} = \sigma^{j_0+j_1+\cdots+j_n} g^{i_0+i_1+\cdots+i_m},$$

whenever exponents i_μ are in $\mathbf{N}$ and j_ν in $\mathbf{Z}$.

Let i be an integer in $\mathbf{Z}$. A word u in Σ^+ is an abelian square iff $\sigma^i(u)$ is an abelian square. Therefore, it is easy to see that, for every word w in Σ^*,

(6) $$\sigma^i(w) \text{ is a-2-free} \quad \text{iff} \quad w \text{ is a-2-free.}$$

Let v be an a-2-free word in Σ^* and g a morphism of the form (2). By (6), $\sigma^i(g(v))$ is a-2-free iff $g(v)$ is a-2-free. Consequently,

(7) $$\sigma^i g \ (= g\sigma^i) \text{ is a-2-free} \quad \text{iff} \quad g \text{ is a-2-free.}$$

3 Existence

We show that the following prefix preserving endomorphism $g_{85}\colon \Sigma^* \to \Sigma^*$, with $\Sigma = \{a, b, c, d\}$, not only generates an a-2-free ω-word but is also an a-2-free morphism, i.e., $g_{85}(w)$ is a-2-free for every a-2-free w in Σ^*. The morphism g_{85} being of the same form as g in (2), we have only to give $g_{85}(a)$ in order to define it. A hyphen - indicates concatenation of words.

(8) $$g_{85}(a) = \begin{array}{l} abcacdcbcdcadcdbdabacabadbabcbdbcbacbcdcacb\text{-} \\ abdabacadcbcdcacdbcbacbcdcacdcbdcdadbdcbca \end{array}$$

Here $|g_{85}(a)| = 85$ and the Parikh vector of $g_{85}(a)$ is $\psi(g_{85}(a)) = (19, 21, 27, 18)$.

Theorem 1. *The endomorphism* g_{85} *defined in* (8) *is a-2-free.*

Proof. Let u in Σ^+ be a word such that an abelian square is a subword of $g_{85}(u)$. Let w be one of the shortest subwords of u such that $g_{85}(w) = \alpha P_1 P_2 \beta$ for some words

α, β in Σ^* and P_1, P_2 in Σ^+ with $\psi(P_1) = \psi(P_2)$. We show that in this situation an abelian square is necessarily a subword of w, too. For this purpose we prove that there remains for consideration only finitely many cases, which can be checked with the aid of a computer within a reasonable time. We examined these cases carefully three times using different computing environments: COMMON LISP was used as a programming language on a 68020-based UNIX workstation and on a 80386-based MS-DOS system, and the language C was used on a 80386-based MS-DOS system. Also different methods were used: precalculated lists of Parikh vectors (described below) in connection with LISP, and in connection with C we checked the relevant subwords of every reduced $g_{85}(w)$ (for the term *reduced* see the text below).

We divide in (12) the different possibilities concerning $g_{85}(w) = \alpha\, P_1P_2\, \beta$ above into five main cases. These cases were chosen with intend to make the required computer programs simple. In the sequel, let x, y, z, t be letters in Σ and v_1, v_2 be words in Σ^* such that $|v_1| = |v_2|$. Consider the situation $g_{85}(w) = \alpha\, P_1P_2\, \beta$ presented above. Recalling that w in $SW(u)$ was of minimal length it is seen that $g_{85}(w)$ has one of the following structures:

(9) $g_{85}(v_1v_2) = P_1P_2,\quad g_{85}(x\, v_1v_2\, z) = \alpha\, P_1P_2\, \beta \quad \text{with} \quad |\alpha| = |\beta|,$

$g_{85}(x\, v_1\, y\, v_2) = \alpha\, P_1P_2,\quad g_{85}(v_1\, y\, v_2\, z) = P_1P_2\, \beta,\quad g_{85}(x\, v_1\, y\, v_2\, z) = \alpha\, P_1P_2\, \beta,$

$g_{85}(x\, t\, v_1\, y\, v_2\, z) = \alpha\, P_1P_2\, \beta \quad \text{with} \quad g_{85}(x\, t\, v_1) \text{ in } PREF(\alpha\, P_1) \quad \text{or}$

$g_{85}(x\, v_1\, y\, v_2\, t\, z) = \alpha\, P_1P_2\, \beta \quad \text{with} \quad g_{85}(v_2\, t\, z) \text{ in } SUFF(P_2\, \beta).$

Let w_1 and w_2 be words obtained from words v_1 and v_2 as follows:

(10) $w_1 = a^{i_0}\, b^{i_1}\, c^{i_2}\, d^{i_3} \quad \text{with} \quad i_\mu = \#_\mu(v_1) - \min(\#_\mu(v_1), \#_\mu(v_2)) \quad \text{and}$

$w_2 = a^{j_0}\, b^{j_1}\, c^{j_2}\, d^{j_3} \quad \text{with} \quad j_\mu = \#_\mu(v_2) - \min(\#_\mu(v_1), \#_\mu(v_2)), \quad \mu = 0,1,2,3.$

Note that here $i_\mu = 0$ or $j_\mu = 0$ for every $\mu = 0,1,2,3$. Informally speaking, w_1 and w_2 are reduced words of v_1 and v_2 obtained by firstly rearranging the occurrences of letters in v_1 and v_2 in alphabetical order and then cancelling, for every letter x in Σ, occurrences that appear in both words v_1 and v_2, i.e., w_1 and w_2 are obtained by erasing the underlined subwords of

$$v_1' = \underline{a^{k_0}}\, a^{i_0}\, \underline{b^{k_1}}\, b^{i_1}\, \underline{c^{k_2}}\, c^{i_2}\, \underline{d^{k_3}}\, d^{i_3} \quad \text{and} \quad v_2' = \underline{a^{k_0}}\, a^{j_0}\, \underline{b^{k_1}}\, b^{j_1}\, \underline{c^{k_2}}\, c^{j_2}\, \underline{d^{k_3}}\, d^{j_3},$$

where $k_\mu + i_\mu = \#_\mu(v_1)$ and $k_\mu + j_\mu = \#_\mu(v_2)$ for every $\mu = 0,1,2,3$. Moreover, there does not exist a letter x in Σ occurring in both w_1 and w_2, since $i_\mu = 0$ or $j_\mu = 0$ in (10). Obviously,

(11) $|w_1| = |w_2| \quad \text{and} \quad \psi(w_1) - \psi(w_2) = \psi(v_1) - \psi(v_2).$

Any words w' and w'' in Σ^* are referred to as *reduced*, and the pair (w', w'') as a *reduced pair*, if there does not exist any letter x in Σ that occurres in both w' and w''.

Below we consider the reduced cases of $g_{85}(w)$ in (9). Instead of the abelian square P_1P_2 we now write R_1R_2. Illustration of these cases is as follows:

(12)

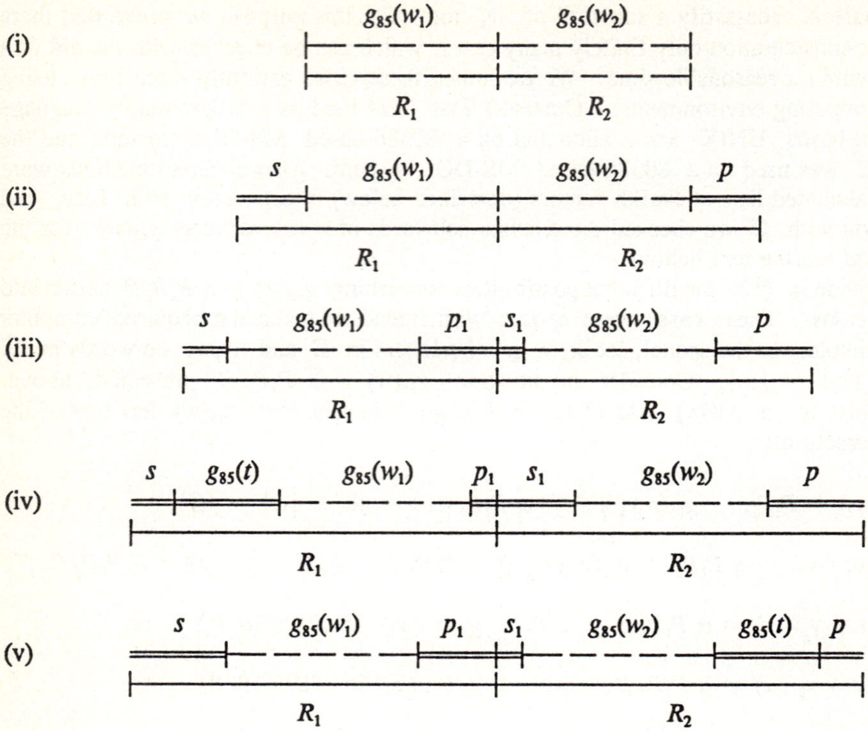

where in accordance with (9) $\alpha s = g_{85}(x)$, $p\beta = g_{85}(z)$, $p_1 \neq \lambda$, $s_1 \neq \lambda$. Moreover, in greater detail: in (12, ii) $0 < |s| = |p| < |g_{85}(a)|$; in (12, iii) $0 \le |s| \le |g_{85}(a)|$, $0 \le |p| \le |g_{85}(a)|$; in (12, iv) $0 < |s| < |g_{85}(a)|$, $0 < |p| \le |g_{85}(a)|$; and in (12, v) $0 < |s| \le |g_{85}(a)|$, $0 < |p| < |g_{85}(a)|$.

Let $\psi_D(w', w'')$ denote the difference vector $\psi(g_{85}(w')) - \psi(g_{85}(w''))$, where w' and w'' are any words in Σ^*. In the cases (ii) - (v) of (12) the existence of an abelian square R_1R_2 implies that the difference vectors

(13) $D_{ii} = \psi(p) - \psi(s)$, $D_{iii} = \psi(s_1 p) - \psi(s p_1)$, $D_{iv} = \psi(s_1 p) - \psi(s\, g_{85}(t)\, p_1)$ and

$$D_v = \psi(s_1\, g_{85}(t)\, p) - \psi(s\, p_1)$$

are equal to $\psi_D(w_1, w_2)$, where w_1 and w_2 are the reduced words corresponding each case (ii) - (v) in question. Note, for example, that in (12.ii) $\psi(R_1) = \psi(R_2)$ means that $\psi(s\, g_{85}(w_1)) = \psi(g_{85}(w_2)\, p)$, i.e., $\psi(s) + \psi(g_{85}(w_1)) = \psi(g_{85}(w_2)) + \psi(p)$ giving $D_{ii} = \psi(p) - \psi(s) = \psi(g_{85}(w_1)) - \psi(g_{85}(w_2)) = \psi_D(w_1, w_2)$. Using precalculated lists of Parikh vectors for prefixes (*preflist.x*) and suffixes (*sufflist.x*) of each $g_{85}(x)$, with x in Σ, one may quickly check with the aid of a computer that

(14) $m_{ii} = 9$, $m_{iii} = 15$, $m_{iv} = 18$ and $m_v = 19$,

where $m_\mu = \max\{|D_{\mu,k}| \mid D_\mu$ goes through all the values in the case μ (all combinations for s, p_1, s_1, p and t are included - but neither xt nor tz is a square) and $k = 0,1,2,3\}$

($D_{\mu,k}$ is the $k+1$ th component of $D_\mu = (D_{\mu,0}, D_{\mu,1}, D_{\mu,2}, D_{\mu,3})$). In order to explain the precalculated lists *preflist.x* and *sufflist.x* mentioned above we give two examples:

preflist.a = ((1,0,0,0) (1,1,0,0) (1,1,1,0) ... (19,21,27,18))

sufflist.a = ((19,21,27,18) (18,21,27,18) (18,20,27,18) ... (1,0,0,0))

The reader may compare these lists with $g_{85}(a)$ in (8).

Considering the absolute values of components of the difference vectors $\psi_D(w_1, w_2)$, with w_1, w_2 as above, let $m(w', w'') = \max\{|(\psi_D(w', w''))_k| \mid k = 0,1,2,3\}$, where w' and w'' are any words in Σ^*. Straightforward calculation (ignoring here firstly (12) and (14)) shows that concerning all possible reduced pairs (w', w'') in the case $1 \le |w'| = |w''| \le 4$, $m(w', w'') \le 19$ the difference vector $\psi_D(w', w'')$ can have only the values:

(15)						
$\psi_D(a, b)$	=	(1, 2, 6, -9)	$\psi_D(ab, cd)$	=	(-11, -5, 11, 5)
$\psi_D(a, c)$	=	(-8, 3, 8, -3)	$\psi_D(ac, bd)$	=	(7, -7, 7, -7)
$\psi_D(aa, bb)$	=	(2, 4, 12, -18)	$\psi_D(aab, ccd)$	=	(-19, -2, 19, 2)
$\psi_D(aa, bc)$	=	(-7, 5, 14, -12)	$\psi_D(aac, bbd)$	=	(8, -5, 13, -16)
$\psi_D(aa, bd)$	=	(-1, -4, 15, -10)	$\psi_D(aad, bcc)$	=	(-13, 14, 13, -14)
$\psi_D(aa, cc)$	=	(-16, 6, 16, -6)	$\psi_D(aacc, bbdd)$	=	(14, -14, 14, -14)
$\psi_D(aa, cd)$	=	(-10, -3, 17, -4)			

and, in addition, the values obtained by cyclic permutation of the vectors $\psi_D(w', w'')$ in (15) and their opposite vectors $-\psi_D(w', w'')$, i.e., in the case $1 \le |w'| = |w''| \le 4$, $m(w', w'') \le 19$, with reduced w' and w'', the difference vector $\psi_D(w', w'')$ can have at most 72 (as one can check) different values that are among the vectors $\pm\psi_D(\sigma^i(w'), \sigma^i(w''))$; $i = 0,1,2,3$; where (w', w'') goes through the reduced pairs expressed by (15). In the end of the proof it will be shown that for every reduced pair (w', w''), with $|w'| = |w''| \ge 5$, one has $m(w', w'') \ge 21$, implying, by (13) and (14), that the case $|w_1| = |w_2| \ge 5$ is contradictory in (12).

Let $|w_1| = |w_2| \le 4$ in (12). Taking (13), (14) and (15) into account one may now, with the aid of a computer, list all the cases that might exist in (12) (note that also the possibility $w_1 = w_2 = \lambda$ is relevant in the cases (ii) - (v) of (12)). The list of the cases is rather long, but, on the other hand, it is straightforward to check that it contains only those cases in which all of the structures of $g_{85}(w) = \alpha P_1P_2 \beta$, presented in (9), obtained by rearranging the letters of w_1 and w_2, or by rearranging the letters and adding some letters to w_1 and w_2 (the same letters to both), are such that w contains an abelian square. In other words, in every case, the pair (w_1, w_2) is not a reduced pair of any pair (v_1, v_2) such that $g_{85}(w) = \alpha P_1P_2 \beta$ has one of the structures of (9) and w is a-2-free. This means that if $|w_1| = |w_2| \le 4$, then $g_{85}(w)$ is a-2-free whenever w is a-2-free. We give two detailed examples of the cases concerning (iii) and (iv) in (12):

(16) $$g_{85}(a\, w_1\, a\, w_2\, b) = \alpha\, R_1R_2\, \beta;$$

with $w_1 = b$, $w_2 = a$, $|\alpha| = 72$, $|\beta| = 1$, $|R_1| = |R_2| = 176$; and

(17) $$g_{85}(a\, b\, w_1\, c\, w_2\, d) = \alpha\, R_1R_2\, \beta;$$

with $w_1 = cd$, $w_2 = ab$, $|\alpha| = 13$, $|\beta| = 5$, $|R_1| = |R_2| = 331$. The reader may easily get convinced that in (16) and (17) the replacement of the subwords w_1 and w_2 of $a\,w_1\,a\,w_2\,b$ and $a\,b\,w_1\,c\,w_2\,d$ by any v_1 and v_2 such that (w_1, w_2) is a reduced pair of (v_1, v_2) gives only words that are not a-2-free. It turns out that the computer program will list only those cases in which $|w_1| = |w_2| \le 2$. This makes the cases very straightforward to check (we assume that the reader does not want to leave the whole work to the computer). Moreover, it is possible to reduce the case in (17) to a shorter one by considering the reduced pair (cd, a) instead of the pair $(b\,w_1, w_2) = (b\,cd, ab)$. Permitting this kind of reduction leaves only the cases $|w_1| = |w_2| \le 1$ to check!

Let $|w_1| = |w_2| \ge 5$ in (12). We show that in this case always $m(w_1, w_2) \ge 21$. This contradiction with (13) and (14) finally makes the proof complete. We write in short Δ to denote $\psi_D(w_1, w_2)$. Without loss of generality we may assume that

$$(18) \qquad w_1 = a^{i_0}\, b^{i_1}\, c^{i_2}\, d^{i_3} \quad \text{with} \quad i_0 = \max\{i_0, i_1, i_2, i_3\}.$$

Indeed, the other cases concerning $\max\{i_0, i_1, i_2, i_3\}$ are readily obtained from the case in (18) by considering the words $\sigma^k(w_1)$ and $\sigma^k(w_2)$; with $k = 1,2,3$; instead of w_1 and w_2, since $m(\sigma^k(w_1), \sigma^k(w_2)) = m(w_1, w_2)$. Firstly, recall that

$$(19) \qquad \begin{aligned} \psi(g_{85}(a)) &= (19, 21, 27, 18) \\ \psi(g_{85}(b)) &= (18, 19, 21, 27) \\ \psi(g_{85}(c)) &= (27, 18, 19, 21) \\ \psi(g_{85}(d)) &= (21, 27, 18, 19), \end{aligned}$$

where the shading of the columns is used to indicate that we use these columns in order to get a lower bound for $m(w_1,w_2)$. Note also that $|\Delta_k|$; with $k = 0,1,2$ or 3; refers to the difference concerning the occurrences of a letter $\sigma^k(a)$, i.e., it refers to the $k{+}1$ th shaded column of (19). If $i_0 = 2$ in (18), then w_1 contains three different letters and, consequently, $w_2 = (\sigma^\mu(a))^j$, for some $\mu \in \{1,2,3\}$ and $j = |w_1| \ge 5$, implying that $|\Delta_\nu| > j\cdot|21 - 27| \ge 30$, where $\nu \equiv (2 + \mu) \pmod 4$ and $\nu \in \{0,1,2,3\}$. A similiar situation is obtained if $i_0 = |w_1|$. Thus we may assume that in (18) one has $3 \le i_0 < |w_1|$, i.e.,

$$(20) \qquad w_1 \in a^{i_0}\,(b^+ \cup c^+ \cup d^+) \quad \text{with} \quad i_0 = \max\{i_0, i_1, i_2, i_3\} \ge 3.$$

Consider separately the three cases of (20).

Let $w_1 = a^{i_0}\, b^{i_1}$. Then $w_2 = c^{j_2}\, d^{j_3}$ with $j_2 + j_3 \ge 5$. Thus $|\Delta_2| = |i_0\cdot 27 + i_1\cdot 21 - j_2\cdot 19 - j_3\cdot 18| > i_0\cdot(27 - 19) + i_1\cdot(21 - 19) \ge 3\cdot(27 - 19) + 2\cdot(21 - 19) = 28$.

Let $w_1 = a^{i_0}\, c^{i_2}$. Then $w_2 = b^{j_1}\, d^{j_3}$ with $j_1 + j_3 \ge 5$. If $j_1 < j_3$, then $|\Delta_2| = |i_0\cdot 27 + i_2\cdot 19 - j_1\cdot 21 - j_3\cdot 18| \ge 3\cdot(27 - 18) + 2\cdot(19 - 21) = 23$. On the other hand, if $j_1 \ge j_3$, then $|\Delta_3| = |i_0\cdot 18 + i_2\cdot 21 - j_1\cdot 27 - j_3\cdot 19| \ge |3\cdot(18 - 27) + 3\cdot(21 - 19)| = 21$.

Let $w_1 = a^{i_0}\, d^{i_3}$. Then $w_2 = b^{j_1}\, c^{j_2}$ with $j_1 + j_2 \ge 5$. If $j_1 = 1$, then $j_2 \ge 4$ and $|\Delta_0| = |i_0\cdot 19 + i_3\cdot 21 - j_1\cdot 18 - j_2\cdot 27| \ge |3\cdot(19 - 27) + (21 - 27) + (21 - 18)| = 27$. On the other hand, if $j_1 \ge 2$, then $|\Delta_3| = |i_0\cdot 18 + i_3\cdot 19 - j_1\cdot 27 - j_2\cdot 21| \ge |2\cdot(18 - 27) + (18 - 21) + 2\cdot(19 - 21)| = 25$.

Thus we see that if $|w_1| = |w_2| \ge 5$ in (12), then always $m(w_1, w_2) \ge 21$, since $m(w_1, w_2) \ge |\Delta_k|$ for every $k = 0,1,2,3$. Recalling (13), (14) and the result concerning the case $|w_1| = |w_2| \le 4$, we obtain that $g_{85}(w)$ is of the form $\alpha\, P_1 P_2\, \beta$ only if w contains an abelian square. This proves Theorem 1. □

Minimality of $|g(a)|$

We study only the case $\Sigma = \{a, b, c, d\}$ and endomorphisms $g\colon \Sigma^* \to \Sigma^*$ of the form 2). In Lemma 5 below it is stated that, for a fixed length of $g(a)$, there exists an a-2-free OL sequence generated by g only if that is the case for g with

$$g(a) = a\,b\,c\,w\,a \quad \text{for some } w \text{ in } \Sigma^+. \tag{21}$$

onsequently, for every fixed length of $g(a)$, when searching candidates for a-2-free ndomorphisms (of the form (2)) over Σ or a-2-free DOL sequences generated by g, t will be sufficient to consider only the morphisms g of the form (21), i.e., only those vords $g(a)$ which begin with *abc* and end with a. The proof of Lemma 5 is based on our auxiliary results presented below. Due to space limitation only some outlines for the roofs are given. After Lemma 5, we discuss in detail the method with the help of which he morphism g_{85} in (8) was found.

Lemma 1. *Let* g_a *and* g *be endomorphisms of the form* (2) *and let* $g_a(a) = g(x)$ *or the letter* x *such that* $first(g(x)) = a$. *Then*
i) g_a *is a-2-free iff* g *is a-2-free;*
ii) *the DOL sequence* $S(G_a)$ *generated by* $G = \{\Sigma, g_a, \sigma^k(\alpha_0)\}$ *is a-2-free for every* k *in* $\{0,1,2,3\}$ *iff the DOL sequence* $S(G)$ *generated by* $G = \{\Sigma, g, \alpha_0\}$ *is a-2-free.*

The following lemma tells us that it is not possible to construct an a-2-free ω-word γ over Σ in such an attractive way that after starting with a letter, say a, every second letter of γ would be in a certain subset of Σ, say in $\{a,c\}$, and the remaining letters in between would be in the complement set $\{b,d\}$, i.e., every prefix of γ of even length would be in $(\{a,c\}\,\{b,d\})^*$. For example, the following effort to start such a γ will fail:

$$a\,b\,c\,d\,c\,b\,a\,b\,c\,b\,a\,d\,a\,b\,a\,d\,c\,d\,a\,b\,a\,d\,a\,\ldots\,.$$

Lemma 2. *Let* Σ_e *and* Σ_o *be disjoint two letter alphabets such that* $\Sigma_e \cup \Sigma_o = \Sigma = \{a,b,c,d\}$. *There does not exist any a-2-free* ω*-word* γ *over* Σ *such that*

$$\gamma = x_0 x_1 x_2 \cdots \quad \textit{with} \quad x_{2i} \in \Sigma_e \;\; \textit{and} \;\; x_{2i+1} \in \Sigma_o \;\; \textit{for every} \;\; i \in \mathbb{N}. \tag{22}$$

One may prove Lemma 2 by studying different possibilities for the subwords of γ. In this approach no computer is needed. Another proof is obtained by using a computer. One finds out that there exist only finitely many words that could be prefixes of an a-2-free γ of the form (22). Indeed, it turns out that a word in $\{ac, bd, ca, db\}$, i.e., a word of the form $x\,\sigma^2(x)$, $x \in \Sigma$, is a subword of every a-2-free word $w \in \Sigma^+$ of length 42. This way of proving Lemma 2 was found by Matti K. Sinisalo, University of Oulu. Moreover, he found out that a word in $\{ab, bc, cd, da\}$, i.e., a word of the form $x\,\sigma(x)$, $x \in \Sigma$, is a subword of every a-2-free word $w \in \Sigma^+$ of length 45.

By utilizing Lemma 1 and Lemma 2 (or the results of Sinisalo) one may prove

Lemma 3. *Let* g *be an endomorphism of the form* (2) *and assume that the DOL sequence* $S(G)$ *generated by* $G = \{\Sigma, g, \alpha_0\}$ *is a-2-free. Then* $first(g(a)) = last(g(a))$.

Informally, the proof of Lemma 3 is obtained by considering the subwords w of length 2 ocurring in the words of $S(G)$ and by studying the image words $g(w)$ and $g^2(w)$. The same holds for the proof of the following

Lemma 4. *Let* g *be a given endomorphism of the form* (2) *such that, for some* α_0 *in* Σ^+, *there exists an a-2-free DOL sequence* $S(G)$ *generated by* $G = \{\Sigma, g, \alpha_0\}$. *Then*

(23) $$SW_2(g(\{a,b,c,d\})) = \cup_{i=0}^{3} \sigma^i(\{ab,ac,ad\}).$$

By using a computer one may check that there does not exist any candidate for g if $|g(a)| < 10$. Similarly, it is seen that the set $SUFF_{10}(g(a))\, PREF_{10}(g(\{b,c,d\}))$ is a-2-free only if $g(a)$ is of the form $a\,b\,c\,w\,a$, $a\,d\,c\,w\,a$, $a\,w\,c\,b\,a$ or $a\,w\,c\,d\,a$ for some w in Σ^+. These facts and the previous lemmas enable us to prove

Lemma 5. *Let* g *and* g_1 *be endomorphisms of the form* (2) *such that* $|g(a)| = |g_1(a)|$ *and*

(24) $$g_1(a) = a\,b\,c\,w\,a \quad \textit{for some } w \textit{ in } \Sigma^+.$$

Then there exists an a-2-free DOL sequence obtained by the iteration of g *only if the same holds true for* g_1, *too.*

In what follows we explain how the a-2-free morphism g_{85} in (8) was found and why the image word $g_{85}(a)$ is of minimal length as regards not only a-2-free morphisms (of the form (2)) but also morphisms (of the considered form) that generate an a-2-free DOL sequences.

By the result of Lemma 5, we now see that when searching an endomorphism g of the form (2) that generates an a-2-free DOL sequence, one does not lose any possibilities, if one assumes that $g(a)$ is of the form $a\,b\,c\,w\,a$. Let this be the form of $g(a)$ below. Now one may check by a computer that there does not exist any morphism g of the form (2) such that $|g(a)| \leq 20$ and the set $g(\{ab,ac,ad\})$ is a-2-free. This can be done, e.g. letting, for every fixed length of $g(a) \leq 20$, the word $w \in \Sigma^+$ run, in alphabetical order, through all a-2-free combinations in the word $g(a) = a\,b\,c\,w\,a$ and testing, for every w, whether the words $g(a)\, pref_3(g(b)) = a\,b\,c\,w\,a\,b\,c\,d$, $g(ab)$, $g(ac)$ and $g(ad)$ are a-2-free. Using $g(a)\, pref_3(g(b))$ instead of $g(a)$ or $g(ab)$ alone helps in finding an abelian square before the construction of the word $g(ab) = g(a)\,b\,c\,d\,\sigma(w)\,b$, where $\sigma(w)$, or, in fact, some suffix of $\sigma(w)$, has to be computed for every new w. After this, one can compute that there exist only 413 words $u = ps$ such that $|p| = |s| = 20$ and $g(a)$ is of the form $p\,w\,s$; with abc in $PREF(p)$ and $last(s) = a$; and the set

$$SUFF_{20}(g(a))\; PREF_{20}(g(\{b,c,d\}))$$

is a-2-free. Below we denote by p_i and s_j the words such that $g(a)$ is of the form $p_i\, w\, s_j = abc\, w'\, a$ and $|p_i| = i$, $|s_j| = j$. Next, from these 413 words $p_{20}s_{20}$, one sees that there remain for consideration only 9 words $p_{10}s_{10}$; recall that the case $|g(a)| \leq 20$ was already checked. By using these 9 words $p_{10}s_{10}$, one may check that there does not exist any morphism g such that $21 \leq |g(a)| \leq 40$ and the set $g(\{ab,ac,ad\})$ is a-2-free. Here too, in order to save computer time, the program should firstly fix the length of $g(a)$, then choose the word $p_{10}s_{10}$ and after that, for every proper canditate w, firstly check the a-2-freeness of $g(a)\ pref_{10}(g(b)) = p_{10}\, w\, s_{10}\, \sigma(p_{10})$ (we tried also some other methods but in general without a success in making the computation notably faster). Next, one can reduce the number of the words $p_{20} s_{20}$ in the list that above contained 413 of them. Studying the list of words $p_{30} s_{30}$ such that the words $s_{30}\, \sigma^i(p_{30})$; $i = 1,2,3$; are a-2-free, i.e., the set $SUFF_{30}(g(a))\, PREF_{30}(g(\{b,c,d\}))$ is a-2-free, (the list is too large to be useful as such), one finds that no more than 95 words $p_{20} s_{20}$ are possible for the desired $g(a) = p_{20}\, w\, s_{20}$. By using these 95 words $p_{20} s_{20}$, one may now check that there does not exist any morphism g such that $41 \le |g(a)| \le 60$ and the set $g(\{ab,ac,ad\})$ is a-2-free. Moreover, one may, by using the reduced list of 95 words $p_{20} s_{20}$, produce the list of possible words $p_{23} s_{23}$. After many reduction experiments this list was found to contain no more than 367 words. We also found that the length 46 of the words $p_i s_j$ was nearly optimal for future calculations (at least in our computing environments). This seems to be due to the fact that there remain too many possibilities for longer words $p_i s_i$; with $i > 23$. Therefore, when using these long words $p_i s_i$, one may not get any proper speed-up for the search of promising g. One could perhaps get some advantage by using words $p_i s_j$; with $i < j$; since there seems to be more different prefixes than suffixes in the lists containing the words $p_i s_i$. For instance, by studying the prefixes and suffixes in the non-reduced list of words $p_{30} s_{30}$; with a-2-free $s_{30}\, \sigma^i(p_{30})$ for $i = 1,2,3$; one finds that there remain only 10 words $p_{10} s_{20}$ giving for $g(a) = p_{10}\, w\, s_{20}$ the following possibilities:

$$\begin{aligned} g(a) &= abcacdc\ u_1\ \ w\ u_2\ u_3, \\ g(a) &= abcacdcbcd\ w\ bcdcacdcbdcdadbdcbca, \text{ or} \\ g(a) &= abcbdadcbc\ w\ u_4\ acdadbdcdbdabdbca; \end{aligned}$$

with u_1 in $\{acb, bca, bcd\}$, u_2 in $\{abc, bac\}$, $u_3 = babdbcdbdadbdcbca$, u_4 in $\{bad, cbc, dbc\}$ and w in Σ^+. Note here that our choice abc as the prefix of $g(a)$ creates an asymmetric situation because of the permutation σ in (2).

Next, one may check all the cases for morphisms g with $61 \le |g(a)| \le 85$. Here the computer time seems to grow proportionally to m^k; with $1.4 < m < 1.5$ and $k = |g(a)|$. We found out that not until in the case $|g(a)| = 64$ the set $g(\{ab,ac,ad\})$ is a-2-free, i.e., the set $\bigcup_{i=0}^{3} \sigma^i(\{ab,ac,ad\})$ is a-2-free. Going further, let us say that g passes *Test n*, if $g(u)$ is a-2-free for every a-2-free u in Σ^+ of length $\le n$. It turns out that not until in the case $|g(a)| = k$ there exists g passing *Test n*, where k and n form pairs (k, n) as follows: (64, 2), (71, 3), (83, 4) and (85, 5). Moreover, all these morphisms g, except g_{85} considered in the previous chapter (the case (85, 5)), are such that they do not generate any a-2-free D0L sequence; this is quite easy to check by Lemma 4 and by the reason that not very many morphisms g pass *Test* 3 (we found about one hundred of them in the case $71 \le |g(a)| \le 100$; only part of the cases were tested for $|g(a)| > 85$). Including g_{85}, only two morphisms g passed *Test* 4! The first one gives the pair (83,4) above and is as follows:

$$\begin{aligned} g_{83}(a) \;\; = \;\; & abcacdcbcdcadbdcadabaca\underline{badbab}cbdbcbacbcdca- \\ & cbacadabdbcdcacdbcbacbcdcacdcbdcdadbdcbca \end{aligned}$$

This morphism does not pass *Test* 5, since, e.g. $g_{83}(acbad)$ is not a-2-free. Moreover, for the underlined subword $u = badbab$ of $g_{83}(a)$, $g_{83}(u)$ is not a-2-free. Consequently, g_{83} does not generate any a-2-free D0L sequence either. The Parikh vector for $g_{83}(a)$ is $\psi(g_{83}(a)) = (19, 20, 26, 18)$.

We carried out the search for g in the case $|g(a)| \le 73$ using two programming languages LISP and C, which ran in parallel and executed the same tests twice on one 68020-based UNIX workstation and on 80286- and 80386-based MS-DOS systems. In the case $74 \leq |g(a)| \leq 85$ only MS-DOS (mainly 80286-based) systems and C language were used. These computer tests were performed in the summer 1990 at the Rovaniemi Institute of Technology. It is assumed that the same tests could be now (spring 1992) repeated by using, say 10, modern microcomputers in a few days. In addition, we let the computers check, for certain prefixes and suffixes of $g(a)$, the cases $86 \leq |g(a)| \leq 100$. In spite of the fact that the time for this checking was as long as in the case $|g(a)| \leq 85$, we did not find any more morphisms g that would pass *Test* 4. Consequently g_{85} presented in (8) is of rare type. Excluding renaming letters, mirror images of $g_{85}(a)$ and powers of g_{85}, we do not know any other endomorphism over a four letter alphabet that is a-2-free or generates an a-2-free D0L sequence.

Acknowledgements

I want to thank the personnel at the Rovaniemi Institute of Technology for its support and for the inspiring atmosphere that prevailled during the course of this work. Special thanks are due to Kari Tuovinen, who made the C language computer programs for producing and testing words and, in particular, translated the original LISP code into the C code, with the help of which the morphism g_{85} in (8) was finally found. The financial support of the Jenny and Antti Wihuri Foundation is also gratefully acknowledged.

References

[1] J. BERSTEL, Some recent results on squarefree words, *Lecture Notes in Comput. Sci.* 166, Springer-Verlag, Berlin - Heidelberg - New York - Tokyo, 1984, 14-25.

[2] R. CORI and M.R. FORMISANO, Partially abelian squarefree words, *RAIRO Inform. Théor. et Appl.* 24, 1990, 509-520.

[3] M. CROCHEMORE, Sharp characterizations of squarefree morphisms, *Theoret. Comput. Sci.* 18, 1982, 221-226.

[4] F.M. DEKKING, Strongly non-repetitive sequences and progression-free sets, *J. Combin. Theory Ser. A* 27, 1979, 181-185.

[5] V. DICKERT, Research topics in the theory of free partially commutative monoids, *Bull. Europ. Assoc. Theoret. Comput. Sci.* 40, 1990, 479-491.

[6] R.C. ENTRINGER, D.E. JACKSON and J.A. SCHATZ, On nonrepetitive sequences, *J. Combin. Theory Ser. A* 16, 1974, 159-164.

[7] P. ERDÖS, Some unsolved problems, *Magyar Tud. Akad. Mat. Kutató Int. Közl.* 6, 1961, 221-254.

[8] V. KERÄNEN, *On the k-freeness of morphisms on free monoids*, Ann. Acad. Sci. Fenn. Ser. A I Math. Dissertationes 61, 1986.

[9] V. KERÄNEN, On the k-freeness of morphisms on free monoids, *Lecture Notes in Comput. Sci.* 247, Springer-Verlag, Berlin - Heidelberg - New York - London - Paris - Tokyo, 1987, 180-188.

[10] M. LECONTE, A characterization of power-free morphisms, *Theoret. Comput. Sci.* 38, 1985, 117-122.

[11] M. LECONTE, Kth power-free codes, *Lecture Notes in Comput. Sci.* 192, Springer-Verlag, Berlin - Heidelberg - New York - Tokyo, 1985, 172-187.

[12] M. LOTHAIRE, *Combinatorics on words*, Addison-Wesley, Reading, Massachusetts, 1983.

[13] P.A. PLEASANTS, Non-repetitive sequences, *Proc. Cambridge Phil. Soc.* 68, 1970, 267-274.

[14] G. ROZENBERG and A. SALOMAA, *The mathematical theory of L systems*, Academic Press, New York - London - Toronto - Sydney - San Fransisco, 1980.

[15] A. SALOMAA, *Jewels of formal language theory*, Computer Science Press, Rockville, Maryland, 1981.

[16] A. THUE, Über unendliche Zeichenreihen, *Norske Vid. Selsk. Skr. I. Mat. Nat. Kl. Christiania* 7, 1906, 1-22.

POLYNOMIAL SIZE TEST SETS FOR CONTEXT-FREE LANGUAGES

Juhani Karhumäki

Department of Mathematics, Turku University, Finland

Wojciech Plandowski, Wojciech Rytter

Institute of Informatics, Department of Mathematics and Informatics,
Warsaw University, ul. Banacha 2, 00-913 Warszawa, Poland
e-mail: rytter@plearn.bitnet

Abstract. We prove that each context-free language has a polynomial size test set. This improves the doubly exponential upper bound obtained in [ACK] and single exponential bound from [KRJ]. An efficient algorithm to find test sets for context-free languages is also presented. The basic tools in the proof are graph-theoretical properties of test sets and periodicities in strings.

1. Introduction

A test set for a language L is any subset F of L such that for each homomorhisms h, g if they agree on F then they agree on L. The homomorphisms h, g agree on the language R iff for each word x in R we have h(x)=g(x). Test sets are related to equivalence of simple translations on languages and to systems of equations on words, see [Ka]. One of the most important problems in the theory of formal languages was the Ehrenfeucht conjecture: each language has a finite test set.

The conjecture was proved to be true in an existential sense, see [AL] : each language has a finite test set. However for some families of recursive languages there is no algorithm to find such test sets. For context-free languages (cfl's, in short) the construction of test sets was proved to be effective, see [ACK]. However the complexity was too high. The upper bound for the size of test sets for context-free languages given in [ACK] was doubly exponential, later it was improved to single exponential in [KRJ]. The main result of this paper is the proof that polynomial size test sets are sufficient and that they can be found by a polynomial time algorithm. The size of the problem is the size of the context-free grammar (of the cfg, in short). For simplicity assume that the considered cfg's are in Chomsky normal form and the size n is the number of productions of a given cfg.

The paper has the following structure. Small test sets are constructed for a sequence of three languages, starting from a finite language L_4. The test set for a simpler language is later used for more complicated language.

We prove, using the technical key lemma, that each linear cfl has a polynomial size test set F. We show that for every cfl L we can find a linear cfl L' which is a test set for L. Then we take the test set F of the test set of L. This gives polynomial size test sets for cfl's.

The crucial role in the paper plays the language L_4 generated by the following grammar:

$$A_4 \to a_4 A_3 \underline{a}_4 \quad A_4 \to b_4 A_3 \underline{b}_4$$
$$A_3 \to a_3 A_2 \underline{a}_3 \quad A_3 \to b_3 A_2 \underline{b}_3$$
$$A_2 \to a_2 A_1 \underline{a}_2 \quad A_2 \to b_2 A_1 \underline{b}_2$$
$$A_1 \to a_1 \underline{a}_1 \quad A_1 \to b_1 \underline{b}_1 \ .$$

The convention is that the capital letters are here nonterminals and small letters are terminals. The starting nonterminal is A_4.

The language L_4 consists of 16 words. The crucial point is the fact that any subset of L_4 of size 15 is a test set for L_4. It is enough to prove it for one such subset. Denote by T_4 the language L_4 without the word $b_4b_3b_2b_1\underline{b}_1\underline{b}_2\underline{b}_3\underline{b}_4$. The following technical fact is the key point in the proof about the polynomial size test sets for cfl's. We postpone its proof, it is given in the last section.

Lemma 1 (key lemma)
The language $T_4 = L_4 - \{\ b_4b_3b_2b_1\underline{b}_1\underline{b}_2\underline{b}_3\underline{b}_4\ \}$ is a test set for L_4.

The proof is long and complicated. Its complicateness is understandable in view of the fact that a version of the key lemma in [ACK] follows directly from our key lemma. This version and the key lemma in [ACK] are syntactically almost the same and they have the same power: they can be used essentially in the same way to prove the main result in [ACK]. The proof of the key lemma in [ACK] took about seven pages. To acquaint the reader with the properties of L_4 we show how it follows from Lemma 1. We formulate the following version of the key lemma from [ACK]:

(#) R'_4 is a test set for R_4,

where R_4 is the language generated by the grammar

$$A_4 \to A_3 \quad A_4 \to b_4 A_3 \underline{b}_4 \qquad A_3 \to A_2 \quad A_3 \to b_3 A_2 \underline{b}_3$$
$$A_2 \to A_1 \quad A_2 \to b_2 A_1 \underline{b}_2 \qquad A_1 \to \varepsilon \quad A_1 \to b_1 \underline{b}_1 \ .$$

and R'_4 is the language R_4 without the word $b_4b_3b_2b_1\underline{b}_1\underline{b}_2\underline{b}_3\underline{b}_4$.

If we consider only homomorphisms which erase letters a_i and $\underline{a}_i$ then we obtain the fact (#) as a simple consequence of Lemma 1. The key lemma in [ACK] trivially follows from (#), since it has stronger assumption than (#) and the same thesis.

2. Construction of test sets for linear cfl's

Let L be a linear cfl generated by a cfg G = (N, T, P, S). N is the set of nonterminals, T is the set of terminals and P is the set of productions, S is the starting symbol. The cfg is linear if the right side of each production contains at most one nonterminal. Each production of G is of the form A->xBy, or A->x, where x, y are strings of terminal symbols and B is a nonterminal.

A graph theoretical approach is used. We construct a multigraph named graph(G) whose edges correspond to productions of the grammar G and nodes correspond to nonterminal symbols. One additional sink node t is added which does not correspond to any nonterminal. The constructed graph has also one distinguished source node S corresponding to the starting nonterminal.

The production A->xBy corresponds to an edge from A to B labelled (x,y). The production A->x corresponds to the edge from A to t labelled (u,v), where u, v are any words such that x=uv. Many distinct edges leading from A to B are possible. Hence the size of the graph measured as the number of edges equals the size of the grammar.
With each path π of graph(G) we associate a word named $[\pi]$ defined as follows:

if $(x_1,\underline{x}_1)$, $(x_2,\underline{x}_2)$, ..., $(x_k, \underline{x}_k)$ is a sequence of labels of consecutive edges of the path π, then $[\pi] = x_1\, x_2..x_k\, \underline{x}_k..\underline{x}_2\, \underline{x}_1$.

Fact

$L(G) = \{\ [\pi] : \pi$ is a path from S to t in graph(G)$\}$.

For each node of the graph construct a directed spanning tree named Tree(v) rooted at v. Its nodes are all nodes accessible from v. Tree(v) is a simple graph, there is at most one directed edge between any pair of nodes.

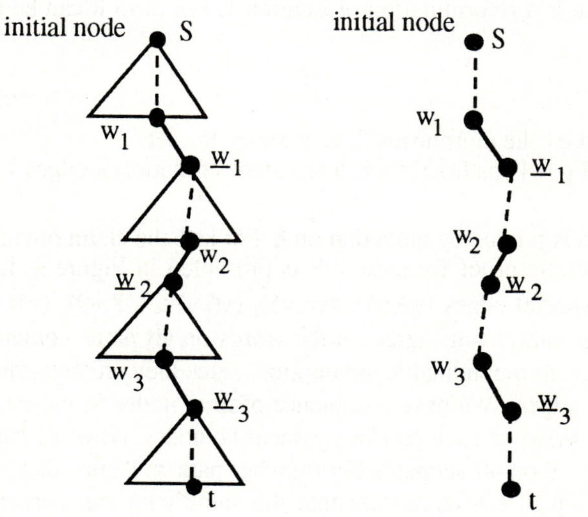

Fig.1. The structure of path $((w_1,\underline{w}_1), (w_2,\underline{w}_2), (w_3,\underline{w}_3))$. S is a source node and t is the sink

Let $\lambda = (w_1,\underline{w}_1),\ (w_2,\underline{w}_2),\ ...,\ (w_k,\underline{w}_k)$ be any sequence of edges of graph(G) not necessarily forming a path, where w_i, $\underline{w}_i$ are endnodes of the edges. Define path(λ) as a path π from S to t which has the following structure. The path π goes from S by a path in Tree(S), then it goes through $(w_1,\underline{w}_1)$, then follows in Tree($\underline{w}_1$) to w_2, then traverses $(w_2,\underline{w}_2)$ and continues in Tree($\underline{w}_2$). At the end it traverses $(w_k,\underline{w}_k)$ and goes in Tree($\underline{w}_k$) from $\underline{w}_k$ to the sink node t, see Figure 1 for the case k=3. For some sequences λ such a path can be undefined (nil). The edges of λ are called special edges.

Denote by F_k the set { [path(λ)] : λ is a sequence of at most k edges in Graph(G)}.

Lemma 2
Assume G is a linear context-free grammar generating L. Then there is a test set F for L of polynomial size. F can be constructed in polynomial time.
Proof.

Let $F=F_6$. It is enough to prove that F_6 is a test set for the language L generated by a linear cfg G. Let h, g be two homomorphisms which agree on F_6. The main point is the application of Lemma 1.

Assume we have a sequence $v_1, v_2, ..., v_m$ of $m \geq 5$ nodes and for each $i<m$ we have two subpaths from v_i to v_{i+1}. Let D_m be the multiset consisting of 2^{m-1} paths from v_1 to v_m formed by combining these subpaths in all possible ways (if two distinct combinations of subpaths give the same path then we consider the resulting path as two distinct copies of the same path). Let D'_m be a set resulting by removing a single path from D_m. Then the following claim is an easy consequence of Lemma 1.

Claim 1

If h, g agree on all words corrresponding to paths in D'_m then h, g agree for all words corresponding to paths in D_m, for $m \geq 5$.

For m=5 this claim is a reformulation of Lemma 1. For m>5 it can be easily proved by induction on m.

Claim 2

For each $k \geq 1$ the morphisms h, g agree on the set

$$F_k = \{ [\text{path}(\lambda)] : \lambda \text{ is a sequence of at most k edges } \}.$$

The claim is proved by induction on k. For $k \leq 6$ the claim obviously holds, since $F = F_6$. The schematic proof for case k=7 is presented in Figure 2. Let π be the path determined by 7 special edges (v2,v3), (v4,v5), (v6,v7), (v8,v9), (v10,v11), (v12,v13) and (v14,v15). The morphisms agree on the words on all paths containing at least one dashed subpath, due to the inductive assumption, since they are determined by a smaller number of special edges. We have a sequence of m=5 nodes S, v5, v9, v13, t. We have also two subpaths between each pair of consecutive nodes. Now we can apply Claim 1. The morphisms agree on all subpaths but (maybe) path π. Hence they agree also on the path π due to Claim 1. (Observe here that the same path can correspond to distinct sequences of special edges.) This completes the proof of Claim 2.

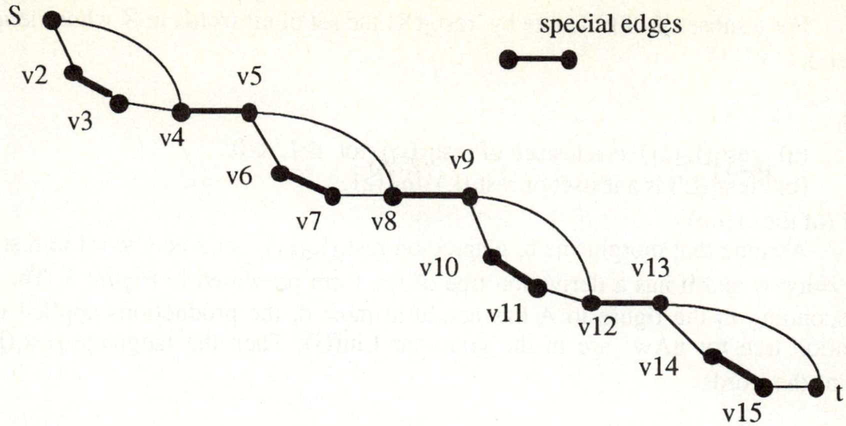

Fig.2.

The thesis follows directly from Claim 2. This completes the proof of the lemma.

3. Construction of test sets for general cfl's

We need the following technical lemma, whose easy proof is omitted.

Lemma 3
If morphisms h, g agree on words u_1v_1, u_1v_2 and u_2v_1 then they agree on u_2v_2 .

Theorem 1 (main result)
Assume G is a context-free grammar generating L. Then there is test set F for L of polynomial size with respect to the size of G. The set F can be constructed in polynomial time.

Proof.

Let L be a cfl generated by a cfg G = (N, T, P, S) in Chomsky normal form and without useless nonterminals. Define a linear cfg Lin(G) which generates a subset L' of L. L' is a linear cfl. For each nonterminal A let w_A be a shortest terminal word generated from A.

The grammar Lin(G) results from G by replacing each production A -> BC by three productions A-> Bw_C , A -> w_BC , A -> w_Bw_C , where B, C are nonterminals. The productions of the form A -> w are unchanged if w is a terminal word.

The height of the node in the derivation tree is its distance (number of edges on the path to) from the root, the height of the root is zero. Let L_d be the language consisting of all words x which have a derivation tree such that all productions corresponding to nodes at height at most d are in the grammar Lin(G) and all other productions are in Lin(G) or G. Let L_0 = L.

For a subset X of L denote by $rest_d(X)$ the set of all words in X whose length is at most d.

Claim

(a) $rest_i(L_{d+1})$ is a testset of $rest_i(L_d)$, for $i \geq 1$, $d \geq 0$

(b) $rest_i(L')$ is a testset of $rest_i(L)$, for $i \geq 1$.

Proof (of the claim)

Assume that morphisms h, g agree on $rest_i(L_{d+1})$. Let z be a word in $rest_i(L_d)$ Then z=uvxw and it has a derivation tree of the form presented in Figure 3. The node corresponding in the figure to A has height at most d, the productions applied in the derivation tree for uAw are in the grammar Lin(G). Then the language $rest_i(L_{d+1})$ contains the words

$$uw_Bxw\ ,\ uvw_Cw\ ,\ uw_Bw_Cw.$$

Hence morphisms h, g agree on these words. Then they agree on z = uvxw , due to Lemma 3. This completes the proof of the point (a).

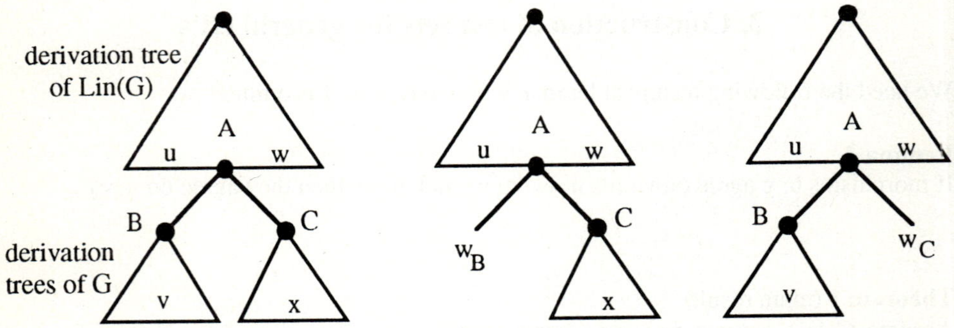

Fig.3

It follows from the point (a) that $rest_d(L') = rest_d(L_d)$ is a test set for $rest_d(L) = rest_d(L_0)$. This completes the proof of the claim.

The testsets have a property that if a language H_i is a test set of a language W_i then the union of the languages H_i is a testset of the union of the languages W_i. Taking H_i=$rest_i(L')$, W_i=$rest_i(L)$ we get the thesis of the theorem.

The next theorem is a simple consequence of Theorem 1.

Theorem 2

The problem of testing equality of homomorphisms on context free languages is solvable in polynomial time. It is also computable in polylogarithmic parallel time with a polynomial number of processors.

4. Sketch of the proof of the key lemma

The proof is similar to the proof of the key lemma in [ACK]. Its idea is to show that certain systems of equations on words imply other equations. The key tool is the periodicity in words, similarly as in [ACK]. Instead of showing that two words are identical it is enough to show that they have the same period and the same length. The proofs of facts related to periodicities are omitted. Useful properties of periodic and primitive words are used, see [Lo].

We use some auxiliary languages. Denote by L_i the language consisting of words generated from the nonterminal A_i, for i=1, 2, 3, 4. Also denote

$\mu = b_3 b_2 b_1 \underline{b}_1 \underline{b}_2 \underline{b}_3$.

Let us fix morphisms h, g. Then we write x" instead of h(x), and x' instead of g(x) for each word denoted by x or y, and for single letters a_i , $\underline{a}_i$, b_i , and $\underline{b}_i$. This shortens the presentation. The assumption of the lemma is equivalent to the following two conditions:

(*) $a''_4 x'' \underline{a}''_4 = a'_4 x' \underline{a}'_4$ for each x in L_3

(**) $b''_4 x'' \underline{b}''_4 = b'_4 x' \underline{b}'_4$ for each x in $L_3 - \{\mu\}$.

Observe that $b_4 b_3 b_2 b_1 \underline{b}_1 \underline{b}_2 \underline{b}_3 \underline{b}_4 = b_4 \mu \underline{b}_4$. The key lemma can be now stated as follows:

Lemma 1' (reformulation of the key lemma)

If (*) and (**) hold then $b''_4 \mu'' \underline{b}''_4 = b'_4 \mu' \underline{b}'_4$.

The general idea of the proof of Lemma 1' is to reduce the problem to simpler ones involving only a single morphism. Properties of the periodicities in strings are used to solve these problems. Denote $T_3 = L_3 - \{\mu\}$

Lemma 4

Let p, q, v, $\underline{v}$, w, $\underline{w}$, and z, $\underline{z}$ be fixed words. Then

(a) if $x'' \in (pq)^* p$ for each $x \in T_3$ then $\mu'' \in (pq)^* p$;

(b) if $vx'' = x''\underline{v}$ for each $x \in T_3$ then $v\mu'' = \mu''\underline{v}$.

(c) if $wx''\underline{w} = zx''\underline{z}$ for each $x \in T_3$ then $w\mu''\underline{w} = z\mu''\underline{z}$.

The technical proof of (a) is given in the full version. Point (b) follows directly from the point (a) due to the fact that equation $vx''=x''\underline{v}$ implies that v is a period of x". Point (c) is a simple consequence of (b). Observe that (b) is a special case of (c).

Lemma 5

(a) Let u, $\underline{u}$, q, α, β be fixed words such that q is not a suffix of α and no nonempty prefix of q is a prefix of β (periodicity q breaks at β). Then

if $ux''\underline{u} \in \alpha q^* \beta$ for each $x \in T_3$ then $u\mu''\underline{u} \in \alpha q^* \beta$,

(b) Assume that $vx''=x'\underline{v}$ for each $x \in L_3$, $zx''=x'\underline{z}$ for all $x \in T_3$. Then $z\mu'' = \mu'\underline{z}$.

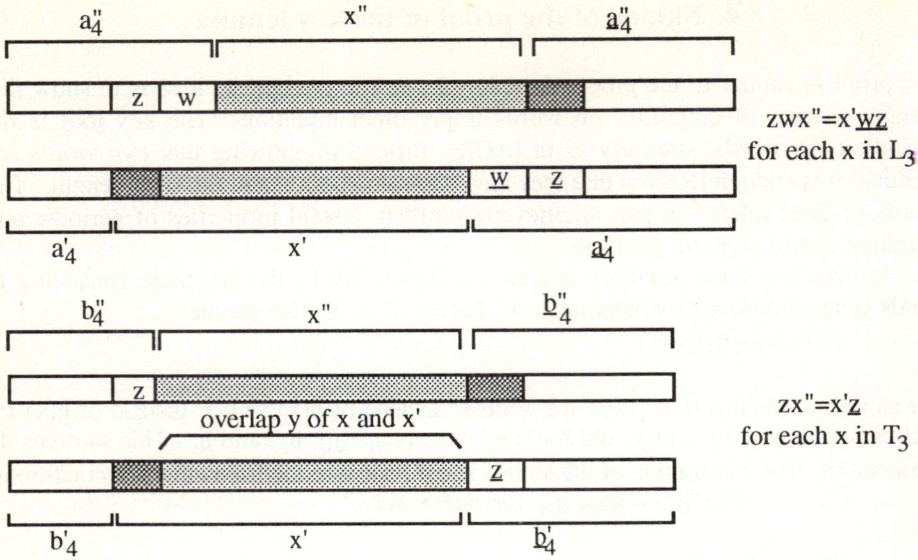

Fig.4. Relation of point (b) to periodicities: x"=zy, where y has period w .

In the proof we need only point (a). However the proof of point (b) is easier. We provide here some informal arguments for the relation between points (b) and periodic structure of words. We consider the basic case, when z is a prefix of v (v=zw, for some w) and z is a suffix of v. Let y, z, w, z and w be as shown in Figure 4, they are related to Case 4 in the proof. Hence wy=yw . This fact implies that all y (corresponding to all elements x of T_3) have the same period w and |zx"|=|x'z| for all x in L_3. zx" and x'z have the same periodic structure. Roughly speaking point (a) implies that $z\mu$" and μ'z have also the same periodic structure. This, together with equality $|z\mu''|=|\mu'z|$ implies that $z\mu''= \mu'z$.

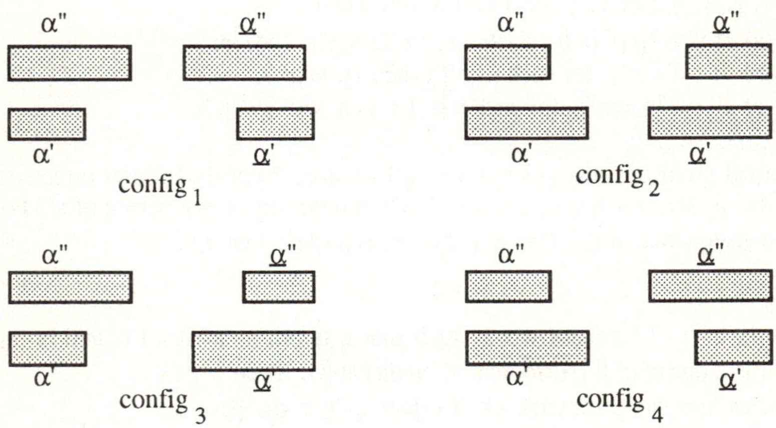

Fig.5.

Let $\leq_p$ be the relation "to be a prefix of" and let $\leq_s$ be the relation "to be a sufix of". We define four conditions for a 4-tuple of words α'', $\underline{\alpha}''$, α', $\underline{\alpha}'$ see Figure 5.

$$\begin{aligned}
config_1(\alpha'', \alpha', \underline{\alpha}'', \underline{\alpha}') &= (\alpha' \leq_p \alpha'') \text{ and } (\underline{\alpha}' \leq_s \underline{\alpha}'');\\
config_2(\alpha'', \alpha', \underline{\alpha}'', \underline{\alpha}') &= (\alpha'' \leq_p \alpha') \text{ and } (\underline{\alpha}'' \leq_s \underline{\alpha}');\\
config_3(\alpha'', \alpha', \underline{\alpha}'', \underline{\alpha}') &= (\alpha' \leq_p \alpha'') \text{ and } (\underline{\alpha}'' \leq_s \underline{\alpha}');\\
config_4(\alpha'', \alpha', \underline{\alpha}'', \underline{\alpha}') &= (\alpha'' \leq_p \alpha') \text{ and } (\underline{\alpha}' \leq_s \underline{\alpha}'').
\end{aligned}$$

The following crucial fact follows directly from the definitions:

if the conditions (*), (**) are satisfied then
$config_i(a''_4, a'_4, \underline{a}''_4, \underline{a}'_4)$ and $config_j(b''_4, b'_4, \underline{b}''_4, \underline{b}'_4)$ hold for some $1 \leq i, j \leq 4$.

Assume conditions (*), (**) are satisfied. The proof of the key lemma proceeds by considering several cases depending on integers i, j in the fact above. Fortunately several pairs of them are symmetric. Our goal is to prove that $b''_4\mu''\underline{b}''_4 = b'_4\mu'\underline{b}'_4$. In each case we have certain type of periodicities. There are essentially 4 cases to be considered.

Case 1 { $config_1(a''_4, a'_4, \underline{a}''_4, \underline{a}'_4)$ holds: $(a'_4 \leq_p a''_4)$ and $(\underline{a}'_4 \leq_s \underline{a}''_4)$}

We have the situation presented in Figure 6. Hence $x' = ux''\underline{u}$ for each x in L_3, in particular $\mu' = u\mu''\underline{u}$. Now we know that $b''_4x''\underline{b}''_4 = b'_4x'\underline{b}'_4 = b'_4ux''\underline{u}\underline{b}'_4$ for each x in T_3. It is sufficient to prove that $b''_4\mu''\underline{b}''_4 = b'_4u\mu''\underline{u}\underline{b}'_4$.

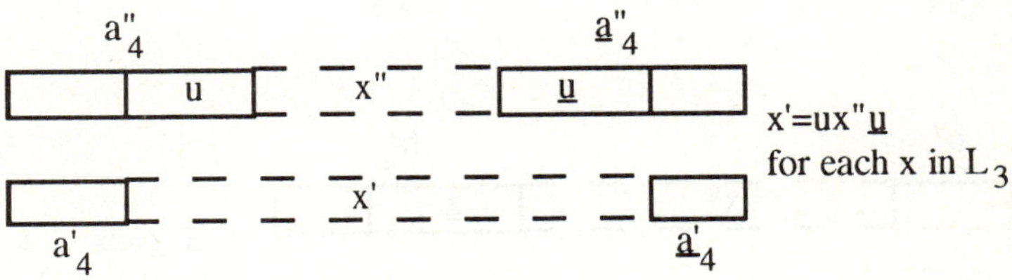

Fig.6.

However the following implication follows directly from Lemma 4(c) :

if $b''_4x''\underline{b}''_4 = b'_4ux''\underline{u}\underline{b}'_4$ for each x in T_3 then $b''_4\mu''\underline{b}''_4 = b'_4u\mu''\underline{u}\underline{b}'_4$.
Hence $b''_4\mu''\underline{b}''_4 = b'_4\mu'\underline{b}'_4$, since $\mu' = u\mu''\underline{u}$. This completes the proof for this case.

Case 2 {$config_1(b''_4, b'_4, \underline{b}''_4, \underline{b}'_4)$ holds}

In this case we have the situation presented in Figure 7. $x' = ux''\underline{u}$ for each x in T_3. On the other hand we know that $a''_4x''\underline{a}''_4 = a'_4x'\underline{a}'_4$ for each x in L_3, this implies that $a''_4x''\underline{a}''_4 = a'_4ux''\underline{u}\underline{a}'_4$ for each x in T_3 and $a''_4\mu''\underline{a}''_4 = a'_4\mu'\underline{a}'_4$. However Lemma 4(c) implies that $a''_4\mu''\underline{a}''_4 = a'_4u\mu''\underline{u}\underline{a}'_4$.

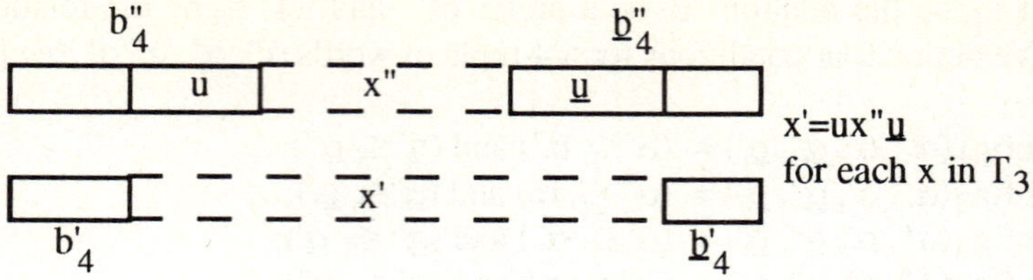

Fig.7.

Now the equalities $a''_4\mu''\underline{a}''_4 = a'_4\mu'\underline{a}'_4$ and $a''_4\mu''\underline{a}''_4 = a'_4 u\mu''\underline{u}\underline{a}'_4$ imply that $\mu' = u\mu''\underline{u}$. Hence $b''_4\mu''\underline{b}''_4 = b'_4\mu'\underline{b}'_4$, due to the equality $\mu' = u\mu''\underline{u}$. This completes the proof for this case.

Case 3 {config$_4$(a''_4, a'_4, $\underline{a}''_4$, $\underline{a}'_4$) and config$_3$(b''_4, b'_4, $\underline{b}''_4$, $\underline{b}'_4$) hold simultaneously}

We have the situation presented in Figure 8. Hence $zux'' = zx'\underline{u} = x''\underline{z}\underline{u}$. For each x in T_3 we have the equation $zux'' = x''\underline{z}\underline{u}$. Lemma 4(b) implies that also $zu\mu'' = \mu''\underline{z}\underline{u}$. The last equation, together with $u\mu'' = \mu'\underline{u}$ implies that $z\mu'\underline{u} = \mu''\underline{z}\underline{u}$. We cut off the suffix $\underline{u}$ and we obtain the equation $z\mu' = \mu''\underline{z}$. It implies the required equation $b''_4\mu''\underline{b}''_4 = b'_4\mu'\underline{b}'_4$.

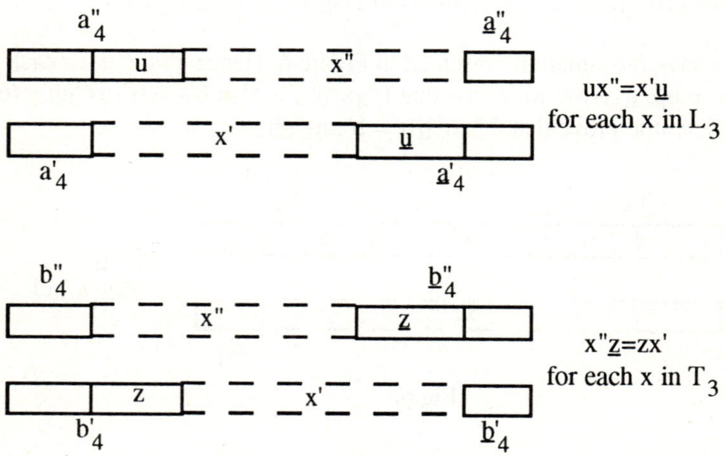

Fig.8.

Case 4 { config$_4$(a''_4, a'_4, $\underline{a}''_4$, $\underline{a}'_4$) and config$_4$(b''_4, b'_4, $\underline{b}''_4$, $\underline{b}'_4$) }

We have the situation $vx''=x'\underline{v}$ for all x in L_3 and $zx''=x'\underline{z}$ for all x in T_3 , for some strings v, $\underline{v}$, z and $\underline{z}$, see Figure 9. See also Figure 4, where a basic subcase is presented.

Then, due to Lemma 5(c) $z\mu'' = \mu'\underline{z}$. It implies directly the required equation $b''_4\mu''\underline{b}''_4 = b'_4\mu'\underline{b}'_4$, This ends the proof for this case.

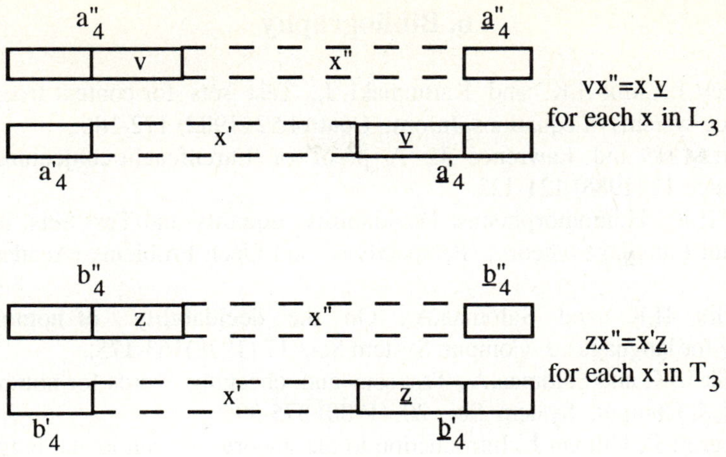

Fig.9.

Two remaining cases config$_2$(a"$_4$, a'$_4$, $\underline{a}$"$_4$, $\underline{a}$'$_4$) and config$_2$(b"$_4$, b'$_4$, $\underline{b}$"$_4$, $\underline{b}$'$_4$) can be considered analogously as Case 1 and Case 2, respectively. The case (config$_3$(a"$_4$, a'$_4$, $\underline{a}$"$_4$, $\underline{a}$'$_4$) and config$_4$(b"$_4$, b'$_4$, $\underline{b}$"$_4$, $\underline{b}$'$_4$) hold simultaneously) is symmetric to Case 3. Other cases are symmetric to Case 4. This completes the proof of the key lemma.

5. Final remarks

We have proved that test sets for context free languages are of polynomial size. The degree of the corresponding polynomial depends on the number of special edges in selected paths. Hence the polynomial is n^6, as we take all subsets of 6 edges, and the edges one-to-one correspond to productions. The time complexity of computing test sets is of the same order. It could be interesting to lower the degree of the corresponding polynomials. The assumption about Chomsky normal form was not essential. Our estimation for the size of test sets is essentially the same for general cfl's and for linear cfl's. Is that possible to improve the size of test sets for linear grammars? The crucial point in the proof was that the language L_4 has a test set T_4 of size 15. We are able to prove that the number 15 can be replaced by 12. However we do not know how to use it to improve the size of test sets for general cfl's. It is also interesting to note that the key lemma does not hold if we replace the number 4 by 3. We are able to prove that T_3 is not a test of L_3. On the basis of this result we can prove that the lower bound for the size of test sets for cfl is $\Omega(n^3)$. This shows the nontriviality of our key lemma. It is related to a similar fact in the key lemma of [ACK], in which 4 pumpings are sufficient and 3 pumpings are not.

6. Bibliography

[ACK] Albert,J.,Culik II,K. and Karhumaki J., Test sets for context free languages and algebraic systems of equations, Inform. Control 52 (1982) 172-186.,

[AL] Albert,M.H. and Lawrence J., A proof of Ehrenfeucht conjecture, Theoret. Comp. Science 41 (1985) 121-123,

[Cu] Culik II,K., Homomorphisms: Decidability, Equality and Test Sets, in R. Book (ed.), Formal Language Theory, Perspectives and Open Problems (Academic Press, 1980).,

[CuS1] Culik II,K. and Salomaa,A., On the decidability of homomorphism equivalence for languages. J. Comput. System Sci. 17 (1978)163-175.,

[CuS2] Culik II,K. and Salomaa,A., Test sets and checking words for homomorphism equivalence, J. Comput. System Sci. 20 (1980) 375-395.,

[HU] Hopcroft J., Ullman J., Introduction to the theory of automata, languages and computation, Addison Wesley (1979),

[Ka1] Karhumaki,J., The Ehrenfeucht conjecture: a compactness claim for finitely generated free monoids, Theoret.Comp.Science 29 (1984) 285-308,

[Ka2] Karhumaki,J., On recent trends in formal language theory, Lect. Notes in Comp. Science 267 (1987) 136-162,

[KRJ] Karhumaki J., Rytter W. and Jarominek S., Efficient construction of test sets for regular and context free languages, MFCS'91, Lect.Notes in Comp.Science (1991)

[Lo] Lothaire, Combinatorics on words, Academic Press (1985)

Quasi-Deterministic 0L Systems

Taishin Y. NISHIDA

School of Computer and Cognitive Sciences
Chukyo University
101 Tokodate Kaizucho, Toyota 470-03, Japan
Email: nishida@sccs.chukyo-u.ac.jp

Abstract. A new variation of 0L systems, called quasi-deterministic 0L systems, is defined. A quasi-deterministic 0L system is a 0L system which generates a bounded set of words at every step, where "bounded" means that the cardinality of the set is always less than a given constant. The family of languages generated by quasi-deterministic 0L systems is properly included by the family of 0L languages and properly includes that of D0L languages. A 0L system is effectively determined whether it is quasi-deterministic or not.

Introduction

Among many stimulative issues concerning L systems [3, 5, 8, 9, 10], it is still attractive to investigate the cardinality of the sets which are generated by an L system in exactly n steps. Dassow has noticed this problem first and he has introduced the notion of the differentiation function of DT0L systems which is the mapping from the steps to the cardinality [1]. He has proved that it is undecidable whether or not two given DT0L systems have the same differentiation function. Similarly, the differentiation function of 0L systems is defined and it is undecidable whether or not two given 0L systems have the same differentiation function. These results mean that a differentiation function cannot be expressed in a closed formula which consists of a finite number of arithmetic operations on a variable corresponding to the steps and a finite number of constants.

The main aim of this paper is to prove that it is decidable whether a differentiation function of a 0L systems is bounded or not. The term "bounded" means that there is an integer C such that the cardinality of the set of words generated in n steps is less than C for every n. If a 0L system is bounded, then it generates a sequence of finite sets of words whose cardinalities are less than C, while a D0L system generates a sequence of singleton sets of a word. This observation gives us an insight that the bounded 0L systems form a new family of 0L systems which is close to the family of D0L systems. So we call them *quasi-deterministic 0L systems* or *D'0L systems* for short.

In this paper we give the definition of D'0L systems. Then we show that the class of D'0L languages (the class of languages which are generated by all D'0L systems) includes that of D0L languages and is included by that of 0L languages. Furthermore, we show that both inclusions are proper. Next, we prove that whether a given 0L system is D'0L or not is effectively decidable. This will be done by classifying the letters appearing in the axiom of the 0L system. Almost all proofs are omitted because of the limit of pages. They will be found in [6].

1 Preliminaries

Let Σ be a finite alphabet. The *length* of a word s over Σ is denoted by $|s|$. If V is any subset of Σ, $|s|_V$ denotes the number of occurrences of letters of V in s. The empty word over Σ is denoted by 1.

Let u and v be two words over Σ. The word v is said to be a *prefix* of u if $u = vt$ for some $t \in \Sigma^*$ and a *suffix* of u if $u = tv$ for some $t \in \Sigma^*$. The word v is said to be a *subword* of u if $u = svt$ for some $st \in \Sigma^*$.

For any set X let card(X) denote the cardinality of X. The empty set is denoted by $\emptyset$. Let x be an element of a set. We denote by x the singleton set $\{x\}$, whenever no confusion occurs.

We denote by $\mathbf{N}$ the set of nonnegative integers and by $\mathbf{N}_+$ the set of positive integers.

Definition [4]. A *0L system* G is a triple $\langle \Sigma, \tau, w \rangle$, where Σ is a finite alphabet, τ is a finite substitution on Σ^*, and w is a word in Σ^* called the *axiom* of G. If τ is a homomorphism on Σ^* rather than a substitution, then G is called *deterministic 0L system* or *D0L system* for short. □

Let $G = \langle \Sigma, \tau, w \rangle$ be a 0L system. The language generated by G is defined by $\tau^*(w)$. A language L is called a 0L language if it is generated by some 0L system.

Let w be a word over Σ and let L be a subset of Σ^*. We denote by alph(w) the set of all and only letters of Σ which actually appear in w, and by alph(L) the set of all and only letters appearing in the words of L. Let τ be a substitution on Σ^*. The *alphabetical projection* of τ, denoted by ψ_τ, is defined as follows:

$$\psi_\tau(x) = \mathrm{alph}(\tau(x)) \text{ for } x \in \Sigma \text{ and}$$

$$\psi_\tau(V) = \mathrm{alph}(\tau(V)) \text{ for } V \subseteq \Sigma.$$

Property 1.1. *For every positive integer k, $\psi_\tau^k = \psi_{\tau^k}$.* □

The reflective and transitive closure of ψ_τ is denoted by Ψ_τ. By Property 1.1, we have $\Psi_\tau(V) = \mathrm{alph}(\tau^*(V))$ for any $V \subseteq \Sigma$.

Let τ be a substitution on Σ^* and let w_1 and w_2 be words over Σ. If $w_1 \in \tau^+(w_2)$, then w_1 is said to be a *descendant* of w_2 and w_2 is said to be an *ancestor* of w_1. Let $w_n \in \Sigma^*$ be a descendant of $w_0 \in \Sigma^*$. A sequence of words $(w_0, w_1, \ldots, w_n)$ is called a *derivation process* from w_0 to w_n if $w_{i+1} \in \tau(w_i)$ for every $i = 0, \ldots, n-1$.

Let $u = a_1 \cdots a_k$ $a_i \in \Sigma$ and $v = b_1 \cdots b_l$ $b_i \in \Sigma$ be two words such that $v \in \tau^n(u)$. A letter a_i is said to be an *ancestor* of a letter b_j if $v = v'sb_jtv''$, $v' \in \tau^n(a_1 \cdots a_{i-1})$, $sb_jt \in \tau^n(a_i)$, and $v'' \in \tau^n(a_{i+1} \cdots a_k)$. On the same conditions, b_j is called a *descendant* of a_i. Let $\Delta = (w_0, \ldots, w_n)$ be a derivation process from w_0 to w_n where $w_0 = a_{01}a_{02} \cdots a_{0l_0}$, $w_i = a_{i1} \cdots a_{il_i}$, and $w_n = a_{n1}a_{n2} \cdots a_{nl_n}$ $a_{kj} \in \Sigma$ and let a_{0j_0} be an ancestor of a_{nj_n}. Then a sequence of letters $(a_{0j_0}, a_{1j_1}, \ldots, a_{nj_n})$ is said to be the *derivation line* from a_{0j_0} to a_{nj_n} in Δ if a_{ij_i} is an ancestor of $a_{(i+1)j_{i+1}}$ for every $i = 0, \ldots, n-1$.

2 Quasi-Deterministic 0L Systems

Now we give the central notion of this paper.

Definition. A 0L system $G = \langle \Sigma, \tau, w \rangle$ is said to be *quasi-deterministic* if there exists a positive integer k such that $\mathrm{card}(\tau^n(w)) < k$ for every nonnegative integer n. □

We abbreviate a quasi-deterministic 0L system to a D'0L system. The next property is directly follows from the definition.

Property 2.1.
(i) Every D0L system is a D'0L system.
(ii) Every finite 0L system, i.e., a 0L system which generates a finite language, is a D'0L system. □

Let us denote by $\mathcal{L}$(D0L) (resp. $\mathcal{L}$(D'0L), $\mathcal{L}$(0L)) the family of all languages generated by all D0L (resp. D'0L, 0L) systems. The following relations obviously hold.

$$\mathcal{L}(\mathrm{D0L}) \subseteq \mathcal{L}(\mathrm{D'0L}) \subseteq \mathcal{L}(\mathrm{0L}).$$

Since there are 0L languages which are not generated by any D0L systems, at least one of the above two inclusions is proper. Indeed, the next lemma and theorem show that both inclusions are proper.

Lemma 2.2. *A U0L system $G = \langle \{a\}, \tau, a^n \rangle$ (i.e., a 0L system over a unary alphabet) is a D'0L system if and only if G is a D0L system (i.e., $\tau(a) = a^k$ for some $k \geq 0$) or G is a finite 0L system (i.e., $\tau(a) = 1$, $\tau(a) = a$, or $\tau(a) = \{a, 1\}$).* □

Theorem 2.3. $\mathcal{L}(\mathrm{D0L}) \subset \mathcal{L}(\mathrm{D'0L}) \subset \mathcal{L}(\mathrm{0L})$.

Proof. The infinite language $\{a^{2i} | i \geq 0\}$ generated by the 0L system $\langle \{a\}, \tau(a) = \{aa, 1\}, aa \rangle$ cannot be generated by any D0L system. The finite language $\{a^i | 0 \leq i \leq n\}$ generated by the 0L system $\langle \{a\}, \tau(a) = \{a, 1\}, a^n \rangle$ cannot be generated by any D0L system. Then the theorem follows from Lemma 2.2. □

Although Theorem 2.3 gives an interesting characterization for the family of D'0L languages, the definition of D'0L systems does not say how to decide whether a given 0L system is a D'0L system or not. For any 0L system G, it is easily decided whether $L(G)$ is finite (cf. [9]) and whether G is a D0L system or not is trivial. But for an infinite nondeterministic 0L system G, it is not so easy to decide whether G is a D'0L system or not; see the following example.

Example 2.1. Let $G_1 = \langle \Sigma, \tau_1, a \rangle$ and $G_2 = \langle \Sigma, \tau_2, a \rangle$ be 0L systems where $\Sigma = \{a, b, c\}$, $\tau_1(a) = \{ab, cb\}$, $\tau_1(b) = b$, $\tau_1(c) = cb$, $\tau_2(a) = \{ab, cb\}$, $\tau_2(b) = b$, and $\tau_2(c) = bc$. Then, using inductions on n, we can prove $\tau_1^n(a) = \{ab^n, cb^n\}$ and $\tau_2^n(a) = ab^n \cup \{b^i c b^{n-i} | 0 \leq i < n\}$ for every $n > 0$. That is, G_1 is a D'0L system and G_2 is not. □

Now we introduce a new notion.

Definition. Let τ be a substitution on Σ^*. A word w in Σ^* is said to be *k-divergent* for τ if $\mathrm{card}(\tau^n(w)) \leq k$ for every $n \geq 0$ and $\mathrm{card}(\tau^m(w)) = k$ for some $m \geq 0$. If there is no such an integer k, then w is called ∞-*divergent* for τ. A word w is called *finitely divergent* for τ if w is k-divergent for some $k \in \mathbf{N}_+$. □

Note that a 0L system $G = \langle \Sigma, \tau, w \rangle$ is a D'0L system if and only if the axiom w is finitely divergent for τ.

We show that there is a subset of Σ^* such that for every word w in the subset w is finitely divergent if and only if $\mathrm{card}(\tau^*(w))$ is finite.

Definition [7]. A word w over Σ is said to be *repeatable* for τ if it is a descendant of itself, i.e., $w \in \tau^+(w)$. Let $G = \langle \Sigma, \tau, w \rangle$ be a 0L system. The set of repeatable words generated by G is called the *repeatable language* and denoted by $P(\tau, w)$: $P(\tau, w) = \{u \in \tau^*(w) | u \in \tau^+(u)\}$. □

Theorem 2.4 is the fundamental theorem for repeatable words and repeatable languages.

Theorem 2.4 (Theorem 4.1 and Theorem 5.1 in [7]). *The set of repeatable words for a finite substitution is a regular set. The class of repeatable languages equals to the class of context free languages.* □

The next theorem states a relation between repeatable words and ∞-divergent words.

Theorem 2.5. *Let w be a repeatable word. Then w is ∞-divergent if and only if* $\mathrm{card}(\tau^*(w)) = \infty$.

Proof. Only if part is a contraposition of Property 2.1 (ii).

Let w be a repeatable word such that $\mathrm{card}(\tau^*(w)) = \infty$. Since $w \in \tau^i(w)$ for some positive integer i, we have

$$\tau^{in+j}(w) = \tau^{in+j}(w) \cup \tau^{i(n-1)+j}(w) \cup \cdots \cup \tau^{i+j}(w)$$

for any j $(0 \le j < i)$. Let $L(n)$ be the set

$$L(n) = \tau^{in}(w) \cup \tau^{in+1}(w) \cup \cdots \cup \tau^{i(n+1)-1}(w)$$

for every $n \in \mathbf{N}_+$. Then

$$\lim_{n \to \infty} L(n) = \tau^*(w) - (\bigcup_{\nu=0}^{i-1} \tau^{\nu}(w)).$$

If there is a positive integer N such that for every $k \ge 0$ $\mathrm{card}(\tau^k(w)) < N$, then $\mathrm{card}(L(n)) < iN$. Because i is a fixed integer, this contradicts the assumption $\mathrm{card}(\tau^*(w)) = \infty$. □

Now the remaining task is to decide whether a word w which is not repeatable for τ and which satisfies $\mathrm{card}(\tau^*(w)) = \infty$ is finitely divergent or not. This will be done in the following sections.

3 Classification of Letters

Let k be a positive integer or ∞. We extend the multiplication and the order relation on N_+ to those on the set $N_+ \cup \infty$ by setting:

$$k \cdot \infty = \infty \cdot k = \infty \text{ and } k < \infty \text{ for every } k \neq \infty.$$

Then the following property becomes obvious.

Property 3.1. *Let τ be a finite substitution on Σ^*. Let w and w' be k-divergent and k'-divergent words for τ, respectively.*
(i) Then ww' is m-divergent with $\max(k, k') \leq m \leq kk'$.
(ii) If $w' \in \tau^*(w)$, then $k' \leq k$. □

Property 3.1 (i) shows that a word $w = a_1a_2 \cdots a_l$ is finitely divergent for τ if and only if every a_i ($i = 1, 2, \ldots, l$) is finitely divergent for τ. Now we consider the properties of letters occurring in a finitely divergent word. If a letter $x \in \Sigma$ is a k-divergent word, then x is called a ***k*-divergent letter**. A 1-divergent letter is called a ***deterministic letter***. A letter $x \in \Sigma$ is deterministic if and only if $\tau(a)$ is a singleton for every $a \in \Psi_\tau(x)$.

3.1 Self-embedding Letters

We adopt, throughout this paper and unless otherwise stated, τ a finite substitution on Σ^*, m the cardinality of Σ, and α the maximum length of words in $\tau(\Sigma)$, respectively; i.e., $m = \text{card}(\Sigma)$ and $\alpha = \max_{w \in \tau(\Sigma)} |w|$.

Definition. A letter x in Σ is said to be *vital* if the empty word 1 is not a descendant of x. A letter which always generates 1 is called *mortal*. A letter x in Σ is said to be *transient* if x does not occur any descendant of x, i.e., $x \notin \text{alph}(\tau^+(x))$. A letter which is not transient is called *persistent*. A persistent letter x is said to be *self-embedding* if x generates no repeatable words, i.e., $P(\tau, x) = \emptyset$. The set of vital, mortal, transient, persistent, and self-embedding letters are denoted by V, M, T, E, and S, respectively. □

We summarize vital, mortal, transient, persistent, and self-embedding letters as follows:
Vital letters $V = \{x \in \Sigma | 1 \notin \tau^+(x)\}$,
Mortal letters $M = \{x \in \Sigma | \tau^m(x) = 1\}$,
Transient letters $T = \{x \in \Sigma | x \notin \text{alph}(\tau^+(x))\}$,
Persistent letters $E = \Sigma - T = \{x \in \Sigma | x \in \text{alph}(\tau^+(x))\}$, and
Self-embedding letters $S = \{x \in E | P(\tau, x) = \emptyset\}$.

We note that all mortal letters are transient and there may be nonvital persistent letters, i.e., $M \subseteq T$ and $(\Sigma - V) \cap E \neq \emptyset$. Some persistent letters are not always "persistent". Indeed a persistent letter x may have a descendant $u \in \tau(x)$ such that x does not occur any descendant of u, i.e., $x \notin \text{alph}(\tau^*(u))$.

From the definition of self-embedding letters, the next property is shown.

Property 3.2. *Let x be a self-embedding letter. Then*
(i) x is vital,
(ii) $|st|_V \geq 1$ for every word $sxt \in \tau^+(x)$, and
(iii) $\text{card}(\tau^*(x)) = \infty$. □

Since any repeatable language is context free and the emptyness problem for context free languages is decidable, self-embedding letters are effectively determined.

Proposition 3.3. *Let x be a letter in $E - S$. Then x is ∞-divergent if and only if* $\mathrm{card}(\tau^*(x)) = \infty$.

Proof. We prove if part only because only if part is obvious.

Since x is not self-embedding, there is a word w in $P(\tau, x)$. If $\mathrm{card}(\tau^*(w)) = \infty$ for some $w \in P(\tau, x)$, then x is ∞-divergent by Theorem 2.5.

Next, let us assume that $\mathrm{card}(\tau^*(w)) < \infty$ for every w in $P(\tau, x)$. Since $\mathrm{card}(\tau^*(x)) = \infty$, there exists $sxt \in \tau^i(x)$ such that $|st|_V > 0$ for some i. Then there exist three sequences of words $(s = s_0, s_1, \ldots, s_n, \ldots)$, $(t = t_0, t_1, \ldots, t_n, \ldots)$, and $(w_1, \ldots, w_n, \ldots)$ which satisfy $s_n \in \tau^i(s_{n-1})$, $t_n \in \tau^i(t_{n-1})$, $|s_n t_n| > 0$, $w_1 \in \tau^i(x)$, $w_n \in \tau^i(w_{n-1})$, and w_n is repeatable for every $n \in \mathbf{N}_+$. Let

$$X_n = \{s_{n-1} \cdots s_0 x t_0 \cdots t_{n-1}, s_{n-1} \cdots s_1 w_1 t_1 \cdots t_{n-1}, \ldots, w_n\}.$$

Then we have $\tau^{in}(x) \supset X_n$ for every $n \in \mathbf{N}_+$. Since every w_n belongs to a finite set, $\lim_{n\to\infty} \mathrm{card}(X_n) = \infty$. □

3.2 Attendant, Compatible, and Null Letters

Henceforth we assume that x is a nondeterministic finitely divergent letter in S.

Now we give an illustrative example, which will help us to explore the structure of finitely divergent words.

Example 3.1. Let τ be a substitution over $\{a, b, c, d, e, f, f_1\}$ given by

$$\tau(a) = \{eae, ebe, eefee\},\ \tau(b) = \{ebe, ecf_1de\},\ \tau(c) = \{ec, ee\},$$

$$\tau(d) = \{de, ee\},\ \tau(e) = ee,\ \tau(f) = f,\ \tau(f_1) = f.$$

Then obviously

$$\tau^1(a) = \{eae, ebe, eefee\}$$

$$\tau^2(a) = \{e^3ae^3, e^3be^3, e^3cf_1de^3, e^4fe^4\}$$

$$\tau^3(a) = \{e^7ae^7, e^7be^7, e^7cf_1de^7, e^7cfde^7, e^8fde^7, e^7cfe^8, e^8fe^8\}$$

$$\vdots$$

holds. Thus it is easily seen that for every $i \geq 3$ $\tau^i(a)$ has the following "center-embedded" structure

$$\tau^i(a) = e^{2^i-1}Ae^{2^i-1}$$

where $A = \{a, b, cf_1d, cfd, efd, cfe, efe\}$. □

We first give a formal definition for such a letter "e" in this example.

Definition. A letter $y \in E \cap \Psi_\tau(x)$ is said to be *attendant* with respect to x if there exists $sx't \in \tau^+(x)$ such that $y \in \mathrm{alph}(st)$ and $x \in \Psi_\tau(x')$. □

Property 3.4. *Any attendant letter is deterministic and vital.* □

Next we define those letters which appear in the "center" of the center-embedded structure in Example 3.1.

Definition. A letter $y \in S \cap \Psi_\tau(x)$ is said to be *compatible* with respect to x if y is not attendant with respect to x. A compatible letter y with respect to x is called *strongly compatible* if $x \in \Psi_\tau(y)$. Otherwise, y is called *weakly compatible.* □

We note that, as a direct consequence of Property 3.4, x itself is not attendant and is strongly compatible.

A compatible letter y is said to be *two-sided* if there exists $syt \in \tau^+(y)$ such that both s and t contain some attendant letters. Otherwise, y is called *one-sided.* A one-sided letter y is said to be *right-sided* (resp. *left-sided*) if there exists $syt \in \tau^+(y)$ such that only t (resp. s) contains attendant letters. We note that there is no two- and one-sided letter nor left- and right-sided letter.

The following two lemmas make basic properties of compatible letters clear.

Lemma 3.5. *For any $sx't \in \tau^+(x)$, if $x \in \Psi_\tau(x')$ then*
(i) *x' is the only one occurrence of strongly compatible letter with respect to x in $sx't$ and*
(ii) *$\tau^m(st)$ is deterministic.* □

Lemma 3.6. *Let w be a word in $\tau^+(x)$ with no occurrence of strongly compatible letter with respect to x.*
(i) *There are at most two occurrences of weakly compatible letters in w.*
(ii) *If x is one-sided, then there is at most one occurrence of one-sided weakly compatible letter in w.*
(iii) *If w contains two weakly compatible letters with respect to x, i.e., $w = sy_1ty_2u$ where y_1 and y_2 are weakly compatible letters with respect to x, then y_1 is left-sided and y_2 is right-sided.* □

A persistent letter which is not attendant nor compatible may occur in some word $w \in \tau^+(x)$.

Definition. A persistent letter $y \in \Psi_\tau(x)$ is said to be a *null* letter with respect to x if $\mathrm{card}(\tau^*(y)) < \infty$ and there exists a positive integer k such that $|w|_y < k$ for every $w \in \tau^*(x)$. □

We denote by C_x, A_x, and N_x the set of all compatible letters, attendant letters, and null letters with respect to x, respectively. The next proposition ensures that the persistent letters in $\Psi_\tau(x)$ are effectively classified.

Proposition 3.7. $\Psi_\tau(x) \cap E = C_x \cup A_x \cup N_x$ *and* $C_x \cap A_x = A_x \cap N_x = N_x \cap C_x = \emptyset$.

Proof. The latter half is obvious from the definitions.

Let y be a persistent letter in $\Psi_\tau(x)$ such that y is not compatible nor attendant. Then y cannot be self-embedding, that is, $P(\tau, y) \neq \emptyset$ and hence $\mathrm{card}(\tau^*(y)) < \infty$ by Proposition 3.3. Since y is not attendant, there exist a positive integer p and a sequence of words $(x, s_1xt_1, \ldots, s_ixt_i, \ldots)$ such that $s_1xt_1 \in \tau^p(x)$, $s_ixt_i \in \tau^p(s_{i-1}xt_{i-1})$, and $|s_it_i|_y = 0$ $(i = 1, 2, \ldots)$. Hence there must exist a constant k such that $|w|_y < k$ for every $w \in \tau^*(x)$ for otherwise x would be ∞-divergent. □

In Example 3.1, a is two-sided strongly compatible, b is two-sided weakly compatible, c and d are one-sided weakly compatible, e is attendant, and f is null with respect to a; while f_1 is a transient letter.

Using the characterization given in this section, we will show that $\tau^i(x)$ has a center-embedded structure if and only if x is a finitely divergent letter (Theorem 4.1). Before doing this, we make it clear what happen when a strongly compatible letter vanishes in some word $w \in \tau^+(x)$.

3.3 Vanishing of strongly compatible letters

It is proved, in the previous subsection, that there is a word w in $\tau^+(x)$ such that w has no occurrence of strongly compatible letter. Let $\Delta = (w_0 = x, w_1, \ldots, w_n = w)$ be a derivation process for such a word w. Then a strongly compatible letter occurs in every word $w_0, \ldots, w_i$ and no strongly compatible letter occurs in every word $w_{i+1}, \ldots, w_n$ for some integer $i > 0$. In this case, there are an occurrence x' of a strongly compatible letter in w_i and a subword u_{i+j} of w_{i+j} such that $u_{i+j} \in \tau^j(x')$ and that $x \notin \text{alph}(\tau^*(u_{i+j}))$ for $j = 1, \ldots, n-i$. We call $\Delta_t = (x', u_{i+1}, \ldots, u_n)$ *transient process* from x' in Δ.

Definition. Let $w \in \tau^+(x)$ be a word which has no occurrence of strongly compatible letter with respect to x. We call w a *transient word* with respect to x if there exist a derivation process $\Delta = (x, w_1, \ldots, w_n = w)$ and a transient process $\Delta_t = (x', u_{i+1}, \ldots, u_n)$ in Δ such that there is an occurrence of a transient letter a_n in u_n and that all letters but x' in the derivation line $(x', a_{i+1}, \ldots, a_n)$ in Δ_t are transient. □

Property 3.8. *Let $w \in \tau^n(x)$ be a transient word with respect to x for some $n > 0$ and let w_i be the last word in the derivation process $(x, \ldots, w_i, \ldots, w)$ such that w_i has an occurrence of a strongly compatible letter with respect to x. Then $n - i \leq \text{card}(T)$.* □

Let y be a weakly compatible letter with respect to x. Since y itself is strongly compatible with respect to y, there may be a transient word $w \in \tau^+(x)$ with respect to y. In this case, w is, in general, not a transient word with respect to x. For example, in Example 3.1, for every $i \geq 3$ $e^{2^i-1}cf_1de^{2^i-1}$ is a transient word with respect to b but it is not a transient word with respect to a.

Let $w \in \tau^*(x)$ be a word which is not transient with respect to any compatible letter. We consider the eight cases which are illustrated in Table 1, according to the number of compatible letters and null letters occurring in w and whether x is one-sided or two-sided. There are no cases corresponding to the – sign in Table 1 because of Lemma 3.6.

x is	number of occurrences of compatible letters in w				
	two	one		none	
		one-sided	two-sided	with null letters	without null letters
one-sided	–	case 1	–	case 2	case 3
two-sided	case 4	case 5	case 6	case 7	csae 8

Table 1

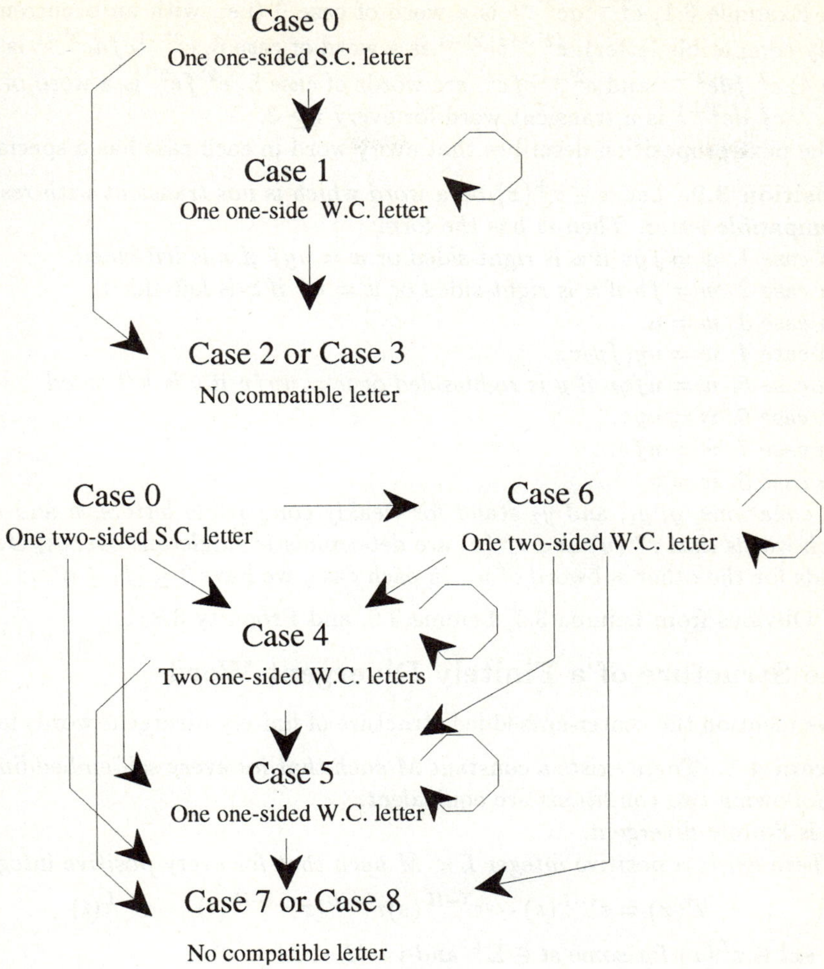

Figure 1. Transitions caused by transient processes. Case 0 stands for the words which has an occurrence of a strongly compatible letter. S.C. and W.C. mean "strongly compatible" and "weakly compatible", respectively.

Transient processes cause transitions between these cases. The possible transitions are illustrated in Figure 1. We note that a derivation process (x', u) is a transient process even if sut is not a transient word as long as it satisfies the following conditions:

(i) $(x, \ldots, s'x't', sut)$ is a derivation process from x to sut,

(ii) $x \notin \mathrm{alph}(\tau^*(u))$, and

(iii) x' is strongly compatible with respect to x.

We also note that the transitions occur at most m times.

In Example 3.1, $e^{2^i-1}ae^{2^i-1}$ is a word of case 0 (i.e., with an occurrence of a strongly compatible letter), $e^{2^i-1}be^{2^i-1}$ is a word of case 6, $e^{2^i-1}cfde^{2^i-1}$ is a word of case 4, $e^{2^i}fde^{2^i-1}$ and $e^{2^i-1}cfe^{2^i}$ are words of case 5, $e^{2^i}fe^{2^i}$ is a word of case 7 and $e^{2^i-1}cf_1de^{2^i-1}$ is a transient word for every $i \geq 3$.

The next proposition describes that every word in each case has a special form.

Proposition 3.9. *Let $w \in \tau^+(x)$ be a word which is not transient with respect to any compatible letter. Then w has the form:*

In case 1. $w = fyv$ if x is right-sided or $w = uyf$ if x is left-sided.
In case 2. $w = fv$ if x is right-sided or $w = uf$ if x is left-sided.
In case 3. $w = u$.
In case 4. $w = uy_1fy_2v$.
In case 5. $w = ufyv$ if y is right-sided or $w = uyfv$ if y is left-sided.
In case 6. $w = uyv$.
In case 7. $w = ufv$.
In case 8. $w = u$.

In the equations, y, y_1, and y_2 stand for weakly compatible letters, u and v stand for such words that $\tau^m(u)$ and $\tau^m(v)$ are deterministic singletons over $A_x \cup M$, and f stands for the other subword of w. In each case, we have $0 \leq |f| \leq \alpha^m$.

Proof. Obvious from Lemma 3.5, Lemma 3.6, and Property 3.8. □

4 The Structure of a Finitely Divergent Word

Now we mention the center-embedded structure of finitely divergent words formally.

Theorem 4.1. *There exists a constant M such that for every self-embedding letter x the following two conditions are equivalent.*
(i) x is finitely divergent.
(ii) There exists a positive integer $L < M$ such that for every positive integer i

$$\tau^i(x) = \tau^{i-L}(s) \cdots \tau^{i-lL}(s)\tau^{i-lL}(x)\tau^{i-lL}(t) \cdots \tau^{i-L}(t)$$

where $sxt \in \tau^L(x)$ for some $st \in \Sigma^+$ and $i - lL < M$.

Then the following algorithm decides whether a self-embedding letter is finitely divergent or not:

1 For every $0 < i \leq M$, find the set A_i such that $\tau^i(x) = s_iA_it_i$ where s_i and t_i are the longest common prefix and suffix, respectively.
2 If there are A_i and A_j $(i \neq j)$ such that $A_i = A_j$, then x is finitely divergent. Otherwise x is infinitely divergent.

In order to prove Theorem 4.1, the following claim must be established for every word of cases 0 to 8 which are considered in the previous section.

Claim. *There is a positive integer M_0 such that for every word $w \in \tau^i(x)$ of case X there exists a positive integer $p < M_0$ satisfying*

$$w = \tau^{i-p}(s) \cdots \tau^{i-lp}(s)w'\tau^{i-lp}(t) \cdots \tau^{i-p}(t)$$

where $w' \in \tau^{i-lp}(x)$, $sxt \in \tau^p(x)$, and $i - lp < M_0$.

In this paper, we only deal with the words of case 0. The other cases are discussed in [6].

Now we assume that every word in $\tau^*(x)$ contains a strongly compatible letter.

Lemma 4.2. *Let $s_1x't_1$ and $s_2x't_2$ be words in $\tau^n(x)$ where n is a nonnegative integer and x' is a strongly compatible letter with respect to x. Then $\tau^{2m-1}(s_1x't_1) = \tau^{2m-1}(s_2x't_2)$.*

Proof. It is sufficient to prove $\tau^{2m-1}(s_1) = \tau^{2m-1}(s_2)$. First observe that $\tau^{2m-1}(s_i)$ ($i = 1,2$) is a singleton. If $\tau^i(s_1) \neq \tau^i(s_2)$ for all $i \geq 2m-1$, then x is ∞-divergent. Therefore, for some i, $\tau^i(s_1) = \tau^i(s_2)$. This implies $\tau^{2m-1}(s_1) = \tau^{2m-1}(s_2)$ since all words in $\tau^m(s_1)$ and $\tau^m(s_2)$ are deterministic (see Theorem 3 in [2]). □

Since $\psi_\tau^i(x)$ is a subset of a finite set Σ for every $i \geq 0$, there exist nonnegative integers $n < 2^m$ and $k \leq 2^m$ such that for every integer $n' \geq n$

$$\psi_\tau^{n'}(x) = \psi_\tau^{n'+k}(x).$$

The integers k and n are called the *period* and *index* of x, respectively. The next lemma shows that the period is less than the cardinality of Σ and the index is less than the square of the cardinality of Σ, i.e., $k < m$ and $n < m^2$.

Lemma 4.3. *For every positive integer p satisfying $sxt \in \tau^p(x)$ for some $st \in \Sigma^+$*

$$\psi_\tau^{ip}(x) = \psi_\tau^{(i+1)p}(x)$$

holds for every $i \geq m$.

Proof. We first note that $\psi_\tau^{ip}(x) = \psi_\tau^{(i+1)p}(x)$ implies $\psi_\tau^{ip}(x) = \psi_\tau^{(i+j)p}(x)$ for every $j \geq 0$. Since $x \in \psi_\tau^p(x)$, we have the ascending sequence

$$\psi_\tau^p(x) \subseteq \psi_\tau^{2p}(x) \subseteq \cdots \subseteq \psi_\tau^{ip}(x) \subseteq \psi_\tau^{(i+1)p}(x) \subseteq \cdots.$$

Because $\psi_\tau^{ip}(x) \subseteq \Sigma$ for every $i \geq 0$, there are at most m different $\psi_\tau^{ip}(x)$'s. □

Let p be the minimum positive integer such that $sxt \in \tau^p(x)$ for some $st \in \Sigma^+$. Then, by Lemma 4.3, the upper bounds of period and index become p and mp, respectively.

Proposition 4.4. *Let k be the period of x and let p be the minimum positive integer satisfying $sxt \in \tau^p(x)$ for some $st \in \Sigma^+$. Then for any nonnegative integer i*

$$\tau^i(x) = \tau^{i-p}(s)\cdots\tau^{i-lp}(s)\tau^{i-lp}(x)\tau^{i-lp}(t)\cdots\tau^{i-p}(t),$$

where $mp + 2m - 1 \leq i - lp < mp + p + 2m - 1$.

Proof. If $i < mp + p + 2m - 1$, then the proposition obviously holds. Now we must prove $\tau^i(x) = \tau^{i-p}(s)\tau^{i-p}(x)\tau^{i-p}(t)$ for every $i \geq mp+p+2m-1$. Let $j = i-(2m-1)$. Then, since $j \geq mp + p$ and p is a multiple of k, $\psi_\tau^j(x) \cap C_x = \psi_\tau^{j-p}(x) \cap C_x$. Any word in $\tau^j(x)$ has the form uyv for some $y \in C_x$. Since $y \in \psi_\tau^{j-p}(x)$, there exists $u'yv' \in \tau^{j-p}(x)$ such that $\tau^{j-p}(s)u'yv'\tau^{j-p}(t) \in \tau^{j-p}(sxt) \subseteq \tau^j(x)$. Then by Lemma 4.2, we have $\tau^{2m-1}(uyv) = \tau^{2m-1}(\tau^{j-p}(s)u'yv'\tau^{j-p}(t))$. Thus we have

$$\tau^i(x) \subseteq \tau^{i-p}(s)\tau^{i-p}(x)\tau^{i-p}(t).$$

Because

$$\tau^i(x) \supseteq \tau^{i-p}(s)\tau^{i-p}(x)\tau^{i-p}(t)$$

always holds for any substitution,

$$\tau^i(x) = \tau^{i-p}(s)\tau^{i-p}(x)\tau^{i-p}(t)$$

is proved. Then the proposition follows immediately. □

Although we get a rather small upper bound for this specific case, the upper bound M for Theorem 4.1 becomes

$$M = 2^{m^{\alpha^m}+1}.$$

I expect that this terrible number is a sham upper bound and there is much smaller one.

References

1. Jürgen Dassow: Eine neue Funktion für Lindenmayer-Systeme, *Elektronische Informationsverarbeitung und Kybernetik* **12**, 515–521 (1976)
2. A. Ehrenfeucht and G. Rozenberg: Simplifications of Homomorphisms, *Information and Control* **38**, 298-309 (1978)
3. G. T. Herman and G. Rozenberg: *Developmental systems and languages*, North Holland (1975)
4. Aristid Lindenmayer: Developmental systems without cellular interactions, their languages and grammars, *Journal of Theoretical Biology* **30**, 455–484 (1971)
5. A. Lindenmayer and G. Rozenberg (eds): *Automata, Languages, Development*, North Holland (1976)
6. T. Nishida: Quasi-deterministic 0L systems, *SCCS Technical Report 92-1-01*, Chukyo University (1992)
7. T. Nishida and Y. Kobuchi: Repeatable words for substitution, *Theoretical Computer Science* **53**, 319–333 (1987)
8. G. Rozenberg and A. Salomaa (eds): *L systems*, Lecture Notes in Computer Science 15, Springer Verlag (1974)
9. G. Rozenberg and A. Salomaa: *The mathematical theory of L systems*, Academic Press (1980)
10. G. Rozenberg and A. Salomaa (eds): *The book of L*, Springer Verlag (1986)

On Growing Context-Sensitive Languages

Gerhard Buntrock[*] Krzysztof Loryś[†]

Abstract

Growing context-sensitive grammars (GCSG) are investigated. The variable membership problem for GCSG is shown to be NP-complete. This solves a problem posed in [DW86]. It is also shown that the languages generated by GCSG form an abstract family of languages and several implications are presented.

Introduction

is well known that the class of languages generated by context-sensitive grammars is ual to NSPACE(n) and that, even for fixed grammars, the membership problem can PSPACE-complete. On the other hand the context-free grammars are known to have, r many applications, too weak derivative power. While many modifications extending ntext-free grammars have been studied, only a few papers concern some restricted verons of context-sensitive grammars. Book ([Boo73]) investigated context-sensitive gramars with certain restrictions on the allowed context or on the use of rewriting rules and owed that some of them cause the grammars to generate only context-free languages. n the other hand when we put the restriction only on the length of derivations to be linr then the context-sensitive grammars are still very powerful, namely they can generate P-complete languages ([Boo78]).

In [DW86] Dahlhaus and Warmuth considered grammars with only growing rules, i.e. les whose right side is longer than the left side. This restriction naturally bounds the ngth of derivations by the length of derivated words. They showed that such growing rammars have the membership problem solvable in polynomial time and even contained the class of problems logspace-reducible to context-free languages. We denote this lass with LOGCFL and refer to [Sud78, Ven91]. With a careful examination of their roof one can get a stronger result: the membership problem for growing grammars is ontained in 1-LOGCFL, the class of problems reducible to context-free languages via ogspace reductions in which the input is read in one-way mode. The class 1-LOGCFL vas introduced by Lautemann ([Lau88]) and by its automata characterization it is known o be a proper subclass of LOGCFL ([Bra77, Lau88]).

Dahlhaus and Warmuth have posed a problem to determine the complexity of the 'variable" membership problem for growing grammars, i.e. the problem in which not only

[*]Institut für Informatik, Universität Würzburg, Am Exerzierplatz 3, D-8700 Würzburg, Germany (e-nail: bunt@informatik.uni-wuerzburg.de).

[†]Instytut Informatyki, Uniwersytet Wroclawski, ul. Przesmyckiego 20, 51-151 Wroclaw, Poland (e-mail: orys@plwruw11.bitnet). This research was supported by the Alexander-von-Humboldt Foundation while he author visited Würzburg Universität.

the tested word but also a growing grammar is a parameter. We solve this problem proving (via generic reduction) its NP-completeness. The reduction is of low complex namely one-way log space. For such one-way reductions see [HIM78, All88].

Having a very elegant and simple definition growing grammars appear at the first lo to generate not an interesting class of languages. In particular, this class (we den it by GCSL) seemed to be not closed under such basic operations as intersection w regular sets or inverse homomorphism. We refute this opinion and show that GCSL for an abstract family of languages. This allows us to give a characterization of GCSL means of weighted grammars. With each such a grammar there is an associated functi determining weights of all terminal and nonterminal symbols, and by a natural extensi also weights of strings. If we restrict grammars to have only productions with the weig of the right side greater than the weight of the left side then they derive exactly GCS Since such weighted grammars allow derivations to have a linear length, it is often mu easier to check a membership of a language in the class GCSL via constructing a weight grammar than via constructing a growing one.

We believe that GCSL is an important class of languages and deserves to be thorough studied. With our paper we hope to increase an interest in it. In the last section we gi some questions which seem to be worth of considerations.

2 Preliminaries

We assume the reader to be familiar with elementary formal language theory and comp tational complexity theory, and use the notations standard according to [HU79]. Here w recall only the notions which are basic for our purposes.

A grammar is a quadruple $\langle N, T, S, P\rangle$, where N and T are finite, disjoint alphabe of nonterminal and terminal (respectively) symbols, $S \in N$ is a start symbol and P is finite set of productions of the form $\alpha \to \beta$ with $\alpha, \beta \in (N \cup T)^*$. The grammars wit $|\alpha| \leq |\beta|$ for all productions from P are called *context-sensitive.*

By replacing the above inequality by a sharp one we obtain growing context-sensiti grammars.

Definition 1 *A context-sensitive grammar* $G = \langle N, T, S, P\rangle$ *is* growing *if* S *does n appear on the right side and* $|\alpha| < |\beta|$ *for any* $(\alpha \to \beta), \alpha \neq S$ *from* P. *We denote th growing grammars by* GCSG *and the class of languages generated by such grammars b* GCSL.

By a *growth ratio* of a production $\alpha \to \beta$ we mean the ratio $\frac{|\alpha|}{|\beta|}$. It is often convenien to deal with GCSG in the following form, which productions, excepting those with S o the left side, have a normalized growth ratio.

Definition 2 *We call a growing context-sensitive grammar* $G = \langle N, T, P, S\rangle$ k-normal *all productions have one of the following form:*

(1) $S \to \beta$, *where* $\beta \in T^*$ *and* $|\beta| \leq k$

(2) $S \to \beta$, *where* $\beta \in (T \cup N)^k$

3) $\alpha \to \beta$, where $\alpha, \beta \in (T \cup N)^*$, $|\alpha| = k$ and $|\beta| = k+1$

oposition 3 *For each language $L \in$ GCSL there exists a number k_o such that for all $\geq k_o$ there exists a k-normal growing context-sensitive grammar G_k such that $L = L(G_k)$.*

oof. Let $G = (N, T, P, S)$ be a growing context-sensitive grammar. We set k_o to be ıal $\max\{|\beta| : (\alpha \to \beta) \in P \text{ for some } \alpha\}$.

The productions of G_k are divided into two subsets P_0 and P_1. To P_0 we put all ›ductions $S \to \beta$ of the form 1 and 2 (from Definition 2) such that β can be derived ›m S in the grammar G. The subset P_1 we create in two stages. At the beginning we t into it all productions $\alpha \to \beta$ from P with $\alpha \neq S$, and then we normalize the length the left and the right side of each of them.

- "Normalization of α": each $\alpha \to \beta$ in P_1 we replace by productions $\alpha_1 \alpha \alpha_2 \to \alpha_1 \beta \alpha_2$ for each $\alpha_1 \alpha_2 \in (T \cup N)^*$ such that $|\alpha_1 \alpha \alpha_2| = k$.
- "Normalization of β": each production $a_1 a_2 \ldots a_k \to b_1 b_2 \ldots b_k b_{k+1} \ldots b_{k+c}$ we replace by productions

$$\begin{array}{lcll} a_1 a_2 \ldots a_k & \to & b_1 B_2 \ldots B_{k+1}, & \\ B_i B_{i+1} \ldots B_{k+i-1} & \to & b_i B_{i+1} \ldots B_{k+i} & \text{for } i = 2, \ldots, c-1, \\ B_c \ldots B_{k+c-1} & \to & b_c \ldots b_{k+c}, & \end{array}$$

  where B_i (for $i = 2, \ldots, k+c-1$) are unique new nonterminals.

is clear that G_k and G are equivalent. □

The restriction imposed on the growth ratio to be greater than one for all productions GCSG implies that all derivations in these grammars are not longer than the derivated ord. This requirement is relaxed in quasi-growing grammars.

efinition 4 *A grammar $G = \langle N, T, S, P \rangle$ is* quasi-growing *if there exists a function $: (N \cup T) \to \mathbb{N}$ (here $\mathbb{N}$ stands for the set of positive integers) such that for all productions $\alpha \to \beta) \in P$ it holds $f(\alpha) < f(\beta)$, where $f(a_1 \ldots a_n) = \sum_{i=1}^{n} f(a_i)$ for $a_1 \ldots a_n \in (N \cup T)^*$. he set of quasi growing grammars we denote by* QGG.

A similar concept of weight functions was already used in string rewriting ([Jan88]). It easy to see that derivations in QGG have length only linearly depending on the length f derivated words. However, in Theorem 19 we show that quasi-growing grammars derive xactly GCSL.

For context-sensitive grammars there are some well known simple normal forms. The ollowing form was proposed by Cremers ([Cre73]):

efinition 5 *A context-sensitive grammar is in* Cremers' normal form *if all productions ıre of the forms*

$$\begin{array}{rcl} AB & \longrightarrow & CD \\ A & \longrightarrow & CD \\ A & \longrightarrow & a \end{array}$$

where A, B, C, D are nonterminals and a is a terminal.

Obviously such a normal form is not growing. From Theorem 19 it follows that every QGG G there exists an equivalent grammar G' in Cremers' normal form; howev the size of G' can be much bigger (even exponentially) than the size of G. The ne proposition shows that for any QGG with only length-nondecreasing productions such normal form can be efficiently obtained.

Proposition 6 *For every* QGG *with only length-nondecreasing productions an equival quasi-growing grammar in Cremers' normal form can be constructed in polynomial tim*

Proof. Let $G = \langle N, T, S, P\rangle$ be a quasi-growing grammar with only length-nondecreasi productions and let f be a corresponding weight function. We can assume that G do not contain productions of the form $\alpha \to \beta$ with $|\alpha| = |\beta| = 1$. A grammar in Creme normal form we obtain by replacing all productions from P which are not in a good for by an equivalent sequence of productions.

Let suppose that $a_1...a_k \to b_1...b_l$ is a production from P, $k \geq 2$. We differentia between two cases:

(1) $l \geq k+2$

(2) $l = k$ or $l = k+1$

In the first case we take the following sequence of productions:

$$\begin{array}{lcll}
a_1a_2 & \to & dA_2 & \\
A_ia_{i+1} & \to & dA_{i+1} & \text{for } i = 2, ..., k-1 \\
A_k & \to & B_kB_{k+1} & \\
dB_i & \to & B_{i-1}b_i & \text{for } i = 2, ..., k-1 \\
dB_2 & \to & b_1b_2 & \\
B_i & \to & b_iB_{i+1} & \text{for } i = k+1, ..., l-2 \\
B_{l-1} & \to & b_{l-1}b_l &
\end{array}$$

The symbols d, A_i (for $i = 2, ..., k$) and B_i (for $i = 2, ..., l-1$) are new nonterminals wit the following weights:

$$\begin{array}{rcl}
f(d) & = & 1 \\
\text{for } i = 2, ..., k: & & \\
f(A_i) & = & f(a_1...a_i) \\
f(B_i) & = & f(b_1...b_i) - 2(i-1) \\
\text{and for } i = k+1, ..., l-1: & & \\
f(B_i) & = & f(b_i...b_l) - l + i
\end{array}$$

Note that, without loss of generality, we can assume that $f(\alpha) \leq f(\beta) - (|\alpha| + |\beta| - 2)$ fo each production $\alpha \to \beta$ from P. If f does not fulfill this condition then instead of f one can take the function $f' = c \cdot f$, where $c = max\{|\alpha| + |\beta| - 2 : \alpha \to \beta \in P\}$.

Now it is easy to check that all above productions are growing with respect to the function f.

For the second case a corresponding sequence of productions can be obtained in an analogous way, so we omit this construction as well as an easy proof of the equivalence of such obtained grammar and grammar G.

From this construction follows that the number of rules will be linear bounded in the length of the encoded grammar G. Hence we see that easily a polynomial time-bounded algorithm can be obtained. □

At the end of this section we recall the notion of *abstract family of languages.*

finition 7 *A* family of languages *is a collection of languages containing at least one* *ıempty language. An* abstract family of languages *(*AFL*) is a family of languages* *sed under union, concatenation, iteration, intersection with regular languages, ε-free* *nomorphism and inverse homomorphism.*

It is well known that the "AFL operations" are not independent. In Section 4 we will ke use of the following proposition (see [GGH69]).

oposition 8 *Let $\mathcal{F}$ be a family of languages. If $\mathcal{F}$ is ε-free and closed under con-* *enation, ε-free homomorphism and inverse homomorphism, then $\mathcal{F}$ is closed under* *ersection with regular sets.*

The complexity of the membership problem for variable grammars

ıe variable membership problem consists in determining, for given a word w and a gram- ır G, whether $w \in L(G)$. It is known that for context-sensitive grammars this problem PSPACE-complete. The open problem posed in [DW86] was to determine its complex- for growing context-sensitive grammars; more precisely, Dahlhaus and Warmuth asked ıether this problem can be solved in polynomial time or whether it is NP-complete. ıe polynomial time algorithm, given by them, for the membership problem with fixed ammars, adapted for the problem with variable grammars, would work in superpoly- mial time because the degree of the polynomial bounding its complexity is dependent the grammar and can be arbitrary large. We give here a strong argument that this oblem presumably is intractable, showing that it is NP-complete under a weak type of ducibility, namely one-way logspace (see [HIM78, All88] for more information about this pe of reducibility).

Let $\bar{G}$ denote a word coding a grammar G. Our problem consists in determining the mplexity of the language

$$L_{GCSL} = \{\langle \bar{G}, w\rangle : G \text{ is a growing context-sensitive grammar and } w \in L(G)\}.$$

As a function coding grammars can be taken any reasonable one, we will not specify . The same concerns also a function coding configurations of Turing machines that will e used later in the proof of NP-hardness of the language L_{GCSL}.

roposition 9 $L_{GCSL} \in \mathrm{NP}$.

roof. Let $\langle \bar{G}, w\rangle \in L_{GCSL}$ and let c be the maximal length of the code of a single ymbol from G. Clearly, $c < |\bar{G}|$. Since w is derived in no more than $|w|$ steps and the ngth of the longest phrase in any derivation of w is bounded by $c|w| < |\bar{G}| \cdot |w|$, it is asy to construct a nondeterministic algorithm which simulates derivations in G in time $v|^2 \cdot |\bar{G}|^2 < |\langle \bar{G}, w\rangle|^4$. □

Now we show how, for each polynomial-time bounded nondeterministic Turing macl M, one can construct growing context-sensitive grammars $G_{w,M}$ such that each $L(G_w$ is over one-letter alphabet and it is nonempty if and only if M accepts w. Moreover this case the length of the shortest word in $L(G_{w,M})$ is bounded by a polynomial in |

Let T_1 be an alphabet in which configurations of Turing machines are coded. assume that $\#, [,] \notin T_1$. For simplicity, from now on, we do not distinguish betw objects like configurations and productions and their codes. We define

$$Seq_{w,M} = \{ \ [l_1\#k_2][l_2^R\#k_3^R][l_3\#k_4][l_4^R\#k_5]...[l_{p-1}^R\#k_{p-1}^R][l_{p-1}\#k_p] :$$

$\forall_{2\leq i \leq p-1}$ k_i and l_i are M's configurations such that l_i is a successor of
l_1 is the initial configuration of M when w is an input, k_p is a predeces
of an accepting configuration and p is an even natural number $\}$.[1]

The following proposition can be easily seen.

Proposition 10 *For any Turing machine M and a word w the language $Seq_{w,M}$ can generated by a context-free grammar $F_{w,M} = \langle N_1, T_1 \cup \{\#, [,]\}, S, P_{w,M}\rangle$. Moreover, any fixed M there exists a GSM [2] which having w as an input produces a code of $F_{w,M}$*

A key part of the grammar $G_{w,M}$ is a subset H_k ($k = |w|$) of its productions whi roughly speaking, allows checking whether corresponding subwords are equal. More p cisely, starting from a word $[l_1\#k_2][l_2^R\#k_3^R][l_3\#k_4][l_4^R\#k_5]...[l_{p-1}^R\#k_{p-1}^R][l_{p-1}\#k_p] \in Seq_u$ and using productions from H_k one can derive a terminal word if and only if for all i wi $1 \leq i < p$ it holds $l_i = k_{i+1}$. Note that H_k is common for all grammars $G_{w,M}$ with $|w| =$

Let $N_2 = T_1 \cup \bar{T_1} \cup \{[,], \#, C\}$ be an alphabet of nonterminal symbols such th $\bar{T_1} = \{\bar{a} : a \in T_1\}$ and let T_2, a terminal alphabet, consist of only one letter d. For eve $a, b \in T_1$, to the system H_k belong the following productions:

$$\begin{array}{llll}
(1) & [a \rightarrow & [C\bar{a} & \\
(2) & C\bar{a}d^i b \rightarrow & d^{i+1}bC\bar{a}, & \forall_{0\leq i<k} \\
(3) & C\bar{a}\#d^{4i}a \rightarrow & \#d^{4(i+1)}, & \forall_{0\leq i<k} \\
(4) & [d^i \rightarrow & d^{i+1}[, & \forall_{0\leq i<k} \\
(5) & [\# \rightarrow & dd[& \\
(6) & [] \rightarrow & ddd & \\
(7) & d \rightarrow & dd. &
\end{array}$$

Proposition 11 *H_k consists of $k(r^2+r+1)+r+3$ productions, where $r = |T_1|$.*

Proposition 12 *Let $u = [u_1\#v_1][u_2\#v_2] \ldots [u_s\#v_s]$, $u_i, v_i \in T_1^*$ for $i = 1, \ldots, s$. If terminal word can be derived from u using production from H_k then for all i with $1 \leq i \leq$ it holds $u_i = v_i$.*

Proof.(sketch) The only way of removing $\#$ and $[$ from a phrase, being derived, is t apply the productions (5) and (6), respectively. Before this can be done, each letter of u has to be changed via production (1) into its counterpart from T_2 and then transporte via production (2) to $\#$, where it is compared with the corresponding letter of v_i. Sinc there is no production changing the order of transported letters it is clear that $u_i = v_i$. □

[1]By l^R we denote the reversal of the string l.

[2]General sequential machine, i.e. a deterministic finite one-way automaton with output, see[HU79] pag 272 for an exact definition

oposition 13 *From each word* $[w\#w]$, ($w \in T_1^*, |w| \leq k$), *system* H_k *derives* d^l, *for* $, l \geq \frac{(|w|+9)(|w|+2)}{2}$.

oof. Let $w = a_1a_2...a_n$, $n \leq k$. One can easily show by induction that for all i, $i \leq n$

$$[w\#w] \stackrel{*}{\Longrightarrow} d^{\frac{i^2+i-2}{2}}[a_i d^{i-1} a_{i+1} d^{i-1} ... d^{i-1} a_n \# d^{4(i-1)} a_i a_{i+1} ... a_n].$$

us finally

$$[w\#w] \stackrel{*}{\Longrightarrow} d^{\frac{n^2+3n}{2}}[\#d^{4n}] \stackrel{*}{\Longrightarrow} d^{\frac{n^2+3n}{2}+2}[d^{4n}] \stackrel{*}{\Longrightarrow} d^{\frac{(n+9)(n+2)}{2}}.$$

w, for $l > \frac{(n+9)(n+2)}{2}$, the words d^l can be derived by use of production (7). □

eorem 14 *The variable membership problem for growing context-sensitive Grammars . the language* L_{GCSL}, *is* NP*-complete under one-way logspace reducibility.*

oof. By Proposition 9 it is sufficient to show that L_{GCSL} is one-way logspace hard for ?.

Let M be a nondeterministic Turing machine working in time $p(n)$. Without loss of nerality we can assume that there exists a polynomial $p'(n)$ which bounds the length of des of the configurations entered by M within $p(n)$ steps. Let N_1, S and $P_{w,M}$ be from $_{w,M}$ (see Proposition 10) and let N_2 and $H_p(|w|)$ be as described above. We set

$$G_{w,M} = \langle N_1 \cup N_2, \{d\}, S, P_{w,M} \cup H_{p(|w|)} \rangle$$

d

$$t_M(w) = \langle G_{w,M}, d^{(p'(|w|)+9)(p'(|w|)+2)p(|w|)} \rangle.$$

All productions from $H_{p(|w|)}$ are polynomially bounded in length and have very simple rm. This, together with Proposition 10, gives that t_M can be constructed to be a one-way gspace transducer.

Since all productions from $H_{p(|w|)}$ contain no symbol from N_1 and $P_{w,M}$ consists of ntext-free rules, each derivation in $G_{w,M}$ can be transformed into an equivalent one in hich no production from $H_{p(|w|)}$ is applied before any production from $P_{w,M}$.

From this and Proposition 13 it is easy to see that $w \in L(M) \Leftrightarrow L(G_{w,M})$ is nonempty $\Leftrightarrow d^{(p'(|w|)+9)(p'(|w|)+2)p(|w|)} \in L(G_{w,M})$. □

Let us note that the languages $G_{w,M}$ can be changed so they would be either empty r would contain at most one word from $\{d\}^*$. Thus the variable membership problem for CSG remains NP-complete even for some restricted subclasses of GCSL.

Closure properties

1 this section we show that the class GCSL is closed under all six AFL operations. A ontrivial part of the proof (Proposition 17) concerns the closure under inverse homoorphism. As a consequence we obtain a characterization of GCSL by quasi-growing rammars (Theorem 19).

Proposition 15 *The class* GCSL *is closed under union, concatenation, iteration, sub tution and* ε*-free homomorphism.*

Proof. By standard methods using the fact that for any language $L \in$ GCSL there exi a growing context-sensitive grammar G generating L such that terminal symbols app only on the right side of productions.

To prove that GCSL is an AFL it is now sufficient, by Propositions 15 and 8, to sh that it is closed under inverse homomorphism.

We start with defining some notions which will be useful in the proof of this fa Following [Loe70, DW86] with each derivation in $G = \langle N, T, S, P\rangle$ we associate a plar acyclic graph called a *derivation graph.* The vertices in such a graph are labeled corresponding symbols and productions used in the derivation. The label of a verte we denote by $l(v)$. For a derivation $\alpha_1 \overset{*}{\underset{G}{\Rightarrow}} \alpha_2 \overset{*}{\underset{G}{\Rightarrow}} \ldots \overset{*}{\underset{G}{\Rightarrow}} \alpha_k$ the derivation gra $D_k = (V_k, E_k)$ is defined inductively. At each step the sources of V_i correspond to α_1 a sinks correspond to α_i.

(1) if $\alpha_1 = a_1a_2...a_r$ then $D_1 = (\{v_j : j = 1, ..., r \text{ and } l(v_j) = a_j\}, \emptyset)$.

(2) if $\alpha_{i-1} = w\alpha w' \Longrightarrow w\beta w' = \alpha_i$ and $\alpha = a_1a_2...a_r$, $\beta = b_1b_2...b_s$ then

- $V_i = V_{i-1} \cup \{v^i, v^i_1, ..., v^i_s\}$, where v^i and v^i_j are new vertices, v^i is labeled by t production $(\alpha \to \beta)$ and $l(v^i_j) = b_j$ for $j = 1, ..., s$,
- $E_i = E_{i-1} \cup \{(v^i, v^i_j) : j = 1, ..., s\} \cup \{(u_j, v^i) \colon j = 1, ..., s\}$ where $u_1, u_2, ..., u_r$ V_{i-1} correspond to α.

Let D be a derivation graph corresponding to $z \overset{*}{\underset{G}{\Rightarrow}} u$. We say that a subgraph $F \subseteq$ is *autonomous* if:

(1) considered as an undirected graph F is connected,

(2) all predecessors of F's vertices also belong to F,

(3) all sinks of F are labeled by productions.

Let $v_1, ..., v_s$ be the sequence of all sources of F and let $v'_1, ..., v'_{s'}$ be the sequence of a successors of F's sinks (remind that there is a natural order among the sources and amon the sinks due to planarity of D). The word $l(v_1)...l(v_s)$ we call *a base* of F and the wor $l(v'_1)...l(v'_{s'})$ we call *a derivative* of F. It is easy to see that the base is a subword of and that $D \setminus F$ is a derivation graph of some derivation $z' \overset{*}{\underset{G}{\Rightarrow}} u$, where z' is created b replacing the base with the derivative.

If v is a node labeled by a production then by $I(v)$ we denote the subgraph of D induced by v and the set of its ancestors. It is easy to see that $I(v)$ is autonomous for an v.

By *a size of autonomous subgraph* we mean the number of its vertices labeled b productions.

Proposition 16 *Let D be a derivation graph for some derivation in k-normal growin context-sensitive grammar.*

) *For any vertex $v \in D$ the subgraph $I(v)$ has a base of length not greater than $(2h-1)\cdot k - 2(h-1)$, where h is the depth of v.*

) *One of the following conditions is fulfilled:*

(a) *there exists an autonomous subgraph F in D of the size at least k and the base not longer than $4k(k-1)$*

(b) *all autonomous subgraphs of D have size smaller than k.*

oof. We omit an easy proof of the part *(1)*.

To prove *(2)* let us suppose that there exists an autonomous subgraph $F \subseteq D$ of the at least k and the base longer than $4k(k-1)$. If F contains a vertex v of the depth hen $I(v)$ satisfies the conditions of the point 1. So let us suppose that all vertices F have depth not greater than $k-1$. We say that a sink of F is maximal if none ts successors belong to F. Let $v_1, ..., v_r$ be succeeding maximal sinks of F. Let now $= \bigcup_{j=1}^{i} I(v_j)$. From *(1)* it follows that the first F_i' of the size at least k has the base longer than $4k(k-1)$. □

oposition 17 *The class* GCSL *is closed under inverse homomorphism.*

oof. Let $L \subseteq \Sigma^*$ be a growing context-sensitive language, let $h : \Delta^* \to \Sigma^*$ and let max$\{|h(a)|: a \in \Sigma\}$.

Since GCSL is closed under substitution it is sufficient to prove the theorem for ε-free Namely, if $\Delta_1 = \{a \in \Delta : h(a) = \varepsilon\}$ and $\Delta_2 = \Delta \setminus \Delta_1$ are not empty, then $h_1 = h|_{\Delta_2}$ -free and $h^{-1}(L) = S(h_1^{-1}(L))$, where S is a substitution such that $S(a) = \Delta_1^* a \Delta_1^*$ for h $a \in \Delta_2$.

By Proposition 3 there exists a k-normal grammar $G = \langle N, \Sigma, P, S\rangle$ generating L for ne $k > 2l$. Now we define a new growing context-sensitive grammar $G' = \langle N', \Delta, P', S'\rangle$ l then we show that $L(G') = h^{-1}(L)$.

We set $N' = N_1 \cup N_2$, where $N_1 = (N \cup \Sigma)^k$ and $N_2 = \bigcup_{i=1}^{l-1} \bigcup_{j=1}^{k-1} \Sigma^i \times (N \cup \Sigma)^j$. The set of productions is divided into four subsets P_i for $i = 0, .., 3$. In the description of these sets we use the following natural projection from N' onto N^*: $C(\langle a_1, a_2, ..., a_j\rangle) =$ $a_2 ... a_j$ for each $\langle a_1, a_2, ..., a_j\rangle \in N'$.

Now we define

$P_0 = \{S' \to w : w \in \Delta^*, |h(w)| \leq 4k(k-1) \text{ and } S \overset{*}{\underset{G}{\Rightarrow}} h(w)\}$,
$P_1 = \{S' \to \beta : \beta \in (N')^{4(k-1)} \text{ and } S \overset{*}{\underset{G}{\Rightarrow}} C(\beta)\}$,
$P_2 = \{\alpha \to \beta : \alpha \in (N')^{4(k-1)}, \beta \in (N')^{4(k-1)+1} \text{ and } C(\alpha) \overset{*}{\underset{G}{\Rightarrow}} C(\beta)\}$
and
$P_3 = \{t_1\alpha \to wt_2 : t_1, t_2 \in N_2 \cup \{\varepsilon\}, \alpha \in (N')^m, m \leq 4(k-1), w \in \Delta^*$ and $h(w)C(t_2)$ is derivable in G from $C(t_1\alpha)$ in less than k steps. $\}$

Let us note that $|wt_2| > |t_1\alpha|$ for all $(t_1\alpha \to wt_2) \in P_3$, so G' is growing.

One can easily prove by induction on the length of derivation that all phrases derived m S' have form $\alpha_1 w_1 \alpha_2 w_2 ... w_{s-1}\alpha_s$, where all α's are from $(N')^*$, all w's are from Δ^* d $S \overset{*}{\underset{G}{\Rightarrow}} C(\alpha_1)h(w_1)C(\alpha_2)h(w_2)...h(w_{s-1})C(\alpha_s)$.

Therefore $w \in L(G') \Rightarrow \exists_{u \in L} h(w) = u$.

Thus it remains to prove that $h^{-1}(L) \subseteq L(G')$.

Let $w \in h^{-1}(L)$ and let $u \in L$ be such that $h(w) = u$. If $|u| \leq 4k(k-1)$ then u derived by a production from P_0. So let assume that $|u| > 4k(k-1)$ and let u_1 be s that $S \overset{*}{\underset{G}{\Rightarrow}} u_1 \overset{*}{\underset{G}{\Rightarrow}} u$ and $|u_1| = 4k(k-1)$. Of course, there exists $z_1 \in (N')^{4(k-1)}$ such t $C(z_1) = u_1$, so $S' \rightarrow z_1$ is in P_1. Let F be a derivation graph for $u_1 \overset{*}{\underset{G}{\Rightarrow}} u$. If F contains autonomous subgraph of the size at least k (by Proposition 16 we can assume that its b is not longer than $4k(k-1)$), then repeating an appropriate number of times an operat of removing a maximal sink from the subgraph we can obtain a subgraph $F' \subseteq F$ of size k. Clearly, F' corresponds to a sequence of k productions which applied to the b of F derive from the word u_1 a word u_2 such that $u_2 \overset{*}{\underset{G}{\Rightarrow}} u$. By the definition of the P_2 we have that $z_1 \underset{G'}{\Rightarrow} z_2$, where $C(z_2) = u_2$. Applying this consideration recursively u_2 and to the graph $F \setminus F'$ we can finally show that there exist words $u_x \in (N \cup \Sigma)^*$ a $z_x \in (N')^*$ such that

(1) $S' \overset{*}{\underset{G'}{\Rightarrow}} z_x$,

(2) $C(z_x) = u_x$,

(3) $u_x \overset{*}{\underset{G}{\Rightarrow}} u$ and all autonomous subgraphs of the corresponding derivation graph are of size $< k$.

Since the autonomous subgraphs of H have basis not longer than $k + (k-1)(k-2)$, t words u_x and u can be divided into subwords $u_x^1, ..., u_x^r$ and $u_1, ..., u_r$, respectively, su that $\forall_{1 \leq i \leq r} k \leq |u_x^i| \leq 2k + (k-1)(k-2)$ and u_i is derived from u_x^i in less than k ste Now for each $i = 1, ..., r$ we define $w_i, s_i, z_{j_i}, z'_{j_i}, z''_{j_i}, t_i$ as follows:

- w_i is the maximal prefix of w such that $h(w_i)$ is a prefix of $u_1 u_2 ... u_i$,
- s_i is the suffix of $u_1 u_2 ... u_i$ such that $u_1 u_2 ... u_i = h(w_i) s_i$,
- if $z_x = z_1 z_2 ... z_y$, where $z_1, ..., z_y \in N'$, then $C(z_{j_i}) = z'_{j_i} z''_{j_i}$ and $u_x^1 ... u_x^i = C(z_1 ... z_{j_i-}$
- $t_i \in N_2 \cup \{\varepsilon\}$ is such that $C(t_i) = s_i z''_{j_i}$ (note that $t_r = \varepsilon$).

One can observe that the following derivation is possible due to productions from I

$$z_x \underset{G'}{\Rightarrow} w_1 t_1 z_{j_1+1} ... z_y \underset{G'}{\Rightarrow} w_1 w_2 t_2 z_{j_2+1} ... z_y \overset{*}{\underset{G'}{\Rightarrow}} w_1 w_2 ... w_r = w$$

Now we are ready to state the main theorems of this section:

Theorem 18 *The class* GCSL *forms an* AFL.

Proof. By the propositions 8, 15, and 17 GCSL is closed under all six AFL operation

eorem 19 *The languages producible by* QGG*s and* GCSG*s are the same.*

oof. Each GCSG can be seen as a QGG simply with a constant weight function.

Let L be generated by a quasi-growing grammar $G = \langle N, T, S, P \rangle$ with weight function Without loss of generality we can assume that S does not appear on the right side of production and that $f(S) = 1$.

Let $c \notin N \cup T$ and let h be a homomorphism such that $h(a) = ac^{f(a)-1}$ for each $N \cup T$. Then $G' = \langle N \cup \{c\}, T, S, P' \rangle$, where $P' = \{h(\alpha) \rightarrow h(\beta) : (\alpha \rightarrow \beta) \in P\}$ is a wing context-sensitive grammar and $L(G) = h^{-1}(L(G'))$. □

Conclusions and open problems

disadvantage of the class of growing context-sensitive languages is that, being a subset linear context-sensitive languages, it does not contain some very simple languages like $:= \{ww^Rw : w \in \Sigma^*\}$ ([Gla64]). It would be interesting to find a natural extension of ntext-sensitive grammars which on the one hand would allow to derive such languages t on the other hand would not lead the membership problem outside LOGCFL.

Another interesting problem concerns the intersection hierarchy for GCSL, i.e. the rarchy of classes $\mathrm{GCSL}^i = \{L : \exists_{L_1, L_2, \ldots, L_i \in \mathrm{GCSL}}\ L = \bigcap_{j=1}^{i} L_j\}$, for $i \geq 1$. It is easy see that the language $\tilde{L}$, defined above, separates the first two levels of the hierarchy d that $\bigcup_{j=1}^{\infty} \mathrm{GCSL}^j \subseteq \mathrm{LOGCFL}$. An analogous hierarchy for context-free languages was own to be infinite ([LW73]). We conjecture that it is also infinite for GCSL.

The complexity of GCSL is still not well examined. In contrary to the context-free nguages GCSL contain nonregular tally languages. What is their complexity? Can one ow that tally languages from GCSL are included in NSPACE($\log n$)? Also the derivative wer of GCSG is not exactly established. Even some very low complexity classes are not own to be included in GCSL. An example of such a class can be the class of languages cognized by one-way deterministic Turing machines in real-time and logarithmic space.

Acknowledgements

anks are due to Hans-Jörg Burtschick, Jaime Caro, Diana Emme, Ulrich Hertrampf, brecht Hoene, Klaus-Jörn Lange, Heribert Vollmer, and Klaus Wagner for some hints d fruitful discussions on the subject of this paper.

We are gratefull also to Manfred Warmuth who brought to our attention that the riable membership problem was already solved in two technical reports [NK87, CH89].

References

[All88] Eric W. Allender. Isomorphisms and 1-L reductions. *Journal of Computer and Sys Sciences*, 36:336–350, 1988.

[Boo73] Ronald V. Book. On the structure of context-sensitive grammar. *International Jour of Computer and Information Sciences*, 2:129–139, 1973.

[Boo78] Ronald V. Book. On the complexity of formal grammars. *Acta Informatica*, 9:171–1 1978.

[Bra77] Franz-Josef Brandenburg. On one-way auxiliary pushdown automata. In *3rd GI C ference on theoretical computer science*, volume 48 of LNCS, pages 132–144. Spring 1977.

[CH89] Sang Cho and Dung T. Huynh. Uniform membership for deterministic growing conte sensitive grammars is P-complete. Technical Report UTDCS-5-89, Computer Scie Department, University of Texas at Dallas, Richardson, Texas 75083, February 1989

[Cre73] A.B. Cremers. Normal forms for context-sensitive grammars. *Acta Informatica*, 3:59– 1973.

[DW86] Elias Dahlhaus and Manfred K. Warmuth. Membership for growing context-sensit grammars is polynomial. *Journal of Computer and System Sciences*, 33:456–472, 198

[GGH69] Seymour Ginsburg, Sheila Greibach, and John Hopcroft. *Studies in Abstract Famil of Languages*. Memoir #87, American Mathematical Society, 1969.

[Gla64] A.W. Gladkij. On the complexity of phrase structure grammars. *Algebra i Logika Se* 3:29–44, 1964. (in Russian).

[HIM78] Juris Hartmanis, Neil Immerman, and Stephen R. Mahaney. One-way log-tape red tions. In *19th* Annual Symposium on Foundations of Computer Science (FOCS), pa 65–71, 1978.

[HU79] John E. Hopcroft and Jeffrey D. Ullman. *Introduction to Automata Theory, Langua and Computation*. Addison-Wesley Series in Computer Science. Addison-Wesley, 197

[Jan88] Matthias Jantzen. *Confluent string rewriting*. Springer, 1988.

[Lau88] Clemens Lautemann. One pushdown and a small tape. In Klaus W. Wagner, editor, *D Siefkes zum 50. Geburtstag*, pages 42–47. Technische Universität Berlin and Universit Augsburg, 1988.

[Loe70] Jacques Loecks. The parsing of general phrase-structure grammars. *Information a Control*, 16:443–464, 1970.

[LW73] Leonard Y. Liu and Peter Weiner. An infinite hierarchy of intersections of context-fr languages. *Mathematical Systems Theory*, 7:185–192, 1973.

[NK87] Paliath Narendran and Kamela Krithivasan. On the membership problem for so grammars. COINS Technical Report CAR-TR-267 & CS-TR-1787 & AFOSR-86-009 Center for Automation Research, University of Maryland, College Park, MD 207 March 1987.

[Sud78] Ivan H. Sudborough. On the tape complexity of deterministic context-free language *Journal of the Association for Computing Machinery*, 25:405–414, 1978.

[Ven91] H. Venkateswaran. Properties that characterize LOGCFL. *Journal of Computer a System Sciences*, 43:380–404, 1991.

Numeration Systems, Linear Recurrences, and Regular Sets

(Extended Abstract)

Jeffrey Shallit*
Department of Computer Science
University of Waterloo
Waterloo, Ontario, Canada N2L 3G1
shallit@graceland.waterloo.edu

Abstract

A numeration system based on a strictly increasing sequence of positive integers $u_0 = 1, u_1, u_2, \ldots$ expresses a non-negative integer n as a sum $n = \sum_{j=0}^{i} a_j u_j$. In this case we say the string $a_i a_{i-1} \cdots a_1 a_0$ is a *representation* for n.

If the lexicographic ordering on the representations is the same as the usual ordering of the integers, we say the numeration system is *order-preserving*. In particular, if $u_0 = 1$, then the *greedy representation*, obtained via the greedy algorithm, is order-preserving. We prove that, subject to some technical assumptions, if the set of all representations in an order-preserving numeration system is regular, then the sequence $u = (u_j)_{j \geq 0}$ satisfies a linear recurrence. The converse, however, is not true.

The proof uses two lemmas about regular sets that may be of independent interest. The first shows that if L is regular, then the set of lexicographically greatest strings of every length in L is also regular. The second shows that the number of strings of length n in a regular language L is bounded by a constant (independent of n) iff L is the finite union of sets of the form xy^*z.

1 Introduction

Let Σ be a finite or infinite alphabet. A *numeration system* is a map $r : \mathbb{N} \rightarrow \Sigma^*$ that assigns a string $r(n)$, called the *representation* of n, to a nonnegative integer n. Sometimes not every integer have a representation, in which case r is a partial function, rather than a function.

*Supported in part by a grant from NSERC.

Common examples of numeration systems include radix-k representation (for k an integer ≥ 2), Fibonacci representation, factorial representation, etc.

In this paper we consider only numeration systems constructed as follows: given a strictly increasing sequence of positive integers $u_0, u_1, u_2, \ldots$ we try to express n as a non-negative integer linear combination of the u_j, say $n = \sum_{0 \leq j \leq i} a_j u_j$. If we can write n in this manner, we say n is u-*representable* (or *representable* if the sequence u is clear from the context) and one representation of n is the string $a_i a_{i-1} \cdots a_1 a_0$. We define the mapping val from strings back to integers as follows:

$$\mathrm{val}(a_i a_{i-1} \cdots a_1 a_0) = \sum_{0 \leq j \leq i} a_j u_j. \quad (1)$$

We say such a representation is *normal* if $a_i \neq 0$. In this paper, we consider only normal representations.

Note that if

$$\gcd(u_0, u_1, \ldots) = g,$$

then every sufficiently large multiple of g is representable. In this paper, we only consider numeration systems whose digits are bounded by some fixed constant.

Since, in general, there may be many normal representations with bounded digits for a number n, we must identify *one* of these representations in order to specify the map r. This can be done in a variety of ways; for example, one could choose $r(n)$ to be the *lexicographically greatest* representation for n, among all nonnegative linear combinations of terms of the sequence u. The lexicographic order we use is defined as follows: if $x = x_1 x_2 \cdots x_i$ and $y = y_1 y_2 \cdots y_j$ are strings, we say x is *lexicographically greater* than y, and write $x > y$, if $i > j$, or if $i = j$ and there exists an integer k, $1 \leq k \leq i$, such that $x_1 = y_1, x_2 = y_2, \ldots, x_{k-1} = y_{k-1}$, but $x_k > y_k$.[1]

Another possible choice for $r(n)$ is the *greedy representation* $\mathrm{gr}(n)$. We define $\mathrm{gr}(n)$ for a positive integer n as follows: let i be the largest index such that $u_i \leq n$. Then successively set $a_i \leftarrow \lfloor n/u_i \rfloor$, $n \leftarrow n - a_i u_i$, and $i \leftarrow i - 1$ until $i < 0$. If $n = \sum_{0 \leq j \leq i} a_j u_j$, then the greedy representation for n is the string $a_i a_{i-1} \cdots a_1 a_0$, and we say n is *greedily representable*. If not, then the representation for n is undefined. The greedy representation for 0 is defined to be ϵ, the empty string. The set of greedy representations for all greedily representable integers is written $G(u)$.

It is easy to see that every non-negative integer is greedily representable iff $u_0 = 1$. If $u_0 \neq 1$, it is possible for a number to be representable, but not

[1] Note that this ordering, called "radix order" by Berstel [2, p. 76] is a well-order (i.e. each nonempty set has a least element), and it differs from the lexicographic order defined by other authors [15, p. 64]; [2]. It is, however, the natural order for our problem.

greedily representable. For example, consider expressing 4 in the numeration system $(u_0, u_1, u_2, \ldots) = (2, 3, 5, \ldots)$. Note that if $u_0 = 1$, then the greedy representation is in fact the lexicographically greatest representation.

A desirable property of any numeration system is that the mapping r that sends an integer n to its representation be *order-preserving.* More precisely, we require that for integers m, n in the domain of r, we have $m > n$ iff $r(m) > r(n)$. It is easy to see that the greedy representation is order-preserving.

Given a numeration system r, based on the sequence u, we define $R(r, u) = R(u)$ to be the set of its representations for all non-negative integers. More formally,

$$R(u) = \sum_{\substack{n \geq 0 \\ n \in \mathrm{Dom}\ r}} r(n). \tag{2}$$

Example 1.

Let $u_j = F_{j+2}$, where F_j is the jth Fibonacci number. Then it can be shown that the set of greedy representations $G(u)$ is

$$\epsilon + 1(0 + 01)^*.$$

(For more on Fibonacci representations, see [3].)

Suppose we are given an increasing sequence of positive integers $(u_j)_{j \geq 0}$, and suppose there is a constant C such that every non-negative integer n has a representation $n = \sum_{0 \leq j \leq i} a_j u_j$ with $0 \leq a_j \leq C$. Let $r(n)$ be defined as the *lexicographically least* representation among all such representations for n.

It is not difficult to show that this r is order-preserving. An example is as follows:

Example 2.

Let $u_j = 2^j$, and consider expressing integers n as $n = \sum_{0 \leq j \leq i} a_j u_j$ with $0 \leq a_j \leq 2$. Now define $r(n)$ to be the lexicographically least representation of this form for n; for example, $r(13) = 221$. Then it is easy to see that $r(n)$ is a string of 1's and 2's; in fact, $R(u)$, the set all such representations, is $(1 + 2)^*$. This corresponds to the well-known observation that every non-negative integer can be written uniquely in base 2, using only the digits 1 and 2. Note that this numeration system is *not* obtained via the greedy algorithm.

The main result of this paper is Theorem 6, which proves the following: suppose the numeration system r is such that the set $R(r, u)$ is regular. Then, subject to some technical conditions on r, the sequence u must satisfy a linear recurrence with integer constant coefficients.

The proof depends on two lemmas about regular sets, which may be of independent interest.

Remarks on the literature.

The point of view we will adopt in this paper is similar to that of Frougny, who has written extensively on this topic. See [8, 9, 10, 11, 12, 13].

In [20], I proved that the set of greedy representations is regular for the numeration systems with bounded digits considered by Fraenkel [7].

2 Lexicographically Largest Strings

Suppose L is a regular language over a finite alphabet Σ. Suppose Σ has a total ordering; for example, suppose $\Sigma = \{0, 1, 2, \ldots, k-1\}$. If $x, y \in \Sigma^n$, we say $x > y$ if x is lexicographically greater than y. More precisely, we say $x > y$ if there exists an integer i, $1 \leq i \leq n$, such that $x_1 = y_1, x_2 = y_2, \ldots, x_{i-1} = y_{i-1}$, but $x_i > y_i$. Then we informally define $B(L)$ as follows: it is the union, over all $n \geq 0$, of the lexicographically largest string of length n in L. More formally, define

$$B(L) = \bigcup_{n \geq 0} \{x \in L \cap \Sigma^n \ : \ \forall \ (y \in L \cap \Sigma^n) \ \ x \geq y\}. \tag{3}$$

For example, $B(1(0+01)^*) = (10)^*(\epsilon + 1)$.

Lemma 1 *If L is regular, then so is $B(L)$.*

Proof.

We show that $L - B(L)$, the relative complement of $B(L)$ in L, is a regular set. The result will then follow, since regular sets are closed under complement, and $B(L) \subseteq L$.

Let $M = (Q, \Sigma, \delta, q_0, F)$ be a deterministic finite automaton accepting L. (See [14] for the basic notions about automata and the notational conventions that we use here.) The idea is to accept $L - B(L)$ with a nondeterministic finite automaton. We use two "fingers" to mimic the behavior of M on input w: the first "finger" imitates M precisely. The second "finger" nondeterministically simulates M on all possible inputs of length $|w|$, trying to find some string in L that is lexicographically greater than w. If we succeed, and w is accepted by M, then we have found a string in L that is not lexicographically greatest, and so we accept.

For space considerations, the details are omitted. ■

Corollary 2 *If L is regular, then so is the set*

$$S(L) = \bigcup_{n \geq 0} \{x \in L \cap \Sigma^n \ : \ \forall \ (y \in L \cap \Sigma^n) \ \ x \leq y\}$$

of *lexicographically smallest strings of every length in* L.

3 Bounded Regular Sets

In this section we prove that if L is a regular language such that the number of strings of length n in L is bounded by a constant (independent of n), then L is the finite union of sets of particularly simple form. More formally, we have

Lemma 3 *The following two statements are equivalent:*

(i) $L \subseteq \Sigma^*$ *is regular and there exists a constant* c *such that* $|L \cap \Sigma^n| \leq c$ *for all* $n \geq 0$

(ii) L *is the finite union of sets of the form* xy^*z, *where* $x, y, z \in \Sigma^*$.

Proof.

$(ii) \Rightarrow (i)$: Suppose

$$L = \sum_{i=1}^{c} x_i y_i^* z_i.$$

Then for each $n \geq 0$, $x_i y_i^* z_i$ contains at most one string of length n. Hence L contains at most c strings of length n.

$(i) \Rightarrow (ii)$: Let q be a non-trivial regular expression denoting L. (When it is necessary to make the distinction, we will use the notation $L(q)$ to represent the language denoted by the regular expression q.)

The result is clearly true for $L = \emptyset$. Thus we may assume that the regular expression q does not contain $\emptyset$. We will show the following two "reduction" steps (recalling that the letters X, Y, W denote finite sets):

(a) If q contains a subexpression of the form t^*, then $L(t^*)$ can be rewritten in the form $X + Yz^*$, and

(b) If q contains a subexpression of the form $s_1^* t s_2^*$, where t contains no star, then $L(s_1^* t s_2^*)$ can be rewritten in the form Wz^*X.

The implication $(i) \Rightarrow (ii)$ will then follow.

(a) Suppose r contains a subexpression of the form t^*. Clearly if $|L(t)| \leq 1$, then t^* is already of the form $X + Yz^*$. Suppose $|L(t)| = 2$, say $L(t) = \{x, y\}$. Choose a positive integer m sufficiently large such that the linear Diophantine equation

$$a|x| + b|y| = m \tag{4}$$

has $\geq c + 1$ solutions (a, b) in non-negative integers. (For example, it suffices to choose $m = c \operatorname{lcm}(|x|, |y|)$.) Then by the hypothesis, we must have

$$x^a y^b = x^{a'} y^{b'}$$

for some distinct pairs (a, b), (a', b') satisfying (4), for otherwise t^* and hence L would contain $\geq c+1$ strings of length n, for some n.

Without loss of generality we may assume $a \geq a'$, $b \leq b'$. Then

$$x^{a-a'} = y^{b'-b}.$$

By [15, Prop. 1.3.1] there exists a string z and integers i, j such that $x = z^i$, $y = z^j$. Hence

$$(x+y)^* = (z^i + z^j)^* = X + z^{ij}(z^{\gcd(i,j)})^*$$

for some finite set X. Thus we can replace the t^* in q by a set of the form $X + Yz^*$.

If $|L(t)| > 2$, we can repeat the argument above on pairs to obtain that *each* element of $L(t)$ is a power of some string z. Thus, there is some (finite or infinite) set S such that

$$L(t) = \sum_{i \in S} z^{a_i}.$$

Set $g = \gcd(a_1, a_2, \ldots)$. Then there is a finite subset of the a_i, say $b_1 \leq b_2 \leq \ldots \leq b_k$, such that $g = \gcd_{1 \leq j \leq k} b_j$. Choose m sufficiently large such that eg is a non-negative integer linear combination of the b_i for all $e \geq m$; by a theorem of Brauer [4, Corollary to Thm. 1], we may in fact choose

$$m = (b_1 - 1)(b_k - 1).$$

Let

$$X = \{z^d \in L(t^*) \ : \ d < gm\};$$

note that X is a finite set. I claim that

$$L(t^*) = X + z^{gm}(z^g)^*.$$

For let $z^d \in L(t^*)$; if $d < gm$, then $z^d \in X$, by construction. Otherwise $d \geq gm$; clearly d must be a multiple of g and hence $z^d \in z^{gm}(z^g)^*$.

It remains to see that $z^{gm}(z^g)^* \subseteq L(t^*)$. We know that if $e \geq m$, then eg is a nonnegative integer linear combination of the b_i. Hence every string z^{eg} with $e \geq m$ must be in $(z^{b_1} + z^{b_2} + \cdots + z^{b_k})^*$, and hence in $L(t^*)$. This completes the proof of (a) and shows incidentally that L is of star-height 1. Hence by a theorem of Cohen [5, Lemma 3.1], we may assume that

$$L = Z_1 + Z_2 + \cdots + Z_j,$$

where each set Z_i can be written in the form

$$w_1 x_1^* w_2 x_2^* \cdots w_k x_k^* w_{k+1}. \tag{5}$$

Part (b) follows in a similar fashion, using [15, Prop. 1.3.4]. For the details, please consult the final paper. ■

After the author proved Lemma 3, other proofs were shown to him by S. Yu, F. Fich, and J. Sakarovitch.

4 Two Lemmas on Linear Recurrences

In this section we provide two useful lemmas about linear recurrences.

Suppose we are given a sequence whose even and odd-numbered terms each satisfy a linear recurrence with integer coefficients, but not the same one, such as

$$\begin{aligned} A_{2n} &= 2A_{2n-1} + A_{2n-2} \\ A_{2n-1} &= 3A_{2n-2} + A_{2n-3} + A_{2n-4} - A_{2n-5}. \end{aligned} \tag{6}$$

Can we then conclude that the sequence A_n itself satisfies a linear recurrence with integer coefficients?

The answer is yes, as the following lemma shows:

Lemma 4 *Let k, d be positive integers, with $d \geq k$, and let $M = [M_{ij}]$ be a $k \times d$ matrix of integers such that for all n sufficiently large we have*

$$\begin{aligned} A_{kn} &= M_{11}A_{kn-1} + M_{12}A_{kn-2} + \cdots + M_{1d}A_{kn-d} \\ A_{kn-1} &= M_{21}A_{kn-2} + M_{22}A_{kn-3} + \cdots + M_{2d}A_{kn-d-1} \\ &\vdots \\ A_{kn-k+1} &= M_{k1}A_{kn-k} + M_{k2}A_{kn-k-1} + \cdots + M_{kd}A_{kn-k-d+1}. \end{aligned}$$

Then the sequence A_n itself satisfies a linear recurrence with constant coefficients.

Proof.

Note that d is the maximum degree of the characteristic polynomials for the subsequences $A_{kn}, A_{kn+1}, \ldots, A_{kn+k-1}$.

By successively substituting the relations for $A_{kn-1}, A_{kn-2}, \cdots$ in the relation for A_{kn}, etc., we can find a $d \times d$ matrix $P = [P_{ij}]$ such that for all n sufficiently large, we have

$$\begin{aligned} A_{kn} &= P_{11}A_{kn-k} + P_{12}A_{kn-k-1} + \cdots + P_{1d}A_{kn-k-d+1} \\ A_{kn-1} &= P_{21}A_{kn-k} + P_{22}A_{kn-k-1} + \cdots + P_{2d}A_{kn-k-d+1} \\ &\vdots \\ A_{kn-d+1} &= P_{d1}A_{kn-k} + P_{d2}A_{kn-k-1} + \cdots + P_{dd}A_{kn-k-d+1}. \end{aligned}$$

Let $f(X)$ be the characteristic polynomial of P. Then each of the sequence

$$A_{kn},\ A_{kn-1},\ \ldots,\ A_{kn-k+1}$$

satisfies the same linear recurrence, namely, the one whose characteristic polynomial is $f(X)$ (at least for n sufficiently large). Thus for n large enough A_n satisfies the linear recurrence whose characteristic polynomial is $f(X^k)$ ■

Our second lemma concerns integer linear recurrences with rational coefficients.

Lemma 5 *Suppose the infinite integer sequence $u_0, u_1, u_2, \ldots$ satisfies a linear recurrence relation*

$$u_n = a_1 u_{n-1} + a_2 u_{n-2} + \cdots a_d u_{n-d},$$

for all $n \geq 0$, where each a_i is a rational number. Then u_n satisfies a (possibly different) linear recurrence relation where all the a_i are actually integers.

Proof.

This follows immediately from a lemma of Fatou [6] (also see [18, Part VIII, Problem 156]). ■

5 Proof of the Main Result

In this section, we prove the result mentioned in the introduction. The details of the proof are a little messy, so it may be helpful to first give the proof in the case of ordinary base-3 representation. In this case, the set of representations $R(u)$ is $\epsilon + (1+2)(0+1+2)^*$. It is easy to see that $B(R(u))$, the set of lexicographically greatest representations, is 2^*. Since

$$\text{val}(1\overbrace{00\cdots0}^{k}) = 1 + \text{val}(\overbrace{22\cdots2}^{k}),$$

it follows that

$$u_k = 1 + 2u_{k-1} + 2u_{k-2} + \cdots + 2u_0. \tag{7}$$

Similarly, we also have

$$u_{k+1} = 1 + 2u_k + 2u_{k-1} + \cdots + 2u_0. \tag{8}$$

Subtracting (7) from (8), we see

$$u_{k+1} - u_k = 2u_k,$$

and hence $u_{k+1} = 3u_k$.

We now state and prove the main result of the paper:

Theorem 6 *Let $u_0, u_1, \ldots$ be a strictly increasing sequence of non-negative integers, and let r be a numeration system based on u. Let $R(u)$ be as in Eq. (2). Suppose that*

(a) $\gcd(u_0, u_1, \ldots) = g$, and $r(kg)$ is defined for all sufficiently large integers k;

(b) for all sufficiently large n, there exists a representation in $R(u)$ of length n;

and

(c) r is order-preserving.

If $R(u)$ is regular, then the sequence $u = (u_n)_{n\geq 0}$ satisfies a linear recurrence with integer constant coefficients.

Note that hypothesis (b) is satisfied by both the greedy representation and the lexicographically greatest representation, for in these cases we have the representation of u_n is $1\,0^n$. In fact, if $u_0 = 1$, then *all three* hypotheses are satisfied when r is the greedy representation.

Proof.

Let $g = \gcd(u_0, u_1, \ldots)$. Then by hypothesis (a), there exists an integer C such that $r(n)$ is defined for all $n \geq C$ with $g \mid n$.

If $L = R(u)$ is regular, then by Lemma 1, the set $B(L)$ of lexicographically greatest strings of every length in L is also regular. By hypothesis (b) of the theorem, there exists C' such that $B(L)$ exactly one string of length j for each $j \geq C'$. Thus Lemma 3 applies and so

$$B(L) = \sum_{i=1}^{k} x_i y_i^* z_i. \tag{9}$$

Similarly, by Corollary 2, the set $S(L)$ can also be written in the form

$$\sum_{i=1}^{k'} x'_i y'^*_i z'_i.$$

To avoid unnecessary complication, we first show how the proof goes in the case $S(L) = 1\,0^*$. At the end of the proof we sketch what needs to be done when $S(L) \neq 1\,0^*$. (Note that in fact $S(L) = 1\,0^*$ for the most common case, when r is the greedy representation.)

The main idea of the proof is as follows: let $w \in B(L)$ be sufficiently long such that $\text{val}(w) > \max(C, C')$. Let $v \in S(L)$ be such that $|v| = |w| + 1$. Then by hypothesis (c) (i.e., r is order-preserving), we must have $\text{val}(v) = \text{val}(w) + g$. Since $S(L) = 10^*$, we have

$$u_{|w|} = \text{val}(w) + g. \tag{10}$$

Now let

$$g' = \mathrm{lcm}_{\substack{1 \le j \le k \\ |y_j| \ne 0}} |y_j|.$$

By replacing y_i with $y_i^{g'/|y_i|}$, adding extra terms to x_i and z_i, and renaming we can rewrite (9) such that $|y_i| = g'$ for all i for which $|y_i| \neq 0$.

Consider a particular term of the sum in (9), say xy^*z. From (10) we have

$$u_{j|y|+|xz|} = \mathrm{val}(xy^j z) + g.$$

We also have

$$u_{(j+1)|y|+|xz|} = \mathrm{val}(xy^{j+1} z) + g.$$

Subtracting, we see

$$u_{(j+1)|y|+|xz|} - u_{j|y|+|xz|} = \mathrm{val}(xy^{j+1}z) - \mathrm{val}(xy^j z)$$

$$= \mathrm{val}(xy0^{j|y|+|z|}) - \mathrm{val}(x0^{j|y|+|z|}),$$

and the last expression on the right is a linear combination of terms of u such that the smallest coefficient which is possibly non-zero corresponds to $u_{j|y|+|z|}$. Hence for j sufficiently large, the subsequence $u_{(j+1)g'+i}$ can be expressed as a non-trivial integer linear combination of the $g' + |x|$ previous terms of u, and the particular linear combination depends only on the value of $i \pmod{g'}$.

By Lemma 4, we can write u itself as a linear recurrence. This completes the proof in the case where $r(n) = 1\,0^n$.

It remains to discuss what needs to be done to handle the more general case, when $S(L)$ is not necessarily equal to $1\,0^*$. The same techniques apply, except that since the leading coefficient of the recurrence we obtain is no longer necessarily 1, we see that u is a linear recurrence with *rational* coefficients. Then by Lemma 5, we see that the linear recurrence actually has *integer* coefficients. This completes the proof. ■

6 Linear Recurrences and Non-Regular Sets

After seeing the main theorem, one immediately wonders if the converse is true. It is not, as the following theorem shows:

Theorem 7 *Suppose $u_j = (j+1)^2$ for $j \ge 0$. Then the set $G(u)$ of greedy representations is not a regular set.*

Note that u_j, of course, satisfies the linear recurrence $u_{j+3} = 3u_{j+2} - 3u_{j+1} + u_j$ for all $j \ge 0$.

Proof.

Let $G(u)$ be the set of greedy representations, and assume it is regular. Then $G(u) \cap 10^*10^*$ would also be a regular set. However, it is easy to see that

$$\begin{aligned} G(u) \cap 10^*10^* &= \{10^a10^b : u_{b+a+2} > u_{b+a+1} + u_b\} \\ &= \{10^a10^b : b^2 < 2a+4\}, \end{aligned}$$

and this set is evidently not regular. ■

It is an open problem to give a sufficient condition for the regularity of $R(u)$.

7 Notes and Acknowledgments

An earlier version of Lemma 3 dealt with the case $c = 1$. I am grateful to E. Bach for having suggested the stronger formulation given in this paper. I would also like to thank A. Lubiw for general discussion and for suggesting an appropriate way to weaken the hypotheses in Theorem 6. C. Stewart pointed out the lemma of Fatou.

After this manuscript was accepted for ICALP '92, I learned from E. Bach of the paper of T. H. M. Smits [21], in which the result of Lemma 4 independently appears.

A. Salomaa has also kindly informed me that Corollary 2 and Lemma 3 have independently appeared in unpublished manuscripts of Păun, Salomaa, et al. [1, 16, 17].

J. Sakarovitch has kindly informed me that Corollary 1 can be deduced from Théorème 2 of his paper [19].

References

[1] M. Andraşiu, J. Dassow, G. Păun, and A. Salomaa. Language-theoretic problems arising from Richelieu cryptosystems. *Theoret. Comput. Sci.*, to appear.

[2] J. Berstel. *Transductions and Context-Free Languages.* Teubner, Stuttgart, 1979.

[3] J. Berstel. Fibonacci words – a survey. In G. Rozenberg and A. Salomaa, editors, *The Book of L*, pp. 13–27. Springer-Verlag, 1986.

[4] A. Brauer. On a problem of Frobenius. *Amer. J. Math.* **64** (1942), 299–312.

[5] R. S. Cohen. Star height of certain families of regular events. *J. Comput. System Sci.* **4** (1970), 281–297.

[6] P. Fatou. Séries trigonométriques et séries de Taylor. *Acta Math.* **30** (1906), 335–400.

[7] A. S. Fraenkel. Systems of numeration. *Amer. Math. Monthly* **92** (1985), 105–114.

[8] C. Frougny. Fibonacci numeration systems and rational functions. In *Math. Found. Comp. Sci.*, Vol. 233 of *Lecture Notes in Computer Science*, pp. 350–359. Springer-Verlag, 1986.

[9] C. Frougny. Linear numeration systems of order two. *Inform. and Comput.* **77** (1988), 233–259.

[10] C. Frougny. Linear numeration systems, θ–developments and finite automata. In B. Monien and R. Cori, editors, *STACS 89*, Vol. 349 of *Lecture Notes in Computer Science*, pp. 144–155. Springer-Verlag, 1989.

[11] C. Frougny. Systèmes de numération lineaires et automates finis (Thèse d'État). Technical Report 89-69, Laboratoire Informatique Théorique et Programmation, Université P. et M. Curie, Université Paris VII, September 1989.

[12] C. Frougny. Fibonacci representations and finite automata. *IEEE Trans. Inform. Theory* **37** (1991), 393–399.

[13] C. Frougny. Representations of numbers and finite automata. *Math. Systems Theory* **25** (1992), 37–60.

[14] J. Hopcroft and J. Ullman. *Introduction to Automata Theory, Languages, and Computation.* Addison-Wesley, 1979.

[15] M. Lothaire. *Combinatorics on Words*, Vol. 17 of *Encyclopedia of Mathematics and Its Applications.* Addison-Wesley, 1983.

[16] G. Păun and A. Salomaa. Closure properties of slender languages. Manuscript, 1991.

[17] G. Păun and A. Salomaa. Thin and slender languages. Manuscript, 1991.

[18] G. Pólya and G. Szegö. *Problems and Theorems in Analysis*, Vol. II. Springer-Verlag, Berlin, 1976.

[19] J. Sakarovitch. Deux remarques sur un théorème de S. Eilenberg. *RAIRO Informatique Théorique* **17** (1983), 23–48.

[20] J. Shallit. A generalization of automatic sequences. *Theoret. Comput. Sci.* **61** (1988), 1–16.

[21] T. H. M. Smits. Conditional recurrent sequences. *Indag. Math.* **2** (1991), 359–365.

The equality problem for rational series with multiplicities in the tropical semiring is undecidable

Daniel KROB

LACIM and CNRS(Institut Blaise Pascal; LITP) [1]

0 Introduction

The tropical semiring is the semiring denoted by $\mathcal{M}$ which has support $\mathbb{N} \cup \{+\infty\}$ and operations $a \oplus b = \min\{a, b\}$ and $a \otimes b = a + b$. It was first introduced in the context of cost minimization in Operations Research. However it appeared that $\mathcal{M}$ plays in fact a central role in several decision problems concerning rational languages (see [8] for a survey of the tropical semiring theory and of its applications). For instance, I. Simon showed that the finite power property for recognizable languages can be reduced to the limitedness problem for the tropical semiring (cf [8]).

One of the main open questions in the theory of the tropical semiring was to see if it is possible to decide whether two given $\mathcal{M}$-rational series are equal or not (cf [8, 9]). We offer here an answer to this problem since we show in this paper that the equality problem for $\mathcal{M}$-rational series over an alphabet with at least two letters is *undecidable*. One should notice that most people thought that a decision procedure existed (cf [8] for instance) and our result is indeed based on a rather surprising encoding of a 10th Hilbert problem.

It is also interesting to precise the structure of the proof of our undecidability result. Indeed it appears that we use as a main tool the tropical "ring" $\mathcal{Z} = (\mathbb{Z} \cup \{+\infty\}, \min, +)$ which is just the extension of $\mathcal{M}$ to arbitrary integers. The importance of $\mathcal{Z}$ comes from the equivalence with respect to decidability of the equality problems for $\mathcal{M}$ and $\mathcal{Z}$. According to this result, we can reduce our problem to showing that the equality problem for $\mathcal{Z}$-rational series over an alphabet with at least two letters is undecidable. To prove this last result, we show in fact that the decidability of the equality

[1] Mailing adress : Université de Rouen; Faculté des Sciences (Informatique); 76134 Mont Saint-Aignan Cedex - FRANCE

problem for $\mathcal{Z}$ is equivalent to the decidability of the local inequality problem for $\mathcal{Z}$. Using this equivalence and a reduction to a 10th Hilbert problem, we prove then the undecidability of the equality problem for $\mathcal{Z}$-rational series over an alphabet with at least two letters. Hence this allows us to obtain the undecidability of the same problem for $\mathcal{M}$-rational series. Moreover our methods give us also immediately other decidability and undecidability results for connected problems. In particular, we solve also another open question (cf [9]) by showing that the equality problem for rational series over an alphabet with at least two letters and with multiplicities in the semiring $\mathcal{N} = (\mathbb{N} \cup \{-\infty\}, \max, +)$ is undecidable.

1 Preliminaries

The *tropical "ring"* is the commutative semiring denoted by $\mathcal{Z}$ which has $\mathbb{Z} \cup \{+\infty\}$ as support, whose addition $\oplus$ is defined by $a \oplus b = \min\{a, b\}$ and whose product $\otimes$ is given by $a \otimes b = a + b$. The operations of $\mathbb{Z}$ are extended to $\mathcal{Z}$ in the usual natural way and the units for $\oplus$ and $\otimes$ are respectively $+\infty$ and 0. The *tropical semiring* is the subsemiring of $\mathcal{Z}$ denoted by $\mathcal{M}$ which has $\mathbb{N} \cup \{+\infty\}$ as support. Let us also introduce the "dual" semiring $\mathcal{N}$ of $\mathcal{M}$ which is the semiring whose support is $\mathbb{N} \cup \{-\infty\}$, whose addition $\oplus$ is given by $a \oplus b = \max\{a, b\}$ and whose product $\otimes$ is defined by $a \otimes b = a + b$. Finally let us consider the subsemiring $\mathcal{Z}^-$ of $\mathcal{Z}$ whose support is $\mathbb{Z}^- \cup \{+\infty\}$. Note that $\mathcal{Z}^-$ is clearly isomorphic to $\mathcal{N}$, an effective isomorphism beeing obtained by the mapping $x \rightarrow -x$ from $\mathcal{Z}^-$ into $\mathcal{N}$.

We refer to [2] for all generalities concerning series and rational series with multiplicities in an arbitrary semiring K. We will denote here by $K \ll A \gg$ the K-algebra of series over A with multiplicities in K and by $K\mathrm{Rat}(A)$ the K-algebra of K-rational series. Let us also recall that a K-representation of order n of a free monoid A^* is just a monoid morphism from A^* into the monoid of square matrices of order n with entries in K. Then a K-automaton of order n is a triple (I, μ, T) where μ is a K-representation of order n of A^* and where I and T are respectively a row and a column vector of order n with entries in K (see [2] for more details).

Let us now precise some notions concerning K-rational series that we will use in the sequel. First we will denote by $\underline{L}$ the characteristic series of any language $L \subset A^*$ which is the series of $K \ll A \gg$ defined by

$$\forall\, w \in A^*,\ (\underline{L}|w) = \begin{cases} 1_K & \text{if } w \in L \\ 0_K & \text{if } w \notin L \end{cases}$$

Note that $\underline{L}$ is always a K-rational series when L is a rational language (cf [2]). We also denote here as usually by $S \odot T$ the Hadamard product of two series S, T which is the series defined by $(S \odot T|w) = (S|w)(T|w)$ for every $w \in A^*$. We recall that $S \odot T$ is a K-rational series when S and T are K-rational series (see [2] for more details). Finally the constant K-rational series whose every coefficient is equal to k, will be always denoted by k.

Let us now recall the following result which is folklore (it is in fact a general property of positive semirings).

ROPOSITION 1.1 : Let S be a rational series of $\mathcal{Z}\mathrm{Rat}(A)$. Then the set

$$\{ w \in A^*,\ (S|w) = +\infty \}$$

a constructible rational language of $\mathrm{Rat}(A)$.

roof : Let π be the morphism of semirings from $\mathcal{Z}$ into the boolean semiring $\mathcal{B}$ efined by $\pi(+\infty) = 0$ and $\pi(z) = 1$ for every $z \in \mathcal{Z} - \{+\infty\}$. We also denote by π its atural extension as an algebra morphism from $\mathcal{Z} \ll A \gg$ into $\mathcal{B} \ll A \gg$. Then we ave

$$\{ w \in A^*,\ (S|w) = +\infty \} = \{ w \in A^*,\ (\pi(S)|w) = 0 \}$$

ur result follows now since $\pi(S)$ is clearly a constructible $\mathcal{B}$-recognizable series. ∎

'ote : It follows also from proposition 1.1 that it is decidable whether a recognizable eries of $\mathcal{Z}\mathrm{Rat}(A)$ is equal to $+\infty$ or whether it has a coefficient equal to $+\infty$.

Finally let us recall the following result of Adler (cf [1, 5]) :

'HEOREM 1.2 : Every diophantine equation is equivalent to an equation of the orm

$$P(x_1, \ldots, x_n) = 1$$

here P is an homogeneous polynomial of degree 4 of some $\mathbb{Z}$-algebra $\mathbb{Z}[x_1, \ldots, x_n]$.

2 Some relations between decidability problems

Let K be a totally ordered semiring. Let us then consider the four problems of equality, inequality, local inequality and local equality for K-rational series over A :

$$P, Q \in \mathrm{KRat}(A), \quad P = Q \ ? \qquad (Eq)$$

$$P, Q \in \mathrm{KRat}(A), \quad P \leq Q \ ? \qquad (Ineq)$$

$$P, Q \in \mathrm{KRat}(A), \quad \exists\, w \in A^*,\ (P|w) \leq (Q|w) \ ? \qquad (LocalIneq)$$

$$P, Q \in \mathrm{KRat}(A), \quad \exists\, w \in A^*,\ (P|w) = (Q|w) \ ? \qquad (LocalEq)$$

In general, these problems are not connected. [2] However it appears that the three first above problems are equivalent with respect to decidability when K is the tropical "ring" or semiring.

PROPOSITION 2.1 : Let $K = \mathcal{Z}$ or $K = \mathcal{M}$. Then the three following assertions that deal with decidability problems for K-rational series, are equivalent :

1. The equality problem (Eq) is decidable.
2. The inequality problem $(Ineq)$ is decidable.
3. The local inequality problem $(LocalIneq)$ is decidable.

---

[2] For instance, when $K = \mathbb{N}$, the equality problem is decidable and the inequality problem is undecidable (see [6]).

Proof : The fact that assertion 2 implies assertion 1 is immediate since we have

$$P = Q \iff \begin{cases} P \leq Q \\ Q \leq P \end{cases}$$

The fact that assertion 1 implies assertion 2 follows also immediately from th relation

$$P \leq Q \iff P = P \oplus Q = \min(P, Q)$$

Let us now show that assertion 3 implies assertion 2. Hence let P, Q be two K rational series where $K = \mathcal{M}$ or $K = \mathcal{Z}$. Then the set $I = \{ w \in A^*,\ (Q|w) = +\infty$ is rational and constructible according to proposition 1.1. Let us now consider th series $\overline{P}$ defined by

$$\overline{P} = (P \odot \underline{A^* - I}) \oplus \underline{I} = \begin{cases} 0 & \text{if } w \in I \\ (P|w) & \text{if } w \notin I \end{cases}$$

which is clearly K-rational. We can define in the same way the series $\overline{Q}$. Then w have

$$\begin{aligned} P \leq Q &\iff \overline{P} \leq \overline{Q} \\ &\iff \forall\, w \in A^*,\ (\overline{P}|w) \leq (\overline{Q}|w) \\ &\iff \forall\, w \in A^*,\ (\overline{P}|w) < (\overline{Q}|w) + 1 \end{aligned}$$

this last equivalence coming from the fact that $\overline{Q}$ has no value equal to $+\infty$. It follow immediately from these relations that we have

$$P \leq Q \iff \neg(\exists\, w \in A^*,\ (\overline{P}|w) \geq (\overline{Q} \odot 1|w))$$

Hence it follows clearly from this last equivalence that assertion 3 implies assertion 2

Let us now show that assertion 2 implies assertion 3. Thus let P, Q be two K rational series. According to proposition 1.1, the set $I = \{ w \in A^*,\ (Q|w) = +\infty$ is an effective rational language. Hence we can decide if I is empty or not. If I i non-empty, the local inequality problem has obviously a positive answer since every word $w \in I$ satisfies to (*LocalIneq*). On the other hand, if I is empty, we have

$$\begin{aligned} (\mathit{LocalIneq}) &\iff \neg(\forall\, w \in A^*,\ (P|w) > (Q|w)) \\ &\iff \neg(\forall\, w \in A^*,\ (P|w) \geq (Q|w) + 1) \\ &\iff \neg(P \geq Q \odot 1) \end{aligned}$$

Note that the second above equivalence follows from the fact that $I = \emptyset$. Hence it follows immediately from these last equivalences that assertion 2 implies assertion 3 This ends the proof of our proposition. ∎

PROPOSITION 2.2 : Let $K = \mathcal{Z}$ or $K = \mathcal{M}$. Then the decidability of the local equality problem (*LocalEq*) for K-rational series implies the decidability of the local inequality problem (*LocalIneq*) for K-rational series.

Proof : It is immediate since we clearly have

$$\exists\, w \in A^*,\ (P|w) \leq (Q|w) \iff \exists\, w \in A^*,\ (P|w) = (P \oplus Q|w) = \min((P|w), (Q|w))$$

Hence our proposition is proved. ∎

3 Undecidability of the equality problem for $\mathcal{Z}$

This section is devoted to the proof of the undecidability of the equality problem for $\mathcal{Z}$-rational series over alphabets with at least two letters. This result implies in fact the undecidability of the same problem for $\mathcal{M}$-rational series as we will see later.

THEOREM 3.1 : Let A be any alphabet with at least 2 letters. Then the equality problem is undecidable for $\mathcal{Z}$-rational series over A.

Proof : We should first notice that the decidability of the equality problem for K-rational series on an arbitrary semiring K and over a k-letter alphabet A_k is equivalent to the decidability of the same problem for a l-letter alphabet A_l when $k, l \geq 2$.

Indeed, it suffices to use an adapted encoding of A_k^* over A_l^* in order to prove this result. For instance, let a, b be two letters of A_l and let σ be the monoid morphism from A_k^* into A_l^* defined by $\sigma(a_i) = a^i b$ for every $a_i \in A_k = \{a_1, \ldots, a_k\}$. Then we can also denote by σ its extension as a K-algebra morphism from $K \ll A_k \gg$ into $K \ll A_l \gg$. It is easy to see that σ is injective and preserves rationality. Hence deciding whether two K-rational series E, F of $K \ll A_k \gg$ are equal, is equivalent to deciding whether the two K-rational series $\sigma(E), \sigma(F)$ of $K \ll A_l \gg$ are equal. This proves our claim.

Our undecidability proof is based on a reduction to Adler's restriction of Hilbert's tenth problem (see theorem 1.2). Let now $P(x)$ be an homogeneous polynomial of degree 4 in several indeterminates of $\mathbb{Z}[x]$. [3] By distinguishing all variables, it is easily seen that the equation $P(x) = 1$ can be transformed in an equivalent way as a system of the form :

$$\left\{ \begin{array}{l} \sum_{i=1}^{p} p_i x_1^{(i)} x_2^{(i)} x_3^{(i)} x_4^{(i)} = 1 \\ \forall\, (i, j, k, l) \in K,\ x_k^{(i)} = x_l^{(j)} \end{array} \right.$$

where K is some subset of $[1,p]\times[1,p]\times[1,4]\times[1,4]$ and where $(p_i)_{i=1,\ldots,p}$ is a family of integers of $\mathbb{Z}$. But this last system can also be transformed into the single equation

$$-\left| \sum_{i=1}^{p} p_i x_1^{(i)} x_2^{(i)} x_3^{(i)} x_4^{(i)} - 1 \right| - \sum_{(i,j,k,l)\in K} \left| x_k^{(i)} - x_l^{(j)} \right| = 0 \qquad (HD)$$

Hence, according to Adler's theorem (cf theorem 1.1) and to the undecidability of Hilbert's tenth problem (cf [5]), it follows from our reduction process that it is undecidable to see whether an equation of the form (HD) has a solution in positive integers.

Let now $A = \{a, b, c, d, e\}$ be a five letter alphabet. According to proposition 2.1 and to our first remark, it suffices to show that the local inequality problem for $\mathcal{Z}$-rational series over A is undecidable in order to prove our theorem. We will show this fact by a suitable encoding of equations (HD) in terms of $\mathcal{Z}$-rational series. But let us now give some lemmas that will allow us to construct this encoding.

[3] Here x denotes of course a vector of variables $x = (x_1, \ldots, x_n)$.

LEMMA 3.2 : Let $k \in \mathbb{Z}$. Then the series $Coeff(k)$ defined by

$$(Coeff(k)|w) = \begin{cases} k\,n_1 n_2 n_3 n_4 & \text{if } w = (((a^{n_1}b)^{n_2}c)^{n_3}d)^{n_4} \text{ with } (n_1,n_2,n_3,n_4) \in \mathbb{N}^4 \\ 0 & \text{if } w \notin (((a^*b)^*c)^*d)^* \end{cases}$$

is a $\mathcal{Z}$-rational series of $\mathcal{Z}\mathrm{Rat}(a,b,c,d)$.

Proof : Using the $\mathcal{Z}$-representation μ of order 1 of A^* defined by

$$\mu(a) = (k) \quad \text{and} \quad \forall\, \alpha \in \{b,c,d\},\ \mu(\alpha) = (0)$$

it can easily be shown that the series

$$S(k,a) = \sum_{w \in A^*} k\,|w|_a\, w$$

is $\mathcal{Z}$-rational. Let now L be the rational language $L = (((a^*b)^*c)^*d)^*$. It is then easy to see that $Coeff(k) = (\underline{L} \odot S(k,a)) \oplus \underline{A^* - L}$ which is hence a $\mathcal{Z}$-rational series.

LEMMA 3.3 : Let $k \in [1,4]$ and $\epsilon \in \{-1,+1\}$. Then the series $Var(k,\epsilon)$ defined by

$$(Var(k,\epsilon)|w) = \begin{cases} \epsilon\, n_k & \text{if } w = (((a^{n_1}b)^{n_2}c)^{n_3}d)^{n_4} \text{ with } (n_1,n_2,n_3,n_4) \in \mathbb{N}^4 \\ 0 & \text{if } w \notin (((a^*b)^*c)^*d)^* \end{cases}$$

is a $\mathcal{Z}$-rational series of $\mathcal{Z}\mathrm{Rat}(a,b,c,d)$.

Proof : Suppose first that $k = 4$. Then, arguing as in lemma 3.2, it is easy to see that

$$\sum_{w \in A^*} \epsilon\,|w|_d\, w$$

is a $\mathcal{Z}$-rational series. Hence it can be clearly shown that $Var(4,\epsilon)$ is a $\mathcal{Z}$-rational series by using the same argument than in lemma 3.2.

We will suppose now that $k \in [1,3]$. Then let us denote $a_1 = a, a_2 = b, a_3 = c, a_4 = d$ and let us consider the $\mathcal{Z}$-representation μ of A^* of order 2 defined by $\mu(a_i) = Id_2$ for every $i \notin \{k, k+1\}$ and by

$$\mu(a_k) = \begin{pmatrix} \epsilon & +\infty \\ +\infty & 0 \end{pmatrix} \quad \text{and} \quad \mu(a_{k+1}) = \begin{pmatrix} +\infty & 0 \\ +\infty & 0 \end{pmatrix}$$

The reader will then easily check that we have for every $n \in \mathbb{N}$

$$\mu(a_k^n\, a_{k+1}) = \begin{pmatrix} +\infty & \epsilon n \\ +\infty & 0 \end{pmatrix}$$

which is an idempotent matrix. Hence it follows immediately from this last relation that we have for every $(n_1,n_2,n_3,n_4) \in \mathbb{N}^4$

$$(\,0\ \ +\infty\,)\ \mu(\,(((a^{n_1}b)^{n_2}c)^{n_3}d)^{n_4}\,) \begin{pmatrix} +\infty \\ 0 \end{pmatrix} = \epsilon\, n_k$$

Thus, if $S(k,\epsilon)$ denotes the $\mathcal{Z}$-rational series defined by $(S(k,\epsilon)|w) = \mu(w)_{1,2}$ for every $w \in A^*$, we clearly have $Var(k,\epsilon) = (S(k,\epsilon) \odot \underline{L}) \oplus \underline{A^* - L}$ where L denotes the rational language $(((a^*b)^*c)^*d)^*$. It follows immediately that $Var(k,\epsilon)$ is a $\mathcal{Z}$-rational series. This ends the proof of our lemma. ∎

EMMA 3.4 : Let M be a square matrix of order n and let $p \in \mathbb{N}$. Let us then enote by $E_{n,p}$ and $\mathcal{N}(M,p)$ the square matrices of order $n\,p$ defined by

$$E_{n,p} = \begin{array}{c} \\ n \\ n \\ \vdots \\ n \end{array} \begin{array}{c} \begin{array}{cccc} n & \dots & n & n \end{array} \\ \begin{pmatrix} +\infty & \dots & +\infty & Id_n \\ Id_n & \dots & +\infty & +\infty \\ \vdots & \ddots & \vdots & \vdots \\ +\infty & \dots & Id_n & +\infty \end{pmatrix} \end{array} \quad \text{and} \quad \mathcal{N}(M,p) = \begin{array}{c} \\ n \\ n \\ \vdots \\ n \end{array} \begin{array}{c} \begin{array}{cccc} n & n & \dots & n \end{array} \\ \begin{pmatrix} M & +\infty & \dots & +\infty \\ +\infty & Id_n & \dots & +\infty \\ \vdots & \vdots & \ddots & \vdots \\ +\infty & +\infty & \dots & Id_n \end{pmatrix} \end{array}$$

et now $(N_i)_{i=1,\dots,p}$ be a family of square matrices of order n. Then we have

$$\mathcal{N}(N_1,p)\,E_{n,p}\,\mathcal{N}(N_2,p)\,E_{n,p}\,\dots\,\mathcal{N}(N_N,p)\,E_{n,p} \;=\; \begin{array}{c} \\ n \\ n \\ \vdots \\ n \end{array} \begin{array}{c} \begin{array}{cccc} n & n & \dots & n \end{array} \\ \begin{pmatrix} N_1 & +\infty & \dots & +\infty \\ +\infty & N_2 & \dots & +\infty \\ \vdots & \vdots & \ddots & \vdots \\ +\infty & +\infty & \dots & N_p \end{pmatrix} \end{array}$$

roof : It is an easy verification that we leave to the reader. ■

Let us now consider an equation of the form (HD). Note that we will use in the equel the notations used in the definition of this equation. Let us then introduce the ational language C over $A = \{a,b,c,d,e\}$ defined by

$$C = (\,(((a^*b)^*c)^*d)^*\,e\,)^p = \prod_{i=1}^{p} (((a^*b)^*c)^*d)^*\,e$$

ny word w of C can described as follows

$$w = w(\underline{n}) = (((a^{n_1^{(1)}}b)^{n_2^{(1)}}c)^{n_3^{(1)}}d)^{n_4^{(1)}}\,e\,\dots\,(((a^{n_1^{(p)}}b)^{n_2^{(p)}}c)^{n_3^{(p)}}d)^{n_4^{(p)}}\,e$$

vhere $\underline{n}$ denotes the vector $(n_1^{(1)},n_2^{(1)},n_3^{(1)},n_4^{(1)},\ \dots\ ,n_1^{(p)},n_2^{(p)},n_3^{(p)},n_4^{(p)})$ of $\mathbb{N}^{4p}$. Let ow (I,μ,T) be an $\mathcal{Z}$-automaton of order n that recognizes the series $Coeff(k)$ of emma 3.2. Then we can define a $\mathcal{Z}$-representation ν of A^* order $n\,p$ as follows

$$\forall\,\alpha \in \{a,b,c,d\},\ \nu(\alpha) = \mathcal{N}(\nu(\alpha),p) \quad \text{and} \quad \nu(e) = E_{n,p}$$

where we took the notations of lemma 3.4. Then, according to lemma 3.2 and to lemma 3.4, it is easy to see that we have

$$\begin{pmatrix} \dots & +\infty & \overset{i\,\downarrow}{I} & +\infty & \dots \end{pmatrix} \nu(w) \begin{pmatrix} \vdots \\ +\infty \\ T \\ +\infty \\ \vdots \end{pmatrix} \overset{i}{\leftarrow} \;=\; \begin{cases} k\,n_1^{(i)}n_2^{(i)}n_3^{(i)}n_4^{(i)} & \text{if } w = w(\underline{n}) \in C \\ 0 & \text{if } w \notin C \end{cases}$$

where each symbol $+\infty$ denotes in fact a block of order n. Hence it follows immediately that the series $Pol_i(k)$ defined by $(Pol_i(k)|w) = 0$ when $w \notin C$ and by

$$(Pol_i(k)|w) = k\,n_1^{(i)}n_2^{(i)}n_3^{(i)}n_4^{(i)}$$

when $w = w(\underline{n}) \in C$ is a $\mathcal{Z}$-rational series. Hence the series $Pol^+ = Pol_1(p_1)$ $Pol_2(p_2) \odot \ldots \odot Pol_n(p_n) \odot -1$ is $\mathcal{Z}$-rational and we clearly have

$$(Pol^+|w) = \begin{cases} \sum_{i=1}^{p} p_i\, n_1^{(i)} n_2^{(i)} n_3^{(i)} n_4^{(i)} - 1 & \text{when } w = w(\underline{n}) \in C \\ -1 & \text{when } w \notin C \end{cases}$$

In the same way, the series $Pol^- = Pol_1(-p_1) \odot Pol_2(-p_2) \odot \ldots \odot Pol_n(-p_n) \odot 1$ $\mathcal{Z}$-rational and we have

$$(Pol^-|w) = \begin{cases} \sum_{i=1}^{p} - p_i\, n_1^{(i)} n_2^{(i)} n_3^{(i)} n_4^{(i)} + 1 & \text{when } w = w(\underline{n}) \in C \\ 1 & \text{when } w \notin C \end{cases}$$

Hence the series $Pol = Pol^- \oplus Pol^+$ is $\mathcal{Z}$-rational and we clearly have

$$(Pol|w) = -\,|\sum_{i=1}^{p} p_i\, n_1^{(i)} n_2^{(i)} n_3^{(i)} n_4^{(i)} - 1|$$

when $w = w(\underline{n})$ is in C and $(Pol|w) = -1$ when $w \notin C$. Arguing as above, bu now with lemma 3.3 instead of lemma 3.2, it is not difficult to see that the serie $Var(i,j,k,l)$ defined by $(Var(i,j,k,l)|w) = 0$ when $w \notin C$ and by

$$(Var(i,j,k,l)|w) = -\,|n_i^{(k)} - n_j^{(l)}| \quad \text{when } w = w(\underline{n}) \in C$$

is $\mathcal{Z}$-rational for every $i, j \in [1,p]$ and $k, l \in [1,4]$. Hence the series HD defined by

$$HD = Pol \odot \bigodot_{(i,j,k,l)\in K} Var(i,j,k,l)$$

is a $\mathcal{Z}$-rational series. Moreover it is easily checked that $(HD|w) = -1$ if $w \notin C$ an that

$$(HD|w) = -\,|\sum_{i=1}^{p} p_i\, n_1^{(i)} n_2^{(i)} n_3^{(i)} n_4^{(i)} - 1| - \sum_{(i,j,k,l\in K} |n_i^{(k)} - n_j^{(l)}|$$

when $w = w(\underline{n}) \in C$. It follows then immediately that the diophantine equatio (HD) has a solution in positive integers if and only if there exists a word $w \in A^*$ suc that $(HD|w) \geq 0$. Hence it follows from a previous remark that the local inequalit problem for $\mathcal{Z}$-rational series over A is undecidable. Thus, according to our reductio work, this ends our proof. ■

Note : Using the same kind of ideas as in the above proof, it can be shown tha any diophantine equation of degree k can be encoded as a local inequality problem fo $\mathcal{Z}$-rational series over an alphabet with $k+1$-letters.

As an immediate corollary of the previous theorem, we obtain according to propo sitions 2.1 and 2.2 :

COROLLARY 3.5 : Let A be an alphabet with at least 2 letters. Then the equality inequality, local equality and local inequality problems are all undecidable questions for $\mathcal{Z}$-rational series over A.

Undecidability of the equality problem for $\mathcal{M}$

.1 Reduction of decidability problems

In this section, we show that the decidability for $\mathcal{M}$ (resp. $\mathcal{N}$) of any problem onsidered in section 2 is equivalent to the decidability of the same problem for $\mathcal{Z}$. et us now first prove this equivalence for $\mathcal{M}$ and $\mathcal{Z}$.

HEOREM 4.1 : Let A be an arbitrary alphabet. Then the equality problem, ıe inequality problem, the local equality problem or the local inequality problem for 1-rational series over A is decidable if and only if the same problem is decidable for -rational series over A.

'roof : Since all the proofs are the same, we shall only show here the equivalence etween the decidability of the equality problems for $\mathcal{M}$ and $\mathcal{Z}$. Clearly we just ave then to prove that the decidability of the equality problem in $\mathcal{M}$ implies the ecidability of the same problem in $\mathcal{Z}$.

Let then R and S be two $\mathcal{Z}$-rational series over the alphabet A. According to the Kleene-Schützenberger theorem, R and S are $\mathcal{Z}$-recognizable series. Let us now conider two $\mathcal{Z}$-automata (I, μ, T) and (J, ν, F) of order m and n recognizing respectively R and S. Let us then consider for every $k \in \mathbb{Z}$ the new vectors $I(k), J(k), T(k), F(k)$ nd the new $\mathcal{Z}$-representations μ_k and ν_k of A^* defined by

$$\forall\, a \in A, \quad \mu_k(a) = (\,\mu(a)_{i,j} + k\,)_{1 \le i,j \le m}\ ,\ \ \nu_k(a) = (\,\nu(a)_{i,j} + k\,)_{1 \le i,j \le n}$$

$$I(k) = (\,I_i + k\,)_{i=1,\dots,m}\ ,\ \ J(k) = (\,J_i + k\,)_{i=1,\dots,n}$$

$$T(k) = (\,T_i + k\,)_{i=1,\dots,m}\ ,\ \ F(k) = (\,F_i + k\,)_{i=1,\dots,n}$$

t is easy to see that we have

$$I(k)\, \mu_k(w)\, T(k) = I\, \mu(w)\, T + 2k + k|w|$$

$$J(k)\, \nu_k(w)\, F(k) = J\, \nu(w)\, F + 2k + k|w|$$

or every word $w \in A^*$. Let now R_k and S_k be the two series recognized respectively y the automata $(I(k), \mu_k, T(k))$ and $(J(k), \nu_k, F(k))$. It follows immediately from the bove computations that we have $S = T$ iff $S_k = T_k$ for any fixed $k \in \mathbb{Z}$. But it is easy o see that T_k and S_k are $\mathcal{M}$-recognizable series when k is greater than

$$-\min_{i,j}\ (\ \mu(a)_{i,j}, \nu(a)_{i,j}, I_i, J_i, T_i, F_i\ )\ \in \mathbb{Z} \cup \{-\infty\}.$$

Hence we showed that the equality of two $\mathcal{Z}$-recognizable series is equivalent to the equality of two $\mathcal{M}$-recognizable series. This ends proving that the decidability of the equality problem for $\mathcal{M}$-rational series implies the decidability of the same problem for $\mathcal{Z}$-rational series. Our theorem is then proved. ■

Using the same method as in the previous theorem, we can also get the following result that shows our equivalence result for $\mathcal{Z}$ and $\mathcal{N}$.

THEOREM 4.2 : Let A be an arbitrary alphabet. Then the equality problem, the inequality problem, the local equality problem or the local inequality problem for $\mathcal{N}$-rational series over A is decidable if and only if the same problem is decidable for $\mathcal{Z}$-rational series over A.

Proof : Since all the proofs are similar, we shall also only show here the equivalence between the equality problems for $\mathcal{N}$ and $\mathcal{Z}$. It suffices then clearly to prove that the decidability of the equality problem for $\mathcal{N}$ implies the decidability of the same problem for $\mathcal{Z}$. But since $\mathcal{N}$ is effectively isomorphic to $\mathcal{Z}^-$ (cf section 1), we have just to prove that the decidability of the equality problem for $\mathcal{Z}^-$ implies the decidability of the same problem for $\mathcal{Z}$.

Let us now take the same notations that in the proof of theorem 4.1. It is easy to see that S_k and R_k are $\mathcal{Z}^-$-recognizable series when k is less than

$$M = -\max_{i,j} \left(\mu(a)_{i,j}, \nu(a)_{i,j}, I_i, J_i, T_i, F_i \right) \in \mathbb{Z}$$

where the above maximum is only taken over the values which are not equal to $+\infty$ (when every value involved in the above maximum is equal to $+\infty$, we set $M = +\infty$). Hence, arguing as in the proof of the previous theorem, it follows that the equality of two $\mathcal{Z}$-rational series is equivalent to the equality of two $\mathcal{Z}^-$-rational series. This ends therefore the proof of our theorem. ∎

4.2 Undecidability of the equality problem for $\mathcal{M}$ and $\mathcal{N}$

As immediate corollaries of the previous results, we get the following undecidability results. Observe that they solve in particular the open problems we speak of in our introduction.

COROLLARY 4.3 : Let A be an alphabet with at least two letters. Then the equality problem, the inequality problem, the local equality problem and the local inequality problem are all undecidable problems for $\mathcal{M}$ or $\mathcal{N}$-rational series over A.

Note : Our undecidability result implies in particular that no "equality theorem" in the sense of Eilenberg (cf [6] th. VI.8.1) can hold in $\mathcal{M}$. However it is interesting to notice that this can also be directly proved. Indeed, if we consider the following family of matrices

$$M_n = \begin{pmatrix} 1 & n \\ 1 & 0 \end{pmatrix}$$

which is indexed by $n \in \mathbb{N}$, it is easy to prove that we have

$$S_n = \sum_{k=0}^{+\infty} (M_n)^k_{1,1}\, a^k = \sum_{k=0}^{n} k\, a^k \; + \sum_{k=n+1}^{+\infty} (n+1)\, a^k$$

It follows from this computation that the two distinct series S_n and S_{n+1} (both associated to a $\mathcal{M}$-representation of order 2) coincide up to the order n. Hence we obtain effectively that no "equality theorem" (cf [6] th. VI.8.1) is possible for the tropical semiring, even in fact with one-letter series.

Moreover we can also obtain as an application of our methods the following decida-ility results which make complete our study of the decidability of the four problems onsidered in section 2 for $\mathcal{Z}$, $\mathcal{M}$ and $\mathcal{N}$.

OROLLARY 4.4 : The equality problem, the inequality problem, the local equa-ty problem and the local inequality problem are decidable for $\mathcal{M}$, $\mathcal{N}$ or $\mathcal{Z}$-rational eries over a *one-letter* alphabet $A = \{a\}$.

roof : According to theorems 4.1, 4.2 and to propositions 2.1 and 2.2, it suffices o show that the local equality problem is decidable for $\mathcal{M}$-rational series over a one-etter alphabet in order to prove our corollary. Using now a classical embedding due o C. Choffrut, of $\mathcal{M}$ into the semiring $Rat(b^*)$ of rational languages over a one-letter lphabet $\{b\}$ (cf [3] or [4]), this last problem appears in fact as an intersection problem or special kinds of rational languages of $Rat(a^* \times b^*)$. The decidability of our result ollows now from the decidability of the intersection problem for $Rat(a^* \times b^*)$ (see [7] or instance). ■

Note : Using a fine study of the iterated power of a square matrix with entries in $\mathcal{Z}$, t can also be shown directly that the equality problem for one-letter $\mathcal{Z}$-rational series s decidable.

Acknowledgements

I want to thank here professor I. Simon for the personal papers he communicated ne and also professor S. Grigorieff who indicated me the crucial reference [5]. Finally I nust also thank the "LAboratoire de Combinatoire et d'Informatique Mathématique" (LACIM; Université du Québec à Montréal) for its kind support during the preparation of the final version of this paper.

References

[1] ADLER A., *A reduction of homogeneous diophantine problems*, Journ. of London Math. Soc. (2), **3**, pp. 446-448, 1971

[2] BERSTEL J., REUTENAUER C., *Rational series and their languages*, Springer Verlag, 1986

[3] CHOFFRUT C., *Séries rationnelles d'image finie*, Technical Report 79-6, LITP, Paris, 1979

[4] CHOFFRUT C., *Rational relations and rational series*, Theor. Comput. Sci., 1992 (To appear)

[5] DAVIS M., MATIJASEVIC Y., ROBINSON J., *Hilbert's tenth problem, diophan-tine equations : positive aspects of a negative solution*, Proceedings of Symposi in Pure Mathematics, Vol. 28, pp. 323-378, 1976

[6] EILENBERG S., *Automata, languages and machines*, Vol. A, Academic Press 1974

[7] GIBBONS A., RYTTER W., *On the decidability of some problems about rational subsets of free partially commutative monoids*, Theor. Comp. Sci., 48, pp. 329-337 1986

[8] SIMON I., *Recognizable sets with multiplicities in the tropical semiring*, [in "MFCS'88 Proceedings"], Lect. Notes in Comput. Sci., **324**, p. 107-120, Springer-Verlag, 1988

[9] SIMON I., *Some open problems for automata with multiplicities*, Private communication

SEMI-COMMUTATIONS and RATIONAL EXPRESSIONS*

M.CLERBOUT - M.LATTEUX - Y.ROOS
LIFL, CNRS UA 369
University of Lille Flandre Artois
59655 VILLENEUVE D'ASCQ Cédex, FRANCE

and

W.ZIELONKA
LABRI, CNRS URA 1304
University of BORDEAUX I
33405 TALENCE Cédex, FRANCE

Abstract

We extend the notion of concurrent iteration defined on trace languages by E.Ochmanski to languages closed under a semi-commutation relation which is the non-symmetric version of partial commutation relation: we iterate strongly connected components of words. For a given semi-commutation relation, this leads to the definition of a new family of rational languages closed under the semi-commutation. We give a necessary and sufficient condition for a language closed under a semi-commutation relation to be a rational language and we proved that the equality between the families of rational languages and recognizable languages closed under a semi-commutation relation is true if and only if the semi-commutation is symmetric (i.e a partial commutation).

To describe concurrent behaviours in an asynchronous distributed system, Mazurkiewicz introduced the notion of traces [Maz77]: in a concurrent system of processes, the order in which independent actions are executed does not matter, so different sequential behaviours which differ only in the order of independent actions will be identified under the concept of trace.
For a fixed independent relation, the set of traces is a monoid called the free partially commutative monoid. These monoids were first studied in order to solve combinatoric problems of rearrangement. Since they provide a good framework to describe parallelism, research on traces has shown a great development over the last years, and several surveys deal with them ([AR86],[Die90],[Per89]).

---

*This work has been partially supported by the EBRA working group 3166 ASMICS and by the PRC "Mathematiques et Informatique".

Semi-commutations, which have been introduced by M.Clerbout in [Cle84] represent both a natural generalization of partial commutations and a new tool for studying synchronizing problem: they model the problem of "producer consumer" . In a sequence of such actions which respects property at each time we produced more than we consumed', it is always possible to substitute the sequence *consume produce* by *produce consume* but the reverse is not always possible. So actions may be "semi-independent". It follows the notion of semi-trace and semi-commutative monoid. The most important properties of semi-commutations can be found in [Lat90].

The main results of research in this area concern the family of recognizable sets of trace monoids, since they describe the behaviour of finite states systems. On one side, W.Zielonka introduced asynchronous automata which allowed executions of independent actions, and he proved that a subset of a trace monoid is recognizable if and only if it is accepted by some finite asynchronous automaton [Zie87]. On the other side, the problem was to find an analogous to Kleene's Theorem.

In the case of trace theory, the two main results are due to Y.Metivier who proved in [Met86] that if R is a recognizable connected trace language R* remains recognizable, and for the second one, to E.Ochmanski [Och85] he introduced a new iteration on trace language which concerns the connected components of the trace language. This leads to the definition of rational trace languages and E.Ochmanski generalized the Kleene's theorem when proving the equality between recognizable trace languages and rational trace languages.

If the result of Y.Metivier was generalized to semi-commutation at the same time by M.Clerbout/M.Latteux in [CL87], it appeared that the result of E.Ochmanski could not be extended easily. The problem is the interpretation of semi-trace recognizable languages (in the semi-trace monoid). In the case of partial commutation, a trace language is recognizable if and only if the union of traces of this language is a recognizable set of the free monoid. It is not true for semi-commutations (see [Och89]).

Here, our point of view is to study languages which are a union of semi-traces (they are closed under a given semi-commutation) rather than semi-trace languages. We defined for these languages a generalization of concurrent iteration: we iterate independently strongly connected components of words and naturally we construct the family of "rational" languages closed under union, product (with respect to the semi-commutation) and concurrent iteration.

We give in this paper a necessary and sufficient condition for a language closed under a semi-commutation to be rational (as explained above) and we proved that the equality between the family of "rational" languages and the family of recognizable languages closed under the semi-commutation holds if and only if the semi-commutation is symmetric, that is a partial commutation.

Preliminaries

Let X be a totally ordered alphabet. As usual, X^* denotes the free monoid over X.

For a word w of X^*, we denote by

$|w|$ the length of w,

$alph(w)$ the set of letters of X which occur in w,

$\Pi_Y(w)$ the projection of w over the subalphabet Y of X, i.e the image of w by the morphism Π_Y defined by: $\forall x \in X$, if $x \in Y$ then $\Pi_Y(x) = x$, else $\Pi_Y(x) = \varepsilon$.

If $L \subseteq X^*$ is a language, we denote by **IF(L)** the set of the iterating factors of L, that is the set $\{u \in X^* |\ \exists x, y \in X^*,\ xu^*y \subseteq L\}$.

A **semi-commutation relation** θ over X is an irreflexive relation defined on X, which is often called an independence relation.

With each semi-commutation relation θ we associate a rewriting system $S_\theta =< X, P >$ which is named a semi-commutation system, in which P is the set of rules $\{xy \to yx,\ (x,y) \in \theta\}$.

We also associate with each semi-commutation relation a **semi-commutation function** f_θ defined on X^* by: $\forall w \in X^*, f_\theta(w) = \{u \in X^* |\ \ w \overset{*}{\underset{P}{\Longrightarrow}} u\}$.

Instead of writing $w \overset{*}{\underset{P}{\Longrightarrow}} u$, we shall denote by $w \overset{*}{\underset{\theta}{\Longrightarrow}} u$ the fact that $u \in f_\theta(w)$.

By extension, if L is a language, i.e a subset of X^*, $f_\theta(L) = \bigcup_{w \in L} f_\theta(w)$.

We say that a language $L \subseteq X^*$ is **θ-closed** if and only if $f_\theta(L) = L$.

With each semi-commutation relation θ, we associate its **dependence graph** or **non-commutation graph** denoted by $(X, \bar{\theta})$ where X is the vertices set and $\bar{\theta} = \{(x,y) | x \neq y, (x,y) \notin \theta\}$ is the edges set.

For a given graph (X, θ), we define an equivalence relation on the set of vertices: $x \in X$ and $y \in X$ are equivalent if there exist an oriented way between x and y and an oriented way between y and x. The subgraph induced by an equivalent class for this relation is a **strongly connected component** of $(X, \bar{\theta})$.

If Y is a subset of X, we denote by $(Y, \bar{\theta})$ the subgraph of the non-commutation graph induced by Y .

If u is a word of X^*, we shall say that u **is strongly connected** if $(alph(u), \bar{\theta})$ is strongly connected.

An important case is when the relation θ is symmetric and we say that θ is a **partial commutation relation.** Note that here the non-commutation graph is undirected and then we shall speak of 'connected graph or word'.

We denote by $\mathcal{REC}(X^*)$ the set of recognizable languages of X^*.

We denote by $\mathcal{REC}_\theta(X^*)$ the set of recognizable θ-closed languages that is the set

$$\{L \subseteq X^* |\ L = f_\theta(L) \ and \ L \in \mathcal{REC}(X^*)\}$$

Just as in the case of free monoid the problem is to find a rational characterization of this family and an equivalent version of the Kleene Theorem. I the case of partial commutation, E.Ochmanski introduced a new iteration fc θ-closed languages, so it seems natural to generalize this operation for sem commutation. Let us recall the following operations on θ-closed languages:

- *θ-union:* $R_1 \cup_\theta R_2 = f_\theta(R_1 \cup R_2) = R_1 \cup R_2$
- *θ-concatenation:*$R_1 ._\theta R_2 = f_\theta(R_1.R_2)$
- *θ -iteration: if* $\forall u \in R$, $(alph(u), \bar{\theta})$ *is strongly connected, then* $R^{*_\theta} = f_\theta(R^*)$

Note that θ-iteration as defined here is a partial operator.

An important result is the following one:

Theorem 1.1 [CL87]– *If R and S are two languages of $\mathcal{REC}_\theta(X^*)$ then $R \cup_\theta S$ and $R ._\theta S$ are in $\mathcal{REC}_\theta(X^*)$. Furthermore, if $\forall u \in R$, u is strongl connected, i.e $(alph(u), \bar{\theta})$ is strongly connected, then R^{*_θ} is in $\mathcal{REC}_\theta(X^*)$.*

So let us define $\mathcal{R}_\theta(X^*)$ as the smallest family of languages containin elementary languages that are $\emptyset$, $\{\varepsilon\}$ and $\{\{x\} |\ x \in X\}$ and closed unde θ-union, θ-concatenation and θ-iteration. Due to the previous result, we hav

$$\mathcal{R}_\theta(X^*) \subseteq \mathcal{REC}_\theta(X^*)$$

In the case of partial commutations, the equality of these two families i an important result due to E.Ochmanski:

Theorem 1.2 [Och85]– *If θ is a partial commutation then $\mathcal{REC}_\theta(X^*) = \mathcal{R}_\theta(X^*)$.*

To answer the question if the equality holds in the general case of semi-commutation, we give in this paper a necessary and sufficient condition for a θ-closed language to be in $\mathcal{R}_\theta(X^*)$ and then we shall be able to conclude. For that, we use the following notions introduced in [Och85]:

In the following, we will denote by θ_s the symmetric part of the semi-commutation relation θ ($\theta_s = \theta \cap \theta^{-1}$).

For $u \in X^*$, $Lex(u)$ denotes the smallest element of $f_{\theta_s}(u)$ for the lexicographic ordering of X^*. For $L \subseteq X^*$, $Lex(L) = \cup_{u \in L} Lex(u)$ and LEX is the set $\{Lex(u) |\ u \in X^*\}$.

For $R \subseteq X^*$, $MAX_\theta(R) = \{u \in R |\ \forall v \in R,\ u \in f_\theta(v) \Rightarrow v \in f_\theta(u)\}$.

It is easy to see that if R is a θ -closed language,

$$MAX_\theta(R) = f_{\theta_s}(MAX_\theta(R)) \quad and \quad R = f_\theta(MAX_\theta(R))$$

e following properties of these sets have already been proved or are easy to ove.

Lemma 1.3 [Och85]– *LEX is a recognizable set of X^* and each factor a word of LEX is also a word of LEX.*

Lemma 1.4 [Och85]– *If θ is a partial commutation over X, and if R is θ-closed language, then $Lex(R) = R \cap LEX$ and $R = f_\theta(Lex(R))$. oreover, $R \in \mathcal{REC}_\theta(X^*) \iff Lex(R) \in \mathcal{REC}(X^*)$.*

Lemma 1.5– *If $R \in \mathcal{REC}_\theta(X^*)$, then $MAX_\theta(R) \in \mathcal{REC}(X^*)$.*

Lemma 1.6 [Och85]– *Let θ be a partial commutation defined on X and $\in X^*$. If $u^* \subseteq LEX$ then u is connected.*

. A necessary and sufficient condition to be in $\mathcal{R}_\theta(X^*)$

Let θ be a semi-commutation and let $R \subseteq X^*$ be a θ-closed language. θ is symmetric, i.e in the case of partial commutation, then Ochmanski has roved that $R \in \mathcal{REC}_\theta(X^*)$ if and only if $Lex(R) \in \mathcal{REC}(X^*)$. imilarly, in the case of semi-commutation we should like to find a language $\subseteq R$ such that $R \in \mathcal{REC}_\theta(X^*)$ if and only if $L \in \mathcal{REC}(X^*)$. Since $_\theta(MAX_\theta(R)) = R$, the language $Lex(MAX_\theta(R))$ seems to be a good candi-ate.
Jnfortunately, $Lex(MAX_\theta(R)) \in \mathcal{REC}(X^*)$ is not a sufficient condition to onclude that $R \in \mathcal{R}_\theta(X^*)$: if $X = \{a,b\}$, $\theta = \{(b,a)\}$ and $R = D_1'^* =$ $w \in \{a,b\}^*, |w|_a = |w|_b,$ *and* $w = uv \Rightarrow |u|_a \geq |u|_b\}$, we get $MAX_\theta(R) =$ $ab)^* \in \mathcal{REC}(X^*)$ but $R \notin \mathcal{R}_\theta(X^*)$ since $R \notin \mathcal{REC}_\theta(X^*)$. As we shall see, the eason of it is that ab is an iterating factor of $MAX_\theta(R)$ which is not strongly onnected.

So, we can give the following condition for a θ-closed language R to be n $\mathcal{R}_\theta(X^*)$, and we are going to prove that it is a necessary and sufficient ondition:

Condition 1–
1- The language $K = Lex(MAX_\theta(R))$ is in $\mathcal{REC}(X^*)$
2- Each iterating factor of K is strongly connected.

To show that *Condition 1* is sufficient, we prove the following proposition:

Proposition 2.1– *Let K be a language of $\mathcal{REC}(X^*)$ such that $K = Lex(K)$, $K = MAX_\theta(K)$ and for each u in $IF(K)$, u is strongly connected.*

Then $f_\theta(K) \in \mathcal{R}_\theta(X^*)$.

Sketch of Proof
The proof is carried on by structural induction on K. It uses the followir elementary results:
If $L = MAX_\theta(L)$ then $\forall L' \subseteq L$, $L' = MAX_\theta(L')$,
if $L = Lex(L)$ then $\forall L' \subseteq L$, $L' = Lex(L')$,
if $L = MAX_\theta(L)$ and $L = L_1.L_2$ then $L_1 = MAX_\theta(L_1)$ and $L_2 = MAX_\theta(L_2$
if $L = Lex(L)$ and $L = L_1.L_2$ then $L_1 = Lex(L_1)$ and $L_2 = Lex(L_2)$. □

It is now easy to prove that **Condition 1** is a sufficient condition for θ-closed language to be in $\mathcal{R}_\theta(X^*)$:

Proposition 2.2– *Let R be a θ-closed language. If $K = Lex(MAX_\theta(R$ is in $\mathcal{REC}(X^*)$ and if each iterating factor of K is strongly connected, then is in $\mathcal{R}_\theta(X^*)$.*

Proof
It is clear that $K = Lex(K)$ and since $K \subseteq MAX_\theta(R)$, $K = MAX_\theta(K$ (Lemma 1.5). So, with Proposition 2.1, we get $f_\theta(K) \in \mathcal{R}_\theta(X^*)$.
And $f_\theta(K) = f_\theta(f_{\theta_s}(Lex(MAX_\theta(R))) = f_\theta(MAX_\theta(R)) = R$, so $R \in \mathcal{R}_\theta(X^*$
□

We are now going to prove that each language R of $\mathcal{R}_\theta(X^*)$ verifies Condition which is in fact equivalent with **Condition 1**:

Condition 2–
1- The language $K = MAX_\theta(R)$ is in $\mathcal{REC}(X^*)$
2- Each iterating factor u of K such that $u^* \subseteq LEX$ is strongly connected.

We immediately show that **Condition 2** is at least stronger that **Condition 1**:

Lemma 2.3– *Each θ-closed language R which verifies Condition 2 verifie also Condition 1*

Proof
It is clear that if we set $K' = Lex(K)$, then K' is in $\mathcal{REC}(X^*)$ (We know tha $K = MAX_\theta(R)$ is a θ_s-closed language and we use Lemma 1.4).
Moreover, if u is an iterating factor of K', then u is also an iterating factor o K since $K' \subseteq K$. And we can find two words x and y such that $xu^*y \subseteq K' \subseteq LEX$. As LEX is closed under factor (Lemma 1.3), we get $u^* \subseteq LEX$ so u is strongly connected. □

We already know that if a language R is in $\mathcal{R}_\theta(X^*)$, so it is in $\mathcal{REC}_\theta(X^*$

:e Theorem 1.1), and when using Lemma 1.5, we can conclude that $MAX_\theta(R)$ in $\mathcal{REC}(X^*)$. But it is rather more difficult to establish the property of the :rating factors of $MAX_\theta(R)$. So we now present some results concerning the :rating factors.

Lemma 2.4– *Let K_1 and K_2 be two languages of $\mathcal{REC}(X^*)$ and u be an :rating factor of $K_1 \cup K_2$. Then we can find a positive integer k and $i \in \{1,2\}$ ıch that u^k is an iterating factor of K_i.*

Proof

here exist two words x, y of X^* such that $xu^*y \subseteq K_1 \cup K_2$. So, at least ıe of the sets $xu^*y \cap K_i$, $i = 1,2$, is infinite. Let $M = (X, Q, q_0, \delta, F)$ be deterministic automaton which recognizes K_i. Since $xu^*y \cap K_i$ is infinite, e can find two different positive integers n and m such that $\delta(q_0, xu^n) =$ $q_0, xu^m)$, $\delta(q_0, xu^n y) \in F$ and $\delta(q_0, xu^m y) \in F$.
$n < m$, then $xu^n(u^{m-n})^*y \subseteq K_i$ and u^{m-n} is an iterating factor of K_i. □

Lemma 2.5– *Let θ be a partial commutation defined on X, K_1, K_2 be vo θ-closed recognizable languages and $u \in X^*$ a connected word (for θ) which : an iterating factor of $K_1 ._\theta K_2$. Then there exists $k \geq 1$ such that u^k is an erating factor of K_1 or K_2.*

Sketch of Proof:

.s u is an iterating factor of $K_1 ._\theta K_2$, we can find two words x, y such that $n \in \mathbb{N}, xu^n y \in K_1 ._\theta K_2$.
.et us choose an integer n such that $n \geq N_1 + N_2 + |u| + 2$.
o for each value $n \geq 0$ there exist two words $w_1 \in K_1$ and $w_2 \in K_2$ such that $u^n y \overset{*}{\underset{\theta}{\Longleftrightarrow}} w_1 w_2$.
3y a Lemma due to Cori and Perrin ([CP85]), we can find words $x', x'', y', y'', u_i',$ u_i'' for $i \in \{1, 2, ..., n\}$ such that

$$w_1 = x' u_1' u_2' ... u_n' y'$$
$$w_2 = x'' u_1'' u_2'' ... u_n'' y''$$
$$x' x'' \overset{*}{\underset{\theta}{\Longleftrightarrow}} x,\ y' y'' \overset{*}{\underset{\theta}{\Longleftrightarrow}} y$$
$$u_i' u_i'' \overset{*}{\underset{\theta}{\Longleftrightarrow}} u,\ \forall i \in \{1, 2, ..., n\}$$
$$\forall i \in \{1, 2, ..., n\}, \forall x \in alph(u_i''), \forall y \in alph(u_{i+1}' ... u_n'),\ (x, y) \in \theta$$

The aim of the proof is to show that as u is connected, for at most P :uccessive indices i (P is bounded by the length of $|u|$), u_i' and u_i'' are both ıon-empty words. This fact leads us to choose a value n big enough to assure hat u^k is an iterating factor of K_1 or K_2. □

The following lemma is a generalization of the previous one, in the case of :oncatenation of more than two languages, and is easily proved by induction.

Lemma 2.6– *Let θ be a partial commutation defined on X, $K_1, K_2, ..., K$ be n recognizable θ-closed languages and $u \in X^*$ be a connected word (for which is an iterating factor of $K_1 ._\theta K_2 \ ... \ ._\theta K_n$. Then there exist a positi integer k and $i \in \{1, 2, ..., n\}$ such that u^k is an iterating factor of K_i.*

The next lemma, due to Yves Roos, characterizes $f_\theta(u^*)$ for any connecte word u.

Lemma 2.7 [Roo89]– *Let θ be a semi-commutation relation and u be strongly connected word such that $\forall x \in alph(u)$, $|u|_x = 1$.*
Then for each factor w of $f_\theta(u^*)$ such that there exists a letter y which satisfie $|w|_y > 2|u|$, we have: $\forall x \in alph(u)$, $|w|_x \geq 1$.

If this property is proved in [Roo89] for words with at most one occur rence of each letter, its remains true for any word: it is sufficient to numbe occurrences of the same letter (then the condition becomes $|w|_y \geq 2|u||u|_y$).

We are now ready to prove the necessary condition (Condition 2) for θ-closed language to be in $\mathcal{R}_\theta(X^*)$.

Proposition 2.8– *Let R be a θ-closed language. If $R \in \mathcal{R}_\theta(X^*)$ then satisfies the two following assertions:*
1- The language $K = MAX_\theta(R)$ is in $\mathcal{REC}(X^*)$
2- Each iterating factor u of K such that $u^* \subseteq LEX$ is strongly connected.

Proof
Theorem 1.1 enables us to state that if R is in $\mathcal{R}_\theta(X^*)$, then it is in $\mathcal{REC}_\theta(X^*$ and then we conclude with Lemma 1.5 that $MAX_\theta(R) \in \mathcal{REC}(X^*)$. So th first assertion is true.
We prove the second one by structural induction.

⊙If $R = \emptyset$ or $R = \{\varepsilon\}$ or $R = \{x\}$ for some letter $x \in X$, the result i obvious.

⊙If $R = R_1 \cup_\theta R_2$, where R_1, R_2 are two languages of $\mathcal{R}_\theta(X^*)$, then fo an iterating factor u of $MAX_\theta(R)$ with $u^* \subseteq LEX$, we can find two word $x, y \in X^*$ such that $xu^*y \subseteq MAX_\theta(R) \subseteq MAX_\theta(R_1) \cup MAX_\theta(R_2)$ and these two languages are in $\mathcal{REC}(X^*)$.
So with Lemma 2.4, we can find a positive integer k such that u^k is an iterating factor of one of the two languages $MAX_\theta(R_1)$ or $MAX_\theta(R_2)$.
And $(u^k)^* \subseteq u^* \subseteq LEX$, so by induction hypothesis, we conclude that u^k and u are strongly connected.

⊙If $R = R_1 ._\theta R_2$ where R_1, R_2 are two languages of $\mathcal{R}_\theta(X^*)$, and if u is an iterating factor of $MAX_\theta(R)$ with $u^* \subseteq LEX$, we can find two words

$y \in X^*$ such that $xu^*y \subseteq MAX_\theta(R) \subseteq MAX_\theta(R_1).{}_{\theta_s}MAX_\theta(R_2)$ and these o θ_s-closed languages are in $\mathcal{REC}(X^*)$.
y Lemma 1.6 u is connected for θ_s and with Lemma 2.6, we get that u^k is iterating factor of one of these languages, and just as above, we conclude at u is strongly connected.

⊙If $R = R_1^{*\theta}$, where R_1 is a language of $\mathcal{R}_\theta(X^*)$ such that each of its ords is strongly connected, then $K = MAX_\theta(R) \subseteq (MAX_\theta(R_1))^{*\theta_s}$.
et us set $K_1 = MAX_\theta(R_1)$. K_1 is a θ_s-closed language.
et u be an iterating factor of K such that $u^* \subseteq LEX$. So we can find two ords $x, y \in X^*$ such that $xu^*y \subseteq K$, and we know (Lemma 1.6) that u is $_s$-connected.
/e have to distinguish two cases:

1) There exists a value N such that $xu^*y \subseteq f_{\theta_s}((K_1 \cup \{\varepsilon\})^N)$.
hen due to Lemma 2.6, we conclude that u^k is an iterating factor of K_1 and at, by induction hypothesis, u is strongly connected.

2) If we cannot find such a value, it means that

$$\forall n \in \mathbb{N}, \exists p, q > n,\ xu^py \in f_{\theta_s}((K_1 \setminus \{\varepsilon\})^q)$$

o we can find q words $u_1, u_2, ..., u_q$ of $K_1 \setminus \{\varepsilon\}$ such that $xu^py \overset{*}{\underset{\theta_s}{\Longleftrightarrow}} u_1u_2...u_q$.
ince u is connected for θ_s, when choosing a sufficiently great value for n we btain the following fact by applying Lemma 2.7:

$$(1) \qquad \exists i, j, k \in \{1, 2, ..., q\},\ alph(u) = alph(u_i...u_j) = alph(u_{j+1}...u_k)$$

uppose that u is not strongly connected. Then there exist two subsets B, C f $alph(u)$ such that $B \cup C = alph(u)$, $B \times C \subseteq \theta$ but $C \times B \cap \theta = \emptyset$.
s each word of K_1 is strongly connected ($K_1 \subseteq R_1$),

$$\forall i \in \{1, 2, ..., q\},\ alph(u_i) \subseteq B \text{ or } alph(u_i) \subseteq C$$

et $u_{i_1}, u_{i_2}, ..., u_{i_l}$ be the words of $\{u_i, u_{i+1}, ..., u_k\}$ whose letters are in B $i_1 < i_2 < ... < i_l$) and $u_{j_1}, u_{j_2}, ..., u_{j_m}$ the other ones ($alph(u_{j_\alpha}) \subseteq C$).
rom (1), we are sure that $\{i_1, i_2, ..., i_l\} \cap \{i, ..., j\}$, $\{i_1, i_2, ..., i_l\} \cap \{j+1, ..., k\}$, $j_1, j_2, ..., j_m\} \cap \{i, ..., j\}$ and $\{j_1, j_2, ..., j_m\} \cap \{j+1, ..., k\}$ are all non-empty.
Then

$$u_{i_1}u_{i_2}...u_{i_l}u_{j_1}u_{j_2}...u_{j_m} \overset{+}{\underset{\theta}{\Longrightarrow}} u_i...u_j...u_k$$

out the converse is not true. So $u_1...u_q$ is not in $MAX_\theta(R)$ since it has a actor which is not maximal, hence xu^*y is not in $MAX_\theta(R)$ since $xu^*y \overset{*}{\underset{\theta_s}{\Longleftrightarrow}}$ $u_1u_2...u_q$.
The hypothesis was impossible: u is strongly connected. □

When putting Proposition 2.2 , Proposition 2.8 and Lemma 2.3 togethe we obtain

Proposition 2.9– *Let θ be a semi-commutation over X and $R \in X^*$ a θ-closed language. Then the following three assertions are equivalent:*
(A) $R \in \mathcal{R}_\theta(X^*)$
(B)
1- The language $K = Lex(MAX_\theta(R))$ is in $\mathcal{REC}(X^*)$ and
2- Each iterating factor of K is strongly connected
(C)
1- The language $K = MAX_\theta(R)$ is in $\mathcal{REC}(X^*)$ and
2- Each iterating factor u of K such that $u^* \subseteq LEX$ is strongly connected.

3. Does the equality $\mathcal{R}_\theta(X^*) = \mathcal{REC}_\theta(X^*)$ hold?

We have already seen that $\mathcal{R}_\theta(X^*) \subseteq \mathcal{REC}_\theta(X^*)$: it's a corollary of Theorem 1.1.
Unfortunately we are going to disprove the converse inclusion with the following corollary of Proposition 2.9:

Theorem 3.1–*Let θ be a semi-commutation relation over X. We have $\mathcal{R}_\theta(X^*) = \mathcal{REC}_\theta(X^*)$ if and only if θ is a symmetric relation.*

Proof
When θ is a partial commutation, the result is due to E.Ochmanski [Och85]. In the other cases, it suffices to find a language of $\mathcal{REC}_\theta(X^*)$ which does not satisfy the properties announced in Proposition 2.9.
Since θ is not symmetric, there exist two letters a, b such that $(b, a) \in \theta$ and $(a, b) \notin \theta$. Let $R = (ba)^*a(a+b)^*$. It is clear that $R = f_\theta(R)$. Note that $(a+b)^* \subseteq LEX$ and $MAX_\theta(R) = R \setminus R'$ where $R' = \{xaby |\ xbay \in R\}$. We get $MAX_\theta(R) = (ba)^*a(a^* + b + bb + bba + bbb^+a^*)$.
Now note that although ba is an iterating factor of $MAX_\theta(R)$, it is not strongly connected.□

We finish with a remark on the definition of $\mathcal{R}_\theta(X^*)$.
In the Ochmanski theory of partial commutation, the iteration of θ-closed languages is defined in a way that can be easily generalized to semi-commutations. For $u \in X^*$, let $(X_1, X_2, ..., X_n)$ be the partition of the graph $(alph(u), \bar{\theta})$ into strongly connected components.
Let us set

$AS(\varepsilon) = \{\varepsilon\}$
$AS(u) = \{\Pi_{X_i}(u) |\ X_i \textit{ is a strongly connected component of } (alph(u), \bar{\theta})\}$

and, for any language $R \subseteq X^*$, $AS(R) = \bigcup_{u \in R} AS(u)$.

Then the "concurrent iteration" of a language R: $R^{*'_{\theta_s}}$ is defined by

$$R^{*'_\theta} = f_\theta((AS(R))^*)$$

we denote by $\mathcal{R}'_\theta(X^*)$ the smallest family containing elementary subsets X^* and closed under union, θ-concatenation and the iteration defined just oove, Ochmanski proved that in the symmetric case $\mathcal{R}_\theta(X^*) = \mathcal{R}'_\theta(X^*)$. 'e are going to prove that this equality is also true in the case of semi-ommutations. It is due to the two following results:

Lemma 3.2– *Let θ be a semi-commutation defined over X. $R \subseteq X^*$ is in $\mathcal{REC}_\theta(X^*)$, $AS(R)$ is also in $\mathcal{REC}_\theta(X^*)$.*

Sketch of Proof

/e proved at first that if $u \in X^*$, $AS(f_\theta(u)) = f_\theta(AS(u))$. So $f_\theta(AS(R)) = S(f_\theta(R)) = AS(R)$ and we get that $AS(R)$ is a θ-closed language. 'he second part of the result ($AS(R)$ is a recognizable set) is obtained by roving that the function AS is a rational transduction. □

Lemma 3.3– *Let θ be a semi-commutation defined on X. f $u \in X^*$, and if P is the number of strongly connected components of alph$(u), \bar{\theta})$, $u \in f_\theta((AS(u))^P)$. Then there exists a constant N bounded by he cardinality of X such that $\forall u \in X^*$, $u \in f_\theta((AS(u) \cup \{\varepsilon\})^N)$.*

Now we prove the same kind of result that in Lemma 3.2 for the family $\mathcal{R}_\theta(X^*)$.

Lemma 3.4– *Let θ be a semi-commutation relation over X.*

$$R \in \mathcal{R}_\theta(X^*) \Longrightarrow AS(R) \in \mathcal{R}_\theta(X^*)$$

Proof

3y Lemma 3.2, we know that $AS(R) \in \mathcal{REC}_\theta(X^*)$, so $MAX_\theta(AS(R))$ is in $\mathcal{REC}(X^*)$ (Lemma 1.5).
We prove now that each iterating factor u of $MAX_\theta(AS(R))$ such that $u^* \subseteq LEX$ is strongly connected, with an induction on R.

⊙ If R is an elementary subset of X^*, the result is obvious.

⊙ If $R = R_1 \cup R_2$ with $R_1, R_2 \in \mathcal{R}_\theta(X^*)$, then

$$\begin{aligned} MAX_\theta(AS(R)) &= MAX_\theta(AS(R_1) \cup AS(R_2)) \\ &\subseteq MAX_\theta(AS(R_1)) \cup MAX_\theta(AS(R_2)) \end{aligned}$$

From Lemma 2.4, we can find some $k > 0$ and $i \in \{1,2\}$ such that u^k is a iterating factor of $MAX_\theta(AS(R_i))$. So, when using the induction hypothesis we deduce that u is strongly connected.

⊙ If $R = R_1 ._\theta R_2$ with $R_1, R_2 \in \mathcal{R}_\theta(X^*)$:
For any word w of $AS(R)$ we can find two words $u_1 \in R_1$, $u_2 \in R_2$ and strongly connected component B of $(alph(u_1u_2), \bar{\theta})$ such that

$$\begin{aligned} w &\in \Pi_B(f_\theta(u_1u_2)) \\ &\subseteq f_\theta(\Pi_B(u_1)\Pi_B(u_2)) \end{aligned}$$

It is clear that for $i \in \{1,2\}$, $B \cap alph(u_i)$ is a union of strongly connected components of $(alph(u_i), \bar{\theta})$ and $AS(\Pi_B(u_i)) \subseteq AS(u_i) \cup \{\varepsilon\}$, so we have

$$\begin{aligned} w &\in f_\theta(\Pi_B(u_1)\Pi_B(u_2)) \\ &\subseteq f_\theta(f_\theta((AS(u_1) \cup \{\varepsilon\})^N) f_\theta((AS(u_2) \cup \{\varepsilon\})^N)) \ (Lemma\ 3.3) \\ &\subseteq f_\theta(f_\theta((AS(R_1) \cup \{\varepsilon\})^N) f_\theta((AS(R_2) \cup \{\varepsilon\})^N)) \\ &= f_\theta((AS(R_1) \cup \{\varepsilon\})^N) ._\theta f_\theta((AS(R_2) \cup \{\varepsilon\})^N) \end{aligned}$$

Let us set for $i = 1,2$, $R'_i = R_i \cup \{\varepsilon\}$.
So we get $AS(R_1 ._\theta R_2) \subseteq f_\theta((AS(R'_1))^N) ._\theta f_\theta(AS(R'_2))^N)$ and

$$\begin{aligned} MAX_\theta(AS(R_1 ._\theta R_2)) &\subseteq MAX_\theta(f_\theta((AS(R'_1))^N) ._\theta f_\theta((AS(R'_2))^N)) \\ &\subseteq MAX_\theta(f_\theta((AS(R'_1))^N)) ._{\theta_s} MAX_\theta(f_\theta((AS(R'_2))^N)) \end{aligned}$$

Hence, if u is an iterating factor of $MAX_\theta(AS(R_1 ._\theta R_2))$ such that $u^* \subseteq LEX$ (which implies that u is connected for θ_s), using Lemma 2.5, we can find a value k such that u^k is an iterating factor of $MAX_\theta(f_\theta((AS(R'_1))^N))$ or of $MAX_\theta(f_\theta((AS(R'_2))^N))$. In these two cases, we have for $i \in \{1,2\}$

$$\begin{aligned} MAX_\theta(f_\theta((AS(R'_i))^N)) &\subseteq MAX_\theta(AS(R'_i) ._\theta AS(R'_i) \ldots ._\theta AS(R'_i)) \ (lemma\ 3.3) \\ &\subseteq MAX_\theta(AS(R'_i)) ._{\theta_s} \ldots ._{\theta_s} MAX_\theta(AS(R'_i)) \end{aligned}$$

and applying Lemma 2.6, we deduce that there exists $i \in \{1,2\}$ such that $u^{k'}$ is an iterating factor of $MAX_\theta(AS(R_i))$. And by induction hypothesis, we conclude that u is strongly connected.

⊙ If $R = R_1^{*\theta}$ with $R_1 \in \mathcal{R}_\theta(X^*)$ where each word of R_1 is strongly connected, then $AS(R_1) = R_1$. So u is an iterating factor of $MAX_\theta(R)$ with $u^* \subseteq LEX$, and since $R \in \mathcal{R}_\theta(X^*)$, u is strongly connected.

Then we deduce from Proposition 2.9 that $AS(R)$ is in $\mathcal{R}_\theta(X^*)$. □

Proposition 3-5– *If θ is a semi-commutation relation on X,*

$$\mathcal{R}_\theta(X^*) = \mathcal{R}'_\theta(X^*)$$

Proof

he inclusion $\mathcal{R}_\theta(X^*) \subseteq \mathcal{R}'_\theta(X^*)$ is clear.
'e prove the reverse inclusion with an induction.
et R be a language of $\mathcal{R}'_\theta(X^*)$.

If R is an elementary subset of X^*, it is also in $\mathcal{R}_\theta(X^*)$.

If R is the union or the θ-concatenation of two languages R_1, R_2 of $'_\theta(X^*)$, then by induction hypothesis R_1 and R_2 are in $\mathcal{R}_\theta(X^*)$ and so is R.

If $R = R_1^{*'_\theta}$ for some language R_1 of $\mathcal{R}'_\theta(X^*)$, then by induction hypoth-sis, $R_1 \in \mathcal{R}_\theta(X^*)$. And by definition $R = f_\theta((AS(R_1))^*)$.
is clear that all words of $AS(R_1)$ are strongly connected, so it remains to rove that $AS(R_1)$ is in $\mathcal{R}_\theta(X^*)$, and it is true with Lemma 3.4. □

ℓeferences

[AR88] J.Aalbersberg and G.Rozenberg. *Theory of traces.* Theoretical Computer Science 60:1-82, 1988.

[Cle84] M.Clerbout. *Commutations partielles et Familles de langages.* Thesis, University of Lille, 1984.

[CL87] M.Clerbout and M.Latteux. *On Semi-commutations.* Information and Computation 73:59-74, 1987.

[CP85] R.Cori and D.Perrin. *Automates et commutations partielles.* RAIRO Informatique 19:21-32, 1985.

[Die90] V.Diekert. *Combinatorics on Traces.* Lecture Notes In Computer Science, 454

[Lat90] M.Latteux. *On semi-commutations.* In M.Ito, editor, Proceedings of the International Colloquium on Words, Languages and Combinatorics, Kyoto 1990, World Scientific, Singapore, 1990.

[Maz77] A.Mazurkiewicz. *Concurrent program schemes and their interpretations.* DAIMI Rep. PB 78, Aarhus University, Aarhus, 1977.

[Met86] Y.Metivier. *Une condition suffisante de reconnaissabilité dans un monoïde partiellement commutatif*. RAIRO Informatique Théorique et Applications,20:121-127, 1986.

[Och85] E.Ochmanski. *Regular Behaviour of Concurrent Systems.* Bulletin of EATCS 27:56-67, Oct 85.

[Och89] E.Ochmanski. *Semi-Commutations for Place/transition systems.* Bulletin of EATCS 38:191-198, June 1989.

[Per89] D.Perrin. *Partial Commutations.* In Proceedings of the 16th International Colloquium on Automata, Languages and Programming (ICALP'89), Lecture Notes in Computer Science 372:637-651.

[Roo89] Y.Roos. *Contribution à l'étude des fonctions de commutation partielle.* Thesis, University of Lille, 1989.

[Zie87] W.Zielonka. *Notes on asynchronous automata.* RAIRO Informatique Théorique et Applications, 21:99-135, 1987.

New Results Concerning Synchronized Finite Automata

Oscar H. Ibarra and *Nicholas Q. Trân*

Department of Computer Science
University of California
Santa Barbara, CA 93106

Abstract

We establish a tight hierarchy of two-way synchronized finite automata having only universal states on the number of allowed processes, i.e. $\mathcal{L}(\text{2SUFA}(k\text{-proc})) \subsetneq \mathcal{L}(\text{2SUFA}(k+1\text{-proc}))$, by studying reduction functions made by two-way deterministic finite automata with a one-way write-only output tape. Consequently for every $k > 1$, $\mathcal{L}(\text{2SUFA}(k\text{-proc}))$ has a complete set under 2DFA reductions and is not closed under union or intersection. In contrast, we show that the corresponding hierarchy collapses for unary alphabets.

Next we study multihead synchronized finite automata having only universal states. We show that $\bigcup_{k=1}^{\infty} \mathcal{L}(\text{2SUFA}(k\text{-heads})) = \bigcup_{k=1}^{\infty} \mathcal{L}(\text{2UFA}(k\text{-heads})) = \text{NLOG}$, and $\bigcup_{k=1}^{\infty} \mathcal{L}(\text{1SUFA}(k\text{-heads})) = \bigcup_{k=1}^{\infty} \mathcal{L}(\text{1UFA}(k\text{-heads}))$. As a result, we establish a tight hierarchy on the number of heads of the two-way machines, i.e. $\mathcal{L}(\text{2SUFA}(k\text{-heads})) \subsetneq \mathcal{L}(\text{2SUFA}(k+1\text{-heads}))$, and a weaker hierarchy of the one-way machines, i.e. $\mathcal{L}(\text{1SUFA}(k\text{-heads})) \subsetneq \mathcal{L}(\text{1SUFA}(2k+1\text{-heads}))$.

Finally, we show that any reasonable class of synchronized pushdown or counter machines accepts all r.e. languages.

1 Introduction

Synchronized alternating Turing machines (SATM) were introduced in [Hro86] to study the effect of allowing processes of an alternating Turing machine to communicate via synchronization. Informally a synchronized alternating machine is an alternating machine with a special subset of internal states called synchronizing states, each of them associated with a synchronizing symbol. If during the course of computation some process enters a synchronizing state, then it has to wait until all other processes enter either an accepting state or a synchronizing state with the same synchronizing symbol. When this happens, all processes are allowed to continue their computation; otherwise, the machine is said to have a deadlock. A computation is successful if all processes terminate in accepting states and no deadlocks occur.

It turns out that synchronization is a natural yet powerful concept with many interesting properties [Wie87, Slo87, Slo88a, Slo88b, HKRS89, DHK+89, IT90]. For example, [DHK+89] showed the relationship $\text{SA-SPACE}(s(n)) = \bigcup_{c\geq 1} \text{SA-TIME}(c^{s(n)}) = \bigcup_{c\geq 1} \text{DSPACE}(c^{s(n)})$ to hold for $s(n) \geq \log(n)$. This implies that the deterministic hierarchy LOGSPACE $\subseteq$ PTIME $\subseteq$ PSPACE $\subseteq$ EXPTIME $\subseteq$ EXPSPACE $\subseteq \ldots$, shifts by two levels and then alternately collapses and separates when synchronization is introduced, i.e. SA-LOGSPACE = SA-PTIME = PSPACE, SA-PSPACE = SA-EXPTIME = EXPSPACE, etc. Recall that alternation shifts this hierarchy by only one level [CKS81].

Similarly, synchronization shifts the computational power of finite automata by two levels in the Chomsky hierarchy. It was shown in [HKRS89] that $\mathcal{L}$(2SAFA) = NSPACE(n), i.e. two-way synchronized alternating finite automata recognize exactly the class of context-sensitive languages. This result was extended in [IT90] to $\mathcal{L}$(1SAFA) = NSPACE(n), and later in [HRS89] to $\mathcal{L}$(1SAFA(k-heads)) = $\mathcal{L}$(2SAFA(k-heads)) = NSPACE(n^k). Hence $\bigcup_{k=1}^{\infty}$ $\mathcal{L}$(SAFA(k-heads)) = PSPACE.

Since most languages considered feasibly computable are context-sensitive, synchronized finite automata emerge as an attractive model for studying feasible parallel computation. Many studies have sought to place restrictions on synchronized finite automata to make them more realistic models of real-world parallel computers. One way is to restrict the number of processes available (which corresponds to the number of processors) to a constant. [HKRS89] nicely characterized one-way and two-way synchronized finite automata with a constant number of processes in terms of multihead nondeterministic finite automata, and thus obtain tight hierarchies of these machines on the number of processes. It was shown there that for any $k \geq 1$,

(i) $\mathcal{L}$(1SAFA(k-proc)) = $\mathcal{L}$(1NFA(k-heads))
$\mathcal{L}$(2SAFA(k-proc)) = $\mathcal{L}$(2NFA(k-heads)), and therefore

(ii) $\mathcal{L}$(1SAFA(k-proc)) $\subsetneq$ $\mathcal{L}$(1SAFA($k+1$-proc))
$\mathcal{L}$(2SAFA(k-proc)) $\subsetneq$ $\mathcal{L}$(2SAFA($k+1$-proc)).

Another way to restrict the power of synchronized finite automata is to allow only universal states so that they reflect the deterministic nature of real-world computers. It is natural to ask whether similar characterizations of $\mathcal{L}$(1SUFA(k-proc)) and $\mathcal{L}$(2SUFA(k-proc)) can be found. [IT90] showed that in terms of deterministic multihead finite automata, no similar nice characterizations exist, since $\mathcal{L}$(1DFA(2-heads)) $-$ $\bigcup_{k=1}^{\infty}$ $\mathcal{L}$(2SUFA(k-proc)) $\neq \emptyset$, and $\bigcup_{k>0}^{\infty}$ $\mathcal{L}$(2SUFA(k-proc)) $\subsetneq$ $\mathcal{L}$(2DFA(2-heads)). However, a tight hierarchy of 1SUFA(k-proc) was obtained directly with techniques in Kolmogorov complexity theory. In short, [IT90] showed that for $k \geq 2$,

(i) $\mathcal{L}$(1SUFA(k-proc)) $\subsetneq$ $\mathcal{L}$(1DFA(k-heads))
$\mathcal{L}$(2SUFA(k-proc)) $\subsetneq$ $\mathcal{L}$(2DFA(k-heads)) and

(ii) $\mathcal{L}$(1SUFA($k-1$-proc)) $\subsetneq$ $\mathcal{L}$(1SUFA(k-proc)),

and posed as an open question whether the corresponding hierarchy of 2SUFA(k-proc) is proper and tight.

In this paper we prove that $\mathcal{L}$(1SUFA($k+1$-proc)) $-$ $\mathcal{L}$(2SUFA(k-proc)) $\neq \emptyset$ for $k \geq 1$ by studying reductions made by two-way deterministic finite automata with a write-only output tape. A tight hierarchy of 2SUFA(k-proc) follows from this result, as well as the corollary that for every $k \geq 2$, $\mathcal{L}$(2SUFA(k-proc)) has a complete set under 2DFA reductions and is not closed under intersection or union.

Next we show that when restricted to unary alphabets, $\bigcup_{k=1}^{\infty}$ $\mathcal{L}$(2SUFA(k-proc)) is exactly the class of unary regular languages. This characterization implies that the hierarchies of 1SUFA(k-proc) and 2SUFA(k-proc) collapse; in contrast, while the hierarchy of 1SAFA(k-proc) collapses, the hierarchy of 2SAFA(k-proc) is proper and tight, since $\mathcal{L}$(2NFA(k-heads)) $\subsetneq$ $\mathcal{L}$(2NFA($k+1$-heads)) hold even for unary languages [Mon80].

We also consider multihead synchronized finite automata. As mentioned earlier [HRS89] showed that $\mathcal{L}$(1SAFA(k-heads)) = $\mathcal{L}$(2SAFA(k-heads)) = NSPACE(n^k) and hence $\bigcup_{k=1}^{\infty} \mathcal{L}$(SAFA($k$-heads)) = PSPACE. We consider a restricted version of multihead synchronized finite automata when only universal states are allowed and show that for these machines, synchronization can be replaced by another set of heads, i.e. $\mathcal{L}$(1SUFA(k-heads)) $\subseteq$ $\mathcal{L}$(1UFA($2k$-heads)) and $\mathcal{L}$(2SUFA(k-heads)) $\subseteq$ $\mathcal{L}$(2UFA($2k$-heads)). From this, we obtain the following characterizations: $\bigcup_{k=1}^{\infty} \mathcal{L}$(2SUFA($k$-heads)) = $\bigcup_{k=1}^{\infty} \mathcal{L}$(2UFA($k$-heads)) = NLOG, and $\bigcup_{k=1}^{\infty} \mathcal{L}$(1SUFA($k$-heads)) = $\bigcup_{k=1}^{\infty} \mathcal{L}$(1UFA($k$-heads)). As another corollary, we obtain a tight hierarchy of the two-way machines on the number of heads, i.e. $\mathcal{L}$(2SUFA(k-heads)) $\subsetneq$ $\mathcal{L}$(2SUFA($k+1$-heads)), and a weaker hierarchy of the one-way machines, i.e. $\mathcal{L}$(1SUFA(k-heads)) $\subsetneq$ $\mathcal{L}$(1SUFA($2k+1$-heads)).

Finally, we consider the power of synchronized pushdown machines. We show that synchronization dramatically enhances their power, because any class of synchronized pushdown machines, restricted in any reasonable way, still accepts all r.e. languages.

The rest of this paper is organized as follows. Section 2 gives definitions relating to synchronized alternating Turing machines and 2DFA reductions. Section 3 states a technical result and uses it to obtain the tight hierarchy of 2SUFA(k-proc), the complete sets under 2DFA reductions, and the non-closure properties under union and intersection. Section 4 gives the results concerning multihead synchronized automata with only universal states, and Section 5 discusses the power of synchronized pushdown automata. Section 6 concludes the paper with a sketch of a proof of the technical result.

2 Definitions

A *2DFA transducer* is a 2DFA T with a two-way read-only input tape, a one-way write-only output tape, and Σ and Π as its input and output alphabets respectively. $L(T) \subseteq \Sigma^*$ denotes the language accepted by 2DFA T, $T(w) \in \Pi^*$ denotes the word generated by T on input w, and $K(T) \subseteq \Pi^*$ denotes $\{T(w) : w \in \Sigma^*\}$.

Let $L_1 \subseteq \Sigma^*$ and $L_2 \subseteq \Pi^*$. We say L_1 is *2DFA-reducible* to L_2 ($L_1 \leq_{\text{2DFA}} L_2$) if there is some 2DFA transducer T such that $T(w) \in L_2$ iff $w \in L_1$.

A formal definition of synchronized alternating Turing machines appeared in [IT90]; we only give a brief description here. An SATM M is an alternating Turing machine whose internal state set is a subset of $Q \times (S \cup \{\epsilon\})$, where S is the *synchronizing alphabet*. Hence each path in a computation tree of M is associated with a *synchronizing sequence*, obtained by concatenating the internal states' second components of the nodes along the path. M accepts x if there is an accepting computation subtree T of M on x in the usual sense for ATM as defined in [CKS81], and furthermore, there is a string s in S^* such that every synchronizing sequence of T is a prefix of s.

1DFA(k-heads), 1NFA(k-heads), 2DFA(k-heads), 2NFA(k-heads), 1UFA(k-

eads), 1SUFA(k-heads), 2UFA(k-heads), 2SUFA(k-heads) denote the one-way and wo-way versions of deterministic, nondeterministic, universal states only, and ynchronized with only universal states k-head finite automata. 1SUFA(k-proc), SAFA(k-proc), SUFA(k-proc), SAFA(k-proc) denote the one-way and two-way ver-ions of universal states only and general synchronized alternating finite automata vhose computation trees on any input w have at most k leaves. 1NPDA(k-heads) lenotes one-way k-head pushdown automata; SA-k-reversal-c-CA, SA-AUX-PDA, A-AUX-SA, SA-AUX-NESA denote the synchronized versions of c-counter au-omata with finite reversal, auxiliary pushdown automata, auxiliary stack automata, nd auxiliary nonerasing stack automata respectively (see [HU79] for formal defini-ions of these machines).

For each w in $\Gamma^n = \{w_1\#w_2\#\ldots\#w_n : w_i \in \{0,1\}^*\}$, $w(i)$ denotes its ith ubword, and $w(i,j)$ denotes the jth symbol of $w(i)$.

3 A Tight Hierarchy of 2SUFA(k-proc)

We begin by stating a technical result, which we will use to obtain our results about 2SUFA(k-proc).

Definition 1 *For each* $n \geq 1$, *let*

$$L_n = \{0^{e_1}\#0^{e_2}\#\ldots\#0^{e_n} : e_1 = \max_{1\leq i\leq n} e_i\}$$

$$K_n = \{x_1\#x_2\#\ldots\#x_n : x_i's \in \{0,1\}^* \ \&\ \exists i \leq n\ \forall j \leq n\ [x_j \textit{ is a prefix of } x_i]\}.$$

Theorem 1 *(Technical Result)* $L_{n+1} \not\leq_{2\text{DFA}} K_n$ *for* $n \geq 1$.

Intuitively, the above theorem says that a 2DFA cannot compress information about $n+1$ integers into n binary words. Since successful synchronization by a finite automata with only universal states and k processes on an input x is the same as reducing x to a string in K_k, a tight hierarchy on k for $\bigcup_{k=1}^{\infty}$ $\mathcal{L}$(2SUFA(k-proc)) follows from Theorem 1. But in fact we can show even more.

Theorem 2 $\mathcal{L}$*(1SUFA(*$k+1$*-proc))* $-$ $\mathcal{L}$*(2SUFA(k-proc))* $\neq \emptyset$ *for* $k \geq 1$.

Proof: It is easy to see that L_{k+1} can be accepted by a 1SUFA($k+1$-proc) M. On input x, M splits into $k+1$ processes. Each process p_i produces a sequence of synchronizing symbols 0^{l_i}, where l_i is the length of the ith segment of x (we can assume without loss of generality that x has the correct form.) The first process also produces an extra synchronizing symbol 1 at the end of its sequence. Clearly $x \in L_{k+1}$ iff M accepts x.

On the other hand, if there is some 2SUFA(k-proc) N that accepts L_{k+1}, then N can be converted into a 2DFA transducer N' that reduces L_{k+1} to K_k as follows. First we observe that N has at most k processes, and furthermore on all inputs, each process of N can be made to halt due to its deterministic nature and the form of strings in L_{k+1}. Then we can construct a halting 2DFA N' that uses a stack of constant size in its finite control to simulate sequentially each of the k processes of N. N' outputs the corresponding synchronizing symbol whenever the simulated process enters a synchronizing state. When a process terminates, N' outputs a $\#$ before simulating another process, unless there is no more left.

If at any time N' detects that the process being simulated is looping, then N outputs k $\#$'s and stops.

Clearly, $x \in L_{k+1}$ iff N' on x outputs a word in K_k, and hence $L_{k+1} \leq_{\text{2DFA}} \mathrm{K}_k$ But this contradicts Theorem 1. ∎

Corollary 1 *For $k \geq 1$, the following are true:*

(i) $\mathcal{L}(\text{1SUFA}(k\text{-proc})) \subsetneq \mathcal{L}(\text{1SUFA}(k+1\text{-proc}))$

(ii) $\mathcal{L}(\text{2SUFA}(k\text{-proc})) \subsetneq \mathcal{L}(\text{2SUFA}(k+1\text{-proc}))$

(iii) K_k is complete under 2DFA reductions for $\mathcal{L}(\text{2SUFA}(k\text{-proc}))$

Proof: Follows from Theorem 2. ∎

It was shown in [IT90] that $\mathcal{L}$(1SUFA(k-proc)) is not closed under Boolean operations for $k \geq 2$, and that $\mathcal{L}$(2SUFA(k-proc)) is not closed under complementation for $k \geq 2$. We are now able to show that

Corollary 2 *$\mathcal{L}(\text{2SUFA}(k\text{-proc}))$ is not closed under intersection or union for $k \geq 2$.*

Proof: Let $k \geq 2$ and

$$R_k = \{0^{e_1}\#0^{e_2}\#\ldots\#0^{e_{k+1}} : e_1 = \max_{1\leq i\leq k} e_i\},$$

$$S_k = \{0^{e_1}\#0^{e_2}\#\ldots\#0^{e_{k+1}} : e_1 = \max_{1\leq i\leq k-1,k+1} e_i\}.$$

Then both R_k and S_k are in $\mathcal{L}$(2SUFA(k-proc)), but $R_k \cap S_k = L_{k+1}$ is not.

Let $U = \{0^{e_1}\#0^{e_2}\#0^{e_3} : e_1 \geq e_2 \text{ or } e_1 \geq e_3\}$. Then U is the union of two languages in $\mathcal{L}$(2SUFA(2-proc)), but $U \not\leq_{\text{2DFA}} K_k$ and hence $U \notin \mathcal{L}$(2SUFA(k-proc)) for any $k \geq 2$. ∎

For the remaining of this section, we restrict our consideration to unary alphabets. We first give a characterization of $\bigcup_{k=1}^{\infty} \mathcal{L}$(2SUFA($k$-proc)).

Theorem 3 *$\mathcal{L}(\text{2SUFA}(k\text{-proc}))$ is the class of regular sets for all $k \geq 1$.*

Proof: Recall that input alphabets are unary. It was shown in [HI68] that $\bigcup_{k=1}^{\infty}$1NPDA(k-heads) accepts only regular languages. Hence it suffices to show that if L is accepted by some 2SUFA(k-proc) M then $\overline{L}$ is accepted by some 1NPDA(k') M'. Without loss of generality, we can assume that M splits into k processes at the start of the computation. Since the input alphabet is unary and processes of M are deterministic, each input head can change direction only at the first or last c squares of its input, and at most t times without going into a loop, for some constants c and t depending only on M. Hence M can be modified so that it never gets into an infinite loop on any input.

If M rejects input x, then either one of its processes finishes in a nonaccepting state, or the synchronizing sequences of two processes differ at some bit. In the ormer case M' needs only to simulate the offending process. Below we show how M' on input x verifies the latter condition. First M' nondeterministically chooses a process to simulate. Since a process of M changes the direction of its input head at most t times and only at the first or last c squares, M' needs at most t heads to perform the simulation. Whenever the process enters a synchronizing state, M' pushes a symbol on its pushdown stack. At some point, M' decides that the last synchronizing symbol witnesses a discrepancy, remembers it in its finite control, and nondeterministically chooses another process to simulate. This time M' pops a symbol from its pushdown stack every time the second process enters a synchronizing state. M' accepts iff there is indeed a discrepancy between the two synchronizing sequences. Note that M' uses at most $2t$ one-way input heads. ∎

Corollary 3 *Let the input alphabet be unary. For $k \geq 1$, the following are true:*

(i) $\mathcal{L}$*(1SUFA(k-proc))* $=$ $\mathcal{L}$*(1SUFA(*$k+1$*-proc))*

(ii) $\mathcal{L}$*(2SUFA(k-proc))* $=$ $\mathcal{L}$*(2SUFA(*$k+1$*-proc))*

(iii) $\mathcal{L}$*(1SAFA(k-proc))* $=$ $\mathcal{L}$*(1SAFA(*$k+1$*-proc))*

(iv) $\mathcal{L}$*(2SAFA(k-proc))* $\subsetneq$ $\mathcal{L}$*(2SAFA(*$k+1$*-proc))*

Proof: (i) and (ii) follow from Theorem 3. (iii) follows from the fact that a 1SAFA(k-proc) can be simulated by a 1NPDA(k-heads) and from [HI68]. (iv) follows from the characterization of 2SAFA(k-proc) in [HKRS89] and from [Mon80]. ∎

4 Synchronized Multihead Finite Automata

[HRS89] showed that $\mathcal{L}$(1SAFA(k-heads)) $=$ $\mathcal{L}$(2SAFA(k-heads)) $=$ NSPACE(n^k), and hence $\bigcup_{k=1}^{\infty} \mathcal{L}$(SAFA($k$-heads)) $=$ PSPACE. It would be interesting to see if similar characterizations can be obtained for multihead synchronized finite automata with only universal states. We now show that synchronization can be replaced by a set of extra heads.

Theorem 4 *For every $k \geq 1$,*

(i) $\mathcal{L}$*(1SUFA(k-heads))* $\subseteq$ $\mathcal{L}$*(1UFA(*$2k$*-heads))* *and*

(ii) $\mathcal{L}$*(2SUFA(k-heads))* $\subseteq$ $\mathcal{L}$*(2UFA(*$2k$*-heads))*

and hence

(i) $\bigcup_{k=1}^{\infty} \mathcal{L}$*(1SUFA(k-heads))* $=$ $\bigcup_{k=1}^{\infty} \mathcal{L}$*(1UFA(k-heads))*

(ii) $\bigcup_{k=1}^{\infty} \mathcal{L}$*(2SUFA(k-heads))* $=$ $\bigcup_{k=1}^{\infty} \mathcal{L}$*(2UFA(k-heads))* $=$ *NLOG*

Proof: We want to show that for every 1SUFA(k-heads) M there is a 1UFA($2k$-heads) M' that accepts the same language. M' uses k heads to simulate M faithfully and keeps the other k heads on the left end of the input until M enters a universal state with b processes. Then M' enters a universal state with $\binom{b}{2}+b$ processes. Each of the first $\binom{b}{2}$ processes simulates a pair of the b processes of M and verifies that the pair will not cause a deadlock via their synchronizing sequences, using the extr k heads. The other b processes of M' continue to simulate the original b processe of M faithfully, with the extra k heads still on the left end of the input. The nex time these b processes enter a universal state, they repeat the verification proces described above. It is easy to see that M' accepts the same language as M.

Now we show that for every 2SUFA(k-heads) M there is a 2UFA($2k$-heads) M that accepts the same language. The simulation for this case is easier. M' uses k heads to simulate M faithfully and the other k heads as a counter of synchronizin states entered so far. Every time a process of M enters a synchronizing state, say the ith one, M' restarts the computation of M on the input from the beginning and verifies that all i^{th} synchronizing symbols are the same. Since every process can make at most n^k moves without looping, it is clear that k heads is enough tc simulate the counter. It is clear that M' and M accept the same language.

From Theorem 4, we obtain a weak hierarchy of 1SUFA(k-heads):

Corollary 4 *$\mathcal{L}$(1SUFA(k-heads)) $\subsetneq$ $\mathcal{L}$(1SUFA($2k+1$-heads)) for every $k \geq 1$.*

Proof: It was shown in [SMI+85] that $\mathcal{L}$(1UFA(k-heads)) $\subsetneq$ $\mathcal{L}$(1UFA($k+1$-heads)) for $k \geq 1$. Hence we have $\mathcal{L}$(1SUFA(k-heads)) $\subseteq$ $\mathcal{L}$(1UFA($2k$-heads)) $\subsetneq$ $\mathcal{L}$(1UFA($2k+1$-heads)) $\subseteq$ $\mathcal{L}$(1SUFA($2k+1$-heads)) for every $k \geq 1$. ∎

On the other hand, we are able to obtain a tight hierarchy of 2SUFA(k-heads).

Lemma 1 *There exists some $j \geq 2$ such that $\mathcal{L}$(2UFA(k-heads)) $\subsetneq$ $\mathcal{L}$(2UFA($kj+1$-heads)).*

Proof: By Immerman's technique and diagonalization. ∎

From this lemma, we can show the following theorem using the method of translation (see [Iba73], for example).

Theorem 5 *For $k \geq 1$,*

(i) $\mathcal{L}$(2UFA(k-heads)) $\subsetneq$ $\mathcal{L}$(2UFA($k+1$-heads))

(ii) $\mathcal{L}$(2SUFA(k-heads)) $\subsetneq$ $\mathcal{L}$(2SUFA($k+1$-heads))

5 Synchronized Pushdown Automata

We have seen the accepting power of nondeterministic multihead pushdown automata is very limited when the alphabet is unary. It is interesting to note that the accepting power of synchronized pushdown automata is on the other end of the computational spectrum, namely, synchronized pushdown automata accept all recursively enumerable sets. In fact, the following theorem shows that even for a severely restricted model of pushdown automata, the claim still holds.

Theorem 6 *Every r.e. language can be accepted by a synchronized alternating one-reversal one-counter machine.*

'roof: It suffices to show that every deterministic two-counter machine M can be imulated by a synchronized alternating one-reversal one-counter machine M', since eterministic two-counter machines can recognize all r.e. languages [HU79]. At the eginning M' splits into five processes h, p_1, p_2, q_1 and q_2; the difference of p_1 and $_2$ represents the content of counter c_1 of M; the difference of q_1 and q_2 represents he content of the other counter c_2. Process h represents the state and input head osition of M.

Every move by M is simulated by M' as follows. Process h first guesses whether he counters c_1 and c_2 are empty, and then produces the synchronizing symbol $q, a, s_1, s_2]$, where q is its current state, a is the symbol under its input head, and $_1$ and $s_2 \in \{0, 1\}$ denote whether the counters are empty. Each of p_1 and p_2 tries o guess and produce the symbol $[q, a, s_1, s_2]$. Furthermore, p_1 and p_2 need to verify hat s_1 reflects correctly the state of counter c_1, which together they represent. Let ay that they have to verify that c_1 is not empty. The next paragraph explains how his can be done. To verify that c_1 is empty, a slightly modified method can be used.

To perform the verification, p_1 spawns an identical process u_1, and p_2 spawns an identical process u_2. Process u_1 then produces a special synchronizing symbol B and then deterministically produces a special synchronizing symbol l for every symbol it pops off its own counter until it becomes empty. Finally, u_1 produces a special synchronizing symbol E and halts in an accepting state. Similarly, process u_2 produces a B and then produces an l for every symbol it pops off its own counter, until at some point u_2 nondeterministically decides that it has produced the same number of l as u_1, and its own counter is still not empty. It then produces an E and halts in an accepting state. Meanwhile, the other processes h, p_1, p_2, q_1, q_2 guess and produce a sequence of synchronizing symbols of the form Bl^iE for some $i \geq 0$; their counters remain the same during this period.

Next, each of q_1 and q_2 tries to guess and produce the symbol $[q, a, s_1, s_2]$, and furthermore they need to verify that s_2 reflects correctly the state of counter c_2, which together they represent. Again, the verification is as described above for p_1 and p_2.

After s_1 and s_2 have been verified, h, p_1, p_2, q_1, q_2 update their configurations according to the next move of M dictated by $[q, a, s_1, s_2]$. Process h moves its head and updates its internal state to reflect the new state and input head position of M. To increment counter c_1, p_1 does nothing while p_2 increments its counter. To decrement counter c_1, p_1 increments its counter while p_2 does nothing. Counter c_2 can be decremented or incremented with the same method. If q is an accepting state, then h, p_1, p_2, q_1, q_2 halt in accepting states.

Now, it is clear that there is an accepting computation tree of M' (in which all processes accept and synchronizing sequences of h, p_1, p_2, q_1, q_2 and of the processes they spawn are prefixes of a common string) iff all processes accept, all corresponding guesses of synchronizing symbols $[a, q, s_1, s_2]$ match, and synchronizing subsequences $B0^iE$ generated by all processes when verifying the validity of an s_1 or s_2 have the same length. This occurs iff the synchronizing symbols $[a, q, s_1, s_2]$ reflect correctly the configurations of M during an accepting computation, iff M accepts its input. Furthermore, the only processes of M' that make a reversal on its counter are those spawned to verify the status an s_1 or s_2, and they make exactly one reversal.

Hence, every r.e. language can be recognized by a one-counter one-reversal synchronized counter machine. ∎

Corollary 5 *For every $k, c \geq 1$ and every space-bound function $s(n)$, SA-k-reversal-c-CA, SA-AUX-PDA(s(n)), SA-AUX-SA(s(n)), SA-AUX-NESA(s(n)) accept the class of r.e. languages.*

6 Proof of the Technical Result

Our objective is to show that $L_{n+1} \not\leq_{\mathrm{2DFA}} K_n$ for $n \geq 1$. We begin with the proof of the trivial case when $n = 1$.

Lemma 2 $L_2 \not\leq_{\mathrm{2DFA}} K_1$.

Proof: If there is some 2DFA transducer T that reduces L_2 to K_1 then we can modify T to obtain a 2DFA T' which accepts L_2 whenever T generates a word in $\{0,1\}^* = K_1$. But this means L_2 must be regular, a contradiction. ∎

The proof is more involved when $n \geq 2$, but the main idea behind it is as follows. Suppose there is some DFA T that reduces L_{n+1} to K_n, so if $x \notin L_{n+1}$ then $T(x) \notin K_n$. This means that two segments of $T(x)$ differ at some bit c, and we call c a *conflict point* of $T(x)$. Now, if we "pump" the *ith* segment of x for some $i \geq 2$ to obtain x' and this only "shifts" the conflict point of $T(x)$ to another conflict point of $T(x')$, then we say that c is an *i-invariant* conflict point of $T(x)$. Intuitively, it means that the conflict point c of $T(x)$ is relatively independent of the *ith* segment of x.

But we show that if T reduces L_{n+1} to K_n, then there is some word $x \notin L_{n+1}$ such that $T(x)$ has n conflict points $c_1, c_2, \ldots, c_n$, where each c_i is the first conflict point between the involved segments of $T(x)$, and each c_i is k-invariant for all k except 1 and i. However no $T(x)$ can have this property, so the assumption that there is a DFA T reducing L_{n+1} to K_n is false.

In the following, let T be a 2DFA transducer of q states with $\Sigma = \Pi = \{0, 1, \#\}$ and $a \geq q+1$. We will be concerned with inputs in the special form $x = 0^{a+e_1 q!}\#0^{a+e_2 q!}\#\ldots\#0^{a+e_n q!} \in \Gamma^n$. We first establish a Normal Form Theorem below for output words $T(x)$, where $x = 0^{a+e_1 q!}\#0^{a+e_2 q!}\#\ldots\#0^{a+e_n q!}$.

Theorem 7 *(Normal Form) Suppose T reduces L_n to $K_{n'}$ for some $n, n' \geq 1$. Then on input $x = 0^{a+e_1 q!}\#0^{a+e_2 q!}\#\ldots\#0^{a+e_n q!}$, where $e_k \geq 0$ for $1 \leq k \leq n$, T generates $w = w(1)\#w(2)\#\ldots\#w(n') \in \Gamma^{n'}$. Furthermore, each $w(k)$, $1 \leq k \leq n'$ has the form $w(k) = c_1 v_1^{f_1} c_2 v_2^{f_2} \ldots c_m v_m^{f_m} c_{m+1}$, where $f_l \in \{e_s : 1 \leq s \leq n\}$ for $1 \leq l \leq m$, and m, $\{c_l : 1 \leq l \leq m+1\}, \{v_l \neq \epsilon : 1 \leq l \leq m\}$ depend only on T, a, and n.*

From now on, T, x, and y will be as defined in Theorem 7; furthermore, since the proof of Theorem 7 shows that each e_i is independent of $|v_1|, |v_2|, \ldots, |v_m|$, we may assume that $e_i > 2L(y)$ for each i, where $L(y) = 2\sum_{1 \leq k \leq m} |v_k| + 1$.

Now, since Theorem 7 guarantees that $y \in \Gamma^{n'}$, $y \notin K_{n'}$ only when two of its subwords differ at some bits. When this is true, we say y has a *conflict point*.

Suppose we change the value of some e_i of x to obtain x'; will $y' = T(x')$ have a conflict point too, and how does it relate to conflict points of y ? The next definition gives us some tools to deal with this question and its consequences.

Definition 2 *Suppose* $y(i) = c_1 v_1^{e_k} c_2 v_2^{e_k} \dots y(i,j) \dots c_s v_s^{e_k} c_{s+1}$, *where* $v_1, \dots, v_s$ *are the loops generated by* $0^{a+e_k q!}$ *(called k-loops) in* $y(i)$ *for some* i, j, *and* k. *Define*

$$l(y,i,j,k) = \begin{cases} \sum_{1 \le t \le s} |v_t|, & \text{if } c_{s+1} = w_1 y(i,j) w_2 \\ \sum_{1 \le t \le s} |v_t|, & \text{if } v_s^{e_k} = v_s^{L(y)} w_1 y(i,j) w_2 \\ \sum_{1 \le t < s} |v_t| & \text{if } v_s^{e_k} = w_1 y(i,j) w_2 v_s^{e_k - L(y)}. \end{cases}$$

If $x \notin L_n$, *then there are some* i_1 *and* i_2, $1 \le i_1 < i_2 \le n$ *and* j *such that* $y(i_1,j) \ne y(i_2,j)$. *We say* j *is a* conflict point *of* y *(between subwords* i_1 *and* i_2*). We say* j *is* k-invariant *if* $l(y,i_1,j,k) = l(y,i_2,j,k)$, *and* k-variant *otherwise.*

The next theorem describes the nature of conflict points for inputs of the special form.

Theorem 8 *There exist* i_1 *and* i_2 *such that for all* $x' = 0^{a+2eq!}\#0^{a+(2e+1)q!}\#0^{a+e_3q!}\#\dots\#0^{a+e_nq!} \in \Gamma^n$ *where* $2L(y) < e \le e_i$ *for* $3 \le i \le n$, $y' = T(x')$ *has a conflict point between subwords* i_1 *and* i_2. *Furthermore, the first such conflict point is 1-variant, 2-variant, and k-invariant for all other* k.

Theorem 8 still holds when the second segment of x is swapped with the kth segment for $k \ge 3$. We are now ready to prove the general case of the technical result.

Theorem 9 $L_{n+1} \not\le_{\text{2DFA}} K_n$ *for* $n \ge 2$.

Proof: Fix an $n \ge 2$ and suppose there is some 2DFA transducer T that reduces L_{n+1} to K_n. Let $x = 0^{a+2eq!}\#0^{a+(2e+1)q!}\#\dots\#0^{a+(2e+1)q!} \in \Gamma^{n+1}, y = T(x)$, and $e > 2L(y)$. Then it follows from Theorem 8 that for each $k \ge 2$ there exist i_k and j_k such that the first conflict point c_k of y between i_k and j_k is 1-variant, k-variant, and k'-invariant for all other k'. We say c_k is only 1-variant and k-variant. There are such n conflict points.

Partition $\{1,2,\dots,n\}$ into subsets $C_1, C_2, \dots, C_m$, such that if i and j are in the same subset then $c_i = c_j$. For each such subset S construct a graph G whose vertices are i_s and j_s and whose edges connect i_s and j_s for all $s \in S$. G must be acyclic; otherwise there is a cycle $v_1, v_2, \dots, v_l$, and without loss of generality, subwords v_1 and v_2 are only 1-variant, 2-variant, v_2 and v_3 are only 1-variant, 3-variant, ..., v_l and v_1 are only 1-variant, $(l+1)$-variant, with respect to the conflict point representing S. But the path from v_1 to v_l implies that they are $l+1$-invariant, a contradiction.

It is impossible to place such n conflict points on n subwords: we start with the largest conflict points, say those in C_1, and select subwords i_1 and j_s for all $s \in C_1$. $|C_1| + p_1$ subwords must be selected, where p_1 is the number of connected components of the graph for C_1, since each conflict point contributes a vertex to a unique component, except the first of that component, which contributes two. Afterward, all selected subwords in a component are effectively the same because $c_1, c_2, \ldots, c_{|C_1|}$ are the *first* conflict points for those subwords. Hence after the first $|C_1|$ conflict points have been selected, only $n - |C_1| - p_1 + p_1 = n - |C_1|$ subwords remain for $n - |C_1|$ conflict points. Continue this process until only $|C_m|$ subwords remain for the last set C_m. But we need at least $|C_m| + 1$ subwords.

Hence $L_{n+1} \not\leq_{\text{2DFA}} K_n$ for $n \geq 2$.

References

[CKS81] A. K. Chandra, D. K. Kozen, and J. Stockmeyer, *Alternation*, J ACM, 28 (1981), pp. 114–133.

[DHK+89] J. Dassow, J. Hromkovič, J. Karhumäki, B. Rovan, and A. Slobodová, *On the power of synchronization in parallel computations*, in Proc. 14th MFCS'89, Lecture Notes in Computer Science 379, Springer-Verlag, 1989, pp. 196–206.

[HI68] M. A. Harrison and O. H. Ibarra, *Multi-tape and multi-head pushdown automata*, Information and Control, 13 (1968), pp. 433–470.

[HKRS89] J. Hromkovič, J. Karhumäki, B. Rovan, and A. Slobodová, *On the power of synchronization in parallel computations*, tech. report, Comenius University, Bratislava, Czechoslovakia, Department of Theoretical Cybernetics and Institute of Computer Science, 1989.

[Hro86] J. Hromkovič, *How to organize the communication among parallel processes in alternating computations.* Manuscript, January 1986.

[HRS89] J. Hromkovič, B. Rovan, and A. Slobodová, *Deterministic versus nondeterministic space in terms of synchronized alternating machines*, tech. report, Comenius University, Bratislava, Czechoslovakia, Department of Theoretical Cybernetics and Institute of Computer Science, 1989.

[HU79] J. Hopcroft and J. Ullman, *Introduction to Automata Theory, Languages, and Computation*, Addison-Wesley, Reading, Mass., 1979.

[Iba73] O. Ibarra, *On two-way multihead automata*, Journal of Computer and System Sciences, 7 (1973), pp. 28–36.

[IT90] O. H. Ibarra and N. Q. Trần, *On space-bounded synchronized alternating Turing machines.* To appear in *Theoretical Computer Science*, 1990.

[Mon80] B. Monien, *Two-way multihead automata over a one-letter alphabet*, RAIRO Inform. Theor., 14 (1980), pp. 67–82.

[Slo87] A. SLOBODOVÁ, *On the power of communication in alternating computations.* Student Research Papers Competition, Section Computer Science (in Slovac), Comenius University, Bratislava, Czechoslovakia, April 1987.

[Slo88a] ——, *On the power of communication in alternating machines*, in Proc. 13th MFCS'88, Lecture Notes in Computer Science 324, Springer-Verlag, 1988, pp. 518–528.

[Slo88b] ——, *Some properties of space-bounded synchronized alternating Turing machines with only universal states*, in Proc. 5th IMYCS'88, Lecture Notes in Computer Science 381, Springer-Verlag, 1988, pp. 102–113.

[SMI+85] S. SAKURAYAMA, H. MATSUNO, K. INOUE, I. TAKANAMI, AND H. TANIGUCHI, *Alternating one-way multihead Turing machines with only universal states*, Trans. IECE Japan E, 68 (1985), pp. 705–711.

[Wie87] J. WIEDERMANN, *On the power of synchronization*, tech. report, VUSEI-AR, Bratislava, Czechoslovakia, 1987.

A Greibach Normal Form for Context-free Graph Grammars

Joost Engelfriet

Department of Computer Science, Leiden University
P.O.Box 9512, 2300 RA Leiden, The Netherlands

Abstract

Every context-free hypergraph grammar that generates a language of bounded degree can be transformed into an equivalent one that has the apex property, i.e., that cannot "pass" nodes from nonterminal to nonterminal. This generalizes Double Greibach Normal Form of context-free grammars. Moreover, it provides a natural grammatical characterization of the context-free hypergraph languages of bounded degree. For grammars with the apex property it is not possible to put a bound on the number of nonterminals in the right-hand sides of the productions.

Introduction

Context-free graph grammars are of importance because they provide a general means of studying graph properties of a recursive nature. Unfortunately, unlike the case of ordinary string grammars, there does not seem to be a single natural notion of context-free grammar for graphs. Thus, many different types of context-free graph grammar have been proposed in the literature, of different generating power. Natural and useful requirements for a graph grammar to deserve the predicate 'context-free' are formulated in [6]. For several of the most interesting of these grammar types it has been shown recently that, although they generate different classes of graph languages, they generate the same graph languages of bounded degree (where a graph language is of bounded degree if there is a fixed upper bound on the degree of all nodes of all graphs in the language). This suggests the existence of a single natural class of context-free graph languages of bounded degree. In this paper we provide another argument for this thesis by showing that the apex graph grammars (introduced in [16] and studied, e.g., in [15,17]) generate that same class of context-free graph languages of bounded degree, and moreover, as opposed to the other grammar types, do not generate any graph languages of unbounded degree. Thus, apex graph grammars form a natural grammatical characterization of "the" class of context-free graph languages of bounded degree. The specific feature of an apex graph grammar is that, during derivation, nodes of the generated graph cannot be "passed" from one nonterminal to another. This property generalizes Greibach Normal Form for context-free grammars [22] and even Double Greibach Normal Form [30] (in which it is required that no right-hand side of a production begins or ends with a nonterminal, cf., e.g., Problem 6 in Section 4.6 of [26]).

The grammar types we have been referring to implicitly are the context-free hypergraph grammars (CFHG grammars, also called hyperedge replacement systems or HR grammars; see [1,23,24,25]) on the one hand, and several types of NLC-like graph grammars (see [20]) on the other hand, in particular the B-edNCE grammars [18] and the C-edNCE grammars [4,10] (that are more powerful than the CFHG grammars, but contain the same graph languages of bounded degree, see Theorem 7 of [19] and [28,5,13,9]). Apex graph grammars are defined by putting a natural static restriction on either CFHG grammars or edNCE grammars. It is shown in [19] that in both cases the same class of graph languages is generated. Thus, to prove our result, we could choose between two formalisms: the CFHG grammar or the edNCE grammar. We have decided for the CFHG grammar, for two reasons. First, CFHG grammars are (in our opinion) easier to understand and visualize than edNCE grammars (but note that, of course, in general edNCE grammars are more powerful). Second, since CFHG grammars can also be used to generate (directed) hypergraphs, we can prove our result for the more general case of hypergraphs. Thus, we will show that every CFHG hypergraph language of bounded degree can be generated by an apex CFHG grammar. Note however that, for hypergraphs, the existence of "the" class of context-free hypergraph languages of bounded degree has not yet been argued as convincingly as for graphs.

In the case of context-free grammars in Double Greibach Normal Form one may assume additionally that there are at most two nonterminals in the right-hand side of each production. For apex CFHG grammars we show that the number of nonterminals in right-hand sides of productions gives rise to a hierarchy that is strict at each level: the set of n-ary trees can be generated with n nonterminals in the right-hand sides, but not with less.

We also provide a grammatical characterization of the class of linear hypergraph languages of bounded degree (where linear means, as usual, that there is at most one nonterminal in each right-hand side). This class is generated by the quasi-linear apex CFHG grammars and equals the substitution closure of the linear apex hypergraph languages (in the case of strings, these languages are called "quasi-rational" in [2]). For quasi-linear apex CFHG grammars two nonterminals in the right-hand sides do suffice. The number of substitutions gives rise to a hierarchy strict at each level.

1. Preliminaries

This section discusses hypergraphs and context-free hypergraph grammars. These grammars are studied, e.g., in [1, 7, 8, 11, 12, 19, 23, 24, 25, 27, 28, 29, 31].

$\mathbf{N} = \{0, 1, 2, \ldots\}$ and, for $n, m \in \mathbf{N}$, $[n, m] = \{i \in \mathbf{N} \mid n \leq i \leq m\}$. For a set A, A^* denotes the set of all finite sequences of elements of A, and $\#A$ its cardinality. For sets A and B, $A \subseteq B$ denotes inclusion and $A \subset B$ denotes proper inclusion. A ranked alphabet is a finite set Σ with a mapping rank: $\Sigma \to \mathbf{N}$.

Let Σ be a ranked alphabet. A *hypergraph over* Σ is a tuple $H = (V, E, \text{nod}, \text{lab}, \text{ext})$, where V is the finite set of nodes, E is the finite set of (hyper)edges, nod: $E \to V^*$ is the incidence function, lab: $E \to \Sigma$ is the labeling function, and ext $\in V^*$ is the sequence of external nodes. It is required that for every $e \in E$, nod(e) has length rank(lab(e)). Moreover, to simplify the presentation, we assume here that the sequences nod(e) and ext do not contain repetitions of nodes (i.e., our hypergraphs are "loop-free" and "identification-free", cf. [12]). In case all symbols of Σ have rank 2, and ext is empty, H is an ordinary (directed, edge-labeled) graph.

If nod(e) $= (v_1, v_2, \ldots, v_r)$, $r \in \mathbf{N}$, then r is said to be the *rank* of e, denoted rank(e); thus, rank(e) = rank(lab(e)). The node v_i is also denoted by nod(e, i), and

e and v_i are *incident*. Similarly, if ext $= (v_1, v_2, \ldots, v_m), m \in \mathbf{N}$, then m is sai to be the *rank* of H, denoted rank(H), and v_i is denoted ext(i). The *degree* of node is the number of edges that are incident with it. A set L of hypergraphs of *bounded degree* if there exists $b \in \mathbf{N}$ such that all nodes of all hypergraphs in have degree $\leq b$. For a given hypergraph H, its components are denoted by V_H, E_H, nod_H, lab_H, and ext_H, respectively.

Context-free hypergraph grammars are based on the operation of *substitutio* of hypergraphs. Let K and H be hypergraphs over Σ, and let e be an edge of K with rank(e) = rank(H). Then $K[H/e]$, the result of substituting H for e in K is obtained by taking the disjoint union of K and H, identifying $\text{nod}_K(e,i)$ with $\text{ext}_H(i)$ for every $i \in [1, \text{rank}(e)]$, and removing edge e. Note that since $\text{nod}_K(e)$ and ext_H are both sequences of distinct nodes, no other nodes are identified.

A *context-free hypergraph grammar* (abbreviated CFHG grammar) is a tupl $G = (\Sigma, \Delta, P, S)$, where Σ is a ranked alphabet, $\Delta \subseteq \Sigma$ is the alphabet of termina symbols (and $\Sigma - \Delta$ is the alphabet of nonterminal symbols), P is the finite set o productions, and $S \in \Sigma - \Delta$ is the initial nonterminal; every production in P i of the form $X \rightarrow H$ where $X \in \Sigma - \Delta$, H is a hypergraph over Σ, and rank(X) = rank(H). An edge e of a hypergraph over Σ is said to be a *nonterminal edge* i lab(e) is nonterminal, and a *terminal edge* otherwise. Application of a productio $X \rightarrow H$ of G is defined as follows. Let K be a hypergraph over Σ and let e be a nonterminal edge of K. Then $X \rightarrow H$ is applicable to e if $\text{lab}_K(e) = X$, and the result of application is the hypergraph $K[H/e]$. We write $K \Rightarrow K'$, where K' is (isomorphic to) $K[H/e]$. The *hypergraph language generated* by G, denoted $L(G)$ is the set of hypergraphs H over Δ such that sing(S) $\Rightarrow^* H$, where sing(S) is the hypergraph ($[1, r]$, $\{e\}$,nod,lab,ext) with r = rank(S), nod(e) = ext = $(1, 2, \ldots, r)$ and lab(e) = S. A hypergraph H over Σ such that sing(S) $\Rightarrow^* H$ is a *sentential form* of G; note that rank(H) = rank(S). We denote by CFHG the class of hypergraph languages generated by CFHG grammars.

CFHG grammars have "all" the nice properties of ordinary (string generating) context-free grammars (cf. [6]). In fact, the latter can be viewed as a special case of the former, as follows. An ordinary alphabet Σ is viewed as a ranked alphabet in which every symbol has rank 2. For a string $w = \sigma_1 \cdots \sigma_m$ over Σ ($m \geq 1$), the *string graph* gr(w) is the hypergraph (V, E, nod, lab, ext) with $V = [0, m], E = [1, m]$, nod(i) = $(i-1, i)$, lab(i) = σ_i, and ext = $(0, m)$. So, the nodes of gr(w) are the positions to the left and right of the symbols of w, and its external nodes are the left-most and the right-most of these positions. Through this representation of strings by graphs of rank 2, we can now view every ordinary (λ-free) context-free grammar as a CFHG grammar of which all symbols have rank 2 and all right-hand sides of productions are string graphs. String substitution is simulated faithfully by hypergraph substitution of the corresponding string graphs.

As an example of a CFHG grammar, consider $G = (\Sigma, \Delta, P, S)$ where $\Sigma = \{S, T, \ell, r, b\}$ and $\Delta = \{\ell, r, b\}$, with rank(S) = 0, rank(T) = 3, and rank(ℓ) = rank(r) = rank(b) = 2, and P consists of the three productions drawn in Fig.1. In pictures of hypergraphs (such as the right-hand sides of these productions) we use the following *pictorial conventions*. A node is indicated by a fat dot, as usual for graphs, and an edge e is indicated by a square box containing lab(e), with a line between e and nod(e, i) labeled by i, for each $i \in$ [1,rank(e)]. These lines are called the "tentacles" of the edge e. An edge e with $\mathrm{rank}(e) = 2$ is also drawn as a directed line from $\mathrm{nod}(e, 1)$ to $\mathrm{node}(e, 2)$, with label $\mathrm{lab}(e)$, as usual for edges of graphs. The external node $\mathrm{ext}(i)$ is indicated by label i, for every $i \in [1, r]$, where r is the rank of the hypergraph. The CFHG grammar G of Fig.1 generates the hypergraph language $L(G)$ that consists of all "binary trees with a linked frontier", such as the one in Fig.2. Such a graph is an ordered binary tree (with its edges labeled ℓ and r, for 'left' and 'right') of which any two consecutive leaves (in the left to right order) are connected by two edges, labeled b and incident with an intermediate node. Note that $L(G)$ is of bounded degree with bound 3.

CFHG contains many hypergraph languages of unbounded degree, such as the set of all "stars" and the set of all graphs of tree-width $\leq k$ (for fixed k).

2. Generalized Greibach Normal Form

Let $X \rightarrow H$ be a production of a CFHG grammar and let e be an edge of H. Suppose that $\mathrm{ext}_H(i) = \mathrm{nod}_H(e, j)$. Then we say that the node $\mathrm{ext}_H(i)$ is *passed* to e (from tentacle i to tentacle j). Also, when $X \rightarrow H$ is applied to an edge f of hypergraph K (with label X), the node $\mathrm{nod}_K(f, i)$ is said to be passed from f to e during application; this is because intuitively this node is the same as the node $\mathrm{nod}_{K[H/f]}(e, j)$. Thus, during a derivation, nodes are passed from nonterminal edges to terminal or nonterminal edges.

A CFHG grammar is *apex* if for every production $X \rightarrow H$, no external node of H is incident with a nonterminal edge. In other words, a CFHG grammar is apex if nonterminal edges do not pass nodes to each other. This means that the terminal part of the sentential form grows at the "outside" (like knitting a sweater). Thus, the derivation process of an apex grammar is particularly easy to understand and visualize. Note that an ordinary context-free grammar, viewed as a CFHG grammar as discussed in Section 1, is apex if and only if, for every production $X \rightarrow \mathrm{gr}(w)$, w does not begin or end with a nonterminal symbol. Thus, the apex CFHG grammar generalizes the Double Greibach Normal Form for context-free grammars.

We denote by A-CFHG the class of hypergraph languages generated by apex CFHG grammars (first considered in [19]). The main result of this paper is stated next. It does not only generalize Double Greibach Normal Form to CFHG grammars that generate a language of bounded degree, but also provides a grammatical characterization of the CFHG languages of bounded degree.

Theorem 1. A-CFHG *is equal to the class of* CFHG *hypergraph languages of bounded degree.*

One direction of this result (that an apex CFHG grammar generates a language of bounded degree) is easy: if the maximal degree of the nodes in the right-hand sides of the productions is m, then the maximal degree of the nodes in the generated hypergraphs is m^2. The other direction is non-trivial. This will be illustrated in the following examples, in which we will also indicate the main idea of the proof. The proof itself is omitted ([14] is entirely devoted to the proofs of Theorems 1 and 3).

(1) To get some feeling for the non-triviality of the result the reader is challenged to find an apex CFHG grammar generating all "binary trees with a linked frontier", i.e., equivalent with the (non-apex) CFHG grammar G of Fig.1. Note that, in G,

each intermediate node is generated by an application of the second production of Fig.1, and is then passed down along both the right-most slope of the tree generated by the left-most T-labeled edge in the production and the left-most slope of the tree generated by the right-most T-labeled edge. An apex CFHG grammar equivalent with G is discussed in [14]; it has more than hundred productions.

(2) Let us now consider an easy example. Fig.3 shows a graph that we will call a "grasshopper". The "grass" is a chain of a-labeled edges and the hops of the grasshopper are represented by the b-labeled edges. We assume that the hops are of length at least 2 (measured by the number of a-labeled edges). A CFHG grammar generating all grasshoppers, in the way suggested by Fig.3, is given in Fig.4. As can be seen from the second production, the grammar is not apex: whenever the grasshopper "takes off" from a node of the chain, this node is passed from nonterminal edge to nonterminal edge until the grasshopper "lands" again. The main idea in the proof of Theorem 1 is that an apex grammar can be obtained by "folding" parts of the graph (or, more precisely, by "folding" parts of derivations of the given non-apex grammar); in this way the nodes that are connected by a "long" edge (such as a hop of the grasshopper) can be generated at the same derivation step and need not be passed along. In this case, to obtain an apex CFHG grammar we "fold" each part of the chain of grass that is hopped over by the grasshopper. The result is shown in Fig.5, and the apex CFHG grammar suggested by this figure is given in Fig.6.

(3) More complicated is the case of two grasshoppers. An example of a "2-grasshopper" is given in Fig.7. As before, the "grass" is a chain of a-labeled edges. The hops of one grasshopper are represented by the b-labeled edges and the hops of a second grasshopper by the c-labeled edges. A non-apex CFHG grammar for the set of all 2-grasshoppers is not difficult to write. The reader may consider the problem of finding an equivalent apex grammar. In this case the graph has to be folded twice: first one should fold the pieces of grass where both grasshoppers are "in the air", and then one has to fold the pieces of graph where one grasshopper is in the air while the other makes several jumps. In the case of 3 grasshoppers the graph has to be folded 3 times, etcetera. A related example is the language of all graphs of cutwidth $\leq k$ (for some fixed k). It is easy to write a non-apex CFHG grammar for this language (cf. [15]), but hard to find an apex CFHG grammar.

The details of the folding construction in [14] are quite tricky. After (iterated) folding, a CFHG grammar G is obtained of which every external node is incident with at least one terminal edge or at least two nonterminal edges. The apex property is then obtained by (iteratively) applying the productions of G to its own right-hand sides. Boundedness of the degree of the generated language is needed in this last step only.

The folding idea is, implicitly, already present in the proof of the Double Greibach Normal Form in (the last section of) [30], but for ordinary context-free grammars it suffices to fold once; thus, the case of graphs is essentially more difficult than the case of strings. We note here that our construction in the proof of Theorem 1 actually transforms an ordinary context-free grammar into an ordinary context-free grammar (in Double Greibach Normal Form).

For every context-free grammar and every $k \in \mathbf{N}$ there is an equivalent context-free grammar in Double Greibach Normal Form such that the distance between any two nonterminals in a sentential form is at least k. This is also true for apex CFHG grammars (see [17], where it is proved for apex edNCE grammars).

3. The Number of Nonterminals

Ordinary context-free grammars can be brought into Double Greibach Normal Form with the additional property that each right-hand side of a production contains at most two nonterminals. We will show that this is not true for apex CFHG grammars: the number of nonterminal edges in the right-hand sides gives rise to a hierarchy, strict at each level. For $k \geq 1$, we denote by kA-CFHG the class of hypergraph languages that can be generated by apex CFHG grammars of which each right-hand side of a production has at most k nonterminal edges. To show that kA-CFHG $\subset$ $(k+1)$A-CFHG, we will prove that $(k+1)$A-CFHG grammars can generate graph languages of higher "density" than kA-CFHG grammars. The notion of "density" that we use refines the notion of (bounded) degree.

Let L be a set of undirected unlabeled graphs, and let f be a function $\mathbf{N} \to \mathbf{N}$. We say that L has $f(n)$ *density* if, for every $n \in \mathbf{N}$, every graph g in L, and every node v of g, $\#\{w \mid w$ is a node of g, there is a path from v to w in g of length at most $n\} \leq f(n)$. In other words, at most $f(n)$ nodes can be reached from a given node in n steps. Clearly, if L is of unbounded degree, then there is no such function; if L is of bounded degree with bound b, with $b \geq 3$, then L has $O((b-1)^n)$ density.

For a (directed edge-labeled) graph H, we denote by und(H) the underlying undirected unlabeled graph. For a set L, und$(L) = \{\text{und}(H) \mid H \in L\}$.

Lemma 1. *Let $k \geq 2$, and let L be a set of graphs. If $L \in k$A-CFHG, then und(L) has $O(k^n)$ density.*

The proof of this lemma is based on the fact that for an apex CFHG grammar G, the generated language has, modulo a constant, the same density as the set of derivation trees of G. And since a derivation tree of a kA-CFHG grammar has degree $\leq k+1$, that density is $O(k^n)$. Let us consider this in some more detail.

Let $G = (\Sigma, \Delta, P, S)$ be a CFHG grammar, and assume that, for every production $X \to H$ of G, the nonterminal edges of H are given a fixed (arbitrary) order. A *derivation tree* of G is a directed ordered tree t of which each node is labeled by a production of G, defined recursively as follows: the root of t is labeled by a production $X \to H$ of G, and if the nonterminal edges of H are labeled $X_1, \ldots, X_m$ (in the given order, $m \geq 0$), then t has m direct subtrees $t_1, \ldots, t_m$ such that t_i is a derivation tree, the root of which is labeled by a production of G with left-hand side X_i. A derivation tree is *complete* if its root is labeled by a production with left-hand side S. The set of all complete derivation trees of G is denoted $D(G)$. It should be clear how to define the hypergraph yield(t) generated by $t \in D(G)$. A conceptually simple way of computing yield(t) is as follows (cf. [11]). Consider for every node n of t an isomorphic copy H_n of the right-hand side of the production that labels n, in such a way that all these copies are disjoint. Then yield(t) is obtained from the disjoint union H of all H_n by first identifying certain nodes of H and then removing all its nonterminal edges. The nodes of H are identified as follows: if $n[i]$ is the i-th child of node n of t, and e_i is the i-th nonterminal edge of H_n (in the given order), then $\text{ext}_{H_{n[i]}}(j)$ should be identified with $\text{nod}_{H_n}(e_i, j)$ for every $j \in [1, r]$, where r is the rank of e_i. Fig.8 shows a complete derivation tree t of the gramma of Fig.6, such that yield(t) is the graph of Fig.5 (or Fig.3). The nodes of t are th big ovals. Rather than showing the label of node n, we have inscribed it with H_n with the nonterminal edges already removed and the non-external nodes encircled. The dotted lines show the identifications that have to be made to obtain yield(t).

For a complete derivation tree t and a node v of yield(t), we denote by tr(v) the node n of t such that v is a non-external node of H_n. Intuitively, the node v of yield(t) is generated by node tr(v) of t. In Fig.8, tr(v) = n for every node v of yield(t) that is encircled within n. Now consider an edge e of yield(t) and two nodes v and w of yield(t) that are both incident with e. In general, due to the passing of nodes by nonterminal edges, tr(v) and tr(w) can be very far apart in t. However, as one can easily see, in the case of an apex CFHG grammar, either tr(v) = tr(w), or tr(v) is a child of tr(w), or vice versa (cf. Fig.8). This makes the mapping tr into an "emulation", defined as follows (see, e.g., [3]).

Let g and g' be two undirected graphs, and let f be a mapping from the nodes of g to the nodes of g'; f is an *emulation* if, for every edge $\{v, w\}$ of g, either $f(v) = f(w)$ or $\{f(v), f(w)\}$ is an edge of g'. Moreover, for $k \in \mathbf{N}$, we say that f is a *k-emulation* if $\#(f^{-1}(v)) \leq k$ for every node v of g', i.e., at most k nodes of g are emulated on one node of g'.

For a derivation tree t we denote by und(t) the underlying undirected unlabeled (unordered) graph.

Lemma 2. *Let G be an apex* CFHG *grammar that generates graphs. Then there exists $k \in \mathbf{N}$ such that for every complete derivation tree t of G there is a k-emulation from und(yield(t)) to und(t).*

Proof. Let k be the maximal number of non-external nodes in the right-hand side of a production of G. Then, for given $t \in D(G)$, the mapping tr is a k-emulation from und(yield(t)) to und(t). □

For two sets L and L' of undirected graphs, *L can be emulated on L'* if there exists $k \in \mathbf{N}$ such that for every graph $g \in L$ there exist a graph $g' \in L'$ and a k-emulation from g to g'. Thus, according to Lemma 2, for every (graph generating) apex CFHG grammar G, und($L(G)$) can be emulated on und($D(G)$).

Lemma 3. *Let L and L' be sets of undirected graphs such that L can be emulated on L'. If L' has $O(f(n))$ density, then so has L.*

The proof of this lemma should be clear. To be precise, if k is the emulation constant, and L' has $h(n)$ density, then L has $k \cdot h(n)$ density (note that a path of length n is mapped by the emulation into a path of length at most n).

Lemma 1 now follows from Lemma's 2 and 3: for a kA-CFHG grammar G, und($D(G)$) has $O(k^n)$ density, and since und($L(G)$) can be emulated on und($D(G)$), und($L(G)$) also has $O(k^n)$ density.

Theorem 2. *For $k \geq 2$, kA-CFHG $\subset$ $(k+1)$A-CFHG.*

Proof. It is easy to see that the set L_{k+1} of $(k+1)$-ary trees is in $(k+1)$A-CFHG. As an example, Fig.9 shows the productions of a 3A-CFHG grammar generating all ternary trees. But und(L_{k+1}) clearly does not have $O(k^n)$ density: for every n there is a tree t_n in L_{k+1} such that at least $(k+1)^n$ nodes can be reached from its root in steps, and $(k+1)^n$ grows faster than $c \cdot k^n$, for any c. Hence, by Lemma 1, L_{k+1} is not in kA-CFHG. □

The proper inclusion 1A-CFHG ⊂ 2A-CFHG will be shown in the next section.

The Quasi-Linear Case

An important special case of the context-free graph grammar is the linear graph grammar (see, e.g., [15]). A CFHG grammar is *linear* if each right-hand side of a production contains at most one nonterminal edge. We denote by LIN-CFHG the class of hypergraph languages generated by linear CFHG grammars.

We will present (in Theorem 3) a grammatical characterization of the LIN-CFHG hypergraph languages of bounded degree, similar to Theorem 1. The folding construction in the proof of Theorem 1 does not preserve linearity (cf. the grammars of Figs.4 and 6). However, it does preserve quasi-linearity, which we now define. A CFHG grammar $G = (\Sigma, \Delta, P, S)$ is *quasi-linear* if there exists a mapping level: $(\Sigma - \Delta) \to (\mathbf{N} - \{0\})$ such that for every production $X \to H$ in P

(1) level(lab$_H(e)$) $\leq$ level(X) for every nonterminal edge e of H, and

(2) there is at most one nonterminal edge e of H such that level(lab$_H(e)$) = level(X).

Note that every linear CFHG grammar is quasi-linear, with level(X) = 1 for every nonterminal X. The grammar of Fig.6 is quasi-linear with level(S) = level(Y) = 2 and level(Z) = 1. The grammar of Fig.1 is not quasi-linear: the second production cannot satisfy (1) and (2) above for any number level(T).

We denote by Q-CFHG and Q-A-CFHG the classes of hypergraph languages generated by quasi-linear (quasi-linear apex, respectively) CFHG grammars. For $k \geq 1$ we denote by kQ-A-CFHG the class of hypergraph languages generated by Q-A-CFHG grammars with level(X) $\leq k$ for every nonterminal X. Note that 1Q-A-CFHG = 1A-CFHG = LIN-A-CFHG.

It can easily be shown that Q-A-CFHG is the closure under substitution of LIN-A-CFHG (and similarly for Q-CFHG). Thus quasi-linearity generalizes the notion of a quasi-rational string language (see Section VII of [2]).

Lemma 4. Q-CFHG = LIN-CFHG.

Proof (sketch). It can be shown that every quasi-linear CFHG grammar G is derivation bounded, i.e., there exists $b \geq 1$ such that for every hypergraph $H \in L(G)$ there is a derivation $S \Rightarrow H_1 \Rightarrow H_2 \Rightarrow \ldots \Rightarrow H$ with $\#\{e \in E_{H_i} \mid e$ is a nonterminal edge of $H_i\} \leq b$. The easy proof is similar to the corresponding result for context-free grammars (see [21,2]). The simulation of a derivation bounded CFHG grammar by a linear CFHG grammar is similar to the one proved for edNCE graph grammars in [15]. The idea is that b nonterminal edges can be simulated by one new nonterminal edge with as many tentacles as the b edges together. □

Theorem 3. Q-A-CFHG *is equal to the class of* LIN-CFHG *hypergraph languages of bounded degree.*

Proof. The folding construction in the proof of Theorem 1 transforms a linear grammar into a quasi-linear apex grammar (see [14]). The other half of the proof follows from Lemma 4. □

For quasi-linear apex grammars the number of nonterminals in the right-hand sides of the productions can be restricted to two; thus, Q-A-CFHG ⊆ 2A-CFHG. The proof of this result is straightforward; as in the proof of Lemma 4, it uses the idea of simulating several nonterminal edges by one.

Lemma 5. *Let $k \geq 2$. For every kQ-A-CFHG grammar an equivalent kQ-A-CFHG grammar G can be constructed such that for each production of G the number of nonterminal edges in its right-hand side is at most 2, and, moreover, their labels are of different level.*

We now show that the quasi-linear A-CFHG languages form a strict hierarchy with respect to level, and that they are properly contained in 2A-CFHG. We use density arguments similar to those in the previous section.

Theorem 4. *For $k \geq 1$, kQ-A-CFHG $\subset$ $(k+1)$Q-A-CFHG.*
And Q-A-CFHG $\subset$ 2A-CFHG.

Proof. For every $k \geq 1$, let G_k be the A-CFHG grammar with nonterminal symbols $X_1, \ldots, X_k$, all of rank 1, initial nonterminal X_1, and the productions shown in Fig.10 (with $i \in [2, k]$). Clearly, G_k is a kQ-A-CFHG grammar, with level$(X_i) = i$. $L(G_1)$ is the set of all chains, and, roughly speaking, $L(G_k)$ is the set of all "combs" with their "teeth" in $L(G_{k-1})$. A graph in und$(L(G_3))$ is shown in Fig.11. Now the following two facts are easy to show by induction on k, using that $\frac{1}{k+1} \cdot n^{k+1} \leq \sum_{i=1}^{n} i^k \leq n^{k+1}$: und$(L(G_k))$ has $O(n^k)$ density, and und$(L(G_{k+1}))$ does not have $O(n^k)$ density, because n^{k+1} grows faster than $c \cdot n^k$ for any constant c. Next, it can be shown, for a set L of graphs, that if $L \in$ kQ-A-CFHG, then und(L) has $O(n^k)$ density. In fact, we may assume that $L = L(G)$ for a kQ-A-CFHG grammar G satisfying Lemma 5. For such a grammar it is easy to see that its derivation trees look like the graphs in $L(G_k)$, more precisely, und$(D(G)) \subseteq$ und$(L(G_k))$. Hence by the above, und$(D(G))$ has $O(n^k)$ density. Lemma's 2 and 3 now imply that und$(L(G))$ has $O(n^k)$ density.

This shows that $L(G_{k+1})$ is in $(k+1)$Q-A-CFHG but not in kQ-A-CFHG. Moreover, the set of binary trees is in 2A-CFHG but not in Q-A-CFHG because it does not have polynomial density (2^n grows faster than any polynomial). It was in fact already shown in [15] that the set of binary trees is not in LIN-CFHG (and hence not in Q-A-CFHG by Theorem 3). □

Thus, the class A-CFHG of CFHG languages of bounded degree consists of a double hierarchy: LIN-A-CFHG $\subset$ 2Q-A-CFHG $\subset$ 3Q-A-CFHG $\subset \ldots$ Q-A-CFHG $\subset$ 2A-CFHG $\subset$ 3A-CFHG $\subset \ldots$ A-CFHG. Strictness of this hierarchy could be shown because the graph languages in kQ-A-CFHG have $O(n^k)$ density, and those in kA-CFHG have $O(k^n)$ density. This was due to the close connection between apex grammars and the notion of emulation.

Acknowledgment. I am grateful to Linda Heyker and George Leih for their help with the proof of Theorem 1.

References

1. M.Bauderon, B.Courcelle; Graph expressions and graph rewritings, Math. Systems Theory 20 (1987), 83-127
2. J.Berstel; "Transductions and Context-free Languages", Teubner, 1979
3. H.L.Bodlaender; The complexity of finding uniform emulations on paths and ring net-

works, Inform. Comput. 86 (1990), 87-106
4. F.J.Brandenburg; On polynomial time graph grammars, Proc. STACS 88, Lecture Notes in Computer Science 294, Springer-Verlag, 1988, pp. 227-236
5. F.J.Brandenburg; The equivalence of boundary and confluent graph grammars on graph languages of bounded degree, in "Rewriting Techniques and Applications", Lecture Notes in Computer Science 488, Springer-Verlag, 1991, pp. 312-322
6. B.Courcelle; An axiomatic definition of context-free rewriting and its application to NLC graph grammars, Theor. Comput. Sci. 55 (1987), 141-181
7. B.Courcelle; On using context-free graph grammars for analyzing recursive definitions, in "Programming of future generation computers II", Elsevier, Amsterdam, 1988, pp. 83-122
8. B.Courcelle; The monadic second-order logic of graphs. I: Recognizable sets of finite graphs, Inform. Comput. 85 (1990), 12-75.
9. B.Courcelle, J.Engelfriet; A logical characterization of the sets of hypergraphs defined by hyperedge replacement grammars; Report 91-41, University of Bordeaux, June 1991
10. J.Engelfriet; Context-free NCE graph grammars, Proc. FCT '89, Lecture Notes in Computer Science 380, Springer-Verlag, 1989, pp. 148-161
11. J.Engelfriet, L.M.Heyker; The string generating power of context-free hypergraph grammars, JCSS 43 (1991), 328-360
12. J.Engelfriet, L.M.Heyker; The term-generating power of context-free hypergraph grammars and attribute grammars, to appear in Acta Informatica
13. J.Engelfriet, L.M.Heyker; Hypergraph languages of bounded degree, Report 91-01, University of Leiden, January 1991
14. J.Engelfriet, L.M.Heyker, G.Leih; Context-free graph languages of bounded degree are generated by apex graph grammars, Report 91-16, Leiden University, August 1991
15. J.Engelfriet, G.Leih; Linear graph grammars: power and complexity, Inform. Comput. 81 (1989), 88-121
16. J.Engelfriet, G.Leih, G.Rozenberg; Apex graph grammars and attribute grammars, Acta Informatica 25 (1988), 537-571
17. J.Engelfriet, G.Leih, G.Rozenberg; Nonterminal separation in graph grammars, Theor. Computer Science 82 (1991), 95-111
18. J.Engelfriet, G.Leih, E.Welzl; Boundary graph grammars with dynamic edge relabeling, JCSS 40 (1990), 307-345
19. J.Engelfriet, G.Rozenberg; A comparison of boundary graph grammars and context-free hypergraph grammars, Inform. Comput. 84 (1990), 163-206
20. J.Engelfriet, G.Rozenberg; Graph grammars based on node rewriting: an introduction to NLC graph grammars, in "Graph Grammars and their Application to Computer Science", Lecture Notes in Computer Science 532, Springer-Verlag, 1991, pp.12-23
21. S.Ginsburg, E.H.Spanier; Derivation bounded languages, JCSS 2 (1968), 228-250
22. S.A.Greibach; A new normal-form theorem for context-free phrase structure grammars, JACM 12 (1965), 42-52
23. A.Habel; Hyperedge replacement: grammars and languages, Ph.D.Thesis, Bremen, 1989
24. A.Habel, H.-J.Kreowski; Some structural aspects of hypergraph languages generated by hyperedge replacement, Proc. STACS '87, LNCS 247, 1987, pp. 207-219
25. A.Habel, H.-J.Kreowski; May we introduce to you: hyperedge replacement, in "Graph-Grammars and their Application to Computer Science", Lecture Notes in Computer Science 291, Springer-Verlag, 1987, pp. 15-26
26. M.A.Harrison; "Introduction to Formal Language Theory", Addison-Wesley, 1978
27. C.Lautemann; Decomposition trees: structured graph representation and efficient algorithms, in Proc. CAAP '88, Lecture Notes in Computer Science 299, 1988, pp. 28-39

28. C.Lautemann; Efficient algorithms on context-free graph languages, Proc. ICALP '88, Lecture Notes in Computer Science 317, Springer-Verlag, 1988, pp. 362-378
29. T.Lengauer, E.Wanke; Efficient analysis of graph properties on context-free graph languages, in Proc. ICALP '88, LNCS 317, 1988, pp. 379-393
30. D.J.Rosenkrantz; Matrix equations and normal forms for context-free grammars, JACM 14 (1967), 501-507
31. W.Vogler; On hyperedge replacement and BNLC graph grammars, in: Graph-Theoretic Concepts in Computer Science WG '89, LNCS 411, 1989, pp. 78-93

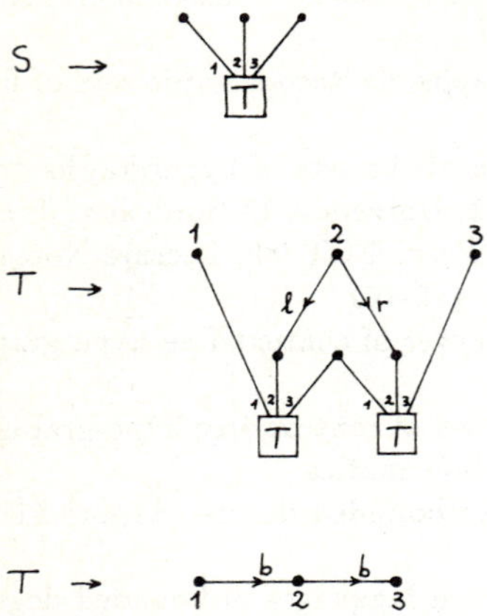

Figure 1. A CFHG grammar generating binary trees with a linked frontier

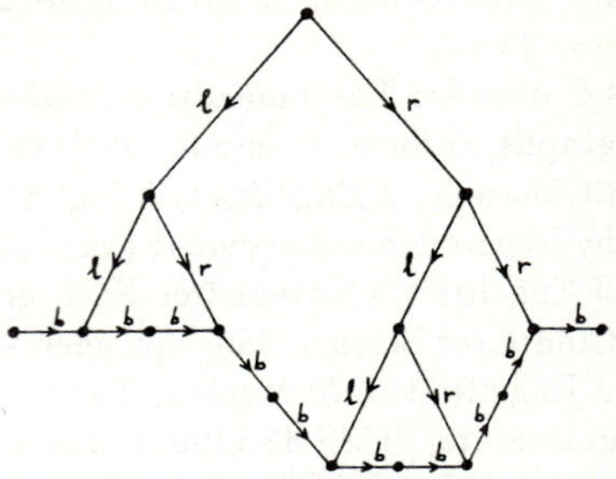

Figure 2. A binary tree with a linked frontier

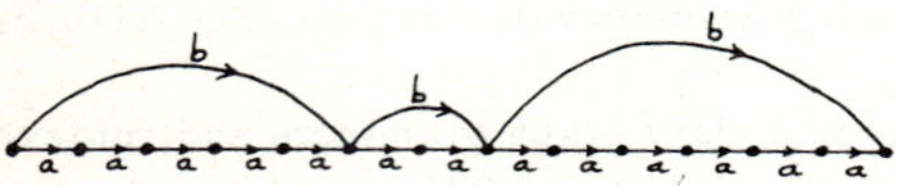

Figure 3. A grasshopper

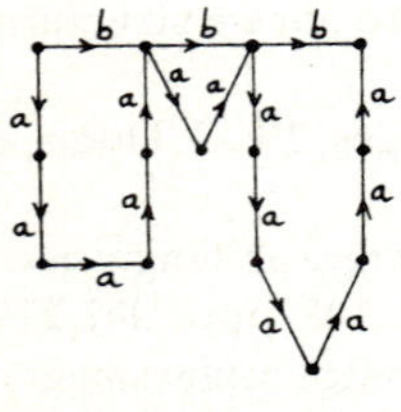

Figure 5. A "folded" grasshopper

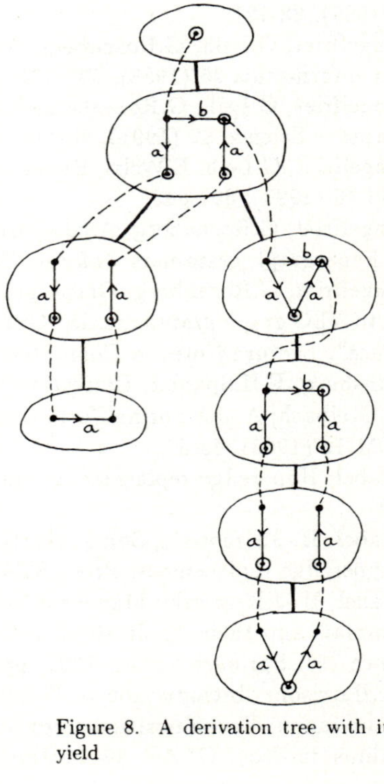

Figure 8. A derivation tree with its yield

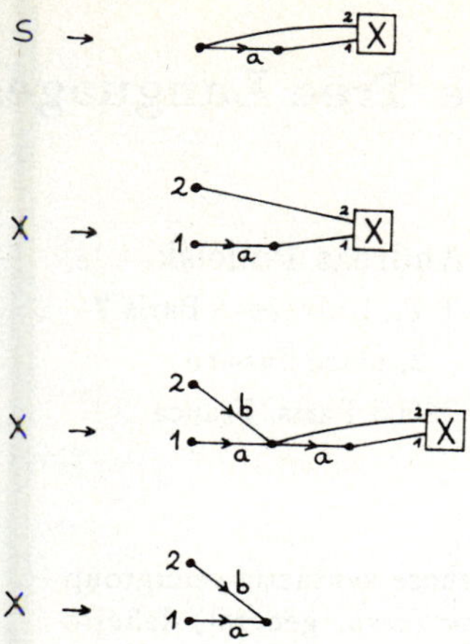

Figure 4. A CFHG grammar generating grasshoppers

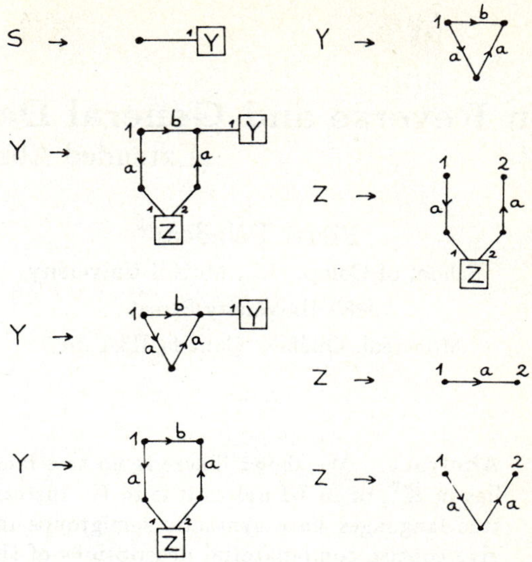

Figure 6. An apex CFHG grammar generating "folded" grasshoppers

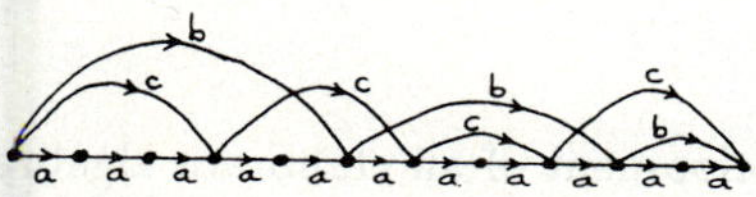

Figure 7. Two grasshoppers

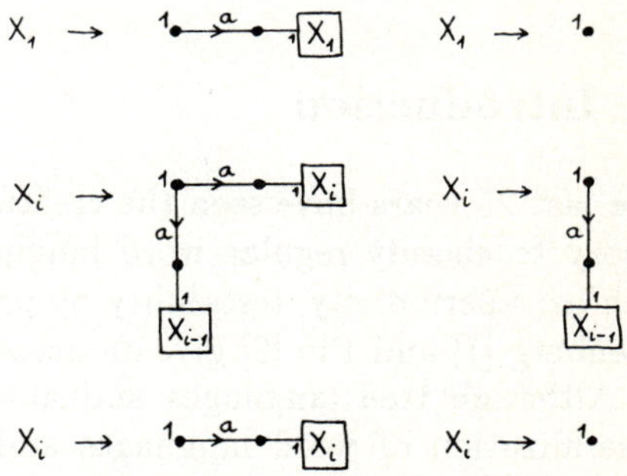

Figure 10. A quasi-linear apex CFHG grammar

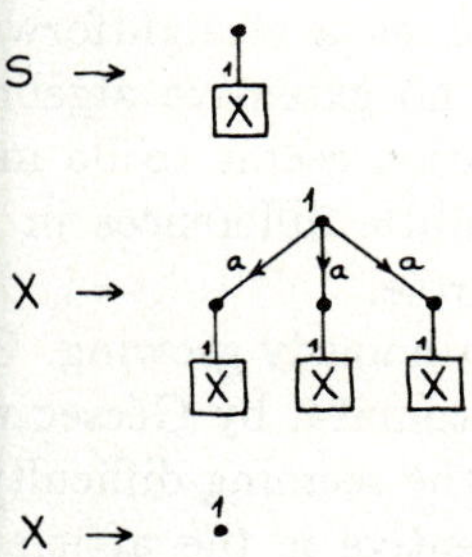

Figure 9. An apex CFHG grammar generating ternary trees

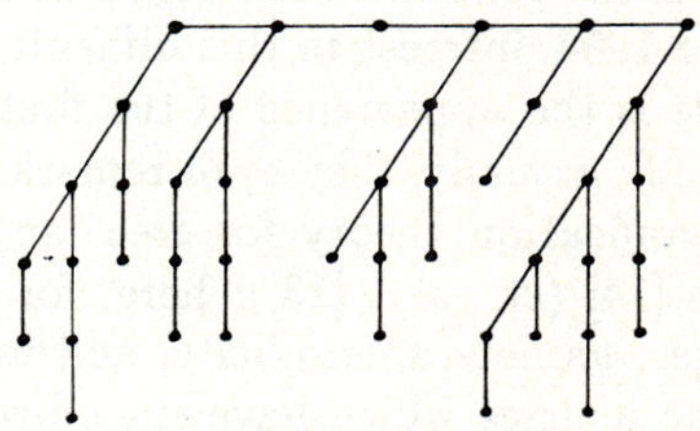

Figure 11. A graph in und($L(G_3)$)

On Reverse and General Definite Tree Languages
(Extended Abstract)

Pierre Péladeau*
School of Comp. Sci., McGill University
3480 University Street
Montréal, Québec, Canada H3A 2A7

Andreas Podelski
L.I.T.P., Université Paris 7
2, place Jussieu
75005 Paris, France

Abstract. We show: There is no tree language whose syntactic semigroup lies in $\mathbf{K^r}$, or in $\mathbf{LI}$ unless it is in $\mathbf{K}$. Instead, reverse (resp. general) definite tree languages have syntactic semigroups in $\mathbf{K^r} \vee \mathbf{J_1}$ (resp. $\mathbf{LI} \vee \mathbf{J_1}$). We give concise combinatorial descriptions of the tree languages whose syntactic semigroups are in $\mathbf{K^r} \vee \mathbf{J_1}$ (resp. $\mathbf{LI} \vee \mathbf{J_1}$), in terms of the properties reverse (resp. general) J_1-definite. These properties are more general than the properties reverse (general) definite as defined by Heuter. Finally we show that they are decidable.

1 Introduction

The last 25 years have seen the creation and development of an elaborate algebrai theory to classify regular word languages according to their structural properties such as: aperiodicity, testability by prefix, suffix, subword, or factor. The books b Eilenberg [1] and Pin [9] give an account of numerous deep results on this subject.

Although tree languages and automata on trees appear as a straightforwar generalization of word languages and automata on words, no extensive algebrai classification of tree languages has emerged yet. The situation seems to be mor involved in the tree than in the word case; furthermore, subtle differences in th ways *how* the generalization proceeds start to play a major role.

Since 1984, interest in this difficult subject has been continuously growing. On incentive is the appearence of the first textbook on tree automata, by Gécsec an Steinby [2]; actually, they even remark on the lack—as well the seeming difficulty— of a classification theory for tree languages. Another incentive is the article b Thomas [14] (cf. also [13]); here, for the first time, a structural property of tre languages, namely aperiodicity, and its connections to logic, are studied in depth From the articles which have since been devoted to the subject we cite here [3, 4 5, 6, 7, 8, 11]. In our view, the overall goal of these works are clean, systemati algebraic characterizations of structural/combinatorial properties of sets of trees possibly, these will allow to derive decision algorithms. Ultimately, the method developed shall lead to a foundational theory connecting algebraic and algorithmi aspects of tree structures.

*Supported by an NSERC postdoctoral fellowship.

Recently, Steinby [11] has proposed a promising approach; he considers syntactic gebras in order to develop a theory of varieties of tree languages. This work actually sumes work started in [10]. Leading into an orthogonal direction, and in fact the e this article pursues, Thomas [13] introduced the notion of a syntactic monoid of tree language. He defines this notion for the purpose of characterizing aperiodicity d showing its decidability. Nivat and Podelski [6, 7] have initiated a systematic udy of the connections between recognizability in the word and in the tree case vhich is: recognizability by finite monoids and by finite algebras, respectively). hey introduce the notion of the (free) tree monoid and the embedding of a tree nguage L into a subset $\widehat{L}$ of this free monoid. Then, they obtain the notion of the ntactic monoid of a tree language through the standard one of a word language.[1]

This close connection between recognizable tree languages and finite monoids ggests the *direct* use of the classification theory of word languages for the study tree languages. This idea has already been applied in order to characterize the rieties of finite/cofinite, definite, aperiodic, label-counting (modulo or threshold) ee languages by exactly the same semigroup/monoid varieties as in the word case. he proofs being rather straightforward (cf. [7]), these results are encouraging. The aracterization of some other properties, however, turn out to be more involved. his is the case for the properties considered in this paper: reverse definite and neral definite.

Heuter [3] calls a tree language *definite* if membership of a tree is determined by s root tree of fixed height and shows that this property of trees is decidable. Nivat d Podelski [7] then looked for an algebraic characterization of these tree languages d found that a tree language L is definite iff its syntactic semigroup $S(L)$ is such at all its idempotents are left zeros (i.e. $S(L)$ belongs to the variety $\mathbf{K}$). In the ord case the syntactic semigroup of a languages belongs to $\mathbf{K}$ iff it is characterized y its prefix of fixed length [9, page 48]. So the straightforward generalization from e word case works for the definite tree languages.

The other semigroup varieties considered in this paper are $\mathbf{K^r}$, $\mathbf{LI}$ and $\mathbf{J_1}$. A emigroup belongs to $\mathbf{K^r}$ (resp. $\mathbf{LI}$) if all its idempotents are right zeros (resp. left r right zeros); and a semigroup belongs to J_1 if it is idempotent and commutative. Ve recall that in the word case the syntactic semigroup of a language belongs to $\mathbf{K^r}$ resp. $\mathbf{LI}$) iff it is characterized by its suffix (resp. and prefix) of fixed lenght. And hat the syntactic monoid of a language belongs to $\mathbf{J_1}$ iff it is characterized by its et of letters (see [9]).

The extension of the property reverse definite from the word to the tree case onsidered by Heuter [4], consists of taking the set of frontier trees of bounded eight as a test set: A tree language is reverse definite iff membership of a tree epends only on the set of all its subtrees of bounded height. Now, the naïve guess hat again the variety characterization of this property in the word case (which is, $\mathbf{K^r}$) holds also in the tree case, is wrong. In fact, we show that *no* tree language an have a syntactic semigroup which belongs to $\mathbf{K^r}$.

We can show that the syntactic semigroups of all reverse definite tree languages ie in the variety $\mathbf{K^r} \vee \mathbf{J_1}$. The converse, however, is not true. Now, the question s: What are the other tree languages with this algebraic property (their syntactic

[1] This notion is only slightly different from the one of [13]; the difference is significant, however, or characterizations other than aperiodicity.

semigroup lying in $\mathbf{K^r} \vee \mathbf{J_1}$? That is: Is there a concise combinatorial description of all these tree languages? The answer is: Yes. We define the following property which we call, for its algebraic characterization, reverse J_1-definite: A tree language L is reverse J_1-definite if the membership of a tree in L is determined by the set of frontier trees of a fixed height, plus the set of "side trees" of bounded height (frontier trees stemming "sidewise from the tree"), plus the set of interior labels as test set. This notion can be motivated by the evaluation model of a frontier-to-root tree automaton (in order to assemble the different "suffix-informations," it reads the whole tree, and thus checks all interior labels).

The situation for the property general definite is analogous: A tree language is general definite iff membership of a tree depends only the set of all its subtrees of bounded height and furthermore on the root tree [4]. Again, the variety characterization of this property in the word case (which is, **LI**) fails to hold in the tree case, and actually no tree languages can have a syntactic semigroup in $\mathbf{LI} \setminus \mathbf{K}$. The syntactic semigroups of all general definite tree languages lie in the variety $\mathbf{LI} \vee \mathbf{J_1}$ but the converse is not true. The definition of general J_1-definite is analogous (the membership of a tree in the tree language L depends furthermore on the root tree of bounded height).

The main technical contributions of this paper are the results that (1) a tree language L is reverse J_1-definite if and only if its syntactic semigroup is in the variety $\mathbf{K^r} \vee \mathbf{J_1}$, and (2) L is general J_1-definite if and only if its syntactic semigroup is in the variety $\mathbf{LI} \vee \mathbf{J_1}$.

We hereby obtain decision algorithms for these two properties, since we also show the decidability of the two conditions on the syntactic semigroups. From [8] we know that these are constructed from given tree automata like in the word case, as it is done by existing systems, e.g. at the University of Aachen and at the LITP in Paris.

These results constitute the first non-trivial combinatorial characterizations of tree languages whose syntactic semigroups lie a in given variety.

2 Preliminaries

For simplicity of exposition we only consider binary trees. A tree t over a finite alphabet A is a mapping $t : supp(t) \longrightarrow A$ where $supp(t) \subseteq \{1,2\}^*$ is a set which is finite and closed under prefixes. The empty tree is denoted by λ (that is, $supp(\lambda) = \emptyset$)). We will also use the notation $a(t_1, t_2)$ for the tree whose root is labeled by 'a', whose left child is the tree t_1 and whose right child is the tree t_2. The set of all trees over the alphabet A is denoted T_A. For the purposes of this paper, where we consider semigroup varieties, it is useful to say that the empty tree does *not* belong to T_A. The height of a tree t, denoted $h(t)$, is the length of the longest path from the root to a leaf. We will use the notation T_A^k and $T_A^{<k}$ to denote the set of all trees of height k and $< k$, respectively. The *outer frontier* $Fr_+(t)$ of a tree $t \in T_A$ is the set of nodes not in $supp(t)$ and whose predecessor is in $supp(t)$; i.e. $Fr_+(t) = \{wi \in \{1,2\}^* \mid i \in \{1,2\},\ w \in supp(t), \text{ and } wi \notin supp(t)\}$. A *pointed-tree* is a pair (t, w) with $t \in T_A$ and $w \in Fr_+(t)$. Pointed trees correspond to the special-trees of [3, 14], and are denoted S_A. Pointed trees form a semigroup with the product defined by $(t, w) \cdot (t', w') = (t'', ww')$ where t'' is the tree obtained by ticking" the tree t' to the tree t in the place w. Let Γ be the set of pointed-trees $, w)$ such that $|w| = 1$ (i.e. $w \in \{1, 2\}$). Notice that the cardinality of Γ is infinite. $^+$ with the pointed tree product then becomes a free semigroup which is isomorphic S_A [6, 8]. Given a tree language $L \subseteq T_A$ we define $\widehat{L} \subseteq S_A$ to be the set of all ointed-trees (t, w) such that $t \in L$. The *syntactic semigroup* of a tree language $\subseteq T_A$ is the syntactic semigroup of the "word" language $\widehat{L} \subseteq \Gamma^+$. It turns out at the syntactic semigroup of a tree language $L \subseteq T_A$ is finite iff L is recognized y a finite tree-automaton [6].

act 2.1 *[8] One can effectively compute the syntactic semigroup of a regular tree nguage.*

ne way to compute the syntactic semigroup of a word language is through its inimal automaton. An *automaton* is a 5-tuple $\mathcal{A} = (Q, A, \delta, i, F)$ where Q is a nite set of states, A is a finite alphabet, $\delta : Q \times A \to Q$ is a transition function, $\in Q$ is the initial state and $F \subseteq Q$ is the set of accepting states. An automaton $\mathcal{A}$ ecognizes a language $L(\mathcal{A}) \subseteq A^+$ in the canonical way. Each word $w \in A^+$ defines function from Q into Q. The *transition semigroup* $S(\mathcal{A})$ of the automaton $\mathcal{A}$ is the emigroup generated by the functions thus defined as w ranges over A^+. Given a egular language $L \subseteq A^+$, there is a standard algorithm (see [9] eg.) that allows us compute, from the regular expression for L, an automaton recognizing L which minimal in some well defined sense. The transition semigroup of this automaton the syntactic semigroup of the language L.

The *root-tree of height k* of a tree $t \in T_A$ is the tree $r_k(t)$ with: $supp(r_k(t)) = w \in supp(t) : |w| \leq k\}$, and the function $r_k(t)$ coincides with t on its support. iven a node $w \in supp(t)$ the *frontier-tree* $t \uparrow w$ is the subtree of t rooted at the ode w. That is, $supp(t \uparrow w) = \{v : wv \in supp(t)\}$, and $t \uparrow w(v) = t(wv)$ for all $\in supp(t \uparrow w)$. The set of all frontier-trees of height k will be denoted by $F_{=k}(t)$. iven also $i \in \{1, 2\}$ the *side-tree* $t \uparrow_i w$ is the pointed tree — actually a letter of Γ - (t', i) where t' is the tree obtained from the frontier tree $t \uparrow w$ by removing the ontier-tree rooted at node wi (notice that the frontier tree rooted at wi may be he empty tree). The set of all side-trees of height k (resp. $\leq k$) will be denoted by $_{=k}(t)$ (resp. $S_{\leq k}(t)$). Finally, the *alphabet* of a tree $t \in T_A$, denoted $A(t)$, is the set f labels which appear on the nodes of t.

xample 2.2

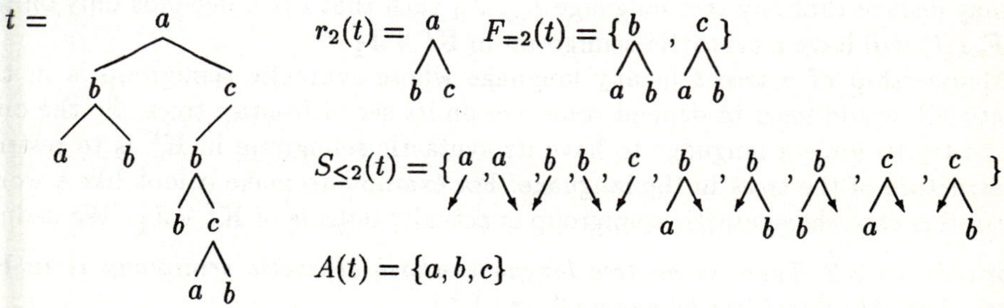

$A(t) = \{a, b, c\}$

□

3 Reverse Definite Tree Languages

In this section we will give an example of a simple reverse definite tree languag and show that its syntactic semigroup belongs to the variety $\mathbf{K^r} \vee \mathbf{J_1}$. We will the deduce that no tree language can have a syntactic semigroup in $\mathbf{K^r}$. Finally, we wi give the combinatorial description of the tree languages whose syntactic semigrou belongs to the variety $\mathbf{K^r} \vee \mathbf{J_1}$.

Example 3.1 Let $A = \{a, b\}$ and consider the tree language $L = \{t \in T_A \mid t$ has leaf with the label 'a'$\}$. The minimal automaton recognizing $\widehat{L}$ is

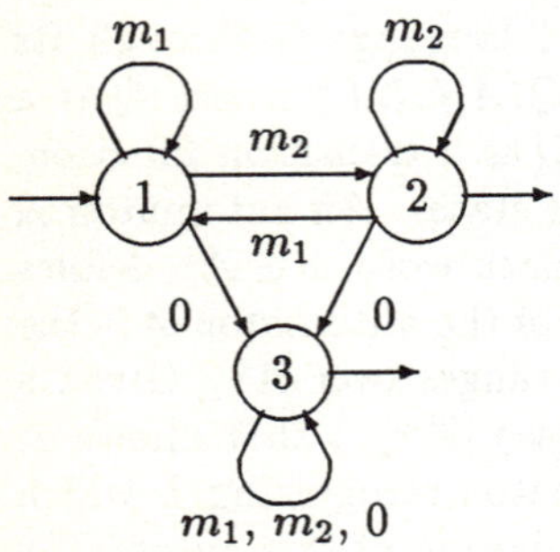

Transition semigroup

	m_1	m_2	0
m_1	m_1	m_2	0
m_2	m_1	m_2	0
0	0	0	0

D-class structure

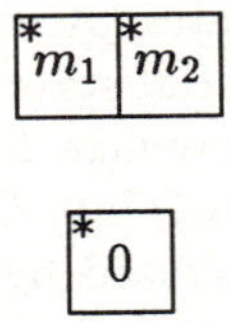

$*$ indicates idempotent

In this automaton

$$
\begin{aligned}
m_1 &= \{(a(\lambda,t),1),\ (a(t,\lambda),2) \mid t \notin L\} \cup \{(b(\lambda,t),1),\ (b(t,\lambda),2) \mid t \notin L \text{ or } t = \lambda \\
m_2 &= \{(a(\lambda,\lambda),1),\ (a(\lambda,\lambda),2)\} \\
0 &= \{(c(\lambda,t),1),\ (c(t,\lambda),2) \mid c \in A \text{ and } t \in L\}
\end{aligned}
$$

Notice that the input alphabet to this automaton is actually infinite, but becaus the automaton itself is finite this input alphabet can be split up into finitely man equivalence classes.

It is easy to verify that this transition semigroup belongs to the variety $\mathbf{K^r} \vee \mathbf{J_1}$ but not to the variety $\mathbf{K^r}$. □

Membership of a tree in the language of the above example depends only on th existence of a leaf (i.e. a frontier tree of height 1) with label a. From this exampl we may deduce that any tree language $L \subseteq T_A$ such that $t \in L$ depends only on th set $F_{=k}(t)$ will have a syntactic semigroup in $\mathbf{K^r} \vee \mathbf{J_1}$.

Membership of a tree t in any language whose syntactic semigroup is in th variety $\mathbf{K^r}$ would need to depend somehow on its set of frontier trees. So the onl way to try to force a language to have its syntactic semigroup in $\mathbf{K^r}$ is to restrict the structure of the trees in the language. For example to make it look like a word. But in this case the syntactic semigroup is actually outside of $\mathbf{K^r} \vee \mathbf{J_1}$. We deduce

Proposition 3.2 *There is no tree language whose syntactic semigroup is in* $\mathbf{K^r}$ *(other than the trivial tree languages* $\emptyset$ *and* Γ^+*).*

Fact 3.3 *A language* $L \subseteq \Gamma^+$ *is recognized by a semigroup in* $\mathbf{K^r} \vee \mathbf{J_1}$ *iff* L *is a boolean combination of languages of the form* $\Gamma^* R$ *and* $\Gamma^* S \Gamma^*$ *with* $R \subseteq \Gamma^+$ *and* $S \subseteq \Gamma$.

In the above fact, R and S need not be finite or otherwise restricted.

Definition 3.4 *Let $k \geq 1$, and define $\sim_k$ to be the congruence on Γ^+ given by $u \sim_k v$ iff u and v have the same suffix of lenght k, and u and v have the same alphabet.*

Notice that because the cardinality of Γ is infinite, the congruence $\sim_k$ is of infinite index.

Fact 3.5 *$L \subseteq \Gamma^+$ is recognized by a semigroup in $\mathbf{K^r} \vee \mathbf{J_1}$ implies L is a union of $\sim_k$-classes for some $k \geq 1$.*

Definition 3.6 *Let $\sim_\wedge$ be the equivalence relation defined on Γ^+ by $u \sim_\wedge v$ iff there exists a $t \in T_A$ such that $u \in \hat{t}$ and $v \in \hat{t}$.*

Definition 3.7 *Let $\equiv_k$ be the equivalence relation defined on T_A by $t \equiv_k t'$ iff there exists $u_0 \sim_k u_1 \sim_\wedge u_2 \sim_k \cdots \sim_\wedge u_{n-1} \sim_k u_n$ with $u_0 \in \hat{t}$ and $u_n \in \hat{t'}$.*

Definition 3.8 *Let Δ_k be the equivalence relation defined on T_A by $t\ \Delta_k\ t'$ iff $F_{=k}(t) = F_{=k}(t')$, $S_{\leq k}(t) = S_{\leq k}(t')$, and $A(t) = A(t')$. A tree language L is called* reverse J_1-definite *if it is the union of Δ_k -classes, for some $k \geq 1$.*

We will now show that $\equiv_k$ and Δ_k are actually the same equivalence relation.

Theorem 3.9 *$t \equiv_k t'$ iff $t\ \Delta_k\ t'$ for all $t, t' \in T_A$ and $k \geq 1$.*

The proof of the "only if" part is easy. In order to show the other direction we will need a normal form representative for the class $[t]_{\equiv_k}$.

Proposition 3.10 *Given a tree $t \in T_A$, there exists a tree $t' \in [t]_{\equiv_k}$ in the following normal form $t' = s(a_1(t_1,\lambda),2)(a_2(t_2,\lambda),2)\cdots(a_n(t_n,\lambda),2)t_{n+1}$ where*

(1) *$s \in S_{\leq k}(t)^*$ is a word containing at least one copy of each side-tree in $S_{\leq k}(t)$,*

(2) *$F_{=k}(t) =$ the set of trees $\{t_1,\ldots,t_{n+1}\}$,*

(3) *$A(t) = A(t')$.*

Proof. We start by building the word s. Pick any node $w \in supp(t)$ such that $(t(w)(\lambda, t{\uparrow}w2), 1) \in S_{\leq k}(t)$ or $(t(w)(t{\uparrow}w1, \lambda), 2) \in S_{\leq k}(t)$. We may suppose w.l.o.g. that we are in the first case. If $t{\uparrow}w1$ is of height $\geq k$ then there is a pointed tree $s = s_0(t(w)(\lambda, t{\uparrow}w2), 1)s_1$ in $\hat{t}$ with $|s_1| \geq k$. Therefore $s = s_0(t(w)(\lambda, t{\uparrow}w2), 1)s_1 \equiv_k s' = (t(w)(\lambda, t{\uparrow}w2), 1)s_0s_1$. If $t{\uparrow}w1$ is of height $< k$ then there is a pointed tree $s = s_0(t(w)(\lambda, t{\uparrow}w2), 1)s_1$ in $\hat{t}$ with $|s_1| < k$. Therefore $s = s_0(t(w)(\lambda, t{\uparrow}w2), 1)s_1 \sim_k s' = (t(w)(\lambda, t{\uparrow}w2), 1)s_0(t(w)(\lambda, t{\uparrow}w2), 1)s_1$. We then mark this side-tree as done.

Notice that this procedure has not introduced any new side-trees of height $\leq k$. So repeating it on s' and so on for each side-tree in t will terminate with a tree $t'' = st'''$ with $s \in S_{\leq k}(t)$ and $t''' \in T_A$ such that there exists no node w in $supp(t''')$ with $height(t'''{\uparrow}w1) < k$ and $height(t'''{\uparrow}w2) \geq k$ (or vice-versa). This second fact implies that t''' may be seen as a complete binary tree all of whose "leaves" are trees of height exacty k.

It remains to show how to write $t''' = (a_1(t_1,\lambda),2)\cdots(a_n(t_n,\lambda),2)t_{n+1}$ with $a_i \in$ k-alpha(t) and $t_i \in F_{=k}(t)$.

Let $a, b \in A$ and $t_1, t_2, t_3 \in T_A^k$. Let furthermore $s_2 \in \widehat{t_2}$ such that $|s_2| = k$. Then $(a(\lambda, t_3), 1)(b(t_1, \lambda), 2)s_2 \sim_k (b(t_1, \lambda), 2)(a(\lambda, t_3), 1)s_2$. Applying this trick recursively on t''' produces the desired normal form. ■

Notice that $S_{\leq k}(t') = S_{\leq k}(t)$, $F_{=k}(t') = F_{=k}(t)$ and $A(t') = A(t)$. It remains to show that the sets $S_{\leq k}(t)$, $F_{=k}(t)$ and $A(t)$ characterize the normal form. By this we mean that we are able to go from any normal form $t' \in [t]_{\equiv_k}$ to any other normal form $t'' \in [t]_{\equiv_k}$ provided $S_{\leq k}(t') = S_{\leq k}(t'')$, $F_{=k}(t') = F_{=k}(t'')$ and $A(t') = A(t'')$.

Clearly for any $s' \in S_{\leq k}(t)^*$ containing at least one copy of each side-tree in $S_{\leq k}(t)$: $s'(a_1(t_1, \lambda), 2) \cdots (a_n(t_n, \lambda), 2)t_{n+1}$ is in the same $\equiv_k$-class as t.

The following two lemmas imply that we may permute any of the t_i's.

Lemma 3.11 *Let $a \in A$ and $t_1, t_2 \in T_A^k$, then $a(t_1, t_2) \equiv_k a(t_2, t_1)$.*

Proof. Let $s_1 \in \widehat{t_1}$ and $s_2 \in \widehat{t_2}$ such that $|s_1| = |s_2| = k$. Then
$(a(\lambda, t_2), 1)s_1 \sim_k (a(\lambda, t_2), 1)(a(\lambda, t_2), 1)s_1 \rightsquigarrow (a(\lambda, t_2), 1)(a(t_1, \lambda), 2)s_2 \sim_k$
$(a(t_1, \lambda), 2)(a(\lambda, t_2), 1)s_2 \rightsquigarrow (a(t_1, \lambda), 2)(a(t_2, \lambda), 2)s_2 \sim_k (a(t_2, \lambda), 2)(a(t_1, \lambda), 2)s_2 \rightsquigarrow$
$(a(t_2, \lambda), 2)(a(\lambda, t_2), 1)s_1 \sim_k (a(\lambda, t_2), 1)(a(t_2, \lambda), 2)s_1 \rightsquigarrow (a(\lambda, t_2), 1)(a(\lambda, t_1), 1)s_2 \sim_k$
$(a(\lambda, t_1), 1)(a(\lambda, t_2), 1)s_2 \rightsquigarrow (a(\lambda, t_1), 1)(a(t_2, \lambda), 2)s_2 \sim_k (a(t_2, \lambda), 2)(a(\lambda, t_1), 1)s_2 \rightsquigarrow$
$(a(t_2, \lambda), 2)(a(t_2, \lambda), 2)s_1 \sim_k (a(t_2, \lambda), 2)s_1$. ■

Lemma 3.12 *Let $a, b \in A$, $t_1, t_2 \in T_A^k$ and $t_3 \in T_A$, then $a(t_1, b(t_2, t_3)) \equiv_k a(t_2, b(t_1, t_3))$.*

Proof. From Lemma 3.11 $a(t_1, b(t_2, \lambda)) \equiv_k a(b(t_2, \lambda), t_1)$. Let s_1 and s_2 be as in the proof of Lemma 3.11. Then $(a(\lambda, t_1), 1)(b(\lambda, \lambda), 1)s_2 \sim_k (b(\lambda, \lambda), 1)(a(\lambda, t_1), 1)s_2 \rightsquigarrow$ $(b(\lambda, \lambda), 1)(a(t_2, \lambda), 2)s_1 \sim_k (a(t_2, \lambda), 2)(b(\lambda, \lambda), 1)s_1$. ■

We now show that we can introduce any label from the set $A(t)$ into the set $\{a_1, \ldots, a_n\}$ without changing the pointed tree s or the set $F_{=k}(t) = \{t_1, \ldots, t_{n+1}\}$. This can be deduced from the following lemma.

Lemma 3.13 *Let $a, b \in A$, $t_1, t_2 \in T_A^k$ and $t_3 \in T_A$, then*
(a) $a(t_1, b(t_2, t_3)) \equiv_k a(t_1, a(t_2, b(t_2, t_3)))$, and
(b) $a(b(t_2, t_3), t_1) \equiv_k a(a(t_2, b(t_2, t_3)), t_1)$.

Proof. Along the lines of the proofs of Lemmas 3.12 & 3.11. ■

We can then combine these lemmas to show how to get to any other normal form t'' such that $S_{\leq k}(t') = S_{\leq k}(t'')$, $F_{=k}(t') = F_{=k}(t'')$ and $A(t') = A(t'')$.

Proof of Theorem 3.9. For the first implication, if $r \rightsquigarrow s$ then $r, s \in \widehat{t}$ for some $t \in T_A$ and obviously Δ_k is reflexive. Similarly if $r \sim_k s$ with $r \in \widehat{t}$ and $s \in \widehat{t'}$ then $t \, \Delta_k \, t'$ because $\sim_k$ does not change any of the sets $F_{=k}(t)$, $S_{\leq k}(t)$ or $A(t)$.

For the other implication we must show that only the sets $F_{=k}(t)$, $S_{\leq k}(t)$ and $A(t)$ are immutable in the normal form. This is certainly clear for the set $S_{\leq k}(t)$.

Lemmas 3.11 to 3.12 imply that we can permute any of the trees t_i. By Lemma 3.13 we may introduce any letter of $A(t)$ in the set $\{a_1, \ldots, a_n\}$. Since we can also permute any two $(a_i(t_i, \lambda), 2)$ and $(a_j(t_j, \lambda), 2)$ and that we may also double $(a_i(t_i, \lambda), 2)$ to $(a_i(t_i, \lambda), 2)(a_i(t_i, \lambda), 2)$, and fuse $(a_i(t_i, \lambda), 2)(a_i(t_i, \lambda), 2)$ to $(a_i(t_i, \lambda), 2)$, we conclude that we may obtain any tree t'' in the desired normal form provided $S_{\leq k}(t') = S_{\leq k}(t'')$, $F_{=k}(t') = F_{=k}(t'')$ and $A(t') = A(t'')$. ■

We can now give the combinatorial description of the tree languages whose syntactic semigroup belongs to $\mathbf{K^r} \vee \mathbf{J_1}$.

Theorem 3.14 *The syntactic semigroup of a tree language $L \in T_A$ belongs to the variety $\mathbf{K^r} \vee \mathbf{J_1}$ iff L is a union of Δ_k -classes for some $k \geq 1$.*

The proof of this theorem is a direct consequence of Theorem 3.9 and the two following propositions.

Proposition 3.15 *If the syntactic semigroup of a tree language $L \subseteq T_A$ belongs to $\mathbf{K^r} \vee \mathbf{J_1}$ then L is a union of $\equiv_k$-classes for some $k \geq 1$.*

Proof. Immediate from Fact 3.5 and the definitions of $\sim_L$ and of the syntactic semigroup of $L \subseteq T_A$. ■

Note *This is a general principle. If $\sim$ is a congruence on Γ^+ and $L \subseteq T_A$ is a tree language whose syntactic semigroup is S such that there exists a morphism $\phi : \Gamma^+/\sim \to S$, then L is a union of $\equiv$-classes, where $\equiv$ is defined as the closure of $\sim$ and $\sim_L$.*

Proposition 3.16 *For any tree $t \in T_A$ and any $k \geq 1$, the syntactic semigroup of the equivalence class $[t]_{\Delta_k}$ belongs to the variety $\mathbf{K^r} \vee \mathbf{J_1}$.*

Proof. Let $L = \{t \in T_A \mid t' \in F_{=k}(t)\}$ for some $t' \in T_A^k$. Then

$$\widehat{L} = \Gamma^*\{(a(\lambda, t), 1), (a(t, \lambda), 2) \mid a \in A, t \in L\}\Gamma^* \cup \Gamma^*\widehat{t'}.$$

So the syntactic semigroup of L belongs to $\mathbf{K^r} \vee \mathbf{J_1}$.

Let $L = \{t \in T_A \mid t' \in S_{\leq k}(t)\}$ for some $t' \in \Gamma$ with $h(t') \leq k$. We may suppose, w.l.o.g, that $t' = (b(\lambda, t''), 1)$ for some $b \in A$ and $t'' \in T_A^{<k}$. Then

$$\begin{aligned} \widehat{L} &= \Gamma^*\{(a(\lambda, t)1), (a(t, \lambda), 2) \mid a \in A, t \in L\}\Gamma^* \cup \\ &\quad \Gamma^*(b(\lambda, t''), 1)\Gamma^* \cup \Gamma^*\{(b(t, \lambda), 2) \mid t \in T_A\}\widehat{t''}. \end{aligned}$$

So again the syntactic semigroup of L is in the variety $\mathbf{K^r} \vee \mathbf{J_1}$.

Let $L = \{t \in T_A \mid b \in A(t)\}$ for some $b \in A$. Then

$$\begin{aligned} \widehat{L} &= \Gamma^*\{(a(\lambda, t)1), (a(t, \lambda), 2) \mid a \in A, t \in L\}\Gamma^* \cup \\ &\quad \Gamma^*\{(b(\lambda, t)1), (b(t, \lambda), 2) \mid t \in L \cup \lambda\}\Gamma^*. \end{aligned}$$

So the syntactic semigroup of L is again in the variety $\mathbf{K^r} \vee \mathbf{J_1}$.

Since we can do all boolean operations on our languages and still have the syntactic semigroup stay in the same variety, we get the desired result. ■

4 General Definite Tree Languages

In this section we show analogous results for "general definite" tree languages. We give an example (which can be generalized) of a general definite tree language whose syntactic semigroup is in the variety $\mathbf{LI} \vee \mathbf{J_1}$, and show that no tree language can have a syntactic semigroup in $\mathbf{LI} \setminus \mathbf{K}$. We then give the combinatorial description of the tree languages whose syntactic semigroups belong to the variety $\mathbf{LI} \vee \mathbf{J_1}$.

The proofs for this section are along the same lines as the ones of the previous section, although the one for Lemma 4.10 is much more involed. In order to meet space requirements, they are left to the journal version.

Example 4.1 Let $A = \{a, b\}$ and consider the tree language $L = \{t \in T_A \mid$ the root of t is labeled 'a' and t has a leaf with the label 'b'$\}$. The minimal automaton recognizing $\widehat{L}$ is

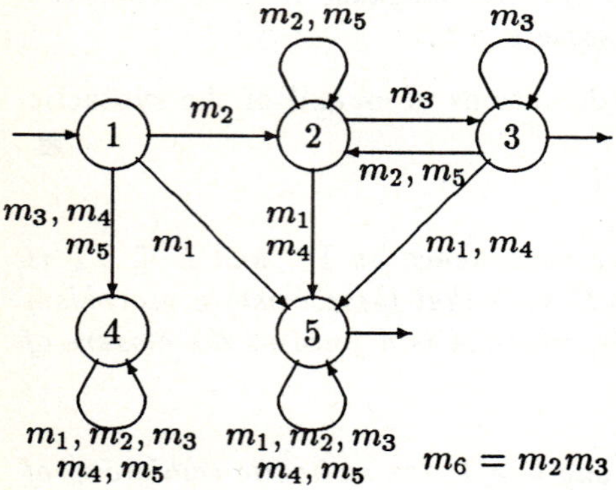

$m_6 = m_2 m_3$

Transition semigroup

	m_1	m_2	m_3	m_4	m_5	m_6
m_1	m_1	m_1	m_1	m_1	m_1	m_1
m_2	m_1	m_2	m_6	m_1	m_2	m_6
m_3	m_4	m_5	m_3	m_4	m_5	m_3
m_4	m_4	m_4	m_4	m_4	m_4	m_4
m_5	m_4	m_5	m_3	m_4	m_5	m_3
	m_1	m_2	m_6	m_1	m_2	m_6

D-class structure

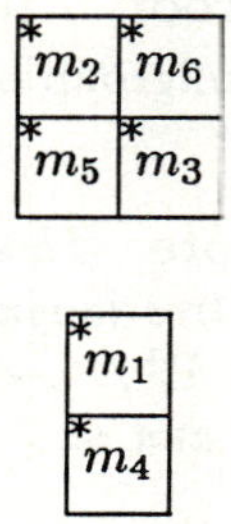

In this automaton

$$
\begin{aligned}
m_1 &= \{(a(\lambda, t), 1), (a(t, \lambda), 2) \mid t \text{ contains a 'b'}\} \\
m_2 &= \{(a(\lambda, t), 1), (a(t, \lambda), 2) \mid t \text{ contains no 'b' or } t = \lambda\} \\
m_3 &= \{(b(\lambda, \lambda), 1), (b(\lambda, \lambda), 2)\} \\
m_4 &= \{(b(\lambda, t), 1), (b(t, \lambda), 2) \mid t \text{ contains a 'b'}\} \\
m_5 &= \{(b(\lambda, t), 1), (b(t, \lambda), 2) \mid t \text{ contains no 'b'}\}
\end{aligned}
$$

One can verify that this semigroup belongs to the variety $\mathbf{LI} \vee \mathbf{J_1}$ but not to the variety $\mathbf{LI}$. □

As for the variety $\mathbf{K^r}$, membership of a tree t in a tree language whose syntactic semigroup lies exactly in the variety $\mathbf{LI}$ will have to depend somehow on its root tree and on its set of frontier trees of some fixed height. So from the above example we deduce

Proposition 4.2 *There is no tree language whose syntactic semigroup is in* $\mathbf{LI} \setminus \mathbf{K}$.

Fact 4.3 *A language* $L \subseteq \Gamma^+$ *is recognized by a semigroup in* $\mathbf{LI} \vee \mathbf{J_1}$ *iff* L *is a boolean combination of languages of the form* $R\Gamma^*$, $\Gamma^* R$ *and* $\Gamma^* S \Gamma^*$ *with* $R \subseteq \Gamma^+$ *and* $S \subseteq \Gamma$.

efinition 4.4 *Let $k \geq 1$, and define $\approx_k$ to be the congruence on Γ^+ given by $\approx_k v$ iff u and v have the same prefix and the same suffix of length k, and u and have the same alphabet.*

Notice that as for $\sim_k$, the congruence $\approx_k$ is of infinite index.

act 4.5 *$L \subseteq \Gamma^+$ is recognized by a semigroup in* $\mathbf{LI} \vee \mathbf{J_1}$ *implies L is a union of $\approx_k$-classes for some $k \geq 1$.*

efinition 4.6 *Let $\cong_k$ be the equivalence relation defined on T_A by $t \cong_k t'$ iff there xists $u_0 \approx_k u_1 \sim u_2 \approx_k \cdots \sim u_{n-1} \approx_k u_n$ with $u_0 \in \widehat{t}$ and $u_n \in \widehat{t'}$.*

efinition 4.7 *Let Λ_k be the equivalence relation defined on T_A by $t \; \Lambda_k \; t'$ iff $_k(t) = r_k(t')$, and $t \; \Delta_k \; t'$. A tree languages L is* general J_1-definite *if it is the nion of Λ_k -classes, for some $k \geq 1$.*

As in the case of reverse definite tree languages, we show that $\cong_k$ and Λ_k are ctually the same equivalence relation.

heorem 4.8 *$t \cong_k t'$ iff $t \; \Lambda_k \; t'$ for all $t, t' \in T_A$ and $k \geq 1$.*

The proof of the "only if" part is easy. In order to show the other direction we will need a normal form representative for the class $[t]_{\cong_k}$.

roposition 4.9 *Given a tree $t \in T_A$, there exists a tree $t' \in [t]_{\cong_k}$ in the following normal form: $r_k(t) = r_k(t')$ and for every $w \in Fr_+(r_k(t))$ with $|w| = k+1$: $t' \uparrow w = s_w(a_{w,1}(t_{w,1}, \lambda), 2)(a_{w,2}(t_{w,2}, \lambda), 2) \cdots (a_{w,n_w}(t_{w,n_w}, \lambda), 2) t_{w,n_w+1}$, where*

(1) *$s_w \in S_{\leq k}(t)^*$ for each w, and each side-tree in $S_{\leq k}(t)$ appears in some s_w or in $r_k(t)$,*

(2) *$F_{=k}(t) =$ the set of trees $\bigcup_w \{t_{w,1}, \ldots, t_{w,n_w+1}\}$,*

(3) *$A(t) = A(t')$.*

Notice that $r_k(t') = r_k(t)$, $S_{\leq k}(t') = S_{\leq k}(t)$, $F_{=k}(t') = F_{=k}(t)$ and $A(t') = A(t)$. It remains to show that these sets characterize the normal form. This can be done in a similar way as for $\cong_k$ except that here we must worry about how to permute letters in $A(t)$, side-trees in $S_{\leq k}(t)$, and trees in $F_{\leq k}(t)$ from one factor-tree rooted at some node w to another rooted at some other node w', where $|w| = |w'| = k+1$.
The following lemma shows how to do this in the case $k = 1$.

Lemma 4.10 *Let $a, b, c \in A$, then $a(b, c) \cong_1 a(c, b)$.*

The general case ($k > 1$) may be deduced from the above lemma. This shows how to permute trees in $F_{=k}(t)$. Labels and side-trees can also be done provided they appear in a factor-tree of height $\geq k$. And we saw how to introduce more copies of the elements in $A(t)$, $S_{\leq k}(t)$ and $F_{=k}(t)$ in Section 3.

So we have shown that the sets $r_k(t)$, $A(t)$, $S_{\leq k}(t)$, $F_{=k}(t)$ characterize the normal form.

We will now show

Theorem 4.11 *The syntactic semigroup of a tree language $L \subseteq T_A$ belongs to the variety* $\mathbf{LI} \vee \mathbf{J_1}$ *iff L is general J_1-definite.*

As for the reverse definite case, the proof of this theorem is a direct consequence of Theorem 4.8 and the two following propositions.

Proposition 4.12 *If the syntactic semigroup of a tree language $L \subseteq T_A$ belongs to* $\mathbf{LI} \vee \mathbf{J_1}$ *then L is a union of $\cong_k$-classes for some $k \geq 1$.*

Proposition 4.13 *For any tree $t \in T_A$ and any $k \geq 1$, the syntactic semigroup of the equivalence class $[t]_{\Lambda_k}$ belongs to the variety* $\mathbf{LI} \vee \mathbf{J_1}$.

5 Decidability

In this section we show that the properties discussed in this paper are decidable.

We denote by $E(S)$ the set of all idempotents of the semigroup S.

The proofs of the following theorems are not very difficult and are defered to the journal version.

Theorem 5.1 $S \in \mathbf{K^r} \vee \mathbf{J_1}$ *iff $E(S)$ is an ideal and each regular D-class of S is a semigroup which belongs to* $\mathbf{K^r}$ *(i.e. $E(S)$ is an ideal and each regular D-class is idempotent and L-trivial).*

Corollary 5.2 *For a tree language $L \in T_A$, it is decidable if the syntactic semigroup of L belongs to* $\mathbf{K^r} \vee \mathbf{J_1}$, *and therefore it is also decidable if L is a union of Δ_k-classes.*

Theorem 5.3 $S \in \mathbf{LI} \vee \mathbf{J_1}$ *iff $E(S)$ is an ideal and each regular D-class of S is a semigroup which belongs to* $\mathbf{LI}$ *(i.e. $E(S)$ is an ideal and each regular D-class is idempotent and H-trivial).*

Corollary 5.4 *For a tree language $L \in T_A$, it is decidable if the syntactic semigroup of L belongs to* $\mathbf{LI} \vee \mathbf{J_1}$, *and therefore it is also decidable if L is a union of Λ_k-classes.*

6 Conclusion and Further Research

We have shown: There is no tree language whose syntactic semigroup lies in $\mathbf{K^r}$, or in $\mathbf{LI}$ unless it is in $\mathbf{K}$; instead, reverse (resp. general) definite tree languages have syntactic semigroups in $\mathbf{K^r} \vee \mathbf{J_1}$ (resp. $\mathbf{LI} \vee \mathbf{J_1}$). We then gave concise combinatorial descriptions of the tree languages whose syntactic semigroups are in $\mathbf{K^r} \vee \mathbf{J_1}$ (resp. $\mathbf{LI} \vee \mathbf{J_1}$), in terms of the properties reverse (resp. general) J_1-definite. These properties are more general than the properties reverse (general) as defined by Heuter [4]. Finally we showed that they are decidable.

The fact that the two properties of Heuter [4] cannot be captured by the syntactic semigroup approach is a shortcoming of this algebraic approach to classifying regular tree languages (i.e., through their syntactic semigroups or monoids), showing that this approach by itself is not sufficient; it might be complemented by the theory of Steinby[11].

The natural next step is to consider the variety of locally testable tree languages. In the word case, a language is called locally testable iff membership of a word depends only on its prefix, its suffix and its set of factors of fixed length. This is the case iff its syntactic semigroup is locally idempotent and commutative (lies in the variety $\mathbf{LJ_1}$) (for references, see [9]). The natural extension to the tree case is to consider the tree languages which are characterized by their root tree, their frontier trees, their side trees and their factor trees of bounded height. One can show that the syntactic semigroups of these tree languages are in $\mathbf{LJ_1}$. We have reasons to think the converse is also true. Notice that the property general J_1-definite is a special case of locally testable, namely of the factor trees being of height 1. We may therefore hope that our proof techniques and results extend to the locally testables. In this case the normal form, and the combinatorial lemmas leading to it will obviously be more involved.

References

1. S. Eilenberg, *Automata, Languages and Machines*, Vol. B, Academic Press, 1976.
2. F. Gécsec, M. Steinby. *Tree Automata.* (Akademiai Kiado, Budapest, 1984).
3. U. Heuter, Definite tree languages, *Bulletin of the EATCS* **35**, 1988, 137–142.
4. U. Heuter, Generalized definite tree languages, *Proc. 14th MFCS*, Lecture Notes in Computer Science **379**, Springer, 1989, 270–280.
5. U. Heuter, *Zur Klassifizierung reguläler Baumsprachen.* Dissertation, University of Aachen (1989), Aachen, FRG.
6. M. Nivat and A. Podelski, Tree monoids and recognizable sets of trees, in H. Aït-Kaci and M. Nivat eds., *Resolution of Equations in Algebraic Structures*, Vol. 1, Academic Press, 1989.
7. N. Nivat and A. Podelski, Definite tree languages (cont'd), *Bulletin of the EATCS* **38**, 1989, 186–190.
8. A. Podelski, *Monoïdes d'Arbres et Automates d'Arbres*, Thèse de Doctorat, Université Paris 7, 1989.
9. J. E. Pin, *Variétés de langages formels*, Masson, Paris, 1984, and *Varieties of Formal Langages*, Plenum, London, 1986.
10. M. Steinby, Syntactic algebras and varieties of recognizable sets, in *Les arbres en algèbre et en programmation, 4ième Coll. Lille (Proc. Coll., Lille 1979)* Université de Lille, 226–240.
11. M. Steinby, "A Theory of Tree Language Varieties," in: Nivat, M., and Podelski, A., eds., *Trees, Automata and Languages*, Elsevier Publishers, 1991.
12. J.W. Thatcher, J.B. Wright, "Generalized Finite Automata Theory with an Application to a Decision Problem of Second Order Logic", *Math.Syst. Th.* **2** (1968).
13. W. Thomas, On noncounting tree languages, *Grundl. Theor. Informatik (Proc. 1. Intern. Workshop, Paderborn 1982)*, 234–242.
14. W. Thomas, Logical aspects in the study of tree languages, *Proc. Ninth Coll. on Trees in Algebra and in Programming*, Cambrige Univ. Press, 1984, 31–49.

Reductions to Sets of Low Information Content

(Extended Abstract)

V. Arvind[*] *Y. Han*[†] *L. Hemachandra*[†]

J. Köbler[‡] *A. Lozano*[§] *M. Mundhenk*[‡] *M. Ogiwara*[¶]

U. Schöning[‡] *R. Silvestri*[//] *T. Thierauf*[**]

Abstract. In this paper we study the complexity of sets that reduce to sparse sets (and tally sets), and the complexity of the simplest sparse sets to which such sets reduce. We show even with respect to very flexible reductions that NP cannot have sparse hard sets unless P = NP; an immediate consequence of our results is: If any NP-complete set conjunctively reduces to a sparse set, then P = NP. We also show that any set A that reduces to some sparse set (via various types of reductions) in fact reduces by the same type of reduction to a sparse set that is simple relative to A. We give a complete characterization of the sets of low instance complexity in terms of reductions to tally sets; it follows that if P $\neq$ NP, then no set of low instance complexity can be complete for NP with respect to disjunctive reductions or conjunctive reductions.

---

*Department of Computer Science and Engineering, Indian Institute of Technology, Delhi, New Delhi 110016, India. Work done while visiting Universität Ulm. Supported in part by an Alexander von Humboldt postdoctoral research fellowship.

†Department of Computer Science, University of Rochester, Rochester, NY 14627, USA. Supported in part by the National Science Foundation under research grant CCR-8957604.

‡Abteilung für Theoretische Informatik, Universität Ulm, Oberer Eselsberg, D-W-7900 Ulm, Germany. Supported in part by the DAAD through Acciones Integradas 1991, 313-AI-e-es/zk.

§Department of Software (L.S.I.), Universitat Politècnica de Catalunya, Pau Gargallo 5, E-08028 Barcelona, Spain. Work supported in part by ESPRIT-II Basic Research Actions Program of the EC under Contract No. 3075 (project ALCOM) and by the DAAD through Acciones Integradas 1991, 313-AI-e-es/zk.

¶Department of Computer Science and Information Mathematics, University of Electro-Communications, 1-5-1, Chofugaoka, Chofu-si, Tokyo 182, Japan. Work done in part while visiting SUNY–Buffalo. Supported in part by the National Science Foundation under research grant CCR-9002292.

// Dipartimento Di Matematica, Università degli Studi di Roma, "La Sapienza," 00198 Rome, Italy. Work done in part while visiting the University of Rochester. Supported in part by Ministero della Pubblica Istruzione through "Progetto 40%: Algoritmi, Modelli di Calcolo e Strutture Informative."

**Abteilung für Theoretische Informatik, Universität Ulm, Oberer Eselsberg, D-W-7900 Ulm, Germany. Work done in part while visiting the University of Rochester. Supported in part by a DFG Post-doctoral Stipend and by the DAAD through Acciones Integradas 1991, 313-AI-e-es/zk.

Overview

This paper is concerned with three basic questions about sparse sets, and a related question about sets of low instance complexity:

Question 1 With respect to what types of reductions might NP have hard or complete sparse sets?[1]

Question 2 If a set A reduces to a sparse set, does it follow that A is reducible to some sparse set that is "simple" relative to A?

Question 3 With respect to what types of reductions might NP have hard or complete sets of low instance complexity, and, relatedly, what is the structure of the class of sets with low instance complexity?

With respect to the first and third questions, intuitively one would expect that even with respect to flexible reductions NP is unlikely to have complete sets whose information content is low. With respect to the second question, one might intuitively feel that the structure imposed on a set by the fact that it reduces to a sparse set makes it plausible that we can indeed find a simple sparse set that can masquerade as the original sparse set. These two intuitions are in many ways certified by the current literature, and by the results of this paper.

The rest of this section summarizes the results of this paper and compares them with previous work.

With regard to Question 1:

- We show that if any NP-complete set conjunctively reduces to a sparse set,[2] then P = NP. This result, which has been obtained independently by Ranjan and Rohatgi [RR], extends Mahaney's Theorem [Mah82] and is incomparable with the strongest previously known result, which is due to Ogiwara and Watanabe: if any NP-complete set bounded truth-table reduces to a sparse set then P = NP [OW91].

  We obtain the above result as a corollary to a more general result we establish regarding the impossibility (if P $\neq$ NP) of NP-complete sets reducing to sparse sets, even via quite flexible reductions.

- One might ask whether in the above-mentioned result of Ogiwara and Watanabe the bounded truth-table case is optimal, or whether it can be

[1] For the reductions we will discuss, the question of sparse hard sets is equivalent to asking what types of reductions might reduce many-one complete sets for NP to some sparse set; we will often use this formulation.

[2] A conjunctive reduction from A to B means that there is a Turing machine with oracle B that accepts A, and the Turing machine's acceptance mechanism is that it accepts if and only if every string it queries is a member of B [LLS75].

extended by making the bound on the number of queries bigger than constan for example, to any function that is $\omega(\log n)$. We show that there are relativize worlds, having tally NP-complete sets with respect to such reductions, in whic the boolean hierarchy is infinite. This provides relativized upper bounds t the work of Ko, Orponen, Schöning, and Watanabe [KOSW86,Orp90,OKSW of Ogiwara and Watanabe, and of the current paper. Our result strengthen the results of Homer and Longpré [HL91], who independently of the work c this paper showed that there are relativized worlds in which there are spars sets that are NP-complete with respect to such bounds and yet (relativized P $\neq$ NP.

With regard to Question 2:

- In the context of recent comparisons between equivalence and reducibility to sparse sets [AHOW,GW91], it is interesting to know for various reductio classes to sparse sets how easy a sparse set can be relative to the set that i reduced to it. We show that any set A that disjunctively reduces (respectively disjunctive bounded truth-table reduces, 2-truth-table reduces) to a sparse set in fact disjunctively reduces (respectively, disjunctive bounded truth-table reduces, 2-truth-table reduces) to a sparse set that is in $\mathrm{P}^{\mathrm{NP}^A}$ (respectively $\mathrm{P}^{\mathrm{NP}^A[\log]}$ (sic), $\mathrm{P}^{\mathrm{NP}^A[\log]}$ (sic)).[3]

With regard to Question 3:

- We completely characterize the sets of low instance complexity (that is, the class IC[log, poly] [KOSW86,Orp90,OKSW]) in terms of reductions to tally sets: IC[log, poly] is exactly the class of sets that both disjunctively and conjunctively reduce to tally sets.
- We show that Orponen's result [Orp90] on 1-truth-table-complete sets for NP applies to disjunctive and conjunctive reductions: If P $\neq$ NP and A is $\leq_c^p$-hard or $\leq_d^p$-hard for NP, then $A \notin$ IC[log, poly].

In this extended abstract all proofs are omitted. Complete versions of this paper are available as technical reports of the University of Ulm (Ulmer Informatik-Bericht Nr. 91-08) and of the University of Rochester (in preparation).

2 Notation

A set T is said to be a *tally* set if $T \subseteq 0^*$. A set S is said to be *sparse* if it has at most polynomially many elements at each length: S is sparse if and only if for some

[3] The [log] means that there is an $\mathcal{O}(\log n)$ bound on the number of calls made to the NP^A oracle (see [Wag90]).

Name	Notation
k disjunctive (truth-table or Turing)	$\leq^p_{k\text{-}d}$
bounded (truth-table or Turing)	$\leq^p_b$
$f(n)$ truth-table	$\leq^p_{f(n)\text{-}tt}$
conjunctive (truth-table or Turing)	$\leq^p_c$
disjunctive (truth-table or Turing)	$\leq^p_d$
bounded conjunctive (truth-table or Turing)	$\leq^p_{bc}$
bounded disjunctive (truth-table or Turing)	$\leq^p_{bd}$

Table 1: Reductions

polynomial p it holds that $(\forall n)[||S \cap \Sigma^n|| \leq p(n)]$. We use TALLY and SPARSE to represent, respectively, the classes of tally and sparse sets. Tally and sparse sets have come to play a large role in modern complexity theory (see, e.g., the surveys [Mah86, Mah89,HOW]).

The reductions discussed in this paper are polynomial-bounded reductions defined by Ladner, Lynch, and Selman [LLS75] (see Table 1).

We will use the following notation to describe downward closures of classes under various reductions; the interrelations among R^p_r(SPARSE) classes have been systematically studied by Book and Ko [BK88] and Ko [Ko89].

Notation 2.1 [BK88,AHOW] For any reducibility $\leq^p_r$ and any class of sets $\mathcal{C}$, let $\mathrm{R}^p_r(\mathcal{C}) = \{A \mid (\exists B \in \mathcal{C})[A \leq^p_r B]\}$.

3 Sets Reducing to Sparse Sets

The study of sparse complete sets was started by the conjecture of L. Berman and J. Hartmanis [BH77] that there are no sparse NP-complete sets; they were motivated to make this conjecture since, if it fails, then not all NP-complete sets are polynomial-time isomorphic (and at that time they conjectured that all NP-complete sets were polynomial-time isomorphic, though recent work has dimmed hopes on that issue [JY85,KMR89]).

The first result along the lines of their sparseness conjecture was P. Berman's proof that P = NP if some subset of 0^* is NP-complete [Ber78]. This result was quickly followed by Fortune's proof that if there is a sparse coNP-complete set, then P = NP [For79]. Finally, Mahaney showed that P = NP if any NP-complete set many-one reduces to a sparse set [Mah82].

Although Mahaney obtained the complete collapse of the polynomial hierarchy in the case of many-one reducibility, possible collapses in the case of more flexible reducibilities have remained an active research area. For the case of Turing reductions, it is known that the existence of sparse Turing-complete sets for NP would collapse the polynomial hierarchy to $\mathrm{P}^{\mathrm{NP}[\log]}$ [Kad89], and the existence of sparse Turing-hard sets for NP would collapse the polynomial hierarchy to Σ_2^p [KL80]; both of these results are known to be essentially optimal with respect to relativizable proof techniques [Kad89,Hel86].

As just noted, for the cases of many-one and Turing reductions, the consequences of sparse NP-complete sets are well-understood. However, with respect to the reductions whose strength lies between Turing and many-one reductions, the question of extending Mahaney's many-one result has proved considerably more challenging. For the case of bounded truth-table reductions, Ukkonen [Ukk83] generalized Berman's result [Ber78] by showing that if there is a tally bounded truth-table hard set for NP, then P = NP. Yesha [Yes83] generalized Fortune's result [For79] by showing that if there is a sparse bounded positive truth-table hard set for coNP, then P = NP. Yesha also (partially) generalized Mahaney's Theorem [Mah82] by showing that if there is a sparse bounded positive truth-table complete set for NP, then P = NP. These results regarding bounded truth-table reductions have been recently subsumed by Ogiwara and Watanabe ([OW91], see also [Wat88] and [JY90, footnote 9]), who showed that if there is a sparse bounded truth-table hard set for NP, then P = NP.

For the case of conjunctive reductions, Ukkonen [Ukk83] and Yap [Yap83] generalized Fortune's result [For79] by showing that if there is a sparse conjunctive hard set for coNP, then P = NP. Yap [Yap83] also (partially) generalized Mahaney's Theorem [Mah82] by showing that if there is a sparse set that is both conjunctive and disjunctive complete for NP, then P = NP. However, in the decade since Mahaney's Theorem, it has remained an open question whether his result can be extended to the case of conjunctive reductions.

We show that if there is a sparse set that is conjunctive hard for NP, then P = NP (Corollary 3.4). This result is incomparable to the result of [OW91] on bounded truth-table reductions to sparse sets. In fact, we prove that $\mathrm{NP} \subseteq \mathrm{R}^p_{bc}(\mathrm{R}^p_{1\text{-}tt}(\mathrm{R}^p_c(\mathrm{SPARSE})))$ implies P = NP. (See Figure 1; note that $\mathrm{R}^p_b(\mathrm{R}^p_c(\mathrm{SPARSE})) = \mathrm{R}^p_{bd}(\mathrm{R}^p_{bc}(\mathrm{R}^p_{1\text{-}tt}(\mathrm{R}^p_c(\mathrm{SPARSE}))))$.

Definition 3.1 [OW91] For a set A in NP, let $W \in \mathrm{P}$ and q be a polynomial such that $A = \{\, x \mid (\exists\, w \in \Sigma^{q(|x|)})\, [\langle x, w\rangle \in W]\,\}$. For $x \in A$, let $w_{max}(x) = \max\{\, w \in \Sigma^{q(|x|)} \mid \langle x, w\rangle \in W\,\}$. Then $Left(A) = \{\, \langle x, w\rangle \mid x \in A,\ w \in \Sigma^{q(|x|)}$ and $w \leq$

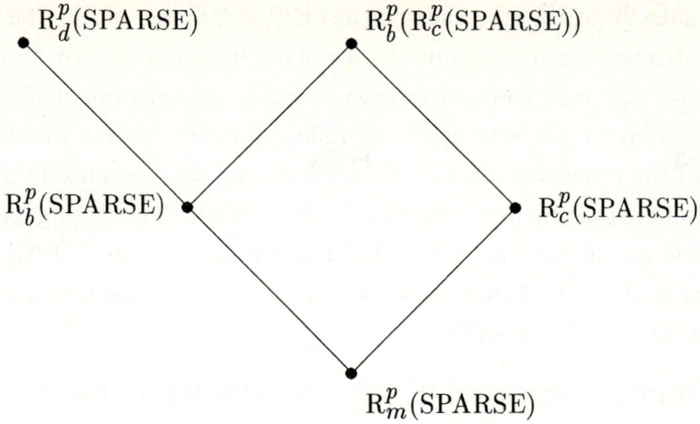

Figure 1: Inclusion structure of some reduction classes to sparse sets (see [Ko89, AHOW]). All inclusions indicated are proper [BK88,Ko89,Gav92].

$w_{max}(x)\}$ is the *left set of* A.[4]

Theorem 3.2 If $A \in$ NP and $Left(A) \in R^p_{bc}(R^p_{1\text{-}tt}(R^p_c(\text{SPARSE})))$, then $A \in$ P.

Corollary 3.3 If $A \in$ NP and $Left(A) \in R^p_c(\text{SPARSE})$, then $A \in$ P.

Of course, from Corollary 3.3, it immediately follows that Mahaney's Theorem generalizes to the case of conjunctive reductions. Corollary 3.4 has been obtained independently by Ranjan and Rohatgi [RR].

Corollary 3.4 If any NP-complete set conjunctively reduces to a sparse set, then P = NP.

Indeed, $Left(A) \leq^p_m A$ for any NP-complete set A, and thus the above theorems apply to the case when some NP-complete (or NP-hard) set reduces (via the reductions named above) to some sparse set.

Recently, Buhrman, Longpré, and Spaan [BLS92] have proved the surprising result that SPARSE $\subseteq R^p_c(\text{TALLY})$, which implies that $R^p_c(\text{SPARSE}) = R^p_c(\text{TALLY})$ and $R^p_b(R^p_c(\text{SPARSE})) = R^p_b(R^p_c(\text{TALLY}))$. Using this result, it can be shown [AKM] that if NP $\subseteq R^p_b(R^p_c(\text{SPARSE}))$ then P = NP, which unifies Theorem 3.2 with the result of Ogiwara and Watanabe [OW91].

We now discuss applications of the above results to the classes UP, PP, and $C_=P$.

Since for every set $A \in$ UP it holds that $Left(A)$ is in UP, it also follows that if UP is contained in $R^p_{bc}(R^p_{1\text{-}tt}(R^p_c(\text{SPARSE})))$ then UP = P. This strengthens the

[4]The left set tacitly depends on the particular witness relation chosen.

results of Watanabe [Wat91], who showed that if P $\neq$ UP then there exists a set in UP that does not many-one polynomial-time reduce to any sparse set.

Consider the set $\{\langle x, m\rangle \mid$ there are at least m satisfying assignments for $x\}$, which has properties similar to left sets and is complete for PP. Under the assumption that this set is in the reduction class of Theorem 3.2 we can compute in polynomial time a set of numbers that includes #SAT(x), the number of satisfying assignments of formula x. Now we can use the result of Cai and Hemachandra [CH91] and Toda (see [ABG90]) that P = PP if there is an FP function that computes on input x a set of numbers that includes #SAT(x).

Theorem 3.5 If PP is contained in $\mathrm{R}^p_{bc}(\mathrm{R}^p_{1\text{-}tt}(\mathrm{R}^p_c(\mathrm{SPARSE})))$ then P = PP.

By the fact that there exist complete sets in $\mathrm{C_{=}P}$ that are one word decreasing self-reducible [OL], we can show that if $\mathrm{C_{=}P}$ is contained in $\mathrm{R}^p_{bc}(\mathrm{R}^p_{1\text{-}tt}(\mathrm{R}^p_c(\mathrm{SPARSE})))$ then $\mathrm{C_{=}P} = \mathrm{P}$. In [AKM] it is shown that for $K \in \{\mathrm{UP}, \mathrm{PP}, \mathrm{C_{=}P}\}$, the assumption that K is contained in $\mathrm{R}^p_b(\mathrm{R}^p_c(\mathrm{SPARSE}))$ implies $\mathrm{P} = K$.

By applying a different proof technique we obtain results similar to Corollary 3.4 for the classes $\mathrm{Mod}_k\mathrm{P}$, $k \geq 2$ [CH90,BGH90].

Definition 3.6 A set L is *perfectly one word decreasing self-reducible* if there exist a function $f \in \mathrm{FP}$ and polynomials p and q such that

1. $L \subseteq \{\langle x, y, i\rangle \mid |y| = p(|x|) \text{ and } 1 \leq i \leq q(|x|)\}$,
2. $(\forall x, i : 1 \leq i \leq q(|x|))[f(\langle x, 0^{p(|x|)}, i\rangle) = \chi_L(\langle x, 0^{p(|x|)}, i\rangle)]$, and
3. $(\forall x)(\forall y \in \Sigma^{p(|x|)} - \{0^{p(|x|)}\})(\exists k)(\forall i : 1 \leq i \leq q(|x|))[f(\langle x, y, i\rangle) = \langle x, \mathit{pred}(y), (i+k) \bmod q(|x|)\rangle$ and $(\langle x, y, i\rangle \in L$ iff $f(\langle x, y, i\rangle) \in L)]$.

Theorem 3.7 If a perfectly one word decreasing self-reducible set L is conjunctively reducible in polynomial time to a sparse set then L is in P.

Since there exist complete sets in the classes $\mathrm{Mod}_k\mathrm{P}, k \geq 2$, that are perfectly one word decreasing self-reducible (in fact, these are essentially the strictly one word decreasing self-reducible complete sets for $\mathrm{Mod}_k\mathrm{P}$ that are defined in [OL]), we have the following corollary.

Corollary 3.8 If $k > 1$ and some $\mathrm{Mod}_k\mathrm{P}$-complete set polynomial-time conjunctively reduces to a sparse set, then $\mathrm{Mod}_k\mathrm{P} = \mathrm{P}$.

For nearly near-testable sets [HH91], the class of sets that have "implicit" polynomial-time membership tests, a similar result holds.

Theorem 3.9 If a nearly near testable set is conjunctively reducible in polynomial time to a sparse set then it is in P.

Finally, we show that the "P = NP if for some k there is a sparse k-tt hard set for NP" result of Ogiwara and Watanabe [OW91] cannot be improved to $\omega(\log n)$-bounded truth-table reductions by any relativizable proof technique, even with respect to the weakened conclusion that the boolean hierarchy [CGH+88] collapses. This extends work of Longpré and Homer (done independently of this work) that shows that for the same truth-table bounds, P = NP results are precluded in some relativized worlds [HL91] (see also [IM89,CGH+89,Kad89] for earlier results on oracles that limit the extent of the collapse one gets from assumptions of NP having sparse hard sets[5]).

Theorem 3.10 For each non-increasing, polynomial-time computable function $f(n) = \omega(\log n)$, there are sets A and T such that T is a tally set that is $\leq_{f(n)\text{-}tt}^{p,A}$-complete for NP^A, and yet the boolean hierarchy relative to A does not collapse.

4 Reductions to Simple Sparse Sets

In this section, we explore the second question mentioned in the introduction:

> If a set A reduces to a sparse set, does it follow that A is reducible to some sparse set that is "simple" relative to A?

Earlier work along these lines has been done both for the case (which is also the case of this paper) of reductions less flexible than Turing reductions [AHOW] and for the case of Turing reductions [GW91]. (Note that Theorems 4.2 and 4.3 are incomparable.)

Theorem 4.1 If $A \leq_{bd}^{p} S$ for some sparse set S, then there is a sparse set $\widehat{S}$ such that $A \leq_{bd}^{p} \widehat{S}$ and $\widehat{S} \in \mathrm{P}^{\mathrm{NP}^A[\log]}$.

Theorem 4.2 If $A \leq_{2\text{-}tt}^{p} S$ for some sparse set S, then there is a sparse set $\widehat{S}$ such that $A \leq_{2\text{-}tt}^{p} \widehat{S}$ and $\widehat{S} \in \mathrm{P}^{\mathrm{NP}^A[\log]}$.

Theorem 4.3 For $k = 2$ and $k = 3$: If $A \leq_{k\text{-}d}^{p} S$ for some sparse set S, then there is a sparse set $\widehat{S}$ such that $A \leq_{k\text{-}d} \widehat{S}$ and $\widehat{S} \in \mathrm{P}^{\mathrm{NP} \oplus A}$.

Also, we have the following results for unbounded disjunctive and unbounded conjunctive reductions.

Theorem 4.4 If $A \leq_{d}^{p} S$ for some sparse set S, then there is a sparse set $\widehat{S}$ such that $A \leq_{d}^{p} \widehat{S}$ and $\widehat{S} \in \mathrm{P}^{\mathrm{NP}^A}$.

[5] Kadin's [Kad89] Turing-hardness relativized claim in fact applies equally well to the truth-table case.

Theorem 4.5 If $A \leq_c^p S$ for some sparse set S, then there is a sparse set $\widehat{S}$ suc that $A \leq_c^p \widehat{S}$ and $\widehat{S} \in \mathrm{NP}^A$.

5 Low Instance Complexity and Polynomial-Tim Reductions

Instance complexity, as defined by Ko, Orponen, Schöning, and Watanab [KOSW86,Orp90,OKSW], is a notion of the complexity of specific instances of problem—a topic that standard complexity theory is ill-suited to study (as an single instance is trivial). In this paper, we are primarily concerned with sets c "low" instance complexity: IC[log,poly] (introduced in [KOSW86]).

We show that the sets of low instance complexity are intimately related to th study of reductions to sets of low information content. In particular, we sho that a set A is of low instance complexity if and only if it both conjunctively an disjunctively reduces to tally sets.

Theorem 5.1 A is in IC[log,poly] if and only if there exist tally sets T_0 and T such that $A \leq_c^p T_0$ and $A \leq_d^p T_1$.

Using the above characterization of IC[log,poly] it is easy to see that IC[log,poly is closed under bounded truth-table reductions (a different proof that explicitl constructs programs appears in [OKSW]):

Theorem 5.2 [OKSW] IC[log,poly] is closed downward under $\leq_b^p$-reductions.

Since IC[log,poly] $\subseteq \mathrm{R}_c^p$(SPARSE), the following corollary is an easy consequenc of the above theorem and Corollary 3.4.

Corollary 5.3 [OKSW] If $\mathrm{P} \neq \mathrm{NP}$ and A is $\leq_b^p$-hard for NP, then $A \notin$ IC[log, poly].

Using Theorem 5.1 and the fact that NP has neither $\leq_c^p$-hard nor $\leq_d^p$-hard tall sets unless $\mathrm{P} = \mathrm{NP}$, it follows that NP has neither $\leq_c^p$-hard nor $\leq_d^p$-hard sets in IC[log,poly] unless $\mathrm{P} = \mathrm{NP}$.

Corollary 5.4 If $\mathrm{P} \neq \mathrm{NP}$ and A is $\leq_d^p$-hard or $\leq_c^p$-hard for NP, then $A \notin$ IC[log, poly].

Finally, using Theorem 3.10 and the fact that every tally set is in IC[log, poly], we conclude that for truth-tables of size $\omega(\log n)$, no analog of Corollary 5.3 can be roved by any relativizable proof technique.

Acknowledgments

or helpful comments, conversations, and suggestions, on this work, and on the udy of kernel constructibility and committability that led us to this work, we are rateful to E. Allender, L. Fortnow, W. Gasarch, A. Hoene, R. Schuler, A. Selman, . Torán, and O. Watanabe. We thank J. Balcázar, M. Hermo, and R. Schuler for iscussions from which Theorem 5.1 resulted.

References

BG90] A. Amir, R. Beigel, and W. Gasarch. Some connections between bounded query classes and non-uniform complexity. In *Proceedings of the 5th Structure in Complexity Theory Conference*, pages 232–243. IEEE Computer Society Press, July 1990.

AHOW] E. Allender, L. Hemachandra, M. Ogiwara, and O. Watanabe. Relating equivalence and reducibility to sparse sets. *SIAM Journal on Computing*. To appear. Preliminary version appears as [AHOW91].

AHOW91] E. Allender, L. Hemachandra, M. Ogiwara, and O. Watanabe. Relating equivalence and reducibility to sparse sets. In *Proceedings of the 6th Structure in Complexity Theory Conference*, pages 220–229. IEEE Computer Society Press, June/July 1991.

AKM] V. Arvind, J. Köbler, and M. Mundhenk. Bounded truth-table and conjunctive reductions to sparse and tally sets. In preparation.

Ber78] P. Berman. Relationship between density and deterministic complexity of NP-complete languages. In *Proceedings of the 5th International Colloquium on Automata, Languages, and Programming*, pages 63–71. Springer-Verlag *Lecture Notes in Computer Science #62*, 1978.

BGH90] R. Beigel, J. Gill, and U. Hertrampf. Counting classes: Thresholds, parity, mods, and fewness. In *Proceedings of the 7th Annual Symposium on Theoretical Aspects of Computer Science*, pages 49–57. Springer-Verlag *Lecture Notes in Computer Science #415*, February 1990.

BH77] L. Berman and J. Hartmanis. On isomorphisms and density of NP and other complete sets. *SIAM Journal on Computing*, 6(2):305–322, 1977.

BK88] R. Book and K. Ko. On sets truth-table reducible to sparse sets. *SIAM Journal on Computing*, 17(5):903–919, 1988.

BLS92] H. Buhrman, L. Longpré, and E. Spaan, 1992. Personal Communication.

CGH+88] J. Cai, T. Gundermann, J. Hartmanis, L. Hemachandra, V. Sewelson, K. Wagner, and G. Wechsung. The boolean hierarchy I: Structural properties. *SIAM Journal on Computing*, 17(6):1232–1252, 1988.

CGH+89] J. Cai, T. Gundermann, J. Hartmanis, L. Hemachandra, V. Sewelson, K. Wagner, and G. Wechsung. The boolean hierarchy II: Applications. *SIAM Journal on Computing*, 18(1):95–111, 1989.

CH90] J. Cai and L. Hemachandra. On the power of parity polynomial time. *Mathematical Systems Theory*, 23(2):95–106, 1990.

[CH91] J. Cai and L. Hemachandra. A note on enumerative counting. *Informatio Processing Letters*, 38(4):215–219, 1991.

[For79] S. Fortune. A note on sparse complete sets. *SIAM Journal on Computin* 8(3):431–433, 1979.

[Gav92] R. Gavaldà. On conjunctive and disjunctive reductions to sparse sets Manuscript, January 1992.

[GW91] R. Gavaldà and O. Watanabe. On the computational complexity of sma descriptions. In *Proceedings of the 6th Structure in Complexity Theor Conference*, pages 89–101. IEEE Computer Society Press, June/July 1991.

[Hel86] H. Heller. On relativized exponential and probabilistic complexity classes *Information and Control*, 71:231–243, 1986.

[HH91] L. Hemachandra and A. Hoene. On sets with efficient implicit membershi tests. *SIAM Journal on Computing*, 20(6):1148–1156, 1991.

[HL91] S. Homer and L. Longpré. On reductions of NP sets to sparse sets. In *Proceedings of the 6th Structure in Complexity Theory Conference*, pages 79–88. IEEE Computer Society Press, June/July 1991.

[HOW] L. Hemachandra, M. Ogiwara, and O. Watanabe. How hard are sparse sets? In preparation (will appear in the Proceedings of the 7th Structure in Complexity Theory Conference).

[IM89] N. Immerman and S. Mahaney. Relativizing relativized computations. *Theoretical Computer Science*, 68:267–276, 1989.

[JY85] D. Joseph and P. Young. Some remarks on witness functions for non-polynomial and non-complete sets in NP. *Theoretical Computer Science* 39:225–237, 1985.

[JY90] D. Joseph and P. Young. Self-reducibility: Effects of internal structure on computational complexity. In A. Selman, editor, *Complexity Theory Retrospective*, pages 82–107. Springer-Verlag, 1990.

[Kad89] J. Kadin. $P^{NP[\log n]}$ and sparse Turing-complete sets for NP. *Journal of Computer and System Sciences*, 39(3):282–298, 1989.

[KL80] R. Karp and R. Lipton. Some connections between nonuniform and uniform complexity classes. In *Proceedings of the 12th ACM Symposium on Theory of Computing*, pages 302–309, April 1980.

[KMR89] S. Kurtz, S. Mahaney, and J. Royer. The isomorphism conjecture fails relative to a random oracle. In *Proceedings of the 21st ACM Symposium on Theory of Computing*, pages 157–166. ACM Press, May 1989.

[Ko89] K. Ko. Distinguishing conjunctive and disjunctive reducibilities by sparse sets. *Information and Computation*, 81(1):62–87, 1989.

[KOSW86] K. Ko, P. Orponen, U. Schöning, and O. Watanabe. What is a hard instance of a computational problem? In *Proceedings of the 1st Structure in Complexity Theory Conference*, pages 197–217. Springer-Verlag *Lecture Notes in Computer Science #223*, June 1986.

[LLS75] R. Ladner, N. Lynch, and A. Selman. A comparison of polynomial time reducibilities. *Theoretical Computer Science*, 1(2):103–124, 1975.

[Mah82] S. Mahaney. Sparse complete sets for NP: Solution of a conjecture of Berman and Hartmanis. *Journal of Computer and System Sciences*, 25(2):130–143, 1982.

[Mah86] S. Mahaney. Sparse sets and reducibilities. In R. Book, editor, *Studies in Complexity Theory*, pages 63–118. John Wiley and Sons, 1986.

[Mah89] S. Mahaney. The isomorphism conjecture and sparse sets. In J. Hartmanis, editor, *Computational Complexity Theory*, pages 18–46. American Mathematical Society, 1989. Proceedings of Symposia in Applied Mathematics #38.

[OKSW] P. Orponen, K. Ko, U. Schöning, and O. Watanabe. Instance complexity. *Journal of the ACM*. To appear.

[OL] M. Ogiwara and A. Lozano. On one-query self-reducible sets. *Theoretical Computer Science*. To appear. Preliminary version appears as [OL91].

[OL91] M. Ogiwara and A. Lozano. On one-query self-reducible sets. In *Proceedings of the 6th Structure in Complexity Theory Conference*, pages 139–151. IEEE Computer Society Press, June/July 1991.

[Orp90] P. Orponen. On the instance complexity of NP-hard problems. In *Proceedings of the 5th Structure in Complexity Theory Conference*, pages 20–27. IEEE Computer Society Press, July 1990.

[OW91] M. Ogiwara and O. Watanabe. On polynomial-time bounded truth-table reducibility of NP sets to sparse sets. *SIAM Journal on Computing*, 20(3):471–483, June 1991.

[RR] D. Ranjan and P. Rohatgi. On randomized reductions to sparse sets. In *Proceedings of the 7th Structure in Complexity Theory Conference*. IEEE Computer Society Press. To appear.

[Ukk83] E. Ukkonen. Two results on polynomial time truth-table reductions to sparse sets. *SIAM Journal on Computing*, 12(3):580–587, 1983.

[Wag90] K. Wagner. Bounded query classes. *SIAM Journal on Computing*, 19(5):833–846, 1990.

[Wat88] O. Watanabe. On $\leq^p_{1-tt}$ sparseness and nondeterministic complexity classes. In *Proceedings of the 15th International Colloquium on Automata, Languages, and Programming*, pages 697–709. Springer-Verlag *Lecture Notes in Computer Science #317*, July 1988.

[Wat91] O. Watanabe. On intractability of the class UP. *Mathematical Systems Theory*, 24:1–10, 1991.

[Yap83] C. Yap. Some consequences of non-uniform conditions on uniform classes. *Theoretical Computer Science*, 26:287–300, 1983.

[Yes83] Y. Yesha. On certain polynomial-time truth-table reducibilities of complete sets to sparse sets. *SIAM Journal on Computing*, 12(3):411–425, 1983.

UP and the Low and High Hierarchies: A Relativized Separation [*†]

Ming-Jye Sheu and Timothy J. Long[‡]

sheu-m@cis.ohio-state.edu, long@nmsu.edu
Department of Computer and Information Science
The Ohio State University
Columbus, Ohio 43210

Abstract. The low and high hierarchies within NP were introduced by Schöning in order to classify sets in NP. It is not known whether the low and high hierarchies include all sets in NP. In this paper, we provide evidence that the class UP may fall in the gap between the low and high hierarchies. Using the circuit lower bound techniques of Håstad and Ko, we construct an oracle set relative to which UP is not in any level of the low and high hierarchies. Since it is known that UP^A is low for PP^A for all sets A, our result also shows that the interaction between UP and PP is crucial for the lowness of UP for PP; changing the base class to any level of the polynomial-time hierarchy destroys the lowness of UP.

1 Introduction

The low and high hierarchies in NP were first introduced by Schöning in order to analyze internal structure of NP [Sch83]. A set $A \in$ NP is said to be in level k of the low hierarch $\Sigma_k^P(A) = \Sigma_k^P$; and a set $B \in$ NP is said to be in level k of the high hierarchy if $\Sigma_{k+1}^P = \Sigma_k^P($ Intuitively, when a set A is in level k of the low hierarchy, then the information content of with respect to the $\Sigma_k^P()$ operator, is so low that $\Sigma_k^P(A)$ has no more power than Σ_k^P. On other hand, when a set B is in level k of the high hierarchy, then the information content of with respect to the $\Sigma_k^P()$ operator, is so high that $\Sigma_k^P(B)$ includes all of Σ_{k+1}^P. The low and hi hierarchies provide a structure for classifying NP sets A based on the first level, of either hierarc to which A belongs. There are numerous examples of sets in the low hierarchy including spa sets in NP, NP sets with polynomial-time probabilistic algorithms, and NP sets with small s circuits. On the other hand, sets in the high hierarchy are complete sets for NP with respect

---

*This work was supported in part by NSF Grant CCR-8909071

†Part of this work was done while both authors were visitors in the the Department of Computer Science at N Mexico State University.

‡The second author is currently a visiting faculty member in the Department of Computer Science at New Mex State University.

ous polynomial-time reducibilities. For example, level 0 of the high hierarchy is the class of complete sets for NP and level 1 of the high hierarchy is the class of $\leq_T^{SN}$-complete sets for NP 83].

The low and high hierarchies demonstrate an interesting structure within NP: the low hierarchy ts at P and grows "upward" toward the complete sets for NP, while the high hierarchy starts complete sets for NP and grows "downward" toward P. One of the important open questions ut this structure is what are the sets in NP that are in neither the low hierarchy nor the high archy. In another words, what are the sets in NP that fall into the "gap" between the low and hierarchies? The existence of a gap between the low and high hierarchy is strongly connected he structure of the polynomial-time hierarchy. In [Sch83], Schöning showed that there exist between the low and high hierarchies if and only if the polynomial time hierarchy is infinite. wever, it is not clear what type of NP sets fall into the gap between the low and high hierarchy. The relation between the gap and the low and high hierarchies is not clear. Note that, if the ynomial time hierarchy collapse to level k, then the low and high hierarchies also collapse to l k. Thus, if there is no gap between the low and high hierarchies, which implies the polynomial rarchy collapes, then both the low and high hierarchy collapse. On the other hand, it is possible t there is a gap between the low and high hierarchies and the low and high hierarchies have er infinitely many or finitely many distinct levels.

The goal of this paper is to investigate whether the class UP includes sets that fall in the gap ween the low and high hierarchies for NP. Recall that UP is the class of sets that are recognized nondeterministic polynomial-time Turing machines that, for all inputs, have either zero or one epting computations. Not only is UP an important class in complexity theory, but it also has ong connections to fundamental theoretical issues in cryptography [GS88].

Because the high hierarchy for NP consists of sets that are complete for NP with respect to ious reducibilities, it seems very unlikely that any sets in UP would also belong to the high rarchy. In fact, there is an oracle relative to which UP is not in level 1 of the high hierarchy. In ng80], Angluin constructed an oracle set A such that $\Delta_2^{P,A} \not\subseteq \mathrm{PP}^A$. If there is a set $L \in \mathrm{UP}^A$ at is $\leq_T^{SN}$-complete for NP^A, then, $\Sigma_2^{P,A} \subseteq \mathrm{NP}^L \subseteq \mathrm{PP}^L \subseteq \mathrm{PP}^{\mathrm{UP}^A} = \mathrm{PP}^A$. This contradicts the t that $\Delta_2^{P,A} \not\subseteq \mathrm{PP}^A$. Note that $\mathrm{PP}^{\mathrm{UP}^A} = \mathrm{PP}^A$ follows from the fact that UP^A is low for PP^A for sets A. This is just a relativization of the recent result by Köbler, Schöning, Toda, and Toran at UP is low for PP[KSTT89]. We extend this result by constructing an oracle set A relative to ich UP is not in any level of the high hierarchy.

On the other hand, considering the fact that UP is low for PP, it is not entirely unreasonable think that UP might belong to the low hierarchy for NP; that is, that each set in UP is low for me level of the polynomial-time hierarchy. To the contrary, in this paper we give evidence that P is not in the low hierarchy by constructing a recursive oracle set A such that UP^A is not in the w hierarchy for NP^A.

Two very interesting observations can be made from our result. First, relative to the constructed oracle set A, UP^A contains sets that fall into the gap between the low and high hierarchies for N Second, since UP^A is not low for NP^A, or for any level of the polynomial-time hierarchy relativ A, and since UP^B is low for PP^B for all sets B, this shows that the lowness of UP for PP depe critically on the interaction between the base class PP and the oracle class UP; replacing the class with any level of the polynomial-time hierarchy destroys the lowness of UP.

Our construction of an oracle set A relative to which UP is not in any level of the high hierar uses an interesting result from Hartmanis and Hemachandra [HH87]. They show that if a nonde ministic polynomial time oracle machine N, relative to every oracle set and on every input str has at most one accepting computation, then N behaves like a deterministic polynomial time or machine with SAT added to the oracle set. Using this property, we are able to show that

1. $\exists A\ \forall k \geq 0\ [\Sigma_{k+1}^P(A) \not\subseteq \Sigma_k^P(\mathrm{UP}(A))]$.

Our construction of an oracle set A relative to which UP is not in any level of the low hierar follows the circuit lower bound techniques from Yao [Yao85], Håstad [Has87], and Ko [Ko Most importantly, we show that Håstad's switching lemma also holds for a new type of rand restriction: restrictions that satisfy the "U property", where a restriction satisfies the "U proper if it assigns at most one variable the value of 1 in each block of a partition of the variables. Us such restrictions, we are able to show that

2. $\exists A\ \forall k \geq 0\ [\Sigma_k^P(\mathrm{UP}(A)) \not\subseteq \Sigma_k^P(A)]$.

Some of the immediate corollaries include:

3. $\exists B\ [\mathrm{BPP}^{\mathrm{UP}(B)} \not\subseteq \mathrm{BPP}^B]$: UP is not low for BPP.
4. $\exists C\ [\#\mathrm{P}^{\mathrm{UP}(C)} \not\subseteq \#\mathrm{P}^C]$: UP is not low for #P.
5. $\exists D\ [\mathrm{UP}(D) \not\subseteq \mathrm{BPP}(D)]$: UP is not contained in BPP.
6. $\exists E\ [\mathrm{UP}(E) \not\subseteq \mathrm{P}/poly(E)]$: UP does not have polynomial size circuits.

Corollary 4 is particularly interesting since UP is known to be low for PP, which implies th UP is low for $\mathrm{P}^{\#\mathrm{P}} = \mathrm{P}^{\mathrm{PP}}$. Our results indicate that UP is unlikely to be low for #P.

2 Definitions and Preliminaries

Sets in the polynomial-time hierarchy relative to A have been characterized in terms of polynomi bounded quantification over predicates in $\mathrm{P}(A)$ [Sto76, Wra76]. We adopt definitions of Σ_k^P predicates from [Ko89]. The classes of the relativized polynomial-time hierarchy (PH(A)) can defined as follows:

$$\Sigma_k^P(A) = \{L \mid (\exists \Sigma_k^{P,1}\text{-predicate } \tau)[x \in L \Leftrightarrow \tau(A; x)]\},$$
$$\Pi_k^P(A) = \text{co-}\Sigma_k^P(A), \text{ for } k \geq 0.$$

The low and high hierarchies in NP were first introduced by Schöning [Sch83].

finition 2.1 *[Sch83]*

. *The low hierarchy in* NP *is* $\{L_k^{P,\Sigma} \mid k \geq 0\}$, *where* $L_k^{P,\Sigma} = \{A \in \mathrm{NP} \mid \Sigma_k^P(A) \subseteq \Sigma_k^P\}$.

. *The high hierarchy in* NP *is* $\{H_k^{P,\Sigma} \mid k \geq 0\}$, *where* $H_k^{P,\Sigma} = \{A \in \mathrm{NP} \mid \Sigma_{k+1}^P \subseteq \Sigma_k^P(A)\}$.

known that $L_k^{P,\Sigma} \subseteq L_{k+1}^{P,\Sigma}$ and $H_k^{P,\Sigma} \subseteq H_{k+1}^{P,\Sigma}$ for all $k \geq 0$.

UP is the class of sets that are acceptable by nondeterministic polynomial-time Turing machines t have, on all inputs, either zero or one accepting computations. The class FewP is the class of that are acceptable by nondeterministic polynomial-time Turing machines that have at most ynomially many accepting computations.

A circuit is a rooted tree, where each non-leaf node is labeled with a gate (AND or OR), and an unlimited number of children. Each leaf is labeled with a constant 0, a constant 1, a variable or a negated variable $\bar{x}$. The dual circuit $\hat{C}$ of a circuit C is defined inductively as follows: the l of a constant or a variable is its negation, the dual of a circuit C which is an OR (or, AND) children $C_i, 1 \leq i \leq n$, is the AND (or, OR) of $\hat{C}_i$, $1 \leq i \leq n$.

A circuit computes a function of its variables. Each variable is represented by v_z for some string Σ^*. Let V be the set of variables in a circuit C. A *restriction* ρ of C is a mapping from V to $1,*\}$. Let $C\lceil_\rho$ denote the circuit C' obtained from C by replacing each variable x with $\rho(x) = 0$ 0 and each y with $\rho(y) = 1$ by 1. For any set $A \subseteq \Sigma^*$, the natural assignment ρ_A on all variables $z \in \Sigma^*$, is: $\rho_A(v_z) = 1$ if $z \in A$ and $\rho_A(v_z) = 0$ if $z \notin A$.

Random restrictions of circuits were first introduced by Furst, Saxe and Sipser [FSS81]. Here use restrictions that are motivated by, but not identical to, the restrictions used by Håstad s87]. We define two probability spaces of restrictions $\hat{R}_{q,\mathcal{B}}^+$ and $\hat{R}_{q,\mathcal{B}}^-$. Let $\mathcal{B} = \{B_j\}_{j=1}^r$ be a tition of the variables in a circuit C and let q be a value between 0 and 1. To define a random triction ρ in $\hat{R}_{q,\mathcal{B}}^+$, for each variable $x \in B_j$, let $\rho(x) = *$ with probability q and let $\rho(x) = 1$ with bability $1-q$. The probability space $\hat{R}_{q,\mathcal{B}}^-$ is defined analogously by changing 1 to 0. Next, for h $\rho \in \hat{R}_{q,\mathcal{B}}^+$, define a restriction $g'(\rho)$ such that independently for each B_j, $1 \leq j \leq r$, let $s_j = *$ h probability q and let $s_j = 0$ with probability $1-q$. Then, let V_j be the set of all variables in which are given value $*$ by ρ. $g'(\rho)$ deterministically selects one variable y in V_j and gives value to y and value 1 to all other variables in V_j. For ρ from $\hat{R}_{q,\mathcal{B}}^-$, $g'(\rho)$ is defined analogously by erchanging the roles of 0 and 1.

One important property of the restriction $\rho g'(\rho)$ is that at most one variable is assigned * or 0 each block B_j, if ρ is a random restriction from $\hat{R}_{q,\mathcal{B}}^+$, and at most one variable is assigned * or n each block B_j, if ρ is a random restriction from $\hat{R}_{q,\mathcal{B}}^-$. We call this property the *U property*.

Circuits can be used to describe the computation of oracle machines on input strings. We call a pth-$(k+1)$ circuit a Σ_k-circuit if it has alternating OR and AND gates, starting with a top OR te. A $\Sigma_k(m)$-circuit is a Σ_k-circuit that (1) the number of gates at level 1 to level $k-1$ is bounded 2^m, and (2) fanins of level $k+1$ are $\leq m$. A depth-$(k+1)$ circuit is a Π_k-circuit ($\Pi_k(m)$-circuit) if it is the dual circuit of a Σ_k-circuit ($\Sigma_k(m)$-circuit). Now we can translate $\Sigma_k^{P,1}$-predicate circuits.

Proposition 2.2 *[FSS81, Ko89, SL92] Let $k \geq 1$. For every $\Sigma_k^{P,1}$-predicate τ there is a polyno q such that for every x, there exists a $\Sigma_k(q(|x|))$-circuit $C_{\tau,x}$ having the property that for any A, $C_{\tau,x}\lceil_{\rho_A}= 1$ if and only if $\tau(A;x)$ is true. Also, for each variable v_z in $C_{\tau,x}$, $|z| \leq q(|x|)$.*

Ko [Ko91] introduced a family of functions f_k^n, $n \geq 1, k \geq 1$, computed by special type circuits.

Definition 2.3 *[Ko91] For each $n \geq 1$ and $k \geq 1$, a C_k^n circuit is a depth-k circuit such that*

1. *C_k^n has alternating OR and AND gates, with a top OR gate,*
2. *all the fanins of C_k^n are exactly n, except the bottom fanins which are exactly $\sqrt{n}$, and*
3. *each leaf of C_k^n is a unique positive variable.*

We let f_k^n denote the function computed by C_k^n.

The switching lemma from Håstad [Has87] has been an important tool in proofs construct oracle sets that separate the polynomial-time hierarchy from PSPACE and separate every le of the polynomial-time hierarchy [Has87, Ko89]. The following lemma states that the switch lemma also holds for random restrictions with the *U property*.

Lemma 2.4 (Switching Lemma) *Let G be an* AND *of* OR*s with bottom fanin $\leq t$, let $\mathcal{B} = \{B$ be a partition of the variables in G, and let q be a number between 0 and 1. Then, for a rand restriction ρ from $\hat{R}^+_{q,\mathcal{B}}$, $Pr(G\lceil_{\rho g'(\rho)}$ is not equivalent to a circuit of* OR *of* AND*s with bottom fa $\leq s$) is bounded by α^s, where $\alpha < 6qt$. The above probability also holds with $\hat{R}^+_{q,\mathcal{B}}$ replaced by $\hat{R}^-_{q,\mathcal{B}}$ or with G being an* OR *of* AND*s circuit to be converted to an* AND *of* OR*s circuit.*

Lemma 2.5 is an application of the switching lemma. Lemma 2.6 shows that a random restricti with the *U property* transforms a circuit computing f_k^n to a circuit that contains a subcirc computing an f_{k-1}^n function.

Lemma 2.5 *For arbitrary $k \geq 2$, let C be a $\Sigma_k(r)$-circuit (or a $\Pi_k(r)$-circuit), where $r \leq t$. V be the set of the variables in C and let $\mathcal{B} = \{B_j\}$ be a partition of the variables in V. $q = 1/(12 \cdot t)$ and let ρ be a random restriction from $\hat{R}^+_{q,\mathcal{B}}$ or $\hat{R}^-_{q,\mathcal{B}}$. The probability that C, af applying the random restriction $\rho g'(\rho)$, can be converted to an equivalent $\Sigma_{k-1}(r)$-circuit (or $\Pi_{k-1}(r)$-circuit, respectively) is greater than 2/3.*

Lemma 2.6 *For each $k \geq 2$, there exists an integer n_k such that the following holds for all $n \geq n$ Let C be a C_k^n circuit computing an f_k^n function, let $q = n^{-1/3}$, and let ρ be a random restricti from $\hat{R}^+_{q,\mathcal{B}}$ if k is even or from $\hat{R}^-_{q,\mathcal{B}}$ if k is odd. The probability that $C\lceil_{\rho g'(\rho)}$ contains a subcirc computing an f_{k-1}^n function is greater than 2/3.*

UP is not in the low hierarchy

his section, we will construct an oracle set A relative to which UP is not in any level of the low rarchy in NP. That is, we show that there exist an oracle set A and a set $L_k \in \mathrm{UP}(A)$ for each 1 such that $\Sigma^P_{k-1}(L_k) \not\subseteq \Pi^P_k(A)$, implying that $\Sigma^P_k(L_k) \not\subseteq \Sigma^P_k(A)$. Therefore, UP($A$) is not in el k of the relativized low hierarchy for all $k \geq 1$.

Let $e(0) = 1$ and $e(n+1) = 2^{2^{e(n)}}$ for all $n > 0$. Then, for arbitrary set A, define sets U^A_k for h $k \geq 0$ as follows:

$$x \in U^A_k \Rightarrow x \text{ is of the form } 0^n w 0^j,\ 0 \leq j < k+2,\ |w| = kn,\ |x| = e(<2k,m>) - n$$
$$\text{for some } m, \text{ and there exists exactly one } y,\ |y| = n, \text{ such that } xy \in A,$$
$$x \notin U^A_k \Rightarrow x \text{ is not of the form } 0^n w 0^j,\ 0 \leq j < k+2,\ |w| = kn,\ |x| = e(<2k,m>) - n$$
$$\text{for some } m, \text{ or } x \text{ is of this form and for all } y,\ |y| = n,\ xy \notin A.$$

w for any fixed $k \geq 0$, when k is even define

$$L^A_k = \{0^n \mid (\exists m)(\exists j, 0 \leq j < k+2)[(k+2)n+j = e(<2k,m>)] \text{ and}$$
$$(\exists y_1, |y_1| = n)(\forall y_2, |y_2| = n)\cdots(\forall y_k, |y_k| = n)0^n y_1 y_2 \ldots y_k 0^j \in U^A_k\},$$

d when k is odd, define

$$L^A_k = \{0^n \mid (\exists m)(\exists j, 0 \leq j < k+2)[(k+2)n+j = e(<2k,m>)] \text{ and}$$
$$(\exists y_1, |y_1| = n)(\forall y_2, |y_2| = n)\cdots(\exists y_k, |y_k| = n)0^n y_1 y_2 \ldots y_k 0^j \notin U^A_k\}.$$

is clear that $L^A_k \in \Sigma^P_k(U^A_k)$ for $k > 0$ and $L^A_0 = U^A_0$.

Our goal is to construct an oracle set A so that $U^A_k \in \mathrm{UP}(A)$ and $L^A_k \notin \Pi^P_{k+1}(A)$ for all $k \geq 0$. is shows that, for all $k \geq 0$, $U^A_k \notin L^{P,\Sigma}_{k+1}$ relative to A and thus UP(A) is not contained in any el of the low hierarchy relative to A.

The construction of set A is by stages. Let $\sigma^{k+1}_1, \sigma^{k+1}_2, \cdots$ be an effective enumeration of $\Sigma^{P,1}_{k+1}$-edicates. We divide each goal $L^A_k \notin \Pi^P_{k+1}(A)$ into an infinite number of requirements:

$$R_{k,i} : (\exists n)[0^n \in L^A_k \Leftrightarrow \sigma^{k+1}_i(A; 0^n) \text{ is true and } U^A_k \in \mathrm{UP}(A)].$$

requirements $R_{k,i}$ are satisfied for all i, then $L^A_k \in \Sigma^P_k(\mathrm{UP}(A))$ and $L^A_k \notin \Pi^P_{k+1}(A)$.

For arbitrary fixed k, $k \geq 0$, let n and m be a pair of numbers such that $(\exists j, 0 \leq j < k+2)[(k+$ $n+j = e(<2k,m>)]$. Let C_0 be the following circuit: C_0 is a (k+1)-level circuit having alternating R and AND gates with a top OR gate. The fanin at each level is 2^n and the variables at the bottom el of C_0 are the variables of the form $0^n y_1 y_2 \ldots y_k 0^j y$, where $|y_1| = |y_2| = \cdots = |y_k| = |y| = n$. so, if k is even then the variables in C_0 are all positive and if k is odd then the variables in C_0 e all negative. In order to ensure that U^A_k is in UP(A), we need to place some restriction on e assignment of variables in C_0. Let circuit $C^v_{k'}$ compute an $f^v_{k'}$ function. Consider the following striction on assignments of variables appearing in $C^v_{k'}$:

The U condition for $C^v_{k'}$: let $G_1, G_2, \ldots$ be the bottom OR gates of $C^v_{k'}$ if k' is odd or the bottom AND gates of $C^v_{k'}$ if k' is even. Let $V_1, V_2, \ldots$ be the set of variables in $G_1, G_2, \ldots$, respectively. If k' is odd, then the assignment of variables in V_i, for all i, must have at most one variable assigned a 1. If k' is even, then the assignment of variables in V_i, for all i, must have at most one variable assigned a 0.

Note that, if k is even, the bottom OR gates of C_0 consist of unique positive variables, and contains a $C^{2^n}_{k+1}$ subcircuit. Thus, for an assignment of variables satisfying the U condition for C_0 correctly computes the predicate "$0^n \in L^A_k$" and U^A_k is in UP(A). When k is odd, the bott AND gates of C_0 consist of unique negative variables. For each negative variable $\bar{x}_j$ in C_0, replace it by a positive variable y_j. Let this new circuit be C'_0. C'_0 then contains a $C^{2^n}_{k+1}$ subcirc computing an $f^{2^n}_{k+1}$ function. It is easy to see that an assignment that satisfies the U condition C'_0 assigns value 1 to at most one variable x_j and assigns 0 to all other variables x_l. Therefore, any assignment satisfying the U condition for C'_0, C_0 correctly computes the predicate "$0^n \in L$ and U^A_k is in UP(A).

Let C_i be the $\Pi_{k+1}(p_i(n))$-circuit corresponding to the negation of the predicate $\sigma^{k+1}_i(A; 0$ Our goal is to find an assignment of strings to A and $\overline{A}$ corresponding to the U condition for C_0 k is even, or the U condition for C'_0, if k is odd, such that C_0 and C_i output different values un this assignment. The requirement $R_{k,i}$ can be rewritten as

$R'_{k,i} : (\exists x = 0^n)[C_0\lceil_{\rho_A} = 1 \Leftrightarrow C_i\lceil_{\rho_A} = 0]$, where ρ_A satisfies
the U condition for C_0 if k is even or for C'_0 if k is odd.

For each $k \geq 1$, let $s_k(v)$ be the minimum r such that some $\Pi_k(r)$-circuit can compute an function when that the assignment of variables in the circuit C^v_k for f^v_k satisfies the U condition C^v_k. The following lower bound on $s_k(v)$ implies that the requirement $R'_{k-1,i}$ can be satisfied.

Theorem 3.1 *For all $k \geq 1$, for sufficiently large v, $s_k(v) \geq \delta v^{1/3}$, where $\delta = 1/12$.*

Proof. The proof is by induction on k. The base of $k = 1$ is done by the traditional counti argument. Our inductive hypothesis is that for some fixed $k' \geq 1$, if C is a $\Pi_{k'}(r)$-circuit wi $r < \delta v^{1/3}$, then there is an assignment σ of variables in the circuit $C^v_{k'}$ computing $f^v_{k'}$ such that satisfies the U condition for $C^v_{k'}$ and such that $C^v_{k'}\lceil_\sigma \neq C\lceil_\sigma$.

For the induction step, let $k = k'+1 \geq 2$. Consider a C^v_k circuit that computes f^v_k. Let $q = v^{-}$ and let $\mathcal{B} = \{B_j\}$, where B_j is the set of variables of the bottom OR gate G_j of C^v_k if k is odd, is the set of variables of the bottom AND gate G_j of C^v_k if k is even.

Let C be a $\Pi_k(r)$-circuit with $r < \delta v^{1/3}$. In order to apply the inductive hypothesis, we ne to show that after applying a random restriction ρ from $\hat{R}^+_{q,\mathcal{B}}$, if k is even, or from $\hat{R}^-_{q,\mathcal{B}}$, if k is od to C^v_k and C, with high probability, $C_k^v\lceil_{\rho g'(\rho)}$ contains a subcircuit computing an f_{k-1}^v function, and with high probability, $C\lceil_{\rho g'(\rho)}$ is equivalent to a $\Pi_{k-1}(r)$-circuit.

(1) is proved by Lemma 2.6 and part (2) is proved by Lemma 2.5.

Now by the inductive hypothesis, there is an assignment σ such that $C_k^v\lceil_{\rho g'(\rho)\sigma} \neq C\lceil_{\rho g'(\rho)\sigma}$. It ains to show that the assignment $\rho g'(\rho)\sigma$ satisfies the U condition for C_k^v.

Consider the case when k is even. Recall that ρ is a restriction from $\hat{R}_{q,B}^+$. For each j, the iction $\rho g'(\rho)$ chooses at most one variable from B_j and assigns 0 or $*$ to it and assigns 1 to all r variables in B_j. If the variable in B_j was assigned $*$ by $\rho g'(\rho)$, then it will be assigned either 0 by σ. In either case, there is at most one variable in B_j being assigned 0 in the assignment $\rho)\sigma$. Thus the assignment $\rho g'(\rho)\sigma$ satisfies the U condition for C_k^v. The case when k is odd is ed similarly. Thus the assignment $\rho g'(\rho)\sigma$ satisfies the U condition for C_k^v. This completes the f of Theorem 3.1. □

Now we can start our construction of A. Initially, let $A(1) = \emptyset$ and let $\overline{A}(1) = \{x \mid |x| \neq$ $2k, m>), \forall k, m \geq 0\}$. In each stage α, we will define sets $A(\alpha+1)$ and $\overline{A}(\alpha+1)$ such that $+1) \cap \overline{A}(\alpha+1) = \emptyset$ and that $A(\alpha+1)$ is an extension of $A(\alpha)$ and $\overline{A}(\alpha+1)$ is an extension of).

At stage α, $\alpha = <2k+1, m>$, we do nothing and let $A(\alpha+1) = A(\alpha)$ and $\overline{A}(\alpha+1) = \overline{A}(\alpha)$. Go ext stage.

At stage α, $\alpha = <2k, m>$, we will try to satisfy $R'_{k,i}$ for the least i such that $R'_{k,i}$ is not yet sfied. Note that for each $k \geq 0$, our diagonalization area for $R'_{k,i}$ will be $\{x \mid |x| = e(<2k, m>)\}$ some m, since all strings not of length $e(n)$, $n = <2k, m>$ for all $k, m \geq 0$, have been assigned to y default. Thus, the diagonalization areas for $R'_{k,i}$ and $R'_{k',j}$ will be disjoint for $k \neq k'$ or $i \neq j$.

Let $n = \lfloor e(\alpha)/(k+2) \rfloor$ and $j = e(\alpha) - (k+2)n$. If n is such that either Theorem 3.1 does not hold $s_{k+1}(2^n)$ or $p_i(n) \geq 1/12 \cdot 2^{n/3}$, then let $A(\alpha+1) = A(\alpha)$ and $\overline{A}(\alpha+1) = \overline{A}(\alpha) \cup \{x \mid |x| = e(\alpha)\}$. to next stage.

Otherwise, let C_0 be the circuit corresponding to the predicate $0^n \in L_k^A$ and let C_i be the uit corresponding to the negation of the predicate $\sigma_i^{k+1}(A; 0^n)$. Replace all the variables v_y in such that $y \in A(\alpha) \cup \overline{A}(\alpha)$ with 1 if $y \in A(\alpha)$ and with 0 if $y \in \overline{A}(\alpha)$. The remaining circuit is a $\Pi_{k+1}(p_i(n))$ circuit having variables of length $e(\alpha)$. There are two cases to consider. In the e that k is even, C_0 contains a $C_{k+1}^{2^n}$ subcircuit computing an $f_{k+1}^{2^n}$ function. By Theorem 3.1, re is a restriction $\rho_{B,B'}$ that satisfies the U condition for C_0 and that $C_0\lceil_{\rho_{B,B'}} \neq C_i\lceil_{\rho_{B,B'}}$, where $B' = \emptyset$ and the restriction $\rho_{B,B'}$ assigns 1 to variables in B, 0 to variables in B', and * to all er variables. Let $A(\alpha+1) = A(\alpha) \cup B$ and $\overline{A}(\alpha+1) = \overline{A}(\alpha) \cup \{x \notin B \mid |x| = e(\alpha)\}$. Cancel i and to the next stage.

In the case that k is odd, we need to replace all of the negative variables $\bar{x}_j$ in C_0 with positive iables y_j. If $\bar{x}_j$ also appears in C_i, replace it there with y_j, and if x_j appears in C_i, replace it re with $\bar{y}_j$. The resulting circuit C'_0 contains a $C_{k+1}^{2^n}$ subcircuit computing an $f_{k+1}^{2^n}$ function. By

Theorem 3.1, there is a restriction $\rho_{B,B'}$ that satisfies the U condition for C_0' and that $C_0' \lceil_{\rho_B}$ $C_i \lceil_{\rho_{B,B'}}$. We let $A(\alpha+1) = A(\alpha) \cup \{y \mid y \text{ is not a replaced variable and } y \in B\} \cup \{x_j \mid y_j (= \bar{x}_j) \in$ and let $\overline{A}(\alpha+1) = \overline{A}(\alpha) \cup \{x \notin A(\alpha+1) \mid |x| = e(\alpha)\}$. Cancel i and go to the next stage. completes the description of the stages.

Now let $A = \cup_{\alpha=1}^{\infty} A(\alpha)$. We claim that all the requirements $R_{k,i}$ will eventually be satisfied see this, assume that i is the least i such that $R_{k,i}$ is not yet satisfied. Assume that each of $i' < i$, is satisfied by stage $\alpha = <2k, m>$, for some m. Then $R_{k,i}$ will be satisfied in the first s $\alpha' = <k, m'>$ such that $m' > m$, Theorem 3.1 holds for $s_{k+1}(2^n)$, and $p_i(n) < 1/12 \cdot 2^{n/3}$, w $n = \lfloor e(\alpha')/(k+2) \rfloor$.

Theorem 3.2 *There exists an oracle set A such that for all $k \geq 0$,* $\mathrm{UP}(A) \not\subseteq L_k^{P,\Sigma}(A)$.

An interesting consequence of Theorem 3.2 is that UP is not contained in the following cla APT, R, BPP, P/*poly*. This implies that UP is not low for any of these classes.

Corollary 3.3 *There exists an oracle set A relative to which* UP *is not contained in, and thus low for, the following classes:* APT, R, BPP, P/*poly*.

Finally, there is an oracle relative to which UP is not low for #P.

Corollary 3.4 *There exists an oracle set A such that* $\mathrm{UP}(A)$ *is not low for* $\#\mathrm{P}(A)$.

4 UP is not in the high hierarchy

In this section we construct an oracle set A relative to which, for all $k \geq 1$, $\Sigma_k^P(A) \not\subseteq \Sigma_{k-1}^P(\mathrm{UP}($ As a result, there is an oracle set A relative to which UP is not in any level of the high hierar Our proof is based on an interesting observation from Hartmanis and Hemachandra [HH87]: nondeterministic polynomial time oracle machine N is a *robustly categorical* machine (mach that, for every oracle, have at most one accepting computation for each input), then N beh like a P oracle machine with SAT added to the oracle set. We will make use of this property separate $\Sigma_k^P(A)$ from $\Sigma_{k-1}^P(\mathrm{UP}(A))$, for all $k \geq 1$.

For arbitrary set A, define sets $\tilde{L}_k^A$ for each $k \geq 1$ as follows:

$$\tilde{L}_k^A = \{1^n \mid (\exists m)(\exists j, 0 \leq j < k+1)[(k+1)n + j = e(<2k+1, m>) \text{ and}$$
$$(\exists y_1, |y_1| = n)(\forall y_2, |y_2| = n) \cdots (Q_k y_k, |y_k| = n) 1^n y_1 y_2 \ldots y_k 1^j \in A]\}$$

where Q_k is $\exists$ if k is odd and Q_k is $\forall$ if k is even. Clearly $\tilde{L}_k^A \in \Sigma_k^P(A)$. Our goal is to construct oracle set A such that $\tilde{L}_k^A \notin \Sigma_{k-1}^P(\mathrm{UP}(A))$, for all $k \geq 1$, implying that $\mathrm{UP}(A)$ is not in any leve the high hierarchy. The construction of A will depend on two crucial technical lemmas. The fi is from Sheu and Long [SL92] and shows that there is an oracle A such that $\tilde{L}_k^A \notin \Sigma_{k-1}^P(A \oplus \mathrm{SA}$ for all $k \geq 1$.

ma 4.1 *[SL92] For arbitrary but fixed $k \geq 1$, let σ be a $\Sigma_{k-1}^{P,1}$ predicate with polynomial bound 'here is a constant n_{k-1} such that if $n \geq n_{k-1}$ and $p(n) < \delta \cdot 2^{n/5}$, where $\delta = 1/48$, then there s an oracle set A such that $1^n \in \tilde{L}_k^A \Leftrightarrow \sigma(A \oplus \mathrm{SAT}; 1^n)$ is false.*

'he second technical lemma is based on ideas from Hartmanis and Hemachandra [HH87].

ma 4.2 *Let NM be an arbitrary nondeterministic polynomial time oracle machine whose ing time is bounded by $p(|n|)$ on each input of length n. Let $U = \{x \mid \forall B[NM^B(x)$ has ost 1 accepting path $]\}$. Then, there exists a deterministic polynomial time oracle machine depending only on NM, such that for all B and for all $x \in U$, $NM^B(x)$ accepts if and only $^{\mathrm{SAT}\oplus B}(x)$ accepts. Furthermore, there exists a polynomial q, depending on p, such that the ing time of $M^{\mathrm{SAT}\oplus B}$ is bounded by $q(|x|)$ on input $x \in U$.*

Iow we are ready to construct the oracle set A. The construction is by stages. To diagonalize (UP(A)), we enumerate $\Sigma_{k-1}^P(\mathrm{NP}(A))$. For machine $NM_i(A)$ not acting like a UP(A) machine, an skip $NM_i(A)$ without diagonalization. For machine $NM_i(A)$ acting like a UP(A) machine, $_i$ is "categorical" and is replaced by $\mathrm{P}^{\mathrm{SAT}\oplus A}$ to reduce the diagonalization to the case of ma 4.1. More formally, let $M_1, M_2, \ldots$ be an effective enumeration of deterministic polynomial oracle machines and let $N_1, N_2, \ldots$ be an effective enumeration of nondeterministic polynomial oracle machines. Let polynomial $p_i(n) = n^i + i$ bound the running time of M_i and N_i on any t string of length n. For arbitrary but fixed $k \geq 1$, and for all $l \geq 1$, define the predicate $4; x)$ as follows

$$\sigma_l^k(A;x) \equiv (\exists y_1, |y_1| \leq p_i(|x|))(\forall y_2, |y_2| \leq p_i(|x|)) \cdots (Q_{k-1} y_{k-1}, |y_{k-1}| \leq p_i(|x|))$$
$$[M_i^{N_j^A}(x y_1 y_2 \ldots y_{k-1}) \text{ accepts}]$$

re Q_{k-1} is $\exists$ if k is even and Q_{k-1} is $\forall$ if k is odd, and $l = <i,j>$. Note that, $\sigma_1^k, \sigma_2^k \cdots$ is an ctive enumeration of $\Sigma_{k-1}^P(\mathrm{NP}(A))$.

Our goal of $\Sigma_k^P(A) \not\subseteq \Sigma_{k-1}^P(\mathrm{UP}(A))$ for all $k \geq 1$ is divided into an infinite number of requirets:

: (1) $l = <i,j>$ and there exist an n and A such that, M_i, in the computation of $\sigma_l^k(A; 1^n)$, makes a query "$y \in L(N_j, A)$" and $N_j^A(y)$ has more than one accepting computation, or

(2) $l = <i,j>$ and there exist an n and A such that $1^n \in \tilde{L}_k^A \Leftrightarrow \sigma_l^k(A; 1^n)$ is false.

Initially, let $A(1) = \emptyset$ and let $\overline{A}(1) = \{x \mid |x| \neq e(<2k+1, m>), \forall k, m \geq 0\}$. In each stage α, we define sets $A(\alpha+1)$ and $\overline{A}(\alpha+1)$ such that $A(\alpha+1) \cap \overline{A}(\alpha+1) = \emptyset$ and such that $A(\alpha+1)$ is an ension of $A(\alpha)$ and $\overline{A}(\alpha+1)$ is an extension of $\overline{A}(\alpha)$. Note that at each stage α, $\alpha = <2k+1, m>$, diagonalization area is $\{x \mid |x| = e(\alpha)\}$.

At stage α, $\alpha = <2k, m>$, we do nothing and let $A(\alpha+1) = A(\alpha)$ and $\overline{A}(\alpha+1) = \overline{A}(\alpha)$. Go to next stage.

At stage α, $\alpha = <2k+1, m>$, we will try to satisfy the requirement $\tilde{R}_{k,l}$, where $l = <i,j>$ is least l for which $\tilde{R}_{k,l}$ is not yet satisfied. Let $n = \lfloor e(\alpha)/(k+1) \rfloor$ and let $p_{M_i}(n) = p_i(n+(k-1)\cdot p$ be the running time of M_i on input $xy_1y_2\cdots y_{k-1}$, where $|x| = n$ and $|y_r| \leq p_i(n)$ for all $1 \leq r$ Let $p_l(n) = p_j(p_{M_i}(n))$. Thus, whether $\sigma_l^k(A; 1^n)$ is true depends only on the membership of str in A of length $\leq p_l(n)$. There are three cases.

Case 1: If $p_l(n) \geq 2^{e(\alpha)}$ or $n < n_{k-1}$, where n_{k-1} is the constant from Lemma 4.1, or p_{M_i} $q(p_{M_i}(n)) \geq 1/48 \cdot 2^{n/5}$, where q is the polynomial from Lemma 4.2, then let $A(\alpha+1) =$ and $\overline{A}(\alpha+1) = \overline{A}(\alpha) \cup \{x \mid |x| = e(\alpha)\}$. Go to next stage.

Case 2: Otherwise, try to find a valid extension A' of $A(\alpha)$ and $\overline{A'}$ of $\overline{A}(\alpha)$ such that, in computation of $\sigma_l^k(A'; 1^n)$, M_i makes a query "$y \in L(N_j, A')$?" and $N_j^{A'}(y)$ has more t one accepting computation. If there are such A' and $\overline{A'}$, then let $A(\alpha+1) = A'$ and $\overline{A}(\alpha+$ $\overline{A}(\alpha) \cup \{x \notin A' \mid |x| = e(\alpha)\}$. Since $L(N_j, A')$ is not in UP(A'), we can cancel l and go to next stage.

Case 3: If for all valid extensions A' of $A(\alpha)$ and $\overline{A'}$ of $\overline{A}(\alpha)$, M_i, in the computation of $\sigma_l^k(A';$ does not make a query "$y \in L(M_j, A')$?" where $N_j^{A'}(y)$ has more than one accepting c putation, then by Lemma 4.2 there is a deterministic polynomial time algorithm with or SAT $\oplus A'$ that computes $w \in L(N_j, A')$ for every query w made by M_i. Thus, there is a terministic oracle machine M^* with oracle SAT $\oplus A'$ such that on input $w = 1^n y_1 y_2 \ldots y$ $|y_r| \leq p_i(n)$ for all r, $1 \leq r < k$, $M^{*(\text{SAT}\oplus A')}$ accepts w if and only if $M_i^{N_j^{A'}}$ accepts Note that the running time of $M^{*(\text{SAT}\oplus A')}$ on inupt w is bounded by $p_{M_i}(n) \cdot q(p_{M_i}($ since there are at most $p_{M_i}(n)$ queries, each of length at most $p_{M_i}(n)$. Also note t $\sigma_l^k(A'; 1^n) \in \Sigma_{k-1}^P(\text{SAT} \oplus A')$ for any valid extension A' of $A(\alpha)$. By Lemma 4.1, th is a valid extension A' and $\overline{A'}$ of $A(\alpha)$ and $\overline{A}(\alpha)$ such that $1^n \in L_k^{A'} \Leftrightarrow \sigma_l^k(A'; 1^n)$ is false. $A(\alpha) = A'$ and $\overline{A}(\alpha+1) = \overline{A}(\alpha) \cup \{x \notin A' \mid |x| = e(\alpha)\}$. Cancel l and go to the next stag

Let $A = \cup_{\alpha=1}^{\infty} A(\alpha)$. We claim that all the requirements $\tilde{R}_{k,l}$ will eventually satisfied. Assu that $l = <i,j>$ is the least i such that $\tilde{R}_{k,l}$ is not yet satisfied. Assume that each of $\tilde{R}_{k,l'}$, $l' <$ satisfied by stage $\alpha = <2k+1, m>$. Then $\tilde{R}_{k,l}$ will be satisfied in the first stage $\alpha' = <2k+1, m'>$ s that $m' > m$, $p_l(n) < 2^{e(\alpha')}$, $n \geq n_{k-1}$, and $p_{M_i}(n) \cdot q(p_{M_i}(n)) < 1/48 \cdot 2^{n/5}$, where $n = \lfloor e(\alpha')/(k+$

Theorem 4.3 $\exists A\ \forall k \geq 1\ [\Sigma_k^P(A) \not\subseteq \Sigma_{k-1}^P(\text{UP}(A))]$.

Corollary 4.4 *There exists an oracle set A relative to which* UP *is not in any level of the low high hierarchies.*

Ko [Ko91] has construced an oracle relative to which both the low and high hierarchies infinite. We combine our proof with Ko's and show that there is an oracle A relative to which t low and high hierarchies have infinitely many distinct levels and UP is not in any level of the l or high hierarchy.

orem 4.5 *There exists an oracle set A such that for all* $k \geq 0$, $\mathrm{UP}(A) \not\subseteq L_k^{P,\Sigma}(A)$, $\mathrm{UP}(A) \not\subseteq$ (A), $L_k^{P,\Sigma}(A) \neq L_{k+1}^{P,\Sigma}(A)$, *and* $H_k^{P,\Sigma}(A) \neq H_{k+1}^{P,\Sigma}(A)$.

:nowledgement

authors would like to thank J. Ramachandran for valuable discussions.

erences

;80] D. Angluin. On counting problems and the polynomial-time hierarchy. *Theoretical Computer Science*, 12:161–173, 1980.

;81] M. Furst, J. Saxe, and M. Sipser. Pairty, circuits, and the polynomial-time hierarchy. In *Proc. 22th Annual IEEE Symposium on Foundations of Computer Science*, pages 260–270, 1981.

38] J. Grollmann and A. Selman. Complexity measures for public-key cryptosystems. *SIAM J. Comput.*, 11(2), April 1988.

;87] J. D. Hastad. *Computational limitations for small-depth circuits.* PhD thesis, Massachusetts Institute of Technology, 1987.

87] J. Hartmanis and L. Hemachandra. One-way functions, robustness, and non-isomorphism of NP-complete sets. In *Proc. Second Annual IEEE Structure in Complexity Theory Conference*, pages 160–173, 1987.

39] K. Ko. Relativized polynomial time hierarchies having exactly *k* levels. *SIAM J. Comput.*, 18(2):392–408, April 1989.

91] K. Ko. Separating the low and high hierarchies by oracles. *Information and Computation*, 90(2):156–177, Feburary 1991.

TT89] J. Köbler, U. Schöning, S. Toda, and J. Torán. Turing machines with few accepting computations and low sets for PP. In *Proc. 4th Structure in Complexity Theory Conference, IEEE*, 1989.

.83] U. Schöning. A low and a high hierarchy within NP. *J. Comput. System Sci.*, 27:14–28, 1983.

)2] M. Sheu and T. Long. The extended low hierarchy is an infinite hierarchy. In *Proc. 9th Annual Symposium on Theoretical Aspects of Computer Sceince*, pages 187–198, 1992.

76] L. Stockmeyer. The polynomial-time hierarchy. *Theor. Comput. Sci.*, 3:1–22, 1976.

a76] C. Wrathall. Complete sets and the polynomial hierarchy. *Theor. Comput. Sci*, 3:23–33, 1976.

)85] A. Yao. Separating the polynomial-time hierarchy by oracles. In *Proc. 26th IEEE Symp. on Foundations of Computer Science*, pages 1–10, 1985.

Analytic Analysis of Algorithms

Philippe FLAJOLET*
Algorithms Project
INRIA, Rocquencourt
F–78150 Le Chesnay (France)

Abstract. *The average case analysis of algorithms can avail itself of the development of synthetic methods in combinatorial enumerations and in asymptotic analysis. Symbolic methods in combinatorial analysis permit to express directly the counting generating functions of wide classes of combinatorial structures. Asymptotic methods based on complex analysis permit to extract directly coefficients of structurally complicated generating functions without a need for explicit coefficient expansions.*

Three major groups of problems relative to algebraic equations, differential equations, and iteration are presented. The range of applications includes formal languages, tree enumerations, comparison–based searching and sorting, digital structures, hashing and occupancy problems.

These analytic approaches allow an abstract discussion of asymptotic properties of combinatorial structures and schemas while opening the way for automatic analysis of whole classes of combinatorial algorithms.

Introduction

In elementary cases, the average case analysis of a combinatorial algorithm follows a simple pattern. First, set up *recurrences* depending upon the structure of the algorithm that relate the complexity on the collection of all inputs of size n to the complexity on inputs of a smaller size $k < n$. Next solve the recurrences explicitly by algebraic manipulations, whenever possible. Conclude by an asymptotic evaluation often based on basic real analysis, for instance approximating discrete sums by integrals. These classical techniques are reviewed for instance in [24, 46, 51, 54, 55, 79].

In this section, we re-examine the analysis of Quicksort. First we recall the usual analysis by means of recurrences. Next, we sketch an alternate derivation by means of generating functions. This provides a simple illustration of the leading theme of this paper: *Generating functions are central to combinatorial enumerations and the average–case analysis of algorithms.*

Part I deals with general methods. Various theories have been developed that furnish direct correspondences between combinatorial structures and generating functions, as explained in Section 1. Asymptotic methods based on complex analysis discussed in Section 2, then permit to extract coefficients directly from the generat-

*Author's electronic mail address: `flajolet@inria.fr`

ng function itself. In this way, wide classes of problems receive satisfactory solutions n asymptotic form.

Part II presents a panorama of some recent investigations where generating functions have been instrumental in arriving at results barely accessible to elementary methods. Three major groups of problems relating to algebraic equations, differential equations, and functional equations are presented. Applications in the average–case analysis of algorithms concern a variety of domains: formal languages, tree enumerations, comparison based searching and sorting, digital structures, hashing and occupancy.

Part III surveys some recent approaches to the analysis of combinatorial schemas, as well as related studies in the automated analysis of some well defined classes of combinatorial problems.

Quicksort and recurrences. The traditional example of Quicksort, the sorting algorithm used in the Unix system, is now discussed. The structure of the algorithm (see [70, Chap. 9]) is well known.

Sorting n elements reduces partioning the n elements with respect to the first element in the file and then sorting the two resulting subgroups of sizes K and $n-1-K$, with K depending on the actual data input to the algorithm. The structure of the algorithm is as follows:

```
procedure QuickSort(l,r : integer);
{sorts the part T[l..r] of a global array T[1..n]}
if r>l then
     i:=Partition(l,r);
     {a[l] is to be placed at position i}
     Quicksort(l,i-1);
     Quicksort(i+1,r)
```

When applied to data in random order, i.e. to a *random* permutation of size n, the random variable K assumes each of its possible values $K \in [0..n-1]$ with equal likelihood. Let $\bar{Q}_n$ be the expected number of comparisons of Quicksort when applied to n random data. The recurrence is based on the recursive structure of the algorithm,

$$\bar{Q}_n = p_n + \sum_{k=0}^{n-1} \pi_{n,k} [\bar{Q}_k + \bar{Q}_{n-1-k}]. \tag{1}$$

There $\pi_{n,k}$ is the probability that the partitioning stage splits the file into two subfiles of sizes k and $n-1-k$, so that $\pi_{n,k} = \frac{1}{n}$ because of our earlier observation on the random variable K. The quantity p_n represents a 'toll' incurred at each recursive call which is the cost (measured in the number of comparisons) for partitioning, and with some standard implementations, we may take $p_n = (n-1)$. Knuth [55, p. 120] explains how to manipulate such recurrences. Reducing summations, and solving a linear equation of order 1, we arrive at an exact solution,

$$\bar{Q}_n = 2(n+1)H_{n+1} - 4n - 2 \qquad \text{with} \qquad H_n = \sum_{j=1}^{n} \frac{1}{j}.$$

Approximating the discrete sum by an integral,

$$H_n \equiv \sum_{j=1}^{n} \frac{1}{j} \sim \int_1^n \frac{dt}{t} \equiv \log n,$$

produces the asymptotic form of the cost,

$$\bar{Q}_n \sim 2n \log n. \tag{2}$$

In this way, Quicksort is proved to be about 40% off from the information theoretic optimum of $n \log_2 n$, which is the final conclusion sought.

Quicksort and generating functions. There is an alternative approach to this problem which in such a simple case seems to be a mere variant of the analysis, but nonetheless reveals some important features of the approaches we plan to discuss here.

Introduce the *generating function* (GF) of the mean values

$$Q(z) = \sum_{n=0}^{\infty} \bar{Q}_n z^n, \tag{3}$$

and set similarly $p(z) = \sum_{n \geq 0} p_n z^n$. Then, the equation corresponding to the recurrence (1) is

$$Q(z) = p(z) + 2 \int_0^z Q(t) \frac{dt}{1-t}. \tag{4}$$

(This could be checked by multiplying in Eq. (1) by z^n and summing over n, employing the usual rules of generating function manipulations [44].) The differential equation, corresponding to (4),

$$\frac{d}{dz} Q(z) = \frac{d}{dz} p(z) + 2 \frac{Q(z)}{1-z},$$

is solved by the usual techniques,

$$Q(z) = \frac{1}{(1-z)^2} \int_0^z \frac{d}{dt} \{p(t)\} (1-t)^2 dt. \tag{5}$$

This integral transform expresses the global cost of the algorithm in terms of the local cost incurred at each recursive call. In the particular case of Quicksort, this leads to the solution: we have $p(z) = z^2/(1-z)^2$, and, carrying out the integration,

$$Q(z) = 2 \frac{\log(1-z)^{-1}}{(1-z)^2} - \frac{2z}{(1-z)^2}. \tag{6}$$

If we expand $Q(z)$, we retrieve again the form of $\bar{Q}_n$ that involves the harmonic numbers.

The solution expressed by (6) can be used to produce direct asymptotic results from the generating function itself, without any need for explicit expansions. The ey observation is that it suffices to examine the generating function *locally* near ts *singularity* at $z = 1$ and apply systematic translation mechanisms described in Section 2. Letting $[z^n]f(z)$ denote the coefficient of z^n in the generating function $f(z)$, a single rule

$$[z^n]\,\frac{1}{(1-z)^2}(\log(1-z)^{-1}) \sim n\log n$$

vill give us the $2n\log n$ result directly.

The generating function approach is the one that leads to higher level generalzations applicable to more complicated algorithms and cost measures.

First, in a suitable framework, the structure of the equation is seen to be a *direct translation* of the specification of the algorithm, and we discuss such aspects n Section 1. The rules exemplified here concern sequential execution and recursive descent into smaller subfiles (with the suitable probability distribution $\pi_{n,k} = 1/n$). Informally, the two rules are

$$\begin{aligned} &\text{Sequential execution: }\mathtt{F;\ G} &&\Longrightarrow\quad F(z)+G(z)\\ &\text{Recursive descent with }\mathtt{F} &&\Longrightarrow\quad \int_0^z F(t)\,\frac{dt}{1-t}.\end{aligned}$$

Such correspondences cover a wide range of problems: *Generating functions of wide classes of combinatorial structures and algorithms can be determined from formal specifications.* The character of these correspondences is systematic enough that we may even use computer algebra programmes to compute the generating function equations automatically, a fact that is explored in Section 1 and further discussed in Part III.

Next, the translation from the local singular behaviour of a function to the asymptotics of its coefficients is a powerful mechanism. General rules valid under simple conditions (analytic continuation) apply, like for instance, the relation

$$[z^n]\,\frac{1}{(1-z)^\alpha}(\log(1-z)^{-1})^k \sim \frac{n^{\alpha-1}}{\Gamma(\alpha)}(\log n)^k.$$

If the partitioning cost of Quicksort becomes of the form $\sqrt{n}$ (due to parallel execution perhaps) a direct asymptotic analysis is still feasible (and easy!), despite the fact that the explicit coefficient expressions become more involved. The general principle is the following: *Generating functions need only be studied locally near their singularities.* Again, this systematic process presented in Section 2 can be subjected to automatic analysis.

Quadtrees. The generating function approach therefore allows for a unified discussion of a whole range of related problems. As a further illustration, consider an analogous two–dimensional problem, namely the analysis of path length in standard quadtrees [26, 49]. In such a tree, there are two successive descents (the first one in a half plane based on the x–coordinate, the second one in a quadrant determined by the y–coordinate). Accordingly, the single integral of (5) gets replaced by a double integral,

$$Q(z) = p(z) + 4\int_0^z \left[\int_0^t Q(u)\,\frac{du}{(1-u)}\right]\frac{dt}{t(1-t)}. \tag{7}$$

The associated differential equation is now of order 2. Thanks to a relation to speci[al] hypergeometric functions, we still have an explicit solution available,

$$Q(z) = \frac{(1+2z)}{(1-z)^2} \int_0^z \frac{(1-t)^3}{t(1+2t)^2} \left[\int_0^t \frac{(1+2v)}{(1-v)^2} \frac{d}{dv} \{ v(1-v) \frac{d}{dv} p(v) \} \, dv \right] dt.$$

Given the principles of singularity analysis, it suffices to examine locally near $z =$ the effect of this integral expression, viewing it as "singularity transformer".

Complex analysis asymptotics render the analysis of such GF's really simpl[e] although coefficients soon turn out to have intractable expressions. For quadtree[s] in higher dimensions $d > 2$, the integral equation is of order d and does not eve[n] admit of closed form solutions any more. However, singularities of such direct an[d] inverse operators can still be studied by appealing to the classification of singularitie[s] of linear differential systems, and complete asymptotic solutions are available. Thi[s] illustrates a further feature of the theory: *Generating functions may be analyzed eve[n] in cases where they admit of no closed form.*

In this way, the cost of partial match and exact match queries in quadtrees of al[l] dimensions has been precisely quantified [26]. It is found that an exact match quer[y] in a quadtree of size n and dimension d has cost asymptotic to $\frac{2}{d}\log n$.

Part I: Methods

1 Symbolic Methods in Combinatorial Analysis

> Laplace discovered the remarkable correspondence between set theoretic operations and operations on formal power series and put it to use with great success to solve a variety of combinatorial problems.
>
> — G.-C. Rota

Early practitioners of combinatorial analysis often realized that certain types o[f] counting problems would invariably lead to definite types of generating functions The systematization of these scattered observations had to wait a bit, however. First, in the late 1950's, Chomsky and Schützenberger discovered that enumerative problems described by regular languages or finite automata lead to rational generating functions, while algebraic functions correspond to (unambiguous) context free languages. Then, Rota and his school on one side, Foata and Schützenberger on an other side, came to general frameworks that would 'explain' such correspondences. Later Joyal [50] with the theory of species, as well as Goulden and Jackson [43] produced frameworks of comparable power. In a remarkable thesis, Greene [45] developed a notion of labelled grammars with a focus on order constraints and analysis of algorithms. Flajolet and Steyaert developed rules initially specialized to trees from which a 'complexity calculus' could be derived for a wide class of algorithms [22, 24, 36, 37, 75]. This was later extended into a much more general system [32] to be discussed in Part III.

We propose now to explain the major principle of a symbolic approach to the derivation of generating functions.

. *Trains.*

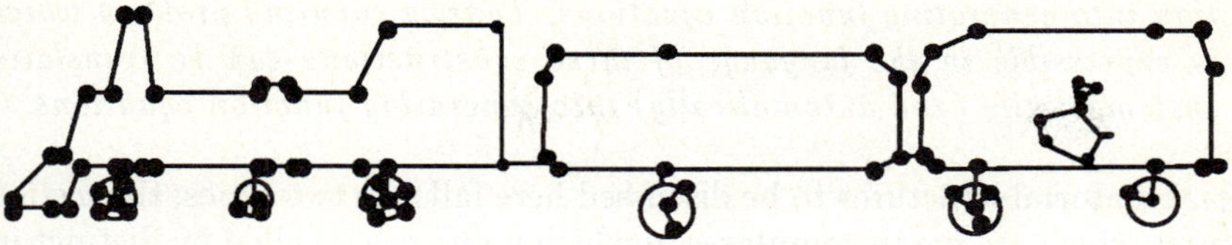

. Formal specification.

$$\begin{cases} \mathit{train} = (\mathit{locomotive} * \mathit{wagons}) \\ \mathit{wagons} = \text{sequence}(\mathit{wagon}) \\ \mathit{locomotive} = \text{sequence}(\mathit{slice}) \\ \mathit{slice} = (\mathit{upper} * \mathit{lower}) \text{ union } (\mathit{upper} * \mathit{lower} * \mathit{wheel}) \\ \mathit{wagon} = (\mathit{locomotive} * \mathit{passengers}) \\ \mathit{passengers} = \text{set}(\mathit{passenger}) \\ \mathit{passenger} = (\mathit{head} * \mathit{belly}) \\ \mathit{wheel} = \text{cycle}(\mathit{wheel\_element}) \\ \mathit{head} = \mathit{belly} = \text{cycle}(\mathit{passenger\_element}) \\ \mathit{upper} = \mathit{lower} = \mathit{wheel\_element} = \mathit{passenger\_element} = \mathit{point}. \end{cases}$$

3. *Generating function equations (excerpts).*

```
train(z):=locomotive(z)*wagons(z);
wagons(z):=Q(wagon);                        % Q(u):=1/(1-u);
locomotive(z):=slice*Q(slice);
slice(z):=upper*lower+upper*lower*wheel;
wheel(z):=center*L(wheel_element);      % L(u):=log(1/(1-u));
.........................
wheel_element(z):=z;
passenger_element(z):=z;
```

4. *Explicit generating function.*

```
                          2    3
                         z  - z  ln(1 - z)
------------------------------------------------------------------
                         /   2    3                            2 \
      2    3             |  (z  - z  ln(1 - z)) exp(ln(1 - z) ) |
(1 - z  + z  ln(1 - z)) |1 - ----------------------------------|
                         |          2    3                      |
                         \     1 - z  + z  ln(1 - z)           /
```

Figure 1. The example of 'random trains' illustrates the power of symbolic methods in combinatorial analysis. We define a complex combinatorial structure [§1] that is formed with sequences, cycles, and sets. A formal specification [§2] is easily set up. From it, generating functions are computable systematically, and a system of equations is compiled from the specification [§3]. The generating function is then solved explicitly [§4].
Currently, the analysis of this problem can be achieved automatically. A system, Lambda–Upsilon–Omega ($\Lambda\Upsilon\Omega$), has been designed by B. Salvy and P. Zimmermann jointly with the author [32, 69, 84]. It does the analysis and via an implementation of complex asymptotic methods and singularity analysis, it is also able to find automatically the asymptotic form of the coefficients: The number of trains of size n satisfies the estimate

$$\frac{\text{train}_n}{n!} \sim 0.07097007911 \cdot 1.930298068^n \; .$$

Principle. *A number of set–theoretic constructions like union, cartesian product, sequence set, cycle set, power set, substitution have direct tanslation into generating function equations. Thus, a counting problem which is expressible in the language of these constructions can be translated systematically (and automatically) into generating function equations.*

Combinatorial structures to be discussed here fall into two types; the *well–labelle*[d] structures which are graph complexes in which nodes are labelled by distinct intege[rs] (from 1 to n when the structure comprises n nodes) and unlabelled ones. Exampl[es] of labelled structures are labelled trees, permutations (when viewed as collectio[ns] of labelled cyclic graphs), etc. Unlabelled trees, formal languages are examples [of] unlabelled objects.

Given a class $\mathcal{F}$ of combinatorial structures, we let $\mathcal{F}_n$ denote the collection [of] objects of size n, and set $F_n = \text{card}(\mathcal{F}_n)$. The *ordinary generating function* (OGF[)] and *exponential generating function* (EGF) are defined respectively to be

$$F(z) = \sum_{n\geq 0} F_n z^n \qquad \text{and} \qquad \hat{F}(z) = \sum_{n\geq 0} F_n \frac{z^n}{n!}. \qquad (8$$

A combinatorial construction is *admissible* if it admits a translation into generatin[g] functions.

The following two theorems are well known under one form or the other. The[y] embody a powerful collection of combinatorial constructions. For detailed definitions the reader is referred to modern treatments of the subject [15, 42, 43, 72, 74, 81] o[r] to the paper [32] where a similar system of notations is developed.

Theorem 1 (Admissible constructions for OGF's) *For unlabelled structures the constructions of union, cartesian product, sequence, cycle, set, multiset, substitution are admissible. The translations into ordinary generating functions are give[n] by the following table*

Construction	Translation (OGF)
$\mathcal{F} = \mathcal{G} \cup \mathcal{H}$	$F(z) = G(z) + H(z)$
$\mathcal{F} = \mathcal{G} \times \mathcal{H}$	$F(z) = G(z) \cdot H(z)$
$\mathcal{F} = \texttt{sequence}(\mathcal{G}) = \mathcal{G}^*$	$F(z) = \frac{1}{1-G(z)}$
$\mathcal{F} = \texttt{set}(\mathcal{G})$	$F(z) = \exp(G(z) - \frac{1}{2}G(z^2) + \frac{1}{3}G(z) - \cdots)$
$\mathcal{F} = \texttt{multiset}(\mathcal{G})$	$F(z) = \exp(G(z) + \frac{1}{2}G(z^2) + \frac{1}{3}G(z) + \cdots)$
$\mathcal{F} = \texttt{cycle}(\mathcal{G})$	$F(z) = \log(1 - G(z))^{-1} + \cdots$
$\mathcal{F} = \mathcal{G}[\mathcal{H}]$	$F(z) = G(H(z))$

Theorem 2 (Admissible constructions for EGF's) *For labelled structures, the constructions of union, partitional product, sequence, cycle, set, substitution are admissible. The translations into exponential generating functions are given by the following table*

Construction	Translation (EGF)
$\mathcal{F} = \mathcal{G} \cup \mathcal{H}$	$\hat{F}(z) = \hat{G}(z) + \hat{H}(z)$
$\mathcal{F} = \mathcal{G} * \mathcal{H}$	$\hat{F}(z) = \hat{G}(z) \cdot \hat{H}(z)$
$\mathcal{F} = \texttt{sequence}(\mathcal{G}) = \mathcal{G}^*$	$\hat{F}(z) = \frac{1}{1-\hat{G}(z)}$
$\mathcal{F} = \texttt{set}(\mathcal{G})$	$\hat{F}(z) = \exp(\hat{G}(z))$
$\mathcal{F} = \texttt{cycle}(\mathcal{G})$	$\hat{F}(z) = \log(1 - \hat{G}(z))^{-1}$
$\mathcal{F} = \mathcal{G}[\mathcal{H}]$	$\hat{F}(z) = \hat{G}(\hat{H}(z))$

2 Complex Analysis and Asymptotics

Es ist eine Tatsache, daß die genauere Kenntnis des Verhaltens einer analytischen Funktion in der Nähe ihrer singulären Stellen eine Quelle von arithmetischen Sätzen ist.

— E. Hecke

Complex analytic methods permit to represent coefficients of generating functions and many combinatorial sums as integrals of an analytic function in the complex plane. The choice of a suitable contour of integration often leads to highly non trivial asymptotic results. A thorough review of theses techniques appears in [38]. Other excellent references are [5, 16, 48, 66].

The first part of this section is devoted to singularity analysis techniques which make it possible to derive estimates on the coefficients of generating functions starting from Cauchy's formula,

$$f_n \equiv [z^n]f(z) = \frac{1}{2i\pi} \oint f(z) \frac{dz}{z^{n+1}}. \tag{9}$$

Singularity analysis. Most functions occurring in combinatorial enumeration problems are built by operators from standard functions that exist over the whole of the complex plane. They thus tend to exist in larger areas of the complex plane. The method of *singularity analysis* is well suited to extracting coefficients of functions lying in a class that enjoys interesting closure properties.

Definition 1 A function analytic at the origin is *star continuable* iff it has a finite number of singularities $\zeta_j = \rho e^{i\theta_j}$ on its circle of convergence $|z| = \rho$ and if for some $\epsilon < \frac{\pi}{2}$ and $\eta > 0$ it is continuable in $|\mathrm{Arg}(ze^{-i\theta_j} - \rho)| > \epsilon$ and $|z| \leq \rho + \eta$.

An *algebraic–logarithmic* element is a formal series

$$F(u) = \sum_{k=0}^{\infty} c_k(\log(\frac{1}{u}))u^{\alpha_k},$$

where the α_k satisfy $\Re(\alpha_1) \leq \Re(\alpha_2) \leq \cdots$, $\Re(\alpha_k) \to \infty$ and each $c_k(x)$ is a polynomial.

A function is *algebraic–logarithmic* iff it is star continuable and near each singularity ζ_j it admits an asymptotic expansion $f(z - \zeta_j) \sim F_j(z - \zeta_j)$, where $F_j(u)$ is an algebraic–logarithmic element.

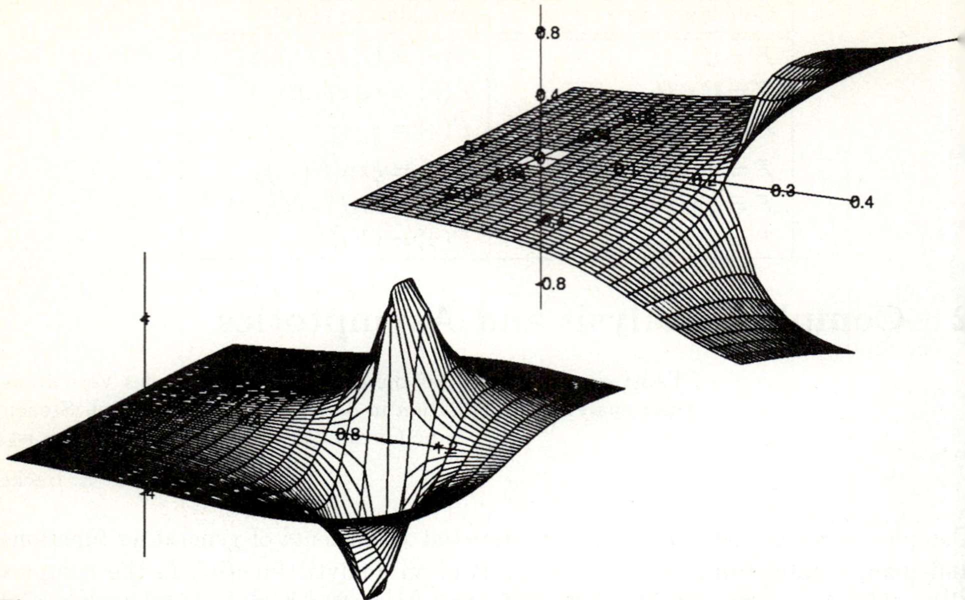

Figure 2. A display of the imaginary parts of two generating functions,

$$f(z) = \frac{1 - \sqrt{1-4z}}{2z} \qquad \text{and} \qquad g(z) = \frac{1}{1-z}.$$

The function $f(z)$ [top] is the ordinary generating function of binary trees with a singularity at $\rho = 1/4$ which is a branch point of the $\sqrt{\ }$ type. The function $g(z)$ [bottom] is the exponential generating function of permutations with a singularity at $\rho = 1$ of a polar type. The singularities are reflected at the level of coefficients,

$$[z^n]f(z) \sim \frac{4^n}{\sqrt{\pi n^3}} \qquad \text{and} \qquad [z^n]g(z) = 1.$$

The first theorem summarizes a variety of results known since about the time of Hadamard [19, 78].

Theorem 3 (Analytic Closure Theorem) *Star continuable functions and algebraic-logarithmic functions are closed under sum, product, Hadamard product*[1]*, integration, and differentiation.*

Star continuable functions and algebraic–logarithmic functions thus enjoy rich closure properties. There is a direct relation between singular expansions and coefficient expansions (see Fig. 2 for an illustration), and the coefficients of algebraic–logarithmic functions can be determined systematically.

Theorem 4 (Singularity analysis) *An asymptotic form of coefficients of algebraic–logarithmic functions is obtained by termwise translation of coefficients*

[1] The Hadamard product of $f(z)$ and $g(z)$ is their term by term product $f \odot g(z) = \sum_n f_n g_n z^n$.

elements, using the rules ($\alpha \notin \{0,-1,-2,\ldots\}$)

$$[z^n](1-z)^{-\alpha}(\log\frac{1}{1-z})^{\beta} \sim \frac{n^{\alpha-1}(\log n)^{\beta}}{\Gamma(\alpha)}\left[1+\frac{C_1}{1!}\frac{\beta}{\log n}+\frac{C_2}{2!}\frac{\beta(\beta-1)}{\log^2 n}+\cdots\right],$$

$$\text{where} \quad C_j = \Gamma(\alpha)\,\frac{d^j}{ds^j}\frac{1}{\Gamma(s)}\bigg|_{s=\alpha}, \quad \Gamma(s)=\int_0^{\infty} e^{-t}t^{s-1}dt.$$

This theorem [29] allows for a vast number of generalizations based on terminating expansions, partial expansions involving $O(.)$ or $o(.)$ error terms, as well as iterated logarithms or functions of slow variation. It is inspired by the classical method of Darboux in asymptotic analysis and by Tauberian theorems. Its principle however relies on integration using contours of the Hankel type.

The applications are numerous. Most of the analysis carried out by Darboux's method can usually be conducted in a transparent way using this method. Instances are uniform tree models and the so–called simple families of trees [64], and generally enumeration problems expressible in terms of context–free languages.

Saddle point integrals. Given an analytic function $f(z)$, a *saddle point* is a point ζ which cancels the derivative, $f'(\zeta)=0$. This terminology is due to the topography of the modulus of the function near ζ which resembles the inside part of a saddle. When computing a complex integral

$$\frac{1}{2i\pi}\int_C e^{g(z)}\,dz,$$

it is often the case that the contour C can be deformed so as to traverse a saddle point, with the value at the saddle point being a local maximum for the modulus. Under suitable conditions, the integral is concentrated near the saddle point. In that case, a local expansion of the analytic integrand holds, and one finds the approximation

$$\frac{1}{2i\pi}\int_C e^{g(z)}\,dz \sim \frac{1}{\sqrt{2\pi g''(\zeta)}}e^{g(\zeta)}.$$

When the method applies, it is often said that the integral is (saddle point) *admissible*. Most notably, the method is useful for the computation of coefficients of whole classes of entire functions [67]. In that case, we should take $g(z)=\log f(z)-(n+1)\log z$, where $f(z)$ is the function to be analyzed.

Theorem 5 (Saddle point coefficient asymptotics) *For an admissible function $f(z)$, one has*

$$[z^n]f(z) \sim \frac{f(\zeta)}{\sqrt{2\pi C}\zeta^{n+1}} \quad \text{where } C=\frac{d^2}{dz^2}\log\frac{f(z)}{z^{n+1}}\bigg|_{z=\zeta},$$

where $\zeta=\zeta_n$ is the smallest real root of $\frac{d}{dz}\log\frac{f(z)}{z^{n+1}}$.

This classical method originating in applied mathematics is discussed in d Bruijn's book [16], and in the context of coefficient extraction in [67, 81]. A revie appears in [69]. Typical applications are to fast growing functions like

$$e^{z/(1-z)},\ e^{z},\ e^{e^{z}-1},\ e^{z+z^{2}/2},\ldots.$$

This covers problems related to increasing subsequences in permutations, set part tions, involutions, etc.

Saddle point methods also serve to analyze coefficients of large powers of func tions. This is one way of establishing the central limit theorem (together with er ror bounds), a fact well explained in [46]. There are also numerous applications t hashing and occupancy problems. Difficult questions present themselves with highe dimensional saddle point problem. Gardy has obtained several general results o urn models [39, 40] and attained a precise quantification of the size of relationa algebra operations applied to collections of random points, using two–dimensiona saddle point techniques. McKay and his collaborators have pushed the analysis t situations where a counting problem of a large size n leads to an n–dimensiona complex integral. For instance,

$$\frac{1}{(2i\pi)^{n}}\oint\oint\cdots\oint\prod_{1\le i<j\le n}(1+z_{i}z_{j})\,\frac{dz_{1}dz_{2}\cdots dz_{n}}{z_{1}^{k+1}z_{2}^{k+2}\cdots z_{n}^{k+1}},$$

gives the number of k–regular graphs, and the integral can be estimated using ar n–dimensional saddle point integral while the dimension of the space n tends tc infinity! See [63] for these promising techniques.

Part II: Classes of Applications

In this part, we explain how the methods introduced can be put to use in order to analyze whole *classes* of problems relative to combinatorial structures and algorithms. We limit ourselves to a descriptive inventory that tries to put in perspective a vast body of literature.

3 Algebraic Functions and Implicit Functions

Regular languages can be specified either by regular expressions or by finite automata. The corresponding GF's either appear as built from the variable z by means of rational operations ($+$, $\times$, quasi–inverse $Q(y)=(1-y)^{-1}$) or as components of linear systems of equations (over $\mathbb{Z}[x]$). At any rate, they are rational [7]. Examples are

$$\frac{z}{1-\dfrac{z}{1-\dfrac{z}{1-\dfrac{z}{1-z}}}},\quad \frac{1-z^{k+1}}{1-2z+z^{k+1}},\quad \frac{z^{4}}{1-4z+4z^{2}+z^{3}-3z^{4}+2z^{5}},$$

epresenting the OGF of plane unlabelled trees (Dyck words) of height at most 4, of inary strings with no runs of more than k consecutive ones, and of binary strings ontaining the pattern 0100.

An immediate consequence of the partial fraction decomposition of rational functions is the following.

Theorem 6 (Rational Asymptotics) *The coefficients of a rational function of $\mathbb{Q}(z)$ are a finite linear combination of 'exponential polynomials' of the form*

$$\lambda \omega^n n^k, \tag{10}$$

with λ, ω algebraic numbers and k an integer.

By grouping the ω's in order of decreasing modulus, this has the character of an asymptotic expansion. In this way, a counting problem relative to a single regular language normally poses no difficulty and falls into a decidable class of problems.

Interesting problems arise from consideration of parameterized sets of rational functions, like trees of bounded height [17], longest runs of ones and carry propagation [56], or occurrences of prefixes of an infinite pattern sequence [65]. The analysis of the clustering of dominant roots, which will have accumulation points, has made it possible to analyze the expected height of trees, and the expected time for carry propagation in binary adders.

Context free languages lead to polynomial nonlinear equations, provided the grammar is unambiguous or we count words with their multiplicities. Thus, the generating function of a context free language is algebraic. This is the famous Chomsky–Schützenberger theorem. By standard elimination theory, such a function satisfies a single polynomial equation,

$$P(z, f(z)) = 0. \tag{11}$$

Near a singularity, an algebraic function admits an expansions into fractional powers of the form $(1 - z/\rho)^{p/q}$, which is also called a Puiseux expansion. The method of singularity analysis applies well to Puiseux elements.

Theorem 7 (Algebraic Asymptotics) *The coefficients of a $\mathbb{Q}(z)$-algebraic function are asymptotic to a sum of 'algebraic elements' of the form*

$$\frac{\lambda}{\Gamma(r/s+1)} \omega^n n^{r/s}, \tag{12}$$

where λ, ω are algebraic numbers, and the exponent r/s is a rational number.

This furnishes a generalized density theorem for context free languages and was used in [23] in order to establish the inherent ambiguity of several context free languages.

Similar singular expansions involving fractional powers also hold for functions implicitly defined by equations of the form $\Phi(z, f(z)) = 0$, where Φ, analytic function of two complex variables, need no longer be a polynomial. By the implicit function theorem, singularities are almost invariably of the $\sqrt{\ }$ type, so that coefficients involve the rational exponent $r/s = -3/2$. Examples are

$$y = z(1 + y^t), \quad y = ze^y, \quad e^y - 2y - 1 + z = 0.$$

The first equation counts regular (plane, unlabelled) t–ary trees; the second one labelled non plane trees whose number is n^{n-1}, a famous result due to Cayley; the third one Schröder's partition systems [15, p. 224].

Implicit functions. Meir and Moon have developed in a series of papers (see e.g., [64]) a general theory of statistics on 'simple families of trees' corresponding to equations of the form $y - z\phi(y) = 0$. Under general conditions, the singularities are again of the $\sqrt{\ }$ type, which leads again to asymptotic forms

$$\lambda\omega^n n^{-3/2}.$$

Using this theory, it is found that various families of trees share common features for instance, path length is of order $n^{3/2}$ on average.

Variations around this theme have led to the analysis of a large number of tree algorithms in the context of symbolic manipulations. We shall cite here: pattern-matching [76, 2], simplification [10], unification [1], common subexpression factorization [34], and term rewriting techniques [11]. See also [9] for an interesting survey.

Another important case of application is to the Cayley function $y = ze^y$. This function shows up in the enumeration of labelled trees, random mappings [28], in the analysis of hashing with linear probing [55], in union find problems [57], in caching algorithms, etc.

A further extension concerns the enumeration of unlabelled non plane trees that involves the operators of Theorem 1 (see the set and multiset constructions). The GF of such graphical trees satisfies a functional relation known to Cayley,

$$f(z) = z \exp\left(f(z) + \frac{1}{2}f(z^2) + \frac{1}{3}f(z^3) + \cdots\right).$$

Although no closed form is available for this GF, it can still be subjected to the same treatment as implicitly defined functions, a general fact discovered by Pólya in his famous 1937 paper [68]: Once it has been recognized that $f(z)$ has radius of convergence $\rho < 1$, the terms in the sum involving $f(z^2), f(z^3)$, etc, being analytic near $z = \rho$, can be treated as 'known' perturbations for all asymptotic purposes. In this way, it is also possible to analyze the GF's

$$f(z) = z + \frac{1}{2}f^2(z) + \frac{1}{2}f(z^2), \quad f(z) = \cfrac{1}{1 - \cfrac{z}{1 - \cfrac{z^2}{1 - \cfrac{z^4}{1 - \cdots}}}},$$

relative to non plane unlabelled binary trees (Otter, 1948) and to structurally isomeric alcohols $C_nH_{2n+1}OH$ without asymmetric carbon atoms. The asymptotic forms found are

$$2.9557 \cdot 0.4399^n n^{-3/2}, \quad 0.3187 \cdot 2.4832^n n^{-3/2}, \quad 0.3067 \cdot 1.6813^n,$$

for Cayley's graphical trees, binary trees and alcohols, respectively [47].

In summary, functions defined implicitly tend to have singularities like those of algebraic functions, involving fractional exponents. This is reflected by the asymptotics of their coefficients of the form $\omega^n n^{-r/s}$. Such a property also holds for many functions satisfying finite and infinite functional equations involving terms like $f(z^2), f(z^3)$ provided that their radius of convergence is < 1.

4 Holonomic Functions and Differential Equations

When discussing the analysis of Quicksort, we have encountered a particular case of the general *probabilistic divide and conquer* schema

$$f_n = e_n + \sum_{k=0}^{n-1} \pi^*_{n,k} f_k. \tag{13}$$

There f_n is the sequence to be analyzed, e_n is a fixed toll sequence, and the $\pi^*_{n,k}$ are proportional to the splitting probabilities $\pi_{n,k}$ that express the chances that a task of size n involve a subtask of size $k < n$.

In comparison based sorting and searching, it is often the case that the $\pi_{n,k}$ involves some rational combination of n and k, a fact also well accounted for by Greene's treatment of 'min-rooting' operators [45]. Naturally occurring examples of $\pi_{n,k}$'s are

$$\frac{1}{n}, \ \frac{2(n-k)}{n(n+1)}, \ \frac{1}{n}[H_n - H_k], \ \frac{2(k-1)(n-k)}{n(n-1)(n-2)},$$

that arise in Quicksort (or binary search trees), fully specified search in 2-d quadtrees, partial match queries in 2-d quadtrees, and median–of–three Quicksort [55, p. 609]. In that case the translation into generating functions leads to integral equations of which (4,7) are typical. This in turn leads to generating functions satisfying differential equations with rational (or equivalently polynomial) coefficients. (Only in simpler cases is the equation an Euler equation that admits elementary solutions.)

Functions satisfying differential equations with polynomial coefficients are sometimes called $\mathcal{D}$–finite and their coefficient sequences which satisfy recurrences with polynomial (in n) coefficients are then called $\mathcal{P}$–recursive. These notions are formalized by the concept of *holonomy* introduced in this range of problems by Zeilberger.

Definition 2 A series $f(z_1, z_2, \ldots, z_r) \in \mathbb{C}[[z_1, z_2, \ldots, z_r]]$ is said to be *holonomic* iff the infinite collection of its partial derivatives

$$\frac{\partial^{j_1}}{\partial z^{j_1}} \frac{\partial^{j_2}}{\partial z^{j_2}} \cdots \frac{\partial^{j_r}}{\partial z^{j_r}} f(z_1, z_2, \ldots, z_r)$$

span a finite dimensional vector space over the field of rational fractions $\mathbb{C}(z_1, z_2, \ldots, z_r)$.

A sequence $f_{n_1,n_2,\ldots,n_r}$ is holonomic iff its generating function $f(z_1, z_2, \ldots, z_r) = \sum_{n_1,n_2,\ldots,n_r} f_{n_1,n_2,\ldots,n_r} z_1^{n_1} z_2^{n_2} \cdots z_r^{n_r}$ is holonomic.

The major closure theorem here is due to Stanley, Lipschitz, and Zeilberger [59, 60, 73, 83].

Theorem 8 (Holonomic Closure) *Holonomic functions are closed under sums, products, Hadamard products, diagonals, algebraic substitutions, integration, differentiation, direct and inverse Laplace transforms.*

Coefficients sequences enjoy the corresponding closure properties. For instance closure under sum, product, convolution, summation, multiplication and division by polynomials in n. Many combinatorial quantities that are expressible as multiple summations of multinomial coefficients with linear constraints (these are sometime called 'multihypergeometric') are in particular holonomic. Using a theory of holonomic symmetric functions, Gessel [41] has established that the generating functions of k–regular graphs and k–Latin rectangles are holonomic. With the closure theorem, Massazza [62] has shown that context-free languages with linear constraints on the number of occurrences of letters have holonomic generating functions. Finally the cost sequences for usual variants of Quicksort and comparison–based search are clearly holonomic.

A major interest of holonomic sequences is that the identities they satisfy form a decidable class. In the 1–dimensional case, the corresponding asymptotic properties are also in essence decidable. The spirit of the available result is captured by the following informally stated theorem.

Theorem 9 (Holonomic Asymptotics) *A holonomic sequence f_n is asymptotic to a sum of elements of the form*

$$\lambda(n!)^{r/s}e^{Q(n^{1/m})}\omega^n n^\alpha(\log n)^k,$$

where r, s, m, k are integers, Q is a polynomial and λ, ω, α are complex numbers.

This theorem is originally due to Birkhoff and his students, and we refer to the useful discussion that Wimp and Zeilberger gave in [82]. The original proof is based on a direct treatment of difference equations in the complex plane.

In our perspective, this theorem relates to the classification of singularities of linear differential equations. The theory of linear differential equations with analytic coefficients [80], distinguishes for solutions of such equations two cases, the regular case and the irregular case. Singular expansions of solutions are then locally composed of elements of one of two types,

$$(1-z/\rho)^{\alpha'}(\log(1-z/\rho))^{k'}, \quad (1-z/\rho)^{\alpha'}(\log(1-z/\rho))^{k'}\exp(\tilde{Q}(1-z/\rho)^{1/m'}).$$

The method of singularity analysis and the method of saddle point integrals are applicable each in one of the two cases. The resulting forms found for coefficients are exactly the ones stated in the theorem on holonomic asymptotics.

The analysis of regular singularities has given results on various multidimensional search problems in k-d trees and quadtrees [30, 26]. For instance the expected cost of a partial match query in a quadtree (alternatively a k-d-tree) when a proportion of $\frac{1}{2}$ or $\frac{2}{3}$, of the coordinates is of the order of

$$n^{(\sqrt{17}-3)/2} \quad \text{and} \quad n^{\theta-1} \text{ with } \theta = \left(\frac{109}{27}+\sqrt{\frac{1320}{81}}\right)^{1/3} + \left(\frac{109}{27}-\sqrt{\frac{1320}{81}}\right)^{1/3}.$$

uch algebraic numbers in the exponents (!) are typical of $\mathbb{Q}(z)$ holonomic functions.

Irregular singularities occur more seldom, a clear example being the number of in-reasing subsequences of a random permutation of n elements which was determined y Pittel and Lifschitz [58] and is asymptotic to

$$\frac{1}{2\sqrt{e\pi}} n^{-1/4} e^{2\sqrt{n}}.$$

n particular, the longest increasing subsequence of the random permutation is roved to have length $O(\sqrt{n})$. (It is actually known to be asymptotic to $2\sqrt{n}$.)

In summary recurrences involving rational functions and summations can usually e treated by means of the theory of singularity of linear differential equations pplied to GF's and combined with singularity analysis or saddle point techniques. n essence, such problems fall into a decidable class.

Functional Equations and Iteration

Ve confine our discussion to *linear functional equations* of the form

$$f(z) = a(z) + b(z) f(\sigma(z)), \tag{14}$$

here $f(z)$ is the unknown function, and a, b, σ are explicitly known. The behaviour epends on the iteration structure of $\sigma(z)$. The formal solution to (14) is found by teration,

$$f(z) = \sum_{k=0}^{\infty} a(\sigma^{\langle k \rangle}(z)) \frac{B(\sigma^{\langle k \rangle}(z))}{B(z)} \qquad \text{with } B(z) = \prod_{j=0}^{\infty} B(\sigma^{\langle j \rangle}(z)), \tag{15}$$

$\sigma^{\langle k \rangle}(z)$ being the k-th iterate of σ.

In the functional equation of (14), everything depends crucially on the dynamics f the iterates of σ. In a few important cases, the iterates are explicit, and one eneral method available relies on the Mellin transform, some of whose uses are ecalled below. In the case of non explicit iterates, singularity analysis or saddle oint techniques have to be applied.

Mellin transforms. Mellin transforms constitute another set of techniques based n complex analysis methods. The Mellin transform of a real function $f(t)$ is defined by

$$f^*(s) = \int_0^{\infty} f(t) t^{s-1} dt.$$

A harmonic sum is a function of the form $\sum_k \lambda_k \varphi(\mu_k x)$, where $\varphi(x)$ is a base function, and the λ_k, μ_k play the rôles of amplitudes and frequencies. Using the inversion theorem for Mellin transforms and simple functional properties, we arrive at a method to evaluate harmonic sums asymptotically.

Theorem 10 (Mellin summation of harmonic sums) *Under conditions* *meromorphic continuation and smallness at* $\pm i\infty$,

$$\sum_k \lambda_k \varphi(\mu_k x) \sim \pm \sum_{\zeta \in H} \operatorname{Res} \left[\sum_k \frac{\lambda_k}{\mu_k^s} \cdot \int_0^\infty \varphi(t) t^{s-1} dt \right]_{s=\zeta},$$

where H *is a left half plane (resp. a right half plane), the sum of residues is over pole* ζ *in the half plane, the sign is* $+$ *(resp.* $-$*), and the expansion applies as* $x \to 0$ *(resp.* $x \to +\infty$*).*

The method of Mellin transforms in disguise lies at the heart of the proof of th prime number theorem (with additional difficulties resulting from the occurrence c the Riemann zeta function!). It is well suited to sums that occur in the analysis c algorithms, like

$$\sum_k 2^{-k} \varphi(x 2^{-k}), \quad \sum_k v_2(k) \varphi(kx), \quad \sum_k d(k) \varphi(kx),$$

where $\varphi(x)$ is often an exponential function e^{-x}, e^{-x^2}, and the coefficients are eithe powers of two or elementary arithmetic functions (e.g., the dyadic valuation $v_2(k$ or the divisor function $d(k)$).

This transform was first introduced in our range of problems by de Bruijn, Knut and Rice [17] for the purpose of analyzing the height of general plane trees (th expression involves the divisor function) which appears to be $\sim \sqrt{\pi n}$. It has als found many uses in the analysis of digital structures (prefix trees, tries, digital searc trees, suffix trees), and in a large number of related areas (protocols, probabilisti counting, carry propagation, dynamic hashing).

Explicit iterations. The analysis of digital tries furnishes an example of the situation where the iteration of $\sigma(z)$ is explicit. The recurrence of expected path lengt in tries is of a new probabilistic divide-and-conquer type,

$$f_n = n - \delta_{n,1} + 2 \sum_{k=0}^{n} \pi_{n,k} f_k \qquad \text{with } \pi_{n,k} = \frac{1}{2^n} \binom{n}{k}.$$

The corresponding EGF satisfies

$$f(z) = z(e^z - 1) + 2e^{z/2} f(\frac{z}{2}).$$

The equation is solved by iteration, after which the solution can be expanded. Th curious phenomenon occurring here is the presence of minute fluctuations in th behaviour of coefficients [55, p. 131]: The expected path length f_n of a trie of size n satisfies an estimate $n \log_2 n + nP(\log_2 n)$ where $P(u)$ has amplitude less than 10^{-5} Such periodicities are traceable to complex poles in a Mellin transform. For instance the transform of $\psi(x) = e^{-x} f(x)$ is

$$\psi^*(s) = \frac{\Gamma(s+1)}{1 - 2^{-s-1}},$$

hich has rightmost poles at $s = -1 + 2ik\pi/\log 2$. The Mellin summation formula hen leads to a Fourier series that expresses these fluctuations, and this in turn is eflected by corresponding fluctuations in the coefficients.

By now, a large number of applications have been given of this analysis technique nd some of its variants (like Rice's integrals). Here, we only refer to [33, 55, 79]. Aplications have been given in the area of trie searching and radix exchange sort [55], ynamic and extendible hashing [20], communication protocols [21], probabilistic ounting and estimation algorithms [27], quadtries and multidimensional searchng [30], suffix trees, Patricia trees and pattern matching in strings (see e.g. [77]), igital trees [31]. Surprising connections with identities of Ramanujan relative to nodular forms were uncovered by Kirschenhofer and Prodinger on the occasion of ariance analysis of digital structures [52, 53].

Divide–and–conquer algorithms also lead to explicitly solvable iterations, espeially when Mellin transforms are used. The recurrence,

$$f_n = e_n + f_{\lfloor n/2 \rfloor} + f_{\lceil n/2 \rceil},$$

vith e_n a known toll sequence, is typical. Let $f(z)$ be the corresponding OGF; the unctional equations is

$$f(z) = e(z) + (2 + z + \frac{1}{z})f(z^2).$$

Take the Mellin transform of $f(e^{-t})$. This is equivalent to considering the Dirichlet eries $\phi(s) = \sum_n f_n n^{-s}$. The series involves a denominator which resembles that of ries, being $1 - 2^{-s}$. Again, this introduces complex poles and fluctuations. However, n this case, we are lead to some fractal function expressing these fluctuations. A ypical case is the analysis of Mergesort [25], for which the expected cost is found to e of the form $\sim n \log_2 n - nQ(\log_2 n)$ for some fractal and periodic function $Q(u)$.

The method is applicable to wide classes of divide and conquer recurrences which re almost invariably found to give rise to periodic fluctuations involving fractals.

Implicit iterations. When the iterates $\sigma^{\langle j \rangle}(z)$ admit of no simple explicit form, one often has to resort to an analysis of individual terms in the sum (14), normally by the battery of complex analysis techniques examined so far. Odlyzko [66] considered the equation

$$f(z) = z + f(z^2 + z^3)$$

that arises when counting balanced 2-3-trees. What is needed is the behaviour of the iterates $\sigma(z) = z^2 + z^3$ near the dominant fixed point of σ, which is equal to the inverse of golden ration, $\varphi = (1 + \sqrt{5})/2$. A delicate analysis of this *singular iteration* problem eventually leads to the number of trees which is $\sim \frac{\varphi^n}{n} R(\log n)$ for some smooth periodic function $R(u)$.

A similar problem arises when analyzing the expected height of binary trees. The iteration

$$y_0(z) = 0; \quad y_h(z) = 1 + z(y_{h-1}(z))^2,$$

gives the OGF of trees whose height is bounded by h. The fixed point y_∞ is the OGF of all binary trees, i.e., the Catalan GF, $\frac{1-\sqrt{1-4z}}{2z}$. The y_h are polynomials of degree 2^{h-1}. Singularity analysis requires investigating the convergence of the y_h ne the singularity 1/4 of the fixed point y_∞. The iteration then becomes singular: f $|z| < \frac{1}{4}$, we have exponential convergence; for $z > 1/4$, there is a double exponenti divergence; at $z = \frac{1}{4}$, there is slow convergence of order $O(h^{-1})$. A fine analysis th reveals a logarithmic singularity for the GF of heights, and the expected height a binary tree with n nodes is found to be asymptotic to $2\sqrt{\pi n}$. Analogous resul hold for any simple family of trees in the sense of Meir and Moon.

At the moment, a complete classification of the various cases of (14) is sti lacking. Some cases appear to involve the theory of analytic iteration and son divergent series. We nonetheless have a number of useful and general tools availab in the form of Mellin transforms and iteration theory of analytic functions.

Part III: Combinatorial Schemas and Automatic Analysis

We have exposed here a few general theorems in symbolic combinatorics an complex asymptotics. They make it possible to approach the analysis of entire classe of problems in combinatorial enumerations and the analysis of algorithms. We sha be brief by necessity here and refer to our paper [32] for a more detailed discussio of the implications of these general methods.

Structure theorems and schemas. A first observation suggested by the result of Part II is that *certain combinatorial and analytic mechanisms can only lead t certain designated types of asymptotic behaviours.* We have seen that regular an context–free structures only lead to exponential polynomials and algebraic asymp totic elements respectively. The probability of a gambler's ruin in $2n$ stages is as sociated to a context free language, and accordingly this probability is asymptoti to $1/\sqrt{\pi n}$, a typical algebraic element. In contrast, logarithms cannot occur in thi range of problems.

The tools presented here are general enough that a large number of problem can be fitted into classes whose asymptotic properties are decidable. Our exam ple of 'trains' in Fig. 1 falls into the category of elementary iterative structures large subset of which has decidable asymptotic properties [32]. Structure theorem describe classes of syntactically specified combinatorial structures and algorithm whose asymptotic properties are decidable and expressible by a well characterize class of formulæ. Roughly speaking, such theorems seem to exist for all structure specified by the constructions of Theorems 1 and 2 excepting substitution. The fu programme of making them explicit is however a delicate task that is yet to b completed.

The relation between structural combinatorics and asymptotic form can b pushed further in order to include results on probability distributions (and not sim ply counting as a function of the size n). The approach is to be contrasted to th 'stochastic' approach that deals directly with continuous limit models like branch ing processes or Brownian motion. The stochastic approach has been successful i lving a number of problems that had resisted a more analytic attack. Most no-ably in this category, we find works of Aldous (diffusion processes and the height f digital search trees [3]), Devroye (branching processes and the height of digital earch trees [18]), Louchard (Brownian motion and interpolation search or dynamic nalysis of algorithms, see e.g. [61]), as well as others.

Our perspective is different, and it attaches itself to the area of multivariate com-lex asymptotics. Multiple inversions are then needed in order to recover coefficients om functions, and the methods draw upon a combination of complex asymptotic echniques which have been presented here, as well as theorems in analytic probabil-y, notably continuity theorems for characteristic functions or moment generating inctions. For instance the analytic scheme $F(z,u) = \exp(uC(z))$ expresses the fact hat an $\mathcal{F}$–structure is built of components of type $\mathcal{C}$. Under wide conditions, the umber of $\mathcal{C}$–components in a random $\mathcal{F}$–structure will obey a law that is Gaussian n the asymptotic limit, as $n \to \infty$. Thus, a common schema covers a variety of eemingly unrelated phenomena. In this way, we find Gaussian limit laws for the umber of cycles in a random permutation, the number of factors of a random poly-omial over $GF(q)$, or the number of components in a random mapping of large ize [35]. First results along these lines were derived by Bender, Canfield and Rich-nond [4, 6, 8]. A classification of some major schemas and their associated laws is iven in Soria's thesis [71]. Even for a structure as complicated as random trains, t is the case that all probability distributions of various components can be char-cterized in their asymptotic form: Non–classical laws as well as standard laws like he Gaussian, geometric and Poisson laws appear in such a structure.

utomatic Analysis The approach of finding general decidable asymptotic prop-rties of combinatorial structures has been prolonged. Flajolet, Salvy and Zimmer-nann [32] have designed a system called Lambda–Upsilon–Omega ($\Lambda\Upsilon\Omega$) that im-lements a number of decision procedures on combinatorial structures like the ones iscussed here. The kernel specification language consists of the constructions of nion, product, sequence, sets, multisets and cycles described in Section 1. The $\Lambda\Upsilon\Omega$ ystem also makes provisions for specifying traversal algorithms on the structures.

A first component of $\Lambda\Upsilon\Omega$ implements the automatic computation of counting generating functions and complexity descriptors that are cost generating functions. Zimmermann [84] has developed the necessary theory which builds on the principles of our Section 1, and he has also found a number of extensions most notably to boolean procedures [85], to some forms of composition, and to exact counting.

After a solving phase, the (usually complicated) generating functions produced need to be subjected to an automatic analysis of their coefficients. Salvy [69] has developed a collection of efficient decision algorithms in the style of computer algebra in order to manipulate general asymptotic scales in Hardy fields and apply transfers from functions to coefficients in the style of the methods of Section 2.

In its current stage, $\Lambda\Upsilon\Omega$ consists of some 20,000 instructions written largely in the computer algebra system Maple. It has provided so far more than 50 different analyses of combinatorial problems and algorithms related to regular languages, finite automata, random walks, term tree, rewriting systems, random mappings, and miscellaneous combinatorial problems. The system can assist experts in the analysis of combinatorial problems specifiable in this language and in a few case it has produced automatically results that had been published in the literature, s that its level of 'competence' is to be considered as reasonably good.

As a final conclusion, we have tried to present a global view of the analysis classes of combinatorial objects, not unlike in spirit to what had been done earli for formal languages and zero–one laws in combinatorics [12, 13, 14]. General resul of an almost 'logical' nature relate *combinatorial structure* and *asymptotic form*. I a number of such cases, decision procedures can be found for asymptotic combina torics. Their development and their implementation within computer algebra is fascinating new area of investigation.

Acknowledgements. The author is grateful to Kevin Compton, Philippe Duma Helmut Prodinger, and Michèle Soria for detailed and constructive comments on th manuscript.
This work was supported in part by the ESPRIT III Basic Research Action Pro gramme of the E.C. under contract ALCOM II (#7141).

References

[1] ALBERT, L., CASAS, R., FAGES, F., TORRECILLAS, A., AND ZIMMERMANN, P. Av erage case analysis of unification algorithms. In *Proceedings of STACS'91* (1991) C. Choffrut and M. Jantzen, Eds., Lecture Notes in Computer Science, pp. 196–213

[2] ALBERT, L., AND FAGES, F. Average case analysis of the Rete pattern–matching algo rithm. In *Automata, Languages and Programming* (1988), T. Lepistö and A. Salomaa Eds., vol. 317 of *Lecture Notes in Computer Science*, Springer Verlag. Proceedings o 15th ICALP Colloquium, Tempere, Finland, July 1988.

[3] ALDOUS, D., AND SHIELDS, P. A diffusion limit for a class of randomly growing binar trees. *Probability Theory and Related Fields 79*, 4 (1988), 509–542.

[4] BENDER, E. A. Central and local limit theorems applied to asymptotic enumeration *Journal of Combinatorial Theory 15* (1973), 91–111.

[5] BENDER, E. A. Asymptotic methods in enumeration. *SIAM Review 16*, 4 (Oct. 1974) 485–515.

[6] BENDER, E. A., AND RICHMOND, L. B. Central and local limit theorems applied t asymptotic enumeration II: Multivariate generating functions. *Journal of Combinato rial Theory,* Series A *34* (1983), 255–265.

[7] BERSTEL, J., AND REUTENAUER, C. *Les séries rationnelles et leurs langages.* Masson Paris, 1984.

[8] CANFIELD, E. R. Central and local limit theorems for the coefficients of polynomials of binomial type. *Journal of Combinatorial Theory,* Series A *23* (1977), 275–290.

[9] CASAS, R., DIAZ, J., AND MARTINEZ, C. Statistics on random trees. In *Automata Languages, and Programming* (1991), J. Leach Albert et al., Ed., vol. 510 of *Lectur Notes in Computer Science*, pp. 186–203. Proceedings of the 18th ICALP Conference Madrid, July 1991.

[10] CASAS, R., FERNÁNDEZ CAMACHO, M.-I., AND STEYAERT, J.-M. Algebraic sim plification in computer algebra: An analysis of bottom-up algorithms. *Theoretica Computer Science 74*, 74 (1990), 273–298.

[11] CHOPPY, C., KAPLAN, S., AND SORIA, M. Complexity analysis of term rewriting systems. *Theoretical Computer Science 67* (1989), 261–282.

[12] COMPTON, K. J. A logical approach to asymptotic combinatorics. I. First order properties. *Advances in Mathematics 65* (1987), 65–96.

3] COMPTON, K. J. A logical approach to asymptotic combinatorics. II. Second-order properties. *Journal of Combinatorial Theory,* Series A *50* (1987), 110–131.

4] COMPTON, K. J. 0–1 laws in logic and combinatorics. In *Proceedings NATO Advanced Study Institute on Algorithms and Order* (Dordrecht, 1988), I. Rival, Ed., Reidel, pp. 353–383.

5] COMTET, L. *Advanced Combinatorics.* Reidel, Dordrecht, 1974.

6] DE BRUIJN, N. G. *Asymptotic Methods in Analysis.* Dover, 1981. A reprint of the third North Holland edition, 1970 (first edition, 1958).

7] DE BRUIJN, N. G., KNUTH, D. E., AND RICE, S. O. The average height of planted plane trees. In *Graph Theory and Computing* (1972), R. C. Read, Ed., Academic Press, pp. 15–22.

18] DEVROYE, L. Branching processes in the analysis of the heights of trees. *Acta Informatica 24* (1987), 277–298.

19] DIENES, P. *The Taylor Series.* Dover, New York, 1958. A reprint of the first Oxford University Press edition, 1931.

20] FAGIN, R., NIEVERGELT, J., PIPPENGER, N., AND STRONG, R. Extendible hashing: A fast access method for dynamic files. *A.C.M. Trans. Database Syst. 4* (1979), 315–344.

21] FAYOLLE, G., FLAJOLET, P., AND HOFRI, M. On a functional equation arising in the analysis of a protocol for a multiaccess broadcast channel. *Advances in Applied Probability 18* (1986), 441–472.

22] FLAJOLET, P. *Analyse d'algorithmes de manipulation d'arbres et de fichiers*, vol. 34–35 of *Cahiers du Bureau Universitaire de Recherche Opérationnelle.* Université Pierre et Marie Curie, Paris, 1981. 209 pages.

23] FLAJOLET, P. Analytic models and ambiguity of context-free languages. *Theoretical Computer Science 49* (1987), 283–309.

24] FLAJOLET, P. Mathematical methods in the analysis of algorithms and data structures. In *Trends in Theoretical Computer Science*, E. Börger, Ed. Computer Science Press, Rockville, Maryland, 1988, ch. 6, pp. 225–304. (Lecture Notes for *A Graduate Course in Computation Theory*, Udine, 1984).

25] FLAJOLET, P., AND GOLIN, M. Mellin transforms and asymptotics: The mergesort recurrence. Report, Institut National de Recherche en Informatique et en Automatique, January 1992. 11 pages.

26] FLAJOLET, P., GONNET, G., PUECH, C., AND ROBSON, J. M. The analysis of multidimensional searching in quad-trees. In *Proceedings of the Second Annual ACM-SIAM Symposium on Discrete Algorithms* (Philadelphia, 1991), SIAM Press, pp. 100–109.

[27] FLAJOLET, P., AND MARTIN, G. N. Probabilistic counting algorithms for data base applications. *J. Comput. Syst. Sci. 31*, 2 (Oct. 1985), 182–209.

[28] FLAJOLET, P., AND ODLYZKO, A. M. Random mapping statistics. In *Advances in Cryptology* (1990), J.-J. Quisquater and J. Vandewalle, Eds., vol. 434 of *Lecture Notes in Computer Science*, Springer Verlag, pp. 329–354. Proceedings of EUROCRYPT'89, Houtalen, Belgium, April 1989.

[29] FLAJOLET, P., AND ODLYZKO, A. M. Singularity analysis of generating functions. *SIAM Journal on Discrete Mathematics 3*, 2 (1990), 216–240.

[30] FLAJOLET, P., AND PUECH, C. Partial match retrieval of multidimensional data. *Journal of the ACM 33*, 2 (1986), 371–407.

[31] FLAJOLET, P., AND RICHMOND, B. Generalized digital trees and their difference–differential equations, Apr. 1991. 15 pages. INRIA Research Report 1423. To appear in *Random Structures and Algorithms.*

[32] FLAJOLET, P., SALVY, B., AND ZIMMERMANN, P. Automatic average-case analysis of algorithms. *Theoretical Computer Science, Series A 79*, 1 (February 1991), 37–109.

[33] FLAJOLET, P., AND SEDGEWICK, R. Digital search trees revisited. *SIAM Journal o[n] Computing 15*, 3 (Aug. 1986), 748–767.

[34] FLAJOLET, P., SIPALA, P., AND STEYAERT, J.-M. Analytic variations on the commo[n] subexpression problem. In *Automata, Languages, and Programming* (1990), M. S. Pa[-]terson, Ed., vol. 443 of *Lecture Notes in Computer Science*, pp. 220–234. Proceeding[s] of the 17th ICALP Conference, Warwick, July 1990.

[35] FLAJOLET, P., AND SORIA, M. Gaussian limiting distributions for the number o[f] components in combinatorial structures. *Journal of Combinatorial Theory,* Series A *53* (1990), 165–182.

[36] FLAJOLET, P., AND STEYAERT, J.-M. A complexity calculus for classes of recursiv[e] search programs over tree structures. In *Proceedings of the 22nd Annual Symposium o[n] Foundations of Computer Science* (1981), IEEE Computer Society Press, pp. 386–393

[37] FLAJOLET, P., AND STEYAERT, J.-M. A complexity calculus for recursive tree algo[-]rithms. *Mathematical Systems Theory 19* (1987), 301–331.

[38] GAMKRELIDZE, R. V., Ed. *Analysis I, Integral Representations and Asymptotic Meth[-]ods*, vol. 13 of *Encyclopedia of Mathematical Sciences*. Springer Verlag, 1989.

[39] GARDY, D. *Bases de données et allocations aléatoires: Quelques analyses de perfor[-]mance.* Doctorate in sciences, Université de Paris–Sud, 1989.

[40] GARDY, D. Méthode de col et lois limites en analyse combinatoire. *Theoretical Com[-]puter Science 92*, 2 (1992), 261–280.

[41] GESSEL, I. M. Symmetric functions and P–recursiveness. *Journal of Combinatori[al] Theory,* Series A *53* (1990), 257–285.

[42] GESSEL, I. M., AND STANLEY, R. P. Algebraic enumeration. Preprint, 1989. To appear as a chapter in the *Handbook of Combinatorics*,R. Graham, M. Grötschel and L. Lovász, Eds.

[43] GOULDEN, I. P., AND JACKSON, D. M. *Combinatorial Enumeration.* John Wiley, New York, 1983.

[44] GRAHAM, R., KNUTH, D., AND PATASHNIK, O. *Concrete Mathematics.* Addison Wesley, 1989.

[45] GREENE, D. H. *Labelled formal languages and their uses.* PhD thesis, Stanford University, June 1983.

[46] GREENE, D. H., AND KNUTH, D. E. *Mathematics for the analysis of algorithms.* Birkhauser, Boston, 1981.

[47] HARARY, F., AND PALMER, E. M. *Graphical Enumeration.* Academic Press, 1973.

[48] HENRICI, P. *Applied and Computational Complex Analysis.* John Wiley, New York, 1977. 3 volumes.

[49] HOSHI, M., AND FLAJOLET, P. Page usage in quadtree indexes. Report 1434, Institut National de Recherche en Informatique et en Automatique, May 1991. 19 pages. Accepted for publication in *BIT.*

[50] JOYAL, A. Une théorie combinatoire des séries formelles. *Advances in Mathematics 42*, 1 (1981), 1–82.

[51] KEMP, R. *Fundamentals of the Average Case Analysis of Particular Algorithms.* Wiley-Teubner, Stuttgart, 1984.

[52] KIRSCHENHOFER, P., AND PRODINGER, H. On some applications of formulæ of Ramanujan in the analysis of algorithms. *Mathematika 38* (1991), 14–33.

[53] KIRSCHENHOFER, P., PRODINGER, H., AND SZPANKOWSKI, W. On the variance of the external path length in a symmetric digital trie. *Discrete Applied Mathematics 25* (1989), 129–143.

[54] KNUTH, D. E. *The Art of Computer Programming,* vol. 1: Fundamental Algorithms. Addison-Wesley, 1968. Second edition, 1973.

5] Knuth, D. E. *The Art of Computer Programming*, vol. 3: Sorting and Searching. Addison-Wesley, 1973.

6] Knuth, D. E. The average time for carry propagation. *Indagationes Mathematicae 40* (1978), 238–242.

7] Knuth, D. E., and Schönhage, A. The expected linearity of a simple equivalence algorithm. *Theoretical Computer Science 6* (1978), 281–315.

8] Lifschitz, V., and Pittel, B. The number of increasing subsequences of the random permutation. *Journal of Combinatorial Theory,* Series A *31* (1981), 1–20.

9] Lipshitz, L. The diagonal of a D-finite power series is D-finite. *J. Algebra 113* (1988), 373–378.

0] Lipshitz, L. D-finite power series. *J. Algebra 122* (1989), 353–373.

1] Louchard, G. The Brownian motion: a neglected tool for the complexity analysis of sorted table manipulation. *RAIRO Theoretical Informatics 17* (1983).

2] Massazza, P. Holonomic functions and their relation to linearly constrained languages. Manuscript, 1991. 14 pages. Based on the authors's Ph. D. Thesis, University of Milan, 1991.

3] McKay, B. D. The asymptotic numbers of regular tournaments, eulerian digraphs and eulerian oriented graphs. *Combinatorica 10*, 4 (1990), 367–377.

4] Meir, A., and Moon, J. W. On the altitude of nodes in random trees. *Canadian Journal of Mathematics 30* (1978), 997–1015.

5] Odlyzko, A. M. Enumeration of strings. In *Combinatorial Algorithms on Words* (1985), A. Apostolico and Z. Galil, Eds., vol. 12 of *NATO Advance Science Institute Series.* Series F: Computer and Systems Sciences, Springer Verlag, pp. 205–228.

6] Odlyzko, A. M. Asymptotic enumeration methods. Preprint, Mar. 1992. To appear as a chapter in the *Handbook of Combinatorics*,R. Graham, M. Grötschel and L. Lovász, Ed.

67] Odlyzko, A. M., and Richmond, L. B. Asymptotic expansions for the coefficients of analytic generating functions. *Aequationes Mathematicae 28* (1985), 50–63.

68] Pólya, G., and Read, R. C. *Combinatorial Enumeration of Groups, Graphs and Chemical Compounds.* Springer Verlag, New York, 1987.

69] Salvy, B. *Asymptotique automatique et fonctions génératrices.* Ph. D. thesis, École Polytechnique, 1991.

70] Sedgewick, R. *Algorithms*, second ed. Addison–Wesley, Reading, Mass., 1988.

71] Soria-Cousineau, M. *Méthodes d'analyse pour les constructions combinatoires et les algorithmes.* Doctorat d'état, Université de Paris–Sud, Orsay, July 1990.

72] Stanley, R. P. Generating functions. In *Studies in Combinatorics,* M.A.A. Studies in Mathematics, Vol. 17. (1978), G.-C. Rota, Ed., The Mathematical Association of America, pp. 100–141.

73] Stanley, R. P. Differentiably finite power series. *European Journal of Combinatorics 1* (1980), 175–188.

74] Stanley, R. P. *Enumerative Combinatorics*, vol. I. Wadsworth & Brooks/Cole, 1986.

75] Steyaert, J.-M. *Structure et complexité des algorithmes.* Doctorat d'état, Université Paris VII, Apr. 1984.

76] Steyaert, J.-M., and Flajolet, P. Patterns and pattern-matching in trees: an analysis. *Information and Control 58*, 1–3 (July 1983), 19–58.

77] Szpankowski, W. Patricia tries again revisited. *Journal of the ACM 37*, 4 (1990), 691–711.

78] Titchmarsh, E. C. *The Theory of Functions*, second ed. Oxford University Press, 1939.

[79] VITTER, J. S., AND FLAJOLET, P. Analysis of algorithms and data structures. I *Handbook of Theoretical Computer Science*, J. van Leeuwen, Ed., vol. A: Algorithm and Complexity. North Holland, 1990, ch. 9, pp. 431–524.

[80] WASOW, W. *Asymptotic Expansions for Ordinary Differential Equations.* Dover 1987. A reprint of the John Wiley edition, 1965.

[81] WILF, H. S. *Generatingfunctionology.* Academic Press, 1990.

[82] WIMP, J., AND ZEILBERGER, D. Resurrecting the asymptotics of linear recurrences *Journal of Mathematical Analysis and Applications 111* (1985), 162–176.

[83] ZEILBERGER, D. A holonomic approach to special functions identities. *Journal o Computational and Applied Mathematics 32* (1990), 321–368.

[84] ZIMMERMANN, P. *Séries génératrices et analyse automatique d'algorithmes.* Thèse d Doctorat, École Polytechnique, Palaiseau, France, 1991.

[85] ZIMMERMANN, P. Analysis of functions with a finite number of return values, Feb 1992. 12 pages.

How to Count Quickly and Accurately:

A Unified Analysis of Probabilistic Counting and Other Related Problems

Peter KIRSCHENHOFER* Helmut PRODINGER*
Technical University of Vienna

Wojciech SZPANKOWSKI†
Purdue University, West Lafayette

Abstract. *We consider a class of probabilistic counting algorithms parameterized by an integer $d \geq 0$ that estimate the number of elements N in a large set. Our algorithms generalize an idea of Flajolet and Martin who limited themselves to the case $d = 0$. As noted by Brassard and Bratley "it is far from obvious how to carry out a more precise analysis of the unbiased estimate of N ...". We present a novel and complete analysis of these new counting algorithms that – to the best of our knowledge – cannot be obtained by an extension of the analysis by Flajolet and Martin. We present results concerning the average value, the variance and the limiting generating function of an estimate of N. Moreover, our novel approach is not limited to probabilistic counting algorithms, and it can be applied in the investigation of several other "splitting algorithms" such as selecting the loser within a group of people, estimating the number of questions necessary to identify the number of distinct objects, searching algorithms based on digital tries, approximate counting, electing d finalists in a contest (cf. polling system), and so forth.*

1 Introduction

Flajolet and Martin in their paper [2] proposed an algorithm in order to estimate the cardinality of large (multi)sets of data. This algorithm is probabilistic in nature and works as follows. In order to estimate the cardinality N of a (multi)set $\mathcal{M}$ every element $x \in \mathcal{M}$ is hashed into a binary string of size m (the choice of m is easy, and $m = 5 + \log N$ suffices [1]). The bitwise OR-composition of modified hashed strings is used to build the so called *bitmap* and to obtain the estimate R_N of the

*Department of Algebra and Discrete Mathematics, Technical University of Vienna, Wiedner Hauptstrasse 8–10, A-1040 Vienna, Austria. This work was supported by Fonds zur Förderung der Wissenschaftlichen Forschung Project P7497-TEC

†Department of Computer Science, Purdue University, West Lafayette, IN 47907, U.S.A. This research was supported by the NSF Grant INT-8912631, and in part by AFOSR grant 90-0107, by the NSF grant CCR-8900305, and by grant R01 LM05118 from the National Library of Medicine.

cardinality N of $\mathcal{M}$. More precisely, the position of the leftmost zero in the *bitma* string approximates $\log_2 N$ (for details see below and [2], [1]).

The original algorithm turns out to have the variance of the estimator R_N ap proximately to 1.257, which is rather large for practical implementations. Therefor a modification of the basic algorithm was also proposed in [2] but this new algorithn is more complex and costly.

In our paper, we propose for the instance of sets without replications a new esti mate $R_{N,d}$ of N that is parameterized by an integer d. The value $d = 0$ correspond to Flajolet-Martin's basic algorithm. Our algorithm has smaller variance (apart forn fluctuations the value tending to $1/12 = 0.083$ as $d \to \infty$). The new idea behind ou algorithm is to *count* the number of 1's up to $d+1$ in every position of the *bitmap* As an estimate of N we propose to take the leftmost position that contains the valu smaller than $d+1$. More formally, our algorithm works as follows. We assume tha there exists a carefully implemented hashing function $hash: \mathcal{M} \to \{0,1\}^m$ wher m is chosen as in [2]. Let y be the image of $x \in \mathcal{M}$ over the $hash(\cdot)$ function, tha is, $y = hash(x)$. The procedure below constructs another string called *bitmap* tha is used to estimate N. As in [2], we define the function $\rho(y)$ for a binary string y a the leftmost position that contains value 1.

```
procedure BITMAP
for i = 0 to m − 1 do bitmap(i) = 0 ;
for x ∈ M do
    begin
    y = hash(x) ;
    index := ρ(hash(x))
    bitmap(index) = min{bitmap(index) + 1, d + 1};
    end;
```

Note that the resulting string *bitmap* is a $d+1$-ary string. For example, for $d =$ 3 we may have *bitmap* = 444444444434100000. As the estimate $R_{N,d}$ we take th position of the leftmost value smaller than $d+1$ (e.g., in the example above we hav $R_{N,3} = 11$). More precisely we define:

$$R_{N,d} = \min\{k : \; bitmap(k) < d+1 \text{ and for all } 0 \le i < k \;\; bitmap(i) = d+1\} \,. \quad (1)$$

It is easy how our algorithm works. The *bitmap* string is a sequence of integers from the set $\{0, 1, \ldots, d+1\}$. If $bitmap(i) = k < d+1$, then exactly k strings y have at the ith position the leftmost one. Moreover, $bitmap(i) = d+1$ means that *at leas* $d+1$ strings y contain the leftmost one at position i.

We now summarize our main results concerning the probabilistic behavior of $R_{N,d}$. In particular, we prove that $ER_{N,d} = \log_2 N - 1 - C_d + P(\log_2 N) + O(\sqrt{N})$ where C_d is a constant such that $C_d = \log_2 d - 0.5 + O(d^{-1/2+\epsilon})$ as $d \to \infty$, and $P(x)$ is a periodic fluctuation of period 1. More importantly, we show that the varianc

: $R_{N,d}$ fulfills, neglecting periodic fluctuations, var$R_{N,d} \to 1/12 = 0.08333\ldots$ as $\to \infty$. Interestingly enough the quantity 1/12 is not uncommon in such a situation nd known as Sheppard's correction (cf. [9]) which says that the discretization by self introduces some fluctuations with the variance equal to 1/12. Finally, in the ext section we make some comments concerning the distribution and the limiting enerating function for $R_{N,d}$.

The reader might have noticed that for $d \geq 1$ our algorithm does not necessarily eglect multiplicities. To be more precise: Only multiplicities greater than d will be gnored by the algorithm. Therefore for data base applications a priori knowledge bout the expected multiplicities of the different elements is necessary. In this paper e confine our analysis on the quantity $R_{N,d}$, which is a good estimator provided here are no multiplicities.

Literature on the probabilistic counting is rather scanty. It was initiated by the aper of Flajolet and Martin [2] who analyzed the case $d = 0$ of our algorithm. 'heir analysis is based on an application of the inclusion-exclusion rule, and – to he best of our knowledge – cannot be extended to $d > 0$. This is also evident rom the analysis of Greenberg, Flajolet and Ladner [6] who analyzed a conflict esolution algorithm which resembles the case $d = 1$ of our analysis. A simplifica-ion of Flajolet-Martin's analysis is presented in Kirschenhofer and Prodinger [11]. 'inally, a different estimator for the probabilistic counting algorithm was analyzed ı Szpankowski and Rego [20] who investigated the maximum over N geometrically istributed random variables. The authors of [20] considered the *last* bit that is qual to one in the *bitmap*. Unfortunately, in this case the variance is even bigger nd equal to 3.507.

It must be stressed that the novelity of our paper also lies in the area of the math-matical analysis of algorithms. In fact, our technique of analysis can be applied to everal other algorithms. As shown in the section below, we reduce our probabilistic :ounting algorithm to a problem on *(digital) trees*. Evaluating a parameter of such tree, we obtain the following functional equation

$$g(z) = \beta a(z/2)g(z/2) + b(z)\,. \tag{2}$$

n general, this equation does not have an explicit finite solution unless $a(z)$ is a very articular function. For example, $a(z) = e^z$ was discussed in Knuth [12] (cf. also 19]), and Szpankowski [18], [19] (cf. also [12]) investigated the case $a(z) = e^z + 1$. For these two cases explicit final solutions of (2) were obtained, but it is rather iopeless to expect the existence of a useful closed form solution for general $a(z)$ and (z). An asymptotic solution to (2) was first suggested in Jacquet and Szpankowski 8] who analyzed the case $a(z) = 1 + (1+z)e^{-z}$. Recently, Flajolet and Richmond 4] investigated asymptotically the above equation with $a(z) = 2z^b/(1+z)^b$ for some nteger $b > 0$. Here, analyzing probabilistic counting algorithms, we must investigate (2) with $a(z) = 1 - e_d(z)e^{-z}$ where $e_d(z) = \sum_{i=0}^{d} z^i/i!$.

2 Main Results

In this section we present our main results. We start with another description of th probabilistic counting algorithm which will slightly differ from the one presented i Section 1, but it helps us to establish recurrence equations necessary for the analysis

Our new description goes as follows. Let us consider an empty *bitmap* string, tha is, with all positions filled by zeros. Assume that N objects (e.g., data, persons, etc. can randomly insert (hit) a 1 at any position of the *bitmap*, however, probability c hitting the $j \geq 0$ position is equal to 2^{-j-1}. In terms of the probabilistic counting this means that the probability of the occurrence of the pattern like $0^j 1 \cdots$ is equal t 2^{-j-1} since 0 and 1 are equally likely (0^j denotes j consecutive zeros). Every objec can hit only one time. In addition, we count the number of hits in any position c the *bitmap*, but we count the number of hits only up to some value $d+1$, wher d is a given parameter. In other words, the *bitmap* is a $d+1$-ary string. Clearly the *bitmap* will contain a lot of $d+1$'s at the begining of the string. This is due t the fact that the probability of a hit decreases exponentially fast with the increas of the position in the *bitmap*. The parameter of interest is the length $R_{N,d}$ of th longest run of $d+1$'s in the front of the *bitmap*. A precise definition of $R_{N,d}$ is give in (1).

Let $F_N(z) = Ez^{R_{N,d}}$ be the probability generating function of the estimato $R_{N,d}$. That is, $[z^j]F_N(z)$ (the jth coefficient at $F_N(z)$) is the probability that th *bitmap* is $(d+1)^j c \dots$ where $c < d$, and $(d+1)^j$ denotes j consecutive values "$d+1$" To derive the recurrence for $F_N(z)$, one may represent the action of our algorithm a a splitting process in a *digital tree*. Imagine N persons flipping a fair coin, and thos who get 1 (a hit) move to the right while the rest to the left. The process continue until less than $d+1$ persons flip a 1. Imagine that at the first step k person flip 1. Then, the contribution to the generating function $F_N(z)$ is as follows: i $k \leq d$ then we need to add $z^0 \sum_{k \leq d} \binom{N}{k} 2^{-N}$, otherwise the contribution become $z \sum_{d+1 \leq k \leq N} \binom{N}{k} 2^{-N} F_{N-k}(z)$. In short, we obtain the following recurrence

$$F_N(z) = \sum_{k=0}^{d} \binom{N}{k} 2^{-N} + z \sum_{k=0}^{N-d-1} \binom{N}{k} 2^{-N} F_k(z) , \qquad (3$$

which is valid for all $N \geq 0$.

Using (3) we can derive recurrences for the first and second moments of $R_{N,d}$ namely: $ER_{N,d} = F'_N(1)$ and var $R_{N,d} = F''_N(1) + F'_N(1) - (F'_N(1))^2$. In particular let $L(z)$ and $W(z)$ be the exponential generating function for $F'_N(1)$ and $F''_N(1$ respectively. That is, $L(z) = \sum_{N \geq 0} F'_N(1) z^N / N!$ and $W(z) = \sum_{N \geq 0} F''_N(1) z^N / N!$ Define also $\widetilde{L}(z) = e^{-z} L(z)$ and $\widetilde{W}(z) = e^{-z} W(z)$. Then, (3) implies

$$\widetilde{L}(z) = f_d(z/2) \widetilde{L}(z/2) + f_d(z/2) \qquad (4$$

$$\widetilde{W}(z) = f_d(z/2) \widetilde{W}(z/2) + 2\widetilde{L}(z) f_d(z/2) \qquad (5$$

here $f_d(z) = 1 - e_d(z)e^{-z}$ and $e_d(z) = 1 + z^1/1! + \cdots + z^d/d!$.

Both functional equations fall into the general equation (2) which we repeat elow

$$g(z) = \beta a(z/2)g(z/2) + b(z)\ . \tag{6}$$

here $\beta \leq 1$. Iterating (6), we obtain a general solution in the form of (cf. [8])

$$g(z) = \sum_{n=0}^{\infty} \beta^n b(z2^{-n}) \prod_{k=1}^{n} a(z2^{-k}) \tag{7}$$

f $g(0) = 0$. Define

$$\varphi(z) = \prod_{j=0}^{\infty} a(z2^j)\ , \tag{8}$$

rovided the infinite product in (8) converges. Then, the general solution (7) can e rewritten as

$$g(z)\varphi(z) = \sum_{n=0}^{\infty} \beta^n b(z2^{-n})\varphi(z2^{-n})\ . \tag{9}$$

The last form is the clue to our asymptotic solution of the functional equation 7), and in particular to solve (4) and (5). In Section 3 we present our detailed lerivations.

Now we are ready to present our main results.

Theorem 1 *Consider the probabilistic counting algorithms described above, and in particular the quantity* $R_{N,d}$.

(i) *The average value of* $R_{N,d}$ *becomes asymptotically as* $N \to \infty$

$$ER_{N,d} \sim \log_2 N - 1 - C_d + P_1(\log_2 N) + O(N^{-1/2+\varepsilon}) \tag{10}$$

where $P_1(x)$ *is a periodic function with period 1 and mean zero, and with* $L = \log 2$ *we have*

$$\begin{aligned} C_d &= \frac{1}{L^2}\int_0^{\infty} e^{-x} e_d(x)\varphi(2x)\frac{\log x}{x}dx \\ &\sim \log_2 d - 0.5 + O(d^{-1/2+\varepsilon}) \qquad as \quad d \to \infty\ . \end{aligned} \tag{11}$$

(ii) *The variance of* $R_{N,d}$ *is for* $N \to \infty$

$$\mathrm{var} R_{N,d} \sim D_d - C_d^2 + P_2(\log_2 N) + O(N^{-1/2+\varepsilon}),$$

where

$$D_d - C_d^2 \sim \frac{1}{12} + O(d^{-1/2+\varepsilon}) \qquad as \quad d \to \infty\ . \tag{12}$$

In particular

$$\begin{aligned} D_d &= \frac{1}{L^3}\int_0^\infty e_d(x)e^{-x}\varphi(2x)\frac{\log^2 x}{x}dx \\ &\sim -C_d - 1/6 + \log_2^2 d + O(d^{-1/2+\varepsilon}) \quad as \quad d\to\infty\ , \end{aligned} \tag{13}$$

and $P_2(x)$ is a periodic function with period 1. ∎

Using numerical integration, we computed for the first few values of d the constants $ER_{N,d}-\log_2 N - P_1(\log_2 N)$ and var $R_{N,d}-P_2(\log_2 N)$. They are displayed in the table below. We note that the variance decreases substantially for initial values of d ($d\le 2$) and then the decrease slows down. From the table, one may conclude that $d\ge 5$ is a good practical choice.

Table 1: Performance of the algorithm for small values of d (neglecting fluctuations).

d	$ER_{N,d}-\log_2 N$	var $R_{N,d}$
0	-0.37	1.26
1	-1.40	0.78
2	-2.00	0.59
3	-2.42	0.49
4	-2.74	0.43
5	-3.02	0.38
6	-3.24	0.35
7	-3.43	0.32

Our technique also allows to deal with the limiting distribution for $R_{N,d}$ and the limiting generating function. For this, we introduce the Poisson transform of $F_N(z)$ as

$$G(z,x) = e^{-x}\sum_{N=0}^{\infty} F_N(z)\frac{x^N}{N!}\ . \tag{14}$$

Then, after some algebra, one proves that $H(z,x)=G(z,x)-1$ satisfies the following recurrence

$$H(z,x) = 2f_d(x/2)H(z,x/2) + (z-1)f_d(x/2)\ , \tag{15}$$

which falls into our general functional equation (6). In particular, as above we can write a solution for $H(z,x)$ as

$$\varphi(x)H(z,x) = (z-1)\sum_{n=0}^{\infty} z^n\varphi(x2^{-n-1})\ , \tag{16}$$

where $\varphi(x)$ is defined in (8).

With the help of (16), we are able to derive the limiting generating function for $F_N(z)$ as N becomes large.

Theorem 2 *Consider the probabilistic counting algorithm as in Theorem 1. For large N the limiting generating function becomes*

$$F_N(z) \sim z^{\log_2 N} \varphi^*(-\log_2 z)/L + P_3(\log_2 N) + O(N^{-1/2+\epsilon}) \,. \tag{17}$$

where $\varphi^*(s)$ is the Mellin transform of $\varphi(x)$ defined in (8). ∎

In addition, in Section 3, we will present a conjecture (28) concerning the limiting distribution for $R_{N,d}$. Details and more precise statements are left to a journal version of the paper.

For the reader's convenience we want to point out that the definition and basic properties of the *Mellin integral transform* are included in the paper [17] by U. Schmid *in this volume.*

Remarks

(i) *Special Cases.* As mentioned before, the case $d = 0$ was analyzed by Flajolet and Martin [2]. In addition, Greenberg *et al.* [6] investigated the function $\varphi(x)$ for $d = 1$ in the analysis of a conflict resolution algorithm for multiple channels. It must be stressed that the analysis of Flajolet and Martin [2] is based on the inclusion-exclusion rule, and seems to be very unlikely extendable to $d \geq 1$. This is partially evident from the analysis in [6]. We note that our analysis is based on a quite different approach, namely, functional equations, and can be proved very successful for some other algorithms (cf. Remark (iii)).

(ii) *Convergence in Probability.* Theorem 1 implies that $R_{N,d}$ converges *in probability* (pr.) to $ER_{N,d}$. More precisely, by Theorem 1 and Chebyshev's inequality we have

$$\Pr\{|R_{N,d}/ER_{N,d} - 1| > \varepsilon\} \leq \frac{\mathrm{var} R_{N,d}}{\varepsilon^2 E^2 R_{N,d}} \sim \frac{const}{\log_2^2 N} \to 0$$

as $N \to \infty$. This proves the announced convergence. It is not clear, however, whether one can extend this convergence to the *almost sure* convergence.

(iii) *Other Problems.* As mentioned in the abstract, our technique might be applied to analyze several other algorithms that are based on the "splitting process" described at the beginning of this section. More precisely, a splitting algorithm divides randomly N objects according to some rule. A good representation of such a splitting process is a digital tree (i.e., a trie [12]). For example, consider a multiaccess system with a large number of users. The following "conflict resolution" algorithm found to be very successful for some multiaccess systems (cf. [6], [8]). Let N be the size of a conflict, that is, the number of users that simultaneously sent packets. To split it, every user flips a fair coin, and those who get 1 "move right" in a digital tree represented this process, while the others "move left". This splitting process continues until all users sent successfully their packets (i.e., subtrees containing conflicting users are of size one). This algorithm can be analyzed in a unified manner by our approach through the functional equation of type (6) and its solution (9 Here is another example that generalizes PATRICIA tries (cf. [12], [19]). Conside N infinite strings that are built over a V-ary alphabet. Using a trie construction a in Knuth [12] (see also [19]) we can split all strings in such a manner that no tw of them will share the common prefix. A trie built in such a manner may contai some internal nodes that are unary. To avoid such a waste of storage, one ma compress this trie to obtain a PATRICIA trie. In the PATRICIA all internal node have degree greater or equal to two. But a generalization of a PATRICIA trie migh be also explored by imposing that all internal nodes have degree of size greater c equal to $d \leq V$. What is the search time (i.e., the length of a path to an externa node) in such a trie? What about the average height, etc.? The analysis of suc a data structure seems to be a nontrivial task, and our approach can be applie to obtain some asymptotics. Finally, several other problems fall into the "splittin process" scenario. We mention here: selecting the loser [15], estimating the numbe of questions necessary to identify the number of distinct objects [14], selecting finalists in a contest where questions are answered randomly, and so forth.

3 Analysis

In this section we present a sketch of proof for Theorem 1. We also offer som remarks regarding Theorem 2.

We start with the average value $ER_{N,d}$. Its modified exponential generatin function $\widetilde{L}(z)$ satisfied the functional equation (4) which we repeat below

$$\widetilde{L}(z) = f_d(z/2)\widetilde{L}(z/2) + f_d(z/2) \quad (18$$

where $f_d(z) = 1 - e_d(z)e^{-z}$ and $e_d(z) = 1 + z^1/1! + \cdots + z^d/d!$. Iterating thi equation, as in (6) – (9), we finally obtain

$$\widetilde{L}(z)\varphi(z) = \sum_{n=0}^{\infty} \varphi(z2^{-n-1}) , \quad (19$$

where

$$\varphi(z) = \prod_{j=0}^{\infty} f_d(z2^j) , \quad (20$$

or, equivalently,

$$\widetilde{L}(z) = \sum_{n=0}^{\infty} \prod_{j=1}^{n+1} f_d\left(z2^{-j}\right) .$$

Observe that, by d'Alembert's criterion, the final expression converges absolutel for all complex numbers z.

The idea of our analysis is to obtain an asymptotic behavior of $\widetilde{L}(z)$ as $z \to \infty$ and then by *depoissonization* [8] to evaluate the coefficients $ER_{N,d}$ of $L(z)$. Tc obtain the asymptotics of $\widetilde{L}(z)$ we apply the Mellin transform technique (cf. [5] and [3]). We shall use the following depoissonization lemma proved in Rais, Jacquet and Szpankowski [16]. (An older and less versatile version of the depoissonization lemma may be found in Jacquet and Szpankowski [8].)

Depoissonization Lemma *Let S_θ be a cone $S_\theta = \{z : |\arg z| < \theta\ ,\ 0 < \theta < \pi/2\}$. If for $z \in S_\theta$ and $z \to \infty$ the following holds for all u in a compact set $\mathcal{U}$*

$$|D(z,u)| < \beta_1 |z|^\varepsilon \tag{21}$$

for some $\beta_1, \varepsilon > 0$, and for $z \notin S_\theta$ and all $u \in \mathcal{U}$

$$|D(z,u)e^z| < \beta_2 |z|^\varepsilon e^{\alpha |z|} \tag{22}$$

for some $0 < \alpha < 1$ and a constant $\beta_2 > 0$, then for large n uniformly in $u \in \mathcal{U}$ the generating function $D_n(u) = [z^n/n!]D(z,u)e^z$ satisfies

$$D_n(u) = D(n,u) + O(n^{-1/2+\varepsilon}) \tag{23}$$

with $\varepsilon < 1/2$. ■

Using the above idea, we first apply the Mellin transform to (19). Let $Q_1(z) = \widetilde{L}(z)\varphi(z)$ and $Q_1^*(s)$ be its Mellin transform. Since the sum in (19) is a harmonic sum, we have (cf. [5])

$$Q_1^*(s) = \frac{2^s}{1-2^s}\varphi^*(s) = \left(\frac{2^s}{1-2^s}\right)^2 \Phi^*(s) \tag{24}$$

where $\varphi^*(s)$ is the Mellin transform of $\varphi(z)$, and $\varphi^*(s) = \Phi^*(s)/(2^{-s}-1)$. Note that $\Phi^*(s)$ is an entire function (cf. [6]). Using the inverse of the Mellin transform (by computing the residues of $Q_1^*(s)$) and $\varphi(z) \sim 1 + O(z^{-M})$, any $M > 0$, we obtain for $z \to \infty$

$$\widetilde{L}(z) \sim \log_2 z \cdot \frac{\Phi^*(0)}{L} - \frac{\Phi^*(0)}{L} - \frac{\Phi^{*\prime}(0)}{L^2} + P(z) \tag{25}$$

where

$$\Phi^*(0) = \int_0^\infty e^{-x} e_d(x)\varphi(2x)\frac{dx}{x}\ ,$$

$$\Phi^{*\prime}(0) = \int_0^\infty e^{-x} e_d(x)\varphi(2x)\frac{\log x}{x}dx\ ,$$

and $P(z)$ is a fluctuating function with a small amplitude.

We need to evaluate $\Phi^*(0)$ and $\Phi^{*\prime}(0)$. The former is rather easy, and by properties of the Mellin transform we find out that $\Phi^*(0) = L = \log 2$. The latter quantity is harder to estimate. To do so, we first present another evaluation of $\Phi^*(0)$ which can be extended to obtain $\Phi^{*\prime}(0)$. Note that $\varphi(2x) - \varphi(x) = e^{-x}e_d(x)\varphi(2x)$. Define also a function $1(x)$ as follows

$$1(x) = \begin{cases} 1 & \text{if } x \geq 1 \\ 0 & \text{if } x < 1\ . \end{cases}$$

Then, we can write

$$
\begin{aligned}
\Phi^*(0) &= \int_0^\infty (\varphi(2x)-\varphi(x))\frac{dx}{x} \\
&= \int_0^\infty (\varphi(2x)-1(2x))\frac{dx}{x} + \int_0^\infty (1(2x)-1(x))\frac{dx}{x} + \int_0^\infty (1(x)-\varphi(x))\frac{dx}{x} \\
&= \int_0^\infty (1(2x)-1(x))\frac{dx}{x} = \int_{1/2}^\infty \frac{dx}{x} = L\,.
\end{aligned}
$$

In a similar manner we can deal with $\Phi^{*\prime}(0)$. This time, however, the calculation is more involved. Nevertheless, we obtain the following

$$
\begin{aligned}
\Phi^{*\prime}(0) &= -L^2/2 - L\int_0^\infty (1(x)-\varphi(x))\frac{dx}{x} \\
&= \int_0^1 \varphi(x)\frac{dx}{x} + \int_1^\infty (\varphi(x)-1)\frac{dx}{x} \\
&\sim O(d^{-1/2}) - \int_1^\infty e_d(x)e^{-x}\frac{dx}{x} \qquad d\to\infty
\end{aligned}
$$

where the last integral can be analyzed using the *incomplete gamma* function. In particular, we can prove that for $d\to\infty$ we have $C_d = \frac{1}{L^2}\Phi^{*\prime}(0) = \log_2 d - 1/2 + O(d^{-1/2+\epsilon})$. Finally, using the Depoissonization Lemma we establish Theorem 1(i).

The proof of Theorem 1(ii) concerning the variance of $R_{N,d}$ is more intricate but it proceeds along the same lines as the above derivation. In particular, we deal now with the functional equation (5) which we repeat below

$$
\widetilde{W}(z) = f_d(z/2)\widetilde{W}(z/2) + 2\widetilde{L}(z)f_d(z/2)\,. \tag{26}
$$

Using the idea from Section 2 (cf. (6) – (9)) we find immediately an explicit solution to (26), namely

$$
\widetilde{W}(z)\varphi(z) = 2\sum_{n=0}^{\infty}\widetilde{L}(z2^{-n-1})\varphi(z2^{-n-1})\,,
$$

which becomes, after inserting the solution (19) for $\widetilde{L}(z)$,

$$
\widetilde{W}(z)\varphi(z) = 2\sum_{j=1}^{\infty}\sum_{n=1}^{\infty}\varphi(z2^{-n-j})\,. \tag{27}
$$

To establish asymptotics of $\widetilde{W}(z)$, and also $F_N''(1)$, we proceed as before through the Mellin transform and depoissonization. Let $Q_2(z)$ and $Q_2^*(s)$ denote respectively the RHS of (27) and its Mellin transform. Then,

$$
Q_2^*(s) = 2\left(\frac{2^s}{1-2^s}\right)^2\varphi^*(s) = 2\left(\frac{2^s}{1-2^s}\right)^3\Phi^*(s)
$$

where $\Phi^*(s)$ is defined as before. Computing residues of the above, one immediately proves for $z\to\infty$

$$
Q_2(z) \sim \log_2^2 z - (3+2C_d)\log_2 z + 2 + 3C_d + D_d + P(\log_2 z)
$$

where C_d was defined before, and

$$D_d = \frac{1}{L^3}\Phi^{*\prime\prime}(0) = \frac{1}{L^3}\int_0^\infty e_d(x)e^{-x}\varphi(2x)\frac{\log^2 x}{x}dx\ .$$

We can analyze D_d in the same manner as before. After some algebra we arrive at

$$\begin{aligned} D_d &\sim C_d - 1/6 + 2/L^2 \int_1^\infty e_d(t)e^{-t}\frac{\log t}{t}dt \\ &\sim C_d - 1/6 + \log_2^2 d + O(d^{-1/2+\epsilon}) \end{aligned}$$

as $d \to \infty$. After applying the Depoissonization Lemma, we complete the proof of Theorem 1(ii).

The limiting generating function $\lim_{N\to\infty} F_N(z)$ can be analyzed in the same manner as above but this time the solution (16) should be used to derive Theorem 2.

With regard to the limiting distribution the Depoissonization Lemma applied to the bivariate function $H(z,x)$ expressed as in (16) strongly suggests

$$\varphi\left(2^{-t-1-\{\log_2 N+t\}}\right) \tag{28}$$

as an approximation for $\Pr\{R_{N,d} - \log_2 N > t\}$ for t fixed and $N \to \infty$. (In the expression (28) $\{x\}$ denotes the fractional part of x.) The fluctuating term is due to the fact that a discrete parameter is approximated by a continuous function. Detailed derivations are left to a journal version of the paper.

Acknowledgment. The first two authors want to thank M. Drmota and P. Grabner for interesting discussions.

References

[1] G. Brassard and P. Bratley, *Algorithmics. Theory and Practice*, Prentice Hall, Englewood Cliffs, 1988.

[2] P. Flajolet and G.N. Martin, Probabilistic Counting Algorithms for Data Base Applications, *J. Comp. Syst. Sci.*, 31, 182-209, 1985.

[3] P. Flajolet and A. Odlyzko, Singularity Analysis of Generating Functions, *SIAM J. Disc. Methods*, 3, 216-240, 1990.

[4] P. Flajolet and B. Richmond, Generalized Digital Trees and Their Difference-Differential Equations, *Random Structures and Algorithms*, 3, 1992, in press.

[5] P. Flajolet, M. Régnier and R. Sedgewick, Some Uses of the Mellin Transform Techniques in the Analysis of Algorithms, in *Combinatorial Algorithms on Words*, Springer NATO ASI Ser. F12, 241-254, 1985.

[6] A. Greenberg, P. Flajolet and R. Ladner, Estimating the Multiplicities of Conflicts to Speed Their Resolution in Multiple Access Channels, *JACM*, 34, 289-325, 1987.

[7] P. Henrici, *Applied and Computational Complex Analysis*, vol. 2, John Wiley & Sons, 1977.

[8] P. Jacquet and W. Szpankowski, Ultimate Characterizations of the Burst Response of an Interval Searching Algorithm: A Study of a Functional Equations, *SIAM J. Computing*, 18, 777-791, 1989.

[9] M. Kendall and A. Stuart, *The Advanced Theory of Statistics*, 4th Ed. vol. 1, Charles and Griffin and Co. Ltd., London (1977).

[10] P. Kirschenhofer and H. Prodinger, On the Analysis of Probabilistic Counting, *Lecture Notes in Mathematics*, 1452 (Eds. E. Hlawka and R. Tichy), 117-120, Springer Verlag, 1990.

[11] P. Kirschenhofer and H. Prodinger, Approximate Counting: An Alternative Approach, *Informatique Théorique et Applications/Theoretical Informatics and Applications*, 25, 43-48, 1991.

[12] D.E. Knuth, *The Art of Computer Programming. Sorting and Searching*, vol. 3., Addison-Wesley, Reading, MA 1973.

[13] R. Morris, Counting Large Numbers of Events in Small Registers, *Comm. ACM*, 21, 840-842, 1978.

[14] B. Pittel and H. Rubin, How Many Random Questions Are Necessary to Identify n Distinct Objects?, *J. Combinatorial Theory, Ser. A*, 55, 292-312, 1990.

[15] H. Prodinger, How to Select a Loser, *Discrete Mathematics*, to appear.

[16] B. Rais, P. Jacquet, and W. Szpankowski, Limiting Distribution for the Depth in Patricia Tries, *SIAM Journal on Discrete Mathematics*, to appear.

[17] U. Schmid, The Average CRI-length of a Tree Collision Resolution Algorithm in Presence of Multiplicity-Dependent Capture Effects, *this Volume.*

[18] W. Szpankowski, Solution of a Linear Recurrence Equation Arising in the Analysis of Some Algorithms, *SIAM J. Alg. Disc. Methods*, 8, 233-250, 1987.

[19] W. Szpankowski, Patricia Tries Again Revisited, *JACM*, 37, 691-711, 1990.

[20] W. Szpankowski and V. Rego, Yet Another Application of a Binomial Recurrence. Order Statistics, *Computing*, 43, 401-410, 1990.

The Average CRI-Length of a Tree Collision Resolution Algorithm in Presence of Multiplicity-Dependent Capture Effects

Ulrich Schmid

Technical University of Vienna
Department of Automation 183/1
Treitlstraße 3, A-1040 Vienna
E-Mail: s@auto.tuwien.ac.at, Phone: 0222/58801-8189

Abstract. We investigate the average length L_n of a collision resolution interval of a simple tree collision resolution algorithm in presence of capture effects: In opposition to the common model, we assume a (non-zero) probability ac^n, $c < 1$ that exactly one of the packets involved in a collision of multiplicity n survives. Our analysis is based on the application of Mellin transform techniques to ordinary generating functions, which yields an asymptotic expansion of the ordinary generating function $L(z)$ of L_n near its dominant singularity $z = 1$. The application of a simple transfer lemma eventually provides the required asymptotic expansion of L_n as $n \to \infty$.

1 Introduction

In the past, a large amount of theoretical (and practical) work has been devoted to the analysis of distributed algorithms suitable for controlling the transmission activities of stations (i.e., transmitting/receiving units) sharing a single broadcast communication channel. Applications of such algorithms may be found in certain computer networks, e.g., in wide area networks based on (satellite) radio channels. Starting from the famous ALOHA system of the Univerity of Hawaii in the late 1970's, a number of different algorithms offering much better characteristics (e.g., throughput, stability, ...) have been developed, cf. [8] for a nice survey.
The appropriate analysis usually relies on a model similar to the following:

(1) A (infinite) population of identical transmitters is supposed to have access to a common time-slotted, noiseless, collision-type broadcast channel, without any centralized channel arbitration mechanism.
(2) Data are transmitted in the form of fixed size packets, which fit into one time slot.
(3) For the whole population of stations, new packets are generated according to a Poisson process with a fixed rate λ.
(4) If $n \geq 2$ stations transmit in the same slot, a collision of multiplicity n occurs and all packets involved in that collision are completely lost. The channel feedback (e.g., collision/no collision) is supposed to be identical for all stations.

A very important class of collision resolution algorithms are *tree algorithms*, see [10] for details. We restrict ourselves to the so-called Q-ary tree algorithm with the obvious blocked access protocol, which works as follows:

No collision resolution: If the whole system is in an idle state, each transmitter having generated a new packet transmits it in the very next slot. Obviously, if there are two or more simultaneous transmission attempts, a collision occurs and a *collision resolution interval* (CRI) is initiated.

Collision resolution: Each transmitter involved in the (initial) collision flips a fair Q-sided coin with values from $1, 2, \ldots, Q$. That value determines the (relative) number of the slot where the packet should be transmitted again. For example, all stations with 1 flipped transmit in the very next slot. If a new collision occurs, it is resolved immediately by the same method, thereby defering the treatment of transmitters having other values (≥ 2) flipped. Note that such stations have to keep track with those "intermediate" resolution processes, e.g., to add Q to their previous relative slot number in order to determine the new one.
On the other hand, all transmitters not involved in the initial collision (which has terminated the idle state of the whole system) remain blocked, that is, inactive until the whole system is idle again. They may contend for the idle slot following a collision resolution interval, probably forcing a new initial collision.

A crucial parameter for the performance of a collision resolution algorithm (of this type) is the (average) length of a collision resolution interval. This random variable denotes the number of slots necessary for resolving an initial collision of multiplicity n, and is clearly independent from the packet generating process due to the blocked access protocol.
Our intention is the derivation of the average CRI-length L_n for our simple tree algorithm when *capture effects* are present. In many real communication systems, the "strongest" of the actually colliding packets is sometimes able to capture the receiver and is therefore received without error. That is, in case of a collision there is a non-zero probability that exactly one of the packets involved survives. To handle this subject, we have to change (4) of the usual model stated above.
In [11], we used a very simple approach for modelling capture: In case of a collision, we assumed a fixed probability $1-p$ that exactly one packet survives. The major difficulty with this idea is the unrealistic assumption of a capture probability which is independent of the multiplicity of the collision. For example, in a radio network, the signal of a dominating packet has to be more powerful than the sum of the signals of the other packets in order to capture the receiver. Thus, the probability of capture decreases fast with the multiplicity of a collision. A suitable choice in this respect (which has the advantage of being mathematically tractable, too) is the assumption of a geometric capture probability ac^n for some $c < 1$. Obviously, $1-ac^n$ is the probability that all packets involved in a collision of multiplicity n get lost.
The outline of the paper is as follows: Section 2 contains the derivation of a functional equation for the ordinary generating function of L_n, which is attacked by Mellin transform techniques in Section 3. Section 4 is devoted to some computations concerning the major term of the asymptotic expansion of L_n as $n \to \infty$ and closes with our major theorem. At last, some conclusions are appended in Section 5.

Functional Equations

According to [10], a collision resolution interval of a tree algorithm with the obvious blocked access protocol may be modelled as a Q-ary tree. Each slot is represented by a node in the tree; the root of the tree corresponds to the initial collision slot. Since the coin-flipping process splits the set of transmitters involved in a particular collision into exactly Q subsets, each node corresponding to a collision slot has exactly Q successors. Empty slots or slots where only a single transmission occurs form the leafs in our tree.

It is possible to adopt this correspondence to suit our needs. We consider Q-ary trees with two types of nodes, C-nodes (Capture-nodes, representing a successful transmission of a packet in a collision slot, or a single transmission), and NC-nodes (NonCapture-nodes, representing a collision slot with the destruction of all packets involved, or an empty slot). Each node is additionally labelled with the multiplicity of the corresponding conflict; 0 is the label for empty slots, 1 for a slot used by a single transmitter.

If we examine the trees generated by the application of these rules, we obtain the following properties:

(1) For each collision resolution interval of a conflict of multiplicity n, there exists a unique tree representation of the resolution process with exactly n C-nodes. The appropriate CRI-length L_n is reflected by the total number of nodes in the tree.

(2) Traversing the tree in preorder, we obtain the traffic on the channel; each C-node represents a slot with a successful transmission.

(3) Leafs correspond to empty slots (label 0) or to single transmission slots (label 1).

(4) Internal nodes correspond to collision slots of multiplicity equal to their label $n \geq 2$. The sum of the labels of the Q successors equals n in case of a NC-node and $n-1$ in case of a C-node.

Such trees may be viewed as a certain "mixture" of digital search trees and radix search tries, cf. [7] for a survey. Therefore, it is not too surprising that we succeded with an analysis based on a novel technique for investigating characteristic parameters of such trees proposed in [6].

We start our detailled derivations by providing the basic recurrence for the average total number of nodes in a tree with exactly n C-nodes, i.e., the average CRI-length L_n, $n \geq 2$:

$$L_n = 1 + (1 - ac^n) \sum_{\Sigma i_l = n} \binom{n}{i_1, \ldots, i_Q} Q^{-n} \sum_{k=1}^{Q} L_{i_k} + ac^n \sum_{\Sigma i_l = n-1} \binom{n-1}{i_1, \ldots, i_Q} Q^{1-n} \sum_{k=1}^{Q} L_{i_k}$$

This is justified by the following straightforward facts: First, the root node, i.e., the (initial) collision slot, contributes 1 to the sum above. The second term corresponds to the non-capture case, where a (root) node with label n has to have Q successors with label $i_1, \ldots, i_Q$ and $\sum_k i_k = n$; note the appropriate multinomial splittin probability. Finally, the third term covers the case of an actual capture, where th label-sum of the Q successors has to be $n-1$. By virtue of the simple identity

$$\sum_{\Sigma i_l = n} \binom{n}{i_1, \ldots, i_Q} Q^{-n} \sum_{k=1}^{Q} L_{i_k} = Q \sum_{i=0}^{n} \binom{n}{i} Q^{-i}(1-1/Q)^{n-i} L_i,$$

we may rewrite the initial recurrence in order to obtain

$$\begin{aligned} L_n = 1 + Q(1-ac^n) \sum_{i=0}^{n} \binom{n}{i} Q^{-i}(1-1/Q)^{n-i} L_i \\ + Qac^n \sum_{i=0}^{n-1} \binom{n-1}{i} Q^{-i}(1-1/Q)^{n-1-i} L_i \qquad \text{for } n \geq 2; \end{aligned} \tag{2.1}$$

the initial values are $L_1 = 1$ and $L_0 = 1$.
It is convenient to proceed with some well-known techniques from the formal operato calculus. Let $\mathcal{D}$ denote the ordinary differential operator, $\mathcal{I}$ the identity operator an $\mathcal{N}$ the 0-substitution operator, all with respect to a variable t. Now, if $l(t)$ denote the *exponential generating function* (EGF) of L_n, it is clear that $L_n = \mathcal{N}\mathcal{D}^n * l(t)$ the star characterizes the application of the operator(s) on the left hand side to th function on the right.
Rewriting our recurrence relation (2.1) accordingly, we obtain for $n \geq 2$

$$L_n = 1 + Q(1-ac^n)\mathcal{N}(1-1/Q+\mathcal{D}/Q)^n * l(t) + Qac^n\mathcal{N}(1-1/Q+\mathcal{D}/Q)^{n-1} * l(t)$$

Multiplying both sides by $z^{n-1}/(n-1)!$ and summing up for $n \geq 2$, we find by usin the fact $\mathcal{N}e^{x\mathcal{D}} * f(t) = f(x)$

$$\begin{aligned} l'(z) - L_1 = e^z - 1 + Q\mathcal{N}(1-1/Q+\mathcal{D}/Q)\left[e^{z(1-1/Q+\mathcal{D}/Q)} - \mathcal{I}\right] * l(t) \\ - Qac\mathcal{N}(1-1/Q+\mathcal{D}/Q)\left[e^{zc(1-1/Q+\mathcal{D}/Q)} - \mathcal{I}\right] * l(t) \\ + Qac\mathcal{N}\left[e^{zc(1-1/Q+\mathcal{D}/Q)} - \mathcal{I}\right] * l(t) \end{aligned}$$

and therefore

$$\begin{aligned} l'(z) = (Q-1)e^{z(1-1/Q)} l(z/Q) + e^{z(1-1/Q)} l'(z/Q) \\ + ace^{zc(1-1/Q)} l(zc/Q) - ace^{zc(1-1/Q)} l'(zc/Q) + e^z - Q. \end{aligned} \tag{2.2}$$

Introducing the *Poisson generating function* (PoGF) of L_n, namely

$$h(z) = \sum_{n \geq 0} h_n \frac{z^n}{n!} = l(z)e^{-z}, \tag{2.3}$$

which implies the inverse pair

$$h_n = \sum_{k=0}^{n} \binom{n}{k} (-1)^{n-k} L_k \qquad \text{and} \qquad L_n = \sum_{k=0}^{n} \binom{n}{k} h_k, \tag{2.4}$$

e have $h_0 = 1$, $h_1 = 0$ and $e^{-z}l'(z) = h(z)+h'(z)$. Thus, multiplying equation (2.2) y e^{-z} yields a simpler functional equation for $h(z)$:

$$h(z) + h'(z) = Qh(z/Q) + h'(z/Q) - ace^{-z(1-c)}h'(zc/Q) + 1 - Qe^{-z} \qquad (2.5)$$

'he major idea for treating this equation, as proposed in [6], is the transition to *rdinary generating functions* (OGFs). If

$$L(z) = \sum_{n\geq 0} L_n z^n \qquad \text{and} \qquad H(z) = \sum_{n\geq 0} h_n z^n$$

enote the appropriate OGFs, (2.4) implies the following relation:

$$\begin{aligned} L(z) &= \sum_{n\geq 0}\sum_{k=0}^{n}\binom{n}{k}h_k z^n = \sum_{k\geq 0} h_k \sum_{n\geq k}\binom{n}{k} z^n = \sum_{k\geq 0} h_k z^k \sum_{n\geq 0}\binom{n+k}{k} z^n \\ &= \frac{1}{1-z} H\Big(\frac{z}{1-z}\Big) \end{aligned} \qquad (2.6)$$

Translating the functional equation (2.5) for the EGF into the corresponding one or the OGF is simple. We use the Borel-transformation in a more or less "formal" manner; alternatively, one might extract the coefficients and sum up as well. Providing

$$\int_0^\infty e^{-t}h'(\alpha t)\,dt = -h_0/\alpha + \frac{1}{\alpha}\int_0^\infty e^{-t}h(\alpha t)\,dt = \frac{H(\alpha) - h_0}{\alpha}$$

and

$$\int_0^\infty e^{-t}e^{-\beta t}h'(\alpha t)\,dt = \frac{1}{1+\beta}\int_0^\infty e^{-s}h'\Big(\frac{\alpha s}{1+\beta}\Big)\,ds = \frac{H\big(\frac{\alpha}{1+\beta}\big) - h_0}{\alpha},$$

we obtain by substituting $z = zt$ in (2.5), multiplying by e^{-t} and integrating

$$H(z) + \frac{H(z) - h_0}{z} = QH(z/Q) + \frac{H(z/Q) - h_0}{z/Q} - ac\frac{H\big(\frac{zc/Q}{1+z(1-c)}\big) - h_0}{zc/Q} + 1 - \frac{Q}{1+z}.$$

Introducing

$$I(z) = H(z) - h_1 z - h_0 = H(z) - 1, \qquad (2.7)$$

cf. equations (2.4) and (2.1), some straightforward algebraic manipulations finally yield

$$I(z) = QI(z/Q) - \frac{aQ}{1+z} I\Big(\frac{zc/Q}{1+z(1-c)}\Big) + \frac{Qz^2}{(1+z)^2}. \qquad (2.8)$$

In order to show that the functional equation (2.8) has indeed a suitable analytic solution, it is possible to employ similar techniques as in [3] to prove the following

Lemma 2.1 (Existence of an Analytic Solution). *Let $t(z)$, $\gamma_1(z)$, $\gamma_2(z)$, $\sigma_1(z)$, $\sigma_2(z)$ denote analytic functions defined on a (possibly open) domain $\mathcal{D}$ of the complex plane. Let H be the semigroup of substitutions generated by σ_1 and σ_2 under the operation of composition of functions: The identity of H is $\varepsilon = \varepsilon(z) = z$, any member $\sigma = \sigma(z) \in H$ can be written in the form $\sigma = \sigma_{i_1}\sigma_{i_2}\dots\sigma_{i_n}$, $|\sigma| = n \geq$ and $i_k \in \{1,2\}$ for $1 \leq k \leq n$.*
If (1) $\sigma_1(z)$ and $\sigma_2(z)$ are such that $\sigma_i(\mathcal{D}) \subseteq \mathcal{D}$ (and therefore $\tau\sigma(\mathcal{D}) \subseteq \tau(\mathcal{D})$ for any $\tau, \sigma \in H$) and $|\sigma_i(z)| \leq |z|$ for $i = 1,2$, and (2) for some $m \geq 0$ arbitrary but fixed

$$\alpha_1 = \sup_{\substack{z\in\mathcal{D}\\|\sigma|=m}} \left|\gamma_1(\sigma(z))\right|, \qquad \alpha_2 = \sup_{\substack{z\in\mathcal{D}\\|\sigma|=m}} \left|\gamma_2(\sigma(z))\right|$$

satisfy the contraction property $\alpha_1 + \alpha_2 < 1$, then the functional equation

$$J(z) = t(z) + \gamma_1(z)J(\sigma_1(z)) + \gamma_2(z)J(\sigma_2(z))$$

has an analytic solution

$$J(z) = \sum_{\sigma\in H} [\gamma_1, \gamma_2]^{\sigma}(z)\, t(\sigma(z))$$

on $\mathcal{D}$, where

$$[\gamma_1,\gamma_2]^{\sigma}(z) = \gamma_{i_1}(\sigma_{i_2}\dots\sigma_{i_n}(z)) \cdots \gamma_{i_{n-1}}(\sigma_{i_n}(z)) \cdot \gamma_{i_n}(z) \quad \text{for } |\sigma| = n,\ n \geq 1,$$
$$[\gamma_1,\gamma_2]^{\varepsilon}(z) = 1.$$

Lemma 2.1 is not directly applicable to the functional equation (2.8) since the required contraction property is violated. However, remembering definition (2.7), it is obvious that $I(z) = z^2 J(z)$. Rewriting (2.8) accordingly, we obtain

$$J(z) = Q^{-1}J(z/Q) - \frac{ac^2/Q}{(1+z)(1+z(1-c))^2} J\left(\frac{zc/Q}{1+z(1-c)}\right) + \frac{Q}{(1+z)^2}.$$

It is easily checked that this equation is treatable by Lemma 2.1 on

$$\mathcal{D} = \{z : |\mathrm{Arg}(z+2/3)| \leq \psi\} \cap \{z : |\mathrm{Arg}(z+1)| \leq \phi\} \tag{2.9}$$

for some $\psi > \phi > \pi/2$; the reason for that (somewhat artificial) region will become clear in the following section. Note that $\gamma_2(z) < 1/Q$ ($Q \geq 2$) is only valid for c small enough ($\gamma_2(z)$ attains its maximum on $\mathcal{D}$ at $z = -2/3$). However, using the "feature" $m \geq 1$ in our Lemma, it is possible to handle arbitrary $c < 1$ as well.
Thus, Lemma 2.1 states the existence of a solution $J(z)$ (and hence $I(z)$), which is analytic on $\mathcal{D}$. Moreover, the fact that the coefficients of $I(z)$ are uniquely defined (which is immediately apperarent from the functional equation (2.5); cf. the recurrence relation (4.4), too) implies that this solution is uniquely determined.
In addition, by a simple majorization of (2.8) divided by z, it is easily seen that

$$I(z) = O(z) \qquad \text{for } z \to +\infty, \tag{2.10}$$

i.e., that the order of $I(z)$ for $z \to \infty$ is smaller than its order as $z \to 0$.

3 Mellin Transform Techniques

As we shall see, equation (2.8) is tractable by Mellin transform techniques. The well-known Mellin transform is a powerful tool in asymptotic analysis, and is applicable to a wide variety of problems; see **[2]** for the very complete theory and **[12]** for application-oriented details. First, we supply a short summary of appropriate theorems (without any proof):

DEFINITION 3.1 (MELLIN-TRANSFORM). *The Mellin-transform of a continuous, real valued function $f(x)$ is the complex valued function $f^*(s)$ defined by*

$$f^*(s) = \mathcal{M}[f(x); s] = \int_0^\infty f(x) x^{s-1}\, dx,$$

provided that the integral is absolutely convergent in the region $\alpha < \Re(s) < \beta$. This region is called the fundamental strip of $f^*(s)$ and is denoted by $\langle \alpha, \beta \rangle$.

THEOREM 3.2 (EXISTENCE AND ANALYTICITY). *If there are two complex numbers α, β with $\Re(-\alpha) < \Re(-\beta)$ and the property*

$$f(x) = \begin{cases} O(x^\alpha), & \text{for } x \to 0 \\ O(x^\beta), & \text{for } x \to \infty, \end{cases}$$

e.g., if the order of $f(x)$ near zero is larger than the order near infinity, then the transform $f^*(s)$ exists on the fundamental strip $\langle -\alpha, -\beta \rangle$, and is analytic within the whole region.

LEMMA 3.3 (TRANSFORM OF HARMONIC FUNCTIONS). *If the transform $f^*(s)$ of the function $f(x)$ exists on the fundamental strip $\langle \alpha, \beta \rangle$, we obtain for a real $c > 0$*

$$\mathcal{M}[f(cx); s] = c^{-s} f^*(s) \qquad \text{within the fundamental strip } \langle \alpha, \beta \rangle.$$

THEOREM 3.4 (ASYMPTOTIC EXPANSION). *If $f^*(s)$ exists on the fundamental strip $\langle \alpha, \beta \rangle$, and satisfies certain smallness-conditions towards $i\infty$ for $\beta \leq \Re(s) \leq M$ (to the right of the fundamental strip), a pole of the transform*

$$f^*(s) \sim \sum_{n=0}^{N} \frac{d_{n,k}}{(s - b_k)^{n+1}} \qquad \text{for } \beta \leq \Re(b_k) < M \text{ and } s \to b_k$$

translates into a term of the asymptotic expansion for $x \to \infty$ according to

$$\sum_{n=0}^{N} \frac{-d_{n,k}}{n!} (-\log x)^n x^{-b_k} \qquad \text{for } x \to \infty.$$

Moreover, the complete expansion of $f(x)$ evaluates to

$$f(x) = \sum_k \text{Term resulting from } b_k + O(x^{-M}) \qquad \text{for } x \to \infty.$$

The classical theory of the Mellin transform is based on real functions $f(x)$. However our theorems (especially Theorem 3.4) are valid for (certain) complex functions $f(z$ as well. There are a number of strikingly elegant applications (cf. [**5**] or [**6**], fo example) which demonstrate the power of this approach. We, too, shall use thi technique to attack the functional equation (2.8).
First of all, equations (2.7) and (2.10) in conjunction with Theorem 3.2 guarante that the Mellin transform $I^*(s)$ of our desired solution $I(z)$ exists. Abbreviating

$$T(z) = \frac{1}{1+z} I\Big(\frac{zc/Q}{1+z(1-c)}\Big) = \begin{cases} O(z^2) & \text{for } z \to 0, \\ O(z^{-1}) & \text{for } z \to \infty; \end{cases}$$

the appropriate fundamental strip of $T^*(s)$ is $\langle -2, 1\rangle$ by virtue of Theorem (3.2). Regarding the last term of (2.8), we have

$$\int_0^\infty \frac{z^{s+1}}{(1+z)^2}\, dz = B(s+2, -s) = \Gamma(s+2)\Gamma(-s)$$

according to [**1**, p.258]; $B(z, w)$ denotes the Beta function. Since $\Gamma(s+2)\Gamma(-s)$ is analytic within the strip $-2 < \Re(s) < 0$, we obtain the fundamental strip $\langle -2, 0\rangle$. Now, using Theorem (3.3), the Mellin transform of $I(z)$ evaluates to

$$I^*(s) = \frac{Q\Gamma(s+2)\Gamma(-s) - aQT^*(s)}{1 - Q^{1+s}}; \tag{3.1}$$

the appropriate fundamental strip is obviously $\langle -2, -1\rangle$.
According to Theorem (3.4), the desired asymptotics for $z \to \infty$ are trivially connected with the singularities of the transform to the right of the fundamental strip. The "first" singularities encountered are simple poles at $-1+\chi_k$, $\chi_k = 2k\pi i/\log Q$ for all integers k, which are caused by the vanishing denominator. Applying de l'Hospital's rule, we find

$$\frac{1}{1-Q^{1+s}} = -\frac{1}{\log Q} \cdot \frac{1}{s+1-\chi_k} + O(1) \qquad \text{for } s \to -1+\chi_k$$

and therefore

$$I^*(s) = -\frac{Q}{\log Q}\big(\Gamma(1+\chi_k)\Gamma(1-\chi_k) - aT^*(-1+\chi_k)\big) \cdot \frac{1}{s+1-\chi_k} \qquad \text{for } s \to -1+\chi_k.$$

The next singularity is a simple pole at $s = 0$, caused by $\Gamma(s)$. The expansion is simple:

$$I^*(s) = -\frac{Q}{Q-1} \cdot \frac{1}{s} + O(1) \qquad \text{for } s \to 0.$$

Further singularities lie on the vertical strip $\Re(s) = 1$. But, since we have to expect a singularity caused by the (not explicitely known) function $T^*(s)$, there is no easy way to state the appropriate expansion. However, to obtain an asymptotic expansion of $I(z)$ up to $O(z^{-1+\varepsilon})$ for $z \to \infty$, it suffices that there are no additional singularities within the strip $0 < \Re(s) \le 1 - \varepsilon$ for some $\varepsilon > 0$.

t is not hard to show that $I^*(s)$ is indeed small towards $\pm i\infty$ within the strip -3/2 $\le \Re(s) \le 1-\varepsilon$, as required by Theorem 3.4: $I^*(s)$ involves $\Gamma(s)$, which is vell-known to be small for large imaginary parts of s, and $T^*(s)$, whose fundamental trip covers the region of interest (because smallness-conditions are always satisfied within the fundamental strip).
Therefore, we may apply Theorem (3.4) and obtain

$$I(z) = \frac{Q(1-aT^*(-1))}{\log Q} \cdot z + \frac{Q}{\log Q} \cdot zP(\log_Q z) + \frac{Q}{Q-1} + O(z^{-1+\varepsilon}) \qquad (3.2)$$

for $z \to \infty$; $\varepsilon > 0$ denotes an arbitrary small positive constant, and

$$P(u) = \sum_{k \neq 0} \big(\Gamma(1+\chi_k)\Gamma(1-\chi_k) - aT^*(-1+\chi_k)\big)e^{-2k\pi iu}$$

is a periodic function with period 1, mean 0 and small amplitude, caused by poles with non-zero imaginary part.
Note that the term $T(z)$, which is not explicitely expressible since it depends on the unknown function $I(z)$, contributes a constant factor to the major term of the desired asymptotics of $I(z)$ for $z \to \infty$! We shall provide an asymptotic expansion of this factor as $c \to 0$ in Section 4.

However, using relation (2.6) and (a weaker form of) expansion (3.2) yield

$$\begin{aligned} L(z) &= \frac{1}{1-z} I\Big(\frac{z}{1-z}\Big) + \frac{1}{1-z} \\ &= \frac{Q(1-aT^*(-1))}{\log Q} \frac{1}{(1-z)^2} + \frac{Q}{\log Q} \frac{1}{(1-z)^2} P\Big(\log_Q \frac{z}{1-z}\Big) + O\Big(\frac{1}{1-z}\Big) \end{aligned}$$

for $z \to 1$. In addition, since $z^{-\chi_k} = 1 + O(z-1)$ for $z \to 1$ uniformly in k and therefore $P\big(\log_Q \frac{z}{1-z}\big) = P\big(\log_Q \frac{1}{1-z}\big) + O(z-1)$, we eventually find

$$L(z) = \frac{Q(1-aT^*(-1))}{\log Q} \frac{1}{(1-z)^2} + \frac{Q}{\log Q} \frac{1}{(1-z)^2} P\Big(\log_Q \frac{1}{1-z}\Big) + O\Big(\frac{1}{1-z}\Big)$$

for $z \to 1$.
The desired coefficient $L_n = [z^n]L(z)$ is most easily obtained by a simple transfer lemma, cf. [4] for a very complete treatment of the subject. For applicability, we need the analyticity of $L(z)$ in a closed domain $\Delta(\phi, \eta) = \big\{z : |z| \le 1+\eta, |\operatorname{Arg}(z-1)| \ge \phi, z \ne 1\big\}$ (an indented disk with radius $1+\eta$, sparing out the singularity $z = 1$). This, however, is a straightforward consequence of Relation (2.6) and the region of analyticity of $I(z)$ according to equation (2.9), as may be seen by considering the image of $\Delta(\phi, \eta)$ under the mapping $z/(1-z)$.
Thus, we obtain

$$L_n = \frac{Q(1-aT^*(-1))}{\log Q} n + \frac{Q}{\log Q} np(\log_Q n) + O(1) \qquad \text{for } n \to \infty, \qquad (3.3)$$

where

$$p(u) = \sum_{k\neq 0}\Big(\chi_k\Gamma(-1-\chi_k) - \frac{aT^*(-1-\chi_k)}{(1+\chi_k)\Gamma(1+\chi_k)}\Big)e^{2k\pi iu}.$$

Note that we restricted ourselves to the computation of the major terms of the asymptotic expansion for $L(z)$ and hence L_n in order to preserve the essentials. Nevertheless, it would be easy to extend the derivations above to obtain a more accurate expression up to a remainder term of order $O(n^{-(1-\varepsilon)})$.

4 Evaluation of a Mellin Integral

At last, we shall derive an asymptotic expression for $T^*(-1)$ as $c \to 0$, thereby providing a means for computing the major term of L_n numerically. Our point of application is the Mellin integral

$$T^*(-1) = \int_0^\infty I\Big(\frac{zc/Q}{1+z(1-c)}\Big)\frac{z^{-2}}{1+z}\,dz. \tag{4.1}$$

The fact that the argument of $I(.)$ in the integrand above remains small, i.e., $O(c)$, over the whole path of integration, enables us to operate with the first few terms of the Taylor expansion

$$I(z) = H(z) - h_0 - h_1 z = \sum_{k\geq 0} h_{k+2}z^{k+2}$$

instead of $I(z)$; remember definition (2.7). That is, using the substitution

$$t = \frac{zc}{1+z(1-c)},$$

we obtain

$$\begin{aligned} T^*(-1) &= \int_0^{\frac{c}{1-c}} I(t/Q)\frac{c-t(1-c)}{c+tc}\frac{c\,dt}{t^2} = cQ^{-2}\int_0^{\frac{c}{1-c}} \frac{I(t/Q)}{(t/Q)^2}\frac{1+t-t/c}{1+t}\,dt \\ &= cQ^{-2}\sum_{k\geq 0} h_{k+2}Q^{-k}\int_0^{\frac{c}{1-c}}\frac{1+t-t/c}{1+t}t^k\,dt. \end{aligned}$$

Since it is not hard to establish that

$$\int_0^{\frac{c}{1-c}}\frac{1+t-t/c}{1+t}t^k\,dt = \sum_{m\geq 0}\binom{m+k}{m}\frac{c^{m+k+1}}{(k+1)(m+k+2)},$$

some algebraic manipulations eventually yield

$$T^*(-1) = cQ^{-1}\sum_{l\geq 1}\frac{c^l}{l(l+1)}\sum_{k=1}^{l}\binom{l}{k}h_{k+1}Q^{-k}. \tag{4.2}$$

What remains to do is to compute the first few h_k. This is accomplished by extracting the coefficients from the functional equation (2.5): For $n \geq 1$, we find

$$h_{n+1}(1-Q^{-n}) + h_n(1-Q^{-(n-1)}) = \\ - Q(-1)^n - ac(-1)^n \sum_{k=0}^{n} \binom{n}{k} h_{k+1}(-c/Q)^k (1-c)^{n-k}.$$

Multiplying the equation above with $(-1)^{m-n}$ and summing up for $1 \leq n \leq m$, $m \geq 1$ while remembering $h_1 = 0$ yields a telescoping sum on the left hand side and therefore

$$\begin{aligned} h_{m+1}(1-Q^{-m}) &= -mQ(-1)^m - ac\sum_{n=1}^{m}(-1)^m \sum_{k=1}^{n}\binom{n}{k} h_{k+1}(-c/Q)^k(1-c)^{n-k} \\ &= mQ(-1)^{m+1} + ac(-1)^{m+1}\sum_{k=1}^{m} h_{k+1}(-c/Q)^k \sum_{n=k}^{m}\binom{n}{k}(1-c)^{n-k} \\ &= mQ(-1)^{m+1} + ac(-1)^{m+1}\sum_{k=1}^{m} h_{k+1}(-c/Q)^k t_{m,k}(c). \qquad (4.3) \end{aligned}$$

Evaluating $t_{m,k}(c)$ involves some standard identities concerning binomial coefficients and yields

$$t_{m,k}(c) = \sum_{j=0}^{m-k}(-c)^j \binom{k+j}{j}\binom{m+1}{m-k-j}.$$

Using $t_{m,m}(c) \equiv 1$ for all $m \geq 1$, some straightforward algebraic manipulations of equation (4.3) establish

$$h_{m+1} = (-1)^{m+1}\frac{mQ - aQ\sum_{k=2}^{m} h_k(-c/Q)^k t_{m,k-1}(c)}{1-Q^{-m}(1-ac^{m+1})}. \qquad (4.4)$$

For example, we most easily get

$$\begin{aligned} h_2 &= \frac{Q}{1-Q^{-1}(1-ac^2)} = \frac{Q}{1-Q^{-1}} + O(c^2) \\ h_3 &= -\frac{2Q+O(c^2)}{1-Q^{-2}(1-ac^3)} = -\frac{2Q}{1-Q^{-2}} + O(c^2); \end{aligned}$$

for $c \to 0$. Remembering equation (3.3) and using the above values in equation (4.2) finally establishes our major result

THEOREM 4.1. *The average CRI-length L_n of the Q-ary tree algorithm with obvious blocked access protocol in presence of multiplicity-dependent capture with probability ac^n is*

$$L_n = \frac{Q(1-aT^*(-1))}{\log Q} n + \frac{Q}{\log Q} np(\log_Q n) + O(1) \qquad \text{for } n \to \infty.$$

The constant $T^*(-1)$ is approximated by

$$T^*(-1) = \frac{c^2}{2Q(1-Q^{-1})} + \frac{c^3}{3Q(1-Q^{-1})} - \frac{c^3}{3Q^2(1-Q^{-2})} + O(c^4) \qquad \text{for } c \to 0,$$

the function $p(u)$ is periodic with periode 1, has very low amplitude, mean 0, an[d] its Fourier expansion is given by

$$p(u) = \sum_{k \neq 0} \left(\chi_k \Gamma(-1-\chi_k) - \frac{aT^*(-1-\chi_k)}{(1+\chi_k)\Gamma(1+\chi_k)} \right) e^{2k\pi i u} \qquad \text{with } \chi_k = \frac{2k\pi i}{\log Q}.$$

5 Conclusions

Using Mellin transforms of ordinary generating functions in conjunction with transfe[r] lemmas, we analyzed the average CRI-length of a Q-ary tree algorithm in presence o[f] multiplicity-dependent capture effects. Relying on a geometric capture probabilit[y] of ac^n, $c < 1$ in case of a collision of multiplicity n, our investigations extend th[e] somewhat unrealistic assumption of a constant probability $a = 1 - p$ (i.e., $c = 1$[)] pursued in [**11**].
However, we should admit that even our improved approach is not able to cove[r] all possible varieties of capture effects sufficiently. For instance, in communicatio[n] networks without a central receiver, capture is often a local phenomenon, i.e., con[-]cerns not all receivers in the same way. Our assumption of all stations agreeing o[n] the channel feedback is therefore too optimistic. The consideration of local phe[-]nomenons, however, is not at all simple and requires a substantial redesign of th[e] model and the whole analysis.

References

1. M. Abramowitz, I. A. Stegun, "Handbook of Mathematical Functions," Dover Publications Inc., New York, 1972.
2. G. Doetsch, "Handbuch der Laplace-Transformation," Birkhauser-Verlag, Basel, 1955.
3. G. Fayolle, Ph. Flajolet, M. Hofri, On a Functional Equation Arising in the Analysis of a Protocol for a Multi-Access Broadcast Channel, Adv. Appl. Prob. 18 (1986), 441–472.
4. Ph. Flajolet, A. Odlyzko, Singularity Analysis of Generating Functions, SIAM J. Discr. Math 3 (1990), 216–240.
5. Ph. Flajolet, H. Prodinger, Register allocation for unary-binary trees, SIAM Journal on Computing 15(3) (1986), 629–640.
6. Ph. Flajolet, B. Richmond, Generalized Digital Trees and their Difference-differential Equations. (Manuscript)
7. Ph. Flajolet, R. Sedgewick, Digital Search Trees revisited, SIAM Journal on Algorithms 15(3) (August 1986), 748-767.
8. R. G. Gallager, A Perspective on Multiaccess Channels, IEEE Trans. on Inf. Th. IT-31, No. 2 (1985), 124–142.
9. Ph. Jacquet, Wojciech Szpankowski, Ultimate Characterizations of the Burst Response of an Interval Searching Algorithm: A Study of a Functional Equation, SIAM J. Comput. 18, No 4 (August 1989), 777-791.
10. P. Mathys, Ph. Flajolet, Q-ary Collision Resolution Algorithms in Random-Access Systems with Free or Blocked Channel Access, IEEE Trans. on Inf. Th. IT-31, No. 2 (1985), 217–243
11. U. Schmid, On a Tree Collision Resolution Algorithm in Presence of Capture. To appear in RAIRO, (1992).
12. J. S. Vitter, Ph. Flajolet, Average Case Analysis of Algorithms and Data Structures, Handbook of Theoretical Computer Science (J. van Leeuwen, ed.) (1990), North Holland.

Polynomial Hash Functions Are Reliable

(Extended Abstract)

M. Dietzfelbinger[1] J. Gil[2] Y. Matias[3] N. Pippenger[2]

[1] Universität–GH–Paderborn[†]
[2] The University of British Columbia[‡]
[3] University of Maryland and Tel Aviv University[§]

Abstract. Polynomial hash functions are well studied and widely used in various applications. They have gained popularity because of certain performances they exhibit. It has been shown that even linear hash functions are expected to have such performances. However, quite often we would like the hash functions to be reliable, meaning that they perform well with high probability; for some certain important properties even higher degree polynomials were not known to be reliable. We show that for certain important properties linear hash functions are not reliable. We give indication that quadratic hash functions might not be reliable. On the positive side, we prove that cubic hash functions are reliable. In a more general setting, we show that higher degree of the polynomial hash functions translates into higher reliability. We also introduce a new class of hash functions, which enables to reduce the universe size in an efficient and simple manner. The reliability results and the new class of hash functions are used for some fundamental applications: improved and simplified reliable algorithms for perfect hash functions and real-time dictionaries, which use significantly less random bits, and tighter upper bound for the program size of perfect hash functions.

1 Introduction

Polynomial hash functions (introduced in [27], studied in some detail in [19]) are widely used in a variety of applications including: dictionaries (e.g., [2, 4, 21, 27]), simulation of parallel machines with shared memory on bounded degree networks (e.g., [11, 13, 22, 23]), cryptography [28], and pseudo random number generation (e.g., [3, 20]). It is well known that if polynomials of higher degree are used then a better performance can be guaranteed. However, higher degree polynomials are more expensive to use, both in terms of representation and in evaluation time.

We present results indicating that in certain cases linear hash functions (provably) exhibit a certain undesirable behavior with much larger probability than higher degree polynomial hash functions. The effects we demonstrate are most influential when reliable performance is required (such as in the case of real-time dictionaries [5] and in parallel computation [7]). Yet, even the most basic algorithms that use hashing (such as hashing with open addressing [12] or perfect hashing [6]) should suffer from these effects.

On the positive side we demonstrate *reliable* performance of polynomials of degree 3. Namely, we show that for any given input set, a randomly chosen degree-3 polynomial hash function "performs well" with very high probability. Such performance was not known even for higher degree polynomials.

We also present a new class $\mathcal{Y}$ of (polynomial) hash functions and show that it is most useful for reducing the size of the universe from which keys are given. The

---

† Fachbereich Mathematik/Informatik and Heinz-Nixdorf-Institute, Universität–GH–Paderborn, D-W-4790 Paderborn, Fed. Rep. Germany. Partially supported by DFG Grant Me 872/1-4.

‡ Department of Computer Science, The University of British Columbia, Vancouver, B.C. V6T 1Z2, Canada.

§ Institute for Advanced Computer Studies, University of Maryland, College Park, MD 20742, USA. Also with the Department of Computer Science, Tel Aviv University, Israel. Partially supported by NSF grants CCR-9111348 and CCR-8906949.

new class $\mathcal{Y}$ enables improved results for the problem of giving a short program :h computes a perfect hash function [18]. The class $\mathcal{Y}$, in conjunction with the resul on reliability of polynomial hash functions, enables us to give algorithms for hashi and real-time dictionary that use substantially less random bits than in prev.o algorithms.

Some proofs were omitted in this extended abstract.

2 Results

Hash functions Carter and Wegman [2] introduced the concept of universal hashi (i.e., hashing which works for *any* input). They suggested linear transformations an polynomials over finite fields (in [27]) for realization of universal classes of hash fun tions. Although universal classes are sufficient for many purposes, they lack ma of the properties truly random functions have. Using explicit and randomized co structions of expanders, Siegel [25] gave a class of hash functions which has mar random like properties and which exhibits high independence. Dietzfelbinger an Meyer auf der Heide [5] described a class of hash functions which have many ra dom like properties. However, selecting a random function from those classes us a large number of random bits. Our results demonstrate that in many applicatio polynomial hash functions can be used reliably instead of these complex classes. W also demonstrate a new class of hash functions that allows for further reduction i the number of random bits without sacrificing reliability.

Properties of polynomial hash functions Universal hashing [2] works as follows: Give a fixed set S of keys from the universe U of all keys, a mapping h from U into th table $\{0, \ldots, m-1\}$ is chosen at random from some suitable "universal" class $\mathcal{H}$ The aim is to obtain a good distribution for the number of collisions between key (A collision is a pair $\{x, y\} \subseteq S, x \neq y$, with $h(x) = h(y)$.)

We establish lower bounds for the probabilities of certain (undesirable) event For example, we show the following for a certain set S of size m. (It is essential t note that these probabilities depend neither on u nor on m.)

$$\mathbf{Prob}(\text{ the number of collisions exceeds } k \cdot m\ ) = \Omega(k^{-2})$$

In many applications of hash functions, it is important to have at most $O(m)$ coll sions. It would be therefore desirable to have that many collisions with high prob ability. The above implies that for linear hash functions, the probability can be a most constant. In contrast, we show that for degree-3 polynomial hash function the number of collisions is indeed $O(m)$ with very high probability. This implies tha such hash functions are "reliable" for many applications. More results on degree- polynomial hash functions concern the number of collisions between two sets of keys They imply that these functions are "reliable" even for the construction of real-tim dictionary.

Hashing and dictionary algorithms Fredman, Komlós and Szemerédi [6] describe a *static hashing* algorithm with the following properties: for an input set $S \subseteq U$ $|S| = s$, $|U| = u$, the algorithm uses $O(s \lg u)$ random bits to construct in $O(s$ expected time an $O(s)$-space lookup table for S; this lookup table supports mem bership queries in $O(1)$ time. We present a reliable version of this algorithm, i.e., on in which the performance is with high probability with fewer random bits. Specifically, the number of random bits used is $\lg \lg u + \lg^{(k)} u + O(k \lg s)$ for any intege $k > 0$. [4] described a dictionary (*dynamic hashing*) in which each instruction ca perform in expected constant time. A real-time dictionary, i.e., a dictionary in which with high probability all instructions execute each in constant time, was described in [5]. This algorithm uses a pseudo-random class of hash functions, which implies $O(s^{\epsilon} \lg u)$ random bits, for any constant ϵ. We show that polynomial hash function can be used for real-time dictionary. The number of random bits required is the same as for the static algorithm (!).

rogram size of perfect hash functions Consider the number of bits required to rep-
sent a perfect hash function (its "program size") mapping a set $S \subset U$, $|S| = s$, to a table of size m. In particular, we are interested in $O(1)$ time 1-probe perfect ash functions [29]. Mehlhorn showed that for $s \leq m < (1-\epsilon)u$ the program size is wer bounded by $\Omega(s^2/m + \lg\lg u - \lg\lg m)$ [17, 18]. Upper bounds for the program ze were given by [6, 9, 18, 15, 26, 24]. Schmidt and Siegel [24] showed that the pro- am size can be $O(\lg\lg u + s)$, within a constant *multiplicative* factor from the lower ound. We give a simple construction for an $O(1)$ time 1-probe perfect hash function ing $\lg\lg u + \lg^{(k)} u + O(s \lg s)$ bits, for any constant $k > 0$. For a universe that is uch larger than the set size, this implies only an additive term of lower order. By ing the construction of [24] we can get program size of $\lg\lg u + \lg^{(k)} u + O(s)$, for y constant $k > 0$.

Basics

1 Hashing

he universe The universe U of possible keys is some finite set (whose cardinality denoted by u throughout). For practicability, one usually imposes some structure U. The following approaches are common:

) Assume $U = \{0, 1, \ldots, u-1\}$ for some prime u; use the structure of U as a finite field under the operations of addition/multiplication modulo u.

) More generally, assume $U = \mathbb{F}_u$, the finite field with u elements, where $u = p^\ell$ for a prime p. Then, $\mathbb{F}_u$ is in particular a ℓ-dimensional vector space $\mathbb{F}_p^\ell$ over $\mathbb{F}_p$.

) Assume $U = \{0,1\}^\ell$, the ℓ-dimensional vector space over the field $\mathbb{F}_2 = \{0,1\}$ [2, 16, 19].

In all cases it is assumed that arithmetical or vector space operations such as ddition, multiplication, or scalar multiplication can be performed in one unit of me. In this paper, we will consider approaches (a) and (b) as well as a variant of): vector spaces over general finite fields.

otation W.l.o.g. we will always assume that some total order $<$ on U is given. For ≥ 1, k-tuples $\langle x_1, \ldots, x_k \rangle$ are denoted by $\mathbf{x}$. For arbitrary subsets $T \subseteq U$ we let $T)_k = \{\mathbf{x} \mid \mathbf{x} = \langle x_1, \ldots, x_k \rangle \in T^k \mid x_1 < \cdots < x_k\}$, a convenient way of denoting e set of all k-element subsets of T. Clearly, $|(T)_k| = \binom{|T|}{k}$.

ash functions A *hash function* h is a mapping $h : U \to M$ for some finite range I. The size $|M|$ of the range will be denoted by m. Two keys $x, y \in U$ *collide* under if $h(x) = h(y)$. It is convenient to use indicator functions for collisions, defined s follows. For $\mathbf{x} = \langle x_1, \ldots, x_k \rangle \in U^k$ we let $\delta^h(\mathbf{x}) = 1$ if $h(x_1) = \cdots = h(x_k)$ nd 0 otherwise. Henceforth we will not distinguish between the sequence $x_1, \ldots, x_k$ nd the vector $\langle x_1, \ldots, x_k \rangle$. More generally, for $S_1, \ldots, S_k \subseteq U$ let $\delta^h(S_1, \ldots, S_k) = \sum_{\mathbf{x} \in S_1 \times \cdots \times S_k} \delta^h(\mathbf{x})$. For $x \in U$ we will write $\delta^h(x, S)$ as a shorthand for $\delta^h(\{x\}, S)$.

Jniversal hashing [2] Let $\mathcal{H} \subseteq {}^U M$ be some class of hash functions. We regard $\mathcal{H}$ as a probability space with the uniform distribution ("pick $h \in \mathcal{H}$ at ran- om"). For $c \geq 1$, the class $\mathcal{H}$ is called *c-universal* if for all $\mathbf{x} = \langle x, y \rangle \in (U)_2$ e have $\mathbf{Prob}(h(x) = h(y)) \leq c/m$. (The case $c = 1$ is particularly nice.) The tandard generalization of this definition [4, 25, 19] runs as follows: The class is $c, k)$*-universal* if for all $\mathbf{x} = \langle x_1, \ldots, x_k \rangle \in (U)_k$ and $i_1, \ldots, i_k \in M$ we have $\mathbf{Prob}(\forall j : 1 \leq j \leq k : h(x_j) = i_j) \leq c/m^k$. We shall use a slightly more stringent efinition [27, 13]: The class $\mathcal{H}$ is (c,k)*-independent* if

a) $\mathbf{Prob}(h(x) = i) \leq c/m$, for all $x \in U$, $i \in M$;

b) for each $\mathbf{x} = \langle x_1, \ldots, x_k \rangle \in (U)_k$, the random variables $h(x_1), \ldots, h(x_k)$ are independent.

Clearly, if $\mathcal{H}$ is (c,k)-independent, then it is (c,k')-independent for $1 \leq k' \leq k$, s well as (c^k, k)-universal and c^2-universal.

Polynomial hash functions For $U = \mathbb{F}_q$ a finite field and $d \geq 1$, let $\mathcal{H}^d$: $\mathcal{H}^d(U) := \{g_\alpha \mid \alpha = \langle \alpha_0, \ldots, \alpha_d \rangle \in U^{d+1}\}$, where for $\alpha \in U^{d+1}$ we let $g_\alpha(x)$: $\sum_{i=0}^{d} \alpha_i x^i$ for $x \in U$. (The computation is carried out in the field U.) These class consist of functions that map U into itself.

Mapping into some smaller range M is effected by using a mapping $\zeta : U \longrightarrow$ with $|\zeta^{-1}(i)|$ about the same for all $i \in M$, i.e., $||\zeta^{-1}(i_1)| - |\zeta^{-1}(i_2)|| \leq 1$ for a $i_1, i_2 \in M$.

If $U = \mathbb{F}_u = \{0, 1, \ldots, u-1\}$ for some prime u, we let $M = \{0, \ldots, m-1\}$ ar $\zeta(x) = x \bmod m$; we write $\mathcal{H}^d_m := \{\zeta \circ g \mid g \in \mathcal{H}^d\}$. (We say that ζ is a modul operation.)

If $U = \mathbb{F}_u$ for some prime power $u = p^\ell$, we use the fact that $\mathbb{F}_u$ is represented b ℓ-dimensional vectors over $\mathbb{F}_p$; and use as ζ a projection from $\mathbb{F}_p^\ell$ onto any ℓ' out the ℓ coordinates (e.g., the last ℓ' ones). Thus, $M = \mathbb{F}_u^{\ell'}$ and $m = u^{\ell'}$, for $1 \leq \ell' \leq$ fixed.

The resulting class is denoted by $\mathcal{H}^d_M$. (We say that ζ is a projection in this case The classes $\mathcal{H}^1_m$ and $\mathcal{H}^1_M$ are called the classes of linear hash functions.

The following fact is the basis of all considerations concerning polynomial has functions.

Fact 3.1 ([27, 19]) *Let $U = \mathbb{F}_u$ be an arbitrary finite field. Then,* (a) *the class $\mathcal{H}^d$ are* $(1, d+1)$*-independent;* (b) *if u is prime, then the classes $\mathcal{H}^d_m$ are* $(c, d+1$ *independent, for $c = c(u, m) = 1 + (m-1)/u < 2$;* (c) *if $u = p^\ell$ is a prime powe $1 \leq \ell' \leq \ell$, $M = \mathbb{F}_p^{\ell'}$, then the classes $\mathcal{H}^d_M$ are* $(1, d+1)$*-independent.*

The input set Let $S \subseteq U$, $|S| = s$, be a set of keys. Then a hash function partitions S into subsets, called *buckets*, $S_i^h := h^{-1}(i) \cap S$, of size $b_i^h = |S_i^h|, i \in M$ Henceforth, the superscript h will be omitted whenever it can be deduced from th context.

We say that h is *r-perfect for S* if $b_i^h \leq r$ for all $i \in M$; h is *perfect for S* if it 1-perfect, i.e., if no two elements of S collide under h.

We will be interested in the quantities $B_k := B_k^h := \sum_{i \in M} \binom{b_i^h}{k}$, for $k \geq 1$, an $A_k := A_k^h := \sum_{i \in M} (b_i^h)^k$, for $k \geq 1$. We will study the probabilistic behavior of B under the assumption that S is fixed.

Upper bounds on B_k and on A_k are one of the main tools used in the constructio of hashing algorithms [5, 7]. Also note that bounds on B_k and A_k imply bound on the maximum bucket size and on the number of buckets that exceed a certai threshold. The analysis of B_2 is of central interest. It was observed in [2] that if h i used in a simple chained hashing scheme for storing the set S, then the number c memory accesses for performing a sequence of instructions in which each element c S is searched for once is proportional to $s + B_2$. The hashing algorithm of [6] relie heavily on the analysis of $\mathbf{E}(B_2)$. It is clear [2, 4] that $B_2 = \sum_{\mathbf{x} \in (S)_2} \delta(\mathbf{x})$, and more generally, $B_k = \sum_{\mathbf{x} \in (S)_k} \delta(\mathbf{x})$, for $k \geq 1$. We note the following properties o polynomials of bounded degree.

Fact 3.2 ([4]) *Let S be fixed. Let h be selected at random from $\mathcal{H}^d_M$. Then there ar constants $c_1 = c_1(k)$, $c_2 = c_2(k)$ such that the following holds:*

(a) For all $i \in M$, $\mathbf{Prob}(b_i \geq k)$ *is bounded above by $c_1 \cdot e^{k-1}/k^k$ if $k \leq d$ an by $c_1 \cdot e^{d-1} \cdot k^{-d}$ if $k > d$.*

(b) For all $x \in S$, $\mathbf{Prob}(\delta(x, S) \geq k)$ *is bounded above by $c_1 \cdot e^{k-1}/k^k$ if $k \leq d-$ and by $c_1 \cdot e^{d-2}/k^{(d-1)}$ if $k > d-1$.*

(c) $\mathbf{E}(B_k) \leq c_2 s^k / m^{k-1}$.

These statements also hold if any $(c, d+1)$-universal class is used.

Worst case inputs As can be seen from Fact 3.2, the higher the degree of the polynomials used, the better the performance that can be guaranteed. However, high degree polynomials use more random bits and have larger evaluation time. Experiments reported by [23] (concerning shared-memory emulation) and [21] (concerning implementations of hash tables on secondary storage) indicate that for the purposes universal classes are used for in these applications linear hash functions perform just as well as higher order polynomials or even as random functions. We shall address the question of the worst case performance of linear (and other polynomial) hash functions. Our example for demonstrating bad behavior will an input set S which is an *arithmetical progression* in U, i.e., $S = \{\alpha + i\beta \mid i = 0, 1, \ldots, s-1\}$ for some $\alpha, \beta \in U$. In particular, when $U = \mathbb{F}_p$ for a prime p, we will use $S = \{0, 1, \ldots, s-1\}$.

3.2 Probability and randomized algorithms

It will be convenient to use the following terminology.

Definition 3.3 *An event $A = A(n)$ is* n-dominant *(or for short* dominant*) if* $\mathrm{Prob}(A) = 1 - O(n^{-\epsilon})$ *for some $\epsilon > 0$. We then say that A occurs with* n-dominant *probability.*

Chebyshev's inequality can be written with this notation as follows.

Fact 3.4 *Let ω be a nonnegative random variable satisfying* $\mathbf{Var}(\omega) \le (\mathbf{E}(\omega))^{2-\delta}$ *for some constant $\delta > 0$. Then for any fixed $\epsilon > 0$, the event $(1-\epsilon)\mathbf{E}(\omega) \le \omega \le (1+\epsilon)\mathbf{E}(\omega)$ is $\mathbf{E}(\omega)$-dominant.*

When the expectation is small, the following fact is handy.

Fact 3.5 (Markov) *Let ω be a nonnegative random variable satisfying* $\mathbf{E}(\omega) < 1$. *Then $\omega < 1$ with $(1/\mathbf{E}(\omega))$-dominant probability.*

Definition 3.6 *An algorithm is* solid *if it solves problem instances of size n within the stated complexity bounds with n-dominant probability.*

Note that if failure of a solid algorithm can be detected, then its success probability can be amplified by means of repetitions to $1 - n^{-k}$ for *any* constant k. Consider an algorithm A for a given problem of size n that consists of poly(n) (possibly dependent) procedures, such that each procedure succeeds with n-dominant probability, and for each procedure a failure can be detected. Then Algorithm A is solid.

Fact 3.7 *Let $\omega_1, \ldots, \omega_n$ be pairwise independent random variables, and let $\omega = \sum_{i=1}^{n} \omega_i$. Then* (a) $\mathbf{Var}(\omega) = \sum_{i=1}^{n} \mathbf{Var}(\omega_i)$; (b) *if, moreover, $0 \le \omega_i \le 1$ for $1 \le i \le n$, then $\mathbf{Var}(\omega) \le \mathbf{E}(\omega)$, with equality occurring if and only if $\omega \equiv 0$.*

We will also be using a technique called 2-independent sampling (or, more generally, k-independent sampling) by [3], see also [13, 1, 10, 14]. The technical basis of this method is given in the following fact.

Fact 3.8 ([3, 13]) *Let $\omega_1, \ldots, \omega_n$ be 0-1-valued random variables with* $\mathbf{E}(\omega_i) = \rho$ *for $1 \le i \le n$, where $\rho n \ge 1$, and let $\omega = \sum_{i=1}^{n} \omega_i$.* (a) *If $\omega_1, \ldots, \omega_n$ are pairwise independent, then* $\mathbf{Prob}(\omega = 0) \le (1-\rho)/\rho n$. (b) *If $\omega_1, \ldots, \omega_n$ are k-wise independent (i.e., every subset $\{\omega_{i_1}, \ldots, \omega_{i_k}\}$ is independent) then* $\mathbf{Prob}(\omega = 0) \le \lfloor k/2 \rfloor^{k+1} \left(\frac{1-\rho}{\rho n}\right)^{\lfloor k/2 \rfloor}$.

4 Polynomial Hash Functions

4.1 Reliability of cubic polynomials

In the following, we assume that U is a finite field $\mathbb{F}_u$. We first consider the random variable $B_2 = \sum_{i \in M} \binom{b_i}{2}$, for a fixed input set $S \subseteq U$ of size s, and for h chosen randomly from $\mathcal{H}_M^3$ (or, from $\mathcal{H}_M^d$ for some $d \ge 3$). Recall the role of the function ζ (modulo-function or projection).

Definition 4.1 (a) $K_\zeta := \{x \in U \mid \zeta(x) = 0 \text{ or } \zeta(-x) = 0\}$. (b) *For* $\mathbf{x} = \langle x_1, x_2\rangle \in (U)_2$, $g \in \mathcal{H}^d$ *let* $\delta'(\mathbf{x})$ *be 1 if* $g(x_1) - g(x_2) \in K_\zeta$ *and 0 otherwise.* (c) $B_2' := \sum_{\mathbf{x}\in(S)_2} \delta'(\mathbf{x})$.

Lemma 4.2 *Assume* $d \geq 1$. *Then* (a) $\delta(\mathbf{x}) \leq \delta'(\mathbf{x})$, *for all* $\mathbf{x} \in (U)_2$. (b) $B_2 \leq B_2'$. (c) $\mathbf{E}(\delta'(\mathbf{x})) = |K_\zeta|/u$ *for all* $\mathbf{x} \in (U)_2$. (d) $1/m \leq |K_\zeta|/u \leq 3/m$. (e) $\mathbf{E}(B_2) \leq \mathbf{E}(B_2') = \binom{s}{2} \cdot |K_\zeta|/u \leq 3s^2/2m$. *In case* ζ *is a projection, the inequalities in (a) (b) are equalities; (d) becomes* $|K_\zeta|/u = 1/m$; *(e) becomes* $\mathbf{E}(B_2) = \binom{s}{2}/m$.

We need the following lemma.

Lemma 4.3 *The family* $\delta'(\mathbf{x})$, $\mathbf{x} \in (U)_2$ *is pairwise independent for all* $d \geq 3$.

From now on, let $S \subseteq U$ with $|S| = s$ be fixed, and assume $h \in \mathcal{H}_m^d$ is randomly chosen.

Lemma 4.4 *Assume* $d \geq 3$. *Then the following holds.* (a) $\mathbf{Var}(B_2') < \mathbf{E}(B_2')$. (b) $\mathbf{E}(B_2') < s^2|K_\zeta|/u$ *with* (s^2/m)*-dominant probability.*

Proof. *(a)* By Lemma 4.3, the family $\delta'(\mathbf{x}), \mathbf{x} \in (S)_2$, is pairwise independent. Thus Fact 3.7 applies. *(b)* Because of (a), we may apply Fact 3.4, noting that $\mathbf{E}(B_2') \leq s^2|K_\zeta|/2u$, by Lemma 4.2(e). ∎

Corollary 4.5 $B_2 \leq s^2 \cdot |K_\zeta|/u$ *with* $(s^2 \cdot |K_\zeta|)/u$*-dominant probability. (Recall that* $|K_\zeta|/u = 1/m$ *in case* ζ *is a projection and* $1/m \leq |K_\zeta|/u \leq 3/m$ *in case* $\zeta(x) = x \bmod m$.)

Theorem 4.6 *Let* $S \subseteq U$ *be fixed and let* $h \in \mathcal{H}_M^d$ *be random,* $d \geq 3$. *Then* $B_2 < 3|S|^2/m$ *with* $|S|^2/m$*-dominant probability.*

Note If $|S|^2/m$ is small, this theorem is not very useful. If $|S|^2/m < 1$, a simple application of Markov's inequality yields $B_2 < |S|^2/m$ (i.e., $B_2 = 0$, or h is perfect) with $m/|S|^2$-dominant probability, see [6, 4].

Remark 4.7 *In the analysis of* B_2' *above, only* differences $g(x_1) - g(x_2)$ *play a role, not the actual values* $g(x)$. *Thus the constant term* α_0 *in the polynomial* $g(x) = \sum_{i=0}^{d} \alpha_i x^i$ *is inessential for the argument; so the lemma is still true if* $\mathcal{H}^d$ *is replaced by* $\mathcal{H}^{d,1}$ = *the set of polynomials over* U *of degree* $\leq d$ *with zero constant term.*

In applications of polynomial hash functions to dynamic hashing the analysis of the following situation is of crucial importance: Assume the hash function h has been applied for S already, forming buckets S_i, $i \in M$. Suppose now that a new set R of r keys arrives, which are added to the already existing buckets. What is the total number of keys (from R and S together) in affected buckets? For $h \in \mathcal{H}_M^d$ random, we denote by $h(R)$ the set $\{i \in M \mid \exists y \in R : h(y) = i\}$. Thus, we are interested in the quantity $\sum_{i\in h(R)} b_i$. Obviously, $\delta(S, R) \geq \sum_{i\in h(R)} b_i$. The following lemma shows that this number is proportional to r with high probability.

Lemma 4.8 (The disturbance lemma) *Assume* $d \geq 3$, *and* S, $R \subseteq U$ *with* $S \cap R = \emptyset$. *Let* $h \in \mathcal{H}_M^d$ *be random. Then* $\delta(S, R) \leq 6sr/m$ *with* (sr/m)*-dominant probability. (In case* ζ *is a projection the right hand side may be replaced by* $2sr/m$.)

Proof. We have $\delta(S, R) = \sum_{x\in S, y\in R} \delta(\langle x, y\rangle) \leq \sum_{x\in S, y\in R} \delta'(\langle x, y\rangle) =: \delta'(S, R)$. Thus, it suffices to establish the claim for $\delta'(S, R)$. By a trivial variation of Lemma 4.3 (using that $S \cap R = \emptyset$) we see that the family $\delta'(\langle x, y\rangle)$, $x \in S$, $y \in R$, is pairwise independent. Thus, Fact 3.7 applies to $\delta'(S, R)$; thus $\delta'(S, R) \leq 2\mathbf{E}(\delta'(S, R))$ with $\mathbf{E}(\delta'(S, R))$-dominant probability. It remains to observe that $\mathbf{E}(\delta'(S, R)) = \sum_{x\in S, y\in R} \mathbf{E}(\delta'(\langle x, y\rangle)) = sr|K_\zeta|/u$. The quantity $|K_\zeta|/u$ has been determined above. ∎

4.2 Sums of higher powers of bucket sizes

Without proofs, we list the following generalizations of the results of Section 4.1.

Theorem 4.9 *Assume $d+1 \geq 2k$. Then the following holds.*

(a) If $s \leq m$ then $B_k = O(s^k/m^{k-1})$ with (s^k/m^{k-1})-dominant probability.
(b) If $s \geq m$ then $B_k = O(s^k/m^{k-1})$ with (s^2/m)-dominant probability.

Theorem 4.10 *Assume $d+1 \geq 2k$.*

(a) If $s \leq m$, then $A_k = O(\mathbf{E}(A_k))$ with (m^{k-1}/s^{k-2})-dominant probability.
(b) If $s \geq m$, then $A_k = O(\mathbf{E}(A_k))$ with (s^2/m)-dominant probability.

4.3 Higher moments of B_2

We want to study the effect of using polynomials of higher degree on the reliability of the estimate $B_2 = O(s^2/m)$. In Section 4.1 we saw that this is true with probability $1 - O(m/s^2)$. Throughout the following, we assume $d+1 \geq k \geq 4$, where k is even, and that a set $S \subseteq U$ is given. We study $\sigma^k(B'_2) := \mathbf{E}\left((B'_2 - E(B'_2))^k\right)$, the kth central moment of B'_2 and aim at using the following classical fact.

Fact 4.11 *Let $k \geq 2$ be even. Assume X is a nonnegative random variable for which $\sigma^k(X) = \mathbf{E}\left((X - E(X))^k\right)$ exists. Then $\mathbf{Prob}(X \geq 2\mathbf{E}(X)) \leq \sigma^k(X)/\mathbf{E}(X)^k$.*

We obtain the following:

Theorem 4.12 *If $d+1 \geq k \geq 2$ and k is even, and $m \leq s^2$, then $\mathbf{Prob}(B'_2 \geq 2\mathbf{E}(B'_2)) = O((m/s^2)^{k/2})$.*

4.4 A refinement of the disturbance lemma

For the implementation of the real-time dictionary in Section 7 we need a refinement of Lemma 4.8. As before, we consider two disjoint sets R and S of keys, with $|S| = s$ and $|R| = r$. For the construction of the real-time dictionary, we need to show that the quantity $\sum_{i \in h(R)} (b_i)^2$ is close to its expectation with r-dominant probability. Note that $\sum_{i \in h(R)} (b_i)^2 \leq \delta((S)_2, R)$. As usual, we define a more manageable random variable. We generalize δ' to inputs from $S \times S \times R$ as follows: $\delta'(\langle x_1, x_2, y\rangle) = \delta'(\langle x_1, x_2\rangle)\delta'(\langle x_2, y\rangle)$ and define $\delta'((S)_2, R) := \sum_{\langle x_1, x_2, y\rangle \in (S)_2 \times R} \delta'(\langle x_1, x_2, y\rangle)$. It is easily seen that $\delta((S)_2, R) \leq \delta'((S)_2, R)$.

Lemma 4.13 *Assume $d \geq 5$, and $h \in \mathcal{H}_M^d$ is chosen at random. Let $S, R \subseteq U$ with $S \cap R = \emptyset$. Then $\delta'((S)_2, R) \leq 9s^2r/m^2$ with $(sr/m)/(1 + m/s + r/s)$-dominant probability.*

Corollary 4.14 *Assume $r \leq s = \Theta(m)$. Then $\sum_{i \in h(R)} (b_i)^2 = O(r)$ with r-dominant probability.*

4.5 Limitations of linear and quadratic hash functions

In this section we demonstrate that in several respects linear and quadratic hash functions, i.e., the classes $\mathcal{H}_M^1$ and $\mathcal{H}_M^2$ are not as well suited for applications as higher degree polynomials. First, Theorem 4.15 shows that the method we used for proving that $\mathcal{H}_M^3$ is reliable will in general not work for linear and quadratic polynomials—thus, $(c,2)$-independence and $(c,3)$-independence of a hash class are not sufficient for proving reliability via Chebyshev's inequality. Then, in Theorem 4.17, we identify some of the limitations of polynomial hash functions: aspects in which the class $\mathcal{H}_M^1$ actually is not as reliable as $\mathcal{H}_m^3$.

Theorem 4.15 *Let $U = \mathbb{F}_u$ for $u = p^\ell$; let $M = S = \mathbb{F}_p$, and let $\zeta : U \to M$ be given as the projection $\mathbb{F}_p^\ell \ni (\alpha_{\ell-1}, \ldots, \alpha_0) \mapsto \alpha_0$. Then the following holds:*

(a) If h is chosen at random from $\mathcal{H}_M^1$ then $\mathbf{Var}(B_2) = \Omega(s^3)$.
(b) If h is chosen at random from $\mathcal{H}_M^2$ then $\mathbf{Var}(B_2) = \Omega(s^2)$.

(Recall from Lemma 4.2 that $\mathbf{E}(B_2) = O(s)$*, and that* δ *and* δ' *coincide in th situation.)*

Remark 4.16 *The proof of Theorem 4.15 heavily relies on the structure of* $\mathbb{F}_u$ *an extension field of* $\mathbb{F}_p$*, and on the special choice of the projection* ζ*. It is not clea whether other choices of* ζ *and the situation where* ζ *is a modulo operation lead similar effects or not.*

The next theorem shows that for some inputs linear hash functions perform provably worse than high degree polynomials.

Theorem 4.17 *Suppose that* $S = \{0, 1, \ldots, s-1\}$*,* $2s(s-1) < u$*,* $m = s$*, and th* $h \in \mathcal{H}_M^1$ *is randomly chosen. Then the following assertions hold, for* $1 \le k \le s$*.*

(a) $\forall i \in M : \mathbf{Prob}(b_i \ge k) = \Omega(k^{-3})$*;*

(b) $\forall x \in S : \mathbf{Prob}($ *x collides with at least k keys* $y \in S - \{x\}$ $) = \Omega(k^{-2})$*;*

(c) $\mathbf{Prob}($ *at least s/2 keys collide with at least k other keys* $) = \Omega(k^{-2})$*;*

(d) $\mathbf{Prob}(A_2 \ge k \cdot s) = \Omega(k^{-2})$*;*

(e) $\mathbf{E}(A_3) = \Omega(s \lg s)$*.*

If these results are compared with the upper bounds from Fact 3.2 for $d \ge$ one sees that the classes $\mathcal{H}^d$ perform provably better than $\mathcal{H}^1$ with respect to th distribution of the b_i and the number of keys a single key collides with; with respec to the quantity A_3 a visible difference already appears for $d = 2$, i.e., for quadrati polynomials. Also, the lower bounds given in Theorem 4.17 should be compared wit the upper bounds $O(u^{-2})$ (in (a)) and $O(u^{-1})$ (in (b)–(d)) provided by Fact 3.2 fo the case $d = 2$.

Theorem 4.17(d) implies that linear hash functions are provably unreliable, an therefore cannot replace the degree-3 polynomials in the applications described i later sections.

5 A New Class of Hash Functions

In this section we present a new class of hash functions. The functions are define in a simple manner and are easy to use. We use these functions for the purpos of reducing the universe size of a given set of keys. The universe reduction is don efficiently in time as well as in space. The analysis is rather elementary.

Definition 5.1 *Let the universe be* $U = \mathbb{F}_p^\ell$*. Then the class* $\mathcal{Y}$ *of hash function from* U *into* $\mathbb{F}_p$ *is defined by* $\mathcal{Y} = \{\langle 1, t, t^2, \ldots, t^{\ell-1}\rangle : t \in \mathbb{F}_p\}$*. For* $\mathbf{t} \in \mathcal{Y}$*,* $\mathbf{t} = \langle 1, t, t^2, \ldots, t^{\ell-1}\rangle$*, the operation of* $\mathbf{t}$ *on* $\mathbf{x} \in U$*,* $\mathbf{x} = \langle x_0, x_1, \ldots, x_{\ell-1}\rangle$*, is defined a the inner product* $\mathbf{t} \cdot \mathbf{x}$*.*

The functions in $\mathcal{Y}$ map the universe U of size $u = p^\ell$ into the range $\mathbb{F}_p$ o size p. The evaluation of a function from $\mathcal{Y}$ can be done in constant time. Each such function can be specified using $\lg p = \lg u/\ell$ bits.

Lemma 5.2 *For any* $x, y \in U$*,* $x \ne y$*, and* $\mathbf{t}$ *chosen at random from* $\mathcal{Y}$ $\mathbf{Prob}(\mathbf{t}(x) = \mathbf{t}(y)) \le (\ell-1)/p$*.*

Lemma 5.3 *Let* $S \subset U$*,* $|S| = s$*, be fixed. Then,* (a) *For a random* $\mathbf{t} \in \mathcal{Y}$*,* $\mathbf{Prob}(\mathbf{t}$ *is not injective over* $S) \le \binom{s}{2}(\ell-1)/p$ (b) *If* $p > \binom{s}{2}(\ell-1)$*, then there exists* $\mathbf{t} \in \mathcal{Y}$*, such that* $\mathbf{t}$ *is injective for* S*.* (c) *If* $p > s^{2+\epsilon}\ell$ *for some constant* $\epsilon > 0$*, and* $\mathbf{t} \in \mathcal{Y}$ *is random then* $\mathbf{t}$ *is injective for* S *with s-dominant probability.*

The class $\mathcal{Y}$ can be used for the following problem: Given s distinct keys drawn from a very large universe, find a hash function, represented by a small number of bits, which maps the universe into a smaller one while maintaining key distinctness. Specifically, for a universe of size u, the number of bits required for reducing the universe to $s^{2+\epsilon}$, for any $\epsilon > 0$, is $\lg\lg u + \lg^{k+1} u + k(2+\epsilon)\lg s + O(\lg\lg s)$ for any integer k. (In [6, 18] the number of bits obtained is $O(\lg\lg u + \lg s)$.) This construction is based on applying a $\mathcal{Y}$ type function k times, with carefully chosen p at each application, and then a linear hash function.

Static Hashing (FKS) Scheme

Fredman, Komlós, and Szemerédi [6] presented a simple and elegant sequential algorithm for constructing a static dictionary. Their two-phase algorithm constructs a two-level hashing scheme. A first-level hash function $h \in \mathcal{H}_s^{1,1}$ partitions the input set S into $|S| = s$ buckets $S_i = S \cap h^{-1}(i)$, $i \in M = \{0, \ldots, s-1\}$; this function is constructed in the first phase. For each bucket S_i of size b_i, a private memory block of size $2b_i^2$ is allocated, and a second-level function $h_i \in \mathcal{H}_{2b_i^2}^{1,1}$ maps the elements of S_i one-to-one into this block; these functions are constructed in the second phase. Each of the phases runs in expected linear time. Our aim in this section is to show how the FKS algorithm can be made to run in linear time with high probability, and in addition using very few random bits.

6.1 Reliability

We show that a slight modification suffices to make the FKS algorithm run in linear time with high probability: We only have to choose the first-level function h from $\mathcal{H}_s^3$ instead of $\mathcal{H}_s^{1,1}$. Although only an expected linear time bound is claimed in [6], the second phase need not be altered at all.

First phase We say the first-level function $h : U \to M = \{0, \ldots, s-1\}$ is *good* if $\sum_{i \in M} b_i^2 = O(|S|)$. In [6] the function h is constructed by repeatedly trying functions selected at random from $\mathcal{H}_s^{1,1}$ until one is found that is good. Each trial is successful with probability at least 1/2, and takes $O(|S|)$ time; thus this phase takes expected time $O(|S|)$. (Note that, according to Theorem 4.17(d), for worst-case input sets S the probability that a random h in $\mathcal{H}_s^{1,1}$ is good is bounded below 1, thus this algorithm is not solid.) We suggest that h is chosen at random from $\mathcal{H}_s^3$. By Theorem 4.6 the first function tried is good with probability $1 - O(s^{-1})$. Thus, the first phase takes time $O(s)$ with s-dominant probability.

Second phase In the FKS algorithm, each bucket S_i, $i \in M$, is *independently* treated as follows: A second-level function h_i is chosen at random from $\mathcal{H}_{2b_i^2}^{1,1}$. Such a function has the form $h_\alpha, \alpha \in \mathbb{F}_p$, where $h_\alpha(x) = (\alpha x \bmod p) \bmod 2b_i^2$. This function is injective on S_i with probability at least 1/2; injectiveness can be tested using $O(b_i)$ operations. This is repeated (independently) until an h_i is found that is injective on S_i. Let t_i denote the number of trials needed to find a suitable h_i. This random variable is geometrically distributed, with $\mathbf{Prob}\,(t_i = 1) \geq 1/2$; hence $\mathbf{E}\,(t_i) \leq 2$ and $\mathbf{Var}\,(t_i) \leq 2$.

Let T_i be the number of operations required to find a suitable function h_i. Then $T_i = O(t_i \cdot b_i)$, and hence $\mathbf{E}\,(T_i) = O(b_i)$ and $\mathbf{Var}\,(T_i) = O(b_i^2)$. Let $T := \sum_{i \in M} T_i$ be the total number of operations necessary to find injective functions for all buckets. Then $\mathbf{E}\,(T) = \sum_{i \in M} O(b_i) = O(|S|)$, and, because the random experiments for the different buckets are independent, $\mathbf{Var}\,(T) = \sum_{i \in M} \mathbf{Var}\,(T_i) = \sum_{i \in M} O(b_i^2) = O(|S|)$. (We used that h is good.) Now Chebyshev's inequality immediately implies that $T = O(|S|)$ with $|S|$-dominant probability.

6.2 The FKS scheme with limited randomness

Now we indicate how the static hashing algorithm may be implemented with only $O(\lg \lg u + \lg s)$ random bits, without impairing its reliability.

Preparation Using the method described in Section 5, we reduce the size of the universe to $s^{2+\varepsilon}$, for some arbitrarily fixed $\varepsilon > 0$. That means, we randomly select a function $h_0 : U \to \{0, \ldots, s^{2+\varepsilon} - 1\}$ that is injective on S with $|S|$-dominant probability. The number of random bits required is $\lg \lg u + \lg^{(k)} u + O(\lg s)$, where k is an arbitrary constant. Let $S' = \{h_0(x) \mid x \in S\}$. (Note that a slight problem arises here: if h_0 fails to be injective on S, we must detect this in the sequel.) Let

$p \in [s^{2+\varepsilon}, 2s^{2+\varepsilon}]$ be an arbitrary prime number. All calculations below are in the field $\mathbb{F}_p$.
First phase Choose randomly *one* function $h \in \mathcal{H}_s^3$. (This requires $4\lg p = O(\lg s)$ random bits.) As in Section 6.1 we have that h is good for S' with $|S|$-dominant probability.
Second phase We provide a method for choosing the second-level functions $h_i, i \in S$. First consider a fixed i. The standard algorithm uses an independent sequence of hash functions from $\mathcal{H}^{1,1}_{2b_i^2}$. Each such function is determined by a unique number from the reduced universe, i.e., by $O(\lg s)$ bits. The naive implementation of the algorithm results in selecting $O(s)$ independent random numbers, i.e., $O(s \lg s)$ random bits. It has been already observed that $O(\lg s)$ functions suffice for the second level. This is based on the fact that the same function can be used by all buckets, and will be injective for around half of them. This would lead to $O(\lg^2 s)$ random bits. We may get by with even fewer random bits in the manner proposed by [3]: we only use a 6-wise independent sequence $(h_{i,j})_{j=0,\ldots,p-1}$ of such functions. (E. g., choose one random seed $\beta = (\beta_{i0}, \ldots, \beta_{i5}) \in \mathbb{F}_p^5$, and let $h_{i,j} = h_{\alpha(i,j)}$, where $\alpha(i,j) = \sum_{l=0}^{5} \beta_{il} j^l$, for $j = 0, \ldots, p-1$.) By Fact 3.8 we have that, for all $z \geq 1$,

$$\mathbf{Prob}\,(t_i > z) = \mathbf{Prob}\,(\forall j < z : h_{\alpha(i,j)} \text{ not } 1-1 \text{ on } S_i) = O(1/z^3).$$

It follows that $\mathbf{E}\,(t_i) = O(1)$ and $\mathbf{Var}\,(t_i) \leq \mathbf{E}\,(t_i^2) \leq \sum_{z\geq 1} \mathbf{Prob}\,(t_i^2 \geq z) = \sum_{z\geq 1} O(1/z^2) = O(1)$. This shows how we can achieve $\mathbf{E}\,(T_i) = O(b_i)$ and $\mathbf{Var}\,(T_i) = O(b_i^2)$, using only $6\lg(s^{2+\varepsilon}) + O(1)$ random bits, for one fixed bucket S_i. It remains to show how to get the seeds $(\beta_{i0}, \ldots, \beta_{i5}), i \in M$, for all these sequences without wasting random bits. We modify the idea of 2-independent sampling. Each of the 6 sequences $(\beta_{il})_{i\in M}$ is generated as a 2-independent sequence induced by a random seed consisting of $2\lg p + O(1)$ bits (e. g., use 6 random linear functions over $\mathbb{F}_p$). Altogether, $12\lg p + O(1) = O(\lg s)$ random bits are needed. The random choices concerning the distinct S_i are now pairwise independent; hence the time bounds T_i, $i \in M$, are pairwise independent. This implies, as in Section 6.1, that $\mathbf{Var}\,(T) = \sum_{i\in M} \mathbf{Var}\,(T_i) = \sum_{i\in M} O(b_i^2) = O(|S|)$, and hence that $T = O(|S|)$ with $|S|$-dominant probability. We summarize.

Theorem 6.1 *There is a randomized algorithm for constructing a perfect hash function that runs in $O(s)$ time, succeeds with s-dominant probability, and uses $\lg\lg u + \lg^{(k)} u + O(\lg s)$ random bits, where k is an arbitrary constant.*

Remark We have described an algorithm that, with probability $O(1/s)$, may fail to work properly. Especially if h_0 produces many collisions, the algorithm will not find a suitable hash function. However, it is not hard to see that the methods we introduced are sufficient for designing an algorithm that always constructs a perfect hash function, has expected linear time, finishes in $O(s)$ time with s-dominant probability, and uses $\lg\lg u + O(\lg s)$ random bits (for $s = \lg^{(k)} u$ where k is some constant).

7 Real-Time Dictionary

In [4] a dictionary is described in which each instruction can be performed in constant expected amortized time, and lookups take constant time in the worst case. Afterwards, in [5], a real-time dictionary was described, i.e., a dictionary in which with high probability all instructions execute each in constant time. To achieve this, in [5] a new class of hash functions was designed. This class demonstrates random-like behavior better than that of polynomial hash functions. However, a function drawn from this class, and consequently the algorithm, requires as many as $O(n^\varepsilon \lg u)$ random bits for a dictionary with n keys drawn from a universe of size u. Below we ketch a method how simple polynomial hash functions suffice to get a real-time dic-ionary. This results in a significant reduction in the number of random bits. We get hat such a real-time dictionary can be implemented using $\lg\lg u + \lg^{(k)} u + O(\lg n)$ andom bits, where k is an arbitrary constant. First of all, we reduce the size of the niverse to $O(n^{2+\epsilon})$, just as in the static case, and only deal with keys of length)$(\lg n)$ from here on. The basic structure of the algorithm is the same as in [5]; the nly modification is that the hash functions used have been substituted by simpler lasses. We do not elaborate below on how to manage global "rehash" operations n a way that will guarantee real-time behavior, but assume that the number n of nstructions to be performed is known beforehand. (See also [8] on this issue.)

We give highlights of the dictionary algorithm. A dictionary with capacity of up o n keys consists of a *primary dictionary* of size $O(n)$ and of a *cache* of size $O(n)$. Keys are first inserted into the cache and later moved into the permanent primary lictionary. Two buffers are alternately used as the cache. When the number of eys in the cache becomes n^ϵ, for an appropriately chosen constant $\epsilon < 1$, the other uffer becomes a *new* empty cache for the next n^ϵ keys. The keys in the *old* cache are nserted into the primary dictionary by a background process that runs concurrently vith the current insertions into the new cache. These insertions must be finished by he time the new cache is filled with n^ϵ new keys.

A way of implementing the cache, using polynomial hash functions, is described n [5]. Thus, we only need to establish the following:

Primary Dictionary There is an implementation of a dictionary in $O(n)$ space, such that if the dictionary holds at most n keys then a *batch* of n^ϵ keys can be inserted in $O(n^\epsilon)$ time with n-dominant probability.

The primary dictionary is implemented using our modified FKS scheme. Thus, a polynomial hash function $h \in \mathcal{H}_n^6$ serves as a first level function. This function is selected once when the primary dictionary is first built. With n-dominant probability, this function will be appropriate (i.e., will have the properties that we need) for the entire life-time of the primary dictionary. Let R be the set of keys already stored in the primary dictionary and S be the set of keys in the cache to be inserted into the primary dictionary. Using the (fixed) first-level function, keys in S can collide within themselves or with keys from R. All keys in R that are collided by S are retrieved. Let S^R be the set of keys retrieved from R. By the disturbance lemma (Corollary 4.14) we have $\sum_{i\in J(R,S)}(b_i)^2 = O(|S|)$ with $|S|$-dominant probability, and hence with n-dominant probability. We can now employ the second phase of the static FKS scheme to the retrieved keys together with S and store them in $O(|S|)$ space.

Evidently, the primary dictionary is identical to the hash table built in the static (modified FKS) hashing scheme. Hence, the real-time dictionary requires the same resources as the static hashing scheme.

Theorem 7.1 *The real-time dictionary can be implemented using* $\lg\lg u + \lg^{(k)} u + O(\lg s)$ *random bits where k is an arbitrary constant.*

References

1. N. Alon, L. Babai, and A. Itai. A fast and simple randomized parallel algorithm for the maximal independent set problem. *J. of Alg.*, 7:567–583, 1986.
2. L. J. Carter and M. N. Wegman. Universal classes of hash functions. *JCSS*, 18:143–154, 1979.
3. B. Chor and O. Goldreich. On the power of two-point based sampling. *J. Complexity*, 5:96–106, 1989.
4. M. Dietzfelbinger, A. R. Karlin, K. Mehlhorn, F. Meyer auf der Heide, H. Rohnert, and R. E. Tarjan. Dynamic perfect hashing: Upper and lower bounds. Technical Report 70, Universität Paderborn, Gesamthochschule, Jan. 1991. *Revised Version* of the paper of the same title that appeared in *FOCS '88*, pp. 524–531.

5. M. Dietzfelbinger and F. Meyer auf der Heide. A new universal class of hash functions and dynamic hashing in real time. In *ICALP '90* (LNCS 443), pp. 6–19, July 1990.
6. M. L. Fredman, J. Komlós, and E. Szemerédi. Storing a sparse table with $O(1)$ worst case access time. *J. ACM*, 31(3):538–544, July 1984.
7. J. Gil and Y. Matias. Fast hashing on a PRAM—designing by expectation. In *SODA '91*, pp. 271–280, Jan. 1991.
8. J. Gil, Y. Matias, and U. Vishkin. Towards a theory of nearly constant time parallel algorithms. In *FOCS '91*, pp. 698–710.
9. C. T. M. Jacobs and P. van Emde Boas. Two results on tables. *IPL*, 22(1):43–48, 1986.
10. A. Joffe. On a set of almost deterministic k-independent random variables. *The Annals of Prob.*, 2:161–162, 1974.
11. A. R. Karlin and E. Upfal. Parallel hashing—an efficient implementation of shared memory. In *STOC '86*, pp. 160–168.
12. D. E. Knuth. *Sorting and Searching*, volume 3 of *The Art of Computer Programming*.
13. C. P. Kruskal, L. Rudolph, and M. Snir. A complexity theory of efficient parallel algorithms. *TCS*, 71:95–132, 1990.
14. M. Luby. A simple parallel algorithm for the maximal independent set problem. *SIAM J. Comput.*, 15(4):1036–1053, 1986.
15. H. G. Mairson. The program complexity of searching a table. In *FOCS '83*, pp. 40–75.
16. G. Markowsky, L. J. Carter, and M. N. Wegman. Analysis of a universal class of hash functions. In *MFCS '78*, pp. 345–354, 1978.
17. K. Mehlhorn. On the program size of perfect and universal hash functions. In *FOCS '82*, pp. 170–175.
18. K. Mehlhorn. *Data Structures and Algorithms*. Springer-Verlag, Berlin Heidelberg, 1984.
19. K. Mehlhorn and U. Vishkin. Randomized and deterministic simulations of PRAMs by parallel machines with restricted granularity of parallel memories. *Acta Inf.*, 21:339–374, 1984.
20. N. Nisan. Pseudorandom generators for space-bounded computations. In *STOC '90*, pp. 204–212.
21. M. V. Ramakrishna and P.-A. Larson. File organization using composite perfect hashing. *ACM Trans. Database Syst.*, 14(9):231–263, 1989.
22. A. G. Ranade. How to emulate shared memory. In *FOCS '87*, pp. 185–194.
23. A. G. Ranade, S. N. Bhatt, and S. L. Johnsson. The fluent abstract machine. In *Proc. of the 5th MIT Conference on Advanced Research in VLSI*, pp. 71–93, 1988.
24. J. P. Schmidt and A. Siegel. The spatial complexity of oblivious k-probe hash functions. *SIAM J. Comput.*, 19(5):775–786, 1990.
25. A. Siegel. On universal classes of fast high performance hash functions, their time-space tradeoff, and their applications. In *FOCS '89*, pp. 20–25.
26. C. Slot and P. van Emde Boas. On tape versus core; an application of space efficient hash functions to the invariance of space. In *STOC '84*, pp. 391–400.
27. M. N. Wegman and L. J. Carter. New classes and applications of hash functions. In *FOCS '79*, pp. 175–182.
28. M. N. Wegman and L. J. Carter. New hash functions and their use in authentication and set equality. *JCSS*, 22:265–279, 1981.
29. A. C. Yao. Should tables be sorted? *J. ACM*, 28(3):615–628, July 1981.

Adaptive Pattern Matching*

R.C. Sekar†
Bellcore, Rm 2A-274
445 South Street
Morristown, NJ 07962.
sekar@thumper.bellcore.com.

R. Ramesh
Dept. of Computer Science
University of Texas at Dallas
Richardson, TX 77083
ramesh@utdallas.edu

I.V. Ramakrishnan
Dept. of Computer Science
SUNY at Stony Brook
Stony Brook, NY 11794.
ram@cs.sunysb.edu

Abstract

Pattern matching is an important operation used in many applications such as functional programming, rewriting and rule-based expert systems. By preprocessing the patterns into a DFA-like automaton, we can rapidly select the matching pattern(s) in a single scan of the relevant portions of the input term. This automaton is typically based on left-to-right traversal (of the patterns) or its variants. By *adapting* the traversal order to suit the set of input patterns, it is possible to considerably reduce the space and matching time requirements of the automaton. The design of such adaptive automata is the focus of this paper. In this context we study several important problems that have remained open even for automata based on left-to-right traversals. Such problems include upper and lower bounds on space complexity, construction of optimal dag automata and impact of typing in pattern matching. An interesting consequence of our results is that lazy pattern matching in typed systems (such as ML) is computationally hard whereas it can be done efficiently in untyped systems.

1 Introduction

Pattern matching is a fundamental operation in a number of important applications such as functional and equational programming, term rewriting and theorem proving. In most of these applications, patterns are partially ordered by assigning priorities. For instance, in languages such as ML [5] and Haskell [6], a pattern occurring earlier in the text has a higher priority over those following it. Applications that do not impose priorities can also be handled as a special case of matching with priorities.

The typical approach to pattern matching is to preprocess the patterns into a DFA-like automaton that can rapidly select the patterns that match the input term[1]. The main advantage of such a matching automaton is that all pattern matches can be identified in a *single* scan (i.e., no backtracking) of portions of input term relevant for matching purposes and is done in time that is *independent* of the number of patterns. Fig. 1 shows such a matching automaton constructed on the basis of a left-to-right traversal of patterns. This automaton can be represented by tables or compiled into case statements. Each state of the automaton corresponds to the prefix of the input term seen in reaching that state and is annotated with the set of patterns that can possibly match. For instance, state s_4 corresponds to having inspected the prefix $f(b,a,x)$, where x denotes the subterm that has not yet been examined. This state is annotated with the pattern set $\{1,2,3\}$ since we cannot rule out a match for any of the three patterns on the basis of the prefix $f(b,a,x)$.

Pattern matching automata have been extensively studied for well over a decade. Augustsson [1] and Wadler [14] describe pattern matching techniques based on left-to-right traversal. A drawback with these methods is that they may reexamine symbols potentially several times; so in the worst case they may end up testing each pattern separately against the input term. This drawback is overcome by the methods of Christian, Graf and Schnoebelen, which are again based on left-to-right traversals or variants thereof. However, their space requirements can become exponential in the size of patterns.

One way to improve space and matching time is to engineer a traversal order to suit the set of patterns or the application domain. We refer to such traversals as *adaptive traversals* and automata

*Research partially supported by NSF grants CCR-8805734,9102159,9110055 and NYS S&T grant RDG 90173.

†This research was completed at SUNY, Stony Brook.

[1] Only matches at the root of the input term are identified. Nonlinearity is not considered since even in applications that allow such patterns, most failures are associated with symbol mismatches, as observed in [3].

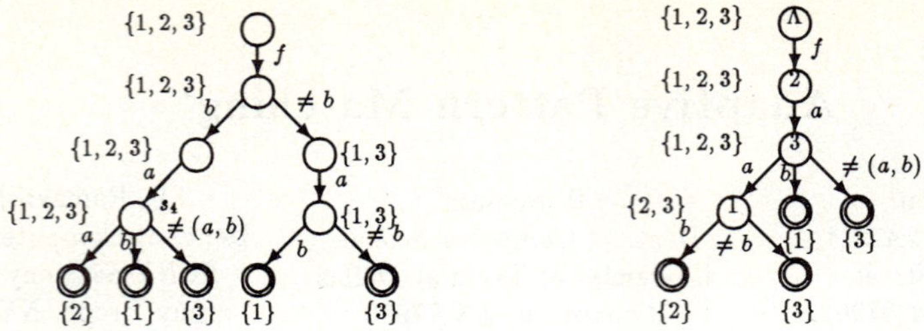

Figure 1: Left-to-right (left) and adaptive automata (right) for $f(x, a, b), f(b, a, a), f(x, a, y)$ with textual order priority. Here x and y denote variables.

based on such traversals as *adaptive automata.* As traversal orders are no longer fixed apriori, an adaptive automaton must specify the traversal order. For instance, in the adaptive automaton shown in fig. 1, each state is labelled with the next argument position to be inspected from that state. Adaptive traversal has two main advantages over a *fixed-order* of traversal such as left-to-right.

• The resulting automaton can be smaller, e.g., the adaptive automaton in fig. 1 has 8 states compared to 11 in the left-to-right automaton. The reduction factor can even become *exponential.*

• Pattern matching requires lesser time with adaptive traversals than fixed-order traversals. Furthermore, fixed-order traversals cannot use the priority information to avoid inspection of symbols irrelevant for determining highest priority matches. For instance, the adaptive automaton announces a match for pattern 1 after examining only the last two arguments of f whereas the left-to-right automaton needs to inspect all three arguments. Observe that examining unnecessary symbols runs counter to the goals of lazy evaluation in the context of functional programs.

The design of adaptive automata is the focus of this paper. In the process, we study several important problems that have remained open even in the context of automata based on fixed-order traversals. These problems include lower and upper bounds on space complexity, construction of optimal dag automata and the impact of typing in pattern matching.

1.1 Summary of Results

We first formalize the concept of adaptive traversal and its special case, namely, fixed traversal orders and then present an algorithm for constructing adaptive automata.

• In section 3 we examine space and matching time complexity. We show that the space requirements of the automata can be quite large by establishing the first known tight exponential lower bounds on size. These results are quite difficult to obtain since proofs must be *independent of traversal order.*

• In section 4 we present several techniques to synthesize traversal orders that can improve space and matching time. We first develop the important concept of a *representative set* which forms the basis of several optimization techniques aimed at avoiding inspection of unnecessary symbols. We present a quadratic-time algorithm for computing representative sets in untyped systems whereas we show that computing these sets is NP–complete for typed systems.

• We then present several powerful strategies for synthesizing traversal orders. Through an intricate example, we show that they all can sometimes increase both space and matching time. We therefore present another strategy based on selecting *indices* that overcomes this drawback. Huet and Levy [7] had established the importance of index in the design of optimal automata for strongly-sequential patterns. Our results extend the applicability of indices even for patterns that are not strongly-sequential.

• In section 4 we synthesize a traversal $S(T)$ from a given traversal T (such as left-to-right) by inspecting index positions whenever possible. Using $S(T)$ in place of T does not affect the termination properties of a functional program. So the programmer can continue to assume T whereas an implementation can benefit from significant improvements in space and time using $S(T)$.

• In section 5 we describe an orthogonal approach to space minimization based on dag representation. By tightly characterizing equivalence of states we directly build an optimal dag automaton. This important problem had remained open [4] even for left-to-right traversals.

• In section 6 we focus on the important problem of index computation in prioritized systems. Laville [9], Puel and Suarez [11] have extended Huet-Levy's [7] index computation algorithm to deal with priorities. However, these algorithms require exponential time in the worst case. In contrast, we present the first polynomial-time algorithm for index computation in untyped systems. Furthermore, we also show that this problem is *co-NP*–complete for typed systems. We therefore present powerful heuristics that can significantly speed up this process in typed systems.

• Our work clearly brings forth the impact of typing in pattern matching. We have shown that several important problems in the context of pattern matching are unlikely to have efficient algorithms in typed systems whereas we have given polynomial time algorithms for them in untyped systems. This clearly demonstrates that typing makes algorithms inherently more complex rather than serve as an optimization that aids in pruning the search space for index computation, as suggested in [9]. Finally, implications of our results are discussed in section 7. Some proof details have been omitted for lack of space and can be found in [13].

2 Preliminaries

In this section we introduce notations that will be used in the rest of this paper. We also present a generic algorithm for adaptive automata construction that forms the basis for the results presented in later sections.

We assume familiarity with the basic concept of a *term.* We use $root(t)$ to denote the symbol appearing at the root of a term t. The notion of a ***position*** is used to refer to subterms in a term as follows. A position is either the empty string Λ that reaches the root or $p.i$ (p is a position and i an integer) which reaches the ith argument of the root of the subterm reached by p. We use t/p to refer to the subterm of t reached by p and $t[p \leftarrow s]$ to denote the term obtained by replacing the subterm t/p by s. A ***substitution*** maps variables to terms. An *instance* $t\beta$ of a term t is obtained by substituting $\beta(x)$ for every variable x in t. If t is an instance of u then we say $u \leq t$ and call u a *prefix* of t. *Fringe* of u is the set of all variable positions in u. We will use x, y and z to denote variables, '_' to denote any unnamed variable and a, b, c, d, f to denote nonvariables. We say that two terms t and s unify iff they possess a common instance. $t \sqcup s$ denotes the least such instance in the ordering given by $\leq$.

A top-down traversal inspects nodes in a term successively and is characterized by a selection function that, having inspected a prefix u, chooses the next position to visit. The *match set* $\mathcal{L}_u$ of a prefix u is the set of patterns that unify with u. Let $\mathcal{P}$ denote the set of fringe positions of u wherein at least one pattern in $\mathcal{L}_u$ has a nonvariable.

Adaptive and Fixed Traversals: An *adaptive* traversal is a top-down traversal wherein the position $p \in \mathcal{P}$ next visited is a function of the prefix u and the given set of patterns $\mathcal{L}$. In a fixed traversal order p is simply a function of $\mathcal{P}$.

Pattern Match: We say that a pattern $l \in \mathcal{L}$ matches t iff $t \geq l$ and $t \not\geq l'$ for any l' with priority higher than that of l. If there is any such l in $\mathcal{L}$ then we say there is a *pattern match* for t. Among patterns of equal priority, we may announce a match for a term t if it is an instance of any one of these patterns.

Ambiguous and Unambiguous Systems: A set of patterns $\mathcal{L}$ is said to be *ambiguous* whenever there is more than one pattern that can match any given term. Otherwise $\mathcal{L}$ is *unambiguous.*

Typed Systems: In typed systems, the set of allowable input terms are constrained by a type discipline. For instance, this constraint may take the form that arguments to a function f must be drawn from the set $\{T, F\}$. In this case, terms $f(T, F), f(F, T), f(T, T)$ and $f(F, F)$ are allowable whereas $f(2, T)$ is not.

Untyped Systems: In untyped systems, there is no such type discipline. Hence the allowable input terms include any term over any alphabet Σ' that is a superset of the alphabet Σ used in the patterns.

Index: A fringe position p for a prefix u is said to be an *index* iff for every term $t \geq u$ for which there is a pattern match, t/p is a nonvariable. Intuitively, index is a position that *must* be examined to determine a pattern match, e.g., any fringe position of u where every pattern in $\mathcal{L}_u$ has a nonvariable is an index.

2.1 Generic Algorithm to build Adaptive Automata

A state v of the automaton remembers the prefix u inspected in reaching v from the start state. Suppose that p is the next position inspected from v. From v there are transitions on all symbols c that are present at p for any $l \in \mathcal{L}_u$. There will also be a transition from v on $\neq$ which will be taken on seeing a

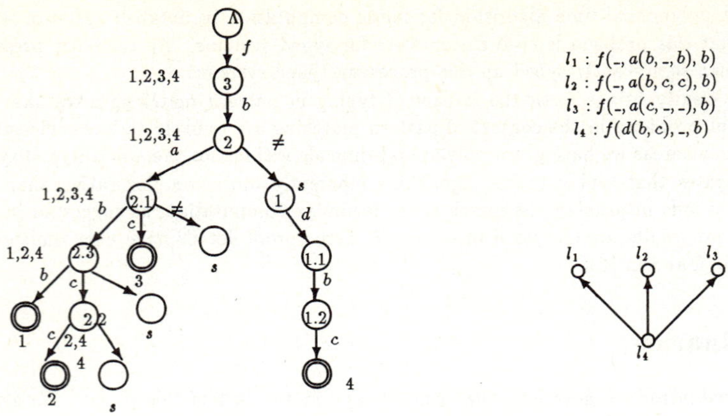

Figure 2: Adaptive automaton constructed by *Build* for patterns shown at top-right, with priorities as in bottom-right. The *pos* field and $\mathcal{L}_u$ are shown with each state. States labelled *s* denote identical subtrees.

symbol different from those on the other edges. A generic algorithm to construct an adaptive automaton is given below. Observe that priorities are implicitly handled by definition of match.

```
Procedure Build(v, u)
1.  Let M denote patterns that match u.
2.  If M ≠ φ then matchset[v] := M.  /* State v announces a match for patterns in M
3.  else
4.        p = select(u) /* select is a function to choose the next position to inspect */
5.        pos(v) = p /* Next position to inspect is recorded in the pos field */
6.        for each symbol for which ∃l ∈ L_u with root(l/p) = c (for some nonvariable c) do
7.              create a new node v_c and an edge from v to v_c labelled c
8.              if c ≠ '_' then Build(v_c, u[p ← c(y_1, ..., y_rank(c))])
9.              else Build(v_c, u[p ←≠])
```

Build is similar to previously known algorithms such as that of Christian [3] and Huet-Levy [7]. The main difference is that *Build* is generic whereas the others prespecified the *select* function. *Build* is invoked by the call $Build(s_0, x)$ where s_0 is the start state of the automaton. It takes two parameters – v specifies a state of the automaton and u specifies the prefix examined in reaching state v. This invocation of *Build* constructs the subautomaton rooted at v. Lines 6, 7 and 8 create transitions based on each symbol that could appear at p for any pattern in $\mathcal{L}_u$. In line 8, *Build* is recursively invoked with the prefix extended to include the symbols seen on the transitions created in line 7. Note that if the symbol is a '_' then the prefix is extended with a new nonvariable symbol '$\neq$' that is not in any pattern (line 9). Therefore the match set of the extended prefix $u[p \leftarrow \neq]$ consists only of those patterns that have a variable at p. Fig. 2 shows a set of patterns along with their priorities and the corresponding automaton constructed by *Build*. The automaton has 25 states since states labelled *s* all have identical subtrees each with 4 states.

Observe that there can be at most S positions of interest (recall that S is the sum of the sizes of all the patterns). Each position, which is an integer sequence, can be encoded as an integer. We omit the details of such an encoding which can be found in implementations such as EQUALS [10]. Using such an encoding, the symbol specified by any position can be accessed without any extra overhead when compared to fixed-traversal orders. Finally, observe that the automata generated by *Build* has a tree structure which is also the representation used by all previous pattern matching methods. Popularity of trees stems from their simplicity and the fact that they readily support case statement representation.

Class of Patterns	Lower bound on space	Upper bound on space	Lower bound on time	Upper bound on time
Unambiguous, no priority	$\Omega(2^n)$	$O(\prod_{i=1}^{n} \|l_i\|)$	$\Omega(a)$	S
Unambiguous, with priority	$\Omega(a^n)$	$O(\prod_{i=1}^{n} \|l_i\|)$	$\Omega(S)$	S
Ambiguous	$\Omega(a^n)$	$O(\prod_{i=1}^{n} \|l_i\|)$	$\Omega(S)$	S

Figure 3: Space and matching time complexity of adaptive automata. Here n, S, a and $|l_i|$ denote respectively the number of patterns, sum of their sizes[2], average size and the size of ith pattern.

3 Space and Matching Time Complexity

We now examine upper and lower bounds (independent of traversal order) on the space and matching time complexity of adaptive tree automata for several classes of patterns. Fig. 3 summarizes our results. Observe that the contribution by depth of the automata (whose upper bound is S) to space is insignificant compared to the exponential lower bound. The major contribution comes from the breadth of the automata. Furthermore, even for patterns without variables (i.e. ground terms) it is hard to reduce the number of states [2] whereas reducing breadth is not. Therefore in the rest of the paper, we will use breadth as the measure of size of the automaton.

We now present the details of lower bound proofs. The proofs for upper bounds can be established along the lines of [4, 12]. The space bounds given in this section are all independent of the traversal order and are established using flat patterns that all have a root symbol f with arbitrarily large arity. It can be shown that for purposes of buiding either the smallest size automaton or one that does matching in the shortest possible time, flat patterns are equivalent to a set of patterns having a common prefix u whose fringe size equals arity of f. (This is because every symbol in the common prefix will have to be inspected on any path to any final state. Then it follows by theorem 4 (section 4.3) that the automaton that inspects all these symbols first is no larger than any other automaton.)

3.1 Unambiguous, Unprioritized Patterns

Consider a set of n nonoverlapping patterns from the alphabet $\{f, a, b\}$ and variables. Since all flat patterns have the same root symbol f, we need only specify the arguments. Therefore n patterns can be represented by a matrix of n rows, where the ith row lists the arguments of f in the ith pattern. Each column has at most one occurrence of a, at most one occurrence of b and the rest are all '_'s. For each pair of patterns l and l', there is at least one column wherein l and l' have different nonvariables and so the system is unambiguous. Fig. 4 shows such a matrix that represents the four patterns $f(a, a, a, \_, \_, \_, a), f(b, \_, \_, a, a, \_, \_), f(\_, b, \_, b, \_, a, \_), f(\_, \_, b, \_, b, b, \_)$. Denote by $S(n)$ the size of the smallest automaton with n such patterns. Now pick any position (i.e., column) to discriminate. If this column contains only one nonvariable then on a positive transition (i.e., transition on a or b) we will once again be left with the same $\mathcal{L}_u$ without any reduction in the problem size. For instance, in fig. 4 if we choose column 7 for inspection then we are left with the problem of building an automaton to match on the basis of the first six positions of each pattern. In other words, we are left with a matrix obtained from that of fig 4 by deleting the last column and hence the problem is still an instance of $S(4)$. On the other hand if the column contains two nonvariables then based on the symbol seen in the column we can now partition the n patterns into two sets each consisting of $n-1$ patterns. These sets must be further discriminated and hence we have two instances of $S(n-1)$, e.g., in the above example, inspecting position 2 results in the pattern sets $\{1,2,4\}$ and $\{2,3,4\}$ as shown in fig. 5. Hence:

$$S(n) = 2 * S(n-1)$$

whose solution is $O(2^n)$

3.2 Unambiguous Prioritized Patterns

To derive lower bound on the size of the automaton for unambiguous prioritized set of patterns, consider the following set of flat patterns with textual order priority:

$$f(c^m(a), x_2, x_3, ..., x_n), f(x_1, c^m(a), x_3, ..., x_n), ..., f(x_1, ..., x_{n-1}, c^m(a))$$

[2]The size of a pattern is the number of nonvariable symbols in it.

$$\begin{bmatrix} a & a & a & \_ & \_ & \_ & a \\ b & \_ & \_ & a & a & \_ & \_ \\ \_ & b & \_ & b & \_ & a & \_ \\ \_ & \_ & b & \_ & b & b & \_ \end{bmatrix}$$

Figure 4: Example Matrix for $n = 4$

$$\begin{bmatrix} a & a & \_ & \_ & \_ & a \\ b & \_ & b & \_ & a & \_ \\ \_ & b & \_ & b & b & \_ \end{bmatrix} \quad \begin{bmatrix} \_ & \_ & a & a & \_ & \_ \\ b & \_ & b & \_ & a & \_ \\ \_ & b & \_ & b & b & \_ \end{bmatrix}$$

Figure 5: Matrices representing $\mathcal{L}_u$ for states reached by transitions on a and b.

Using the set of patterns constructed as above, we obtain a lower bound on space of $\Omega(a^n)$ as follows. Denote by $S(n)$ the size of the automaton for patterns of the above form. It can be shown that the smallest size automaton is obtained by first inspecting all the c's and then the a in the first column, then those in the second column and so on. (This because it can be shown that each position examined in automaton thus obtained is an index. Then it follows by lemma 4 (see section 4.3) that this automaton is no larger than any other automaton.) Now consider the first m states of the automaton. Each state has a $\neq$ branch that leads to a subautomaton for matching $n-1$ patterns. Therefore:

$$S(n) \geq mS(n-1)$$

whose solution is m^n. Noting that m is the average size of patterns, we have the lower bound a^n.

3.3 Ambiguous Patterns

Consider again the set of patterns used in section 3.2. Now assume that there is no priority among these patterns. Observe that the automata A for this set of patterns (which must report all matches since there is no priority relationship among the patterns) can be used as an automaton for the prioritized system considered in the previous section. To do this, on reaching a final state v, we simply announce a match for the pattern with the highest priority among those in *matchset*$[v]$. Hence this automaton cannot be any smaller than that for patterns with priority. Thus we arrive at a lower bound of $\Omega(a^n)$ for ambiguous unprioritized patterns. The same lower bound holds even when priorities are allowed since the automata for unprioritized patterns can be used for matching prioritized patterns in the manner mentioned above.

3.4 Matching Time

One possible measure of matching time is the average root-to-leaf path length of the automaton. This measure can exhibit the following anomaly. Suppose an automaton A has the property that for each path in automaton A', the corresponding path(s) in A is shorter or has the same length. Then clearly A has better matching time but this may not be reflected in the average root-to-leaf path lengths. The reason for this is that there may be several paths in A' that correspond to a single path in A. Suppose that there is one path P of length 2 in A and another path Q of length 7. Also assume that there are two paths in A' corresponding to P with length 3 and one corresponding to Q with length 7. Note that although each path in A is shorter than (or equal in length to) the corresponding path in A', the average path length of A is 4.5 whereas the average path length of A' is only 4.3!

A good measure that does not suffer from such an anomaly is average over the mean matching times for all patterns, where the mean matching time for a pattern l is the average path length of those paths that lead to a matching state for l. This quantity is 4.5 for A whereas it is 5 for A'. Based on this measure, upper and lower bounds on matching time are given in fig. 3. These bounds are established with the same construction used to obtain the space bounds.

4 Synthesizing Traversal Orders

Observe that *Build* does not specify the selection function. We now present several techniques that can be used in any combination to derive selection functions for improving space and matching time.

4.1 Representative Sets

Suppose v is the state reached on seeing prefix u. Observe that there may be patterns in $\mathcal{L}_u$ for which no match is announced at any descendant of v. For instance, consider the patterns in fig. 2 and the prefix $u = f(\_, a(b, \_, b), \_)$. Although $\mathcal{L}_u = \{l_1, l_4\}$ observe that a match for l_4 can be declared only if the 3rd argument of f is b. In such a case we declare a match for the higher priority pattern l_1. Inspecting any position only on behalf of a pattern such as l_4 is wasteful, e.g., inspection of position 1 for u is useless since it is irrelevant for declaring a match for l_1. To avoid inspecting such positions, we identify a *representative set* $\overline{\mathcal{L}_u}$ that is a minimal set consisting only of those patterns for which a match is announced at a descendant of v. In the above example $\overline{\mathcal{L}_u} = \{l_1\}$. Representative sets form the core of several optimizations discussed later. Laville's notion of accessible patterns [9] is the same as our representative set. Our contribution here is that our definition yields a simple algorithm for computing this set. The definition of accessible patterns does not yield such an algorithm and so [9] uses the notion of compatible patterns (which corresponds to our match set $\mathcal{L}_u$) in place of accessible patterns. Observe that we can also use $\mathcal{L}_u$ in place of $\overline{\mathcal{L}_u}$ in all the following optimizations, but doing so may make the optimizations less effective. For instance, our algorithm for computing indices in untyped systems (section 6.1) will not be complete if $\mathcal{L}_u$ is used in place of $\overline{\mathcal{L}_u}$.

To compute $\overline{\mathcal{L}_u}$ note that any pattern l that has the following property need not be present in $\overline{\mathcal{L}_u}$: whenever a match for l is identified for a term t at a descendant of v we can also identify a match for another pattern (of equal or higher priority) for the same term. A precise definition of the property is

$$\forall t \geq u \;\; [t \geq l \Rightarrow \exists l' \in \mathcal{L}_u \; [(t \geq l') \wedge (priority(l') \geq priority(l))]]$$

We can arrive at $\overline{\mathcal{L}_u}$ by repeatedly deleting patterns such as l. However, the above definition does not immediately lend itself to a decision procedure since it refers to an infinite number of instances of u. Although the set of terms t to be considered can be restricted to a finite set, it still does not lead to an efficient procedure since:

Theorem 1 *Computing $\overline{\mathcal{L}_u}$ is NP-complete for typed systems.*

The proof of this therem uses a reduction similar to that used in the proof of theorem 9 (see section 6.2). We omit the proof details here. For untyped systems a more efficient method can be developed by considering only $l \sqcup u$ and not its instances. Specifically,

Theorem 2 *$\overline{\mathcal{L}_u}$ can be computed in $O(nS)$ time in untyped systems by deleting l such that*

$$\exists l' \;\; [(l \sqcup u) \geq l'] \wedge [priority(l') \geq priority(l)]$$

Proof: First we need to show that the two characterizations of the l to be deleted are equivalent. It is easy to see that any l satisfying the second characterization will also satisfy the first one. To show that any l that satisfies the first characterization also satisfies the second, let $l_1, ..., l_k$ be all those patterns such that any instance v of l and u (and hence of $l \sqcup u$) is also an instance of one of $l_1, ..., l_k$. Now consider the term t obtained from $l \sqcup u$ by instantiating each variable by $\neq$. This term cannot be an instance of some l_i unless $l_i \leq (l \sqcup u)$. Observe that this l_i can serve the role of l' in the second characterization. Thus the two characterizations are equivalent. It can be easily seen that computing $\overline{\mathcal{L}_u}$ using the simpler characterization can be done in $O(nS)$ time. ∎

Note that in the above definition, if l and l' have equal priority and $l \sqcup u = l' \sqcup u$, then either one can be retained in $\overline{\mathcal{L}_u}$ and the other discarded. (This choice will not affect the structure of the automaton below the current state, but will only determine whether a match is announced for l or l' in a final state below.) Except for the choice among such patterns, the second definition specifies $\overline{\mathcal{L}_u}$ uniquely. Finally, note that in unambiguous systems $\mathcal{L}_u = \overline{\mathcal{L}_u}$.

4.2 Greedy Strategies

We resort to greedy strategies for implementing function *select* since dynamic programming approach will take exponential time as the subautomata themselves can have exponential size. We begin by listing several strategies to choose a p such that:

1. the number of distinct nonvariables at p in any pattern in $\overline{\mathcal{L}_u}$ is *minimized.*
2. the number of distinct nonvariables at p in any pattern in $\overline{\mathcal{L}_u}$ is *maximized.*
3. the number of patterns having nonvariables at p is maximized.
4. Let $\overline{\mathcal{L}_1}, ..., \overline{\mathcal{L}_r}$ be the match sets of the children of v. Then $max(|\overline{\mathcal{L}_1}|, ..., |\overline{\mathcal{L}_r}|)$ is minimized.
5. $\Sigma_{i=1}^{r} \overline{\mathcal{L}_i}$ is minimized.

For improving space, the rationale for these strategies is as follows. Strategy 1 locally minimizes the breadth. Strategy 2 attempts to distinguish quickly among patterns so as to contain the (potentially exponential) blow-up. Since only patterns with a variable at p are duplicated among the children of v strategy 3 minimizes their number. Strategy 5 carries this one step further by minimizing the number of such duplications. Strategy 4 attempts to minimize the size of the largest subautomaton rooted at any child of v since it determines the size of the automaton in case of blow-up's. For improving time strategy 3 (and also 5) minimizes the number of patterns for which inspection of the symbol at p is unnecessary. For strategies 2 and 4 observe that in the worst case, time spent in matching below v is proportional to sum of sizes of patterns in $\mathcal{L}_u$. By quickly discriminating among patterns, this quantity can be rapidly reduced.

All of the above greedy strategies suffer from the drawback that:

Theorem 3 *For each of the above greedy strategies there exist pattern sets for which automata of smaller size and matching time can be obtained by making a choice different from that given by the greedy strategy.*

Proof: For strategies 2,3,4 and 5 consider the following set of patterns with equal priorities:

$$f(a,a,\_,\_) \quad f(b,\_,a,\_) \quad f(c,\_,\_,a)$$
$$f(\_,b,b,b) \quad f(\_,c,c,c) \quad f(\_,d,d,d)$$

After inspecting the root, all these strategies will choose one of positions 2,3 or 4. It can be shown (by enumerating all possible matching automata for these patterns) that the smallest breadth, number of states and matching time obtainable by this choice are 20, 47 and 4.25 respectively. These figures can be reduced to 15, 45 and 4 respectively by choosing position 1. ∎

The construction of this example is quite intricate. The key idea is to make each of the pattern sets $\{1,4,5,6\}$, $\{2,4,5,6\}$ and $\{3,4,5,6\}$ strongly-sequential[3] whereas any set containing two of the first three patterns and one of the last three is not.

4.3 Selecting Indices

We now propose an important strategy that does not suffer the drawbacks of the above strategies. The key idea is to select a p that is an index for u. We show that this strategy yields automaton of smaller (or same) size and smaller (or same) matching time than obtainable by any other choice. The importance of index was known only in the context of strongly-sequential systems. Our result demonstrates its applicability to patterns that are not strongly-sequential. Specifically, given a subautomata A rooted at v that selects some position p, we construct another subautomata A' that selects an index q and show that size and matching time of A' are no greater than that of A. This construction proceeds through a series of steps each of which interchange the order of positions q and another position seen immediately preceding it. The following theorem (stated without proof) makes the above construction precise.

Theorem 4 *With p, q and v as above suppose that $pos(w) = q$ for every child w of v. Then the order of inspection of p and q can be interchanged without increasing the size or matching time of the automaton. By repeating this interchange we can obtain A' from A.*

It is quite interesting to note that repetition of interchange steps which is permitted only under restrictive conditions, is sufficient to globally rearrange the traversal order in A to get A'. Although the theorem only asserts that size and matching time of A' is no larger than that of A, fig. 6 shows that they can be strictly smaller for A'. We remark that by doing interchange steps as above, size can sometimes be reduced byas much as an exponential factor and time by $O(n)$.

Using arbitrary traversal orders (determined at compile time) is not appropriate in functional programming since termination properties of the program depend on the traversal order. Therefore the programmer must be made aware of the traversal order apriori. Given this constraint, we now show that it is possible to internally change the traversal order in such a way that it does not affect the termination properties and at the same time realize the advantages of adaptive traversal[4].

[3] In strongly-sequential systems, any prefix u with $|\mathcal{L}_u| \geq 1$ must have an index.

[4] We remark that this transformation is applicable only to functional languages with no side-effects.

Figure 6: Example to illustrate size and matching time reduction due to transformation. Figure shows automata for $f(\_, b), f(a, a)$ and $f(\_, a)$ with textual order priority.

Definition 5 (Monotonic Traversals) *Given a prefix u, suppose a traversal order T selects p as the next position to be visited. T is said to be monotonic iff for any prefix $u' \geq u$ (where u'/p a variable), the traversal once again selects p unless u'/p corresponds to a variable position $\forall l \in \overline{\mathcal{L}_{u'}}$.*

Traversals that possess monotonicity property include most known traversal orders such as depth-first and breadth-first (as well as variations of these without left-to-right bias). In particular it includes all fixed-traversal orders mentioned earlier such as left-to-right and right-to-left. It also includes traversals used in implementations of strongly-sequential systems.

Given a monotonic traversal T, we can obtain another traversal $S(T)$ from it by inspecting indices whenever possible without affecting termination properties. Specifically, let $S(T)$ denote the traversal that uses the following strategy to pick the position to inspect for a prefix u.

- If u has indices then arbitrarily select one of them.
- Otherwise, select the position in the fringe of u that will be the first to be visited by T.

Theorem 6 *Size and matching time can never become worse if $S(T)$ is used in place of T. Each path in the automaton using $S(T)$ examines a subset of the positions examined on the corresponding path in T.*

5 Minimizing Space using DAGs

The previous section discussed synthesizing traversal orders in order to improve space and time. We now discuss an orthogonal approach to minimize space based on sharing equivalent states. An obvious way to achieve sharing is to use standard FSA minimization techniques for optimizing the tree automaton constructed by *Build*. A more efficient method is to identify equivalence of two states *without even generating* the subautomata rooted at these states. Observe that the tree automaton may be exponentially larger than the dag automaton, in which case the naive approach uses exponential time and space whereas direct construction can potentially use only polynomial space and polynomial time. This important problem of directly building an optimal automata even when restricted to left-to-right traversals has remained open [4]. We now propose a solution to this problem.

To identify equivalent states, suppose that two prefixes u_1 and u_2 have the same representative set $\overline{\mathcal{L}_u}$ and differ in positions p such that every pattern in $\overline{\mathcal{L}_u}$ has a variable at or above p. Since such positions are irrelevant for determining a match, these two prefixes are equivalent. On the other hand, it can also be shown that if they have different representative sets or differ in any other position then they are not equivalent. Based on this observation, we define the relevant prefix of u as the term obtained by replacing subterms at positions such as p above by the symbol $\neq$. For instance, the prefixes corresponding to different states marked 's' in fig. 2 are different, but they all have the same relevant prefix $f(x, \neq, b)$. By showing that two states are equivalent iff the corresponding relevant prefixes are identical we establish:

Theorem 7 *The automaton obtained by merging states with identical relevant prefixes is optimal.*

Merging equivalent states as described above can substantially reduce the space required by the automata, e.g., the tree automaton in fig. 2 has 25 states which can now be reduced to 16 by sharing. Also recall that for the patterns in fig. 4, parts of the automaton reached by positive transitions alone is exponential. We can show that by sharing states this part of the automaton will become polynomial!

5.1 Impact of DAGs on Space and Matching Time Complexity

Since sharing affects space requirements alone, all the results established so far not relating to spac (such as theorem 6 for time) continue to hold for dags as well. In what follows we discuss the impact dags on some of the results established earlier regarding space. We can show that the upper bound o size of dag automata is $O(2^n S)$ which is much smaller than the corresponding bound $O(\prod_{i=1}^{n} |l_i|)$ fc tree automata. We can also establish a lower bound of $O(2^n)$ for ambiguous patterns. For unambiguou patterns, it is not clear whether the lower bound on size is exponential. For instance, it appears tha the patterns used in the lower bound proof on size of tree automata for unambiguous patterns, posses a polynomial-size dag automaton. Reasoning about lower bounds becomes extremely complicated fo dags since it is difficult to capture behavior of sharing formally. Finally, all the greedy strategies as we as the strategy of selecting indices can, in some cases, increase the space of dag automata. The failur of the index selection strategy further demonstrates the complexity of sharing.

6 Index Computation

Recall that an index for a prefix u is a position on its fringe that must be inspected to announce a matc for any pattern in $\mathcal{L}_u$. In the absence of priorities, the positions that must be inspected to announc a match for a pattern l are exactly those fringe positions wherein l has a nonvariable. With priorities however, we may have to inspect positions wherein l has a variable in order to rule out a match fo higher priority patterns. But it is not obvious which variable position of l must be inspected and so i is not clear how to compute indices in prioritized systems. Therefore Laville [8] proposed an indirec method for index computation that first transforms the prioritized patterns into an equivalent set o unprioritized patterns, which is then used for index computation. For each pattern l, the transformation generates a set $\mathcal{M}_l$ of its instances that are not instances of any higher priority pattern. (For typed systems, only those instances that observe the type discipline are generated.) The transformed system is $\bigcup_{l\in\mathcal{L}} \mathcal{M}_l$.

Puel and Suarez [11] developed a compact representation for the sets $\mathcal{M}_l$ based on the notion o *constrained terms*. These are terms with constraints placed on the substitutions taken by the variables in them. For instance, the constrained term $\{eq(x,y)|(x \neq a) \vee (y \neq b)\}$ denotes the set of terms $eq(t_1, t_2)$ with either $t_1 \neq a$ or $t_2 \neq b$.

A precise semantics of constrained terms is given by first regarding a term t with variables as denoting the set $\mathcal{I}(t)$ of its instances. Now the constrained term $\{t|\varphi\}$ denotes the set of instances of t that also satisfy the constraint formula φ. To define a constraint formula, first define an atomic constraint to be of the form $t \neq s$ or of the form $x \neq s$, where x is a variable in t and all the variables in s are unnamed[5]. In the former case the terms that satisfy the constraint must belong to the set $\mathcal{I}(t) - \mathcal{I}(s)$. In the latter case, the substitution taken by x must not be an instance of s. An arbitrary constraint is obtained by combining atomic constraints using $\vee$ and $\wedge$. A constraint $\varphi \vee \psi$ is satisfied by all (and only) terms that satisfy either φ or ψ. Similarly, the constraint $\varphi \wedge \psi$ is satisfied by all (and only) terms that satisfy φ as well as ψ. Finally, note that in typed systems, there is always an implicit constraint imposed by the type discipline that requires a variable to be substituted only by a term that is of the same type as the variable. Henceforth we may use several methods to simplify constrained terms. These methods are quite intuitive and their correctness readily follows from the above semantics.

Using constrained terms the set $\mathcal{M}_l$ can be represented compactly as

$$\{l|(l \neq l_1) \wedge \cdots \wedge (l \neq l_k)\}$$

where $l_1, \ldots, l_k$ are all the patterns with priority greater than l. The constraints $l \neq l_i$ can be simplified to constraints on the variables in l. To simplify a constraint $l \neq l'$, we first check if the two patterns unify. If not, the constraint simplifies to *true*. Otherwise, let $x_1, ..., x_n$ be all the variables in l that are substituted by nonvariable terms (say $t_1, ..., t_n$) when unifying l and l'. Then the simplified constraint is

$$(x_1 \neq t_1) \vee \cdots \vee (x_n \neq t_n)$$

[5] This is to prevent having constraints such as $x \neq a(y) \wedge y \neq b$ that will complicate the development of the materials in the rest of the section.

Using this procedure, simplification of all the constraints on a pattern l can be done in $O(S)$ time. For illustration, consider the following example that defines equality on an enumerated data type $t = a_1|a_2|\dots|a_n$.

$$\begin{aligned} eq(a_1,a_1) &= true \\ \vdots \quad & \vdots \quad \vdots \\ eq(a_n,a_n) &= true \\ eq(x,y) &= false \end{aligned}$$

Note that no two among the first n patterns unify and so the condition $eq(a_i,a_i) \neq eq(a_j,a_j)$ holds vacuously and hence the first n patterns appear unchanged in the transformed system. The last rule, however, unifies with every other rule and gets translated into

$$eq(x,y)|(x \neq a_1 \vee y \neq a_1) \wedge (x \neq a_2 \vee y \neq a_2) \wedge \dots \wedge (x \neq a_n \vee y \neq a_n) \tag{1}$$

where the condition in the ith disjunct prevents instances of the ith pattern from being recognized as instances of the last rule.

As seen above, the constraints are in CNF and it is not clear how indices can be obtained from them. So Puel and Suarez simplify it into DNF which makes the structure of the terms in the set denoted by a constrained term apparent. For instance, when $n = 2$ the above constrained term simplifies to:

$$\{eq(x,y)|(x \neq a_1 \wedge x \neq a_2) \vee (x \neq a_1 \wedge y \neq a_2) \vee (x \neq a_2 \wedge y \neq a_1) \vee (y \neq a_1 \wedge y \neq a_2)\}$$

Note that in typed systems, constraints such as $x \neq a_1 \wedge x \neq a_2$ will be simplified to $false$ and so the above constraint simplifies to $[x \neq a_1 \wedge y \neq a_2] \vee [x \neq a_2 \wedge y \neq a_1]$.

Once a constrained term $t_c = \{t|\varphi_1 \vee \dots \vee \varphi_m\}$ is in DNF, indices of a prefix u w.r.t. t_c can be easily picked as follows. Firstly, any variable position of u wherein t has a nonvariable symbol is an index for u. Secondly, consider any variable position p wherein t has a variable x, but a constraint on the substitution for x appears in every conjunct φ_i. Note that for any term to be an instance of this constrained term, it must satisfy one of the constraints φ_1 through φ_m. Since x appears in every one of these conjuncts, its substitution must be examined in order to determine whether a term satisfies the constraint. Therefore all such positions are also indices of u.

The above conversion of the constraint on a pattern l from CNF to DNF is very expensive and can take $O(|l|^n)$ time. Therefore Puel and Suarez's algorithm, which is based on such a conversion, has exponential time complexity for *both typed and untyped systems.* Laville's algorithm is also exponential for typed and untyped systems. In contrast, we now present the first polynomial-time algorithm for untyped systems that operates directly on the original patterns.

6.1 Algorithm for Index Computation in Untyped Systems

The index for a prefix is computed in two steps. First we compute the set of indices of the prefix w.r.t. each of the constrained patterns in $\overline{\mathcal{L}_u}$. The intersection of the sets thus computed yields the indices of the prefix w.r.t. $\overline{\mathcal{L}_u}$. We compute the indices of a prefix u w.r.t. to a single constrained pattern l as follows:

> Let $l_1, l_2, \cdots l_k$ be the patterns in $\overline{\mathcal{L}_u}$ that have priority over l and also unify with l. Following two steps specify the indices of u w.r.t. l.
> 1. Each variable x in u such that l has a nonvariable at the position corresponding to x.
> 2. Each variable x that must be the only position to be instantiated in $l \sqcup u$ to determine a match for some higher priority pattern l_j.

We remark that a similar algorithm was suggested by Laville as a heuristic for fast index computation. However his algorithm is not complete for untyped systems. We now illustrate the algorithm on $l_1 = f(a,b,c)$, $l_2 = f(a,\_,\_)$ and $l_3 = f(\_,\_,c)$ (with textual order priority) and the prefix $u = f(x,y,z)$. Observe that x, y and z are all indices of l_1 by step 1. The only index for l_2 is x by step 1. Observe that step 2 does not yield any additional indices for l_2. For l_3, z is an index by step 1 and x is an index by step 2 (with $l_j = l_2$). The intersection of all these positions is x which is therefore an index for u. Index computation takes $O(nS)$ time using this algorithm. We show:

Theorem 8 *The algorithm for computing indices in untyped systems is sound and complete.*

Proof: We need only establish that the above algorithm computes all and only the indices of u w.r.t. the constrained pattern

$$l_c = \{l \mid \underbrace{(l \neq l_1) \wedge \cdots (l \neq l_k)}_{\gamma}\}$$

Clearly, any position computed using step 1 is an index w.r.t. l_c. For any position selected using step 2 observe that the constraint $l \neq l_j$ will simplify to $x \neq t$ for some term t; so the constraint $x \neq t$ will appear in every conjunct in the DNF of γ. All the positions selected in step 2 are thus indices of l_c.

For completeness, we need to show that if a position p in u is not selected by steps 1 or 2 then it is not an index of l_c. This is accomplished by giving an instance of l_c that has a variable at p. Suppose that p is not selected by step 1 or step 2. Then for each l_i, it must be the case that either l_i/p is variable or that there is more than one position in the fringe of $l \sqcup u$ that need to be instantiated for determining a match for l_i. In either case, the disjunct obtained by simplifying $l \neq l_i$ contains at least one literal $(x_i \neq t)$ for some variable x_i different from l/p. Consider the term s obtained from $l \sqcup u$ by substituting $\neq$ for each of the x_i's mentioned above. Note that the substitutions for the variables in $l \sqcup u$ satisfies the constraints in l_c. Hence s is an instance of l and u but still has a variable at p. This proves that p is not an index of u.

6.2 Index Computation in Typed Systems

For typed systems it is very unlikely we can escape exponential complexity since we show:

Theorem 9 *Index computation problem for typed systems is co-NP-complete.*

For intuition into the reason for the complexity gap between typed and untyped systems, we draw an analogy between index computation with constrained terms and the satisfiability problem. Note that the literals in constraints generated by Puel and Suarez's algorithm are all of the form $x \neq t$ where x is a variable and t an arbitrary term. Such constrained terms are analogous to boolean formulas with only negative literals. Such boolean formulas are trivially satisfiable (by a truth assignment that assigns *false* to every literal). Similarly index computation is simple in untyped systems. However, in typed systems, there are implicit positive constraints introduced by the type discipline. For instance, there is an implicit constraint $(x = a_1) \vee \cdots \vee (x = a_n)$ on the constrained term (1). Thus we have a constraint that contains both positive and negative literals. Such a constrained term is analogous to a boolean formula with both positive and negative literals and hence the complexity gap.

Proof: The index selection problem, when posed as a decision problem, takes the form "Does u possess an index w.r.t. pattern set $\mathcal{L}$?" To show that this problem is in *co-NP*, we need to show that the problem of deciding whether p is not an index is in *NP*. To do this, let $p_1, ..., p_r$ be the set of fringe nodes of u. We first guess r instances $t_1, ..., t_r$ of u such that t_i/p_i is a variable for $1 \leq i \leq r$. Then we verify that each t_i is an instance of (at least) one of the patterns in $\mathcal{L}$ and if so we declare that u has no index. All this can clearly be accomplished in polynomial time and hence the problem of determining whether u does not possess an index is in *NP* and so the index selection problem is in *co-NP*.

To show that the problem is *co-NP*-complete, we reduce the complement of satisfiability to this problem. Let $\varphi_1 \wedge \cdots \wedge \varphi_n$ be an instance of satisfiability problem where φ_i is a disjunct of literals of the form x or $\neg x$, $x \in \{x_1, \ldots, x_m\}$. We transform this into an index computation problem in the following system consisting of $n+1$ patterns, with textual order priority. The roots of these patterns is the symbol f (which once again stands for the common prefix shared by all the patterns) with $m+1$ arguments. The last m arguments of f are of a type that consists of nonvariables a and b. The $(n+1)$th pattern is of the form $f(x_0, x_1, ..., x_m)$. To specify the first n patterns, let $t_1, ..., t_n$ be terms that do not unify with each other. Also assume that there is a term t of the same type as $t_1, ..., t_n$ which does not unify with any of these terms. Now we specify the ith pattern (for $1 \leq i \leq n$) as $f(t_i, s_1, .., s_m)$, where s_j is a or b depending upon whether x_j or $\neg x_j$ occurs in φ_i. If neither occur in φ_i then s_j is a variable. Observe that the size of this pattern set is polynomial in the size of $\varphi_1, ..., \varphi_n$. With this construction, we will now show that determining whether $f(x_0, ..., x_m)$ possesses an index is equivalent to determining whether $\varphi_1 \wedge \cdots \wedge \varphi_n$ is *not* satisfiable.

First we transform the above pattern set into a set of constrained patterns. Following the transformation the $(n+1)$th pattern becomes

$$\{f(x_0,...,x_m)|(x_0 \neq t_1 \vee \varphi_1) \wedge \cdots \wedge (x_0 \neq t_n \vee \varphi_n)\}$$

Here we have slightly abused the notation in replacing $x_j \neq a$ by x_j and $x_j \neq b$ (i.e. $x_j \neq a$ by type discipline) by $\neg x_j$. By using heuristic 1, this constraint can be simplified to $(x_0 \neq t_1) \vee (\varphi_1 \wedge \cdots \wedge \varphi_n)$ which shows that $f(x_0,...,x_m)$ has an index (namely x_0) iff the input to the SAT problem is *not* satisfiable. ∎

6.3 Heuristics for Fast Index Computation in Typed System

Suppose the constrained term is of the form

$$\{l|\underbrace{[(x \neq t_1 \vee \varphi_1) \wedge \cdots \wedge (x \neq t_n \vee \varphi_n)]}_{\gamma} \wedge \underbrace{\psi_1 \wedge \ldots \wedge \psi_k}_{\delta}\}$$

where $\psi_1, \psi_2, \ldots \psi_k$ and $\varphi_1, \varphi_2, \ldots \varphi_n$ are disjuncts that do not contain occurrences of x. Suppose that there exists a term t (valid for type of x) such that t does not equal any of $t_1, t_2, \ldots, t_n$. Then:

Rule 1: The indices obtainable from the above constraint are exactly those obtainable from the constrained term

$$\{l|[(x \neq t_1) \vee (\varphi_1 \wedge \varphi_2 \wedge \cdots \varphi_n)] \wedge \delta\}$$

If $t_1, t_2, \ldots, t_n$ are the only terms valid for type of x then:

Rule 2: If each φ_i consists only of a single variable y, then y is an index.

The power of rule 1 is evident from the fact that in order to apply it we need only identify some variable x such that the constraints on x do not simplify to false[6]. Once such an x is identified, we can avoid expanding γ into 2^n terms and instead concern ourselves with just two of the terms in the expansion, namely, $(x \neq t_1)$ and $\gamma' = \varphi_1 \wedge \cdots \wedge \varphi_n$.

To illustrate the power of rule 2 consider the constrained term in (1). Note that $x \neq a_1 \wedge x \neq a_2 \wedge \cdots \wedge x \neq a_n$ simplifies to *false* and the φ_i's (which are of the form $y \neq a_i$) also contain only one variable y and hence y is an index. Interchanging the role of x and y we also note that x is an index. Note that computing these indices has taken time linear in sum of sizes of rules. In contrast, Puel and Suarez's algorithm spends exponential time reducing the constraint to DNF.

Proof of Correctness of Rule 1: Expanding γ and taking the conjunction with δ we get

$$[(x \neq t_1 \wedge \cdots \wedge x \neq t_n) \wedge \delta] \vee [\underbrace{(\varphi_1 \wedge \cdots \wedge \varphi_n)}_{\gamma'} \wedge \delta] \vee [\epsilon \wedge \delta]$$

where ϵ is a disjunction of terms containing at least one constraint on x and some φ's. The first term in this expansion constrains only x. Therefore no variable other than x that appears in any φ_i can be an index unless it appears in every conjunct in DNF of δ. Furthermore, since x apears in all the terms in the above expansion except γ', it is clear that x will be an index iff γ' reduces to false. This is exactly captured by considering only $[(x \neq t_1) \wedge \delta] \vee [\gamma' \wedge \delta]$. ∎

Proof of Rule 2: Once again we expand γ as above and eliminate the term $(x \neq t_1 \wedge \cdots \wedge x \neq t_n)$ which reduces to *false*. Now observe that every other conjunct in the expansion contains at least one φ_i. If each φ_i is of the form $y \neq t_i$ then each conjunct in DNF will constrain y and so y is an index.

7 Concluding Remarks

In this paper we studied pattern matching with adaptive automata. We presented lower and upper bounds on the worst-case size of the automata for several classes of patterns. We then discussed how to improve space and matching time by synthesizing traversal orders. We showed that a good traversal order selects indices whenever possible and uses one of the greedy strategies (especially strategy 4 or 5) otherwise. Although the greedy strategies may sometimes fail, it appears from the complexity of

[6]By saying "φ reduces to false" we mean that φ is not satisfiable, i.e., it is a contradiction.

the counter examples that such failures may be rare. For functional programming, we synthesized traversal $S(T)$ from a monotonic traversal T. Since using $S(T)$ does not affect termination propertie the programmer can assume T whereas an implementation can benefit from significant improvemen in space and matching time.

We also discussed an orthogonal approach to space minimization by sharing equivalent states. Reca that even the index selection strategy may fail to improve space of dag automata. This occurs becaus index selection may adversely affect the way in which descendants of a state can be shared. Since it difficult to predict sharing among descendant states, the possibility of improving space without usin indices does not appear to be practical. So the best approach is to use all the strategies in section and use sharing as an additional source of space optimization over tree automata.

Our work clearly brings forth the impact of typing in prioritized pattern matching. We have show that several important problems in the context of pattern matching are unlikely to have polynomial-tim algorithms for typed systems whereas we have given polynomial-time algorithms for them in untype systems. This raises the question whether it is worthwhile to consider typing for pattern matching. It i not clear how often typing information can be used to find an index (or to determine that a pattern doe not belong to $\overline{\mathcal{L}_u}$) which cannot be found otherwise. On the other hand there is a significant penalty i terms of computational effort for both these problems if we use typing information.

References

[1] L. Augustsson, Compiling Pattern Matching, *FPCA '85, LNCS 201.*

[2] D. Comer and R. Sethi, Complexity of Trie Index Construction, *FOCS '76.*

[3] J. Christian, Fast Knuth-Bendix Completion : Summary, *RTA '89, LNCS 355.*

[4] A. Graf, Left-to-Right Tree Pattern Matching, *RTA '91, LNCS 488.*

[5] R.Harper, R.Milner and M.Tofte, The Definition of Standard ML, *Report ECS-LFCS-88-62, LFCS, University of Edinburgh, 1988.*

[6] P. Hudak et al, Report on Haskell, *Draft Proposed standard circulated by IFIP WG2.8., 1988.*

[7] G. Huet and J.J. Levy, Computations in Nonambiguous Linear Term Rewriting Systems, *Tech. Rep No. 359, INRIA, Le Chesney, France, 1979.*

[8] A. Laville, Lazy Pattern Matching in the ML Language, *FST&TCS '87.*

[9] A. Laville, Implementation of Lazy Pattern Matching Algorithms, *ESOP '88, LNCS 300.*

[10] K. Owen, S. Pawagi, C. Ramakrishnan, I.V. Ramakrishnan, R.C. Sekar, Fast Parallel Implementation of Functional Languages - The EQUALS Experience, *to appear in JLFP '92.*

[11] L. Puel and A. Suarez, Compiling Pattern Matching by Term Decomposition, *LFP '90.*

[12] Ph. Schnoebelen, Refined Compilation of Pattern Matching for Functional Languages, *Science of Computer Programming, 11, pp. 133–159, 1988.*

[13] R.C. Sekar, R. Ramesh and I.V. Ramakrishnan, Adaptive Pattern Matching, *TR 14/91, SUNY, Stony Brook.*

[14] P. Wadler, Efficient Compilation of Pattern Matching, in S.L. Peyton-Jones, Ed, The Implementation of Functional Programming Languages, *Prentice Hall International, 1987.*

andomized interpolation and approximation of sparse polynomials

(Preliminary version)

Yishay Mansour*

Abstract

We present a randomized algorithm that interpolates a sparse polynomial in polynomial time in the bit complexity model. The algorithm can be also applied to approximate polynomials that can be approximated by sparse polynomials (the approximation is in the L^2 norm).

Introduction

consider the problem of interpolating a sparse polynomial with integer coefficients. The interpola- model is the following. The algorithm can query the polynomial P on any input x, and receives value $P(x)$ with some finite accuracy. The accuracy parameter determines the number of signifi- bits that are given from the value of $P(x)$, which enables to truncate $P(x)$ after that number of ificant bits. The running time of the algorithm is the number of bit operations performed.

The parameters of the sparse polynomial are: n – the number of variables, t – an upper bound on number of non-zero coefficients, d – an upper bound on the degree of a single variable and L – an er bound on the absolute value of the largest coefficient. A t-sparse polynomial is a polynomial at most t non-zero coefficients. The number of bits required to express a t-sparse polynomial is $(\log L + n \log d)) = poly(n, t, \log d, \log L)$[1].

Our main result is a randomized algorithm that interpolates a t sparse polynomial in time $poly(n, t,$ $d, \log L)$. In the analysis of the algorithm we count the number of operations performed, but since accuracy that our algorithm requires is very small (logarithmic in the above parameters), the ning time remains $poly(n, t, \log d, \log L)$ in the bit complexity model.

An interesting application of our algorithm is finding an *approximation* by a sparse polynomial. approximation is in the L_2 norm, i.e. the expected error squared is small. The main result is that polynomial P can be approximated by a t-sparse polynomial Q, such that the power of $P - Q$ is nded by ϵ, then our algorithm can find a t-sparse polynomial Q' such that the power of $P - Q'$ ounded by $O(\epsilon)$. (The *power* of a polynomial is the sum of the squares of its coefficients.) Our rithm achieves this by finding only the "large" coefficients of P, rather than finding all of the fficients of P.

IBM - T. J. Watson Research Center, P. O. Box 704, Yorktown Heights NY 10598. Part of the research done while author was at Aiken Computation Laboratory, Harvard University, Cambridge Ma

We denote by $poly(a)$ a function which is bounded by a polynomial in a, i.e. $a^{O(1)}$

Interpolation of a univariate polynomial can be computed in time $O(d \log d)$, using the Fast Fo Transform (FFT). The FFT does not use the information about the sparsity of the polynomial. t fore, its running time is polynomial in d, rather than t and $\log d$. Since the FFT algorithm doe consider the sparsity of the polynomial, for multivariate polynomials where the number of pos coefficients is exponential, its running time is exponential.

Previous works on sparse multi-variate polynomial interpolation were mainly in the algebraic m where each operation requires one unit of time. When considering those works in the bit compl model one has to take into account the number of bits which were given to the polynomial, and the size of the result of the polynomial and the required accuracy on the result. Zippel [Zip79] g probabilistic algorithm that runs in time $poly(n, t, d)$. His algorithm evaluates the polynomial at p of size $\Theta(\log ntd)$ bits and requires $\Theta(d \log ntd + \log L)$ bits, which implies that the running time i bit complexity model is $poly(n, t, d, \log L)$.

Grigoriev and Karpinski [GK87] showed a deterministic NC algorithm to compute a matchin the case that the permanent is sparse. (The permanent can be viewed as a generic multivariate nomial.) Ben-Or and Tiwari [BOT88] gave a deterministic algorithm to compute a sparse multiva polynomial that runs in time $poly(n, t, \log d)$. The algorithm evaluates the polynomial at points o $O(t \log n)$ bits and requires accuracy of $O(td \log n + \log L)$ bits. Therefore, the running time o algorithm in the bit complexity model requires time $poly(n, t, d, \log L)$. Grigoriev, Karpinski and Si [GKS90] showed how to interpolate a rational function with $O(nt^t)$ points, which is independent o degree d ! However in the bit complexity model the running time is $poly(n, t^t, d, \log L)$. Algorithm interpolation of sparse polynomials over finite field appears in [GKS88, Kar89].

The work here extends the technique developed in [GL89, KM91] for the Fourier transform the hypercube. The extension handles a ring of integers modulus an integer (i.e. Z_d). After we s how to handle Z_d, we use the previous techniques and extend the result to handle Z_d^n.

Remark: In this work we are interested in the bit complexity model. In the analysis we use algebraic model and count the number of operations. By guaranteeing that the precision that algorithm requires is small, we make sure that its running time in both models is similar.

The paper is organized as following. Section 2 mentions a few know facts about the Che bounds, discrete Fourier transform and defines the black box model. Section 3 has the interpolatio univariate sparse polynomials with small coefficients, which is extended in Section 4 to handle arbit coefficients. The precision accuracy that the algorithm requires is discuss in Section 5. Section 6 with the approximation of a real value function by sparse polynomials. Section 7 sketches ho extend the case of univariate polynomials to multivariate polynomials.

2 Preliminaries

Chernoff bounds: In the analysis of our randomized algorithms we use Chernoff bounds. following theorem formalizes the bounds in a way which would be most useful for us. (See [HR89]

Theorem 2.1 *(Chernoff bound) Let $X_1, \ldots, X_m$ be i.i.d with $E[X_i] = p$, $|X_i| \le H$ and let S $X_1 + \cdots + X_m$. If $m \ge \frac{2H^2}{\epsilon^2} \ln \frac{2}{\delta}$ then,*

$$Pr\left[\left|\frac{S_m}{m} - p\right| > \epsilon\right] \le \delta$$

plex numbers: The magnitude of a complex number $z = a+ib$ is $||z|| = \sqrt{a^2+b^2}$. Let $\omega_d = e^{\frac{2\pi}{d}i}$ e d-th root of unity. Recall that $||\omega_d^k|| = 1$, for any d and k.

rete Fourier Transform: We list here a few known facts about the Discrete Fourier Transform. ider a polynomial $P(x) = \sum_{j=0}^{d-1} a_j x^j$. Each coefficient of the polynomial satisfies,

$$a_j = \frac{1}{d}\sum_{k=0}^{d-1} P(\omega_d^k)\omega_d^{-kj}.$$

arseval identity relates the sum of the squares of the coefficients to the values of the polynomial llows,

$$\sum_{j=0}^{d-1} a_j^2 = \frac{1}{d}\sum_{k=0}^{d-1} P(\omega_d^k)P(\omega_d^{-k}) = \frac{1}{d}\sum_{k=0}^{d-1} ||P(\omega_d^k)||^2.$$

or any integer k not divisible by d,

$$\sum_{m=0}^{d-1} (\omega_d^k)^m = 0.$$

Black Box Model: Our algorithm assumes that has a black box that evaluates the polynomial he algorithm evaluates the polynomial at roots of unity of the form ω_d^k. The interaction between lgorithm and the black box is the following.

o evaluate the polynomial P at ω_d^k, the algorithm gives the black box the pair (d,k). Denote by $d)$ the output of the black box on input (d,k).

he black box has *precision accuracy* ϵ if for any input (d,k), the absolute error is at most ϵ, i.e.,

$$||P(\omega_d^k) - \rho(d,k)|| \leq \epsilon.$$

et $H = \max_k\{||P(\omega_d^k)||\}$. The number of bits required for precision accuracy ϵ is $O(\log H + \log \frac{1}{\epsilon})$.

Univariate polynomials

his section we discuss interpolation of t-sparse polynomials with integer coefficients. The result is d on a searching technique that we develop to search in the space of the possible coefficients. The ch technique is similar to the one in [GL89, KM91].

Given a polynomial $P(x)$ we split it into two polynomials. One that includes the coefficients of odd ees (denoted by $P_{1,1}$), while the other includes the coefficients of even degrees (denoted by $P_{0,1}$). e that $P_{0,1}(x) = (P(x)+P(-x))/2$ and $P_{1,1}(x) = (P(x)-P(-x))/2$. In a similar way we define (x), where $0 \leq \alpha \leq 2^\ell - 1$ and $0 \leq \ell \leq \log d$, as follows,

$$P_{\alpha,\ell}(x) = \sum_{j:\ e_j \bmod 2^\ell = \alpha} a_{e_j} x^{e_j}$$

The algorithm works as follows. It keeps a list of the current candidates polynomials, $P_{\alpha,\ell}$. The rithm takes a polynomial $P_{\alpha,\ell}$ off the list and tests if it is identically zero. If the polynomial is identically zero no action is taken. (This implies that we terminate the search for coefficien exponents e, $e = \alpha \bmod 2^\ell$, since all those coefficients are zero.) Otherwise we continue recurs searching, i.e. we add to the list of candidates polynomials the polynomials $P_{\alpha+2^\ell,\ell+1}$ and P_ϵ When $\ell = \log d$ then we reached a non-zero coefficient. (Namely the polynomial $P_{\alpha,\log d}(x) = a_\alpha x$

Since there are at most t non-zero coefficients and for each coefficient we perform $\log d$ recu calls, the total number of recursive calls is bounded by $O(t \log d)$.

We are not able to compute the values of the polynomials that we create *exactly*, but rathe are able to approximate them very well. One way to overcome this is to approximate the value polynomial at randomly selected points, and if at some point it is more than 1/2 then we assume it is a non-zero polynomial. Here we chose an alternative way to test for zero polynomials that is b on approximating the power of the polynomial. Since the polynomial has integer coefficients, if a non-zero polynomial then it has power at least one. We approximate the power of the polyno and if the approximation is more than 1/2 we assume that the polynomial is non-zero. We prefe method that is based on approximating the power since it extends more naturally to the case of coefficients and to the case of approximation by sparse polynomials.

3.1 The main procedure

In this subsection we outline the main idea of how to recover the coefficients of a t-sparse polyno $P(x) = \sum_{j=1}^{t} a_{e_j} x^{e_j}$, whose coefficients are integers. The basic strategy is to recover the exponen bit after bit, starting from the least significant bit.

Let d be the smallest power of two greater than $\max_j\{e_j\}$, i.e. $d = 2^{\lceil \log(1+\max_j\{e_j\})\rceil}$. Consider sets $S_\ell = \{e_j \bmod 2^\ell \;:\; 1 \le j \le t\}$, for $0 \le \ell \le \log d$. The set S_ℓ includes the ℓ least significant bi the exponents e_j. Clearly the size of a set S_ℓ is bounded by t, and the set S_0 includes only zero. show how given the set S_ℓ to compute the set $S_{\ell+1}$.

From the definition of S_ℓ it follows that if $\alpha \notin S_\ell$ then $P_{\alpha,\ell}(x) \equiv 0$ while if $\alpha \in S_\ell$ then $P_{\alpha,\ell}(x)$ One step in the construction is testing if a polynomial is identically zero. Assume that we are g such a procedure $ZERO(P_{\alpha,\ell})$ that returns $TRUE$ if and only if $P_{\alpha,\ell} \equiv 0$ (latter we will show ho implement it).

Given the set S_ℓ we construct the set $S_{\ell+1}$ as follows. For each $\alpha \in S_\ell$ we create two inte $\alpha_0 = \alpha$ and $\alpha_1 = \alpha + 2^\ell$. For each α_σ, $\sigma \in \{0,1\}$, we test $ZERO(P_{\alpha_\sigma,\ell})$. If $ZERO(P_{\alpha_\sigma,\ell})$ is FAI we add α_σ to $S_{\ell+1}$. Since the set $S_{\log d}$ includes all the exponents of P, when we recover $S_{\log d}$ we found all the exponents of P.

Claim 3.1 *The algorithm constructs each S_ℓ correctly.*

Proof: The proof is by induction on ℓ. Initially S_0 includes only zero, so it is correct. Assume we constructed S_ℓ correctly. We have to show two properties of $S_{\ell+1}$: that any element added sh have been added and that every element of $S_{\ell+1}$ was added at some point.

An element $\beta = \alpha_\sigma$, $\sigma \in \{0,1\}$, is added to $S_{\ell+1}$ only if $ZERO(P_{\beta,\ell+1})$ is $FALSE$. This im that $\sum_{j:e_j \bmod 2^{\ell+1}=\beta} a_{e_j}^2 \neq 0$. Therefore, there exists some coefficient a_{e_j}, such that $e_j \bmod 2^{\ell+1} =$ which is not zero.

For the second part assume that $\beta \in S_{\ell+1}$. Let $\beta \bmod 2^\ell = \alpha$. By the induction hypothesis $\alpha \in$ When we handle α, in one of the two cases we test $\alpha_\sigma = \beta$. Since $\beta \in S_{\ell+1}$, $ZERO(P_{\beta,\ell+1}$ $ZERO(P_{\alpha_\sigma,\ell+1})$ is $FALSE$. Hence β was added to $S_{\ell+1}$.

emma 3.2 *The number of calls to the procedure ZERO is bounded by* $O(t \log d)$ □

far we have an algorithm that requires an oracle for computing $P_{\alpha,\ell}(x)$, and to the procedure *O*. In the next subsection we show how to approximate $P_{\alpha,\ell}$, and in the subsection after that we how to approximate *ZERO*.

Approximating $P_{\alpha,\ell}(x)$

is subsection we show how to approximate $P_{\alpha,\ell}(x)$. The main tool in the approximation is the ing lemma.

emma 3.3 *For any polynomial* $P(x)$, α *and* ℓ, *such that* $0 \le \alpha \le 2^\ell$, *then*

$$P_{\alpha,\ell}(\omega_d^k) = \frac{1}{2^\ell} \sum_{m=0}^{2^\ell - 1} P\left(\omega_d^{k+m\frac{d}{2^\ell}}\right) \omega_d^{-\alpha m \frac{d}{2^\ell}}$$

roof: The proof is somewhat technical. We basically transform the right hand side to the left side. We start by substituting for $P(x)$ the expression $\sum_{j=0}^{d} a_j x^j$. This means that the right side is,

$$\frac{1}{2^\ell} \sum_{m=0}^{2^\ell-1} \sum_{j=0}^{d-1} a_j \omega_d^{j(k+m\frac{d}{2^\ell})} \omega_d^{-\alpha m \frac{d}{2^\ell}}$$

an change the order of summation and rewrite it as,

$$\sum_{j=0}^{d-1} a_j \omega_d^{jk} \left[\frac{1}{2^\ell} \sum_{m=0}^{2^\ell-1} \left(\omega_d^{(j-\alpha)\frac{d}{2^\ell}} \right)^m \right]$$

ider the expression between the brackets. If $j = \alpha \bmod 2^\ell$, then the expression is one, and wise it is equal to zero. Therefore we can rewrite it as,

$$\sum_{j:\, j \bmod 2^\ell = \alpha} a_j \omega_d^{jk} = P_{\alpha,\ell}(\omega_d^k).$$

□

Ve can use Lemma 3.3 to approximate $P_{\alpha,\ell}$. Define the following random variable,

$$A_m(\alpha,\ell,k) = \frac{1}{m} \sum_{i=1}^{m} P\left(\omega_d^{k+\beta_i \frac{d}{2^\ell}}\right) \omega_d^{-\alpha\beta_i \frac{d}{2^\ell}}$$

e β_i is chosen uniformly in the set $\{0,\ldots,2^\ell - 1\}$. The random variable $A_m(\alpha,\ell,k)$ is an approx-ion for $P_{\alpha,\ell}(\omega_d^k)$ using m sample points. The following lemma, which is an application of Chernoff ds, states how "good" is the approximation.

emma 3.4 *Let* $P(x)$ *be a polynomial such that* $\max_j\{\|P(\omega_d^j)\|\} \le H$. *There exists a constant c,* *that for* $m \ge cH^4 \log \frac{2}{\delta}$ *then, for any k,*

$$Prob\left[\|A_m(\alpha,\ell,k) - P_{\alpha,\ell}(\omega_d^k)\| \ge \frac{1}{16H} \right] \le \delta$$

The main use of Lemma 3.4 would be through the following claim.

Claim 3.5 *Let a and b be complex numbers such that $\|a\| \leq H$. If $\|a-b\| \leq \frac{1}{16H}$ then,*

$$\left|\|a\|^2 - \|b\|^2\right| \leq \frac{1}{4}.$$

3.3 Testing for Zero Polynomials

Our aim is to show that with high probability we can approximate each $ZERO(\alpha, \ell)$ correctly. would imply that we can run the entire algorithm using the approximation, rather than te $ZERO(\alpha, \ell)$. The approximating function is named $APPROX - ZERO$. The main idea would approximate the power of the function. If the polynomial is the zero polynomial, then the pov zero, otherwise the power is at least one (since we assumed that the coefficients are integers).

For a polynomial $P(x) = \sum a_j x^j$, let $power(P) = \sum_j a_j^2$. In order to approximate the power polynomial we define the following random variable

$$B_{m_1,m_2}(\alpha, \ell) = \frac{1}{m_1} \sum_{i=1}^{m_1} \|A_{m_2}(\alpha, \ell, k_i)\|^2$$

where k_i is chosen uniformly in the set $\{0, \ldots, d-1\}$. The random variable $B_{m_1,m_2}(\alpha, \ell)$ is an ap imation of $power(P_{\alpha,\ell})$. The following lemma states how "good" is the approximation.

Lemma 3.6 *Let $P(x)$ be a polynomial such that $H = \max_k\{\|P(w_d^k)\|\}$. There exists a consta such that for $m_1 \geq cH^4 \log\frac{1}{\delta}$ and $m_2 \geq cH^4 \log\frac{m_1}{\delta}$ then,*

$$Prob\left[|B_{m_1,m_2}(\alpha, \ell) - power(P_{\alpha,\ell})| \geq \frac{1}{2}\right] \leq \delta$$

Proof: Using Lemma 3.3 one can show that $\max_k\{\|P_{\alpha,\ell}(\omega_d^k)\|\} \leq H$. The proof of the le uses Chernoff bounds. We first show that choosing m_1 values of k_i's at random guarantees that, probability $1 - \delta/2$,

$$\left|power(\alpha, \ell) - \frac{1}{m_1} \sum_{i=1}^{m_1} \|P_{\alpha,\ell}(\omega_d^{k_i})\|^2\right| \leq \frac{1}{4}.$$

Based on Lemma 3.4 we show that for every k_i, with probability $1 - \delta/(2m_1)$, the random var $A_{m_2}(\alpha, \ell, k_i)$ approximates $P_{\alpha,\ell}(\omega_d^{k_i})$ with error less than $1/16H$. By Claim 3.5 this implies tha difference between $\|A_{m_2}(\alpha, \ell, k_i)\|^2$ and $\|P_{\alpha,\ell}(\omega_d^{k_i})\|^2$ is at most $1/4$.

Now we are ready to define $APPROX-ZERO(\alpha, \ell)$. In order to compute $APPROX-ZERO($ we first compute $B_{m_1,m_2}(\alpha, \ell) = b$, where m_1 and m_2 are chosen according to Lemma 3.6, then i smaller than $1/2$ we return $TRUE$, and otherwise we return $FALSE$. The following theorem s that this is a good approximation.

Theorem 3.7 *For any ℓ and α, $0 \leq \alpha \leq 2^\ell$, then*

$$Prob[APPROX - ZERO(\alpha, \ell) \neq ZERO(\alpha, \ell)] \leq \delta$$

Proof: It is easy to show that if $APPROX - ZERO(\alpha, \ell) \neq ZERO(\alpha, \ell)$. then

$$|B_{m_1,m_2}(\alpha, \ell) - power(P_{\alpha,\ell})| \geq \frac{1}{2}$$

and the theorem follows from Lemma 3.6.

Approximating a single coefficient

ed to approximate each coefficient that we found at the end of the algorithm with an error of at ε. Given an exponent e, we show how to approximate its coefficients, a_e.

fine the random variable $COEF_m(e)$ to be,

$$COEF_m(e) = \left\lfloor \frac{1}{2} + \frac{1}{m}\sum_{i=1}^{m} P(\omega_d^{\beta_i})\omega_d^{-\beta_i e} \right\rfloor$$

β_i is chosen uniformly from $\{0,\ldots,d-1\}$. The random variable $COEF_m(e)$ is an approximation coefficient a_e using m sample points. The following lemma states how close is the random le $COEF_m(e)$ to the coefficient a_e.

emma 3.8 *Let P be a polynomial and $H = \max_j\{\|P(\omega_d^j\|\}$, then for $m \geq 32H^2 \log\frac{2}{\delta}$,*

$$Prob\left[COEF_m(e) \neq a_e\right] \leq \delta$$

roof: If $COEF_m(e) \neq a_e$ then $|COEF_m(e) - a_e| \geq 1/2$, since they are both integers. The claim lemma follows from Chernoff bounds. □

Univariate Interpolation Algorithm: Small Coefficients

an now use the building blocks in this section to show the following result.

heorem 3.9 *Let P be a non-zero univariate polynomial with at most t integer coefficients, each of itude at most L, and $H = \max_k\{\|P(w_d^k)\|\}$. The running time of the algorithm is $O(H^8 t \log d \log^2 \frac{1}{\delta'}$ $\log\frac{1}{\delta}) = poly(n, t, \log d, \log L)$, where $\delta' = O(\frac{1}{H^8 t \log d \log^2\frac{1}{\delta}})$*

roof: There are $O(t\log d)$ recursive calls. Each time in $APPROX - ZERO$ we test $O(H^4 \ln\frac{1}{\delta})$ ent k_i's. For each k_i we use $O(H^4\log\frac{1}{\delta})$ points to approximate it. At the end, for each of the t cients we call the procedure $COEF$ and use $O(H^2\log\frac{1}{\delta})$ points. □

he main drawback of the above result is that the running time depends on H, which could be ortional to the largest coefficient, i.e. L, and not polynomial in bit representation of the coefficients, og L. In the next subsection we show how to overcome this deficiency.

Handling large coefficients

aim of this section is to modify the algorithm of the previous section so that its running time d be polynomial in log L. The technique we use in handling large coefficients is similar in spirit to scaling techniques in graph algorithms. We first find the exponents whose coefficients are "very" . We approximate the coefficients of those exponents and subtract them from the polynomial. we only approximate the coefficients, the new polynomial may still have for those exponents with zero coefficients. The main advantage is that the new polynomial has, in the worst case, the same ber of non-zero coefficients, while the maximum value of a coefficient is smaller.

Consider a t-sparse polynomial $P(x)$, i.e.

$$P(x) = \sum_{j=1}^{t} a_{e_j} x^{e_j}.$$

As before, let the parameter d be the smallest power of two larger than the degree, i.e. $d = 2^{\lceil \log(1+m$ Similarly, let L be the smallest power of two that is larger than the maximum coefficient in abs value, i.e. $L = 2^{\lceil \log(\max_j\{|a_{e_j}|\})\rceil}$.

Our aim, is to perform the interpolation in time that is polynomial in t, $\log d$ and $\log L$. following is the basic idea of the algorithm. Given L, the bound on the largest coefficient, we con the sets $Z_1 = \{e_j : L/2 \leq |a_{e_j}| \leq L\}$ and $Z_2 = \{e_j : L/(8\sqrt{t}) \leq |a_{e_j}| \leq L\}$. We find a set Y, that $Z_1 \subset Y \subset Z_2$. Once we found the set Y (in time polynomial in t, $\log d$ and $\log L$), we approxi each of the coefficients of the exponents in Y, with a absolute error of less than a $L/4$. This in that, with high probability, the difference between the real coefficient and the approximated coeffi is less than $L/4$. For each $e_j \in Y$, denote by γ_{e_j} the approximated value of the coefficient a_{e_j}. In to ensure that the new polynomial is integral we enforce γ_{e_j} be integral. More formally, let Q b polynomial we created,

$$Q(x) = \sum_{e_j \in Y} \gamma_{e_j} x^{e_j}.$$

Consider the polynomial $P'(x) = P(x) - Q(x)$. Assuming that $Z_1 \subset Y \subset Z_2$, and $|\gamma_{e_j} - a_{e_j}| \leq$ the polynomial P' is t-sparse, and the maximum size coefficient in P' is at most $L/2$. This implies in polynomial time we reduced the problem of recovering the coefficients of a polynomial with maxi coefficient L (i.e. $P(x)$), to the problem of recovering the coefficients whose maximum coefficie $L/2$ (i.e. $P'(x)$). Therefore after $O(\log L)$ such iterations the maximum coefficient is $O(\sqrt{t})$. Onc coefficients are of size $O(\sqrt{t})$ the algorithm of Theorem 3.9 would recover all the coefficients.

4.1 Approximating the power

The main idea is that with a sample size that depends on t and is independent of L we can approxi *power*$(P_{\alpha,\ell})$. We start by modifying the approximation of $P_{\alpha,\ell}$.

Lemma 4.1 *Let P be a t-sparse polynomial whose magnitude of each coefficient is bounded There exists a constant c, such that for $m \geq ct^4 \log \frac{2}{\delta}$ then, for any k,*

$$Prob\left[\|A_m(\alpha,\ell,k) - P_{\alpha,\ell}(\omega_d^k)\| \geq \frac{L}{64t}\right] \leq \delta$$

As before, this implies the following about the difference in the magnitude.

Claim 4.2 *Let a and b be complex numbers such that $\|a\| \leq tL$. If $\|a-b\| \leq \frac{L}{64t}$ then,*

$$\left|\|a\|^2 - \|b\|^2\right| \leq \frac{L^2}{16}.$$

We can now claim the result about approximating the power.

mma 4.3 *Let P be a t-sparse polynomial whose magnitude of each coefficient is bounded by L. exists a constant c, such that for $m_1 \geq ct^4 \log \frac{1}{\delta}$ and $m_2 \geq ct^4 \log \frac{m_1}{\delta}$ then,*

$$Prob\left[|B_{m_1,m_2}(\alpha,\ell) - power(P_{\alpha,\ell})| \leq \frac{3L^2}{32}\right] \leq \delta$$

oof: The polynomial P is t-sparse, furthermore, each coefficient is bounded by L. Therefore, agnitude of $P(w_d^k)$ is bounded by tL, and the magnitude of $P_{\alpha,\ell}(w_d^k)$ is bounded by tL. Once we ed the magnitude of $P_{\alpha,\ell}$ we can use the Chernoff bounds to claim that with probability $1-\delta/2$ fference between $power(\alpha,\ell)$ and $\frac{1}{m_1}\sum_{1\leq j\leq m_1}||P_{\alpha,\ell}(\omega_d^{k_j})||^2$, is less than $L^2/32$.

sed on Lemma 4.1, for every k_i, with probability $1-\delta/(2m_1)$ the random variable $A_{m_2}(\alpha,\ell,k_i)$ ximates $P_{\alpha,\ell}(\omega_d^{k_i})$ with error less than $L/(64t)$. By Claim 4.2 this implies that the difference en $||A_{m_2}(\alpha,\ell,k_i)||^2$ and $||P_{\alpha,\ell}(\omega_d^{k_i})||^2$ is at most $L^2/16$. □

w we define the equivalent of the $ZERO$ and $APPROX-ZERO$ procedures. Let $LARGE(P,L)$, P is a polynomial and L an integer, be $TRUE$ if and only if P has some coefficient a_j such that $L/2$. As before, we cannot compute $LARGE(P,L)$ directly, but we can approximate it. We ximate the value of $LARGE(P_{\alpha,\ell},L)$ by $APPROX-LARGE(\alpha,\ell,L)$ which is define as follows: ute $B_{m_1,m_2}(\alpha,\ell)=b$ and if $b\geq\frac{5L^2}{32}$ return $TRUE$, otherwise return $FALSE$.

mma 4.4 *Let P be a t-sparse polynomial whose magnitude of each coefficient is bounded by L. ny α and ℓ, $0\leq\alpha\leq 2^\ell$, then*

$$Prob[APPROX-LARGE(\alpha,\ell,L)=FALSE \text{ and } LARGE(P_{\alpha,\ell},L)=TRUE]\leq\delta$$

roof: Assume that $LARGE(P_{\alpha,\ell},L)$ is $TRUE$. This implies that $power(P_{\alpha,\ell})\geq L^2/4$, since some cient is larger than $L/2$. With probability $1-\delta$

$$|B_{m_1,m_2}(\alpha,\ell)-power(P_{\alpha,\ell})|\geq\frac{3L^2}{32}.$$

mplies that with probability $1-\delta$ we have $B_{m_1,m_2}(\alpha,\ell)\geq\frac{5L^2}{32}$, which is equivalent to $APPROX-$ $GE(\alpha,\ell,L)$ being $TRUE$. □

emma 4.5 *Let P be a t-sparse polynomial whose magnitude of each coefficient is bounded by L. ny α and ℓ, $0\leq\alpha\leq 2^\ell$, then*

$$Prob[APPROX-LARGE(\alpha,\ell,L)=TRUE \text{ and } LARGE(P_{\alpha,\ell},\frac{L}{4\sqrt{t}})=FALSE]\leq\delta$$

roof: Assume that $LARGE(P_{\alpha,\ell},\frac{L}{4\sqrt{t}})$ is $FALSE$. This implies that $power(P_{\alpha,\ell})\leq L^2/64$, since efficient is larger than $\frac{L}{8\sqrt{t}}$. With probability $1-\delta$

$$|B_{m_1,m_2}(\alpha,\ell)-power(P_{\alpha,\ell})|\geq\frac{3L^2}{32}.$$

implies that with probability at most δ we have $B_{m_1,m_2}(\alpha,\ell)\geq\frac{5L^2}{32}$, which is equivalent to $ROX-LARGE(\alpha,\ell,L)=TRUE$. □

he above two lemmas show that if $APPROX-LARGE(\alpha,\ell,L)$ is $TRUE$, then, with high ability, $P_{\alpha,\ell}$ is a non-zero polynomial (and even has a coefficient larger than $\frac{L}{8\sqrt{t}}$), and if $P_{\alpha,\ell}$ has fficient larger than $L/2$, then with high probability $APPROX-LARGE(\alpha,\ell,L)$ is $TRUE$.

ote that the number of recursive calls when we use the procedure $LARGE$ is bounded, as before, og d, since every time $LARGE$ is $TRUE$, we are focusing on at least one of the t coefficients.

4.2 Approximating a single coefficient

The approximation of a single coefficient is done as in the case of small coefficients, in Section 3.4 only difference is that we need to approximate each coefficient that we found with absolute erro than $L/4$.

Lemma 4.6 *There is a constant c, such that for $m \geq ct^2 \log \frac{1}{\delta}$, then*

$$Prob\left[|COEF_m(e) - a_e| \geq L/4\right] \leq \delta.$$

4.3 Putting it all together

The basic building block would be a procedure $FIND(P, L)$ that receives a polynomial P with c cients of magnitude at most L, and returns a list of exponents and their respective coefficients such with high probability this list includes all the coefficients larger than $L/2$ and no coefficient sm than $\frac{L}{8\sqrt{t}}$.

The procedure $FIND(P, L)$ works as the algorithm in Section 3, but uses the subroutine $APPR$ $LARGE$ instead of $APPROX - ZERO$. At the beginning we start with a set Y_0 that has or At phase ℓ, for each element $\alpha \in Y_\ell$, we create $\alpha_0 = \alpha$ and $\alpha_1 = \alpha + 2^\ell$. We add α_σ to $Y_{\ell+1}$ o $APPROX - LARGE(\alpha_\sigma, \ell+1, L)$ is $TRUE$.

Since we are modifying the original polynomial we describe how to compute the output tha black box returns to the algorithm. We have a polynomial $R(x)$ which is initially identically zero. polynomial would be used to "change" the original polynomial P that the algorithm is approximat Each time the algorithm queries the black box on a point ω_d^k, we query the original black box on (d, k) and get a reply $\rho(d, k)$ (which is simply $P(\omega_d^k)$ with finite accuracy), we reply to the algor with $\rho(d, k) - R(\omega_d^k)$.

The main algorithm runs in phases. We maintain that after the i-th phase, with high probab $P(x) - R(x)$ is a polynomial with at most t non-zero coefficients and its maximum coefficient is at $L/2^i$. The i-th phase consists from the following steps: (1) We call $FIND(P - R, L/2^{i-1})$ that re a set Y_i. With high probability the set Y_i includes all the coefficients in the range $L/2^i$ to $L/2^{i-1}$ For each exponent in $e \in Y_i$ we approximate its coefficient a_e by γ_e, such that the difference is at $L/2^{i+1}$. Let $Q_i(x) = \sum_{e \in Y_i} \gamma_e x^e$. (3) We add the polynomial $Q_i(x)$ to the polynomial $R(x)$.

Once the size of the coefficients is $O(\sqrt{t})$ the algorithm uses the algorithm for small coefficients Theorem 3.9) to completely recovers the coefficients, and adds this polynomial to R. At this with high probability, the polynomial $R(x)$ at is identical to the polynomial $P(x)$.

We have established the following theorem.

Theorem 4.7 *There exists a randomized algorithm that interpolates a t-sparse polynomial, w coefficients are integers less than L. The algorithm runs in time polynomial in t, $\log d$ and $\log L$.*

5 Precision accuracy

Our algorithms use complex numbers. When we measure the bit complexity we need to specify accuracy to which we require the complex numbers to be. This in turn also determines the bit plexity of the algorithm, since it determines the size of the numbers that the algorithm operates most of our operations are simply averaging over a set of random inputs, the algorithm requires . low accuracy. The following claim shows how we can bound the precision accuracy.

.aim 5.1 *Adding m numbers with precision accuracy ϵ/m results in an outputs that has an ab- error of at most ϵ. Furthermore, if each of the m numbers is bounded in absolute value by H heir description requires only $O(\log H + \log \frac{m}{\epsilon})$ bits to guarantee a precision accuracy of ϵ in the*

.

ıe values used in $COEF_m(e)$ and $A_m(\alpha, \ell, k)$ are bounded by tL. This implies that using $tL + \log \frac{m}{\epsilon}$) bits (and precision accuracy ϵ/m) in their computation, we introduce an additive of at most $O(\epsilon)$. Since those random variables are used only as an approximation, the additional could be absorbed in the error margins that we assumed before. (Recall that $B_{m_1,m_2}(\alpha, \ell)$ is define gh A_m, and therefore the precision accuracy of B_{m_1,m_2} is determine by the precision accuracy of

Approximation by polynomials

ne that rather than recovering all the coefficients of the polynomial we are interested in recovering e "large" coefficients of the polynomial. This is a very interesting case, since in many cases we regard the other coefficients as "noise" that is not interesting.

he formal scenario is the following. As before we are given a polynomial P as a black box, however, e not guaranteed that it is either t-sparse or has integer coefficients. We assume that there is polynomial Q, with t integer coefficients, such that the power of $P - Q$ is at most ν.

ur algorithm will recover a polynomial Q', such that the power of $P - Q'$ is bounded by $\nu + O(\epsilon)$. number of recursive calls done by the algorithm is bounded by $O((t + \nu) \log d)$. The correctness is d as before (each coefficient larger than one would be reached, with high probability). The running can be bounded by observing that the difference between P and Q can cause the algorithm, for specific value of ℓ, to search at most $O(\nu)$ values that would not lead to a coefficient of size at one. The time to perform a single $APPROX - ZERO$ or $APPROX - LARGE$ remains the

.

Ve made a restriction that the approximating polynomial Q has integer coefficients. This restriction be easily removed in the following way. Assume that there is a polynomial R that has real icients, and the power of $P - R$ is at most ν. Consider the polynomial R' that is achieved by iplying the coefficients of R by t/ϵ, and rounding them to the nearest integer. The difference een $R(x)$ and $R'(x) * (\epsilon/t)$ is bounded by ϵ. This implies that the polynomial $P(x) * (t/\epsilon)$ has a approximation by an integer polynomial R'.

Multi-variate polynomials

technique that we developed for univariate polynomials extends to multi-variate polynomials, and ideas similar to [GL89, KM91] and [Zip79]. In this Section we give a sketch of the ideas of the nsion.

The idea for multi-variate polynomials is the following. As before, We first recover all the "large ficients". We proceed variable by variable, and find the exponents of the "large coefficients". Then we approximate the coefficients, and subtract then from the original polynomial. The main diffe is that we recover the exponents of the interesting coefficients sequentially. We omitt the detail to lack of space.

Acknowledgments

I would like to thank László Babai, Persi Diaconis, Marek Karpinski, Eyal Kushilevitz, Michael R and Prasoon Tiwari, for the discussions that I had with each of them.

References

[BOT88] M. Ben-Or and P. Tiwari. A deterministic algorithm for sparse multivariate polyno interpolation. In *Proceedings of the* 20th *Annual ACM Symposium on Theory of Compu* pages 301–309, May 1988.

[GK87] D. Y. Grigoriev and M. Karpinski. The matching problem for bipartite graphs with po mially bounded permanents is in nc. In 28th *Annual Symposium on Foundations of Com Science, Los Angeles, California*, pages 166–172, October 1987.

[GKS88] D. Y. Grigoriev, M. Karpinski, and M. Singer. Fast parallel algorithms for sparse multiva polynomial interpolation over finite fields. Technical Report Research report no. 8523 University of Bonn, 1988. to appear in Siam J. on Comp.

[GKS90] D. Y. Grigoriev, M. Karpinski, and M. Singer. Interpolation of sparse rational functions w out knowing bounds on exponents. In 31th *Annual Symposium on Foundations of Com Science, St. Louis*, pages 840–846, October 1990.

[GL89] O. Goldreich and L. Levin. A hard-core predicate for all one-way functions. In *Proc. ACM Symposium on Theory of Computing*, pages 25–32. ACM, 1989.

[HR89] Torben Hagerup and Christine Rub. A guided tour to chernoff bounds. *Info. Proc. I* 33:305–308, 1989.

[Kar89] M. Karpinski. Boolean circuit complexity of algebraic interpolation problems. Techr Report TR-89-027, ICSI, 1989.

[KM91] Eyal Kushilevitz and Yishay Mansour. Learning decision trees using the fourier spectr In *Proceedings of STOC '91*, pages 455–464, may 1991.

[Zip79] R. E. Zippel. Probabilistic algorithms for sparse polynomials. In *EUROSAM*, pages 216– Springer Lecture notes in computer science, vol. 72, 1979.

Two Strikes Against Perfect Phylogeny

Hans L. Bodlaender* Mike R. Fellows† Tandy J. Warnow‡

Abstract

One of the major efforts in molecular biology is the computation of phylogenies for species sets. A longstanding open problem in this area is called the Perfect Phylogeny problem. For almost two decades the complexity of this problem remained open, with progress limited to polynomial time algorithms for a few special cases, and many relaxations of the problem shown to be NP-Complete. From an applications point of view, the problem is of interest both in its general form, where the number of characters may vary, and in its fixed-parameter form. The Perfect Phylogeny problem has been shown to be equivalent to the problem of triangulating colored graphs[30]. It has also been shown recently that for a given fixed number of characters the yes-instances have bounded treewidth[45], opening the possibility of applying methodologies for bounded treewidth to the fixed-parameter form of the problem. We show that the Perfect Phylogeny problem is difficult in two different ways. We show that the general problem is NP-Complete, and we show that the various finite-state approaches for bounded treewidth cannot be applied to the fixed-parameter forms of the problem.

Introduction

orically, one of the major efforts in molecular biology has been the computation of phylogenetic s, or *phylogenies*, which describe the evolution of a set of species from a common ancestor. A *ogeny* for the set S of species, is a rooted tree in which the leaves represent the species in S the internal nodes of the tree represent the ancestral species. The computational complexity etermining a most-likely phylogeny for the species set then depends, among other things, on the species set is described. One of the standard models uses *characters* to describe species. :, a character is an equivalence relation on the species set, partitioning the set into the different *acter states.* Under this model, a proposed phylogeny will also assign character states to each e hypothesized species indicated by the internal nodes. The desired property for the phylogeny e following:

> For each state of each character, the set of nodes in the tree having that state should form a connected component.

When the phylogeny has this property, it is said to be *perfect*, and the characters are also said e *perfectly compatible.* The Perfect Phylogeny problem[28] (in short: PP; also known as the racter Compatibility problem[21]) is then as follows.

Perfect Phylogeny: For a given set of characters defining a species set S, does a perfect phylogeny exist?

Department of Computer Science, P.O. Box 80.089, 3508 TB Utrecht, the Netherlands. This author was partially orted by the Esprit II Basic Research Actions Program of the EC under contract no. 3075 (project ALCOM).
Computer Science Department, University of Victoria, Victoria, B.C., Canada.
Department of Mathematics, University of Southern California, Los Angeles, California, USA. This author was ially supported by NSF Grants no. CCR88-13632 and no. DMS90-05833

If the number of characters is a fixed constant k, we call the problem the k-Perfect Phylo problem.

This approach to constructing phylogenies was probably first discussed in the biological li ture in the 1960's (see [13, 58] for two of the earliest papers, and the series of papers by LeQu [38, 39, 40, 41]), but was given its precise mathematical formulation by Estabrook and othe a series of papers beginning in 1972 (see [16, 17, 18, 19]). In 1974, Buneman showed[12] that Perfect Phylogeny problem reduced to a graph-theoretic problem, which we call the *Triangul Colored Graphs* problem (or TCG). A graph is said to be *triangulated* if every induced cycle cont at least four vertices. The Triangulating Colored Graphs problem is:

> **Input:** Graph $G = (V, E)$, coloring $c : V \to Z$.
> **Question:** Does there exist a supergraph $G' = (V, E')$ of G which is properly colored by c and which is triangulated?

If I is the instance of the Perfect Phylogeny problem, and G_I the corresponding instanc the Triangulating Colored Graphs problem, then vertices of G_I correspond to the character st of I, with states of the same character having the same color. Two vertices are adjacent if corresponding character states share a species in common. Thus, the number of colors of T corresponds to the number of characters in the Perfect Phylogeny problem.

In 1990, Kannan and Warnow [30] showed that these two problems were polynomially eq lent. Linear time algorithms for the case of two and three-colored graphs have been found [7 (corresponding to two and three character compatibility), and a polynomial time algorithm for case of quaternary characters has been found [31]). The latter algorithm can be used to const phylogenetic trees from DNA sequences. For the general case, the best that is known is an $O(n$ algorithm to triangulate (if possible) a k-colored graph [45].

In this paper we will show the following:

Theorem A: Perfect Phylogeny is NP-complete.

Theorem B: k–PERFECT PHYLOGENY is not finite-state for bounded treewidth, for $k \geq$

The significance of Theorem B is the following. There are a large number of papers, that s that many problems, that are often combinatorially hard, become linear time solvable on gra with bounded treewidth, given with a suitable tree-decomposition. (The latter can be foun $O(n \log n)$ time [8, 50].) See, amongst others, [1, 4, 5, 6, 11, 14, 29, 32, 33, 54, 59]. The underl technique of all these results is — in a certain sense — the same, and can be described as foll for each node of a rooted tree-representation of the input graph, some information of a cer type is computed. This computation for a node can be done quickly when given the informat computed for the children of the node. In many cases, this information is an element, taken f a finite set. In such a case, we call the problem 'finite state'. By theorem B, such an algori is not possible for k–Perfect Phylogeny for $k \geq 4$. For problems, that like k-Perfect Phylog for fixed k have no growing parameter associated with it, all general techniques to solve them graphs with a given tree-decomposition of constant bounded treewidth can be seen as special c of this finite state concept. (In contrast, problems like Independent Set, with a growing param associated to it, require a generalization of the finite state concept. Here, the 'information' constant size table, with each entry an integer. However, an extension of our arguments show t such approaches also cannot yield linear time algorithms.) (The result also shows, that the gr reduction method from [3] will not work for the problem with $k \geq 4$.)

In contrast, for $k = 2, 3$, k-Perfect Phylogeny is finite state. (For $k = 2$, this is trivial. $k = 3$, it follows from the characterization in [7] that the problem can be formulated in mon second order form, and hence, by the result of Courcelle [14], it is finite state.)

Since a standard tool for molecular biologists involves checking small subsets of characters perfect compatibility, efficient algorithms for small k can be of use.

Preliminary definitions and results

que in a graph $G = (V, E)$ is a subset S of V, such that for all $v, w \in S, (v, w) \in E$. A graph $V, E)$ is triangulated, if and only if it does not contain an induced cycle of length at least four. known [52, 26] that a graph G is triangulated if and only if there exists an linear ordering of ertex set $v_1, v_2, \ldots, v_n$, such that for each i, the neighbors of v_i which follow v_i in the ordering, a clique. Such an ordering is called a *perfect elimination scheme.*

'he following lemma is due to Dirac [15].

ma 1 *Let $G = (V, E)$ be a triangulated graph which is not a complete graph. Then V contains non-adjacent simplicial vertices.*

. graph $G = (V, E)$ with vertex coloring $c : V \to Z$ is c-triangulatable if there exists a rgraph $G' = (V, E')$, $E \subset E'$, which is properly colored by c (thus $(v, w) \in E'$ implies $c(v) \neq$) and which is triangulated. The supergraph G' is said to be a c-triangulation of G.

useful characterization of c-triangulable graphs is with the help of tree-decompositions.

nition A tree-decomposition of a graph $G = (V, E)$ is a pair $(\{X_i \mid i \in I\}, T = (I, F))$ with $\mid i \in I\}$ a collection of subsets of V, and T a tree, such that

$\bigcup_{i \in I} X_i = V$.

For all $(v, w) \in E$, there exists an $i \in I$ with $v, w \in X_i$.

For all $v \in V$, $\{i \in I \mid v \in X_i\}$ forms a connected subtree of T.

treewidth of a tree-decomposition $(\{X_i \mid i \in I\}, T = (I, F))$ is $\max_{i \in I} |X_i| - 1$. The treewidth graph is the minimum treewidth over all possible tree-decompositions of that graph.

sider $G = (V, E)$ with tree-decomposition $(\{X_i \mid i \in I\}, T = (I, F))$. The graph $H = (V, E')$ $(v, w) \in E' \Leftrightarrow \exists i \in I,\ v, w \in X_i$ contains G as a subgraph, has the same treewidth as G, is triangulated. (Define $T_v = \{i \in I \mid v \in X_i\}$, for all $v \in V$. Then $(v, w) \in E'$, if and only $\cap T_w \neq \emptyset$. So H is the intersection graph of subtrees of a tree, hence H is triangulated, see) The following proposition can now easily be observed.

position 1 *A graph $G = (V, E)$ with coloring $c : V \to C$, (C a set of colors), is c-ngulatable, if and only if there exists a tree-decomposition $(\{X_i \mid i \in I\}, T = (I, F))$ of G, that for all $i \in I$, $v, w \in V$: if $v \neq w$, and $v, w \in X_i$, then $c(v) \neq c(w)$.*

In [9] a short proof of the following fact can be found:

position 2 *Let $(\{X_i \mid i \in I\}, T = (I, F))$ be a tree-decomposition of $G = (V, E)$. Let $W \subset V$ n a clique in G. Then there exists an $i \in I$ with $W \subseteq X_i$.*

One can also easily verify the following propositions.

position 3 *Let $(\{X_i \mid i \in I\}, T = (I, F))$ be a tree-decomposition of $G = (V, E)$. Let $i_0 \in I$ be h that X_{i_0} is not a separator of G, i.e., $G[V - X_{i_0}]$ is connected. Then there exists a set $I' \subseteq I$, h that $(\{X_i \mid i \in I'\}, T[I'])$ is a tree-decomposition of G, and i_0 is a leaf of T.*

position 4 *Let $(\{X_i \mid i \in I\}, T = (I, F))$ be a tree-decomposition of $G = (V, E)$. Suppose $x_2, \ldots, x_r$ form a path in G, $x_1 \in X_{i_0}$, $x_2 \in X_{i_1}$. Then for every $i_2 \in I$, that lies on the path n i_0 to i_1 in T: $X_{i_2} \cap \{x_1, x_2, \ldots, x_r\} \neq \emptyset$.*

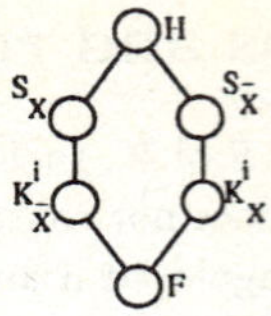

Figure 1: The Decision Component

3 Perfect Phylogeny is NP-complete

This section is devoted to the proof of the following result:

Theorem 1 *Triangulating Colored Graphs is NP-Complete, even when every color is give exactly two vertices.*

As TCG and PP are polynomial equivalent, it directly follows from this result that Per Phylogeny is NP-complete.

That Triangulating Colored Graphs is in NP is obvious: given the colored graph $G = (V,$ present a triangulation $G' = (V, E')$, and in polynomial time we can check that G' is prop colored and triangulated.

We now show that TCG is NP-hard, by a reduction from 3-SAT to TCG.

For a given instance I of 3-SAT, we create a graph, G_I, which consists of decision compon and clause components. We assume that no clause contains both a variable and its complem For each variable X and for each clause i containing either X or $\overline{X}$, we have the decision compor given by figure 1.

We call the variable H the *head*, F is called the *foot*, the variables S_X and $S_{\overline{X}}$ are called *shoulders*, and K_X^i and $K_{\overline{X}}^i$ are called *knees*. For each variable X, we will superimpose the r co of the decision component (corresponding to the r clauses containing X or $\overline{X}$), so that only and $K_{\overline{X}}^i$ are not identified with other vertices. Thus, there will be one vertex H, one vertex and for every variable X, if X or $\overline{X}$ appear in r clauses $i_1, i_2, \ldots, i_r$, then there will be one pai shoulders S_X and $S_{\overline{X}}$, and r pairs of knees, $K_X^{i_1}$, $K_{\overline{X}}^{i_1}$, $K_X^{i_2}$, $K_{\overline{X}}^{i_2}, \ldots, K_X^{i_r}$, $K_{\overline{X}}^{i_r}$.

We assign colors so that every color class consists of exactly two vertices. The head H and F are given the same color, each pair of shoulders, S_X and $S_{\overline{X}}$ is given the same color, and e pair of knees K_X^i and $K_{\overline{X}}^i$ is given the same color.

Note that there are exactly two color-respecting triangulations for the variable component X: you either add the edges in all paths $H - K_X^i - S_X - F$, or you add the edges in all pa $H - K_{\overline{X}}^i - S_{\overline{X}} - F$. Each way of triangulating the graph can be described as adding a *Mark of Zo* in one of two possible orientations. Thus, a triangulation either includes all edges (H, K_X^i) or edges $(H, K_{\overline{X}}^i)$. We will refer to the first orientation as the *positive orientation*, and the second the *negative orientation*. When the triangulation is positively oriented, we will set X to *true*, otherwise we will set X to *false*.

We now describe the clause components. For the ith clause (X, Y, Z) we have the graph gi by Figure 2.

Note that we do not add any new vertices, but only add edges between knees which alre exist. The knees K_X^i, K_Y^i, and K_Z^i are said to be *active*, while the complements $K_{\overline{X}}^i, K_{\overline{Y}}^i$ and are said to be *inactive*. In general, if the literal L appears in the ith clause, then K_L^i is said to active, and its complement $K_{\overline{L}}^i$ is said to be inactive. Thus, for each pair of knees K_L^i and exactly one will be active, and the other inactive.

As G_I can be constructed in polynomial time, given I, NP-hardness of TCG follows from following lemma.

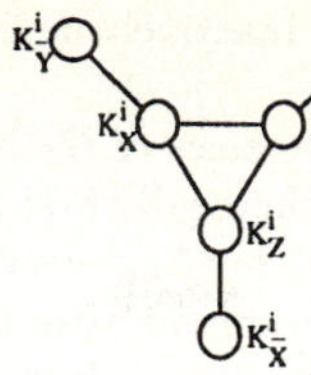

Figure 2: The Clause Component

ma 2 *The 3-SAT instance I is satisfiable if and only if G_I can be triangulated without intro-ng edges between vertices of the same color.*

of: We first show that if G_I has a color-respecting triangulation, then I can be satisfied. So s assume that G_1 is a color-respecting triangulation of G_I. As we mentioned before, G_1 defines s a truth assignment for the variables. We need to show that under this truth assignment each se contains at least one literal, set to true.

uppose that the truth function we derive from G_1 does not satisfy the clause $i = (X, Y, Z)$ in . we assume that the graph G_1 does not contain any of the edges between H and K_X^i, K_Y^i, or We will show that this contradicts G_1 being both properly colored by c and triangulated.

3y our comments earlier, G_1 must contain the Mark of Zorro in one of the two possible orienta-s; since we exclude the edges (H, K_α^i), for $\alpha \in \{X, Y, Z\}$, it must include the negative orienta-s of the Mark of Zorro in the decision components for X, Y and Z. Thus, we assume that each of following edges is in G_1: $(H, K_{\bar{\alpha}}^{i}), (S_{\bar{\alpha}}, K_{\bar{\alpha}}^{i}), (S_{\bar{\alpha}}, F)$, for each variable $\alpha \in \{X, Y, Z\}$ and clause ntaining α. Consider the subgraph G_2 of G_1 induced by the vertex set $\{H, F, S_{\bar{\alpha}}, K_\alpha^i, K_{\bar{\alpha}}^{i} : \alpha$ $X, Y, Z\}\}$. This subgraph G_2 is triangulated, since G_1 is triangulated. However, we will show G_2 does not admit a perfect elimination scheme (which respects the coloring), and hence is triangulated.

ince G_2 is triangulated, it must contain at least two simplicial vertices (see Lemma 1). Because s properly colored, only H can possibly be simplicial (every other vertex is adjacent to two ices of the same color). Therefore, we see that G_2 can not be both triangulated and properly red, contradicting our hypothesis.

Thus, we have shown that a color-respecting triangulation of G_I implies satisfiability of I.

We now show the converse. Suppose I is satisfiable, and that G_I is the graph we derive from I, that f is a satisfying truth assignment for I. We will show that we can triangulate G_I without ing edges between vertices of the same color, using the truth assignment f.

We will assume that we have renamed the variables so that X is always true, and $\overline{X}$ always . We now describe some terminology we use in defining the triangulated supergraph of G_I. all that we distinguish between *active* and *inactive* knees (see our discussion following the nition of the clause component). We now describe another way of distinguishing vertices. If able X is true, we call S_X and K_X^i *true*, thus each S_X is a *true shoulder*, and each K_X^i is a *true* . Similarly, the complements are called *false shoulders* or *false knees*.

To triangulate G_I, add the following edges: the positively oriented Mark of Zorro in each decision ponent. the complete graph on $\{$*true shoulders, true knees*$\}$, and the complete bipartite graph $\{$*true shoulders, false knees*$\}$.

Thus, we have added to the neighbor set of each true shoulder the foot F, the head H, and y knee and every true shoulder. We have added to the neighbor set of each true knee the head nd every true knee as well. It is obvious that this enlarged graph G' is properly colored. We now show it is triangulated by exhibiting a perfect elimination scheme for G'.

Consider the following partition of the vertex set of G_I into five subsets: $S_1 = \{$ False shoulders, ctive false knees$\}$, $S_2 = \{$ The head $H\}$, $S_3 = \{$ Active false knees adjacent to inactive false knees}, $S_4 =$ { Active false knees adjacent to inactive true knees}, and $S_5 =$ { True knees K_X^i, shoulders S_X, and foot F }.

These sets constitute a partition of the vertices of G_I into five pair-wise disjoint sets. We these sets to produce a perfect elimination scheme, by first listing all the vertices in S_1, then t in S_2, and so forth, down to S_5. A tedious, but not very complicated case analysis shows every vertex in simplicial in the graph which remains after the previous vertices have been del We omit this analysis from this version of the paper. It now follows that G' is a properly col triangulated supergraph of G_I. ∎

4 Non-cutset-regularity of the problem with four colo

In [20], Fellows and Abrahamson developed the theory of cutset regularity of graphs. To desc the theory, we first define some terminology used in it.

A t-boundary graph G contains a distinguished ordered subset of t nodes, called $bd(G)$). binary operator $\oplus$ on two t-boundary graphs is defined as follows: $G \oplus H$ is the t-boundaried g obtained by identifying the ith boundary nodes in $bd(G)$ with the ith boundary node in $bd(H)$ each $i = 1, 2, \ldots, t$. For a fixed graph family F, we then define an equivalence relation on th of t-boundary graphs as follows: Two t-boundary graphs X and Y are equivalent ($X \sim_F Y$) if only if for every t-boundary graph Z, $X \oplus Z \in F \iff Y \oplus Z \in F$. The "small" universe U is defined to be the set of all t-boundaried graphs that arise in the parsing of graphs of treew at most t. A graph family F is *t-cutset regular* iff $\sim_F$ has finite index on U^t_{small}.

One of the main results in [20] is the following:

Theorem 2 *(Fellows and Abrahamson [20]) A graph family F is t-finite state if and only if t-cutset regular.*

An important consequence of this result is that, if a graph family is t-finite state, then recogni of this family can be done efficiently, and without computing obstruction sets.

Using this theorem, we can show that the class of triangulatable t-colored graphs is not t-fi state, for $t \geq 4$.

Consider the following two classes of 4-colored 4-boundary graphs:

For $r \geq 2$, let $G_r = (V_r, E_r, B, f)$ with

- $V_r = \{w_1, w_2, w_3, w_4\} \cup \{z_j \mid 1 \leq j \leq 4r\}$,
- $E_r = \{(w_1, z_1), (w_2, z_2), (w_1, z_2), (w_2, z_1), (w_3, z_{4r-1}), (w_4, z_{4r}), (w_3, z_{4r}), (w_4, z_{4r-1})$ $\{z_j, z_{j+1}) \mid 1 \leq j < 4r\}$,
- $B = \{w_1, w_2, w_3, w_4\}$, and
- $f(w_j) = j (1 \leq j \leq 4)$.

Let $c : V_r \to \{1, 2, 3, 4\}$ be the coloring of G_r, defined by $c(w_j) = j$ $(1 \leq j \leq 4)$, and $c(z_j$ $(j+1) \bmod 4 + 1$ $(1 \leq j \leq 4r)$. See Figure 3 for an example.

For $s \geq 2$, let $H_s = (V'_s, E'_s, B'_s, f'_s)$ with

- $V'_s = \{y_j \mid 1 \leq j \leq 4s\}$,
- $E'_s = \{(y_{j_1}, y_{j_2}) \mid 1 \leq j_1, j_2 \leq 4s,\ j_1 \neq j_2,\ |j_1 - j_2| \leq 2\}$,
- $B'_s = \{y_1, y_2, y_{4s-1}, y_{4s}\}$, and
- $f'_s(y_1) = 1$, $f'_s(y_2) = 2$, $f'_s(y_{4s-1}) = 3$, $f'_s(y_{4s}) = 4$.

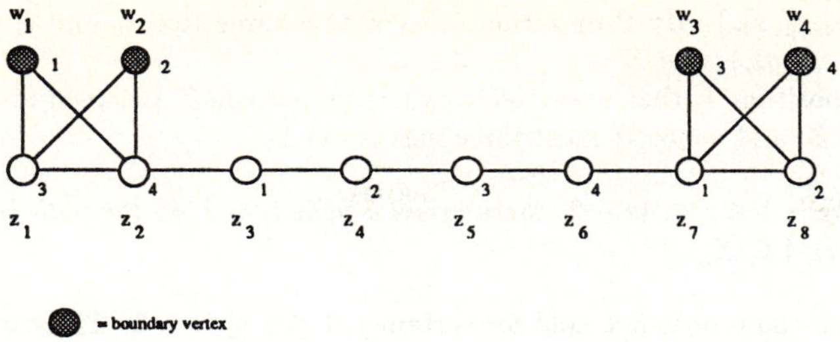

Figure 3: G_2

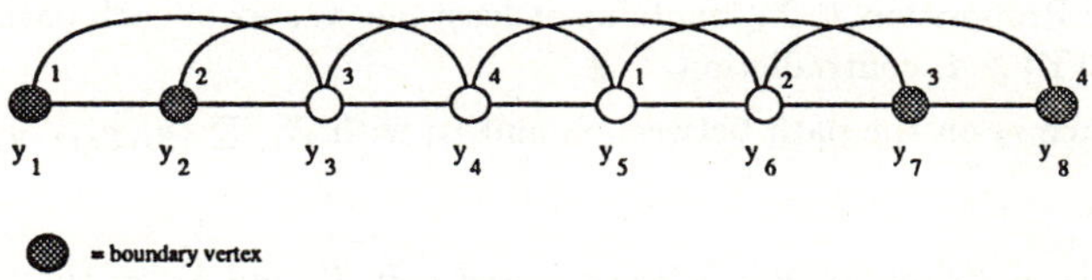

Figure 4: H_2

$c : V_s' \to \{1,2,3,4\}$ be the coloring of G_s, defined by $c(y_j) = (j-1) \bmod 4 + 1$. See Figure 4 n example.

Note that for every $r, s \geq 2$, c is a coloring of $G_r \oplus H_s$.

nma 3 *If $s \leq r$, then $G_r \oplus H_s$ is c-triangulatable.*

of: First add an edge between z_2 and $z_{4(r-s)+3}$. (If $r = s$, then omit this step.) We now ngulate the cycle on the edge $(z_2, z_{4(r-s)+3})$, and the remainder of the graph independently. cycle with edges $(z_2, z_{4(r-s)+3})$, and (z_i, z_{i+1}) for $1 \leq i < 4(r-s)+3$ can be triangulated, as vertices on the cycle contain more than two different colors (see [30], theorem 3.1.) Further, edges $(y_j, z_{j+4(r-s)-1})$ for all j, $2 < j \leq 4s-1$, $(y_j, z_{j+4(r-s)})$ for all j, $2 < j < 4s-1$, and $z_{j+4(r-s)+1})$ for all j, $2 \leq j < 4s-1$. The graph now looks as depicted in Figure 6.

One can easily verify that this graph is triangulated. There are no edges between vertices of same color. ∎

Suppose $s > r$. Let $G = G_r \oplus H_s$. Suppose we have a tree-decomposition $(X_i \mid i \in I\}, T = (I, F))$ G, with for all $i \in I$, for all $v, w \in V$, if $v \neq w$ and $v, w \in X_i$ then $c(v) \neq c(w)$. Let be the triangulated graph $(V, \{(v,w) \mid \exists i,\ v, w \in X_i,\ v \neq w\})$. (See Proposition 1.) From position 2, it follows that there exists an $i_0 \in I$ with $X_{i_0} = \{y_1, y_2, z_1, z_2\}$, and an $i_1 \in I$ with

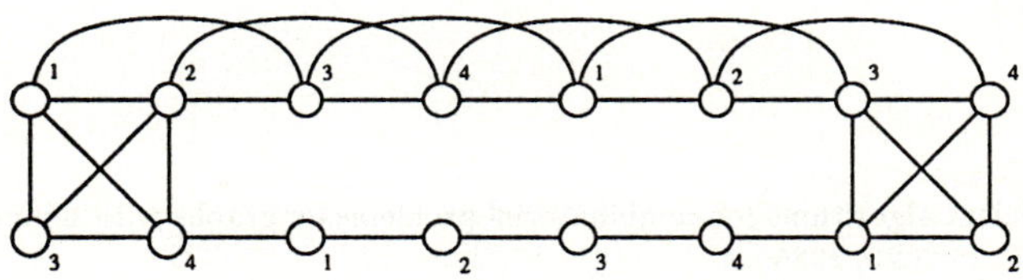

Figure 5: $G_2 \oplus H_2$

$X_{i_1} = \{y_{4s-1}, y_{4s}, z_{4r-1}, z_{4r}\}$. By Proposition 3, we may assume that i_0 and i_1 are leaves fron
Write $Y = \{y_1, y_2, \ldots, y_{4s}\}$, and $Z = \{z_1, z_2, \ldots, z_{4r}\}$.

Note from Proposition 4, that every node i_2 on the path in T between i_0 and i_1 contain least one vertex in Z, and hence at most three vertices in Y.

Claim 1 *For every j, $1 \leq j \leq 4s-2$, there exists a node $i_2 \in I$ on the path between i_0 and i T, with $\{y_j, y_{j+1}, y_{j+2}\} \subseteq X_{i_2}$.*

Proof: Suppose the claim does not hold for certain j, $1 \leq j \leq 4s-2$. There exists a node i_3 with $\{y_j, y_{j+1}, y_{j+2}\} \subseteq X_{i_3}$. By assumption, i_3 lies not on the path from i_0 to i_1 in T. Let i_4 the unique node that lies on each of the paths between i_0 and i_1, i_0 and i_2, and i_1 and i_2. T must exist a vertex $y_{j'} \notin X_{i_4}$, with $j' \in \{j, j+1, j+2\}$. Note that there exist four paths in G from $\{y_j, y_{j+1}, y_{j+2}\}$ to $\{y_1, y_2, y_{4s-1}, y_{4s}\}$ that are vertex disjoint, except that two paths share vertex $y_{j'}$. Now, by Proposition 4, X_{i_4} contains at least one vertex of each path, and, as $y_{j'} \notin$. we have that $|X_{i_4} \cap Y| \geq 4$, contradiction. ∎

Note that for such i_2 on the path between i_0 and i_1, with $X_{i_2} \supseteq \{y_j, y_{j+1}, y_{j+2}\}$, there mus a $z_{j'} \in X_{i_2}$.

Claim 2 *Suppose i_2, i_3 lie on the path between i_0 and i_1 in T, and $X_{i_2} = \{y_{4\alpha+1}, y_{4\alpha+2}, y_{4\alpha+3}, z$ $X_{i_3} = \{y_{4\beta+1}, y_{4\beta+2}, y_{4\beta+3}, z_{j_2}\}$, $0 \leq \alpha, \beta < s$, $\alpha \neq \beta$. Then $c(z_{j_1}) = c(z_{j_2}) = 4$, and $j_1 \neq j_2$.*

Proof: By case analysis. Omitted from this extended abstract. ∎

It follows that there must be at least s different vertices z_j with $c(z_j) = 4$. So $G_r \oplus H_s$ can o be c-triangulatable, when $s \leq r$. Hence we have the following theorem.

Lemma 4 *$G_r \oplus H_s$ is c-triangulatable, if and only if $s \leq r$.*

It follows that every graph G_r must be in a different equivalence class, and hence TCG with f colors and 4-Perfect Phylogeny are not cutset-regular, and hence, by theorem 2 not finite-st Clearly, the same results also hold for a larger number of colors or characteristics.

Theorem 3 *For every $k \geq 4$, k-Perfect Phylogeny, and Triangulating Colored Graphs with k co are not finite state for bounded treewidth.*

With a slightly more complex, but further more or less similar construction one can s that the number of equivalence classes can be exponential in the number of vertices of the gra involved. From this, it follows that not only the problem is not only not finite state, but also t no other linear time table based approach is possible. (For instance, consider Independent Set. the size of independent sets is a parameter that can be $O(n)$ large, it is not finite state. Howe there exists a 'table based' linear time algorithm for the problem, when restricted to graphs, gi with a tree-decomposition of constant bounded treewidth (see e.g. [1].) When this situation occ then the number of equivalence classes is still polynomial. Hence, it cannot occur for the Perf Phylogeny problem.)

References

[1] S. Arnborg. Efficient algorithms for combinatorial problems on graphs with bounded decomposabil – A survey. *BIT*, 25:2–23, 1985.

[2] S. Arnborg, D. Corneil, and A. Proskurowski. Complexity of finding embeddings in a k-tree. *SI J. Alg. Discr. Meth.*, 8:277–284, 1987.

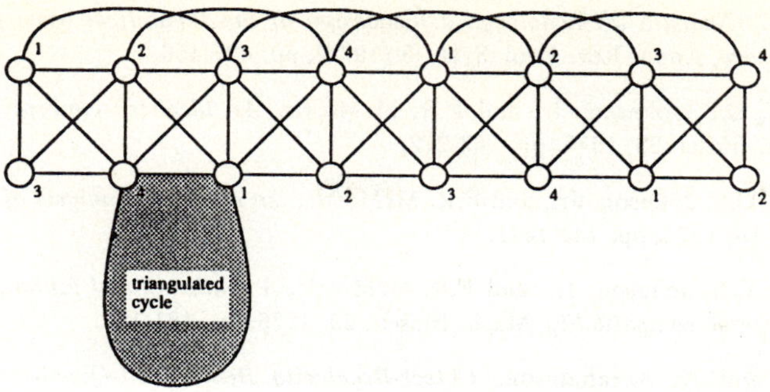

Figure 6: $G_i \oplus H_2$ triangulated

S. Arnborg, B. Courcelle, A. Proskurowski, and D. Seese. An algebraic theory of graph reduction. Technical Report 90-02, Laboratoire Bordelais de Recherche en Informatique, Bordeaux, 1990. To appear in Proceedings 4th Workshop on Graph Grammars and Their Applications to Computer Science.

S. Arnborg, J. Lagergren, and D. Seese. Easy problems for tree-decomposable graphs *J. Algorithms*, 12:308–340, 1991.

S. Arnborg and A. Proskurowski. Linear time algorithms for NP-hard problems restricted to partial k-trees. *Disc. Appl. Math.*, 23:11–24, 1989.

H. L. Bodlaender. Dynamic programming algorithms on graphs with bounded tree-width. In *Proceedings of the 15'th International Colloquium on Automata, Languages and Programming*, pages 105–119. Springer Verlag, Lecture Notes in Computer Science volume 317, 1988.

H. L. Bodlaender and T. Kloks. *A simple linear time algorithm for triangulating three-colored graphs.* Technical Report RUU-CS-91-13, Department of Computer Science, Utrecht University, the Netherlands, 1991. To appear in: Proceedings STACS'92.

H.L. Bodlaender and T. Kloks. Better algorithms for the pathwidth and treewidth of graphs. In *Proceedings 18'th International Colloquium on Automata, Languages and Programming*, pages 544–555. Springer Verlag, Lecture Notes in Computer Science volume 510, 1991.

H. L. Bodlaender and R. H. Möhring, The pathwidth and treewidth of cographs. In *Proceedings 2nd Scandinavian Workshop on Algorithm Theory*, pages 301–309. Springer Verlag Lecture Notes in Computer Science volume 447, 1990.

K. Booth and G. Lueker. Testing for the consecutive ones property, interval graphs, and graph planarity using PQ-tree algorithms. *J. Comp. Syst. Sc.*, 13:335–370, 1976.

R. B. Borie, R. G. Parker, and C. A. Tovey. Automatic generation of linear algorithms from predicate calculus descriptions of problems on recursive constructed graph families. Manuscript, 1988.

P. Buneman. A characterization of rigid circuit graphs. *Discrete Math.* 9:205–212, 1974.

J. Camin and R. Sokal, *A method for deducing branching sequences in phylogeny,* Evolution 19, (1965), pp. 311-326.

B. Courcelle. The monadic second-order logic of graphs I: Recognizable sets of finite graphs. *Information and Computation*, 85:12–75, 1990.

G. A. Dirac. On rigid circuit graphs. *Abh. Math. Sem. Univ. Hamburg*, 25: 71-76, 1961.

[16] G.F. Estabrook, *Cladistic Methodology: a discussion of the theoretical basis for the inductic evolutionary history*, Annu. Rev. Evol. Syst., 3 (1972), pp. 427-456.

[17] G.F. Estabrook, C.S. Johnson, Jr. and F.R. McMorris, *An idealized concept of the true clac character*, Math. Biosci. 23, 1975, pp. 263-272.

[18] G.F. Estabrook, C.S. Johnson, Jr., and F.R. McMorris, *An algebraic analysis of cladistic charac* Discrete Math., 16, 1976, pp. 141-147.

[19] G.F. Estabrook, C.S. Johnson, Jr., and F.R. McMorris, *A mathematical foundation for the ana of cladistic character compatibility*, Math. Biosci., 29, 1976, pp. 181-187.

[20] M. R. Fellows and K. Abrahamson, *Cutset-Regularity Beats Well-Quasi-Ordering for Bou Treewidth.* Manuscript, Nov. 1989.

[21] J. Felsenstein. Numerical methods for inferring evolutionary trees. *The Quaterly Review of Bio* Vol. 57, No. 4, Dec. 1982.

[22] W. M. Fitch and E. Margoliash. The construction of phylogenetic trees. *Science*, 155, 1967.

[23] L. R. Foulds, and R. L. Graham, The Steiner problem in phylogeny is NP-Complete. *Advance Applied Mathematics*, 3:43–49, 1982.

[24] D. R. Fulkerson and O. A. Gross. Incidence matrices and interval graphs. *Pacific J. Mathema* 15:835–855, 1965.

[25] F. Gavril. The intersection graphs of subtrees in trees are exactly the chordal graphs. *J. Combinat Theory series B*, 16:47–56, 1974.

[26] M. C. Golumbic. *Algorithmic Graph Theory and Perfect Graphs.* Academic Press, New York, 19

[27] D. Gusfield. *The Steiner tree problem in phylogeny.* Technical Report 332, Department of Comp Science, Yale University, Sept. 1984.

[28] D. Gusfield. Efficient algorithms for inferring evolutionary trees. *Networks*, 21:19–28, 1991.

[29] A. Habel. *Hyperedge Replacement: Grammars and Languages.* PhD thesis, Univ. Bremen, 1988.

[30] S. Kannan and T. Warnow. Triangulating three-colored graphs. In *Proceedings Second Annual A SIAM Symp. on Discrete Algorithms*, pages 337–343, San Francisco, Jan. 1991. Also to appea SIAM J. on Discrete Mathematics.

[31] S. Kannan and T. Warnow. Inferring evolutionary history from DNA sequences. In *Proceedings Annual Symposium on the Foundations of Computer Science*, pages 362–371, St. Louis, Misso 1990.

[32] J. Lagergren. *Algorithms and Minimal Forbidden Minors for Tree-decomposable Graphs.* PhD the Royal Institute of Technology, Stockholm, Sweden, 1991.

[33] C. Lautemann. Efficient algorithms on context-free graph languages. In *Proceedings of the International Colloquium on Automata, Languages and Programming*, pages 362–378, 1988. Sprir Verlag Lectures Notes in Computer Science volume 317.

[34] C. G. Lekkerkerker and J. Ch. Boland. Representations of a finite graph by a set of intervals on real line, *Fund. Math.* 51:45–64, 1962.

[35] W. J. LeQuesne. The uniquely evolved character concept and its cladistic application, *Syst. Zc* 23:513–517, 1974.

[36] W. J. LeQuesne. The uniquely evolved character concept. *Syst. Zool.*, 26:218–223, 1977.

[37] W.J. LeQuesne, *A method of selection of characters in numerical taxonomy*, Syst. Zool., 18, 201-205, 1969.

W.J. LeQuesne, *Further studies on the uniquely derived character concept,* Syst. Zool., 21, pp. 281-288, 1972.

W.J. LeQuesne, *The uniquely evolved character concept and its cladistic application*, Syst. Zool., 23, pp. 513-517, 1974.

W.J. LeQuesne, *Discussion of preceeding papers*, In G.F. Estabrook (ed.), *Proc. Eighth International Conference on Numerical Taxonomy,*, pp. 416-429. W.H. Freeman, San Francisco, 1975.

W.J. LeQuesne, *The uniquely evolved character concept*, Syst. Zool., 26, pp. 218-223, 1977.

F. R. McMorris. Compatibility criteria for cladistic and qualitative taxonomic characters. In *Proceedings 8th Internatinal Conference on Numerical Taxonomy*, G.F. Estrabrook, ed., pp. 339-415. W.H. Freeman, San Francisco, 1975.

F. R. McMorris. On the compatibility of binary qualitative taxonomic characters. *Bull. Math. Biol.*, 39:133-138, 1977.

F. R. McMorris and C. A. Meacham. Partition intersection graphs. *Ars Combinatorica*, 16-B:135-138, 1983.

F. R. McMorris, T. Warnow, and T. Wimer. *Triangulating colored graphs.* Submitted to Information Processing Letters.

C. A. Meacham and G. F. Estabrook. Compatibility methods in systematics. *Annual Review of Ecology and Systematics*, 16:431–446, 1985.

C. A. Meacham. Evaluating characters by character compatibility analysis. In: T. Duncan and T. F. Stuessy (eds.), *Cladistics: Perspectives on the estimation of evolutionary history*, pp. 152–165. Columbia Univ. Press: New York, 1984.

C. A. Meacham. Theoretical and computational considerations of the compatibility of qualitative taxonomic characters. In: J. Felsenstein (ed.), *Numerical Taxonomy*, pages 304–314. NATO ASI Series, volume G1. Springer-Verlag: Berlin, Heidelberg, 1983.

A. Proskurowski. Separating Subgraphs in k-trees: Cables and Caterpillars. *Discrete Math.*, 49:275–285, 1984.

B. Reed. Finding approximate separators and computing treewidth quickly. Manuscript, 1992. To appear in: Proceedings of the 24'th Annual Symposium on Theory of Computing STOC'92.

N. Robertson and P. D. Seymour. *Graph minors XIII: The disjoint path problem.* Manuscript, September 1986.

D. J. Rose. Triangulated graphs and the elimination process. *J. Math. Anal. Appl.*, 32:597–609, 1970.

D. J. Rose. On simple characterization of k-trees. *Discrete Math.*, 7:317–322, 1974.

P. Scheffler. Linear-time algorithms for NP-complete problems restricted to partial k-trees. Report R-MATH-03/87, Karl-Weierstrass-Institut Für Mathematik, Berlin, GDR, 1987.

] R. R. Sokal and P. H. A. Sneath. *Principles of Numerical Taxonomy.* W.H. Freeman, San Francisco, 1963.

] R. E. Tarjan. *Data Structures and Network Algorithms.* Society for Industrial and Applied Mathematics, Philadelphia, 1983.

] J. R. Walter. *Representations of Rigid Circuit Graphs.* Ph.D. thesis, Wayne State University.

] E. O. Wilson. A Consistency Test for Phylogenies Based upon Contemporaneous Species. *Systematic Zoology*, 14:214–220.

] T. V. Wimer. *Linear algorithms on k-terminal graphs.* PhD thesis, Dept. of Computer Science, Clemson University, 1987.

Disjunctive Systems and L-Domains

Guo-Qiang Zhang
Artificial Intelligence Laboratory
Department of EECS,
The University of Michigan
Ann Arbor, MI 48109

Abstract. Disjunctive systems are a representation of L-domains. They use sequents of the form $X \vdash Y$, with X finite and Y pairwise disjoint. We show that for any disjunctive system, its elements ordered by inclusion form an L-domain. On the other hand, via the notion of stable neighborhoods, every L-domain can be represented as a disjunctive system. More generally, we have a categorical equivalence between the category of disjunctive systems and the category of L-domains. A natural classification of domains is obtained in terms of the style of the entailment: when $|X| = 2$ and $|Y| = 0$ disjunctive systems determine coherent spaces; when $|Y| \leq 1$ they represent Scott domains; when either $|X| = 1$ or $|Y| = 0$ the associated cpos are distributive Scott domains; and finally, without any restriction, disjunctive systems give rise to L-domains.

1 Introduction

Discovered by Coquand [Co90] and Jung [Ju90] independently, L-domains form one of the maximal cartesian closed categories of algebraic cpos. Together with the work of Smyth [Sm83] which shows that the SFP domains of Plotkin [Pl76] is the largest cartesian closed category inside the ω-algebraic cpos, we have a better picture of good categories of domains that may be used in denotational semantics of programming languages.

The primary contribution of this paper is to give a representation of L-domains, based concretely on sets and relations. These concrete structures, which are called *disjunctive systems*, provide a general framework encompassing coherent spaces, Scott domains, and bifinite L-domains. Our representation follows the idea of Scott [Sc82] in his work on information systems, and hence it inherits the benefits of these structures.

Previous work has shown that Scott domains, SFP domains, and stable domains can all be represented by structures similar to information systems ([LaWi84], [Sc82], [Zh90b], [Zh89]). The topic of this paper comes very naturally along this line of research. It adds to our belief that *good categories of domains all have nice concrete representations.*

A disjunctive system is a structure with sequents of the form $X \vdash Y$ on propositions (tokens) about computation. These sequents generalize the usual notion of sequents due to Gentzen: Although X is required to be finite (and non-empty), Y need not be so. A sequent $X \vdash Y$ should read 'conjunction of X's entails disjunction of Y's'. The name 'disjunctive' originates from the work of Johnstone [Jo77]. It reflects the condition that whenever we write $X \vdash Y$, Y must be pairwise disjoint, in the sense that two distinct propositions in Y always contradict each other. In terms of sequents, this is expressed by

$$\forall a, b \in Y.\, a \neq b \Longrightarrow \{a, b\} \vdash \emptyset.$$

Perhaps *exclusive-or* is the most familiar example of 'disjunctiveness'. However, situations like this arise frequently in computation. For example, although a storage can hold different data, it can only store one datum at a time. Essentially, anything which has to do with resources involves the notion of disjunctiveness. Suppose a candy bar, a pack of cigarette, and a cup of coffee cost one dollar each. With one dollar, you can buy a candy bar, a pack of cigarette, or a cup of coffee. However, you can only buy one of them.

The notion of disjunctiveness has appeared already in different areas of theoretical study. It was considered in [Die76] and [Jo77] in the context of category theory. When looking for a topological characterization of stable functions of Berry [Be78], one naturally arrives at stable neighborhoods [Zh], [Zh90a], those open sets which have the disjoint property (see Section 2).

Our purpose in considering disjunctive systems is to study concrete representation of L-domains and its consequences. The concept of *elements* serves as a bridge between disjunctive systems and domains. An element of a disjunctive system is a set x of propositions such that whenever $X \vdash Y$ for some $X \subseteq x$, $Y \cap x$ is non-empty. Ordered by set inclusion these elements form a cpo, with $\emptyset$ the bottom. We are going to show that disjunctive systems represent L-domains, those algebraic cpos in which every principal ideal $\downarrow x$ is a complete lattice. This main result is expressed as the following two theorems, which is generalized to the equivalence of the related categories in Section 6.

Theorem 1.1 *For a disjunctive system $\underline{A}$, the set of elements $|\underline{A}|$ ordered by inclusion is an L-domain.*

Theorem 1.2 *Every L-domain can be represented by a disjunctive system.*

To prove the first theorem a characterization of finite elements is needed. They turn out to be those which are generated by a *finite* set of tokens relative to a given element. The disjoint nature of the sequents makes it true that any *relative intersection* of a collection of compatible elements is again an element. Therefore, the greatest lower bound of any collection of elements exists within every principal ideal. In fact this is the construction with which we get the finite elements, as well as the least upper bounds of compatible sets.

To show that every L-domain can be represented by a disjunctive system, we use the notion of stable neighborhoods. Stable neighborhoods were introduced in [Zh89a] to give a topological characterization of stable functions. They are those Scott open sets whose minimal points are pairwise incompatible. The disjoint property o stable neighborhoods makes them natural candidates for propositions of disjunctiv systems. We show that there is an order preserving, 1-1 correspondence, betwee an L-domain and the disjunctive system constructed from its stable neighborhood.

Since disjunctive systems are based on a limited use of sets, they provide concrete approach to domains. What is more important is, however, that the suggest an interesting classification of domains. Call a cpo an $[m, n]$-domain if i is determined by a disjunctive system whose sequents are of the form $X \vdash Y$ wit $|X| \leq m$ *and* $|Y| \leq n$. Similarly, call a cpo an (m, n)-domain if it is determine by a disjunctive system whose sequents are of the form $X \vdash Y$ with $|X| < m$ o $|Y| < n$. Note that to be able to express disjointness, X should be allowed to hav at least two elements. Note also that for technical advantage, X is required to b non-empty. For convenience let $*$ stand for *no restriction*.

By this classification, coherent spaces [Gi87] are exactly the [2, 0]-domains. Othe existing domains can also be classified in this way. We summarize the classificatio by the following table.

Parameters	**Domain Types**
[2, 0]	Coherent Spaces
[*, 1]	Scott Domains
(2, 1)	Distributive Scott Domains
(*, *)	L-Domains

On the other hand, the classification of domains in this way brings forwar an extremely rich family of domains. Many interesting issues arise. For example What are the [3, 2]-domains? Do they form a cartesian closed category? Note these are follow-up questions which we do not attempt to settle here. Whether this classification generates any interesting category of domains or not remains to be seen.

The general notion of sequent structures and non-deterministic information systems were introduced in [Zh] and [DG90], respectively. The purpose of [Zh] was to give a representation of SFP domains (see also [Zh90b]), and the main purpose of [DG90] was to give a characterization of partial orders representable by non-deterministic information systems. This paper uses a different structure which is not contained in sequent structures or non-deterministic information systems: instead of having $\vdash$ as a relation between *finite subsets* of tokens, we need it to be a relation that relates a finite subset to *any* pairwise disjoint subset. As mentioned at the end, whether this need is absolutely necessary is unknown at the time this paper is finished.

2 L-Domains and Stable Neighborhoods

We assume the reader's familiarity with some basic definitions of domain theory, such as complete partial orders (cpos), finite elements, Scott open sets, algebraic cpos, etc. With respect to a cpo D, we write $\uparrow T$ for the upper closed set (upper set)

$$\{d \in D \mid d \sqsupseteq t \text{ for some } t \in T\},$$

here T is a subset of D. Similarly $\downarrow T$ stands for the set

$$\{d \in D \mid d \sqsubseteq t \text{ for some } t \in T\}.$$

cases where $T = \{x\}$, a singleton set, we just write $\uparrow x$ or $\downarrow x$.

The notion of minimal upper bounds are quite relevant here, so we recall some elated definitions. Let T be a subset of a cpo D. An element d is said to be minimal upper bound of T if d dominates every element in T, and there is no ement strictly below d with this property. The set of minimal upper bounds of T written as $\bowtie T$. We call $\bowtie T$ complete if whenever d is an upper bound of T, d already bigger than some element of $\bowtie T$. A cpo D is said to have *property m* if very subset has a complete set of minimal upper bounds. For a given set T of a po, we define

$$\begin{aligned}
&\mathcal{U}^0(T) = T,\\
&\mathcal{U}^{n+1}(T) = \{x \mid x \in \bowtie S \text{ for some finite subset } S \subseteq \mathcal{U}^n(T)\},\\
&\mathcal{U}^\infty(T) = \textstyle\bigcup_{n \geq 0} \mathcal{U}^n.
\end{aligned}$$

Let D be an algebraic cpo (the non-algebraic ones do not seem to be interesting). t is called an L-domain if for every x in D, the principal ideal $\downarrow x$ is a complete attice. The following characterization of L-domains is useful.

Theorem 2.1 *(Jung) An algebraic cpo D is an L-domain if and only if D has roperty m and $\mathcal{U}^\infty = \mathcal{U}^1$.*

When we show that every L-domain can be represented as a disjunctive system ater in Section 4, stable neighborhoods are used.

Definition 2.1 *Let D be a cpo. u is a stable neighborhood of D if it is a Scott-open set of D such that $u = \uparrow C$, where C is a collection of finite elements of D with the property*

$$(c_1, c_2 \in C \;\&\; \uparrow c_1 \cap \uparrow c_2 \neq \emptyset) \Longrightarrow c_1 = c_2.$$

We write the set of stable neighborhoods of a cpo D as $\mathbf{SN}(D)$. In general, $\mathbf{SN}(D)$ does not necessarily form a topology. It is not closed under finite intersections, neither arbitrary unions. However, stable neighborhoods are closed under *disjoint unions.*

Definition 2.2 *Let D be a cpo. The set of minimal points of a Scott open set u, written as μu, consists of $m \in u$ such that $\forall x \sqsubseteq m.\, x \in u \Longrightarrow x = m$.*

Clearly minimal points of a Scott open set are finite elements. Also, a Scott open set u is a stable neighborhood if and only if μu is pairwise incompatible. As a corollary of Theorem 2.1, we have a characterization of L-domains in terms of stable neighborhoods.

Proposition 2.1 *Let D be an algebraic cpo. D is an L-domain if and only if the collection of stable neighborhoods on D is closed under finite intersections.*

3 Disjunctive Systems

Definition 3.1 *A disjunctive system is a pair*

$$\underline{A} = (A, \vdash)$$

where A is a set of tokens and $\vdash$ is a relation such that whenever $X \vdash Y$, X is non-empty finite subset of A and Y is a pairwise disjoint subset of A in the sen that

$$\forall a, b \in Y.\ a \neq b \Longrightarrow \{a, b\} \vdash \emptyset.$$

Note that the following additional axioms can be put on $\vdash$:

$$(Identity) \qquad a \vdash a,$$

$$(Weakening) \qquad \frac{X' \supseteq X \qquad X \vdash Y \qquad Y \subseteq Y' \qquad Y \in \#A}{X' \vdash Y'},$$

$$(Cut) \qquad \frac{X \vdash Y, a \qquad a, X' \vdash Y' \qquad Y \cup Y' \in \#A}{X, X' \vdash Y, Y'}.$$

Here $\#A$ is the collection of pairwise disjoint subset of A.

When these axioms are satisfied we call a disjunctive system *normal.* However, a far as the representation of domains is concerned, these axioms may not be required From the definition of elements given later, we will see that if any of the axioms i violated, the associated propositions can not appear in any element. In any case every L-domain can be represented by a normal disjunctive system, as can be seen from the proof of Theorem 1.2. Additionally, it seems neater not to require normalit when dealing with the classification of domains determined by disjunctive systems

Note that when we write $X \vdash Y$, X is always a non-empty, finite subsets o propositions, and Y pairwise disjoint. Of course Y can be infinite, or even uncount able, keeping in mind that our intended purpose is to recast various kinds of domain in a general setting, including L-domains, which may have an uncountable numbe of finite elements.

A disjunctive system determines a family of subsets of propositions called it *elements.*

Definition 3.2 *The elements $|\underline{A}|$, of a disjunctive system $\underline{A} = (A, \vdash)$ consist o subsets x of propositions which are closed under entailment:*

$$(X \subseteq x \ \&\ X \vdash Y) \Longrightarrow x \cap Y \neq \emptyset.$$

Given an element x of a disjunctive system $\underline{A}$, we have $|Y \cap x| = 1$ for any entailment $X \vdash Y$ with $X \subseteq x$. This follows from the pairwise disjoint nature of Y. Note for any disjunctive system $\underline{A}$, we have $\emptyset \in |\underline{A}|$. Although this is easy to check, we had to take some care earlier for the definition of the entailment relation. The requirement that X be non-empty in any entailment $X \vdash Y$ is crucial. Imagine if $\emptyset \vdash \emptyset$ were allowed in a disjunctive system $\underline{A}$. According to the definition of an element of a disjunctive system, $|\underline{A}|$ would be empty, contradicting the expectation that $(|\underline{A}|, \subseteq)$ be a cpo. This is not surprising because logically speaking, $\emptyset \vdash \emptyset$ means **true** $\vdash$ **false**, resulting in an inconsistent structure.

Noting that for disjunctive systems the least upper bound of a directed set is the union of elements in that set, we have

Proposition 3.1 *For a disjunctive system $\underline{A}$, $(|\underline{A}|, \subseteq)$ is a complete partial order.*

The following proposition is the key to the construction of finite elements and the proof of the representation theorems.

Proposition 3.2 *Let $\underline{A}$ be a disjunctive system, and let T be a compatible subset of elements of $\underline{A}$. Then $\bigcap T$ is an element of $\underline{A}$.*

Proof Since T is compatible, there is some element x of $\underline{A}$ such that $\forall y \in T.\ y \subseteq x$. Let $X \vdash Y$ be an entailment in $\underline{A}$ such that $X \subseteq \bigcap T$. We have $X \subseteq y$ for every y in T. But such y's are elements of $\underline{A}$. Therefore $Y \cap y = Y \cap x \neq \emptyset$ for all y in T. This implies $Y \cap \bigcap T \neq \emptyset$.

□

Suppose P is any subset of an element x. According to the previous proposition,

$$\bigcap\{y \mid P \subseteq y \subseteq x \ \&\ y \in |\underline{A}|\}$$

is an element of $\underline{A}$. We write $\lceil P \rceil_x$ for such an element. Elements of the form $\lceil P \rceil_x$ with P finite are of special interest to us. They form the basis of a cpo $(|\underline{A}|, \subseteq)$.

Proposition 3.3 *Given a disjunctive system $\underline{A}$, $\lceil P \rceil_x$ is a finite element of $(|\underline{A}|, \subseteq)$ for any finite subset P of an element $x \in |\underline{A}|$.*

Proof Let S be a directed subset in the cpo $(|\underline{A}|, \subseteq)$, and let $\lceil P \rceil_x \subseteq \bigcup S$. (Note that $\bigsqcup S = \bigcup S$ when S is directed.) Clearly $\lceil P \rceil_x = \lceil P \rceil_{\bigcup S}$. Since $P \subseteq \bigcup S$, $P \subseteq z$ for some $z \in S$. Therefore $z \supseteq \lceil P \rceil_{\bigcup S} = \lceil P \rceil_x$.

□

4 Representation Theorems

In this section we present the proofs of the two representation theorems mentioned in the introduction.

Proof of Theorem 1.1: Given a disjunctive system $\underline{A}$, and P, Q, subsets of an element x of the disjunctive system, we have $\lceil P \rceil_x \sqcup \lceil Q \rceil_x = \lceil P \cup Q \rceil_x$. It is then clear that $x = \bigcup\{\lceil P \rceil_x \mid P$ is a finite subset of $x\}$, with $\{\lceil P \rceil_x \mid P$ is a finite subset of $x\}$ directed. Hence $(|\underline{A}|, \subseteq)$ is algebraic. It is an L-domain because by Proposition 3.2, any subset S of a principle ideal $\downarrow x$ has a greatest lower bound, which is the intersection of all elements in S. Again by Proposition 3.2, the least upper bound of any subset S is the element

$$\bigcap\{y \mid \bigcup S \subseteq y \subseteq x \ \&\ y \in |\underline{A}|\}.$$

Hence $\downarrow x$ is a complete lattice.

□

Given an L-domain D, we can associate it with a disjunctive system $\mathcal{I}(D)$ $(A, \vdash)$. Here A is the collection of stable neighborhoods of D excluding D, and $\vdash$ given by

$$X \vdash Y \qquad \text{if} \qquad \bigcap X \subseteq \bigcup Y,$$

with X finite and Y pairwise disjoint (in set theoretic terms).

We now show that every L-domain can be represented as a disjunctive system

Proof of Theorem 1.2: Let D be an L-domain. It is enough to prove th $D \cong |\mathcal{I}(D)|$. Define a mapping $\varphi: (D, \sqsubseteq) \to (|\underline{A}|, \subseteq)$ by letting

$$\varphi(d) = \{u \mid u \in \mathbf{SN}(D) \setminus \{D\} \ \&\ d \in u\}.$$

We show that φ is an order preserving isomorphism between $(D, \sqsubseteq)$ and $(|\underline{A}|, \subseteq)$

The non-trivial part is to show that φ is onto. Let $x \in |\mathcal{I}(D)|$ be a non-botto (otherwise $\varphi(\bot) = x$). We show that $\bigcap x = \uparrow t$ for some $t \in D$. For any $u \in x$, w have $\{u\} \vdash \{\uparrow p \mid p \in \mu u\}$. Therefore, there is some $p_0 \in \mu u$ such that $\uparrow p_0 \in x$, sinc x is an element of $\mathcal{I}(D)$. It follows that

$$\bigcap x = \bigcap \{\uparrow p \mid \uparrow p \in x\}.$$

However, $\{p \mid \uparrow p \in x\}$ is a directed set. To see this, let $p_0, p_1 \in \{p \mid \uparrow p \in x\}$ The fact that D is an L-domain implies that $\bowtie\{p_0, p_1\}$ is a complete set of minim upper bounds of $\{p_0, p_1\}$. Moreover, $\uparrow\bowtie\{p_0, p_1\}$ is again a stable neighborhood, b Proposition 2.1. Hence

$$\{\uparrow p_0, \uparrow p_1\} \vdash \{\uparrow p \mid p \in \bowtie\{p_0, p_1\}\}.$$

It follows that $\uparrow p \in x$ for some $p \in \bowtie\{p_0, p_1\}$. This p is an upper bound of p_0 an p_1.

Therefore $\bigcap x = \uparrow t$, where $t = \bigsqcup\{p \mid \uparrow p \in x\}$. From this we know that for an $u \in \mathbf{SN}(D)$ containing t, u must belong to x. In summary, all the above shows tha $x = \varphi(t)$ for some $t \in D$.

□

5 Classification of Domains

A disjunctive system is called $[m, n]$ if all the sequents involved are of the forn $X \vdash Y$ with $|X| \leq m$ *and* $|Y| \leq n$. It is (m, n) if the sequents are of the forn $X \vdash Y$ with $|X| < m$ *or* $|Y| < n$. The corresponding cpos are called $[m, n]$-domain or (m, n)-domains, respectively. To explain the notation better, note that $[m, n$ suggests a 'closed interval', and (m, n) an 'open interval'.

Consider a disjunctive systems $(A, \vdash)$, where all the entailments $X \vdash Y$ are such that $|X| = 2$ and $|Y| = 0$. Clearly $x \in |\underline{A}|$ if and only if

$$\forall a, b \in x. \{a, b\} \not\vdash \emptyset.$$

Therefore, the elements of such systems have the property that

$$(x \in |\underline{A}| \ \&\ y \subseteq x) \Longrightarrow y \in |\underline{A}|.$$

ence every such a disjunctive system represents a coherent space. Conversely, it an be easily seen that every coherent space can be represented by a disjunctive stem of this kind. We have, therefore,

roposition 5.1 *The* [2, 0]*-domains are exactly coherent spaces.*

Implied by the results of Scott [Sc82], we have

roposition 5.2 *The* [$*$, 1]*-domains are exactly Scott domains.*

Suppose $(A, \vdash)$ is a disjunctive system with the entailments of the form $X \vdash Y$ ich that either $|X| = 1$ or $|Y| = 0$. What can be said about the cpo it represents? is a distributive Scott domain. Let T be a compatible subset of $|\underline{A}|$. It can be een that the union $\bigcup T$ must also be an element. Therefore the elements ordered nder inclusion is a Scott domain. It is distributive because in this case the greatest ower bound of any collection of elements is just the intersection of these elements.

Further more, any distributive Scott domain can be represented by a disjunctive ystem of this kind, even with the condition $|Y| \leq 1$ holding for all entailments $X \vdash Y$. The proof of this claim can be derived from [Zh89]. This leads to the ollowing proposition.

roposition 5.3 *The* (2, 1)*-domains are exactly distributive Scott domains.*

Continuing in this line, we only need that the following relation $\prec$ be well-founded or a disjunctive system $(A, \vdash)$ to represent a dI-domain. For $a, b \in A$, $a \prec b$ if $a \neq b$ nd for some Y containing b, $a \vdash Y$. The well-foundedness of $\prec$, however, does not eem to be expressible in terms of the types of disjunctive systems as it stands: I uspect dI-domains are none of the $[m, n]$-domains or (m, n)-domains.

6 Categorical Equivalence

This section introduces approximable mappings between disjunctive systems. It then shows that the category of disjunctive systems with approximable mappings is equivalent to the category of L-domains.

For technical convenience we work with a particular kind of disjunctive systems — those which are normal (see Section 3) and *expressive*.

Definition 6.1 *A normal disjunctive system* $(A, \vdash)$ *is called expressive if*

$$\{a\} \vdash Y \Longrightarrow \exists b \in Y.\ \{a\} \vdash \{b\},$$

and for every finite subset $X \subseteq A$, *there is some* $Y \subseteq A$ *such that*

$$X \vdash Y \ \&\ \forall b \in Y.\ [\forall a \in X.\ \{b\} \vdash \{a\}].$$

We abbreviate the above statement as $X \rightharpoonup Y$ *(note* $\rightharpoonup$ *is not symmetric).*

Note that from Proposition 2.1 and the proof of Theorem 1.2 we know that the expressiveness condition will not pose a restriction on the expressive power of disjunctive systems. This will be confirmed again later in the categorical equivalence theorem.

We now introduce morphisms on expressive disjunctive systems called approximable mappings. This makes expressive disjunctive systems a category. Approximable mappings show how disjunctive systems are related to one another and they correspond to continuous functions between the associated L-domains.

Definition 6.2 *Let* $\underline{A} = (A, \vdash_{\underline{A}})$, $\underline{B} = (B, \vdash_{\underline{B}})$ *be expressive disjunctive systems. An approximable mapping from* $\underline{A}$ *to* $\underline{B}$ *is a relation* $R \subseteq A \times B$ *which satisfies*

$$\forall S \subseteq^{fin} R \quad \forall X \forall Y. \quad (\pi_1 S \rightharpoonup_{\underline{A}} X \ \& \ \pi_2 S \rightharpoonup_{\underline{B}} Y) \Longrightarrow \forall a \in X \exists b \in Y.\, a\, R\, b.$$

Here π_1 and π_2 are projections to the first and the second component, respectively, and $\subseteq^{fin}$ stands for 'a finite subset of'.

Clearly, when $a \vdash a'$ we have $a' \rightharpoonup \{a, a'\}$. Taking $\{(a', b')\} \subseteq R$, $X = \{a, a'\}$, $Y = \{b'\}$ in the previous definition, we have, for an approximable mapping R,

$$a \vdash a' \ \& \ a'\, R\, b' \Longrightarrow a\, R\, b'.$$

Proposition 6.1 *Expressive disjunctive systems with approximable mappings form a category, written as* **EDIS**.

Proof We check that approximable mappings compose. Other axioms for a category can be checked similarly. Let $\underline{A}$, $\underline{B}$ and $\underline{C}$ be expressive disjunctive systems and $R : \underline{A} \to \underline{B}$ and $S : \underline{B} \to \underline{C}$ be approximable mappings. Let $R \circ S$ be the relational composition. We show that $R \circ S$ is an approximable mapping. Suppose for a finite set I, $\forall i \in I.\ a_i\,(R \circ S)\, c_i$ and $\{a_i \mid i \in I\} \rightharpoonup_{\underline{A}} X$, $\{c_i \mid i \in I\} \rightharpoonup_{\underline{C}} Z$. There exists a $u_i \in B$ such that $a_i\, R\, u_i$, $u_i\, S\, c_i$ for any $i \in I$. Let $\{u_i \mid i \in I\} \rightharpoonup_{\underline{B}} Y$. The existence of such Y follows from the expressiveness of $\underline{B}$. Since R is an approximable mapping, $\forall p \in X \exists q \in Y.\ p\, R\, q$. But for each $q \in Y$ we have $q\, S\, c_i$ for all $i \in I$. Therefore there exists some $r \in Z$ such that $q\, S\, r$, since $q \rightharpoonup q$. Hence $\forall p \in X \exists r \in Z.\ p\,(R \circ S)\, r$.

□

Approximable mappings determine continuous functions via the construction given in the following definition.

Proposition 6.2 *Let* R *be an approximable mapping from* $\underline{A}$ *to* $\underline{B}$. *Define* $f_R : |\underline{A}| \to |\underline{B}|$ *by*

$$f_R(x) = \{ b \in B \mid \exists a \in x \,.\, a\, R\, b \}.$$

Then f_R *is a continuous function from* $|\underline{A}|$ *to* $|\underline{B}|$.

Proof The only difficulty is to show that f_R is well-defined. Let $x \in |\underline{A}|$ and let $R : \underline{A} \to \underline{B}$ be an approximable mapping. To show $f_R(x) \in |\underline{B}|$ let $Y \subseteq^{fin} f_R(x)$ and $Y \vdash_{\underline{B}} Z$. For each $b \in Y$ there is some $a \in x$ such that $a\, R\, b$. Write X for such a collection of a's. Because $\underline{A}$ and $\underline{B}$ are expressive, there are X', Y' such hat $X \rightharpoonup_{\underline{A}} X'$ and $Y \rightharpoonup_{\underline{B}} Y'$. This means we have $X \subseteq x$ and $X \vdash_{\underline{A}} X'$, which mplies $X' \cap x \neq \emptyset$. Now let $u_0 \in X' \cap x$. By Definition 6.2 there is a $v_0 \in Y'$ such that $u_0 \, R \, v_0$. Thus $v_0 \in f_R(x)$. We must also have $v_0 \vdash_{\underline{B}} Z$, which implies, by expressiveness, $v_0 \vdash_{\underline{B}} c$ for some $c \in Z$. Therefore $c \in f_R(x) \cap Z$. Thus f_R is well-defined.

□

To show that the category of expressive disjunctive systems and the category of L-domains (which is written as **L**) are equivalent, we use one of MacLane's results ([Ma71], pp 91). By this result, a functor F determines an equivalence of the categories if it is full and faithful, and each L-domain D is isomorphic to $F(\underline{A})$ for some expressive disjunctive system $\underline{A}$.

Theorem 6.1 EDIS *is equivalent to* **L**.

Proof Let F : **EDIS** $\to$ **L** be the functor given by

$$F(\underline{A}) = |\underline{A}|$$
$$F(R) = f_R.$$

That each L-domain D is isomorphic to $F(\underline{A})$ for some expressive disjunctive system $\underline{A}$ is shown in Theorem 1.2. It remains to show that F is full and faithful. First we show that F is full. Let $\underline{A}$ and $\underline{B}$ be expressive disjunctive systems, and $f : F(\underline{A}) \to F(\underline{B})$ a continuous function. Define a relation $R \subseteq A \times B$ by letting $a \, R \, b$ if $b \in f(\overline{a})$. We check that this relation is an approximable mapping form $\underline{A}$ to $\underline{B}$. Let $\{ (a_i, b_i) \mid i \in I \}$ be a finite subset of R. Assume

$$\{ a_i \mid i \in I \} \rightharpoonup_{\underline{A}} X$$

and

$$\{ b_i \mid i \in I \} \rightharpoonup_{\underline{B}} Y.$$

For any $a \in X$, we have $a \vdash \{a_i\}$ for every $i \in I$. Thus we have $b_i \in f(\overline{a_i}) \subseteq f(\overline{a})$ for any $i \in I$. Now $\{ b_i \mid i \in I \} \vdash_{\underline{B}} Y$. Therefore $f(\overline{a}) \cap Y \neq \emptyset$. This means for some $b \in Y$, $b \in f(\overline{a})$, or $a \, R \, b$.

We now show that the continuous function f_R determined by the above R is actually equal to f. Let $x \in |\underline{A}|$. Suppose $b \in f_R(x)$. By definition there is some $a \in x$, $a \, R \, b$. That is, $b \in f(\overline{a})$. Therefore $b \in f(x)$, by the monotonicity of f. Thus $f_R(x) \subseteq f(x)$. On the other hand, let $b \in f(x)$. By the continuity of f there is some $a \in x$ such that $b \in f(\overline{a})$. Hence $a \, R \, b$ and $b \in f_R(x)$. This means $f(x) \subseteq f_R(x)$. Hence $f = f_R$.

Secondly, we show that F is faithful. Suppose $R, S : \underline{A} \to \underline{B}$ are approximable mappings such that $f_R = f_S$. Let $a \, R \, b$. Then $b \in f_S(\overline{a})$. This means for some $a' \in \overline{a}$, $a' \, S \, b$, which implies $a \, S \, b$. Therefore, $R \subseteq S$. By symmetry, $S \subseteq R$ and hence $R = S$.

□

7 Future Work

There are many interesting issues to be explored about disjunctive systems.

A sub-system relation can be introduced to disjunctive systems using the method presented in [LaWi84], so that recursive equations can be solved by using fixed-point constructions. Technically, we do not anticipate any difficulty for such a treatment. The question is, however: what kind of equations do we have? The equations usually involve constructions on disjunctive systems. However, we have not dealt with constructions such as function space on the category of expressive disjunctive systems. It is not clear to the author at this moment how best function space can be introduced (that is the reason we have not included a sub-system relation).

Secondly, there is the question of whether $(*, *)$-domains and $(*, \infty)$-domains are the same (i.e., L-domains), in other words: whether the infinity of Y in the disjunctive entailment $X \vdash Y$ is absolutely necessary. On the surface, it looks like $(*, \infty)$-domains should be the bifinite L-domains. Clearly, every bifinite L-domain can be represented by a $(*, \infty)$-disjunctive system. The other way round, however, is not true: $(*, \infty)$-disjunctive systems can represent more than just the bifinite L-domains (an example can be found in [DG90]). One can even have an uncountable set of minimal upper bounds for a finite set of finite elements in a $(*, \infty)$-domain. Given *any* non-empty set Q, one can construct the following $(*, \infty)$-disjunctive system. Let the token set be

$$\{a, b\} \cup Q \times \{0\} \cup Q \times \{1\}.$$

The entailment relation is specified by

$$\{a, b\} \vdash \{(q, 0), (q, 1)\} \text{ for all } q \in Q,$$
$$\{(q, 0), (q, 1)\} \vdash \emptyset \text{ for all } q \in Q.$$

There is a 1-1 correspondence between functions

$$f : Q \to \{0, 1\}$$

and elements

$$\{a, b\} \cup \{(q, f(q)) \mid q \in Q\}.$$

Therefore, elements $\{a\}$, $\{b\}$ have a set of minimal upper bounds with cardinality the same as that of the power set of Q.

All this suggests that maybe $(*, \infty)$-domains are L-domains. We have no proof of this at the moment: the proof of Theorem 1.2 cannot be easily adapted to the new situation. It seems some intrinsic construction must be used to get a $(*, \infty)$-disjunctive system for an L-domain. However, it would be even more interesting if one could find an example of an L-domain which is not $(*, \infty)$. What, then, are the $(*, \infty)$-domains?

References

[Be78] Berry, G., Stable models of typed λ-calculi, *Lecture Notes in Computer Science* 62 (1978).

[Die76] Diers, Y, *Catégories Localisables*, thèse de doctorat d'éetat, Paris VI (1976).

[DG90] Droste, M., and Gobel, R., Non-deterministic information systems and their domains, *Theoretical Computer Science*, 75 (1990).

[Co89] Coquand, T., Categories of embeddings. Proceedings of the third annual symposium on logic in computer science, (1988)

[Gi87] Girard, J.-Y., Linear Logic, *Theoretical Computer Science* 50 (1987).

[Gu85] Gunter, C., *Profinite solutions for recursive domain equations*, PhD thesis, Department of Computer Science, Carnegie-Mellon University, (1985).

[Jo77] Johnstone, P.T., A syntactic approach to Diers' localizable categories, *Lecture Notes in Mathematics* 753 (1977).

[Jo82] Johnstone, P.T., *Stone spaces*, Cambridge University Press (1982).

[Ju90] Jung, A., Cartesian closed categories of algebraic cpos, *Theoretical Computer Science* 70 (1990)

[LaWi84] Larsen, K.G., Winskel, G., Using information systems to solve recursive domain equations effectively, *Lecture Notes in Computer Science* 173 (1984).

[Pl76] Plotkin, G., A powerdomain construction, *SIAM J. Computing* 5 (1976).

[Sc82] Scott, D. S., Domains for denotational semantics, *Lecture Notes in Computer Science* 140 (1982).

[Sm83] Smyth, M.B., The largest cartesian closed category of domains, *Theoretical Computer Science* 27, (1983).

[Zh] Zhang, G.Q., *Logic of Domains*, Birkhaueser, Boston (1991).

[Zh89] Zhang, G.Q., DI-domains as information systems, ICALP-1989, Italy. Revised version to appear in *Information and Computation* (1989).

[Zh90a] Zhang, G.Q., Stable neighborhoods, *Theoretical Computer Science* vol 93 (1992).

[Zh90b] Zhang, G.Q., A representation of SFP, to appear in *Information and Computation* (1990).

[Zh91] Zhang, G.Q., A monoidal closed category of event structures, proceedings of the 7-th conference on Mathematical Foundations of Programming Semantics, Pittsburgh, (1991).

Optimal Parallel Algorithms for Periods, Palindromes and Squares (Extended Abstract)

Alberto Apostolico*
Purdue University and
Università di Padova

Dany Breslauer†‡
Columbia University

Zvi Galil‡
Columbia University and
Tel-Aviv University

Summary of results

Optimal concurrent-read concurrent-write parallel algorithms for two problems are presented:

- Finding all the periods of a string. The period of a string can be computed by previous efficient parallel algorithms only if it is shorter than half of the length of the string. Our new algorithm computes all the periods in optimal $O(\log\log n)$ time, even if they are longer. The algorithm can be used to compute all initial palindromes of a string within the same bounds.
- Testing if a string is square-free. We present an optimal $O(\log\log n)$ time algorithm for testing if a string is square-free, improving the previous bound of $O(\log n)$ given by Apostolico [1] and Crochemore and Rytter [12].

We show matching lower bounds for the optimal parallel algorithms that solve the problems above on a general alphabet. The lower bounds for testing if a string is square-free and finding all initial palindromes are derived by a modification of the lower bound for finding the period of a string [7].

*Partially supported by NSF Grant CCR-89-00305, by NIH Library of Medicine Grant R01-LM05118, by AFOSR Grant 90-0107, by NATO Grant CRG 900293 and by the National Research Council of Italy.

†Partially supported by an IBM Graduate Fellowship. Part of the work done while visiting at Università de L'Aquila, L'Aquila, Italy.

‡Partially supported by NSF Grant CCR-90-14605.

1 Introduction

We present optimal CRCW-PRAM algorithms for the problems of finding all periods of a string and testing if a string is square-free. Both solutions are the fastest possible optimal parallel algorithms for these problems over a general alphabet. The two algorithms start with many independent calls to a string matching routine which are performed in parallel and the results of the string matching problems are later combined to give an answer to the problem being solved.

A parallel algorithm is said to be *optimal* if the time-processor product, that is the total number of operations performed, is equal to that of the fastest sequential algorithm. Note that a simple algorithm can compute all periods of a string in constant time if n^2 processors are available. Another simple algorithm can test if a string is square-free using n^3 processors.

A lower bound of $\Omega(\frac{\log n}{\log\log n})$ by Beame and Hastad [4], for computing the parity of n input bits on a CRCW-PRAM with any polynomial number of processors, implies that most interesting problems would take at least that time. However, many problems on strings, including the problems solved in this paper, have trivial CRCW-PRAM algorithms that work in constant time using a polynomial number of processors. This fact suggests that an optimal parallel algorithm that is faster than that lower bound is possible. Our goal is to design fast optimal parallel algorithms.

1.1 Periods

A string $S[1..n]$ has *a period* p if $S[i] = S[i+p]$ for $i = 1 \cdots n-p$. *The period* of S is defined as its shortest period. The period of a string is computed in linear time as a step in Knuth, Morris and Pratt's sequential string matching algorithm [15] and in optimal $O(\log\log n)$ parallel time on a CRCW-PRAM as a step in Breslauer and Galil's string matching algorithm [6]. A recent lower bound by Breslauer and Galil [7] shows that the $O(\log\log n)$ bound is the best possible over a general alphabet, where only comparisons between symbols are allowed. However, Breslauer and Galil's [6] algorithm as well as an algorithm discovered by Vishkin [21] compute the period p only if $p \le \lceil \frac{n}{2} \rceil$; knowing the fact that $p > \lceil \frac{n}{2} \rceil$ is sufficient to obtain good string matching algorithms. We show that given an optimal parallel algorithm for string matching one can compute all the periods, including those which are longer than half of the length of the input string, in the same processor and time bounds of the string matching algorithm. In particular Breslauer and Galil's [6] algorithm can be used to obtain an optimal $O(\log\log n)$ time CRCW-PRAM algorithm that computes the period of a string.

A *palindrome* is a string which reads the same forward and backward. Formally, $S[1..k]$ is a palindrome if $S[i] = S[k+1-i]$ for $i = 1..k$. A string $S[1..n]$ has an initial palindrome of length k if the prefix $S[1..k]$ is a palindrome. We show how our algorithm can be used to detect all initial palindromes of a string in the same time bound. We can prove also that this is the best time bound possible for any optimal parallel algorithm that solves this problem over a general alphabet. The lower bound is obtained by a modification of a lower bound for string matching of Breslauer and Galil [7] and will be described in the full paper [8].

1.2 Squares

A nonempty string of the form xx is called a *square*. A string that does not contain any square is called *square-free*. For example, the strings aa, $abab$ and $baba$ are squares which are contained in the string $baababa$. It is trivial to show that any string of length larger than three on an alphabet of two symbols contains a square. However, there exist strings of infinite length on a three letter alphabet that are square-free as shown by Axel Thue [19, 20] at the beginning of the century. We develop an efficient parallel algorithm that tests if a string is square-free, improving to $O(\log\log n)$ the previous bound of $O(\log n)$, given by Apostolico [1] and Crochemore and Rytter [12]. A version of our algorithm which will be described in the full paper [2] can detect all squares in the same bounds. We prove also that this is the best time bound possible for an optimal parallel algorithm that solves this problem over a general alphabet.

There exist few sequential algorithms to solve this problem. Algorithms by Apostolico and Preparata [3], by Crochemore [9] and by Main and Lorentz [17] find all the squares in a string of length n in $O(n \log n)$ time. Main and Lorentz [17] also show that $O(n \log n)$ comparisons are necessary even to decide if a string is square-free. In another paper, Main and Lorentz [18] show that the latter problem of deciding whether a string is square-free can be solved in $O(n)$ time if the alphabet is finite. Crochemore [10] also gave a linear time algorithm for this problem.

In parallel, an algorithm by Crochemore and Rytter [12] can test if a string is square-free in optimal $O(\log n)$ time. This algorithm uses $O(n^2)$ space. Other algorithms by Apostolico [1] can test if a string is square-free and even detect all the squares in the same time and processor bounds using only linear auxiliary space. The algorithm for testing if a string is square-free is even more efficient in the case of a finite alphabet and achieves the same time bound of $O(\log n)$ using only $\frac{n}{\log n}$ processors. Apostolico's algorithms [1] assume that the alphabet is ordered, an assumption which is not necessary to solve this problem. All these algorithms are designed for the CRCW-PRAM computation model.

All the parallel algorithms mentioned above are optimal since the time-processor product is $O(n \log n)$ which is the best possible in the case of a general alphabet. Apostolico's [1] algorithm for testing square-freeness in case of finite alphabet is also optimal since the time-processor product is $O(n)$, the best running time of a sequential algorithm for this problem.

The algorithm described in this paper is a parallel version of the sequential algorithm of Main and Lorentz [18].

1.3 The CRCW-PRAM model

The algorithms described in this paper are for the concurrent-read concurrent write parallel random access machine model. We use the weakest version of this model called the *common CRCW-PRAM*. In this model, many processors have access to a shared memory. Concurrent read and write operations are allowed at all memory locations. In case that several processors attempt to write simultaneously at the same memory location, we assume they always attempt to write the same value.

Our algorithms use a string matching algorithm as a "black-box" to find all ocurrences of a short string in a longer string. The input to the string matching lgorithm will consist of two strings: *pattern*[1..m] and *text*[1..n] and the output a Boolean array *match*[1..n] that has a *true* value at each position where an ccurrence of the pattern starts in the text. We use the Breslauer and Galil [6] parllel string matching algorithm that takes $O(\log\log n)$ time on a $\frac{n}{\log\log n}$-processor RCW-PRAM. This algorithm is the fastest optimal parallel string matching algoithm on a general alphabet as implied by a lower bound of Breslauer and Galil [7]. a faster string matching algorithm on a finite alphabet exists, it would imply a ster algorithm for finding the periods and for testing if a string is square-free.

We use also an algorithm of Fich, Ragde and Wigderson [13] to compute the inimum of n integers between 1 and n in constant time using an n-processor RCW-PRAM. We use this algorithm, for example, to find the first occurrence of a tring in an other string. After the occurrences are computed by the string matching lgorithm mentioned above, we look for the smallest i such that $match[i] = true$.

Finally, we use the following theorem:

Theorem 1.1 (Brent): Any synchronous parallel algorithm of time t that consists f a total of x elementary operations can be implemented on p processors in $\lceil x/p \rceil + t$ ime.

This theorem can be used for example to slow down a constant time p-processor lgorithm to work in time t using p/t processors. Coming back to the example bove, which finds the first occurrence of one string in an other, we see that the econd step of finding the smallest index of an occurrence takes constant time on n rocessors, while the call to the string matching procedure takes $O(\log\log n)$ time n $\frac{n}{\log\log n}$ processors. By Theorem 1.1 the second step can be slowed down to work n $O(\log\log n)$ time on $\frac{n}{\log\log n}$ processors.

2 Finding the periods

We describe an algorithm that given a string $S[0..n]$ will compute all the periods of S. The output of the algorithm will be a Boolean array $P[1..n]$ such that $P[i] = true$ f and only if i is a period of S. Note that, for the convenience of the presentation, n this section the input string $S[0..n]$ is of length $n+1$ and starts with $S[0]$.

We will prove the following theorem:

Theorem 2.1: There exists an algorithm to compute $P[1..n]$ that takes $O(\log\log n)$ time using $\frac{n}{\log\log n}$ processors.

Corollary 2.2: The period of a string S can be computed in the same time and processor bounds.

Proof: The period of S is the smallest i such that $P[i]$ is *true*. We use the technique of Fich, Ragde and Wigderson [13] to compute the minimum of n integers in the range $1..n$ in constant time using an n-processor CRCW-PRAM. (This step can be slowed down to work in optimal $O(\log\log n)$ time by Theorem 1.1.) □

Corollary 2.3: All initial palindromes of a string S can be computed in the same time and processor bounds.

Proof: Suppose we want to compute all initial palindromes of a string w that does not contain the symbol \$. We present $w\$w^R$ (where w^R is the string w reversed) as input to the algorithm that computes all periods of a string. Each period of this string corresponds to an initial palindrome of w. Two copies of the string $w\$w^R$ are aligned with each other shifted by some offset and the overlapping parts are identical if and only if the overlapping part is an initial palindrome of w. This reduction was used by Fischer and Paterson [14]. □

Example: The string *abaab* has an initial palindrome *aba*. This initial palindrome corresponds to the period *abaab\$ba* of the string *abaab\$baaba*.

Proof of Theorem 2.1:

The algorithm will proceed in independent stages which are all computed simultaneously and described in the next section. In stage number η, $0 \leq \eta < m$, we will compute only $P[n - l_\eta + 1..n - l_{\eta+1}]$; where the sequence $\{l_\eta\}$ is a decreasing sequence defined as $l_0 = n$, $l_{\eta+1} = \lfloor \frac{2}{3} l_\eta \rfloor$ and m is the smallest integer for which $l_m = 0$. Note that each stage is assigned to compute a disjoint part of the output array P and the entire array is covered.

We denote by T_η the time it takes to compute stage number η using P_η processors. The number of operations at stage η will be denoted by $O_\eta = T_\eta P_\eta$. We show later how to implement stage number η in $T_\eta = O(\log\log l_\eta)$ time and $O_\eta = l_\eta$ operations using Breslauer and Galil's [6] parallel string matching algorithm.

Since all stages of our algorithm are executed in parallel the total number of operation performed in all stages is $\sum_\eta O_\eta \leq \sum_\eta \left(\frac{2}{3}\right)^\eta n = O(n)$ and the time is $\max T_\eta = O(\log\log n)$. By Theorem 1.1 the algorithm can be implemented using $\frac{n}{\log\log n}$ processors in $O(\log\log n)$ time. □

2.1 A single stage

We describe a single stage η, $0 \leq \eta < m$, that computes $P[n - l_\eta + 1..n - l_{\eta+1}]$ in optimal $O(\log\log l_\eta)$ time. Note, that since a period p implies that $S[0..n-p] = S[p..n]$, there must be an occurrence of $S[0..l_{\eta+1}]$ starting at each position p which is a period of S in the range computed by this stage.

We start with a call to a string matching algorithm to find all occurrences of $S[0..l_{\eta+1}]$ in $S[n-l_\eta+1..n]$. Let q_i, $i = 1..r$, denote the indices of all these occurrences (all indices are in the string $S[0..n]$, thus $n - l_\eta < q_i \leq n - l_{\eta+1}$).

If there were no occurrences found, the string S has no period in the range computed by this stage and all entries of $P[n - l_\eta + 1..n - l_{\eta+1}]$ can be set to *false*. Otherwise, we continue with another call to a string matching algorithm to find all occurrences of $S[0..l_{\eta+1}]$ in $S[0..l_\eta - 1]$. Let p_i, $i = 1..k$, denote the indices of all these occurrences (note that $p_1 = 0$).

If there was only one occurrence of $S[0..l_{\eta+1}]$ in $S[n - l_\eta + 1..n]$, it can be verified to be a period in $O(l_\eta)$ operations. However, if there are $r > 1$ occurrences, $O(rl_\eta)$ operations may be needed to verify all of them. Luckily the sequences $\{p_i\}$ and $\{q_i\}$ have a "nice" structure as we show in the following lemmas. This structure enables us to proceed efficiently to test which of the q_i's is actually a period of S.

Lemma 2.4 (Lyndon and Schutzenberger [16]): If a string of length m has two periods of length p and q and $p+q \leq m$, then it has also a period of length $\gcd(p, q)$.

Lemma 2.5: If a string $A[1..l]$ has period p and occurs only at positions $p_1 < p_2 < \cdots < p_k$ of a string $B[1..\lceil \frac{3}{2} l \rceil]$, then the p_i's form an arithmetic progression with ifference p.

roof: Assume $k \geq 2$. We prove that $p = p_{i+1} - p_i$ for $i = 1 \cdots k-1$. The string $[1..l]$ has periods p and $q = p_{i+1} - p_i$. Since $p \leq q \leq \lceil \frac{l}{2} \rceil$, by Lemma 2.4 it has lso a period of length $gcd(p,q)$. But p is the shortest period so $p = gcd(p,q)$ and p ust divide q. The string $B[p_i..p_{i+1}+l-1]$ has period p. If $q > p$ then there must e another occurrence of A at position $p_i + p$ of B, a contradiction. □

emma 2.6: The sequences $\{p_i\}$ and $\{q_i\}$ form an arithmetic progression with ifference $\mathcal{P}$, where $\mathcal{P}$ is the period of $S[0..l_{\eta+1}]$.

roof: The sequences p_i and q_i are indices of occurrences of a string of length $_{+1}+1$ in strings of length l_η. Recall that $l_{\eta+1} = \lfloor \frac{2}{3} l_\eta \rfloor$. By Lemma 2.5 the p_i's and $_i$'s form an arithmetic progression with a difference $\mathcal{P}$, the period of $S[0..l_{\eta+1}]$. □

The sequences $\{p_i\}$ and $\{q_i\}$ can be represented using three integers (each): the tart of the sequence, the difference, and the length of each sequence. This repreentation can be easily obtained from the output of the string matching algorithm n constant time and l_η processors.

Some of the q_i's can be ruled out of being periods of S immediately as we show the following lemma.

emma 2.7: If $k < r$ then q_i is not a period of S for $1 \leq i \leq r-k$.

roof: Assume q_i is a period of S and $1 \leq i \leq r-k$. In this case $S[q_i..n] = S[0..n-q_i]$. The string $S[q_i..n]$ has $r-i+1 > k$ occurrences of $S[0..l_{\eta+1}]$, which are $q_i \cdots q_r$. But $S[0..n-q_i]$ has only k occurrences of $S[0..l_{\eta+1}]$; contradiction. □

There might be two reasons why $q_r + \mathcal{P}$ is not included in the $\{q_i\}$ sequence:

1. If $S[q_r + \mathcal{P}..\mathcal{N}] \neq S[0..\mathcal{N} - q_r - \mathcal{P}]$, and $\mathcal{N} = \min(n, q_r + \mathcal{P} + l_{\eta+1})$ we call it **a mismatch**.

2. If there is no mismatch then the only reason that $q_r + \mathcal{P}$ is not in the $\{q_i\}$ sequence is that $q_r + \mathcal{P} + l_{\eta+1} > n$. We call this case **an overflow**.

Lemma 2.8 (a mismatch): If $S[q_r + \mathcal{P}..\mathcal{N}] \neq S[0..\mathcal{N} - q_r - \mathcal{P}]$ then, S has at nost one period in the range computed by this stage. This only possible period may xist if $k \leq r$ and it is q_{r-k+1}.

Proof: By Lemma 2.7 all q_i, $1 \leq i < r-k+1$, are not periods. Assume q_i is a period and $i > r-k+1$, then $S[q_i..n] = S[0..n-q_i]$. Since $r-i+2 \leq k$ and $p_j = (j-1)\mathcal{P}$, also $S[q_r + \mathcal{P}..\mathcal{N}] = S[p_{r-i+2}..\mathcal{N} - q_i]$. By the assumption of a mismatch $S[q_r + \mathcal{P}..\mathcal{N}] \neq S[0..\mathcal{N} - q_r - \mathcal{P}]$. So $S[p_{r-i+2}..\mathcal{N} - q_i] \neq S[0..\mathcal{N} - q_r - \mathcal{P}]$. But $S[p_{r-i+2}..p_{r-i+2} + l_{\eta+1}] = S[0..l_{\eta+1}]$ and also $\mathcal{N} - q_r - \mathcal{P} \leq l_{\eta+1}$; contradiction. □

Lemma 2.9 (an overflow): If $S[q_r + \mathcal{P}..n] = S[0..n - q_r - \mathcal{P}]$ then:

a. If $r < k$ then $q_1, .., q_r$ are periods of S.

b. If $r \geq k$ then $q_{r-k+2}, .., q_r$ are periods of S. In this case q_{r-k+1} can also be a period of S.

Proof: Assume $S[q_r + \mathcal{P}..n] = S[0..n - q_r - \mathcal{P}]$. It is enough to show that q_i is a period of S, for $\max(r-k+2, 1) \leq i \leq r$.

By the definitions of the $\{q_i\}$ and $\{p_i\}$ sequences

$$S[0..p_{r-i+1} + l_{\eta+1}] = S[q_i..q_r + l_{\eta+1}] \quad (1$$

since both substrings are covered by $r-i+1$ occurrences of $S[0..l_{\eta+1}]$. Since $r-i+2 \leq k$ also

$$S[p_{r-i+2}..p_{r-i+2} + l_{\eta+1}] = S[0..l_{\eta+1}]. \quad (2$$

But $n - q_r - \mathcal{P} < l_{\eta+1}$ and $S[q_r + \mathcal{P}..n] = S[0..n - q_r - \mathcal{P}]$. By taking prefixes of (2

$$S[q_r + \mathcal{P}..n] = S[p_{r-i+2}..p_{r-i+2} + n - q_r - \mathcal{P}]. \quad (3$$

By combining equalities (1) and (3), we get that $S[0..n - q_i] = S[q_i..n]$. □

The computation in stage η can be summarized as follows:

1. Compute the $\{q_i\}$ and $\{p_i\}$ sequences.
2. If $k \leq r$, check if q_{r-k+1} is a period of S.
3. If $S[q_r + \mathcal{P}..\mathcal{N}] = S[0..\mathcal{N} - q_r - \mathcal{P}]$ then,
   a. If $r < k$, then $q_1, .., q_r$ are all periods of S.
   b. If $r \geq k$, then $q_{r-k+2}, .., q_r$ are all periods of S.

Lemma 2.10: Stage number η is correct and it takes $O(\log\log l_\eta)$ time and $O(l_\eta)$ operations.

Proof: Correctness of the algorithm follows from Lemmas 2.7, 2.8 and 2.9. The two calls to a string matching algorithm to compute the $\{q_i\}$ and $\{p_i\}$ sequences take $O(\log\log l_\eta)$ time and $O(l_\eta)$ operations if we use Breslauer and Galil's [6] string matching algorithm. The sequences $\{q_i\}$ and $\{p_i\}$ can be represented by three integers which can be computed from the output of the string matching algorithm (which is assumed to be a Boolean vector representing all occurrences) in constant time and $O(l_\eta)$ operations. Steps *2* and *3* can be done also in constant time and $O(l_\eta)$ operations. □

2.2 A lower bound

The algorithm that finds all periods of a string described above achieves the best time bound possible for an optimal parallel algorithm as implied by a lower bounds of Breslauer and Galil [7]. This lower bound can be modified to a lower bound for finding the initial palindromes of a string.

The reduction from periods to initial palindromes of Corollary 2.3 does not imply the immediate translation of the lower bound to this problem. The details of the lower bound will be given in the full paper [8].

3 Looking for squares

We describe an algorithm that tests if a string $S[1..n]$ is square-free in $O(\log\log n)$ time on a $\frac{n \log n}{\log\log n}$-processor CRCW-PRAM. This algorithm is optimal since the time-processor product is $O(n \log n)$, the same as the running time of the fastest sequential algorithm on general alphabet.

We will prove the following theorem:

Theorem 3.1: There exists an algorithm that tests if a string $S[1..n]$ is square-free and takes $O(\log\log n)$ time using $\frac{n \log n}{\log\log n}$ processors.

The algorithm will work in independent stages which are all computed simultaneously. In stage number η, $1 \leq \eta \leq \lceil \log_2 n - 1 \rceil$, we look only for squares xx where $2^\eta - 1 \leq |x| < 2^{\eta+1} - 1$. If a square is found, a global variable will be set to indicate that the string is not square-free. Note that the complete range of possible lengths of x is covered and if there exist a square it will be discovered.

We denote by T_η the time it takes to compute stage number η on P_η processors. The number of operations performed in stage η will be denoted by $O_\eta = T_\eta P_\eta$.

We show later how to implement stage η in $T_\eta = O(\log\log 2^\eta)$ time and $O_\eta = O(n)$ operations using Breslauer and Galil's [6] string matching algorithm. Since there are $O(\log n)$ stages, the total number of operations is $O(n \log n)$. By Theorem 2.1 the complete algorithm can be implemented in $\max T_\eta = O(\log\log n)$ time and $\frac{n \log n}{\log\log n}$ processors.

3.1 A single stage

We describe a single stage η, $1 \leq \eta \leq \lceil \log_2 n - 1 \rceil$, that looks only for squares xx where $2^\eta - 1 \leq |x| < 2^{\eta+1} - 1$. Every comparison to a symbol $S[q]$ out of the range $S[1..n]$ is assumed to be answered as unequal.

Partition the input string $S[1..n]$ into blocks of length l_η, where l_η is defined as $l_\eta = 2^{\eta-1}$. That is, block number k, for $1 \leq k < \frac{n}{l_\eta}$, is $S[(k-1)l_\eta + 1..kl_\eta]$. Stage η consists of sub-stages which are also computed simultaneously. There is a sub-stage for each block of length l_η, in which the algorithms checks if there is a square xx such that $2^\eta - 1 \leq |x| < 2^{\eta+1} - 1$ and the first x contains that block.

Each such sub-stage starts with a call to the string matching algorithm to find all occurrences of the kth block, $S[(k-1)l_\eta + 1..kl_\eta]$, in $S[(k+1)l_\eta..(k+4)l_\eta - 2]$. Let $p_1 < p_2 < \cdots < p_r$, be the indices of these occurrences. Then $(k+1)l_\eta \leq p_i \leq (k+3)l_\eta - 1$ for $1 \leq i \leq r$.

Note, that for each square xx such that $2^\eta - 1 \leq |x| < 2^{\eta+1} - 1$ and the first copy of x contains the block $S[(k-1)l_\eta + 1..kl_\eta]$ there must be an occurrence of $S[(k-1)l_\eta + 1..kl_\eta]$ at position $(k-1)l_\eta + |x| + 1$. This occurrence is included in the $\{p_i\}$ sequence.

Lemma 3.2: For each p_i, we can verify in constant time and $O(l_\eta)$ operations if there is a square xx of length $p_i - (k-1)l_\eta - 1$ that contains the block $S[(k-1)l_\eta + 1..kl_\eta]$.
Proof: Let $l = p_i - (k-1)l_\eta - 1$ be the length of the square we are looking for. For all ζ in the range $kl_\eta < \zeta \leq (k-1)l_\eta + l$ check if $S[\zeta - l] = S[\zeta]$ and if $S[\zeta] = S[\zeta + l]$. Let ζ_L be the largest index in this range such that $S[kl_\eta + 1..\zeta_L] = S[kl_\eta + l + 1..\zeta_L + l]$ and ζ_R be the smallest index such that $S[\zeta_R..(k-1)l_\eta + l] = S[\zeta_R - l..(k-1)l_\eta]$.

Using the algorithm of Fich, Ragde and Wigderson [13] we can find ζ_L and ζ_R in constant time and $O(l_\eta)$ operations.

We show that there is a substring xx of length l containing the block $S[(k-1)l_\eta+1..kl_\eta]$ if an only if $\zeta_L \geq \zeta_R - 1$. Recall that since there is an occurrence of $S[(k-1)l_\eta+1..kl_\eta]$ at position p_i, we know that $S[(k-1)l_\eta+1..kl_\eta] = S[p_i..p_i+l_\eta-1]$

If $\zeta_L \geq \zeta_R - 1$ then $S[\zeta_R - l..\zeta_L] = S[\zeta_R..\zeta_L + l]$. The last equality means that there are squares of length l starting at positions $\zeta_R - l$ up to $\zeta_L - l + 1$.

On the other hand, if there is a square xx such that the first x contains $S[(k-1)l_\eta+1..kl_\eta]$ that starts at position π, then for all ζ in the range $\pi+|x| \leq \zeta \leq (k-1)l_\eta+l$, $S[\zeta-|x|] = S[\zeta]$ and for all ζ in the range $kl_\eta < \zeta < \pi+|x|$, $S[\zeta] = S[\zeta+|x|]$. But $\zeta_L < \zeta_R - 1$ so $S[\zeta_L+1] \neq S[\zeta_L+|x|+1]$ and $S[\zeta_R-|x|-1] \neq S[\zeta_R-1]$; a contradiction to the existence of a square of length $|x|$ at position π. □

We could verify all the p_i's in constant time using Lemma 3.2 but it might take $O(rl_\eta)$ operations if the length of the $\{p_i\}$ sequence is r. Luckily, we do not have to verify all the p_i's. If the length of the $\{p_i\}$ sequences $r > 2$ then there is a square as the following lemma shows:

Lemma 3.3: If the number of occurrences of $S[(k-1)l_\eta+1..kl_\eta]$ in $S[(k+1)l_\eta..(k+4)l_\eta-2]$ is larger than two then S[1..n] has a square. This square is actually shorter than the squares which are supposed to be found by this stage.

Proof: All the occurrences $\{p_i\}$, $1 \leq i \leq r$ satisfy $(k+1)l_\eta \leq p_i \leq (k+3)l_\eta - 1$. If $r \geq 3$ then either $p_2 - p_1 \leq l_\eta$ or $p_3 - p_2 \leq l_\eta$. In this case there is a square of length $p_2 - p_1$ or $p_3 - p_2$ (respectively) starting at position p_1 or p_2 (respectively). □

The computation in each sub-stage of stage η can be summerized as follows:

1. Compute the $\{p_i\}$ sequence.

2. If the $\{p_i\}$ sequence has more than two elements then by Lemma 3.3 the string $S[1..n]$ has a square. This square will be found also by some stage μ, $\mu < \eta$.

3. If the $\{p_i\}$ sequence has at most two elements, check if these elements correspond to squares using the procedure described in Lemma 3.2.

Note that if the $\{p_i\}$ sequence has only a small number of elements we can actually find all the squares efficiently using Lemma 3.2. The complete details of an algorithm that finds all the squares will be described in the full paper [2].

Lemma 3.4: Stage number η is correct and it takes $O(\log\log n)$ time on $\frac{n}{\log\log n}$ processors.

Proof: For correctness we have to show that if the string $S[1..n]$ has a any square xx, $2^\eta - 1 \leq |x| < 2^{\eta+1} - 1$, then some square will be found.

Assume there is such a square. Since $2l_\eta - 1 \leq |x|$, there must be a block of length l_η that is completely contained in the first x. The sub-stage assigned to that block will either find the square xx or conclude that there is a shorter square by Lemma 3.3. In both cases a square has been found. Note that some squares can be detected by several sub-stages simultaneously.

Stage η consists of $\frac{n}{l_\eta}$ independent sub-stages. In each sub-stage, step number 1 takes $O(\log\log l_\eta)$ time and $O(l_\eta)$ operation. Steps number 2 and 3 take constant time and $O(l_\eta)$ operations. Since all sub-stages are computed in parallel the time age η takes is $O(\log\log l_\eta)$ and the total number of operations performed is $\frac{n}{l_\eta}l_\eta =$ (n). □

Theorem 3.1 follows from the last lemma since the $\log n$ stages are executed in arallel.

.2 A Lower Bound

/e prove a lower bound for testing if a string is square-free by a reduction to ie lower bound for string matching by Breslauer and Galil [7]. This lower bound on the number of comparison rounds the algorithm performs when there are p omparisons in each round. This bound holds for the CRCW-PRAM model in case f a general alphabet where the only access an algorithm has to the input strings is y comparison of symbols.

Breslauer and Galil [7] show that an adversary can fool any algorithm which laims to check if a string has a period shorter than half of its length in less than $(\lceil\frac{n}{p}\rceil + \log\log_{\lceil 1+p/n\rceil} 2p)$ rounds of p comparisons each.

We will not go into the complete details of that lower bound. We only use the ict that the adversary of Breslauer and Galil [7] answers the comparisons in each ound in such a way that after $\Omega(\lceil\frac{n}{p}\rceil + \log\log_{\lceil 1+p/n\rceil} 2p)$ rounds it is still possible hat the string $S[1..n]$ has a period shorter than half of its length, or that it does not ave any such period. The string generated by that adversary has also the following roperty: If $S[i] = S[j]$, for $i < j$, then for any integer k, such that $k \equiv i \bmod (j-i)$ nd $1 \leq k \leq n$, $S[k] = S[i]$.

emma 3.5: The string generated by the adversary of Breslauer and Galil has a eriod smaller than half of its length if and only if it has a square.

roof: If the string generated by the adversary has a period which is shorter than alf the length of the string then there is a square starting at the beginning of the tring, which is this period repeated twice.

It remains to show that if there is a square then the string has a period shorter han half of its length. Assume that a square of length $2l$, starts at position i. This neans that $S[i+k] = S[i+l+k]$ for $k = 0..l-1$. By the property of the string enerated by the adversary, the string $S[1..n]$ has a period l, which is smaller than alf the length of S. □

Now, we are ready to prove the lower bound.

Theorem 3.6: Any optimal parallel algorithm that tests if a string $S[1..n]$ is square-ree takes $\Omega(\log\log n)$ time.

roof: Any optimal algorithm performs at most $n\log n$ comparisons in each round ince the fastest sequential algorithm that solves this problem takes $O(n\log n)$ time s shown by Main and Lorentz [17]. Assume an algorithm performs $p = n\log n$ omparisons in each round. By the lower bound of Breslauer and Galil [7] and Lemma 3.5 after $\Omega(\log\log n)$ rounds the adversary can decide if $S[1..n]$ is square-ree or not, fooling any algorithm which terminates in less rounds. □

4 Acknowledgments

We would like to thank Roberto Grossi, Kunsoo Park and Laura Toniolo for comments on early versions of this paper.

References

[1] Apostolico, A. (1991), Optimal parallel detection of squares in strings, *CS-TR-91-026, Purdue and Algorithmica, in press.*

[2] Apostolico, A. and Breslauer D. (1991), An optimal $O(\log\log n)$ time parallel algorithm for detecting all repetitions in a string, *In preparation.*

[3] Apostolico, A. and Preparata, F. P. (1983), Optimal off-line detection of repetitions in a string, *Theoretical Computer Science 22*, 297-315.

[4] Beame, P., and Hastad, J. (1989), Optimal Bound for Decision Problems on the CRCW PRAM, *Journal of ACM 36:3*, 643-670.

[5] Brent, R. P. (1974), The parallel evaluation of general arithmetic expressions, *J. ACM 21*, 201-206.

[6] Breslauer, D. and Galil, Z. (1990), An optimal $O(\log\log n)$ parallel string matching algorithm, *SIAM J. Comput. 19:6*, 1051-1058.

[7] Breslauer, D. and Galil, Z. (1991), A lower bound for parallel string matching, *Proc. 23rd ACM Symp. on Theory of Computation*, 439-443.

[8] Breslauer, D. and Galil Z. (1991), Finding all the periods and initial palindromes of a string in parallel, *manuscript.*

[9] Crochemore, M. (1981), An optimal algorithm for computing the repetitions in a word, *Information Processing Letters 12:5*, 244-250.

[10] Crochemore, M. (1986), Transducer and repetitions, *Theoretical Computer Science 45*, 63-86.

[11] Crochemore, M. and Rytter, W. (1991), Usefulness of the Karp-Miller-Rosenberg algorithm in parallel computations on strings and arrays, *Theoretical Computer Science 88*, 59-82.

[12] Crochemore, M. and Rytter, W. (1991), Efficient Parallel Algorithms to Test Square-freeness and Factorize Strings, *Information Processing Letters 38*, 57-60.

[13] Fich, F. E., Ragde, R. L., and Wigderson, A. (1984), Relations between concurrent-write models of parallel computation, *Proc. 3rd ACM Symp. on Principles of Distributed Computing*, 179-189.

[14] Fischer, M. J. and Paterson, M. S. (1974), String-Matching and other products, *SIAM-AMS proceedings, Vol 7*, 113-125.

[15] Knuth, D. E., Morris, J. H. and Pratt, V. R. (1977), Fast pattern matching in strings, *SIAM J. Comput. 6*, 322-350.

[16] Lyndon, R. C. and Schutzenberger, M. P. (1962), The equation $a^M = b^N c^P$ in a free group, *Michigan Math. J. 9*, 289-298.

[17] Main, G. M. and Lorentz, R. J. (1984), An $O(n \log n)$ algorithm for finding all repetitions in a string, Journal of Algorithms 5, 422-432.

[18] Main, G. M. and Lorentz, R. J. (1985), Linear time recognition of squarefree strings, in *Combinatorial Algorithms on Words*, Edited by A. Apostolico and Z. Galil, 271-278.

[19] Thue, A. (1906), Über unendliche Zeichenreihen, *Norske Vid. Selsk. Skr. Mat. Nat. Kl. (Cristiania)*, Nr. 7, 1-22.

[20] Thue, A. (1912), Über die gegenseitige Lage gleicher Teile gewisser Zeichenreihen, *Norske Vid. Selsk. Skr. Mat. Nat. Kl. (Cristiania)*, Nr. 1, 1-67.

[21] Vishkin, U. (1985), Optimal parallel pattern matching in strings, *Information and Control 67*, 91-113.

Near-perfect Token Distribution

Andrei Z. Broder[1], Alan M. Frieze[2*], E. Shamir[3**], and E. Upfal[4]

[1] DEC Systems Research Center
130 Lytton Ave, Palo Alto, California, USA
[2] Department of Mathematics, Carnegie-Mellon University
Pittsburgh, Pennsylvania, USA
[3] Department of Computer Science, The Hebrew University, Jerusalem, Israel
[4] IBM Almaden Research Center, San Jose, California, USA, and
The Weizmann Institute of Science, Rehovot, Israel.

Abstract. Suppose that n tokens are arbitrarily placed on the n nodes of a graph. At each parallel step one token may be moved from each node to an adjacent node. An algorithm for the *near-perfect token distribution* problem redistributes the tokens in a minimum number of steps, so that, at the end, no more than $O(1)$ tokens reside at each node. (In perfect distribution, at the end, exactly one token resides at each node.)

In this paper we present a simple algorithm that works for all *extrovert* graphs, a new property which we define and study. In terms of connectivity requirements, extrovert graphs are in-between expanders and compressors. Our results lead to an optimal solution for the near-perfect token distribution problem on almost all cubic graphs. The new solution is conceptually simpler than previous algorithms, and applies to graphs of minimum possible degree.

1 Introduction

Suppose that n tokens are arbitrarily placed on the n nodes of a graph, with no more than M tokens at each site. At each parallel step one token may be moved from each node to an adjacent node. An algorithm for the *near-perfect token distribution* problem redistributes the tokens in a minimum number of steps, so that, at the end, no more than $O(1)$ tokens reside at each node. (In perfect distribution, at the end, exactly one token resides at each node.)

Algorithms for the near-perfect token distribution have direct applications to fundamental problems in parallel and distributed systems. We give two examples here: First, consider each node as a processor, each edge as a communication line, and each token as a message. An efficient solution to the near-perfect token distribution problem can be combined with an efficient permutation routing algorithm to give an efficient routing algorithm for an arbitrary communication request. Second, consider each token as an independent process awaiting execution and each node as a processor. Then an efficient algorithm for the near-perfect token distribution problem can be used for workload distribution among the processors.

* A portion of this work was done while the author was visiting DEC SRC. Supported in part by NSF Grant CCR 0089112.

** A portion of this work was done while the author was visiting DEC SRC.

These, and other applications, require a solution in a fully distributed model, here all information processing are done by processors that reside at the nodes of e graph. When the computation is initiated, each process has no information on e status (e.g. number of tokens) of other processors in the system. Information transmitted via messages along the edges of the graph. In each step a processor n either send one token through an edge, or one $O(\log n)$-bit message. Further-ore, most applications require an efficient solution on graphs that are as sparse as ossible.

Peleg and Upfal [4] have shown that any solution to the near-perfect token dis-ibution problem requires $\Omega(M + \log n)$ parallel steps, and this run time can be chieved only on expander graphs. They gave an optimal $O(M + \log n)$ algorithm r *sqrt* (α, β)-*expander* graphs. These are n node, d-regular expander graphs with e following additional strong properties: there exist constants $\alpha > 1$ and $\beta > 0$ ich that any set of vertices X, with $|X| \leq \beta n$, has at least $\alpha |X|$ neighbors outside and there exists an i with $0 < i \leq d$ such that $\alpha > (d + i^2)/(i + 1)$. These two onditions imply that the expansion is at least $\sqrt{d}$.

In this paper we present an optimal solution for almost all cubic graphs, which the best one can hope for from the point of view of the minimum degree required.

Our solution is based on identifying a novel property, called *extrovertness* which eplaces the sqrt-expander property of [5, 4]. Extrovertness is stronger than the xpansion property but weaker than the compression property of [3] (i.e every ex-rovert graph is an expander, and every compressor graph is an extrovert graph). n Section 2 we present an asymptotically optimal solution for any extrovert graph, hich immediately leads to a solution for any graph G such that for some bounded onstant c, the union $\cup_{i \leq c} G^i$ is extrovert. The main result of this paper, which is erhaps of independent significance, is given in Section 3: we show that for almost ll cubic graphs G, the graph $G \cup G^2$ (that is, the graph of all paths of length at ost two in G) is extrovert. (In fact, it can be even shown that for almost all cubic raphs the graph G^2 is extrovert, but the proof is rather intricate and the constants re much worse.)

2 The New Algorithm

Our new approach to the token distribution problem is based on the *extrovertness* roperty:

Definition 1. *A vertex $v \in S \subseteq V$ is called* extrovert *in S if v has strictly more neighbors in $V \setminus S$ than it has in S.*

Definition 2. *A set $S \subseteq V$ is called b-*extrovert *if at least $b\,|S|$ of its vertices are extrovert.*

Definition 3. *A graph G is called (a, b)-*extrovert *if any set $S \subseteq V$ of size at most $a\,|V|$ is b-extrovert.*

Our algorithm gives an optimal solution for any n-node (a, b)-extrovert graph $G = (V, E)$.

Let d be the maximum degree in G and let $\gamma = (2d+1)/(2d+2)$. The algorithm works in $\log_{1/\gamma} M$ phases. (Recall that M is the initial maximum number of tokens in any node.)

Let $t_j(v)$ be the number of tokens at node v at the start of phase j, and let $M_j = \max_v \{t_j(v)\}$. Let $U_j = \{v : t_j(v) \geq M_j/2\}$ denote the set of "heavily loaded" nodes.

Each phase reduces the maximum number of tokens in any node by a factor of γ until no vertex has more than $2/a = O(1)$ tokens. Note that in any phase $|U_j| \leq a\,|V|$ because $M_j > 2/a$ and the total number of tokens is $|V|$.

At the beginning of phase j the algorithm constructs a dag (directed acyclic graph) $D_j = (V_j, E_j)$ with the following properties:

1. $U_j \subseteq V_j \subseteq V$. (All heavily loaded vertices are in the dag.)
2. Each edge in D is the directed version of some edge in G, and each edge in G has at most one directed version in D.
3. For every $v \in U_j$, outdegree$(v) >$ indegree(v).

After the dag is constructed, phase j consists of $M_j/(2d+2)$ iterations; in each iteration each node in V_j sends one token (if it has one) through each of its outgoing edges in D_j.

Lemma 1. $M_{j+1} \leq \gamma M_j$.

Proof. Let v be an arbitrary vertex in G. If v is not in D_j, and therefore not in U_j, then it does not receive (or send) any tokens in phase j, and thus it has less than $M_j/2 < \gamma M_j$ tokens at the end of this phase.

If v is in the dag but not in U_j, it starts the phase with less than $M_j/2$ tokens and it may receive in this phase no more than $dM_j/(2d+2)$ tokens. Thus at the end of the phase it has no more than $(2d+1)/(2d+2)M_j = \gamma M_j$ tokens.

Finally, if $v \in U_j$, it sends more tokens than it receives in each iteration. Thus it loses at least $M_j/(2d+2)$ tokens in phase j and has less than γM_j tokens at the end of the phase. ∎

For a set of vertices $S \subseteq V$ let $\Phi(S) = \{v \in S : v \text{ is extrovert}\}$. Given an (a,b)-extrovert graph we construct the dag D_j, "bottom-up," using the following procedure.

1. Set $W \leftarrow U_j$.
2. Repeat until $W = \emptyset$:
   (a) Connect all vertices in $\Phi(W)$ to all of their G neighbors outside W.
   (b) $W := W \setminus \Phi(W)$.

Lemma 2. *The above procedure constructs a dag with the required properties, and terminates in* $O(\log N)$ *iterations.*

Proof. Clearly when the procedure terminates, all of U_j is in the dag, and each edge in G was used no more than once. When a vertex $v \in U_j$ joins the dag it has more neighbors in the dag than in the remaining W, thus *outdegree*$(v) >$ *indegree*(v) in the final dag. Since the graph G is (a,b)-extrovert, and $|W| \leq |U_j| \leq a\,|V|$, in each iteration $|\Phi(W)| \geq b\,|W|$, and the procedure terminates within $O(\log n)$ iterations. ∎

Corollary 3. *If G is an (a,b)-extrovert graph, then the algorithm solves the near-perfect token distribution problem on G in $O(\log M)$ phases. When phase $j-1$ ends, no vertex has more than $M_j = \gamma^{j-1}M$ tokens and phase j takes $O(\gamma^{j-1}M + \log n)$ parallel steps.*

If we execute the $O(\log M)$ phases sequentially the total run time is $O(M + \log n \log M)$. Peleg and Upfal [4] have shown that all the dags for the last $O(\log\log n)$ phases can be constructed in parallel. Since $\log n$ becomes dominant only when $M_j \leq \log n$, this suffices to reduce the total run-time to $O(M + \log n)$. The same idea can be applied to our algorithm. We briefly sketch the construction here, the reader is referred to [4] for a more detailed description.

The difficulty in constructing the dags in parallel is that in order to construct the dag D_j one needs to know the set U_j, and the set U_j is defined in turn, by the dag D_{j-1}. This difficulty is overcome by replacing the set U_j in the construction, by an overestimate set $\tilde{U}_j$ that contains U_j and is easier to compute. We also compute, for each node, an overestimate of the number of tokens it will have at the end of each phase.

The overestimate sets are constructed *sequentially* as follows: Assume that we have already constructed the set $\tilde{U}_j$, and that each node has a bound on the number of tokens it will have at the end of phase $j-1$. We run the first $c\log\log n$ phases of the construction of D_j, assuming that $U_j = \tilde{U}_j$. Let $\tilde{W}_j$ denote the set W at the end of these $O(\log\log n)$ phases. We fix the number of phase we run so that $|\tilde{W}_j| = O(n/(\log n)^2)$.

Now we compute for each node, a bound on the number of tokens it will have at the end of phase j. If a node v has a neighbor in $\tilde{W}_j$, it assumes that this neighbor will be connected to it in D_j, and will send to it $\gamma^{j-1}M$ tokens. Besides this, if v is in the partially computed dag, it can compute in $O(\log\log n)$ time, a bound on how many tokens it will receive in phase j. Thus, every node which potentially is in the dag D_j, has an estimate on the number of tokens it will receive in phase j. Vertex v is in $\tilde{U}_{j+1}$ if according to this estimate it has more than $\gamma^j M/2$ tokens, at the end of phase j.

These estimates add no more than $d|\tilde{W}_j|M_j$ "dummy" tokens in each phase. We can fix the parameters so that no more than n dummy tokens are created in the whole process, thus, this procedure can be executed until $M_j \leq 4/b$. Once the sets $\tilde{U}_j$ are constructed, the $O(\log\log n)$ dags can be constructed in parallel in $O(\log n)$ parallel steps, with messages of size $O(\log\log n)$ bits. Thus, we prove:

Theorem 4. *The near-perfect token distribution problem can be solved on any (a,b)-extrovert graph G in $O(M + \log n)$ time.*

Let $H(G,c) = \bigcup_{j\leq c} G^i$, for some constant c (i.e., the graph $H(G,c)$ has an edge for any path of length c or less in G). Clearly, a parallel step on the graph $H(G,c)$ can be simulated on the original graph G in constant time. Thus, our result applies to any graph G such that $H(G,c)$ is extrovert for some constant c.

The main result of this paper, which we prove in the next section, is that if G is a random cubic graph, then with high probability the graph $H = G \cup G^2$ is extrovert. Thus we prove:

Theorem 5. *There is an $O(M + \log n)$ solution for the near-perfect token distribu- tion problem on almost all cubic graphs.*

3 Extrovert Graphs via Random Cubic Graphs

We first state three properties that almost all random cubic graphs possess; then w show that if a cubic graph G has these properties, then the graph $H = G \cup G^2$ i extrovert.

In the definition of these properties $G = (V, E)$ is a cubic graph on n vertices S is a subset of V, and $i(S)$ is the number of edges that are contained in S. A constants were chosen for convenience and no attempt was made to optimize them

Property 1. *If $S \subseteq V$ and $|S| \leq (\ln n)/4$ then $i(S) \leq |S|$.*

Property 2. *For any fixed $\epsilon > 0$ there exists a function $a(\epsilon) > 0$ such that fo $S \subseteq V$, if $|S| \leq a(\epsilon)n$ then $i(S) \leq (1 + \epsilon)|S|$.*

Property 3. *There are fewer than $\ln\ln n$ vertices on cycles of length less tha $\ln\ln\ln n$*

We will now prove that almost all cubic graphs have these three properties. W will use the model of Bender and Canfield [1] and Bollobás [2]. Specifically, we wi adopt the notation of [2].

Let $[k]$ stand for the set $\{1, \ldots, k\}$. Let $W = [3n]$ and let $W_i = \{3(i-1)+1, 3(i-1)+2, 3i\}$, for $i = 1, 2, \ldots, n$ be a partition of W into n sets of size 3, called *block* For $w \in W$ we define $\psi(w) = \lceil w/3 \rceil$ so that $w \in W_{\psi(w)}$. The elements of W ar called *points* to distinguish them from the vertices of the graph.

A configuration is a partition of W into $m = 3n/2$ pairs. The set of configuratior is denoted Φ. For $F \in \Phi$ let $\mu(F)$ be the multigraph with vertex set $[n]$ and m edge $\{\{\psi(x), \psi(y)\} : \{x, y\} \in F\}$.

We consider Φ as a probability space in which each $F \in \Phi$ is equally likely. I this setting, for large n, the probability that $\mu(F)$ is a simple graph is at least 1/1 and conditional on this, each cubic graph with vertex set $[n]$ is equally likely. W need only show therefore that almost all $\mu(F)$ have the above three properties.

Proof of Property 1.

$$\begin{aligned}
\mathbf{Pr}(\text{Prop. 1 fails}) &\leq \sum_{k=4}^{(\ln n)/4} \mathbf{E}(\#\text{ of } k\text{-sets with at least } k+1 \text{ edges}) \\
&\leq \sum_{k=4}^{(\ln n)/4} \binom{n}{k} \binom{3k}{k+1} \left(\frac{k}{n}\right)^{k+1} \leq \sum_{k=4}^{(\ln n)/4} \left(\frac{ne}{k}\right)^k 2^{3k} \left(\frac{k}{n}\right)^{k+1} \\
&= \frac{1}{n} \sum_{k=4}^{(\ln n)/4} k(8e)^k = o(1)
\end{aligned}$$

Explanation of second inequality: $\binom{n}{k}$ counts sets of k blocks. Focus now on the block $W_1, W_2, \ldots, W_k$. If vertex set $\{1, 2, \ldots, k\}$ contains more than k edges then there ar

+1 pairs contained in $X = W_1 \cup W_2 \cup \ldots \cup W_k$. We can choose the lowest numbered oint of each pair in at most $\binom{3k}{k+1}$ ways. Having fixed these as $a_1, a_2, \ldots, a_k$ say, e probability that they are each paired in X is at most $(k/n)^{k+1}$, for given the irings of $a_1, a_2, \ldots, a_i$, the probability that a_{i+1} is paired in X is at most k/n. □

roof of Property 2. For definiteness let

$$a(\epsilon) = \left(\frac{(1+\epsilon)^{1+\epsilon}}{2e^{2+\epsilon}3^{1+\epsilon}}\right)^{1/\epsilon}.$$

hen

$$\begin{aligned}\mathbf{Pr}(\text{Prop. 2 fails}) &\le \sum_{k=4}^{a(\epsilon)n} \binom{n}{k}\binom{3k}{(1+\epsilon)k}\left(\frac{k}{n}\right)^{(1+\epsilon)k} \\ &\le \sum_{k=4}^{a(\epsilon)n} \left(\frac{ne}{k}\left(\frac{3ek}{(1+\epsilon)n}\right)^{1+\epsilon}\right)^k \\ &= \sum_{k=4}^{a(\epsilon)n} \left(\frac{e^{2+\epsilon}3^{1+\epsilon}}{(1+\epsilon)^{1+\epsilon}}\left(\frac{k}{n}\right)^{\epsilon}\right)^k = o(1).\end{aligned}$$

Split the range at, say, $k = \sqrt{n}$. If $k \le \sqrt{n}$ then $k/n \le 1/\sqrt{n}$ so the sum up to $= \sqrt{n}$ is $O(1/n^{2\epsilon})$; for $k \ge \sqrt{n}$ the summand is at most 2^{-k}.) □

Proof of Property 3. Say that a cycle is small if it has at most $\ell = \ln\ln\ln n$ vertices. hen

$$\begin{aligned}\mathbf{E}(\#\text{ of vertices on small cycles}) &\le \sum_{k=3}^{\ell} k\binom{n}{k}\frac{1}{2}(k-1)!6^k\left(\frac{1}{3n-2k}\right)^k \\ &\le \sum_{k=3}^{\ell} 2^k = o(\log\log n).\end{aligned}$$

nd so we can use the Markov inequality. (Explanation of first inequality: having hosen k vertices and a cyclic ordering in $\binom{n}{k}\frac{1}{2}(k-1)!$ ways, there are 3^k choices for he points in pairs not in edges of the cycle and 2^{k-1} ways of pairing up the points n the blocks to make the edges of the cycle. The probability that these k pairs exist s at most $(3n-2k)^{-k}$.) □

Theorem 6. *If G is a cubic graph that has properties 1, 2, and 3, then the graph $H = G \cup G^2$ is extrovert.*

Proof. For a set $S \subseteq V$ let

$$\Phi(S) = \{v \in S : v \text{ is extrovert in } S \text{ with respect to } H\}$$

and let $g(S) = |\Phi(S)|$.

From now until the end of the proof all adjacency and neighborhood relations, when not specified otherwise, are with respect to G.

Let ϵ be a sufficiently small constant and let S be a set of vertices with $|S|$: $\frac{1}{4}a(\epsilon)n$. Let $T = \Gamma(S)$ be the set of neighbors of S in G. (Thus $|S \cup T| \leq a(\epsilon)n$ Let $i = i(S)$, the number of edges in S. Now partition S and T as follows: Let

$$S_j = \{v \in S : v \text{ has } j \text{ neighbors in } S\}, \qquad j = 0, 1, 2, 3.$$
$$T_j = \{v \notin S : v \text{ has } j \text{ neighbors in } S\}, \qquad j = 1, 2, 3.$$

Finally let $s = |S|$, $s_j = |S_j|$, and $t_j = |T_j|$.

We first prove that S is extrovert in H under two assumptions:
1. No vertex of S lies on a cycle of length ≤ 4 in G.
2. $|S| \geq s^* > 0$ where s^* is an absolute constant.

These assumptions are removed in the last part of the proof.

Under Assumption 1, each vertex in V has 9 distinct neighbors in H. For v t be extrovert with respect to S in H, it must have at least 5 neighbors in H outsid S. Having two G-neighbors in T_1 is sufficient, but if v has no neighbor in S_3, the one G-neighbor in T_1 is sufficient.

There are two cases in the proof. In the first case, $i \geq (1 - \alpha)s$ for a suitabl chosen $\alpha > 0$. In this case $T_2 \cup T_3$ is quite small, and the proof works readily; I the second case, $i < (1 - \alpha)s$, and the graph induced by S breaks into $k \geq \alpha$ components out of which $k' \geq (1 - 2\alpha)k$ are trees of size at most α^{-2}. Here w estimate the number of extrovert vertices in terms of αk.

We start by deriving a series of relations among the s_j's and the t_j's.

From Property 2 applied to $S \cup T_2 \cup T_3$ we deduce that

$$i + 2t_2 + 3t_3 \leq (1 + \epsilon)(s + t_2 + t_3),$$

or

$$(1 - \epsilon)t_2 + (2 - \epsilon)t_3 \leq (1 + \epsilon)s - i. \qquad (1$$

Let

$$S_2^g = \{v \in S_2 : v \text{ has no neighbor in } S_3\}$$

and let $s_2^g = |S_2^g|$. Observe that if $v \in S_2^g$ is adjacent to $w \in T_1$ then $v \in \Phi(S$ Clearly

$$s_2^g \geq s_2 - 3s_3 \qquad (2$$

Claim 7.

$$s_1 \geq s_3 - 2\epsilon s$$

Proof. Consider the graph G_S induced by S. Let C be a component of G_S with a least one edge. Let C have σ_i vertices of degree i, for $i = 1, 2, 3$, and let $\sigma = |C|$. Du to Property 2, by removing at most $\epsilon\sigma + 1$ edges from C we can obtain a spannin tree C'. Observe that in C'

$$\sigma_1' = \sigma_3' + 2.$$

Now add back the removed edges. Each additions reduces the difference between th number of degree 3 vertices and degree 1 vertices by at most two. Hence

$$\sigma_1 - \sigma_3 \geq \sigma_1' - \sigma_3' - 2(\epsilon\sigma + 1) = -2\epsilon\sigma.$$

Now add up over all the components of S.

Now from equations (2) and Claim 7 we obtain that

$$\begin{aligned} s &= s_0 + s_1 + s_2 + s_3 \\ &\le s_0 + s_1 + s_2^g + 4s_3 \\ &\le s_0 + s_1 + s_2^g + 4s_1 + 8\epsilon s \end{aligned}$$

ence

$$s_0 + 5s_1 + s_2^g \ge (1 - 8\epsilon)s. \tag{3}$$

Now choose α small but large compared with ϵ. There are two cases.

ase 1: $i \ge (1-\alpha)s$.

We deduce from (1) that

$$2t_2 + 3t_3 \le \frac{2}{1-\epsilon}(\alpha + \epsilon)s \le 3\alpha s. \tag{4}$$

The number of edges coming from $T_2 \cup T_3$ into S is $2t_2 + 3t_3$. Delete from S all e vertices hit by such edges. A remaining vertex v has out-of-S neighbors only T_1, thus v has at least two such neighbors if $v \in S_0 \cup S_1$ and at least one such eighbor if $v \in S_2^g$. Hence

$$\begin{aligned} g(S) &\ge s_0 + s_1 + s_2^g - (2t_2 + 3t_3) \\ &\ge \frac{1}{5}(s_0 + 5s_1 + s_2^g) - 3\alpha s \\ &\ge \left(\frac{1-8\epsilon}{5} - 3\alpha\right) s \end{aligned} \tag{5}$$

ase 2: $i < (1-\alpha)s$.

In this case the number of components k of G_S satisfies

$$k \ge \alpha s. \tag{6}$$

he number of components of G_S of size greater than $s/(\alpha k)$ is at most αk. Hence here are at least $(1-\alpha)k$ components of size $< s/(\alpha k) \le \alpha^{-2}$.

laim 8. *If $i < (1-\alpha)s$ then the number of components of S of size $\le \alpha^{-2}$ that re trees is at least $(1-2\alpha)k$.*

roof. We can assume without loss of generality that $G_{S\cup T}$ is connected.

If $s + t \ge A \ln n/\ln\ln n$ then we use Property 3 to assert that at most $o(k)$ of the mall components are not trees.

If $s + t < A \ln n/\ln\ln n$ then at most one of the small components is not a tree; therwise two non-tree components can be connected via T and contradict Property . By Assumption 2

$$s \ge s^* > \alpha^{-2} \tag{7}$$

o $\alpha k \ge \alpha^2 s \ge 1$.

Hence in both cases the number of small components that are not trees is less han αk.

□

Suppose now that the k components have sizes $c_1, c_2, \ldots, c_k$, and that thes components span $i_1, i_2, \ldots, i_k$ edges. We re-write equation (1) as

$$\begin{aligned}(1-\epsilon)t_2 + (2-\epsilon)t_3 &\leq (1+\epsilon)\sum_{1\leq t\leq k} c_t - \sum_{1\leq t\leq k} i_t \\ &= \sum_{1\leq t\leq k} (c_t - i_t) + \epsilon s \leq k + \epsilon s \end{aligned} \quad (8$$

Now we divide the components of s of size $\leq \alpha^{-2}$ that are trees into three classe

- k_1 trees that are isolated vertices.
- k_2 trees that are paths of length ≥ 1.
- k_3 trees that have at least one degree 3 vertex.

We now classify the leaves of these trees. Within these trees there are

- l_1 isolated vertices adjacent to T_1, and thus $k_1 - l_1$ other vertices triply adjacen to $T_2 \cup T_3$.
- l_2 end-of-path leaves adjacent to T_1, and thus $2k_2 - l_2$ other vertices doubl adjacent to $T_2 \cup T_3$.
- l_3 other leaves doubly adjacent to T_1, and thus at least $3k_3 - l_3$ other vertice adjacent to $T_2 \cup T_3$.

Note that $l_1 + l_2 + l_3$ counts extrovert vertices with respect to H. Counting edge out of $T_2 \cup T_3$ we get the inequality

$$2t_2 + 3t_3 \geq 3(k_1 - l_1) + 2(2k_2 - l_2) + (3k_3 - l_3)$$

or

$$3l_1 + 2l_2 + l_3 \geq (3k_1 + 4k_2 + 3k_3) - (2t_2 + 3t_3)$$

and hence

$$l_1 + l_2 + l_3 \geq (k_1 + k_2 + k_3) - \frac{2}{3}(t_2 + 2t_3).$$

Now using Claim 8 and equation (8) and assuming $\epsilon < 1/4$, we obtain that

$$\begin{aligned} l_1 + l_2 + l_3 &\geq (1-2\alpha)k - \frac{2}{3}\frac{1}{1-\epsilon}(k + \epsilon s) \\ &\geq \left(\frac{1}{9} - 2\alpha\right)k - \frac{8\epsilon}{9}s \end{aligned}$$

Hence

$$g(S) \geq ((1-18\alpha)\,\alpha - 8\epsilon)\, s/9 \quad (9$$

Equations (5) and (9) conclude the proof under the two initial assumptions, sinc they show that for any set $S \subseteq V$, if $|S| \leq a(\epsilon)n$, then $g(S)/|S|$ is greater than certain constant. We now proceed to remove the assumptions that we made.

Let b be the lower bound on $g(S)/|S|$ derived above. Suppose first that $|S| \geq$ $(\ln n)/4$. Remove the $k \leq \ln\ln n$ vertices in S that lie on cycles of length $\leq \ln\ln\ln n$

us obtaining S'. All extrovert vertices in S' are extrovert in S with the possible ceptions of vertices at distance 1 or 2 from the removed vertices. (Because the moved vertices are on cycles, there are only $3k$ neighbors.) Therefore

$$g(S) \geq g(S') - 3k \geq (b - o(1))\,|S|\,.$$

Suppose now that

$$\left(8 + \frac{24}{b}\right) \leq |S| \leq \frac{1}{4}\ln n$$

y Property 1, there is at most one cycle in $S \cup \Gamma(S) \cup \Gamma^2(S)$. If this cycle has length ss than 4 and contains vertices of S, remove them to obtain S'. Then, as before

$$g(S) \geq g(S') - 12 \geq b(s-4) - 12 \geq \frac{bs}{2}.$$

It remains to consider the case $2 \leq |S| \leq s^*$, where $s^* = \max\left\{8 + 24/b, \alpha^{-2}\right\}$, to ccommodate also equation (7). We shall argue that S contains at least one extrovert ertex.

Consider the graph K that consists of all edges incident to S in G. (In other vords, K is the graph induced by $S \cup T$ except for the edges whose both endpoints re in T.) Without loss of generality we can assume that K is connected.

Clearly K has the following properties:

- Any $v \in S$ has degree 3 in K.
- Any leaf in K belongs to T_1.
- Any neighbor of a leaf in K belongs to S.
- There is at most one cycle in K.

If K is a tree then there must be a $v \in S$ that is the neighbor of two leaves in K, and therefore is extrovert in S. Similarly if K contains a cycle, then if there is any $v \in S$ that does not belong to the cycle, then there must be a $v' \in S$ that is the neighbor of two leaves in K. (Consider the tree that "hangs" at v away from the cycle.) It remains to consider the case when all the vertices in S belong to a cycle in K. But then all vertices in S are extrovert. □

References

1. Bender, E. A. and Canfield, E. R.: The asymptotic number of labeled graphs with given degree sequences. Journal of Combinatorial Theory B, **24** (1978) 296–307
2. Bollobás, B.: Random graphs, Academic Press, (1985)
3. Pippenger, N.: On networks of noisy gates. 26th Annual Symposium on Foundations of Computer Science (1985) 30–38
4. Peleg, D. and Upfal, E.: The token redistribution problem. SIAM Journal of Computing, **18** (1989) 229–243
5. Peleg, D. and Upfal, E.: The Generalized Packet Routing Problem. Theoretical Computer Science, **53** (1987) 281–293

This article was processed using the LaTeX macro package with LLNCS style

Fast Integer Merging on the EREW PRAM

Torben Hagerup*

Max–Planck–Institut für Informatik,
Im Stadtwald, D-W-6600 Saarbrücken, Germany.
Email: torben@mpi-sb.mpg.de

Mirosław Kutyłowski†

Fachbereich Mathematik–Informatik and Heinz–Nixdorf–Institut,
Universität–GH Paderborn, Postfach 1621, D-W-4790 Paderborn, Germany.
Email: mirekk@uni-paderborn.de

Abstract. We investigate the complexity of merging sequences of small integers on the EREW PRAM. Our most surprising result is that two sorted sequences of n bits each can be merged in $O(\log \log n)$ time. More generally, we describe an algorithm to merge two sorted sequences of n integers drawn from the set $\{0, \ldots, m-1\}$ in $O(\log \log n + \log m)$ time using an optimal number of processors. No sublogarithmic merging algorithm for this model of computation was previously known. The algorithm not only produces the merged sequence, but also computes the rank of each input element in the merged sequence. On the other hand, we show a lower bound of $\Omega(\log \min\{n, m\})$ on the time needed to merge two sorted sequences of length n each with elements in the set $\{0, \ldots, m-1\}$, implying that our merging algorithm is as fast as possible for $m = (\log n)^{\Omega(1)}$. If we impose an additional stability condition requiring the ranks of each input sequence to form an increasing sequence, then the time complexity of the problem becomes $\Theta(\log n)$, even for $m = 2$. Stable merging is thus harder than nonstable merging.

1 Introduction

A PRAM is a synchronous parallel machine with a global memory accessible to all processors. The EREW (*exclusive-read exclusive-write*) PRAM is the weakest member of the PRAM family; it disallows concurrent access to the same memory cell by more than one processor, even for reading purposes. The resulting difficulties of interprocessor communication preclude superfast algorithms for even very simple problems: Lower time bounds of $\Omega(\log n)$ for the EREW PRAM were established by Cook, Dwork and Reischuk [7] for the problem of computing the OR of n bits (this bound holds even for the stronger CREW PRAM, which allows concurrent reading),

*Supported in part by the Deutsche Forschungsgemeinschaft, SFB 124, TP B2, VLSI Entwurfsmethoden und Parallelität, and in part by the ESPRIT II Basic Research Actions Program of the EC under contract No. 3075 (project ALCOM).

†Supported by Deutsche Forschungsgemeinschaft grant ME 872/1–4. Present address: Institute of Computer Science, University of Wrocław, Przesmyckiego 20, PL-51-151 Wrocław, Poland. Email: mirekk@plwruw11.bitnet

Beame, Kik and Kutyłowski [1] for the problem of creating n copies of an input t, and by Snir [11], under certain assumptions, for the problem of *searching*, i.e., locating a value in a sorted table of size n. In fact, except for certain algorithms ven by Snir [11] with running times down to $\Theta(\sqrt{\log n})$, the only EREW PRAM gorithms known to the authors that use $o(\log n)$ time on inputs of size n are either r trivial problems (e.g., compute the OR of n bits, at most one of which is a 1), have the property that each output bit is determined by a small set of input ts that is very easy to compute (e.g., the ruling-set algorithm of Cole and Vishkin]). Against this background one might strongly suspect that merging two sorted quences of length n each also requires $\Omega(\log n)$ time on the EREW PRAM. Indeed, nce merging a sequence of length 1 with a sorted sequence of length n is surely o more difficult than merging two sorted sequences of length n each, such a result llows from the lower bound of Snir mentioned above. Note, however, that the lower ound for merging derived from Snir's result is unsatisfactory for two reasons. First, nir places restrictions on the values that can be stored in memory cells and on the ternal workings of individual processors. Second, his proof uses Ramsey theory and erefore breaks down if the values that can occur in the input sequences are drawn om a restricted domain. As shown by Snir, the restrictions placed on memory cells nd processors are essential in the sense that without them, the searching problem an be solved faster than in $\Theta(\log n)$ time. The same can be seen to hold if the ntries in the table to be searched are drawn from a sufficiently small domain, but either effect is apparent for the more difficult problem of merging. We here answer he question of whether sublogarithmic merging on the EREW PRAM is possible n some circumstances and explore the fine boundary between what can and what annot be done. Since the results obtained turn out to be very sensitive to the exact efinition of merging, we back off for a while to provide such definitions.

A sequence $(x_1, \ldots, x_n)$ of integers is called *sorted* if $x_1 \leq \cdots \leq x_n$. Given two orted sequences X and Y of integers, we denote by $X \cup Y$ the sorted sequence f integers that contains as many occurrences of i, for each integer i, as X and put together. We consider several different merging problems. Common to all f these is that the input consists of two sorted sequences $X = (x_1, \ldots, x_n)$ and $X' = (x_{n+1}, \ldots, x_{2n})$ with elements drawn from a set $\{0, \ldots, m-1\}$, where n and m are positive integers known to all processors; without loss of generality we always ssume that $n \geq 4$ and $m \geq 2$. The *value-merging* problem simply calls for the onstruction of the sequence $X \cup X'$. *Rank-merging*, on the other hand, provides a lescription of how to construct $X \cup X'$ from the input sequences. More precisely, he rank-merging problem is to compute a permutation r of $\{1, \ldots, 2n\}$, called a *rank vector* and represented by the sequence $(r(1), \ldots, r(2n))$, such that $x_{r^{-1}(1)} \leq x_{r^{-1}(2)} \leq \cdots \leq x_{r^{-1}(2n)}$. When wanting to be explicit about the parameters n and m, we speak of (n,m)-value-merging and (n,m)-rank-merging. Furthermore, when the meaning is clear from the context, we will omit the qualifiers "value" and "rank".

A procedure for (n,m)-rank-merging clearly implies a procedure for (n,m)-value-merging with the same resource bounds (up to constant factors). Value-merging at first glance might seem more natural than rank-merging. The latter is needed, however, whenever the actual objects to be merged are records containing other information in addition to the integer keys according to which the records are to be merged, since value-merging the keys alone does not allow the full records to be put in order. The definition of rank-merging given above is precise, but oft cumbersome. In most proofs to follow, we therefore employ a less precise but mo convenient informal terminology that we now introduce. The key point is to consid the input elements not as integers, of which several may be identical, but as pairwi distinct objects with several attributes. An input element x_i has a *value*, the integ x_i, and an *original position*, i, and the goal of rank-merging is to mark each inp element with a *rank* from the set $\{1, \ldots, 2n\}$ such that distinct input elements recei distinct ranks and smaller elements receive smaller ranks. Equivalently, the goal to store the input element of rank i in position i of an output sequence of length $2n$ for $i = 1, \ldots, 2n$. In the formal definition, the rank of x_i of course corresponds t $r(i)$.

Consider the easiest among the merging problems introduced above, that ($(n, 2)$-value-merging, i.e., given two sorted sequences X and X' of n bits each, cor struct the sorted sequence of $2n$ bits that contains exactly as many 0's as X an X' put together. Despite its seeming simplicity, this is a fascinating problem whos complexity is still unresolved. Observe first that it is easy to solve the problem i constant time on a CREW PRAM. If a processor is associated with each bit in X and each such processor inspects its own bit and the bit following it, if any, exactl one processor, detecting the change from 0 to 1 in X, will know the number of 0's i X. The number of 0's in X' can be determined in the same way, and the sum of th two counts can be broadcast to each of $2n$ processors, each of which can then easil produce one output bit. Since only the broadcasting part of this algorithm use concurrent reading, it is also evident that even an EREW PRAM can compute an desired bit of the output in constant time. Computing all output bits faster than i $\Theta(\log n)$ time, however, is a challenging problem, and both authors expended con siderable energy trying to prove that it cannot be done. The lower bound of Beame Kik and Kutyłowski cited above shows that any EREW PRAM algorithm that uses the same general approach as the above CREW PRAM algorithm is doomed to fail since broadcasting the total number of 0's to sufficiently many processors will require $\Omega(\log n)$ time.

Our main result is that not only $(n, 2)$-value-merging, but in fact $(n, 2)$-rank-merging can be done in $O(\log\log n)$ time on an EREW PRAM. An interesting feature of the algorithm is that, intuitively speaking, a significant portion of the computation is carried out by processors operating on incorrect data. A major concern in the design of the algorithm was to prevent such processors from interfering with the useful part of the computation, e.g., by causing concurrent reading or writing. We also extend the algorithm to solve the general (n, m)-rank-merging problem, in which case the running time becomes $O(\log\log n + \log m)$, and reduce the number of processors used to obtain an algorithm that executes $O(n)$ operations, which is optimal. On the other hand, we prove that (n, m)-value-merging requires $\Omega(\log \min\{n, m\})$ time, thereby showing our algorithm to be as fast as possible for $m = (\log n)^{\Omega(1)}$ (assume that we switch to a standard $O(\log n)$-time algorithm for $m \geq n$). Our discovery that integer merging is easier than general merging on the EREW PRAM to some extent parallels what is known for the CREW PRAM. General merging can be done in $O(\log\log n)$ time on the CREW PRAM with an optimal number of processors [9] and there is a corresponding lower bound [4], whereas an algorithm that runs in $O(\log\log\log m)$ time using an optimal number of processors known for the case of input numbers drawn from the set $\{0, \ldots, m-1\}$ [2].

It is frequently desirable to impose an additional restriction on the rank vector obtained by merging two sorted sequences X and X' of length n each. Requiring that $r(1) < \cdots < r(n)$ and $r(n+1) < \cdots < r(2n)$ (informally, the relative order of the elements in X and X' is to be preserved), we obtain the problem of *stable (rank-) merging*. Stable merging clearly is no easier than unadorned rank-merging, which we will also call *unstable merging*. Our fast algorithm solves the unstable merging problem only and cannot be extended to stable merging. In fact, we prove a lower time bound of $\Omega(\log n)$ for stable $(n, 2)$-merging. Stable merging is hence more difficult than unstable merging. Although such a result might seem natural for randomized algorithms (compare, e.g., the integer sorting algorithm of Rajasekaran and Reif [10]), in our deterministic setting it is somewhat unexpected.

2 Lower bounds

In this section we first show that (n, m)-value-merging needs $\Omega(\log \min\{n, m\})$ time on the EREW PRAM. We then give an $\Omega(\log n)$ lower time bound for stable merging. All of our lower bounds allow any number of processors and memory cells, arbitrary local computation, the storage of arbitrary values in memory cells, and nonuniformity. They are derived from the following result, proved by Beame, Kik and Kutyłowski [1] and valid under the same assumptions.

Lemma 2.1 *Every EREW PRAM algorithm that takes as input a single bit x and stores the value of x in n distinct fixed memory cells has a running time of $\Omega(\log n)$.*

Corollary 2.2 *In order to show that an EREW PRAM algorithm runs in $\Omega(\log t)$ time, it suffices to exhibit two inputs that differ in just one component (i.e., the contents of exactly one input cell differ), but whose associated outputs, as computed by the algorithm, differ in t components.*

We first use Corollary 2.2 to show the lower bound on (n, m)-value-merging.

Theorem 2.3 *(n, m)-value-merging requires $\Omega(\log \min\{n, m\})$ time on the EREW PRAM.*

Proof: Let $t = \min\{n, m\}$ and consider the n-element sequences $X = (1, 2, \ldots, t-2, t-1, t-1, \ldots, t-1)$, $Y_1 = (0, t-1, t-1, \ldots, t-1)$ and $Y_2 = (t-1, t-1, \ldots, t-1)$. Viewed as input to a merging algorithm, (X, Y_1) and (X, Y_2) differ in exactly one component, but $X \cup Y_1 = (0, 1, \ldots, t-3, t-2, t-1, \ldots, t-1)$ and $X \cup Y_2 = (1, 2, \ldots, t-2, t-1, t-1, \ldots, t-1)$ differ in their first $t-1$ components. The theorem now follows from Corollary 2.2. □

Section 4 presents an (n, m)-rank-merging algorithm that runs in $O(\log\log n + \log m)$ time. By Theorem 2.3, this is as fast as possible for $m = (\log n)^{\Omega(1)}$. For $m = (\log n)^{o(1)}$ the complexity of the problem is unknown.

In the remainder of this section we consider stable $(n, 2)$-merging. We first show a lower time bound of $\Omega(\log n)$ for the *strictly stable merging* of two sorted sequences $X = (x_1, \ldots, x_n)$ and $X' = (x_{n+1}, \ldots, x_{2n})$, which we define to be the problem of computing a permutation r that satisfies the following condition in addition to those of stable merging: $r(i) < r(j)$ for all integers i and j with $1 \le i \le n < j \le 2n$ such that $x_i = x_j$ (i.e., in the case of ties, elements in X are to be considered smaller). Since this result is subsumed by our next lower bound and mostly serves as a warm-up to the latter, we shall not formulate it as a theorem.

Suppose that an algorithm for strictly stable merging maps the input sequences $X_0 = (0, 0, \ldots, 0)$ and $X' = (0, 0, \ldots, 0)$ to the rank vector r_0, and the input sequences $X_1 = (0, 0, \ldots, 1)$ and X' to the rank vector r_1. Then $(r_0(n+1), \ldots, r_0(2n)) = (n+1, n+2, \ldots, 2n)$, but $(r_1(n+1), \ldots, r_1(2n)) = (n, n+1, \ldots, 2n-1)$. The desired lower bound again follows from Corollary 2.2.

As announced above, we can extend the last result to (not necessarily strict) stable merging. Note how the above proof breaks down in the absence of strict stability.

Theorem 2.4 *Stable $(n, 2)$-merging requires $\Omega(\log n)$ time on the EREW PRAM.*

Proof: Consider $2n+1$ inputs $(X_0, Y_0), \ldots, (X_{2n}, Y_{2n})$ defined as follows: For $i = 0, \ldots, n$, X_i consists of $n - i$ zeros followed by i ones, and Y_i consists of n ones. For $i = 0, \ldots, n$, X_{n+i} consists of n ones, and Y_{n+i} consists of i zeros followed by $n - i$ ones. Given a stable merging algorithm, let r_i be the rank vector computed by the algorithm on the input sequences (X_i, Y_i), for $i = 0, \ldots, 2n$, and let $\Phi_i = \sum_{j=1}^{n} r_i(j)$, that is, Φ_i is the sum of the ranks of the elements of X_i in $X_i \cup Y_i$. In order to gain intuition for the remainder of the proof, consider an experiment consisting of $2n$ steps. In the ith step, for $i = 1, \ldots, 2n$, we replace (X_{i-1}, Y_{i-1}) by (X_i, Y_i) as input to the merging algorithm and observe the resulting change in Φ_i.

Since $\Phi_0 = \sum_{j=1}^{n} j$ and $\Phi_{2n} = \sum_{j=1}^{n}(n+j)$, we have $\Phi_{2n} - \Phi_0 = n^2$. It follows that for some i_0 with $1 \le i_0 \le 2n$, $\Phi_{i_0} - \Phi_{i_0-1} \ge n/2$. Let $t = \sqrt{n/2}$ and consider two possibilities:

Case 1: $\max_{1 \le j \le n}(r_{i_0}(j) - r_{i_0-1}(j)) \le t$. In this case at least t elements of X_i change their rank in step i_0 of the experiment, i.e., $(r_{i_0}(1), \ldots, r_{i_0}(n))$ and $(r_{i_0-1}(1), \ldots, r_{i_0-1}(n))$ differ in at least t components. Since the input sequences (X_{i_0}, Y_{i_0}) and (X_{i_0-1}, Y_{i_0-1}) differ in exactly one component, the lower bound now follows from Corollary 2.2 and the fact that $\log t = \Omega(\log n)$.

Case 2: *For some j with $1 \le j \le n$, $r_{i_0}(j) > r_{i_0-1}(j) + t$.* Note that in a stable merging of two sequences X and Y, the rank of the jth element of X is precisely j plus the number of elements in Y of smaller rank. Hence the element of X_i that increases its rank by more than t in step i_0 of the experiment changes its relative order with respect to more than t elements of Y_i. By another application of the same principle, each of these more than t elements of Y_i changes its rank in the same step of the experiment. The lower bound now follows as in Case 1. □

If even the requirement of stability is relinquished, merging becomes substantially easier, as demonstrated in the following sections.

3 Bitonic ranking

For $k \in \mathbb{N}$ and $b_1, b_2, \ldots, b_k \in \{0, 1\}$, we call a sequence of 0's and 1's a $b_1 b_2 \cdots b_k$-*sequence* if it belongs to the set described by the regular expression $b_1^* b_2^* \cdots b_k^*$, i.e., if consists of zero or more occurrences of b_1, followed by zero or more occurrences of , etc. 010- and 101-sequences are collectively said to be *bitonic*. In this section we ·scribe an algorithm for *bitonic ranking*, i.e., given a bitonic sequence $(x_1, \ldots, x_n)$, ·mpute a permutation r of $\{1, \ldots, n\}$ such that $x_{r^{-1}(1)} \leq \cdots \leq x_{r^{-1}(n)}$. Note, as ·fore, that ranking can be used to sort, i.e., to actually arrange the input elements . sorted order, and that the opposite is true provided that each element is marked ith its original position. Since the concatenation of a sorted bit sequence with the ·verse of another such sequence is bitonic, rank-merging of sorted bit sequences ·duces to bitonic ranking.

'heorem 3.1 *When $n \geq 4$ is a power of 2, a bitonic sequence of length n can be ınked in $O(\log\log n)$ time on an EREW PRAM using $O(n)$ processors and $O(n)$ pace.*

'roof: We show Theorem 3.1 by induction on n, i.e., by giving a recursive algorithm ıat performs as stated. It turns out to be necessary for the inductive argument to ıake an additional assumption about the rank vector r computed by the recursive lgorithm. Recall that every permutation can be written as a *product of cycles*, e., as the composition of a number of cyclic permutations. We inductively assume hat r has a particularly simple cycle structure, namely that it can be written as product of *disjoint transpositions*, i.e., of cyclic permutations of disjoint sets of ize 2.

Given a bitonic input sequence of length n, where $n \geq 4$ is a power of 2, let $= 2^{\lfloor \log n/2 \rfloor}$ and $t = n/s$ and consider the input sequence to be stored in an $s \times t$ *ableau* $\mathcal{A}$ in column-major order, i.e., the first s elements in the sequence are stored, rom top to bottom, in the first column of the tableau, the next s elements are stored n the next column, etc. Number the rows and columns of all tableaus consecutively ·tarting at 1. It is easy to see that each row of $\mathcal{A}$ is bitonic, so that the rows of $\mathcal{A}$:an be sorted by recursive applications of the algorithm (or trivially, if $t = 2$). The ·esulting tableau $\mathcal{B}$ has at most one column containing distinct values (i.e., both)'s and 1's). In the following assume that $\mathcal{B}$ has such a column, called the *critical :olumn*; it will be easy to see that the case in which $\mathcal{B}$ has no critical column is ıandled correctly.

Since the critical column of $\mathcal{B}$ is bitonic, the most natural way to solve the original problem would be first to sort the rows of $\mathcal{A}$ to obtain $\mathcal{B}$, and then to sort the critical column of $\mathcal{B}$, which results in a sorted sequence (stored in column-major order). However, this would give a recurrence relation for the running time $T(n)$ of the algorithm on input of size n of the approximate form $T(n) = 2T(\sqrt{n}) + \Omega(1)$, which solves to $T(n) = \Omega(\log n)$. Since this is clearly too slow, we must proceed differently. Our approach is to sort the critical column of $\mathcal{B}$ *before* $\mathcal{B}$ has been constructed, i.e., while the rows of $\mathcal{A}$ are being sorted! In this way we obtain an algorithm that uses only constant time in addition to a number of recursive calls on arguments of size $O(\sqrt{n})$, all of which can be started simultaneously; the running time is $O(\log\log n)$, as desired.

We cannot actually compute the critical column of $\mathcal{B}$ and store it in a fixed place (this would require $\Omega(\log n)$ time). What we can do, however, is to construct two candidate sequences, one of which will equal the critical column C of $\mathcal{B}$. Note carefully that we take C to be simply a sequence of 0's and 1's; the additional information associated with the elements in the critical column of $\mathcal{B}$, such as their original pos tions, is not available. Begin by constructing the sequence $D = (d_1, \ldots, d_s)$, wher d_i is the number of 0's in the ith row of $\mathcal{A}$ (or $\mathcal{B}$), for $i = 1, \ldots, s$. Computing c is easy since the ith row of $\mathcal{A}$ is bitonic: In the sequence of values there is at mos one change from 0 to 1 and one change from 1 to 0, and d_i is a simple function c the positions of these changes, which can be determined in constant time by lettin each pair of consecutive elements be inspected by a processor. If d is the index c the critical column, then for each $i \in \{1, \ldots, s\}$ we have $d = d_i$ or $d = d_i + 1$. Henc if the parity of d were known, a processor P_i associated with the ith element of I could determine the exact value of d and hence the ith element of C. The parity of being unknown, $P_1, \ldots, P_s$ instead construct an "even candidate" $E = (e_1, \ldots, e_s)$ equal to C if d is even, and an "odd candidate" $U = (u_1, \ldots, u_s)$, equal to C if is odd (see Fig. 1). In precise terms, if d_i is even, then P_i sets $e_i := 0$ and $u_i := 1$ otherwise it sets $e_i := 1$ and $u_i := 0$. Observe that E and U are both bitonic.

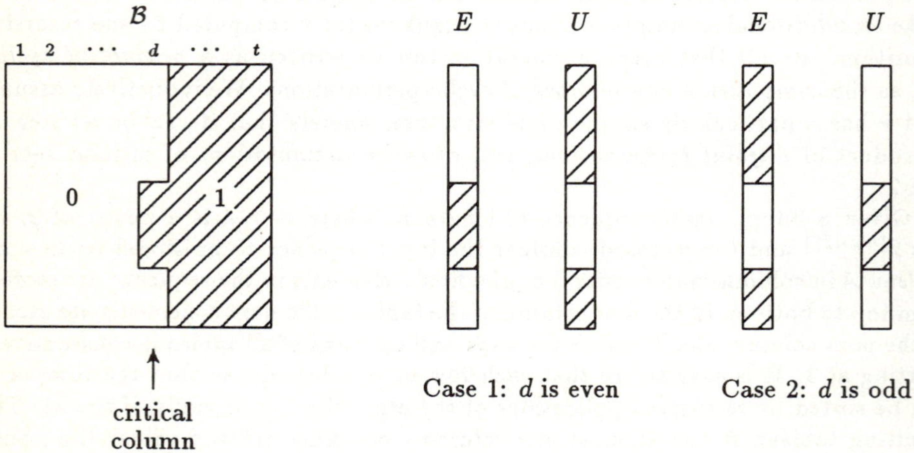

Fig. 1. The candidate critical columns E and U.

Recursively rank both E and U and recall that this produces permutations r_E and r_U of $\{1, \ldots, s\}$ that sort E and U, respectively. Our remaining task is, using constant time after the construction of $\mathcal{B}$, to permute the elements in the critical column of $\mathcal{B}$, now marked with their original positions, according to the proper one of r_E and r_U (i.e., r_E if d is even, and r_U if d is odd). The problem again is that the parity of d is unknown. We might therefore apply both r_E and r_U, arguing that permuting elements in columns other than the critical column does no harm. More precisely, let P_i move the elements in position i of its candidate critical columns in $\mathcal{B}$, i.e., those indexed by d_i and $d_i + 1$ (introduce dummy columns numbered 0 and $t + 1$), according to r_E and r_U, with r_E operating in the column of even index and r_U operating in the column of odd index. However, while this applies one of the permutations r_E and r_U correctly in column d, the other permutation may be applied partly in column $d - 1$ and partly in column $d + 1$. This is because not all processors have the same two candidate critical columns, and it does not lead to the desired result (some positions in the tableau may afterwards hold either 0 or elements). In order to counter this problem, we modify the procedure so that it moves an element to a new position only if the element in that position before the move has a different value (e.g., a 0 may be moved to a position previously holding a 1, but not to one previously holding a 0). This clearly prevents movement of elements outside of the critical column, as desired. Because r_E and r_U are products of disjoint transpositions, according to the inductive hypothesis, the modified procedure still has the effect of permuting the elements within the critical column – we have simply cancelled all interchanges of 0's with 0's and of 1's with 1's). Since the critical column is sorted and no other column is modified, the resulting tableau is sorted.

At this point we have obtained a permutation r that sorts the input sequence. Because an element may be moved both by a transposition within a row and subsequently by a transposition within the critical column, however, r is not necessarily a product of disjoint transpositions. In order to satisfy the inductive assumption, a final constant-time computation described below replaces r by a product r' of disjoint transpositions that is equivalent to r, i.e., results in the same distribution of 0's and 1's.

Let G be an undirected graph with n vertices arranged in an $s \times t$ array and with each vertex corresponding in the natural way to a tableau position. There is an edge between two vertices if and only if elements in the corresponding positions are interchanged during the sorting procedure described above. G has "horizontal" edges, corresponding to interchanges within rows during the construction of $\mathcal{B}$, and "vertical" edges, corresponding to interchanges within the critical column. Since only vertices in the critical column have incident vertical edges and no vertex is incident to more than one horizontal edge or to more than one vertical edge, it is easy to see that no connected component of G contains more than 4 vertices. Call a vertex of G a 01-vertex if the corresponding position initially contains a 0, but after the application of r contains a 1, and define a 10-vertex analogously. Since all "interaction" takes place within connected components of G, the number of 01-vertices within a connected component equals the number of 10-vertices. Now construct a new graph G' from G as follows: Within each connected component of G, replace all edges by new edges that form a perfect matching of the 01-vertices with the 10-vertices. G' in a natural way corresponds to a product of disjoint transpositions, which clearly is equivalent to r and can be taken as r'. Since connected components are treated independently and each component represents a problem of constant size, the construction of G' and r' from G can be carried out in constant time on an EREW PRAM if only we can associate a unique processor with each connected component of G. But since each nontrivial component contains exactly one vertical edge, this is easy: Associate a processor with each vertex of G and assign responsibility for a nontrivial component to the processor associated with the upper endpoint, say, of the vertical edge. This ends the description of the recursive algorithm, which can clearly be executed using $O(n)$ processors and $O(n)$ space. □

4 Optimal merging

In this section we extend the basic algorithm of the previous section in two directions: First we show how the algorithm can be modified to handle input numbers in the set

$\{0, \ldots, m-1\}$, for arbitrary $m \geq 2$. Then we reduce the number of processors used by the modified algorithm in order to obtain an algorithm with optimal speedup.

Lemma 4.1 *(n, m)-rank-merging problems can be solved on an EREW PRAM using $O(\log\log n + \log m)$ time, $O(nm)$ processors and $O(nm)$ space.*

Proof: Without loss of generality assume that $n \geq 4$ and that n is a power of 2 (to achieve this extend each input sequence by a suitable number of dummy elements larger than all other elements). Since nm processors can easily create m copies of the entire input in $O(\log m)$ time, it suffices to show how to compute the ranks of all occurrences of an arbitrary but fixed value $i \in \{0, \ldots, m-1\}$ using $O(\log\log n + \log m)$ time, n processors and $O(n)$ space in addition to a constant number of applications of the algorithm of Theorem 3.1 to inputs of size $O(n)$. This can be done as follows (see Fig. 2):

Given two sorted input sequences X and X' of length n each and with elements in the set $\{0, \ldots, m-1\}$, let Y be the concatenation of X with the reverse of X'. In order to compute ranks for all occurrences of some value $i \in \{0, \ldots, m-1\}$, first derive a 010-sequence V from Y by replacing each value $\leq i$ by 0 and each value $> i$ by 1. Derive another 010-sequence W from Y by replacing each value $< i$ by 0 and each value $\geq i$ by 1. Sort both V and W and form a new 010-sequence Z as the componentwise exclusive-or of the two resulting 01-sequences. For $j = 1, \ldots, 2n$, it is easy to see that Z contains a 1 in position j exactly if sorting Y places an i in position j, i.e., if $j \in R$, where R is the set of ranks to be distributed among the occurrences of i in Y. Marking each 1 in Z with its position in Z and subsequently sorting Z stores the elements of R in the last positions of a sequence S_R. In a similar fashion we store in the last positions of another sequence S_I (constructed as explained below) the elements of the set I of positions that contain an occurrence of i in the (unsorted) sequence Y. Matching corresponding positions in the two sequences S_I and S_R now defines a bijection between I and R, i.e., computes the desired ranks.

In order to compute S_I, first derive from each half of Y a 010-sequence by replacing each occurrence of i by 1 and each occurrence of a value different from i by 0. Mark each 1 in either half with its position in Y, sort the halves separately and concatenate the sorted first half with the reverse of the sorted second half. This creates a new 010-sequence, the sorting of which easily yields S_I. This ends the proof of Lemma 4.1. □

Before deriving an optimal algorithm from Lemma 4.1, we discuss the reasons why this is not as easy as it might seem. The standard way of merging optimally, given a nonoptimal subroutine for merging, is to begin by merging sequences of equally-spaced *representatives*. As a result of this computation, each representative knows its rank in the opposite sequence to within the distance between successive representatives. Each representative then determines its exact rank in the opposite sequence, typically by binary search in the relevant subsequence, which splits the original problem into a collection of small and independent subproblems that are easily solved in parallel. The binary search by the representatives is easy on the CREW PRAM, but on the EREW PRAM requires coordination between representatives. This coordination between representatives can be achieved by means of prefix

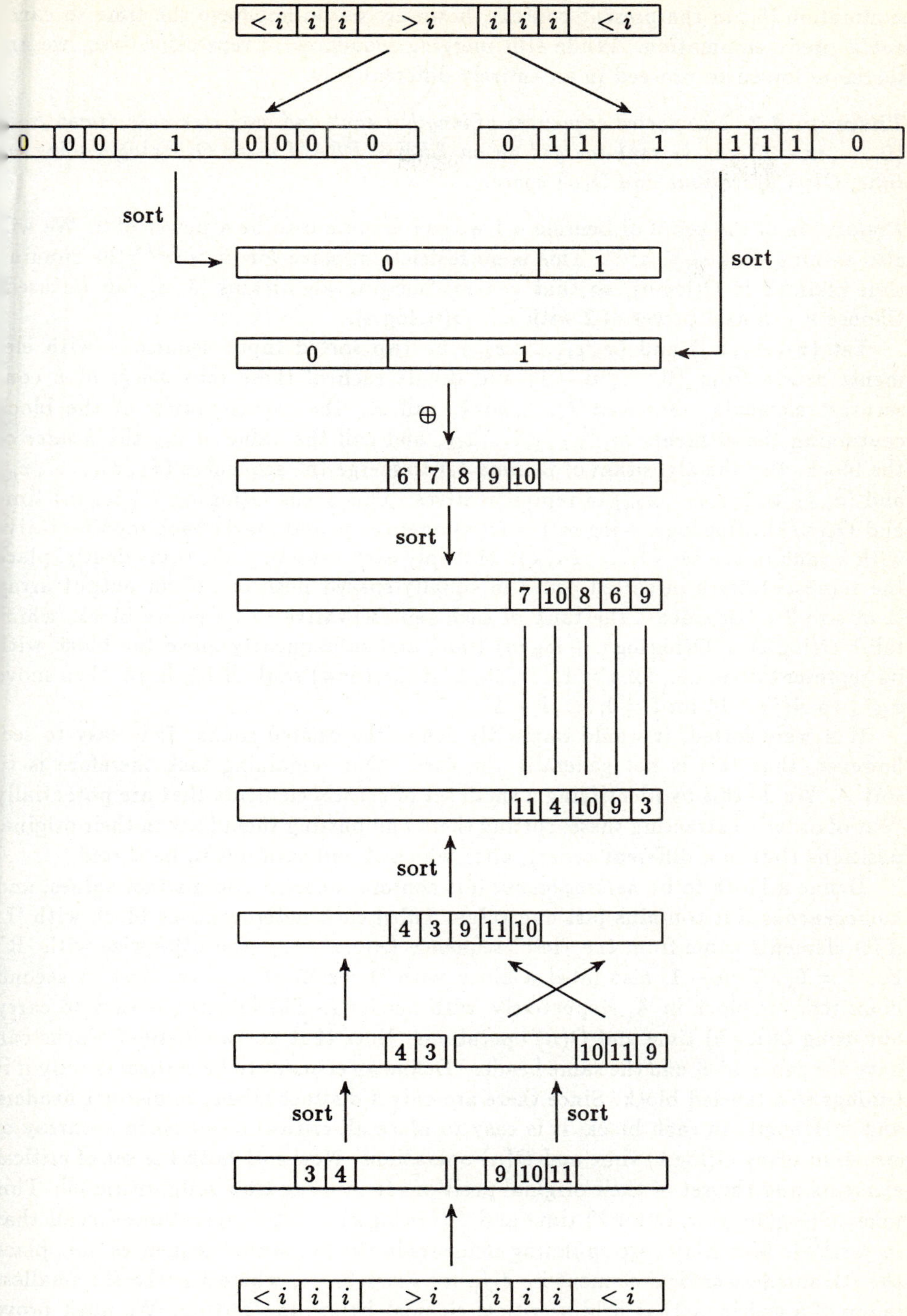

Fig. 2. Computing the ranks of all occurrences of i. Nonzero integers mark key values of 1.

summation [8]; in the present context, however, we do not have the time to carr out a prefix summation. While still merging sequences of representatives, we ar therefore forced to proceed in an entirely different way.

Theorem 4.2 *Two sorted sequences of length* n *each and with elements drawn fror* $\{0,\ldots,m-1\}$ *can be rank-merged on an EREW PRAM using* $O(\log\log n + \log m$ *time,* $O(n)$ *operations and* $O(n)$ *space.*

Proof: As in the proof of Lemma 4.1 we can assume n to be a power of 2. We wil also assume that $m \leq n^{1/3}$. This is no restriction, since for $m > n^{1/3}$ the runnin time claimed is $O(\log n)$, so that general merging algorithms [3, 8] can be used Choose $k \leq n$ as a power of 2 with $k = \Theta(m \log n)$.

Let $(x_1,\ldots,x_n)$ and $(x_{n+1},\ldots,x_{2n})$ be two sorted input sequences with ele ments drawn from $\{0,\ldots,m-1\}$ and divide each of these into *blocks* of k con secutive elements. For $i = 1,\ldots,2n/k$, call x_{ik} the *representative* of the block containing the elements $x_{(i-1)k+1},\ldots,x_{ik}$, and call the value of x_{ik} the *header* o the block. Use the algorithm of Lemma 4.1 to merge the sequences $(x_k, x_{2k},\ldots,x_n$ and $(x_{n+k}, x_{n+2k},\ldots,x_{2n})$ of representatives. This needs $O(\log\log n + \log m)$ tim and $O((n/k)m(\log\log n + \log m)) = O(n)$ operations and marks each representativ with a rank in the set $\{1,\ldots,2n/k\}$. Multiply each rank by k or, equivalently, plac the representatives in sorted order in equally-spaced positions of an output arra A of size $2n$. Broadcast the rank of each representative to its entire block, whicl takes $O(\log k) = O(\log\log n + \log m)$ time, and subsequently move the block witl its representative, i.e., for $i = 1,\ldots,2n/k$, if the (new) rank of x_{ik} is jk, then mov x_{ik-l} to $A[jk-l]$, for $l = 0,\ldots,k-1$.

If A were sorted, it would implicitly define the desired ranks. It is easy to see however, that this is not generally the case. Our remaining task therefore is t sort A. We do this by identifying a small set of *critical* elements that are potentiall "out of order", extracting these, sorting them and putting them back in their origina positions (but in a different order), after which A will turn out to be sorted.

Define a block to be *heterogeneous* if it contains at least two distinct values, and *homogeneous* if it contains just one value. Label each heterogeneous block with 'L if its elements came from the "left" sequence $(x_1,\ldots,x_n)$, and otherwise with 'R' For $i = 0,\ldots,m-1$, also label a block with '1' or '2' if it is the first or second homogeneous block in A, respectively, with header i. The labeling is easy to carry out using $O(\log k)$ time and $O(n)$ operations. Note that no two distinct blocks can have the same label *and* the same header. Define an element to be *critical* exactly if it belongs to a labeled block. Since there are only 4 distinct labels, m distinct headers and k elements in each block, it is easy to place all critical elements in an array of size $4km$ using $O(\log k)$ time and $O(n)$ operations. Now sort both the set of critical elements and the set of their original positions in A using Cole's algorithm [5]. This takes $O(\log(km)) = O(\log k)$ time and $O(km \log k) = O(n)$ operations (recall that $m \leq n^{1/3}$). Match up corresponding elements in the two sorted sequences, i.e., place the ith smallest critical element in $A[j]$, for $i = 1,2,\ldots$, where j is the ith smallest index of a cell in A that held a critical element before the sorting. We must prove that A is now sorted.

Since every element of a heterogeneous block is critical, the subsequence of the sequence stored in A consisting of all noncritical elements is clearly sorted. After he sorting of the critical elements, the same by definition applies to the subsequence consisting of all critical elements. Furthermore, since no noncritical element s preceded by a larger element before the sorting of the critical elements, the same ondition holds after the sorting. The only remaining problem is hence that some ıoncritical element might be followed by a smaller critical element after the sorting. Co see that this is impossible, note that before the sorting, a noncritical element with value i is followed by at most $2(k-1)$ smaller elements (each such element must ›e a nonrepresentative in a block with header $\geq i$ but containing a value smaller han i; each input sequence contributes at most one such block), while it is preceded y at least $2k$ critical elements with value i (those in the first two homogeneous locks with header i); the number of positions available to critical elements before he noncritical element under consideration hence exceeds the number of positions equired by smaller critical elements. □

References

[1] P. Beame, M. Kik and M. Kutyłowski, Information broadcasting by exclusive read PRAMs, submitted.

[2] O. Berkman, J. JáJá, S. Krishnamurthy, R. Thurimella and U. Vishkin, Some triply-logarithmic parallel algorithms, *in* Proc. 31st Annual IEEE Symposium on Foundations of Computer Science (1990), pp. 871–881.

[3] G. Bilardi and A. Nicolau, Adaptive bitonic sorting: An optimal parallel algorithm for shared-memory machines, *SIAM J. Comput.* **18** (1989), pp. 216–228.

[4] A. Borodin and J. E. Hopcroft, Routing, merging, and sorting on parallel models of computation, *J. Comput. Syst. Sci.* **30** (1985), pp. 130–145.

[5] R. Cole, Parallel merge sort, *SIAM J. Comput.* **17** (1988), pp. 770–785.

[6] R. Cole and U. Vishkin, Deterministic coin tossing with applications to optimal parallel list ranking, *Inform. and Control* **70** (1986), pp. 32–53.

[7] S. Cook, C. Dwork and R. Reischuk, Upper and lower time bounds for parallel random access machines without simultaneous writes, *SIAM J. Comput.* **15** (1986), pp. 87–97.

[8] T. Hagerup and C. Rüb, Optimal merging and sorting on the EREW PRAM, *Inform. Process. Lett.* **33** (1989), pp. 181–185.

[9] C. P. Kruskal, Searching, merging, and sorting in parallel computation. *IEEE Trans. Comput.* **32** (1983), pp. 942–946.

[10] S. Rajasekaran and J. H. Reif, Optimal and sublogarithmic time randomized parallel sorting algorithms, *SIAM J. Comput.* **18** (1989), pp. 594–607.

[11] M. Snir, On parallel searching, *SIAM J. Comput.* **14** (1985), pp. 688–708.

Approximation Algorithms for Graph Augmentation

Samir Khuller *   *Ramakrishna Thurimella* †

Abstract

We study the problem of increasing the connectivity[1] of a graph at an optimal cost. Since the general problem is NP-hard, we focus on efficient approximation schemes that come within a constant factor from the optimal. Previous algorithms either do not take edge costs into consideration, or run slower than our algorithm. Our algorithm takes as input an undirected graph $G_0 = (V, E_0)$ on n vertices, that is not necessarily connected, and a set *Feasible* of m weighted edges on V, and outputs a subset *Aug* of edges which when added to G_0 make it 2-connected. The weight of *Aug*, when G_0 is initially connected, is no more than twice the weight of the least weight subset of edges of *Feasible* that increases the connectivity to 2. The running time of our algorithm is $O(m + n \log n)$. We also study the problem of increasing the edge connectivity of any graph G, to k, within a factor of 2 (for any $k > 0$). The running time of this algorithm is $O(nk \log n(m + n \log n))$. We observe that when k is *odd* we can use different techniques to obtain an approximation factor of 2 for increasing edge connectivity from k to $(k + 1)$ in $O(kn^2)$ time.

1 Introduction

Augmenting the connectivity of communication networks is increasingly becoming important provide reliable means of communication. Informally, the problem is the following: we hav "current" network, and we wish to increase the connectivity of the network by adding edg Each *feasible* edge has an associated weight, and we want to achieve the desired connectivity adding the least total weight of edges.

The Weighted Augmentation Problem: Consider a graph $G = (V, E)$ with a weight w_i on edge e_i, where E is the set of feasible edges. We are also given a "current" network, graph $G_0(V, E_0)$ that is a subgraph of G. The goal is to add a minimum weight set of ed *Aug*, to G_0, such that the resulting graph is k-connected for a given k. The edges that we permitted to add, are required to be edges from the graph G (the feasibility graph). If $E_0 =$ and $k = 1$ then the problem is that of finding a minimum weight spanning tree on V. It is w known that this problem can be solved very efficiently. For $k > 1$, the problem turns out to NP-hard. We will study approximation algorithms for the general problem.

Our Results: Let m and n denote the number of edges and the number of vertices, respective of the input graph. In this paper we provide a simple algorithm (for $k = 2$) that produces approximation within a factor of 2 in $O(m + n \log n)$ time. Note that for sparse networks t algorithm is faster by a factor of n^2/m. (The previous algorithm for the problem is due

*Institute for Advanced Computer Studies (UMIACS), The University of Maryland at College Park, Coll Park, MD 20742. This work is partially supported by NSF grants CCR-8906949, CCR-9103135 and CCR-91113

†Department of Mathematics and Computer Science, University of Denver, Denver, CO 80208. Part of t work was done while this author was with UMIACS.

[1]Connectivity refers to both edge and vertex connectivity throughout the paper unless stated explici otherwise.

31] and takes $O(n^2)$ time.) We design algorithms for both edge and vertex connectivity. The dup is obtained not only by using a better subroutine for branchings in directed graphs, but lso doing away with the dynamic programming subroutine (computation of function $DIST$) l by [FJ81]. Since the subroutine $DIST$ used by [FJ81] computes $\Theta(n^2)$ values there is no to make their algorithm more efficient. It should be noted however that our strategy is lar to that of [FJ81].

We also show that an approximation factor of 2 can be obtained to augment the edge nectivity of any graph to being k-edge connected. The running time of this algorithm is $k \log n(m + n \log n))$. This improves the approximation factor of 3 that was given by [FJ81] ncrease the edge connectivity of an unconnected graph to 2, albeit the running time of our me is significantly higher. When k is *odd* we show how to obtain the same approximation or for increasing edge connectivity from k to $(k+1)$ in $O(kn^2)$ time.

nificance of the Results: The subtlety of the construction, and the manner in which it is bined with the branching gives the scheme its simplicity and elegance.

The application of algorithms for finding k-edge disjoint branchings (for directed graphs) to ementing edge connectivity in undirected graphs appears to have gone unnoticed until now. s gives the first polynomial time algorithm for obtaining the approximation factor of 2 for ementing the edge connectivity of *any* graph to *any* k (when the edges have weights). The lge disjoint branching problem is solved by formulating it as a weighted matroid intersection blem [Ed79, G91a]. It remains of interest to see how the solutions to various problems in the a of Matroids will give rise to approximation schemes for NP-hard problems.

evious Work: The problem of increasing the connectivity to 2 was first considered by [ET76], o showed that it was NP-hard. In subsequent work, Frederickson and JáJá [FJ81] showed t the problem remained NP-hard even if G_0 is connected. They also provided approximation orithms to solve the problem with a running time of $O(n^2)$ and produced an approximation hin a factor of 2 of the optimal solution [FJ81] when G_0 is initially connected. For the special e of the edge weights satisfying the triangle inequality they designed algorithms that produce approximation within a factor of $\frac{3}{2}$ of the optimal solution [FJ82].

The more general problem of increasing the edge connectivity for *any* k has not been studied he weighted case to our knowledge. We also observe that the NP-hardness proof of [FJ81] ends to work for any $k > 0$.

lated Work: When the feasible set of edges is the complete graph, and all the edges have t weight, the augmentation can be done optimally to obtain any desired edge connectivity polynomial time. In this setting, there have been a number of papers that have studied ph augmentation problems, starting from the seminal paper by Eswaran and Tarjan who t introduced the problem of making a connected graph biconnected by adding the smallest of edges [ET76]. A linear time implementation was provided by Rosenthal and Goldner G77] (see also [HR91b]). This has recently been extended to triconnecting a graph optimally Hsu and Ramachandran [HR91a]. Subsequent work has led to the generalization of the blem in many ways, and also led to faster algorithms. This was first shown by Watanabe l Nakamura [WN87, Wa88]. Recently much faster algorithms have been obtained by Naor, sfield and Martel [NGM90] and by Gabow [G91b]. A generalization of the problem was sidered by Frank [Fr90] who has shown a more general problem to be solvable in polynomial e, but the algorithm is not very efficient. In the context of planar graphs, [KB91] have given proximation algorithms for the case when edges have to be added, maintaining planarity.

For weighted directed graphs the following problem is solvable in polynomial time [Ed79, '89, G91a]. In a digraph find a minimum weighted set of edges so as to have k-edge disjoint hs from a source to every vertex. Gabow provides the fastest implementation of Edmond's algorithm [G91a]. We use this algorithm to obtain an approximation factor of 2 for the e connectivity augmentation problem in undirected graphs.

2 Preliminaries

A graph is said to be k-vertex (k-edge) connected if it has at least $(k+1)$ vertices (edg and the deletion of any $(k-1)$ vertices (edges) leaves the graph connected. A *branching* directed graph G rooted at a vertex r is a spanning tree of G such that each vertex excep has indegree exactly 1 and r has indegree zero. The *weight* of a branching is the sum of weights of the edges of the branching. A *minimum* weight branching is a branching with least weight. For the definitions of bridges, 2-edge (2-vertex) connected components the read referred to [Ev79]. The 2-vertex connected components of a graph are also referred to as *blo* For a vertex v in a rooted tree Γ, let the components formed by the deletion of v be ca $C_1(v), C_2(v), \ldots, C_{d(v)}(v)$, where $d(v)$ is the degree of v in Γ. If v is not the root, we will ass that $C_1(v)$ is the component that contains the root, and the other components correspond subtrees rooted at the children of vertex v. In a rooted tree, for a vertex u we denote its pa by $p(u)$.

As a matter of notation, we refer to an undirected edge between two vertices x and y (x, y). On the other hand, a directed edge from x to y is denoted by $x \to y$.

We can assume that G_0 is a connected graph. If the graph is not initially connected, fin minimum spanning tree and add the edges of the tree to G_0. This would yield an approximat within a factor of 3 to the problem of finding a least-weight 2-connected spanning subgraph a 2-connected weighted graph. We can also assume that G is 2-connected.

3 Increasing Edge Connectivity from 1 to 2

Notice that we only need to show how to increase the edge connectivity of a tree due to following observation. If we are given G_0 with nontrivial 2-edge connected components, t we can shrink the vertex sets of these components into single vertices, resulting in a tree wh edges are the bridges of G_0. The edges to be retained from *Feasible* are the minimum wei edges that connect vertices of different 2-edge connected components of G_0. (Observe that edges of *Feasible* that connect vertices of the same 2-edge connected component are of no use augmenting G_0. Similarly, among the edges that connect different 2-edge connected compone only the minimum weight edge is of interest.)

From G_0, we will construct a directed graph G^D, and find a minimum weight branching fr a vertex r. In case there is no branching that spans all the vertices, we can show that there no way to increase the connectivity of the network. Using a minimum weight branching of G we can find a set of edges of $G - G_0$ whose addition will increase the connectivity of G_0. can also show that the total weight of the edges added by this technique is bounded by tw the weight of an optimal augmentation.

Algorithm *Find Edge Aug*
Input: Graph G and a spanning tree G_0.
Output: A set of edges $Aug \subseteq E - E_0$.

(1) (*Construct* $G^D = (V, E_D)$)

(a) Pick an arbitrary leaf r and root the tree G_0 at r by directing all the edges towar the root. Denote the resulting tree by Γ.

(b) Add to E_D the directed tree edges of Γ and set their weight to zero.

(c) Consider the edges that belong to $G = (V, E)$ but do not belong to G_0 (edges in $E - E_0$). For each such edge (u, v), if (u, v) is a back edge (i.e., it connects a vertex to one of its ancestors), we add one directed edge to E^D (shown below); otherwise, we add two directed edges to E^D. (We will refer to these directed edges as *images* of (u, v), and we say these directed edges are *generated* by (u, v).)

Suppose that the edge e with weight w, joins vertices u and v belonging to the tree Γ. There are two cases depending on the relative locations of u and v in the tree Γ. (See Figure 1.)

(i) If u is an ancestor of v (the converse is symmetric): then add an edge $u \to v$ in G^D with weight w.

(ii) If neither u nor v is an ancestor of the other: let $t = l.c.a(u, v)$ (least common ancestor in the rooted tree Γ). Add edges $t \to u$ and $t \to v$ in G^D, each with weight w.

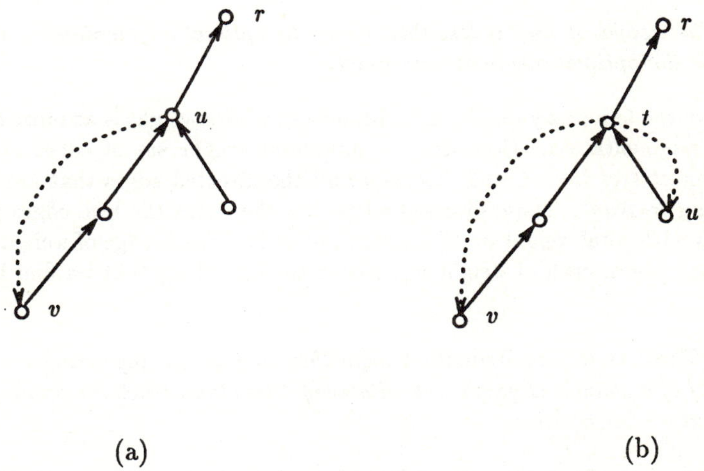

Figure 1: Construction of G^D.

2) Find a minimum weight branching in G^D rooted at r. For each directed edge e that is picked as part of the branching, and that does not belong to the directed tree Γ, add the corresponding edge in $E - E_0$ that generated e. The set of edges added is Aug.

bserve that all edges of $G^D - \Gamma$ are such that they connect a vertex to one of its descendants ı Γ). We now show the following.

emma 3.1 *If G is 2-edge connected, then the directed graph G^D is strongly connected.*

roof: Clearly, all the vertices of G^D can reach the root r using edges from the tree Γ. Further, ; us assume that G^D is not strongly connected. Of all the vertices that cannot be reached from e root, let u be the vertex that is closest to the root in Γ. Clearly, the entire subtree rooted at must consist of unreachable vertices. Since the image of the edge $(u, p(u))$ in G is not a bridge G, there must be another edge (v, s) in G going from a vertex v that is in the subtree rooted at u, to vertex s that is not in this subtree. Such an edge would have generated a directed e from a vertex w to v in G^D where w is an ancestor of v (specifically the least-common-ance of v and s). Since w is a proper ancestor of u, it is reachable from r in G^D. Therefore reachable from r, and hence u as well. Thus we obtain a contradiction.

Lemma 3.2 *If G is 2-edge connected, then the edge connectivity of the graph G_0 together the edges in Aug is at least 2.*

Proof: Assume G is 2-edge connected. Then, by the previous lemma, we can find a minim weight branching in G^D. Next, assume that despite the addition of the edges in *Aug* to G_0, resulting graph has bridges. All such bridges are the tree edges in Γ. Let $(u, p(u))$ be one s edge of Γ that is closest to the root (it does not have to be unique). Since vertices in the subt rooted at u, are reached from r in the branching it must be the case that there is a direc edge $w \rightarrow v$, from a vertex w (ancestor of u) to v in the minimum weight branching. Such edge would have been generated by an edge connecting v to a vertex not in the subtree roo at u. This edge would belong to *Aug* and hence the edge $(u, p(u))$ is not a bridge.

Lemma 3.3 *The weight of Aug is less than twice the optimal augmentation. (Weight(Aug $2C^*$ where C^* is the optimal augmentation weight.)*

Proof: We prove the lemma by exhibiting a branching whose weight is at most twice the wei of the optimal augmentation. Consider the minimum weight set of edges Aug^* that wo increase the connectivity from 1 to 2. Consider all the directed edges that are "generated" edges that belong to Aug^*. These directed edges together with the tree edges yield a stron connected graph with total weight on the edges at most $2C^*$ (each edge of weight w generated most two directed edges, each of weight w). Hence the branching that we find has total wei at most $2C^*$.

Theorem 3.4 *There is an approximation algorithm to find an augmentation to increase edge connectivity of a connected graph to 2 with weight less than twice the optimal augmentati that runs in $O(m + n \log n)$ time.*

Proof: The correctness of the algorithm follows from Lemma 2 and Lemma 3. Since the brid connected components can be found in $O(m+n)$ time [AHU] and a minimum weight branchi can be found in $O(m + n \log n)$ time [GGST86]. Since the least common ancestors for the pairs can be found in $O(m + n)$ time by using the algorithm of Harel and Tarjan [HT84], t theorem follows.

The approximation factor of 2 for the algorithm is tight. In [FJ81] an example is given show that their algorithm can actually result in an approximation that is twice the optim solution. The same example works for our algorithm as well.

4 Increasing Vertex Connectivity from 1 to 2

We can assume w.l.o.g. that G_0 is a connected graph just as in the case of edge connectivit Our overall strategy is similar to the one used in the previous section. That is, we first obta a tree structure Γ of the blocks of G_0, construct a weighted, directed graph G^D using Γ an G. Then find a minimum weight branching in G^D which will indicate the edges of $E - E_0$ th are to be added to increase the connectivity of G_0. We remark that Γ, in the case of verte connectivity, is quite different from that of the previous section.

gorithm *Construct Block Cut Tree*
ut: Connected graph G_0.
tput: Block cut tree Γ of G_0.

.) Let $a_1, a_2, \ldots$ and $B_1, B_2, \ldots$ be the articulation points and blocks of $G_0 = (V, E_0)$, respectively. The vertex set $V(\Gamma)$ is a union of V_a and V_b where $V_a = \{a_1, a_2, \ldots\}$ and $V_b = \{b_i \mid B_i$ is a block of $G_0\}$. Associated with each vertex in $V(\Gamma)$, is a set. For $a_i \in V_a$, $X_i = \{a_i\}$. For $b_i \in V_b$, $Y_i = \{v_j \mid v_j \in V$ and v_j is not a cut vertex in $G_0\}$.

2) The edge set $E(\Gamma)$ consists of edges (a_i, b_j) where a_i is an articulation point that belongs to block B_j.

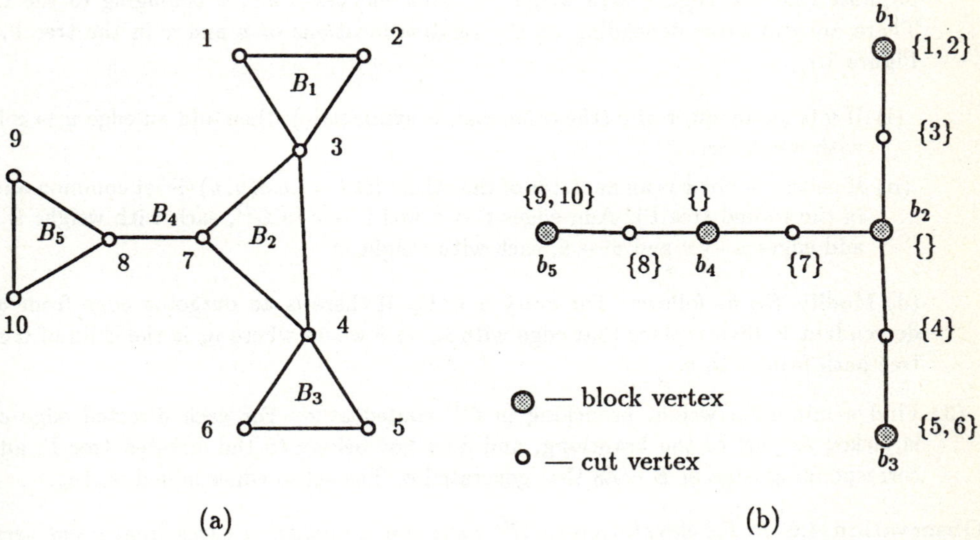

Figure 2: Construction of block cut vertex tree Γ.

Figure 2 illustrates the above construction via an example.

bservation 4.1 *In the block cut tree Γ, each edge is between a vertex in V_a and a vertex in*

.

bservation 4.2 *Consider the sets associated with the vertices of Γ. Each vertex of G_0 belongs exactly one such set.*

In the rest of the section, for a vertex u of V, the vertex of Γ that *corresponds* to u is u u is an articulation point, and b_i otherwise where B_i is the unique block containing u. In e following, by *superimposing* an edge $(x, y) \in G$ on Γ, we mean adding an edge between $b \in V(\Gamma)$ where the associated sets X and Y contain x and y respectively.

lgorithm *Find Vertex Aug*
put: Graph G, a connected subgraph G_0, and a block cut tree Γ of G_0.
utput: A set of edges $Aug \subseteq E - E_0$.

(1) Superimpose all the edges of $E - E_0$ on Γ. Discard all the self-loops. Among the mult edges retain the cheapest edge, discarding the rest.

(2) (*Construct* $G^D = (V, E_D)$)

(a) Pick an arbitrary leaf of Γ to be the root r, and direct all the edges of Γ toward Continue to denote the resulting tree by Γ.

(b) Add to E_D the directed tree edges of Γ and set their weight to zero.

(c) Consider the superimposed edges of $E - E_0$ on Γ. Let (u, v) be one such superimpo edge. If (u, v) is a back edge (i.e. it connects a vertex to one of its ancestors), we add directed edge to E^D (shown below); otherwise, we add four directed edges to E^D. (will refer to these directed edges as *images* of (u, v), and we say these directed edges *generated* by (u, v).)

Suppose that the edge e with weight w, joins vertices u and v belonging to the tree There are two cases depending on the relative locations of u and v in the tree Γ. (Figure 3.)

(i) If u is an ancestor of v (the other case is symmetric): then add an edge $u \rightarrow v$ in with weight w.

(ii) If neither u nor v is an ancestor of the other: let $t = l.c.a(u, v)$ (least common ances in the rooted tree Γ). Add edges $t \rightarrow u$ and $t \rightarrow v$ in G^D, each with weight w. A add edges $u \rightarrow v$ and $v \rightarrow u$, each with weight w.

(d) Modify E_D as follows. For every $u \in V_a$, if there is an outgoing edge from u to descendant v, then replace that edge with $u_v \rightarrow v$ where where u_v is the child of u on t tree path from u to v.

(3) Find a minimum weight branching in G^D rooted at r. For each directed edge e th is picked as part of the branching, and does not belong to the directed tree Γ, add t corresponding edge in $E - E_0$ that generated e. The set of edges added is *Aug*.

Observation 4.3 *In the directed graph G^D there are no outgoing edges from a cut vertex any of its descendants in Γ.*

Observation 4.4 *Consider the components formed on the deletion of a vertex $u \in V_a$ from The edges of G when superimposed on $\Gamma - u$ connect all these components.*

Lemma 4.5 *If G is 2-vertex connected, then the directed graph G^D is strongly connected.*

Proof: Clearly all the vertices of G^D can reach the root r using edges of the tree Γ. Let assume that G^D is not strongly connected. Of all the vertices that cannot be reached from t root, let u be a vertex that is closest to the root in Γ. Clearly, the entire subtree rooted at must consist of unreachable vertices and every proper ancestor of u is reachable from r. T proof is a little involved and we break it into cases.

Case 1: $u \in V_a$.

Since u is not a cut vertex in G, there must be at least one edge connecting a vertex i $C_1(u)$ to some vertex in $C_i(u)$ by Observation 4.4. Let this edge be (v, s) where $s \in C_1($ and $v \in C_i(u)$. Now there are two subcases to consider:

(a) s is an ancestor of v.

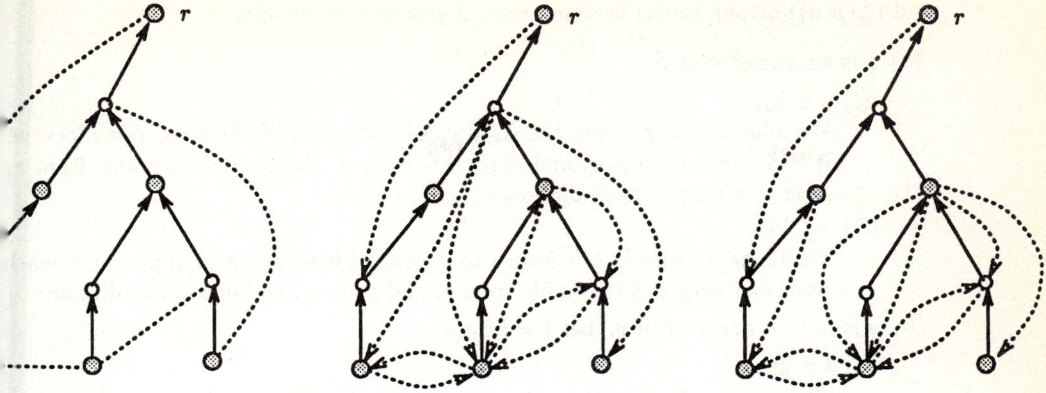

Rooted tree Γ and edges of $E - E_0$ (b) G^D after step 2(c) (c) G^D after step 2(d)

Figure 3: Construction of G^D in the case of vertex connectivity.

(i) $s \in V_a$.
Corresponding to edge (s, v) we added an edge in G^D, from the child s_v of s (on the path from s to v) to v. Since s_v is a block vertex, it is distinct from u. Clearly s_v is an ancestor of u, and hence reachable from r. Thus v is reachable from r and so is u, yielding a contradiction.

(ii) $s \in V_b$.
We add an edge in G^D from s to v. Since s is reachable from r (because it is an ancestor of u), so is v and hence u, yielding a contradiction.

(b) s is not an ancestor of v. Let $t = l.c.a(s, v)$.

(i) $t \in V_a$.
Corresponding to edge (s, v) we added an edge in G^D, from the child t_v of t (on the path from t to v) to v. Since $t_v \in V_b$, it is distinct from u. It is reachable from r and hence v is reachable from r, and so is u, yielding a contradiction.

(ii) $t \in V_b$.
Clearly t is an ancestor of u, hence reachable from r. We added an edge in G^D from t to v, hence v is reachable and u as well, yielding a contradiction.

2: $u \in V_b$.
Consider the cut vertex $p(u)$. Notice that $p(u) \neq r$ since $r \in V_b$. Let the roots of the subtrees $C_1(p(u)), C_2(p(u)), \ldots C_k(p(u))$ be $r_1 (= r), r_2, \ldots r_k$, where k is the degree of $p(u)$. Assume that $C_2(p(u))$ refers to the component containing u (hence $r_2 = u$). Partition the components into two groups as follows. The first group contains all the components whose roots are reachable from r, and the second group contains the rest. (Notice that both the groups are non-empty.) Since G is biconnected there must exist an edge (s, v) where s belongs to a vertex in $C_i(p(u))$ and v belongs to a vertex in $C_j(p(u))$, such that $C_i(p(u))$

and $C_j(p(u))$ belong to the first and second groups respectively.

(a) s is an ancestor of v.

(i) $s \in V_a$.
We added an edge from the child s_v of s to v in G^D. Since s_v is a block ve it is distinct from $p(u)$ and reachable from r. Hence v is reachable from r, so is r_j giving a contradiction.

(ii) $s \in V_b$.
We added an edge in G^D from s to v. Since s is an ancestor of $p(u)$ it is reach from r. Hence v is reachable from r, and so is r_j, giving a contradiction.

(b) s is not an ancestor of v. Let $t = l.c.a(s, v)$.

(i) $t \neq p(u)$.
There is an edge in G^D from either t or t_v, to v. Since both t and t_v are reach from r, so is v and hence r_j, giving a contradiction.

(ii) $t = p(u)$.
Note that r_i is reachable from r. Because of edge (s, v) we generate the follov edges in G^D: $r_i \to s, r_j \to v, s \to v, v \to s$. Hence v is reachable from r, an is r_j, yielding a contradiction.

Lemma 4.6 *If G is 2-vertex connected, then the vertex connectivity of the graph G_0 toge with the edges in Aug is at least 2.*

Proof: Assume that despite the addition of the edges in Aug to G_0, the resulting graph h cut vertex u. We will now show that u is destroyed as a cut vertex in the tree Γ, and hence in Consider the components $C_1(u), \ldots, C_{d(u)}(u)$ in Γ. Partition the components into two gro as follows. The first group contains all the components that get connected to $C_1(u)$ (by an e or a path) when the edges of Aug are superimposed on Γ. The second group contains the r Notice that both the groups are non-empty. Since G^D is strongly connected all the vertices reachable from the root in the minimum weight branching. Since there are no outgoing ed from u to its descendants by Observation 4.3, these must be an edge $s \to v$ in the branch that satisfies the following. This edge has the property that $s \in C_i(u)$ and $v \in C_j(u)$, wh $C_i(u)$ and $C_j(u)$ belong to the first and second groups respectively. The edge that genera $s \to v$ in Aug would connect $C_i(u)$ to $C_j(u)$ in $G_0 + Aug$, yielding a contradiction.

Lemma 4.7 *The weight of Aug is less than twice the optimal augmentation. (Weight(Aug) $2C^*$ where C^* is the optimal augmentation weight.)*

Proof: We prove the lemma by exhibiting a branching whose weight is at most twice the weig of the optimal augmentation. Consider the minimum weight set of edges Aug^* that wou increase the connectivity from 1 to 2. Consider all the directed edges that are "generated" edges that belong to Aug^*. These directed edges together with the tree edges yield a stron connected graph with total weight on the edges at most $4C^*$ (each edge of weight w_i generat at most four directed edges, each of weight w_i). Now pick a minimum weight branching in t graph. Notice that for each cross edge (u, v) (when neither u nor v is an ancestor of the othe even though we generate four directed edges in G^D, no minimum weight branching will use mo than two of these four edges. (Otherwise, it will not be a valid branching.) Hence the branchi that we find has total weight at most $2C^*$.

orem 4.8 *There is an approximation algorithm to find an augmentation to increase the x connectivity of a connected graph to 2 with weight less than twice the optimal augmentation runs in $O(m + n \log n)$ time.*

of: The correctness of the algorithm follows from Lemma 5 and Lemma 6. Since the nnected components can be found in $O(m+n)$ time [AHU] and a minimum weight branching be found in $O(m + n \log n)$ time [GGST86]. Since the least common ancestors for the m s can be found in $O(m + n)$ time by using the algorithm of Harel and Tarjan [HT84], the rem follows. □

Increasing Edge Connectivity to k

his section, we show that it is possible to obtain an approximation factor of 2 for increasing edge connectivity of a graph to any k. The algorithm takes as input an undirected graph $V, E_0)$ on n vertices and a set *Feasible* of m weighted edges on V, and finds a subset *Aug* of es which when added to G_0 make it k-edge connected. The weight of *Aug*, is no more than e the weight of the least weight subset of edges of *Feasible* that increases the connectivity. also observe that the problem is NP-hard (for any k) by extending the proof that was given FJ81] for incrementing 1-connected graphs to 2-connected optimally.

Consider a directed graph G with weights on the edges, and a fixed root r. How does one find *cheapest* directed subgraph H^D that has k-edge disjoint paths from a fixed root r to each ex v ? Gabow [G91a] gives the fastest implementation of a weighted matroid intersection orithm to solve this problem in $O(kn(m + n \log n) \log n)$ time. (See also [Ed79, FT89].)

To solve our problem (approximation algorithm), in the undirected graph G_0 replace each irected edge (u, v) by two directed edges $u \to v$ and $v \to u$ with each edge having weight 0. each edge in the set *Feasible* (u, v) we replace it by two directed edges $u \to v$ and $v \to u$ h weight $w(u, v)$ (the weight of the undirected edge). Call this graph G^D. Now run Gabow's orithm on the graph G^D, asking for k-edge disjoint paths from each vertex to the root. If the ected edge $u \to v$ is picked in H^D and $w(u, v) > 0$ (we can assume all edges of set *Feasible* e weight > 0 else we can always include it in *Aug*) we add (u, v) to *Feasible.* (This is a eralization of the scheme in [KV92], where it was shown only for the case when E_0 is empty.)

mma 5.1 *The graph $E_0 + Aug$ is a k-edge connected graph.*

oof: Suppose there is a cutset C of $(k-1)$ edges in $G_k = G_0 + Aug$. Assume that it separates into components C_1 and C_2. Let r be in C_1. Consider a vertex v in C_2. It cannot be that re were k-edge disjoint paths from r to v in G^D (because of the cutset C). □

eorem 5.2 *The total weight of* Aug *is at most twice the weight of the optimal augmentation.*

oof: Consider the optimal augmentation Aug^*. Consider the following subgraph: add the i-parallel edges corresponding to edges in E_0 and Aug^*. In this graph clearly there are k-edge joint (directed) paths from r to each vertex v (by directing the undirected paths from r to v propriately). Thus the optimal solution to the problem (Aug) must have total cost less than Aug^*. □

6 Increasing Edge Connectivity from k to $k+1$ (for odd k)

The method described in Section 3 can be applied in a more general setting: to find an app imation to the problem of increasing the edge connectivity of a k-edge connected graph $G_k = (V, E_k))$ to $k+1$ when k is odd. The key idea is to use the representation of cuts th due to Karzanov and Timofeev [KT86]. We note that this structure was used by Naor, Gus and Martel [NGM90] to solve the augmentation problem exactly for the unweighted case w *Aug* allows an edge to be added between any pair. The structure itself can be constructe $O(kn^2)$ time. We briefly describe this structure in the following. In this structure each ve of G_k maps to exactly one vertex in the representation, i.e., each vertex of the representa corresponds to a subset (possibly empty) of vertices of G_k. Moreover, the cuts of a graph be represented by a tree when the connectivity of G_k is odd; otherwise, the structure is a "t like" structure. Denote this structure by τ. The cut information about G_k is encoded as foll When k is odd, the number of cuts in a k-edge connected graph is $O(n)$ [DKL76]. Assume t removing a cut (i.e., a set of k edges whose removal disconnects the graph) from G_k result two components H_1 and H_2 where $V(G_k) = V(H_1) \cup V(H_2)$. In τ there exists an edge wh removal results in two subtrees with the following property: the union of the subsets of vertices represented by the vertices of one of the subtrees is $V(H_1)$ and of the other is $V(H$

To increase the connectivity of G_k, construct τ and a set of edges F on the vertices of τ t correspond to the set of edges *Feasible*. That is, for every $(u, v) \in$ *Feasible*, include an edge i between the unique pair of vertices of τ that corresponds to u, v. Now, run the algorithm *F Edge Aug* with $\tau + F$ for G and τ for G_0. At the end of the algorithm, pick the subset of edges (denote it as *Aug'*) of *Feasible* that correspond to the edges of the output *Aug*. It foll from Theorem 3.4 and the fact the τ can be build in $O(kn^2)$ time that the time complexity the entire process is $O(kn^2)$. We establish the correctness of our method.

Lemma 6.1 *The graph $E_k + Aug'$ is a $(k+1)$-edge connected graph.*

Proof: Assume that there is a cutset C of k edges in $E_k + Aug'$. Notice that all the edges C must come from E_k as we assumed that G_k is k-edge connected. Assume that C separa the augmented graph into components H_1 and H_2. The cut tree representation guarantees t there is a tree edge in τ that corresponds to C. Denote the two subtrees resulting from removi this tree edge by τ_1 and τ_2. Since this tree edge is not a bridge in $\tau + Aug$, by Lemma 3.2 th is an edge of *Aug* that connects a vertex of τ_1 to a vertex of τ_2. There is an edge in *Aug'* t corresponds to this edge of *Aug*. But this edge must connect a vertex of H_1 to a vertex of contradicting the assumption that H_1 is separated from H_2 when C is deleted from $G_k + A$
□

The next theorem follows from Lemma 3.3 and the observation that when the edges cor sponding the subset of edges from the optimal augmentation are added to τ, the resulting gra is 2-edge connected.

Theorem 6.2 *The total weight of* Aug *is at most twice the weight of the optimal augmentati*

Acknowledgments: We thank Professors Joseph JáJá and Uzi Vishkin for encouragement a useful discussions. We would also like to thank the referees for useful comments on an earl draft.

References

[AHU] A.V. Aho, J.E. Hopcroft and J.D. Ullman, *The design and analysis of computer algorithms*, Addiso Wesley, 1974.

76] E.A. Dinits, A.V. Karzanov and M.L. Lomosonov, "On the structure of a family of minimal weighted cuts in a graph," *Studies in Discrete Optimization* [In Russian], A.A. Fridman (ear decomposition), Nauka, Moscow, pp. 290-306, 1976.

[90] D. Naor, D. Gusfield and C. Martel, "A fast algorithm for optimally increasing the edge-connectivity," *31st Annual Symposium on Foundations of Computer Science*, pp. 698–707, (1990).

] J. Edmonds, "Matroid intersection," *Annals of Discrete Mathematics*, No. 4, pp. 185–204, (1979).

] S. Even, *Graph Algorithms*, Computer Science Press, Potomac, Md., 1979.

6] K. P. Eswaran and R. E. Tarjan, "Augmentation problems," *SIAM Journal on Computing*, Vol. 5, No. 4, pp. 653–665, (1976).

A. Frank, "Augmenting graphs to meet edge-connectivity requirements," *31st Annual Symposium on Foundations of Computer Science*, pp. 708–718, (1990).

] G. N. Frederickson and J. JáJá, "Approximation algorithms for several graph augmentation problems," *SIAM Journal on Computing*, Vol. 10, No. 2, pp. 270–283, (1981).

] G. N. Frederickson and J. JáJá, On the relationship between the biconnectivity augmentation and traveling salesman problems," *Theoretical Computer Science*, Vol. 19, No. 2, pp. 189–201, (1982).

9] A. Frank and E. Tardos, "An application of submodular flows," *Linear Algebra and its Applications*, 114/115, pp. 320–348, (1989).

a] H. N. Gabow, "A matroid approach to finding edge connectivity and packing arborescences," *23rd Annual Symposium on Theory of Computing*, pp. 112–122, (1991).

b] H. N. Gabow, "Applications of a poset representation to edge connectivity and graph rigidity," *32nd Annual Symposium on Foundations of Computer Science*, pp. 812–822, (1991).

T86] H. N. Gabow, Z. Galil, T. Spencer and R. E. Tarjan, "Efficient algorithms for finding minimum spanning trees in undirected and directed graphs," *Combinatorica*, 6 (2), pp. 109–122, (1986).

1a] T. S. Hsu and V. Ramachandran, "A linear time algorithm for triconnectivity augmentation," *32nd Annual Symposium on Foundations of Computer Science*, pp. 548–559, (1991).

1b] T. S. Hsu and V. Ramachandran, "On finding a smallest augmentation to biconnect a graph," 2nd *Annual International Symposium on Algorithms*, Springer Verlag LNCS 557, pp. 326–335, (1991).

4] D. Harel and R. E. Tarjan, "Fast algorithms for finding nearest common ancestors," *SIAM Journal on Computing*, Vol 13, No. 2, pp. 338–355, (1984).

1] G. Kant and H. Bodlaender, "Planar Graph Augmentation Problems," *1991 Workshop on Algorithms and Data Structures*, pp. 286–298, (1991).

6] A.V. Karzanov and E.A. Timofeev, "Efficient algorithm for finding all minimal edge cuts of a nonoriented graph," *Cybernetics*, pp. 156-162, Translated from *Kibernetika*, No. 2, pp. 8–12, (1986).

2] S. Khuller and U. Vishkin, "Biconnectivity approximations and graph carvings," Technical Report UMIACS-TR-92-5, CS-TR-2825, Univ. of Maryland, January (1992), Also to appear in *24th Annual Symposium on Theory of Computing*, (1992).

[90] D. Naor, D. Gusfield and C. Martel, "A fast algorithm for optimally increasing the edge-connectivity," *31st Annual Symposium on Foundations of Computer Science*, pp. 698–707, (1990).

7] A. Rosenthal and A. Goldner, "Smallest augmentations to biconnect a graph," *SIAM Journal on Computing*, Vol. 6, No. 1, pp. 55–66, (1977).

8] T. Watanabe, "An efficient augmentation to k-edge connect a graph," Tech. Report C-23, Department of Applied Math., Hiroshima University, April 1988.

87] T. Watanabe and A. Nakamura, "Edge-connectivity augmentation problems," *Journal of Comp. and Sys. Sciences*, 35 (1), pp. 96–144, (1987).

Fast Incremental Planarity Testing

Jeffery Westbrook*

Abstract

The *incremental planarity testing* problem is to perform the following operations on a biconnected planar graph G of at most n vertices: test if an edge can be added between two vertices while preserving planarity; add edges and vertices that preserve planarity. Let m be the total number of operations. We present fast data structures for this problem that can be used in conjunction with the previous algorithm of Di Battista and Tamassia to achieve an $O(\alpha(m,n))$ worst-case amortized time per test operation. If the graph is biconnected, a sequence of n additions can be performed in total time $O(m\alpha(m,n))$ worst-case plus $O(n)$ expected time. Our tree data structure is flexible and can answer in $O(1)$ time queries about parents, roots, and nearest common ancestors while performing tree modifications such as inserting nodes, cutting edges, and merging or splitting nodes. If the graph is not biconnected then insertions of edges and vertices require $O(\log n)$ amortized expected time per operation.

The study of graph planarity has a long and rich history. The off-line planarit testing problem, in which one must test if a given graph is planar, has been studied since the nineteen-sixties. If can be solved in $O(n)$ sequential time [3, 12] an $O(\log n)$ time on a CRCW PRAM [17]. Recently Di Battista and Tamassia introduced and studied the *incremental planarity testing* problem [1, 2]: beginning with a biconnected graph G process on-line an intermingled sequence of the following query and update operations.

1. *Test*(u,v): Determine if an edge can be inserted between vertices u and v o while still preserving the planarity of G.

2. *Insert Vertex*(v,e): Split edge e into two edges e_1, e_2 by inserting a new vertex v. Return e_1 and e_2.

3. *Insert Edge*(u,v): Insert a new edge e between u and v. Return e. The insertion must preserve planarity.

The above repertoire of operations is sufficient to construct any biconnected planar graph. A related problem is static on-line testing problem, in which one must preprocess a fixed planar graph G so as to answer on-line a sequence of *Test*(u,v) operations. Di Battista and Tamassia gave an algorithm that performs *Test*(u,v) in

*Department of Computer Science, Yale University, New Haven, CT 06520-2158. Research partially supported by National Science Foundation Grant CCR-9008653.

$O(\log n)$ worst-case time per query, where n is the maximum number of vertices in the graph, and performs the updates in total time $O(n \log n)$, hence amortized time per update operation $O(\log n)$. They solve the static testing problem with $O(n)$ preprocessing time and $O(\log n)$ time per query.

In this paper we give algorithms for incremental planarity testing with nearly constant time per operation. Let m be the total number of operations. We perform

- *Test*(u, v) queries in $O(\alpha(m, n))$ worst-case amortized time per operation.
- Update operations in total time $O(m\alpha(m, n))$ worst-case plus $O(n)$ expected, or $O(\alpha(m, n)$ amortized expected time per update.

The function $\alpha(m, n)$ is the functional inverse of Ackermann's function, and is less than 4 for any practical value of n (see [18]).

Di Battista and Tamassia extended their results to include update operations that allow the construction of general 1 and 0-connected graphs, with the time bounds of $O(\log n)$ per operation. Our results can also be extended to general graphs, handling queries in $O(\alpha(m, n))$ time but still requiring $O(\log n)$ amortized expected time per update. Our techniques solve the static testing problem in worst-case $O(1)$ time per query and $O(n)$ preprocessing time.

Our approach speeds up the algorithm of Di Battista and Tamassia by the use of more sophisticated data structures. The problem is partioned into two subproblems, one involving maintaining a tree decomposition of a graph into its 3-connected components, and the other involving doing tests and updates into a fixed 3-connected embedding. Di Battista and Tamassia solve each subproblem in $O(\log n)$ time per query and $O(\log n)$ amortized time per update.

Our contributions occur in three areas. First, we recast the algorithm of Di Battista and Tamassia in a form that simplifies the data structure operations required. We also state the algorithm without using the conceptual framework of orientations of undirected 2-connected graphs into directed st-graphs. Second, we give $O(\alpha(m, n))$-time algorithms for maintaining a collection of fixed embeddings of a 2-connected planar graphs undergoing edge and vertex insertion and operations that merge one graph into another. We use fast split-find and union-find data structures in combination with dynamic perfect hash tables. Third, to solve the decomposition tree maintenance subproblem we develop a flexible tree data structure that can answer in $O(1)$ time queries about parents, roots, and nearest common ancestors while performing tree modifications such as inserting nodes, cutting edges, and merging or splitting nodes. The structure utilizes the micro-set approach of Gabow and Tarjan [8] but is based on a different tree representation. We note that several groups of researchers have developed fast data structures for maintaining 3-edge-connected and 3-vertex connected component tree decompositions [2, 14, 16], but they did not give algorithms to find nearest common ancestors.

It is straightforward to reduce the union-find problem to the problem of maintaining 3-connected components, and lower bounds of $\Omega(m\alpha(m, n))$ are known for the union-find problem in both pointer machine [15] and cell probe model [6]. Hence a time bound of $O(m\alpha(m, n))$ is the best possible for any algorithm that maintains a decomposition of the 3-connected components of the graph.

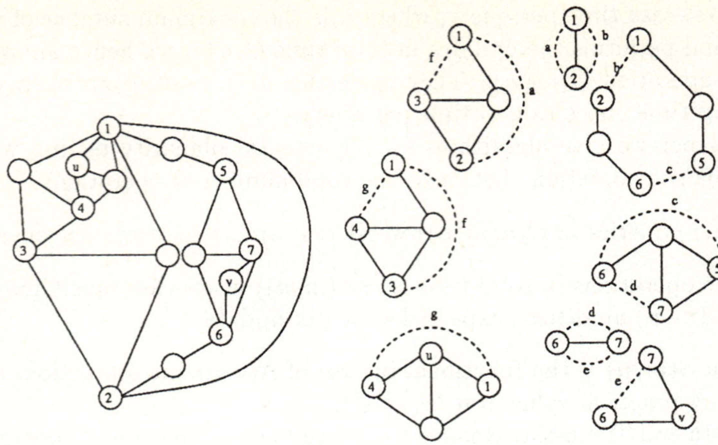

(a) (b)

Figure 1: (a) A planar graph G with a given embedding. Numbered vertices belong t split pairs. (b) Decomposition of graph G into 3-connected components. Dashed line represent virtual edges; each is labeled by the split that produced it.

1 The Decomposition Tree

If a planar graph is 3-connected it has a unique embedding [9] and on-line testin reduces to testing whether an edge can be inserted into a fixed embedding. If th graph is not 3-connected then the set of possible embeddings is determined by the 3 connected components of the graph. We review the decomposition of a 2-connecte multigraph G into its 3-connected components as described in [11]. Let n be th number of vertices and m the number of edges. For convenience we assume $m \geq 3$ Let $\{a, b\}$ be a pair of vertices. Partition the edges of G into classes $E_1, E_2, \ldots, E$ such that two edges are in the same class if and only if they both lie on a commo (not necessarily simple) path containing neither a nor b except as an endpoint. Fo a given j let $E^j = \cup_{i=1}^{j} E_i$. The pair $\{a, b\}$ is called a *separation pair* if there is a j such that $|E^j| \geq 2$ and $|E - E^j| \geq 2$. If a graph contains no separation pair the it is 3-connected.

The 3-connected components are produced by a recursive splitting procedur that takes a separation pair and partitions the edges as above into two subgraphs Each subgraph contains a virtual edge $\langle a, b \rangle$ labeled by i, the number of the split The result, after some massaging, is a collection of component subgraphs. Eac subgraph is either a 3-connected graph; a simple cycle or "polygon"; or two vertices connected by three or more edges, called a "multi-bond". The true edges ar partitioned among the components; a virtual edge with a given label occurs in tw components; a separation pair may occur in multiple components. Let G_1 and G_2 be components that both contain a virtual edge $\langle a, b \rangle$ with label i. To *merge* G_1 and G_2 means to construct the graph induced by $E_1 \cup E_2$ minus the two virtual edges $\langle a, b \rangle$ labeled i. The 3-connected components are unique, have total size $O(m)$, and can be constructed in time $O(m)$ [11]. Figure 1a gives an example of a planar graph and its 3-connected components.

For each 3-connected component create a node of type S, P, or R, depending ı whether the split component is a polygon, a multi-bond, or a 3-connected graph, spectively. Add an edge between two nodes ν and μ if the corresponding compo- ents contain a virtual edge $\langle a, b\rangle$ with the same label. The copy of edge $\langle a, b\rangle$ in is called the location of μ in ν. Root the tree arbitrarily. For each true edge we eate a q node and make it a child of the node whose graph contains the edge. This ves the "SPQR" decomposition tree of Di Battista and Tamassia . For increased mplicity, however, we dispose of P nodes as follows. If the P node ν has a parent node μ then ν is merged into μ. The k-edge multi-bond in the resultant graph of is replaced by a single edge of *multiplicity* k. The children of ν become children f μ associated with the single edge. If the P node has no R node parent then the node is converted to an R node whose graph consists of a single edge of appro- riate multiplicity. For each vertex v create a p node and make it a child of the earest common ancestor of all tree nodes ν corresponding to components in which appears (this ancestor itself contains v). The children of an R node are ordered rbitrarily amongst themselves. the children of an S node are ordered according to rder around the polygon. Children with the same virtual edge (of multiplicity $>$) are ordered arbitrarily among themselves. Figure 4a shows the tree for the graph f Figure 1

Let ν be connected to its parent μ by $\{a, b\}$. These two vertices are called ıe *poles* of ν. Let ω be the grandparent of ν. Node ν is *marked* if there is no mbedding of the component μ such that the location of ν and the location of ω lie n a common face. The *characteristic ancestor triple* of nodes μ and ν, $ca(\mu, \nu)$, is triple $(\omega, \omega_\mu, \omega_\nu)$ such that ω is the nearest common ancestor of μ and ν and ω_μ nd ω_ν are the ancestors of μ and ν respectively, that are children of ω. If $\omega = \nu$ hen $\omega_\nu = \nu$; similarly for μ.

The Test and Update Algorithms

Ve sketch the algorithms because they differ somewhat from those of [2]. The test lgorithm essentially determines whether in each component on the tree path the ocations of u and v can be placed on a common face. This is always true for R and nodes but depends on the embedding of the R nodes.

Let p_u and p_v be the p nodes of u and v, respectively, with $ca(p_u, p_v) = (\omega, \omega_u, \omega_v)$. hen *Test*$(u, v)$ answers TRUE if and only if: 1) there are no marked nodes on the aths from p_u to ω_u and p_v to ω_v other than possibly ω_u and ω_v themselves, and the ocations of ω_u and ω_v in ω share a common face; or 2) the least marked ancestor f p_u, γ, has parent δ such that vertex v is a pole of δ and shares a face with the ocation of γ in δ (Or vice versa).

The insertion of a vertex into an edge e causes the the corresponding q node to e replaced by an S node with children q_1 and q_2.

The insertion of an edge $\langle u, v\rangle$ is more complex. Let $ca(p_u, p_v)= (\nu, \mu_u, \mu_v)$. We ssume the three nodes are distinct. (The other cases are similar.) The nodes on he tree paths from p_u and p_v to ν are transformed in three phases. In phase 1 one f two preliminary transformations, shown in Figure 2, is done to node ν, depending n its type. If neither applies ν is labeled by "$*$". In phase 2, the paths down

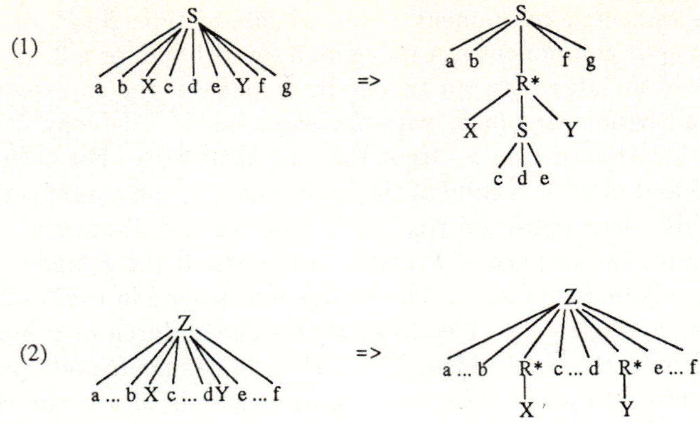

Figure 2: Preliminary transformations. Nodes X and Y are the locations of u and v in th critical node ν. (a) X and Y have different virtual edges in a parent S node. (b) X and share a virtual edge in their parent.

from μ_u and μ_v to the homes of u and v, respectively, are processed using th recursive transformation procedure diagrammed in Figure 3. At each step, th procedure transforms the current node μ according to its type and then recurse on the child of μ that corresponds to the location of v in ν. Each transformatio leaves an intermediate R node labeled "$*$" and assigned a flip orientation. In phas 3, each marked node ν is merged into its parent μ with specified flip orientatio The children of the merged node retain their previous relative order. The set c unmarked children of μ are examined to determine if they become marked in th merged node. Note that a node can be marked only once. If initial transformation was applied, the merging phase produces two nodes that are children of the same node. In this case, these two children are merged into one by equating the poles c the respective graphs. After phase 3, vertices u and v are incident on a common fac of some triconnected graph and the edge is inserted. Figure 4 shows an example.

Lemma 2.1 [1] *The total number of transformation steps over any sequence $O(n)$. There are $O(n)$ marks made.*

To implement the test and update algorithms we need two fast data structures The first supports the following operations on fixed embeddings.

Test whether two vertices are adjacent in a fixed embedding.

Create a new graph consisting of three or four edges in a polygon or multi-bon

Merge two graphs connected by a shared virtual edge, fixing a embedding of th first with respect to the second. The endpoints of the shared edge are the poles c at least one of the graphs.

Insert a true edge between two vertices on the same face.

The second data structure handles the following operations on the decompositio tree:

The query *find mark*(ν), which returns the least marked ancestor of node ν, an *ca*(ν, μ) the characteristic ancestor query.

add leaf(u, v): Make a new leaf v that is a child of u.

demote(x, y): Let v be the parent of x and y, and let $x, x_1, \ldots, x_k, y$ be th

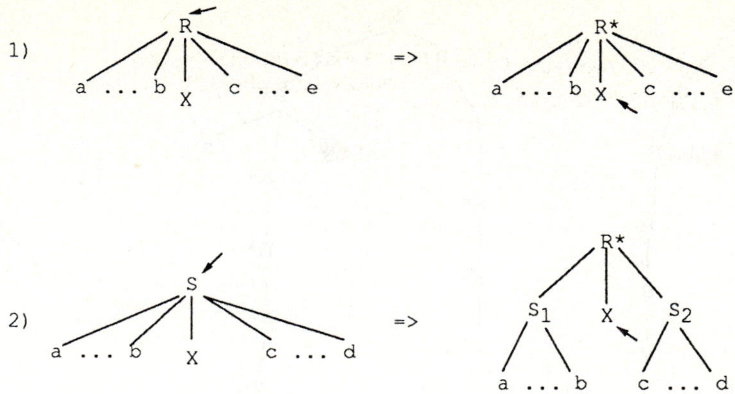

Figure 3: Transformations on nodes along the path from the critical node to the home of a vertex v. The arrow indicates which node is being examined in the current step, and after the transformation which node will be examined in the next processing step. The node X represents the location of v in the current node.

children of x in order from x to y. Create a node node u with parent v and children $x, \ldots, y$ in that order.

mark(u): Mark node u.

merge(u, v): Merge nodes u and v into one, making all children of u and v children of the new node. The new node has parent *parent*(v).

3 Fixed embeddings

In this section we consider the problem of maintaining a collection of 3-connected graphs with fixed embeddings. Each has a pair of poles which are endpoints of a virtual edge. The two faces incident on this polar edge are called the *exterior* faces. A given vertex appears exactly once as a non-pole and an arbitrary number of times as a pole. There are $O(n)$ edges, hence $O(n)$ faces.

For each face choose 2 representative vertices so that the following properties are maintained: (i): a vertex incident on k faces is a representative of at least $k-2$ of them. (ii): if a vertex is a pole it is a representative of all faces on which it is incident Let $\mathcal{W}$ be the set of pairs of face representatives. Call v a *regular* vertex of the at most two incident faces that for which it is not a representative.

Lemma 3.1 *Vertices u and v are incident on a common face if and only if one of the following is true: u and v share a regular face; u is in the name of one of the two regular faces of v; v is in the name of a regular face of u; u and v are a representative pair.*

An initial set of representatives can be chosen in $O(n)$ time using an st-numbering algorithm (see [1]). By inductions we can show that properties i and ii can be maintained throughout any sequence of modification operations. In an edge addition, it is easily seen that we can delete the old representative pair and create two new ones without increasing the number of faces any vertex is regular in. Next consider

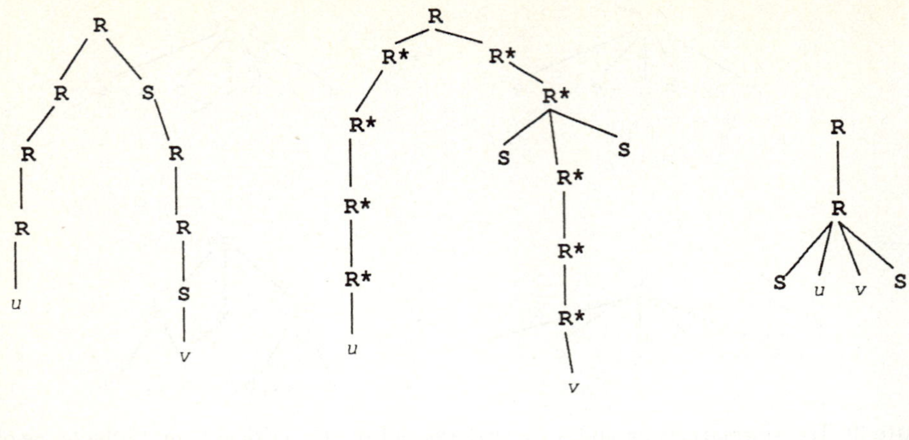

Figure 4: (a) The decomposition tree for the graph of Figure 1a. "q" nodes and "p" node with the exception of u and v are omitted for clarity. (b) The decomposition tree after th first two phases of processing for *Insert Edge*(u, v). (b) The decomposition tree after al processing for *Insert Edge*(u, v). The tree corresponds to the embedding of the graph i Figure 1b.

merging graphs A and B, connected by $\{u, v\}$. A least one graph, say A, has pole $\{u, v\}$. The exterior faces of A are merged into two faces of B. Which is merge with which depends on the specified flip orientation. Two representative pairs ar deleted and regular vertices of A's exterior faces become regular vertices of B's faces Note that the set of edges that go from bounding exterior to interior faces exactly determines the set of nodes in the decomposition tree that must be marked.

The ordered list of vertices and edges around a non-exterior face is maintaine using a version of the split-insert-find data structure of [13]. A similar data structur is described in a subsequent section. This can be used to find information abou the name of a face and the location of representatives around the boundary in $O(1$ time. Edge insertions cause splits and vertex insertions cause inserts. To handl merging of faces we add a level of indirection between the labels on split-find set and actual face labels. The labels on split-find sets are managed with a standar fast union-find data structure [18]. When two faces are merged the sets of labels o split-find sets on that faces are unioned.

The set $\mathcal{W}$ of representative pairs is stored in a dynamic perfect hash table [5] Representative pair lookups can be done in worst-case constant time. Insertions and deletions into the perfect hash table can be done in $O(1)$ amortized expected time The split-find operations have cost $O(1)$.

Lemma 3.2 *The time for m operations on a collection of subgraphs of n total vertices is $O(m\alpha(m, n))$ worst-case time plus $O(n)$ expected time in managing the hash table.*

4 Fast Tree Data Structures

In this section we discuss a tree data structure that answers *parent*(v), *find mark*(v), and *ca*(u, v) queries in a tree subject to *add leaf*(u, v), *demote*(x, y), *mark*(v) and *merge*(u, v) operations. In fact, our data structure will only handle *merge*(u, v) when u is a child of v. This is denoted by *pmerge*(u). Merges between siblings are simulated by managing the sets of merged siblings with a fast disjoint set union algorithm. Such merges involve children associated with the same multiple edge. When a node corresponding to k merged siblings is merged into its parent, then all the original siblings are merged with the parent. Let T' be the version of the decomposition tree T with only resulting from doing only parent merges as described above. We can show as an invariant that two siblings are merged in T if and only if the corresponding nodes are siblings in T'. This invariant allows us to convert characteristic ancestor triples in T' into characteristic ancestor triples in T with two extra ancestor queries. The details are omitted.

We take the maximum value of n as known beforehand. If it is not, then the data structures can be recomputed whenever n doubles. Since recomputation takes linear time, the total overhead is $O(n)$.

Our data structure based on the standard parentheses encoding of an n-node tree. Add a super-root to the tree whose only child is the original root. Now encode the tree by performing a depth-first traversal, writing an open paren, "(", when an edge is descended in the traversal, and a close paren ")" when it is subsequently ascended. The result is a string σ of length $2n$. Denote by $(_v$ and $)_v$ the open and close parens of the edge from node v to its parent. The open and close parens of v enclose exactly all the paren pairs corresponding to the descendents of v. The parent of v is the least pair of parens enclosing those of v.

Following a modification operation, the new tree encoding can be constructed from the old encoding by a constant number of changes to the paren string. Operation *add leaf*(u, v) is done by inserting into σ a new pair "()" corresponding to u immediately after $(_v$. Operation *demote*(u, v) is implemented by inserting "(" immediately before $(_u$ and ")" immediately after $)_v$. Operation *pmerge*(v) is done by removing $(_v$ and $)_v$ from the string.

To perform fast parent and findmark queries, we appropriate the microset techniques used by Gabow and Tarjan [8] and Imai and Asano [13] to solve the split-insert-find list maintenance problem. The string σ is partitioned into blocks of consecutive paren symbols. Each block contains between b^c and and $2b^c$ symbols, where $b = \log\log n$ and c will be chosen later. Call these the level-1 blocks. The string of level-1 blocks is partitioned into level-2 blocks containing between $2^{cb}/b^c$ and $2 \cdot 2^{cb}/b^c$ consecutive level-1 blocks. The parens within each level-1 block are stored in a machine word (assumed to have size $\Omega(\log n)$ bits) which is treated as a micro-memory. Blocks at higher levels are treated as linked lists. By using appropriate lookup tables and relabeling techniques, we can achieve the following. (See [8, 13] for more information.)

Lemma 4.1 *A sequence of* m parent(v), find mark(v), add leaf(u, v), demote(u, v), mark(v), *and* pmerge(v) *operations can be performed in time* $O(m + n)$ *and space* $O(n)$. *With the addition of* merge(u, v), *all operations can be performed in time*

$O(m\alpha(m,n))$.

Now we turn to the processing of characteristic ancestor queries. For simplicity we will talk only of nearest common ancestors; the extension to characteristic ancestors is relatively straightforward. We assume that the relative order of two blocks can determined in $O(1)$ time. This can be achieved by implementing the algorithm of Sleator and Deitz for maintaining order in a list [4] on top of our block partition. Together with an appropriate micro-memory table, this allows us to compute in $O(1)$ whether one node is an ancestor of another. We now describe the basic algorithm for computing $nca(u,v)$ when neither is an ancestor of the other. Assume that $)_u$ precedes $(_v$ in the paren string. Compute the following five nodes: a_1, the nearest common ancestor of all the parens between $)_u$ and $(_v$, including $)_u$ and $(_v$, in the level-1 block containing $)_u$; a_2, the nca of all the parens in the level-1 blocks between $)_u$ and $(_v$ in the same level-2 block as $)_u$, not including the blocks containing $)_u$ and $(_v$; a_3, the nca of the parens in the level-2 blocks between $)_u$ and $(_v$, not including those of $)_u$ and $(_v$; a_4, analogous to a_2 but in level-2 block containing $(_v$; a_5, analogous to a_1 but in the level-1 block containing $(_v$,

Lemma 4.2 *The nearest common ancestor of u and v is one of $a_1, \ldots, a_5$.*

We omit the proof. The values a_1 and a_5 can be computed in $O(1)$ time with a microset lookup table. The least common ancestor among $a_1, \ldots, a_5$ can be found in $O(1)$ time with five ancestor queries. To find the values of a_2, a_3, and a_4 we use the following lemma.

Lemma 4.3 *Let u be the label on block x and v the label on block y. Then the nearest common ancestor of x and y labels a block in between x and y, inclusive.*

Again we omit the proof. For each level i block, $i \geq 2$, we construct a *label tree* which contains one node for each distinct label on a level $i-1$ block in the level i block. The node corresponding to label x is made a child of its least ancestor in the full tree among the other labels. Each label tree is rooted at the NULL label. Each label tree is equipped with an auxiliary data structure, to be described below, that performs characteristic ancestor queries in $O(1)$ time. The values of a_1, a_2, and a_3 are computed by queries within the appropriate three label trees. Hence an nca query can be answered in $O(1)$ time.

The auxiliary tree data structure is subject to the same modifications as the decomposition tree. A demote in the full tree may cause a demote in the label tree. A splitting of a block may cause a demotion of a single label. A parent merge results in a merge in the label tree when the merged nodes are both labels. We use an extension of the data structure designed by Gabow [7] to perform nearest common ancestors with leaf insertions only. The critical ideas are use of a compressed version of the original tree. and the use of a "fat preorder number".

The idea of a compressed tree comes from [10]. Given T, let $s(x)$ be the number of descendents of x. A child w of v is *light* if $2s(w) \leq s(v)$, otherwise *heavy*. Deleting every edge from a light child to its parent leaves a set of disjoint *heavy paths* of length at least 1. The root of a heavy path is called an *apex*. To *compress* a node v means to take T_v, the subtree of T rooted at v, compute sizes, and make every node a child of its least proper ancestor that is an apex. The result is denoted $C(T_v)$. It is easy to see that $C(T)$ has logarithmic depth. We can extract $ca(x,y)$ information in T

om $ca(x,y)$ information in $C(T)$, using the fact that the nearest common ancestor f two nodes x,y on the same heavy path is simply whichever is nearer the apex.

Fat preorder numbers were invented by Gabow [7]. Let U be any tree in which or each vertex v with parent u, $\beta s(v) \leq s(u)$ for some constant $\beta > 1$. Let c nd e be positive integers that satisfy $c \leq \beta^e + 2$. In a fat preorder numbering, ach node v is assigned an interval $[f^*(v)..g^*(v)]$ of the nonnegative integers such hat: 1) the interval has size $c(s(v))^e$; 2) all descendents of v are assigned intervals n the range $[f^*(v) + s(v)^e + 1 \ .. \ g^*(v) - s(v)^e]$; and 3) only descendents of v are ssigned intervals in the range $[f^*(v)..g^*(v)]$. The value $f(v) = f^*(v) + s(v)^e$ is the *at preorder number* of v. To assign intervals, use a simple recursive procedure. The oot gets interval $[0..cn^e]$. Let v be a node assigned interval $[f^*(v)..g^*(v)]$. Starting t $f^*(v) + s(v)^e + 1$, assign consecutive intervals of appropriate size to the children f v, then recursively visit the children. Each node $v \in U$ is given an *ancestor able* $A_v[i]$, $0 \leq i \leq e(1 + \log_\beta n)$, where $A_v[i]$ is the first ancestor b of v that has $c-2)s(b)^e \geq \beta^i$. Given the fat preorder numbers and the ancestor tables, a $ca(x,y)$ uery can be answered in $O(1)$ time [7]. By using the fat preorder scheme on $C(T)$ ve can compute characteristic ancestors in T.

In Gabow's original scheme, a node v is recompressed and reassigned a new ubinterval of its parent interval time its size increases by a factor of α because of eaf additions. One may choose c and e such that v has a big enough interval to ccommodate all requests for new intervals from its children between the times that v itself is recompressed. After *add leaf*(u,v), u is made a child of v in T and of the ppropriate apex in $C(T)$. The weight of each ancestor of u in $C(T)$ is incremented and recompression is done where appropriate.

We can extend this to handle the *pmerge*(v) operation without actually changing the structure of T or $C(T)$. A set of merged nodes forms a subtree that is represented by its root node. The sets of merged nodes are managed with a relabel-the-smaller-half technique. If v is heavy, the heavy path list is adjusted as appropriate. If v is a light child of a heavy node, no change besides relabeling is done. If v is a light child of a light node, the ancestor table of the descendents of v are modified so as to reflect the merge. This is done so that characteristic ancestor triples can be found quickly.

The *demote*(x,y) is the most complex. Let v be the new node introduced by the demote and let u be the parent of x,y. If $2s(v)$ is more than $s(u)$ and u has no other heavy child then make v heavy. Insert it in the appropriate heavy path (or start a new one) and add v it as a child of the corresponding apex. Otherwise, the new children of v were formerly light children of u. Make v light. If all children of v are also light, then recompress v and assign it a new subinterval of u's interval. All descendents of v now correctly contain v in their ancestor tables. If v has a heavy child x then we let v take the place of x in $C(T)$ and add a new single node to represent x as a child of v. This can be done in $O(1)$ time given a level of indirection between ancestor table entries and actual nodes. The remaining demoted nodes are made children of v and are recompressed and assigned new subintervals in the interval of v.

Turning to the correctness of the above algorithm, we first note that it maintains the invariant that for any node v a child of u in $C(T)$, $\beta s(v) \leq s(u)$. Secondly, we note that at all time the ancestor tables are correctly updated. It remains to choose

α, c, and e so that between the times that node v is recompressed, its interval i large enough to satisfy all requests for new subintervals from its children. We als must satisfy $c \leq \beta^e + 2$. The worst case space utilization, $I(n)$, to construct a nod of size n occurs when the node is constructed by repeated demotions. We have

$$I(1) = c, I(n) = \max_{x_0, \ldots, x_k} \left\{ \sum_{i=0}^{k} I(x_i) + cn^e \right\},$$

given $\sum_{i=0}^{k} x_i = n - 1$ and $\alpha x_i \leq n$ for all i. By induction one may show that thi recurrence has solution $I(n) \leq \frac{cn^e}{1 - \frac{1}{\alpha^{e-1}}}$. The choice of $\alpha = 5/4$, $c = 9$, and $e =$ satisfies all constraints.

Lemma 4.4 *A sequence of* m add leaf(u, v), demote(u, v), pmerge(v) *and* ca$(u, v$ *operations in the auxiliary tree can be performed in time* $O(m + n(\log n)^2)$ *where* *is the maximum size of the tree.*

Proof: The relabeling caused by merges takes time $O(n \log n)$. The cost to adjus ancestor tables following a merge, or to recompress/reassign a subtree rooted a v, is $As(v)$, where $A = O(\log n)$. A is the cost of rebuilding the ancestor tabl of a node. Charge the cost of a merge to the light node being merged with it parent. This is a one-time charge. Charge the cost of a recompression/reassignmen to the node causing the recompression, either by increasing in size by a factor o α or by being the new parent of a collection of demoted nodes. The total amoun charged to a node of size n at the time it is recompressed is given by $T(n) = T(n/\alpha) + O(An) + O(A(1 - 1/\alpha)n) = O(An)$.

Call a node live if it is currently a light child or will become a light child at some time in the future. Otherwise call it dead. If a dead node is demoted by a live node, give its charge to the light new. Otherwise, share $O(\alpha k)$ of its charge among the k descendents it received while live. A node is charged $O(A)$ at most once for each light ancestor it had when inserted, hence $O(A \log n)$ total. Any remaining charge is "dead" charge. By managing dead charge carefully, sharing it whenever the node is demoted by another dead node, we can keep the amount of dead charge bounded by a constant times the total size of the dead node's siblings. At the conclusion of the algorithm this cost is shared proportionately by size among its light siblings. Thus in the end each live node u has total charge $O(As(u))$. Hence each level of the final $C(T)$ tree has $O(An)$ total cost per level, and there are $O(\log n)$ levels.

Theorem 4.5 *A sequence of* m add leaf(u, v), demote(u, v), *and* nca(u, v) *operations in the decomposition tree can be performed in time* $O(m\alpha(m, n))$.

Proof: This follows by choosing the block partition sizes such that there are $O(n/(\log n)^2)$ total nodes in label trees.

By adding an operation *Make Vertex*() one may extend the problem incremental planarity testing problem to general 0-connected graphs. In this case one must use auxiliary decompositions of the graph into 1 and 2-connected components. Edges can always be inserted between different components. To handle test queries between vertices in distinct blocks, a second test algorithm is applied to the 2-connected decomposition tree. This algorithm is described in [2] and can be implemented in

(1) time using our tree data structure. When two 1-connected or two connected mponents are combined into one, the data structures for the smaller one are rebuilt om scratch and inserted into those of the larger. The total time for these operations $O(n \log n)$.

References

1] G. D. Battista and R. Tamassia. Incremental planarity testing. In *Proc. 30th IEEE FOCS*, pages 436–441, 1989.

2] G. D. Battista and R. Tamassia. On-line graph algorithms with SPQR-trees. In *Proc. 17th ICALP*, 1990.

3] K. Booth and G. Lueker. Testing for the consecutive ones property, interval graphs, and graph planarity using PQ-tree algorithms. *J. Comput. System Sci.*, 13:335–379, 1976.

4] P. F. Deitz and D. D. Sleator. Two algorithms for maintaining order in a list. In *Proc. 19th ACM STOC*, pages 365–372, 1987.

5] M. Dietzfelbinger, A. Karlin, K. Mehlhorn, F. M. auf der Heide, H. Rohnert, and R. E. Tarjan. Dynamic perfect hashing: upper and lower bounds. In *Proceedings 29th IEEE FOCS*, pages 524–531, 1988.

6] M. L. Fredman and M. E. Saks. The cell probe complexity of dynamic data structures. In *Proc. 21st ACM STOC*, pages 345–354, Seattle, WA, May 1989.

7] H. N. Gabow. Data structures for weighted matching and nearest common ancestors with linking. In *Proc. 1st ACM-SIAM SODA*, pages 434–443, 1990.

8] H. N. Gabow and R. E. Tarjan. A linear-time algorithm for a special case of disjoint set union. *J. Comput. Syst. Sci.*, 30:209–211, 1985.

9] F. Harary. *Graph Theory*. Addison-Wesley, Reading, MA., 1972.

10] D. Harel and R. E. Tarjan. Fast algorithms for finding nearest common ancestors. *SIAM J. Comput.*, 13(2):338–355, 1984.

11] J. Hopcroft and R. E. Tarjan. Dividing a graph into triconnected components. *SIAM J. Comput.*, 2:135–158, 1973.

12] J. Hopcroft and R. E. Tarjan. Efficient planarity testing. *J. ACM*, 21:549–568, 1974.

13] H. Imai and T. Asano. Dynamic orthogonal segment intersection search. *Journal of Algorithms*, 8:1–18, 1987.

14] G. F. Italiano and Z. Galil. Fully dynamic algorithms for edge connectivity problems. In *23rd ACM STOC*, pages 317–327, 1991.

[15] J. A. La Poutré. Lower bounds for the union-find and split-find problems on pointer machines. In *Proc. 22nd ACM Symposium on Theory of Computing*, pages 34–44, 1990.

[16] J. A. La Poutré. On-line maintenance of triconnected components. These proceedings.

[17] V. Ramachandran and J. H. Reif. An optimal parallel algorithm for graph planarity. In *Proc. 30th FOCS*, pages 282–287, 1989.

[18] R. E. Tarjan. Efficiency of a good but not linear set union algorithm. *J. ACM*, 22:215–225, 1975.

Maintenance of Triconnected Components of Graph[s]
(Extended Abstract)

J.A. La Poutré

Department of Computer Science, Princeton University,

Princeton NJ 08544, U.S.A.

and

Department of Computer Science, Utrecht University,

3508 TB Utrecht, The Netherlands

Abstract

In this paper, optimal algorithms and data structures are presented to maintain the triconnected components of a general graph, under insertions of edges in the graph. At any moment, the data structure can answer the following type of query: given two nodes in the graph, are these nodes triconnected. Starting from an "empty" graph of n nodes (i.e., a graph with no edges) the solution runs in $O(n + m.\alpha(m, n))$ total time, where m is the total number of queries and edge insertions. The solution allows for insertions of nodes also.

1 Introduction

Dynamic algorithms are known for a number of graph problems, including computing transitive closures [13, 14, 15, 22], minimal spanning trees [4, 5], planarity testing [2] and shortest paths [1, 23]. We consider the problem of maintaining the triconnected components (also called the 3-vertex-connected components) of general graphs under insertions of edges and nodes. For an undirected graph, k-edge/vertex-connectivity is defined as follows. Two nodes x and y are k-vertex-connected or k-edge-connected if there are k different paths between them that are vertex-disjoint or edge-disjoint, respectively. Computing triconnected components of a graph is an important problem in the theory of algorithms. As was shown by Hopcroft and Tarjan in [12], the triconnected components of a (static) graph can be computed in linear time. Applications include circuit lay-out problems, reliability problems, and planarity testing.

*At Princeton University, the research was supported by a NATO Science Fellowship awarded by N.W.O., and partially supported by DIMACS (Center for Discrete Mathematics and Theoretical Computer Science), a National Science Foundation Science and Technology Center - NSF-STC88-09648. At Utrecht University, the research was partially supported by the ESPRIT Basic Research Actions of the E.C. under contract no. 3075 (project ALCOM).

[26, 19], Westbrook and Tarjan and La Poutré presented optimal algorithms for aintaining 2-edge-connected and 2-vertex-connected components of general graphs nder the insertions of edges in the graph. The solutions have a total running me of $O(n + m.\alpha(m,n))$, starting from an empty graph of n nodes, where α is the verse Ackermann function and m is the number of edge insertions and queries. The lutions allow for insertions of nodes also (in the same time bounds, taking n as the nal number of nodes). As can be observed in [19, 26], although an optimal solution r maintaining 2-edge/vertex-connected components for initially connected graphs simple and can be obtained mainly by using optimal Union-Find algorithms, he problem of optimally maintaining these components for general graphs is much arder and requires involved data structures. In [10], an algorithm was presented r maintaining the 3-edge-connected components of graphs in the special case that he initial graph is connected, taking $O((n+m).\alpha(m,n))$ time. In [19], an optimal lgorithm was presented for maintaining the 3-edge-connected components of general raphs, which takes $O(n+m.\alpha(m,n))$ time. Whereas 2- and 3-edge-connectivity and -vertex-connectivity are equivalence relations on the constituents of a graph (viz., n nodes and on edges respectively), this is not the case for 3-vertex-connectivity. Moreover, any two nodes may both be contained in several 3-vertex-connected omponents.) Di Battista and Tamassia [3] presented an algorithm for maintaining he triconnected components of a graph in the special case in which the initial graph s biconnected. This solution has a running time of $O((n+m).\alpha(m,n))$. They also resented an extension to the general case, which, however, has a running time of $O(n.\log n + m)$.

n this paper, we present optimal algorithms and data structures for maintaining he triconnected components of a general graph under insertions of edges and nodes. The algorithm starts from an empty graph with n nodes (i.e., a graph with no edges nd n nodes), in which edges are inserted one by one, and where at any time for ny two nodes the query that asks whether these nodes are 3-vertex-connected can e answered. The solution has a total running time of $O(n + m.\alpha(m,n))$, where m s the number of edge insertions and queries. The solution allows for insertions of odes also (in the same time bounds, taking n as the final number of nodes).

The abstract is organized as follows. Section 2 contains the preliminaries. Section 3 describes cycle trees. In Section 4, cluster partitions and rooted cluster trees are introduced with which in Section 5, the algorithms are constructed. In this extended abstract, we only present a rough sketch, and we omit details and special cases.

2 Preliminaries

We use the terminology of [11]. Consider an undirected graph. Two nodes x and y are *k-vertex-connected* iff there exist k different vertex-disjoint paths between x and y. If the removal of a set of vertices and edges separates the vertices x and y, then that set is called a *cut set* for x and y.

We call a set S of at least 2 nodes a *2vc-class* if the nodes are 2-vertex-connected and if there does not exist a node not in S that is 2-vertex-connected with the nodes of S (i.e., the class is maximal). We call a set S of at least 3 nodes a *3vc-class* the nodes are 3-vertex-connected and if there does not exist a node not in S tha is 3-vertex-connected with the nodes of S (i.e., the class is maximal). Similarly, w call a set S of 2 nodes that satisfies these conditions a *3vc-pair*. See Figure 1 fc an example of a graph and its 3vc-classes (where the 3vc-classes are indicated b dotted lines.) The *2-vertex-connected components* (or blocks) of a graph G are th subgraphs of G that are induced by the 2vc-classes of nodes. We can also defin 3-vertex-connected components of graphs (where now new edges are added), bu we will not give the definition here. Two nodes in a 2-vertex-connected componen are k-vertex-connected in H iff they are k-vertex-connected in G ($k \geq 1$). On th other hand, for a 3vc-class or 3vc-pair C_3, there exists exactly one 2vc-class C_2 tha contains C_3, and for all other 2vc-classes $C_2' \neq C_2$, $|C_2' \cap C_3| \leq 1$. So, 3vc-classes ar determined by the 2-vertex-connected components in which they are contained.

For a 2-vertex-connected graph $G =< V, E >$ and a set of nodes $S \subseteq V$ we defin graph $G_3(S)$ by $G_3(S) =< S, E(G,S) >$, where for each pair of nodes in S tha are 3-vertex-connected in G, there are 3 parallel edges between them in $E(G,S)$ and where for the other pairs of nodes in S, $E(G,S)$ contains the edge(s) of E with these nodes as end nodes. We call a set $S \subseteq V$ a *coherent set* if $G_3(S)$ i 2-vertex-connected. We have that two nodes x and y in a coherent set S are 3 vertex-connected in G iff they are 3-vertex-connected in $G_3(S)$. In addition, a se $\{u,v\}$ is a cut set for x and y in G iff it is a cut set for x and y in $G_3(S)$.

2.1 Problem Description

The problems that we consider are as follows. Let a graph be given. Then th following operations may be applied: operation *insert*(x,y), i.e., insert an edge (x,y) in the graph, and operation *Is3vc(x, y)*, i.e., output whether x and y are two node in the graph that are 3-vertex-connected, and return the name of the 3-vertex connected component (3vc-class or 3vc-pair) in which they both are contained (i any). We call this problem the *3vc-problem*. In addition, the above collection o operations can be extended with the insertion of a new (isolated) node in the graph (We will consider this operation only in the last steps of our solutions.)

3 Cycle Trees and 2-Vertex-Connected Graphs

In this section, we describe cycle trees and their constituents for splitting up 2-vertex-connected graphs. (A related way of splitting up a graph with a resulting tree can be found in e.g. [25], and a way in terms of *st*-graphs in [3].) Subsequently we describe the use of such trees for the problem of maintaining 3-vertex-connected components (also cf. [3]), and we describe problems that arise for arbitrary graphs.

3.1 Cycle Trees

First, we consider a 2-vertex-connected graph G containing at least three nodes. G can be subdivided into so-called 3vc-classes, cycles, and bars in a way as follows. A pair of nodes $\{u, v\}$ that is a cut pair (i.e., a cut set) of G, and such that u and v are 3-vertex-connected, is called a *3vc-bar* of G. Let $\{u, v\}$ be a 3vc-bar of G. The deletion of u and v from G yields a number of connected components $H_1, .. , H_k$. For each $H_i =< V_i, E_i >$, let H_i' be the subgraph of G induced by the node set $V_i \cup \{u, v\}$, where the edges between u and v are deleted (if any), and where three new edges between u and v are inserted. Then H_i' is 2-vertex-connected, and each 3vc-class of G is contained in exactly one H_i'. Moreover, H_i' and H_j' have exactly the two nodes u and v in common. Finally, each other 3vc-bar of G is contained in exactly one H_i', and, moreover, it is a 3vc-bar of H_i'. This process can be continued on the resulting graphs, until we have obtained graphs that do not contain 3vc-bars. A final graph in this process can be a 3vc-class of nodes of G, or it can be a simple cycle if all multiple edges between two nodes are replaced by one edge. To denote the latter case, we just refer to the graph as a (simple) cycle. We call each pair of consecutive nodes on such a cycle a *cycle bar.* The result graphs are independent of the order of splitting.

In this way, we define the *cycle tree* of G corresponding to this splitting as follows. The nodes of the cycle tree are cycle nodes, class nodes, and bars, where each *cycle node* corresponds to one resulting cycle, each *class node* corresponds to one 3vc-class, and each *bar* is a pair of nodes of G that form a 3vc-bar or a cycle bar. The tree is defined recursively as follows. If all the nodes of G are 3-vertex-connected, and, hence, form one 3vc-class of nodes, then the tree is the class node for that 3vc-class. Otherwise, if G is a simple cycle, then the tree consists of the cycle node for that cycle, together with all bars corresponding to cycle bars of the cycle, where the cycle node has a link to each such bar. These links are ordered according to the order in which the bars occur in the cycle. Otherwise, there is a 3vc-bar $[u, v]$ for G. The deletion of u and v from G yields a number of graphs $H_1', .. , H_k'$ as defined above. The tree for G is constructed as follows. Create the bar $[u, v]$. For each H_i', obtain its cycle tree. If this cycle tree has a bar $[u, v]$, then identify this with the above 3vc-bar $[u, v]$. Otherwise, make a link between $[u, v]$ and the unique 3vc-class in H_i' that contains u and v. This yields a tree that is independent of the order of splitting. We denote the tree by $Cyc(G)$. See Figure 2 for an example of the cycle tree for the graph in Figure 1 (where class nodes, cycle nodes, bars, and links are drawn as dotted ellipses, boxes, heavy lines, and light lines, respectively). The collection of bars, classes, and cycles in $Cyc(G)$ in which a node x of G is contained is a subtree of $Cyc(G)$, called the *subtree induced by x.*

3.2 Insertions and Implementation

Suppose edge (x, y) is inserted in G, where x and y are not 3-vertex-connected. Let $P(x, y)$ be the tree path between the two subtrees induced by x and y in $Cyc(G)$. Then all 3vc-classes on $P(x, y)$, all nodes in bars on $P(x, y)$, and nodes x and y together form one new 3vc-class K resulting from the insertion. Moreover, each cycle s occurring on $P(x, y)$ is split into two new cycles, viz., according to the node in cycle s that are in a bar on $P(x, y)$ or that are x or y.

To compute the path $P(x, y)$ efficiently, the tree $Cyc(G)$ is rooted in some class node or cycle node r. Then, in the rooted cycle tree, the path can be obtained by traversing the two root paths in $Cyc(G)$ starting from the tops of the subtrees induced by x and by y, respectively, such that the traversal steps for both paths are performed in an alternating way, until a common class node, cycle node, or bar is visited by both traversals, or the traversal for x reaches a bar containing y, or vice versa. (This can be checked by means of marking.) Then $P(x, y)$ consists of the two traversed path parts up to the common element. This computation of $P(x, y)$ requires a number of traversal steps that is at most twice the length of $P(x, y)$.

An implementation of a cycle tree has the following main aspects. Each node x has a representative x_s for each cycle s in which it is contained, and it has a representative x_K for each 3vc-class K in which it is contained. The 3vc-classes are implemented as Union-Find structures on these representatives. The name of the set for the 3vc-class is the corresponding class node. Moreover, the cycles are implemented as "Circular" Split-Find structures [20] on these representatives (in the order determined by the cycle). The name of the list for the cycle is the corresponding cycle node. The bars do not occur as such in the implementation. The representatives of nodes for cycles are also used for representing the 3vc-pairs. For an implementation as a rooted tree, we have the following. For each node x in G, there is a pointer $max(x)$ to its representative in a 3vc-class or cycle that is closest to the root. A class node or a cycle node has two pointers to the representatives (for this class or cycle) of the two nodes that are in its father bar in $Cyc(G)$. (This implements the relation between bars and cycles/classes, being the links in the cycle tree.)

Now, edge insertions can be processed by computing $P(x, y)$ in the rooted cycle tree, by performing Unions for joining 3vc-classes and separate nodes, and by performing Splits for splitting cycles into two subcycles. (Here, since bars are not implemented as such, instead of bars, it is checked whether end nodes of bars are visited by both traversals. Moreover, the father F of a bar can be obtained from the end nodes x and y of that bar, by using $max(x)$ and $max(y)$ and by obtaining the related class/cycle nodes from a Find operation; then F is returned for either both the nodes x and y or for only one node, say x, where the father pointer of F points to a representative for y.) A query that asks whether two nodes x and y are 3-vertex-connected can be performed by means of $max(x)$ and $max(y)$ and the classes or cycles containing these representatives, together with the nodes in the father bars of these classes or cycles. (A query that asks whether a nodes x is contained in a given 3vc-class with class name K can be performed similarly.)

3.3 Extensions

The above structure of rooted cycle trees can be used to efficiently process insertions of edges in a 2-vertex-connected graph, while queries are performed from time to time. In [3], a similar optimal solution for such graphs, i.e., for graphs that are 2-vertex-connected, is presented (implemented by SPQR trees). In particular, these ooted trees allow efficient computation of tree paths needed to process edge insertions. However, in a general graph, connected components and 2-vertex-connected components may be joined arbitrarily, and therefore cycle trees (or SPQR trees) or 2-vertex-connected components must be linked from time to time. If we use a straightforward "redirection" technique to maintain the father pointers while trees re linked from time to time, then this takes $O(n.\log n)$ time for the redirections lready. In the sequel, we will deal with this problem.

4 Cluster Partitions and Rooted Cluster Trees

We introduce cluster partitions and rooted cluster trees for 2-vertex-connected graphs, with which our maintenance algorithms for general graphs will be constructed in Section 5.

4.1 Cluster Partitions

For a 2-vertex-connected graph $G =< V, E >$, a cluster partition $CP(G)$ (CP for short) is as follows. The vertex set of G is partitioned into (not necessarily disjoint) sets of at least three nodes, called *clusters*, and to each cluster C, a collection $not3bar(C)$ is associated containing sets of two nodes in C that are 3-vertex-connected in G. The cluster partition must satisfy the following. Every two clusters have at most two nodes in common. Every cluster C is a coherent set, and every node of C occurs in at most two pairs in $not3bar(C)$. For a cluster C, a 3vc-bar $[x, y]$ of $G_3(C)$ is called an *internal bar* of C. The following constraints are satisfied.

1. For every node x and 3vc-class K in which x is contained, there is a cluster that contains at least three nodes of K, amongst which is x.
2. Every cut pair in G is contained in $G_3(C)$ of some cluster C.
3. For $\{x, y\} \in not3bar(C)$, $[x, y]$ is not an internal bar of C.
4. For every pair $\{x, y\}$ of nodes in G, there is at most one cluster C such that $\{x, y\} \subseteq C$ and $\{x, y\} \notin not3bar(C)$.

In addition, an arithmetic constraint holds. There also is a distinction between so-called *minor* or *major* clusters (and, in association, classes), where the union of all the major clusters is a coherent set. We will not describe this here. See Figure 3 for an example, where the clusters are drawn as heavy lines.

We say that a 3vc-class K *occurs* in a cluster C if the cluster contains at least three nodes of the class. We also say that C *contains* a part of K, or just that it contains K (for short). The intersection of such a 3vc-class and cluster is called a *subclass* of that class. The 3vc-classes occurring in at least two clusters are called *multiple* classes. The nodes of G occurring in at least two clusters are called *multiple* nodes or m-nodes. For a cluster C, let $tree(C)$ denote a cycle tree for $G_3(C)$, where cycles occurring both in G and $G_3(C)$ get the same name, and where a 3vc-class in $G_3(C)$ gets the same name as the class in G in which it is contained. It then follows that for a cluster C, $tree(C)$ is a subtree of $Cyc(G)$, and that every cycle node s of $Cyc(G)$ is contained in $tree(C')$ for exactly one cluster C'.

An implementation of a cluster partition has the following main aspects. Each cluster is represented by a new node, called a cluster node. For each m-node x or multiple 3vc-class K, there is a representative x_{CP} or K_{CP} for the cluster partition. For each m-node x that occurs in a cluster C, there is a representative x_C for the node in that cluster. Similarly, for a multiple 3vc-class K, there is a representative K_C for a subclass of K in a cluster C. These representatives form a set in a Union-Find structure with K_{CP} as set name. Furthermore, we have various lists containing representatives (e.g., for $not3bar(C)$).

4.2 Rooted Cluster Trees

Consider a graph G with cycle tree $Cyc(G)$ rooted in a class or cycle r. The formal concept of a rooted cluster tree $RCT(G)$ for G on $CP(G)$ (RCT for short) is as follows. The nodes of RCT are the clusters and the multiple 3vc-classes of G, and the 3vc-bars of $Cyc(G)$. The root of RCT is some cluster in which the root of $Cyc(G)$ occurs. For a cluster C, we denote by $max(C)$ the bar or 3vc-class in $tree(C)$ that is closest to the root r in $Cyc(G)$. The father relation is such that it satisfies the following. If root r of $Cyc(G)$ is a multiple 3vc-class itself, then its RCT-father is the root of RCT. The RCT-father of a *multiple 3vc-class* K with father $[x, y]$ in $Cyc(G)$ is the cluster C in which K occurs and for which $max(C) \neq K$, if it exists; otherwise, it is a cluster containing K and the nodes x and y, or, if K is major, it may be $[x, y]$ itself. The RCT-father of a *cluster* C is the representative for $max(C)$ in RCT, where if C is chosen as the RCT-father of class $max(C)$, then the RCT-father of C is the father of $max(C)$ in $Cyc(G)$ (which is a bar). The RCT-father of a *bar* $[x, y]$ with father F in $Cyc(G)$ is either a cluster C that contains x, y, and F, or, if F is a multiple 3vc-class, it may be F itself. Thus, the father relation defines the links in the tree. We call a link a class link, if it connects a 3vc-class and a cluster, and we call it a cut link otherwise.

For each node x, $clus(x)$ is a maximal cluster in RCT containing x, i.e., it contains x whereas all clusters and bars on its root path do not contain x. Moreover, each cluster C contains a set $Assoc(C)$ of $O(1)$ of its m-nodes. This is such that for every node x in a cluster C, $clus(x) = C$ or $x \in Assoc(C)$.

The implementation has the following main aspects. A class link between cluster C and class K is implemented by the corresponding class representative K_C. Like for cycle trees, bars are not implemented as such. These are omitted, together with the incident edges. Instead, each cut link directed to a bar is represented by a record with pointers to the two m-nodes forming the bar, together with pointers to representatives for these m-nodes. For convenience in this abstract, we describe (and use henceforth) an implementation for rooted cluster trees in which classes are only adjacent to clusters. (Actually, this property can be obtained by extending 3vc-classes with new nodes and by creating new clusters.) Then the above representatives or two m-nodes are the representatives for the cluster that is the begin of the link nd the representatives for a cluster that satisfies the constraints of being the father f the bar. A cluster or a class name a in $RCT(G)$ has a pointer $father(a)$ to its utgoing link. Each m-node x has a pointer to the cluster $clus(x)$ together with pointer to its representative for that cluster. For each cluster C, $Assoc(C)$ is nplemented as a list of representatives in C. A rooted cluster tree can now be onstructed from a cluster partition in linear time.

'he idea of processing insertions is as follows. A path in the structure that is related o the path $P(x,y)$ in $Cyc(G)$ can be obtained as follows. First a path P in $RCT(G)$ s obtained, which is the path between the two subtrees of clusters and class names n $RCT(G)$ that "contain" x or y, respectively. Then, in each cluster C on the path except for possibly two clusters closest to the root), a path in $tree(C)$ is obtained, pecified by the links on P that are incident with C. In this way, an equivalent of $P(x,y)$ in $Cyc(G)$ can be computed, viz., where each class name on $P(x,y)$ may e replaced by a feasible sequence consisting of this class name or representatives or this class in clusters. The update is done by performing local updates inside lusters (updating the local tree $tree(C)$ for a cluster C), and by global updates on he cluster tree (joining 3vc-classes).

A query that asks whether two nodes x and y are 3-vertex-connected is performed by inspecting $clus(x)$ and $clus(y)$. If $clus(x)$ contains y, then a local query is performed in $clus(x)$, and similarly with x and y reversed. Otherwise, the father and grandfather of these clusters are used for the queries, which may result in the following intermediate output: a node x, a cluster C_x that contains x, and a subclass k_x in C_x (and similarly for y), such that if x and y are 3-vertex-connected, then k_x and k_y are subclasses of some class K that contains both x and y. Then a local query inside cluster C_x is performed, computing whether x is contained in k_x, and similarly for y. (Actually, representatives of these nodes are then returned which also are representatives for class K, if K exists.) For the local queries in a cluster C, the (information of) the local cycle tree $tree(C)$ is used.

5 Resulting Algorithms

To solve the 3vc-problem for general graphs, we maintain the 2-vertex-connected components of a graph, and we maintain the 3-vertex-connected components in each 2-vertex-connected component. For maintaining 2-vertex-connected components, we use an optimal solution [19, 26]. Each node has a representative for each 2-vertex-connected component in which it is contained. Now, the representatives of these nodes for a 2-vertex-connected component play the role of the nodes described in the previous section. The query whether two nodes are 2-vertex-connected also returns their representatives for the common 2-vertex-connected component in which they are contained (see [19, 20]).

To obtain the optimal time bound of $O(n+m.\alpha(m,n))$ for m operations on n nodes, we have the following approach for an implementation using address calculation. We develop a generalisation of the concept of microsets as presented in [9], that allow joining of microsets and queries within the α-bound. We call it *dynamic microsets*. We will not describe it here. (The concept of microsets was also used for e.g. obtaining a linear-time Split-Find algorithm in [9].) For "small" 2-vertex-connected components (i.e., of size $O(\log\log n)$), a cluster tree consists of one cluster only, and the cluster is a dynamic microset. The dynamic microset for a cluster C encodes the structure of $tree(C)$ into one machine word. Now, if two or more small 2-vertex-connected components are joined, this is done by joining the dynamic microsets as long as the resulting component is still small. The queries and updates inside a 2-vertex-connected component can be performed by using the encoded information in the microsets.

A larger 2-vertex-connected component is represented by a cluster tree with clusters of larger size, where the cluster tree is used for a coherent part of this component only (which corresponds to a "super" cluster), and by parts (i.e., subtrees) of the (global) cycle tree for the remaining part of the 2-vertex-connected component (which can be seen as parts of the cycle tree that are attached to the super cluster). (E.g., for "intermediate" 2-vertex-connected components of $O(\log n)$ nodes there are altogether $O(\frac{\log n}{\log\log n})$ clusters of size $O(\log\log n)$, and, similarly, for the remaining "large" components of size $O(n)$, there are altogether $O(\frac{n}{\log n})$ clusters of size $O(\log n)$.) For each such cluster C, the local cycle tree $tree(C)$ is implemented as a rooted cycle tree (by means of the representatives of the nodes for cluster C) where Union-Find structures are used for implementing 3vc-classes and Split-Find structures are used for implementing cycles.

Then, if the insertion of an edge causes that two or more 2-vertex-connected components of similar size are joined, we have the following. The occurring 2-vertex-connected components each contain exactly 2 nodes ("connection nodes") that are the nodes in common with the other 2-vertex-connected components or that are end nodes of the inserted edge. Then the insertion corresponds to first inserting auxiliary edges between the two connection nodes in each participating 2-vertex-connected component. Then, the 2-vertex-connected components are joined (by the algorithms for 2-vertex-connectivity). The cycle tree for the resulting 2-vertex-connected component now actually consists of the cycle trees of the constituent old 2-vertex-connected components, integrated by a newly arisen cycle consisting of the connection nodes. This is reflected in our algorithms by creating a new cluster for the newly arisen cycle and by linking the rooted cluster trees to it (after some adaptations for the above parts of the global cycle trees, which we will not consider here). These cluster trees are reconstructed w.r.t. the father relation, except for the largest one. (We omit many cases and details in this description.) The complexity for these linkings for e.g. the largest components is $O(l.\log l)$ steps, if there are l "large" clusters of size $O(\log n)$ altogether (where a step includes a Find). Since $l = O(\frac{n}{\log n})$, this yields $O(n)$ steps in total only, and a similar analysis holds for the other components. If the insertion of an edge causes that a "small" 2-vertex-connected component C (of size $O(\log\log n)$) is joined with a larger component D (e.g., of size $\Omega(\log n)$), then the cycle tree of C is obtained and attached as a subtree in the representation of D. The other cases are similar, where if e.g. a collection of only "small" 2-vertex-connected components are joined into a component of "intermediate" size (between $\Omega(\log\log n)$ and $O(\log n)$), a new representation consisting of one cluster is created for the resulting 2-vertex-connected component. The insertion of an edge inside a 2-vertex-connected component is performed according to the strategy given in Subsection 4.2 and Section 3 (inside *tree*(C) the algorithms on rooted cycle trees are used). We will not give further details here.

To obtain the optimal time bound for a Pointer Machine, hence, *without* address calculation, we use "contraction cluster partitions" of cluster partitions, in a way somewhat related to the division trees in [19, 20]. (This solution is more involved than the previous solution, once the dynamic microset structure is available.) The idea is to use so-called double cluster partitions consisting of a (base) cluster partition and a contraction cluster partition. Then each cluster in the contraction cluster partition is again partitioned into a "local cluster partition", which is implemented as a rooted cluster tree. All these local cluster partitions together form the base cluster partition. Since, in general, it is not possible to contract a 2-vertex-connected component into an existing contraction cluster of another 2-vertex-connected component if these components are joined (like this is possible for linking trees in [19, 20]), these double cluster partitions are more involved than the division trees of [19, 20]. We use these double cluster partitions to build a layered structure corresponding to Ackermann values, with a technique related to [7, 8, 16, 19, 20]. I.e., there are at most i layers numbered decreasing from i, with $i = O(n, n)$. Each layer consists of a cluster partition, and the cluster partitions of layer $j+1$ and j ($j < i$) form a double cluster partition. Then a similar approach for answering queries as described above is used, where now a "local" query in a cluster C in some layer j of the structure corresponds to a recurrent call in the cluster partition of cluster C in layer $j+1$, if it exists (i.e., if $j < i$), and to a "normal" local call on *tree*(C) otherwise (i.e., if $j = i$). We will not give further details here.

In this way, we obtain the following theorem (where the insertion of an edge is called *nonessential* if the end nodes of the new edge are already 3-vertex-connected).

Theorem 5.1 *There exists a data structure and algorithms that solve the 3vc-problem such that the following holds. The total time that is needed starting from an empty graph with n nodes is $O(m.\alpha(m,n))$ (where m is the number of edge insertions and queries), whereas the fth operation can be performed in $O(\alpha(f,n))$ worst-case time if it is a query or a nonessential insertion. The initialisation can be performed in $O(n)$ time, and the entire structure takes $O(n)$ space.*

The above theorem can be augmented to allow insertion of new nodes in the graph with a time complexity of $O(n + m.\alpha(m,n))$. Then n, m and f in the theorem denote the current number at the moment of consideration.

If the *initial graph* is *not empty*, then first the 2- and 3-vertex-connected components are computed, using [24, 12]. The structure is then initialised by creating one cluster for each 2-vertex-connected component, that contains all nodes in that component. In this way, we obtain a solution such that the total time that is needed starting from a graph with n nodes is $O((n_{red}+m).\alpha(m,n))$ (where m is the number of edge insertions and queries, and n_{red} is the number of nodes in the initial graph that are contained in connected components of size at least 2), the initialisation can be performed in $O(n+e)$ time (where e is the number of edges in the initial graph and the entire structure takes $O(n)$ space. (Note that $n_{red} \leq n$.)

6 Concluding Remarks

We have presented a solution for maintaining triconnected components of grapl that runs in $O(n + m.\alpha(m,n))$ time starting from the "empty" graph with n node where m is the number of queries and edge insertions. The solution allows fc insertions of nodes also (in the same time bounds, taking n as the final numbe of nodes). These solutions are optimal for both the Cell Probe Machine and th Pointer Machine. This is proved by a reduction from the Union-Find problem to th maintenance problem (like in [26] for biconnectivity). The optimality then follow by [6, 17].

A modification of the solution for maintaining triconnected components can be use as part of a (deterministic) algorithm for incremental planarity testing in $O(n$ - $m.\alpha(m,n))$ time, where n and m are as above. This improves the results in [2, 27 We refer to [21].

References

[1] G. Ausiello, G.F. Italiano, A. Marchetti Spaccamela, and U. Nanni, Incremental Algorithm for Minimal Length Paths, Proc. 1st Ann. ACM-SIAM Symp. on Discrete Algorithms (SODA 1990, 12-21.

[2] G. Di Battista and R. Tamassia, Incremental Planarity Testing, Proc. 30th Ann. Symp. o Found. of Comp. Sci. (FOCS) 1989, 436-441.

[3] G. Di Battista and R. Tamassia, On-Line Graph Algorithms with SPQR-Trees, Proc. 17th Int. Colloquium on Automata, Languages and Programming (ICALP) 1990, 598-611.

[4] D. Eppstein, G.F. Italiano, R. Tamassia, R.E. Tarjan, J. Westbrook and M. Yung, Mainte nance of a Minimum Spanning Forest in a Dynamic Planar Graph, Proc. 1st Ann. ACM-SIAM Symp. on Discrete Algorithms (SODA) 1990, 1-11.

[5] G.N. Frederickson, Data structures for on-line updating of minimum spanning trees, wit applications, SIAM J. Computing 14 (1985), pp.781-798.

[6] M.L. Fredman and M.E. Saks, The Cell-Probe Complexity of Dynamic Data Structures, Proc 21th Ann. ACM Symp. on Theory of Comput. (STOC) 1989, 345-354

[7] H.N. Gabow, A Scaling Algorithm for Weighted Matching on General Graphs, Proc. 26th Ann. Symp. on Found. of Comp. Sci. (FOCS) 1985, 90-100.

[8] H.N. Gabow, Data Structures for Weighted Matching and Nearest Common Ancestors with Linking, Proc. 1st Ann. ACM-SIAM Symp. on Discrete Algorithms (SODA) 1990, 434-443.

[9] H.N. Gabow and R.E. Tarjan, A Linear Time Algorithm for a Special Case of Disjoint Se Union, J. Comput. Syst. Sci. 30 (1985), 209-221.

[10] Z. Galil and G.F.Italiano, Fully Dynamic Algorithms for Edge Connectivity Problems, Proc 23th Ann. ACM Symp. on the Theory of Computing (STOC) 1991.

[11] F. Harary, Graph Theory, Addison-Wesley Publishing Company, Reading, Mass., 1969.

[12] J. Hopcroft and R.E. Tarjan, Dividing a graph into Triconnected Components, SIAM J Comput. 2 (1973).

[13] G.F. Italiano, Amortized efficiency of a path retrieval data structure, Theoretical Computer Science, 48, (1986), pp. 273-281.

[14] G.F. Italiano, Finding paths and deleting edges in directed acyclic graphs, Information Processing Letters, 28, (1988), pp. 5-11.

[15] J.A. La Poutré and J. van Leeuwen, Maintenance of Transitive Closures and Transitive Reductions of Graphs, In: H. Göttler, H.J. Schneider (Eds.), Graph-Theoretic Concepts in Computer Science 1987, Lecture Notes in Computer Science Vol. 314, Springer-Verlag, Berlin, pp. 106-120.

[16] J.A. La Poutré, New Techniques for the Union-Find Problem, Proc. 1st Ann. ACM-SIAM Symp. on Discrete Algorithms (SODA) 1990, 54-63.

[17] J.A. La Poutré, Lower Bounds for the Union-Find and the Split-Find Problem on Pointer Machines, Proc. 22th Ann. ACM Symp. on Theory of Comput. (STOC) 1990, 34-44.

[18] J.A. La Poutré, J. van Leeuwen and M.H. Overmars, Maintenance of 2- and 3-connected components of graphs, Part I: 2- and 3-edge-connected components, Tech. Rep. RUU-CS-90-26, Utrecht University, 1990.

[19] J.A. La Poutré, Maintenance of 2- and 3-connected components of graphs, Part II: 2- and 3-edge-connected components and 2-vertex-connected components, Tech. Rep. RUU-CS-90-27, Utrecht University, 1990.

[20] J.A. La Poutré, Dynamic Graph Algorithms and Data Structures, Ph.D. Thesis, Utrecht University, 1991.

[21] J.A. La Poutré, Fast Deterministic Algorithms for Incremental Planarity Testing, Princeton University, in preparation.

[22] F.P. Preparata and R. Tamassia, Fully Dynamic Techniques for Point Location and Transitive Closure in Planar Structures, Proc. 29th Ann. Symp. on Found. of Comp. Sci. (FOCS) 1988, pp. 558-567.

[23] H. Rohnert, A dynamization of the all pairs least cost path problem, In: K. Mehlhorn (ed.), 2nd Annual Symposium on Theoretical Aspects of Computer Science 1985, Lecture Notes in Computer Science Vol. 182, Springer-Verlag, Berlin, pp. 279-286.

[24] R.E. Tarjan, Depth-First Search and Linear Graph Algorithms, SIAM J. Comput. 1 (1972) 146-160.

[25] W. Tutte, Connectivity in Graphs, University of Toronto Press, 1966.

[26] J. Westbrook and R.E. Tarjan, Maintaining Bridge-Connected and Biconnected Components On-Line, Tech. rep. CS-TR-228-89, Princeton University, 1989 (to appear in Algorithmica).

[27] J. Westbrook, Fast Incremental Planarity Testing, Proc. 19th Int. Colloquium on Automata, Languages and Programming (ICALP) 1992.

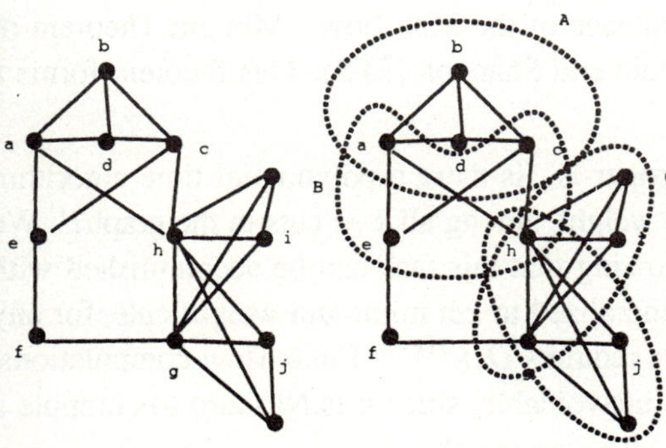

Figure 1

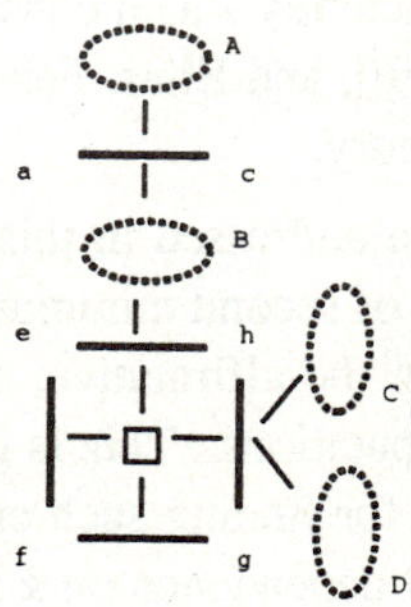

Figure 2

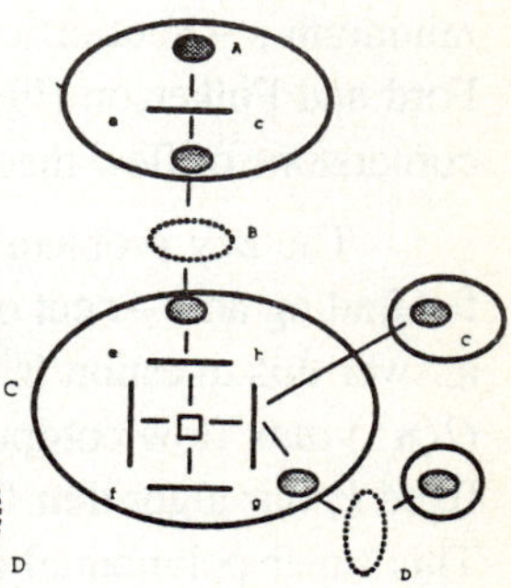

Figure 3

SUBOPTIMAL CUTS: THEIR ENUMERATION, WEIGHT AND NUMBER
(Extended Abstract)

Vijay V. Vazirani

Indian Institute of Technology
New Delhi, India

Mihalis Yannakakis

AT&T Bell Laboratories
Murray Hill, NJ 07974

ABSTRACT: We present (1) an algorithm that enumerates the cuts of a network by increasing weight with polynomial delay, and (2) an algorithm that computes the k-th minimum weight in polynomial time for fixed k. We also show that in the case of undirected networks there are only polynomially many cuts that have the k-th minimum weight for any fixed k (whereas directed networks can have exponentially many different minimum cuts).

1. INTRODUCTION

Minimum cuts have been extensively studied in the past, and by now we seem to have a satisfactory understanding of such (i.e., optimal) cuts, both structurally and algorithmically. In this paper, we extend this theory to suboptimal cuts.

Very early on it was realized that a maximum flow algorithm can be used to find a minimum $s-t$ cut efficiently - as a consequence of the Max flow - Min cut Theorem of Ford and Fulkerson [FF], and Elias, Feinstein and Shannon [EFS]. This theorem forms a cornerstone of flow theory.

The key problem addressed in this paper is, 'is there a polynomial time algorithm for finding an $s-t$ cut of second minimum weight, among all $s-t$ cuts in the graph?' We answer this question in the affirmative, showing that this task can be accomplished with $O(n^2)$ max flow computations. This is generalized to kth minimum weight cuts, for any fixed k; our algorithm for finding such cuts requires $O(n^{2(k-1)})$ max flow computations. This super-polynomial dependence on k is unavoidable, since it is NP-hard to compute a maximum cut in a graph in which each edge has unit weight.

Let us next consider the question of enumeration of cuts in a graph. The problem of enumerating all minimum $s-t$ cuts was solved by Picquard and Querrane [PQ] (and independently by Suurbale in an unpublished report [S]). They obtained a partial order by computing a maximum $s-t$ flow and shrinking strongly connected components in the resulting residual graph, and they showed that ideals in this partial order correspond exactly to minimum $s-t$ cuts. Hence the problem reduces to enumerating ideals in this partial order, which can be easily accomplished with only polynomial delay. (This partial order is central to our paper as well; for instance, it is used for proving the results of the previous paragraph.) The other interesting case previously considered was that of minimum weight cuts in an undirected edge-weighted graph. Such cuts are called edge connectivity cuts. Dinits, Karzanov and Lomosonov showed that a graph can have at most $\binom{n}{2}$ edge connectivity cuts, and they showed how to enumerate them using only $n-1$ max flow computations [DKL]. (See [NV] for the corresponding parallel algorithm).

Using a schema of Lawler [L], we present algorithms for enumerating all cuts by increasing weight in the following situations: s-t cuts in directed and undirected graphs, all cuts in directed and undirected graphs, and ideal cuts (see Section 2 for definition) in a directed acyclic graph. Each algorithm outputs cuts with polynomial delay; in particular, requiring at most n max flow computations to output the next cut. Moreover, averaged over all cuts, each algorithm, except for the one for ideal cuts, requires less than two max flow computations per cut.

As stated above, an undirected graph can have at most $\binom{n}{2}$ minimum cuts. In each of the other four situations, there may be exponentially many minimum cuts. Our last result establishes a bound of $O(n^{3k-1})$ for the number of kth minimum weight cuts in an undirected graph. The best lower bound we have is $\Omega(n^{2k})$; we leave the open problem of closing this gap.

Finally, using these properties, we have a polynomial time algorithm for finding all kth minimum weight cuts in an undirected graph. As a corollary, it follows that the problem of counting the number of kth minimum weight cuts in this situation is in P, for any fixed k. In contrast, for the remaining four situations, the problem of counting the number of minimum weight cuts is #P-complete; for directed and undirected $s-t$ cuts this was established by Provan and Ball [PB], and it is shown here for the other two situations.

Minimum cuts have found numerous applications, and we expect the same for our results as well. One possible application of our results is in studying the reliability and connectivity of networks. For example, consider a network in which each edge has failure probability p, and all failures occur independently. Let f_l denote the number of subsets of l edges that disconnect s from t in the case of the $s-t$ reliability problem (and f_l is defined analogously in the other situations). If the network has n nodes and m edges then, the probability that s is disconnected from t is $P(s,t) = \sum_{l=1}^{m} f_l p^l (1-p)^{m-l}$. If p is very small, then a good approximation is obtained by considering only kth minimum weight cuts, for small k. These applications once again point out the importance of obtaining an efficient approximation algorithm for the problem of counting the number of ideals in a partial order, which is known to be #P-complete.

The rest of this paper is organized as follows. In Section 2 we give definitions and present the algorithms for enumerating cuts in increasing order of weight in the various situations. In Section 3 we prove results on the structure of suboptimal cuts and use them to derive our algorithms for computing the kth minimum weight. In Section 4 we study the number of kth minimum weight cuts in a graph.

2. ENUMERATING ALL CUTS BY INCREASING WEIGHT

We first give some definitions and establish the notation. Let $G = (V,E)$ be a directed or undirected graph. In general, the edges of the graph may have associated weights; i.e., there is a function $wt : E \to Z^+$ assigning a positive weight to each edge. The unweighted case is the same as the special case where all the edges have weight 1. A *cut* of the graph is a partition $C = (S,T)$ of the vertices into two sets S, T; i.e., $S \cup T = V$ and $S \cap T = \emptyset$. The *weight* of the cut (S,T), denoted $wt(S,T)$ is the total weight of the edges that go from S to T; that is, $wt(S,T) = \sum_{\substack{u \in S \\ v \in T}} wt(u,v)$. In the directed case, the order of the sets S, T in the partition is important (only edges directed from S to T count in the weight of the cut), whereas in the undirected case it is not. In the undirected case we identify the cut (S,T) with the cut (T,S). That is, an undirected graph has 2^{n-1} cuts and a directed graph has 2^n cuts.

Let s and t be two vertices of G; s is usually called the *source* and t the *sink* of the cut. An $s-t$ *cut* is a cut (S,T) where s is in the first part S and t in the second T. The weight of the minimum $s-t$ cut is denoted by $wt(s,t)$. Also, for two disjoint subsets X, Y of vertices with $s \in X$ and $t \in Y$, we will denote by $wt(X;Y)$ the minimum weight of a cut (S,T) where $X \subseteq S$ and $Y \subseteq T$. This is the same as the minimum weight of an $s-t$ cut in the graph that is obtained by identifying all vertices of X with s and vertices of Y with t.

There are four problems of enumerating cuts by increasing weight (or of just enumerating the weights), depending on whether the graph is directed or undirected, and on whether we consider all cuts of the graph or only $s-t$ cuts for a given pair of vertices s, t. For a directed acyclic graph G, there is another restricted type of cuts that is often useful to consider: ideal cuts. An *ideal cut* is a cut (S,T) such that there are no edges directed from T to S. In this section we will present enumeration algorithms for these problems.

We say that an enumeration algorithm has *polynomial delay* $p(n)$ if it spends time at most $p(n)$ between successive outputs; thus, the first k outputs are generated in time $kp(n)$. This is an important property because in many cases, we cannot afford and do not really want to enumerate all the solutions (the cuts in our case), but rather we want to produce the most "important" ones in some metric (for instance, in the reliability example above), or in other cases we need to find a solution that satisfies some other complicated side conditions and thus we generate the solutions in order of preference until we find an acceptable one.

1. $s-t$ Cuts in a Graph.

In the case of $s-t$ cuts, an undirected graph can be treated as a special case of a rected graph by replacing each edge by two arcs with opposite directions. Let $= (V,E)$ be a directed graph on $n+2$ vertices with special vertices s, $t \in V$. Number e remaining vertices from 1 to n. We will represent each $s-t$ cut as an n-dimensional ector over 0/1; vertices corresponding to 0's (respectively, 1's) are assumed to be on the me side as s (resp. t). A k-dimensional 0/1 vector, for $k < n$, will represent a partially ecified cut; only the sides of vertices numbered 1 to k are specified.

Consider a full binary tree of height n. Its leaves are named in the standard way by /1 vectors of dimension n, and internal nodes at depth k are represented by 0/1 vectors of imension k. The leaves represent all possible $s-t$ cuts, and internal nodes represent parally specified cuts. Our algorithm will enumerate all $s-t$ cuts, i.e. leaves, by increasing eight.

Let v be an n-dimensional 0/1 vector. For $1 \le k \le n$, let $v = v'bv''$, where v' is the -1 dimensional vector consisting of the first $k-1$ components of v, b is the kth component of v, and v'' is the $n-k$ dimensional vector consisting of the remaining components. y *the immediate child of* v *at depth* k we mean the k-dimensional vector $v'\bar{b}$ where $= 1-b$. Our algorithm uses the following obvious fact:

ACT: Given a partially specified cut, we can find with one max flow computation a minnum weight $s-t$ cut consistent with it.

The algorithm is as follows. We will store in a heap, HEAP, candidate cuts. HEAP s initialized with an arbitrary minimum $s-t$ cut, v. Actually, we will store $(v,1)$, saying hat the immediate child of v at depth 1 is as yet unexplored.

In each iteration, we pick a minimum weight cut from HEAP, say (v,k), and output . The immediate children of v at depths k, $k+1$, ..., n are as yet unexplored. For each ne of them, we find a consistent minimum weight $s-t$ cut, and we insert all these in IEAP. For example, if u is such a cut corresponding to depth $k+l$, then we will store $u,k+l+1)$. Notice that the immediate children of u at depths 1, ..., $k+l$ have already een explored. The algorithm stops when HEAP is empty.

Clearly, in each iteration we will require at most n max flow computations. Morever, the amortized number of max flow computations per cut output that is required over he entire run of the algorithm is no more than 2.

Theorem 2.1. The above algorithm enumerates all $s-t$ cuts in a graph G by increasing veight with polynomial delay.

roof: For any partially specified cut u in $\{0,1\}^k$, $k<n$, let completions(u) denote the set f all $s-t$ cuts that are consistent with u. In the complete binary tree, this will be the set f leaves in the subtree rooted at u.

Consider the situation at any point in the algorithm. Let v_1, ..., v_r be the cuts output o far. Consider the set of all possible immediate children of these vectors, i.e. at all levls from 1 to n. Retain an immediate child iff no vector v_i is consistent with it. By an asy induction, one can prove that the completions of the retained vectors are disjoint, and their union is precisely the set of $s-t$ cuts that have not been output so far.

Now, at this point in the algorithm, HEAP contains the lightest cut in completions(u), for each retained u. Hence, if all $s-t$ cuts have not already been output the algorithm will definitely output a cut at this point; moreover, the cut output will be the lightest remaining cut. □

2.2. All Cuts in an Undirected Graph

Notice that now the partitions $(U,\overline{U})$ and $(\overline{U},U)$ for $U \subseteq V$ represent the same cut. We avoid enumerating cuts twice by assuming that one designated vertex is always on the 0 side. Assume that besides this vertex there are n vertices in the graph, numbered from 1 to n. As before, we will consider a complete binary tree of height n, whose leaves represent cuts.

The enumeration procedure is similar to that in Section 2.1. The only differences are: we will initialize HEAP with an arbitrary minimum cut in G. This is found as follows: for each i, $1 \leq i < n$, identify vertices $1, \ldots, i$, and find a minimum weight cut that separates them from vertex $i+1$. Pick the lightest of these cuts over all i.

Partially specified cuts in which both the 0 side and the 1 side are non-empty are extendable optimally as in case (1) (section 2.1). For a partially specified cut of the form 0^k, the 1 side is empty. We extend this optimally as follows: consider the set of minimum cuts that separate vertices $1, \ldots, i$ from vertex $i+1$ for $k \leq i < n$ that were found during the initialization phase. Pick the lightest of these cuts.

2.3. All Cuts in a Directed Graph

In this case the partitions $(U,\overline{U})$ and $(\overline{U},U)$ represent different cuts. So, we will assume that G has n vertices numbered $1, \ldots, n$.

We initialize HEAP by first computing $2(n-1)$ cuts: for each i, $1 \leq i < n$, find minimum cuts with $1, \ldots, i$ as a source and $i+1$ as a sink, and with $i+1$ as a source and $1, \ldots, i$ as a sink (note that these are different because the graph is directed). Pick the lightest of these cuts.

This time, partially specified cuts of the form 0^k and 1^k are handled in the special way elaborated in case (2). All other partially specified cuts are extended as in case (1).

2.4. Enumerating Ideal Cuts in a DAG.

Let G be a dag (a directed acyclic graph) with two special nodes s and t. Assume that s is the unique smallest element in the partial order specified by the dag, and that t is the unique largest element. An *ideal cut* is specified by an ideal U in the partial order, i.e., a set U of nodes such that all predecessors of the nodes of U belong to U. The cut consists of all edges running from U to $\overline{U} = V-U$. We wish to enumerate all ideal cuts by increasing weight.

Some other enumeration problems can be expressed in this form by choosing the dag G appropriately. One such case is the problem of generating the elements of the

artesian product in sorted order: We are given k sorted (finite) sets $X_1, \ldots, X_k$ of num-
ɜrs. The cost of a tuple v of the Cartesian product $X = X_1 \times \cdots \times X_k$ is the sum of its
omponents. The problem is to enumerate the Cartesian product in order of increasing
ost. Algorithms with polynomial delay for this problem and applications are discussed in
ASY]. This problem is isomorphic to the problem of enumerating ideal cuts in a dag that
onsists of a set of disjoint paths from s to t. Given a collection of sets $X_1, \ldots, X_k$, con-
ruct a dag G that has one $s-t$ path p_i for each set X_i; the path p_i has one arc of weight
$_{ij}$ for each element w_{ij} of X_i. An ideal cut must cut every path p_i at exactly one edge.
hus, there is an one-to-one, weight-preserving correspondence between ideal cuts and
lements of the Cartesian product.

The problem of enumerating in order the ideal cuts of a dag G can be reduced to the
roblem of Section 2.1 as follows. Let $\hat{G}$ be the graph obtained from G by adding for each
rc $u \rightarrow v$ the reverse arc $v \rightarrow u$ with very large weight M, say larger than the sum of the
eights of all the arcs of G. Consider a cut $(U,\overline{U})$. If it is an ideal cut in G then it has the
ame weight $wt(U,\overline{U}) < M$ in $\hat{G}$, while if is not an ideal cut of G then its weight in $\hat{G}$ is
rger than M. Therefore, the ideal cuts of G form an initial segment in a listing of the
uts of $\hat{G}$ by increasing weight.

Thus, we can enumerate the ideal cuts of a dag G by applying the algorithm of Sec-
ion 2.1 to $\hat{G}$. Of course, during the execution of the algorithm we can ignore candidate
uts of $\hat{G}$ that are not ideal. That is, we consider only those leaves of the complete binary
ree that represent valid ideal cuts. Given a partially specified cut, it is easy to check
whether it extends to a valid ideal cut: Suppose the partial cut puts set X of vertices
ogether with s and set Y with t. This partial cut is extendable to a valid ideal cut iff no
ertex in X is larger in the partial order that a vertex in Y.

3. COMPUTING THE WEIGHTS OF THE CUTS IN ORDER

Let G be a (weighted) directed or undirected graph and s, t two special vertices.
Consider the weights of all $s-t$ cuts, and list them (without duplicates) in increasing order
$w_1 < w_2 < w_3 < \cdots$. For any weight w_i, there may be many cuts having that weight.
We want to generate a listing of the weights without having to generate all the cuts that
achieve them. We say that an $s-t$ cut (S,T) is a *kth minimum weight* $s-t$ *cut* if it has the
th weight w_k in the listing. We have analogous definitions for the case in which we do
ot have specified vertices s and t.

The following lemma gives a key property of suboptimal cuts.

Lemma 3.1. Let (S,T) be a suboptimal $s-t$ cut, i.e., $wt(S,T) > wt(s,t)$. Then there
exists a set $X \subseteq S-\{s\}$ and a set $Y \subseteq T-\{t\}$, with $|X| \le 1$ and $|Y| \le 1$, such that
$wt(S,T) \ge wt(X \cup \{s\}; Y \cup \{t\}) > wt(s,t)$.

Proof: Interpreting the weights of the edges as capacities, compute the maximum flow
rom s to t in G, and obtain the residual graph. Shrink maximal strongly connected com-
onents in the residual graph, leaving edges that run between strongly connected compo-
ents (SCCs). These edges define a partial order among the SCCs. It was shown by [PQ]

that ideals in this partial order correspond exactly to minimum $s-t$ cuts in G. Now there are two cases.

Case 1: There is a strongly conencted component containing two vertices s' and such that $s' \in S$ and $t' \in T$. It may be the case that $s' = s$ or that $t' = t$ (but not both Since there is no minimum $s-t$ cut that separates s' and t', it follows tha $wt(\{s,s'\};\{t,t'\}) > wt(s,t)$. Since (S,T) is a cut that separates s, s' from t, t' $wt(S,T) \geq wt(\{s,s'\};\{t,t'\})$.

Case 2: Every strongly connected component contains vertices either entirely from S or entirely from T. In this case, since (S,T) is not a minimum $s-t$ cut, there must b two components, say A and B, such that $A \subseteq S$ and $B \subseteq T$, even though A is smaller tha B in the partial order (the smallest strongly connected component in the partial order is th one containing t). Pick two vertices $s' \in A$ and $t' \in B$. Any minimum $s-t$ cut that put s' on the same side as s must put t' on the same side as well. Therefore $wt(\{s,s'\};\{t,t'\}) > wt(s,t)$. As before, $wt(S,T) \geq wt(\{s,s'\};\{t,t'\})$. This proves th lemma. □

Theorem 3.1. Let (S,T) be a *kth*-minimum weight $s-t$ cut, $k > 1$. Then there are sub sets $X \subseteq S-\{s\}$ and $Y \subseteq T-\{t\}$, where $|X| \leq k-1$ and $|Y| \leq k-1$, such tha $wt(S,T) = wt(X \cup \{s\};Y \cup \{t\})$.

Proof: By Lemma 3.1, there exists an $X_1 \subseteq S-\{s\}$ and a $Y_1 \subseteq T-\{t\}$, where $|X_1| \leq$ and $|Y_1| \leq 1$, such that $wt(S,T) \geq wt(X_1 \cup \{s\};Y_1 \cup \{t\}) > wt(s,t)$. If the first tw terms are equal, we are done. Otherwise, identify the vertex, if any, of X_1 with s, and o Y_1 with t. In the new graph G_1, the minimum $s-t$ cut has weight $wt(X_1 \cup \{s\};Y_1 \cup \{t\}$ and (S,T) is a heavier cut, hence Lemma 3.1 holds again. Every $s-t$ cut of G_1 corre sponds to an $s-t$ cut of G with the same weight. Therefore, (S,T) is now a *j*th-minimum weight cut in G_1 for some j, $1 < j < k$. By repeatedly applying Lemma 3.1 in the same manner, we will get a heavier cut each time and the process will terminate in at most $k-$ steps. The set X of (at most $k-1$) vertices that are identified with s and the set Y of ver tices identified with t satisfy the condition $wt(S,T) = wt(X \cup \{s\};Y \cup \{t\})$. □

Corollary 3.1. Each *k*th-minimum weight $s-t$ cut is a minimum $X \cup \{s\};Y \cup \{t\}$ cut fo some subsets X, Y of cardinality at most $k-1$. □

The corollary suggests the following algorithm. Take all possible pairs of (disjoint) subsets X, Y with at most $k-1$ nodes; there are $O(n^{2(k-1)})$ pairs. For each such pair compute the weight of the minimum cut separating $X \cup \{s\}$ from $Y \cup \{t\}$, and select the k smallest distinct values from this multiset of weights. Thus, we have:

Theorem 3.2. Given a graph G and vertices s, t, we can compute the smallest k weights of $s-t$ cuts using n^{2k} max flow computations. □

There are a number of improvements that one can make to the simple algorithm outlined above, by analyzing further the structure of suboptimal cuts. Instead of computing the indicated minimum cuts for the choices of X and Y from scratch, it is better to compute them incrementally.

emma 3.2. Let H be the residual network with respect to a maximum $s-t$ flow in G. t C be an SCC of H and u, v two vertices of C (possibly $u = s$ or $v = t$ but not vicersa).

) If the weight of the minimum $s-t$ cut in G is z and the weight of the minimum $u-v$ t in C is z', then the minimum cut in G separating s and u from t and v has weight $(\{s,u\};\{t,v\}) = z+z'$.

) Let G' be obtained from G by identifying s with u and v with t. The residual network G' with respect to the maximum $s-t$ flow (i.e., max $\{s,u\}-\{t,v\}$ flow in G) can be tained from the network H by replacing the SCC C by the residual network of C with spect to the maximum $u-v$ flow, and then identifying u with s and v with t. □

emma 3.3. Let H be the residual network with respect to a maximum $s-t$ flow in G. t A, B be two strongly connected components of H such that A is an immediate predessor of B in the partial order (i.e., there is no other SCC C with $A < C < B$), A is not e SCC that contains t, and B is not the SCC that contains s.

) If the weight of the minimum $s-t$ cut in G is z and the total weight of the edges of H om A to B is z', then the minimum cut in G separating s and A from t and B has weight $(\{s\}\cup A;\{t\}\cup B) = z+z'$.

) Let G' be obtained from G by identifying the vertices of A with s and the vertices of B ith t. The residual network of G' with respect to the maximum $s-t$ flow can be tained from the network H by reversing the direction of all the arcs from A to B, and en identifying A with s and B with t. □

We give now an algorithm for listing in increasing order the weights of the $s-t$ its. (There is still some redundancy in the algorithm below, but we prefer to keep the esentation simple.) We assume that the rest of the vertices have been numbered arbitrary 1, 2, ..., n; and let s be 0 and t be $n+1$. If we shrink a set of vertices A that does not clude s and t, then the new vertex gets a number by some convention, say it gets the nallest number in A; if A includes s or t then it stays with the number of s or t. We keep a priority queue Q the candidate weights. Every element z of Q has an associated list (z) of graphs H. Every graph in the list is a graph obtained from the input graph G by rinking possibly some subsets of vertices, then computing a maximum $s-t$ flow in the sulting graph which is a flow of value z, and then computing the residual graph. Of ourse, if we only want the k first weights, then we stop the algorithm at that point (and e may prevent also Q from acquiring more than k elements by dropping larger values).

lgorithm

iput: A graph G and specified vertices s and t.

utput: A listing in increasing order of the weights of the $s-t$ cuts.

ethod: 1. Compute a maximum flow from s to t and the corresponding residual network with respect to the flow. Insert the value z of the flow into Q, and let the list $L(z)$ consist of H.

. While Q is not empty do the following:

$z = extract-min(Q)$.

Output z.

For every graph H in $L(z)$ do the following:

a. For every nontrivial strongly connected component C of H and for every ve tex u $(u \neq t)$ of C do:

Let v be the next vertex of C in the numbering (in case u is the last verte of C, then let v be the first vertex of C if $v \neq s$).

Compute a maximum flow from u to v in C and let z' be its value.

Insert $z+z'$ into Q, if it is not already there.

Replace the SCC C in H by the residual network of the max $u-v$ flov identify u with s and v with t, and append the resulting graph H' to the li $L(z+z')$.

b. Shrink all the strong components of H to obtain an acyclic graph $\hat{H}$.

For every nontransitive arc $A \rightarrow B$ of $\hat{H}$, where $A \neq t, B \neq s$, do:

Let z' be the weight of the arc. Insert $z+z'$ into Q, if it is not already there

In $\hat{H}$ identify vertex A with s and B with t, and append the resulting grap H' to the list $L(z+z')$.

□

Theorem 3.3. The above algorithm outputs correctly the weights of $s-t$ cuts in increa ing order. □

The case where we do not have specified vertices in the graph, but wish to generat the weights of all the cuts in order can be solved by a similar algorithm. The same hold for the case of ideal cuts.

Let us note finally that the exponential dependence on k is to be expected. Th problem of computing the maximum cut in a graph, even a graph without weights, is NP complete. In this case, the maximum weight of the graph is the k-th minimum weight fo a value of $k \leq n^2$. Therefore, if we had an algorithm that worked in time polynomial i both n and k then we could use it to solve the maximum cut problem.

4. THE NUMBER OF CUTS

In this section we examine how many k-th minimum weight cuts (or $s-t$ cuts) ther can be in a graph with n nodes. For $s-t$ cuts it is easy to see that this number can b exponential even for $k = 1$: Take a (directed or undirected) graph that consists of n dis joint paths of length 2 from s to t with all arcs having weight 1. The minimum $s-t$ cu has weight n, and there are 2^n such cuts depending on whether we cut the first or the sec ond edge of each path.

In the case of cuts in a directed graph with no fixed source and sink, the number i still exponential. Take the above graph, and for each arc add the opposite arc with larg weight H, say $H = n+1$; see Figure 1. The minimum cuts in this graph coincide with th 2^n minimum $s-t$ cuts.

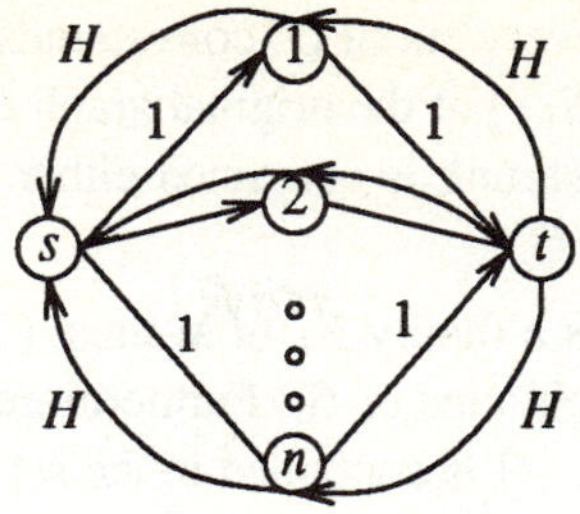

Figure 1

We show that for every fixed k, an undirected graph can have at most a polynomial umber of k-th minimum weight cuts.

heorem 4.1. For each k, an undirected graph has at most $O(n^{3k-1})$ k-th minimum eight cuts. □

First, we need the following lemma.

emma 4.1. Let s, t be two adjacent nodes of an undirected graph G. Consider the maxi-um $s-t$ flow, let H be the residual graph with respect to the flow, and let P be the partial rder (directed acyclic graph) obtained from H by shrinking its strongly connected compo-ents. If the minimum $s-t$ cut has k-th minimum weight in G, then the partial order P has idth at most $2k-1$.

roof Sketch: We outline the proof for the case $k = 2$. Consider the maximum $s-t$ flow, nd let C be a strongly connected component of the residual graph H, other than the com-onents containing s and t. Let f_C be the amount of flow that goes through C (the amount lowing into C is equal to the amount flowing out). The key property is that every edge onnecting C with the rest of the graph must be saturated by the maximum $s-t$ flow in ne of the two directions, going in or going out. This follows from the fact that we started ith an undirected graph, and from the construction of the residual graph. Therefore, the ut in G separating C from the rest of the graph has weight $2f_C$.

Suppose that P has width 4 or more, take an antichain with at least four nodes, con-ider the flows through the corresponding strongly connected components, and let C_1, C_2, C_3, C_4 be the SCCs with the four smallest flows, say f_1 through f_4. The minimum $s-t$ ut has weight $W > f_{1+} \cdots + f_4$; the inequality is strict because of the edge $[s,t]$. The ut that isolates component C_1 has weight $a = 2f_1 < W$. The cut that separates compo-ents C_1 and C_2 from the rest of the graph has weight $2f_1 + 2f_2$ which is strictly larger han a and smaller than W. Therefore, W cannot be the 2nd minimum weight cut, a con-radiction. □

We outline now the proof of Theorem 4.1. Number the nodes of the graph G from 1 o n so that each node j is connected to one of the previous nodes $1, ..., j-1$. As in the last ection, every kth-minimum weight cut (S,T) is a minimum X, Y cut for some sets of odes X, Y of cardinality k or less. However, we do not need $2k$ nodes to represent the ut. Let G' be a graph obtained from G by shrinking some subsets of nodes and let s, t be two distinguished nodes of G'; every cut of G' corresponds to a cut of G. We say that th tuple (G',s,t) *represents* a cut (S,T) of the original graph G if (S,T) is consistent with C (i.e., every set of nodes that is shrunk is contained either in S or in T) and (S,T) corre sponds to a minimum $s-t$ cut in G'.

We will argue that there is a family F_k of at most $(2k-1)!n^k$ tuples (G',s,t) tha represents all kth-minimum weight cuts of G. Furthermore, for each tuple (G',s,t) in F there is an index i such that $\{1,\ldots,i\}$ is contained in the set shrunk into s and $i+1$ belong to the set shrunk into t. Since $i+1$ is adjacent to one of the first i nodes, s and t are adja cent in G'. Since $wt(S;T)$ is the kth-minimum weight in G, the minimum $s-t$ cut ha rank at most k in the shrunk graph G'. From Lemma 4.1 it follows then that there are n more than $\begin{bmatrix} n \\ 2k-1 \end{bmatrix}$ ideals in the corresponding partial order, and thus there are no mor than this number of cuts represented by the tuple (G',s,t). Since F_k has at mos $(2k-1)!n^k$ tuples, it follows that there are at most n^{3k-1} kth-minimum weight cuts.

The family F_k is constructed by an algorithm similar to that of the last section. W maintain again a priority queue Q of weights. For every value z in Q there is an associate list $L(z)$ of directed graphs H with a source s and sink t. Recall that each such graph H i obtained from G by shrinking some sets of nodes, computing a maximum $s-t$ flow in th resulting graph, and then computing the residual network. We can associate with each H tuple (G',s,t) by ignoring the directions of the arcs of H. Now we initialize the priorit queue as follows. For each index i, let G_i be the graph obtained by shrinking the set o nodes $(\{1,\ldots,i\}$ into a node s, and letting t be the node $i+1$. Compute the maximum $s-$ flow in G_i, insert its value z_i into Q and append the residual network of G_i to $L(z_i)$.

The main loop is essentially the same. In the kth iteration, we extract the minimum element w_k of Q and update the data structures in the same way. The family F_k consist of the tuples (G',s,t) associated with the networks H in the list $L(w_k)$. To bound the car dinalities of the sets F_k, consider the ith iteration, and let H be a network in the list $L(w_i)$ In step (a) we insert no more than n new networks in a higher list. By Lemma 4.1, the width of the partial order is at most $2i-1$, thus the degree of the transitive reduction of H is also bounded by this quantity, and therefore step (b) inserts at most $(2i-1)n$ networks into a higher list. It follows that the number f_i of tuples in F_i obeys the recurrence $f_{i+1} \le (2i-1)nf_i$, from which it follows that $f_k \le (2k-1)!n^k$. □

The best lower bound we have is $\begin{bmatrix} n \\ 2k \end{bmatrix}$: Take a cycle where all the edges have weight 1. The kth-minimum weight is $2k$ and is achieved by all partitions (S,T) where S and T alternate k times around the cycle.

By the proof of Theorem 4.1 we can get the following improvement on Theorem 3.2 for the case of all cuts in an undirected graph.

Theorem 4.2. Given an undirected edge-weighted graph G, we can compute the smallest k weights of cuts in G using $O(n^k)$ max flow computations. □

Finally, we discuss the complexity of computing the number of kth minimum weight cuts for fixed k in the various settings. The complexity is polynomial in the case all cuts in undirected graphs. Provan and Ball have shown that the problems of counting the number of ideals in a partial order and the number of minimum weight $s-t$ cuts in rected and undirected graphs are #P-complete. We note also:

heorem 4.3. The following problems are #P-complete: counting the number of miniium weight ideal cuts in a DAG and counting the number of minimum weight cuts in a rected graph.

roof: Reduction from counting the number of ideals in a partial order P. Add a source s nd a sink t and a suitable number of edges from s to the nodes of P and/or from them to t, that for every node of P the in-degree becomes equal to the out-degree. All ideal cuts the resulting DAG have equal (thus, minimum) weight, and their number is equal to the umber of ideals of P. For the second problem add heavy reverse edges. □

cknowledgment: We wish to thank Richard Karp for valuable discussions.

. REFERENCES

ASY] A. V. Aho, T. G. Szymanski, M. Yannakakis, "Sorting the Cartesian Product", *Proc. Princeton Conf. on Information Sciences*, 1980.

DKL] E. A. Dinits, A. V. Karzanov, M. V. Lomosonov, "On the Structure of a Family of Minimal Weighted Cuts in a Graph", *Studies in Discrete Optimization* [in Russian], A. A. Fridman (Ed.), Nauka, Moscow, pp. 290-306, 1976.

EFS] P. Elias, A. Feinstein, C. E. Shannon, "Note on Maximum Flow on a Network", *IRE Trans. Inform. Theory*, pp. 117-119, 1956.

FF] L. R. Ford, D. R. Fulkerson, *Flows in Networks*, Princeton University Press, 1962.

K] A. Kanevsky, "On the Number of Minimum Size Separating Vertex Sets in a Graph and How to Find All of Them", *Proc. 1st Annual ACM-SIAM Symp. on Discrete Algorithms*, 1990.

KT] A. V. Karzanov, E. A. Timofeev, "Efficient Algorithms for Finding All Minimal Edge Cuts of a Nonoriented Graph", *Cybernetics*, pp. 156-162, 1986.

L] E. L. Lawler, "A Procedure for Computing the *K* Best Solutions to Discrete Optimization Problems and its Application to the Shortest Path Problem", *Management Science*, 18(7), pp. 401-405, 1972.

NV] D. Naor, V. V. Vazirani, "Representing and Enumerating Edge Connectivity Cuts in RNC", *Proc. 2nd Workshop on Algorithms and Data Structures*, 1991.

PQ] J. C. Picard, M. Queyrane, "On the Structure of All Minimum Cuts in a Network and Applications", *Mathematical Programming Study* 13, pp. 8-16, 1980.

PB] S. Provan, O. Ball, "The Complexity of Counting Cuts and of Computing the Probability that a Graph is Connected", *SIAM J. on Computing*, 12(4), pp. 777-788, 1983.

S] J. W. Suurballe, "Note on Finding All Minimum Cuts", Bell Laboratories Technical Memorandum, 1974.

Gröbner Bases: An Introduction

Bruno Buchberger
Research Institute for Symbolic Computation
Johannes Kepler Universität, A-4040 Linz, Austria
email: buchberg@risc.uni-linz.ac.at

Gröbner bases are certain finite sets of multivariate polynomials. Many fundamental problems in polynomial ideal theory (commutative algebra, algebraic geometry) can be solved by easy algorithms when
the input polynomials are first transformed into Gröbner bases form.

In this lecture we present, first, the building blocks of Gröbner bases theory:

- the notion of admissible orderings on the set of power products
- the notion of reduction w.r.t. polynomial sets
- the notion of Gröbner bases
- the main theorem that establishes an algorithmic characterization of Gröbner bases
- the main algorithm that transforms arbitrary sets of polynomials into equivalent Gröbner bases.

Then we explain the most important applications of Gröbner bases:

- polynomial ideal membership and radical membership test
- canonical simplification modulo polynomial ideals
- computation in polynomial residue class rings and algebraic number computations
- solvability and solution of polynomial systems of equations
- solution of diophantine equations with polynomial coefficients (syzygy computation)
- computation of the Hilbert function

- computation of algebraic relations
- test for invertible polynomial maps
- implicitization of parametric hypersurfaces.

Besides their intrinsic mathematical interest, most of these problems have ractical relevance in engineering, geometric modeling, robotics etc.

Some of the problems also play a key role in automated mathematics, e.g. utomated geometrical theorem proving.

Implementations of the Gröbner bases method are widely available as a outine constituent of most of the current symbolic computation software ystems.

We will make a few comments on implementation issues and the theoretcal and practical complexity of the method.

Finally, we will view the method in the more general frame of the "critical air / completion" approach and discuss various recent generalizations.

The method was introduced in the author's Ph.D. thesis as early as 965. Recently, quite a few authors dedicated their research effort to an xtensive exploration of its theoretical and algorithmic aspects. A recent ibliography, [Sweedler 90], on the method contains more than 300 items. The first textbook on the method, [Becker, Kredel, Weispfenning 92] will apear soon. There are a number of survey articles available, see for instance Buchberger 85] and [Buchberger 87].

References

Buchberger 85] B. Buchberger: Gröbner Bases: An Algorithmic Method in Polynomial Ideal Theory. In N. K. Bose, editor, *Multidimensional Systems Theory*, pages 184-232. D. Reidel Publishing Company, Dordrecht-Boston-Lancaster, 1985.

Buchberger 87] B. Buchberger: Applications of Gröbner Bases in Non-Linear Computational Geometry. In *Proc. Workshop on Scientific Software*, IMA, Minneapolis, USA. Springer, 1987.

Sweedler 90] M. Sweedler: Bibliography on Gröbner Bases. Technical Report, Computer Science Department, Cornell University, Ithaca, NY, 1990.

[Becker, Kredel, Weispfenning 92] T. Becker, H. Kredel, V. Weispfenning: The Theory of Gröbner Bases. Springer, NY, 1992, to appear.

Buchberger's Algorithm: The Term Rewriter's Point of View

(Extended Abstract)

Reinhard Bündgen
Wilhelm-Schickard-Institut, Universität Tübingen
D-7400 Tübingen, Fed. Rep. of Germany
⟨buendgen@informatik.uni-tuebingen.de⟩

Abstract: We show how Buchberger's algorithm for $\mathbf{Q}[x_1, \ldots, x_r]$ can be simulated using term completion modulo AC. To specify the rational numbers an infinite term rewriting system is needed. However, for the simulation of each particular ideal completion a finite approximation of the infinite rule set is sufficient. This approximation can be constructed during the completion. Then the division operation in $\mathbf{Q}$ reduces to a narrowing procedure which becomes part of the critical pair computation process.

1 Introduction

The similarity between Buchberger's algorithm [Buc65] and the Knuth-Bendix completion procedure [KB70] has long been observed. The first attempt to relate the two completion procedures was published by Loos ([Loo81]) and the latest contributions to that theme the author is aware of ([KKW89], [Pot89]) yield a *unification* of the two procedures. It is our goal to find an even clearer and simpler connection between polynomial- and term completion procedures: We want to describe Buchberger's algorithm in terms of term rewriting systems and their completion.

In our attempt to explain the relationship between the two procedures, we showed already how we can simulate Buchberger's algorithm by term completion modulo associativity and commutativity (AC) for polynomials over finite fields ([Bün91b]) and an extended version of Buchberger's algorithm for integral polynomials ([Bün91a]). In this paper we want to conclude this study by exhibiting the relation between polynomial- and term completion for polynomials over infinite fields. In particular, we will present a simulation of Buchberger's algorithm for polynomials over the rational numbers.

Buchberger's algorithm (in its plain form) is an effective method to compute Gröbner bases for ideal presentations in polynomial rings over *fields in which division is effective*. The main problem with term rewriting specifications is that fields cannot be specified as an equational variety. However it turns out that an *effective division* in the coefficient domain is the essential property for Buchberger's algorithm to work.

Effective division means that for some quotient c_1/c_2, the equation $c_2 \cdot x = c_1$ must have a solution. Therefore if c_1 and c_2 are represented by ground terms t_1 and t_2, then a ground erm t' must exist such that $t_2 \cdot t' \leftrightarrow^*_{\mathcal{R}} t_1$ for a term rewriting system $\mathcal{R}$ specifying the oefficient domain, i.e. $t_2 \cdot x$ and t_1 must be unifiable modulo $\mathcal{R}$. If $\mathcal{R}$ is canonical then nifiers modulo $\mathcal{R}$ can be computed by a narrowing procedure. This narrowing procedure ; then a part of the term completion procedure simulating Buchberger's algorithm.

'or finite fields $GF(q)$, we demanded ([Bün91b]) that the initial algebra (ground term nodel) of the canonical term rewriting system R_q specifying the coefficient domain is somorphic to $GF(q)$ and therefore $t_1 \cdot x$ and t_2 are unifiable modulo R_q for all ground erms t_1 and t_2. Hence division is effective in the term specification of $GF(q)$.

Ve can relax the requirement of effective division somewhat. Namely, it suffices that nly the divisions by leading coefficients generated during the completion are effective. n order to specify the rational numbers by the ground term model of an equational pecification, we need an infinite set of equations which may be thought of as the limit of a series of equation sets $(\mathcal{E}_i)_{i\in\mathbf{N}}$. The initial model of each $\mathcal{E}_i$ is isomorphic to an extension $\mathbf{Z}(i) = \mathbf{Z}\langle\frac{1}{p_1},\dots\frac{1}{p_i}\rangle$ of $\mathbf{Z}$ for distinct prime numbers p_i. It can be shown that for each $\mathcal{E}_i$ there exists a canonical term rewriting system $R(p_1,\dots,p_i)$ which is extensible to a canonical term rewriting specification $R(p_1,\dots,p_i)X$ of polynomials over $\mathbf{Z}\langle\frac{1}{p_1},\dots\frac{1}{p_i}\rangle$. n addition, for each ideal $\mathcal{I}$ in $\mathbf{Q}[x_1,\dots,x_r]$ there is an extension $\mathbf{Z}(i)$ of $\mathbf{Z}$ such that the Gröbner base of $\mathcal{I}$ in $\mathbf{Q}[x_1,\dots,x_r]$ is equal to the Gröbner base of $\mathcal{I}$ in $\mathbf{Z}(i)[x_1,\dots,x_r]$. We show how starting with a specification of $\mathbf{Z}$ the rewrite specification of $\mathbf{Z}(i)$ can be computed incrementally during the simulation of Buchberger's algorithm and thereby present a method to work with an infinite term rewriting system. Figure 1 presents the simple version of Buchberger's algorithm we want to simulate using term completion. *Simplify*, *delete*, *orient*, *collapse*, *compose* and *deduce* are the inference rules taken from of the abstract completion procedure in [Der89].

We assume the reader is familiar with term completion and Buchberger's algorithm (see e.g. [DJ90] and [Buc85]). All missing proofs can be found in [Bün91c].

2 Polynomials as Terms

In this section, we want to specify the abstract data type *polynomial* by a many-sorted canonical term rewriting system. There is a coefficient sort *Cœf*, a sort for power products *Ind* (short for indeterminate) and a sort *Poly* for polynomials. The set of operators together with their signatures and intended meanings is listed in Table 1.

From now on we consider a, b, c, d to be variables of sort *Cœf*; x, y, z to be variables of sort *Ind*; and f, g, h to be variables of sort *Poly* if they occur in terms of our polynomial algebra. The well-known canonical term rewriting system for commutative rings with unit elements ([PS81]) is given in Specification 1.

Specification 1

$$R = \{ \begin{array}{llllllll} 1: & a+0 & \to & a, & 6: & a+-a & \to & 0, \\ 2: & a\cdot 0 & \to & 0, & 6x: & b+(a+-a) & \to & b, \\ 3: & a\cdot 1 & \to & a, & 7: & -(a+b) & \to & -a+-b, \\ 4: & -0 & \to & 0, & 8: & a\cdot -b & \to & -(a\cdot b), \\ 5: & -(-a) & \to & a, & 9: & a\cdot(b+c) & \to & (a\cdot b)+(a\cdot c)\} \end{array}$$

The ground term model of R is isomorphic to the integers because only terms of the form

$G \leftarrow \mathbf{BA}(I, \succ)$

[Buchberger's algorithm. I is a set of polynomials presenting an ideal in $F[x_1, \ldots, x_r]$ where F is a field. $\succ$ is a terminating ordering on polynomials. Then G is the Gröbner base of I.]

(1) [Initialize.] $\mathcal{E} := \{p \leftrightarrow 0 \mid p \in I\}$; $\mathcal{R} := \emptyset$.
(2) [Simplify.] **while** the *simplify*-inference rule applies **do** $(\mathcal{E};\mathcal{R}) := Simplify((\mathcal{E};\mathcal{R}))$.
(3) [Delete.] **while** the *delete*-inference rule applies **do** $(\mathcal{E};\mathcal{R}) := Delete((\mathcal{E};\mathcal{R}))$.
(4) [Stop?] **if** $\mathcal{E} = \emptyset$ **then return** $G = \{l - r \mid l \rightarrow r \in \mathcal{R}\}$ **and stop.**
(5) [Orient.] Let $a \leftrightarrow b \in \mathcal{E}$; $\mathcal{E} := \mathcal{E} \setminus \{a \leftrightarrow b\}$; $l := LT(a-b)/LC(a-b)$; $r := -RED(a-b)/LC(a-b)^a$; $\mathcal{R} := \mathcal{R} \cup \{l \rightarrow r\}$.
(6) [Collapse.] **while** the *collapse*-inference rule applies **do** $(\mathcal{E};\mathcal{R}) := Collapse((\mathcal{E};\mathcal{R}))$.
(7) [Compose.] **while** the *compose*-inference rule applies **do** $(\mathcal{E};\mathcal{R}) := Compose((\mathcal{E};\mathcal{R}))$.
(8) [Deduce.] Compute all critical pairs $\mathcal{P}$ of rules in $\mathcal{R}$. $\mathcal{E} := \mathcal{E} \cup \mathcal{P}^b$; **continue** with step 2. □

[a] The leading term (LT) and the reductum (RED) is computed w.r.t. $\succ$.
[b] Here we identify pairs and equations. Equivalently we could compute the set S of all s-polynomials of $\{r - l \mid l \rightarrow r \in \mathcal{R}\}$ and let $P = \{s \leftrightarrow 0 \mid s \in S\}$.

Figure 1: Algorithm *BA*

op.	domain	range	meaning
0 :		$\rightarrow$ *Coef*	coefficient zero
1 :		$\rightarrow$ *Coef*	coefficient unit element
$-$:	*Coef*	$\rightarrow$ *Coef*	additive inversion for coefficients
$+$:	*Coef* $\times$ *Coef*	$\rightarrow$ *Coef*	coefficient addition (AC)
$\cdot$:	*Coef* $\times$ *Coef*	$\rightarrow$ *Coef*	coeffient multiplication (AC)
I :		$\rightarrow$ *Ind*	x^0, unit-element of *Ind*
X, Y, X_i :		$\rightarrow$ *Ind*	generators of type *Ind*
. :	*Ind* $\times$ *Ind*	$\rightarrow$ *Ind*	concatenation (AC)
M :	*Coef* $\times$ *Ind*	$\rightarrow$ *Poly*	monomial constructor
Ω :		$\rightarrow$ *Poly*	zero polynomial
$\oplus$:	*Poly* $\times$ *Poly*	$\rightarrow$ *Poly*	polynomial addition (AC)
$\odot$:	*Poly* $\times$ *Poly*	$\rightarrow$ *Poly*	polynomial multiplication (AC)

Table 1: Signature $\mathcal{F}$ of *RX*

0, -1, 1, $-1 + -1$, $1 + 1, \ldots$ are irreducible. The canonical rewrite rule specification of polynomials over commutative rings with unit elements is given in Specification 2.

Specification 2 *RX = R ∪ Pol where*

$$
\begin{array}{rlll}
Pol = \{ & 1: & x.I & \to x, \\
& 2: & M(0,x) & \to \Omega, \\
& 3: & M(a,x) \oplus M(b,x) & \to M(a+b,x), \\
& 3x: & f \oplus M(a,x) \oplus M(b,x) & \to f \oplus M(a+b,x), \\
& 4: & M(a,x) \odot M(b,y) & \to M(a\cdot b, x.y), \\
& 4x: & f \odot M(a,x) \odot M(b,y) & \to f \odot M(a\cdot b, x.y), \\
& 5: & f \oplus \Omega & \to f, \\
& 6: & f \odot \Omega & \to \Omega, \\
& 7: & f \odot M(1,I) & \to f, \\
& 8: & f \odot (g \oplus h) & \to (f \odot g) \oplus (f \odot h), \\
& 9: & (f \odot M(a,x)) \oplus (f \odot M(b,x)) & \to f \odot M(a+b,x), \\
& 9x: & g \oplus (f \odot M(a,x)) \oplus (f \odot M(b,x)) & \to g \oplus (f \odot M(a+b,x)), \\
& 10: & f \odot M(a+1,I) & \to f \oplus (f \odot M(a,I)), \\
& 11: & f \oplus (f \odot M(-1,I)) & \to \Omega, \\
& 11x: & g \oplus f \oplus (f \odot M(-1,I)) & \to g, \\
& 12: & f \oplus (f \odot M(-1+a,I)) & \to f \odot M(a,I), \\
& 12x: & g \oplus f \oplus (f \odot M(-1+a,I)) & \to g \oplus (f \odot M(a,I))\}
\end{array}
$$

The ground normal forms of the sort *Ind* are isomorphic to the power products over the *Ind*-constants. The ground normal forms w.r.t. *RX* of sort *Poly* are of the forms

$$\Omega,\ M(k,\vec{X}),\ M(k_1,\vec{X_1}) \oplus \ldots \oplus M(k_n,\vec{X_n})$$

where the $k, k_i \neq 0$, $\vec{X}$, $\vec{X_i}$ are ground terms in normal form and the $\vec{X_i}$ are disjoint within a sum of monomials for all $1 \leq i \leq n$. Thus the canonical form for polynomials produced by *RX* is the distributive normal form which is convenient for Gröbner basis computations. We will denote the isomorphism from polynomials to terms in *RX*-normal forms by ψ.

To obtain a term rewriting specification $(\mathcal{F}^*, R_qX)$ for multivariate polynomials over finite fields $GF(q)$ of order $q = p^r$ (for some prime number p), the specification $(\mathcal{F}, RX)$ must be modified as follows.

1. If $q = p$ then $\mathcal{F}^* = \mathcal{F}$ and R_qX is the result of AC-completing
$$(\{\underbrace{a + \ldots + a}_{p-times} \leftrightarrow 0\}; RX).$$
2. If $q = p^r$ for $r > 1$ then $\mathcal{F}^* = \mathcal{F} \cup \{A\}$ for a new constant A of sort *Cœf*. R_qX is the result of AC-completing $(\psi(m_\alpha(A)) \leftrightarrow 0\}; R_qX)$, where $m_\alpha(x)$ is the minimal polynomial of an algebraic extension $Z_p(\alpha)$ of Z_p which is isomorphic to $GF(q)$. Then $\psi(\alpha) = A$.

Theorem 1 *RX is canonical modulo AC and for each $q = p^r$ where p is prime there exists an AC-canonical term rewriting system R_qX.* □

3 Simulating Polynomial Completion

In this section, we briefly summarize the main results of [Bün91b] and [Bün91a] which demonstrate that Buchberger's algorithm can be simulated by term completion modulo

AC if the coefficient domain consists of the integers or is a finite field.

We first want to describe the reduction relation of a polynomial. Therefore we need t introduce a convenient notation for term representations of polynomials. $\vec{X}, \vec{Y}, \vec{Z}$ or an indexed version of these vectors represents a power product in ground normal form. W will use k, n or any indexed version of these letters as repetition factors of coefficient. Thus ka is a short hand for $\underbrace{a + \ldots + a}_{k\text{-times}}$ if k is positive otherwise it means $\underbrace{-a + \ldots + -a}_{-k\text{-times}}$ k is an abbreviation for $k1$ or $-k(-1)$ respectively. We also write $\bigoplus_{i=1}^{n} M(k_i, \vec{X}_i)$ fo $M(k_1, \vec{X}_1) \oplus \ldots \oplus M(k_n, \vec{X}_n)$ where sums of zero terms denote the constants 0 or respectively. Optional parts in a term will be put in brackets. Thus $a[+b]$ means a c $a+b$. In a rule, optional parts on both sides of the arrow belong together. In our example the reader can easily detect which of several optional terms in a rule or an equation ar related.

Let P be a polynomial. Then $\psi(LT(P)) \rightarrow \psi(RED(P))$ is the *p-rule* associated with P For each p-rule there exists a set of extension rules which describes the reduction relatio associated with P (see Specification 3). If $LC(P) = 1$ then there are two extension rule and otherwise there are eight extension rules. These extension rules are called the *c-rule* associated with P.

Specification 3 (rule types)

$$
\begin{array}{lll}
I.0: & M(1, \vec{X}) & \rightarrow \bigoplus_i M(k_i, \vec{X}_i), \\
I.1: & M(a, \vec{X}) & \rightarrow \bigoplus_i M(k_i a, \vec{X}_i), \\
I.2 & M(a, \vec{X}.x) & \rightarrow \bigoplus_i M(k_i a, \vec{X}_i.x), \\
\\
II.0: & M(k, \vec{X}) & \rightarrow \bigoplus_i M(k_i, \vec{X}_i), \\
II.1: & M(ka, \vec{X}) & \rightarrow \bigoplus_i M(k_i a, \vec{X}_i), \\
II.2: & M(ka, \vec{X}.x) & \rightarrow \bigoplus_i M(k_i a, \vec{X}_i.x), \\
II.3: & M(ka+b, \vec{X}) & \rightarrow M(b, \vec{X}) \oplus \bigoplus_i M(k_i a, \vec{X}_i), \\
II.4: & M(ka+b, \vec{X}.x) & \rightarrow M(b, \vec{X}.x) \oplus \bigoplus_i M(k_i a, \vec{X}_i.x), \\
II.5: & M(-a, \vec{X}) & \rightarrow M((k-1)a, \vec{X}) \oplus \bigoplus_i M((-k_i)a, \vec{X}_i), \\
II.6: & M(-a, \vec{X}.x) & \rightarrow M((k-1)a, \vec{X}.x) \oplus \bigoplus_i M((-k_i)a, \vec{X}_i.x), \\
II.7: & M(-a+b, \vec{X}) & \rightarrow M((k-1)a+b, \vec{X}) \oplus \bigoplus_i M((-k_i)a, \vec{X}_i), \\
II.8: & M(-a+b, \vec{X}.x) & \rightarrow M((k-1)a+b, \vec{X}.x) \oplus \bigoplus_i M((-k_i)a, \vec{X}_i.x).
\end{array}
$$

Buchberger's algorithm transforms a polynomial ideal $\mathcal{I} = (P_1, \ldots, P_k)$ presented by a set of polynomials into a Gröbner basis presenting the same ideal. Using the term rewriting specifications of polynomials this completion algorithm can be simulated by term completion modulo AC:

(1) [Choose coefficient domain] Let CX be RX or $R_q X$.

(2) [Translate Ideal] $\mathcal{E} := \{\psi(P_1) \leftrightarrow 0, \ldots, \psi(P_k) \leftrightarrow 0\}$

(3) [Complete] complete $(\mathcal{E}; CX)$ and call the result CX^*.

(4) [Extract Gröbner base.] Let G be the set of polynomials associated with the c-rules in $CX^* \setminus CX$ □

G is a Gröbner base of $\mathcal{I}$ and the rules in $CX^* \setminus CX$ are c-rules which describe the reduction relation associated with G. The following observations can be made:

1. There exist term orderings which for p-rules simulate polynomial orderings based on lexicographic orderings over power products.
2. If P is in the Gröbner base of $\mathcal{I}$ then all c-rules associated with P are contained in $CX^* \setminus CX$[1].
3. If CX describes a finite field then all c-rules in $CX^* \setminus CX$ are of type I. Note that for every type II rule $M(ka, \vec{X}.[x]) \rightarrow \bigoplus_i M(k_i a, \vec{X}_i)$, the coefficient term ka unifies with 1 modulo $R_q X$. Therefore the rule $M(1, \vec{X}.[x]) \rightarrow \bigoplus_i M(k_i k', \vec{X}_i)$ (where $kk' = 1$ in $GF(q)$) and its associated c-rules will be derived during the completion.
4. Let $c(P)$ be the set of c-rules associated with a polynomial P. Then $CX \cup c(P)$ is confluent and terminating.
5. During the term completion, the analogue of the selection of the leading term of a polynomial in the orientation step of Buchberger's algorithm is a symmetrization procedure in the Abelian group of polynomial addition ($\oplus$).
6. The s-polynomial computation in Buchberger's algorithm is simulated by the computation of critical pairs between two c-rules.

Theorem 2 *The computation of Gröbner bases for multivariate polynomials over the integers or over finite fields (using the lexicographic ordering over power products) can be simulated by term completion modulo associativity and commutativity.* □

Polynomials over Fields

Let $RX \cup C$ be the result of the completion of $(\emptyset; RX \cup I_R)$ where I_R is a set of p-rules. Then all equalities, provable using the term rewriting system $RX \cup C$ hold in all models of $RX \cup C$ and the rules in C, define a quotient in all modes of RX. The variety defined by RX is the set of all polynomial rings over commutative rings with unit elements. This includes all polynomial rings over fields. Let P be a polynomial in $\mathbf{Z}[x_1, \ldots, x_r]$ which is an element of the ideal $\mathcal{I}$ described by the set of p-rules I_R, i.e. $\psi(P)$ reduces to Ω by $RX \cup C$. But then P interpreted as a polynomial in $F[x_1, \ldots, x_r]$ is also an element of the ideal $\mathcal{I}$ in $F[x_1, \ldots, x_r]$ where F is a field.

In Section 3, we already showed how to extend the specification of the coefficient domain in order to specify polynomials over finite fields. Now we want to specialize the specification RX of $\mathbf{Z}[x_1, \ldots, x_r]$ in order to compute Gröbner bases in polynomial rings over an infinite field, namely in $\mathrm{Q}[x_1, \ldots, x_r]$. Let us recall the difference between completion of polynomials over rings and polynomials over fields discovered so far:

Fields: The resulting set of rules describing the Gröbner base consists of type-I c-rules only. Type-II rules can be 'divided' by the leading coefficient of the associated p-rule because for every ground term of sort *Cœf* there exists a multiplicative inverse.

Rings: The canonical rule set for the ideal base can also contain type-II c-rules.

Thus in order to compute Gröbner bases for polynomials over fields, we must be able to make type-II rules monic, i.e. we must 'divide' a type-II rule by its leading coefficient so that the resulting type-I rules makes the original rule confluent. The type-II rule can then be ignored. Note that a type I rule does not superpose with any rule specifying the coefficient domain, and thus no unexpected critical pairs can be created. All other

[1] See [Bün91a] for a description of the special case where G contains a constant polynomial.

aspects of the simulation carry over from the cases discussed in Section 3. In particula observations 1–6 hold.

Example 1 Consider the ideals $\mathcal{I}_1 = (x^2y - x^2 + 2xy, y^2 - y + 1)$ and $\mathcal{I}_2 = (x^2y -$ $2y, xy^2 - x)$ translated to the following sets of p-rules:

$$\begin{array}{llll} I_1^R = \{ & 1: & M(1,X.X.Y) & \rightarrow & M(1,X.X) \oplus M(-1+-1,X.Y), \\ & 2: & M(1,Y.Y) & \rightarrow & M(1,Y) \oplus M(-1,I) \}, \end{array}$$

$$I_2^R = \{\ 1:\ M(1,X.X.Y) \rightarrow M(1+1,Y),\ \ 2:\ M(1,X.Y.Y) \rightarrow M(1,X)\}$$

for which the following sets of c-rules can be deduced:

$$\begin{array}{llll} G_1^R = \{ & 1: & M(a,Y.Y) & \rightarrow & M(a,Y) \oplus M(-a,I), \\ & 2: & M(a,Y.Y.x) & \rightarrow & M(a,Y.x) \oplus M(-a,x), \\ & 3,4: & M(a,X.X[.x]) & \rightarrow & M(a+a,X.Y[.x]) \oplus M(-a+-a,X[.x])\}, \end{array}$$

$$\begin{array}{lll} G_2^R = \{ & 1,2: M(a,X.Y.Y[.x]) & \rightarrow M(a,X[.x]), \\ & 3,4: M(a,X.X[.x]) & \rightarrow M(a+a,Y.Y[.x]), \\ & 5-8: M(a+a[+b],Y.Y.Y[.x]) & \rightarrow M(a+a,Y[.x])[\oplus M(b,Y.Y.Y[.x])], \\ & 9-12: M(-a[+b],Y.Y.Y[.x]) & \rightarrow M(-a+-a,Y[.x]) \oplus M(a[+b],Y.Y.Y[.x])\} \end{array}$$

All rules in G_1^R are of type I. Therefore $G_1 = \{y^2 - y + 1, x^2 - 2xy - 2x\}$ is a Gröbne basis of $\mathcal{I}_1$ in $\mathbf{Q}[x_1, \ldots, x_r]$. G_2^R contains rules of type II, the leading coefficient of th associated p-rule being 1+1. Therefore $G_2 = \{xy^2 - x, x^2 - 2y^2, 2y^3 - 2y\}$ is not necessaril a Gröbner basis of $\mathcal{I}_2$ in $\mathbf{Q}[x_1, \ldots, x_r]$. Suppose $R'X$ is a canonical extension of RX suc that there exists a ground term H such that $H \cdot (1+1) \rightarrow^*_{R'X} 1$, then $H + H \rightarrow 1$ mus be derivable from $R'X$ and thus also the rule $M(1,Y.Y.Y) \rightarrow M(1,Y)$ and its associate c-rules which makes the rules $G_2^R.5$ – $G_2^R.12$ confluent. Then

$$\begin{array}{llll} G_2^{R'} = \{ & 1,2: & M(a,X.Y.Y[.x]) & \rightarrow & M(a,X[.x]), \\ & 3,4: & M(a,X.X[.x]) & \rightarrow & M(a+a,Y.Y[.x]), \\ & 5,6: & M(a,Y.Y.Y[.x]) & \rightarrow & M(a,Y[.x])\} \end{array}$$

is a Gröbner base of $\mathcal{I}_2$ in $\mathbf{Q}[x_1, \ldots, x_r]$ because no new critical pairs can be created. □

5 A Canonical TRS for $\mathbf{Z}\langle 1/n_1, \ldots, 1/n_m \rangle$

In this section, we present a canonical term rewriting system for extension rings of Z These extensions can be specified as finitely presented commutative rings. General result on the completion of finitely presented rings can be found in [LC86].

We begin with $\mathbf{Z}\langle 1/n \rangle$, that is the commutative ring which is generated by 0, 1 and $1/n$ where $n(1/n) = 1$. In this extension of Z, every power of a prime factor of n has a inverse. In order to specify $\mathbf{Z}\langle 1/n \rangle$ for $n \geq 2 \in \mathbf{N}$ by a term rewriting system, we add a new constant $\frac{1}{n}$ of sort *Cœf* to $\mathcal{F}$ and complete

Specification 4 $\mathcal{E}_{R(n)} = \{ \underbrace{\tfrac{1}{n} + \ldots + \tfrac{1}{n}}_{n\text{-times}} \leftrightarrow 1\} \cup \mathcal{E}_R$ *where*

$$\mathcal{E}_R = \{a + 0 \leftrightarrow a, a + -a \leftrightarrow 0, a \cdot 1 \leftrightarrow a, a \cdot (b + c) \leftrightarrow (a \cdot b) + (a \cdot c)\}.$$

he resulting canonical term rewriting system is

pecification 5

$$
\begin{array}{lllll}
R(n) = R \cup \{ & \alpha: & \underbrace{\tfrac{1}{n}+\ldots+\tfrac{1}{n}}_{n\text{-times}} & \rightarrow & 1, \\
& \alpha x: & a+\underbrace{\tfrac{1}{n}+\ldots+\tfrac{1}{n}}_{n\text{-times}} & \rightarrow & a+1, \\
& \beta: & \underbrace{(a\cdot\tfrac{1}{n})+\ldots+(a\cdot\tfrac{1}{n})}_{n\text{-times}} & \rightarrow & a, \\
& \beta x: & b+\underbrace{(a\cdot\tfrac{1}{n})+\ldots+(a\cdot\tfrac{1}{n})}_{n\text{-times}} & \rightarrow & b+a, \\
& \gamma: & -\tfrac{1}{n} & \rightarrow & -1+\underbrace{\tfrac{1}{n}+\ldots+\tfrac{1}{n}}_{(n-1)\text{-times}}, \\
& \delta: & -(a\cdot\tfrac{1}{n}) & \rightarrow & -a+\underbrace{(a\cdot\tfrac{1}{n})+\ldots+(a\cdot\tfrac{1}{n})}_{(n-1)\text{-times}} \}.
\end{array}
$$

heorem 3 *For every $n \geq 2 \in \mathbf{N}$, the term rewriting system $R(n)$ is AC-canonical.* □

iven a finite set $S \subset \mathbf{Q}$, we can construct a canonical term rewriting system for an xtension $\mathbf{Z}'$ of $\mathbf{Z}$ such that each element of S has an inverse in $\mathbf{Z}'$. The idea is to ompute $R(n)$ for a sufficiently large n. However, this is not satisfactory for the following easons:

- Assume we have constructed a term rewriting system R_S w.r.t. which every element in S has an inverse. If we now want to obtain such a term rewriting system for a set $S \cup \{s\}$, $s \in \mathbf{Q}$, we must start all over again.
- AC-matching and AC-unification is extremely expensive for large, non-linear terms.

herefore we want to present canonical term rewriting systems for extensions of $\mathbf{Z}$ which an be extended incrementally. Clearly, to find the inverse of any number n, only the nverses of its prime factors must be known. Following this idea, we want to construct anonical term rewriting systems for multiple extensions $\mathbf{Z}\langle 1/n_1, \ldots 1/n_m\rangle$ of $\mathbf{Z}$ where he n_i are distinct prime numbers. To get the canonical specification of such an extension ve must complete

pecification 6 $\mathcal{E}_{R(n_1,\ldots,n_m)} = \{n_1\frac{1}{n_1} \leftrightarrow 1, \ldots, n_m\frac{1}{n_m} \leftrightarrow 1\} \cup \mathcal{E}_R$.

This results in

pecification 7 $R(n_1,\ldots,n_m) = R(n_1) \cup \ldots \cup R(n_m) \cup$

$$\cup \{\ \varepsilon_{ij} : \tfrac{1}{n_i}\cdot\tfrac{1}{n_j} \rightarrow -1+\underbrace{\tfrac{1}{n_i}+\ldots+\tfrac{1}{n_i}}_{u_{ij}\text{-times}} + \underbrace{\tfrac{1}{n_j}+\ldots+\tfrac{1}{n_j}}_{v_{ij}\text{-times}} \mid 1 \leq i < j \leq m,$$

$$v_{ij}n_i + u_{ij}n_j = n_i n_j + 1, 0 < u_{ij} < n_i, 0 < v_{ij} < n_j \textit{ minimal } \}$$

$$\cup\{\ \varepsilon_{ij}x : a\cdot\tfrac{1}{n_i}\cdot\tfrac{1}{n_j} \rightarrow -a+\underbrace{(a\cdot\tfrac{1}{n_i})+\ldots+(a\cdot\tfrac{1}{n_i})}_{u_{ij}\text{-times}} + \underbrace{(a\cdot\tfrac{1}{n_j})+\ldots+(a\cdot\tfrac{1}{n_j})}_{v_{ij}\text{-times}} \mid$$

$$1 \leq i < j \leq m, v_{ij}n_i + u_{ij}n_j = n_i n_j + 1, 0 < u_{ij} < n_i, 0 < v_{ij} < n_j \textit{ minimal } \}.$$

Theorem 4 *For every set of distinct prime numbers $\{n_1, \ldots, n_m\}$, the term rewriting system $R(n_1, \ldots, n_m)$ is canonical modulo AC.*

6 Completion in $\mathbf{Z}\langle 1/n_1, \ldots 1/n_m\rangle[x_1, \ldots, x_r]$

The term rewriting system $R(n_1, \ldots, n_m)X = R(n_1, \ldots, n_m) \cup Pol$ specifies the polynomial ring over $\mathbf{Z}\langle 1/n_1, \ldots 1/n_m\rangle$.

Theorem 5 *For all sets $\{n_1, \ldots, n_m\}$ of distinct prime numbers the term rewriting system $R(n_1, \ldots, n_m)X$ is canonical modulo AC.*

In the context of completion with $R(n_1, \ldots, n_m)X$, we have to consider a new rule type. Before defining this type , we introduce some notation. We use $\vec{A}$, $\vec{B}$ or any indexed version of these to denote a term of sort *Cœf* in ground normal form. $\vec{A}a$ (where a is coefficient variable) denotes the $R(n_1, \ldots, n_m)$-normal form of $(\vec{A} \cdot a)$. Now the following rule types can be defined:

Specification 8 (rule types)

$$\begin{array}{llll} III.0 : M(\vec{A}, \vec{X}) & \rightarrow \bigoplus_i M(\vec{A}_i, \vec{X}_i), & III.1 : M(\vec{A}, \vec{X}.x) & \rightarrow \bigoplus_i M(\vec{A}_i, \vec{X}_i.x), \\ III.2 : M(\vec{A}a, \vec{X}) & \rightarrow \bigoplus_i M(\vec{A}_i a, \vec{X}_i), & III.3 : M(\vec{A}a, \vec{X}.x) & \rightarrow \bigoplus_i M(\vec{A}_i a, \vec{X}_i.x). \end{array}$$

Rules of type III.0 are called p-rules too. Type-III rules include the type-II.0, -II.1 and -II.2 rules. Let $\vec{A}$ be such that there exists a ground term $\vec{A}'$ and $\vec{A} \cdot \vec{A}' \rightarrow^*_{R(n_1,\ldots,n_m)} 1$. Then by fairness and completeness of the procedure *COMPLETE* for all of the type-III rules above, the following type-I rules can be derived:

$$M(a, \vec{X}[.x]) \rightarrow \bigoplus_i M((\vec{A}_i \cdot \vec{A}')a), \vec{X}_i[.x]).$$

The original type-III rules are confluent w.r.t. the derived type-I rule. Therefore every type-III rule with an invertible coefficient term in its left-hand side can be transformed into a type-I rule.

Otherwise if $\vec{A}$ and $R(n_1, \ldots, n_m)X$ are such that there is no multiplicative inverse of $\vec{A}$ in the set of $R(n_1, \ldots, n_m)X$-ground normal forms then for every type-III rule a type-I rule

$$M(ka, \vec{X}[.x]) \rightarrow \bigoplus_i M(\vec{B}_i a, \vec{X}_i[.x])$$

can be derived from the completion of the type-III rule and $R(n_1, \ldots, n_m)X$ where $\psi^{-1}(\vec{A}) = k_1/k_2$ for k_1, k_2 relative prime, $k = k_1 k_2$ and $\vec{B}_i = \psi^{-1}(k_2\psi(\vec{A}_i))$.

7 Computing Gröbner Bases in $\mathbf{Q}[x_1, \ldots, x_r]$

Given a specification $R(n_1, \ldots, n_m)X$, we can compute a Gröbner base for an ideal associated with a set of p-rules if every irreducible type-III rule created during the completion can be made 'monic', i. e. the leading coefficients of all these type-III p-rules have their inverse elements in $T(\{0, 1, \frac{1}{n_1}, \ldots, \frac{1}{n_m}, -, +, *\})$. In this case the completion procedure is the same as in the case of finite fields.

To compute Gröbner bases for arbitrary ideals using term completion, we need an infinite rewrite system specifying $\mathbf{Z}\langle 1/n_1, \ldots, 1/n_k, \ldots\rangle[x_1, \ldots, x_r]$ where '$1/n_1, \ldots, 1/n_k, \ldots$' denotes a sequence of all inverted primes. Of course this is not practicable; but to complete any given ideal, only divisions by a finite number of primes must be defined. It is hardly possible to predict which set $\{n_1, \ldots, n_m\}$ of prime numbers must be chosen such that the specification $R(n_1, \ldots, n_m)X$ can be used to compute the Gröbner base for a given set of polynomials. We propose two scenarios to overcome these difficulties:

Lucky guess: This is not a very intelligent solution, but it is of course possible to guess the right set of primes for which the completion succeeds. determine the leading coefficients computed during the completion. If there is a heuristic to determine the leading coefficients computed during the completion, *lucky guess* would be a very good solution. Also if an upper bound for the coefficients is known, 'guessing' all primes less than or equal to this bound yields an effective algorithm.

Incremental approximation of $\mathbf{Q}[x_1, \ldots, x_r]$: In this case we start the completion with $\mathcal{R} = RX$ as the specification of the polynomial domain and a set of p-rules describing the ideal. Whenever we encounter an irreducible type-II c-rule

$$M(ka, \vec{X}[.x]) \quad \rightarrow \quad \bigoplus_i M(\vec{A}_i a, \vec{X}_i[.x])$$

which cannot be made monic, we set $\mathcal{F} := \mathcal{F} \cup \{\frac{1}{n_i}\}$ and

$$\mathcal{R} := COMPLETE(\ (\{\tfrac{1}{n_i} + \ldots + \tfrac{1}{n_i} \leftrightarrow 1\};\ \mathcal{R}), \succ)$$

where n_i is the smallest prime number which divides k and has not been introduced to $\mathcal{R}$. The factorization of k can actually be simulated by term rewriting methods (AC-matching). Let *EXTEND* be a procedure which performs this update of $\mathcal{F}$ and $\mathcal{R}$. Now we can formulate the completion algorithm *RPCOMPLETE* in Figure 2 for ideal completion in $\mathbf{Q}[x_1, \ldots, x_r]$.

RPCOMPLETE simulates completion with an infinite set of rules. It is correct because the resulting canonical term rewriting system is $\mathcal{R}^* = R(n_1, \ldots, n_m)X \cup C$, where $\{n_1, \ldots, n_m\}$ is a set of prime numbers and C is a set of type-I c-rules. Type-I c-rules do not superpose with any rule of sort *Cœf*. Thus no critical pairs between a rule in C and a rule in $R(n_1, \ldots, n_k, \ldots)X \setminus R(n_1, \ldots, n_m)X$ exist.

8 Conclusion

Together with the result in [Bün91b] we showed that Buchberger's algorithm can be simulated by term completion modulo AC. A key outcome of our work is that division of polynomials by leading coefficients reduces to a unification problem in the term rewriting simulation. The solution of this unification is part of the term completion process. Therefore the initial algebra semantics of the term rewriting system representing the polynomial ring plays a central rôle.

We do not propose to replace Buchberger's algorithm by a term completion procedure. On the contrary, Buchberger's algorithm is in general much more efficient than its simulations by term completion. However, we argue that term completion is a good means to study algebraic completion procedures. The structure of term rewriting systems and term completion procedures is easy and well-understood. Many complicated features of

$\mathcal{R}^* \leftarrow$ **RPCOMPLETE**$(I_E, \succ)$

[Rational polynomial AC-term completion with 'Buchberger strategy'.
I_E is a set of ground equations of sort *Poly* and $\succ$ is a terminating term ordering which includes the reverse lexicographical ordering on multivariate polynomials described by ground terms. Then $\mathcal{R}^* = R(n_1, \ldots, n_m)X \cup C$ is a canonical term rewriting system where $\{n_1, \ldots, n_m\}$ is a set of prime numbers and C is a set of type-I c-rules. The polynomials associated with C describe a Gröbner base in the domain of rational polynomials of the ideal described by I_E.]

(1) [Initialize.] $\mathcal{F} := \{0, 1, -, +, \cdot, ., M, \oplus, \odot\}$; $\mathcal{R} := RX$; $FX := RX$; $\mathcal{E} := I_E$; $C := \emptyset$.

(2) [Simplify.] **while** the *simplify*-inference rule applies **do** $(\mathcal{E}; \mathcal{R}) := Simplify((\mathcal{E}; \mathcal{R}))$.

(3) [Delete.] **while** the *delete*-inference rule applies **do** $(\mathcal{E}; \mathcal{R}) := Delete((\mathcal{E}; \mathcal{R}))$.

(4) [Stop?] **if** $\mathcal{E} = \emptyset$ **then return** $\mathcal{R}^* = \mathcal{R}$ **and stop**.

(5) [Orient.] Let $a \leftrightarrow b \in \mathcal{E}$; $\mathcal{E} := \mathcal{E} \setminus \{a \leftrightarrow b\}$;

(5.1) [Select leading term.]
Transform $a \leftrightarrow b$ into a rule $l \rightarrow r$ of type I, II or III.

(5.2) [Extend rule.]
$\mathcal{R} := COMPLETE((\emptyset; FX \cup \{l \rightarrow r\}), \succ)$; $C := C \cup (\mathcal{R} \setminus FX)$;

(5.3) [Trivial ideal.] **if** $f \rightarrow \Omega \in C$ **then return** $\mathcal{R}^* = \mathcal{R}$ **and stop;**

(5.4) [Extend Specification.]
if C contains a rule of the form $M(ka, \vec{X}) \rightarrow RHS$ **then**
{ $EXTEND(k, \mathcal{F}, FX;\ \mathcal{F}, FX)$[a]; **continue with step 5.2** }.

(6) [Collapse.] $\mathcal{R} := FX \cup C$; **while** the *collapse*-inference rule applies **do** $(\mathcal{E}; \mathcal{R}) := Collapse((\mathcal{E}; \mathcal{R}))$.

(7) [Compose.] **while** the *compose*-inference rule applies **do** $(\mathcal{E}; \mathcal{R}) := Compose((\mathcal{E}; \mathcal{R}))$; $C := \mathcal{R} \setminus FX$.

(8) [Deduce.] Compute all critical pairs $\mathcal{P}$ between two c-rules in C which are associated with different p-rules; $\mathcal{E} := \mathcal{E} \cup \mathcal{P}$; **continue** with step 2. □

[a] In analogy to the ALDES syntax, the parameters in front of the ';'-sign are input parameters and those following the ';'-sign are output parameters.

Figure 2: Algorithm *RPCOMPLETE*

algebraic completion (e. g. symmetrization or complicated rewrite relations for the generalized Buchberger's algorithm for polynomials over rings, see also [Bün91c]) can be easily explained in the setting of term completion. Therefore term completion provides an appropriate framework to present, explain and illustrate various algebraic completion procedures. Buchberger's algorithm is highly optimized w. r. t. the algebraic domain it is supposed to work for. Translating it into a term completion procedure elucidates which parts of the original procedure are essential for completion (the parts that are invariant w. r. t. different term completion strategies, -orderings and coefficient representations) and which features are included for efficiency reasons only (e. g. the left-hand sides of rules contain only monomials, or computing with polynomials in distributive normal form only).

'herefore it should be much easier to modify or generalize an existing algebraic comple-ion procedure given a term completion presentation. Once a sound modified completion rocedure has been found, it can still be optimized and adapted to its input data.

) Acknowledgements

wish to thank Prof. Loos for initiating this work and many discussions and Prof. Kuechlin or helpful comments.

References

Buc65] Bruno Buchberger. *Ein Algorithmus zum Auffinden der Basiselemente des Restklassenringes nach einem nulldimensionalen Polynomideal.* PhD thesis, Universität Innsbruck, 1965.

Buc85] Bruno Buchberger. Gröbner bases: an algorithmic method in polynomial ideal theory. In N. K. Bose, editor, *Recent Trends in Multidimensional Systems Theory*, chapter 6, Reidel, 1985. (Also Report CAMP-83.29, U. Linz, 1983).

Bün91a] Reinhard Bündgen. Completion of integral polynomials by AC-term completion. In Stephen M. Watt, editor, *International Symposium on Symbolic and Algebraic Computation*, pages 70 – 78, 1991. (Proc. ISSAC'91, Bonn, Germany, July 1991).

Bün91b] Reinhard Bündgen. Simulating Buchberger's algorithm by Knuth-Bendix completion. In Ronald V. Book, editor, *Rewriting Techniques and Applications (LNCS 488)*, Springer-Verlag, 1991. (Proc. RTA'91, Como, Italy, April 1991).

[Bün91c] Reinhard Bündgen. *Term Completion Versus Algebraic Completion.* PhD thesis, Universität Tübingen, D-7400 Tübingen, Germany, May 1991. (reprinted as report WSI 91–3).

[Der89] Nachum Dershowitz. Completion and its applications. In H. Aït-Kaci and M. Nivat, editors, *Resolution of Equations in Algebraic Structures*, chapter 2, Academic Press, 1989.

[DJ90] Nachum Dershowitz and Jean-Pierre Jouannaud. Rewrite systems. In Jan van Leeuven, editor, *Formal Models and Semantics*, chapter 6, Elsevier, 1990.

[KB70] Donald E. Knuth and Peter B. Bendix. Simple word problems in universal algebra. In J. Leech, editor, *Computational Problems in Abstract Algebra*, Pergamon Press, 1970. (Proc. of a conference held in Oxford, England, 1967).

[KKW89] A. Kandri-Rody, D. Kapur, and F. Winkler. Knuth-Bendix procedure and Buchberger algorithm – a synthesis. In *International Symposium on Symbolic and Algebraic Computation*, pages 55–67, 1989. (Proc. ISSAC'89; in the proceedings only Winkler is listed as author).

[LC86] Philippe Le Chenadec. *Canonical Forms in Finitely Presented Algebras.* Pitman, London, 1986.

[Loo81] Rüdiger Loos. Term reduction systems and algebraic algorithms. In Jörg H. Siekmann, editor, *GWAI-81 (IFB 47)*, pages 214–234, Springer-Verlag, 1981. (Proc. German Workshop on AI, Bad Honnef, 1981).

[Pot89] Loïc Pottier. *Algorithmes de complétion et généralison en logique du premier ordre.* PhD thesis, Université Nice — Sophia Antipolis, Parc Valrose, Fac. Sciences, F-06034 Nice, France, February 1989.

[PS81] G. Peterson and M. Stickel. Complete sets of reductions for some equational theories. *Journal of the ACM*, 28:223–264, 1981.

Completion of Rewrite Systems with Membership Constraints

Hubert Comon*

Abstract. We consider a constrained equational logic where the constraints are membership conditions $t \in \underline{s}$ where $\underline{s}$ is interpreted as a regular tree language. Our logic includes a fragment of second order equational logic (without λ-expressions) where second order variables range over regular sets of contexts. The problem with constrained equational logics is the failure of the critical pair lemma. This is the reason why we propose new deduction rules for which the critical pair lemma is restored. Computing critical pairs requires however to solve some constraints in a second-order logic with membership constraints. This is the most difficult result of the paper: we give a terminating set of transformation rules for these formulas, which decides the existence of a solution.

Since an order-sorted signature is nothing but a bottom-up tree automaton, order-sorted equational logic falls into the scope of our study; our results show how to perform order-sorted completion without regularity and without sort decreasingness. It also shows how to perform unification in the order-sorted case, with some higher-order variables (without any regularity assumption).

Introduction

Constraints have been introduced in logic programming languages because they provide with adequate notations for some particular computation domains. They also allow to use tailored decision procedures and therefore to enhance efficiency. This also holds true for equational logic: constraints enhance expressiveness since some (infinite) sets of equations can be represented by a single constrained equation. On the efficiencey side, it is possible to express control strategies, such as ordered strategies, and to avoid computing large sets of most general unifiers (see [15]). However, the problem with the constrained version of equational logic is the failure of the critical pair lemma. For example, consider the rewrite system

$$\mathcal{R} = \left\{ \begin{array}{rrcl} x \in \{a\} : & f(x) & \rightarrow & b \\ & a & \rightarrow & b \end{array} \right.$$

where $x \in \{a\}$ represents some sort constraint on the variable x. (Actually, it could be any constraint provided that $x \mapsto b$ does not satisfy it.) Then $f(a) \rightarrow b$ using the first rule and $f(a) \rightarrow f(b)$ using the second rule. $f(b)$ and b are both irreducible, which means that $\mathcal{R}$ is not locally confluent. On the other hand, there is no critical pair (in the usual sense).

*CNRS and LRI, Bat. 490, Université de Paris Sud, 91405 ORSAY cedex, France. E-mail comon@lri.lri.fr

This problem did show up some years ago in the framework of *order-sorted equational logic* (see e.g. [14]) which may be viewed as a logic with membership constraints [19]. There are two ways to overcome the failure of the critical pair lemma:

1. To design restrictions under which the lemma holds. This is what has been done for order-sorted equational logic: the rewrite systems are supposed to be *sort decreasing* in [14, 20] and similar conditions are assumed in Toyama's conditional systems [21]. For ordering constraints, some conditions have been recently proposed [17].

2. To extend the notion of critical pair. This solution amounts to add deduction rules in the completion procedure. These new deduction rules have to be designed in such a way that they specialize to the classical deduction system for unconstrained equations. (For example, adding deduction rules for superpositions inside variables would not be a good solution).

We choose this second solution in our paper, considering the case of membership constraints. This gives in particular a solution to order-sorted completion without any assumption on the signature (usually regularity is required) nor any assumption on the rewrite system: we do not require the sort-decreasingness of the rules. We also believe that our technique can be applied to other constraints systems. Given a variable which is constrained by a formula $\phi(x)$, the idea is to guess terms t such that $\phi(t)$ is valid but for which t can be rewritten in t' in such a way that $\phi(t')$ is no longer valid hence resulting in a failure of the critical pair lemma. Of course, there may exist infinitely many such t and we need to introduce second order variables in order to express these infinite sets. Let us give an example.

$$\mathcal{R} = \left\{ \begin{array}{lrcl} x \in f^+(g(f^*(a))) : & f(x) & \rightarrow & a \\ & g(x) & \rightarrow & h(x,x) \end{array} \right.$$

where we used a regular expression for the membership condition, instead of a sort as it is usually done.[1] In this formulation, the critical superpositions between all instances of the first rule and the second rule become apparent: $a \leftarrow f(f^n(g(f^k(a)))) \rightarrow f(f^n(h(f^k(a), f^k(a))))$ for every $n \geq 1, k \geq 0$. Now, the problem is to express finitely all terms on the right side (which are all irreducible). Terms $f^k(a)$ will be represented by a variable x, with the constraint $x \in f^*(a)$. Contexts $f^n(\cdot)$ will be represented by a second-order variable X with the constraint $X \in f^+$. Then we get

$$x \in f^*(a) \wedge X \in f^+ : \;\; a = f(X(h(x,x)))$$

Additional problems will arise with these second-order variables. First, we have to consider some other kind of critical pairs (this problem is well known in higher-order equational logic, see e.g. [18] for recent developments). Then, and that is the main problem, we have to solve unification problems that arise in further critical pairs computations. These unification problems may involve second-order variables, and second-order unification is undecidable [10]. Fortunately, it turns out that the fragment of second-order logic that we have to consider here leads to decidable

[1] Order-sorted signatures are bottom-up tree automata. Therefore it is equivalent to give the regular expression itself or the signature.

unification problems. That is what we show in section 3, and that is the mos difficult result of our paper. This decidability property is also the reason why w believe that our technique improves over [11, 16, 21] where unification cannot b decidable.

In [3], the authors propose to enrich the term structure by (roughly) adding in teger exponents. However, they need to forbid nested integer exponents in order t keep the decidability of unification. We do not have such limitations. On the othe hand we have other limitations, which shows that our results are complementary t those of [3]. Similarly, the result of W. Farmer that monadic second-order unifica tion is decidable [7] cannot help us because we do not have such assumptions o the signature. Finally, the work of A. Megrelis [1] should be pointed out because precisely, he avoids additional hypothesis by adding new deduction rules. Unfor tunately, his logic is currently too rich for expecting a confluence criterion. Thi means that completion will always run forever in his framework. Nevertheless, tak ing advantage of Megrelis results, C. Kirchner and H. Kirchner are designing a nev notion of reduction and new inference rules which lead to a completion procedure fo order-sorted rewrite systems without any sort-decreasing condition. We believe tha such a technique will give results close to ours in the case of membership constraints

We present the logic in section 1, we give the deduction rules in section 2 an prove a critical pair lemma. Then, in section 3, we show how to solve combinatio of equations and constraints which occur during the completion procedure.

1 The Constrained Logic

1.1 Syntax

1.1.1 Terms and Contexts

$T(F,\mathcal{X},\mathcal{CX})$ is the set of terms built on the finite set F of function symbols each of which is given with a fixed arity, the (disjoint) infinite set $\mathcal{X}$ of *variable symbol* which are considered as having an arity 0, and the (disjoint) infinite set of *contex variable symbols* $\mathcal{CX}$ which are considered as having arity 1. Elements of $\mathcal{CX}$ are denoted using upper case letters: $X, X_1, X', Y, Z, \ldots$ whereas elements of $\mathcal{X}$ are denoted using lower case letters: $x, x_1, x', y, z \ldots$ As usual, $T(F,\mathcal{X})$ (resp. $T(F)$) denotes the subset of terms which do not contain context variables (resp. variables)

Example 1.1 $f(X(x), Y(X(a))) \in T(F,\mathcal{X},\mathcal{CX})$ if $Y, X \in \mathcal{CX}$, $x \in \mathcal{X}$ and $a, f \in F$ have respectively arity 0 and arity 2.

If t is a term, $Var(t)$ is the set of variables (both large and small ones) occurring in t. A *context* is a term in $T(F)$ with a "hole". More precisely, it is a pair (t,p) where $t \in T(F)$ and p is a position of t. Such a context is also denoted $t[\cdot]_p$. We use concatenation of contexts and terms in the usual way: the concatenation of the context $t[\cdot]_p$ and the term u is written $t[u]_p$. $\mathcal{C}(F)$ is the set of all contexts. $t[\cdot]_p \in \mathcal{C}(F)$ may also be viewed as the mapping $\lambda u.t[u]_p$ from $T(F)$ to $T(F)$. Finally, $\mathcal{C}(F,\mathcal{X},\mathcal{CX})$ is the set of second order (monadic) terms $t[X]_p$ where $X \in \mathcal{CX}$ and $t \in T(F,\mathcal{X},\mathcal{CX})$. $X(Y) \in \mathcal{C}(F,\mathcal{X},\mathcal{CX})$ is also written $X \cdot Y$ or $\lambda x.X(Y(x))$.

.1.2 Sort Expressions and Context Expressions

et Q be a finite set of *sort symbols*, disjoint from $F \cup \mathcal{X} \cup \mathcal{CX}$. $\bot_S, \top_S, \epsilon, \top_C, \bot_C$ re also particular symbols, which do not belong to $Q \cup \mathcal{X} \cup \mathcal{CX} \cup F$. Then, the set $\mathcal{E}$ of *sort expressions* and the set $\mathcal{CE}$ of *context expressions* are defined as the least ets which satisfy:

- $Q \cup \{\top_S, \bot_S\} \subseteq \mathcal{SE}$ and $\{\epsilon, \top_C, \bot_C\} \subseteq \mathcal{CE}$.
- $\mathcal{SE}$ and $\mathcal{CE}$ are closed under the boolean operations $\wedge, \vee, \neg$
- $f(q_1, \ldots, q_n) \in \mathcal{SE}$ if $f \in F$ and $q_1, \ldots, q_n \in \mathcal{SE}$
- $f(q_1, \ldots, q_n)[C]_i \in \mathcal{CE}$ if $1 \leq i \leq n$, $q_1, \ldots, q_n \in \mathcal{SE}$ and $C \in \mathcal{CE}$
- $C \cdot q \in \mathcal{SE}$ if $C \in \mathcal{CE}$ and $q \in \mathcal{SE}$. $C \cdot C' \in \mathcal{CE}$ and $C^* \in \mathcal{CE}$ if $C, C' \in \mathcal{CE}$

.1.3 Formulas

The atomic formulas of the constraint language are:

- "$t[X]_p \in C$" where $t[X]_p \in \mathcal{C}(F, \mathcal{X}, \mathcal{CX})$
- "$t \in q$" where $t \in T(F, \mathcal{X}, \mathcal{CX})$ and $q \in \mathcal{SE}$
- "$s = t$" where $s, t \in T(F, \mathcal{X}, \mathcal{CX})$
- "$X = t[Y]_p$" where $X, Y \in \mathcal{CX}$ and $t[Y]_p \in \mathcal{C}(F, \mathcal{X}, \mathcal{CX})$.

nd the logical connectives $\wedge, \vee, \exists x, \exists X$. In addition $\bot$ is the empty disjunction and $\top$ is the empty conjunction. Let $\mathcal{F}$ be the set of all such formulas. A *membership onstraint* is a formula $\phi \in \mathcal{F}$ which does not contain any equation. A *membership olved form* is a conjunction of constraints of the form $X \in C$ and $x \in q$ where X nd x are variables occurring only once.

.1.4 Constrained Rules and Constrained Equations

An *equation* is a pair of terms $s, t \in T(F, \mathcal{X}, \mathcal{CX})$, which is denoted $s = t$. s, t cannot e contexts in our logic. Note also that we make no difference between $s = t$ and $= s$. A *constrained equation* is a pair $(\phi, s = t)$ where ϕ is a membership constraint nd $s = t$ is an equation. Such a pair is written $\phi : s = t$. A *constrained rule* is a onstrained equation in which the order on s, t is relevant. It is written $\phi : s \rightarrow t$. (We do not need to assume here that $Var(t) \subseteq Var(s)$.) As we will see in section 3, we may always assume that the constraints are membership solved forms.

1.2 Semantics

1.2.1 Sort Expressions and Context Expressions

Let $\mathcal{A}$ be a finite bottom-up tree automaton with states Q (see [9] for definitions and basic results on tree languages). Then, we define the semantics mappings $[\![\cdot]\!]_{\mathcal{A}}$ from $\mathcal{SE}$ into $2^{T(F)}$ (subsets of $T(F)$) and from $\mathcal{CE}$ into $2^{\mathcal{C}(F)} \subseteq 2^{T(F)^{T(F)}}$ (subsets of mappings from $T(F)$ to $T(F)$) as follows:

- $[\![q]\!]_{\mathcal{A}}$ is the language recognized by $\mathcal{A}$ in the state q, if $q \in Q$. [2]
- $[\![\bot_S]\!]_{\mathcal{A}} = \emptyset$, $[\![\top_S]\!]_{\mathcal{A}} = T(F)$, $[\![\top_C]\!]_{\mathcal{A}} = \mathcal{C}(F)$, $[\![\bot_C]\!]_{\mathcal{A}} = \emptyset$, $[\![\epsilon]\!]_{\mathcal{A}} = \{id_{T(F)}\}$ where $id_{T(F)}$ is the identity function on $T(F)$.
- $[\![\cdot]\!]_{\mathcal{A}}$ is a morphism from $(\mathcal{SE}, \wedge, \vee, \neg)$ into $(2^{T(F)}, \cap, \cup, \bar{\cdot})$ where $\overline{A}$ is the complement of A in $T(F)$. This means, for example, that $[\![q \wedge q']\!]_{\mathcal{A}} = [\![q]\!]_{\mathcal{A}} \cap [\![q']\!]_{\mathcal{A}}$.
- $[\![\cdot]\!]_{\mathcal{A}}$ is a morphism from $(\mathcal{CE}, \wedge, \vee, \neg)$ into $(2^{\mathcal{C}(F)}, \cap, \cup, \bar{\cdot})$
- $[\![f(q_1, \ldots, q_n)]\!]_{\mathcal{A}} = \{f(t_1, \ldots, t_n) \in T(F) \mid t_1 \in [\![q_1]\!]_{\mathcal{A}}, \ldots, t_n \in [\![q_n]\!]_{\mathcal{A}}\}$
- $[\![f(q_1, \ldots, q_n)[C]_i]\!]_{\mathcal{A}} = \{\lambda x.f(t_1, \ldots, t_n)[u[x]_p]_i \mid \lambda x.u[x]_p \in [\![C]\!]_{\mathcal{A}}, \forall i, t_i \in [\![q_i]\!]_{\mathcal{A}}\}$
- $[\![C \cdot q]\!]_{\mathcal{A}} = \{t[u]_p \in T(F) \mid u \in [\![q]\!]_{\mathcal{A}}, t[\cdot]_p \in [\![C]\!]_{\mathcal{A}}\}$
- $[\![C \cdot C']\!]_{\mathcal{A}} = \{\lambda x.t[u[x]_p]_q \in \mathcal{C}(F) \mid t[\cdot]_q \in [\![C]\!]_{\mathcal{A}}, u[\cdot]_p \in [\![C']\!]_{\mathcal{A}}\}$
- $[\![C^*]\!]_{\mathcal{A}} = \bigcup_{n \geq 0} [\![C^n]\!]_{\mathcal{A}}$ where $C^0 = \{id_{T(F)}\}$ and $C^n = C \cdot C^{n-1}$ for $n \geq 1$.

1.2.2 Solutions of Formulas

A *ground assignment* is a mapping σ from $\mathcal{X} \cup \mathcal{CX}$ to $T(F) \cup \mathcal{C}(F)$ such that $\sigma|_{\mathcal{X}}$ is a mapping from $\mathcal{X}$ to $T(F)$ and $\sigma|_{\mathcal{CX}}$ is a mapping from $\mathcal{CX}$ to $\mathcal{C}(F)$. Such an assignment is extended into an F-algebra morphism from $T(F, \mathcal{X}, \mathcal{CX})$ to $T(F)$.

A *solution* of an equation $s = t$ is a ground assignment σ such that $s\sigma \equiv t\sigma$ ($\equiv$ is the identity on $T(F)$). A *solution* of a membership constraint $t \in q$ (resp. $X \in C$) w.r.t. the automaton $\mathcal{A}$ is a ground assignment σ such that $t\sigma \in [\![q]\!]_{\mathcal{A}}$ (resp. $X\sigma \in [\![C]\!]_{\mathcal{A}}$). These definitions of solutions are lifted to formulas in $\mathcal{F}$ in a straightforward way. $[\![\phi]\!]_{\mathcal{A}}$ denotes the set of all solutions of $\phi \in \mathcal{F}$ with respect to the automaton $\mathcal{A}$.

1.2.3 Constrained Equations

Let $\mathcal{A}$ be an automaton with states Q, then $[\![\phi : s = t]\!]_{\mathcal{A}} = \{s\sigma = t\sigma \mid \sigma \in [\![\phi]\!]_{\mathcal{A}}\}$. This means that constrained equations are nothing but the denotation of infinite sets of ground equations. A similar definition holds for rewrite rules. All definitions of reduction relation, equality step, ... for rewrite systems carry over the constrained case, replacing constrained equations (or rules) by the corresponding (infinite) sets of equations (or rules) they denote. Thus, we will use the notations of [6]. For example, $\underset{E,\mathcal{A}}{\longleftrightarrow}$ (in general, we will omit these subscripts) is the relation $\underset{[\![E]\!]_{\mathcal{A}}}{\longleftrightarrow}$.

2 Deduction on Constrained Equations

This section is devoted to the completion of constrained equations. As shown in introduction, we have to design new deduction rules which restore the critical pair lemma.

---

[2] Then $[\![q]\!]_{\mathcal{A}}$ is a regular tree language: that is why we use sometimes a regular expression instead of a sort symbol in the examples.

Deduce 1

$$\frac{\phi : l \rightarrow r \qquad \psi : g \rightarrow d}{\phi' : l[d]_p\sigma = r\sigma}$$

If $\begin{cases} 1a.\ p \text{ is a non variable position of } l \\ 1b.\ \phi \wedge \psi \wedge l|_p = g \mapsto^* \phi' \wedge \sigma \end{cases}$

Deduce 2

$$\frac{x \in q \wedge \phi : l \rightarrow r \qquad \psi : g \rightarrow d}{\phi' : r\{x \mapsto X_1(d)\} = l\{x \mapsto X_1(d)\}}$$

If $X_1(g) \in q \wedge \psi \wedge \phi \wedge X_1(d) \notin q \mapsto^* \phi'$ is satisfiable

Deduce 3

$$\frac{X_1 \in C \wedge \phi : l \rightarrow r \qquad \psi : g \rightarrow d}{\phi' : l' = r\{X_1 \mapsto X_3 \cdot X_2\}\sigma}$$

If $\begin{cases} 3a.\ X_3 \cdot X_2 \in C \wedge X_2(w) = g \wedge \phi \wedge \psi \mapsto^* \phi' \wedge \sigma \\ 3b.\ l|_p \equiv X_1(w) \\ 3c.\ l' \text{ is } l \text{ in which } \textit{all} \text{ occurrences of } X_1(w) \text{ are replaced with } X_3(d) \text{ and } \sigma \text{ is applied.} \end{cases}$

Figure 1: Deduction Rules

.1 Critical pair lemma and Deduction Rules

'he deduction rules are given in figure 1.

)educe 1 is the classical deduction rule. We adopt the convention that a substitution may also be viewed as a solved set of equations. Therefore, a substitution may appear as part of a constraint. The constraint $\phi' \wedge \sigma$ is moreover assumed to be in *solved form*. This means that ϕ' is a membership solved form. Finally, $\mapsto^*$ is the reduction of a constraint to one of its solved forms.

)educe 2 corresponds intuitively to the superposition between a left hand side and a sort expression: x is assigned to $X_1(g)$, which has to belong to q. Condition $X_1(d) \notin q$ states that there is not already a trivial rewrite proof of the conclusion; we require that $x \in q \wedge \phi : l \rightarrow r$ cannot be applied on l if we replace x with $X_1(d)$, thus commuting the rewrite steps. This condition is not necessary, but it is very useful in pruning unnecessary deductions.

)educe 3 corresponds intuitively to superpositions "inside" a second-order variable X_1: X_1 is split into the two second-order variables X_3 and X_2 ($X_1 = X_3 \cdot X_2$). Then, the superposition takes place at X_2. This is the meaning of condition 3*a*. where we have to solve $g = X_2(w)$. Then we get l' and r' by simplification of the extrema of the critical peak

$$r\sigma \leftarrow l[X_3(X_2(w))]_p\sigma \rightarrow l[X_3(d)]_p\sigma$$

We indeed know that $X_2(w)\sigma$ can be reduced by $\psi : g \to d$: we apply thi rule as long as possible on the right side of the peak. We get l' (condition 3c.

The critical pair lemma states that, for "non-commuting peaks" there is a de duction rule leading to a simpler proof:

Lemma 2.2 *Let $s, t, u \in T(F)$ and R be a set of constrained rules where the con straints are membership solved forms. If $s \underset{\mathcal{R}}{\longleftarrow} t \underset{\mathcal{R}}{\longrightarrow} u$, then either there is a rewrit proof (in $\mathcal{R}$) of $s \overset{*}{\underset{\mathcal{R}}{\longleftrightarrow}} u$ (i.e. a proof $s \overset{*}{\underset{\mathcal{R}}{\longrightarrow}} \overset{*}{\underset{\mathcal{R}}{\longleftarrow}} u$) or there is a deduction rule L such that $\mathcal{R} \vdash_D \phi : s' = t'$ and a proof $s \overset{*}{\underset{\mathcal{R}}{\longrightarrow}} \underset{\phi:\, s'=t'}{\longleftrightarrow} \overset{*}{\underset{\mathcal{R}}{\longleftarrow}} u$.*

Note that some simplification steps have been incorporated in the deductior rules, for technical reasons which will become apparent in section 3.

2.2 Other rules and Refutation Completeness

Up to now, the rewrite relation is only defined on ground terms. If we want to desigr simplification rules, we have to lift the it to non-ground terms. Let a *constrainec term* be a pair $\phi : s$ where ϕ is a solved membership constraint and s is a term.

Definition 2.3 *Given a set $\mathcal{R}$ of constrained rules, the rewrite relation $\to_{\mathcal{R}}$ is de fined on constrained terms by:*

$$\phi : s \xrightarrow[\psi:l\to r]{p,\theta} \phi : s' \quad \text{iff} \quad s|_p \equiv l\theta \text{ and } \theta \wedge \phi \models \psi \text{ and } s' \equiv s[r\theta]_p$$

The rules **Delete**, **Simplify**, **Compose** and **Collapse** are now identitical to clas sical rules, except that the notion of simplification is replaced with the above one.

We also use the classical **Orient** rule. We do not address here the problem of orderings on second order terms. We could use, e.g. a recursive path ordering where any second order variable is not comparable with any other symbol.

Finally, we rely on [6] for the notions of *fairness, completion procedure,...* Because of our critical pair lemma, we get results similar to those of [6], in particular the refutation completeness theorem for completion procedures that do not fail.

Example 2.4 Let us show how our completion procedure works on the example given in introduction.

$$\mathcal{R} = \left\{ \begin{array}{lrcl} x \in f^+(g(f^*(a))) : & f(x) & \to & a \\ & g(x) & \to & h(x, x) \end{array} \right.$$

$$\mathcal{R} \vdash_{\textbf{Deduce 2}} \quad x \in f^*(a) \wedge X \in f^+ : \ a = f(X(h(x, x)))$$

but it is also possible to apply **Deduce 2**, overlapping the first rule in itself, which leads to the equation

$$Y \in f^+ : \ Y(a) = a$$

Orienting the two last equations into rules, we get a convergent system because the only deduction rules that are applicable lead to renamings of the last rules.

3 Constraints Solving and Unification

We reach now the second difficulty: how to solve constraints? This is not obvious since our logic involves both higher order variables and membership constraints. However, it turns out that the problems which can be generated during the completion procedure are all solvable.

3.1 Restricting the constraints that have to be considered

The terms occurring while running the completion procedure enjoy the following property: if $X \in \mathcal{CX}$ occurs twice in t at positions p and q, then $t|_p \equiv t|_q$. Let $WT(F, \mathcal{X}, \mathcal{CX})$ be the set of terms which satisfy the above property and which may be viewed as DAGS which are linear w.r.t. the second-order variables. Assuming that, when a simplification rule is applied at an inner position of $X(w)$ where $X \in \mathcal{CX}$, then it is also applied at the same position for every other occurrence of $X(w)$ (which corresponds to a fair control), then the above property is kept during completion as shown by the following lemma.

Lemma 3.5 *If E (resp.$\mathcal{R}$) is a set of constrained equations (resp. rules) $\phi : s = t$ (resp. $\phi : s \rightarrow t$) where $s, t \in WT(F, \mathcal{X}, \mathcal{CX})$, and if $E, \mathcal{R} \vdash E', \mathcal{R}'$, then every term occurring in $E', \mathcal{R}'$ also belongs to $WT(F, \mathcal{X}, \mathcal{CX})$.*

Now, looking back at the constraints we have to solve for the deduction rules of the completion procedure, it turns out that we only have to consider constraints of the form $\phi \wedge s = t$ where ϕ is a membership constraint and $s, t \in WT(F, \mathcal{X}, \mathcal{CX})$ or membership constraints alone. Using as long as possible a *variable abstraction* rule, it is possible to restrict our attention to *unification problems* which are existentially quantified disjunctions of formulas of the form:

$$s_1 = t_1 \wedge \ldots \wedge s_n = t_n \wedge X_1 = U_1 \wedge \ldots \wedge X_m = U_m \wedge M$$

where M is a membership constraint, $s_1, \ldots, s_n \in T(F, \mathcal{X})$, each term t_i either belongs to $T(F, \mathcal{X})$ or has the form $Y_i(t_i')$ where $t_i' \in T(F, \mathcal{X})$, each variable in $\mathcal{CX}$ occurs at most once in $s_1, \ldots, s_n, t_1, \ldots, t_n$, $X_1, \ldots, X_m \in \mathcal{CX}$, $U_1, \ldots, U_m \in T(F, \mathcal{X}, \mathcal{CX})$, each variable of $\mathcal{CX}$ occurs at most once in each U_i, $X_1, \ldots, X_m$ do not occur in $s_1, \ldots, s_n, t_1, \ldots, t_n$ and $X_1 = U_1 \wedge \ldots \wedge X_m = U_m$ is in DAG solved form (i.e. each variable occurs at most once as a member of an equation and the extended occur check fails. See [13] for more details.)

3.2 Transformation Rules

We first consider decomposition rules for both constraints and equations yThese rules are given in figure 2 (except for the so-called "flexible/flexible" case of [12] which is combined with a merge rule and displayed as the last rule of figure 3). Decomposition of membership constraints rely on a number of technical constructions of tree automata (for the first time, we use the fact that sort expressions are interpreted as regular tree languages). The construction of f^{-1} was already given in [4] The other constructions (C^{-1}, the extension of f^{-1} to context expressions and split) are similar. They are given in [5] and omitted here.

$$\frac{f(t_1,\ldots,t_n) = f(u_1,\ldots,u_n)}{t_1 = u_1 \wedge \ldots \wedge t_n = u_n}$$

$$\frac{f(t_1,\ldots,t_n) \in q}{\bigvee_{(q_1,\ldots,q_n)\in f^{-1}(q)} t_1 \in q_1 \wedge \ldots \wedge t_n \in q_n} \qquad \frac{f(t_1,\ldots,t_n)[U]_i \in C}{\bigvee_{(q_1,\ldots,C_i,\ldots,q_n)\in f^{-1}(C)} U \in C_i \wedge \bigwedge_{i\neq j} t_j \in}$$

$$\frac{X(s) \in q \wedge X \in C}{\bigvee_{(C',q')\in C^{-1}(q)} X \in C' \wedge s \in q'} \qquad \frac{X(u) \in C}{\bigvee_{(C_0,C_1)\in \mathrm{split}(C)} X \in C_0 \wedge u \in C_1}$$

$$\frac{f(s_1,\ldots,s_n) = X(s)}{\bigvee_{1\leq i\leq n} \begin{array}{r}(X = f(s_1,\ldots,s_n)[\cdot]_i \wedge s = s_i)\\ \vee(\exists Y : X = f(s_1,\ldots,s_n)[Y]_i \wedge s_i = Y(s) \wedge Y \in \neg\epsilon)\end{array}}$$

Figure 2: Decomposition Rules

Merge rules are described in figure 3. We also use classical clash, occur-check, coalesce rules and other simplification rules which are described in detail in [5]. The definition of a *solved variable* is an extension (in the same spirit) of the definition given in [13]. We also need an extension of the occur-check ordering which is more interesting.

3.2.1 Congruence Closure and Occur-Check Relation

The *strict equality* $=_{S,\phi}$ defined by a conjunction ϕ of equations and membership constraints is the least equivalence relation on first-order terms (in $T(F, \mathcal{X}, \mathcal{CX})$) containing all equations $s = t$ occurring in ϕ. This relation (also known as *congruence closure*) has been shown decidable in polynomial time [8].

The *strict occur-check relation* $>_\phi$ is the least transitive relation on equivalence classes modulo $=_{S,\phi}$ such that $c >_\phi c'$ if there are terms $s \in c$ and $s' \in c'$ such that s' is a (strict) subterm of s. $\geq_\phi$ is the union $=_{S,\phi} \cup >_\phi$. This relation $\geq_\phi$, restricted to variables is the classical occur-check relation [13].

3.2.2 Merge Rules

Now, let us elaborate a little bit on the last merge rule. This rule is actually a number of merges followed by decompositions. It solves the "flexible/flexible" case. First let us explain the notations. We use $t[t_1]_{p_1}\ldots[t_n]_{p_n}$ to denote the simultaneous replacement of terms $t_1,\ldots,t_n$ at uncomparable positions $p_1,\ldots,p_n$ of t. We also denote $\mathrm{Inv}(\Phi, \{1,\ldots,n\})$ any set of representatives of the classes under the equivalence relation $\sim_\Phi$ defined on $\{1,\ldots,n\}$ by $i \sim_\Phi j$ iff $\Phi(i) = \Phi(j)$.

$$\frac{x \in q \wedge x \in q'}{x \in q \wedge q'} \qquad\qquad \frac{X \in C \wedge X \in C'}{X \in C \wedge C'}$$

$$\frac{x = s \wedge x \in q}{x = s \wedge s \in q} \text{ If } s \not>_\phi x \text{ and } s \notin \mathcal{X} \qquad\qquad \frac{X = t[Y]_p \wedge X \in C}{X = t[Y]_p \wedge t[Y]_p \in C} \text{ If } X \not\equiv Y$$

$$\frac{x = f(s_1, \ldots, s_m) \wedge x = g(t_1, \ldots, t_n)}{x = f(s_1, \ldots, s_m) \wedge f(s_1, \ldots, s_m) = g(t_1, \ldots, t_n)}$$

$$\frac{x = f(s_1, \ldots, s_m) \wedge x = X(t)}{x = f(s_1, \ldots, s_m) \wedge f(s_1, \ldots, s_m) = X(t)}$$

$$\frac{x = X_1(s_1) \wedge \ldots \wedge x = X_n(s_n)}{\begin{array}{l} \displaystyle\bigvee_{i=1}^{n} \exists Z_1, \ldots, Z_n : \ x = X_i(s_i) \wedge \bigwedge_{j \neq i} s_i = Z_j(s_j) \wedge X_j = X_i \cdot Z_j \\ \displaystyle\bigvee_{\substack{n \geq m \geq 2,\ \Phi \text{ surjective} \\ \Phi : \{1, \ldots, n\} \to \{1, \ldots, m\} \\ f \in F,\, k = a(f) \geq m \\ 1 \leq i_1 < \ldots < i_m \leq k}} \exists Z_0, \ldots, Z_n, \exists x_1, \ldots x_k : x = Z_0(f(x_1, \ldots, x_k)) \bigwedge_{q=1}^{m} \bigwedge_{\Phi(j) = i_q} x_{i_q} = Z_j(s_j) \\ \displaystyle\bigwedge_{\substack{1 \leq j \leq n \\ \{j_1, \ldots, j_m\} = \mathrm{Inv}(\Phi, \{1, \ldots, n\})}} X_j = Z_0(f(x_1, \ldots, x_k)[Z_{j_1}(s_{j_1})]_{i_{\Phi(j_1)}} \ldots [Z_{j_m}(s_{j_m})]_{i_{\Phi(j_m)}} [Z_j]_{i_{\Phi(j)}}) \end{array}}$$

All these Merge rules assume that there is no positive occur-check. The last rule assumes that x is a maximal variable w.r.t. $>_\phi$ among the unsolved first order variables. Moreover, $x = X_1(s_1) \ldots, x = X_n(s_n)$ are all equations of ϕ whose one member is x.

Figure 3: Merge rules

The rule may look complicated and some restrictions may seem unnecessary. Let us give some informal explanations of both the rule and the restictions on its application. For simplicity, assume $n = 2$. There are actually two large disjunctions in the target of the rule. These disjunctions correspond to two cases: for every solution $x\sigma \equiv X_1\sigma[s_1\sigma]_{p_1} \equiv X_2\sigma[s_2\sigma]_{p_2}$, either p_1 and p_2 are comparable (this is the first disjunction) or they are not comparable (this is the second disjunction). Consider the first case. With only this case, we can simplify the rule into

$$\begin{array}{rl} x = X_1(s_1) \wedge x = X_2(s_2) \vdash & (\exists Y : \ x = X_1(s_1) \wedge s_1 = Y(s_2) \wedge X_2 = X_1 \cdot Y) \\ & \vee(\exists Y : \ x = X_2(s_2) \wedge s_2 = Y(s_1) \wedge X_1 = X_2 \cdot Y) \end{array}$$

each of the two conjunctions corresponding either to $p_1 \leq p_2$ or to $p_2 \leq p_1$. If we have only unary function symbols, then this rule alone is sufficient. But there is still a difficulty. Indeed, we cannot use here size arguments (as in classical unification methods [13]). For, consider the problem $x = X(s) \wedge x = Y(t)$ where $X(s)$ has more nodes than $Y(t)$ (hence $X(s)$ is greater w.r.t. the size). If we merge the two equations, keeping the smallest one, we get $x = Y(t) \wedge Y(t) = X(s)$ which leads by decomposition to the two problems: $\exists Z : x = Y(t) \wedge X = Y \cdot Z \wedge t = Z(s$ and $\exists Z : x = Y(t) \wedge Y = X \cdot Z \wedge s = Z(t)$. But this last problem is no longer a unification problem because Y occurs both as a member of an equation and inside an equation between first order terms. This situation is avoided in our set of rules because we choose which equation is kept (among $x = X(s)$, $x = Y(t)$) *after* the decomposition step. However, this means that we will have to give new termination arguments, based on the occur check relation we defined above, instead of size.

The last disjunction corresponds to (some) uncomparable positions of the holes in a solution (see [5] for a discussion).

3.2.3 Correctness Results

Lemma 3.6 *Occur-check, Coalesce, Clash, Elimination rules as well as all the rules given in figures 2 and 3 preserve the set of solutions. They also do transform unification problems into unification problems.*

3.3 Termination and Completeness

This is the main results of our paper: the set of rule is terminating and the irreducible formulas are solved forms.

Definition 3.7 *A formula* $\phi \in \mathcal{F}$ *is in* DAG solved form *if it is either* $\top$, $\bot$, *or else it has the following form:*

$$\exists \vec{z}, \exists \vec{Z} : x_1 = t_1 \wedge \ldots \wedge x_n = t_n \wedge X_1 = u_1[Z_1]_{p_1} \wedge \ldots \wedge X_r = u_r[Z_r]_{p_r} \\ \wedge y_1 \in q_1 \wedge \ldots \wedge y_m \in q_m \wedge Y_1 \in C_1 \wedge \ldots \wedge Y_k \in C_k$$

where

- $x_1, \ldots, x_n \in \mathcal{X}$ *and each* x_i *has only once occurrence in* $x_i = t_i \wedge x_{i+1} = t_{i+1} \wedge \ldots \wedge Y_k \in C_k$
- $X_1, \ldots, X_r \in \mathcal{CX}$ *and, for all* i, X_i *has only one occurrence in* $X_i = u_i[Z_i]_{p_i} \wedge \ldots \wedge Y_k \in C_k$
- $t_1, \ldots, t_n \in T(F, \mathcal{X}, \mathcal{CX})$ *and* $u_1[Z_1]_{p_1}, \ldots, u_r[Z_r]_{p_r} \in \mathcal{C}(F, \mathcal{X}, \mathcal{CX})$
- $y_1, \ldots, y_m \in \mathcal{X}$ *and* $Y_1, \ldots, Y_k \in \mathcal{CX}$ *are distinct variables*
- $[\![q_1]\!]_{\mathcal{A}}, \ldots, [\![q_m]\!]_{\mathcal{A}}, [\![C_1]\!]_{\mathcal{A}}, \ldots, [\![C_k]\!]_{\mathcal{A}}$ *are not empty.*

Lemma 3.8 *DAG solved forms which are distinct from* $\bot$ *are satisfiable.*

Theorem 3.9 *The rules of figures 2 and 3 together with Clash, Occur-check, Coalesce and Elimination Rules do terminate on unification problems. Every formula, which is irreducible with respect to these rules is a finite disjunction of DAG solved forms.*

The proof of this result is interesting by itself (see [5]); it consists in interpreting the formulas in an algebra of terms which is ordered by the *associative path ordering* [2].

Acknowledgment I thank Jean-Pierre Jouannaud, in particular for the idea of extending the ocur-check relation to the quasi-ordering $\geq_\phi$ on ground terms.

References

[1] Aristide. A logic of semi-functions, inclusion and equality. the setting. Research Report 89-R-058, Centre de Recherche en Informatique, Nancy, France, 1989.

[2] L. Bachmair and D. A. Plaisted. Termination orderings for associative-commutative rewriting systems. *J. Symbolic Computation*, 1(4):329–349, Dec. 1985.

[3] H. Chen and J. Hsiang. Logic programming with recurrence domains. In *Proc. ICALP 91, LNCS*, Madrid, 1991.

[4] H. Comon. Equational formulas in order-sorted algebras. In *Proc. ICALP, Warwick*. Springer-Verlag, July 1990.

[5] H. Comon. Completion of rewrite systems with membership constraints. Research Report 699, LRI, Univ. Paris-Sud, Sept. 1991.

[6] N. Dershowitz and J.-P. Jouannaud. Rewrite systems. In J. van Leeuwen, editor, *Handbook of Theoretical Computer Science*, volume B. North-Holland, 1990.

[7] W. Farmer. A unification algorithm for second order monadic terms. *Annals of Pure and Applied Logic*, 39:131–174, 1988.

[8] J. Gallier, P. Narendran, D. Plaisted, S. Raatz, and W. Snyder. An algorithm for finding canonical sets of ground rewrite rules in polynomial time. Draft, Jan. 1988.

[9] M. Gécseg and M. Steinby. *Tree Automata*. Akademia Kiadó, Budapest, 1984.

[10] W. D. Goldfarb. Note on the undecidability of the second-order unification problem. In *Theoretical Computer Science*, pages 225–230, 1981.

[11] B. Gramlich. Unification of term schemes. theory and applications. Research Report SR-88-18, SEKI, Univ. Kaiserslautern, Germany, 1988.

[12] G. Huet. *Résolution d'équations dans les langages d'ordre* $1, 2, \ldots \omega$. Thèse d'Etat, Univ. Paris 7, 1976.

[13] J.-P. Jouannaud and C. Kirchner. Solving equations in abstract algebras: A rule-based survey of unification. In J.-L. Lassez and G. Plotkin, editors, *Computational Logic: Essays in Honor of Alan Robinson*. MIT-Press, 1991.

[14] C. Kirchner, H. Kirchner, and J. Meseguer. Operational semantics of OBJ-3. In *Proc. 15th ICALP, Tampere, LNCS 317*. Springer-Verlag, July 1988.

[15] C. Kirchner, H. Kirchner, and M. Rusinowitch. Deduction with symbolic constraints. *Revue Française d'Intelligence Artificielle*, 4(3):9–52, 1990.

[16] H. Kirchner. Schematization of infinite sets of rewrite rules generated by divergent completion process. *Theoretical Comput. Sci.*, 67(2–3):303–332, Oct. 1989.

[17] R. Nieuwenhuis and A. Rubio. Basic superposition is complete. In *Proc. ESOP*, 1992.

[18] T. Nipkow. Higher order critical pairs. In *Proc. IEEE Symp. on Logic in Comp. Science*, Amsterdam, 1991.

[19] G. Smolka. *Logic programming over polymorphically order-sorted types*. PhD thesis, Univ. Kaiserslautern, May 1989.

[20] G. Smolka, W. Nutt, J. A. Goguen, and J. Meseguer. Order-Sorted Equational Computation. In H. Aït-Kaci and M. Nivat, eds, *Resolution of Equations in Algebraic Structures, Volume 2*, chapter 10. Academic Press, 1989.

[21] Y. Toyama. Confluent term rewriting systems with membership conditions. In *Proc. 1st Int. Workshop on Conditional Term Rewriting Systems, Orsay, LNCS 308*, 1988.

A New Metric Between Polygons, and How to Compute it

(extended abstract)

Günter Rote

Technische Universität Graz, Institut für Mathematik,
Steyrergasse 30, A-8010 Graz, Austria
electronic mail: `rote@ftug.dnet.tu-graz.ac.at`

Abstract. The bounded Lipschitz distance can be used as a distance measure between polygons or between functions. It has several properties which make it attractive to use from the viewpoint of applications. There are several alternative definitions of the metric, some of which are interesting in their own right. We also give an algorithm for computing the metric, and we discuss possible implementations.

1 Introduction

1.1 Problem statement

For a given function $f\colon [0,1] \to \mathbb{R}$ we want to compute its *bounded Lipschitz norm*

$$\|f\|_{\mathrm{BL}} = \max\Big\{ \int_0^1 f(x)g(x)\,dx : |g(x)| \le M,\ |g(x) - g(y)| \le |x - y| \Big\}$$

Here, M is a scaling parameter of the norm whose meaning will be explained later. practice, f could be given by a sequence of values $(f_1, \ldots, f_n)$. In that case, the bounc Lipschitz norm looks as follows:

$$\|(f_1, \ldots, f_n)\|_{\mathrm{BL}} = \max\Big\{ \sum_{i=1}^{n} f_i g_i : |g_i| \le m,\ |g_{i+1} - g_i| \le 1 \Big\}$$

For the new parameter m we have $m = M \cdot n$.

1.2 Why is it interesting to compute the distance between functions?

As defined above, the bounded Lipschitz norm is a *norm* on the vector space of sequences or of all piecewise continuous functions. Of course, a norm gives rise tc *metric* $d_{\mathrm{BL}}(f,g)$ in the usual way:

$$d_{\mathrm{BL}}(f,h) = \|f - h\|_{\mathrm{BL}}.$$

This concept of *distance* between functions is where the applications are.

ern recognition: Functions of one parameter. In pattern recognition, it is often re-ed to classify an "image" of an object by deciding which one of several given patterns s most like it. In many cases, the image and the patterns can be specified as one-metric functions, or, more concretely, as sequences of measured values. For example, vending machine that accepts money bills, an inserted bill is scanned by moving rough a beam of light, and the absorption of the light as it passes through the bill corded. The resulting light-and-dark pattern must be compared to each of several ible patterns.

ern recognition: Shapes. A shape might be given as a simple polygon or, more gener-as a simply connected region. There are several ways to represent this as a function ne variable. If the "image" polygon and the pattern polygons are *convex* bodies C, one possibility is the support function h: $[0, 2\pi) \to \mathbb{R}$

$$h(\alpha) = \max\{\, x\cos\alpha + y\sin\alpha : (x, y) \in C\,\}.$$

ther possibility is to parameterize the boundary by arc length

$$\begin{pmatrix} x(t) \\ y(t) \end{pmatrix}, \quad \text{for } 0 \le t \le L,$$

$\dot{x}(t)^2 + \dot{y}(t)^2 = 1$, and use the tangent direction α: $[0, L] \to [0, 2\pi)$ as the function characterizes the curve:

$$\begin{pmatrix} \dot{x}(t) \\ \dot{y}(t) \end{pmatrix} = \begin{pmatrix} \cos\alpha(t) \\ \sin\alpha(t) \end{pmatrix},$$

s approach has for example been taken by Arkin et al. (1990), who considered the L_1 L_2 norms of the resulting funtion $\alpha(t)$.

Other distances, including the well-known Hausdorff distance, have been investigated his context, see Atallah (1983) and Cox et al. (1989).

ality control. In the production of artificial fabric, one aspect of quality is that the ric should be as smooth and uniform as possible. A real piece of fabric will always ibit irregularities. To monitor the quality, the fabric is passed under a light beam and thickness is continuously measured. Now it is important to assign a numerical value t "measures" the deviation of the measured thickness distribution of a piece of fabric n the ideal uniform thickness. In this application, it is important that this distance ue is computed fast, because data arrive continuously at a high rate.

In what respect is the proposed distance different from other metrics?

ou take any L_p norm,

$$\left(\int_0^1 |f(x)|^p \, dx \right)^{1/p},$$

your favourite value of p, this norm will not be able to distinguish between functions t have "the same values distributed in a different way over the domain". For example,

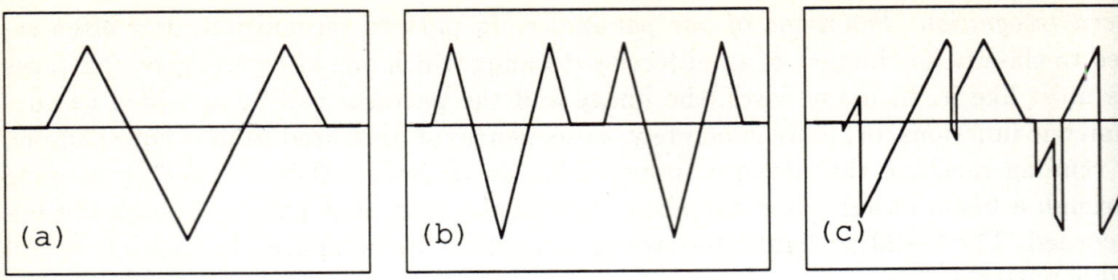

Fig. 1. Three functions which have the same L_p metrics.

in figure 1a, figure 1b, and figure 1c, the L_p norms will always be identical, although functions look completely different. The same holds for the $L_{\max}$ or L_∞ norm,

$$\sup_{0\le x\le 1} |f(x)|$$

Figure 1a has one big "hole", wherease figure 1b has two small holes. It is nat that small oscillations are more tolerable than big systematic irregularities, and sev small holes should be better than one big hole of the same total "area". This is certa true for the quality control of fabrics, at least.

Another distance measure is the *discrepancy*,

$$\max_{0\le a\le b\le 1} \left| \int_a^b f(x)\, dx \right|,$$

or the maximum sum of a contiguous subsequence of $(f_1, \dots, f_n)$. The discrepancy d distinguish between big holes and small holes. However, contrary to the bounded Lipsc norm, the discrepancy is influenced only by the biggest hole, and the smaller holes not play any role at all. The situation is similar to the $L_{\max}$ norm.

The Hausdorff distance between polygons C and D,

$$\max \left\{ \max_{x\in C} \min_{y\in D} d(x,y),\ \max_{y\in D} \min_{x\in C} d(x,y) \right\},$$

suffers from the same limitation.

To my knowledge, the bounded Lipschitz norm is the only distance measure betw functions that is both sensitive to locality on the x-axis and maintains a global vie the function on the whole interval. In other words, it distinguishes between big holes small holes, and it still takes care of every hole, big and small.

1.4 Examples

We will illustrate the bounded Lipschitz norm of a function and not of a sequence, beca this exhibits the essential properties of the bounded Lipschitz distance more clearly more intuitively.

Figure 2a shows a function f and the corresponding function g which optimizes integral in (1). In general, g will try to follow the trend of f, being as large as poss

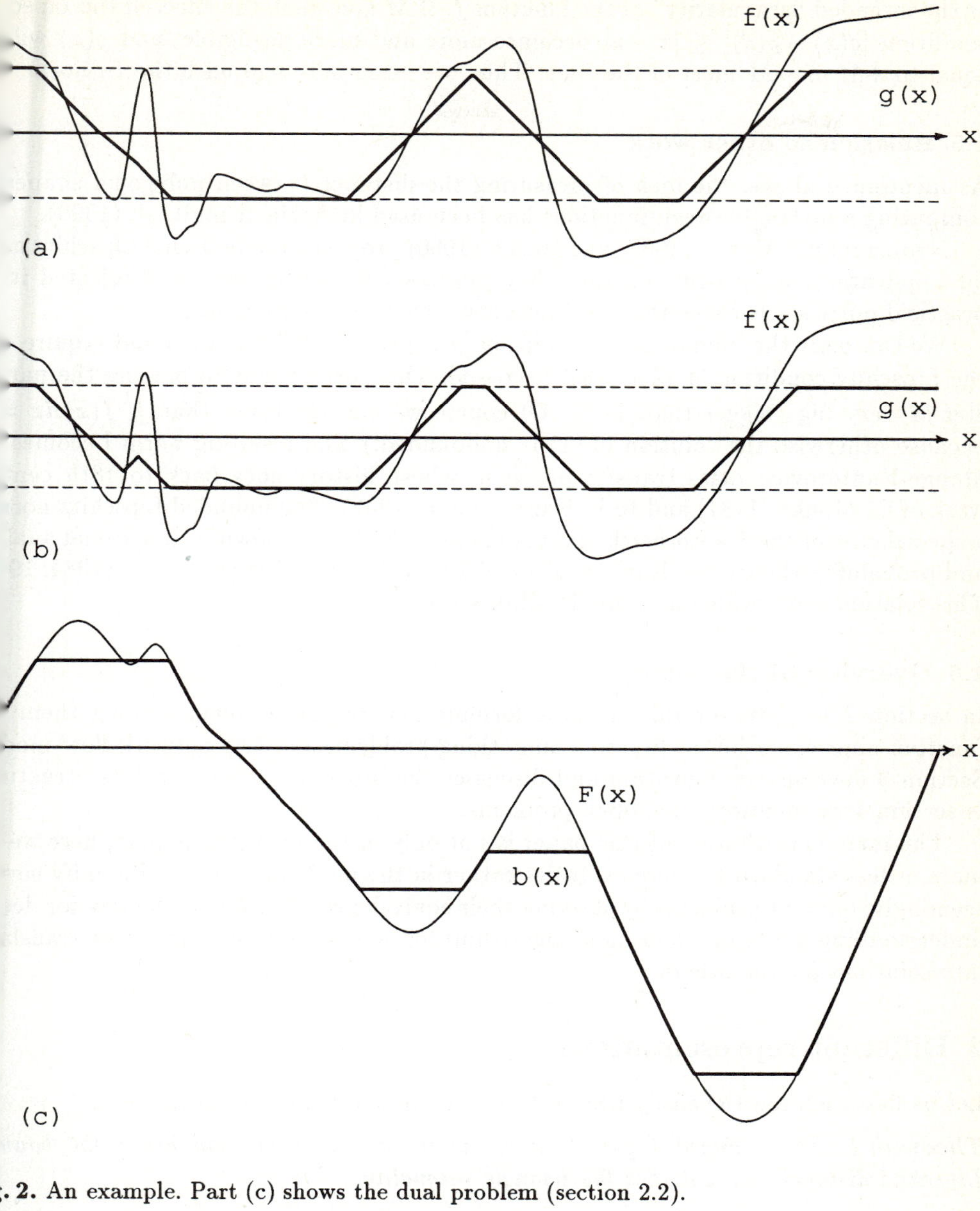

;. 2. An example. Part (c) shows the dual problem (section 2.2).

en f is positive and as small as possible when f is negative. We see that g cannot follow oo closely in the left part because f wiggles too fast, but the big wave in the right t is fully taken into account. Figure 2b shows what happens when the parameter M in constraint $|g(x)| \leq M$ is decreased. Previously, the wave in the middle was too short $g(x)$ to follow it completely, but now there is no longer a difference between the big ht wave and the middle wave. Thus M is something like the "precision" of the norm or the intended "granularity" of the function f. If M goes to 0, the effect of the Lipsc condition $|g(x) - g(y)| \leq |x - y|$ becomes more and more negligible, and $g(x)$ wil equal to $+M$ or $-M$ most of the time. Thus, we essentially approach the L_1 norm.

1.5 Relation to other work

As mentioned above, the idea of measuring the distance between polygonal shape computing a metric between functions has been used in Arkin et mult. al. (1990).

Neunzert and Wetton (1987) and Hackh (1990) proposed the bounded Lipschitz n for applications in quality control. They discussed its advantages, but rejected it practical purposes because they did not know how to compute it fast.

We can omit the boundedness condition $|g(x)| \leq 1$ in definition (1) and require the Lipschitz condition $|g(x) - g(y)| \leq |x - y|$. This corresponds to making the par eter M very big (bigger than 1/2). (Of course we have to insist that $\int_0^1 f(x)\,dx =$ because otherwise the solution of (1) is unbounded.) The resulting norm becomes Monge-Kantorovich mass transfer problem, whose history goes back to 18th cent work of G. Monge (1781) and to L. Kantorovitch (1942). The bounded Lipschitz nor a special case of the Kantorovich-Rubinstein norm which is known in functional anal and probability theory, see Kantorovich and Rubinshteĭn (1958) or Rachev (1984, 19 This relation is described in more detail in section 2.3.

1.6 Overview of the paper

In section 2 we give several alternate formulations of the problem; among them function approximation or function smoothing problem, and two network flow mod Section 3 develops an algorithm and discusses the possible options for data struct Is section 4 we mention some open problems.

The main contribution of this paper is not only in the algorithmic part; here we more or less standard techniques. It lies rather in the modeling aspect: We unify sev seemingly different problems by showing their equivalence. This opens the way for dee understanding of the problems, and algorithms for one of the problems can be transla into solutions for the others.

2 Different representations

Let us first mention the easily proved fact that we really have a norm:

Theorem 1. *The bounded Lipschitz norm is in fact a norm, and hence the boun Lipschitz distance* d_{BL} *satisfies the triangle inequality.*

2.1 A network flow model

It is possible to translate the restrictions in (2) into a minimum cost network flow mo as shown in figure 3. The flow on the arc from A_{i-1} to A_i is the variable g_i and i restricted between the capacities $-m \leq g_i \leq +m$. The cost on this arc is f_i. On the a between A_i and B the flow can be at most 1 in either direction, except for A_0 and where the flow is unrestricted. This models the Lipschitz condition. The cost on th arcs is 0.

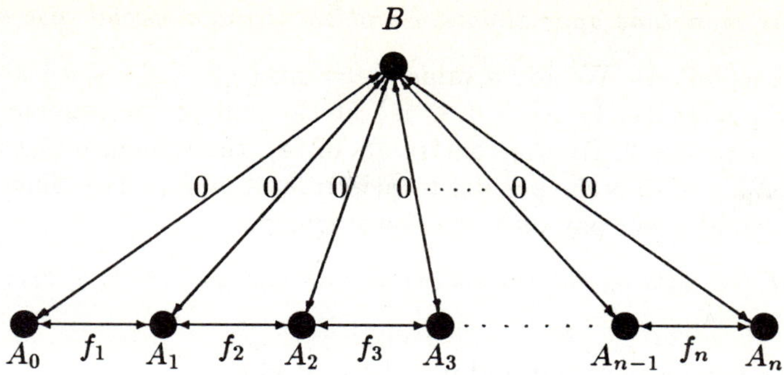

.3. The network flow model.

The dual problem: smoothing a function

e linear programming dual of problem (2), after some transformation, can be written follows: Let F_i denote the partial sums of the sequence f_i:

$$F_k := f_1 + f_2 + \cdots + f_k \tag{3}$$

en we have

$$\|(f_1, \ldots, f_n)\|_{\mathrm{BL}} = \min\Big\{ m \cdot \sum_{i=1}^{n} |b_i - b_{i-1}| + \sum_{i=0}^{n} |b_i - F_i| : b_0 = F_0,\, b_n = F_n \Big\}. \tag{4}$$

is problem is of interest in itself because it can be interpreted as a problem of smoothing unction which is given by a sequence $(F_0, F_1, \ldots, F_n)$. We are looking for a "smooth" uence $(b_0, b_1, \ldots, b_n)$ that approximates the given one. The right sum in the above pression (4) is the approximation error, and the left sum is a penalty term that keeps sequence smooth. By varying the parameter m one can control the desired degree smoothness. Figure 2c shows the dual problem to the example of figure 2b. (This is continuous version of the problem again.) The graph of $b(x)$ is shown as a thick rve overlaid over the graph of $F(x) := \int_0^x f(t)\,dt$. The reader may notice a correspon- nce between the ranges where $b(x)$ follows $F(x)$ or where it remains constant, and the nges in figure 2a where $g(x)$ is at its upper or lower bound or where it has slope ± 1, spectively. This is nothing but a manifestation of complementary slackness in linear ogramming duality.

A similar curve fitting problem has been considered by Ohmura et al. (1990) and by ai (1991):

$$\min\Big\{ \sum_{i=1}^{n} |b_i - F_i| : b_1 \le b_2 \le \cdots \le b_n \Big\}. \tag{5}$$

re a given sequence must be approximated by a monotone sequence.

Theorem 2. *The monotone approximation problem* (5) *is a special case of* (4).

This can be seen as follows. We add a value $F_0 := \min\{ F_i : 1 \le i \le n \}$ at the beginn and a value $F_{n+1} := \max\{ F_i : 1 \le i \le n \}$ at the end of the sequemce, and we $n := n + 1$ and $m := n + 2$. By the constraints of (4), the sequence (b_i) must rise fr $b_0 = F_0$ to $b_n = F_n$, and thus the penalty term is at least $m(F_n - F_0)$. Since the multip m is so large it would never pay off to go down again:

Lemma 3. *With the data as defined above, we will always have $b_i \le b_{i+1}$ in an opti solution of* (4).

Thus the penalty term is constant and may be neglected.

Imai gave an $O(n \log n)$ algorithm which uses mergeable heaps. Our algorithm wh we give in the next section is very simple and has the same complexity, and, wh appropriately specialized, it uses just plain heap operations. We also give a few ot algorithms whose complexity depends on the size of the F_i values.

2.3 Another network flow model — the Monge-Kantorovich mass transfe problem

A different reformulation leads to the following dual definition:

$$\|(f_1, \ldots, f_n)\|_{\mathrm{BL}} = \min\Big\{ \sum_{i=1}^{n} \sum_{j=1}^{n} c_{ij} x_{ij} : \sum_{j=1}^{n} x_{ij} - \sum_{j=1}^{n} x_{ji} = f_i, \text{ for } i = 1, 2, \ldots, n; \ \ x_{ij} \ge 0$$

where $c_{ij} := \min\{|i - j|, 2m\}$. This is a mininum-cost network flow problem with no $C_1, C_2, \ldots, C_n$. The original problem (2) is the dual problem to this. The restricti are the flow conservation equations. The supply/demand at node C_i is f_i. We assu $\sum_{i=1}^{n} f_i = 0$. By routing the flow x_{ij} on arc (i, j) over the vertices $C_{i+1}, C_{i+2}, \ldots, C_j$ if $|i - j| \le 2m$, and via an auxiliary vertex D otherwise, we can correctly model the co c_{ij} with the arcs shown in the network of figure 4. The costs on the arcs between C and C_i are 1, in both directions, and the costs on the arcs between C_i and D are m, both directions. The flow must be nonnegative, but there are no upper capacities.

This model has a nice direct interpretation: f_i is the excess or the lack of material location i. We have to transport the material to balance the excesses and the dema in the cheapest possible way. Without the additional vertex D, this is the original m transfer problem that goes back to G. Monge (1781) and to L. V. Kantorovitch (194 In our case, the price that we have to pay is proportional to the distance, except that only have to pay a flat rate of $2m$ for very long distances.

The cost coefficients c_{ij} can be generalized to be an arbitrary metric on the point $\{1, 2, \ldots, n\}$ (or on the interval $[0, 1]$ in the continuous formulation), and unter suita conditions the equivalence between the different dual formulations still holds. The resu ing general metric between functions is known as the Kantorovich-Rubinstein dista or the Wasserstein distance, depending on the formulation, see Rachev (1984, 1991).

Note that the network in both network flow models of figure 3 and figure 4 is seri parallel. Booth (1990) has given an $O(E \log E)$ algorithm for a mininum-cost network fl

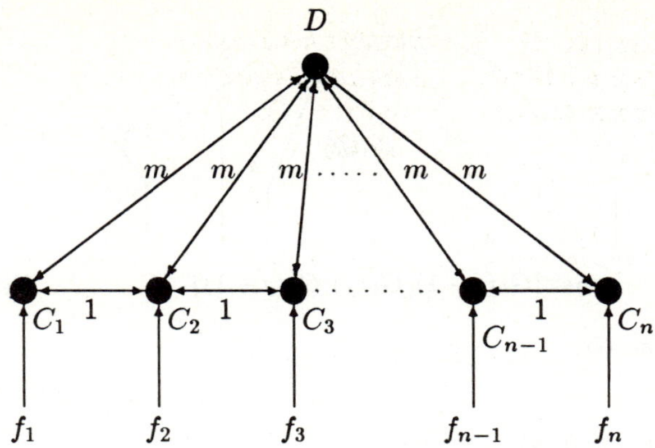

;. 4. Another network flow model.

a series-parallel network with E arcs. When applied to our networks, the algorithm .ncides for the two networks; the algorithm in section 3 can actually be derived as a ecial case and simplified version of Booth's algorithm.

Algorithms

1 Dynamic programming

e can solve the linear program in (2) by considering all possible values $j = -m, \ldots, m$ g_k, successively for $k = 1, \ldots n$. Define

$$G_k(j) := \max\Big\{ \sum_{i=1}^{k} f_i g_i : |g_i| \le m,\ |g_{i+1} - g_i| \le 1, g_k = j \Big\}.$$

ıen we can set up the recursion

$$G_{k+1}(j) = f_{k+1} \cdot j + \max\{G_k(j-1), G_k(j), G_k(j+1)\}, \text{ for } -m \le j \le m, \qquad (6)$$

ıere we take $G_k(-m-1) = G_k(m+1) = -\infty$ for the "out-of-boundary" values.

mma 4. $G_k(j)$ is a concave function of j.

Thus, we can represent the function $G_k(j)$ by the "start value" $S := G_k(-m)$ and e decreasing sequence of increments $E_j := G_k(j+1) - G_k(j)$. Let us see how G_{k+1} is rived from G_k. The maximum of the three expressions in (6) can be obtained by taking ree copies of the graph of the function G_k: one which is shifted one unit to the left, one hich is shifted one unit to the right, and an unshifted one. Then we take the pointwise aximum, see figure 5. The result will have the same sequence of increments E_j as the iginal G_k, except that two horizontal pieces $E_j = 0$ are inserted in the middle, and ıe first and the last entry is deleted. Adding the linear function $f_{k+1} \cdot j$ means that we ıve to add the value f_{k+1} to each E_j. We also have to compute the new start value $= G_{k+1}(-m)$ from $G_k(-m)$. The following algorithm summarizes this procedure.

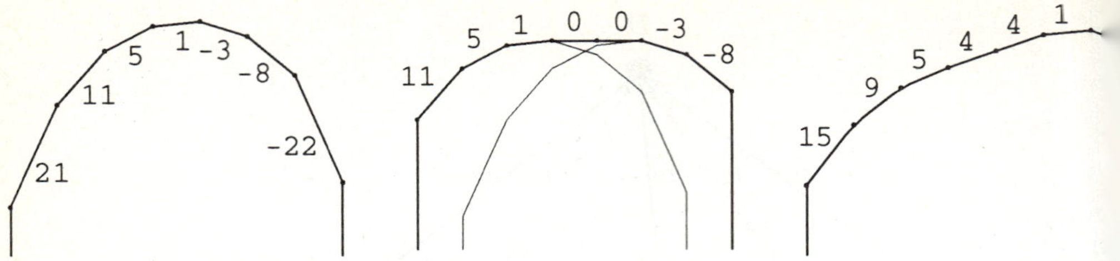

$G_k(j)$ $\max\{G_k(j-1), G_k(j), G_k(j+1)\}$ G_{k+1}

Fig. 5. Deriving G_{k+1} from G_k.

> (∗ Start with $G_0(j) \equiv 0$: ∗)
> Let $E = (0, 0, \ldots, 0)$; ($2m$ zeros)
> $S := 0$;
> **for** k := 1 **to** n **do**
>   (∗ compute G_k: ∗)
>   insert 0,0 into the decreasing sequence E (at the correct position);
> $S := S +$ the largest (i. e., first) element of E;
>   remove the largest and the smallest element from E;
>   add f_k to all elements of E;
> $S := S + (-m) \cdot f_k$;
> **end for;**
> add all positive elements of E to S; this is the result.

Instead of adding a constant to all elements of E, which costs too much time, we c also add it to a separate value F which we maintain; we represent the "true value" of . by $E'_j + F$. The resulting algorithm is as follows:

> Let $E' = (0, 0, \ldots, 0)$; ($2m$ zeros)
> $S := 0$; $F := 0$;
> **for** k := 1 **to** n **do**
>   (∗ compute G_k: ∗)
>   insert $-F$,$-F$ into E';
> $S := S + F +$ the largest element of E';
>   remove the largest and the smallest element from E';
> $F := F + f_k$;
> $S := S + (-m) \cdot f_k$;
> **end for;**
> for all elements e' of E' with $e' + F > 0$, add $e' + F$ to S; this is the result.

3.2 Data structures

Clearly, there is really no need to represent E' as a sorted list. The amount of work constant per iteration of the loop, except for the operations of the set E'. Let us discu necessary operations on E':

insert
find and remove the smallest and largest element.

have $2n$ operations of each type. E' has always $2m$ or $2m+2$ elements. The elements t are inserted are the partial sums F_i of (3).

Depending on the size of the universe $\hat{F} := \max_k |F_k|$, there are several choices of a structures.

dinary priority queues. This leads to $O(n \log m)$ time and $O(m)$ storage if we simply heaps. (Actually, we have to use max-min heaps, see Atkinson et al. 1986.) Since list E' is initialized with zeros, we can in fact achieve $O(n \log \min\{m, n\})$ time and $\min\{m, n\})$ space. By using a balanced search tree augmented with appropriate in-mation fields, we can solve the problem on-line with $O(\log m)$ time per data point: mediately after we have read the next value f_i, we can output the bounded Lipschitz rm of the sequence so far.

bsets of a small universe. The remaining possibilities are more "data-dependent", be-use they are only useful when the data are integers or $\hat{F}$ is small. In applications, $\hat{F}$ will t be too large, because it does not make sense to measure the image with a very high ecision. If $\hat{F}$ is large or the data are "real numbers", this approach can still be used if e is satisfied with an approximate value, because the data can be scaled and rounded small integers. Using the appropriate data structures (see Mehlhorn and Näher 1990) can achieve $O(m)$ storage and $O(n \log\log \hat{F})$ time. If the f_i are very small it might en be preferable to use a much simpler sequential search, which would lead to a time $O(\sum_i |f_i|)$ with $O(\hat{F})$ storage.

e off-line min approach. If we can afford to sort the values F_i by some method of stribution sort or radix sort, using $O(n+\hat{F})$ time or $O(n \log_n \hat{F})$ time, respectively, we n then use the algorithm for the off-line-min problem (cf. the three Americans' book, o et al. 1974).

This leads to a very simple algorithm with a complexity of $O(n\alpha(m, m))$, where $m, n)$ is the extremely slowly growing inverse function of the Ackermann function, Tarjan and van Leeuwen (1984). Using the improvement of Gabow and Tarjan (1985), ich makes the algorithm slightly more complicated, we can even achieve $O(n)$ time. ll these times are in addition to sorting.)

However, there is a slight catch: Usually, in the off-line min problem, we are given a quence of INSERT and DELETE-MIN operations. We are given the whole sequence in vance. We start by looking at the smallest element that was inserted and find out at at time it was deleted: this is the nearest DELETE-MIN that follows the respective SERT. We match off this INSERT and this DELETE-MIN, look at the second-smallest ement, and so on.

In our case, we also have DELETE-MAX operations, and these might interfere with e DELETE-MIN's: We cannot be sure whether an element for which we are looking r the corresponding DELETE-MIN operation should not really have been removed by a DELETE-MAX. There is no easy way to decide this, except by carrying out operations in sequence, which is just what we want to avoid.

We solve this dilemma by cutting the sequence of operations into *blocks* of length 4 i. e., we have $2m$ INSERT's, m DELETE-MIN's and m DELETE-MAX's.

Lemma 5. *Let x be the median of the $2m$ elements that are initially in the list bef the a block starts. Then all elements which are removed by DELETE-MIN's in the bl are smaller than x, and all elements which are removed by DELETE-MAX's in the bl are larger than x.*

It follows that we can process the DELETE-MIN's and DELETE-MAX's in a bl separately without caring about interference.

4 Open questions and further research

The algorithm can be extended compute the norm of a piecewise linear functions. a function with n linear pieces, it takes $O(n \log n)$ time. In addition, $O(n)$ equati involving at most $O(n)$ square roots must be solved. More precisely, if the function f s sign changes, the algorithm needs $O(n \log n + s^2)$ time. The details will be given i later version of this paper.

A question for further research is the minimization of the bounded Lipschitz dista under various transformations of the functions, like scalings of coordinates or cyclic ro tions of the definition interval. This is of interest when comparing two polygonal shap as described in the introduction. The arbitrary choice of the starting point for the para eterization of the boundary or a rotation of the polygons should not have an influe on their distance measure.

The definition of the bounded Lipschitz norm in section 2.3 can be extended natura to two- and higher-dimensional arrays (functions of more than one variable). Already two dimensions, it is an open question whether the bounded Lipschitz norm can computed more efficiently than by solving a network flow problem. Even with $m =$ i. e., with the distance coefficients $c((x_i, y_i), (x_j, y_j)) = |x_i - x_j| + |y_i - y_j|$, there is simple way to compute the metric. (Setting $m = \infty$ corresponds to deleting the sup vertex D in figure 4.)

Another open question is whether the heap operations in section 3.2 are really nec sary if nothing is assumed about the values F_i. Since we are not interested in the *seque* in which the elements are removed from the data structure but only in the *set* of eleme which are removed by DELETE-MAX and DELETE-MIN operations (actually, only their *sum*), it is not clear whether $\Omega(n \log \min\{m, n\})$ is a lower bound for the proble

References

A. V. Aho, J. E. Hopcroft, and J. D. Ullman: *The Design and Analysis of Computer Algorith* Addison-Wesley, 1974.

E. Arkin, L. P. Chew, D. P. Huttenlocher, K. Kedem, and J. S. B. Mitchell: An efficien computable metric for comparing polygonal shapes, in: *SODA'90, Proc. 1st ACM-SIA Symp. Discrete Algorithms, San Francisco, January 1990*, Society for Industrial and Appli Mathematics, Philadelphia 1990, pp. 129–137.

J. Atallah: A linear time algorithm for the Hausdorff distance between convex polygons, *Inform. Process. Lett.* **17** (1983), 207–209.

D. Atkinson, J.-R. Sack, N. Santoro, T. Strothotte: Min-max heaps and generalized priority queues, *Commun. ACM* **29** (1986), 996–1000.

Booth: *Flows in Series-Parallel Networks*, Ph. D. dissertation, Department of Computer Science, Princeton University, 1990.

Cox, H. Maitre, M. Minoux, C. Ribeiro: Optimal matching of convex polygons, *Pattern Recognition Letters* **9** (1989), 327–334.

N. Gabow and R. E. Tarjan: A linear-time algorithm for a special case of disjoint set union, *J. Comput. Syst. Sci.* **30** (1985), 209–221.

Hackh: Quality control of artificial fabrics, Forschungsbericht, Arbeitsgruppe Technomathematik, Universität Kaiserslautern, 1990.

Imai: A geometric fitting problem of two corresponding sets of points on a line, *IEICE Trans. Fund. Electron. Commun. Comput. Sci.* **E 74** (1991), 665–667 plus Hiroshi Imai's picture on p. 668.

Kantorovitch: On the translocation of masses, *Comptes Rendues (Doklady) de l'Académie des Sciences de l'URSS* **37** (1942), 199–201.

V. Kantorovich and G. Sh. Rubinshteĭn: On a space of completely additive functions (in Russian), *Vestnik Leningrad. Univ.* **13**, 7 (1958), 52–59.

Mehlhorn and S. Näher: Bounded ordered dictionaries in $O(\log\log N)$ time and $O(n)$ space. *Inform. Proc. Lett.* **35** (1990), 183–189.

Monge: Mémoire sur la théorie des déblais et des remblais, *Mémoires de l'Académie Royale des Sciences*, Paris, 1781, 666–704 and pl. XVIII–XIX; see also: Sur les déblais et les remblais, *Histoire de l'Académie Royale des Sciences*, Paris, 1781, 34–38.

Neunzert and B. Wetton: Pattern recognition using measure space metrics, Forschungsbericht Nr. 28, Arbeitsgruppe Technomathematik, Universität Kaiserslautern, 1987.

Ohmura, S. Wakabayashi, J. Miyao, and N. Yoshida: Improvement of one-dimensional placement in VLSI design (in Japanese), *Trans. Inst. Electron. Commun. Eng. Japan* **J73-A**, 11 (1990), 1858-1866.

T. Rachev: The Monge-Kantorovich mass transfer problem and its stochastic applications (in Russian), *Teor. Veroyatn. Primen.* **29** (1984), 625–653; English translation: *Theory Prob. Appl.* **29** (1984), 647–676.

T. Rachev: *Probability Metrics and the Stability of Stochastic Models*, Wiley, Chichester 1991.

E. Tarjan and J. van Leeuwen: Worst-case analysis of set union algorithms, *J. Assoc. Comput. Mach.* **31** (1984), 245–281.

is article was processed using the LaTeX macro package with LMAMULT style

On Nearest-Neighbor Graphs

Michael S. Paterson * *F. Frances Yao* †

Abstract

The "nearest neighbor" relation, or more generally the "k nearest neighbors" relation, defined for a set of points in a metric space, has found many uses in computational geometry and clustering analysis, yet surprisingly little is known about some of its basic properties. In this paper, we consider some natural questions that are motivated by geometric embedding problems. We derive bounds on the relationship between size and depth for the components of a nearest-neighbor graph and prove some probabilistic properties of the k-nearest-neighbors graph for a random set of points.

1 Introduction

Neighborhood-preserving mappings from a set of points in space to a regular data array are useful for speeding up many computations in physical simulation. For example, complicated many-body interactions between particles can often be approximated by the dominant forces exerted by near neighbors. If the data for nearby particles is stored with close address indices in an array, one can take advantage of fast vectorized operations in contiguous memory to perform such computations. The question is, therefore, whether such neighborhood-preserving mappings from a point set to a regular array always exist. In this paper, we attempt to answer this question by first studying the performance of a known mapping scheme, called the monotone logical grid (MLG), and then establishing some combinatorial properties of the nearest-neighbor graph that are relevant to geometric embedding.

The outline of the extended abstract is as follows. In Section 2, we introduce some basic concepts and notation. In Section 3, we describe the MLG mapping and analyze its performance. We give in Section 4 some examples of point sets which are hard to embed in a grid. In Section 5, we prove a polynomial upper bound on the size of a component of the NNG in terms of its diameter. Some concluding remarks and open problems are given in Section 6.

---

*Department of Computer Science, University of Warwick, Coventry, CV4 7AL, England. The work was done while this author was visiting Xerox Palo Alto Research Center. He is partially supported by the ESPRIT BRA Program of the EC under contract 7141 (ALCOM II).

†Xerox Palo Alto Research Center, 3333 Coyote Hill Road, Palo Alto, CA 94304, USA.

Preliminaries

this section, we introduce some notation and basic facts about the nearest-ighbor graph.

Let $V = \{v_1, v_2, \ldots, v_n\}$ be a set of points in R^ℓ. The *nearest neighbor* of v_i is point v_j, $j \neq i$, with minimum Euclidean distance from v_i. To make the nearest ighbor unique we choose the point v_j with maximum index in case of ties, and note it by $\mathrm{nn}(v_i)$. For any v, we define the directed edge $e(v) = \langle v, nn(v) \rangle$. The arest-neighbor graph of V, denoted by NNG(V), is the directed graph $\langle V, E \rangle$ where $= \{e(v) | v \in V\}$. It is easy to verify that the graph NNG(V) has the following operties:

- Along any directed path in NNG(V), the edges have non-increasing lengths.
- The only cycles in NNG(V) are 2-cycles. For $|V| \geq 2$, each weakly connected mponent C of NNG(V) contains exactly one 2-cycle. This pair of vertices is called ie *bi-root* of C.
- For a point set V in two dimensions, NNG(V) is a planar graph. Furthermore, vo edges incident into a vertex v must meet at an angle of at least 60°, hence the -degree of a vertex is at most 6.
-) NNG(V), when considered as an undirected graph, is a subgraph of DT(V) (the elaunay graph of V) and can be constructed in asymptotically the same time as T(V). Thus, the NNG of n points can be constructed in $O(n \log n)$ time for two imensions (see [5]), in $O(n^{4/3+\epsilon} + k \log n)$ time for three dimensions where k is the mplexity of DT(V) (see [6]), and in $O(n^{\lfloor (\ell+1)/2 \rfloor})$ time for dimensions $\ell > 3$ (see]).

We can generalize NNG(V) to k-NNG(V), the *k-nearest-neighbors* graph of V, y introducing k edges from a vertex to its k nearest neighbors. At present these raphs have not been studied extensively, and few of their combinatorial properties re known.

3 Monotone Logical Grid

oris[2] proposed a data structure, called the *Monotone Logical Grid* (MLG), as a ay for storing and indexing a set of points in R^ℓ. The MLG maps a set S of n oints in R^ℓ to an ℓ-dimensional array A of size $n^{\frac{1}{\ell}} \times \cdots \times n^{\frac{1}{\ell}}$. When $\ell = 1$, the ILG is simply a sorted linear array. For $\ell \geq 2$, we first sort the points into $n^{\frac{1}{\ell}}$ uckets of equal size by the values of their x_1-coordinates. Then the i-th bucket is tored recursively as an $(\ell - 1)$-dimensional MLG in the subarray of A corresponding $x_1 = i$.

It was observed in [2] that, on average, the MLG seems to preserve near-neighbor elations rather well. In other words, points that are close in real space tend to have lose addresses in the MLG array. However, these claims were based on experimental data and no precise statements or proofs were given. By analyzing the performanc of MLG for a random point set, we can indeed make the following statement i support of the empirical results. The proof is omitted here.

Theorem 1 *Let $V = \{v_1, v_2, \ldots, v_n\}$ be n points chosen independently and un formly from a square of size $m \times m$ where $n = m^2$, and let k be fixed. For any i, th probability that, in an MLG for V, all k nearest neighbors of v_i lie within x-offs $O(\sqrt{k})$, and y-offset $O(\sqrt{k} + \sqrt{m})$ from the address of v_i is $1 - o(1)$.*

In spite of this good expected performance, it is not hard to construct exampl where the MLG leads to large dilation for almost all edges in NNG(V).

Example. (See Figure 1.) Let V consist of $2m$ subsets $V_1, V_2, \ldots, V_{2m}$, where eac subset has $m/2$ points, and let ϵ be chosen such that $1/(2m) > \epsilon > 0$. For $i \geq 1$, th points of V_{2i} are equally spaced on the vertical segment $(i-\epsilon)\times[i-\frac{1}{2}, i]$, and the poin of V_{2i+1} are equally spaced on the adjacent vertical segment $(i+\epsilon)\times[i-\frac{1}{2}, i]$. For V_1 we place its points uniformly on the interval $\epsilon \times [0, \frac{1}{2}]$). Thus, for $1 \leq i \leq m-1$, th points of V_{2i} and V_{2i+1} are matched pairwise as nearest neighbors. However, unde the MLG mapping, the points of V_{2i} will have y-indices $m/2 + 1, m/2 + 2, \ldots, m$ while the points of V_{2i+1} will have y-indices $1, 2, \ldots, m/2$. Thus the y-offset in th MLG for nearest neighbors is $m/2$ for all points except those in V_1 and V_{2m}.

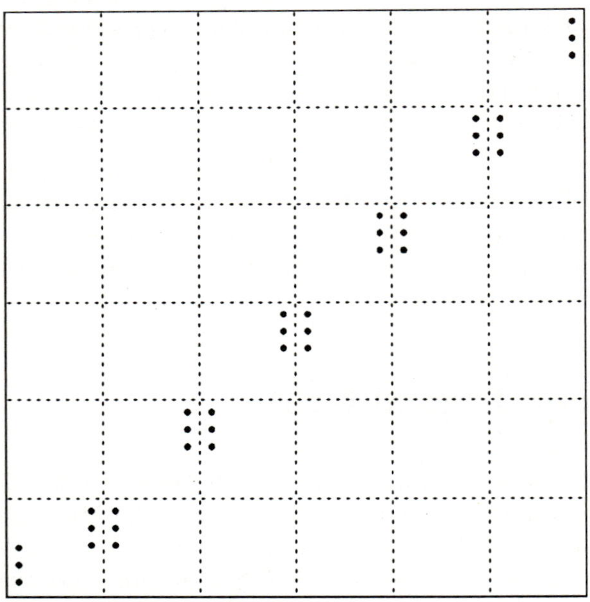

Figure 1: Monotone Logical Grid with large dilation

One natural strategy for embedding NNG(V) into a grid is to embed the in dividual components of NNG(V) separately. This could certainly work well if al components of NNG(V) were small. At present, we have little knowledge regard ng the distribution of component size for a random NNG(V). However, one can rove the following result on the expected number of components in NNG(V) (cf], Lemma 1).

heorem 2 *The expected number of components in NNG(V) for a uniform random oint set V is at least* $0.31 \times |V|$. □

4 Dilation of Embeddings for the NNG

he *dilation* of an embedding of a k-NNG in a graph G is the maximum over all dges $\langle u, v \rangle$ in the k-NNG of the path length in G between the images of u and v. It s easy to see that any embedding of a rapidly branching tree in a finite-dimensional rray must have a large dilation. We begin by constructing a 2-NNG which contains uch a tree.

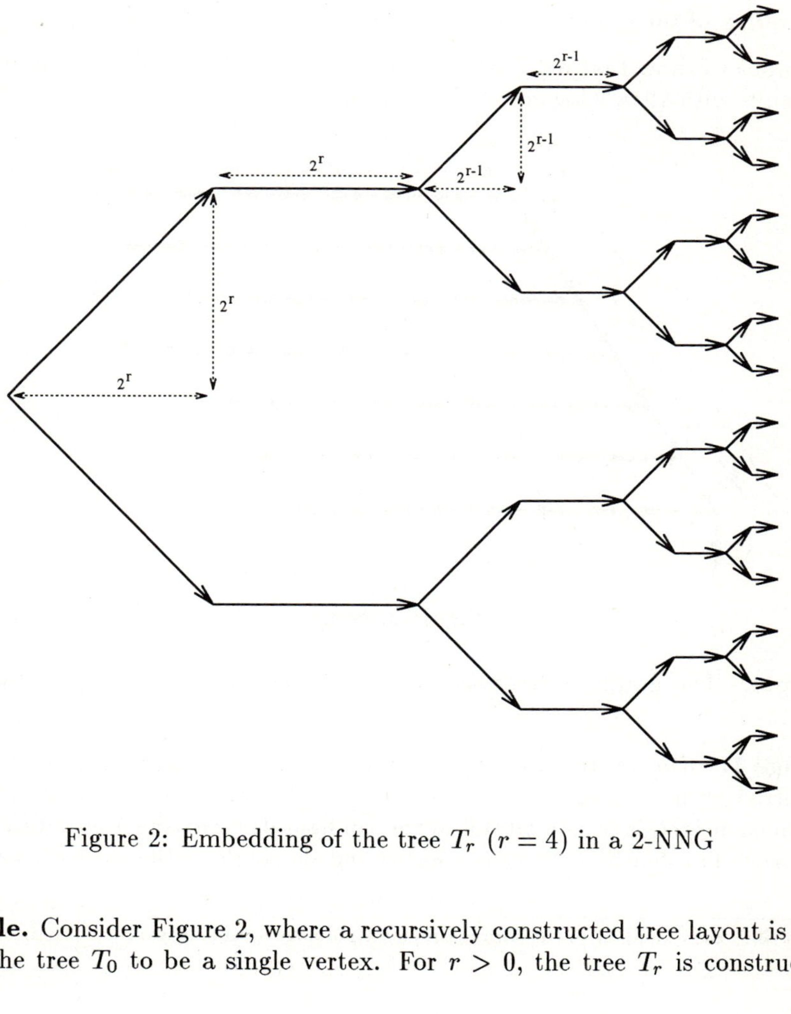

Figure 2: Embedding of the tree T_r ($r = 4$) in a 2-NNG

Example. Consider Figure 2, where a recursively constructed tree layout is shown. Define the tree T_0 to be a single vertex. For $r > 0$, the tree T_r is constructed as follows. The root of T_r has two children, w_1 and w_2, placed with x-offset 2^r an y-offsets 2^r and -2^r respectively from the root. Each w_i is connected by a horizont edge of length 2^r to the root of a copy of T_{r-1}. It is clear from the diagram th the edges from any vertex v of T_r go to the nearest q neighbors of v for some $q \leq$ Thus T_r is a subgraph of the 2-NNG for the underlying point set.

Theorem 3 *For all n, there exists a planar set V of n vertices such that any em bedding of 2-NNG(V) in an ℓ-dimensional array has dilation $\Omega(n^{1/\ell}/\log n)$. particular, for $\ell = 2$ we obtain a dilation of $\tilde{\Omega}(n^{1/2})$.* [1]

Proof. For any set of n elements of a ℓ-dimensional array, almost all pairs elements are Manhattan-distance $\Omega(n^{1/\ell})$ apart. For the 2-NNG in the Example w constructed, the maximum (undirected) path length in the graph between any tw vertices is $O(\log n)$. The result follows.

For an NNG, we define the *depth* $d(v)$ of a vertex v to be the path length in th NNG from v to a vertex in the bi-root. The *depth* of an NNG is the maximum dept of any vertex of the graph.

In order to show that a 1-NNG may require considerable dilation, we construc an example with $\Omega(m^5/\log m)$ vertices and depth $O(m)$.

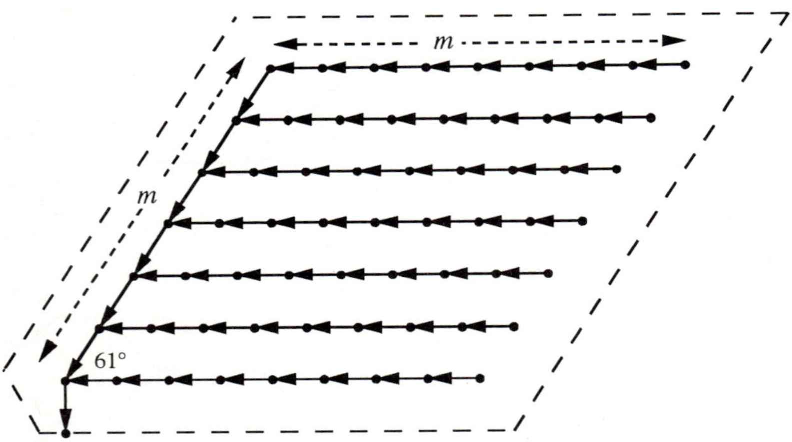

Figure 3: Stage 1

Example. The graph is described in four stages. Let m be any suitably larg integer.

Stage 1. $\Theta(m^2)$ vertices are arranged in a diamond array as shown in Figure The dotted outline indicates an *exclusion zone*. No vertex outside this zone can b the nearest neighbor of any vertex in the configuration, except the indicated 'roo of the tree. The depth of the tree is $\Theta(m)$ and the angle at the corner near the roo is 61°.

[1] The tilde accent indicates the suppression of a polylogarithmic factor.

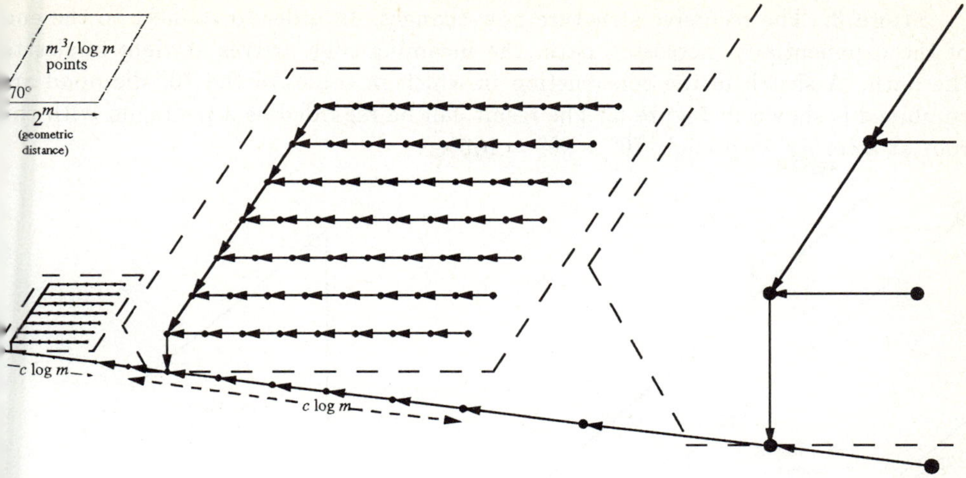

Figure 4: Stage 2

Stage 2. In this stage a sequence of $\Theta(m/\log m)$ of the diamonds from Stage 1 are connected using a chain of edges of exponentially growing length. The result is a configuration of $\Theta(m^3/\log m)$ vertices, which can be enclosed in a diamond shape as shown with the root next to a corner with angle 70°. See Figure 4.

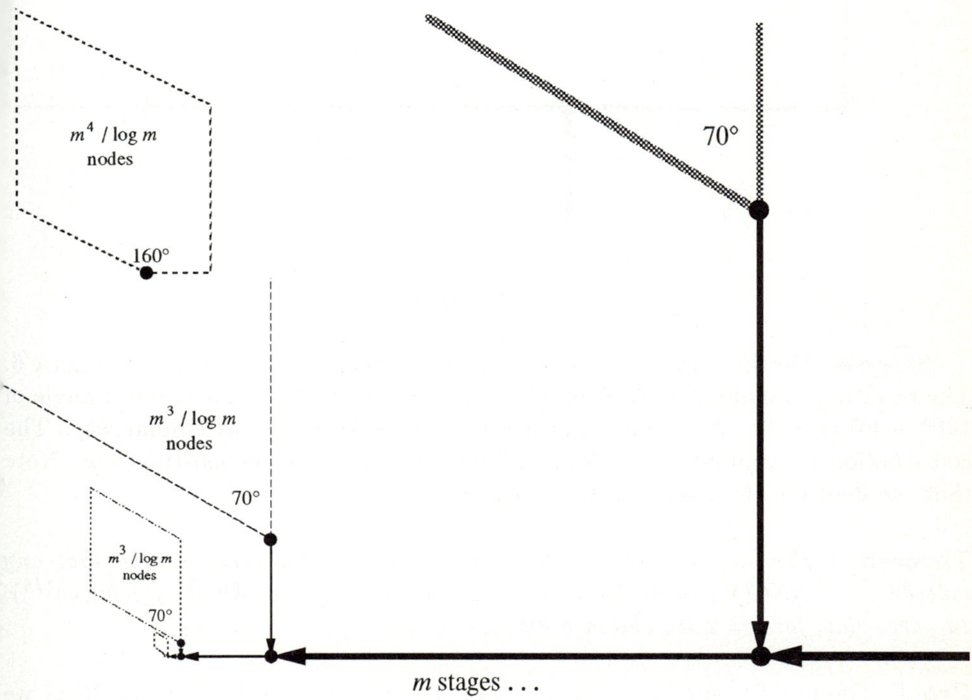

Figure 5: Stage 3

Stage 3. The recursive structure now changes. In order to connect to the end of the exponentially increasing path, the incoming edge arrives at right angles to the path. A sketch of the construction in which m copies of the $70°$ diamond are combined is shown in Figure 5. The result can be regarded as a pentagon with the root at a corner with angle $70° + 90° = 160°$.

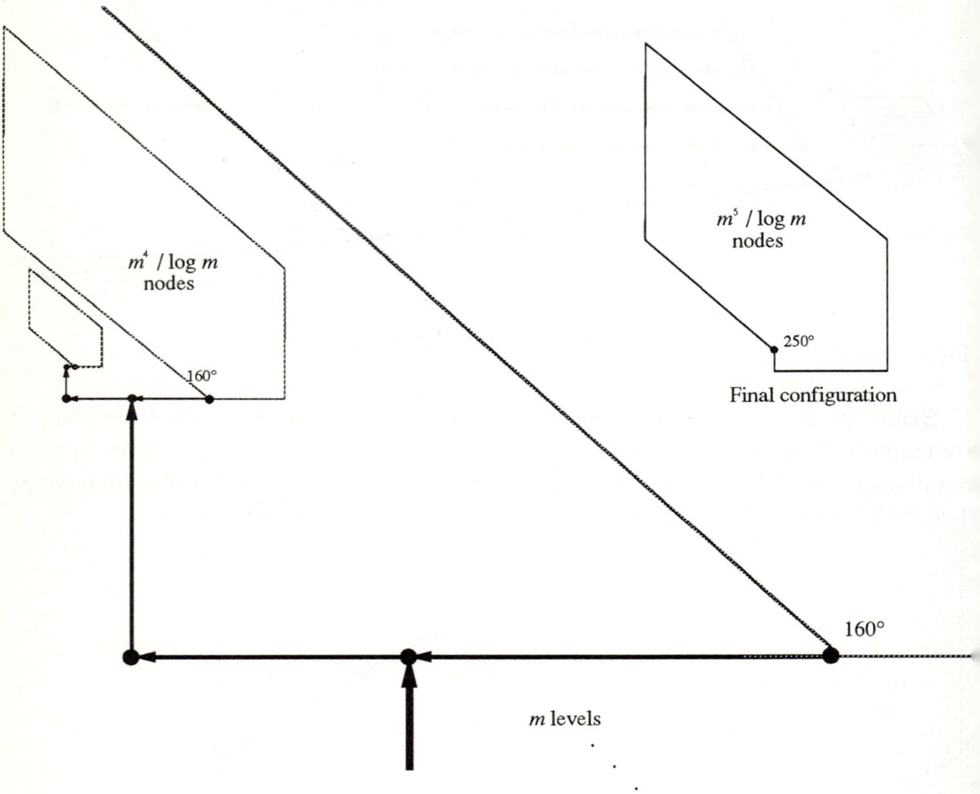

Figure 6: Stage 4

Stage 4. The final stage is similar to the previous one and is sketched in Figure 6. The resulting configuration of $\Theta(m^5/\log m)$ vertices has its root at a reflex angle of $160° + 90° = 250°$. We cannot continue to another stage in any similar way. The construction is terminated by adding a final vertex close to the existing root. Note that the depth of the NNG constructed is $\Theta(m)$.

Theorem 4 *For any n, there exists a planar set V of n vertices such that any embedding of NNG(V) in an ℓ-dimensional array has dilation $\Omega(n^{1/\ell}/(n\log n)^{1/5})$. In particular, for $\ell = 2$ we obtain a dilation of $\tilde{\Omega}(n^{0.3})$.*

Proof. The proof is similar to the proof of the previous theorem. In the NNG we constructed, the maximum (undirected) path length in the graph between any two vertices is $O(m) = O((n\log n)^{1/5})$. The result follows. □

Size and Depth of the NNG

In this section we show that the $O(D^5)$ size for an NNG of depth D, which we obtained above, cannot be increased to an exponential size such as we obtained for -NNG's. We prove an upper bound of $O(D^9)$.

It is convenient to define the depth $d(e)$ of an edge $e = e(v)$ to be $d(v)$. The length of an edge $e = \langle u, v\rangle$ is denoted $||e||$ or $||u - v||$. For any $v \in V$ define $C(v)$ to be the open circle with center v and radius $||e(v)||/2$.

Lemma 1 *For all v_i, v_j, $C(v_i) \cap C(v_j) = \emptyset$.*

Proof. If $C(v_i) \cap C(v_j) \neq \emptyset$, then

$$||v_i - v_j|| < ||e(v_i)||/2 + ||e(v_j)||/2 \leq \max\{||e(v_i)||, ||e(v_j)||\}.$$

This inequality contradicts the nearest-neighbor property. □

Theorem 5 *If $G = \langle V, E\rangle$ is a connected NNG with depth D then $|V| = O(D^9)$.*

Proof. Let $R = \{z_1, z_2\}$ be the bi-root of G. We will normalize distances by taking $||z_1 - z_2|| = 1$. Let O be the midpoint of the edge (z_1, z_2). For $i \geq 0$, define B_i to be the open disk with radius 8^i and center O. For any vertex v, define $b(v)$ to be the minimal i such that B_i contains all vertices on the directed path from v to R, and let $S_i = \{v | b(v) = i\}$. The proof continues with three lemmas.

Lemma 2 $|S_i| = O(D^2)$.

Proof. Suppose $v \in S_i$. Since v or some vertex on the path from v to R is at a distance at least 8^{i-1} from the origin, and each edge on this path is of length at most $||e(v)||$, we have $||e(v)|| \geq (8^{i-1} - 1/2)/d(v) \geq (8^{i-1} - 1/2)/D$. The disk $C(v)$ is contained within a circle of radius $8^i + ||e(v)||/2$ centered at O, and $||e(v)|| \leq \min\{||v - z_1||, ||v - z_2||\} < 8^i + 8^{-i-1}$. By Lemma 1 the disks $C(v)$ are disjoint for all $v \in S_i$ and so a comparison of areas yields the inequality

$$((8^{i-1} - 1/2)/2D)^2 \cdot |S_i| \leq ((8^i + (8^i + 8^{-i-1})/2)^2.$$

Hence, $|S_i| = O(D^2)$. □

Lemma 3 *For all $i > 0$, there are at most six directed edges from $\bigcup_{j>i} S_j$ to $\bigcup_{j<i} S_j$.*

Proof. Any such edge comes from outside B_i to inside B_{i-1}. Consider any pair of edges $\langle v_1, w_1\rangle$, $\langle v_2, w_2\rangle$ such that vertices v_1, v_2 are outside B_i, while w_1, w_2 are within B_{i-1}, and $\angle v_1 O v_2 = \psi$ say. Without loss of generality we can assume that $||v_1|| \geq ||v_2||$. Suppose that $||v_1|| = \theta t$ and $||v_2|| = t$ where $\theta \geq 1$.

As w_1 is a nearest neighbor of v_1, we have

$$1 \leq \frac{||v_1 - v_2||}{||v_1 - w_1||} < \frac{t\sqrt{1+\theta^2 - 2\theta\cos\psi}}{\theta t - 8^{i-1}} \leq \frac{\sqrt{1+\theta^2-2\theta\cos\psi}}{\theta - 1/8},$$

since $t \geq 8^i$. Therefore

$$0 > \theta - 1/8 - \sqrt{1+\theta^2 - 2\theta\cos\psi} \geq 7/8 - \sqrt{2(1-\cos\psi)}$$

since the middle expression is monotonically increasing in θ. Hence $\cos\psi < 79/128$ and so $\psi > 2\pi/7$. The lemma follows since, if there were seven or more edges from $\bigcup_{j>i} S_j$ to $\bigcup_{j<i} S_j$, then there would be two outer vertices v_1, v_2 subtending an angle of $2\pi/7$ or less at the origin. □

The *extension* operation on a set E of edges consists of deleting one edge $e \in E$ and replacing it by the set of predecessor edges of e. We associate with any set E of edges the *characteristic* of E, denoted $\chi(E)$, which is the 7-tuple $\langle d_1, \ldots, d_7\rangle$ where d_i is the depth of the edge with the ith smallest depth among edges of E. If E has fewer than seven edges then some of the last components of $\chi(E)$ take the value ∞. A simple property of extensions is stated in the following lemma.

Lemma 4 *If the set E' is the result of applying one or more extension operations to E then $\chi(E') \succeq \chi(E)$, where '$\succeq$' denotes lexicographic order.* □

Define $E_i = \{\langle v, v'\rangle | b(v) \geq i, b(v') \leq i\}$. Since E_{i+1} can be obtained by applying a sequence of extension operations to E_i, we have $\chi(E_{i+1}) \succeq \chi(E_i)$.

If $\chi(E_{i+1}) = \chi(E_i)$ then every edge represented in $\chi(E_{i+1})$ must also have been a member of E_i. By Lemma 3 there can be at most six edges in $E_i \cap E_{i+1}$, and so in this case the characteristic of E_{i+1} has at most six finite components. Hence each edge of E_{i+1} is represented in $\chi(E_{i+1})$. The equality of the characteristics therefore implies that $E_i = E_{i+1}$. Suppose $v \in S_i$, i.e. $b(v) = i$. Then $e(v) \in E_i = E_{i+1}$, and so $b(v) \geq i+1$. This contradiction proves that $S_i = \emptyset$ when $\chi(E_{i+1}) = \chi(E_i)$.

Thus at each step in the sequence $\chi(E_1), \chi(E_2), \ldots$, either the characteristic increases in the lexicographic order, or else remains the same in which case S_i is empty. Since the characteristic can take at most $O(D^7)$ values, by Lemma 2 the total number of vertices in G is $O(D^7 \cdot D^2)$, and the Theorem is proved. □

Recently, Eppstein (private communication) was able to refine the above argument to obtain an improved $O(D^6)$ upper bound.

6 NNG in Higher Dimensions

We can prove a generalization of Theorem 5, that, in any fixed dimension ℓ, the size of a connected NNG is bounded above by a polynomial in its depth D. The upper bound is of the form $O(D^{\ell+\tau(\ell)})$ where $\tau(\ell)$ is the number of disjoint unit hyperspheres which can touch a given unit hypersphere, i.e., the so-called *kissing* umber in dimension ℓ. Conway and Sloane [4] give a fascinating history of the kiss-g number question. Currently, the exact value of $\tau(\ell)$ is known only for dimensions $= 1, 2, 3, 8, 24$. For general ℓ, there are exponential upper bounds and lower bounds $^{\ell} \leq \tau(\ell) \leq \beta^{(\ell)}$ where $\alpha \approx 2^{0.207}$ and $\beta \approx 2^{0.401}$.

For the purpose of proving our upper bound, it is more convenient to use the ollowing equivalent definition of the kissing number: $\tau(\ell)$ is the number of points hat can be placed on the surface of sphere in ℓ dimensions, so that the angular eparation between any two points is at least $60°$.

heorem 6 *If $G = \langle V, E\rangle$ is a connected NNG with depth D in ℓ dimensions then* $V| = O(D^{\ell+\tau(\ell)})$.

roof. The discussions in Section 5 generalize from dimension two to higher dimen-ions without any difficulty, once we change "circles" to "hyperspheres", "disks" to balls", and the "characteristic" of an edge set from a 7-tuple to a $(\tau(\ell)+1)$-tuple. he statements of Lemmas 1 and 4 remain the same, while Lemma 2 now states that $S_i| = O(D^{\ell})$, and Lemma 3 states that the number of directed edges from $\bigcup_{j>i} S_j$ o $\bigcup_{j<i} S_j$ is at most $\tau(\ell)$ for all $i > 0$. The proofs of the lemmas go through with ittle modification. □

7 Conclusion and Open Problems

Ve have explored some of the combinatorial properties of k-NNG's. Whereas for ≥ 2, a k-NNG can have components of size exponential in the diameter, for $k = 1$ he size is limited by a polynomial. We have also constructed NNG's which require lilation $n^{\Omega(1)}$ for embeddings into an array of any finite dimension.

For uniform random distributions of points in the plane, Theorem 2 gives the xpected number of connected components of the NNG. It would be of interest to stimate some further properties of the NNG, such as the distribution of component izes and the expected size of the largest component. Similar questions arise for he k-NNG and, in particular, we would like to know the minimum value of k (as a unction of n) for which the k-NNG can be expected to have a 'giant' component. An understanding of the probabilistic properties of the k-NNG might be exploited n designing embeddings with improved computational characteristics.

Our knowledge of the combinatorial properties of NNG's is still very limited. We know of no graph-theoretic characterization of NNG's and suspect that deciding, for a given graph, whether or not it is the NNG for some set of points may be a hard problem. So far we know even less about k-NNG's for $k > 1$.

The gap between the exponent of 5 in our construction in Section 4 and the exponent of 9 in Theorem 5 has been partially closed by Eppstein's improvement. Detailed estimates for the corresponding exponents in higher dimensions have not been made.

An understanding of the relationship between the dimension of a space and the combinatorial or statistical properties of the k-NNG's would be of use in classificatio theory. For example, if a set of points were representable as a fairly well-behave distribution over a manifold of unknown dimension, could we estimate the dimensio of the manifold from statistical properties of the k-NNG?

Acknowledgement

The authors wish to thank Marshall Bern, Dan Greene and Shang-Hua Teng fc helpful discussions.

References

[1] M. Bern, "Two probabilistic results on rectilinear Steiner trees," *Algorithmic* **3**, 1988, 191-204.

[2] J. Boris, "A vectorized 'near neighbors' algorithm of order N using a monotoni logical grid," *Journal of Computational Physics* **66**, 1986, 1-20.

[3] B. Chazelle, "An optimal convex hull algorithm and new results on cuttings," *Proc. 32nd IEEE Symp. Foundations of Computer Science*, 1991, 29-38.

[4] J. H. Conway and N. J. A. Sloane, *Sphere Packings, Lattices and Groups* Springer-Verlag, New York, 1988.

[5] H. Edelsbrunner, *Algorithms in Combinatorial Geometry.* Springer-Verlag 1987.

[6] J. Matoušek, "Reporting points in halfspaces," *Proc. 32nd IEEE Symp. Foundations of Computer Science*, 1991, 207-215.

A tail estimate for Mulmuley's segment intersection algorithm*

JIŘÍ MATOUŠEK[†] RAIMUND SEIDEL[‡]

Abstract

We analyze the distribution of the running time of Mulmuley's randomized algorithm for computing the intersections of n given segments in the plane. Its expectation has been known to be $O(n \log n + k)$, where k is the number of intersections. We show that for values of k not too close to n ($k \geq Cn \log^{15} n$ for a large enough constant C), the running time is sharply concentrated around the expected value; e.g., the probability that the expected value is exceeded more than twice is $O(n^{-c})$, where c can be an arbitrarily large constant (its choice determines the value of C needed in the assumption). Our proof uses an isoperimetric inequality for permutations.

Introduction and statement of results

computational geometry many of the algorithms developed in the last few years have n randomized algorithms. Their performance analysis usually produces just some upper ind on their expected running times (the expectation is over internal random choices de by the algorithm, and the bound holds for any specific input; in this sense it is a st-case bound), and various elegant methods have been developed for such a kind of lysis (see for instance [CS89], [Sei91]).

Knowing the expected running time, the next natural question seeks a bound for the bability that the expected running time will be exceeded by a substantial amount. Such stions have also been considered in several papers (e.g. [MSW92],[AGMKW91]). In this er we prove another result of this kind, namely for Mulmuley's segment intersection orithm. This is an elegant randomized algorithm that, given a set of n segments in the ne, computes all their intersection in (optimal) expected time $O(n \log n + k)$, where k he number of intersections. We will describe this algorithm in section 2. For its tail imates, we prove the following:

eorem 1.1 *For any constant c, there is a constant C, so that the probability that the nning time of Mulmuley's segment intersection algorithm deviates from its expected value more than $CK^{1/2}n^{1/2}\log^{7.5} n$ (where $K = k + n$) is bounded by $O(n^{-c})$. In particular,*

---

*This research by the first author has been supported by International Computer Science Institute at keley, CA, USA. Research by the second author was supported by NSF Presidential Young Investigator ard CCR-9058440.

[†]Department of Applied Mathematics, Charles University, Malostranské nám. 25, 118 00 Praha 1, choslovakia. Email: MATOUSEK@CSPGUK11.BITNET

[‡]Computer Science Division, University of California, Berkeley, CA 94720 USA. Email: lel@cs.berkeley.edu

for the number of intersections k exceeding $n \log^{15} n$ *by a suitable multiplicative cons factor, the probability of surpassing the expected running time more than twice is* $O(n^{-}$

Remarks: Our interest in the tail analysis of this particular algorithm was kin by the observation that this algorithm appeared to be impervious to previous meth of obtaining tail estimates. In particular the so-called Chernoff bounds did not ap Moreover, even the standard trick of repeatedly stopping the algorithm if it runs " long" and restarting it (see [AGMKW91]) does not apply. For this method requires the expected running time be konwn in advance, which is usually not the case with algorithm, since k, the number of intersections, is usually not known in advance.

Some of the logarithmic factors in Theorem 1.1 could be removed by refining the pre methods (and making the presentation somewhat more complicated). However, it se that the present methods are not suitable for obtaining good tail estimates for all value K. We have to leave such estimates as an interesting open problem.

For our proof, we use the so-called isoperimetric inequality. This tool is well-know combinatorics and other branches of mathematics, but only recently it has been app in computational geometry (by T. Hagerup, [Hag90]). This allows to deal with the rat complicated interactions of the segments in Mulmuley's algorithm. Hopefully the b method as well as some of the special tricks introduced in this paper will allow to get g tail estimates for many more randomized algorithms.

We should mention at this point that Mehlhorn, Sharir, and Welzl [MSW92] have recer managed to give a tail estimate for the space requirement of Mulmuley's algorithm

2 Mulmuley's algorithm

Let S be a set of n straight line segments in the plane. We are interested in finding intersecting pairs of segments in S.

Non-trivial deterministic algorithm for solving this problem were given by Bentley Ottmann [BO79], Chazelle [Cha86], and Chazelle and Edelsbrunner [CE88]. The last of th methods achieved optimal $O(k+n \log n)$ worst case running time. Rather simple randomi algorithms with $O(k + n \log n)$ expected running time were independently developed Mulmuley [Mul88] and Clarkson and Shor [CS89].

Here we briefly describe Mulmuley's algorithm and some aspects of the analysis of running time. For a complete analysis of its expected running time the reader is referred [Mul88] or [Sei91].

For the sake of ease of presentation let us assume we are dealing with a set S o segments that is non-degenerate in the sense that no two segments of S have the sa endpoint, no three segments intersect in a common point, no two segment endpoints h the same x-coordinate, and that no segment endpoint lies in the relative interior of so other segment. As usual such non-degeneracy could be simulated by standard perturbati methods [Ede87, pp. 185], or also the algorithm could easily be modified so that none these assumptions are necessary.

Mulmuley's algorithm does more than just determine which pairs of segments in S int sect. It constructs what we call the *trapezoidal decomposition* (or trapezoidation) induced S. This decomposition $T(S)$ can be intuitively defined as follows: First draw a sufficien large axis-parallel rectangle frame F that contains in its interior all segments of S. N draw all segments of S in the rectangle F. Finally, from each intersection point and fr each segment endpoint draw its *vertical extensions*, i.e. start drawing two vertical rays, c g up, the other going down, that extend until they hit a segment of S or the boundary (see Figure 1). Thus F is decomposed into trapezoids that each have two vertical sides of which can have length 0). Using a sweep argument it is not hard to prove that $\mathcal{T}(S)$ ains exactly $3(n+k)+1$ trapezoids; thus, when viewed as a planar graph, $\mathcal{T}(S)$ has $+ n)$ faces, edges, and vertices.

et us define for any subset $R \subset S$ the trapezoidal decomposition $\mathcal{T}_S(R)$ in a similar way ollows: Draw all segments of R in the rectangle frame F, and for each intersection point gments in R as well as for each endpoint of a segment in S (note: *in* S) draw its vertical nsions. But now the rays that form the vertical extensions extend only until they hit a nent of R or the boundary of the rectangle F (see Figure 2). The decomposition $\mathcal{T}_S(R)$ itions F into $2n+r+3k_R+1$ trapezoids, where $r = |R|$ and k_R is the number of pairs tersecting segments in R, and thus $\mathcal{T}_S(R)$ as a planar graph has $O(n+k_R)$ faces, edges, vertices.

Mulmuley's algorithm for computing $\mathcal{T}(S) = \mathcal{T}_S(S)$ is very simple: First compute $\mathcal{T}_S(\emptyset)$ then insert the segments of S into the trapezoidation in random order to compute $R)$ for $R = \emptyset, \{s_1\}, \{s_1, s_2\}, \ldots, \{s_1, s_2, \ldots, s_n\}$, where $s_1, \ldots, s_n$ is an enumeration of segments of S in a random order.

We still have to specify how a segment s is introduced into the current trapezoidation. rder to do this we need to give more detail about the representation of trapezoidations. muley chooses a somewhat idiosyncratic representation that essentially works as follows: sider $\mathcal{T}_S(R)$, and consider some segment $s \in R$. Assume s is intersected by i other nents in R. Thus s is partitioned into $i+1$ *pieces.* Conceptually Mulmuley makes each e of each segment in R into a narrow zero-width face, and obtains this way from $\mathcal{T}_S(R)$ ane graph that we denote by $G_S(R)$. Figure 3 should make the idea clear. Note that $R)$ has the same number of vertices as $\mathcal{T}_S(R)$ and thus it also has complexity $O(n+k_R)$. note that in the graph $G_S(R)$ faces that correspond to trapezoids of $\mathcal{T}_S(R)$ have at most edges around them, only the zero-width faces can have more than a constant number of es around them.

How does one now introduce segment s into $G_S(R')$ to obtain $G_S(R)$, where $R' = R \backslash \{s\}$? orks in two phases (that of course could be combined into one). In the first phase one rmines which faces and edges of $G_S(R')$ are intersected by s. In the second phase new faces of $G_S(R)$ are created. This involves splitting the faces of $G_S(R')$ that are rsected by s, creating the zero-width faces for the pieces of s, introducing the vertical ensions from the intersection points of s with the other segments of R, and merging s that are separated by vertical edges that are not part of vertical extensions any more ause of the introduction of s. We omit the details of the second phase.

The first phase can simply be done as follows: Since the two endpoints of s are vertices $G_S(R')$ already, we can determine in constant time, say, the leftmost face of $G_S(R')$ that ntersected by s. Now we "thread" s through $G_S(R')$ in the usual way, similar to the remental line arrangement construction algorithm [EOS86, CGL85]: We walk along the ment s; assume we just entered a face f through some edge e; we determine through ich edge the segment s leaves f and which new face the segment enters by simply testing edges of f (say, in clockwise order starting after e). Now we have reached a new face, and repeat. Obviously, this procedure can also tell when we have reached the right endpoint s.

What is the time necessary for introducing segment s? It is not hard to see that the t of phase one dominates the cost of phase two. Thus it suffices to consider just the cost phase one. This cost is clearly the sum of the degrees of all the faces of $G_S(R')$ that

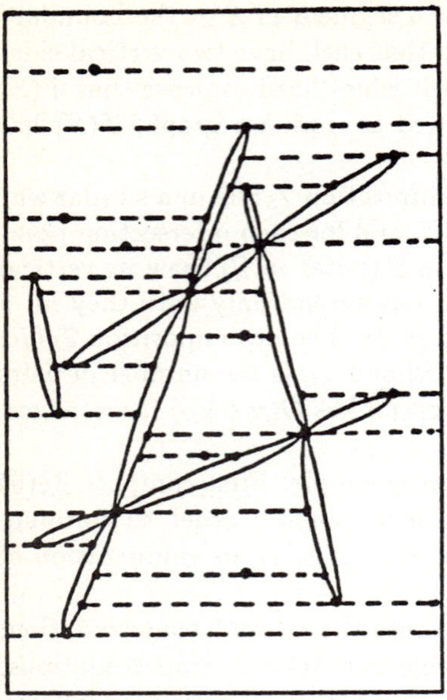

Figure 3: $G_S(R)$ with its zero-width faces

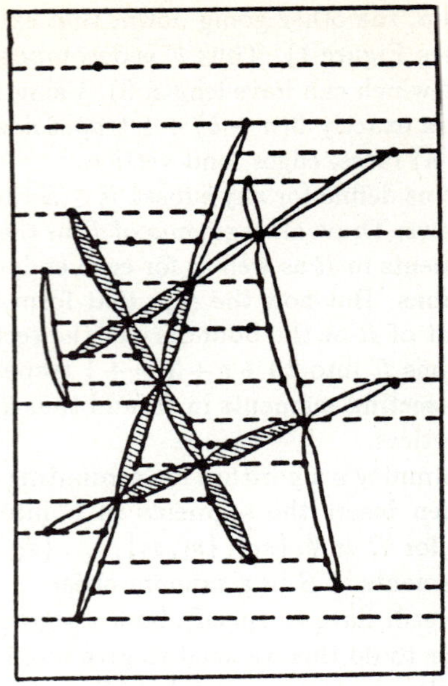

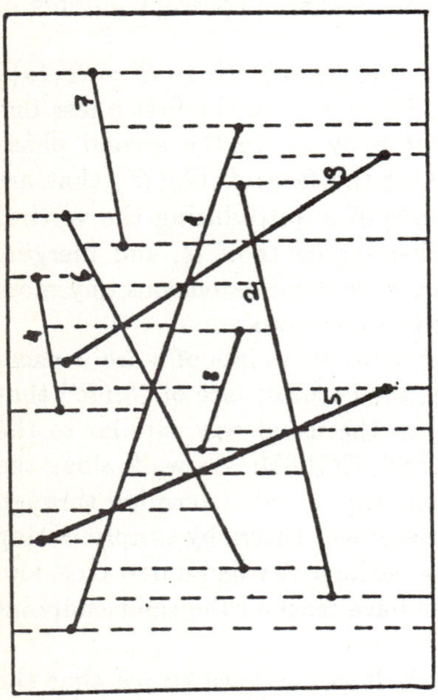

Figure 1: Trapezoidation $\mathcal{T}(S)$

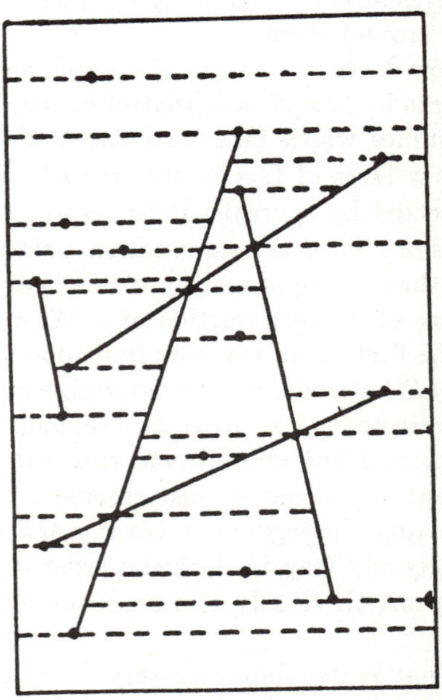

intersected by s (here the degree of a face f of $G_S(R')$, for short $deg(f, G_S(R'))$, is the ber of edges of $G_S(R')$ incident to f).

The cost of inserting s can also (maybe more conveniently) be expressed in terms of R): It is $\sum_{f \in \mathcal{P}_R(s)} deg(f, G_S(R))$, where $\mathcal{P}_R(s)$ is the set of all zero-width faces of R) that either derive from pieces of s in $G_S(R)$ or from those pieces of other segments that are incident to intersection points with s (see Figure 4). In other words, the cost serting s to obtain $G_R(S)$ is proportional to the number of vertical extension of $G_R(S)$ abut on a piece in $\mathcal{P}_R(s)$.

This finishes our description of Mulmuley's algorithm.

The tail estimate

Isoperimetric inequality

proof is based on the so-called *concentration of measure* phenomenon. This proof nique has been used with great success in the geometry of Banach spaces (see e.g. book [MS86]), and later on in random graphs and other branches of combinatorics. mples and much more background material can be found e.g. in a survey paper by lobás [Bol87], to which we also refer for the results quoted below. One of the tools ch this theory offers are *isoperimetric inequalities*. A basic example of an isoperimetric uality is the following:

Consider the set Ω of all permutations on $[n] = \{1, \ldots, n\}$, equipped by the standard form probabilistic measure μ (i.e. every sigleton set has measure $1/|\Omega|$), and define the ance d of two permutations $\pi_1, \pi_2 \in \Omega$ by

$$d(\pi_1, \pi_2) = |\{i \in [n];\ \pi_1(i) \neq \pi_2(i)\}|.$$

$A \subseteq \Omega$ be some set of "good" permutations with $\mu(A) \geq 1/2$, and for a number t, let denote the set of all permutations in distance at most t from A (permutations "close to od ones"). The isoperimetric inequality says that there is only a very small fraction of mutations which are not close to good ones; quantitatively

$$\mu(\Omega \setminus A_t) \leq \exp(-\frac{(t - \sqrt{6n})^2}{8n}),$$

s the interesting values of t are of the order $\sqrt{n}$ and more. Such a kind of inequalities are ved using martingales, and sometimes one also uses martingales directly. We will use an perimetric inequality to show that the running time of Mulmuley's algorithm is sharply centrated around its expected value. The basic scheme of our argument is as follows:

Let S be the input set of n segments with some implicit canonical ordering ($s_1, \ldots, s_n$ is enumeration of the segments of S in that ordering). Let $T(\pi)$ denote the running time Mulmuley's algorithm when the segments are processed in the order $s_{\pi(1)}, s_{\pi(2)}, \ldots, s_{\pi(n)}$. Markov's inequality, for at most half of the permutations $\pi \in \Omega$ the running time $T(\pi)$ ceeds the expected value more than twice, so A will be the set of all permutations π for ich $T(\pi) \leq 2ET$. Now if we know that changing the order of at most t segments increases running time by a small enough amount (e.g., by at most ET), the set A_t will contain ly permutations for which the running time does not exceed the expectation too much. en the isoperimetric inequality should guarantee that the permutations which are not in occur with a very small probability.

This program cannot be realized quite as stated, since it is not difficult to constr an example of a segment set and of a permutation under which the Mulmuley's algori behaves nicely but by exchanging the order of just two segments, it becomes very bad (we need to change the order of at least about $\sqrt{n}$ segments).

We use two tricks to overcome these difficulties. First, we will not consider all per tations, but only those which are "locally nice", and we show by a separate argument t most of the permutations are locally nice. Second, we use a different metric on the se permutations, depending on the properties of the considered set of segments. This g good tail estimates when the number k of intersections of the segments is not too small. give tail estimates also for the case when k is close to n ($k = n \log n$, say) remains as open problem.

Let us begin the proof. Let $I(s)$ denote the number of intersections of a segment s with other segments of S plus one. For a subset $X \subseteq [n]$, we will write $I(X)$ for $\sum_{i \in X} I($ We put $K = I(S) = 2k + n$. For technical reasons, we will suppose that the canon ordering is chosen is such a way that $I(s_1) \geq I(s_2) \geq \ldots \geq I(s_n)$. For $\pi_1, \pi_2 \in \Omega$ we $\Delta(\pi_1, \pi_2) = \{i \in [n];\ \pi_1(i) \neq \pi_2(i)\}$. We define a metric ρ on Ω as follows:

$$\rho(\pi_1, \pi_2) = \sum_{i \in \Delta(\pi_1, \pi_2)} \sqrt{|I(s_i)|}.$$

We now need an isoperimetric inequality for our metric probability space (Ω, μ, ρ). may use a general inequality due to Schechtman (see [Bol87]), we only need to estim a parameter called *size* of our space. Let us give the necessary definitions (only in s a generality as we need them; see [Bol87] for a general formulation): Let (Ω, μ, ρ) b finite metric probability space with a uniform probability measure, and let $\mathcal{P}_0, \mathcal{P}_1, \ldots,$ be a sequence of partitions of Ω, such that $\mathcal{P}_0$ is the trivial partition, $\mathcal{P}_m$ is the discr partition and every $\mathcal{P}_i$ is a refinement of $\mathcal{P}_{i-1}$. A sequence $c_1, \ldots, c_m$ of real numbers i *decomposition sequence* of (Ω, μ, ρ) if for every $A, B \in \mathcal{P}_i$ which are contained in the sa part $C \in \mathcal{P}_{i-1}$ there exists a bijection $\varphi : A \to B$ such that $\rho(x, \varphi(x)) \leq c_i$ for all $x \in$ If there exists a decomposition sequence $c_1, \ldots, c_m$ with $\sum_{i=1}^{m} c_i^2 = s$, the *size* of $(\Omega, \mu, \rho$ at most s. The following isoperimetric inequality was established by Schechtman:

Theorem 3.1 *Let (Ω, μ, ρ) be a metric probability space of size at most s and let $A \subseteq \Omega$ a set of measure $\alpha > 0$. Then for $t \geq \sqrt{2s \ln(1/\alpha)}$, it is*

$$\mu(\Omega \setminus A_t) \leq \exp\left(-\frac{(t - \sqrt{2s\ln(1/\alpha)})^2}{2s}\right).$$

Lemma 3.2 *The size of the above defined probability metric space (Ω, μ, ρ) is at most 4*

Proof: We set $m = n - 1$, and we define a sequence $\mathcal{P}_0, \ldots, \mathcal{P}_m$ of partitions of Ω by

$$\mathcal{P}_i = \big\{\{\pi \in \Omega;\ \pi(1) = x_1, \ldots, \pi(i) = x_i\};\ x_1, \ldots, x_i \in [n] \text{ distinct}\big\}.$$

Let $C \in \mathcal{P}_{i-1}$, $C = \{\pi \in \Omega;\ \pi(1) = x_1, \ldots, \pi(i-1) = x_{i-1}\}$ and let $A \in \mathcal{P}_i$ be the of permutations of C with $\pi(i) = x_i$ and B the set of permutations of C with $\pi(i) =$ Let us set $i' = \pi^{-1}(x_i')$. Since for $j \in [i-1]$ it is $\pi(j) = x_j \neq x_i'$, it must be $i' \geq i$. define a mapping $\varphi : A \to B$ as follows: if π' denotes the image $\varphi(\pi)$, we set $\pi'(i) =$ $\pi'(i') = x_i$ and $\pi'(j) = \pi(j)$ for $j \neq i, i'$. Obviously φ is a bijection between A and B, a we have $\rho(\pi, \varphi(\pi)) \leq \sqrt{I(s_i)} + \sqrt{I(s_{i'})} \leq 2\sqrt{I(s_i)}$ (since we assumed that $I(s_i) \geq I($ for $i' \geq i$). Hence $c_i = 2\sqrt{I(s_i)}$ defines a decomposition sequence, and the bound for size is $\sum_{i=1}^{n-1} c_i^2 = 4K$. □

Locally nice permutations

ɔrder to show that the change in the running time of Mulmuley's algorithm is small ugh under a small change of the permutation, we have to restrict ourselves to a subset ll permutations.

Let r be a segment (not necessarily in S), let x be a point of r, and let $\pi \in \Omega$ be a mutation ordering the segments of S. We say that a segment $s \in S$ is *visible from x ng r under π* if s intersects r and no segment $s' \in S$ intersecting r between the points x $r \cap s$ precedes s in the ordering defined by π.

We define the set of *interesting segments* as follows: It consists of all segments of S, vertical lines passing through the endpoints of the segments of S and all vertical lines sing through intersections of segments of S.

For an integer ℓ we say that a permutation $\pi \in \Omega$ is *locally ℓ-nice* if for every interesting ment r and for every point $x \in r$, the number of segments visible from x along r under π t most ℓ. Let C_1 be a large enough constant to be fixed later. We say that a permutation Ω is *locally nice* if it is locally ℓ-nice with $\ell = C_1 \log n$. We let Ω_0 denote the set of all ally nice permutations.

mma 3.3 *The complement of Ω_0 has measure at most $O(n^{-C_2})$, where the constant C_2 be made arbitrarily large by a suitable choice of the constant C_1.*

oof: Consider the following game (which is similar in spirit to games considered by lmuley [Mul88]): There are n distinct points ordered along a horizontal line that are noved according to some permutation π. The score of the game is the number of times t a current leftmost point was removed. If we consider π to be drawn from Ω according the uniform probability measure, then the score of our game becomes a random variable which can be represented as a sum $X_n + X_{n-1} + \ldots + X_2 + X_1$ of 0-1 random variables, ere $X_i = 1$ iff when i points are remaining the leftmost one is removed next. Note that X_i's are independent and that $\mathrm{E}X_i = 1/i$. Thus $\mathrm{E}X = H_n \approx \log n$, moreover Chernoff's hnique for bounding the tail of sums of independent random variables applies [1] and lds that

$$\Pr(X > c \log n) = O(n^{-(1+c(\log c - 1))}) = O(n^{-C_2}).$$

Now observe that counting the segments $s \in S$ that are visible from a point x along a ment r under permutation π amounts to playing and scoring the game outlined above ice (once in each of the two directions on r away from x). Moreover, if more than $C_1 \log n$ ments are visible from x along r under π, then in at least one of the two directions along he score of our game according to π would have to be at least $(C_1/2)\log n$. Although general r will not be intersected by each of the n segments of S (especially on each side point x) we will assume so by placing bogus intersection points arbitrarily on r for such n-intersecting segments of S. This can only increase the score. Thus the probability at x sees more than $c \log n = (C_1/2)\log n$ segments along r in one direction only is $n^{-(1+c(\log c-1))})$, and hence the measure of the set of permutations that are "bad" for x ɔng r, is $O(2n^{-(1+c(\log c-1))})$.

Since on any segment r there can be at most $n+1$ combinatorially different points x, d since we are considering only $O(n^2)$ interesting segments, the measure of the set of rmutations that are bad for some x on some r (i.e. not locally nice permutations) is $n^3 n^{-(1+c(\log c-1))}) = O(n^{-C_2})$. □

3.3 Bounding the change in the running time

We are now aiming at an estimate of the following kind: If $\pi, \sigma \in \Omega_0$ are two locally n permutations with $\rho(\pi, \sigma) \leq t$, then $|T(\pi) - T(\sigma)|$ is bounded by a suitable function of n and t. It turns out that the contributions of a change in the permutation to a change the running time are of two kinds: one comes from the intersections on the segments wh order has beed changed, and one comes from the intersections of other segments seen by segments with changed order.

Let $\Delta \subseteq [n]$ and $\pi \in \Omega$. Imagine that we are inserting the segments of S one by one, a whenever we insert a segment s whose index belongs to Δ, we mark all vertices of all fa with s on their boundary in the arrangement (not trapezoidation!) of the already inser segments. We define $Z(\pi, \Delta)$ as the total number of vertices marked during this proc (even if a vertex has been marked several times, we count it only once).

Lemma 3.4 *Let $\pi, \sigma \in \Omega_0$ be two locally nice permutations, and let $\Delta = \Delta(\pi, \sigma)$. Then*

$$|T(\pi) - T(\sigma)| = O((I(\Delta) + Z(\pi, \Delta) + Z(\sigma, \Delta)) \log^4 n).$$

Proof: Let π/Δ be the $(n - |\Delta|)$-permutation obtained from π by removing all eleme whose index is in Δ. Define σ/Δ analogously. Then $\pi/\Delta = \sigma/\Delta$. Thus to prove the lem it suffices to show that

$$|T(\pi) - T(\pi/\Delta)| = O((I(\Delta) + Z(\pi, \Delta)) \log^4 n).$$

In other words we want to bound the difference of the cost of inserting all segments in t order dictated by permutation π and the cost of inserting the segments according to π b with the segments whose index is in Δ omitted.

At the end of section 2 we argued that the cost of inserting segment s to obtain trapez dation $\mathcal{T}_S(R)$ is proportional to the number of abutments between vertical extensions a segment pieces in $\mathcal{P}_R(s)$ (the pieces of s and other segments' pieces that touch s).

When considering $T(\pi)$, where π is locally ℓ-nice, how often can a vertical extension charged with such an abutment? Clearly $O(\ell^2)$ times, since the vertical extension can "shortened" because of a newly inserted segment at most ℓ times, and since for each of the segments the segment piece hit by the vertical extension can be shortened at most 2ℓ tim

It follows that the contribution of a vertical extension to $T(\pi)$ for π locally nice $O(\log^2 n)$.

Now consider $T(\pi/\Delta)$. It is not hard to see that π locally ℓ-nice implies that $\pi\Delta$ locally ℓ^2-nice. By the arguments above we can therefore conclude that the contribution a vertical extension to $T(\pi/\Delta)$ is $O(\log^4 n)$.

To prove our desired bound it now suffices to observe that the only vertical extensic that could possible contribute to $|T(\pi) - T(\pi/\Delta)|$ are the $2I(\Delta)$ ones from endpoints intersection points of segments with index in Δ, and vertical extension from the mark vertices that are counted by $Z(\pi, \Delta)$.

□

3.4 Complexity of many faces in a dynamic segment arrangeme

In view of Lemma 3.4, we need to estimate the quantity $Z(\pi, \Delta)$, which is a question abo the complexity of many faces in a dynamically evolving arrangement of segments.

We begin by recalling a result of Edelsbrunner et al. on the complexity of many faces . static arrangement of segments. Let us consider an arrangement of n segment with k wise intersections. Let $f(n,m,k)$ denote the maximum possible complexity of m distinct s in such an arrangement.

eorem 3.5 [AEGS90]

$$f(n,m,k) = O(m^{2/3}k^{1/3} + n\alpha(\frac{k}{n}) + n \min(\log\frac{k}{n}, \log m)).$$

is bound is tight in most situations, see [AEGS90]. We will use a somewhat less precise und $f(n,m,k) = O(m^{2/3}k^{1/3} + n\log n)$, which simplifies the calculations a little.

We now want to pass to a dynamical setting. Imagine that segments of a certain set of n ments are inserted one by one. At some moment, all vertices of some face of the current angement are marked, and this happens m times in all. We are interested in the total nber of marked vertices.

If the order of insertions is arbitrary, it is not hard to come up with an example in ich $\Omega(mn)$ vertices are marked (for $k = mn$). We will thus restrict ourselves to inserting segments in orders which are locally nice permutations according to the definition from tion 3.2. Let $g(n,m,k)$ denote the maximum possible number of marked vertices under s assumption.

mma 3.6 $g(n,m,k) = O(m^{2/3}k^{1/3}\log n + n\log^3 n)$.

Proof: We will use a divide-and-conquer approach, and our proof is very similar to mbination lemma" used e.g. by Edelsbrunner et al. [EGS90]. Let us choose one point each of the marked faces, and let P be the set of m these points. Let S_1 be the set of ments inserted in the first $\lfloor n/2 \rfloor$ steps, let us call them the *blue* segments, and let S_2 the remaining – *red* – segments. Let P_1 be the points of P whose corresponding faces ve been marked before inserting the first red segments, and P_2 the remaining points of Let us put $m_1 = |P_1|$, $m_2 = |P_2|$. Let k_1 be the number of intersections among the e segments and k_2 the number of intersections among the red segments. We will derive ecurrence relation for $g(n,m,k)$.

Let us consider all the marked vertices. We may assume that no three segments intersect a common point, so there are at most two segments participating at every vertex. We 1 neglect vertices which are endpoints of segments (since there only $2n$ of them). The naining vertices are naturally classified as blue-blue, blue-red and red-red. The number marked red-red vertices is clearly bounded by $g(\lceil n/2 \rceil, m_2, k_2)$. The number of marked ie-blue vertices is at most $g(\lfloor n/2 \rfloor, m_1, k_1) + f(\lfloor n/2 \rfloor, m_2, k_1)$.

It remains to bound the number of red-blue vertices. Imagine that before inserting the st red segment, we place one observer on every blue edge seen by a point of P_2 (i.e. ntained in the boundary of a face with some point of P_2 in it, in the arrangement of blue rtices). Then we insert the red segments one by one. When inserting a red segment, a ue edge may get partitioned into two edges. If only one of these new edges is seen by ints of P_2, the observer living on that edge goes into the part visible from P_2. If both rts are visible, the old observer goes into one part and a new observer is placed into the cond part.

Every observer always looks along his current blue edge in both directions (and cannot see through segments). By our assumption on the insertion order of the segm every observer sees at most $O(\log n)$ red-blue vertices created during his observation. the other hand, every marked red-blue vertex is seen by some observer. We thus wan bound the total number of observers. In the beginning, there are at most $f(\lfloor n/2 \rfloor, m_2,$ observers. Let us bound the number of observers placed during the insertion of red segme Imagine that every red segment is not inserted at once, but rather in pieces correspond to edges of the current arrangement. The need for a new observer on a divided blue segm arises when both parts are visible from some point of P_2. This may happen if for at le one edge e of the two red edges adjacent to the dividing vertex either (i) e does not parti the face containing it into two faces, or (ii) e partitions its containing face into two fa but both these new faces contain a point of P_2.

In case (i), either e is the first or the last edge along the red segment, or the numbe connected components of the union of the current segments decreases by one, hence case can occur for at most $2n$ red edges, contributing at most $4n$ observers in total.

In case (ii) the partitioning of the set P_2 is strictly refined, so case (ii) can occur fo most $m_2 - 1$ red edges, contributing less than $2m$ observers. Hence the total numbe observers is at most $f(n,m,k) + O(n+m) = O(f(n,m,k))$.

This together with the bound for $f(n,m,k)$ gives us a recurrence

$$g(n,m,k) \leq g(\lfloor n/2 \rfloor, m_1, k_1) + g(\lceil n/2 \rceil, m_2, k_2) + O(m^{2/3}k^{1/3}\log n + n\log^2 n),$$

and it is not difficult to verify that this implies the Lemma. □

By a simple trick, we can now pass to bounding the quantity $Z(\pi, \Delta)$ defined in preceding section.

Lemma 3.7 *For a locally nice permutation π, it is*

$$Z(\pi,\Delta) = O(|\Delta|^{2/3}k^{1/3}\log n + (n + I(\Delta))\log^3 n).$$

Proof: Let $T = \{s_i \in S;\ i \in \Delta\}$ and $U = S \setminus T$. When proving the above bound $Z(\pi,\Delta)$, we can disregard all marked vertices which are segment endpoints or are incid to segments of T. We now consider the arrangement of $S = U \cup T$, and for every segm $s \in U$, we remove a very small piece of s around every its intersection with a segm of T. In this way, every segment of U is partitioned into several segments, obtainin new set U' from U with $|U'| \leq n + I(\Delta)$. Now every segment of T lies in a single face the arrangement of U', and so the number of marked vertices in the arrangement of U' bounded by $g(|U'|, |\Delta|, k) \leq g(n + I(\Delta), |\Delta|, k)$ and we apply the previous Lemma. □

3.5 End of the proof

Now it remains to put the auxiliary results together. We set $\delta = C_3 n^{-C_2}$, where C_2 is constant appearing in the bound of Lemma 3.3 and C_3 is another constant, large enou to guarantee $\delta > 4\mu(\Omega \setminus \Omega_0)$. Let A be the set of all permutations $\pi \in \Omega_0$ for wh $T(\pi) \leq (1+\delta)\mathrm{E}T$. By Markov's inequality $\mu(A \cup (\Omega \setminus \Omega_0)) \geq \delta/(1+\delta) \geq \delta/2$, a by Lemma 3.3, $\alpha = \mu(A) \geq \delta/4 = C_3 n^{-C_2}/4$. Hence $\ln(1/\alpha) = O(\log n)$ and by isoperimetric inequality (1) and by Lemma 3.2, we get

$$\mu(\Omega \setminus A_t) \leq \exp\left(-\frac{(t - O(\sqrt{K}\log n))^2}{8K}\right) + O(n^{-C_2}).$$

n we set t to $C_4\sqrt{K\log n}$, where C_4 is another large enough constant, we get $\mu(\Omega\backslash A_t) =$ $^{-c})$, where c can be made arbitrarily large by the choice of the other constants.

et $\pi \in A_t$ and let $\sigma \in A$ satisfy $\rho(\pi,\sigma) \le t$. For $\Delta = \Delta(\pi,\sigma)$, we thus have

$$\sum_{i\in\Delta} \sqrt{I(s_i)} \le t,$$

h implies $|\Delta| \le t$ and we show that then also $I(\Delta) \le t\sqrt{n}$. Indeed, we know that $\le n$ for all $s \in S$, and considering a sum $\sum_{i=1}^n x_i^2$ under restrictions $0 \le x_i \le \sqrt{n}$, $_i \le t$, we see that it is maximized by setting a maximum possible number of the x_i's to maximum value and the others to zero. In this case $\sum_{i=1}^n x_i^2 \le n(t/\sqrt{n}) = t\sqrt{n}$.

Let us substitute this into the bound of Lemma 3.4. We know that $t \ge \sqrt{n}$, so $n+I(\Delta) \le$ $\overline{n}$, and from Lemmas 3.4 and 3.7 we get

$$|T(\pi) - T(\sigma)| = O(t^{2/3}K^{1/3}\log^5 n + tn^{1/2}\log^7 n).$$

ally by the substitution $t = O(\sqrt{K\log n})$, we get

$$|T(\pi) - T(\sigma)| = O(K^{2/3}\log^{16/3} n + K^{1/2}n^{1/2}\log^{7.5} n) = O(K^{1/2}n^{1/2}\log^{7.5} n),$$

ch concludes the proof of Theorem 1.1. □

eferences

MKW91] H. Alt, L. J. Guibas, K. Mehlhorn, R. M. Karp, and A. Wigderson. A Method for Obtaining Randomized Algorithms with Small Tail Probabilities. *Algorithmica*, submitted.

GS90] B. Aronov, H. Edelsbrunner, L.J. Guibas, and M. Sharir. The number of edges of many faces in a line segment arrangement. Tech. Report 90-46, DIMACS Center, Piscataway, NJ, 1990. *Combinatorica*, to appear.

)79] J. L. Bentley and T. A. Ottmann. Algorithms for reporting and counting geometric intersections. *IEEE Transactions on Computers*, 28:643–647, 1979.

187] B. Bollobás. *Martingales, isoperimetric inequalities and random graphs*, pages 113–139. Math. Soc. J. Bolyai, 1987.

88] B. Chazelle and H. Edelsbrunner. An optimal algorithm for intersecting line segments in the plane. In *Proc. 29. IEEE Symposium on Theory of Computer Science*, pages 590–600, 1988.

L85] B. Chazelle, L. Guibas, and D. T. Lee. The power of geometric duality. *BIT*, 25(1), 1985.

a86] B. Chazelle. Reporting and counting segment intersections. *Journal of Computer and System Sciences*, 32:156–182, 1986.

89] K. L. Clarkson and P. Shor. New applications of random sampling in computational geometry II. *Discrete & Computational Geometry*, 4, 1989.

le87] H. Edelsbrunner. *Algorithms in combinatorial geometry*. Springer, 1987.

[EGS90] H. Edelsbrunner, L. Guibas, and M. Sharir. The complexity and construct of many faces in arrangements of lines and segments. *Discrete & Computatic Geometry*, 5:161–196, 1990.

[EOS86] H. Edelsbrunner, J. O'Rourke, and R. Seidel. Constructing arrangement hyperplanes with applications. *SIAM Journal on Computing*, 15:341–363, 19

[Hag90] T. Hagerup. An application of martingales to computational geometry. Lect at Computational geometry workshop, Dagstuhl, FRG, October 1990.

[1] T. Hagerup and C. Rüb. *A Guided Tour of Chernoff Bounds.* Inform. P Letters 33 (1989/90), pp 305–308.

[MSW92] K. Mehlhorn, M. Sharir, and E. Welzl. *Tail estimates for the space complexit randomized incremental algorithms.* To appear in Proc. 3rd ACM-SIAM Sy on Discrete Algorithms, 1992.

[MS86] V. D. Milman and G. Schechtman. *Asymptotic theory of finite dimensic normed spaces.* LNM 1200, Springer-Verlag, 1986.

[Mul88] K. Mulmuley. A fast planar partition algorithm: Part I. In *Proc. 29. IE Symposium on Theory of Computer Science*, pages 33–43, 1988.

[Sei91] R. Seidel. Backwards analysis of randomized geometric algorithms. Internatic Computer Science Institute (ICSI) Berkeley, Report TR-92-014. Also to app in **New Trends in Discrete and Computational Geometry**, J. Pach (e Springer Verlag.

Lower Bounds on the Complexity of Simplex Range Reporting on a Pointer Machine

EXTENDED ABSTRACT

Bernard Chazelle
Department of Computer Science
Princeton University

Burton Rosenberg
Department of Mathematics and
Computer Science
Dartmouth College

Abstract

We give a lower bound on the following problem, known as *simplex range reporting:* Given a collection P of n points in d-space and an arbitrary simplex q, find all the points in $P \cap q$. It is understood that P is fixed and can be preprocessed ahead of time, while q is a query that must be answered on-line. We consider data structures for this problem that can be modeled on a pointer machine and whose query time is bounded by $O(n^{\delta} + r)$, where r is the number of points to be reported and δ is an arbitrary fixed real. We prove that any such data structure of that form must occupy storage $\Omega(n^{d(1-\delta)-\varepsilon})$, for any fixed $\varepsilon > 0$. This lower bound is tight within a factor of n^{ε}.

1 Introduction

Given a set of n points in d-space, precompute a data structure capable of counting or reporting all points inside an arbitrary query simplex. This problem, known as *simplex range searching*, has been extensively studied in recent years [4, 5, 7, 8, 10, 13, 16, 18, 19]. On the practical side, the problem relates to fundamental questions in computer graphics, (e.g., hidden surface removal), while theoretically it touches on some of the most central issues in algorithm design and combinatorial geometry (e.g, derandomization, geometric graph separation, k-sets). In spite of all the attention, however, only recently have optimal or quasi-optimal solutions been discovered. If m is the amount of storage available, it is possible to achieve a query time of roughly $n/m^{1/d}$, where "roughly" means that an extra factor of the form n^{ε} [4] or $(\log n)^{O(1)}$ [13, 14] must be added to the complexity bound. What allows us to brand these solutions quasi-optimal is an (almost) matching lower bound established in the arithmetic model of computation [2]. This lower bound is very general and holds for any realistic computing model, but it is limited to the case where searching is interpreted as *counting* or more generally computing the cumulative weight of weighted points inside the query.

This has left open the question of proving the optimality of the known algorithm in the *reporting* case: this is the version of the problem where the points inside th query must be found and reported one by one. To date, only the case of orthogona range reporting has been satisfactorily resolved [3]. To prove lower bounds in th counting case is difficult enough, but the difficulty is compounded in the reportin case, because of the possibility for the algorithm to amortize the search over th output. This design paradigm, known as *filtering search* [1], is based on the idea tha if many points must be reported then the search can be slowed down proportionatel which is then likely to result in a slimmer data structure.

We look here at the typical case where a query time of the form $O(n^\delta + r)$ i sought, where r is the size of the output and δ is any fixed constant. We sho that on a pointer machine any data structure with a query time of that form mus be of size $\Omega(n^{d(1-\delta)-\varepsilon})$, for any fixed $\varepsilon > 0$. This lower bound is quasi-optima Despite the apparent restrictions we place on the model, we must mention tha the overwhelming majority of data structures proposed in the literature for rang searching fall in that category. The magnitude of our lower bound is striking. I says, for example, that in E^{20} to achieve a query time even as inefficient as $O(\sqrt{n}+r$ still requires approximately n^{10} storage!

The proof rests on a combination of graph-volumic and integral-geometric argu ments. The next section defines the model and proves a technical lemma regardin the spread of information across the data structure. Section 3 contains the proof c the main result. Concluding thoughts are given in Section 4.

2 The Complexity of Navigation on a Pointer Ma- chine

We assume some familiarity with the *pointer machine model* [17]. As in [3] the dat structure is modeled as a directed graph $G = (V, E)$ of outdegree at most 2. Le $P = \{p_1, \ldots, p_n\}$ be a set of n points in E^d. To each node v of the data structure an integer $f(v)$ is attached. If $f(v) = i$ is not zero, then node v is associated wit point p_i. A *query* q is a simplex in E^d, and the algorithm must report all points i $P \cap q$. When presented with q, the algorithm begins at a starting node and, afte following pointers across the data structure, terminates with a working set $W(q$ that is required to contain the answer, namely,

$$\{\, i \mid p_i \in q \,\} \subseteq \{\, f(v) \mid v \in W(q) \,\}.$$

The *size* of the data structure G is n, the number of nodes in the graph. Note tha our model accommodates static as well as self-adjusting data structures.

A data structure G is termed (a, δ)-*effective* if for any query q, we have $|W(q)| \leq$ $a(|P \cap q| + n^\delta)$. A collection of queries $Q = \{q_i\}$ is called (c, k, δ)-*favorable* if fo all i, $|P \cap q_i| > n^\delta$ and for all $i_1 < \cdots < i_k$, $|P \cap q_{i_1} \cap \cdots \cap q_{i_k}| < c$. We want t show that if δ is small, an (a, δ)-effective data structure must be large. Using th following result, which generalizes a lemma given in [3], we can lower-bound a dat structure's size by exhibiting a (c, k, δ)-favorable set of queries.

Lemma 2.1 *For any fixed* $a, \delta > 0$ *and* $c \geq 2$*, if* G *is* (a,δ)*-effective and* Q *is* (c, k, δ)*-favorable, then*

$$|V| > |Q| n^{\delta}/(3(k-1)2^{8ac^2}),$$

for n *large enough.*

Proof: We exploit the fact that the data structure can quickly answer a large number of very different queries to show that the data structure is itself large. More precisely, we look at the *c-sets* of V,

$$V^{(c)} = \{ W \subseteq V \mid |W| = c \}.$$

Recall that a tree is *rooted* if its edges are directed and the root is the only node with no incoming edge. Given any subset $W \subseteq V$, we define the *diameter* of W in G, denoted $\Lambda_G(W)$, as the minimum number of edges in any rooted tree that spans W and is a subgraph of G. It is ∞ if no such tree exists. This definition applies to any directed graph, in particular to subgraphs of G. Below we shall need Λ_T, where T is a rooted tree and a subgraph of G.

The number of c-sets in G of diameter smaller than r is bounded by,

$$\begin{aligned}\left|\{ W \in V^{(c)} \mid \Lambda_G(W) < r \}\right| &\leq \left|\{ (z, W) \in V \times V^{(c)} \mid \forall w \in W,\ d(z,w) < r \}\right| \\ &\leq |V| 2^{rc},\end{aligned}$$

because of the limitation on the outdegree of G. Suppose now that query q is presented to the algorithm. Fix a rooted tree $T' \subseteq G$ which contains exactly the vertices of $W(q)$. Because the algorithm reaches all the nodes in $W(q)$, such a tree exists. We can select from $W(q)$ a subset W that contains exactly one $w \in W$ with $f(w) = i$ for each $p_i \in P \cap q$.

Let T be the Steiner minimal tree of W inside of T'. Note that $\Lambda_T(Z) \geq \Lambda_G(Z)$ for any $Z \subseteq G$. Embed the tree T in the plane and number the vertices of W in a natural order around the border of T. Then, $W = w_1, w_2, \ldots, w_s$, where $s = |P \cap q|$, and,

$$\sum_{j=1}^{s-1} \Lambda_T(\{w_j, w_{j+1}\}) \leq 2|T|.$$

Consider the c-sets,

$$W_i = \{ w_i, \ldots, w_{c+i-1} \},\ \ i = 1, \ldots, s-c+1.$$

It is clear that,

$$\Lambda_T(W_i) \leq \sum_{j=i}^{c+i-2} \Lambda_T(\{w_j, w_{j+1}\}).$$

Summing over all i,

$$\sum_{i=1}^{s-c+1} \Lambda_T(W_i) \leq (c-1) \sum_{j=1}^{s-1} \Lambda_T(\{w_j, w_{j+1}\}) \leq 2(c-1)|T|.$$

Since, $|T| \leq |W(q)|$, if we assume that G is (a,δ)-effective and Q is $(c,k,\delta$ favorable (thus $|P \cap q| > n^{\delta}$):

$$\sum_{i=1}^{s-c+1} \Lambda_T(W_i) < 4a(c-1)\,|P \cap q|,$$

for n large enough. By Markov's inequality,

$$\left|\{\, i \mid \Lambda_T(W_i) \geq 8a(c-1)\,\}\right| \leq |P \cap q|/2,$$

and therefore,

$$\left|\{\, i \mid \Lambda_T(W_i) < 8a(c-1)\,\}\right| \geq |P \cap q|/2 - c + 1 > |P \cap q|/3.$$

Because $\Lambda_T(W_i) \geq \Lambda_G(W_i)$, this is also a lower bound on the number of c-sets wit diameter in G less than $8a(c-1)$.

This argument is valid for any q in Q. Since $|P \cap q_{i_1} \cap \cdots \cap q_{i_k}| < c$, for ap propriate indices $i_1 < \cdots < i_k$, a small c-set will be counted at most $k-1$ times Thus,

$$\left|\{\, w \in V^{(c)} \mid \Lambda_G(w) < 8a(c-1)\,\}\right| > |Q|\ |P \cap q|/(3(k-1)) > |Q|\,n^{\delta}/(3(k-1))$$

for large enough n.

In view of the upper bound given at the beginning of this proof, the result follow easily. □

3 A Lower Bound for Simplex Range Reporting

According to the discussion of the previous section, any algorithm for solving simple range reporting in time $O(n^{\delta}+r)$ can be modeled as an (a,δ)-effective data structure for some suitable constant a. The desired lower bound follows, therefore, from the construction of a set P of n points along with a $(c, \lceil \log n \rceil, \delta)$-favorable query set Q of size $\Omega(n^{d(1-\delta)-\delta\varepsilon})$, for any fixed $\varepsilon > 0$.

Let $q \in E^d$ be any nonzero vector in Euclidean d-space. The hyperplane $H_q \subset E^d$ is defined by

$$H_q = \{\, x \in E^d \mid \langle x, q \rangle - |q|^2 = 0 \,\}.$$

For any real $\mu > 0$, the *slab* $H_{q,\mu}$ is the set of all points within distance μ from H_q

$$H_{q,\mu} = \{\, x \in E^d \mid\ |\langle x, q \rangle - |q|^2| \leq \mu |q| \,\}.$$

The point q is the *defining point* of the slab $H_{q,\mu}$. Although our final result is stated for a collection of simplex queries, the query set we construct is a collection of slabs. Once a favorable query set has been constructed, using slabs for queries we can replace the slabs by very long flat simplices using elementary perturbation techniques.

We note a shift in our notation. The set Q is represented as a set of points. The collection of queries is actually,

$$\{ H_{q,\mu} \,|\, q \in Q \}.$$

Let $C_d = [0,1]^d$ be the unit d-cube in E^d. We construct a favorable query set in two steps. First we position the slabs so that their arrangement has certain geometric properties: their intersection with C_d must be large, but their k-wise intersections with each other must be small. Next, n points are thrown at random into C_d and we verify that with high probability the slabs are favorable for this point set.

Further on we shall demonstrate that a sufficient condition for any k of the slabs to intersect in a small volume is that any k of the defining points have a large convex hull. This relates to Heilbronn's problem [11, 12, 15]: what is the largest area, over all point-sets $P = \{p_1, \ldots, p_m\} \subset C_2$, of the smallest triangle with vertices in P? Here we require that the convex hull of k points in d dimensions should have volume $\Omega(1/m)$. This can be achieved if $k \geq \log m$:

Theorem 3.1 (Chazelle[2]) *For any $d > 1$ there exists a constant $c > 0$ such that a random set of m points in C_d has, with probability greater than $1 - 1/m$, the property that the convex hull of any $k \geq \log m$ of these points has volume greater than ck/m.*

Hence a random point set is likely to be "good" for the construction of a favorable query set.

Let Q_0 be a random set of m points uniformly distributed in C_{d-1}. Theorem 3.1 assures that with high probability any $k \geq \log m$ points will "enclose" a large volume. Embed Q_0 in C_d via the map:

$$(x_1, \ldots, x_{d-1}) \mapsto \frac{1}{2}(x_1 + 1, \ldots, x_{d-1} + 1, 2).$$

Fixing a real $(0 < \mu < 1)$, each image point gives rise to $\Theta(1/\mu)$ new points by the maps,

$$x \mapsto 2\mu i x,$$

where i ranges over all integers such that $1/2 \leq 2\mu i \leq 3/4$. To be precise, $i \in [i_0, i_1]$ with $i_0 = \lceil 1/(4\mu) \rceil$ and $i_1 = \lfloor 3/(8\mu) \rfloor$. Let Q be the image of Q_0 under the composition of these two maps. That is, the image of Q_0 under the map (figure 1):

$$\begin{array}{llll} \phi(x,z): & C_{d-1} \times [i_0, i_1] & \longrightarrow & C_d \\ & (x_1, \ldots, x_{d-1}, i) & \mapsto & \mu i(x_1 + 1, \ldots, x_{d-1} + 1, 2). \end{array}$$

Lemma 3.1 *Assume μ goes to zero with increasing m. Then,*

1. *Q is a set of size $\Theta(m/\mu)$.*
2. *For all $q \in Q$ the slabs $H_{q,\mu}$ have an intersection with C_d of volume $\Theta(\mu)$.*
3. *Any $k = \lceil \log m \rceil$ of these slabs have an intersection of volume,*

$$O(\mu^d m (\log m)^{d-2}).$$

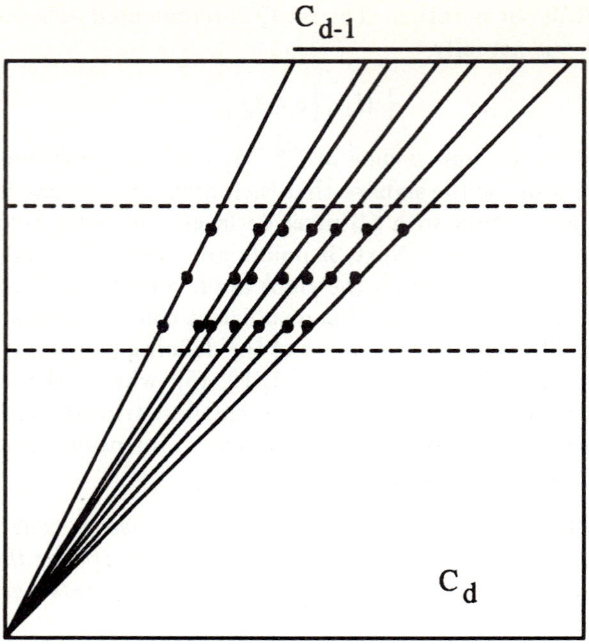

Figure 1: Building the query set.

Proof: The first claim is trivial. The second follows from the fact that each coordinate of any $q \in Q$ is in the interval $[1/4, 3/4]$. So a ball of radius $1/4 - \mu$ and center q intersects H_q in a hyperdisk D which lies entirely inside C_d. The cylinder of height 2μ and cross section D at its midpoint is inside C_d. Here we assume, by increasing m if necessary, that $\mu \ll 1/4$. This gives the lower bound on the volume of $H_{q,\mu} \cap C_d$. The upper bound follows by placing a sufficiently large ball around q, say of radius $\sqrt{d}$, so as to contain the piece of $H_{q,\mu}$ that lies in C_d.

The third claim is substantiated as follows. Let $H_{q_1,\mu}, \ldots, H_{q_k,\mu}$ be the $k = \lceil \log m \rceil$ slabs, where the q_i are all distinct. If q_i and q_j are collinear with the origin, the intersection is empty. If they are not, let $p_1, \ldots, p_k$ be the points in C_{d-1} which gave rise to $q_1, \ldots, q_k$. The convex hull of the p_i has $(d-1)$-dimensional volume at least $c_1 k/m$ for some appropriate positive constant c_1. Triangulate the convex hull using $O(k^{d-1})$ simplices and choose one among the simplices of largest $(d-1)$-volume. After renumbering, the vertices of this simplex are $p_1, \ldots, p_d$ and it has volume at least $c_2/(k^{d-2}m)$. We conclude that,

$$|\det(q_1, \ldots, q_d)| \geq c_3/(k^{d-2}m),$$

where the q_i have been renumbered according to the same pattern as the p_i and c_2 and c_3 are positive constants depending only on the dimension. The lemma follows from the next result. □

Lemma 3.2 *Given $k = \lceil \log m \rceil$, from every set $q_1, \ldots, q_k \subset Q$ a subset $q_{i_1}, \ldots, q_{i_d}$*

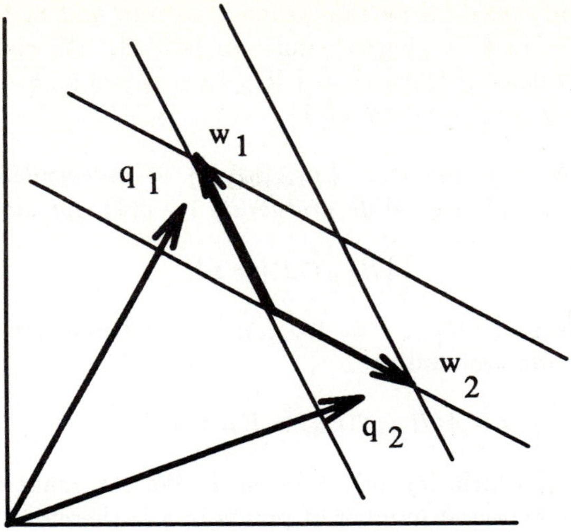

Figure 2: Intersection parallelotope.

can be selected such that,

$$Vol\left(H_{q_1,\mu}\cap\cdots\cap H_{q_k,\mu}\right) \le Vol\left(H_{q_{i_1},\mu}\cap\cdots\cap H_{q_{i_d},\mu}\right) = O\left(\mu^d m(\log m)^{d-2}\right).$$

Proof: We can still assume that no two q_i's are collinear with the origin. The first inequality is trivial. In general, let $q_1,\ldots,q_d$ be linearly independent vectors. The polytope $H_{q_1,\mu}\cap\cdots\cap H_{q_d,\mu}$ is a translate of the parallelotope defined by d vectors w_j where,

$$\langle w_j, q_i\rangle = \begin{cases} 2\mu\,|\,q_i\,| & \text{if } i=j \\ 0 & \text{otherwise.}\end{cases}$$

To be more precise,

$$H_{q_1,\mu}\cap\cdots\cap H_{q_d,\mu} = \left\{\textstyle\sum_{i=1}^d \alpha_i w_i \,\middle|\, 0\le\alpha_i\le 1,\ i=1,\ldots,d\right\} + x_o,$$

where x_o is the unique point of E^d satisfying,

$$\langle x_o, q_i\rangle - |\,q_i\,|^2 = -\mu\,|\,q_i\,|$$

for all $i = 1,\ldots,d$ (figure 2). Denote by $[w]$ the matrix $(w_1,\ldots,w_d)$, by $[q]$ the matrix $(q_1,\ldots,q_d)$, and by Λ the diagonal matrix with $\Lambda_{ii} = |\,q_i\,|$. Note that $\det[w]$ is the volume of the parallelotope $H_{q_1,\mu}\cap\cdots\cap H_{q_d,\mu}$. From $[w]^T[q] = (2\mu)^d\Lambda$ we have

$$\det[w]\det[q] = (2\mu)^d\,|\,q_1\,|\cdots|\,q_d\,|.$$

Recall that from the set $q_1,\ldots,q_k\subset Q$ we can select d vectors such that $|\det[q]|\ge c_3/(k^{d-2}m)$, and $\sqrt{d}/4\le|\,q_i\,|\le 3\sqrt{d}/4$. This gives the bound. □

We finish the proof of the lower bound with a probabilistic analysis of the interaction between the query set Q and n points chosen randomly in the unit cube C_d. For any real $0<\delta<(d-1)/d$ and any fixed $\varepsilon>0$, set:

$$\mu = 1/(\tau n^{1-\delta}),\quad m = n^{d(1-\delta)-1-\varepsilon},$$

where τ depends only on d. Note that μ tends to zero and m tends to infinity as n tends to infinity. Set $k = \lceil \log m \rceil$ and $c = \lceil d^2/\varepsilon \rceil$. We claim that with high probability the collection of slabs $H = \{ H_{q,\mu} \mid q \in Q \}$ is (c, k, δ)-favorable for the point set P, where Q is as in Lemma 3.1.

Lemma 3.3 *Let the n points* $P = \{p_1, \ldots, p_n\}$ *be independently and uniformly distributed in the unit cube* C_d. *With probability* $1 - o(1)$, *for all* $q \in Q$,

$$|H_{q,\mu} \cap P| > n^{\delta}.$$

Proof: The points $p_i \in H_{q,\mu}$, $i = 1, \ldots, n$, are independent Bernoulli random variables with common probability,

$$p = \mathrm{Vol}(H_{q,\mu} \cap C_d) > K\mu = K/(\tau n^{1-\delta}),$$

for an appropriate K which depends only on d. We can make τ small enough so that $np > 2n^{\delta}$. The expected number of points in q is therefore $E(|H_{q,\mu} \cap P|) = np > 2n^{\delta}$. The Chernoff bound [6, 9] states that, for $X = \{x_1, \ldots, x_n\}$ a Bernoulli random variable where $x_i = 1$ with probability p and $x_i = 0$ with probability $1 - p$

$$\mathrm{Prob}\left(\sum_{i=1}^{n} x_i \leq (1-\kappa)np\right) \leq \left(\frac{e^{-\kappa}}{(1-\kappa)^{1-\kappa}}\right)^{np},$$

for $0 < \kappa < 1$. Therefore, the probability that $|H_{q,\mu} \cap P| \leq np/2$ is less than $(2/e)^{np/2}$. Taking the disjunction over all $q \in Q$,

$$\begin{aligned}
\mathrm{Prob}\,(\exists q \in Q \text{ s.t. } |H_{q,\mu} \cap P| \leq np/2) &\leq |Q|\,\mathrm{Prob}\,(|H_{q,\mu} \cap P| \leq np/2) \\
&< (m/\mu)\,(2/e)^{np/2} \\
&< n^{d(1-\delta)-\delta-\varepsilon}(2/e)^{n^{\delta}}.
\end{aligned}$$

It is not difficult to see that this probability goes to 0 as n goes to infinity. Therefore, with probability approaching 1, every $H_{q,\mu}$ has more than $np/2 > n^{\delta}$ points in it. □

Lemma 3.4 *Let P be a set of n random points chosen uniformly in the unit cube* C_d. *With probability* $1 - o(1)$, *for all distinct* $q_1, \ldots, q_k \in Q$,

$$|H_{q_1,\mu} \cap \cdots \cap H_{q_k,\mu} \cap P| < c.$$

Proof: The events $p_i \in H_{q_1,\mu} \cap \cdots \cap H_{q_k,\mu}$, for $i = 1, \ldots, n$, are independent Bernoulli random variables with common probability,

$$p = \mathrm{Vol}(H_{q_1} \cap \cdots \cap H_{q_k}) < K\mu^d m (\log m)^{d-2},$$

for an appropriate constant K, (Lemma 3.1(3)), We again refer to the Chernoff bound: for any positive real κ,

$$\mathrm{Prob}\left(\sum_{i=1}^{n} x_i \geq (1+\kappa)np\right) \leq \left(\frac{e^{\kappa}}{(1+\kappa)^{1+\kappa}}\right)^{np},$$

hus if $np < 1$ then for any integer $b \geq 1$,

$$\text{Prob}\left(\sum_{i=1}^{n} x_i \geq b\right) \leq \left(\frac{enp}{b}\right)^b .$$

The expected number of points in $H_{q_1,\mu} \cap \cdots \cap H_{q_k,\mu}$ is less than 1 for n sufficiently large, hence,

$$\text{Prob}\left(\,|\,H_{q_1,\mu} \cap \cdots \cap H_{q_k,\mu} \cap P\,| \geq c\,\right) \leq \left(\frac{K'(\log m)^{d-2}}{cn^{\varepsilon}}\right)^c ,$$

where K' is a positive constant. Recall from Lemma 3.2 that the upper bound on the volume of a k-wise intersection of query slabs is derived by considering a subset of only d of them. Therefore,

$$\text{Prob}\left(\,\exists q_1, \ldots, q_k \in Q \text{s.t.}\,|\,H_{q_1,\mu} \cap \cdots \cap H_{q_k,\mu} \cap P\,| \geq c\,\right) \\ \leq \binom{|Q|}{d}\left(\frac{K'(\log m)^{d-2}}{cn^{\varepsilon}}\right)^c ,$$

which goes to 0 as n increases. □

What has been shown is the existence of a collection H of $\Theta(\,n^{d(1-\delta)-\delta-\varepsilon}\,)$ slabs and a set of n points P such that H is $(\,\lceil d^2/\varepsilon \rceil, \lceil\,(d(1-\delta)-1-\varepsilon)\log n\,\rceil, \delta\,)$-favorable with respect to P. We can now apply Lemma 2.1 and derive,

Theorem 3.2 *Simplex reporting on a pointer machine in E^d with a query time of $O(\,n^{\delta}+r\,)$, where r is the number of points reported and $0 < \delta \leq 1$, requires space $\Omega(\,n^{d(1-\delta)-\varepsilon}\,)$, for any fixed $\varepsilon > 0$.*

4 Conclusion

Our bound implies that if the search time is to be in $O(\,(\log n)^b + r\,)$, for b arbitrarily large, then the space must be in $\Omega(\,n^{d-\varepsilon}\,)$ for all fixed $\varepsilon > 0$. We believe that this can be improved to $\Omega(\,n^d/\text{polylog}(n)\,)$, but leave this as an open problem. Finally, we would like to approach the question of halfspace range searching, where we expect similar techniques to give a bound of $\Omega(\,n^{\lfloor d/2 \rfloor-\varepsilon}\,)$ for polylog-time queries.

Acknowledgments

The authors would like to thank Laszlo Lovàsz for helpful discussions. They acknowledge the National Science Foundation for supporting this research in part under Grant CCR–9002352.

References

[1] B. Chazelle. Filtering search: A new approach to query-answering. *SIAM Journal on Computing*, 15(3):703–724, 1986.

[2] B. Chazelle. Lower bounds on the complexity of polytope range searching. *Journal of the American Mathematical Society*, 2(4):637–666, 1989.

[3] B. Chazelle. Lower bounds for orthogonal range searching: I. The reporting case. *Journal of the ACM*, 37(2):200–212, April 1990.

[4] B. Chazelle, M. Sharir, and E. Welzl. Quasi-optimal upper bounds of simplex range searching and new zone theorems. In *The Proceedings of the Sixth Annual Symposium on Computational Geometry*, pages 23–33, 1990.

[5] B. Chazelle and E. Welzl. Quasi-optimal range seaching in spaces of finite Vapnik Chervonenkis dimension. *Discrete and Computational Geometry*, 4(5):467–489, 1988.

[6] H. Chernoff. A measure of asymptotic efficiency for tests of a hypothesis based on the sum of observations. *The Annals of Mathematical Statistics*, 23:137–142, 1952.

[7] H. Edelsbrunner, D. G. Kirkpatrick, and H. A. Maurer. Polygonal intersection searching. *Information Procesing Letters*, 14(2):74–79, 1982.

[8] H. Edelsbrunner and E. Welzl. Halfplanar range seach in linear space and $O(n^{0.695})$ query time. *Information Processing Letters*, 23:289–293, 1986.

[9] P. Erdös and J. Spencer. *Probabilistic Methods in Combinatorics.* Academic Press, New York, 1974.

[10] D. Haussler and E. Welzl. ε-nets and simplex range queries. *Discrete & Computational Geometry*, 2(3):237–256, 1987.

[11] J. Komlós, E. Szemerédi, and J. Pintz. On Heilbronn's triangle problem. *Journal of the London Mathematical Society*, 24(2):385–396, 1981.

[12] J. Komlós, E. Szemerédi, and J. Pintz. A lower bound for Heilbronn's problem. *Journal of the London Mathematical Society*, 25(2):13–24, 1982.

[13] J. Matoušek. Efficient partition trees. In *Proceedings of the Seventh Annual Symposium on Computational Geometry*, pages 1–9, 1991.

[14] J. Matoušek. Range searching with efficient hierarchical cuttings. manuscript, 1991.

[15] W. O. J. Moser. Problems on extremal properties of a finite set of points. *Discrete Geometry and Convexity*, 440:52–64, 1985.

[16] M. Paterson and F. F. Yao. Point retrieval for polygons. *Journal of Algorithms*, 7:441–447, 1986.

[17] R. E. Tarjan. A class of algorithms which require nonlinear time to maintain disjoint sets. *Journal of Computing System Science*, 18:110–127, 1979.

[18] E. Welzl. Partition trees for triangle counting and other range searching problems. In *Proceedings Fourth Annual Symposium on Computational Geometry*, pages 23–33, 1988.

[19] D. E. Willard. Polygon retrieval. *SIAM Journal on Computing*, 11:149–165, 1982.

Infinitary Logic for Computer Science

Phokion G. Kolaitis*[1] and Moshe Y. Vardi[2]

[1] Computer and Information Sciences
University of California, Santa Cruz
Santa Cruz, CA 95064
kolaitis@cse.ucsc.edu
[2] IBM Research
Almaden Research Center
San Jose, CA 95120-6099
vardi@ibm.com

Abstract

Infinitary logic $L_{\infty\omega}$ extends first-order logic by allowing infinitary conjunctions and disjunctions (i.e., conjunctions with an infinite number of conjuncts and disjunctions with an infinite number of disjuncts). One usually thinks of infinitary logic as a fairly esoteric logic, which is not of much interest in computer science. Surprisingly, a certain fragment $L^{\omega}_{\infty\omega}$ of $L_{\infty\omega}$ turns out to be of great interest in computer science. This fragment is obtained by restricting formulas to contain a finite number of distinct variables, though the formulas can be of infinite length, The advantage of dealing with $L^{\omega}_{\infty\omega}$ is that its the expressive power can be completely characterized in game-theoretic terms. We will describe applications of this logic to the study of 0-1 laws and the expressive power of database query languages.

1 Introduction

In recent years the model theory of finite structures [Fag90] has been a meeting point for research in computer science, combinatorics, and mathematical logic. Results and techniques from finite-model theory have found interesting applications to several other areas, including database theory [CH82, Var82] and complexity theory [Ajt83, Gur84, Imm86].

One of the distinguishing features of finite-model theory is the fact that first-order logic has severely limited expressive power on finite structures (cf. [Fag75, AU79, Gai82]). In view of that fact, there is an emphasis in finite-model theory on logics that go beyond first-order logic. One way to increase the limited expressive power of first-order logic is by adding fixpoint constructs. Such extensions of first-order logic have been the subject of extensive study – focusing on their expressive power, their relationship to complexity classes, in particular to P and PSPACE, and their asymptotic probability properties (cf. [AV89, AV91, Cha88, CH82, Imm86, KV87, Var82]). Of particular importance among these extensions are *positive-fixpoint logic*,

* During the preparation of this paper this author was partially supported by NSF Grant CCR-9108631

which is obtained from first-order logic by adding least fixpoints of positive first-order formulas, and *partial-fixpoint logic*, which is obtained from first-order logic by adding fixpoints of general first-order formulas.

Since first-order logic has a finitary syntax, a different way to increase its expressive power is to allow for infinitary formation rules. One of the most powerful logics resulting this way is the infinitary logic $L_{\infty\omega}$ which allows for arbitrary disjunctions and conjunctions. This logic has been studied extensively on infinite structure (cf. [BF85]), but it is too powerful to be of interest or use in finite-model theory, because every class of finite structures that is closed under isomorphisms is definable in $L_{\infty\omega}$.

A more interesting extension from the perspective of finite-model theory is the infinitary logic $L^{\omega}_{\infty\omega}$, which consists of all formulas of $L_{\infty\omega}$ with a finite number of distinct variables. More formally, $L^{\omega}_{\infty\omega}$ is the union $\bigcup_{k=1}^{\infty} L^{k}_{\infty\omega}$, where $L^{k}_{\infty\omega}$ is the collection of all formulas of $L_{\infty\omega}$ with k variables. The infinitary logic $L^{\omega}_{\infty\omega}$ was studied originally on infinite structures [Bar77], but it turned out to have several uses in finite-model theory [KV90a, KV90b]. In fact, $L^{\omega}_{\infty\omega}$ could be said to underlie much of the work on lower bounds for expressibility on finite structures in [Imm82, IR87, LM89, CFI89], although its use there is rather implicit.

One of the reasons for the importance of $L^{\omega}_{\infty\omega}$ in finite-model theory is the fact that from the expressive power standpoint $L^{\omega}_{\infty\omega}$ constitutes a proper extension of fixpoint logics [KV90a]. Thus, $L^{\omega}_{\infty\omega}$ can be used to derive both positive and negative results about fixpoint logics. On the positive side, structural results about $L^{\omega}_{\infty\omega}$ transfer to similar results for fixpoint logics, while, on the negative side, lower-bound results for expressibility in $L^{\omega}_{\infty\omega}$ yield, a fortiori, lower bounds for expressibility in fixpoint logics (but not vice-versa). The main advantage of $L^{\omega}_{\infty\omega}$ is that the expressive power of the infinitary logics $L^{k}_{\infty\omega}$ with k variables, $k \geq 1$, has a clean and precise characterization in terms of certain k-pebble games between two players on a pair of structures (cf. [Bar77, Imm82, KV90a]). In contrast, fixpoint logic is not known to possess a similar property.

In this paper we review two applications of $L^{\omega}_{\infty\omega}$ in computer science. The first concerns 0-1 laws [KV90a] and the second concerns expressibility results for database query languages [KV90b].

We start by reviewing the application to 0-1 laws. In general, if C is a class of finite structures over some vocabulary and if P is a property of some structures in C, then *the asymptotic probability* $\mu(P)$ on C is the limit as $n \to \infty$ of the fraction of the structures in C with n elements which satisfy P, provided that the limit exists. We say that P *is true almost everywhere* on C in case $\mu(P)$ is equal to 1. If $\mu(P) = 0$, then we say that P *is false almost everywhere*. It turns out that many interesting properties on the class G of all finite graphs are either true almost everywhere or false almost everywhere. It is, for example, well known and easy to prove that μ(connectivity)=1, μ(rigidity)=1, while μ(planarity)=0 and μ(l-colorabilty)=0, for $l \geq 2$ [Bol79]. On the other hand, properties such as "evenness" (i.e., the property of the domain consisting of an even number of elements) do not have asymptotic probabilities. The study of asymptotic probabilities can be viewed as the foundations of the theory of average-case complexity, since it focuses on the average-case properties of mathematical structures (cf. [ACV92]).

Fagin [Fag76] and Glebskii et al. [GKLT69] were the first to establish a fascinating connection between logical definability and asymptotic probabilities. More specifically, they showed that if C is the class of all finite structures over some relational vocabulary and if P is any property expressible in first-order logic, then $\mu(P)$ exists and is either 0 or 1. This result, which is known as *the 0-1 law for first-order logic*, became the starting point of a series of investigations aiming in discovering the relationship between expressibility in a logic and asymptotic probabilities. The survey by Compton [Com88] contains an eloquent account of developments in this area (see also [Gur92] for an expository introduction).

One of the reasons for the limited expressive power of first-order logic is the absence of any fixpoint mechanism. Thus, an important direction of investigation pursued the study of 0-1 laws for extensions of first-order logic that allow for fixpoint constructs. Talanov [Tal81] showed that the 0-1 holds for first-order logic augmented with a *transitive closure* operator. This result was extended by Talanov and Knyazev [TK86], and, independently, by Blass, Gurevich and Kozen [BGK85] who proved that a 0-1 law holds for *positive-fixpoint logic*.[3] The latter results were extended further in [KV87], where a 0-1 law was established for a logic that is equivalent to partial fixpoint logic.

We show here that all these results are subsumed by a 0-1 law for $L^{\omega}_{\infty\omega}$ [KV90a]. This result reveals a boundary of 0-1 laws for fragments of $L_{\infty\omega}$, since "evenness" is expressible as a countable disjunction of first-order sentences (a disjunction, however, which involves infinitely many distinct variables). Unlike the previous proofs of 0-1 laws for fixpoint logics, the proof described here is in the spirit of "pure" finite-model theory; the proof uses the game-theoretic characterization of $L^{\omega}_{\infty\omega}$ and it does not appeal to "infinitistic" arguments.

Our second application involves the expressive power of database query languages. The study of rule-based query languages has been a focal area of research in database theory during the past few years [Ull89]. A significant part of this research has concentrated on the language Datalog, which, in its purest form, is the language of function-free and negation-free Horn clauses. As a query language, Datalog can be viewed as the data sublanguage of general logic programming. Datalog programs always express queries that are computable in time bounded by a polynomial in the size of the underlying database. Moreover, in spite of its syntactical simplicity, Datalog can capture queries that are complete for polynomial time computations, such as the *path systems* query [Coo74]. The exact expressive power of Datalog is completely understood in certain cases. More specifically, if only *ordered* databases are considered, then a query is expressible by a Datalog program if and only if it is computable in polynomial time. This result is implicit in [Imm86, Var82] and stated explicitly in [Pap85, BG87].

The picture, however, changes dramatically on general (unordered) databases. In this case, as pointed out first in [CH85], there are natural polynomial time properties that are not expressible by any Datalog program. This follows easily from the fact that in general Datalog programs compute queries that are *strongly monotone*, i.e., they are preserved by adding tuples to database predicates (EDBs), or by adding

[3] To be precise, Talanov and Knyazev's result was in terms of a certain iterative extension of first-order logic, which in particular includes positive-fixpoint logic. The study of 0-1 laws for iterative extensions of first-order logic was further pursued by Knyazev [Kny89].

elements to the universe of the database, or by identifying elements of the universe of the database (i.e., collapsing multiple elements into a single element). Thus, if a query is not strongly monotone, then there is no Datalog program that expresses it.

One natural way to increase the expressive power of Datalog, without changing the semantics, is to allow equalities $u = v$ and inequalities $u \neq v$ in the bodies of the rules. After all, these constructs are quite natural in a database query language. This extension of Datalog, which we will denote by Datalog($\neq$), was considered first by Shmueli [Shm87], who showed that it is more expressive than Datalog, and then was investigated further in [GMSV87, AG89], where it was shown that Datalog($\neq$) differs from Datalog in more aspects than just expressive power. If a query is computed by a Datalog($\neq$) program, then it has the property that it is *monotone*, i.e., it is preserved by adding new elements to the universe of the database and by adding tuples to the database predicates. As a result, there is no Datalog($\neq$) program that expresses the complement of the transitive closure query, or any other polynomial time query that lacks the above monotonicity property.

The preceding remarks establish that the Datalog($\neq$) queries and, a fortiori, the Datalog queries, form a proper subset of the polynomial time queries. So far, however, only modest progress has been made towards a complete understanding of which monotonic queries are expressible in Datalog($\neq$) [LM89, ACY91]. We show here that the expressive power of Datalog($\neq$) can be studied by considering it as a fragment of L^ω [KV90b], the latter being the existential negation-free fragment of $L^\omega_{\infty\omega}$. We demonstrate the usefulness of our approach by considering the class of *fixed subgraph homeomorphism* queries. Here we have a fixed pattern graph H and we have to determine whether it can be homeomorphically embedded in a given directed graph. The complexity of fixed subgraph homeomorphism queries was completely classified by Fortune et al. [FHW80] in terms of two dichotomies. They defined a class C of pattern graphs and showed that an H-subgraph homeomorphism query is (1) polynomial for pattern graphs H in C, (2) NP-complete for pattern graphs H in the complement $\overline{C}$ of C, and (3) polynomial for acyclic input graphs. Of course, the two dichotomies, pattern graphs in C vs. pattern graphs in $\overline{C}$ and general input graphs vs. acyclic input graphs, are proper only if P $\neq$ NP.

Without using any complexity-theoretic assumptions, we show here that the two dichotomies are indeed proper in terms of expressibility in Datalog($\neq$). More specifically, on the positive side we show that the H-subgraph homeomorphism query is expressible in Datalog($\neq$) for every pattern graph H in the class C, and we also prove that if only acyclic graphs are allowed as inputs, then the H-subgraph homeomorphism query is expressible in Datalog($\neq$) for every pattern graph H. On the negative side we demonstrate that if the pattern graph H is in the complement $\overline{C}$ of C, then the fixed subgraph homeomorphism query is *not* expressible in L^ω and, a fortiori, is not expressible in Datalog($\neq$) as well. The proof is an interesting illustration of how the game-theoretic characterization of L^ω can be used to translate a complexity lower-bound proof to an expressibility lower-bound proof. One of the main difficulties in game-theoretic proofs of lower bounds is finding the right structures on which to play the game. The idea underlying our proof is to play the game on structures generated by the reduction in the NP-hardness result of [FHW80].

In summary, we believe that two examples reviewed in this paper supports our claim that the infinitary logic $L^\omega_{\infty\omega}$ is of fundamental importance in finite-model theory. We refer the reader to [AVV92, DLW91, KV92] for further investigations of the relationship between fixpoint logics, $L^{\omega}_{\infty\omega}$, and its game-theoretical characterization.

2 Infinitary Logics and Fixpoint Logics

2.1 Infinitary Logics with a Fixed Number of Variables

Different infinitary logics arise by allowing for infinite disjunctions and conjunctions or by allowing for infinite strings of quantifiers, or by allowing for both at the same time. We consider the infinitary logic $L_{\infty\omega}$, which is the extension of first-order logic that results by allowing infinite disjunctions and conjunctions in the syntax, while keeping the quantifier strings finite (cf. [BF85]). To illustrate the gain in expressive power, recall the well-known fact that the property "there is an even number of elements" is not expressible by any first-order sentence on finite structures. Let ρ_n be a first-order sentence stating that there are exactly n elements. Then the infinitary sentence $\bigvee_{n=1}^{\infty} \rho_{2n}$ asserts that "there is an even number of elements".

We now define formally the syntax of the infinitary logic $L_{\infty\omega}$.

Definition 1. Let σ be a vocabulary consisting of finitely many relational and constant symbols and let $\{v_1, \ldots, v_n\}$ be a countable set of variables. The class $L_{\infty\omega}$ of *infinitary formulas* over σ is the smallest collection of expressions such that

- it contains all first-order fomulas over σ;
- if φ is an infinitary formula, then so is $\neg\varphi$;
- if ψ is an infinitary formula and v_i is a variable, then $(\forall v_i)\varphi$ and $(\exists v_i)\varphi$ are also infinitary formulas;
- if Ψ is a set of infinitary formulas, then $\bigvee \Psi$ and $\bigwedge \Psi$ are also infinitary formulas

∎

The semantics of infinitary formulas is a direct extension of the semantics of first-order logic, with $\bigvee \Psi$ interpreted as a disjunction over all formulas in Ψ and $\bigwedge \Psi$ interpreted as a conjunction.

In general, infinitary formulas, even infinitary sentences, may have an infinite number of distinct variables. We now focus attention on fragments of $L_{\infty\omega}$ in which the total number of variables is required to be finite. Variables, however, may have an infinite number of occurrences in such formulas.

Definition 2. Let k be a positive integer.

- The *infinitary logic with k variables*, denoted by $L^k_{\infty\omega}$, consists of all formulas of $L_{\infty\omega}$ with at most k distinct variables.
- The infinitary logic $L^{\omega}_{\infty\omega}$ consists of all formulas of $L_{\infty\omega}$ with a finite number of distinct variables. Thus, $L^{\omega}_{\infty\omega} = \bigcup_{k=1}^{\infty} L^k_{\infty\omega}$.
- We write $L^k_{\omega\omega}$ for the collection of all first-order formulas with at most k variables.

∎

The family $L^{\omega}_{\infty\omega}$ of the infinitary languages $L^{k}_{\infty\omega}$, $k \geq 1$, was introduced first by Barwise [Bar77], as a tool for studying positive-fixpoint logic on infinite structures.

We now give some examples that illustrate the expressive power of infinitary logic with a fixed number of variables.

Example 1. Paths and Connectivity
Assume that the vocabulary σ consists of a single binary relation E and let $p_n(x,y)$ be a first-order formula over σ asserting that there is a path of length n from x to y. The obvious way to write $p_n(x,y)$ requires $n+1$ variables, namely

$$(\exists x_1 \dots \exists x_{n-1})(E(x,x_1) \wedge E(x_1,x_2) \wedge \dots \wedge E(x_{n-1},y)).$$

It is well known, however, that each $p_n(x,y)$ is equivalent to a formula in $L^3_{\omega\omega}$, i.e. a first-order formula with at most three distinct variables x,y,z. To see this, put

$$p_1(x,y) \equiv E(x,y)$$

and assume, by induction on n, that $p_{n-1}(x,y)$ is equivalent to a formula in $L^3_{\omega\omega}$. Then

$$p_n(x,y) \equiv (\exists z)[E(x,z) \wedge (\exists x)(x = z \wedge p_{n-1}(x,y))].$$

It follows that "connectivity" is a property of graphs expressible in $L^3_{\infty\omega}$, since it is given by the formula $(\forall x \forall y)(\bigvee_{n=1}^{\infty} p_n(x,y))$. Similarly, the property "there is no cycle" is also in $L^3_{\infty\omega}$, since it is definable by $(\forall x)(\bigwedge_{n=1}^{\infty} \neg p_n(x,x))$. Notice that if P is *any* set of positive integers, then the property "x and y are connected by a path whose length is a number in P" is expressible in $L^3_{\infty\omega}$ via the formula $\bigvee_{n \in P} p_n(x,y)$. It follows that $L^{\omega}_{\infty\omega}$ can express non-recursive properties on finite graphs.
∎

Properties such as "connectivity" and "there is no cycle" are also known to be expressible in positive-fixpoint logic. We consider next extensions of first-order logic with fixpoint formation rules and compare the resulting logics to $L^{\omega}_{\infty\omega}$.

2.2 Fixpoint Logics

Let σ be a vocabulary, let S be an n-ary relation symbol not in σ, let $\varphi(x_1,\dots,x_n,S)$ be a first-order formula over the vocabulary $\sigma \cup \{S\}$, and let $\mathbf{D}$ be a finite structure over σ. The formula φ gives rise to an operator $\Phi(S)$ from n-ary relations on the universe D of $\mathbf{D}$ to n-ary relations on D, where

$$\Phi(T) = \{(a_1,\dots,a_n) : \mathbf{D} \models \varphi(a_1,\dots,a_n,T)\},$$

for every n-ary relation T on D.

Every such operator $\Phi(S)$ generates a sequence of *stages* that are obtained by iterating $\Phi(S)$. We will be interested here in the relationship between the stages of the operator and its *fixpoints*.

Definition 3. Let $\mathbf{D}$ be a finite structure over the vocabulary σ.

- The stages Φ^m, $m \geq 1$, of Φ on $\mathbf{D}$, are defined by the induction:

$$\Phi^1 = \Phi(\emptyset), \ \Phi^{m+1} = \Phi(\Phi^m).$$

- We say that a relation T on D is a *fixpoint* of the operator $\Phi(S)$ (or, of the formula φ) if $\Phi(T) = T$.

■

Intuitively, one would like to associate with an operator $\Phi(S)$ the "limit" of its stages. This is possible only when the sequence Φ^m, $m \geq 1$, of the stages "converges", i.e., when there is an integer m_0 such that $\Phi^{m_0} = \Phi^{m_0+1}$ and, hence $\Phi^{m_0} = \Phi^m$, for all $m \geq m_0$. Notice that in this case Φ^{m_0} is a fixpoint of $\Phi(S)$, since $\Phi^{m_0} = \Phi^{m_0+1} = \Phi(\Phi^{m_0})$. The sequence of stages, however, may not converge. In particular, this will happen if the formula $\varphi(\mathbf{x}, S)$ has no fixpoints. Thus, additional conditions have to be imposed on the formulas considered in order to ensure that the sequence of stages converges.

A formula $\varphi(x_1, \ldots, x_n, S)$ is *positive in* S if every occurrence of S in φ is within an even number of negations. Positivity is a natural syntactic condition that guarantees convergence. Indeed, if $\varphi(\mathbf{x}, S)$ is positive in S, then the associated operator Φ is *monotone* (i.e., if $T_1 \subseteq T_2$, then $\Phi(T_1) \subseteq \Phi(T_2)$) and, as a result, the sequence Φ^m, $m \geq 1$, of stages is increasing. If $\mathbf{D}$ is a finite structure with s elements, then every stage Φ^m has at most s^n elements and, consequently, there is an integer $m_0 \leq s^n$ such that $\Phi^{m_0} = \Phi^m$ for every $m \geq m_0$. Thus, the sequence of stages of $\varphi(\mathbf{x}, S)$ converges to Φ^{m_0}. Moreover, it is easy to verify that Φ^{m_0} is the *least fixpoint* of $\varphi(\mathbf{x}, S)$, i.e., it is a fixpoint of φ with the property that $\Phi^{m_0} \subseteq T$ for every fixpoint T of φ. We write φ^∞ or Φ^∞ to denote the least fixpoint of φ.

Positive-fixpoint logic is first-order logic augmented with the *least fixpoint* formation rule for *positive* formulas. The canonical example of a formula of positive-fixpoint logic is provided by the least fixpoint $\varphi^\infty(x, y)$ of the first-order formula

$$E(x, y) \vee (\exists z)(S(x, z) \wedge S(z, y)).$$

In this case $\varphi^\infty(x, y)$ defines the *transitive closure* of the *edge* relation E. It follows that *connectivity* is a property expressible in positive-fixpoint logic, but, as is well known (cf. [Fag75, AU79]), not in first-order logic.

The theory of positive-fixpoint logic on infinite structures was developed in Moschovakis [Mos74]. Chandra and Harel [CH82] were the first to focus attention on the collection FP of properties expressible in positive-fixpoint logic on finite structures (*positive-fixpoint properties*). Since that time, positive-fixpoint logic has been studied extensively on finite structures and this has resulted to a thorough understanding of its expressive power (cf. [Cha88] for a survey of results in this area). We should remark that often in the literature positive-fixpoint logic is referred to as simply *fixpoint logic*.

How can we obtain logics with fixpoint constructs that are more expressive than positive-fixpoint logic? A more powerful logic results if one iterates arbitrary first-order operators, until a fixpoint is reached (which may never happen). In this case we may have non-terminating computations, unlike positive-fixpoint logic, where the iteration is guaranteed to converge.

Definition 4. Let σ be a vocabulary, let S be an n-ary relation symbol not in σ, let $\varphi(x_1, \ldots, x_n, S)$ be a first-order formula over the vocabulary $\sigma \cup \{S\}$, let D be a finite structure over σ, and let Φ^m, $m \geq 1$, be the sequence of stages of the associated operator $\Phi(S)$.

If there is an integer m_0 such that $\Phi^{m_0} = \Phi^{m_0+1}$, then we put $\varphi^\infty = \Phi^\infty = \Phi^{m_0}$; otherwise, we set $\varphi^\infty = \Phi^\infty = \emptyset$. In the former case we say that φ *converges* on $\mathbf{D}$, and in the latter case we say that φ *diverges* on $\mathbf{D}$. We call φ^∞ the *partial-fixpoint* of φ on $\mathbf{D}$.

Partial-Fixpoint Logic is first-order logic augmented with the partial-fixpoint formation rule for arbitrary first-order formulas. ∎

Partial-fixpoint logic on finite structures has been investigated by Abiteboul and Vianu [AV89]. In particular, they established that the class PFP of partial-fixpoint properties on finite structures coincides with the class of properties expressible in the language RQL *(ranked query language)*, introduced by Chandra and Harel [CH82] and studied also in [KV87] under the name *iterative logic.* The latter is an extension of first-order logic obtained by adding **while** looping as an iteration construct.

Notice that for every first-order formula $\varphi(x_1, \ldots, x_n, S)$, if $\mathbf{D}$ is a finite structure with s elements, then either the sequence Φ^m, $m \geq 1$, of stages converges or it cycles. Which of the two is the case can be determined by carrying out at most 2^{s^n} iterations of Φ. Thus, the computation of the partial-fixpoint requires space polynomial in the size s of the structure $\mathbf{D}$, since we only have to store one stage at a time and compute the next stage, while making sure that the current level of iteration has not exceeded 2^{s^n}. Notice also that if $\varphi(x_1, \ldots, x_n, S)$ is a positive in S formula, then the partial-fixpoint of φ coincides with the least fixpoint of φ, because the sequence of stages converges. It follows that partial-fixpoint logic is an extension of positive-fixpoint logic. It was shown in [AV91] that the two logic have the same expressive power if and only if P=PSPACE.

The key observation underlying this paper is that positive-fixpoint logic and partial-fixpoint logic can be viewed as fragments of $L^\omega_{\infty\omega}$.

Theorem 5 [KV92]. *Let $\varphi(x_1, \ldots, x_m, S)$ be an $L^k_{\omega\omega}$- formula over the vocabulary $\sigma \cup \{S\}$, where S is an m-ary relation, $m \leq k$. Let $\Phi(S)$ be the operator associated with φ. Then, for every $l \geq 1$, the stage $\Phi^l(x_1, \ldots, x_k)$ of Φ is definable by an $L^k_{\omega\omega}$-formula on all finite structures over σ. As a result, the partial fixpoint $\varphi^\infty(x_1, \ldots, x_k)$ of $\varphi(x_1, \ldots, x_k, S)$ is definable by a formula of $L^k_{\infty\omega}$ on all finite structures over σ.*

Note, however, that $L^\omega_{\infty\omega}$ is more expressive than partial-fixpoint logic, since $L^\omega_{\infty\omega}$ can express non-recursive properties.

2.3 $L^k_{\infty\omega}$-Equivalence and Pebble Games

Two structures are *equivalent* in some logic L if no sentence of L distinguishes them. This is a central concept in every logic and plays an important role in model theory. We discuss here equivalence in the infinitary logics with a fixed number of variables.

Definition 6. Let $\mathbf{A}$ and $\mathbf{B}$ be two structures over the vocabulary σ and let k be a positive integer.

- We say that **A** is $L^k_{\infty\omega}$*-equivalent* to **B**, and we write $\mathbf{A} \equiv^k_{\infty\omega} \mathbf{B}$, if **A** and satisfy the same sentences of $L^k_{\infty\omega}$.
- We say that **A** is $L^k_{\omega\omega}$*-equivalent* to **B**, and we write $\mathbf{A} \equiv^k_{\omega\omega} \mathbf{B}$, if **A** and satisfy the same sentences of $L^k_{\omega\omega}$.

∎

The connection between definability in $L^k_{\infty\omega}$ and the equivalence relation $\equiv^k_{\infty}$ is described by the following two propositions.

Proposition 7 [KV90a]. *Let σ be a vocabulary, let $\mathcal{F}$ be the class of all fini structures over σ, and let k be a positive integer. Then every equivalence class of th equivalence relation $\equiv^k_{\infty\omega}$ on $\mathcal{F}$ is definable by a sentence of $L^k_{\infty\omega}$.*

Proposition 8 [KV90a]. *Let C be a class of finite structures over the vocabulary and let k be a positive integer. Then the following statements are equivalent:*

1. *The class C is $L^k_{\infty\omega}$-definable, i.e. there is a sentence φ of $L^k_{\infty\omega}$ such that fo any finite structure* **A** *over σ we have that*

$$\mathbf{A} \in C \Longleftrightarrow \mathbf{A} \models \varphi.$$

2. *If* **A** *and* **B** *are finite structures over σ such that $\mathbf{A} \in C$ and $\mathbf{A} \equiv^k_{\infty\omega} \mathbf{B}$, the $\mathbf{B} \in C$.*

It is known that $\equiv^k_{\infty\omega}$-equivalence can be characterized in terms of an *infinitar k-pebble game.* This game was implicit in Barwise [Bar77] and was described in deta in Immerman [Imm82].

Definition 9. Assume that **A** and **B** are two structures over the vocabulary σ an let $c_1, \ldots, c_l$ and $d_1, \ldots, d_l$ be the interpretations of the constant symbols of σ on and **B**, respectively.

The *k-pebble game* between Players I and II on the structures **A** and **B** has th following rules:

Player I chooses one of the two structures **A** and **B** and places a pebble on on of its elements. Player II responds by placing a pebble on an element of the othe structure. Player I chooses again one of the two structures and the game continue this way until k pebbles have been placed on each structure.

Let a_i and b_i, $1 \leq i \leq k$, be the elements of **A** and **B** respectively picked b the two players in the i-th move. Player I wins the game at this point if one of th following two conditions holds:

- Two of the pebbles are on the same element of one of the structures, while th corresponding two pebbles are on different elements of the other structure, i.e $a_i = a_j$ and $b_i \neq bj$ (or $a_i \neq a_j$ and $b_i = b_j$), for some i and j such tha $1 \leq i < j \leq k$.
- The previous condition fails and the mapping h with

$$h(a_i) = b_i,\ 1 \leq i \leq k,$$

and

$$h(c_j) = d_j,\ 1 \leq j \leq l,$$

is *not* an isomorphism between the substructures of **A** and **B** with universes

$$\{a_1, \dots, a_k\} \cup \{c_1, \dots, c_l\}$$

and

$$\{b_1, \dots, b_k\} \cup \{d_1, \dots, d_l\}$$

respectively.

both conditions fail, then Player I removes some corresponding pairs of pebbles d the game resumes until again k pebbles have been placed on each structure. ∎

Player II wins the game if he can continue playing "forever", i.e. if Player I can ever win a round of the game. The concept of a *winning strategy* for Player II in the -pebble game is formalized in terms a family of partial isomorphisms; see [KV90a] r details.

The crucial connection between k-pebble games and $L^k_{\infty\omega}$-equivalence is provided y the following result.

heorem 10 [Bar77, Imm82, KV90a]. *Let* **A** *and* **B** *be two finite structures ver the vocabulary σ, and let k be a positive integer. The following are equivalent:*

1. $\mathbf{A} \equiv^k_{\omega\omega} \mathbf{B}$.
2. $\mathbf{A} \equiv^k_{\infty\omega} \mathbf{B}$.
3. Player II has a winning strategy for the k-pebble game on **A** *and* **B**.

As a consequence of Proposition 8 and Theorem 10, we get a game-theoretic haracterization of definability in the logics $L^k_{\infty\omega}$, $k \geq 1$, for classes of finite strucures.

roposition 11 [KV90a]. *Let C be a class of finite structures over the vocabulary and let k be a positive integer. Then the following statements are equivalent:*

1. The class C is $L^k_{\infty\omega}$-definable.
2. If **A** *and* **B** *are finite structures over σ such that $\mathbf{A} \in C$ and Player II has a winning strategy for the k-pebble game on* **A** *and* **B**, *then $\mathbf{B} \in C$.*

The preceding results provide tools for establishing that certain properties are ot expressible in infinitary logic with a finite number of variables. More specifically, n order to establish that a property Q is not expressible by any formula of $L^\omega_{\infty\omega}$ on inite structures it is enough to show that for any $k \geq 1$ there are finite structures $\mathbf{A}_k$ and $\mathbf{B}_k$ such that $\mathbf{A}_k \models Q$, $\mathbf{B}_k \not\models Q$, and Player II has a winning strategy for he k-pebble game on $\mathbf{A}_k$ and $\mathbf{B}_k$. Moreover, Proposition 11 guarantees that this nethod is also *complete*, i.e., if Q is not expressible in $L^k_{\infty\omega}$, then such structures $\mathbf{A}_k$ and $\mathbf{B}_k$ must exist.

3 0-1 Laws for Infinitary Logics

Let σ be a vocabulary consisting of finitely many relation symbols *only* and let C b the class of all finite structures over σ with universe an initial segment $\{1, 2, \ldots, n$ of the integers for some $n \geq 1$. If P is a property of (some) structures in C, the the *(labeled) asymptotic probabilty* $\mu(P)$ on C is defined to be equal to the limit a $n \to \infty$ of the fraction of structures in C of cardinality n which satisfy P, provide this limit exists. If L is a logic, we say that the 0-1 *law holds for* L in case $\mu(P$ exists and is equal to 0 or 1 for every property P expressible in the logic L.

In the past, 0-1 laws for various logics L were proved by establishing first *transfer theorem for* L of the following kind:

> *There is an infinite structure* **R** *over the vocabulary* σ *such that for any property* P *expressible in* L *we have:*
>
> $$\mathbf{R} \text{ satisfies } P \iff \mu(P) = 1 \text{ on } C.$$

This method was discovered by Fagin [Fag76] in his proof of the 0-1 law for firs order logic on finite structures. It was also used later in [BGK85] to establish th 0-1 law for positive-fixpoint logic and in [KV87] to show that the 0-1 law holds fo partial-fixpoint logic.

It turns out that there is a countable structure **R** over the vocabulary σ that sa isfies the above equivalence for all these logics. Moreover, this structure **R** is uniqu up to isomorphism. We call **R** the *countable random structure over the vocabular* σ. The random structure **R** is characterized by an infinite set of *extension axiom* which, intuitively, assert that every *type* can be *extended* to any other possible typ The precise definitions are as follows.

Definition 12. Let σ be a vocabulary consisting of relation symbols only.

- If $\mathbf{x} = (x_1, \ldots, x_m)$ is a sequence of distinct variables, then a *type* $t(\mathbf{x})$ *in th variables* $\mathbf{x}$ *over* σ is the conjunction of all the formulas in a *maximally consiste* set S of equalities $x_i = x_j$, inequalities $x_i \neq x_j$, atomic formulas in the variable $\mathbf{x}$, and negated atomic formulas in the variables $\mathbf{x}$.
- Let z be a variable that is different from all the variables in $\mathbf{x}$. We say that *type* $t(\mathbf{x}, z)$ *extends the type* $s(\mathbf{x})$ if every conjunct of $s(\mathbf{x})$ is also a conjunct o $t(\mathbf{x}, z)$.
- With each pair $s(\mathbf{x})$ and $t(\mathbf{x}, z)$ of types such that t extends s we associate first-order *extension axiom* $\tau_{s,t}$ stating that

$$(\forall \mathbf{x})(s(\mathbf{x}) \to (\exists z) t(\mathbf{x}, z)).$$

∎

Let T be the set of all extension axioms. The theory T was studied by Gaifma [Gai64], who showed, using a *back and forth* argument, that any two countable model of T are isomorphic (T *is an* ω*-categorical theory*). Fagin [Fag76] realized that th extension axioms are relevant to the study of probabilities on finite structures an proved that on the class C of all finite structures over a finite vocabulary σ

$$\mu(\tau_{s,t}) = 1$$

r any extension axiom $\tau_{s,t}$. The equivalence between truth on **R** and almost sure uth on C (and consequently the 0-1 law for first-order logic on finite structures) llows from these two results by an application of the compactness theorem of athematical logic.

We now show that the 0-1 law holds for the infinitary logic $L^{\omega}_{\infty\omega}$ on finite strucres using the game-theoretic characterization of $L^{\omega}_{\infty\omega}$. Unlike the proofs that are ised on the transfer property, our proof here does note employ any "infinitistic" ethods; the proofs does not assume the 0-1 law for first-order logic, it does not volve the random structure **R** or any other infinite structure, and it does not make se of compactness or of any of its consequences. Moreover, the 0-1 law is derived rectly without establishing a transfer theorem first. The 0-1 law for $L^{\omega}_{\infty\omega}$ on the ne hand subsumes all earlier ones in [BGK85, Fag76, KV87, Tal81, TK86], and on e other hand provides a unifying treatment of 0-1 laws for first-order logic and its tensions with fixpoint operators or infinitary syntax.

Let k be a fixed positive integer. If **A** is a finite structure, then we write [**A**] for *the uivalence class* of **A** with respect to the equivalence relation $\equiv^{k}_{\infty\omega}$. In what follows e will show that there is a tight connection between 0-1 laws and the asymptotic robabilities of equivalence classes [**A**]. Actually, this turns out to be a general fact at holds for arbitrary probability measures.

So far all the results discussed here are about the *uniform* probability measures on , i.e., all structures with n elements carry the same probability. There is, however, well developed study of random structures under *variable* probability measures. his started with the work of Erdös and Rényi [ER60] and is presented in detail in ollobás [Bol85]. In general, for each $n \geq 1$ one has a probability measure pr_n on all ructures in C with n elements, where pr_n may be a non-uniform distribution. The *symptotic probability* $pr(P)$ of a property P (*relative to the probability measures* r_n, $n \geq 1$) is defined by $pr(P) = \lim_{n\to\infty} pr_n(P)$, provided this limit exists. If L is logic, then we say that a *0-1 law holds for L relative to the measure pr* if for every entence ψ of L the asymptotic probability $pr(\psi)$ exists and is either 0 or 1. Notice at, strictly speaking, pr is not a probability measure, because it is not countably dditive (it is, however, finitely additive).

We now establish necessary and sufficient condition for the existence of 0-1 laws r $L^{k}_{\infty\omega}$ under arbitrary probability measures.

heorem 13 [KV90a]. *Let K be a class of finite structures over the vocabulary σ, t k be a positive integer, and let pr_n, $n \geq 1$, be a sequence of probability measures n the structures in K with n elements. Then the following are equivalent:*

1. *The 0-1 law holds for the infinitary logic $L^{k}_{\infty\omega}$ relative to the measure pr.*
2. *There is an equivalence class C of the equivalence relation $\equiv^{k}_{\infty\omega}$ such that $pr(C) =$ 1.*

We now return to the uniform measure μ on $L^{k}_{\infty\omega}$ and give a proof of the 0-1 law r $L^{k}_{\infty\omega}$ using the preceding Theorem 13 and the characterization of $\equiv^{k}_{\infty\omega}$ in terms f pebble games.

heorem 14 [KV90a]. *Let C be the class of all finite structures, let k be a positive nteger, and let θ_k be the conjunction of all extension axioms with at most k variables.*

If **A** *is a finite structure that is a model of* θ_k*, then* $\mu([\mathbf{A}]) = 1$*. As a result, the 0-1 law holds for* $L^k_{\infty\omega}$ *relative to the uniform measure on* C*.*

Proof Sketch: If **A** and **B** are both models of θ_k, then it is easy to verify that Player II has a winning strategy in the k-pebble game on **A** and **B**. Intuitively, the winning strategy for Player II is provided by the elements of **A** and **B** witnessing the extension axioms with at most k variables.

It follows from Theorem 10 that if **A** and **B** are both models of θ_k, then $\mathbf{A} \equiv^k_\infty \mathbf{B}$. Since $\mu(\theta_k) = 1$ [Fag76], it follows that $\mu([\mathbf{A}]) = 1$, for any finite structure **A** that is a model of θ_k. The 0-1 law now follows by Theorem 13. ∎

4 Datalog($\neq$) and Infinitary Logic

Datalog is a fragment of positive-fixpoint logic that has attracted a lot of attention [Ull89]. We consider here Datalog($\neq$), which extends Datalog by allowing inequalities [AG89, GMSV87, Shm87]. A Datalog($\neq$) program is a finite collection of rules of the form

$$t_0 \leftarrow t_1, t_2, \ldots, t_l.$$

The expression t_0 is the *head* of the rule, while $t_1, \ldots, t_l$ form the *body of the rule*. The head of the rule is an atomic formula $S(x_1, \ldots, x_n)$, where S is a relational symbol. The expressions in the body can be equalities $x_j = x_k$, inequalities $x_j \neq x_k$, or atomic formulas $R(x_1, \ldots, x_m)$. Notice that negated atomic formulas are not allowed in a Datalog($\neq$) program. The relational symbols appearing in the heads of the rules form the *intensional database predicates* (IDBs), while the others are the *extensional database predicates* (EDBs). One of the IDB predicates is designated as the *goal* predicate of the program. (A Datalog program is a Datalog($\neq$) program such that the body of every rule consists entirely of atomic formulas, i.e., no equalities or inequalities are allowed.)

A Datalog($\neq$) program is evaluated by repeatedly applying its rules and deriving *facts* (i.e., ground atomic formulas) till no more facts can be derived. In what follows, π^∞ will denote the semantics of a Datalog($\neq$) program π. Chandra and Harel [CH85] were the first to point out that Datalog and its extensions are related to the existential fragment of fixpoint logic. More precisely, their results imply that Datalog($\neq$) is equivalent to the negation-free existential fragment of positive-fixpoint logic, while Datalog is equivalent to the negation and inequality-free existential fragment of positive-fixpoint logic. To see that, assume first for simplicity that π is a Datalog($\neq$) program with a single IDB predicate S of arity r. We can assume without loss of generality that all the heads are of the form $S(y_1, \ldots, y_r)$ and that the bodies of the rules do not share any variable other than $y_1, \ldots, y_r$. Let the bodies be $\varphi_1, \ldots, \varphi_m$, and let $x_1, \ldots, x_t$ be the variables used in these bodies. Let φ_π denote the formula $(\exists x_1, \ldots, x_t) \bigvee_{i=1}^m \varphi_i$. Then π^∞ is equivalent to the positive-fixpoint formula φ_π^∞. If the Datalog($\neq$) program π has more than one IDB predicates, then it can be viewed as a system of monotone operators, but it is known that in the context of positive-fixpoint logic a system of operators can be collapsed to a single operator [Mos74].

Example 2. Let π be the Datalog($\neq$) program

$$T(x,y,w) \leftarrow E(x,y), w \neq x, w \neq y$$
$$T(x,y,w) \leftarrow E(x,z), T(z,y,w), w \neq x.$$

Then φ_π is the formula

$$(E(x,y) \land w \neq x \land w \neq y) \lor (\exists z)(E(x,z) \land T(z,y,w) \land w \neq x).$$

On any graph $\mathbf{G} = (V, E)$ the least fixpoint φ_π^∞ expresses the query "is there a path from x to y that does not go through w?". ∎

The above discussion points out the connection between Datalog($\neq$) and the negation-free existential fragment of $L^\omega_{\infty\omega}$.

Definition 15. Let L^k be the collection of all formulas of $L^k_{\infty\omega}$, $k \geq 1$, that are obtained from atomic formulas, equalities, and inequalities using infinitary disjunctions, infinitary conjunctions, and existential quantification only. We also put

$$L^\omega = \bigcup_{k=1}^{\infty} L^k$$

and call this logic the *existential negation-free fragment of* $L^\omega_{\infty\omega}$. ∎

Our next result shows that L^ω is powerful enough to express every Datalog($\neq$) query.

Theorem 16 [KV90b]. *If π is a Datalog($\neq$) program, then π^∞ is definable by a formula of L^ω.*

The results above imply that, in order to show that a particular query Q is not expressible in Datalog($\neq$), it is enough to establish that Q is not definable in L^ω. This requires an extension of the framework developed in Section 2.3.

Definition 17. Let k be a positive integer and let $\mathbf{A}$ and $\mathbf{B}$ be two structures over the vocabulary σ. We write $\mathbf{A} \preceq^k \mathbf{B}$ to denote that every sentence of L^k that is true in $\mathbf{A}$ is also true in $\mathbf{B}$. ∎

The connection between expressibility in L^k and the relation $\preceq^k$ is described by the following proposition.

Proposition 18 [KV90b]. *Let C be a class of finite structures over the vocabulary σ and let k be a positive integer. Then the following statements are equivalent:*

1. *The class C is L^k-definable, i.e. there is a sentence φ of L^k such that for any finite structure $\mathbf{A}$ over σ we have that*

$$\mathbf{A} \in C \Longleftrightarrow \mathbf{A} \models \varphi.$$

2. *If $\mathbf{A}$ and $\mathbf{B}$ are finite structures over σ such that $\mathbf{A} \in C$ and $\mathbf{A} \preceq^k \mathbf{B}$, then $\mathbf{B} \in C$.*

The relation $\preceq^k$ can be characterized in terms of an asymmetric version of th k-pebble games of Section 2.3.

Definition 19. Let σ be a vocabulary consisting of relational symbols and constar symbols. Let **A**, **B** be two structures over σ and let $c_1, \ldots, c_l$ and $d_1, \ldots, d_l$ be th interpretations of the constant symbols of σ on **A** and **B**, respectively.

The *existential k-pebble game* between Players I and II on the structures **A** an **B** is the k-pebble games between Players I and II on the structures **A** and **B** wher the pebbles of Player I are always placed on **A** while the pebbles of Player II ar always placed on **B**.

Let a_i and b_i, $1 \leq i \leq k$, be the elements of **A** and **B** respectively picked b the two players in the i-th move. Player I wins the game at this point if one of th following two conditions holds:

- Two of the pebbles are on the same element of **A** while the corresponding tw pebbles are on different elements of the other structure, i.e., $a_i = a_j$ and $b_i \neq b$ for some i and j such that $1 \leq i < j \leq k$.
- The previous condition fails and the mapping h with

  $$h(a_i) = b_i, \ 1 \leq i \leq k,$$

  and

  $$h(c_j) = d_j, \ 1 \leq j \leq l,$$

  is *not* a one-to-one homomorphism[4] between the substructures of **A** and **B** wit universes

  $$\{a_1, \ldots, a_k\} \cup \{c_1, \ldots, c_l\}$$

  and

  $$\{b_1, \ldots, b_k\} \cup \{d_1, \ldots, d_l\}$$

  respectively.

Otherwise, the game continues. ∎

Player II wins the game if he has a winning strategy that allows him to continu playing "forever", i.e., if Player I can never win a round of the game. A forma definition of a winning strategy can be given in terms a family of partial one-to-on homomorphisms; see [KV90b] for details.

We give next an example that illustrate existential k-pebble games and revea that the relation "Player II wins the existential k-pebble game on **A** and **B**" is no a symmetric relation on pairs of structures

[4] An *one-to-one homomorphism* from a structure **A** into a structure **B** is a one-to-one mapping h from the domain of **A** into the domain of **B** such that the constants of **A** are mapped to the corresponding constants of **B** and if $(w_1, \ldots, w_n)$ is a tuple in a relation R of **A**, then $(h(w_1), \ldots, h(w_n))$ is a tuple in the relation R of **B**.

Example 3. Paths of different length

Assume that the vocabulary σ consists of a single binary relation symbol E. Let $\mathbf{A}$ be a graph that is a directed path with m vertices and let $\mathbf{B}$ be a directed path with n vertices, where $n > m \geq 2$. Assume that $\mathbf{A}$ has vertices $\{a_1, \ldots, a_m\}$ and edges (a_i, a_{i+1}), $1 \leq i < m$, while $\mathbf{B}$ has vertices $\{b_1, \ldots, b_n\}$ and edges (b_j, b_{j+1}), $1 \leq j < n$.

It is quite clear that Player II wins the existential k-pebble game on $\mathbf{A}$ and $\mathbf{B}$, for any $k \geq 1$. His strategy is to simply "copy" on $\mathbf{B}$ the moves of Player I on $\mathbf{A}$. More precisely, whenever Player I places a pebble on a node a_i of $\mathbf{A}$, Player II responds by placing a pebble on the node b_i of $\mathbf{B}$.

In contrast, we claim that Player I can win the existential 2-pebble game on $\mathbf{B}$ and $\mathbf{A}$ (and, hence, Player I can also win the existential k-pebble game on $\mathbf{B}$ and $\mathbf{A}$ for any $k \geq 2$). Player I begins by placing his two pebbles on the nodes b_1, b_2 of $\mathbf{B}$. Player II has to respond by placing his two pebbles on two consecutive nodes of $\mathbf{A}$; actually, his best option is to place the two pebbles on a_1, a_2. Player I now removes the pebble from b_1 and places it on b_3. Player II's only option is to remove the pebble from a_1 and place it on a_3. Player I can continue playing this way, moving along the path on $\mathbf{B}$ and forcing Player II to move along the path on $\mathbf{A}$. After a while, Player II has two pebbles on b_{m-1}, b_m and Player I has his corresponding pebbles on a_{m-1}, a_m. Since $n > m$, Player I can win the game in his next move by removing the pebble from b_{m-1} and placing it on b_{m+1}. At this point, Player II can no longer maintain a one-to-one homomorphism. ∎

It turns out that $\preceq^k$ can be characterized in terms of the existential k-pebble game according to the following crucial result.

Theorem 20 [KV90b]. *Let k be a positive integer and let $\mathbf{A}$, $\mathbf{B}$ be two finite structures over the vocabulary σ. Then the following two statements are equivalent:*

1. $\mathbf{A} \preceq^k \mathbf{B}$.
2. *Every first-order sentence of the infinitary logic L^k that is true on $\mathbf{A}$ is also true on $\mathbf{B}$.*
3. *Player II has a winning strategy for the existential k-pebble game on $\mathbf{A}$ and $\mathbf{B}$.*

As a consequence of Proposition 18 and Theorem 20, we obtain next a game-theoretic characterization of definability in the logic L^k for classes of finite structures.

Theorem 21 [KV90b]. *Let k be a positive integer and let $\mathcal{C}$ be a class of finite structures over the vocabulary σ. Then the following two statements are equivalent:*

1. *The class $\mathcal{C}$ is L^k-definable, i.e., there is a sentence ψ of L^k such that for any finite structure $\mathbf{A}$ over σ we have that $\mathbf{A} \in \mathcal{C} \iff \mathbf{A} \models \psi$.*
2. *If $\mathbf{A}$ and $\mathbf{B}$ are finite structures over σ such that $\mathbf{A} \in \mathcal{C}$ and Player II has a winning strategy for the existential k-pebble game on $\mathbf{A}$ and $\mathbf{B}$, then $\mathbf{B} \in \mathcal{C}$.*

Theorem 21 provides a tool for showing that certain queries are not expressible in the existential negation-free fragment L^ω of $L^\omega_{\infty\omega}$ on finite structures and, *a fortiori*, not expressible in Datalog($\neq$) on finite structures. More specifically, in order to prove that a query Q is not expressible in L^ω on finite structures it suffices to show that for every $k \geq 1$ there are two structures $\mathbf{A}_k$ and $\mathbf{B}_k$ such that $\mathbf{A}_k$ satisfies Q, $\mathbf{B}_k$ do not satisfy Q, and Player II has a winning strategy for the existential k-pebble gam on $\mathbf{A}_k$ and $\mathbf{B}_k$. Theorem 21 also guarantees that this method is *complete*. Namel if Q is not expressible in L^ω, then such structures $\mathbf{A}_k$ and $\mathbf{B}_k$ must exist for eve $k \geq 1$.

5 Expressibility of Fixed Subgraph Homeomorphism Queries

Let $H = (V_H, E_H)$ be a fixed directed graph with nodes $v_1, \ldots, v_l$ and edges $e_1, \ldots,$ Let $G = (V, E, s_1, \ldots, s_l)$ be a directed graph with l distinguished points $s_1, \ldots,$ such that $s_i \neq s_j$ for $i \neq j$. We say that H is *homeomorphic to the distinguish subgraph of* G if G contains m pairwise node-disjoint simple paths $p_1, \ldots, p_m$[5] suc that if the edge e_i in H is from v_j to $v_{j'}$, then p_i is a path from s_j to $s_{j'}$. Th *H-subgraph homeomorphism* query is to determine, for a fixed graph H, whether is homeomorphic to the distinguished subgraph of an input graph G. We use th term *pattern graphs* to refer to such graphs H. In what follows we assume that pa tern graphs have no isolated nodes, since if they do, then we can remove all isolate nodes and obtain a fixed subgraph homeomorphism query that is equivalent to th original one.

Fortune et al. [FHW80] classified the computational complexity of fixed subgrap homeomorphism queries in terms of two different dichotomies that are described i what follows. The first dichotomy is expressed in terms of the collection C of a directed graphs with a distinguished node called the *root* and with the property tha either the root is the head of every edge or the root is the tail of every edge (note tha the root can be both the head and the tail of an edge, when the graph contains a sel loop from the root to itself). Fortune et al. [FHW80] showed that the H-subgrap homeomorphism query can be answered in polynomial time if the pattern graph is in C, but it is NP-complete if the pattern graph H is in the complement $\overline{C}$ C. For the second dichotomy Fortune et al. [FHW80] considered the restriction the subgraph homeomorphism problem to acyclic graphs. They showed that if onl acyclic graphs are allowed as inputs, then the H-subgraph homeomorphism quer can be answered in polynomial time for every pattern graph H.

Notice that these two dichotomies yield a computational distinction only P $\neq$ NP. In this section, without using any complexity theoretic assumptions, w establish that the two dichotomies are indeed proper in terms of expressibility i Datalog($\neq$). More specifically, on the positive side we show that the H-subgrap homeomorphism query is expressible in Datalog($\neq$) for every pattern graph H i the class C. We also prove that if only acyclic graphs are allowed as inputs, then th H-subgraph homeomorphism query is expressible in Datalog($\neq$) for every patter graph H. Finally, on the negative side we demonstrate that if the pattern graph is in the complement $\overline{C}$ of C, then the fixed subgraph homeomorphism query is *no* expressible in L^ω and, a fortiori, is not expressible in Datalog($\neq$) as well. The res of the present section is devoted to the description of these results.

[5] Two simple paths are node-disjoint if they have no point in common, except that end points may be equal.

As mentioned earlier, Fortune et al. [FHW80] exhibited polynomial-time algorithms for certain cases of the fixed subgraph homeomorphism problem. We strengthen here these results by establishing that in these cases the fixed subgraph homeomorphism problem is actually expressible in Datalog($\neq$).

Theorem 22 [KV90b]. *Let H be a pattern graph in the class C. Then the H-subgraph homeomorphism query is expressible in Datalog($\neq$).*

Proof Sketch: Fortune et al. show that for pattern graphs H in C the H-subgraph homeomorphism problem can be reduced to a network flow problem, where we want to know whether the input graph, viewed as an appropriate directed network with node capacities, can carry a flow greater or equal than the outdegree k of the root (or the indegree k of the root if the root is the tail of every edge). This reduction establishes that the H-subgraph homeomorphism query is decidable in polynomial time for pattern graphs H in C. Using the Max-Flow Min-Cut Theorem (cf. [Bol79]) and the fact that k is fixed, we can express this query by a Datalog($\neq$) program. ▮

Theorem 23 [KV90b]. *Let H be a fixed graph. Then there is a Datalog($\neq$) program π_H that expresses the H-subgraph homeomorphism problem for acyclic graphs.*

Proof Sketch: Consider a pattern graph H with nodes $v_1, \ldots, v_k$. We denote the edge between v_i and v_j, if it exists, by e_{ij}. Here the input graph G has distinguished distinct nodes $s_1, \ldots, s_k$, and we have to determine whether H is homeomorphic to the distinguished subgraph of G. Consider the following pebble game between two players on a graph G. (We note that this game is different from the existential pebble game described in Section 4. In particular, the current game is played on a single structure, while the games in Section 4 were played on pairs of structures). To each edge e_{ij} in H there corresponds a pebble p_{ij}. A position in the game consists of a placement of a subset of the pebbles on the nodes of G. Initially, all pebbles are placed on G: each pebble p_{ij} is placed on s_i. A round of the game consists of the following:

- Player I points to one of the pebbles, say p_{ij}, that is placed on a node u.
- Player II moves p_{ij} to a node v such that there is an edge from u to v, there is no other pebble on v, and v is not a distinguished node of G except perhaps s_j. If v is s_j, then Player II removes p_{ij} from G.

Player I (resp., II) wins if II (resp., I) cannot make a move. Thus, Player II wins if and only if all pebbles have been removed from G. It can be shown that Player II has a winning strategy if and only if H is homeomorphic to the distinguished subgraph of G (the proof is almost identical to the proof of Lemma 4 in [FHW80], which relates homeomorphism in acyclic graphs to a certain *single*-player pebble game). It remains to show that the existence of a winning strategy for Player II can be expressed in Datalog($\neq$). ▮

It should be pointed out that the preceding positive results presented in Theorem 22 and Theorem 23 depend on the use of inequalities in Datalog($\neq$) programs in a crucial way. Indeed, fixed subgraph homeomorphism queries are not expressible in Datalog (without inequalities), since such queries are not strongly monotonic.

We focus now on fixed subgraph homeomorphism queries in which the patter graph H is in the complement $\overline{C}$ of the class C and show that they are not expressibl in L^ω. The main idea behind the proof is to convert a complexity lower-bound proo to an expressibility lower-bound proof. For this, we analyze the NP-hardness proo of the fixed subgraph homeomorphism queries for pattern graphs H in $\overline{C}$ and extrac certain graphs on which we play existential k-pebble games. We then apply the game theoretic characterization of L^ω given by Theorem 21 to obtain the lower bound o the expressibility of the fixed subgraph homeomorphism queries.

Recall that the class C consists of all directed graphs with a distinguished nod and with the property that either this distinguished node is the head of every edg or it is the tail of every edge. It follows that the complement $\overline{C}$ of the class C is th collection of all directed graphs G such that G contains at least one of the followin subgraphs:

1. A graph H_1 that consists of two disjoint edges, i.e., H_1 has four distinct node s_1, s_2, s_3, s_4 and two disjoint edges (s_1, s_2), (s_3, s_4).
2. A graph H_2 that is a path of length 2 through three distinct nodes, i.e., H_2 ha three distinct nodes s_1, s_2, s_3 and edges (s_1, s_2), (s_2, s_3),
3. A graph H_3 that is a cycle of length 2, i.e., H_3 has two distinct nodes s_1, s_2 an edges (s_1, s_2), (s_2, s_1).

Notice that these three graphs give rise to the following natural fixed subgrap homeomorphism queries:

1. Given a directed graph G and four distinct nodes s_1, s_2, s_3, s_4, are there tw node-disjoint simple paths from s_1 to s_2 and from s_3 to s_4?
2. Given a directed graph G and three distinct nodes s_1, s_2, s_3, is there a simpl path from s_1 to s_3 that goes through s_2?
3. Given a directed graph G and two distinct nodes s_1 and s_2, is there a simpl cycle containing both s_1 and s_2?

We wish to establish that if the pattern graph H is in the class $\overline{C}$, then the fixe subgraph homeomorphism query with pattern H is not expressible in L^ω. The fol lowing lemma and the preceding description of the class $\overline{C}$ imply that it is enough t establish this only for the subgraph homeomorphism queries with pattern H_1, H_2 and H_3.

Lemma 24 [KV90b]. *Assume that F_1 and F_2 are two directed graphs such that F is a subgraph of F_2. If the fixed subgraph homeomorphism query with pattern F_1 i not expressible in L^ω, then the fixed subgraph homeomorphism query with pattern F is not expressible in L^ω as well.*

Our next goal is to show that the fixed subgraph homeomorphism query wit pattern H_1 (i.e., "are there two node-disjoint simple paths between two given pair of distinct vertices?") is not expressible in L^ω. Fortune et al. [FHW80] showed tha the SATISFIABLITY problem with instance a Boolean formula φ reduces to the H_1- subgraph homeomorphism query with input the graph $(G_\varphi, s_1, s_2, s_3, s_4)$. In othe words, they showed that a Boolean formula φ is satisfiable if and only if in the graph G_φ there are two node-disjoint simple paths from s_1 to s_2 and from s_3 to s_4

Our expressibility lower bound will be proved by playing existential k-pebble games on appropriate graphs that we will extract from the NP-hardness proof. Since the winning strategy for Player II in these games will turn out to be quite complicated, we introduce first certain k-pebble games on Boolean formulas that will be used as an auxiliary device to describe the moves of Player II in the existential k-pebble games.

Usually, a *a truth assignment* is understood to be a mapping from a set $\{x_1, \ldots, x_k\}$ of variables to the set $\{\textbf{true}, \textbf{false}\}$ of truth values. In what follows, we will be considering "extended" truth assignments in which we keep track of the truth values assigned to *literals*, i.e., both variables x_i and negated variables $\overline{x}_i$, $1 \leq i \leq k$. The understanding is that if $\overline{x}_i$ is assigned value **true**, then x_i is assigned value **false** at the same time, and vice versa. Similarly, if $\overline{x}_i$ is assigned value **false**, then x_i is assigned value **true** at the same time, and vice versa.

Definition 25. Let k be a positive integer and let φ be a Boolean formula in conjunctive normal form with variables $x_1, \ldots, x_m$ and clauses $c_1, \ldots, c_n$. The *k-pebble game between Players I and II on the formula* φ has the following rules:

Player I moves first by placing a pebble either on one of the literals $x_1, \overline{x}_1, \ldots, x_m, \overline{x}_m$ or on one of the clauses $c_1, \ldots, c_n$ of the formula φ. If the pebble is placed on a literal, then Player II has to assign a truth value to it; otherwise, Player II has to select a literal from the clause pebbled by Player I and assign the value **true** to it. The game continues this way until each Player has made k moves. If at some point during these moves a literal has been assigned value both **true** and **false** by Player II, then Player I wins the game; otherwise, Player I removes some of his pebbles and the game resumes until again he has placed k-pebbles and Player II has responded.

We say that *Player II wins the k-pebble game on the formula* φ if he has a *winning strategy* that allows him to continue playing the game "forever". ∎

If the formula φ is satisfiable, then Player II wins the k-pebble game for every $k \geq 1$, by playing according to a satisfying assignment for φ. On the other hand, if φ is an unsatisfiable Boolean formula with k variables, then Player I wins the $(k+1)$-pebble game on φ. Indeed, Player I places one pebble on each positive literal of φ during his first k moves, forcing this way Player II to determine a truth assignment for all the literals of φ. Since φ is unsatisfiable, there is a clause c that is not satisfied by this truth assignment. Player I wins the $(k+1)$-pebble game on φ by placing his last pebble on the clause c. It should be pointed out that there are unsatisfiable formulas with k variables for which Player I can win the game with as few as 2 pebbles. Consider, for example, the formula

$$x_1 \wedge x_2 \wedge \ldots \wedge x_k \wedge (\overline{x}_1 \vee \overline{x}_2 \vee \ldots \vee \overline{x}_k).$$

Player I wins the 2-pebble game on the above formula, as follows. In his first move Player I places a pebble on the clause $(\overline{x}_1 \vee \overline{x}_2 \vee \ldots \overline{x}_k)$. If Player II responds by making the literal $\overline{x}_i$ true, then Player I wins by placing his second pebble on the clause consisting of the literal x_i.

Let φ_k, $k \geq 1$, be the *complete Boolean formula on the variables* $x_1, \ldots, x_k$. This is the only formula in conjunctive normal form that has 2^k distinct clauses, each with k distinct literals. For example, φ_2 is the formula

$$(x_1 \vee x_2) \wedge (\overline{x}_1 \vee x_2) \wedge (\overline{x}_1 \vee \overline{x}_2) \wedge (\overline{x}_1 \vee \overline{x}_2).$$

Notice that each φ_k, $k \geq 1$, is an unsatisfiable formula and, consequently, Player wins the (k+1)-pebble on φ_k. In contrast, Player II wins the k-pebble game on th formula φ_k by playing according to the following strategy. Suppose that Player places a pebble on a literal y. If the truth value of y is determined at present, the Player II maintains this truth value; otherwise, Player II assigns a truth value (**tru** or **false**) to y arbitrarily. Suppose that Player I places a pebble on a clause c of φ. This means that at most $(k-1)$ pairs of literals of the form $(x_i, \overline{x}_i)$ have their trut values determined at present. Thus, c contains at least one literal whose truth valu is not determined at present. Player II wins by selecting such a literal from c an assigning value **true** to it.

Theorem 26 [KV90b]. *Let H_1 be a graph consisting of two disjoint edges. The the fixed subgraph homeomorphism query with pattern H_1 is not expressible in th infinitary logic L^ω.*

Proof Sketch: We have to show that there is no $k \geq 1$ such that the fixed subgrap homeomorphism query with pattern H_1 is expressible in the infinitary logic L^k Theorem 21 implies that it is enough to show that for every $k \geq 1$ we can find graph A_k with distinguished nodes w_1, w_2, w_3, w_4 and a graph B_k with distinguishe nodes s_1, s_2, s_3, s_4 such the following hold:

1. In the graph A_k there are two node-disjoint simple paths from w_1 to w_2 an from w_3 to w_4;
2. In the graph B_k there is no pair of node-disjoint simple paths from s_1 to s_2 an from s_3 to s_4;
3. Player II has a winning strategy for the existential k-pebble game on the graph $(A_k, w_1, w_2, w_3, w_4)$ and $(B_k, s_1, s_2, s_3, s_4)$.

The graph A_k will consist of just two node-disjoint simple paths. The lengths o these paths will be determined later on. Let φ_k be the complete Boolean formula o the variables $x_1, \ldots, x_k$ and let G_{φ_k} be the graph associated with it in the reductio of the SATISFIABILITY problem to the fixed subgraph homeomorphism query wit pattern H_1. We take $(B_k, s_1, s_2, s_3, s_4)$ to be the graph $(G_{\varphi_k}, s_1, s_2, s_3, s_4)$, wher s_1, s_2, s_3, s_4 are the nodes introduced at the very end of the construction of G_{φ_k} Since φ_k is an unsatisfiable formula, there is no pair of node-disjoint simple path from s_1 to s_2 and from s_3 to s_4. The length of the paths in A_k is determined by th length of certain paths in B_k.

The proof of the theorem will be complete once we establish that Player II has winning strategy in the existential k-pebble game on the graphs $(A_k, w_1, w_2, w_3, w_4$ and $(B_k, s_1, s_2, s_3, s_4)$. The basic idea is that Player II views each move of Player I a a move in a k-pebble game on the formula φ_k that is played simultaneously with th existential k-pebble game on the graphs $(A_k, a_1, a_2, a_3, a_4)$ and $(B_k, s_1, s_2, s_3, s_4)$ Player II keeps track of the current configuration of the k-pebble game on φ_k an uses his winning strategy for this game to determine his moves on the existentia k-pebble game. ∎

The lower-bound proof for the patterns H_2 and H_3 is similar to the proof for the attern H_1.

orollary 27. *Let H be a pattern graph in $\overline{C}$. Then the fixed subgraph homeomorphism query with pattern H is not expressible in the infinitary logic L^ω.*

eferences

CV92] S. Abiteboul, K. Compton, and V. Vianu. Queries are easier than you thought (probably). In *Proc. 11th ACM Symp. on Principles of Database Systems*, 1992. To appear.

CY91] F. Afrati, S. S. Cosmadakis, and M. Yannakakis. On Datalog vs. polynomial time. In *Proc. 10th ACM Symp. on Principles of Database Systems*, 1991.

G89] M. Ajtai and Y. Gurevich. DATALOG vs. first-order logic. In *Proc. 30th IEEE Symp. on Foundations of Computer Science*, pages 142–146, 1989.

jt83] M. Ajtai. Σ_1^1 formulae on finite structures. *Ann. of Pure and Applied Logic*, 24:1–48, 1983.

U79] A. V. Aho and J. D. Ullman. Universality of data retrieval languages. In *Proc. 6th ACM Symp. on Principles of Programming Languages*, pages 110–117, 1979.

V89] S. Abiteboul and V. Vianu. Fixpoint extensions of first-order logic and Datalog-like languages. In *Proc. 4th IEEE Symp. on Logic in Computer Science*, pages 71–79, 1989.

AV91] S. Abiteboul and V. Vianu. Generic computation and its complexity. In *Proc. 23rd ACM Symp. on Theory of Computing*, pages 209–219, 1991.

AVV92] S. Abiteboul, Moshe Y. Vardi, and V. Vianu. Fixpoint logics, relational machines, and computational complexity. In *Proc. 7th IEEE Symp. on Structure in Complexity Theory*, 1992. To appear.

Bar77] J. Barwise. On Moschovakis closure ordinals. *Journal of Symbolic Logic*, 42:292–296, 1977.

BF85] J. Barwise and S. Feferman, editors. *Model-Theoretic Logics*. Springer-Verlag, 1985.

BG87] A. Blass and Y. Gurevich. Existential fixed-point logic. In E. Börger, editor, *Computation Theory and Logic. Lecture Notes in Computer Science 270*, pages 20–36, 1987.

BGK85] A. Blass, Y. Gurevich, and D. Kozen. A zero–one law for logic with a fixed point operator. *Information and Control*, 67:70–90, 1985.

Bol79] B. Bollobas. *Graph Theory*. Springer-Verlag, 1979.

Bol85] B. Bollobas. *Random Graphs*. Academic Press, 1985.

CFI89] J. Cai, M. Fürer, and N. Immerman. An optimal lower bound on the number of variables for graph identification. In *Proc. 30th IEEE Symp. on Foundations of Computer Science*, pages 612–617, 1989.

CH82] A. Chandra and D. Harel. Structure and complexity of relational queries. *Journal of Computer and System Sciences*, 25:99–128, 1982.

CH85] A. Chandra and D. Harel. Horn clause queries and generalizations. *Journal of Logic Programming*, 1:1–15, 1985.

Cha88] A. Chandra. Theory of database queries. In *Proc. 7th ACM Symp. on Principles of Database Systems*, pages 1–9, 1988.

Com88] K. J. Compton. 0-1 laws in logic and combinatorics. In I. Rival, editor, *NATO Adv. Study Inst. on Algorithms and Order*, pages 353–383. D. Reidel, 1988.

Coo74] S. A. Cook. An observation of time-storage trade-off. *Journal of Computer and System Sciences*, 9:308–316, 1974.

[DLW91] A. Dawar, S. Lindell, and S. Weinstein. Infinitary logic and inductive definab- ity over finite structures. Research report, Univ. of Pennsylvania, 1991.

[dR87] M. de Rougemont. Second-order and inductive definability on finite structure *Zeitschrift für Mathematische Logik und Grundlagen der Mathematik*, 33:47–6 1987.

[ER60] P. Erdös and A. Rényi. On the evolution of random graphs. *Publ. Math. In Hungar. Acad. Sci.*, 7:17–61, 1960.

[Fag75] R. Fagin. Monadic generalized spectra. *Zeitschrift für Mathematische Log und Grundlagen der Mathematik*, 21:89–96, 1975.

[Fag76] R. Fagin. Probabilities on finite models. *Journal of Symbolic Logic*, 41:50–5 1976.

[Fag90] R. Fagin. Finite-model theory—a personal perspective. In S. Abiteboul a P. Kanellakis, editors, *Proc. 1990 International Conference on Database Theor* pages 3–24. Springer-Verlag Lecture Notes in Computer Science 470, 1990. T appear in *Theoretical Computer Science*.

[FHW80] S. Fortune, J. Hopcroft, and J. Wyllie. The directed homeomorphism proble *Theoretical Computer Science*, 10:111–121, 1980.

[Gai64] H. Gaifman. Concerning measures in first-order calculi. *Israel Journal of Mat ematics*, 2:1–18, 1964.

[Gai82] H. Gaifman. On local and nonlocal properties. In J. Stern, editor, *Logic Coll quium '81*, pages 105–135. North Holland, 1982.

[GKLT69] Y. V. Glebskii, D. I. Kogan, M. I. Liogonki, and V. A. Talanov. Range and d gree of realizability of formulas in the restricted predicate calculus. *Cybernetic* 5:142–154, 1969.

[GMSV87] H. Gaifman, H. Mairson, Y. Sagiv, and M. Y. Vardi. Undecidable optimizatic problems for database logic programs. In *Proc. 2nd IEEE Symp. on Logic Computer Science*, pages 106–115, 1987.

[Gur84] Y. Gurevich. Toward logic tailored for computational complexity. M. M. Ricther et al., editor, *Computation and Proof Theory, Lecture Notes Mathematics 1104*, pages 175–216. Springer-Verlag, 1984.

[Gur92] Y. Gurevich. Zero-one laws: The logic in computer science column. *Bulletin the European Association for Theoretical Computer Science*, 1992.

[Imm82] N. Immerman. Upper and lower bounds for first-order expressibility. *Journ of Computer and System Sciences*, 25:76–98, 1982.

[Imm86] N. Immerman. Relational queries computable in polynomial time. *Informatio and Control*, 68:86–104, 1986.

[Kny89] V. V. Knyazev. Iterative extensions of first-order logic. *Mathematical Problem in Cybernetics*, pages 123–130, 1989. MR 91a:03062.

[KV87] Ph. G. Kolaitis and M. Y. Vardi. The decision problem for the probabilities higher-order properties. In *Proc. 19th ACM Symp. on Theory of Computin* pages 425–435, 1987.

[KV90a] Ph. G. Kolaitis and M. Y. Vardi. 0-1 laws for infinitary logics. In *Proc. 5t IEEE Symp. on Logic in Computer Science*, pages 156–167, 1990.

[KV90b] Ph. G. Kolaitis and M. Y. Vardi. On the expressive power of Datalog: too and a case study. In *Proc. 9th ACM Symp. on Principles of Database System* pages 61–71, 1990. Full version appeared in IBM Research Report RJ8010 March 1991.

[KV92] Ph. G. Kolaitis and M. Y. Vardi. Fixpoint logic vs. infinitary logic in finite model theory. In *Proc. 7th IEEE Symp. on Logic in Computer Science*, 1992 To appear.

[LM89] V. S. Lakshmanan and A. O. Mendelzon. Inductive pebble games and the ex pressive power of DATALOG. In *Proc. 8th ACM Symposium on Principles of Database Systems*, pages 301–310, 1989.

[Mos74] Y. N. Moschovakis. *Elementary Induction on Abstract Structures.* North Holland, 1974.

[Pap85] C. H. Papadimitriou. A note on the expressive power of Prolog. *Bulletin of the EATCS*, 26:21–23, 1985.

[Shm87] O. Shmueli. Decidability and expressiveness aspects of logic queries. In *Proc. 6th ACM Symposium on Principles of Database Systems*, pages 237–249, 1987.

[Tal81] V. A. Talanov. Asymptotic solvability of logical formulas. *Combinatorial-Algebraic Methods in Applied Mathematics*, pages 118–126, 1981. MR 85i:03081, ZBL 538.03007.

[TK86] V. A. Talanov and V. V. Knyazev. The asymptotic truth of infinite formulas. In *Proceedings of All-Union Seminar on Discrete and Applied Mathematics and it Applications*, pages 56–61, 1986. MR 89g:03054.

[Ull89] J. D. Ullman. *Database and Knowledge-Base Systems, Volumes I and II.* Computer Science Press, 1989.

[Var82] M. Y. Vardi. The complexity of relational query languages. In *Proc. 14th ACM Symp. on Theory of Computing*, pages 137–146, 1982.

This article was processed using the LaTeX macro package with LLNCS style

Characterization of Temporal Property Classes*

Edward Chang [†] Zohar Manna [†] Amir Pnueli [‡]

Abstract. This paper presents two novel characterizations of the classes of properties of reactive systems in terms of their expression by temporal logic. The first family of characterizations concerns the safety-progress classification, which describes a hierarchy within the set of temporal properties. Previous characterizations of this hierarchy depended critically on the use of past temporal operators. The characterization presented here identifies the future formulas that belong to each class. This characterization is shown to be complete.

The second characterization concerns the safety-liveness classification, which partitions temporal properties into the classes of safety and liveness. While automata-theoretic and temporal logic characterizations of the safety class have been known for some time, a complete characterization of the liveness class by temporal logic remained open. This paper provides such a characterization.

1 Introduction

Reactive systems are systems that are expected to maintain an ongoing interactio with their environment, rather than to produce some final result on termination. A reactive system may be viewed as a generator of *computations*, i.e., infinite sequence of states or events. A (temporal) *property* is a set of such computations. A program P *satisfies* the property Π if every computation of P belongs to Π. In [MP90] we presen a hierarchy of temporal properties, the *safety-progress* hierarchy, which correspond to the first several layers of the Borel topological hierarchy. The main interest in this hierarchy is that properties belonging to the same class have many features in common, and in particular each class can be associated with a characteristic proo rule used to verify that a given program satisfies a property of this class. Since proof rules for the lower classes are simpler than those for the higher classes, i is a worthwhile endeavor to identify, for a given property, the lowest class in the hierarchy to which it belongs.

Several characterizations of the hierarchy are presented in [MP90]. The tempora characterization is based on *canonical formulas*, which are boolean combinations o

*This research was supported in part by the National Science Foundation under grants CCR 89-11512 and CCR-89-13641, by the Defense Advanced Research Projects Agency under contrac NAG2-703, by the United States Air Force Office of Scientific Research under contract AFOSR-90 0057, and by the European Community ESPRIT Basic Research Action project 3096 (SPEC).

[†]Department of Computer Science, Stanford University, Stanford, CA 94305

[‡]Department of Computer Science, Weizmann Institute, Rehovot, Israel

rmulas of the form $\Box p$ and $\Box \Diamond p$, where p is an arbitrary *past* formula. This stricted form avoids the use of other future operators such as the *until* operator $\mathcal{U}$. follows that, in order to determine the minimal class to which a formula belongs, should be transformed first to canonical form. While this is always possible, the neral transformation may lead to an exponential blowup. Furthermore, certain operties are more naturally expressed by future formulas, and the need to translate past-oriented specifications may be awkward and unnatural.

In this paper, we remedy these drawbacks of the past-based characterization by resenting a considerably wider syntactic characterization of the safety-progress hiarchy using *standard formulas.* This characterization is applicable to arbitrary mporal formulas, including those that use the *until* and *unless* (*waiting-for*) opertors. Consequently, without any preliminary transformation, the presented characrization provides an upper bound on where a formula lies in the hierarchy.

The second classification for which we present a temporal logic characterization the partition of properties into *safety* and *liveness* classes. The fact that temporal roperties naturally partition into two disjoint classes was first observed by Lamport [Lam77]. Semantic characterizations of the safety and liveness classes were given [Lam85] and [AS85], respectively.

Since the safety class belongs to both the safety-liveness partition and the safetyrogress hierarchy, it possesses a (canonical) temporal logic characterization. This haracterization was presented in [LPZ85]. An important step in the temporal charcterization of the safety-liveness partition was taken by [Sis85], which provided a omplete future-based characterization of safety properties. For liveness, only partial haracterizations were given.

In this paper we improve on these results by presenting a complete temporal logic haracterization of the liveness class, as well as a simpler proof of the completeness f the temporal logic characterization of the safety class.

In Section 2 we sketch the previously known results about the safety-progress lassification and its characterization by canonical formulas. In Section 3 we present he new characterization by standard formulas. Section 4 deals with the safetyiveness classification and its temporal characterization.

2 Safety-Progress: Previous Results

In this section we present the previously known facts about the safety-progress hierarchy as given in [MP90].

2.1 The Language-Theoretic View

First, we present the language-theoretic characterization of the safety-progress hierarchy.

We consider a set Σ of *states*, which we also refer to as *letters.* We refer to an infinite sequence of letters as an (infinite) *word*, and to a finite sequence of letters as a *finite word*.

Let Σ^* denote the set of all finite words, Σ^+ denote the set of all nonempty finite words, and Σ^ω denote the set of all (infinite) words. A *language* is a subset of Σ^ω, i.e., a set of words. A *finitary language* is a subset of Σ^+, i.e., a set of finite nonempty words. In the abstract setting considered here, a *property* is a set of infinite words, i.e., a language.

The *length* $|\sigma|$ of a finite word $\sigma : a_1, \ldots, a_j$ is j. The length of an infinite word is defined to be ∞. We write $\widehat{\sigma} \prec \sigma$ (equivalently $\sigma \succ \widehat{\sigma}$) to denote that $\widehat{\sigma}$ is a *prefix* of σ.

We introduce four operators A, E, R, and P that construct languages from finitary languages. Let Φ be a finitary language.

- The language $A(\Phi)$ consists of all words σ such that *every* prefix of σ is in Φ. For example, $A(a^+b^*) = a^\omega + a^+b^\omega$.
- The language $E(\Phi)$ consists of all words σ such that *some* prefix of σ is in Φ. For example, $E(a^+b^*) = a^+b^* \cdot \Sigma^\omega$. In fact, it is true for every finitary language Φ that $E(\Phi) = \Phi \cdot \Sigma^\omega$.
- The language $R(\Phi)$ consists of all words σ such that *infinitely many* prefixes of σ are in Φ. For example, $R(\Sigma^* b) = (\Sigma^* b)^\omega$. The words of this language are precisely those that have infinitely many occurrences of b.
- The language $P(\Phi)$ consists of all words σ such that *all but finitely many* prefixes of σ are in Φ. For example, $P(\Sigma^* b) = \Sigma^* b^\omega$. The words of this language are precisely those that, from a certain point on, have only the letter b.

Based on these four operators we define six classes of languages. A language $\Pi \subseteq \Sigma^\omega$ is defined to be

- A *safety* language if $\Pi = A(\Phi)$ for some finitary Φ.
- A *guarantee* language if $\Pi = E(\Phi)$ for some finitary Φ.
- An *m-obligation* language if $\Pi = \bigcap_{i=1}^{m} \Big(A(\Phi_i) \cup E(\Psi_i)\Big)$ for some finitary languages $\Phi_i, \Psi_i, i = 1, \ldots, m$.
- A *response* language if $\Pi = R(\Phi)$ for some finitary Φ.
- A *persistence* language if $\Pi = P(\Phi)$ for some finitary Φ.
- An *m-reactivity* language if $\Pi = \bigcap_{i=1}^{m} \Big(R(\Phi_i) \cup P(\Psi_i)\Big)$ for some finitary languages $\Phi_i, \Psi_i, i = 1, \ldots, m$.

It follows, for example, that the languages $a^\omega + a^+b^\omega$, $a^+b^* \cdot \Sigma^\omega$, $(\Sigma^* b)^\omega$, and $\Sigma^* b^\omega$ are safety, guarantee, response, and persistence languages, respectively.

In the following, let κ stand for one of the classes: *safety, guarantee, obligation, response, persistence*, or *reactivity*.

We also refer to a κ-language as a *κ-property*. We refer to the classification induced by the language-theoretic characterization as the *semantic* characterization, in contrast to the syntactic characterizations of the temporal formulas or automata that specify properties of a given class.

2.2 Temporal Logic

In this section we introduce the language of temporal logic. The language of temporal logic is constructed inductively from a set of propositions $\mathcal{P}$. We assume an interpretation I such that, for each state $s \in \Sigma$, $I(s) \subseteq \mathcal{P}$ specifies the propositions that are true in s.

Every proposition is a formula; if p and q are formulas, then the following are formulas:

$$\neg p \qquad p \vee q \qquad \bigcirc p \qquad p\,\mathcal{U}\,q \qquad \ominus p \qquad p\,\mathcal{S}\,q$$

A formula whose only logical symbols are $\neg$ or $\vee$ is called a *state formula*. A formula which does not use $\bigcirc$ or $\mathcal{U}$ is called a *past formula*. A formula which does not use $\ominus$ or $\mathcal{S}$ is called a *future* formula.

For a state $s \in \Sigma$ and a proposition $p \in \mathcal{P}$, we define $s \models p$ *iff* $p \in I(s)$. A computation $\sigma = s_1, s_2, \ldots$ *satisfies formula* p *at* $j > 0$, denoted $(\sigma, j) \models p$, as follows:

$(\sigma,j) \models p$ *iff* $s_j \models p$, where p is a proposition

$(\sigma,j) \models \neg p$ *iff* $(\sigma,j) \not\models p$

$(\sigma,j) \models p \vee q$ *iff* $(\sigma,j) \models p$ or $(\sigma,j) \models q$

$(\sigma,j) \models \bigcirc p$ *iff* $(\sigma,j+1) \models p$

$(\sigma,j) \models p\,\mathcal{U}\,q$ *iff* $(\sigma,k) \models q$ for some $k \geq j$, and $(\sigma,i) \models p$ for all i, $j \leq i < k$

$(\sigma,j) \models \ominus p$ *iff* $j > 1$ and $(\sigma,j-1) \models p$

$(\sigma,j) \models p\,\mathcal{S}\,q$ *iff* $(\sigma,k) \models q$ for some $k \leq j$, and $(\sigma,i) \models p$ for all i, $k < i \leq j$

A computation σ is a *model* for p, denoted $\sigma \models p$, exactly when $(\sigma,1) \models p$. Formulas p and q are said to be *equivalent* if they have the same models. Formulas p and q are *congruent* if, for every computation σ and every position $j > 0$, $(\sigma,j) \models p$ *iff* $(\sigma,j) \models q$. A formula p is *satisfiable* if p has at least one model. A formula p is *valid* if every computation is a model for p.

For a formula p, we define the *property specified by* p, denoted $Spec(p)$, to be the set of all models of p. A property Π is said to be *specifiable by temporal logic* if there exists a formula p such that $\Pi = Spec(p)$.

The following derived temporal operators are also used below:

$p \Rightarrow q = \Box(p \rightarrow q)$	– "entails"		
$\Diamond p = true\ \mathcal{U}\, p$	– "eventually"	$\diamondminus p = true\ \mathcal{S}\, p$	– "earlier"
$\Box p = \neg \Diamond \neg p$	– "always"	$\boxminus p = \neg \diamondminus \neg p$	– "always in the past"
$p\,\mathcal{W}\,q = p\,\mathcal{U}\,q \vee \Box p$	– "waiting-for"	$p\,\mathcal{B}\,q = p\,\mathcal{S}\,q \vee \boxminus p$	– "back-to"

The operators $\mathcal{W}$ and $\mathcal{B}$ are also called *unless* and *weak since*, respectively.

2.3 Characterization by Canonical Formulae

We define a formula to be

- *A canonical safety formula* if it has the form $\Box p$,
- *A canonical guarantee formula* if it has the form $\Diamond p$,

- *A canonical m-obligation formula* if it has the form $\bigwedge_{i=1}^{m}(\Box p_i \vee \Diamond q_i)$,
- *A canonical response formula* if it has the form $\Box \Diamond p$,
- *A canonical persistence formula* if it has the form $\Diamond \Box p$,
- *A canonical m-reactivity formula* if it has the form $\bigwedge_{i=1}^{m}(\Box \Diamond p_i \vee \Diamond \Box q_i$

where $p, p_1, \ldots, p_m, q_1, \ldots, q_m$, are arbitrary past formulas.

The following theorem establishes the connection between the semantic chara terization and the characterization by canonical formulas.

Theorem 1 *A property specifiable by temporal logic is a κ-property iff it is speci able by a canonical κ-formula.*

It is not too difficult to show that a property Π specifiable by a canonical formula is a κ-property. For example, if Π is specifiable by the canonical safety fo mula $\Box p$, then it is equal to $A(\Phi)$, where Φ is the set of all finite words $\sigma : s_1, \ldots,$ such that p holds at position k of σ.

The other direction is more involved and requires the construction of an automaton that accepts exactly the models of the formula specifying the property

In Figure 1, we present a diagram that displays the six classes of the hiera chy with their respective characterizations in the language-theoretic and tempor logic frameworks. A line connecting two classes in the diagram represents stri containment of the lower class in the higher one.

3 Safety-Progress: Characterization by Standard Formulas

In this section we introduce a new characterization of properties based on the defi nition of *standard κ-formulas.*

Definition 2 *Standard κ-formulas* are defined as follows.

a. *standard safety formulas*

- Every past formula is a standard safety formula.
- The negation of a standard guarantee formula is a standard safety formula.
- If p and q are standard safety formulas, then so are $p \vee q$, $p \wedge q$, $\bigcirc p$, $\Box p$, an $p \mathcal{W} q$.

b. *standard guarantee formulas*

- Every past formula is a standard guarantee formula.
- The negation of a standard safety formula is a standard guarantee formula.
- If p and q are standard guarantee formulas, then so are $p \vee q$, $p \wedge q$, $\bigcirc p$, $\Diamond p$ and $p \mathcal{U} q$.

c. *standard obligation formulas*

- Every standard safety and standard guarantee formula is a standard obligatio formula.

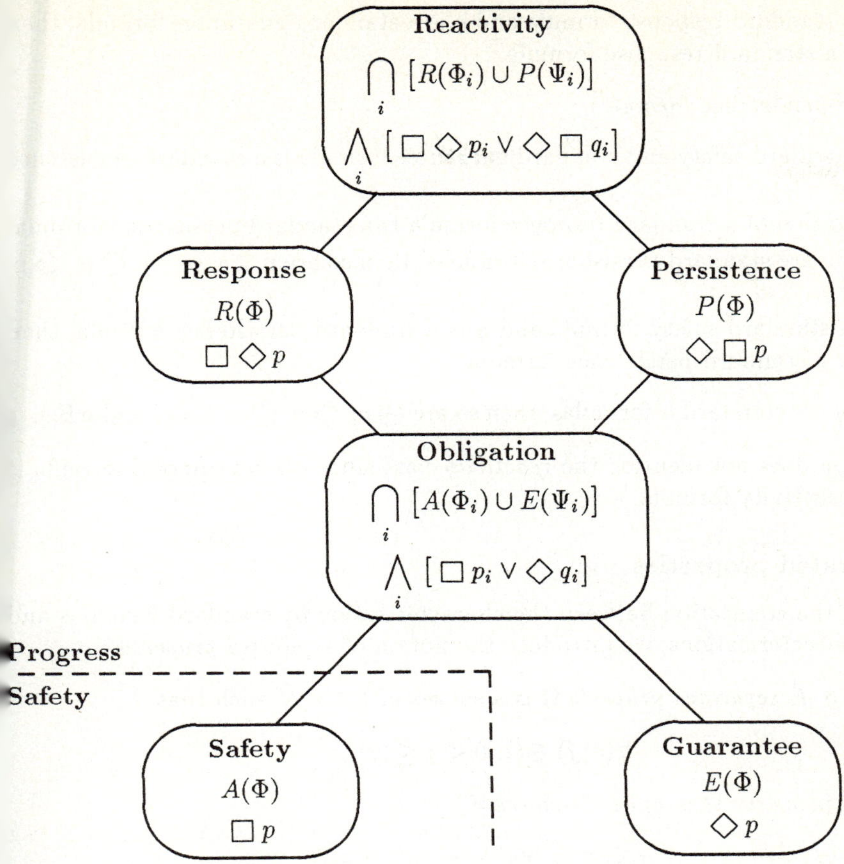

Figure 1: Inclusion Relations between the Classes

- If p and q are standard obligation formulas, then so are $\neg p$, $p \vee q$, $p \wedge q$, and $\bigcirc p$.
- If p is a standard obligation formula and q is a standard guarantee formula, then $p \mathcal{U} q$ is a standard obligation formula.
- If p is a standard safety formula and q is a standard obligation formula, then $p \mathcal{W} q$ is a standard obligation formula.

d. *standard response formulas*

- Every standard safety and standard guarantee formula is a standard response formula.
- The negation of a standard persistence formula is a standard response formula.
- If p and q are standard response formulas, then so are $p \vee q$, $p \wedge q$, $\bigcirc p$, $\Box p$, and $p \mathcal{W} q$.

- If p is a standard response formula and q is a standard guarantee formula, then $p\,\mathcal{U}\,q$ is a standard response formula.

e. *standard persistence formulas*

- Every standard safety and standard guarantee formula is a standard persistence formula.
- The negation of a standard response formula is a standard persistence formula.
- If p and q are standard persistence formulas, then so are $p \vee q$, $p \wedge q$, $\bigcirc p$, $\diamondsuit p$ and $p\,\mathcal{U}\,q$.
- If p is a standard safety formula and q is a standard persistence formula, then $p\,\mathcal{W}\,q$ is a standard persistence formula.

f. If p and q are standard κ-formulas, then so are $\ominus p$, $\diamondminus p$, $\boxminus p$, $p\,\mathcal{S}\,q$, and $p\,\mathcal{B}\,q$. ■

The definition does not mention the *reactivity* class since every temporal formula is a standard reactivity formula.

3.1 Separated properties

To establish the connection between the characterization by standard formulas and the other characterizations, we introduce the notion of *separated properties*.

Definition 3 A *separated property* Ω is a subset of $\Sigma^+ \times \mathcal{N}$ such that

$$\forall \langle \sigma, j \rangle \in \Omega.\ 0 < j \leq |\sigma|$$

A separated property Ω is *upwards-closed* if

$$\forall \langle \sigma, j \rangle \in \Omega.\ \forall \sigma' \succ \sigma.\ \langle \sigma', j \rangle \in \Omega$$

Similarly, a separated property Ω is *downwards-closed* if

$$\forall \langle \sigma, j \rangle \in \Omega.\ \forall \sigma' \prec \sigma.\ j \leq |\sigma'| \rightarrow \langle \sigma', j \rangle \in \Omega \ ■$$

Theorem 4 We can characterize standard *safety, guarantee, response*, and *persistence* formulas as follows.

a. If p is a standard safety formula, there exists a downwards-closed separated property Ω such that

$$(\sigma, j) \models p \quad \textit{iff} \quad \text{for every } \sigma' \prec \sigma.\ j \leq |\sigma'| \rightarrow \langle \sigma', j \rangle \in \Omega$$

b. If p is a standard guarantee formula, there exists an upwards-closed separated property Ω such that

$$(\sigma, j) \models p \quad \textit{iff} \quad \text{for some } \sigma' \prec \sigma.\ \langle \sigma', j \rangle \in \Omega$$

d. If p is a standard response formula, there exists a separated property Ω such that

$$(\sigma, j) \models p \quad \textit{iff} \quad \text{for infinitely many } \sigma' \prec \sigma.\ \langle \sigma', j \rangle \in \Omega$$

e. If p is a standard persistence formula, there exists a separated property Ω such that

$$(\sigma, j) \models p \quad \textit{iff} \quad \text{for all but finitely many } \sigma' \prec \sigma.\ \langle \sigma', j \rangle \in \Omega$$

efinition 5 *The finitary property Φ induced by a separated property Ω is*

$$\{\sigma | \langle \sigma, 1 \rangle \in \Omega\}$$

orollary 6 *Every standard κ-formula specifies a κ-property.*

roof: The finitary properties Φ induced by the separated properties Ω described Theorem 4 clearly establish this corollary. ∎

orollary 7 *Every standard obligation formula specifies an obligation property.*

Note that Corollaries 6 and 7 establish one direction of the connection between e standard and the semantic characterizations. The other direction follows trivially y observing that every canonical κ-formula is a standard κ-formula.

.2 Completeness of the Future Fragment

'he standard formulas classification restricted to future formulas is complete. This stated in the following theorem.

'heorem 8 *A property Π that is specifiable by temporal logic is a κ-property if and nly if it is specifiable by a future standard κ-formula.*

'roof: We will outline the proof which appears in the full version of this paper. By 'heorem 1 it suffices to show that every canonical κ-formula is equivalent to some tandard κ-formula.

Recall that canonical formulas are boolean combinations of formulas of the form $\Box p$ and $\Box \Diamond p$, where p is an arbitrary past formula. We say that a finite sequence f states $\sigma : s_1, \ldots, s_j$ *end-satisfies* p, denoted $\sigma \dashv p$, if $(\sigma \cdot \sigma', j) \models p$ for any $\sigma' \in \Sigma^\omega$; ince p is a past formula, the choice of σ' is irrelevant.

We use a translation derived from [SPH84] to obtain, for every past formula p, a tar-free expression α such that $\sigma \dashv p$ exactly when σ is in the language of α. Then Zuc86] provides a translation from star-free expressions to finitary future temporal ogic, in which formulas are interpreted over finite state sequences and the predicate *ast*, which holds only at the last state of a sequence, is introduced. Thus we have a ranslation from every past formula p to a finitary future formula φ such that $\sigma \dashv p$ f and only if $\sigma \models \varphi$.

It is a property of the translation that an infinite sequence σ is a model of $\varphi\langle true / last\rangle$, the future formula obtained by replacing *last* by *true* in φ, exactly when there is some prefix of σ that satisfies φ. Therefore $\varphi\langle true / last\rangle$ is equivalent to $\Diamond p$, and since $\varphi\langle true / last\rangle$ is a standard guarantee formula, this shows the case for $\kappa = guarantee$. The case for $\kappa = safety$ follows by duality.

For the case of $\kappa = response$ (and $\kappa = persistence$, by duality), we construct the *limit of* φ, a future formula which accepts the set of infinite states sequences with infinitely many prefixes satisfying φ; these are also exactly the models of $\Box \Diamond p$. This construction is given in [Zuc86] and, after some manipulation, yields a standard response formula. ∎

3.3 A Verification Strategy Using Standard Formulas

In this section we will briefly describe how the standard formulas characterizatio may be used in verification. We represent programs as *fair transition systems*. I this model, a program P consists of the following components: V (a finite set c *state variables*), Σ (a set of states, each of which is an interpretation over V), $\mathcal{T}$ (finite set of *transitions*), and Θ (an assertion characterizing all initial states).

Every transition $\tau \in \mathcal{T}$ is associated with a *transition relation* $\rho_\tau(V, V')$, whic may refer to unprimed and primed versions of the state variables. The purpose c ρ_τ is to express a relation between a state s and its successor s'; we define s' t be a *τ-successor* of s if $\langle s, s' \rangle \models \rho_\tau(V, V')$, where $\langle s, s' \rangle$ is the joint interpretatio which interprets $x \in V$ as $s(x)$ and x' as $s'(x)$. A transition τ is *enabled on a stat* s if $En(\tau)$, given by $(\exists V')\rho_\tau(V, V')$, holds. Consider an infinite sequence of state $\sigma: s_1, s_2, \ldots$. A transition τ is *enabled at position k* if τ is enabled on s_k. A transition τ is *taken at position k* if s_{k+1} is a τ-successor of s_k. The sequence σ is *computation of P* if it satisfies the following requirements:

- *Initiality* — State s_1 is *initial*, i.e., $s_1 \models \Theta$
- *Consecution* — For every $j > 0$, there is some $\tau \in \mathcal{T}$ such that s_{j+1} is a τ-successor of s_j
- *Justice* — For every $\tau \in \mathcal{T}$, if τ is continually enabled beyond some position j, then τ is taken at some position $k > j$.

Given a temporal logic specification φ, P *satisfies* φ if every computation of P i a model of φ. One approach to verification, i.e., proving that P satisfies φ, is t determine where φ belongs in the safety-progress hierarchy, choose a suitable proo rule for properties of that class, translate φ to an equivalent specification φ' fo which the proof rule is applicable, and then perform the verification.

Consider, for example, a system in which a server process must respond to re quests by a client process, unless the server is able to observe that the client mus eventually achieve some goal. This can be specified by:

$$\varphi: \quad (request \rightarrow \Diamond\, response)\, \mathcal{W}\, (\Diamond\, goal)$$

According to the standard formulas characterization, φ is a response property, an we might hope to apply rule C-RESPONSE. This rule states that, in order to show $\Box \Diamond p$, it suffices to find intermediate assertions φ_i and helpful transitions τ_i such that the following conditions hold. Every $(\neg p)$-position entails that one of the φ holds. The final φ_0 entails p. Every transition from a φ_i-position results in a φ_j position with $j \leq i$, and the helpful transition τ_i must result in a φ_j-position with $j < i$. Finally, φ_i entails that the helpful transition is enabled, so by the justic requirement, τ_i is eventually taken.

Before applying rule C-RESPONSE, we must first translate φ to its equivalen canonical form $\Box \Diamond p$. This can be achieved by

$$p: \quad \ominus\!\!\!\Diamond\, goal \;\vee\; (\neg request)\, \mathcal{B}\, response$$

and we may the proceed with the verification.

C-RESPONSE

Find intermediate assertions $\varphi_0, \ldots, \varphi_k$
and helpful transitions $\tau_1, \ldots, \tau_k$:

R1. $\neg p \Rightarrow \bigvee_{i=0}^{k} \varphi_i$

R2. $\varphi_0 \Rightarrow p$

For every $i > 0$:

R3. $(\varphi_i \;\wedge\; \rho_\tau) \Rightarrow \bigvee_{j \leq i} \varphi_j'$ for every $\tau \in \mathcal{T}$

R4. $(\varphi_i \;\wedge\; \rho_{\tau_i}) \Rightarrow \bigvee_{j < i} \varphi_j'$

R5. $\varphi_i \Rightarrow En(\tau_i)$

$\Box \Diamond p$

There are two principal drawbacks to this approach. First, in translating to anonical form, we obtain a formula that has less intuitive meaning than the original . Second, we are obligated to work with past formulas during the verification. In articular, the premises of rule C-RESPONSE are entailments.

Consider instead the following strategy. Starting with the original specification , we attempt to replace subformulas of φ by state formulas. In the example given bove, for instance, it would suffice to identify state formulas p and q such that the rogram P satisfies the following:

(1) $\quad p \;\Rightarrow\; request \rightarrow \Diamond\, response$

(2) $\quad q \;\Rightarrow\; \Diamond\, goal$

(3) $\quad p\, \mathcal{W}\, q$

Ve can further simplify (1) if we identify a state formula r such that P satisfies:

(1a) $\quad r \;\Rightarrow\; \Diamond\, response$

(1b) $\quad p \;\Rightarrow\; request \rightarrow r$

The formulas that we obtain through this procedure, such as (1a), (1b), (2), and (3), have a simple form, and we can provide straightforward proof rules for verifying them. Rule WAITING, for example, can be used for formulas (1b) and (3), and a

WAITING

Find strengthening assertion φ:

W1. $p \rightarrow (r \vee \varphi)$

W2. $\varphi \rightarrow q$

W3. $(\varphi \;\wedge\; \rho_\tau) \rightarrow (r \;\vee\; \varphi)$ for every $\tau \in \mathcal{T}$

$p \Rightarrow q\, \mathcal{W}\, r$

simpler version of rule C-RESPONSE can be used for (1a) and (2). In this manner we can verify complex specifications piecemeal without having to translate the original specification into any particular form, and our proof rules yield temporal conclusions from first-order premises.

As an example, consider the following program:

local x, y, z : **integer where** $x = y = 0, z > 0$

$$P1 :: \left[\begin{array}{l} \ell_0\text{: }\textbf{while } y = 0 \textbf{ do} \\ \quad \left[\begin{array}{l} \ell_1\text{: }\textbf{if } x = 2 \textbf{ then} \\ \quad \left[\begin{array}{l} \ell_2\text{: } x := 1 \\ \ell_3\text{: } x := 0 \end{array}\right] \end{array}\right] \\ \ell_4\text{: } x := 1 \\ \ell_5\text{: } x := 0 \end{array}\right] \quad || \quad P2 :: \left[\begin{array}{l} m_0\text{: }\textbf{loop forever do} \\ \quad \left[\begin{array}{l} m_1\text{: } x := 2 \\ m_2\text{: }\textbf{if } x = 1 \textbf{ then} \\ \quad m_3\text{: } y := 1 \\ m_4\text{: }\textbf{if } y = 1 \textbf{ then} \\ \quad m_5\text{: } z := 0 \end{array}\right] \end{array}\right]$$

The "goal" of this program is to set $z = 0$. It is clear that this happens exactly $y = 1$ eventually holds. Process $P2$ signals "request" by setting $x = 2$, and proces $P1$ will "respond" by setting $x = 1$. If $P2$ observes a response before $P1$ changes t $x = 0$, then $P2$ will set $y = 1$. We can prove

$$(x = 2 \rightarrow \diamondsuit(x = 1)) \ \mathcal{W} \ \diamondsuit(z = 0)$$

by taking p, q, and r to be:

$$\begin{array}{ll} p: & (y = 0) \ \wedge \ (z > 0) \ \wedge \ at\_\ell_{0\ldots3} \\ q: & (y = 1) \ \wedge \ (z > 0) \\ r: & at\_\ell_{0\ldots3} \end{array}$$

4 Safety-Liveness: A New Characterization

First, consider the language-theoretic characterization of the safety and livene properties, as formulated in [AS85].

- A property Π is a *safety* property if it equals $A(\Phi)$ for some finitary langua Φ.
- A property Π is a *liveness property* if every finite word $\sigma \in \Sigma^+$ is a pref of some infinite word σ' belonging to Π. That is, every finite word can l extended to an infinite word in Π.

The temporal characterization of these two classes is based on the followi definition.

- A *canonical safety formula* is any formula of the form $\square p$, where p is son past formula.
- A *canonical liveness formula* is any formula of the form $\diamondsuit \bigvee_{i=1}^{k}(p_i \wedge \bigcirc f_i$ where $p_1 \ldots, p_k$ are past formulas such that $\square(\bigvee_{i=1}^{k} p_i)$ is valid, and $f_1, \ldots,$ are satisfiable future formulas.

The definition of canonical safety formula is identical to the one used for tl safety-progress hierarchy, and is repeated here for completeness.

To establish the connection between the characterization by canonical tempor formulas and the semantic characterization, we use the following separation theore due to Gabbay [Gab87]. The following version of this theorem has been adapted t fit our notation.

Theorem 9 (Separation) *Every temporal formula is congruent to a formula of the form* $\bigvee_{i=1}^{k}(p_i \wedge \bigcirc f_i)$*, where* $p_1 \ldots, p_k$ *are past formulas and* $f_1, \ldots, f_k$ *are future formulas.*

Theorem 10 *A property specifiable by temporal logic is a* λ*-property iff it is specifiable by a canonical* λ*-formula, where* λ *is either* safety *or* liveness.

Proof: We consider separately the cases of safety and liveness. Let Π be a property specified by the canonical safety formula $\Box\, p$. Taking the finitary language Φ to be the set of all finite words $s_1, \ldots, s_k$ satisfying the past formula p at position k, it is not difficult to see that $\Pi = A(\Phi)$.

In the other direction, assume that the property Π is specified by the temporal formula φ and is known to be a safety property. In this case there exists a finitary language Φ such that $\Pi = A(\Phi)$. Consider the formula $\psi : \diamondminus(\mathit{first} \wedge \varphi)$, where *first* stands for the formula $\neg \ominus \mathit{true}$ and is true precisely at position 1 of every word.

The formula ψ has the property that, for every word σ and position $j > 0$, $(\sigma, j) \models \psi$ iff $(\sigma, 1) \models \varphi$ iff $\sigma \in \Pi$. Applying the separation theorem to ψ, we obtain a congruent formula $\bigvee_{i=1}^{k}(p_i \wedge \bigcirc f_i)$. Without loss of generality, we may assume that every f_i is satisfiable; otherwise we can drop the conjunction $p_i \wedge \bigcirc f_i$ from the formula. Define the formula $\chi : \Box(\bigvee_{i=1}^{k} p_i)$, which is obviously a canonical safety formula. We will show that a word σ satisfies χ iff it belongs to Π. This will show that Π is specifiable by a canonical safety formula.

Assume first that $\sigma \in \Pi$. In that case it satisfies φ which implies that ψ holds at all positions of σ. This means that $\bigvee_{i=1}^{k}(p_i \wedge \bigcirc f_i)$, and therefore $\bigvee_{i=1}^{k} p_i$, holds at all positions of σ. It follows that $\chi : \Box(\bigvee_{i=1}^{k} p_i)$ holds at position 1 of σ.

Next, consider a word σ that satisfies $\chi : \Box(\bigvee_{i=1}^{k} p_i)$. Consider an arbitrary position $j > 0$. Obviously, for some $i = 1, \ldots, k$, p_i holds at position j. Since f_i is satisfiable, there exists a word $\widehat{\sigma} : \widehat{s}_1, \widehat{s}_2, \ldots$, satisfying f_i. Consider the word $\sigma' : s_1, \ldots, s_j, \widehat{s}_1, \widehat{s}_2, \ldots$, obtained by concatenating $\widehat{\sigma}$ to the end of the j-prefix of σ. Clearly, $p_i \wedge \bigcirc f_i$ holds at position j of σ'. It follows that $\bigvee_{i=1}^{k}(p_i \wedge \bigcirc f_i)$, and therefore ψ, holds at position j. Consequently, σ' belongs to Π and, therefore, all of its prefixes, including $s_1, \ldots, s_j$, belong to Φ. Since j was an arbitrarily chosen position, this shows that all prefixes of σ belong to Φ, and therefore σ belongs to Π.

Next, consider the case of liveness. In one direction, let Π be a property specifiable by the canonical liveness formula $\Diamond \bigvee_{i=1}^{k}(p_i \wedge \bigcirc f_i)$. Let $\sigma : s_1, \ldots, s_j$ be an arbitrary finite word. Since $\Box(\bigvee_{i=1}^{k} p_i)$ is valid, $\bigvee_{i=1}^{k} p_i$ must hold at all positions, including position j of σ. This means that $(\sigma, j) \models p_i$ for some $i = 1, \ldots, k$. Since f_i is satisfiable, there exists a word $\widehat{\sigma} : \widehat{s}_1, \widehat{s}_2, \ldots$, satisfying f_i. Consider the word $\sigma' : s_1, \ldots, s_j, \widehat{s}_1, \widehat{s}_2, \ldots$, obtained by concatenating $\widehat{\sigma}$ to the end of σ. Clearly, $p_i \wedge \bigcirc f_i$ holds at position j of σ'. It follows that $\bigvee_{i=1}^{k}(p_i \wedge \bigcirc f_i)$ holds at position j, implying that σ' satisfies $\Diamond \bigvee_{i=1}^{k}(p_i \wedge \bigcirc f_i)$. Consequently, σ' belongs to Π, and we have shown that any finite word can be extended to a word in Π.

In the other direction, assume that the property Π is specified by the temporal formula φ and is known to be a liveness property. Again we construct the formula $\psi : \diamondminus(\mathit{first} \wedge \varphi)$ and, using the separation theorem, the formula $\chi : \bigvee_{i=1}^{k}(p_i \wedge \bigcirc f_i)$ congruent to ψ. Since φ specifies Π, it follows that a word σ is in Π iff it satisfies

ψ iff it satisfies $\diamondsuit \chi = \diamondsuit \bigvee_{i=1}^{k}(p_i \wedge \bigcirc f_i)$. We will show that $\square(\bigvee_{i=1}^{k} p_i)$ is valid thus showing that $\diamondsuit \chi$ is a canonical liveness formula specifying Π.

Let $\sigma : s_1, \ldots, s_j$ be an arbitrary finite word. Since Π is a liveness property there exists an infinite word extending σ, $\sigma' : s_1, \ldots, s_j, s_{j+1}, \ldots$, which belongs to Π. Since φ specifies Π, $\bigvee_{i=1}^{k}(p_i \wedge \bigcirc f_i)$ and, therefore, $\bigvee_{i=1}^{k} p_i$ must hold at all positions of σ', including position j. As $\bigvee_{i=1}^{k} p_i$ is a past formula, this implies that it holds at position j of σ. This shows that $\bigvee_{i=1}^{k} p_i$ holds at an arbitrary position of an arbitrary finite sequence, so $\square(\bigvee_{i=1}^{k} p_i)$ is valid. ∎

References

[AS85] B. Alpern and F.B. Schneider. Defining liveness. *Info. Proc. Lett.*, 21:181–185, 1985.

[AS87] B. Alpern and F.B. Schneider. Recognizing safety and liveness. *Dis Comp.*, 2:117–126, 1987.

[Gab87] D. Gabbay. The declarative past and imperative future. In B. Banieqba H. Barringer, and A. Pnueli, editors, *Temporal Logic in Specification*, Le Notes in Comp. Sci. 398, pages 407–448. Springer-Verlag, 1987.

[Lam77] L. Lamport. Proving the correctness of multiprocess programs. *IEE Trans. Software Engin.*, 3:125–143, 1977.

[Lam83] L. Lamport. What good is temporal logic. In R.E.A. Mason, editor, *Pro IFIP 9th World Congress*, pages 657–668. North-Holland, 1983.

[Lam85] L. Lamport. *Distributed Systems — Methods and Tools for Specification*, chapter Basic Concepts, pages 19–30. Lec. Notes in Comp. Sci. 190 Springer-Verlag, 1985.

[LPZ85] O. Lichtenstein, A. Pnueli, and L. Zuck. The glory of the past. In *Proc. of the Workshop on Logics of Programs*, Lec. Notes in Comp. Sci. 193, pages 196–218. Springer-Verlag, 1985.

[MP89] Z. Manna and A. Pnueli. Completing the temporal picture. In *Proc. 16th Int. Colloq. Aut. Lang. Prog.*, Lec. Notes in Comp. Sci. 372, pages 534–558 Springer-Verlag, 1989. To appear in Theoretical Computer Science.

[MP90] Z. Manna and A. Pnueli. A hierarchy of temporal properties. In *Proc. 9th ACM Symp. Princ. of Dist. Comp.*, pages 377–408, 1990.

[Sis85] A.P. Sistla. On characterization of safety and liveness properties in temporal logic. In *Proc. 4th ACM Symp. Princ. of Dist. Comp.*, pages 39–48 1985.

[SPH84] R. Sherman, A. Pnueli, and D. Harel. Is the interesting part of process logic uninteresting. *SIAM J. Comp.*, 13:825–839, 1984.

[Zuc86] L. Zuck. *Past Temporal Logic.* PhD thesis, Weizmann Institute, 1986.

Lazy Lambda Calculus: Theories, Models and Local Structure Characterization

(Extended Abstract)

C.-H. Luke Ong[1]

Department of Information Systems and Computer Science
National University of Singapore, Lower Kent Ridge Road, Singapore 0511

Abstract

Lambda Calculus is commonly thought to be the basis for functional programming. However, there is a fundamental mismatch between the "standard" theory of *sensible* Lambda Calculus and the practice of *lazy* evaluation which is a distinctive feature of functional programming. This paper proposes modification of a number of key notions in the sensible theory along the lines of laziness. Starting from the *strongly unsolvables* as the meaningless terms, we define and investigate properties of *lazy* (or *weakly sensible*) λ-theories, *lazy* λ-models and a number of *lazy* behavioural preorders on λ-terms. In the second part, we show that all these notions have a natural place in a class of *lazy* PSE-*models*. A major result of this paper is a new *local structure theorem* for lazy PSE-models. This characterizes the ordering between denotations of λ-terms in the model by a new lazy behavioural preorder.

This paper is concerned with the Lambda Calculus as a foundation for lazy functional rogramming. The untyped Lambda Calculus as studied in [Bar84] is a proof system with quations as the only kind of judgement. The language of the system has a simple syntax overned by the following *BNF* rule:

$$\Lambda \ni M \quad ::= \quad x \quad | \quad (MM) \quad | \quad (\lambda x.M)$$

here x ranges over Var, a denumerable collection of variables. A key notion of the Lambda alculus is *solvability*. Recall that a λ-term M is *solvable* if there are λ-terms $N_1, \cdots, N_n$ for ome $n \in \omega$ such that $MN_1 \cdots N_n = \mathbf{I} \stackrel{\text{def}}{=} \lambda x.x$ (or more suggestively, $M\vec{X} = \mathbf{I}$, the equation n meta-variables $\vec{X}$ is solvable). A term is *unsolvable* if it is not solvable. Wadsworth (see .g. Ch. 8 *ibid.*) showed that:

$$\begin{array}{llll} M \text{ is solvable} & \Leftrightarrow & M \text{ has a } \textit{head normal form}, & \text{(proof-theoretic)} \\ & \Leftrightarrow & M \text{ converges under } \textit{head reduction}. & \text{(operational)} \end{array}$$

his characterization lays the foundation for the *sensible theory*, according to which the neaning of a λ-term is based on *head normal forms* (p. 41 *ibid.*) via the notion of *Böhm ree.* Since unsolvable terms diverge under head reduction (p. 171 *ibid.*), they are identified. his then, is the commonly accepted foundation for functional programming languages.

However, most functional languages are *lazy* — they do not evaluate "under an abstracion": programs are evaluated only up to *weak head normal form* (whnf) (see [PJ87]). So ve have a fundamental mismatch between theory and practice. To see that the theory really loes need to be modified, let $\mathbf{\Omega} \equiv (\lambda x.xx)(\lambda x.xx)$, the standard unsolvable term. In the ensible theory, $\lambda x.\mathbf{\Omega}$ and $\mathbf{\Omega}$, being both unsolvable are considered equal (as meaningless rograms). However, in the lazy regime, they should have different meaning: the former is weak head normal form (and so, has a meaning) whereas the latter is a non-terminating rogram.

[1]Current address: Trinity College, Cambridge CB2 1TQ, England; Email: chlo1@cl.cam.ac.uk

How should classical notions in the standard Lambda Calculus like Böhm tree, sensible λ-theory and λ-model be modified along the lines of lazy computation? Are there mathematical models for such lazy notions? This paper provides some answers to these questions.

1 A Theory of Lazy Lambda Calculus

We define a minimal lazy functional programming language to serve as our paradigm. First some notations.

Notations We denote ω as $\{0,1,\cdots\}$ and **Seq** as the collection of finite sequences of *positive* integers. The empty sequence is denoted $\langle\,\rangle$. For any sequences $\alpha \equiv \langle a_1,\cdots,a_n \rangle$ and $\beta \equiv \langle b_1,\cdots,b_m \rangle$, we write concatenation of sequences as $\alpha * \beta$. $\alpha \leqslant \beta \stackrel{\text{def}}{=} [n \leqslant m$ and $a_i = b_i$ for $1 \leqslant i \leqslant n]$. The *length* of α, denoted $|\alpha|$ is n. For any sets A, B, we write $A \subseteq_{\text{fin}} B$ ($A \subset_{\text{fin}} B$) to mean A is a finite (proper) subset of B. We shall often use the "vector" notation for sets as well as for λ-terms: $\vec{M} \equiv M_1,\cdots,M_n$ for $n \geqslant 1$; $|\vec{M}|$, the *length* of $\vec{M}$ is n. $\vec{A} \subseteq_{\text{fin}} \vec{B}$ means $[|\vec{A}| = |\vec{B}|$ and $A_i \subseteq_{\text{fin}} B_i$ for each $1 \leqslant i \leqslant n]$; $\vec{A} \subset_{\text{fin}} \vec{B}$ means $[\vec{A} \subseteq_{\text{fin}} \vec{B}$ and for some i, $A_i \subset_{\text{fin}} B_i]$. A *cpo* $\langle D, \sqsubseteq \rangle$ is a set D partially ordered w.r.t. $\sqsubseteq$ with a least element $\bot$ and that every ω-increasing sequence has a least upper bound in D. For cpo's D and E, we denote $[D \to E]$ as the cpo of Scott-continuous functions from D to E. Our standard reference for the Lambda Calculus is [Bar84] and we adopt all definitions and notational conventions therein with the notable exception of substitution: we write $M[N/x]$ to mean "in M, substitute N for all occurrences of x". $\Lambda(\underline{A})$ is the collection of λA-*terms*, *i.e.* λ-terms generated from elements of A regarded as *formal constants* (which are written $\underline{a} \in \Lambda(\underline{A})$ for $a \in A$); we write $\Lambda(\underline{A})^o$ for the collection of *closed* λA-terms. $\lambda\beta \vdash M = N$ or simply $M = N$ means that M and N are β-*convertible*. We shall assume known standard combinators like $\mathbf{Y}, \mathbf{K}, \mathbf{S}, \mathbf{\Omega}$ *etc.*

Definition 1.1 We define **lazy reduction** $\Downarrow$ as the least binary relation over Λ^o satisfying the following rules: (we read $M \Downarrow K$ as "M reduces or computes lazily to weak head normal form K")

$$\frac{}{\lambda x.P \Downarrow \lambda x.P} \qquad\qquad \frac{M \Downarrow \lambda x.P \quad P[N/x] \Downarrow L}{MN \Downarrow L}$$

The first rule stipulates a *weak* reduction scheme (*i.e.* no reduction under an abstraction) and the second corresponds to a call-by-name parameter passing mechanism. Note that $\Downarrow$ is a partial function. In [Abr89,AO92], the programming language $\lambda\ell \stackrel{\text{def}}{=} \langle \Lambda^o, \Downarrow \rangle$ is called the **Lazy Lambda Calculus.** *Programs* of the language $\lambda\ell$ are closed terms, *values* are weak head normal forms (whnf) which, in the absence of constants, are just the abstractions. It is possible to show that $\Downarrow$ corresponds precisely to a reduction strategy that contracts the leftmost redex at each step *provided* the redex is not a subterm of an abstraction. We write $M\Downarrow \stackrel{\text{def}}{=} \exists K.M \Downarrow K$, to be read "$M$ converges"; and $M\Uparrow \stackrel{\text{def}}{=} \neg[M\Downarrow]$ "M diverges".

In the lazy operational setting exemplified by $\lambda\ell$, a term is meaningful iff it has a whnf (*i.e.* β-convertible to an abstraction), which by the Standardisation Theorem, is equivalent to $M\Downarrow$. Since not all unsolvables have whnf, it makes sense to characterize those that do not; and these are precisely the meaningless (and non-terminating) terms in the lazy setting.

Definition 1.2 For $n \in \omega$, we say that M has **order** n, denoted $M \in \mathbf{O}_n$, if n is the largest i such that for variables $x_1,\cdots,x_i$ and for some N, we have $M = \lambda x_1 \cdots x_i.N$. So M has order 0 if M is not β-convertible to any abstraction. For example, $x\vec{M}, \mathbf{\Omega} \in \mathbf{O}_0$. We say that has **order** ∞, denoted $M \in \mathbf{O}_\infty$ if for any $n \in \omega$, however large, there is some N such that $= \lambda x_1 \cdots x_n.N$. For example, $\mathbf{YK} \in \mathbf{O}_\infty$ — think of $\mathbf{YK}$ informally as the "solution" to e equation (in meta-variable X) $\mathsf{X} = \lambda x.\mathsf{X}$. M has **proper order** 0, denoted $M \in \mathbf{PO}_0$, if $\in \mathbf{O}_0$ and that $\neg[\exists \vec{N}.M = x\vec{N}]$. We say that a term is **strongly unsolvable** just in case has proper order 0. A term is **weakly solvable** if it is β-convertible to an abstraction.

In contrast to the sensible approach, the meaningful terms in a lazy computational setting e the *weakly solvables*. We can show: for $M \in \Lambda^o$, M is weakly solvable iff $M\Downarrow$; and at M is strongly unsolvable iff $M\Uparrow$. Having pinned down the strongly unsolvables as e meaningless terms in the lazy regime, we characterize the unsolvables according to how eeply nested" an abstraction they are (convertible to) in the following sense: define, for $\geqslant 1$, $\mathbf{PO}_n \stackrel{\text{def}}{=} \{ M : \exists N \in \mathbf{PO}_0.M = \lambda x_1 \cdots x_n.N \}$; $\mathbf{PO}_\infty \stackrel{\text{def}}{=} \mathbf{O}_\infty$.

heorem 1.3 (Characterization) *The collection* $\{ \mathbf{PO}_n : n \in \omega \cup \{ \infty \} \}$ *is a partition the unsolvables. For* $n \in \omega \cup \{ \infty \}$, *we say that* M *has* **order of unsolvability** n *just in se* $M \in \mathbf{PO}_n$. □

zy λ-Theory Recall that a *sensible* λ-theory is one that equates all unsolvables. As we ve seen in the example of $\mathbf{\Omega}$ and $\lambda x.\mathbf{\Omega}$, such a notion does not fit the lazy computational tting. In light of the above characterization (due to Longo), we propose the notion of a y or *weakly sensible λ-theory*. For each $n \in \omega \cup \{ \infty \}$, define $\mathcal{PO}_n$ to be the collection equations of the form $M = N$ such that $M, N \in \mathbf{PO}_n \cap \Lambda^o$. For any set of equations tween closed terms $\mathcal{T}$, let $\mathsf{Th}(\mathcal{T})$ be the theory obtained by adding $\mathcal{T}$ as new axioms to e axioms and rules of $\lambda\beta$. Define $\mathcal{PO} \stackrel{\text{def}}{=} \mathsf{Th}(\cup_{n \in \omega \cup \{\infty\}} \mathcal{PO}_n)$. For any λ-theory $\mathcal{T}$, we write $+(P = Q)$ for $\mathsf{Th}(\mathcal{T} \cup \{ P = Q \})$. Note that $\mathcal{PO}$ is consistent because $\bigcup_{n \in \omega \cup \{\infty\}} \mathcal{PO}_n \subset \mathcal{K}$ the λ-theory obtained by identifying all unsolvables, and that $\mathcal{K}$ is consistent (see [Bar84, 411]).

efinition 1.4 Let $\mathcal{T}$ be a λ-theory. We say that $\mathcal{T}$ is **zero sensible** if $\mathcal{T}$ equates all the rongly unsolvables *i.e.* $\mathsf{Th}(\mathcal{PO}_0) \subseteq \mathcal{T}$. $\mathcal{T}$ is **finitely sensible** if $\mathcal{T}$ equates all unsolvables finite orders. $\mathcal{T}$ is **pre-lazy** if $\mathcal{PO} \subseteq \mathcal{T}$. A **fully lazy λ-theory** $\mathcal{T}$ is a λ-theory which uates any two unsolvable terms iff they have the same order of unsolvability, *i.e.* for any $, n \in \omega \cup \{ \infty \}$, for any $M \in \mathbf{PO}_m$ and $N \in \mathbf{PO}_n$, $\mathcal{T} \vdash M = N$ iff $m = n$.

Note that any sensible λ-theory is zero- and finitely sensible as well as pre-lazy but not lly lazy; any fully lazy λ-theory is zero- and finitely sensible and pre-lazy but not sensible. e present some results in lazy λ-theory and refer the reader to [Ong88b, Ch. 2] for proofs.

roposition 1.5 *(i) Any zero sensible λ-theory which equates all* $\mathbf{PO}_\infty$*-terms is pre-lazy.*
ii) Any pre-lazy λ-theory is either fully lazy or finitely sensible.
iii) Any pre-lazy λ-theory which does not equate $\mathbf{\Omega}$ *and* $\lambda x.\mathbf{\Omega}$ *is fully lazy.* □

What are some examples of fully lazy λ-theories? In what follows, we study two such eories induced from lazy behavioural preorders: *Longo-tree preorder* and *applicative simation*. Since the Böhm tree λ-theory $\mathcal{BT}$ [Bar84, p. 425] is sensible, it has to be rected for our purpose. We propose *Longo trees* as the replacement for Böhm trees in e lazy regime. Intuitively, Longo trees are "fully lazy" variants of Böhm trees. Define $_{\mathrm{L}} \stackrel{\text{def}}{=} \{ \top \} \cup \{ \lambda\vec{x}.\bot : \vec{x} \subseteq \mathsf{Var} \} \cup \{ \lambda\vec{x}.y : \vec{x}, y \subseteq \mathsf{Var} \}$. Let $\leqslant_{\mathrm{L}}$ be the least partial order over $_{\mathrm{L}}$ with $\top$ as the top element and that $\lambda\vec{x}.\bot \leqslant_{\mathrm{L}} \lambda\vec{x}.s$ for any $s \in \Sigma_{\mathrm{L}} - \{ \top \}$. Note that $\bot$ is e least element and that $\lambda\vec{x}.y$ with $\vec{x}, y \subseteq \mathsf{Var}$ are the maximal elements. We denote head duction as $\twoheadrightarrow_{\mathrm{h}}$.

Definition 1.6 (Informal) The **Longo tree** of a λ-term M, $\text{LT}(M)$, is a Σ_{L}-labelled tre defined inductively as follows:

$$\text{LT}(M) \stackrel{\text{def}}{=} \begin{cases} \lambda x_1 \cdots x_n.\bot & \text{if } M \in \mathbf{PO}_n \text{ for } n \in \omega; \\ \top & \text{if } M \in \mathbf{PO}_\infty; \end{cases}$$

else if $M \twoheadrightarrow_{\text{h}} \lambda\vec{x}.yM_1 \cdots M_m$ for $m \geqslant 0$, then

$\text{LT}(M) \stackrel{\text{def}}{=}$

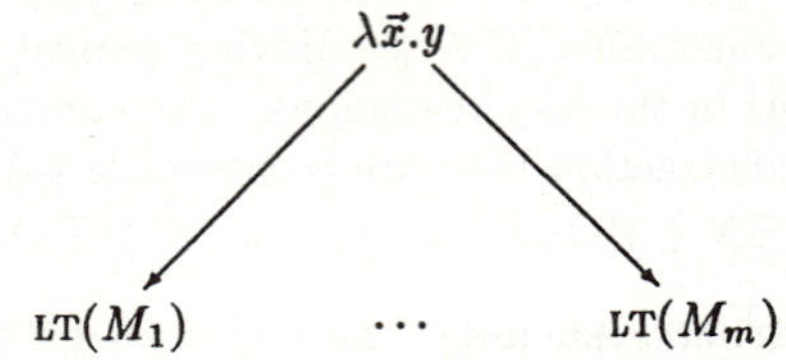

We define a partial ordering $\subseteq_{\text{L}}$ on Longo trees informally as follows: $L_1 \subseteq_{\text{L}} L_2$ iff L_2 obtained from L_1 by

- replacing some occurrence(s) of $\bot$ in L_1 by the Longo tree of some λ-term, and / or
- replacing some occurrence(s) of $\lambda\vec{x}.\bot$ in L_1 by $\top$.

This definition is the version presented in [Lon83]. We write $M \sqsubseteq_{\text{LT}} N \stackrel{\text{def}}{=} \text{LT}(M) \subseteq_{\text{L}} \text{LT}(N$ It is easy to show that $\sqsubseteq_{\text{LT}}$ is a preorder on Λ. We call the induced λ-theory the **Long λ-theory**, denoted $\mathcal{LT}$ *i.e.* $\mathcal{LT} \vdash M = N \stackrel{\text{def}}{=} [M \sqsubseteq_{\text{LT}} N \;\;\&\;\; N \sqsubseteq_{\text{LT}} M]$. The Longo λ-theory trivially fully lazy. We establish an important property of the Longo-tree preorder as follow (see [Ong88b, Theorem 2.3.1.1] for a proof):

Theorem 1.7 (Precongruence) *The Longo-tree preorder $\sqsubseteq_{\text{LT}}$ is a pre-congruence: for terms M and N, whenever $M \sqsubseteq_{\text{LT}} N$ then for any context $C[\,]$, $C[M] \sqsubseteq_{\text{LT}} C[N]$.* ∎

What is (a characterization of) the *maximal consistent* fully lazy λ-theory?

Definition 1.8 We define a binary relation $\sqsubseteq^{\text{B}}$ on Λ^o called **applicative simulation** a $M \sqsubseteq^{\text{B}} N \stackrel{\text{def}}{=} \forall \vec{P} \subseteq \Lambda^o.M\vec{P}\Downarrow \implies N\vec{P}\Downarrow$. It is easy to see that $\sqsubseteq^{\text{B}}$ is a preorder; in fact, it is *precongruence* (we have two proofs in [AO92], see also [Ong92]). By abuse of notation, w define $\lambda\ell$ to be the induced λ-theory of $\sqsubseteq^{\text{B}}$. A **fully lazy preorder** $\sqsubseteq$ over Λ^o is defined t be a preorder that induces a fully lazy λ-theory and that for any $\Theta \in \mathbf{PO}_0$ and $\Upsilon \in \mathbf{PO}_\infty$ $\Theta \sqsubseteq \lambda x.\Theta \sqsubseteq \lambda x_1x_2.\Theta \sqsubseteq \cdots \sqsubseteq \lambda x_1 \cdots x_n.\Theta \sqsubseteq \cdots \sqsubseteq \Upsilon$.

We can show that all the strongly unsolvables are identified in $\lambda\ell$, similarly for $\mathbf{PO}_\infty$ terms. Hence, by Proposition 1.5(i), $\lambda\ell$ is pre-lazy. Since $\mathbf{PO}_0$ and $\mathbf{PO}_1$ are not provabl equal in $\lambda\ell$, by (iii) of the same proposition, the λ-theory $\lambda\ell$ is fully lazy. In fact, we can als show that the *preorder* $\sqsubseteq^{\text{B}}$ is fully lazy. The λ-theory $\lambda\ell$ is the *canonical* fully lazy λ-theor in the following sense:

Proposition 1.9 *$\lambda\ell$ is the maximal consistent fully lazy λ-theory i.e. for any $M, N \in \Lambda$ such that $\lambda\ell \nvdash M = N$, either $\lambda\ell + (M = N)$ is inconsistent or it is not fully lazy.* ∎

As an immediate consequence, we have $\mathcal{LT} \subset \lambda\ell$ (to see that the inclusion is strict, jus consider $\lambda x.xx = \lambda x.x(\lambda y.xy)$). This may be strengthened to $\sqsubseteq_{\text{LT}} \subset \sqsubseteq^{\text{B}}$ (see Theorem 3.12 A very general theory of full abstraction for the Lazy Lambda Calculus based on the notio of applicative simulation has been developed; we direct the reader to [AO92] for details.

azy λ-Models All the well-known λ-models *e.g.* D_∞ [Sco72], $\mathcal{P}(\omega)$ [Plo72,Sco76], $\mathbb{T}^\omega$ 'lo78] and $\mathcal{B}$ (*Böhm tree model*, see e.g. [Bar84, Ch. 18]) are sensible; and so these concrete odels are of no use to our purpose. What about the very definition of a λ-*model*? Are .e three equivalent formulations of a λ-model *viz.* environmental, functional and first-order, ecific and accurate enough to capture a model of the Lazy Lambda Calculus? In the lazy gime, the denotations of Ω and $\lambda x.\Omega$ cannot be the same: the former is divergent and .e latter is a *value* and must have a well-defined denotation. We build such distinction of enotations into the notion of a lazy λ-model at the *definitional* level.

A *quasi-applicative structure with divergence* **(q-aswd)** is a structure $\langle A, \cdot, \Uparrow \rangle$ such that s a binary operation on the set A, $\Uparrow \subset A$ (we write $a \in \Uparrow$ as $a\Uparrow$: "a diverges") satisfying .e following *left-strictness* axiom: $\forall a \in A.[a\Uparrow \Rightarrow \forall x \in A. a \cdot x\Uparrow]$. We define $\Downarrow \stackrel{\text{def}}{=} (A - \Uparrow)$. A *iasi-applicative transition system* **(q-ats)** is a structure $\langle A, \mathsf{eval} \rangle$ such that eval is a proper rtial function from A to A^A *i.e.* $\mathsf{dom}(\mathsf{eval}) \subset A$. A q-aswd (q-ats) is **pointed** if $\Uparrow$ (or $A - \mathsf{dom}(\mathsf{eval})$) respectively) is a singleton set. Given a q-aswd $\mathcal{A} = \langle A, \cdot, \Uparrow \rangle$, we define an sociated partial function $\mathsf{eval} : A \rightharpoonup A^A$ by

$$\mathsf{eval}(a) \stackrel{\text{def}}{=} \begin{cases} x \mapsto a \cdot x & \text{if } a\Downarrow; \\ \text{undefined} & \text{else.} \end{cases}$$

he structure $\mathcal{A}^{q-ats} \stackrel{\text{def}}{=} \langle A, \mathsf{eval} \rangle$ is clearly a q-ats. Conversely, given a q-ats $\mathcal{A} = \langle A, \mathsf{eval} \rangle$ ith a distinguished element $\bot \in (A - \mathsf{dom}(\mathsf{eval}))$, we define an associated application

$$a \cdot b \stackrel{\text{def}}{=} \begin{cases} f(b) & \text{if } \mathsf{eval}(a) = f; \\ \bot & \text{else,} \end{cases}$$

ıd define $\Uparrow \stackrel{\text{def}}{=} (A - \mathsf{dom}(\mathsf{eval}))$. It is easy to see that the left-strictness axiom is satisfied. /e define the structure $\mathcal{A}^{q-aswd} \stackrel{\text{def}}{=} \langle a, \cdot, \Uparrow \rangle$.

Let us recall some standard notions: let $\langle A, \cdot \rangle$ be an **applicative structure** (*i.e.* $\cdot$ is binary operation on A) and $X \subseteq A$. For any $X \subseteq A$, an endofunction f on A is X-*epresentable* if there is an $d \in X \subseteq A$ such that $f(a) = d \cdot a$ for any $a \in A$; we call such d a *epresentative* of f. An A-representable function is simply called a **representable** function. /e denote $[A \to_\Downarrow A]$ as the class of all $\Downarrow$-representable endofunctions on A.

efinition 1.10 We define **environment** and **functional lazy λ-models** as structures tisfying the following conditions with explanatory comments in the remark thereafter:

Environment Lazy λ-Model $\mathcal{A} = \langle A, \cdot, \Uparrow, [\![-]\!]^{\mathcal{A}}_{-} \rangle$	**Functional Lazy λ-Model** $\mathcal{A} = \langle A, \mathsf{Fun}, \mathsf{Gr}, [\![-]\!]^{\mathcal{A}}_{-} \rangle$
• $\langle A, \cdot, \Uparrow \rangle$ is a q-aswd,	• $\langle A, \mathsf{Fun} \rangle$ is a q-ats $(\langle A, \cdot, \Uparrow \rangle \stackrel{\text{def}}{=} \langle A, \mathsf{Fun} \rangle^{q-aswd})$,
• $[\![-]\!]^{\mathcal{A}}_{-}$ *preserves* application,	• $A \stackrel{\mathsf{Fun}}{\rightharpoonup} [A \to_\Downarrow A] \stackrel{\mathsf{Gr}}{\to} A$, $\mathsf{range}(\mathsf{Gr}) \subseteq \Downarrow$ & $\mathsf{Fun} \circ \mathsf{Gr} = 1_{[A \to_\Downarrow A]}$,
• $\mathcal{A} \models (\beta)$,	
• $\mathcal{A} \models (\xi)$,	• $[\![-]\!]^{\mathcal{A}}_{-} : \Lambda(\underline{A}) \times A^{\mathsf{Var}} \to A$ satisfying:
• $\forall M \in \Lambda^o. M\Downarrow \implies [\![M]\!]^{\mathcal{A}}_{\rho}\Downarrow$.	- $[\![-]\!]^{\mathcal{A}}_{-}$ *preserves application*,
	- $[\![\lambda x.M]\!]^{\mathcal{A}}_{\rho} \stackrel{\text{def}}{=} \mathsf{Gr}(d \mapsto [\![M]\!]^{\mathcal{A}}_{\rho[x \mapsto \underline{d}]})$.

Remark By $[\![-]\!]_{-}^{\mathcal{A}}$ *preserving application* is meant: $\forall M, N \in \Lambda(\underline{A}), \forall \rho \in A^{\mathsf{Var}}, \forall a \in A,$

$$\begin{array}{lcl} [\![\underline{a}]\!]_\rho^{\mathcal{A}} & \stackrel{\text{def}}{=} & a, \\ [\![x]\!]_\rho^{\mathcal{A}} & \stackrel{\text{def}}{=} & \rho(x), \\ [\![MN]\!]_\rho^{\mathcal{A}} & \stackrel{\text{def}}{=} & [\![M]\!]_\rho^{\mathcal{A}} \cdot [\![N]\!]_\rho^{\mathcal{A}}. \end{array}$$

The axiom (ξ): $\forall M, N \in \Lambda(\underline{A}).\forall \rho \in A^{\mathsf{Var}}.\forall a \in A.[[\![M]\!]_{\rho[x\mapsto \underline{a}]}^{\mathcal{A}} = [\![N]\!]_{\rho[x\mapsto \underline{a}]}^{\mathcal{A}} \Rightarrow [\![\lambda x.M]\!]_\rho^{\mathcal{A}}$ $[\![\lambda x.N]\!]_\rho^{\mathcal{A}}]$. In the functional model, we require $(d \mapsto [\![M]\!]_{\rho[x\mapsto \underline{d}]}^{\mathcal{A}}) \in [A \to_{\Downarrow} A]$. No that $[A \to_{\Downarrow} A]_\perp$ is a *retract* of A. The application $\cdot$ therein is that which is defined $\langle A, \mathsf{Fun}\rangle^{q-aswd}$. We use $\mathcal{A} \models M = N$ and $\mathcal{A} \models M \sqsubseteq N$ with their standard meaning, s e.g. [Bar84].

Proposition 1.11 *There are bijective translations*[2] *between pointed environment lazy models and pointed functional lazy λ-models.*

It should be clear that a lazy λ-model is, a *fortiori*, a λ-model. We have presented a ve rough sketch of the model theory of the Lazy Lambda Calculus, leaving out all notions n needed in the sequel. The reader is directed to [Ong88b, Ch. 3] for a complete treatmer A **fully lazy λ-model** is defined to be a lazy λ-model whose λ-theory (*i.e.* collection equations valid in the model) is fully lazy. Examples: $\mathbb{D}$, the initial solution to the doma equation $D = [D \to D]_\perp$ in the category of cpo's and continuous functions, is a fully la λ-model. We refer the reader to [Ong88b, Ch. 4] for analysis of its properties. Anoth example of a fully lazy λ-model is the *lazy* PSE*-model*, to which we now turn.

2 Plotkin-Scott-Engeler Algebras and Models

We introduce a family of combinatory algebras called *Plotkin-Scott-Engeler* (PSE) *algebra* First formalized by Engeler [Eng81], PSE-algebras may be seen as a "code-free" variant the well-known *graph model* $\mathcal{P}(\omega)$ due independently to Plotkin [Plo72] and Scott [Sco7 Let B be a non-empty set satisfying

(io) $$\beta \subseteq_{\text{fin}} B \ \&\ b \in B \iff (\beta; b) \in B;$$

where the meta-variable β ranges over *finite* sets (similarly for $\beta_0, \beta_1, \cdots, \gamma_i, \cdots$). We ca $(\beta; b)$ an **io-pair. Plotkin-Scott-Engeler** (PSE) **algebra** is the applicative structu $\langle \mathcal{P}(B), \cdot \rangle$ where $\mathcal{P}(B)$ is the powerset of B; application is defined as follows: for $d, e \in \mathcal{P}(B$ $d \cdot e \stackrel{\text{def}}{=} \{ b : (\beta; b) \in d \ \&\ \beta \subseteq_{\text{fin}} e \}$. In this paper, we shall focus on the class of *free* PSE*-algebra i.e.* those freely generated from a *base set* A.

Definition 2.1 Let A be a non-empty set such that none of its element is an io-pair. W define inductively a family of sets $\{ B_n : n \in \omega \}$ as follows:

$$B_0 \stackrel{\text{def}}{=} A; \quad B_{n+1} \stackrel{\text{def}}{=} B_n \cup \{ (\beta; b) : \beta \subseteq_{\text{fin}} B_n \ \&\ b \in B_n \}$$

and define $B \stackrel{\text{def}}{=} \bigcup_{n\in\omega} B_n$. B thus defined clearly satisfies the axiom (**io**). We define D the **free** PSE**-algebra** generated from base set A, as the applicative structure $\langle \mathcal{P}(B), \cdot$

[2]There is also bijective translations to and from *pointed lazy first-order lazy λ-model* whose definition have omitted. See [Ong88b, Ch. 3] for details.

n the following, we shall often confuse the applicative structure D_A with its underlying et $\mathcal{P}(B)$. The intuition behind the io-pairs and application should be clear: $(\beta;b)$ is an atomic" instruction giving output b whenever the input contains β. Since we assume A not o contain any io-pair, we have $A\cdot d=\varnothing$ for any $d\in D_A$. Note that for any free algebra D_A nd for any $b\in B$, there are $\beta_1,\cdots,\beta_n\subseteq_{\text{fin}} B$ and $a\in A$ such that $b=(\beta_1;\cdots(\beta_n;a)\cdots)$.

We denote $[A\rightarrow_{\mathrm{r}} A]$ as the set of representable endofunctions on A. We record some lementary topological results of PSE-algebras. Let B be a non-empty set satisfying (io). $\mathcal{P}(B),\subseteq\rangle$ is a complete algebraic lattice. The basis for the Scott topology on $\mathcal{P}(B)$ is given y sets of the form $\{d\in\mathcal{P}(B):\beta\subseteq_{\text{fin}} d\}$ with β ranging over finite subsets of B. Moreover, $\in[(\mathcal{P}(B))^n\rightarrow\mathcal{P}(B)]$ iff f is continuous argument-wise iff $f(\vec{d})=\bigcup\{f(\vec{\beta}):\vec{\beta}\subseteq_{\text{fin}}\vec{d}\}$. For $\geqslant 1$, $[(\mathcal{P}(B))^n\rightarrow_{\mathrm{r}}\mathcal{P}(B)]=[(\mathcal{P}(B))^n\rightarrow\mathcal{P}(B)]$.

Any PSE-algebra $\langle D,\cdot\rangle$ is a combinatory algebra which may be λ-*expanded* into a λ-odel (see *e.g.* [Lon83]). λ-expansion is specfied by a *graph function* Gr which, given any epresentable endofunction f, selects a unique representative $\mathsf{Gr}(f)$ from REP(f), the set of epresentatives of f.

- The *standard* λ-expansion is specified by the graph function: $\mathsf{Gr}^{\mathrm{st}}(f)\stackrel{\text{def}}{=}\{(\beta;b):\beta\subseteq_{\text{fin}} B\ \&\ b\in f(\beta)\}$. $\langle D_A,\mathsf{Fun},\mathsf{Gr}^{\mathrm{st}},[\![-]\!]^{\mathrm{st}}_{-}\rangle$ is a *functional λ-model* (see [Bar84]), which we shall call the **standard** PSE**-model**. Note that Fun : $D_A\rightarrow[D_A\rightarrow_{\mathrm{r}} D_A]$ is the induced map from the applicative structure $\langle D_A,\cdot\rangle$.
- The second λ-expansion is applicable only to free PSE-algebras generated from a base set A, say. The defining graph function is $\mathsf{Gr}^{\mathrm{lazy}}(f)\stackrel{\text{def}}{=}\mathsf{Gr}^{\mathrm{st}}(f)\cup A$. The structure $D_A^{\mathrm{lazy}}\stackrel{\text{def}}{=}\langle D_A,\mathsf{Fun},\mathsf{Gr}^{\mathrm{lazy}},[\![-]\!]^{\mathrm{lazy}}_{-}\rangle$ is a *lazy functional λ-model* which we call the *lazy free* PSE*-model generated from A* or simply **lazy** PSE**-model**. Note that $\langle D_A,\mathsf{Fun}\rangle$ is just $\langle D_A,\cdot\rangle^{q-ats}$ and that D_A^{lazy} is *pointed* as a lazy λ-model.

Notation Let $P(\beta_1,\cdots,\beta_n)$ be a predicate with each meta-variable β_i ranging over finite ubsets of B. We write $[\mu\vec{\beta}.P(\vec{\beta})]$ to be the least $\vec{\beta}\subseteq_{\text{fin}}\vec{B}$ such that $P(\vec{\beta})$ holds, to wit, $P([\mu\vec{\beta}.P(\vec{\beta})])$ holds, and if $\vec{\delta}\subset_{\text{fin}}[\mu\vec{\beta}.P(\vec{\beta})]$ then $P(\vec{\delta})$ does not hold.

Let B be a set satisfying the axiom (io). We say that an element x of $\mathcal{P}(B)$ is **saturated** if whenever $(\beta;b)\in x$ then $(\gamma;b)\in x$ for any finite superset γ of β. Not every element in REP(f) s saturated. For any $f\in[D_A\rightarrow D_A]$, consider $X_f\stackrel{\text{def}}{=}\{([\mu\beta.b\in f(\beta)];b):\exists\beta\subseteq_{\text{fin}}B.b\in f(\beta)\}$. t is easy to see that X_f is a representative of f and that X_f is *not* saturated.

Longo [Lon83] established a semantic characterization of unsolvability in lazy PSE-models. He defined a family of sets $\{A_{(n)}:n\in\omega\}$ as follows: $A_{(0)}\stackrel{\text{def}}{=}\varnothing$, and for $n\geqslant 1$, $A_{(n)}\stackrel{\text{def}}{=}$ $[\![\lambda x_1\cdots x_n.\varnothing]\!]^{\mathrm{lazy}}$. By abuse of notation, we define $A_{(\infty)}\stackrel{\text{def}}{=}B$. Note that $[\![\lambda x_1\cdots x_n.A]\!]^{\mathrm{st}}=$ $\{(\beta_n;\cdots(\beta_1;a)\cdots):a\in A\ \&\ \beta_i\subseteq_{\text{fin}} B\}$.

Lemma 2.2 *(i) For $n\in\omega$, $A_{(n+1)}=[\![\lambda x_1\cdots x_n.A]\!]^{\mathrm{st}}\cup[\![\lambda x_1\cdots x_{n-1}.A]\!]^{\mathrm{st}}\cup\cdots\cup A$.*
(ii) For any $d\in D_A$ and any $n\in\omega$, $A_{(n+1)}\cdot d=A_{(n)}$.
(iii) For any $M\in\Lambda$ and any $n\in\omega$, $[\![\lambda x_1\cdots x_n.M]\!]^{\mathrm{lazy}}=[\![\lambda x_1\cdots x_n.M]\!]^{\mathrm{st}}\cup A_{(n)}$. □

As a corollary of the preceding lemma, we have:

Theorem 2.3 (Longo) *For any $M\in\Lambda$ and any $n\in\omega\cup\{\infty\}$,*

(i) $M\in\mathbf{PO}_n\iff[\![M]\!]^{\mathrm{lazy}}=A_{(n)}$;
(ii) $M\in\mathbf{O}_n\iff[[\![M]\!]^{\mathrm{lazy}}\supseteq A_{(m)}\Leftrightarrow m\leqslant n]$. □

The theorem says that D_A^{lazy} is a fully lazy λ-model. So far, we have seen two valuation of λ-terms in the combinatory algebra D_A, namely $[\![-]\!]_-^{\text{lazy}}$ and $[\![-]\!]_-^{\text{st}}$. We now introduce third, non-standard valuation.

Definition 2.4 Let $f \in [(D_A)^n \to D_A]$ where $n \in \omega$. Define $[\![\lambda x_1 \cdots x_n.f(x_1,\cdots,x_n)]\!]^*$ as

$$\{(\gamma_1;\cdots(\gamma_n;c)\cdots) \in [\![\lambda x_1 \cdots x_n.f(x_1,\cdots,x_n)]\!]^{\text{st}} : \vec{\gamma} \equiv [\mu\vec{\beta}.(c \in f(\vec{\beta}))]\}.$$

Contrast this with the definition of $\lambda^-\vec{x}.f(\vec{x})$ in [Lon83, p. 186]. Note that any b $[\![\lambda x_1 \cdots x_n.f(x_1,\cdots,x_n)]\!]^*$ has the form $(\beta_1;\cdots(\beta_m;a)\cdots)$ where $a \in A$ and $m \geqslant n$. It is easy to show that for any $d_1,\cdots,d_n$ in D_A, we have $[\![\lambda x_1 \cdots x_n.f(x_1,\cdots,x_n)]\!]^* d_1 \cdots d_n = f(d_1,\cdots,d_n)$.

Proposition 2.5 *Let $f \in [D_A \to D_A]$ and $n \geqslant 2$. Suppose*

$$(\neg\varnothing) \qquad \forall d_1.\cdots.\forall d_{n-1}.\exists d_n.f(d_1,\cdots,d_{n-1},d_n) \neq \varnothing.$$

If $0 \leqslant p < n$, then $[\![\lambda x_1 \cdots x_n.f(x_1,\cdots,x_n)]\!]^* d_1 \cdots d_p$ does not *contain any $a_0 \in A$ and it is* not *saturated.* □

Definition 2.6 For $m \geqslant 2$, we define $\mathbf{P}_m \stackrel{\text{def}}{=} \lambda x_0 \cdots x_m.x_m x_o \cdots x_{m-1}$ called **Böhm permutator** of degree m. Define $\mathbf{p}_m \stackrel{\text{def}}{=} [\![\mathbf{P}_m]\!]^* \in D_A$. These permutators play a pivotal role in the "Böhm-out" technique. Since $\mathbf{p}_m$ satisfies condition $(\neg\varnothing)$ in the preceding proposition, we have an immediate corollary.

Corollary 2.7 *Let x_i, d_i range over D_A. Then, $\mathbf{p}_m x_0 \cdots x_m = x_m x_o \cdots x_{m-1}$. Further, for $l \leqslant m$, $\mathbf{p}_m d_1 \cdots d_l$ is not saturated and does not contain any $a_0 \in A$.* □

3 Lazy PSE-Ordering and Local Structure Characterization

We introduce a new behavioural preorder on λ-terms $\lesssim_{\text{PSE}}$ which we call the *lazy Plotkin Scott-Engeler preorder*. The least λ-terms w.r.t. $\lesssim_{\text{PSE}}$ are the strong unsolvables and the top elements are just the unsolvables of infinite order. The λ-theory induced by $\lesssim_{\text{PSE}}$ is just the Longo λ-theory.

- The lazy PSE-ordering has a *sound* interpretation in a class of *lazy* λ-models (including D_A^{lazy} and $\mathbb{D}$) satisfying certain *monotonicity* conditions.
- We prove a *local structure theorem* for $D_A^{\text{lazy}} = \langle D_A, \mathsf{Fun}, \mathsf{Gr}^{\text{lazy}}, [\![-]\!]_-^{\text{lazy}} \rangle$ lazy PSE-models: $\forall M, N \in \Lambda$,
$$M \lesssim_{\text{PSE}} N \iff D_A^{\text{lazy}} \models M \subseteq N.$$

The preceding result strengthens an (apparently unsubstantiated) claim of Longo (see second para. before Remark 3.9 in *op. cit.*). To the best of our knowledge, this is the first local structure theorem for "non-sensible" (lazy) λ-models; it complements local structure characterizations of well-known *sensible* λ-models *viz.* $\mathcal{P}(\omega)$ and D_∞ in [Hyl76] and $\mathbb{T}^\omega$ in [BL80].

The lazy PSE-ordering is defined via a family of inductively defined behavioural preorders indexed over ω. For $m \in \omega$, we define $M \in \mathcal{F}_m \stackrel{\text{def}}{=} \exists N \in \Lambda.M = \lambda x_1 \cdots x_m.N$, and $\mathcal{F}_\infty \stackrel{\text{def}}{=} \mathbf{PO}_\infty$.

efinition 3.1 (Lazy PSE-ordering) For any λ-terms M and N,

- $M \sqsubseteq^{0}_{\mathrm{PSE}} N$ always holds.
- $M \sqsubseteq^{k+1}_{\mathrm{PSE}} N \stackrel{\mathrm{def}}{=}$

$$
\begin{array}{lll}
[1] & M \in \mathbf{PO}_0 \quad \text{or} & \\
[2] & N \in \mathbf{PO}_\infty \quad \text{or} & \\
[3a] & M \in \mathbf{PO}_n, n \geqslant 1 \quad \Longrightarrow & N \in \mathcal{F}_n, \quad \& \\
[3b] & M \twoheadrightarrow_{\mathrm{h}} \lambda x_1 \cdots x_n.yM_1 \cdots M_m \quad \Longrightarrow & N \twoheadrightarrow_{\mathrm{h}} \lambda x_1 \cdots x_{n+l}.yN_1 \cdots N_{m+l} \quad \& \\
 & & \forall 1 \leqslant i \leqslant m. M_i \sqsubseteq^{k}_{\mathrm{PSE}} N_i \quad \& \\
 & & \forall 1 \leqslant j \leqslant l. x_{n+j} \sqsubseteq^{k}_{\mathrm{PSE}} N_{m+j};\ m, n, l \geqslant 0.
\end{array}
$$

- Finally, $M \sqsubseteq_{\mathrm{PSE}} N \stackrel{\mathrm{def}}{=} \forall k \in \omega. M \sqsubseteq^{k}_{\mathrm{PSE}} N$.

It is easy to see that $\sqsubseteq_{\mathrm{PSE}}$ is a preorder. It follows immediately from the definition of $\sqsubseteq_{\mathrm{PSE}}$ hat the least and greatest elements w.r.t. $\sqsubseteq_{\mathrm{PSE}}$ are the $\mathbf{PO}_0$- and $\mathbf{PO}_\infty$-elements respectively. n fact, we can show that $\sqsubseteq_{\mathrm{PSE}}$ is a *fully lazy preorder* and that $\lambda x_1 \cdots x_n.\Theta \sqsubseteq_{\mathrm{PSE}} \lambda x_1 \cdots x_n.M$ or any $M \in \Lambda$ and any $\Theta \in \mathbf{PO}_0$. That the preorder $\sqsubseteq_{\mathrm{PSE}}$ should have the $\mathbf{PO}_\infty$-terms s top elements might seem rather bizarre at first sight, since not just $\mathbf{PO}_\infty$ but the whole f the unsolvables are collapsed onto the bottom element in practically all the λ-models we ave experience with. We have an informal operational justification: think of $M \sqsubseteq_{\mathrm{PSE}} N$ s "N having more 'lazy information' than M". In the lazy regime, convergence (to an bstraction) is the only kind of "information quantum" observable. It follows that the more eeply-nested an abstraction a term is β-convertible to, the more information it contains. o extrapolate this argument further, terms which have the greatest amount of information re those which are convertible to unboundedly deeply nested abstractions — unsolvables of nfinite order. Intuitively, the lazy PSE-ordering is characterized by the following:

- the $\mathbf{PO}_\infty$-terms are the top (not just maximal) elements,
- $M \sqsubseteq_{\mathrm{PSE}} N$ if $M \sqsubseteq_{\mathrm{LT}} N'$ where N is some *finite η-expansion* (see [Bar84, page 232]) of N'.

roposition 3.2 *The λ-theory induced by the lazy PSE-ordering is precisely the Longo -theory.* □

The first major result on lazy PSE-ordering is the following Soundness Theorem. We say hat D has **lazy approximable application** ([Ong88b, Ch. 2]) if $\langle D, \cdot \rangle$ is an applicative tructure with a continuous application $\cdot$, and that there is a continuous map from $D \times \omega$ to D: $(d,n) \mapsto d_n$ satisfying the following: for any $d, e \in D$, $m, n \in \omega$,

$$
\begin{array}{llll}
\text{(i)} & d = \bigsqcup_{n \in \omega} d_n, & \text{(iv)} & d_{n+1} \cdot e \sqsubseteq (d \cdot e_n)_n, \\
\text{(ii)} & d_0 = \bot, & \text{(v)} & (d_n)_m = d_{\min(m,n)}. \\
\text{(iii)} & \bot \cdot d = \bot, & &
\end{array}
$$

or simplicity, we abbreviate $\mathcal{M} \models N \sqsubseteq M$ as $N \sqsubseteq M$.

heorem 3.3 (Soundness) *Let $\langle D, \sqsubseteq \rangle$ be a cpo which has a top element $\top$ and lazy pproximable application. Suppose $\mathcal{M} = \langle D, \mathsf{Fun}, \mathsf{Gr}, [\![-]\!]_{-} \rangle$ is a lazy λ-model satisfying he following* **monotonicity conditions***:*

(1) the graph function Gr *is monotonic w.r.t.* $\sqsubseteq$; *(3) for any* $\Upsilon \in \mathbf{PO}_\infty$, $\mathcal{M} \models \Upsilon = \top$;
(2) $\mathcal{M}$ *is a fully lazy* λ*-model;* *(4)* $\mathcal{M} \models (\eta^-)$;

where $(\eta^-) \stackrel{\text{def}}{=} \forall d \in D.\mathcal{M} \models \underline{d} \sqsubseteq \lambda x.\underline{d}x$. *Then,* $M \precsim_{\text{PSE}} N$ *implies* $\mathcal{M} \models M \sqsubseteq N$.

A second major result of this paper is the following:

Theorem 3.4 (Local Structure) *For any* $M, N \in \Lambda$, $M \precsim_{\text{PSE}} N$ *iff* $D_A^{\text{lazy}} \models M \sqsubseteq N$.

The "$\Rightarrow$" direction follows from the Soundness Theorem. The rest of this section is devote[d] to a (sketched) proof of the *completeness* part "$\Leftarrow$" of the theorem.

We formalize two notions: the *occurrence* of a subterm of a λ-term, and the number [of] descendents a subterm of a λ-term has (with reference to the Longo tree of the term). L[et] $\alpha, \beta \in \mathbf{Seq}$ and $i \geqslant 1$. Define M_α, the *subterm of* M *at occurrence* α inductively as follow[s]:

$$M_{\langle\rangle} \stackrel{\text{def}}{=} M;$$
$$M_{\langle i\rangle * \beta} \stackrel{\text{def}}{=} \begin{cases} (M_i)_\beta & \text{if } M \twoheadrightarrow_{\text{h}} \lambda x_1 \cdots x_n.yM_1 \cdots M_m, 1 \leqslant i \leqslant m \text{ and } n \geqslant 0\ ; \\ \text{undefined} & \text{else.} \end{cases}$$

If M_α is defined and $M_\alpha \twoheadrightarrow_{\text{h}} \lambda x_1 \cdots x_n.yM_1 \cdots M_m$ for some $m \geqslant 0$, we say the **branchin[g] factor of** M_α is m. If, for example, $M_\alpha \twoheadrightarrow_{\text{h}} \lambda\vec{x}.y$ then the branching factor of M_α is 0. Thi[s] corresponds precisely to the number of descendents at node α of LT(M). If M_α is undefine[d] then we say the branching factor of M at α is 0.

A key idea of the completeness proof is that of a α-M-N *substitutor* for $\alpha \in \mathbf{Seq}$ an[d] $M, N \in \Lambda$, a notion which is borrowed from Barendregt's proof of the local structure theore[m] of $\mathcal{P}(\omega)$ in [Bar84, Ch. 19].

Definition 3.5 Let $M, N \in \Lambda, \alpha \in \mathbf{Seq}$ and $p \geqslant 1$. We say that p is **large enough fo[r]** α-M-N if p is larger than the branching factors of M_β and of N_β for all $\beta \leqslant \alpha$. An α-M-[N] **substitutor** for variables $y_0, \cdots, y_n$ is a substitution (mapping any L to $L^{+m} \in \Lambda(\underline{D_A})$) whose action on any term $L \in \Lambda$ is defined as

$$L^{+m} \stackrel{\text{def}}{=} L[\underline{\mathbf{p}_m}y_0/y_0] \cdots [\underline{\mathbf{p}_m}y_n/y_n]$$

where $\mathbf{p}_m$ is as defined in Definition 2.6, with the proviso that m is any number large enoug[h] for α-M-N. Whenever m is understood, we shall write $+_m$ simply as $+$ and $\underline{\mathbf{p}_m} \in \Lambda(\underline{D_A})$ as $\mathbf{p}_m$.

Lemma 3.6 *(i) Let* $M = \lambda x_1 \cdots x_m.yN_1 \cdots N_l$ *with* y *distinct from* $x_1, \cdots, x_m$. *Then, fo[r]* $r > l$, $D_A^{\text{lazy}} \models A_{(q)} \sqsubseteq M^{+_r}$ *iff* $q \leqslant m$.
(ii) Let $M = yM_1 \cdots M_m$ *and* $N = zN_1 \cdots N_n$. *If* $y \not\equiv z$ *or* $m \neq n$ *then* $D_A^{\text{lazy}} \not\models M \sqsubseteq N$. □

Proposition 3.7 *If* $M \not\precsim_{\text{PSE}}^1 N$ *then* $D_A^{\text{lazy}} \not\models M^+ \sqsubseteq N^+$ *for any* $\langle\,\rangle$*-*M*-*N *substitutor* $+$. □

Observation 3.8 Observe that if $M \precsim_{\text{PSE}}^1 N$ according to clauses [1], [2] or [3a] (but no[t] [3b]) in Definition 3.1, then for all $k \geqslant 1$, $M \precsim_{\text{PSE}}^k N$, *i.e.* $M \precsim_{\text{PSE}} N$. Hence, if $M \precsim_{\text{PSE}}^k N$ and $M \not\precsim_{\text{PSE}}^{k+1} N$ for some $k \in \omega$, then we can deduce from the definition of $\precsim_{\text{PSE}}^k$ that $M \twoheadrightarrow_{\text{h}}$ $\lambda x_1 \cdots x_n.yM_1 \cdots M_m$ and $N \twoheadrightarrow_{\text{h}} \lambda x_1 \cdots x_{n+l}.yN_1 \cdots N_{m+l}$ for some $n, m, l \geqslant 0$; and

(1) *either* for some $1 \leqslant i \leqslant m$, $M_i \not\precsim_{\text{PSE}}^k N_i$

(2) *or for some* $1 \leqslant j \leqslant l$, $x_{n+j} \not\sqsubseteq^k_{\text{PSE}} N_{m+j}$.

Suppose (1) above is true, say, $M_i \not\sqsubseteq^k_{\text{PSE}} N_i$, then we must have $M_i \sqsubseteq^{k-1}_{\text{PSE}} N_i$. One can ontinue in this fashion, and arrive at $P \not\sqsubseteq^1_{\text{PSE}} Q$, where P and Q are *homologous* subterms f M and N respectively *i.e.* the occurrence of P in M is the same as the occurrence of Q in '. Roughly speaking, P and Q are the "largest" homologous pair (of respective subterms f M and N) which "causes" $M \not\sqsubseteq_{\text{PSE}} N$. We formalize the intuition as follows:

)efinition 3.9 Let α range over **Seq**. Suppose $M \not\sqsubseteq_{\text{PSE}} N$. We define $M \not\sqsubseteq_{\text{PSE}} N$ @ α ıductively as follows:

- If $M \not\sqsubseteq^1_{\text{PSE}} N$, then $M \not\sqsubseteq_{\text{PSE}} N$ @ $\langle\,\rangle$.
- Suppose $M \sqsubseteq^k_{\text{PSE}} N$ and $M \not\sqsubseteq^{k+1}_{\text{PSE}} N$ for $k \geqslant 1$. Then, with reference to the notations in the preceding observation, $M \not\sqsubseteq_{\text{PSE}} N$ @ $\langle i \rangle * \beta$ or $\langle m+j \rangle * \beta$ according as $M_i \not\sqsubseteq_{\text{PSE}} N_i$ @ β or $x_{n+j} \not\sqsubseteq_{\text{PSE}} N_{m+j}$ @ β respectively. Note that by definition, $[M \sqsubseteq^k_{\text{PSE}} N$ and $M \not\sqsubseteq^{k+1}_{\text{PSE}} N]$ iff $M \not\sqsubseteq_{\text{PSE}} N$ @ α where $|\alpha| = k$.

'heorem 3.10 *If* $M \not\sqsubseteq_{\text{PSE}} N$ @ α *then* $D^{\text{lazy}}_A \not\models M^+ \subseteq N^+$ *for any* α-M-N *substitutor* $+$. □

The (completeness part of the) local structure theorem is an immediate corollary: just onsider the trivial α-M-N substitution that performs no substitution at all. As another orollary, since $\cdot$ and $\lambda x.-$ are monotonic operations in the lazy PSE-model, if $[\![M]\!]^{\text{lazy}} \subseteq$ $N]\!]^{\text{lazy}}$ then $[\![C[M]]\!]^{\text{lazy}} \subseteq [\![C[N]]\!]^{\text{lazy}}$ for any context $C[\,]$. Now, applying the local structure heorem, we get an important property of the preorder $\sqsubseteq_{\text{PSE}}$:

'heorem 3.11 (Precongruence) *For* $M, N \in \Lambda$, *if* $M \sqsubseteq_{\text{PSE}} N$ *then* $C[M] \sqsubseteq_{\text{PSE}} C[N]$ *for ny context* $C[\,]$. □

'onclusions and Related Work The relationship between the three lazy behavioural reorders are as follows:

'heorem 3.12 (Comparison) $\sqsubseteq_{\text{LT}} \subset \sqsubseteq_{\text{PSE}} \subset \sqsubseteq^{\text{B}}$. □

'or a proof of the inclusion, see Propositions 4.6.6.2 and 4.6.6.4 in [Ong88b]; to see that the nclusions are strict: let $U \stackrel{\text{def}}{=} \lambda x.x(x\Upsilon\Theta)\Upsilon$ and $V \stackrel{\text{def}}{=} \lambda x.x(\lambda y.x\Upsilon\Theta y)\Upsilon$ for any $\Theta \in \mathbf{PO}_0$ and $\Upsilon \in \mathbf{PO}_\infty$. We have: $U \not\sqsubseteq_{\text{LT}} V$ and $V \not\sqsubseteq_{\text{LT}} U$; $U \sqsubseteq_{\text{PSE}} V$ and $V \not\sqsubseteq_{\text{PSE}} U$; $U \sqsubseteq^{\text{B}} V$ and $V \sqsubseteq^{\text{B}} U$ (proof f the last pair of assertions is non-trivial, see *e.g.* [Ong88b, Theorem 4.5.1.1] or [AO92]). 'urther directions of research in the Lazy Lambda Calculus are reported in [Ong88b]:

- We investigate the full abstraction problem in the Lazy Lambda Calculus focussing on the Lazy Lambda Calculus with *convergence testing* $\lambda\ell_C$ and *Labelled Lazy Lambda Calculus* $\lambda\ell_\omega$ in Ch. 4 *ibid*. A general method for constructing fully abstract models which are *retracts* of $\mathbb{D}$ for a class of sufficiently expressive variants of the language $\lambda\ell$ is developed. A number of non-full abstraction results are obtained. We refer the reader to [Ong88a] for a summary and [AO92] for a systematic treatment.
- In Ch. 5 *ibid*., a proof system for the Lazy Lambda Calculus (with *convergence testing*) based on Scott's logic of existence which is *correct* w.r.t. the fully lazy λ-theory $\lambda\ell$ is introduced. The proof system is given a *sound* and *complete* interpretation in *partial categories*.
- In [Ong92], we study a system of Concurrent Lambda Calculi and establish a general pre-congruence theorem for *applicative bisimulation*.

In this paper, we have presented some key ideas of a theory of the *Lazy Lambda Calculu* Our approach reflects actual programming practice and our analysis parallels that of tl sensible Lambda Calculus. Much of the mathematical elegance of the sensible theory, w believe, is matched by the lazy theory.

Acknowledgements Samson Abramsky initiated the study of the Lazy Lambda Calcul in [Abr89]. The work described here formed part a PhD thesis written under his supervisio I am indebted to him for his kindness and help. I thank Anthony Dekker for helpful commen on a draft of this paper.

References

[Abr89] S. Abramsky. The lazy lambda calculus. In David Turner, editor, *Research Topi in Functional Programming*, Addison-Wesley, 1989.

[AO92] S. Abramsky and C.-H. L. Ong. Full abstraction in the lazy lambda calculu *Information and Computation*, 105 pp., 1992. To Appear.

[Bar84] H. P. Barendregt. *The Lambda Calculus: Its Syntax and Semantics.* Nortl Holland, revised edition, 1984.

[BL80] H. Barendregt and G. Longo. Equality of λ-terms in the model $\mathbb{T}^\omega$. In *To H. Curry: Essays on Combinatory Logic, Lambda Calculus and Formalism*, Academ Press, 1980.

[Eng81] E. Engeler. Algebras and combinators. *Algebra Universalis*, 13:389–392, 1981.

[Hyl76] J. M. E. Hyland. A syntactic characterization of the equality in some models the lambda calculus. *J. London Math. Soc. (2)*, 12:361–370, 1976.

[Lon83] G. Longo. Set-theoretical models of lambda calculus: theories, expansions an isomorphisms. *Annals of Pure and Applied Logic*, 24:153–188, 1983.

[Ong88a] C.-H. L. Ong. Fully abstract models of the lazy lambda calculus. In *Proc. 29t Conf. Foundations of Computer Science*, pages 368 – 376, IEEE Computer Societ Press, 1988.

[Ong88b] C.-H. L. Ong. *The Lazy Lambda Calculus: An Investigation into the Foundatio of Functional Programming.* PhD thesis, University of London, 1988.

[Ong92] C.-H. L. Ong. The concurrent lambda calculus 1: a general pre-congruence thec rem for applicative bisimulation. 1992. Unpublished Manuscipt, 27pp.

[PJ87] S. L. Peyton Jones. *The Implementation of Functional Programming Language* Prentice-Hall, 1987.

[Plo72] G. D. Plotkin. *A Set-Theoretical Definition of Application.* Technical Report MIP R-95, School of A.I., Univ. of Edinburgh, 1972.

[Plo78] G. D. Plotkin. $\mathbb{T}^\omega$ as a universal domain. *J. Computer and System Science* 17:209–236, 1978.

[Sco72] D. S. Scott. Continuous lattices. In F. W. Lawvere, editor, *Toposes, Algebrai Geometry and Logic*, pages 97–136, Springer-Verlag, Berlin, 1972. Lecture Note i Mathematics 274.

[Sco76] D. S. Scott. Data types as lattices. *SIAM J. Computing*, 5:522–587, 1976.

Logic Programming Semantics Made Easy

Els Laenens
Dirk Vermeir

Department of Computer Science
University of Antwerp (UIA)
Universiteitsplein 1, B2610 Wilrijk, Belgium

Carlo Zaniolo

MCC
3500 West Balcones Center Drive,
Austin, Texas 78759-6509, USA

ABSTRACT

We propose a new model-theoretic semantics for logic programs, called pure semantics, based on the notions of unfounded set and assumption set. The pure semantics emerges from the observation that major logic programming semantics have the following feature in common: given an 'intended model' M, the set of negative literals in M corresponds exactly with the greatest unfounded set w.r.t. the set of positive literals in M. In other words, a model contains redundant information as its negative part can be described in function of its positive part. Thus, the total models and the partial models of programs can now be characterized by a set of positive literals. Based on this idea, we develop the pure semantics for logic programs. The result is a remarkably simple semantics that unifies previous approaches and explains how partial model semantics follows from a weaker closed world assumption.

. Introduction

'his note presents a new model-theoretic semantics for logic programs, called pure emantics, based on the notions of unfounded set[Gel88a] and assumption set [Lae90a, Lae90b]. 'he logic programs under consideration are general logic programs[Gel89a] (or seminega- ive programs[Lae90c]), i.e. sets of rules of the form

$$L_1 , L_2 , .. , L_n \rightarrow A$$

vhere A is an atom and $L_1 , .. , L_n$ are literals (atoms or negated atoms).

Given a rule r, $H(r)$ denotes the head of r and $B(r)$ denotes the set of all literals in th body of r. A term, atom, rule is *ground* if it is variable-free.

Given a logic program P, the Herbrand Universe of P (denoted H_P) is the set of all poss ble ground terms. The Herbrand Base of P (denoted $B[P]$) is the set of all possib ground atoms whose predicate symbols occur in P and whose arguments are elements c H_P. A *ground instance* of a rule r in P is a rule obtained from r by replacing every vari able X in r by $\phi(X)$, where ϕ is a mapping from the set of all variables occurring in P t H_P. The set of all ground instances of all rules in P is denoted by $ground(P)$. Any subse of $B[P] \cup Not\ B[P]$ is called an *interpretation* of P if it is consistent, i.e. there are n two literals A and B in X such that $A = Not\ B$.

Notation. Let S be a set of ground literals. The set of atoms that occur positively, resp negatively, in S is denoted $pos(S)$, resp. $neg(S)$. $Not\ S$ denotes the set $\{Not\ s \mid s \in S\}$
□

Definition [Gel88a]. Let P be a logic program and I an interpretation of P. $X \subseteq B[P]$ is a *unfounded set* of P w.r.t. I if for each $p \in X$ and for each $r \in ground(P)$ with $H(r)=p$ one of the following holds:

1. $B(r) \cap Not\ I \neq \emptyset$, i.e. r is *blocked* in I
2. $B(r) \cap X \neq \emptyset$.

The *greatest unfounded set* of P w.r.t. I, denoted $U(I)$, is the union of all sets that ar unfounded sets of P w.r.t. I. $U(I)$ is easily seen to be an unfounded set. □

The pure semantics emerges from the observation that major logic programming seman tics - such as the well-founded semantics [Gel88a], the stable partial model semantics [Sac90a] the three-valued stable semantics[Prz90a] - have the following feature in common: given a 'intended model' M, the set of negative literals in M corresponds exactly with the greates unfounded set w.r.t. the set of positive literals in M, i.e. $neg(M)=U(pos(M))$. In othe words, a model contains redundant information as its negative part can be described i function of its positive part. So we can follow the line of the stable mode semantics[Gel88b] in the sense that a model is a set of atoms rather than a set of literals Based on this idea, we next develop the pure semantics for logic programs. The result i a remarkably simple semantics that unifies previous approaches and explains how partia model semantics follows from a weaker closed world assumption.

The proofs of all results presented in the sequel can be found in [Lae91a].

2. The Pure Semantics

Definition. Let P be a logic program. A *positive interpretation* of P is any subset o $B[P]$. □

Intuitively, an unfounded set w.r.t. an interpretation I contains atoms that are known to be non-inferrable from I and from any extension of I. The next definition (which resembles e definition of unfounded sets) introduces assumption sets. Assumption sets differ from ıfounded sets in that the atoms they contain are not inferrable from the interpretation at and but might be inferrable from extensions of it. As an example, consider the logic pro-ram consisting of two rules $Not\ p \rightarrow q$ and $Not\ q \rightarrow p$; and let I be the empty set. he greatest unfounded set w.r.t. I is empty whereas $\{p,q\}$ is an assumption set as nei-er p nor q is inferrable from I. Formally, this amounts to the following definition.

efinition. Let P be a logic program and let I be a positive interpretation of P. $\subseteq B[P]$ is an *assumption set* of P w.r.t. I if for each p in X and for each rule r in $round(P)$ with $H(r)=p$, one of the following holds:

. $B(r) \cap Not\ I \neq \varnothing$

. $B(r) \cap X \neq \varnothing$

. $B(r) \not\subseteq I \cup Not\ U(I)$, i.e. r is *non-applicable* in I

he *greatest assumption set* of P w.r.t. I, denoted $A(I)$, is the union of all sets that are ssumption sets of P w.r.t. I. $A(I)$ is easily seen to be an assumption set.

is said to be *assumption-free* iff $A(I) \cap I=\varnothing$. □

ince the notion of assumption set relaxes the notion of unfounded set by adding an alter-ative condition (3), we conclude:

roposition 1. Given a logic program P and a positive interpretation I of P, every nfounded set of P w.r.t. I is an assumption set of P w.r.t. I. □

et us take a closer look at the new condition in the above definition. Traditionally, a ule r is applicable in a (partial) interpretation I if $B(r) \subseteq I$ which means that the atoms ccurring positively in r's body are *true* w.r.t. I, i.e. $pos(B(r)) \subseteq I$, and that the atoms ccurring negatively in its body are *false* w.r.t. I, i.e. $neg(B(r)) \subseteq Not\ I$. It is our feel-ng that this (traditional) definition of false atoms (i.e. an atom p is false w.r.t. I iff $Iot\ p \in I$) obscures what falsity is really about. Indeed, it is essential to remember that n atom is not false because its negation is in the interpretation, but that the negation of n atom is in the interpretation because the atom is found to be false. An excellent illus-ration of this is the fixpoint computation of the well-founded model in[Gel88a] where each teration adds, besides the positive atoms as inferred using the immediate consequence ransformation, also the negation of each atom in the greatest unfounded set which is the et of ground atoms that are already known to be false. This suggests that the greatest nfounded set w.r.t. an interpretation I is actually the set of all atoms that are false w.r.t. I. t follows that the classical definition of falsity only makes sense when $neg(I)=U(I)$. So ondition 3 in the definition of assumption set uses a more exact definition of rule appli-ability, based on the essential meaning of falsity: an atom p is false w.r.t. I iff $p \in U(I)$. s we shall see later, this tantamounts to a more conservative application of the CWA.

ntuitively, the greatest assumption set w.r.t. a positive interpretation I is the set of all toms that are non-inferrable from I. In view of this fact, whenever $I \cap A(I) \neq \varnothing$ we now that I contains 'assumptions', i.e. atoms for which no proper motivation of their resence in I exists. Accordingly, $\overline{A(I)}=B[P]-A(I)$ contains all atoms that are inferr-ble from I and consequently I is not deductively closed whenever $\overline{A(I)}-I \neq \varnothing$. It is evident that we want a model to be deductively closed and assumption-free at the sam time. This leads us to the introduction of pure models.

Definition. Let P be a logic program and M a positive interpretation of P. M is a *pur (partial) model* of P iff its complement, $\overline{M}=B[P]-M$, is the greatest assumption set w.r. M, i.e. $\overline{M}=A(M)$. □

The following equivalence is useful in the proofs.

Proposition 2. Let P be a logic program and let M be a positive interpretation. M is pure model iff it is assumption-free and its complement $\overline{M}$ is an assumption set w.r.t. M □

Since pure models are assumption-free, they are - according to Proposition 1 - also fre from 'unfounded literals'.

Proposition 3. Given a logic program P and a pure model M of P. The $M \cap U(M)=\varnothing$. □

The pure semantics is universal in the sense that it captures the meaning of every logi program.

Theorem 1. Every logic program has at least one pure model.

Proof (sketch). $pos(W_P^{\infty}(\varnothing))$[Gel88a] is a pure model of P. □

3. Examples

Consider the logic program P_1.

Not p $\rightarrow$ *q*
Not q $\rightarrow$ *p*

Let $M_1=\varnothing$. Then $U(M_1)=\varnothing$. $\overline{M_1}=B[P_1]=\{p, q\}$ is the greatest assumption set w.r. M_1 as no rule for p or q is applicable in M_1. Hence, M_1 is a pure model of P_1.

Let $M_2=\{p\}$. Then $U(M_2)=\{q\}$. So $\overline{M_2}=\{q\}$ is the greatest unfounded set w.r.t. M and since every unfounded set is an assumption set (Proposition 1), we find that $\overline{M_2}$ is a assumption set. Moreover, $\{p,q\}$ is not an assumption set as the rule *Not q* $\rightarrow$ *p* doe not satisfy any of the conditions in the definition of assumption set: *Not q* $\rightarrow$ *p* is appl cable in M_2, not blocked in M_2 and its body does not contain any 'assumptions'. So, $\overline{M}$ is the greatest assumption set w.r.t. M_2 which means that M_2 is a pure model of P_1. I much the same way, one can show that also $M_3=\{q\}$ is a pure model. And that's i $M_4=\{p, q\}$ is not a pure model as the greatest assumption set w.r.t. M_4 is M_4 itself.

Consider P_2 consisting of only one rule.

Not p $\rightarrow$ *p*

The empty set is the only pure model of P_2 since $B[P_2]=\{p\}$ is an assumption set w.r. $\varnothing$ (as $U(\varnothing)=\varnothing$) and also w.r.t. $\{p\}$.

e following logic program P_3 (two copies of P_1)

$Not\ p \rightarrow q$
$Not\ q \rightarrow p$
$Not\ a \rightarrow b$
$Not\ b \rightarrow a$

s nine pure models: $\varnothing$, $\{a\}$, $\{b\}$, $\{p\}$, $\{q\}$, $\{p,a\}$, $\{p,b\}$, $\{q,a\}$, and $\{q,b\}$.

onsider P_4.

$Not\ b \rightarrow a$
$Not\ a \rightarrow b$
$Not\ c \rightarrow a$
$Not\ a, Not\ b \rightarrow c$

his program has two pure models: $\{a\}$ and the empty set.

onsider P_5 (basically two copies of P_4).

$Not\ b \rightarrow a$
$Not\ a \rightarrow b$
$Not\ c \rightarrow a$
$Not\ a, Not\ b \rightarrow c$

$Not\ p \rightarrow q$
$Not\ q \rightarrow p$
$Not\ r \rightarrow q$
$Not\ q, Not\ p \rightarrow r$

$a, q \rightarrow s$

$_5$ has four pure models: the empty set, $\{a\}$, $\{q\}$ and $\{a, q, s\}$.

. Relation to Other Approaches

he stable model semantics[Gel88b] is not universal. Some logic programs such as P_2 do ot have any stable model. The main difference between the stable and the pure semantics s that - although both are based on positive interpretations - they are actually two-valued nd three-valued respectively. Indeed, a rule r is applicable in a stable model M if $os(B(r)) \subseteq M$ and $neg(B(r)) \cap M=\varnothing$, i.e. $B(r) \subseteq M \cup Not\ \overline{M}$; whereas r is appli- able in a pure model M if $B(r) \subseteq M \cup Not\ U(M)$. In other words, the set of atoms that re considered false w.r.t. M is $\overline{M}$ in the case of a stable model M and $U(M)$ in the case of pure model M. So $B[P]-(M \cup U(M))=\overline{M}-U(M)$ is the set of atoms that are ndefined w.r.t. a pure model M. It is then intuitively clear that the pure models that are wo-valued or total (i.e. $M \cup U(M)=B[P]$ or $\overline{M}-U(M)=\varnothing$) are exactly the stable nodels.

efinition. Given a logic program P, a positive interpretation I of P is called *total* iff $\cup\ U(I)=B[P]$. □

Theorem 2. The total pure models of a logic program P are identical to the stabl models[Gel88b] of P.

Proof (sketch). This theorem follows immediately from Theorem 4 (b) (which is prove independently). □

Further relationships between the pure model semantics and other model-theoreti approaches will be described in terms of 'equivalences' between interpretations and posi tive interpretations.

Definition. Let P be a logic program. We say that a positive interpretation A of P i *equivalent* with an interpretation I of P iff $I = A \cup Not\ U(A)$. □

Theorem 3. Let P be a logic program. The pure models of P are equivalent with th three-valued stable models[Prz90a] of P, i.e.

(a) If M is a three-valued stable model of P, then $pos(M)$ is a pure model of P; and

(b) If M is a pure model of P, then $M \cup Not\ U(M)$ is a three-valued stable model o P.

Proof (sketch). (a) One can show[Lae91a] that three-valued stable models of P are fixpoint of W_P and that assumption-free fixpoints of W_P are equivalent with pure models of P. Sc if $pos(M)$ is assumption-free, then we can infer that it is a pure model of P. It is easy t see that $M - A(pos(M))$ is a model of P/M. But, by definition, M is the least model o P/M so $M \cap A(pos(M)) = \emptyset$. Hence, $pos(M)$ is indeed assumption-free and part (a) o the theorem follows.

(b) It is straightforward to show that $M \cup Not\ U(M)$ is a three-valued model[Prz89a] o $P/(M \cup Not\ U(M))$.

Suppose that M' is the least three-valued model of $P/(M \cup Not\ U(M)$. The $M' \subseteq M \cup Not\ U(M)$. One can show that for each $p \in M - M'$ and for each rule r i $ground(P)$ with $H(r) = p$ either $B(r) \not\subseteq M \cup Not\ U(M)$, or $B(r) \cap Not\ M \neq \emptyset$, o $B(r) \cap (M - M') \neq \emptyset$. In other words, $M - M'$ is an assumption set w.r.t. M. So $M - M'$ must be empty as M is a pure model and thus assumption-free. Therefore $M = pos(M')$ Moreover, as M' is a three-valued stable model of P, it is a fixpoint of W_P and thu $M' = pos(M') \cup Not\ U(pos(M'))$ [Lae91a]. So, we find $M' = M \cup Not\ U(M)$ whic proves part (b) of the theorem. □

This result is significant, since it clarifies the significance of the three-valued model pro posed in[Prz90a] as follows:

- complications, such as three-valued logic and negative literals in the model, ar unnecessary, and, in fact obscure, the fundamental conceptual shift described next.
- the transition from total stable models to partial stable models tantamounts to th relaxation of the CWA, whereby only atoms that belong to the greatest unfounde set are now assumed to be false.

We now consider for a given logic program P the partially-ordered set of pure models o P where " $\subseteq$ " (set inclusion) is the partial order, and turn our attention to its maxima nd minimal elements.

Definition. Let P be a logic program and $< \mu , \subseteq >$ the partially ordered set where μ denotes the set of pure models of P. □

As an example, the next figures depict the partially ordered sets of pure models of P_3 and P_5.

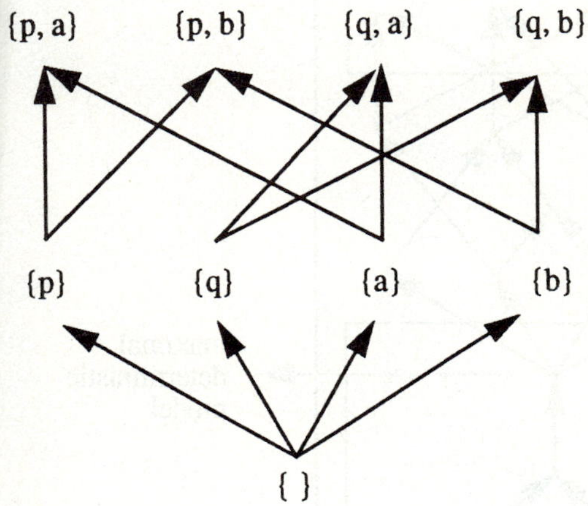

Fig. 1.

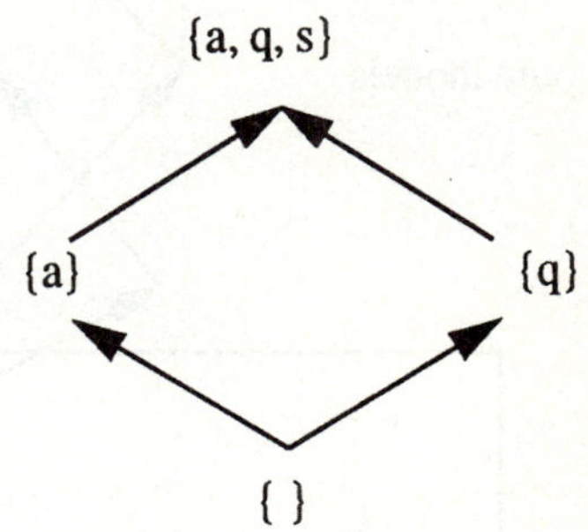

Fig. 2.

Theorem 4. Let P be a logic program.

(a) P has a unique minimal pure model that is equivalent with the well-founded partial model[Gel88a] of P;

(b) The maximal pure models of P are equivalent with the stable partial models[Sac90a] or regular models[You90a] of P;

(c) The greatest lowerbound in $< \mu , \subseteq >$ of all maximal pure models of P exists and is equivalent with the maximal deterministic model[Sac90a] of P.

Proof (sketch). (a) One can show (see[Lae91a]) that the assumption-free fixpoints of W_P are equivalent with the pure models of P, from which it follows immediately that the well-founded model, i.e. the least fixpoint of W_P, is equivalent with the unique minimal pure model.

(b) It is easy to show that the pure models of P are equivalent with the strongly-founded[Sac90a] models of P. Therefore, as stable partial models are maximally strongly founded (Proposition 3 in[Sac90a]), it follows that they are equivalent with the maximal pure models.

Since both regular models and stable partial models are three-valued models, it is straightforward to see that the regular models for P are exactly the stable partial models for P. Indeed, the notions justifiable for regular models and founded for stable models are equivalent and clearly the principle of minimal undefinedness for regular models corresponds to the maximality of stable partial models.

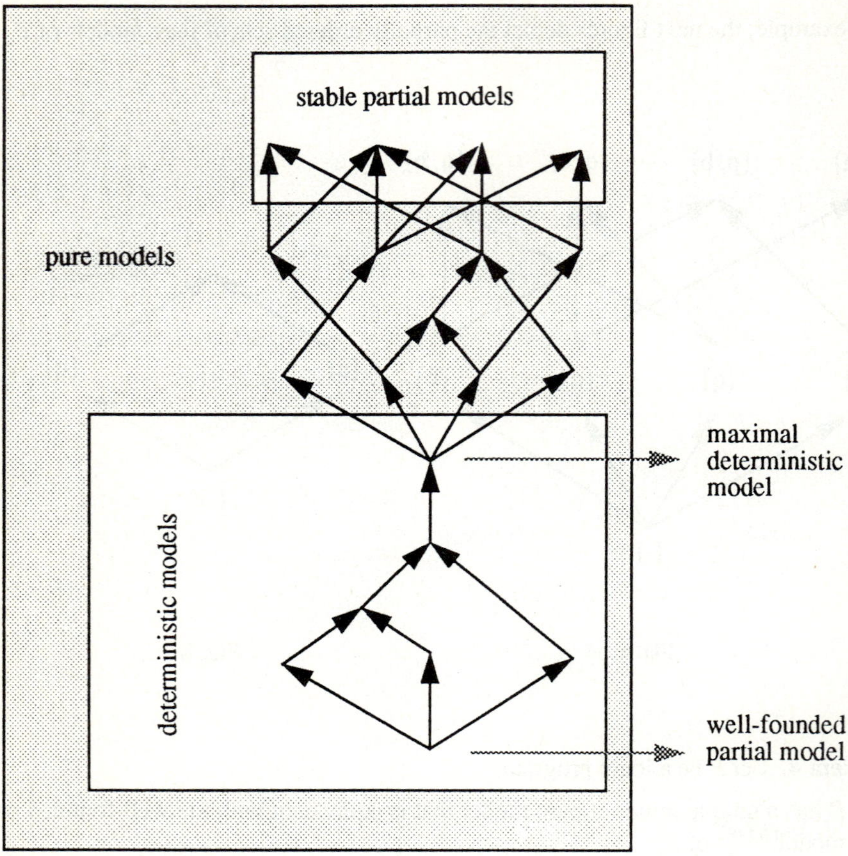

Fig. 3.

(c) In[Lae91a] we show that the greatest lower bound in $< \mu , \subseteq >$ of all maximal pure models of P exists, by proving that for every two pure models, M_1 and M_2, that are contained in the deterministic set[Sac90a], (i.e. the intersection of all maximal pure models), $pos(W_P^\infty(M_1 \cup M_2))$ is again a pure model that contains $M_1 \cup M_2$ and which is a subset of the deterministic set. So, the glb G of all maximal pure models is the maximal pure model that is contained in the deterministic set. Since the pure models of P are equivalent with the strongly-founded models of P, G is equivalent with the maximal strongly-founded model that is contained in the deterministic set, which is exactly the definition of the maximal deterministic model. □

o, $< \mu , \subseteq >$ provides us with a dag structure (directed acyclic graph) of which

- the root, i.e. unique node without incoming arcs, represents the well-founded model,
- the leafs, i.e. nodes without outgoing arcs, represent the stable partial models,
- the glb of the leafs represents the maximal deterministic model, and
- the 'ancestors' of the glb of the leafs represent the deterministic models.

This is depicted in Figure 3.

From Figures 1 and 2 we learn for instance that:

- P_3 has four stable partial models; its unique deterministic model is the well-founded model;
- P_5 has a single stable partial model which therefore coincides with its maximal deterministic model; its well-founded model is a different deterministic model;
- etc.

5. Conclusion

Much of the current research aims to provide semantics which is universal and deals effectively with incompleteness. The contribution of this paper is to simplify and unify previous approaches by eliminating the need for three-valued logic and negated literals in defining partial models. Moreover, it shows that the conceptual transition from total model semantics to partial model semantics boils down to a simple relaxation of the CWA.

The 'purification' of traditional logic programming semantics as presented in this paper also clarifies the meaning of negation as failure. In[Lae92a], the pure semantics for extended logic programs is shown to be a natural extension of the pure semantics for general logic programs.

6. Acknowledgment

The first author would like to thank Natraj Arni for interesting discussions.

References

Gel88a.
A. Van Gelder, K. Ross, and J. S. Schlipf, *The Well-Founded Semantics for General Logic Programs*, 1988. UCSC-CRL-88-16

Gel88b.
M. Gelfond and V. Lifschitz, "The Stable Model Semantics for Logic Programming," in *Proc. of the Intl. Conf. on Logic Programming*, pp. 1071-1079, 1988.

Gel89a.
M. Gelfond and V. Lifschitz, *Logic Programs with Classical Negation*, 1989. Manuscript

Lae90a.
E. Laenens and D. Vermeir, "A Fixpoint Semantics of Ordered Logic," *Journal of Logic and Computation*, vol. 1, no. 2, pp. 159-185, 1990.

Lae90b.

E. Laenens, "Foundations of Ordered Logic," *PhD Thesis, University of Antwerp (UIA)*, 1990.

Lae90c.

E. Laenens, D. Vermeir, and D. Sacca, "Extending logic programming," in *Proceedings of the SIGMOD conference*, pp. 184-193, 1990.

Lae91a.

E. Laenens, D. Vermeir, and C. Zaniolo, "Pure models for logic programs: a simplification and unification of logic programming semantics," *Technical report 91-31*, 1991.

Lae92a.

E. Laenens, *A universally consistent semantics for extended logic programs*, Submitted for puclication, 1992.

Prz89a.

T. Przymusinski, "Three-Valued Formalizations of Non-Monotonic Reasoning and Logic Programming," *Proc. 1st Int. Conference on Principles of Knowledge Representation and Reasoning*, pp. 341-349, 1989.

Prz90a.

T. Przymusinski, "Well-founded semantics coincides with three-valued stable semantics," *Fundamenta Informaticae*, vol. 13, pp. 445-463, IOS Press, 1990.

Sac90a.

D. Sacca and C. Zaniolo, "Stable models and Non-determinism for logic programs with negation," *Proc. ACM Symp. on Principles of Database Systems*, 1990.

You90a.

Jia-Huai You and Li Yan Yuan, "Three-Valued Formalization of Logic Programming: Is It Needed?," in *Proc. of the PODS'90 conference*, pp. 172-182, 1990.

On the Complexity of Dataflow Analysis of Logic Programs

Saumya K. Debray
Department of Computer Science
University of Arizona
Tucson, AZ 85721, USA

Abstract

It is widely held that there is a correlation between complexity and precision in dataflow analysis, in the sense that the more precise an analysis algorithm, the more computationally expensive it must be. The details of this correspondence, however, appear to not have been explored extensively. This paper reports some results on this tradeoff in the context of Horn logic programs. A formal notion of the "precision" of an analysis algorithm is proposed, and this is used to characterize the worst case computational complexity of a number of dataflow analysis algorithms with different degrees of precision.

1 Introduction

It is widely held that there is a correlation, in dataflow analysis, between precision and complexity: at one extreme, there are very efficient algorithms that rarely give useful information about any program; at the other extreme are highly precise algorithms that may sometimes be very expensive computationally. Most algorithms for dataflow analysis tend to fall in between these extremes: some are generally quite efficient but may fail to be very precise under some circumstances, while others tend to be fairly precise but may not always be very efficient.

The details of this tradeoff, however, appear to not have been explored extensively. Some researchers have given complexity results for dataflow analyses for other language families (e.g., Jones and Muchnick [18] and Myers [28] for imperative languages, Hudak and Young for functional languages [15]). However, these have typically addressed the complexity of particular analyses, rather than explore the correlation between precision and complexity of dataflow analysis in general. In the logic programming context, a variety of dataflow analysis algorithms have been proposed, with varying degrees of precision, and worst-case complexities ranging from linear time [10] to exponential in the input size [4, 13, 14, 16, 21, 26, 27, 36]. These results do not, however, really give us an adequate understanding of *intrinsic* tradeoffs between precision and complexity, in the sense of how expensive any "sufficiently precise" analysis must be.

A problem that arises immediately is that of making the notion of an analysis being "sufficiently precise" sufficiently precise. In the next section, we offer one way to address this problem. The remainder of the paper uses this notion of "sufficient precision" to examine classes of dataflow analysis algorithms of varying degrees of precision and obtain characterizations of their worst case complexities. We consider two broad classes of dataflow analyses of logic programs in this paper. The first class, which distinguishes between different function symbols and constants, is important in the context of type inference [13, 14, 24, 36] and sharing analysis [2, 25]. The second class, which does not distinguish between different function symbols but focusses instead on groundness of variables and on dependencies and aliasing between them, is important in the context of compile-time parallelization and optimization of logic programs [3, 9, 16, 21, 26, 27]. Of the various dataflow analyses we have seen proposed in the literature, the vast majority tend to belong to one of these two classes. There have also been proposals that combine aspects of both these classes of dataflow analyses [17]. For these classes of analyses, we have the following specific results:

1. Precise analysis of programs that contain no aliasing, where there are only two distinct constants and no function symbols of nonzero arity, is EXPTIME-complete.

   The problem is PSPACE-hard under the additional constraint that there is no recursion.

   The problem is NP-complete if, in addition to the constraints listed above, we require also that the maximum arity of any predicate symbol be $O(1)$; it remains NP-complete even if the maximum arity is restricted to 3.

2. If we allow function symbols of nonzero arity, then precise analysis of programs that contain no recursion or aliasing is PSPACE-hard, even if the maximum arity of any function or predicate symbol is restricted to 2, and the maximum number of literals in the body of any clause is restricted to 2.

3. Precise groundness and alias analyses of programs that contain no function symbols, where there are no failed execution branches at runtime, is EXPTIME-complete. This addresses one aspect of the worst-case complexity of a groundness analysis proposed by Marriott, Søndergaard and Jones [21].

   The problem is co-NP-complete if, in addition to the above constraints, we require also that there be no recursion, and that the maximum arity of any predicate be $O(1)$; it remains co-NP-complete even if the maximum arity of any predicate is restricted to 6.

   The problem remains co-NP-complete even if each clause is allowed to have at most one success pattern: this addresses the worst-case complexity of any algorithm whose precision is similar to that of a groundness analysis proposed by Codish, Dams and Yardeni [4].

4. Analysis algorithms with polynomial-time worst case complexities can be obtained if dependencies between variables can be ignored or if such dependencies can be assumed to be transitive, and if the arity of each predicate as well as the number of distinct calling and success patterns for any predicate is $O(1)$ [9, 10].

It should be noted, moreover, that the techniques proposed here have applicability beyond just the class of definite Horn programs. Our approach can also be sed to separately examine independent sources of complexity in an analysis algo-thm, and study their contributions to the overall complexity of the algorithm: this illustrated, in Section 4, via a groundness analysis algorithm due to Marriott, øndergaard and Jones [21]. Our results indicate, somewhat surprisingly, that even or classes of programs whose syntactic structure is extremely simple, the worst case omplexity of precise dataflow analysis can be exponential in the program size for a ide variety of dataflow analyses. The implication is that the complexity cannot be ny better when considering the entire Prolog language, even for relatively modest ataflow analyses.

To the best of our knowledge, this paper is the first attempt to explore this orrelation in a systematic way. While we have restricted ourselves to analysis of ogic programs, the technique we use, based on the notion of exactness sets, does ot depend on logic programs in any way, and is equally applicable to flow analyses f other language families.

We assume that the reader is acquainted with the basics of logic programming nd Prolog, in particular with the operational semantics: details may be found in 19]. Details of constructions and proofs have been omitted due to space constraints: he interested reader is referred to [11].

2 Preliminaries

2.1 What is the "Precision" of an Analysis Algorithm?

A natural characterization of the precision of a flow analysis algorithm is relative o some other algorithm: the first is more precise than the second if and only if for very program and every input, the results obtained using the first algorithm are niformly more precise than those obtained from the second. Cousot and Cousot se such a characterization to show that the space of all abstract interpretations of ny given language, when ordered according to precision, form a complete lattice, nd thereby infer the existence of a "most precise" abstract interpretation [6, 7]. Because this expresses the precision of an analysis in terms of that of other analysis lgorithms, we refer to this notion of precision as *relative precision*.

Such a characterization is unsuitable for our needs, however, because it is not mmediately obvious how it can be used to obtain insights into the intrinsic com-putational costs associated with analyses of different degrees of precision. What we need, rather, is a way to characterize the precision of an analysis in a way that loes not require any reference to any other analysis algorithm. Our approach is to onsider the precision of an analysis to be given by the class of programs for which he analysis is exact, i.e., where the analysis is *sound* (anything that can happen at untime is predicted by the analysis) and *minimal* (anything that is predicted by he analysis can happen at runtime). We refer to the class of programs for which an nalysis algorithm A gives exact results as the *exactness set* of A, written $\Xi(A)$. In he language of abstract interpretation, $\Xi(A)$ for an analysis algorithm A is the class of programs for which the concretization of the fixpoint computed by A coincides with the collecting semantics of the program.[1] The following result is not difficult to see:

Proposition 2.1 *Given two analysis algorithms A and B, if A is relatively more precise than B, then* $\Xi(B) \subseteq \Xi(A)$. ■

The converse, however, need not hold in general.

2.2 Some Simple but Interesting Classes of Programs

Certain classes of programs are especially interesting for examining the tradeoff between precision and complexity of dataflow analyses. Intuitively, they attempt to express assumptions that are common in dataflow analysis algorithms. The following list enumerates the main classes of programs we consider, and the motivation for considering them.

Failure-free programs : This class contains programs that do not contain failed execution branches at runtime. This class is considered in order to deal with the complexity and precision of dataflow analyses, such as [3, 9, 16, 21, 26, 27], that ignore the possibility of failure.

Function-free programs : This class consists of programs that do not contain any function symbols or constants. It is motivated by a variety of groundness and alias analyses used in compile-time optimization and parallelization, e.g., [3, 21, 26, 27], which focus on groundness and aliasing between variables, and ignore the distinction between different function symbols.

Recursion-free programs : This class consists of programs that contain no recursion. It is motivated by the desire to study the complexity and precision of dataflow analysis algorithms that sacrifice precision at recursive calls for efficiency reasons, e.g., [9].

Alias-free programs : This consists of programs that do not, at any point during execution, alias together two distinct variables. It is motivated by the fact that tracking aliasing may be expensive, so that we may be interested in discussing the extent to which an analysis problem may be simplified by not having to track aliasing precisely (see, for example, [10])

Bounded-arity programs : This is a family of sets of programs: for each $n \geq 0$, the class of n-bounded arity programs consists of programs where every predicate and function symbol has arity at most n. In analyzing their algorithms, some researchers have argued that in most programs encountered in practice, the arities of predicates do not increase as program size increases (see, for example, [9, 10, 30, 34]). Bounded-arity programs are used to study the complexity/precision behavior of various dataflow analyses under this assumption.

[1] Note that this is notion of "exactness" is strictly stronger than that proposed in, e.g., [20].

Analysis of Programs with Nonzero-Arity Function Symbols

his section considers the analysis of programs that contain function symbols of on-zero arity. Information about constants and function symbols is of interest in ie context of type inference [13, 14, 24, 36] and sharing analysis [2, 25]. It is well nown that Horn programs containing both recursion and nonzero-arity function ymbols are undecidable: indeed, Tärnlund shows that even the class of alias-free orn programs containing at most one body literal per clause, and with one binary inction symbol and one constant, is Turing-complete [33]. This immediately implies ie following:

heorem 3.1 *Exact analysis of alias-free Horn programs is recursively unsolvable. he problem remains unsolvable even if we restrict ourselves to one binary function ymbol and one constant, and one body literal per clause.* ∎

The first class of decidable Horn programs we consider is the class of alias-free ecursion-free programs: our intention is to investigate the complexity of dataflow nalysis algorithms that strive to not sacrifice precision except when aliasing or reursion is involved. We show that in this case, precise dataflow analysis is PSPACE-ard. To this end, we first give a procedure that, given a quantified Boolean formula , generates, using logarithmic space, a logic program P_φ that can be used to esablish the truth of φ. Given any quantified Boolean formula φ, the program P_φ onstructed in this manner is recursion-free, has bounded arity (the arity of every redicate and function symbol is bounded by 2), each clause contains at most two ody literals, and is alias-free. This leads immediately to the following result:

heorem 3.2 *Exact analysis of recursion-free, alias-free, bounded-arity, boundediteral Horn programs is PSPACE-hard. The problem remains PSPACE-hard even if ie arity of each function and predicate symbol is restricted to at most 2, and there re at most 2 literals in the body of any clause.*

roof By reduction from the *Quantified Boolean Formulae* (QBF) problem, which PSPACE-complete [32]. ∎

This result is significant because various authors, for example [20], have argued iat flow analysis algorithms that are exponential in the number of literals in a lause body may be tolerable in practice because in most commonly encountered rograms, the number of literals in clause bodies is small, and can reasonably be ssumed to be $O(1)$. This result shows that depending on how precise an analysis trives to be, it may have a bad worst-case complexity even if we assume that the umber of literals per clause is $O(1)$.

4 Analysis of Programs without Nonzero-Arity Functio Symbols

The construction used in Theorem 3.2 relies on the existence of a binary functio symbol, which is used to represent truth assignments as lists of truth values. W next examine the extent to which disallowing function symbols with nonzero arit simplifies dataflow analysis:

Theorem 4.1 *Exact analysis of alias-free Horn programs containing at least tw distinct constants, and no function symbols of nonzero arity, is EXPTIME-complet*

Proof By reduction from the *Recursive Monotone Boolean Formulae* (RMBF problem, shown by Hudak and Young to be EXPTIME-complete [15]. ■

Thus, precise analysis of programs can be expensive even if the analysis algorith does not try to deal with compound terms and aliasing. Now it may not be unrea sonable for an analysis algorithm to try to be precise for non-recursive program but to surrender some precision when dealing with recursive calls. To examine th complexity of such analyses, we next consider the case where the exactness sets c the analyses do not contain recursive programs. We have the following result:

Theorem 4.2 *Exact analysis of alias-free recursion-free Horn programs containin at least two distinct constants and no function symbols of nonzero arity is PSPAC hard.*

Proof By reduction from QBF. ■

In the construction used in the proof of this theorem, the maximum arity of pred icates increases as the number of quantifiers increases. In the literature on dataflo analysis of logic programs, some researchers have assumed that the maximum arit of predicates in any program is bounded, i.e., $O(1)$. We next consider the effect o imposing this additional restriction on the complexity of analyses:

Theorem 4.3 *Precise analysis of recursion-free, alias-free, bounded-arity Horn pro grams, whose only function symbols are two distinct constants, is NP-complete.*

Proof By reduction from the *Satisfiability Problem for Boolean Formulae* (SAT) which is known to be NP-complete [5]. ■

5 Precise Analysis of Function-free Programs

As seen from the discussion of the previous section, precise analysis of program containing function symbols can be quite difficult. In this section, we conside precise analysis of programs that do not contain any function symbols and do no have any failed execution branches. This is motivated by a variety of groundness an aliasing analyses that have been proposed in the literature in recent years, primaril the context of compiler parallelization and code optimization of logic programs , 9, 16, 21, 26, 27]. While the syntactic structure of programs in this class is very stricted, the analysis problem is, somewhat surprisingly, not as simple as one might xpect.

For the complexity results of this section, we rely on aliasing to mimic the evalation of monotone Boolean formulae. The essential idea here is to associate, with ch proposition φ, two logic variables $X_{\varphi L}$ and $X_{\varphi R}$. A truth value of **true** is enoted by having $X_{\varphi L}$ and $X_{\varphi R}$ aliased together, while a truth value of **false** is enoted by having these variables not aliased. We have the following result:

heorem 5.1 *Exact alias analysis of function-free failure-free Horn programs is XPTIME-complete.*

roof By reduction from RMBF. ∎

We next consider what happens to the complexity of dataflow analysis if precision sacrificed only at recursive calls. To this end, we focus on recursion-free programs:

heorem 5.2 *Exact alias analysis for recursion-free, function-free, failure-free, ounded-arity programs is co-NP-complete.*

roof The proof is by reduction from the *Inequivalence Problem for Monotone oolean Formulae*, which is co-NP-complete [1]. ∎

It turns out that even if all we have is an analysis that only gives information bout groundness of different argument positions at entry and exit from a predicate, is possible to extract enough aliasing information from this to obtain the following esults:

heorem 5.3 *Exact groundness analysis for function-free, failure-free Horn prorams is EXPTIME-complete.* ∎

heorem 5.4 *Exact groundness analysis for recursion-free, function-free, failureree, bounded-arity programs is co-NP-complete.* ∎

While most groundness and alias analyses proposed in the logic programming litrature are not precise enough to include the class of recursion-free, function-free, ailure-free, bounded-arity programs in their exactness sets, we know of at least one nalysis algorithm, proposed by Marriott, Søndergaard and Jones [21], that appears o be precise enough to give exact results for this class of programs [22]. This analsis requires the solution of a co-NP-hard problem at each iteration to determine whether a fixpoint has been attained (the analysis manipulates propositional formuae containing the connectives $\wedge$, $\vee$ and $\Leftrightarrow$, and checking to see whether a fixpoint as been attained involves deciding the equivalence of two such formulae: this is a roper generalization of the equivalence problem for monotone Boolean formulae, and therefore is co-NP-hard (see [1])). However, this source of inefficiency is ab sent in recursion-free programs, and one may reasonably inquire after the worst-cas complexity of this analysis if this source of inefficiency is removed. It follows fro Theorem 5.4 that unless $P = NP$, their algorithm can be exponential in the size the input in the worst case. This illustrates, interestingly, that there are at least tw independent sources of complexity in this algorithm, a fact that is by no means ob vious from the description of the algorithm. This demonstrates how our technique can be used to separate out different sources of complexity in an analysis algorithm

6 Bounding the Number of Calling and Success Patterns

The description of the arguments in a procedure call in a logic program, in terms c abstract domain elements, is referred to as a *calling pattern*, while the descriptio of the arguments on return from the call is referred to as a *success pattern* (for mor precise definitions of these terms, see [10]). Since a procedure can be called fro many different points in a program, there may be more than one calling pattern for it and since it may be defined by a number of different clauses, any given calling patter may correspond to more than one success pattern. One possible reason an analysi may be imprecise is that it may not keep track of all the different calling and/o success patterns it encounters for a predicate, instead "lubbing" them together t obtain a conservative approximation. Such a strategy is used in a number of analyses for example [3, 8, 23]: it does not appear too unreasonable if we believe that fo most programs encountered in practice, predicates have specific argument position consistently used with the same mode, type, etc. (see, for example, [12]). In thi section, we examine whether such a strategy can lead to fundamental improvement in the worst case behavior of analysis algorithms.

If an analysis approximates a set of calling and/or success patterns by their leas upper bound in the abstract domain, there will be a loss in precision in genera However, if every predicate in the program under consideration has at most on calling pattern and one success pattern, then there is no loss of precision due to thi approximation. The first restriction we consider is described by Codish et al [4] who describe a bottom-up analysis that maintains at most one success pattern pe program clause (since the analysis is a bottom-up one, there is no notion of callin patterns). The worst case complexity of the resulting analysis is $O(2^n)$, where n i the size of the program [4]. We show that unless $P = NP$, it may not be possible t do better than this:

Theorem 6.1 *Exact alias analysis of recursion-free, function-free, failure-free bounded arity programs, where each clause has at most one success pattern for an calling pattern, is co-NP-complete.*

Proof Identical to that of Theorem 5.2. ∎

If, however, variable dependency information is used in a fairly limited way, it is possible to obtain analysis algorithms with polynomial-time worst case complexity One of the properties that complicates the handling of sharing between variables is that sharing is not transitive in general: the fact that a program variable X ma share variables with a program variable Y at runtime, and that Y may share variables with a program variable Z, does not imply that X shares with Z. The representation and processing of sharing and dependency information can be simplified considerably, at the cost of some precision, by assuming that sharing is transitive. Define the class of *transitive-sharing* programs to be those programs where all sharing between variables is transitive. Then, the results of [9] imply the following:

Theorem 6.2 *Let A be any analysis algorithm such that $\Xi(A)$ is contained in the class of function-free, failure-free, bounded-arity, transitive-sharing programs where each predicate has at most one calling and one success pattern. Then, there is an equivalent analysis algorithm whose worst-case time complexity is polynomial in the size of the input program.* ∎

If aliasing can be disregarded entirely, analyses can be carried out quite efficiently: it is shown, in [10], that abstract domains that allow aliasing to be ignored admit sound flow analysis algorithms whose worst case complexity, assuming $O(1)$ calling and success patterns per predicate, is linear in program size. Define a program to satisfy the *variable-free calls* property if none of the calls arising during the execution of that program contain variables. We have the following result [10]:

Theorem 6.3 *Let A be any analysis algorithm such that $\Xi(A)$ is contained in the class of alias-free, failure-free, bounded-arity programs that satisfy the variable-free calls property and where each predicate has at most one calling and one success pattern. Then, there is an equivalent algorithm whose worst-case time complexity is linear in the size of the input program.* ∎

Analyses satisfying these requirements include the *rigid type* analysis of Janssens [17], and Sato and Tamaki's depth-abstraction analysis [31].

7 Conclusions

While it is generally believed that there is a tradeoff between complexity and precision of flow analysis algorithms, in the sense that "sufficiently precise" analyses must also be correspondingly expensive, little work appears to have been done on quantifying this tradeoff. This paper takes a step towards formally addressing this tradeoff in the context of logic programming. We offer a characterization of "sufficient precision" of dataflow analyses. We consider the implications, with regard to the precision/complexity tradeoff, of some of the different ways in which various analysis algorithms may sacrifice precision. Our results, indicate, somewhat surprisingly, that even for classes of programs whose syntactic structure is extremely simple, the worst case complexity of precise dataflow analysis can be exponential in the program size for a wide variety of dataflow analyses. The implication is that the complexity cannot be any better when considering the entire Prolog language, even for relatively modest dataflow analyses of logic programs.

8 Acknowledgements

I am grateful to Steve Mahaney and Sampath Kannan for patiently instructin me in the elements of complexity theory. Discussions with Nevin Heintze an Niels Jørgensen were very helpful in crystallizing the ideas presented here. Manue Hermenegildo, Dean Jacobs, Anno Langen, Kim Marriott, and Harald Søndergaar graciously tolerated, and responded to, innumerable questions regarding the preci sion of their flow analysis algorithms. Pascal Van Hentenryck pointed out a numbe of errors in an earlier version of this paper. This work was supported in part by th National Science Foundation under grant number CCR-8901283.

References

[1] P. A. Bloniarz, H. B. Hunt III, and D. J. Rosenkrantz, "Algebraic Structure with Hard Equivalence and Minimization Problems", *J. ACM* vol. 31 no. 4 Oct. 1984, pp. 879-904.

[2] M. Bruynooghe, "Compile time Garbage Collection", *Proc. IFIP Working Con ference on Program Transformation and Verification*, Elsevier, 1986.

[3] J.-H. Chang, A. M. Despain, and D. DeGroot, "AND-Parallelism of Logic Pro grams Based on A Static Data Dependency Analysis", in *Digest of Papers Compcon 85*, IEEE Computer Society, Feb. 1985, pp. 218-225.

[4] M. Codish, D. Dams, and Y. Yardeni, "Abstract Unification and a Bottom up Analysis to Detect Aliasing in Logic Programs", Technical Report CS90 10, Department of Applied Mathematics and Computer Science, Weizman Institute of Science, Rehovot, Israel, May 1990.

[5] S. A. Cook, "The Complexity of Theorem-proving Procedures", *Proc. Thir ACM Symposium on Theory of Computing*, pp. 151-158.

[6] P. Cousot and R. Cousot, "Abstract Interpretation: A Unified Lattice Model fo Static Analysis of Programs by Construction or Apporoximation of Fixpoints" *Proc. Fourth ACM Symposium on Principles of Programming Languages*, 1977 pp. 238-252.

[7] P. Cousot, and R. Cousot, "Systematic Design of Program Analysis Frame works", Proc. Sixth ACM Symposium on Principles of Programming Languages 1979, pp. 269-282.

[8] S. K. Debray and D. S. Warren, "Automatic Mode Inference for Logic Pro grams", *J. Logic Programming* vol. 5 no. 3 (Sept. 1988), pp. 207-229.

[9] S. K. Debray, "Static Inference of Modes and Data Dependencies in Logic Pro grams", *ACM Transactions on Programming Languages and Systems* vol. 11 no. 3, July 1989, pp. 419-450.

[10] S. K. Debray, "Efficient Dataflow Analysis of Logic Programs", *J. ACM* (to appear).

11] . S. K. Debray, "On the Complexity of Dataflow Analysis of Logic Programs", Technical Report TR 91-27, Dept. of Computer Science, The University of Arizona, Tucson, AZ 85721, USA, Dec. 1991.

12] W. Drabent, "Do Logic Programs Resemble Programs in Conventional Languages?", *Proc. IEEE Symposium on Logic Programming*, San Francisco, CA, Aug. 1987, pp. 389-396.

13] N. Heintze and J. Jaffar, "A Finite Presentation Theorem for Approximating Logic Programs", Research Report RC-16089, IBM T. J. Watson Research Center, Yorktown Heights, August 1990. (Preliminary version appeared in *Proc. Seventeenth ACM Symposium on Principles of Programming Languages*, Jan. 1990.)

14] N. Heintze and J. Jaffar, "A Decision Procedure for a Class of Set Constraints", Technical Report CMU-CS-91-110, School of Computer Science, Carnegie Mellon University, Pittsburgh, February 1991. (Preliminary version appeared in *Proc. Fifth IEEE Symposium on Logic in Computer Science.*)

15] P. Hudak and J. Young, "Higher-Order Strictness Analysis in the Untyped Lambda Calculus", *Proc. Thirteenth ACM Symp. on Principles of Programming Languages*, Jan. 1986.

16] D. Jacobs and A. Langen, "Accurate and Efficient Approximation of Variable Aliasing in Logic Programs", *Proc. North American Conference on Logic Programming*, Cleveland, Ohio, Oct. 1989, pp. 154-165.

17] G. Janssens, "Deriving Run-time Properties of Logic Programs by means of Abstract Interpretation", PhD Dissertation, Dept. of Computer Science, Katholieke Universiteit Leuven, Belgium, March 1990.

18] N. D. Jones and S. S. Muchnick, "Complexity of Flow Analysis, Inductive Assertion Synthesis, and a Language due to Dijkstra", in *Program Flow Analysis: Theory and Applications*, eds. S. S. Muchnick and N. D. Jones, Prentice Hall, 1981, pp. 381-393.

[19] J. W. Lloyd, *Foundations of Logic Programming*, Springer Verlag, 1984.

[20] H. Mannila and E. Ukkonen, "Flow Analysis of Prolog Programs", *Proc. Fourth IEEE Symposium on Logic Programming*, San Francisco, CA, Sept. 1987.

[21] K. Marriott, H. Søndergaard and N. D. Jones, "Denotational Abstract Interpretation of Logic Programs", Manuscript, Dept. of Computer Science, University of Melbourne, May 1990.

[22] K. Marriott and H. Søndergaard, personal communication, March 1991.

[23] C. S. Mellish, "Some Global Optimizations for a Prolog Compiler", *J. Logic Programming* vol. 2 no. 1 (Apr. 1985), pp. 43-66.

[24] P. Mishra, "Towards a Theory of Types for Prolog", *Proc. IEEE Symposium on Logic Programming*, Atlantic City, 1984, pp. 289-298.

[25] A. Mulkers, W. Winsborough and M. Bruynooghe, "Analysis of Shared Data Structures for Compile-Time Garbage Collection in Logic Programs", *Proc. Seventh International Conference on Logic Programming*, Jerusalem, Israel, June 1990, pp. 747-764.

[26] K. Muthukumar and M. Hermenegildo, "Determination of Variable Dependence Information at Compile Time through Abstract Interpretation", *Proc. North American Conference on Logic Programming*, Cleveland, Ohio, Oct. 1989, pp. 166-185.

[27] K. Muthukumar and M. Hermenegildo, "Combined Determination of Sharing and Freeness of Program Variables through Abstract Interpretation", *Proc. Eighth International Conference on Logic Programming*, Paris, June 1991, pp. 49-63.

[28] E. W. Myers, "A Precise Inter-procedural Data Flow Algorithm", *Proc. Eighth ACM Symposium on Principles of Programming Languages*, 1981, pp. 219-230.

[29] R. Ramakrishnan and S. Sudarshan, "Top-down vs. Bottom-up Revisited", *Proc. 1991 International Symposium on Logic Programming*, San Diego, Oct. 1991, pp. 321-336. MIT Press.

[30] Y. Sagiv and J. D. Ullman, "Complexity of a Top-Down Capture Rule", Technical Report STAN-CS-84-1009, Dept. of Computer Science, Stanford University, Stanford, CA, July 1984.

[31] T. Sato and H. Tamaki, "Enumeration of Success Patterns in Logic Programs", *Theoretical Computer Science 34* (1984), pp. 227-240.

[32] L. J. Stockmeyer and A. R. Meyer, "Word Problems requiring Exponential Time", *Proc. Fifth ACM Symposium on Theory of Computing*, 1973, pp. 1-9.

[33] S.-Å. Tärnlund, "Horn Clause Computability", BIT **17**, pp. 215-226.

[34] J. D. Ullman, "The Complexity of Ordering Subgoals", *Proc. Seventh ACM Symposium on Principles of Database Systems*, Austin, TX, March 1988, pp. 74-81.

[35] Ullman, J.D., "Bottom-Up Beats Top-Down for Datalog", In *Proceedings of the ACM Symposium on Principles of Database Systems*, pages 140-150, Philadelphia, PA, 1989.

[36] E. Yardeni and E. Shapiro, "A Type System for Logic Programs", in *Concurrent Prolog: Collected Papers*, vol. 2, ed. E. Shapiro, pp. 211-244.

Comparison of Abstract Interpretations[1]

Agostino Cortesi — Dept. of Mathematics, University of Padova, ITALY

Gilberto Filè — Dept. of Mathematics, University of Padova, ITALY

William Winsborough — Dept. of Computer Science, Pennsylvania State. Univ., USA

Abstract

The central idea of the technique of Abstract Interpretation is that the analysis of a program consists of executing it on a special (abstract) domain D of values in which each operation μ, used during the normal execution, is interpreted as a corresponding operation μ_D on D. More formally, an (abstract) domain is a complete lattice that enjoys a Galois insertion into the concrete domain. An interpretation consists of a domain and a collection of monotone operations over that domain. Due to the success of the technique, several interpretations with similar purposes have been proposed. In the classical theory of abstract interpretation, the only way we have to compare two interpretations is to show that one abstracts the other. The weakness of this type of comparison is that it does not allow us to compare various interpretations with respect to some, but not all, of the information they express. This paper presents new notions that enable us to make such precise comparisons. These notions are applicable to the comparison of abstract interpretations for all kinds of programming languages and with all sorts of analysis frameworks. We show that our new notions form a natural extension to the existing notion of one interpretation abstracting another. We demonstrate the utility of our technique by using it to compare abstract interpretations for analysis of logic programs with respect to their ability to infer variable groundness.

Introduction

he technique of abstract interpretation [CCou 77] is a successful framework for con-ructing analyses of the run-time behavior of programs. The information inferred such analyses can be used by compilers to improve the quality of their generated de. The central idea of the technique is that the analysis of a program consists of ecuting it on a special (abstract) domain D of values in which each operation μ, ed during the normal execution, is interpreted as a corresponding operation μ_D D. In our theory, a domain is a complete lattice that enjoys a Galois insertion to the concrete domain. An interpretation consists of a domain and a collection of onotone operations over that domain. The operations are presumed to obey stan-rd safety properties with respect to their concrete analogs. Due to the success

[1]Work supported by CNR, P. F. Sistemi Informatici e Calcolo Parallelo, grant 91.000898.PF69

of the technique, several abstract interpretations with similar purposes have bee proposed [Abr 87]. Such interpretations can be compared according to the precisic and the cost [Debr 91] of the induced analyses. In the present study, we address th former criterion.

In the classical theory of abstract interpretation [CCou 77], the only way we hav to compare two interpretations is to show that one abstracts the other. Thus, a information expressed in the domain is treated as equally relevant. The weakness this type of comparison is that it does not allow us to compare various interpretatio with respect to some, but not all, of the information they express. For example, v are interested in comparing the precision with which variable groundness in log programs is inferred by using domains that also express other information, suc as variable independence. This paper presents new notions that enable us to mak such precise comparisons. These notions are applicable to the comparison of abstra interpretations for all kinds of programming languages and with all sorts of gener analysis constructions (e.g., a collecting semantics).

Assume that we have a set of *subject* interpretations that we want to compar Our comparison technique works as follows.

1. The part of the subject domains with respect to which we wish to do th comparison is formalized as a *reference* domain $\mathcal{R}$.

2. For each *subject* domain D, a function f_D is constructed that maps each valu of D to its corresponding value in $\mathcal{R}$. We formalize the intuition that th function *completely extracts* from the values of D their content expressible i $\mathcal{R}$ and demonstrate the existence of such a function.

3. We say that one subject domain element $d_1 \in D_1$ *$\mathcal{R}$-abstracts* another $d_2 \in D$ if $f_{D_1}(d_1) \sqsupseteq_{\mathcal{R}} f_{D_2}(d_2)$.

4. We say that subject domain D_1 *$\mathcal{R}$-abstracts* subject domain D_2 if for eac individual $d_1 \in D_1$ there exists $d_2 \in D_2$ such that $f_{D_1}(d_1) = f_{D_2}(d_2)$.

5. We say that one subject interpretation $\mathbf{D}_1 = \langle D_1, \mu_1^1, \ldots, \mu_1^k \rangle$ *$\mathcal{R}$-abstracts* an other subject interpretation $\mathbf{D}_2 = \langle D_2, \mu_2^1, \ldots, \mu_2^k \rangle$ if

   (a) D_1 $\mathcal{R}$-abstracts D_2, and

   (b) if $d_1 \in D_1$ $\mathcal{R}$-abstracts $d_2 \in D_2$, then for each i, $\mu_1^i(d_1)$ $\mathcal{R}$-abstract $\mu_2^i(d_2)$.

The significance of showing that $\mathbf{D}_1$ $\mathcal{R}$-abstracts $\mathbf{D}_2$ is the following. For an generic analysis construction, a simple inductive argument shows that, for any give program, the information computed by using $\mathbf{D}_1$ *$\mathcal{R}$-abstracts* that computed by usin $\mathbf{D}_2$. In other words, $\mathbf{D}_1$ computes the information expressed by $\mathcal{R}$ at least as we as $\mathbf{D}_2$.

The idea underlying our technique is to compare different interpretations by pro jecting them into a (typically more abstract) reference domain and comparing thei behavior under this (abstracted) view. In this way we leave out of the compariso irrelevant information expressed in the subject domains. As this intuition suggest turns out that when the reference domain $\mathcal{R}$ is abstracted by the subject domains, -abstraction collapses to the classical notion of abstraction.

We demonstrate the utility of our technique by using it to compare two abstract terpretations for analysis of logic programs with respect to their ability to infer hen variables are bound to ground terms. Groundness information can be used the compiler to improve unification code and to avoid the overhead of preparing r backtracking that will never occur. The interpretations we consider are `Prop` [aSo 89] and `Sharing` [JaLa 89]. In the classical theory, it is possible to show only at `Prop` and `Sharing` are incomparable (see Section 3). Using our new theory, e show that, with respect to their ability to compute groundness, `Prop` is strictly ore powerful than `Sharing`. Moreover we prove that a subset of `Prop` precisely aracterizes the groundness computation of `Sharing`.

The organization of this paper is as follows. Section 2 presents our new theory. he rest of the article describes the application of this theory to the comparison of e interpretations `Prop` and `Sharing` for the groundness analysis of logic programs. ection 3 contains the definitions of `Sharing` and `Prop`, the proof of the incomparability of `Prop` and `Sharing` (which is the only statement one can make about these o interpretations using the classical notions of [CCou 77]) and the results that the w theory of $\mathcal{R}$-abstraction yields. In Section 4 we discuss in general the choice of e reference domain $\mathcal{R}$.

Theory

et X and Y be complete lattices and $\gamma_{XY} : X \longrightarrow Y$ and $\alpha_{YX} : Y \longrightarrow X$ be onotonic functions. The 4-tuple $G_{YX} = (\gamma_{XY}, Y, X, \alpha_{YX})$ is a *Galois insertion* if $y \in Y : y \sqsubseteq_Y \gamma_{XY}(\alpha_{YX}(y))$ and $\forall x \in X : x = \alpha_{YX}(\gamma_{XY}(x))$, where $\sqsubseteq_Y$ is the ordering 1 Y. Observe that the last condition implies that γ_{XY} is one-to-one. The functions $_{XY}$ and α_{YX} are called the concretization and the abstraction function, respectively. is said to abstract Y.

otice that, for the sake of clarity, the following notation is used: f_{TS} is a function om T to S.

emma 2.1 ([JoNie 91]) *Let $(\gamma_{XY}, Y, X, \alpha_{YX})$ be a Galois insertion.*

$$\forall y \in Y \text{ and } \forall x \in X \qquad \alpha_{YX}(y) \sqsubseteq_X x \Leftrightarrow y \sqsubseteq_Y \gamma_{XY}(x)$$

efinition 2.2 *Let D be a complete lattice. An interpretation is a tuple* $\mathbf{D} =$ $D, \mu^1_D, \ldots, \mu^k_D\rangle$ *where $\mu^1_D, \ldots, \mu^k_D$ are monotone functions of type $D^n \longrightarrow D$.*

1 general, the operations μ^i_D may take some other arguments besides elements of . However, as they are irrelevant here, we omit them.

efinition 2.3 (Classical Abstraction) *An interpretation $\mathbf{D} = \langle D, \mu^1_D, \ldots, \mu^k_D\rangle$ bstracts an interpretation $\mathbf{C} = \langle C, \mu^1_C, \ldots, \mu^k_C\rangle$ if*

1. *there is a Galois insertion $(\gamma_{DC}, C, D, \alpha_{CD})$ and*
2. $\forall c_1 \ldots c_n \in C : \forall d_1 \ldots d_n \in D : \forall j \in [1, k]$:

$$(\,\forall i \in [1, n] : c_i \sqsubseteq_C \gamma_{DC}(d_i)\,) \Rightarrow \mu^j_C(c_1, \ldots, c_n) \sqsubseteq_C \gamma_{DC}(\mu^j_D(d_1, \ldots, d_n)).$$

In what follows we will be considering domains D, L and $\mathcal{R}$ that enjoy Galc insertions into a common concrete domain, C. In the following two lemmas, t relation between such domains is studied. Assume in the following that G_{CL} $(\gamma_{LC}, C, L, \alpha_{CL})$, $G_{CD} = (\gamma_{DC}, C, D, \alpha_{CD})$, and $G_{C\mathcal{R}} = (\gamma_{\mathcal{R}C}, C, \mathcal{R}, \alpha_{C\mathcal{R}})$ are Galc insertions.

Lemma 2.4 *Consider the domains D and L. Let $\varepsilon_{DL} = \alpha_{CL} \circ \gamma_{DC}$ and ε_{LD} $\alpha_{CD} \circ \gamma_{LC}$. The following holds:*

i) ε_{DL} *and* ε_{LD} *are monotone;*

ii) $\forall d \in D.\ \ \varepsilon_{LD}(\varepsilon_{DL}(d)) \sqsupseteq_D d$;

iii) $\forall l \in L.\ \ \varepsilon_{DL}(\varepsilon_{LD}(l)) \sqsupseteq_L l$.

proof: Condition *i)* is straightforward. Let $d \in D$ and $l \in L$.

$$\begin{array}{rcll} \varepsilon_{LD}(\varepsilon_{DL}(d)) & = & \alpha_{CD}(\gamma_{LC}(\alpha_{CL}(\gamma_{DC}(d)))) & \\ & \sqsupseteq_D & \alpha_{CD}(\gamma_{DC}(d)) & \textit{as } G_{CL} \textit{ is a Galois insertion} \\ & = & d & \textit{as } G_{CD} \textit{ is a Galois insertion.} \end{array}$$

The proof that $\varepsilon_{DL}(\varepsilon_{LD}(l)) \sqsupseteq_L l$ is analogous. □

Functions such as ε_{DL} and ε_{LD} of Lemma 2.4 are very important for our theory, shown below.

Definition 2.5 *We say that a monotone function $\varepsilon_{DL} : D \longrightarrow L$* completely e tracts *the L-information from D if $\forall d \in D : \varepsilon_{DL}(d) = \alpha_{CL}(\gamma_{DC}(d))$.*

Observe that the function $\varepsilon_{DL} : D \longrightarrow L$ defined above completely extracts the information from D in the following sense $\forall d \in D : \varepsilon_{DL}(d) = \sqcap_L \{l \in L \mid \gamma_{LC}(l) \sqsupseteq \gamma_{DC}(d)\}$. Thus $\varepsilon_{DL}(d)$ is the smallest element $l \in L$ such that $\gamma_{LC}(l) \sqsupseteq_C \gamma_{DC}(d)$. other words, $\varepsilon_{DL}(d)$ is the best approximation in L of the information expressed d.

Notice also that Lemma 2.4 shows that the complete extraction functions betwe L and D satisfy conditions very close to those of a Galois insertion. To get a Galc insertion it would suffice to turn a $\sqsupseteq$ sign of *ii)* or *iii)* into an $=$ sign. The followi result states a sufficient and necessary condition for obtaining a Galois insertic between L and D.

Lemma 2.6 *Let $\alpha_{DL} = \alpha_{CL} \circ \gamma_{DC}$ and $\gamma_{LD} = \alpha_{CD} \circ \gamma_{LC}$. The following conditions a equivalent.*

i) $(\gamma_{LD}, D, L, \alpha_{DL})$ *is a Galois insertion.*

ii) $\gamma_{LC}(L) \subseteq \gamma_{DC}(D)$

proof:

$i \Rightarrow ii)$ Let $l \in L$. Define $d_l = \gamma_{LD}(l)$. We will show that $\gamma_{DC}(d_l) = \gamma_{LC}(l)$. Obser the following two facts:

a) $\gamma_{DC}(d_l) \sqsupseteq_C \gamma_{LC}(l)$. This follows from the definition of d_l and the fact that $(\gamma_{DC}, C, D, \alpha_{CD})$ is a Galois insertion.

b) $\alpha_{DL}(d_l) = l$. This follows immediately from the definition of d_l and hypothesis *i*). Thus, by the definition of α_{DL}, also $\alpha_{CL}(\gamma_{DC}(d_l)) = l$.

By Lemma 2.1, $\alpha_{CL}(\gamma_{DC}(d_l)) \sqsubseteq_L l \Leftrightarrow \gamma_{DC}(d_l) \sqsubseteq_C \gamma_{LC}(l)$.
By *b*), $\alpha_{CL}(\gamma_{DC}(d_l)) \sqsubseteq_L l$. Thus $\gamma_{DC}(d_l) \sqsubseteq_C \gamma_{LC}(l)$, and it follows by *a*) that $\gamma_{DC}(d_l) = \gamma_{LC}(l)$.
This proves that $\gamma_{LC}(L) \subseteq \gamma_{DC}(D)$.

$\Leftarrow$ *ii*) By Lemma 2.4, we need to prove only that $\alpha_{DL}(\gamma_{LD}(l)) = l$.

$$\begin{aligned} \alpha_{DL}(\gamma_{LD}(l)) &= \alpha_{CL}(\gamma_{DC}(\alpha_{CD}(\gamma_{LC}(l)))) \\ &= \alpha_{CL}(\gamma_{DC}(\alpha_{CD}(\gamma_{DC}(d)))) && \textit{for } d \in D \textit{ s.t. } \gamma_{LC}(l) = \gamma_{DC}(d) \\ &= \alpha_{CL}(\gamma_{DC}(d)) && \textit{since } G_{CD} \textit{ is Galois ins.} \\ &= \alpha_{CL}(\gamma_{LC}(l)) \\ &= l && \textit{since } G_{CL} \textit{ is Galois ins.} \quad \Box \end{aligned}$$

efinition 2.7 (Coherency) *A Galois insertion* $G_{DL} = (\gamma_{LD}, D, L, \alpha_{DL})$ *is co-*
rent (with respect to fixed and implicit G_{CL} *and* G_{CD}*) if* $\alpha_{DL} = \alpha_{CL} \circ \gamma_{DC}$ *and*
$_D = \alpha_{CD} \circ \gamma_{LC}$. *Similarly, if* $(\alpha_{CD} \circ \gamma_{LC}, D, L, \alpha_{CL} \circ \gamma_{DC})$ *is a Galois insertion, we*
y L coherently abstracts *D*.

ssume that **L** abstracts **D** according to Definition 2.3. Intuitively, this means that l operations on D can be simulated on L. We want to extend this notion so as to press the fact that L can simulate a subset of D. To this end we introduce the tion of $\mathcal{R}$-abstraction where, intuitively, $\mathcal{R}$ is a domain that represents the part of that L can abstract. From this intuition one can already understand that if $\mathcal{R}$ is stracted both by L and by D, then $\mathcal{R}$-abstraction gives back classical abstraction ee Theorem 2.10). First the definition of $\mathcal{R}$-abstraction is given at the domain vel and then it is lifted to the (abstract) interpretation level.

efinition 2.8 *Let* $\varepsilon_{D\mathcal{R}} : D \longrightarrow \mathcal{R}$ *and* $\varepsilon_{L\mathcal{R}} : L \longrightarrow \mathcal{R}$ *completely extract the* *-information from D and L, respectively.*
r $l \in L$ *and* $d \in D$*, we say that the element l* $\mathcal{R}$-abstracts *the element d if* $_{\mathcal{R}}(d) \sqsubseteq_{\mathcal{R}} \varepsilon_{L\mathcal{R}}(l)$.
he domain L $\mathcal{R}$-abstracts *the domain D if* $\varepsilon_{L\mathcal{R}}(L) \subseteq \varepsilon_{D\mathcal{R}}(D)$, *i.e. if* $\forall l \in L : \exists d \in$ $: \varepsilon_{L\mathcal{R}}(l) = \varepsilon_{D\mathcal{R}}(d)$.

he concept of $\mathcal{R}$-abstraction is further extended to interpretations, by considering so the operations defined on the domains.

efinition 2.9 ($\mathcal{R}$-Abstraction) *Let* D, L, *and* $\mathcal{R}$ *be domains and consider* $\mathbf{D} =$ $), \mu^1_D, \ldots, \mu^k_D\rangle$ *and* $\mathbf{L} = \langle L, \mu^1_L, \ldots, \mu^k_L\rangle$. *Let* $\varepsilon_{D\mathcal{R}} : D \longrightarrow \mathcal{R}$ *and* $\varepsilon_{L\mathcal{R}} : L \longrightarrow \mathcal{R}$ *be* *e two monotone functions that completely extract the* $\mathcal{R}$*-information from D and* *, respectively. We say that* **L** $\mathcal{R}$-abstracts **D** *if the following conditions hold:*

1. *The domain L* $\mathcal{R}$*-abstracts the domain D.*

2. *Let μ_D and μ_L be any corresponding operations defined on D and L, respe tively, of the following types. $\mu_D : D^n \longrightarrow D$ and $\mu_L : L^n \longrightarrow L$. F $d_1 \ldots d_n \in D$, $l_1 \ldots l_n \in L$.*
If $\varepsilon_{DR}(d_1) = \varepsilon_{LR}(l_1)$, $\ldots$, and $\varepsilon_{DR}(d_n) = \varepsilon_{LR}(l_n)$ then $\mu_L(l_1, \ldots, l_n)$ 𝓡 abstracts $\mu_D(d_1, \ldots, d_n)$.

Given two abstract interpretations **L** and **D**, if **L** $\mathcal{R}$-abstracts **D** and **D** $\mathcal{R}$-abstrac **L**, we say that **L** and **D** are *$\mathcal{R}$-equivalent*. On the other hand, if **L** $\mathcal{R}$-abstracts but **D** does not $\mathcal{R}$-abstract **L**, we say that **L** *properly $\mathcal{R}$-abstracts* **D**.

In what follows, we show that the definition of $\mathcal{R}$-abstraction generalizes that classical abstraction (cf. Definition 2.3). In fact, the classical notion of abstracti is obtained from the notion of $\mathcal{R}$-abstraction when $\mathcal{R}$ is abstracted by the two inte pretations under consideration. In particular, it is true that **L** C-abstracts **D** if a only if **L** coherently abstracts **D** (recall that C is the concrete domain).

Theorem 2.10 *Let* **D**, **L** *be interpretations and assume that* **D** *and* **L** *both cohe ently abstract $\mathcal{R}$. Then* **L** *$\mathcal{R}$-abstracts* **D** *if and only if* **L** *coherently abstracts* **D**.

proof:

$\Rightarrow$) As usual, a Galois insertion of a domain X into a domain Y is G_{YX} $(\gamma_{XY}, Y, X, \alpha_{YX})$. Since L $\mathcal{R}$-abstracts D, it must be that $\gamma_{LR}(L) \subseteq \gamma_{DR}(D$ Hence, by Lemma 2.6, there is a Galois insertion $G_{DL} = (\alpha_{RD} \circ \gamma_{LR}, D, L, \alpha_{RL}$ $\gamma_{DR})$ coherent with G_{RD} and G_{RL}. By simple rewritings one can show th G_{DL} is coherent w.r.t. G_{CD} and G_{CL}. That is, $\alpha_{RD} \circ \gamma_{LR} = \alpha_{CD} \circ \gamma_{LC} = \gamma$ and $\alpha_{RL} \circ \gamma_{DR} = \alpha_{CL} \circ \gamma_{DC} = \alpha_{DL}$.
Consider any $l_1, \ldots, l_n \in L$ and $d_1, \ldots, d_n \in D$ such that $\forall i \in [1..n]$ $d_i \sqsubseteq$ $\gamma_{LD}(l_i)$. Rewriting γ_{LD} one obtains $d_i \sqsubseteq_D \alpha_{RD}(\gamma_{LR}(l_i))$, which, by the fact th $\gamma_{LR}(L) \subseteq \gamma_{DR}(D)$, gives $\gamma_{DR}(d_i) \sqsubseteq_{\mathcal{R}} \gamma_{LR}(l_i)$. Again by the fact that $\gamma_{LR}(L)$ $\gamma_{DR}(D)$, there exist $\hat{d}_1, \ldots, \hat{d}_n$ such that $\gamma_{LR}(l_i) = \gamma_{DR}(\hat{d}_i)$ $\forall i \in [1..n]$. It a property of Galois insertions that $d_i \sqsubseteq_D \hat{d}_i$ $\forall i \in [1..n]$. By $\mathcal{R}$-abstracti of **L** and **D** it follows that $\gamma_{DR}(\mu_D(\hat{d}_1, \ldots, \hat{d}_n)) \sqsubseteq_{\mathcal{R}} \gamma_{LR}(\mu_L(l_1 \ldots, l_n))$. A plying the abstraction function α_{RD} to both sides, by monotonicity of α one obtains $\alpha_{RD}(\gamma_{DR}(\mu_D(\hat{d}_1, \ldots, \hat{d}_n))) \sqsubseteq_D \alpha_{RD}(\gamma_{LR}(\mu_L(l_1 \ldots, l_n)))$, and th $\mu_D(\hat{d}_1, \ldots, \hat{d}_n) \sqsubseteq_D \gamma_{LD}(\mu_L(l_1 \ldots, l_n))$, because G_{RD} is a Galois insertion. F nally, by monotonicity of μ_D, we get $\mu_D(d_1 \ldots, d_n) \sqsubseteq_D \gamma_{LD}(\mu_L(l_1 \ldots, l_n))$, required.

$\Leftarrow$) Since L coherently abstracts D, $G_{DL} = (\gamma_{LD}, D, L, \alpha_{DL})$ is a coherent Galc insertion. By Lemma 2.6 this implies that $\gamma_{LR}(L) \subseteq \gamma_{DR}(D)$ (and also th $\gamma_{LC}(L) \subseteq \gamma_{DC}(D)$). Thus, $\forall l \in L\ \exists \hat{d} \in D$ such that $\gamma_{LR}(l) = \gamma_{DR}(\hat{d})$.
Let $l_1, \ldots, l_n \in L$ and let $\hat{d}_1, \ldots, \hat{d}_n$ be the corresponding elements of D.

$\mu_D(\hat{d}_1,\ldots,\hat{d}_n)$	$\sqsubseteq_D$	$\gamma_{LD}(\mu_L(l_1,\ldots,l_n))$	**L** abstracts **D**
$\mu_D(\hat{d}_1,\ldots,\hat{d}_n)$	$\sqsubseteq_D$	$\alpha_{RD}(\gamma_{LR}(\mu_L(l_1,\ldots,l_n)))$	rewriting γ_{LD}
$\gamma_{DR}(\mu_D(\hat{d}_1,\ldots,\hat{d}_n))$	$\sqsubseteq_{\mathcal{R}}$	$\gamma_{DR}(\alpha_{RD}(\gamma_{LR}(\mu_L(l_1,\ldots,l_n))))$	γ_{DR} monotone
$\gamma_{DR}(\mu_D(\hat{d}_1,\ldots,\hat{d}_n))$	$\sqsubseteq_{\mathcal{R}}$	$\gamma_{LR}(\mu_L(l_1,\ldots,l_n))$	$\gamma_{LR}(L) \subseteq \gamma_{DR}(D)$

We wish to emphasize that this theory is applicable to the comparison of abstract terpretations for all kinds of programming languages and with all sorts of analysis ameworks. In the rest of the paper we apply the theory to the comparison of terpretations for analysis of logic programs with respect to their ability to infer riable groundness.

Application of the Theory

s mentioned in the Introduction, we apply the theory of the $\mathcal{R}$-abstraction deveped in the previous section to the comparison of two abstract interpretations for gic programs, namely `Sharing` [JaLa 89] and `Prop` [MaSo 89,CFWi 91].

Let us introduce some notation. Consider a first order language $\mathcal{L}$ consisting ranked function and predicate symbols and the set of variables *Var*. Let $\mathbf{V}$ be e finite set of variables $\{x_i : i \leq n\}$ for some suitably large n. If t is any term nstructed on $\mathcal{L}$, $Var(t)$ is the set of variables occurring in term t. A *substitution* is a function from $\mathbf{V}$ to terms. *Subst* is the set of the renaming equivalence classes idempotent substitutions. The domain and the range of a substitution are defined $Dom(\sigma) = \{x \in \mathbf{V} : \sigma(x) \neq x\}$ and $Rng(\sigma) = \{x \in \mathbf{V} : \exists y \in Dom(\sigma)$ such that $\in Var(\sigma y)\}$. In the examples we will denote by $a, b, c, \ldots$ constant terms.

Logic programs compute substitutions, and the basic operation is the unificaon. Thus, their concrete interpretation is $\mathbf{C}{=}\langle \wp(Subst), \mathbf{u}_{subst}\rangle$, where $\wp(Subst)$ the powerset of *Subst* and $\mathbf{u}_{subst} : \wp(Subst) \times Subst \longrightarrow \wp(Subst)$ is the conete unification that works as follows. Let $\vartheta \in Subst$, A be the selected atom and $\leftarrow B_1, \ldots, B_n$ be the clause to be used, then if δ is the most general unifier of the uation $(A = H)$, $\mathbf{u}_{subst}(\{\vartheta\}, \delta) = \vartheta \circ \mu$, where μ is the most general unifier of the uation $(A\vartheta = H)$.

Properties of substitutions that are particularly useful for optimization tools e *groundness, covering, sharing* and *independence*. Let $\Sigma \subseteq Subst$, Σ *grounds* a riable x if $\sigma \in \Sigma \Longrightarrow Var(\sigma x) = \emptyset$. A set of variables S_1 *covers* S_2 with respect to if $\sigma \in \Sigma \Longrightarrow \bigcup_{x \in S_1} Var(\sigma x) \supseteq \bigcup_{x \in S_2} Var(\sigma x)$. Two sets of variables S_1 and S_2 e equivalent w.r.t. Σ if both S_1 covers S_2 and S_2 covers S_1 w.r.t Σ. Two variables $, x_2$ *share* with respect to Σ if for some $\sigma \in \Sigma$ $Var(\sigma x) \cap Var(\sigma y) \neq \emptyset$. Conversely, $, x_2$ are *independent* w.r.t. Σ if for all $\sigma \in \Sigma$, $Var(\sigma x) \cap Var(\sigma y) = \emptyset$.

Notice the difference between variable sharing and the other properties. Groundess, covering and independence are "sure" information: they must be true for all $\in \Sigma$. At the contrary, sharing is a "possible" information: one requires that some $\in \Sigma$ shows the sharing. Observe that "possible" sharing gives "sure" independence.

The rest of this section consists of three parts: in the first one, the interpretations `rop` and `Sharing` are briefly recalled. In the second part it is shown the only result bout the relation between `Prop` and `Sharing` that can be proven using the classical otion of abstraction between interpretations. The final part contains the more teresting relationship between `Prop` and `Sharing` yielded by the theory developed the previous section.

The interpretations $\langle \texttt{Prop}, \mathbf{u}_{pr} \rangle$ and $\langle \texttt{Sharing}, \mathbf{u}_{sh} \rangle$

The first domain we consider is Prop [MaSo 89,CFWi 91]. Its elements are (equi alence classes of) propositional formula over $\mathbf{V}$ and the logical connectives $\wedge, \vee, \leftarrow$ Intuitively, the formula $x_1 \wedge x_2$ says that x_1 and x_2 are both ground, $(x_1 \wedge x_2)$ $(x_3 \wedge x_4 \wedge x_5)$ expresses the equivalence of $\{x_1, x_2\}$ and $\{x_3, x_4, x_5\}$. The formu $x_1 \vee x_2$ represents sets of substitutions that ground either x_1 and x_2 or both. We s that the last formula represents "possible" groundness. At the contrary, a formu without disjunctions represents only "sure" groundness and covering information.

Example 3.1 *Let* $\sigma = \{x_1/y_1,\ x_2/h(y_2),\ x_3/g(y_1, y_2), x_4/a\}$. *$\sigma$ is approximated, f instance, by the following element of* Prop: $[((x_1 \wedge x_2) \leftrightarrow x_3) \wedge x_4]_{\equiv}$ *This formul in fact, says that x_4 is ground and that the sets $\{x_1, x_2\}$ and $\{x_3\}$ cover each othe Also the formula $[x_4]_{\equiv}$ approximates σ, but the formula $[x_1 \leftrightarrow x_4]_{\equiv}$ does not.*

Thus, a formula φ approximates all substitutions that present the groundness an covering expressed by φ. The notion of approximation between Prop and $\wp(Subs$ is formalized by a concretization function γ_{pr} and by its adjoint function α_{pr}. F their definition see [CFWi 91], where it is shown that Prop is a complete latti w.r.t. logical implication and $(\gamma_{pr}, \wp(Subst), \texttt{Prop}, \alpha_{pr})$ is a Galois insertion.

The abstract unification $\mathbf{u}_{pr}$ over Prop mimics the concrete unification $\mathbf{u}_{subst}$ an it is as follows. $\mathbf{u}_{pr} : \texttt{Prop} \times Subst \longrightarrow \texttt{Prop}$. $\mathbf{u}_{pr}([\varphi]_{\equiv}, \delta)$ is the logical conjunctic $[\varphi \wedge \varphi_\delta]_{\equiv}$, where $\varphi_\delta = \bigwedge\{x \leftrightarrow Var(t) : x/t \in \delta\} \equiv \alpha_{pr}(\{\delta\})$. Thus the first abstra interpretation we consider is the pair $\langle \texttt{Prop}, \mathbf{u}_{pr} \rangle$.

Let us turn to Sharing [JaLa 89]. The elements of Sharing are sets of se of variables. Let $A = \{Z_1, \ldots, Z_n\} \in \texttt{Sharing}$, where $Z_i \subseteq \mathbf{V}$. The element represents the following information: if $Z_i = \{w_1, \ldots, w_m\}$ and σ is a substitutic approximated by A then the terms $\sigma(w_1), \ldots, \sigma(w_m)$ may share a common variabl In other words, the elements $Z_i \in A$ are sets of variables that can share w.r.t. son substitutions approximated by A. Sharing represents also "sure" groundness an covering: the variable x is ground w.r.t. A if x does not belong to $Var(A)$; S covers S_2 w.r.t. $A = \{Z_1, \ldots, Z_n\}$ if whenever $(S_2 \cap Z_i) \neq \emptyset$ then also $(S_1 \cap Z_i)$ $\emptyset$. Intuitively, one may understand that the elements of Sharing cannot represe "possible" groundness and covering. Clearly, for $A \in \texttt{Sharing}$, for any variable either $x \in Var(A)$ or it does not. A similar reasoning holds for covering.

Example 3.2 (JaLa 89) *Let* $\sigma = \{x_1/f(a, y_1), x_2/y_1, x_3/g(y_2), x_4/h(y_1, y_2), x_5/b$ *The state of* Sharing *that best approximates σ is $A = \{\{x_1, x_2, x_4\}, \{x_3, x_4\}\}$. A ca tures, for instance, the fact that x_2 and x_3 are surely independent and that groundi either x_1 or x_2 grounds the other.*

Thus Sharing is $\wp(\wp(\mathbf{V}))$. Obviously Sharing is a complete lattice w.r.t. set i clusion. Let α_{sh} and γ_{sh} be respectively the abstraction and concretization functio between Sharing and $\wp(Subst)$, it can be shown that $(\gamma_{sh}, \wp(Subst), \texttt{Sharing}, \alpha_{sh})$ a Galois insertion. The definition of the abstract unification $\mathbf{u}_{sh}$ is quite technical an therefore it is omitted. These formal definitions can be found in [JaLa 89,CFWi 91b

Let us briefly discuss the main differences between Prop and Sharing. We have ready observed that both of them represent "sure" groundness and covering. How-ver they differ since, on the one hand, Sharing represents also "sure" independence formation and, on the other hand, Prop represents also "possible" groundness and overing. We will show below that because of these differences, using the classical otion of interpretation abstraction [CCou 77],one can only show that $\langle$Prop, $\mathbf{u}_{pr}\rangle$ nd $\langle$Sharing, $\mathbf{u}_{sh}\rangle$ are not comparable. This answer is unsatisfactory because, as we ave seen, these two interpretations both compute "sure" groundness and covering nd it is important to compare them on this common part. Using the new notion of -abstraction this is possible. In fact, this notion allows us to disregard parts of the terpretations that we want to compare, in order to focus only on the common part. o this aim it is crucial to find a "good" reference domain which should represent his common information. Immediately below, the choice of the reference domain r the comparison of Sharing and Prop is discussed. This problem is considered in s generality in Section 4.

For comparing Sharing and Prop we choose as reference domain a sublattice of rop, which represents only the "sure" groundness and covering information, that common to both the interpretations. This new domain, called Cov, is defined as e set of (equivalence classes of) propositional formula of the form $\bigwedge\{\wedge S_1 \rightarrow \wedge S_2\}$, r $S_1, S_2 \subseteq \mathbf{V}$. Notice that Cov does not contain disjunctions that express, as e remarked above, the "possible" groundness and covering that we want disre-ard. That this restriction yields a loss of expressiveness is shown by the following sult [CFWi 91b].

heorem 3.3 Cov $\subset$ Prop.

he classical incomparability of Prop and Sharing

s mentioned before, using the classical notion of abstraction only the incompara-ility of $\langle$Prop, $\mathbf{u}_{pr}\rangle$ and $\langle$Sharing, $\mathbf{u}_{sh}\rangle$ can be shown.

heorem 3.4 *There is no coherent Galois insertion between* Prop *and* Sharing.

roof: By Lemma 2.6 it is sufficient to prove that γ_{pr}(Prop) and γ_{sh}(Sharing) are comparable.

i) That γ_{pr}(Prop) $\not\subseteq$ γ_{sh}(Sharing) is shown using the formula $[x \vee y]_{\equiv} \in$ Prop. There is no value $A \in$ Sharing such that $\gamma_{sh}(A) = \gamma_{pr}([x \vee y]_{\equiv})$, because Sharing does not express "possible" groundness.

ii) That γ_{sh}(Sharing) $\not\subseteq$ γ_{pr}(Prop) is shown using the element $A = \{\{x\}, \{y\}\} \in$ Sharing. A approximates all substitutions in which x and y are independent. Since Prop does not represent independence, there is no formula $f \in$ Prop such that $\gamma_{pr}(f) = \gamma_{sh}(A)$. □

he results that the new theory yields

ccording to the schema presented in Section 2, in order to compare $\langle$Prop, $\mathbf{u}_{pr}\rangle$ and Sharing, $\mathbf{u}_{sh}\rangle$ w.r.t. Cov we have first to compare the domains Prop and Sharing w.r.t. Cov and then to compare $\mathbf{u}_{pr}$ and $\mathbf{u}_{sh}$. At the domain-level the followin result holds [CFWi 91b].

Theorem 3.5

i) The domain Sharing Cov-*abstracts the domain* Prop

ii) The domain Prop Cov-*abstracts the domain* Sharing

Moreover, the intuitive idea that Cov is the intersection of Sharing and Prop ca be formalized as follows. It is easy to turn the domain Cov into an abstract inte pretation $\langle \texttt{Cov}, \mathbf{u}_{cov}\rangle$, where $\gamma_{cov} = \gamma_{pr}$ and $\mathbf{u}_{cov} = \mathbf{u}_{pr}$. Recall that $\texttt{Cov} \subset \texttt{Prop}$.

Theorem 3.6 $\gamma_{cov}(\texttt{Cov}) = \gamma_{pr}(\texttt{Prop}) \cap \gamma_{sh}(\texttt{Sharing})$

The above theorems state that the two domains Prop and Sharing both com pletely represent the information expressed by Cov, i.e. "sure" groundness and cov ering. However, when one compares also the operations $\mathbf{u}_{sh}$ and $\mathbf{u}_{pr}$, it turns ou that $\langle \texttt{Prop}, \mathbf{u}_{pr}\rangle$ is strictly more powerful than $\langle \texttt{Sharing}, \mathbf{u}_{sh}\rangle$ [CFWi 91b].

Theorem 3.7 *The interpretation* $\langle \texttt{Sharing}, \mathbf{u}_{sh}\rangle$ *properly* Cov-*abstracts the inter pretation* $\langle \texttt{Prop}, \mathbf{u}_{pr}\rangle$.

This result teaches us two things:

1) The "sure" independence of Sharing is not useful for computing groundnes and covering information. In fact, for this task there is no difference betwee using $\langle \texttt{Sharing}, \mathbf{u}_{sh}\rangle$ and $\langle \texttt{Cov}, \mathbf{u}_{cov}\rangle$. This intuition is staded formally as fo lows [CFWi 91b]

   Theorem 3.8 $\langle \texttt{Sharing}, \mathbf{u}_{sh}\rangle$ *and* $\langle \texttt{Cov}, \mathbf{u}_{cov}\rangle$ *are* Cov-*equivalent.*

2) The ability of Prop of representing "possible" groundness and covering is use ful for computing "sure" groundness and covering. The following exampl illustrates this phenomenon.

   Example 3.9 *Consider the formula of* Prop $f = [(x \leftrightarrow y) \vee (x \leftrightarrow z)]_{\equiv}$. *Th formula is not a formula of* Cov. *Moreover the only formula* f' *of* Cov *suc that* $\gamma_{cov}(f') \supseteq \gamma_{pr}(f)$ *is the constant* True*: no sure information is expresse by* f. *Consider now the substitution* $\delta = \{y/z\}$. $\mathbf{u}_{pr}([f]_{\equiv}, \delta) = [f \wedge (y \leftrightarrow z)]_{\equiv} = [(x \leftrightarrow y) \wedge (x \leftrightarrow z) \wedge (y \leftrightarrow z)]_{\equiv}$, *whereas* $\mathbf{u}_{cov}(\text{True}, \delta) = [y \leftrightarrow z]_{\equiv}$.

4 The selection of the reference domain

We want to clarify an important point of our theory for comparing interpretations how does one choose the reference domain with respect to which the comparison c the interpretations is made? For the comparison of $\langle \texttt{Prop}, \mathbf{u}_{pr}\rangle$ and $\langle \texttt{Sharing}, \mathbf{u}_{sh}\rangle$ presented in the previous section, the domain Cov was chosen. Despite the motiva tion given above, this choice may seem surprising, since the purpose was to compar the two domains with respect to groundness; Cov expresses not only groundness, bu lso covering information. The reason for this choice is that both $\langle \texttt{Prop}, \mathbf{u}_{pr}\rangle$ and $\texttt{Sharing}, \mathbf{u}_{sh}\rangle$ use covering for computing groundness. If we had chosen a simpler eference domain — for instance $\wp(\mathbf{V})$ where $\Delta \in \wp(\mathbf{V})$ represents the variables hat are surely ground — the reference domain would have been too weak for our oal in the following sense. That an element A of **Sharing** has the same image in $\wp(\mathbf{V})$ as an element $[f]_{\equiv}$ of **Prop** does not imply that the image in $\wp(\mathbf{V})$ of $_{sh}(A,\delta)$ approximates that of $\mathbf{u}_{pr}([f]_{\equiv},\delta)$. This is because A may contain covering nformation (relevant for computing groundness) that is not expressible in Δ and hus not necessarily present in $[f]_{\equiv}$. Hence $\mathbf{u}_{sh}(A,\delta)$ may compute groundness that $_{pr}([f]_{\equiv},\delta)$ does not compute. In fact, by using $\wp(\mathbf{V})$ as reference domain, one could rove only incomparability results. Thus, if the reference domain is "too abstract," othing interesting can be shown because important information is not captured by he reference domain. On the other hand, if the reference domain is "too concrete," ?-abstraction reduces to classical abstraction, as shown in Theorem 2.10.

It is also important to notice that since Cov contains all the information used n Sharing for computing groundness, as shown by Theorem 3.8, the fact that $\texttt{Sharing}, \mathbf{u}_{sh}\rangle$ properly Cov-abstracts $\langle \texttt{Prop}, \mathbf{u}_{pr}\rangle$ implies a corresponding relation etween analyses constructed by using these two interpretations. Namely, the analsis realized by using $\langle \texttt{Prop}, \mathbf{u}_{pr}\rangle$ is more precise with respect to groundness than hat obtained by using $\langle \texttt{Sharing}, \mathbf{u}_{sh}\rangle$

Thus the method for choosing the reference domain is simply this. If one wants o compare interpretations $\mathbf{D}_1$ and $\mathbf{D}_2$ with respect to some property p, one seeks a eference domain $\mathcal{R}$ that expresses p and any information used to compute p in either $\mathbf{D}_1$ or $\mathbf{D}_2$. In general the choice of the reference domain is not unique. Any domain $\mathcal{R}$ that expresses p and that yields $\mathbf{D}_1$ $\mathcal{R}$-abstracts $\mathbf{D}_2$ is sufficient to demonstrate hat the quality of the p-computation of (an analysis based on) $\mathbf{D}_2$ is at least as recise as that of $\mathbf{D}_1$.

For instance, to compare $\langle \texttt{Prop}, \mathbf{u}_{pr}\rangle$ and $\langle \texttt{Sharing}, \mathbf{u}_{sh}\rangle$ one can use Prop as he reference domain, instead of Cov. It turns out that $\langle \texttt{Sharing}, \mathbf{u}_{sh}\rangle$ properly rop-abstracts $\langle \texttt{Prop}, \mathbf{u}_{pr}\rangle$.

References

[Abr 87] Abramsky S. and Hankin C., "Abstract Interpretation of Declarative Languages." Ellis Horwood, 1987.

[Bruy 91] Bruynooghe, M., "A practical framework for the abstract interpretation of logic programs." *Journal of Logic Programming*, vol.10, 1991.

[CCou 77] Cousot, P. and Cousot, R., "Abstract interpretation: a unified framework for static analysis of programs by construction of approximation of fixpoints." *Proc. Fourth ACM POPL*, 1977.

[CoFi 90] Cortesi, A. and Filè, G., "Abstract interpretation of logic programs: an abstract domain for groundness, sharing, freeness and compoundness analysis." *Proc. ACM Symposium on Partial Evaluation and Semantics Based Program Manipulation.*, P. Hudak and N. Jones (eds.), SIGPLAN NOTICES vol.26, n.11, 1991.

[CFWi 91] Cortesi, A., Filè, G. and Winsborough W., "Prop revisited: Propositional formula as abstract domain for groundness analysis." *Proc. Sixth Annual IEEE Symposium on Logic in Computer Science.* G. Kahn (ed.), pp.322-327, Amsterdam 1991.

[CFWi 91b] Cortesi, A., Filè, G. and Winsborough W., "Comparison of Abstract Interpretations." Tech.Rep. 14/91 Dip. Matematica Pura e Applicata, Università di Padova, 1991.

[Debr 89] Debray, S. K., "Static inference of modes and data dependencies in logic programs." *TOPLAS*, Vol.11, pp.418-450, 1989.

[Debr 91] Debray, S. K., "The mithical Free Lunch (notes on the complexity/precision tradeoff in dataflow analysis of logic programs)", draft, Univ. of Arizona, 1991.

[JaLa 89] Jacobs, D. and Langen, A., "Accurate and efficient approximation of variable aliasing in logic programs." *Proc. NACLP.* Cleveland, 1989.

[JoNie 91] Jones, N.D. and Nielson, F., "Abstract Interpretation: a semantic-based tool for program analysis" draft version, 1991.

[JoSo 87] Jones, N. and Søndergaard, H., "A semantic based framework for the abstract interpretation of Prolog." In *Abstract Interpretation of Declarative Languages.* S. Abramsky and C. Hankin (eds.), Ellis Horwood, 1987.

[MaSo 89] Marriott, K. and Søndergaard, H., "Notes for a tutorial on abstract interpretation of logic programs." NACLP. Cleveland, 1989.

[MaSo 90] Marriott, K. and Søndergaard, H., "Abstract interpretation of logic programs: the denotational approach." *Proc. GULP'90.* A. Bossi (ed.), Padova, 1990.

[Mel 81] Mellish, C. S., "The automatic generation of mode declarations for Prolog programs." *Proc. Workshop on Logic Programming for Intelligent Systems.* Los Angeles, 1981.

A proposed categorical semantics for Pure ML*

Wesley Phoa and Michael Fourman
Laboratory for the Foundations of Computer Science
James Clerk Maxwell Building, The King's Buildings
Edinburgh EH9 3JZ, Scotland

Abstract

This paper outlines an approach to modelling the purely functional fragment of ML: it concentrates on the semantics of the Modules system. Our proposed semantics is *set-theoretic:* types and values are modelled by sets and functions in a *topos*, a categorical model of constructive set theory. *Synthetic domain theory* allows us to make sense of fixed points and recursive domains in a set-theoretic setting, while the notions of *classifying topos* and 'generic' structure provide a useful way of interpreting signatures, functors and sharing, as well as Extended ML specifications. We only give an informal account, concentrating on motivation and examples rather than giving a rigorous formal development—only elementary category theory is used.

Introduction

he process of specifying the formal semantics of Standard ML involved an unusu-ly close interplay between theoretical and practical considerations. The official efinition, [12], was finalised in 1989. But a denotational semantics would be use-l, e.g. for reasoning about programs. Indeed, a draft semantics (by Don Sannella) elped guide the development of the language; but it never reached a definitive form.

A model of ML should be clean and intuitive; the different aspects of the language ould be treated in a unified framework; it should be modular; and we should be ble to extend it easily to handle more sophisticated languages. Finally, it should gitimise familiar, informal, methods of reasoning.

This paper sketches an approach to modelling ML, focussing on the Modules stem. There are at least four ways of interpreting Modules:

- *Operational:* following the official Definition of Standard ML: [12], [2], [21];
- *Type-theoretic:* treating signatures as specifying existential types: [13], [3];
- *Algebraic:* identifying a signature with its class of models: [19], [7]; and
- *Categorical:* using the notion of generic structure, as proposed here.

*Research supported by SERC grant GR/F 31199, 'Formal System Design'; the first author was so supported by an 1851 Research Fellowship

Though our categorical methods look superficially different from the other ap- proaches, they are closely related to the algebraic viewpoint; in fact, some of ou proposals formally resemble ideas from algebraic specification, though the setting w work in is more suited to (higher-order) program logic. There are also connection with the type-theoretic view, not fully understood. But our approach was mor directly inspired by the informal picture of the Modules system presented in [2 and [21].

Although we focus on modules, an important feature of our approach is that th different parts of the language fit together in a single, relatively simple framewor motivated by set theory; to demonstrate this, we have included a brief discussion o the semantics of Core ML.

Our work is based on some rather technical results, but no knowledge of thes will be required: only some familiarity with categories. In fact, the presentation i quite informal. The precise details are being worked out; for example, the use c classifying toposes to model ML polymorphism is described in detail in [17] (in th more general framework of *polynomial categories*).

History and acknowledgements The idea of using classifying toposes to inter pret ML polymorphism and modules was proposed by Fourman some years ago, an independently by Phoa in 1991; this paper is the first place in which it has bee worked out. Walters (with his student Lack) has been developing similar ideas i the context of imperative languages based on distributive categories.

Phoa was supported by an 1851 Research Fellowship. He would also like t thank Bob Walters, Mike Johnson, James McKinna, Dave Berry and especially E Kazmierczak for stimulating discussions, and Carl Gunter, Philippa Gardner and th members of the ML club at the LFCS for their encouragement. Healfdene Gogue and John Power patiently read drafts and made valuable comments. So did th referees.

2 The semantics of Core ML

2.1 Constructive set theory and toposes

Our original intuition is that data types, and more complicated semantic domains are *sets*. But for technical reasons—the nonexistence of fixed points, recursive types polymorphic types and so on—we must work, not with sets, but with domains This can be inconvenient: for example, we must always check that functions ar continuous. It is also not clear how far we can reason naively, or set-theoretically about a domain-theoretic semantics. It does seem clear that most constructions o domains should be specified abstractly, using *category theory*.

Since the mid-1970s, Scott has been suggesting that set theory can after all pro vide a basis for denotational semantics: only we must use, not ordinary (classical set theory, but *constructive* set theory. (In recent years, a lot of progress has bee made: e.g. [18] and [15].) We can combine the set-theoretic and categorical view points by working within a categorical formulation of constructive set theory. Such a formulation is provided by *topos theory:* [8].

A topos is a category which 'behaves like' the category of sets: any topos provides n interpretation of constructive set theory, suitably formulated. Though the details f the definition are not needed in this paper, we reproduce it here for completeness:

)efinition 2.1 A category $\mathcal{E}$ (with pullbacks) has a **subobject classifier** if there ; a "set of truth-values" $\Omega \in \mathcal{E}$ and a map $tt: 1 \to \Omega$ such that for any mono $X' \rightarrowtail X \in \mathcal{E}$, there is a unique map $\chi_{X'}: X \to \Omega$ such that $X' \rightarrowtail X$ is the pullback f tt along $\chi_{X'}$. A category $\mathcal{E}$ is a **topos** if it has finite limits, exponentials and a ubobject classifier. ■

The simplest example of a topos is *Sets*, the usual category of sets. In this aper we will work with $\mathcal{E}\!f\!f$, Hyland's effective topos: [4]. This is the 'universe f recursive mathematics'. For example, Church's thesis, and all the theorems of ecursive mathematics, are true in the effective topos.

.2 Synthetic domains

'here are a number of full subcategories of $\mathcal{E}\!f\!f$ whose objects 'behave like domains': or example, see [18], [15], [1]. We will review the necessary results from [15], which s the most naive of these papers; however, nearly all our remarks apply to the other ategories as well.

There is a simple way to define a certain full subcategory $\mathcal{C}\mathcal{S}ig$ of $\mathcal{E}\!f\!f$, the category f **complete Σ-spaces**. We can regard the objects of $\mathcal{C}\mathcal{S}ig$ simply as sets. Some xamples of complete Σ-spaces are:

- basic data types like $\mathbb{N}$ and 2 (note that in $\mathcal{E}\!f\!f$, $2 \neq \Omega$)
- products $P \times Q$ and sums $P + Q$, where $P, Q \in \mathcal{C}\mathcal{S}ig$
- (total) set-theoretic function spaces $[P \to Q]$
- spaces $[P \rightharpoonup Q]$ of 'computable' partial functions

'or example, it contains an object $[\mathbb{N} \to \mathbb{N}]$ whose elements correspond to total re- ursive functions; and an object $[\mathbb{N} \rightharpoonup \mathbb{N}]$, whose elements 'are' the partial recursive unctions. That is, effectivity is built into the model.

The important facts about complete Σ-spaces are:

- if $P \in \mathcal{C}\mathcal{S}ig$, then there is a *canonical* partial order on P, under which P has suprema of chains;
- the canonical partial order on a function space $[P \to Q]$ or $[P \rightharpoonup Q]$ coincides with the pointwise order; and
- if $P, Q \in \mathcal{C}\mathcal{S}ig$, *every* map $P \to Q$ is monotone and preserves suprema of chains.

This canonical ordering has a simple definition motivated by recursion theory.

These results mean that we can regard a complete Σ-space, *qua* set, as a predo- nain, without having to impose a partial order on it; and domain-theoretic construc- ions, such as function spaces, coincide with the ordinary set-theoretic constructions. 'rom now on we will simply refer to complete Σ-spaces as **predomains**.

Equality types in ML have a semantic counterpart in our model. A predomai P is an **equality predomain** if

$$\forall p, p' \in P. (p = p') \vee (p \neq p')$$

is true in $\mathcal{E}\!f\!f$. This is equivalent to the existence of an equality function $P \times P \to$ It is not a vacuous definition because the logic of $\mathcal{E}\!f\!f$ is constructive. For example, N is an equality predomain; but $[\mathbb{N} \rightharpoonup \mathbb{N}]$ is not, since there is no effective procedu for deciding whether two partial recursive functions are equal.

From now on we write $\mathcal{S} = \mathcal{E}\!f\!f$ to emphasise that nothing which follows depend on the fact that we are working with the effective topos $\mathcal{E}\!f\!f$; it all works mor generally. See [16] for other examples of suitable toposes.

2.3 Semantics of the Core

There are two aspects of ML which we will not consider here. Firstly, we are no concerned with references and exceptions. Secondly, we do not deal with type re construction. So we are really working with XML, a pure, explicitly typed variar of SML: see [13].

The way we interpret Core ML without polymorphic values is standard. Follow ing Plotkin and [20], we do not use domains (with a bottom element) but predomair and partial function spaces. For example

$$[\![\texttt{nat}]\!] = \mathbb{N}, \qquad [\![\sigma \,\texttt{*}\, \tau]\!] = [\![\sigma]\!] \times [\![\tau]\!], \qquad [\![\sigma \,\texttt{->}\, \tau]\!] = [[\![\sigma]\!] \rightharpoonup [\![\tau]\!]]$$

Datatypes are interpreted by recursive domains—domain equations can be solved i the usual way. For example, $[\![\texttt{nat list}]\!]$ is the solution, in the category of pred mains, of the domain equation

$$X \cong 1 + (\mathbb{N} \times X)$$

In the case of 'algebraic' datatypes such as lists, for which it makes sense to solv this equation set-theoretically, it can be shown that the domain-theoretic solutio of this equation coincides with the set-theoretic solution. (This applies, not just t 'flat' predomains, but to predomains in general.) One consequence is that algebrai datatypes constructed from equality predomains are again equality predomains, a ML requires. Another consequence is that the naive unrestricted induction principle for algebraic datatypes are, by definition, sound.

Once we have defined the semantics of closed types using predomains, ML poly morphism can be added transparently: we use *polynomial categories* over $\mathcal{S}$, a in [17]. In this case they are just the classifying toposes to be defined in the nex section: a value whose type contains a type variable is interpreted in the topo $\mathcal{S}[X]$, not $\mathcal{S}$. For example, the polymorphic identity `fn x => x :  'a -> 'a` i interpreted by the arrow

$$\ulcorner 1_X \urcorner : 1 \to [X \rightharpoonup X] \in \mathcal{S}[X]$$

'Real' polymorphic types in $\mathcal{S}$ itself (though they do exist) are not required.

3 Classifying toposes

3.1 Basic definitions

The notion of classifying topos, and of a 'generic' or 'indeterminate' structure of a signature or model of a theory, goes back to the 1970s: see [6, chapter 6]. It can be seen as a special case of a general notion of classifying category. The classifying toposes we'll use are not as general as the ones described in [6]—they can be constructed more directly. However, the following discussion is still slightly oversimplified—for example, we gloss over all the 2-categorical issues that arise.

We will need a notion of morphism between toposes that "preserves all set-theoretic structure and all set-theoretic properties".

Definition 3.1 If $\mathcal{E}, \mathcal{F}$ are toposes, an ***a*-functor** $f^*: \mathcal{E} \to \mathcal{F}$ is a functor that preserves finite limits, exponentials and the subobject classifier, and has a right adjoint (and hence preserves arbitrary colimits). ∎

Suppose $\mathcal{E}, \mathcal{F}$ are constructed out of $\mathcal{S}$ in some way, so that they contain full subcategories of predomains in the same way that $\mathcal{S}$ does. Since constructions on predomains are described set-theoretically, a-functors must preserve all domain theoretic constructions too—such as partial function spaces and solutions of domain equations.

In fact, we will deal with toposes $\mathcal{E}$ that come equipped with an embedding $\mathcal{S} \hookrightarrow \mathcal{E}$ (itself an a-functor). They are constructed by "adding extra objects to $\mathcal{S}$". Such a topos is called an **$\mathcal{S}$-topos**; note that it will contain a full subcategory of predomains, just like $\mathcal{S}$ does. We restrict our attention to toposes equipped with such an embedding, and a-functors $f^*: \mathcal{E} \to \mathcal{F}$ such that

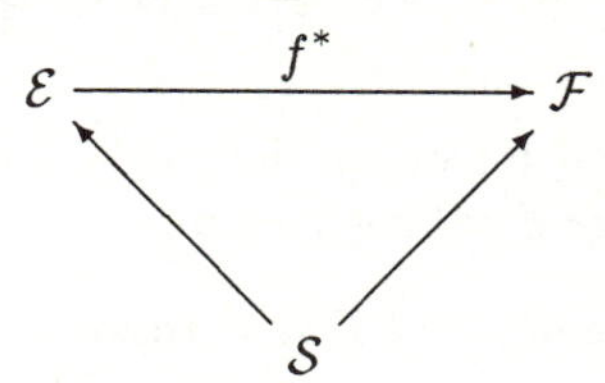

We will refer to f^* as an ***a*-functor over $\mathcal{S}$**. Let **$\mathcal{S}$-Top** denote the category of $\mathcal{S}$-toposes and a-functors over $\mathcal{S}$.

We also need to talk about $\mathcal{E}$-toposes and a-functors over $\mathcal{E}$ for $\mathcal{S}$-toposes $\mathcal{E}$; the definitions are just the same.

A typical way to construct an $\mathcal{E}$-topos is by adding an 'indeterminate' or 'generic' predomain X to $\mathcal{S}$ to obtain a new topos $\mathcal{S}[X]$. This will contain all the old objects of $\mathcal{S}$, but also new ones like

$$2 \times [X \rightharpoonup \mathbb{N}], \qquad [\mathbb{N} \rightharpoonup X\ \mathtt{list}], \ldots$$

Intuitively, X is a new predomain about which we know nothing. "Substituting a real predomain $P \in \mathcal{S}$ for X" will give us an a-functor $\mathcal{S}[X] \to \mathcal{S}$ taking

$$X \mapsto P, \qquad 2 \times [X \rightharpoonup \mathbb{N}] \mapsto 2 \times [P \rightharpoonup \mathbb{N}], \ldots$$

The precise definition of $\mathcal{S}[X]$ is categorical:

Definition 3.2 The **polynomial topos (for predomains) over** $\mathcal{S}$ is a topos $\mathcal{S}[X]$ such that $\mathcal{S} \hookrightarrow \mathcal{S}[X]$, containing a distinguished predomain $X \in \mathcal{S}[X]$, the **generic predomain**, and having the following universal property:

- For every $\mathcal{S}$-topos $\mathcal{E}$ and for every predomain $P \in \mathcal{E}$ there is a unique a-functor $\mathcal{S}[X] \to \mathcal{E}$ over $\mathcal{S}$ taking $X \mapsto P$.

$\mathcal{S}[X]$ is said to **classify** predomains in $\mathcal{S}$, and in $\mathcal{S}$-toposes. ∎

Note that we can adjoin more than one indeterminate predomain to $\mathcal{S}$, to obtain

$$\mathcal{S}[X_1, \ldots, X_n]$$

We can also add a generic equality predomain Y to $\mathcal{S}$ to obtain a polynomial topos $\mathcal{S}[Y]$; the definition is nearly the same, and is left to the reader. We don't discuss equality predomains in the rest of the paper, as they do not, in this paper, introduce any additional problems.

It also makes sense to add an indeterminate element x to a predomain P in $\mathcal{S}$ (or some other topos) to obtain a new topos $\mathcal{S}[x : P]$. Again, "substituting any real element $p \in P$ for x" defines an a-functor $\mathcal{S}[x : P] \to \mathcal{S}$ over $\mathcal{S}$ taking $x \mapsto p$.

Before giving the precise definition, note that if $\mathcal{E}$ is an $\mathcal{S}$-topos, any predomain $P \in \mathcal{S}$ can be regarded as a predomain in $\mathcal{E}$ via the embedding; we call it $P \in \mathcal{E}$ again. Similarly if $\mathcal{F}$ is an $\mathcal{E}$-topos and $P \in \mathcal{E}$ is a predomain, we have a predomain $P \in \mathcal{F}$.

Definition 3.3 Given a $\mathcal{S}$-topos $\mathcal{E}$ and a predomain $P \in \mathcal{E}$, the **polynomial topos (for elements of** P**) over** $\mathcal{E}$ is a topos $\mathcal{E}[x : P]$ such that $\mathcal{E} \hookrightarrow \mathcal{E}[X]$, containing a distinguished element $x: 1 \to P$, the **generic element of** P, and having the following universal property:

- For every $\mathcal{E}$-topos $\mathcal{F}$ and for every element $p: 1 \to P \in \mathcal{F}$, there is a unique a-functor $\mathcal{E}[X] \to \mathcal{F}$ over $\mathcal{E}$ taking $x \mapsto p$.

$\mathcal{E}[x : P]$ is said to **classify** elements of P in $\mathcal{E}$-toposes. ∎

We will sketch the constructions. However, they aren't needed in the rest of the paper—we only ever use the universal property of $\mathcal{S}[X]$ or $\mathcal{S}[x : P]$.

Proposition 3.4 *(Cf. [8, ex. 2, p. 211]) Given a predomain $P \in \mathcal{E}$, $\mathcal{E}[x : P]$ is the slice topos $\mathcal{E}/P$. The embedding $\mathcal{E} \hookrightarrow \mathcal{E}/P$ sends $Q \mapsto \left(\begin{smallmatrix} P\times Q \\ \downarrow \pi_0 \\ P \end{smallmatrix}\right)$, and x is the arrow $1 \cong \left(\begin{smallmatrix} P \\ \downarrow 1_P \\ P \end{smallmatrix}\right) \to \left(\begin{smallmatrix} P\times P \\ \downarrow \pi_0 \\ P \end{smallmatrix}\right)$ arising from the diagonal map $\delta: P \times P \to P$.*

Proposition 3.5 *(Cf. [17, section 4.2]) Recall from [15], [16] that the category of predomains in $\mathcal{S}$ is the* externalisation *(in the sense of [5]) of an internal full subcategory $\mathbb{C} = (C_1 \rightrightarrows C_0)$ in $\mathcal{S}$ arising from a map $f: G \to C_0$ in $\mathcal{S}$: see [6, example 2.38]. Then $\mathcal{S}[X]$ is $\mathcal{S}/C_0$, where $X = \left(\begin{smallmatrix} G \\ \downarrow f \\ C_0 \end{smallmatrix}\right) \in \mathcal{S}/C_0$.*

A similar construction gives the classifying topos $\mathcal{S}[Y]$ for equality predomains.

4 The semantics of the Modules system

4.1 Signatures without substructure components

The ML modules system allows programmers to define *structures*, which are collections of types, values and substructures. Each structure can match one or more *signatures* or 'interfaces:' lists of names for type components, names/types for value components, and signatures for substructures. There are also *functors*, which take structures of one signature to structures of another. The reader is referred to [9], [2] for more details.—From now on, the term 'a-functor' refers to an a-functor *over* $\mathcal{S}$.

Signatures will be interpreted using classifying toposes. We illustrate this via an example. Consider the signature

```
signature SIGA =
  sig
    type t;
    val f : t -> nat
  end;
```

To give a structure of this signature in our model, we must specify a predomain P and an element $f_0 \colon 1 \to [P \rightharpoonup \mathbb{N}]$ in $\mathcal{S}$. But this amounts to giving an a-functor out of a polynomial topos:

$$(\mathcal{S}[X])[x : [X \rightharpoonup \mathbb{N}]] \longrightarrow \mathcal{S}$$

We say that the topos

$$[\![\mathtt{SIGA}]\!] = (\mathcal{S}[X])[x : [X \rightharpoonup \mathbb{N}]]$$

classifies structures of signature SIGA. The particular 'structure'

$$\langle X, x \colon 1 \to [X \rightharpoonup \mathbb{N}]\rangle$$

in $[\![\mathtt{SIGA}]\!]$ is called the **generic** structure of signature SIGA.

Proposition 4.1 *The generic structure of signature SIGA has the following universal property: given any 'real' structure* s : SIGA *in* $\mathcal{S}$*, i.e. any pair* $\langle P \in \mathcal{S}, p \in [P \rightharpoonup \mathbb{N}]\rangle$*, there is a unique a-functor*

$$[\![\mathtt{s}]\!] \colon [\![\mathtt{SIGA}]\!] \longrightarrow \mathcal{S}$$

such that $[\![\mathtt{s}]\!](X) = P$ *and* $[\![\mathtt{s}]\!](x) = p$.

We have assumed that each structure is assigned its principal signature. Actually, signature matching can be modelled using 'instantiation' a-functors between appropriate classifying toposes. The result will generalise to signatures with substructure components. The proof uses only the universal properties of the classifying toposes, and is essentially straightforward.

Proposition 4.2 *If* SIG2 $\geq$ SIG1 *(see [12, 5.9]) then there is an a-functor*

$$\mathrm{inst}^{\mathtt{SIG1}}_{\mathtt{SIG2}} \colon [\![\mathtt{SIG1}]\!] \longrightarrow [\![\mathtt{SIG2}]\!]$$

Given a structure of signature SIG2 *(an a-functor* $[\![\mathtt{SIG2}]\!] \to \mathcal{S}$*), precomposing with* $\mathrm{inst}^{\mathtt{SIG1}}_{\mathtt{SIG2}}$ *yields the corresponding structure of signature* SIG1.

4.2 A general signature is a cocone of toposes

Recall from [2] that structures and signatures can be visualised as directed acylic graphs or **DAG**s. A signature with no substructure components corresponds to a DAG with just one node. More generally, the signatures

```
signature SIGB =
  sig
    structure ba1 : SIGA;
    structure ba2 : SIGA
  end;
```

```
signature SIGC =
  sig
    structure ca : SIGA;
    structure cb : SIGB
    sharing ca = cb.ba1
  end;
```

correspond to the following DAGs:

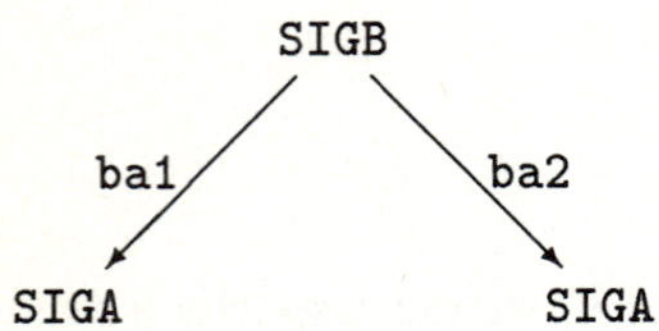

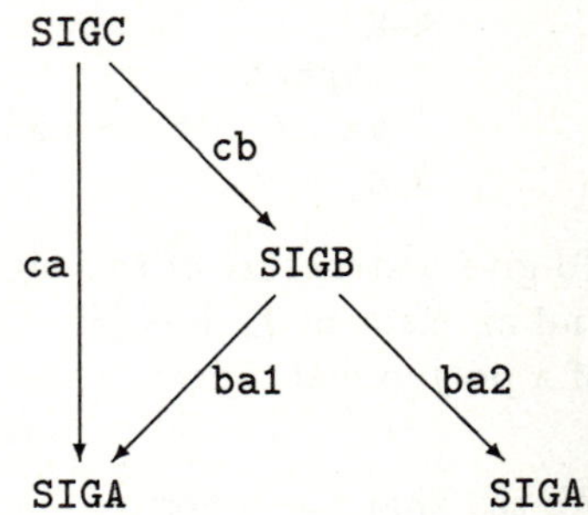

where we have labelled the vertices with the names of the appropriate signatures and the edges with the substructure component names. Note that these DAGs have a distinguished vertex, the **head**, labelled by the main signature. Any other vertex can be reached from the head. The terminal vertices correspond to signatures without substructure components.

[For simplicity, we assume in this paper that if a signature specifies substructure components, it specifies *only* substructure components, not type or value components. We also assume that sharing specifications involve only components of the same signature, so that no admissification (structure unification) is involved. Both assumptions can be dispensed with: the first, simply by spelling out how type and value identifiers are bound to objects and arrows in toposes; the second, by adding an extra vertex for each sharing specification, labelled by a signature computed from a principal admissifier.]

In our model, the interpretation of a signature SIG with DAG G will be a diagram of shape G, *with all the arrows reversed*, in the category $\mathcal{S}$-**Top**. This diagram must be a colimiting cocone, whose head is an $\mathcal{S}$-topos we call

$$[\![\texttt{SIG}]\!]$$

We construct this diagram inductively. For example, we already know what $[\![\texttt{SIGA}]\!]$ is; so we can simply let $[\![\texttt{SIGB}]\!] = [\![\texttt{SIGA}]\!] + [\![\texttt{SIGA}]\!]$. That is, we have a coproduct diagram

$$[\![\texttt{SIGA}]\!] \xrightarrow{[\![\texttt{ba1}]\!]} [\![\texttt{SIGB}]\!] \xleftarrow{[\![\texttt{ba2}]\!]} [\![\texttt{SIGA}]\!]$$

And given this, we can obtain ⟦SIGC⟧ by forming the finite colimit diagram

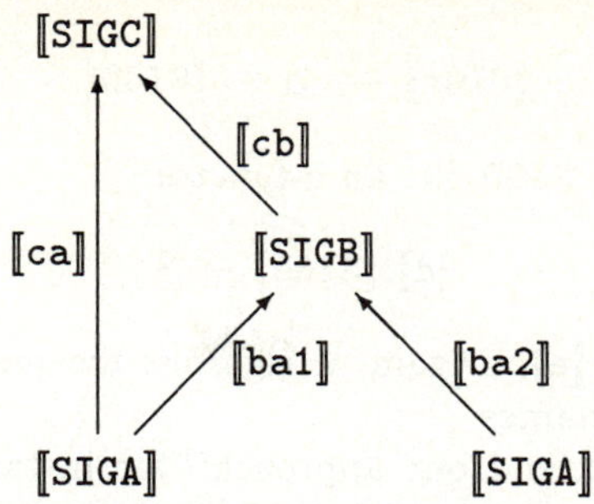

One can show that $\mathcal{S}$-**Top** has the finite colimits we require. (They are in fact easy to construct: all the toposes ⟦SIG⟧ turn out to be slice toposes of $\mathcal{S}$, so we simply use the finite limits in $\mathcal{S}$.)

A structure `c : SIGC` is interpreted as an a-functor

$$[\![\texttt{c}]\!]\colon [\![\texttt{SIGC}]\!] \to \mathcal{S}$$

Because ⟦SIGC⟧ is a colimit, to construct ⟦c⟧ it suffices to construct suitable a-functors ⟦SIGA⟧, ⟦SIGB⟧ $\to \mathcal{S}$, i.e. to interpret the components of c.

From ⟦c⟧ we can recover the components of c. For example,

$$[\![\texttt{c.ca}]\!] = [\![\texttt{c}]\!] \circ [\![\texttt{ca}]\!]\colon [\![\texttt{SIGA}]\!] \longrightarrow [\![\texttt{SIGC}]\!] \longrightarrow \mathcal{S}$$

The commutativity of the diagram enforces the sharing requirement

$$[\![\texttt{c.ca}]\!] = [\![\texttt{c}]\!] \circ [\![\texttt{ca}]\!] = [\![\texttt{c}]\!] \circ [\![\texttt{cb}]\!] \circ [\![\texttt{ba1}]\!] = [\![\texttt{c.cb.ba1}]\!]$$

Here is an example involving non-local sharing. Suppose we have defined a fixed structure `a : SIGA`, and consider the signature

```
signature SIGD =
  sig
    structure da1 : SIGA;
    structure da2 : SIGA
    sharing da1 = a
  end;
```

This signature corresponds to the DAG

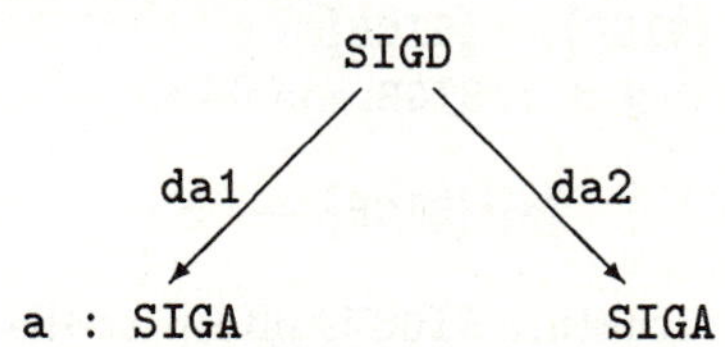

The topos ⟦SIGD⟧ appears in the colimit diagram

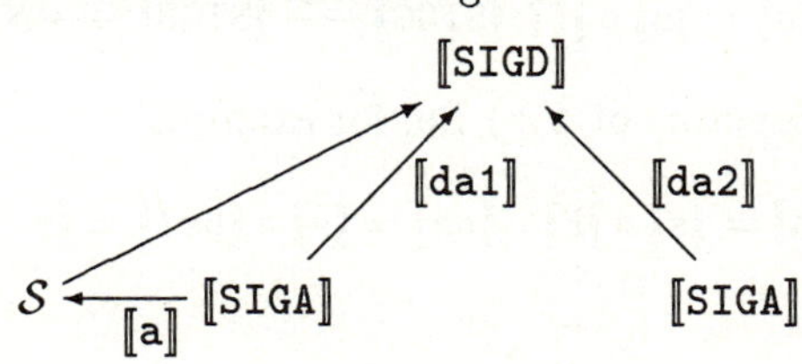

Note that since the diagram commutes, $[\![\mathtt{da1}]\!]\colon [\![\mathtt{SIGA}]\!] \to [\![\mathtt{SIGD}]\!]$ is the composite

$$[\![\mathtt{SIGA}]\!] \xrightarrow{[\![\mathtt{a}]\!]} \mathcal{S} \hookrightarrow [\![\mathtt{SIGD}]\!]$$

If we have a structure `d : SIGD`, i.e. an a-functor

$$[\![\mathtt{d}]\!]\colon [\![\mathtt{SIGD}]\!] \to \mathcal{S}$$

this forces $[\![\mathtt{d.da1}]\!]$ to be $[\![\mathtt{a}]\!]$. Again, a rigorous treament awaits a more careful analysis of the binding of names.

It seems possible to extend our approach: for instance, we hope to describe higher-order modules in our framework (note that the labelled DAGs we drew can be parametrised by objects in $\mathcal{S}$). However, other possible extensions of the modules system, such as recursive definitions, seem more problematic.

4.3 A functor is an a-functor in the wrong direction

Again we illustrate the idea using an example. Consider the functor

```
functor F(sb : SIGB) : SIGC =
  struct
    structure ca = sb.ba1;
    structure cb = sb
  end;
```

`F` will be modelled by an a-functor

$$[\![\mathtt{F}]\!]\colon [\![\mathtt{SIGC}]\!] \longrightarrow [\![\mathtt{SIGB}]\!]$$

We define $[\![\mathtt{F}]\!]$ by using the fact that $[\![\mathtt{SIGC}]\!]$ is the colimit of a cocone.

In more detail, the definition of `F` gives us a-functors

$$\mathrm{id}_{[\![\mathtt{SIGB}]\!]} \circ [\![\mathtt{ba1}]\!] = [\![\mathtt{ba1}]\!] \circ \mathrm{id}_{[\![\mathtt{SIGA}]\!]}\colon [\![\mathtt{SIGA}]\!] \longrightarrow [\![\mathtt{SIGB}]\!]$$

$$\mathrm{id}_{[\![\mathtt{SIGB}]\!]}\colon [\![\mathtt{SIGB}]\!] \longrightarrow [\![\mathtt{SIGB}]\!]$$

which are compatible and so give us a cocone with vertex $[\![\mathtt{SIGB}]\!]$; this in turn induces the required a-functor $[\![\mathtt{F}]\!]\colon [\![\mathtt{SIGC}]\!] \to [\![\mathtt{SIGB}]\!]$.

Now, if we have a structure `s : SIGB`, modelled by an a-functor

$$[\![\mathtt{s}]\!]\colon [\![\mathtt{SIGB}]\!] \longrightarrow \mathcal{S}$$

the structure `u = F(s)` of signature `SIGC` is interpreted as the composite

$$[\![\mathtt{u}]\!] = [\![\mathtt{s}]\!] \circ [\![\mathtt{F}]\!]\colon [\![\mathtt{SIGC}]\!] \longrightarrow [\![\mathtt{SIGB}]\!] \longrightarrow \mathcal{S}$$

(Cf. the instantiation functors of 4.2.) So, for example,

$$[\![\mathtt{u.ca}]\!] = [\![\mathtt{s}]\!] \circ [\![\mathtt{F}]\!] \circ [\![\mathtt{ca}]\!] = [\![\mathtt{s}]\!] \circ [\![\mathtt{ba1}]\!] = [\![\mathtt{s.ba1}]\!]$$

which is what we want.

4.4 Extended ML specifications are classifying toposes

It can be shown that $[\![\texttt{SIG}]\!]$ is always a *slice* topos of $\mathcal{S}$, i.e.

$$[\![\texttt{SIG}]\!] = \mathcal{S}/S$$

for some $S \in \mathcal{S}$. This S is (literally) the "set of structures of signature SIG in $\mathcal{S}$".

Suppose we adjoin some axioms to SIG to obtain an Extended ML signature SIG'. These axioms will define a subset (subobject)

$$S' \rightarrowtail S$$

the "set of structures of signature SIG in $\mathcal{S}$ satisyfing the axioms". Hence we get a new slice topos

$$[\![\texttt{SIG'}]\!] = \mathcal{S}/S' \hookleftarrow \mathcal{S}/S = [\![\texttt{SIG}]\!]$$

and it is not hard to show that this is the classifying topos for structures of signature SIG', in an obvious sense: it contains the generic structure of signature SIG that satisfies the axioms, and, since a-functors preserve the truth of all set-theoretic assertions, any instantiation of this structure via an a-functor $[\![\texttt{SIG'}]\!] \to \mathcal{S}$ will also satisfy the axioms.

When extended to specifications, the polynomial category approach to ML polymorphism gives us a way of defining polymorphic predicates and formulate suitable polymorphic axioms with implicit type quantifiers. We simply work within the logic of the classifying topos, which is set-theoretic. This seems to be a useful way of interpreting, and reasoning about, polymorphic specifications—an aspect of Extended ML that has, up to now, seemed problematic.

5 Some final comments

This paper is a prospectus, not a report on work already completed. But the approach already sheds some light on the 'meaning' of Modules in SML and Extended ML, while retaining a pleasantly close relationship with the official Definition.

Our work does not use the full power of synthetic domain theory. We didn't need 'real' polymorphic types, upper, lower and convex powerdomains, or subtyping and bounded quantification. But the fact that these constructions exist in $\mathcal{E}\!f\!f$ means that our approach applies to languages more sophisticated than ML.

We have tried to show that ML is essentially a simple language, by interpreting it in an intuitively naive framework. We believe that many of the tools previously used to describe aspects of ML are too powerful: ML polymorphism is simpler than System F, and using type theory to model ML Modules is overkill. Both can be captured in a much more elementary way. It is also nice that the use of polynomial categories in both cases reflects the analogy between the *operational* treatments of ML polymorphism and ML Modules, emphasised in [2].

One attractive feature of the strategy sketched here is that it is modular. The interpretations of closed types, of first-order polymorphism, and of the Modules system can be kept quite separate. In fact, the last two make sense in a more general setting, namely that of *categories-with-structure*. However, in this case, classifying categories are constructed differently: cf. [17].

References

[1] P. Freyd, P. Mulry, G. Rosolini, D. S. Scott: Extensional PERs, in: Proc. of 5th Annual Symposium on Logic in Computer Science, 1990.

[2] R. Harper, R. Milner, M. Tofte: A type discipline for program modules, in: Proc. of TAPSOFT '87, Springer LNCS 250, 1987.

[3] R. Harper, J. Mitchell, E. Moggi: Higher-order modules and the phase distinction, in: Proc. 17th ACM Symposium on Principles of Programming Languages, 1990.

[4] J. M. E. Hyland: The effective topos, in: *The L. E. J. Brouwer Centenary Symposium* (ed. A. S. Troelstra, D. van Dalen), North-Holland, 1982.

[5] J. M. E. Hyland, E. P. Robinson, G. Rosolini: The discrete objects in the effective topos, *Proc. Lond. Math. Soc.* (3) 60 (1990) 1–36.

[6] P. T. Johnstone: *Topos Theory*, Academic Press, 1977.

[7] E. Kazmierczak: An institution for Extended ML, talk at the 9th Workshop on Abstract Data Types; report, in preparation.

[8] J. Lambek, P. J. Scott: *Introduction to Higher-Order Categorical Logic*, Cambridge University Press, 1986.

[9] D. MacQueen: Modules for Standard ML, in: *Standard ML* (R. Harper, D. MacQueen, R. Milner), available as ECS-LFCS-86-2, Laboratory for the Foundations of Computer Science, University of Edinburgh.

[10] D. MacQueen: Using dependent types to express modular structure, in: Proc. 13th ACM Symposium on Principles of Programming Languages, 1990.

[11] R. Milner, M. Tofte: *Commentary on Standard ML*, MIT Press, 1990.

[12] R. Milner, M. Tofte, R. Harper: *The Definition of Standard ML*, MIT Press, 1990.

[13] J. Mitchell, R. Harper: The essence of ML, in: Proc. 15th ACM Symposium on Principles of Programming Languages, 1988.

[14] J. Mitchell, G. D. Plotkin: Abstract data types have existential types, *ACM Trans. Programming Languages and Systems,* 10(3):470–502.

[15] W. K.-S. Phoa: Effective domains and intrinsic structure, in: Proc. of 5th Annual Symposium on Logic in Computer Science, 1990.

[16] W. K.-S. Phoa: *Domain Theory in Realizability Toposes*, dissertation, University of Cambridge, November 1990; available as CST-82-91/ECS-LFCS-91-171, Department of Computer Science, University of Edinburgh.

[17] W. K.-S. Phoa: A simple categorical model of first-order polymorphism, submitted.

[18] G. Rosolini: Categories and effective computations, in: Lecture Notes in Computer Science 283, Springer, 1987.

[19] D. T. Sannella, A. Tarlecki: Extended ML, an institution-independent framework for formal program development, in: *Proc. of Category Theory and Computer Programming (Guildford)*, Springer LNCS 240, 1987.

[20] R. D. Tennent, *Semantics of Programming Languages*, Prentice-Hall, 1991.

[21] M. Tofte: *Operational Semantics and Polymorphic Type Inference*, dissertation, 1988; available as CST-52-88/ECS-LFCS-88-54, Department of Computer Science, University of Edinburgh.

What Good Are Digital Clocks?*

Thomas A. Henzinger[†] Zohar Manna[‡] Amir Pnueli[§]

Abstract. Real-time systems operate in "real," continuous time and state changes may occur at any real-numbered time point. Yet many verification methods are based on the assumption that states are observed at integer time points only. What can we conclude if a real-time system has been shown "correct" for integral observations?

Integer time verification techniques suffice if the problem of whether all real-numbered behaviors of a system satisfy a property can be reduced to the question of whether the integral observations satisfy a (possibly modified) property. We show that this reduction is possible for a large and important class of systems and properties: the class of systems includes all systems that can be modeled as timed transition systems; the class of properties includes time-bounded invariance and time-bounded response.

Introduction

'er the past few years, we have seen a proliferation of formal methodologies for soft-re and hardware design that emphasize the treatment of real-time requirements. Any mal approach to real-time systems typically

- assumes C, a mathematical *model of computation*;
- assumes T, a mathematical *model of time*;
- uses a formal language $\mathcal{L}_I$, the *implementation language*, for describing systems and their behavior over time;
- uses a formal language $\mathcal{L}_S$, the *specification language*, for describing timing requirements of systems;
- presents algorithms and/or proof rules that facilitate a formal argument that a particular system meets a particular specification under the assumed semantics of computation and time.

---

*A full version of this paper (including all proofs) is available as a technical report from Cornell Uni-'sity and Stanford University. The research was supported in part by the National Science Foundation der grants CCR-89-11512 and CCR-89-13641, by the Defense Advanced Research Projects Agency der contract NAG2-703, by the United States Air Force Office of Scientific Research under con-ct AFOSR-90-0057, and by the European Community ESPRIT Basic Research Action Project 3096 ᐳEC).

†Department of Computer Science, Cornell University, Ithaca, NY 14853, U.S.A.

‡Department of Computer Science, Stanford University, Stanford, CA 94305, U.S.A.

§Department of Applied Mathematics, The Weizmann Institute of Science, Rehovot, Israel.

The central question "Does the system S, given as an expression of $\mathcal{L}_I$, meet t specification ϕ, given as an expression of $\mathcal{L}_S$, with respect to the semantical assum tion (C, T)?" is called the *real-time verification problem* for the parameters $(C, T, \mathcal{L}_I, \mathcal{L}$ and written as

$$S \overset{?}{\models}_{(C,T)} \phi.$$

While several instances of the real-time verification problem have been studied exte sively [10], comparatively little effort has been invested in (1) justifying the adequa of various semantical assumptions and (2) exploring connections between different a proaches. Both questions are intimately linked with each other. For example, t justification of an integer model of time must relate the model to real-numbered tim

We demonstrate that the comparative analysis of models can be used to justi timing assumptions and to broaden the scope of applicability of verification metho that have been developed for restrictive semantical assumptions. In particular, contrast two important instances of the real-time verification problem — one with *integral* model of time (let T be the integers Z) and one with a *dense* model of time (T be the reals R). To avoid distraction from the issues that concern time, we conside in both cases, a simple trace semantics C of computation.

Our results are twofold. First, we define a sufficient criterion, which we call *digit ability*, for the real-time verification problem to be invariant under the time model:

$$S \models_{\mathsf{R}} \phi \quad \text{iff} \quad S \models_{\mathsf{Z}} \phi$$

for digitizable systems S and digitizable specifications ϕ. Roughly speaking, a set dense-time traces is digitizable iff it contains a trace ρ precisely when it contains integral-time traces that result from observing ρ with fictitious digital clocks. Bo general-purpose real-time programming languages $\mathcal{L}_I$ and real-time specification la guages $\mathcal{L}_S$ all of whose expressions are digitizable have been proposed in the literatu We show that (1) every timed transition system [15] is digitizable and (2) every *qua tative* (time-independent) property and the most common *quantitative* (hard real-tim properties are digitizable. Timed transition systems can model sets of real-time pr cesses that communicate through a shared memory or by message passing in multipr cessing and multiprogramming environments [16]. The class of digitizable properti includes all time-bounded invariance and time-bounded response requirements.

Second, we show that even nondigitizable dense-time properties may benefit fro integer time verification. For any property ϕ that is definable in metric temporal logic [19], we construct a (possibly somewhat weaker) property ϕ^W such that $S \models_{\mathsf{Z}} \phi$ impli $S \models_{\mathsf{R}} \phi^W$ for every timed transition system S. From this result we derive a scheme f approximating dense-time properties conservatively by integral-time properties.[1]

It follows that we can use verification techniques that have been developed for integral-time model to prove (or approximate) properties of real-time systems under dense-time model. This observation is of practical interest for two reasons:

- The dense-time model does not only correspond more closely to the physical reali than the integral-time model, but it is essential for composing systems [1, 6].

[1] Recently it has been brought to the authors' attention that a conservative approximation of re numbered time by integer time was proposed independently by [8].

- Integral-time verification methods are not only significantly simpler than dense-time techniques, but apply to a wider range of real-time properties (because many instances of the dense-time verification problem are undecidable) [2, 4].

addition, a wide variety of integral-time verification methods is readily available the verification of timed transition systems (including model-checking algorithms 4, 12, 25] and proof systems [13, 15, 25]). Our results prescribe the sound application these techniques for proving how systems behave in real-numbered time.

The paper is organized as follows. In Section 2, we introduce both integral-time and ıse-time models for real-time systems. We define the notion of digitizability and, for ;itizable systems and specifications, we reduce the dense-time verification problem to integral-time problem. In Section 3, we briefly review timed transition systems and tric temporal logic. We show that all timed transition systems and an important ss of metric formulas are digitizable. We then proceed to determine which dense-time operty of a timed transition system is established whenever a metric specification is ecked under the assumption of integer time.

Semantics: Discrete Trace Models

is paper studies discrete trace models for real-time systems: *trace* models formalize a stem as the set of its possible behaviors; *discrete* trace models formalize the behavior a system as an infinite sequence of snapshots of the global system state at certain nes. Discrete trace models have obvious limitations:

ace (or *linear-time*) models abstract information about internal decisions of a system. More refined (*branching-time*) models are necessary to study the internal branching structure of a system [5].

screte trace models require the number of snapshots within a behavior to be countable. It follows that the state of a system cannot be observed at every real-numbered point in time. This restriction is adequate for modeling *discrete* processes, which change their state only finitely often between any two points in time. While the behavior of a discrete process can be described completely by an ω-sequence of state changes, more refined models are necessary to study *continuous* processes [21].

spite (or, often, because of) these limitations, discrete trace models have been used ccessfully for the analysis of a wide variety of real-time systems and phenomena that se in practice [1, 3, 15, 17, 20, 23, 25, 27].

Discrete trace models differ significantly in the kinds of systems and phenomena they n capture adequately [5]. These differences are caused by different models of time. e distinguish between *analog-clock* models, which record the precise real-numbered ne of every snapshot (using a fictitious "analog clock"), and *digital-clock* models, ıich approximate the time of every snapshot by an integer (using a fictitious "digital ock").

1 Timed state sequences

rmally, discrete trace models define the behavior of a system as an infinite sequence observations. An *observation* is a pair that consists of a state showing the result of a snapshot and a time-stamp recording the time of the snapshot. An infinite sequer

$$\rho: \quad (\sigma_0, \mathsf{T}_0) \longrightarrow (\sigma_1, \mathsf{T}_1) \longrightarrow (\sigma_2, \mathsf{T}_2) \longrightarrow (\sigma_3, \mathsf{T}_3) \longrightarrow \cdots$$

of observations is called a *timed state sequence* $\rho = (\sigma, \mathsf{T})$; it consists of an infinite quence σ of states σ_i, for $i \geq 0$, and an infinite sequence T of corresponding time-stan $\mathsf{T}_i \in$ *TIME*. At this point, we do not commit to any particular time domain *TIM* we only assume that there is a total ordering $<$ on *TIME* and demand that

Monotonicity (time does not decrease): $\mathsf{T}_i \leq \mathsf{T}_{i+1}$ for all $i \geq 0$. This conditi ensures that the logical order of snapshots is consistent with the temporal ord while permitting adjacent observations with the same time-stamp.[2]

Progress (time progresses): for all $\mathsf{T} \in$ *TIME*, there is some $i \geq 0$ such that $\mathsf{T}_i \geq$ This condition ensures that the time-stamps of observations do not conver thus permitting only finitely many snapshots between any two points in tir This restriction is adequate for modeling discrete processes.

We will consider two types of fictitious clocks. *Analog-clock* (or *dense-time*) mod are discrete trace models that take *TIME* to be the real numbers R; *digital-clock integral-time*) models take *TIME* to be the integers Z.[3] We call a timed state sequer of an analog-clock model *precisely timed*; of a digital-clock model, *digitally timed*. set of precisely timed state sequences is called a *dense-time property*; a set of digita timed state sequences, an *integral-time property*.

We routinely use the following operations on timed state sequences. For any tin state sequence $\rho = (\sigma, \mathsf{T})$, let ρ^- be its state component σ; applied to a set Π timed state sequences, this "untime" operation yields the corresponding set Π^- of st components. By $\rho^i = (\sigma^i, \mathsf{T}^i)$, for $i \geq 0$, we denote the timed state sequence tl results from the timed state sequence $\rho = (\sigma, \mathsf{T})$ by deleting the first i observations particular, $\rho^0 = \rho$). We write TSS_{TIME} for the set of timed state sequences over t time domain *TIME*.

2.2 Dense time: Analog-clock models

Every real-time system S defines a dense-time property, namely $[\![S]\!]_{\mathsf{R}}$, the set of possil behaviors of S in real-numbered time. Similarly, every real-time specification ϕ defi a dense-time property, namely $[\![\phi]\!]_{\mathsf{R}}$, the set of real-time traces that meet the spec cation ϕ. The real-time system S satisfies the specification ϕ (denoted by $S \models_{\mathsf{R}} \phi$) $[\![S]\!]_{\mathsf{R}} \subseteq [\![\phi]\!]_{\mathsf{R}}$. Checking containment of dense-time properties is difficult, often unc cidable, for many implementation and specification languages [1, 3, 4]. In comparis containment of integral-time properties can be tested for transition-system based, a tomata based, and temporal-logic based languages [3, 4, 5, 12]. This is why ma researchers have sacrificed the density of time and given their languages digital-clc interpretations [11, 14, 17, 24, 25, 27].

---

[2] The possibility of several contemporaneous, yet ordered, snapshots allows us to model, if desir simultaneous parallel transitions of a system by nondeterministic interleaving [5].

[3] A third, extremely imprecise, type of fictitious clock results in *untimed* models. If time is imr terial, we can take *TIME* to be any trivial one-element domain. In this case, we may identify a tin state sequence with its state component.

We follow the same path, but rather than being content with solving the simpler egral-time verification problem, we would like to employ integer time techniques to ve the original dense-time verification problem (or some approximation thereof). Our n is to interpret real-time systems and real-time specifications with the analog-clock nantics $\Pi_{\mathsf{R}} \subseteq TSS_{\mathsf{R}}$ as defining an integral-time property $\Pi_{\mathsf{Z}} \subseteq TSS_{\mathsf{Z}}$ such that

nenability: The integral-time verification problem (denoted by $S \models_{\mathsf{Z}} \phi$), namely if $[\![S]\!]_{\mathsf{Z}} \subseteq [\![\phi]\!]_{\mathsf{Z}}$, can be solved.

lequacy: The answer to the integral-time verification problem gives some insight about the dense-time verification problem; ideally, $S \models_{\mathsf{Z}} \phi$ iff $S \models_{\mathsf{R}} \phi$.

.e remainder of this paper presents a digital-clock interpretation of nontrivial im-mentation and specification languages that satisfies both goals of amenability and equacy.

3 Integer time: Digital-clock models

.ere are two deliberations we must take into account when defining an integral-time nantics for real-time systems. First, we want the interpretation of real-time languages be independent of the clock model: if an expression denotes the set of timed state quences that share a certain model-independent characteristics, then it ought to do independently of the time domain. For example, a specification asserting in an alog-clock model the bounded-response property that "every request p is followed by esponse q within 5 time units" should have the same meaning in the corresponding gital-clock model. So if an expression (of an implementation or specification language) fines in the analog-clock model the set Π_{R} of precisely timed state sequences, then *clock-independent integral-time semantics*

$$\mathsf{Z}(\Pi_{\mathsf{R}}) = \Pi_{\mathsf{R}} \cap TSS_{\mathsf{Z}}$$

the subset of digitally timed state sequences in Π_{R}.

Second, in order for a digital-clock model to be adequate, the integral-time se-antics $[\![S]\!]_{\mathsf{Z}}$ of a system S must bear some resemblance to its dense-time seman-:s $[\![S]\!]_{\mathsf{R}}$. At the least, we want both $[\![S]\!]_{\mathsf{R}}$ and $[\![S]\!]_{\mathsf{Z}}$ to satisfy the same qualitative operties, such as the response property that "every request p is followed by a re-onse q." Unfortunately, this is not the case for all real-time systems S and their ock-independent integral-time semantics $[\![S]\!]_{\mathsf{Z}} = \mathsf{Z}([\![S]\!]_{\mathsf{R}})$. Consider, for example, a stem S that changes its state at the times $i + \delta_i$, for $i \in \mathsf{N}$ and a sequence of reals $< \delta_i < 1$ that converges toward 0. Every behavior in $[\![S]\!]_{\mathsf{R}}$ must contain a nonintegral ne-stamp (for any choice of time unit). Consequently, $\mathsf{Z}([\![S]\!]_{\mathsf{R}}) = \emptyset$, which shows at the clock-independent integral-time semantics is a very poor representation for S.) preserve all qualitative properties, we cannot simply "throw away" behaviors with ate changes between integral times, rather we must collect these state changes and cord them at integral times. For this purpose, we introduce the notion of digitizing a nse-time behavior.

igitization

ıppose we observe the behavior of a discrete process with a digital clock. The step-dth of the clock (i.e., the time that elapses between successive clock ticks) is determined by the precision we are interested in; from now on, we normalize the stepwid to 1. For example, if the system behavior

$$\rho: \quad (p, 0.2) \longrightarrow (q, 5.8) \longrightarrow (p, 5.9) \longrightarrow (q, 8) \longrightarrow (p, 8.4) \longrightarrow (q, 8.6) \longrightarrow \cdots$$

is observed by a clock that ticks at all integer times (and shows the integral "clc time" n throughout the real-time interval $[n, n+1)$ for all $n \in \mathsf{Z}$), we obtain t observation sequence

$$[\rho]_0: \quad (p, 0) \longrightarrow (q, 5) \longrightarrow (p, 5) \longrightarrow (q, 8) \longrightarrow (p, 8) \longrightarrow (q, 8) \longrightarrow \cdots.$$

The introduction of a fictitious digital clock suggests an integral-time semantics th identifies two systems iff they cannot be distinguished by the clock. While this integr time semantics preserves all qualitative properties, such as response (since $[\rho]_0^- = \rho^-$ it does not preserve time-bounded response. To see this, consider the two timed st sequences ρ and $[\rho]_0$ given above; while every state p of the digitally timed sequence $[\rho$ is followed by a state q within 5 time units, this is not the case for the precisely tim sequence ρ. It follows that integer time verification of bounded response would not sound.

The suggested digital-clock semantics is not discriminative enough, because we ha fixed the first tick of our digital clock, rather arbitrarily, at dense time 0. If our clo ticks, for example, at times $0.5 + n$ for all integers n, then the bounded-response pro erty in question cannot distinguish between the system behavior ρ and the resulti observation sequence

$$[\rho]_{0.5}: \quad (p, 0) \longrightarrow (q, 6) \longrightarrow (p, 6) \longrightarrow (q, 8) \longrightarrow (p, 8) \longrightarrow (q, 9) \longrightarrow \cdots$$

(neither ρ nor $[\rho]_{0.5}$ satisfy the property that every p is followed by a q within 5 ti units). So let us suppose we have infinitely many different fictitious clocks available f observation — one for every real number $0 \leq \epsilon < 1$. We will henceforth identify digi clocks with reals in the interval $[0, 1)$, assuming that the clock ϵ ticks at all times $\epsilon +$ for $n \in \mathsf{Z}$. For any precisely timed state sequence $\rho = (\sigma, \mathsf{T})$ and digital clock $0 \leq \epsilon <$ let the *ϵ-digitization* $[\rho]_\epsilon = (\sigma, [\mathsf{T}]_\epsilon)$ of ρ be the digitally timed state sequence th results from recording the time-stamps in ρ with the digital clock ϵ:

$$[\rho]_\epsilon: \quad (\sigma_0, [\mathsf{T}_0]_\epsilon) \longrightarrow (\sigma_1, [\mathsf{T}_1]_\epsilon) \longrightarrow (\sigma_2, [\mathsf{T}_2]_\epsilon) \longrightarrow (\sigma_3, [\mathsf{T}_3]_\epsilon) \longrightarrow \cdots,$$

where $[x]_\epsilon = \lfloor x \rfloor$ if $x \leq \lfloor x \rfloor + \epsilon$, otherwise $[x]_\epsilon = \lceil x \rceil$. In words, the ϵ-digitization $[\rho$ is obtained from ρ by rounding, with respect to ϵ, all time-stamps to integers. F any dense-time property Π, the *digitization* $[\Pi]$ of Π is the integral-time property th contains all ϵ-digitizations of sequences in Π:

$$[\Pi] \;=\; \{[\rho]_\epsilon \mid \rho \in \Pi \text{ and } 0 \leq \epsilon < 1\};$$

$[\{\rho\}]$ is abbreviated to $[\rho]$.

Now we can define an alternative integral-time semantics that identifies two syster iff they cannot be distinguished by *any* digital clock: the digitization $[[\![S]\!]_{\mathsf{R}}]$ is the *di itized integral-time semantics* of a real-time system S. We will prove that the digitiz integral-time semantics preserves all qualitative properties as well as many importa quantitative timing properties, including bounded response, for the real-time syster at are admissible in the following sense. Since we want both clock independence *and* equacy, we must restrict ourselves to systems S whose clock-independent integral-ne semantics $\mathsf{Z}([\![S]\!]_{\mathsf{R}})$ and digitized integral-time semantics $[[\![S]\!]_{\mathsf{R}}]$ coincide; that is, are limited to real-time systems whose dense-time behaviors Π satisfy $\mathsf{Z}(\Pi) = [\Pi]$. e call these dense-time properties Π *closed under digitization*, because they are pre-ely the dense-time properties for which $[\Pi] \subseteq \Pi$ (i.e., $\rho \in \Pi$ implies $[\rho] \subseteq \Pi$). We will monstrate that the restriction to systems whose possible behaviors are closed under ;itization is not severe.

verse digitization

rther restrictions on the class of considered real-time systems can guarantee that digital-clock model does not identify any two systems with distinct dense-time haviors. We say that a dense-time property Π is *closed under inverse digitization* iff $\subseteq \Pi$ implies $\rho \in \Pi$ for all $\rho \in TSS_{\mathsf{R}}$. A dense-time property Π is defined to be *gitizable* iff it is closed under both digitization and inverse digitization. Consequently, is digitizable precisely when

$$\rho \in \Pi \quad \text{iff} \quad [\rho] \subseteq \Pi$$

all $\rho \in TSS_{\mathsf{R}}$. Digitization is a bijection between the set of digitizable properties d the set of integral-time properties. Hence our integral-time semantics is one-to-one systems with digitizable dense-time behaviors.[4]

4 Semantic invariance of the verification problem

is not difficult to check that for systems and specifications defining digitizable proper-s, the dense-time verification problem can be reduced to an integral-time verification oblem. In fact, we have a stronger result:

heorem 1 (Dense-time versus integral-time verification) *If the dense-time operty Π is closed under digitization and the dense-time property Ψ is closed under verse digitization, then $\Pi \subseteq \Psi$ iff $\mathsf{Z}(\Pi) \subseteq \mathsf{Z}(\Psi)$.*

For instance, assume that we are given a real-time system S whose dense-time mantics $[\![S]\!]_{\mathsf{R}}$ is digitizable, and a specification ϕ whose dense-time semantics $[\![\phi]\!]_{\mathsf{R}}$ closed under inverse digitization. Then we may check if $S \models_{\mathsf{R}} \phi$ by attempting to ove, instead, that $S \models_{\mathsf{Z}} \phi$, which can be done using verification methods that have en developed for integer time. Either semantic techniques, such as model-checking gorithms, or syntactic techniques, using axioms and proof rules, can be employed to lve the integer time verification problem. Both soundness and completeness results r any integer time method carry over to the corresponding verification problem about al-numbered time.[5]

The following two propositions are helpful for showing that certain dense-time prop-ties are closed under (inverse) digitization. We say that a dense-time property Π is

[4] An alternative restriction that allows the inversion of digitization is provided by the more restricted ass of *synchronous* systems, which change states only at integer times [7, 9].

[5] In fact, the premises of Theorem 1 are needed for *soundness* only; *completeness* (i.e., $\Pi \subseteq \Psi$ plies $\mathsf{Z}(\Pi) \subseteq \mathsf{Z}(\Psi)$) holds for arbitrary dense-time properties Π and Ψ.

qualitative iff $\rho \in \Pi$ implies $\rho' \in \Pi$ for all precisely timed state sequences ρ and ρ' wi identical state components (i.e., $\rho^- = \rho'^-$).

Proposition 1 (Qualitative properties) *Every qualitative property is digitizable.*

Proposition 2 (Boolean operations) *Closure under digitization is preserved by a bitrary unions and intersections of dense-time properties. Closure under inverse digi zation is preserved by arbitrary intersections only. If a dense-time property Π is clos under digitization, then its complement $TSS_{\mathsf{R}} - \Pi$ is closed under inverse digitizatio*

3 Syntax: Transition Systems, Temporal Logic

Let us now turn to a particular implementation language — timed transition systen — and a particular specification language — metric temporal logic. Both languag have been shown to allow a crisp description of a wide variety of real-time systen and timing requirements that are encountered in practice (see [16] for timed transitic systems and [19] for metric temporal logic). Timed transition systems were introduce originally, with an integral-time semantics [15]; metric temporal logic is known to l decidable under an integral-time interpretation and undecidable under a dense-tin interpretation [4]. We use, nonetheless, both languages to define *dense-time* propertie by interpreting them over precisely timed state sequences.

3.1 Timed transition systems

Timed transition systems generalize the conventional model of discrete transition sy tems [18, 26] by imposing real-time constraints on the transitions: with each transitic we associate a minimal and a maximal delay. We only summarize the formal definitic of timed transition systems in this paper; for a more verbose introduction to time transition systems and their applications consult [15, 16].

A *transition system* $S = \langle \Sigma, \Theta, \mathcal{T} \rangle$ consists of three components:

1. a (possibly infinite) set Σ of *states*;
2. a subset $\Theta \subseteq \Sigma$ of *initial states*;
3. a finite set $\mathcal{T}$ of *transitions*. Each transition $\tau \in \mathcal{T}$ is a binary relation on Σ; th is, it defines for every state $\sigma \in \Sigma$ a (possibly empty) set of τ-successors $\tau(\sigma) \subseteq \Sigma$ We say that the transition τ is *enabled* on a state σ iff $\tau(\sigma) \neq \emptyset$.

An infinite sequence σ of states is a *computation* (execution sequence) of the transitio system $S = \langle \Sigma, \Theta, \mathcal{T} \rangle$ iff it satisfies the two requirements of

Initiality: $\sigma_0 \in \Theta$.

Consecution: For all $i \geq 0$ there is a transition $\tau \in \mathcal{T}$ such that $\sigma_{i+1} \in \tau(\sigma_i)$. We sa that τ is *taken* at position i and *completed* at position $i + 1$.

Let $[\![S]\!]$ be the set of computations of S.

We incorporate time into the transition system model by assuming that all trans tions happen "instantaneously," while real-time constraints restrict the times at whic nsitions occur. These timing constraints are classified into two categories: *lower-und* and *upper-bound* requirements. They ensure that transitions occur neither too ly nor too late, respectively. All of our time bounds are natural numbers N; for ational convenience, we assume that $\infty \geq n$ for all $n \in \mathsf{N}$.

A *timed transition system* $S = \langle \Sigma, \Theta, \mathcal{T}, l, u \rangle$ consists of an underlying transition tem $S^- = \langle \Sigma, \Theta, \mathcal{T} \rangle$ as well as

4. a *minimal delay* $l_\tau \in \mathsf{N}$ for every transition $\tau \in \mathcal{T}$;

5. a *maximal delay* $u_\tau \in \mathsf{N} \cup \{\infty\}$ for every transition $\tau \in \mathcal{T}$. We require that $u_\tau \geq l_\tau$ for all $\tau \in \mathcal{T}$.

timed state sequence $\rho = (\sigma, \mathsf{T})$ is a *computation* of the timed transition system $= \langle \Sigma, \Theta, \mathcal{T}, l, u \rangle$ iff the state sequence σ is a computation of the underlying transition tem S^-, and

wer bound: For every transition $\tau \in \mathcal{T}$ and all positions $i \geq 0$ and $j > i$ such that $\mathsf{T}_j < \mathsf{T}_{i+1} + l_\tau$, if τ is taken at position j, then τ is enabled on σ_i. In other words, once enabled, τ is delayed for at least l_τ time units; it can be taken only after being continuously enabled for l_τ time units.

per bound: For every transition $\tau \in \mathcal{T}$ and position $i > 0$ such that τ is enabled on σ_i, there is some position $j > i$ with $\mathsf{T}_j \leq \mathsf{T}_{i-1} + u_\tau$ such that either τ is completed at position j, or τ is not enabled on σ_j. In other words, once enabled, τ is delayed for at most u_τ time units; it cannot be continuously enabled for more than u_τ time units without being completed.[6]

e timed transition system S defines the set $[\![S]\!]_{TIME} \subseteq TSS_{TIME}$ of computations S over the time domain $TIME$. Thus, S defines in the analog-clock model the dense-ne property $[\![S]\!]_{\mathsf{R}}$; in the digital-clock model, S defines the integral-time property $]_{\mathsf{Z}} = \mathsf{Z}([\![S]\!]_{\mathsf{R}})$. The timing constraints of S can be viewed as filters that prohibit tain execution sequences of the underlying untimed transition system S^-, because $]_{TIME} \subseteq [\![S^-]\!]$. Special cases are a minimal delay 0 and a maximal delay ∞ for a nsition τ. While the former does not rule out any computations of S^-, the latter ls to S^- a *weak-fairness* (justice) assumption [22]: τ cannot be continuously enabled hout being taken.

eorem 2 (Timed transition systems) *The set of dense-time computations of a ed transition system is digitizable.*

Proof. (1) To see that the set of computations of a timed transition system is closed der digitization, observe that $\mathsf{T}_j \geq \mathsf{T}_i + l$ implies $[\mathsf{T}_j]_\epsilon \geq [\mathsf{T}_i]_\epsilon + l$ and $\mathsf{T}_j \leq \mathsf{T}_i + u$ plies $[\mathsf{T}_j]_\epsilon \leq [\mathsf{T}_i]_\epsilon + u$ for all $\mathsf{T}_i, \mathsf{T}_j \in \mathsf{R}$, $l, u \in \mathsf{N} \cup \{\infty\}$, and $0 \leq \epsilon < 1$. These servations guarantee that whenever any ϵ-digitization $[\rho]_\epsilon$ of a timed state sequence ρ

[6] The reader familiar with [15] will notice that we have slightly modified the lower-bound and upper-und requirements. This is because in the present paper we do not insist that the computations of a ed transition systems are deterministic (a timed state sequence is called *deterministic* iff with any sition either the state changes or the time increases, but not both). The lower-bound and upper-nd requirements are chosen so that the set of computations of any timed transition system is closed ler a suitable notion of *timed stuttering*, which allows the refinement of timed transition systems by easing the visible portion of the state space. Closure under timed stuttering, its applications, and interaction with closure under digitization and inverse digitization is discussed in [14].

violates a lower-bound requirement for the positions $i < j$, then so does ρ; and whene
ρ satisfies an upper-bound requirement for position i at position $j > i$, then so d
every ϵ-digitization $[\rho]_\epsilon$.

(2) Similarly, to see that the set of computations of a timed transition system closed under inverse digitization, observe that there is, for all $\mathsf{T}_i, \mathsf{T}_j \in \mathsf{R}$ and l, u $\mathsf{N} \cup \{\infty\}$, some $0 \leq \epsilon < 1$ such that $\mathsf{T}_j < \mathsf{T}_i + l$ implies $[\mathsf{T}_j]_\epsilon < [\mathsf{T}_i]_\epsilon + l$, and so $0 \leq \epsilon < 1$ such that $\mathsf{T}_j > \mathsf{T}_i + u$ implies $[\mathsf{T}_j]_\epsilon > [\mathsf{T}_i]_\epsilon + u$. ■

It follows that every timed transition system S defines the same qualitative (timed) property in either clock model (i.e., $[\![S]\!]^-_{\mathsf{R}} = [\![S]\!]^-_{\mathsf{Z}}$).

3.2 Metric temporal logic

Metric temporal logic (MTL for short) is an extension of linear temporal logic [26] th is interpreted over timed state sequences (rather than state sequences). To express ti ing requirements, the temporal operators are replaced with time-constrained versio such as the constrained *eventually* operator $\Diamond_{[2,4]}$ meaning "eventually within 2 t time units." We briefly review the syntax and semantics of MTL; for a more detail introduction to metric temporal logic and its applications consult [4, 19].

Syntax and semantics

Let P be a set of atomic propositions. The *formulas* ϕ of MTL are built from atom propositions by boolean connectives and time-constrained versions of the *until* ope tor $\mathcal{U}$; they are defined inductively as follows:

$$\phi \ ::= \ p \mid \neg\phi \mid \phi_1 \wedge \phi_2 \mid \phi_1 \mathcal{U}_I \, \phi_2,$$

where $p \in P$ and I is an interval of the nonnegative real line both of whose end-poi are natural-number constants.[7] Intervals may be open, halfopen, or closed; emp bounded, or unbounded. We say that all intervals of the forms $[0, n)$ and (m, n), $m \in \mathsf{N}$ and $n \in \mathsf{N} \cup \{\infty\}$, are *open*; all intervals of the forms $[m, n]$ and $[m, \infty)$, $m, n \in \mathsf{N}$, are *closed*. We freely denote intervals by pseudo-arithmetic expressio For example, the expressions $\leq m$ and $> n$ stand for the intervals $[0, m]$ and $(n, \infty$ respectively. The expression $x + I$, where I is an interval and $x \in \mathsf{R}$, denotes the interv $\{x + y \mid y \in I\}$.

The formulas of MTL are interpreted over timed state sequences whose states a taken from some set Σ such that each state in Σ provides an interpretation for t atomic propositions. Let $\rho = (\sigma, \mathsf{T})$ be a timed state sequence with $\sigma_i \in \Sigma$ for all $i \geq$ For any MTL-formula ϕ, the satisfaction relation $\rho \models \phi$ is defined inductively as follow

$\rho \models p$ iff $\sigma_0 \models p$;
$\rho \models \neg\phi$ iff $\rho \not\models \phi$;
$\rho \models \phi_1 \wedge \phi_2$ iff $\rho \models \phi_1$ and $\rho \models \phi_2$;
$\rho \models \phi_1 \mathcal{U}_I \, \phi_2$ iff $\rho^i \models \phi_2$ for some $i \geq 0$ with $\mathsf{T}_i \in \mathsf{T}_0 + I$, and $\rho^j \models \phi_1$ for all $0 \leq j < i$.

[7] The restriction to *constant* time bounds on temporal operators distinguishes our version of M from [19]. This restriction is necessary for the results of this paper as well as the decidability resu of [4]. The use of integer rather than rational time bounds is insignificant, because any formula wi rational constraints can be scaled appropriately.

: an MTL-formula ϕ, let the set $[\![\phi]\!]_{TIME} \subseteq TSS_{TIME}$ of *TIME-models* contain all ıed state sequences ρ over the time domain *TIME* such that $\rho \models \phi$. Thus, ϕ defines the analog-clock model the dense-time property $[\![\phi]\!]_{\mathsf{R}}$; in the digital-clock model, lefines the integral-time property $[\![\phi]\!]_{\mathsf{Z}} = \mathsf{Z}([\![\phi]\!]_{\mathsf{R}})$.

Additional temporal operators can be defined as usual. The defined operators $\Diamond_I \phi$ ·nstrained *eventually*) and $\Box_I \phi$ (constrained *always*) stand for $true\, \mathcal{U}_I\, \phi$ and $\neg\Diamond_I \neg\phi$, pectively. It follows that the formula $\Box_I p$ (or $\Diamond_I p$) is satisfied by a timed state se-ence $\rho = (\sigma, \mathsf{T})$ iff p holds at all states σ_i (at some state σ_i, respectively) with $\in \mathsf{T}_0 + I$. We also define a constrained *unless* operator as the dual of the *until* erator: $\phi_1\, {}_I\mathsf{U}\, \phi_2$ stands for $\neg((\neg\phi_2)\, \mathcal{U}_I\, (\neg\phi_1))$ (intuitively, "either ϕ_1 is true through-; the interval I or ϕ_2 is true at some point before the first time at which ϕ_1 is false I"). We can apply the definition of the *unless* operator to move negations through *til* operators. Hence we may obtain, from any MTL-formula, an equivalent formula, ıtaining both *until* and *unless* operators, in which all negations are in front of atomic ›positions.

We usually suppress the universal interval $[0, \infty)$ as a subscript. Thus the MTL-erators $\Diamond$, $\Box$, and $\mathcal{U}$ coincide with the conventional unconstrained *eventually*, *always*, d *until* operators of linear temporal logic [26]. The standard unconstrained *unless* erator, ϕ_1 *unless* ϕ_2 [22], is definable by the MTL-formula $(\phi_1 \vee \phi_2)\, \mathsf{U}\, \phi_2$.

gitizability

·t every MTL-formula defines a digitizable property. For instance, the R-models of formula $\Diamond_{<1} p$ are not closed under digitization, and the R-models of the disjunction $_{1} p \vee \Diamond_{\geq 2} p$ are not closed under inverse digitization. Yet the most common timing uirements of systems turn out to be digitizable.

Bounded invariance. A *bounded-invariance property* is a dense-time property at can be defined by an MTL-formula of the form

$$\Box(\phi_1 \rightarrow \Box_I \phi_2),$$

ere I is open and contains at least one integer point, and both ϕ_1 and ϕ_2 define alitative properties. A typical example of a bounded-invariance property is best-case rformance, say, a lower bound of $l > 0$ time units on termination: $\Box(p \rightarrow \Box_{<l} \neg q)$ for ›positions p and q that indicate the beginning and the end of execution, respectively.

Bounded response. A *bounded-response property* is a dense-time property that ı be defined by an MTL-formula of the form

$$\Box(\phi_1 \rightarrow \Diamond_I \phi_2),$$

ere I is closed and nonsingular (or $I = [0, 0]$), and both ϕ_1 and ϕ_2 define qualitative ›perties. A typical example of a bounded-response property is best-case performance, /, an upper bound of $u \geq 0$ time units on termination: $\Box(p \rightarrow \Diamond_{\leq u} q)$.

Clearly, MTL-formulas that contain only unconstrained temporal operators define alitative properties, which are digitizable by Proposition 1. Furthermore:

ıeorem 3 (Bounded invariance and bounded response) *Bounded-invariance ›perties and bounded-response properties are digitizable.*

Proof. Similar to the proof of Theorem 2. This is because each lower-bo requirement of a timed transition system is a bounded-invariance condition, and e upper-bound requirement is a bounded-response condition. ■

Now let us put Theorems 2, 3, and 1 together. These results imply that integer ti verification techniques can be used to prove the dense-time correctness of timed tr sition systems with respect to conjunctions of unconstrained MTL-formulas, bound invariance formulas, and bounded-response formulas. An integer time model-check algorithm for checking MTL-formulas over timed transition systems is given in [an integer time proof system for bounded-invariance and bounded-response proper in [15].

In fact, Theorem 1 demands that the requirement specification of a timed transit system be closed under inverse digitization only (rather than being fully digitizab Hence we can enlargen the class of MTL-formulas that can be established by inte time verification techniques. We say that an MTL-formula ϕ is *weakly constrainec* it satisfies the following three conditions:

- Every negation in ϕ is in front of an atomic proposition.
- Every *until* operator in ϕ is constrained by an open interval.
- Every *unless* operator in ϕ is constrained by a closed interval.

All unconstrained MTL-formulas are weakly constrained (since the universal interva both open and closed). But neither bounded-invariance nor bounded-response formu are weakly constrained.

Theorem 4 (Weakly constrained formulas) *The* R*-models of a weakly constra formula of* MTL *are closed under inverse digitization.*

Proof. Show by induction on the structure of a weakly constrained MTL-formul that $[\rho]_0 \models \phi$ implies $\rho \models \phi$ for all $\rho \in TSS_{\mathsf{R}}$. ■

Weakening and strengthening specifications

Theorem 4 can be applied in additional ways. First, from a proof that a timed transit system S satisfies an *arbitrary* MTL-formula ϕ in the digital-clock model, we can in that S satisfies in the analog-clock model all weakly constrained formulas that implied by ϕ. Second, to show that S satisfies a dense-time specification ϕ, it suffi to show that S satisfies in the digital-clock model any weakly constrained formula th implies ϕ.

Let ϕ^W (and ϕ^S) be the weakly constrained MTL-formula that results from MTL-formula ϕ by the following process of *weakening* (*strengthening*, respectively):

1. push negations all the way to the inside;
2. replace the interval I of every *until* operator with the smallest open inter containing I (the largest open, possibly empty, interval contained in I);
3. replace the interval I of every *unless* operator with the largest closed inter contained in I (the smallest closed interval containing I).

Proposition 3 (Weakening and strengthening) *Let ϕ be an* MTL*-formula. Th ϕ implies ϕ^W, and ϕ^S implies ϕ (in either clock model).*

Thus, if a timed transition system S is shown to satisfy an MTL-specification ϕ in ger time, then we can infer that S satisfies the weaker formula ϕ^W in dense time. ilarly, to prove that S satisfies ϕ in dense time, it suffices to show that S satisfies stronger formula ϕ^S in integer time. Naturally, we may not succeed in proving ϕ^S. his case, it is often advisable to approximate ϕ with ϕ^W by refining the granularity he digital-clock model. Suppose we need to show that S satisfies the specification $_{2]} p \vee \Diamond_{=3} p$ in dense time. By strengthening, we obtain $\Diamond_{(1,2)} p$, and suppose this nula is not satisfied by S. Using integer time, we can only establish the weaker erty $\Diamond_{(0,3)} p \vee \Diamond_{(2,4)} p$ (i.e., $\Diamond_{(0,4)} p$). However, by refining the stepwidth of the tal-clock model by a factor of 10 (multiply all delays in S by 10), we can show that atisfies $\Diamond_{(9,21)} p \vee \Diamond_{(29,31)} p$. In this fashion, by decreasing the time between two k ticks further, we can arbitrarily increase our confidence that S indeed satisfies dense-time specification $\Diamond_{[1,2]} p \vee \Diamond_{=3} p$. This is a typical example of how the tal-clock model may be used conservatively to reason about dense time.

:nowledgements. We thank Rajeev Alur for uncountably many stimulating discussions, Moshe Vardi for asking which dense-time properties are proved by integer time methods.

eferences

ALUR, R., AND DILL, D. Automata for modeling real-time systems. In *ICALP 90: Automata, Languages, and Programming*, M. Paterson, Ed., Lecture Notes in Computer Science 443. Springer-Verlag, 1990, pp. 322–335.

ALUR, R., FEDER, T., AND HENZINGER, T. The benefits of relaxing punctuality. In *Proceedings of the Tenth Annual Symposium on Principles of Distributed Computing* (1991), ACM Press, pp. 139–152.

ALUR, R., AND HENZINGER, T. A really temporal logic. In *Proceedings of the 30th Annual Symposium on Foundations of Computer Science* (1989), IEEE Computer Society Press, pp. 164–169.

ALUR, R., AND HENZINGER, T. Real-time logics: complexity and expressiveness. In *Proceedings of the Fifth Annual Symposium on Logic in Computer Science* (1990), IEEE Computer Society Press, pp. 390–401.

ALUR, R., AND HENZINGER, T. Logics and models of real time: A survey. In *Real Time: Theory in Practice*, J. de Bakker, K. Huizing, W.-P. de Roever, and G. Rozenberg, Eds., Lecture Notes in Computer Science 600. Springer-Verlag, 1992.

BARRINGER, H., KUIPER, R., AND PNUELI, A. A really abstract concurrent model and its temporal logic. In *Proceedings of the 13th Annual Symposium on Principles of Programming Languages* (1986), ACM Press, pp. 173–183.

BERRY, G., AND COSSERAT, L. The Esterel synchronous programming language and its mathematical semantics. In *CMU Seminar on Concurrency*, Lecture Notes in Computer Science 197. Springer-Verlag, 1985.

BURCH, J. Approximating continuous time. Presented at the IEEE Computer Society Workshop on VLSI, Orlando, Florida, 1991.

CASPI, P., PILAUD, D., HALBWACHS, N., AND PLAICE, J. Lustre: a declarative language for programming synchronous systems. In *Proceedings of the 14th Annual Symposium on Principles of Programming Languages* (1987), ACM Press.

[10] DE BAKKER, J., HUIZING, K., DE ROEVER, W.-P., AND ROZENBERG, G., Eds. *Real Ti Theory in Practice.* Lecture Notes in Computer Science 600. Springer-Verlag, 1992.

[11] EMERSON, E., MOK, A., SISTLA, A., AND SRINIVASAN, J. Quantitative temporal reas ing. Presented at the First Annual Workshop on Computer-aided Verification, Greno France, 1989.

[12] HAREL, E., LICHTENSTEIN, O., AND PNUELI, A. Explicit-clock temporal logic. In *P ceedings of the Fifth Annual Symposium on Logic in Computer Science* (1990), IE Computer Society Press, pp. 402–413.

[13] HENZINGER, T. Half-order modal logic: how to prove real-time properties. In *Proceedi of the Ninth Annual Symposium on Principles of Distributed Computing* (1990), A Press, pp. 281–296.

[14] HENZINGER, T. *The Temporal Specification and Verification of Real-time Systems.* P thesis, Stanford University, 1991.

[15] HENZINGER, T., MANNA, Z., AND PNUELI, A. Temporal proof methodologies for r time systems. In *Proceedings of the 18th Annual Symposium on Principles of Programm Languages* (1991), ACM Press, pp. 353–366.

[16] HENZINGER, T., MANNA, Z., AND PNUELI, A. Timed transition systems. In *Real Ti Theory in Practice*, J. de Bakker, K. Huizing, W.-P. de Roever, and G. Rozenberg, E Lecture Notes in Computer Science 600. Springer-Verlag, 1992.

[17] JAHANIAN, F., AND MOK, A. Safety analysis of timing properties in real-time syste *IEEE Transactions on Software Engineering SE-12*, 9 (1986), 890–904.

[18] KELLER, R. Formal verification of parallel programs. *Communications of the ACM 1* (1976), 371–384.

[19] KOYMANS, R. Specifying real-time properties with metric temporal logic. *Real-t Systems 2*, 4 (1990), 255–299.

[20] LYNCH, N., AND ATTIYA, H. Using mappings to prove timing properties. In *Proceedi of the Ninth Annual Symposium on Principles of Distributed Computing* (1990), A Press, pp. 265–280.

[21] MALER, O., MANNA, Z., AND PNUELI, A. A formal approach to hybrid systems. In *I Time: Theory in Practice*, J. de Bakker, K. Huizing, W.-P. de Roever, and G. Rozenbe Eds., Lecture Notes in Computer Science 600. Springer-Verlag, 1992.

[22] MANNA, Z., AND PNUELI, A. *The Temporal Logic of Reactive and Concurrent Syste* Springer-Verlag, 1992.

[23] MERRITT, M., MODUGNO, F., AND TUTTLE, M. Time-constrained automata. In *CC CUR 91: Theories of Concurrency*, Lecture Notes in Computer Science. Springer-Verl 1991.

[24] NICOLLIN, X., RICHIER, J.-L., SIFAKIS, J., AND VOIRON, J. ATP: an algebra for tin processes. In *Proceedings of the IFIP WG2.2/2.3 Working Conference on Programm Concepts and Methods* (1990), M. Broy and C. Jones, Eds., Elsevier Science Publish (North-Holland), pp. 415–442.

[25] OSTROFF, J. *Temporal Logic of Real-time Systems.* Research Studies Press, 1990.

[26] PNUELI, A. The temporal logic of programs. In *Proceedings of the 18th Annual Symposi on Foundations of Computer Science* (1977), IEEE Computer Society Press, pp. 46–5

[27] PNUELI, A., AND HAREL, E. Applications of temporal logic to the specification of re time systems. In *Formal Techniques in Real-time and Fault-tolerant Systems*, M. Jose Ed., Lecture Notes in Computer Science 331. Springer-Verlag, 1988, pp. 84–98.

Behavioural Abstraction in TCCS

Faron Moller Chris Tofts
Department of Computer Science
University of Edinburgh

Abstract

In this paper, we extend the theory of TCCS, the authors' Temporal Calculus of Communicating Systems, with a behavioural congruence which abstracts away internal communications. An equational theory for this equivalence is presented, and proven to be complete over the sublanguage of finite sequential agents.

Introduction

ı [MT90] the Temporal Calculus of Communicating Systems TCCS was introduced. 'his calculus is an extension of Milner's Calculus CCS which allows for the description and analysis of the temporal properties of concurrent systems. In this earlier aper, the semantic definition of the language was presented, and an equational ıeory was presented for a semantically defined congruence over the language. In MT91], a notion of relative speed was defined and analyzed. However, these relaons between processes did not abstract away from the internal behaviour of the gents being related.

The observational congruence of CCS [Mil89] is one of the most elegant and seful equivalences defined over process calculi. The ability to ignore silent interal communications between concurrent processes allows for a comparison of an nplementation with an abstract specification of the external behaviour of the sysem under consideration. This abstraction is characterized in CCS by the following hree tau laws, where τ represents an internalized (silent) event, $+$ represents a hoice between computation paths, and juxtaposition represents sequencing.

$$a\tau x = ax \qquad \tau x = \tau x + x \qquad a(x + \tau y) = a(x + \tau y) + ay$$

'he first law tells us that intermediate τ's can be ignored completely; the second hat initial τ's can be ignored up to always allowing them to occur to preempt the nvironment from following a particular choice of computation path; and the third ıw says that a τ can indeed cause the preemption of the environment from following certain choice of computation path.

In the theory of CCS, observation congruence is defined as the most liberal ongruence contained within a naturally defined equivalence relation which ignores ilent actions. Due to the preemptive nature of silent actions in choice contexts, his equivalence is not a congruence, which explains the reason for the form of the definition of the congruence. In this paper, we define a notion of equivalence tha expresses when two *timed* processes have equivalent temporal behaviour modulo th occurrence of unobservable internal communications; we show this relation to be non-trivial extension of Milner's observational equivalence. The equivalence fails t be substitutive over the calculus for reasons similar to the case of CCS, but als due to other interesting temporal effects. We then continue to examine in detail th underlying congruence derived from this equivalence.

In Section 2, we present the syntax of the calculus. The syntax deviates fro [MT90] in some minor ways. Reasons for this deviation are discussed in Section 6. I Section 3, we present the operational semantic definition of the calculus. In Section 4 we define our observational equivalence relation, and prove that it is substitutiv with respect to the full language of TCCS. In Section 5, we describe an equationa axiomatization for the language which is sound and complete for reasoning abou finite sequential processes. This theory includes timed versions of the tau laws c CCS presented above.

Due to space limitations, we only give the barest hints as to the proofs of th results stated in the present paper. The full proofs, as well as an extended analysis c the temporal properties of an alternating bit protocol, are included in the overviev paper [MT92].

2 The Language TCCS

Our language TCCS is a timed extension of CCS, Milner's *Calculus of Commun cating Systems* of [Mil89]. It is a slightly modified (though semantically equivalent version of the language presented in [MT90, Mol91, MT91].

To define the language, we first presuppose a set $\mathcal{L}$ of *atomic action symbol* not containing τ or ε, and we define $Act = \mathcal{L} \cup \{\tau\}$. We assume that $\mathcal{L}$ can b partitioned into two equinumerous sets with a *complementation* bijection $\bar{\cdot}$ define between them extended by $\overline{\overline{a}} = a$. These complementary actions form the basis o the handshake communication method in our calculus, analogous to that of CCS We also presuppose some set $\mathcal{X}$ of *process variables*, and take $\mathcal{T}$ to represent som time domain (that is, a totally ordered set, bounded below by, but not containing 0, the unit of a natural addition operation + defined on $\mathcal{T}$), be it for instance th discrete domain of positive integers, or the dense domain of positive rationals, o the continuous domain of positive reals.

The collection of TCCS expressions, ranged over by P, is then defined by th following BNF expression, where we take $a \in Act$, $X \in \mathcal{X}$, $t \in \mathcal{T}$, and f rangin over *relabelling functions*, those $f : Act \to Act$ such that $\overline{f(a)} = f(\overline{a})$ for $a \neq \tau$ an $f(\tau) = \tau$.

$$P \;::=\; \mathbf{0} \;|\; \underline{\mathbf{0}} \;|\; X \;|\; a.P \;|\; \underline{a}.P \;|\; (t).P \;|\; P+P \;|\; P \mathbin{+\!\!+} P \;|\; P\,|\,P \;|\; P\backslash a \;|\; P[f] \;|\; \mu x.P$$

he intuitive interpretation of these terms can be given as follows.

- $\mathbf{0}$ represents the completely dead process. It can neither perform any computation, nor can it witness the passage of any time. In terms of a concrete machine, it may be thought of as one which is shut down or with its "plug pulled". One of the consequences is that whenever $\mathbf{0}$ is a component of a machine, then the whole machine will be temporally dead; the local temporal deadlock gives rise to a global temporal deadlock.
- $\underline{\mathbf{0}}$ represents the process which can never perform any actions, but rather idles forever.
- X represents the process bound to the variable X in some assumed environment.
- $a.P$ represents the process which can perform the (atomic) action a and evolve into the process P upon so doing. The process cannot progress through time before performing the action a, and the action is assumed to occur instantaneously in time. If we wish to model a system where actions have some amount of duration (for instance, if we wish to talk of the computation of some hardware component with a certain known propagation delay), then we can naturally treat the action as a composite event consisting of the two atomic actions a_s (start event a) and a_f (finish event a), with the two actions separated in time by the appropriate duration t_a. Also note that atomic actions can be *causally* related without one being *temporally* dependent on the other. For instance, if we consider the process $a.b.P$, the action b is causally dependent on the action a in that it cannot have occurred *unless* (notice that we do not say *until*) the action a has occurred; however, the two actions are considered to happen simultaneously in time.
- $\underline{a}.P$ represents the process which can perform the (atomic) action a, but will allow any amount of time to pass before doing the action.
- $(t).P$ represents the process which will do no computation for an amount t of time, but at that point in time will commence behaving as the process P. Thus prefixing P by a composite event a consisting of start and finish actions a_s and a_f respectively with duration t_a would be written as $a_s.(t_a).a_f.P$.
- $P + Q$ represents a choice between the two processes P and Q. The process behaves as the process P or the process Q, with the choice being made at the time of the first action, or else at the occurrence of a passage of time when only one of the operands may allow the time passage to occur. In this latter case, the second *"stopped"* process is dropped from the computation. This operator is referred to as *weak* choice.
- $P \mathbin{+\!\!+} Q$ represents a stronger notion of choice between the two processes P and Q. The process behaves as the process P or the process Q, with the choice being made only at the time of the first action. Thus for instance any initial passage of time must be allowed by both P and Q. This operator is referred to as *strong* choice.

- $P \mid Q$ represents the parallel composition of the two processes P and Q. Each of the processes may do any actions independently, or they may synchronise on complementary actions, resulting in a τ action. Any passage of time must be allowed and recorded by each of P and Q.
- $P\backslash a$ represents the process P with the action $a \in \mathcal{L}$ (as well as its complement action $\overline{a}$) restricted away, that is, not allowed to occur.
- $P[f]$ represents the process P with its actions renamed by the relabelling function f.
- $\mu x.P$ represents the (least fixed point) solution to the equation $x = P$.

Remark 2.1 *The symbolic calculus CCS can be viewed as the subcalculus of TCCS obtained by omitting deadlock* $\mathbf{0}$*, action prefix* $a.P$*, temporal prefix* $(t).P$ *and strong choice* $P +\!\!+ Q$*. In CCS, the interpretation of behaviour is that it occurs within time but that the time is abstracted away; one action occurs after another, but there is no way of expressing or knowing how much time may or must pass before this second action occurs. The nil process of CCS is interpreted by* $\underline{\mathbf{0}}$ *rather than* $\mathbf{0}$*, as time can always pass when it is being ignored. For a similar reason, action prefix is interpreted by* $\underline{a}.P$ *rather than* $a.P$*. There is no place for temporal prefix in the subcalculus for CCS, as the delaying would simply be ignored. Finally, in this situation, where no temporal deadlocking can occur, the two choice operators coincide semantically, so the strong choice operator is dropped.*

As a point of notation, we shall occasionally omit the dot when applying the prefix operators, and also drop trailing $\mathbf{0}$'s (but *not* trailing $\underline{\mathbf{0}}$'s), thus for instance rendering $\underline{a}.\mathbf{0} + (t).b.\underline{\mathbf{0}}$ as $\underline{a} + (t)b\underline{\mathbf{0}}$. Also, we shall allow the prefix operators to take precedence over the concurrency operator, which will take precedence over the recursion operator, which finally in turn will take precedence over the choice operators; outside of this, the binary operators will be left associative, and we shall freely use parentheses to override the precedence rules.

3 The Semantics of TCCS

In this section, we provide a mathematical semantic definition of the language TCCS. The (transitional) semantics of TCCS is presented structurally, outlining what actions and time delays a process can witness. In order to define our semantics however, we must first define a predicate which will allow us to describe when a process may idle for a particular amount of time. This function is defined as follows.

Definition 3.1 *The relation* $\uparrow \subseteq TCCS \times \mathcal{T}$ *(written* $P \uparrow t$ *for* $(P,t) \in \uparrow$*) is defined to be the least relation satisfying the natural deduction style inference rules laid out in Figure 1. These rules are to be read as follows: if the relation(s) above the inference line can be inferred, then we can infer the relation below the line. The substitution*

$$\frac{}{\underline{\mathbf{0}} \uparrow t} \qquad \frac{P \uparrow t}{P + Q \uparrow t} \qquad \frac{P \uparrow t}{P \backslash a \uparrow t}$$

$$\frac{}{\underline{a}.P \uparrow t} \qquad \frac{Q \uparrow t}{P + Q \uparrow t} \qquad \frac{P \uparrow t}{P[f] \uparrow t}$$

$$\frac{}{(t).P \uparrow s}(s \leq t) \qquad \frac{P \uparrow t, \; Q \uparrow t}{P \mathbin{+\!\!+} Q \uparrow t} \qquad \frac{P\left\{{}^{\mu x.P}\!/_{x}\right\}^{\mathcal{G}} \uparrow t}{\mu x.P \uparrow t}$$

$$\frac{P \uparrow t_0}{(t).P \uparrow s}(s \leq t_0 + t) \qquad \frac{P \uparrow t, \; Q \uparrow t}{P \mid Q \uparrow t}$$

Figure 1: Delay capabilities of TCCS terms

perator $P\left\{{}^{Q}\!/_{x}\right\}^{\mathcal{G}}$ *replaces only the* guarded *free occurrences of* x *in* P *with the rm* Q.

nformally, $P \uparrow t$ means that the process term P may idle for time t without per- rming any computation. This semantic relation is identical to (though easier to ork with than) the functional definition of the maximal delay capability of a term s defined in [MT90]

In Figure 2, we present the operational rules for our language. Again they e presented in an identical natural deduction style. Our transitional seman- cs over TCCS then is given by the least relations $\longrightarrow \subseteq \mathrm{TCCS} \times Act \times \mathrm{TCCS}$ nd $\rightsquigarrow \subseteq \mathrm{TCCS} \times \mathcal{T} \times \mathrm{TCCS}$ (written $P \xrightarrow{a} Q$ and $P \overset{t}{\rightsquigarrow} Q$ for $(P, a, Q) \in \longrightarrow$ and $P, t, Q) \in \rightsquigarrow$ respectively) satisfying the rules laid out in Figure 2. Notice that these les respect the informal description of the constructs given in the previous section.

emark 3.2 *The reason that our substitution operator* $P\left\{{}^{Q}\!/_{x}\right\}^{\mathcal{G}}$ *only replaces uarded occurrences of* x *in* P *with* Q *is that we must be sure that no behaviour can inferred from an unguarded variable occurrence. This problem does not arise in e symbolic calculus CCS, where we could equally have used the standard notion f substitution and replaced all freely-occurring* x*'s. However, it does arise when nsidering the passage of time. For instance, if we used the usual definition of bstitution in our semantic definitions, then we would deduce that* $\mu x.x + (1).x \uparrow 1$ *hereas* $\mu x.x + (1).x \not\rightsquigarrow 1$. *The way in which our semantics is defined, unguarded cursive variables cannot contribute toward the semantic behaviour of a term, so in ct a term with unguarded variables is semantically identical to itself where those riables are replaced by the deadlock process* $\mathbf{0}$.

In [MT90] we defined strong $\mathcal{T}$-congruence over the calculus TCCS. This relation oes not abstract away the internal behaviour of a process; all internal communica- ons between agents are observable by the environment. In this paper, we examine

$$\frac{}{a.P \xrightarrow{a} P} \qquad \frac{P \xrightarrow{a} P'}{P \mathbin{+\!\!+} Q \xrightarrow{a} P'} \qquad \frac{P \xrightarrow{a} P',\ Q \xrightarrow{\bar{a}} Q'}{P \mid Q \xrightarrow{\tau} P' \mid Q'}$$

$$\frac{}{\underline{a}.P \xrightarrow{a} P} \qquad \frac{Q \xrightarrow{a} Q'}{P \mathbin{+\!\!+} Q \xrightarrow{a} Q'} \qquad \frac{P \xrightarrow{a} P'}{P\backslash b \xrightarrow{a} P'\backslash b}(a \neq b, \bar{b})$$

$$\frac{P \xrightarrow{a} P'}{P + Q \xrightarrow{a} P'} \qquad \frac{P \xrightarrow{a} P'}{P \mid Q \xrightarrow{a} P' \mid Q} \qquad \frac{P \xrightarrow{a} P'}{P[f] \xrightarrow{f(a)} P'[f]}$$

$$\frac{Q \xrightarrow{a} Q'}{P + Q \xrightarrow{a} Q'} \qquad \frac{Q \xrightarrow{a} Q'}{P \mid Q \xrightarrow{a} P \mid Q'} \qquad \frac{P\left\{{}^{\mu x.P}/_{x}\right\}^{\mathcal{G}} \xrightarrow{a} P'}{\mu x.P \xrightarrow{a} P'}$$

$$\frac{}{\mathbf{\underline{0}} \overset{t}{\rightsquigarrow} \mathbf{\underline{0}}} \qquad \frac{P \overset{t}{\rightsquigarrow} P',\ Q \overset{t}{\rightsquigarrow} Q'}{P + Q \overset{t}{\rightsquigarrow} P' + Q'} \qquad \frac{P \overset{t}{\rightsquigarrow} P'}{P\backslash a \overset{t}{\rightsquigarrow} P'\backslash a}$$

$$\frac{}{\underline{a}.P \overset{t}{\rightsquigarrow} \underline{a}.P} \qquad \frac{P \overset{t}{\rightsquigarrow} P',\ Q \not\rightsquigarrow t}{P + Q \overset{t}{\rightsquigarrow} P'} \qquad \frac{P \overset{t}{\rightsquigarrow} P'}{P[f] \overset{t}{\rightsquigarrow} P'[f]}$$

$$\frac{}{(s+t).P \overset{s}{\rightsquigarrow} (t).P} \qquad \frac{Q \overset{t}{\rightsquigarrow} Q',\ P \not\rightsquigarrow t}{P + Q \overset{t}{\rightsquigarrow} Q'} \qquad \frac{P_i\left\{{}^{\mu x.P}/_{x}\right\}^{\mathcal{G}} \overset{t}{\rightsquigarrow} P'}{\mu x.P \overset{t}{\rightsquigarrow} P'}$$

$$\frac{}{(t).P \overset{t}{\rightsquigarrow} P} \qquad \frac{P \overset{t}{\rightsquigarrow} P',\ Q \overset{t}{\rightsquigarrow} Q'}{P \mathbin{+\!\!+} Q \overset{t}{\rightsquigarrow} P' \mathbin{+\!\!+} Q'}$$

$$\frac{P \overset{s}{\rightsquigarrow} P'}{(t).P \overset{s+t}{\rightsquigarrow} P'} \qquad \frac{P \overset{t}{\rightsquigarrow} P',\ Q \overset{t}{\rightsquigarrow} Q'}{P \mid Q \overset{t}{\rightsquigarrow} P' \mid Q'}$$

Figure 2: Operational Rules for TCCS

the theory of observational congruence over TCCS, where such communications ar hidden. To do so, we shall require the following definitions for τ abstraction.

Definition 3.3 *If $s \in Act^*$ then $\widehat{s} \in \mathcal{L}^*$ is the sequence s with all occurences of deleted.*

Definition 3.4

$$\overset{\epsilon}{\Longrightarrow} \ = \ (\overset{\tau}{\longrightarrow})^* \quad \textit{(also written} \Longrightarrow\textit{)}$$

$$\overset{as}{\Longrightarrow} \ = \ \Longrightarrow\overset{a}{\longrightarrow}\overset{s}{\Longrightarrow}$$

$$\overset{t}{\approx\!\!>} \ = \ \Longrightarrow\overset{t_1}{\rightsquigarrow}\Longrightarrow \cdots \Longrightarrow\overset{t_n}{\rightsquigarrow}\Longrightarrow \quad \textit{where } t = t_1 + \ldots + t_n.$$

Observational Equality

this section, we develop the theoretical background for our equivalence relation hich abstracts away internal silent actions. The theory is based on the analogous amework for CCS, but there are several subtle difficulties which appear in the lculus extended with time.

efinition 4.1 *A binary relation $\mathcal{R}$ over agents is a $\mathcal{T}$-*bisimulation *if whenever* $,Q) \in \mathcal{R}$ *then* $\forall a \in Act$ *and* $\forall t \in \mathcal{T}$*:*

(i) if $P \stackrel{a}{\longrightarrow} P'$ *then* $Q \stackrel{\widehat{a}}{\Longrightarrow} Q'$ *for some* Q' *with* $(P', Q') \in \mathcal{R}$*;*

(ii) if $Q \stackrel{a}{\longrightarrow} Q'$ *then* $P \stackrel{\widehat{a}}{\Longrightarrow} P'$ *for some* P' *with* $(P', Q') \in \mathcal{R}$*;*

(iii) if $P \stackrel{t}{\leadsto} P'$ *then* $Q \stackrel{t}{\approx\!\!>} Q'$ *for some* Q' *with* $(P', Q') \in \mathcal{R}$*;*

(iv) if $Q \stackrel{t}{\leadsto} Q'$ *then* $P \stackrel{t}{\approx\!\!>} P'$ *for some* P' *with* $(P', Q') \in \mathcal{R}$*.*

and Q *are* $\mathcal{T}$-equivalent *or* $\mathcal{T}$-bisimilar, *written* $P \approx Q$*, if* $(P, Q) \in \mathcal{R}$ *for some -bisimulation* $\mathcal{R}$*.*

act 4.2 $\approx$ *is the largest $\mathcal{T}$-bisimulation, and is an equivalence relation.*

As with the case of the symbolic calculus CCS, this relation is not a con- uence over the whole calculus, in particular, over the choice operators. The ual counterexamples are as follows: $a \approx \tau.a$ but $a + b \not\approx \tau.a + b$ and $a \mathbin{+\!\!+} b \not\approx$ $a \mathbin{+\!\!+} b$. There are also other interesting counterexamples to congruicity involv- g the timing constructs. For example, we have that $(s)\tau(t)\mathbf{0} \approx (s+t)\mathbf{0}$, but $)\tau(t)\mathbf{0} \mathbin{+\!\!+} (s+t)aP \not\approx (s+t)\mathbf{0} \mathbin{+\!\!+} (s+t)aP$. In fact, the left hand side will al- ays deadlock after time $s+t$, whereas the right hand side will always perform the event and evolve into P after time $s+t$.

We are thus interested in refining this relation to be a congruence. This relation n be defined as follows.

efinition 4.3 *A binary relation $\mathcal{R}$ over agents is an* (observational) $\mathcal{T}$-congruence *whenever* $(P, Q) \in \mathcal{R}$ *then* $\forall a \in Act$ *and* $\forall t \in \mathcal{T}$*:*

(i) if $P \stackrel{a}{\longrightarrow} P'$ *then* $Q \stackrel{a}{\Longrightarrow} Q'$ *for some* Q' *with* $P' \approx Q'$*;*

(ii) if $Q \stackrel{a}{\longrightarrow} Q'$ *then* $P \stackrel{a}{\Longrightarrow} P'$ *for some* P' *with* $P' \approx Q'$*;*

(iii) if $P \stackrel{t}{\leadsto} P'$ *then* $Q \stackrel{t}{\leadsto} Q'$ *for some* Q' *with* $(P', Q') \in \mathcal{R}$*;*

(iv) if $Q \stackrel{t}{\leadsto} Q'$ *then* $P \stackrel{t}{\leadsto} P'$ *for some* P' *with* $(P', Q') \in \mathcal{R}$*.*

and Q *are* $\mathcal{T}$-congruent *written* $P \cong Q$*, if* $(P, Q) \in \mathcal{R}$ *for some $\mathcal{T}$-congruence* $\mathcal{R}$*.*

Hence, the only refinement required to Definition 4.1 is in the very first action the process. While the process idles, the definition remains restrictive (the final t clauses), whereas after performing an action, the resulting derivative terms ne only be $\mathcal{T}$-bisimilar, according to Definition 4.1.

Fact 4.4 $\cong$ *is the largest* $\mathcal{T}$*-congruence, and is indeed a congruence relation.*

Fact 4.5 $\cong$ *is contained within* $\approx$.

Thus we do indeed have a congruence relation. Moreover, this is the least refi ment we could have made of our original $\mathcal{T}$-bisimulation in order to get a congruen in the sense that $\cong$ is the largest congruence contained in $\approx$. To prove this, we u the following alternate characterization of $\cong$.

Fact 4.6 $P \cong Q$ *iff* $P \mathbin{+\!\!+} R \approx Q \mathbin{+\!\!+} R$ *for all* R.

With this, we can prove our result.

Fact 4.7 $\cong$ *is the largest congruence contained within* $\approx$.

Thus we have formally defined our congruence relation. Definition 4.3 can used as the basis for a partition-based algorithm for efficiently deciding the equal of (finite-state) agents. Such an automated decision procedure is implemented in t Edinburgh Concurrency Workbench [Mol91a], and has been used to analyze seve simple timed systems such as the protocol examined in [MT92]. However, usi this definition by hand can be tedious, so in the next section we consider equation axioms for syntactically proving congruicity between processes.

5 An Equational Theory for TCCS

In this section, we develop a set of equational laws for the calculus TCCS whi are sound with respect to observational congruence. We shall use = to represe derivability in our equational theory.

As a start, we restrict our attention to the finite sequential subset of TCCS, th is, the subcalculus given by the following syntax.

$$P \;::=\; \mathbf{0} \mid \underline{\mathbf{0}} \mid a.P \mid \underline{a}.P \mid (t).P \mid P+P \mid P \mathbin{+\!\!+} P$$

We shall refer to this subcalculus as TCCS_0.

Our equational theory for TCCS_0 is presented in Figure 3. We can straightfo wardly verify that these laws are sound with respect to congruence $\cong$ over the f calculus TCCS, by constructing appropriate observational congruences. Hence get the following result.

$$(s)(t)x = (s+t)x \qquad \underline{\mathbf{0}} = (t)\underline{\mathbf{0}}$$

$$(x+y)+z = x+(y+z) \qquad (x \mathbin{+\!\!+} y) \mathbin{+\!\!+} z = x \mathbin{+\!\!+} (y \mathbin{+\!\!+} z)$$
$$x+y = y+x \qquad x \mathbin{+\!\!+} y = y \mathbin{+\!\!+} x$$
$$x+x = x \qquad x \mathbin{+\!\!+} x = x$$
$$x+\mathbf{0} = x \qquad x \mathbin{+\!\!+} \underline{\mathbf{0}} = x$$

$$(t)x+(t)y = (t)(x+y) \qquad (t)x \mathbin{+\!\!+} (t)y = (t)(x \mathbin{+\!\!+} y)$$

$$ax+(y \mathbin{+\!\!+} \mathbf{0}) = ax \mathbin{+\!\!+} y \qquad \underline{a}x+\underline{b}y = \underline{a}x \mathbin{+\!\!+} \underline{b}y$$
$$\underline{a}x+(t)\underline{b}y = \underline{a}x \mathbin{+\!\!+} (t)\underline{b}y$$

$$x \mathbin{+\!\!+} (y+z) = (x \mathbin{+\!\!+} y)+(x \mathbin{+\!\!+} z)$$

$$\underline{a}x+\underline{\mathbf{0}} = \underline{a}x \qquad \underline{a}x \mathbin{+\!\!+} \mathbf{0} = ax$$
$$\underline{a}x = \underline{a}x+ax \qquad (t)x \mathbin{+\!\!+} \mathbf{0} = \mathbf{0}$$
$$\underline{a}x = \underline{a}x+(t)\underline{a}x$$

$$\Big(\underline{a}x \mathbin{+\!\!+} (t)y\Big)+(t)\Big(\underline{a}x \mathbin{+\!\!+} z\Big) = \underline{a}x \mathbin{+\!\!+} (t)\Big(y+z\Big)$$

$$\underline{a}x = \underline{a}\tau x$$
$$\underline{\tau}x = \underline{\tau}x+x$$
$$\underline{\tau}\Big(x+(\underline{a}y \mathbin{+\!\!+} z)\Big) = \underline{\tau}\Big(x+(\underline{a}y \mathbin{+\!\!+} z)\Big)+\underline{a}y$$
$$\underline{a}\Big(x+(\underline{\tau}y \mathbin{+\!\!+} z)\Big) = \underline{a}\Big(x+(\underline{\tau}y \mathbin{+\!\!+} z)\Big)+\underline{a}y$$

Figure 3: Equational Theory for $TCCS_0$

act 5.1 (Soundness) *$P = Q$ implies $P \cong Q$.*

The tau laws presented are the timed versions of the analogous tau laws for CS as presented in [Mil89]. There are many varieties of tau laws in our callus which are formed by interchanging the two choice operators, as well as the o action prefix operators. Many of these are sound, such as $\tau x = \tau x \mathbin{+\!\!+} x$ and $x \mathbin{+\!\!+} \underline{\tau}y) = a(x \mathbin{+\!\!+} \underline{\tau}y) + ay$. However, several are quickly seen to be unsound, ch as $a\underline{\tau}x = ax$ and $\tau x = \tau x + x$. Each of the sound instances is derivable from ose given in Figure 3.

Next, we show that these laws are complete for reasoning over $TCCS_0$. To do is, we work with the following normal form. For simplicity, we allow ourselves to ite $(0)P$ to be read as P, $(\infty)P$ to be read as $\underline{\mathbf{0}}$, and $P \overset{0}{\leadsto} Q$ to be read as $P \equiv Q$.

Definition 5.2 *A term is in* disjoint saturated normal form (DSNF) *if it is of th form*

$$\sum_{i\in I}(s_i)\Big(\underline{a_i}p_i +\!\!+ (s_i')\mathbf{0}\Big) + (s)\mathbf{0}$$

where each p_i is itself in DSNF, s_i may be 0, each of s and s_i' may be 0 or ∞ $s \geq s_i + s_i'$ for each $i \in I$, if $P \overset{t}{\rightsquigarrow}\overset{a}{\Longrightarrow} P'$ then $P \overset{t}{\rightsquigarrow}\overset{a}{\longrightarrow} P'$, and if $a_i = a_j$, $p_i \approx$ and $s_i \leq s_j$ then $s_i + s_i' < s_j$.

We can show that every term in $TCCS_0$ can be equated to some term in DSNF.

Fact 5.3 *Every term can be equated using the equational theory to a term in DSN*

The importance of terms being expressible in DSNF is shown by the following lemm

Lemma 5.4 *If P is in DSNF and $P \cong P + (s)\Big(\underline{a}p +\!\!+ (s')\mathbf{0}\Big)$ then we have th $P = P + (s)\Big(\underline{a}p +\!\!+ (s')\mathbf{0}\Big)$.*

Using this lemma, we can demonstrate the completeness of the equational theo over $TCCS_0$ by considering congruent terms in DSNF.

Fact 5.5 *If P and Q are in DSNF and $P \cong Q$, then $P = Q$.*

Corollary 5.6 (Completeness) *For p and q in $TCCS_0$, $p \cong q$ implies $p = q$.*

We can add further laws for reasoning completely about terms containing rel belling and restriction, as these operators distribute through the sequential operato as expected (*cf* the analogous laws in CCS as presented in [Mil89]). However, the is a problem with the complete axiomatization of the language of finite agents co taining the parallel construct. The usual approach is via an Expansion Theore which is used to transform any term into a congruent sequential one. However, th cannot be accomplished in general. For instance, the term

$$(1)\underline{a}\mathbf{0} \mid \underline{b}\mathbf{0}$$

has no equivalent sequential term. Naïvely one might expect this term to be equiv lent to the term $b(1)\underline{a}\mathbf{0} + (1)\big(\underline{a}\mathbf{0} \mid \underline{b}\mathbf{0}\big)$; however, this expanded term cannot perfor any action after any fraction of a time unit. In the case of a discrete time doma though, there does exist an Expansion Theorem; this has been presented in [MT9(

In [Che91], a subcalculus of TCCS has been extended with time variables whi allow an Expansion Theorem to be proven for strong congruence over any arbitra time domain. However, the price paid with the introduction of time variables a more complicated calculus with restrictions on the use of the time variables maintain decidability of strong congruence.

Conclusion

e have presented the theory of observational congruence over the calculus TCCS. nis extends the work on the calculus introduced in [MT90]. Our language is one many attempts to provide a concurrent calculus for reasoning about the effects time. Similar works can be found for example in [Mil83, BB91, Che91, HR90, ro90, RR86, NRSV90, Wan90]. We believe that TCCS gives one of the clearest counts of the nature of non-determinism that can occur within temporal systems. hough syntactically it is a simple extension of CCS, its naturality is demonstrated the similarity of the equational extensions required to address the inclusion of ostraction.

The syntax of the calculus has changed somewhat from the original paper [MT90] several respects. Firstly, the arbitrary delay operator $\delta.P$ has been replaced the delaying constructs presented above, namely delayed action prefix $\underline{a}.P$ and ntemporal nil $\underline{\mathbf{0}}$, as these are the only interesting occurrences of temporal delay. ore importantly, the weak and strong choice operators have changed from $\oplus$ and respectively to $+$ and $+\!\!+$ respectively. Though this may lead to confusion, there e several reasons for this recasting of the syntax.

- The choice operator of CCS is more naturally represented in a timing framework by the weak choice operator. In particular, the Expansion Theorem presented in [MT90] is expressed using the weak sum of prefixed terms.
- The use of the $\oplus$ operator conflicted with its common interpretation (as advocated for example by [Hen88]) as internal nondeterminism.
- Most other timed calculi have only one explicit choice operator, generally represented by the $+$ operator, which for the most case more closely resembles the weak choice operator of TCCS. Hence, the new syntax allows for an easier comparison of the various calculi.
- The calculus is implemented within the Edinburgh Concurrency Workbench [Mol91a]. For this reason, some ascii symbol had to replace the $\oplus$ operator for the user interface.

The definition of behavioural abstraction which we have presented seems the tural extension to that of CCS for a timed system. The cases where the equivalence not a congruence show some interesting properties of silent actions with respect temporal behaviour, notably the fact that $(t)\tau(s)P \not\cong (t+s)P$.

The equational theory is a simple and natural extension of the basic equational eory of TCCS, matching ones intuitions in the light of the additions required to e theory of CCS to represent abstraction.

References

[BB91] Baeten, J.C.M. and J.A. Bergstra, *Real Time Process Algebra*, Forma Aspects of Computing, Vol 3, 1991.

[Che91] Chen, L., *Decidability and Completeness in Real Time Processes*, Re search Report No ECS-LFCS-91-185, University of Edinburgh Depart ment of Computer Science, 1991.

[Gro90] Groote, J.F., *Specification and Verification of Real Time Systems i ACP*, Proceedings of the 10th International Symposium on Protoco Specification, Testing and Verification, Ottawa, 1990.

[Hen88] Hennessy, M.C., **Algebraic Theory of Processes**, MIT Press, 1988.

[HR90] Hennessy, M. and T. Regan, *A Temporal Process Algebra* Technical Re port No. 2/90, University of Sussex Department of Computer Scienc April 1990.

[Mil83] Milner, R., *Calculi for Synchrony and Asynchrony*, Theoretical Com puter Science, Vol 25, 1983.

[Mil89] Milner, R., **Communication and Concurrency**, Prentice–Hall Inter national, 1989.

[MT90] Moller, F. and C. Tofts, *A Temporal Calculus of Communicating Sys tems*, Proceedings of CONCUR'90, LNCS 458, Springer-Verlag, 1990.

[MT91] Moller, F. and C. Tofts, *Relating Processes with Respect to Speed*, Pro ceedings of CONCUR'91, LNCS 527, Springer-Verlag, 1990.

[MT92] Moller, F. and C. Tofts, *An Overview of TCCS*, Research Report, Uni versity of Edinburgh Department of Computer Science, 1992.

[Mol91] Moller, F., *Process Algebra as a Tool for Real Time Analysis*, IV Ban Higher Order Workshop, Springer-Verlag, 1991.

[Mol91a] Moller, F., *The Edinburgh Concurrency Workbench (Version 6.0)*, Tech nical Note No LFCS-TN-34, University of Edinburgh Computer Scienc Department, 1991.

[NRSV90] Nicollin, X., J.L. Richier, J. Sifakis and J.Voiron, *ATP: An Algebra fo Timed Processes*, Proceedings of IFIP Working Conference on Program ming Concepts and Methods, North Holland, 1990.

[RR86] Reed, G.M. and A. Roscoe, *A Timed Model for Communicating Sequen tail Processes*, Proceedings of ICALP'86, LNCS 226, Springer-Verlag 1986.

[Wan90] Wang Yi, *Real-time Behaviour of Asynchronous Agents*, Proceedings o CONCUR'90, LNCS 458, Springer-Verlag, 1990.

Timing Petri Nets Categorically*

Carolyn Brown and Doug Gurr
DAIMI, Århus University

Abstract

We establish a precise connection between the two distinct ways in which category theory has been applied to the study of Petri Nets. This connection relates morphisms between nets which consider only their static structure to morphisms on their purely dynamic behaviour, and is based on a CCS-like notion of simulation. We apply these results to the timing of Petri Nets.

Introduction

etri nets [Rei85] are an established and relatively successful model of concurrent ocesses. They admit an appealing operational interpretation, the "Token Game", d their simple graphical presentation makes them a convenient system with which describe or specify concurrent systems, as is evidenced by the successful industrial plication of systems such as Jensen's Petri Net Tool [Jen90]. This paper connects e two distinct ways in which category theory has been applied to Petri Nets.

The first approach, which was pioneered by Winskel [Win84], is to define a tegory whose objects are Petri nets and whose morphisms represent the refinement one net by another. There are several advantages to this categorical approach. The mpositionality of morphisms expresses the compositionality of refinement, thus if $_1$ refines N_0 and N_2 refines N_1 then N_2 must refine N_0. Further, structure in the tegory of nets (for example, products and coproducts) allows us to build composite ts from simpler sub-components. In addition, we can use the logic associated with ch structure to reason about refinements of nets. Finally, a categorical approach ten makes it straightforward to extend and generalise results by modifying the ructures under consideration.

The other approach observes that the token game can be extended to give a belled transition system operational semantics for a net, in which the states are arkings. The transitions are computations, that is, sequential and parallel comsitions of events of the net. In fact, such a transition system yields a category hose objects are markings and whose morphisms are computations: we shall refer this as the behaviour category of the net, since it presents all possible behaviours the net. One then defines a category whose objects are behaviour categories, in [MM88, DMM89]. However, difficulties arise in giving a meaningful computaonal interpretation to the morphisms of such a category, and in obtaining a natural

*Supported by ESPRIT CEDISYS and CLICS and Danish DART grants.

connection between a category whose objects are behaviour categories and a cat gory whose objects are nets. We seek an adjunction or coreflection such as Nielse and Winskel [NW91, Win84] have established between different models of concu rent processes. Finding such an adjunction relating categories of nets to categori of their behaviours has been a long-standing problem.

This paper establishes such a reflection between a category **Behav** whose o jects are behaviour categories and the category **Net** [BGdP91] whose objects a Petri Nets. In addition we give a meaningful computational interpretation to t morphisms in **Behav**. The underlying idea is that a net N_1 refines a net N_0 if N can simulate any behaviour of N_0, in the sense that for any computation in N_0 fro marking M_0 to M_0', there is a corresponding computation in N_1 from a marking M corresponding to M_0 to a marking M_1' corresponding to M_0'. The morphisms b tween behaviour categories reflect precisely this notion of refinement. They consi of a simulation relation S between states of the two nets and a function C from t computations in N_0 to those of N_1 such that whenever $\langle M_0, M_1\rangle \in S$ and $M_0 \xrightarrow{c} M$ then $M_1 \xrightarrow{Cc} M_1'$ and $\langle M_1, M_1'\rangle \in S$. This notion of simulation is closely connect to the definition of bisimulation in the process algebra CCS.

The primary application presented in this paper is to timing nets. Behavio categories provide a suitable framework in which to define Petri Nets incorporati information about the duration of computations. We show that the category **Beha** of behaviour categories gives rise to a category of timed nets **Time** in which t morphisms have an appealing computational interpretation as refinements (in t above sense) which respect specified timing constraints. In the remainder of th paper, we describe the lifting of the categorical structure possessed by the catego **Net** to the categories **Behav** and **Time**, and briefly describe a logical system f reasoning about timed nets.

2 Categories of Nets and Behaviours

We recall some elementary definitions of Petri net theory [Rei85].

Definition 2.1 A *Petri Net*, denoted N, is a 4–tuple $\langle E, B, pre, post\rangle$ where E a B are sets, and *pre* and *post* are functions from $E \times B$ to **N**.

We shall call elements of E ***events*** and elements of B ***conditions***. We shall call *p* and *post* the pre– and post–condition relations of N respectively. With each of t multirelations *pre* and *post* we associate a function of the same name from E to B (the free abelian monoid on B, with unit $\emptyset$) defined by

$$pre(e) = \sum_{b\in B} pre(e,b)b \quad \text{and} \quad post(e) = \sum_{b\in B} post(e,b)b.$$

We call $pre(e)$ the pre–condition set and $post(e)$ the post–condition set of e. *loop-free* Petri net is a net in which every event has disjoint pre– and post–conditi sets. We write **Petri** for the class of all loop-free Petri nets.

The state of a Petri net is described by a finite multiset over B (that is, a multis which is zero on all but finitely many $b \in B$), called a *marking*, which indicates whi conditions hold, and in what multiplicities. The occurrence of an event, call firing, transforms the state of the net by consuming its pre–condition set and oducing its post–condition set. A multiset $\Sigma n_i e_i$ of events can fire *concurrently* the present marking M contains the multiset $\Sigma n_i pre(e_i)$. We denote this firing $\xrightarrow{\Sigma n_i e_i} M - \Sigma n_i pre(e_i) + \Sigma n_i post(e_i)$.

Intuitively, a computation is either an event (possibly idle) or the parallel or quential composition of two computations. The sequential composition of computions c and d occurs when c can fire and $pre(d) = post(c)$: it results in marking $st(d)$. The parallel composition of c and d occurs when they can fire simultanesly, consuming $pre(c) + pre(d)$ and producing $post(c) + post(d)$. The following mma shows that we can represent the behaviour of a net N by a category with jects the markings of N and morphisms the computations of N.

emma 2.2 Let N $= \langle E, B, pre, post\rangle$ be a Petri net. The following data:

- Objects: elements of $B^{\oplus}$,
- Morphisms defined inductively by the following formation rules:

$$\frac{}{pre(e) \xrightarrow{e} post(e)}\ e \in E \qquad \frac{}{b \xrightarrow{id_b} b}\ b \in B^{\oplus}$$

$$\frac{b_0 \xrightarrow{c_0} b_1 \quad b_1 \xrightarrow{c_1} b_2}{b_0 \xrightarrow{c_0;c_1} b_2} \qquad \frac{b_0 \xrightarrow{c} b_1 \quad b_0' \xrightarrow{c'} b_1'}{b_0 + b_0' \xrightarrow{c|c'} b_1 + b_1'}$$

subject to the following equivalences:

(id)	$id_{pre(c)}; c = c; id_{post(c)} = c$	$(ass1)$	$c_0;(c_1;c_2) = (c_0;c_1);c_2$
$(unit)$	$c\vert id_{\emptyset} = c$	$(ass2)$	$c_0\vert(c_1\vert c_2) = (c_0\vert c_1)\vert c_2$
(com)	$c_0\vert c_1 = c_1\vert c_0$		

- composition: given by ;

efine a category $\mathcal{B}$(N). We shall call $\mathcal{B}$(N) the "behaviour category of N".

Remark 2.3 The category $\mathcal{B}$(N) was called $\mathcal{K}$(N) in [DMM89].

Notation 2.4 For notational convenience, we shall write
$M_0 = M + pre(e) \xrightarrow{c|id_M} M_1 = M + post(e)$ as $M_0 \xrightarrow{c} M_1$.

For our purposes, it is convenient to work with a slightly different representation of the behaviour category of a net. Thus we identify the behaviour category of a net N $= \langle E, B, pre, post\rangle$ with the tuple $\langle E^+, B, \partial_0, \partial_1\rangle$, where E^+ is the set of arrows of $\mathcal{B}$(N) and $\partial_0, \partial_1 : E^+ \rightarrow B^{\oplus}$ are the source and target maps respectively of the category $\mathcal{B}$(N). It is clear that $\mathcal{B}$(N) can be recovered from this representation. Conversely, we can recover this representation from $\mathcal{B}$(N) by taking B to be the irreducible elements of the free abelian monoid $B^{\oplus}$.

Note that if we require that $(c_0|c'_0);(c_1|c'_1) = (c_0;c_1)|(c'_0;c'_1)$ (as is done, fo example, in [MM88]) then $|$ is a symmetric monoidal structure on $\mathcal{B}(\mathrm{N})$. This precisely the coherence condition necessary to ensure that $\mathcal{B}(\mathrm{N})$ is an abelian-monoi enriched category. However, this rule is not admissible when we consider the tim taken by events, as the following example shows.

Example 2.5 Consider the computations $(c;d)|(c';d')$ and $(c|c');(d|d')$ in a be haviour category $\mathcal{B}(\mathrm{N})$. Let c and d' have duration 1 and let c' and d have duration Let the duration of the sequential composition of computations be the sum of the durations, and the duration of the parallel composition of computations be the max imum of their durations. The duration of $(c;d)|(c';d')$ is $max\{1+2,2+1\} = 3$, whil the duration of $(c|c');(d|d')$ is $max\{1,2\} + max\{1,2\} = 4$. Thus the interchange la $(c_0|c'_0);(c_1|c'_1) = (c_0;c_1)|(c'_0;c'_1)$ fails for this natural example.

We expect a map between behaviour categories to indicate how to simulate, c implement, the computations of one net in another. An evident first choice of ma between behaviour categories is a structure-preserving functor. This is the approac of [BG90] and also [MM88]. In order to reflect the vicissitudes of real life, we take more subtle (and more complex) notion of morphism between behaviour categorie Our definition is intended to capture the notion of simulation, and so is closel related to the definition of bisimulation in *CCS*.

We first consider a relation which is crucial in what follows[1].

Definition 2.6 Let F be a function from a set B' to a set B. The relation $F^+ \subseteq B^{\oplus} \times B'^{\oplus}$ is given by $\langle M_0, M'_0\rangle \in F^+$ if and only if $M_0F \leq M'_0$.

Lemma 2.7 $G^+ \circ F^+ = (GF)^+$, where $\circ$ denotes relational composition.[2]

Remark 2.8 The above result shows that if id_B is the identity function on B the id_B^+ is the left identity relation for all F^+ arising from functions $F\colon B' \to B$ and th right identity relation for all F^+ arising from functions $F\colon B \to B'$.

Lemma 2.9 The following data:

- objects are the behaviour categories generated by the elements of **Petri**,
- morphisms from $\langle E^+, B, \partial_0, \partial_1\rangle$ to $\langle E'^+, B', \partial'_0, \partial'_1\rangle$ are pairs $\langle C, F\rangle$ where C is a function from E^+ to E'^+ such that $C(c;d) = C(c);C(d)$ and $C(c|d) = C(c)|C(d)$, and F is a function from B' to B such that if $\langle M_0, M'_0\rangle \in F^+$ an $M_0 \xrightarrow{c} M_1$ then $M'_0 \xrightarrow{Cc} M'_1$ and $\langle M_1, M'_1\rangle \in F^+$, and
- composition given by $\langle g, G\rangle\langle f, F\rangle = \langle gf, FG\rangle$

define a category, which we shall denote **Behav**.

Proof: We first show that composition is well-defined. Let $\langle C, F\rangle\colon \langle E_0^+, B_0, \partial_0, \partial_1\rangle \to \langle E_1^+, B_1, \partial'_0, \partial'_1\rangle$ and $\langle D, G\rangle\colon \langle E_1^+, B_1, \partial'_0, \partial'_1\rangle \to \langle E_2^+, B_2, \partial''_0, \partial''_1\rangle$ be morphisms i

[1] This relation was called F-**ok** in [BGdP91].

[2] Given by $R \circ S = \{\langle a,b\rangle \mid \exists c.\langle a,c\rangle \in R \text{ and } \langle c,b\rangle \in S\}$.

hav. Suppose that $\langle M_0, M_2\rangle \in (FG)^+$ and $M_0 \xrightarrow{c} M_0'$. Then taking $M_1 = M_0F$ have $\langle M_0, M_1\rangle \in F^+$. Now $\langle M_0, M_2\rangle \in (FG)^+$ implies that $M_0FG \le M_2$ and $\langle M_1, M_2\rangle \in G^+$. Now $M_1 \xrightarrow{Cc} M_1'$ and $\langle M_1, M_2\rangle \in G^+$ so $M_2 \xrightarrow{DCc} M_2'$ and $\langle M_1', M_2'\rangle \in G^+$ and thus $\langle M_0', M_2'\rangle \in F^+ \circ G^+$. By Lemma 2.7 $F^+ \circ G^+ = (FG)^+$ d so $\langle gf, FG\rangle$ is a morphism from $\langle E_0^+, B_0\rangle$ to $\langle E_2^+, B_2\rangle$ in **Behav**.
ssociativity follows from the associativity of function composition.
ne identity on a behaviour $\langle E^+, B, \partial_0, \partial_1\rangle$ is the pair $\langle id_{E^+}, id_B\rangle$. This is a morism in **Behav** since whenever $\langle M_0, M_0'\rangle \in id_B^+$ and $M_0 \xrightarrow{c} M_1$ then by the finition of evolution, $M_0' \xrightarrow{c} (M_0' + M_1 - M_0)$ which contains M_1. □

he intended interpretation of morphisms in **Behav** is that they indicate how to nplement the computations of one net in another net. It follows from the definition **Behav** and that if $\langle C, F\rangle : \mathcal{B}(\mathrm{N}) \to \mathcal{B}(\mathrm{N}')$ and M_i, M_i' are markings of net N, N':

- if $M_0 \xrightarrow{c} M_1$ then there exist M_0', M_1' such that $\langle M_0, M_0'\rangle \in F^+$ and $M_0' \xrightarrow{Cc} M_1'$, with $\langle M_1, M_1'\rangle \in F^+$,
- if $M_0 \xrightarrow{c} M_1$ and $M_2 \xrightarrow{d} M_3$ then there exist M_i' with $\langle M_0 + M_2, M_0' + M_2'\rangle$, $\langle M_1 + M_3, M_1' + M_3'\rangle \in F^+$ and $M_0' + M_2' \xrightarrow{(Cc)|(Cd)} M_1' + M_3'$,
- if $M_0 \xrightarrow{c} M_1$ and $M_1 \xrightarrow{d} M_2$ then there exist M_0', M_2' such that the pairs $\langle M_0, M_0'\rangle$ and $\langle M_2, M_2'\rangle$ are in the relation F^+ and $M_0' \xrightarrow{(Cc);(Cd)} M_2'$.

hus every computation of N has an implementation in N′. Further, the sequential omposition of two computations in N is implemented by the sequential composition f their implementations in N′, while their parallel composition is implemented by ne parallel composition of their implementations in N′.

Note that although morphisms between behaviour categories are not functors, ney do preserve both sequential and parallel composition of computations. We now ecall the definition of the category **Net** of Petri nets defined in [BGdP91]. We shall nen define a functor $\mathcal{B}$: **Net** $\to$ **Behav** and prove that it has a right adjoint.

emma 2.10 The following data:

- objects are elements of **Petri**,
- a morphism from $\langle E, B, pre, post\rangle$ to $\langle E', B', pre', post'\rangle$ is a pair of functions $\langle f, F\rangle$ with $f: E \to E'$ and $F: B' \to B$ such that

$$\begin{array}{ccc} E\times B' & \xrightarrow{id_E\times F} & E\times B \\ {\scriptstyle f\times id_{B'}}\downarrow & \le & \downarrow{\scriptstyle pre} \\ E'\times B' & \xrightarrow[pre']{} & \mathbf{N} \end{array} \quad \text{and} \quad \begin{array}{ccc} E\times B' & \xrightarrow{id_E\times F} & E'\times B' \\ {\scriptstyle f\times id_{B'}}\downarrow & \ge & \downarrow{\scriptstyle post} \\ E'\times B' & \xrightarrow[post']{} & \mathbf{N}, \end{array}$$

that is, for each $e \in E$ and each $b' \in B'$, we have

$$pre\langle e, Fb'\rangle \geq pre'\langle fe, b'\rangle \quad \text{and} \quad post\langle e, Fb'\rangle \leq post'\langle fe, b'\rangle,$$

- and composition is function composition in each component

define a category, which we shall denote **Net**.

We recall the following important results from [BGdP91]:

Theorem 2.11 Let $\langle f, F\rangle: \mathrm{N} \to \mathrm{N}'$ be a morphism in **Net**. For any $e \in E$, i $\langle M_0, M_0'\rangle \in F^+$ and $M_0 \xrightarrow{e} M_1$ in N, then $M_0' \xrightarrow{fe} M_1'$ and $\langle M_1, M_1'\rangle \in F^+$.

Corollary 2.12 Let $\langle f, F\rangle: \mathrm{N} \to \mathrm{N}'$ be a morphism in **Net**. Let f^+ be the extensio of $f: E \to E'$ to $f^+: E^+ \to E'^+$ given by $f^+(c; d) = (f^+c); (f^+d)$ and $f^+(c|d) =$ $(f^+c)|(f^+d)$. If $c \in E^+$, $\langle M_0, M_0'\rangle \in F^+$ and $M_0 \xrightarrow{c} M_1$ in N, then $M_0' \xrightarrow{f^+c} M$ and $\langle M_1, M_1'\rangle \in F^+$.

Theorem 2.13 Let N be a loop free net, let N′ be any net, let $f: E \to E'$ an $F: B' \to B$, and suppose that for all pairs of markings $\langle M_0, M_0'\rangle \in F^+$, wheneve $pre(e) \subseteq M$ then $pre(f(e)) \subseteq M'$ and whenever $M_0 \xrightarrow{e} M_1$ and $M_0' \xrightarrow{f(e)} M_1'$ the $\langle M_1, M_1'\rangle \in F^+$. Then $\langle f, F\rangle$ is a morphism from N to N′ in **Net**.

The above results enable us to exhibit an adjunction between **Net** and **Behav**.

Proposition 2.14 There is a faithful functor $\mathcal{B}$: **Net** $\to$ **Behav** which takes a ne N to its behaviour category $\mathcal{B}(\mathrm{N})$ and a morphism $\langle f, F\rangle$ to the pair $\langle f^+, F\rangle$, wher f^+ is the extension of f from E to E^+ induced by putting $f^+(c; d) = f^+(c); f^+(d$ and $f^+(c|d) = f^+(c)|f^+(d)$.

Proof: By Corollary 2.12, if $\langle f, F\rangle: \mathrm{N} \to \mathrm{N}'$ in **Net** then $\langle f, F\rangle$ satisfies the condi tion that if $\langle M_0, M_0'\rangle \in F^+$ and $M_0 \xrightarrow{c} M_1$ then $M_0' \xrightarrow{f^+c} M_1'$ and $\langle M_1, M_1'\rangle \in F^+$ Thus $\langle f^+, F\rangle: \mathcal{B}(\mathrm{N}) \to \mathcal{B}(\mathrm{N}')$ in **Behav**. Faithfulness is immediate from the def inition of $\mathcal{B}$. As already observed, the identity on $\langle E^+, B, \partial_0, \partial_1\rangle$ in **Behav** i $\langle id_{E^+}, id_B\rangle = \mathcal{B}(id_{\mathrm{N}})$. In the case of composition, a simple structural inductio on E^+ shows that $(fg)^+ = f^+g^+$, while Lemma 2.7 shows that $F^+ \circ G^+ = (FG)^+$ from which it follows that $\langle (gf)^+, (FG)^+\rangle = \langle g^+, G\rangle\langle f^+, F\rangle$. □

Proposition 2.15 There is a functor $\mathcal{N}$: **Behav** $\to$ **Net** which takes the morphism $\langle C, F\rangle$ to $\langle C, F\rangle$ and the behaviour $\langle E^+, B, \partial_0, \partial_1\rangle$ to the net $\langle E^+, B, pre, post\rangle$ where pre and $post$ are given by $pre(e, b) = \partial_0(e)(b)$ and $post(e, b) = \partial_1(e)(b)$.[3]

Proof: We need only check that, whenever $\langle C, F\rangle: \langle E^+, B, \partial_0, \partial_1\rangle \to \langle E'^+, B', \partial_0', \partial_1'$ then $\langle C, F\rangle$ is a morphism in **Net** from $\langle E^+, B, pre, post\rangle$ to $\langle E'^+, B', pre', post'\rangle$ We must show that, for all $e \in E^+$ and $b' \in B'$, $pre\langle e, Fb'\rangle \geq pre'\langle Ce, b'\rangle$ an $post\langle e, Fb'\rangle \leq post'\langle Ce, b'\rangle$. This is precisely the content of Theorem 2.13. □

[3]Recalling that $\partial_0(e)$ and $\partial_1(e)$ are multisets over B.

mark 2.16 The event set of $\mathcal{N}(\mathcal{B}(\mathrm{N}))$ contains that of N, as it comprises all nputations.

eorem 2.17 $\mathcal{N}$ is right adjoint to $\mathcal{B}$.

oof: We show that there is a natural bijection

$$\mathbf{Behav}(\mathcal{B}(\mathrm{N}), \langle E'^{+}, B', \partial_0', \partial_1' \rangle) \cong \mathbf{Net}(\mathrm{N}, \mathcal{N}(\langle E'^{+}, B', \partial_0', \partial_1' \rangle)).$$

ppose $\langle f, F\rangle : \mathrm{N} \to \mathcal{N}(\langle E'^{+}, B', \partial_0', \partial_1'\rangle)$ in **Net**. To see that $\langle f^{+}, F\rangle : \mathcal{B}(\mathrm{N}) \to$ $'^{+}, B', \partial_0', \partial_1'\rangle)$ is a morphism in **Behav**, we simply check that F^{+} satisfies the operty that if $\langle M_0, M_0'\rangle \in F^{+}$ and $M_0 \xrightarrow{c} M_1$ then there exists a multiset M_1' er B' such that $M_0' \xrightarrow{Cc} M_1'$ and $\langle M_1, M_1'\rangle \in F^{+}$. This follows from Corollary 2.12. nversely, if $\langle C, F\rangle : \mathcal{B}(\mathrm{N}) \to \langle B', E'^{+}, \partial_0', \partial_1'\rangle$ in **Behav** then denoting by $C|_E$ the striction of C to the generators of E^{+}, the pair $\langle C|_E, F\rangle$ is a morphism in **Net** om N to $\mathcal{N}(\langle B', E'^{+}, \partial_0', \partial_1'\rangle)$, by Theorem 2.13.
ese two constructions evidently define the desired natural bijection. □

rollary 2.18 $\mathcal{N}$ **is a reflection (that is, a full and faithful right adjoint).**

is reflection shows that there is a morphism from N to N′ in Net precisely when ery computation of N can be simulated by a computation of N′. This gives us a ecise correspondence between morphisms of nets which consider only their static ucture, and morphisms on their purely dynamic behaviour, and justifies the claim at morphisms in **Net** represent refinements.

Timed Nets

[Gur90], the second author uses a monoid structure to represent the composition durations of sequential programs. We now show that a second monoidal structure n be used to represent the composition of the durations of parallel processes.

Our intention is to time computations using a function from computations to eir durations. We consider durations to form a poset equipped with two binary erations $\cdot$ and $\otimes$, which compose the durations of two computations to give the rations of their sequential and parallel compositions, respectively. We require cer- in natural properties of these operations: first, they should both be associative, nce both sequential and parallel composition are associative. Parallel composition durations should be commutative, as parallel composition is commutative: se- ential composition of durations is not required to be commutative. Finally, we quire that there be a distinguished zero duration which is the unit of both parallel d sequential composition of durations, and is the bottom element of the poset. hese considerations lead us to the following definition.

efinition 3.1 A time object $\mathcal{T}$ is a tuple $\langle \mathbf{T}, \cdot, \otimes, 0, \leq\rangle$ such that $\langle \mathbf{T}, \leq\rangle$ is a rtially ordered set with bottom element 0, $\langle \mathbf{T}, 0, \cdot\rangle$ is a monoid, $\langle \mathbf{T}, 0, \otimes\rangle$ is an elian monoid, both $\cdot$ and $\otimes$ are monotone in each argument with respect to $\leq$, d

$$(m \cdot m') \otimes (n \cdot n') \leq (m \otimes n) \cdot (m' \otimes n') \quad \text{for all } m, m', n, n' \in \mathbf{T}. \qquad (*)$$

Notation 3.2 $\mathcal{T}$ will also denote the underlying category $\langle \mathbf{T}, 0, \cdot \rangle$ of a time objec

The condition $(*)$ expresses the fact that the sequential composition of parallel com- putations cannot be faster than the corresponding parallel composition of sequenti computations. Example 2.5 showed that we cannot expect equality here. Althoug in general there is no notion of ordering on the homsets of $\mathcal{B}(\mathrm{N})$, timing induces suc an ordering. Thus we can distinguish the fastest computations, just as Gehlot ar Gunter [GG90] distinguish those with maximal parallelism.

Remark 3.3 A time object is a poset–enriched, one–object category with a la symmetric monoidal structure.

Example 3.4 The following are examples of time objects:

- $\mathbf{T} = \mathbf{N}$ with the usual ordering, $\cdot$ is addition and $\otimes$ is *max*,
- $\mathbf{T} = \mathbf{R}$ with the usual ordering $\cdot$ is addition and $\otimes$ is *average*, and
- $\mathcal{T}$ an ordered ring.

We make the natural assumption that assigning a duration to each event of a n determines the duration of every computation of the net. The following definitic and proposition show that we can use the behaviour category $\mathcal{B}(\mathrm{N})$ to express th smooth lifting of durations from events to computations.

Definition 3.5 Let $\mathcal{T} = \langle \mathbf{T}, \cdot, \otimes, 0, \leq \rangle$ be a time object. A T–**timed net** is a trip $\langle \mathrm{N}, \mathcal{T}, \tau \rangle$ consisting of a net N and a function $\tau : E \to \mathbf{T}$.

Proposition 3.6 Let $\mathcal{T}$ be a time object and let $\langle \mathrm{N}, \mathcal{T}, \tau \rangle$ be a T–timed net. The τ extends uniquely to a functor $\tau : \mathcal{B}(\mathrm{N}) \to \mathcal{T}$ such that $\tau(c_0 | c_1) = \tau(c_0) \otimes \tau(c_1)$.

We shall call such a functor τ a $\mathcal{T}$–**timing**. Since τ is a functor, we have

$$\tau(c_0; c_1) = \tau(c_0) \cdot \tau(c_1) \quad \text{and} \quad \tau(id_M) = 0 \quad \text{for all } M \in B^{\oplus},$$

and hence τ assigns a unique duration to any computation of N.

Example 3.7

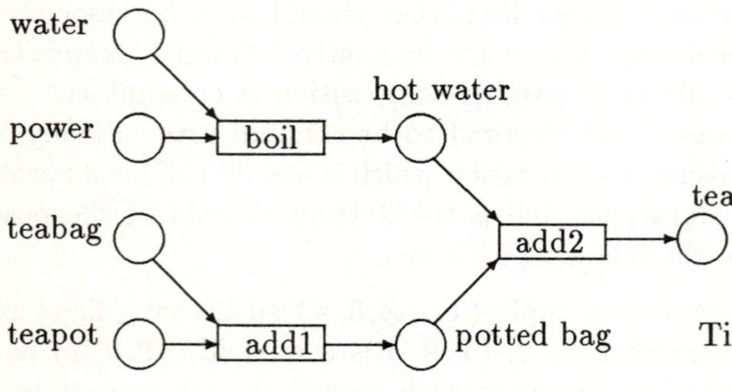

τ is given by:

$\tau(\text{boil}) = 5$

$\tau(\text{add1}) = 1$

$\tau(\text{add2}) = 1$

Time object = $\langle \mathbf{N}, 0, +, max, \leq$

The computation (boil|add1); add2 makes tea efficiently: $\tau((\text{boil}|\text{add1}); \text{add2}) = 6$
The computation boil; add1; add2 makes tea inefficiently: $\tau(\text{boil}; \text{add1}; \text{add2}) = 7$.

Simulation with Time Constraints

: have shown that a Petri net gives rise a category of behaviours, and that a timed ; can be represented by a certain kind of functor from such a category to a time ject. In addition, the adjunction $\mathcal{B} \dashv \mathcal{N}$ shows that a simulation of one net by other is precisely a morphism between their behaviour categories. We now extend : category **Behav** of net behaviours to obtain a category in which the objects are ned nets and the morphisms represent simulations of timed nets. Following the proach described in the introduction, a refinement of one timed net by another uld allow any timed computation of the source net to be simulated by a timed mputation of the target net. We say that a timed computation d simulates another d simulates c in our original sense and furthermore has shorter duration. Thus, a ned specification is a specification of desired computations together with an upper und on the duration of each computation.

In this section, we define a category **Time** of timed nets. In addition, we exhibit tain adjunctions between **Behav** and **Time** which enable us to construct the free "time unconstrained" specification corresponding to a given net.

mma 4.1 The poset structure on T extends to $\mathcal{T}$–timings via, $\tau : \mathcal{B}(\mathrm{N}) \to \mathbf{T} \leq$ $: \mathcal{B}(\mathrm{N}) \to \mathbf{T}$ if and only if for each computation c of N, $\tau(c) \leq \tau'(c)$.

mma 4.2 Let $\langle C, F\rangle$ be a morphism from $\mathcal{B}(\mathrm{N})$ to $\mathcal{B}(\mathrm{N}')$ in **Behav** and let τ' be timing of $\mathcal{B}(\mathrm{N}')$. Then $\tau' C$ is a timing of $\mathcal{B}(\mathrm{N})$.

roposition 4.3 The following data:

- Objects: Pairs $\langle \mathcal{B}(\mathrm{N}), \tau\rangle$ where $\mathcal{B}(\mathrm{N})$ is a behaviour category and τ is a $\mathcal{T}$–timing of $\mathcal{B}(\mathrm{N})$.
- Morphisms: a morphism from $\langle \mathcal{B}(\mathrm{N}), \tau\rangle$ to $\langle \mathcal{B}(\mathrm{N}'), \tau'\rangle$ is a morphism $\langle C, F\rangle$ from $\mathcal{B}(\mathrm{N})$ to $\mathcal{B}(\mathrm{N}')$ in **Behav** such that $\tau' C \leq \tau$.
- Composition: as in **Behav**

efine a category which we shall denote **Time**.

roof: Morphisms are well defined by Lemma 4.2. Identities and associativity fol- w from the corresponding results for **Behav** and the transitivity of $\geq$. □

et $\langle C, F\rangle : \langle \mathcal{B}(\mathrm{N}), \tau\rangle \to \langle \mathcal{B}(\mathrm{N}'), \tau'\rangle$ be a morphism in **Time**. Since $\langle C, F\rangle : \mathcal{B}(\mathrm{N}) \to$ (N') is also a morphism in **Behav**, for each $c \in E^+$, $C(c)$ implements c. Further, ; $\tau'(Cc) \leq \tau(c)$, Cc is no slower than c. Thus $\langle C, F\rangle$ is precisely a refinement which spects upper bounds on the durations of computations.

emark 4.4 *We can also define a category by reversing the inequality in Proposi- on 4.3. Morphisms in this category are refinements which respect lower bounds on e time.*

here is an evident forgetful functor $\mathcal{U}$ from **Time** to **Behav** mapping $\langle C, F\rangle$: $3(\mathrm{N}), \tau\rangle \longrightarrow \langle \mathcal{B}(\mathrm{N}'), \tau'\rangle$ to $\langle C, F\rangle : \mathcal{B}(\mathrm{N}) \longrightarrow \mathcal{B}(\mathrm{N}')$. The following result shows nat $\mathcal{U}$ always has a right adjoint and in certain cases has a left adjoint.

Theorem 4.5 The forgetful functor $\mathcal{U}$: **Time** $\rightarrow$ **Behav** has a right adjoint $\perp$: **Behav** $\rightarrow$ **Time**. Further, if the monoid **T** has a top element $\top_{\mathbf{T}}$ then $\mathcal{U}$ has left adjoint $\top$: **Behav** $\rightarrow$ **Time**.

Proof: We define $\perp(\mathcal{B}(\mathrm{N}))$ to be $\langle\mathcal{B}(\mathrm{N}), \perp\rangle$ where $\perp$ is the constant function wi value 0. Similarly, we define $\top(\mathcal{B}(\mathrm{N}))$ to be $\langle\mathcal{B}(\mathrm{N}), \top\rangle$ where $\top$ is the consta function with value $\top_{\mathbf{T}}$. It is routine to verify that $\perp$ and $\top$ are functors, and

$$\mathbf{Time}(\langle\mathcal{B}(\mathrm{N}), \tau\rangle, \perp(\mathcal{B}(\mathrm{N}'))) \cong \mathbf{Behav}(\mathcal{U}(\langle\mathcal{B}(\mathrm{N}), \tau\rangle), \mathcal{B}(\mathrm{N}'))$$

as $\mathcal{U}(\langle\mathcal{B}(\mathrm{N}), \tau\rangle) = \mathcal{B}(\mathrm{N})$ and $\langle C, F\rangle$ is a morphism from $\mathcal{B}(\mathrm{N})$ to $\mathcal{B}(\mathrm{N}')$ in **Behav** and only if it is a morphism from $\langle\mathcal{B}(\mathrm{N}), \tau\rangle$ to $\langle\mathcal{B}(\mathrm{N}'), \perp\rangle$ in **Time**. The conditi that $\tau C \geq \perp$ being satisfied trivially. Similarly $\top$ is left adjoint to $\mathcal{U}$.

Thus, for any timing $\tau: \mathcal{B}(\mathrm{N}') \rightarrow \mathbf{T}$ and any morphism $\langle C, F\rangle: \mathcal{B}(\mathrm{N}) \rightarrow \mathcal{B}(\mathrm{N}')$ **Behav**, for any computation c of N we have $\top(c) \geq \tau(C(c))$, since $\top(c) = \top_{\mathbf{T}}$ Hence $\langle C, F\rangle$ is a morphism in **Time** from $\top: \mathcal{B}(\mathrm{N}) \rightarrow \mathbf{T}$ to $\tau: \mathcal{B}(\mathrm{N}') \rightarrow \mathbf{T}$. Th $\top(\mathcal{B}(\mathrm{N}))$ represents the free or time–unconstrained specification of the net N.

5 Structure in Behav and Time

The category **Net** has considerable categorical structure [BGdP91]. In particular, has finite products and coproducts and is symmetric monoidal closed. This structu gives rise to a language for combining simple nets to define more complex ones. A important issue is the extent to which this structure lifts to **Behav**, since we expe the behaviour of a compound net to be determined in a precise way by the behavio of its components.

The coproduct in **Behav** is induced by that in **Net**. Thus $\langle E_0^+, B_0, \partial_0, \partial_1\rangle$ $\langle E_1^+, B_1, \partial_0', \partial_1'\rangle = \langle(E_0 + E_1)^+, B_0 \times B_1, \delta_0, \delta_1\rangle$, where δ_0 and δ_1 are given by t evident extension to $(E_0 + E_1)^+$ of the functions given by:

$$\delta_0(e_0, 0) = \Sigma_{B_0}\Sigma_{B_1} pre_0(e_0, b_0)\langle b_0, b_1\rangle \text{ and } \delta_1(e_0, 0) = \Sigma_{B_0}\Sigma_{B_1} post_0(e_0, b_0)\langle b_0, b_1\rangle$$

with similar definitions for $(e_1, 1)$. The injections are $\langle in_i, \pi_i\rangle$. We are able inject E_i^+ into $(E_0 + E_1)^+$ because $E_0^+ + E_1^+ \subseteq (E_0 + E_1)^+$. Thus there is injection of the computations of N_i into the computations of $\mathrm{N}_0 + \mathrm{N}_1$. The copro uct in **Behav** induces a coproduct in **Time**, given by $\langle\mathcal{B}(\mathrm{N}_0), \tau_0\rangle + \langle\mathcal{B}(\mathrm{N}_1), \tau_1\rangle$ $\langle\mathcal{B}(\mathrm{N}_0) + \mathcal{B}(\mathrm{N}_1), \tau_0 + \tau_1\rangle$ where $\tau_0 + \tau_1: (E_0 + E_1 \rightarrow \mathbf{T})$ is given by $(\tau_0 + \tau_1)(e, i)$ $\tau_i(e)$.

Behav does not have products but does have a symmetric monoidal structu which is the product on a subcategory.

Proposition 5.1 The following assignment:

- $\langle E_0^+, B_0, \partial_0, \partial_1\rangle \oslash \langle E_1^+, B_1, \partial_0', \partial_1'\rangle = \langle(E_0 \times E_1)^+, B_0 + B_1, \delta_0, \delta_1\rangle$, where and δ_1 are given inductively by

$$\begin{aligned} &\delta_i(\langle e_0, e_1\rangle) = \partial_i(e_0) + \partial_i'(e_1) && \text{for } \langle e_0, e_1\rangle \in E_0 \times E_1 \\ &\delta_i(\langle c_0, c_1\rangle; \langle c_0', c_1'\rangle) = \partial_i(c_0; c_0') + \partial_i'(c_1; c_1') \\ &\delta_i(\langle c_0, c_1\rangle | \langle c_0', c_1'\rangle) = \partial_i(c_0 | c_0') + \partial_i'(c_1 | c_1'), \end{aligned}$$

- $\langle C, F\rangle \oslash \langle D, G\rangle = \langle C \oslash D, F + G\rangle$, where $C \oslash D$ has an inductive definition analogous to that of δ_i above,

efines a symmetric monoidal structure on **Behav**. Furthermore, $\oslash$ is the categorical roduct on the subcategory consisting of those morphisms in which the map on omputations restricts to a function between the generating sets.

'his subcategory is precisely the image under $\mathcal{B}$ of **Net**. In fact, the categories **Net** nd $\mathcal{B}(\mathbf{Net})$ are isomorphic. The "projections" $\langle \pi_i, in_i\rangle : \mathcal{B}(\mathrm{N}_0) \oslash \mathcal{B}(\mathrm{N}_1) \to \mathcal{B}(\mathrm{N}_i)$ here in_i is the injection of B_i into $B_0 + B_1$ and π_i is the evident extension of the rojection function from $E_0 \times E_1$ to $(E_0 \times E_1)^+$ (analogous to the definition of δ_i bove) are not categorical projections, since the universal property fails. This is ecause the pair $\langle C_0 c, C_1 c\rangle$ is not in general an element of $(E_0 \times E_1)^+$, as is clear if e consider the case where $C_0 c = e_0; e_1$ and $C_1 c = e_0$. However, in the case where $C_i(E) \subseteq E_i$ (that is, each C_i restricts to a map between generating sets) there is a niversal arrow given by $\langle\langle C_0, C_1\rangle, [F_0, F_1]\rangle$.

The monoidal structure $\oslash$ on **Behav** lifts to a monoidal structure $\oslash$ on **Time** ia

$$\langle \mathcal{B}(\mathrm{N}_0), \tau_0\rangle \oslash \langle \mathcal{B}(\mathrm{N}_1), \tau_1\rangle = \langle \mathcal{B}(\mathrm{N}_0) \oslash \mathcal{B}(\mathrm{N}_1), \tau_0 \vee \tau_1\rangle$$

here $(\tau_0 \vee \tau_1)\langle e_0, e_1\rangle = max\{\tau_0 e_0, \tau_1 e_1\}$, and extended to computations in the usual ay. Further, as for **Behav**, the monoidal structure $\oslash$ is the categorical product n the subcategory of **Time** comprising all objects together with those morphisms hich are in $\mathcal{B}(\mathbf{Net})$. This is an interesting result, as the introduction of timing ften makes it impossible to find a categorical product [Gur90].

6 A Proof System for Timing Nets

n this section we briefly describe a calculus for deriving the duration of computations f a given net. The calculus derives sequents of the form $[c: M_0 \to M_1, t]$, whose ntended interpretation in any timed net $\langle N, \tau\rangle$ is that the computation c from narking M_0 to marking M_1 has duration t.

$$\text{cst}\frac{}{\vdash [e: pre(e) \to post(e), \tau(e)]}\, e \in E$$

$$\text{zero}_{sl}\frac{\vdash [c: M_0 \to M_1, t \cdot 0]}{\vdash [c: M_0 \to M_1, t]} \qquad \text{zero}_{sr}\frac{\vdash [c: M_0 \to M_1, 0 \cdot t]}{\vdash [c: M_0 \to M_1, t]}$$

$$\text{seq}\frac{\vdash [c_0: M_0 \to M_1, t_0] \quad \vdash [c_1: M_1 \to M_2, t_1]}{\vdash [c_0; c_1: M_0 \to M_2, t_0 \cdot t_1]} \qquad \text{zero}_p\frac{\vdash [c: M_0 \to M_1, t \otimes 0]}{\vdash [c: M_0 \to M_1, t]}$$

$$\text{par}\frac{\vdash [c_0: M_0 \to M_0', t_0] \quad \vdash [c_1: M_1 \to M_1', t_1]}{\vdash [c_0 | c_1: M_0 + M_1 \to M_0' + M_1', t_0 \otimes t_1]} \qquad \text{com}\frac{[c: M_0 \to M_1, t_0 \otimes t_1]}{[c: M_0 \to M_1, t_1 \otimes t_0]}$$

Example 6.1 Timing Efficient Tea Making (see Example 3.7)

$$\frac{\dfrac{[\mathrm{boil} : (\mathrm{pow} + \mathrm{wat}) \rightarrow \mathrm{hw}, 5] \qquad [\mathrm{ad1} : (\mathrm{bag} + \mathrm{pot}) \rightarrow \mathrm{pb}, 1]}{[\mathrm{boil}|\mathrm{ad1} : (\mathrm{pow} + \mathrm{wat} + \mathrm{bag} + \mathrm{pot}) \rightarrow \mathrm{hw} + \mathrm{pb}, 5]} \qquad [\mathrm{ad2} : (\mathrm{hw} + \mathrm{pb}) \rightarrow \mathrm{tea}, 1]}{[(\mathrm{boil}|\mathrm{ad1}); \mathrm{ad2} : (\mathrm{pow} + \mathrm{wat} + \mathrm{bag} + \mathrm{pot}) \rightarrow \mathrm{tea}, 6]}$$

It is routine to verify that these rules are sound with respect to interpretation in any timed net. Furthermore, we can construct a timed net in which $\tau(c{:}\, M_0 \rightarrow M_1) = t$ if and only if $\vdash [c{:}\, M_0 \rightarrow M_1, t]$. Future work will relate this system to the logic described in [BGdP91] for reasoning about net refinements.

7 Acknowledgements

This work has benefited considerably from the comments and suggestions of Simon Ambler, Valeria de Paiva, John Power, Jørn Schmidt, and Glynn Winskel.

References

[BG90] C. T. Brown and D. J. Gurr, *Timing Petri Nets*, manuscript, December 1990.

[BGdP91] C. T. Brown, D. J. Gurr, and V. C. V. de Paiva, *A Linear Specification Language for Petri Nets*, Technical Report DAIMI PB - 363, DAIMI, Århus University 1991.

[DMM89] P. Degano, J. Meseguer, and U. Montanari, *Axiomatizing Net Computations and Processes*, in *Proc. LICS*, 1989.

[GG90] V. Gehlot and C. Gunter, *A Proof-Theoretic Operational Semantics for True Concurrency*, in *Proc. LICS*, 1990.

[Gur90] D. J. Gurr, *Semantic Frameworks for Complexity*, PhD thesis ECS-LFCS-91-130, University of Edinburgh, 1990.

[Jen90] K. Jensen, *Coloured Petri Nets: a high level language for system design analysis* in G. Rozenberg, editor, *Proc. Advances in Petri nets*, Springer–Verlag, 1990.

[MM88] J. Meseguer and U. Montanari, *Petri Nets are Monoids: A new algebraic foundation for net theory*, Technical Report SRI-CSL-88-3, C.S. Lab., SRI International, January 1988.

[NW91] M. Nielsen and G. Winskel, *Concurrency*, to appear as a chapter in the Handbook of Theoretical Computer Science.

[Rei85] W. Reisig, *Petri Nets: an Introduction*, EATCS Monographs on Theoretical Computer Science Springer–Verlag, 1985.

[Win84] G. Winskel, *Categories of Models for Concurrency*, In S.D. Brookes, A.W Roscoe, and G. Winskel, editors, *Proc. Seminar on Semantics of Concurrency (CMU, Pittsburgh)*, LNCS 197, Springer–Verlag, 1984.

ASYNCHRONOUS CELLULAR AUTOMATA FOR INFINITE TRACES*

Paul GASTIN
Université PARIS 6
LITP, Institut Blaise Pascal
4, place Jussieu
75 252 PARIS CEDEX 05
FRANCE
gastin@litp.ibp.fr

Antoine PETIT
Université PARIS SUD
LRI, URA CNRS 410
Bât. 490
91 405 ORSAY CEDEX
FRANCE
petit@lri.lri.fr

Abstract: The notion of infinite traces has been proposed to generalize both the finite traces defined by Mazurkiewicz and the infinite words. The family of recognizable languages of infinite traces has been introduced by means of recognizing morphisms and characterized by c-rational expressions. In this paper, we provide another characterization of this family using asynchronous cellular automata (which carry the most intuitive idea of finite state concurrent machines). To this purpose, we give effective constructions for c-rational operations on these automata, which are of independent interest.

Introduction

concept of traces has been introduced by Mazurkiewicz [Maz77] as a suitable semantics for -sequential processes. Let us refer, for instance, surveys [Maz86, AR88, Per89] or the ograph [Die90] where also extensive bibliographies on the subject are given.

order to describe the behaviours of non sequential processes which never terminate, the oid ℂ of complex traces has been proposed [Die91]. In this monoid, the subset ℝ of real es is of first importance since it turns out that a complex trace is simply a real trace together n an alphabetic information. In fact, previously to the definition of ℂ, real traces were ady introduced and intensively studied [BMP89, Gas90, Kwi90, GR91].

ognizable languages describe the behaviours of finite state systems and hence form one of basic families of a monoid. As regards infinite traces, the family of recognizable languages been defined by recognizing morphisms in [Gas91] and characterized by c-rational ressions in [GPZ91] and [DGP91]. But, until now, it suffers from lack of characterizations means of finite automata. The aim of this paper is to remedy this deficiency. For lack of ce, we focus on automata on recognizable real trace languages only.

recognizable languages of finite traces, Zielonka introduced *asynchronous* and *nchronous cellular* automata which are major special classes of trace automata [Zie87, 89, CMZ89]. These automata are more suited for the description of concurrent processes e, thanks to their decentralized control, they allow concurrent executions of independent ons.

üchi automaton for infinite words is simply a classical finite word automaton together with eptance conditions for infinite paths [Eil74, Tho90, PP91]. Hence, we will extend

is work has been partly supported by the ESPRIT Basic Research Actions N° 3166 (ASMICS) and N° 3148 MON) and by the PRC C^3 and Math-Info.

asynchronous cellular automata to accept infinite real traces using a suitable Büchi acceptance condition for infinite paths. In order to obtain a class of automata which are suit for infinite traces, we will use a decentralized acceptance condition. We show that these Bü asynchronous cellular automata accept recognizable real trace languages.

Conversely, we prove that each recognizable real trace language is accepted by some Bü asynchronous cellular automaton. To this purpose, we use effective constructions to show the class of Büchi asynchronous cellular automata is closed under union, intersecti concatenation and a restricted infinite iteration.

Therefore, we extend the deep Zielonka's theorem to infinite traces with a characterizatio the family of recognizable real trace languages by means of Büchi asynchronous cell automata.

For lack of space, this paper does not contain all proofs. They can be found in [GP Nevertheless, we emphasize on the fact that all the constructions are given and hence a Bü asynchronous cellular automaton can be obtained, in an effective way, for any recogniza language of real traces.

2. Preliminaries

In this section, we recall definitions and results on traces which are relevant in this paper. more detailed discussions, the reader is referred to [Maz86, AR88, Per89, Die90] as conce finite traces, to [Gas91, GPZ91] as concerns real traces.

2.1. Dependence graphs and traces

In the following (X,D) denotes a finite dependent alphabet, i.e. X is a finite alphabet toget with a reflexive and symmetric dependence relation $D \subseteq X \times X$. We denote by $I = X \times X \setminus D$ independence relation which is the complement of D.

A dependence graph over (X,D) is (an isomorphism class of) a labelled acyclic graph [V,E where V is the (at most) countable set of vertices, $E \subseteq V \times V$ is the set of arcs, $\lambda: V \to X$ the labelling and which satisfies:

i) the induced partial order (V,E^*) is well-founded: there does not exist an infi descending chain i.e. an infinite sequence (x_i) such that $(x_{i+1}, x_i) \in E$ for all i.

ii) edges are between dependent vertices:

$$\forall x,y \in V, \qquad ((\lambda(x),\lambda(y)) \in D \Leftrightarrow x = y \text{ or } (x,y) \in E \text{ or } (y,x) \in E$$

Then, for any a in X, the restriction to the set of vertices with label a is well-ordered. T allows to think of dependence graphs by standard representations where the vertices are p (a,i) with $a \in X$ and i is a countable ordinal. The set of dependence graphs is denoted $\mathbb{G}(X,D)$ or simply by $\mathbb{G}$ when there is no confusion. It is a monoid for the operat $[V_1,E_1,\lambda_1][V_2,E_2,\lambda_2] = [V,E,\lambda]$ where $[V,E,\lambda]$ is the disjoint union of $[V_1,E_1,\lambda_1]$ a $[V_2,E_2,\lambda_2]$ together with new arcs (p_1,p_2) for all vertices $p_1 \in V_1$, $p_2 \in V_2$ such t $(\lambda_1(p_1),\lambda_2(p_2)) \in D$. The neutral element is the empty graph $1 = [\emptyset,\emptyset,\emptyset]$.

The concatenation is generalized to an infinite product as follows. Let (g_i) be any sequence dependence graphs. Then $g = g_1g_2...g_i... \in \mathbb{G}$ is defined as the disjoint union of the g_i with ions new arcs from g_i to g_j for i<j whenever the vertices have dependent labels. Note that catenation and infinite product are associative operations.

ng these operations, we relate words and dependence graphs. We define the mapping $X^\infty \to \mathbb{G}$ by $\varphi(a) = [\{x\},\emptyset,x \mapsto a]$ for all a in X and $\varphi(a_1a_2\ldots) = \varphi(a_1)\varphi(a_2)\ldots$ for all finite nfinite word $a_1a_2\ldots$ in X^∞. Note that φ is not injective, for instance $\varphi(ab) = \varphi(ba)$ iff) $\in$ I. We define an equivalence $\sim_I$ on X^∞ by $u \sim_I v$ if $\varphi(u) = \varphi(v)$ for all u,v in X^∞. It s out that φ is a surjective morphism from the free monoid X^* onto the monoid $\mathbb{FG}(X,D)$ inite dependence graphs. The monoid of finite traces introduced by A. Mazurkiewicz is isely $X^*/\sim = \mathbb{FG}(X,D)$ which is denoted by $\mathbb{M}(X,D)$ or simply by $\mathbb{M}$ throughout.

Infinite traces

similar way, we will define the monoid of infinite traces from the monoid $\mathbb{G}$. But it turns hat $\varphi : X^\infty \to \mathbb{G}$ is neither surjective nor a morphism. The set $\varphi(X^\infty) = X^\infty/\sim$ of dependence hs which can be represented by words is denoted by $\mathbb{R}(X,D)$ or simply by $\mathbb{R}$ when there confusion. Its elements are called real dependence graphs or real traces. $\mathbb{R}$ is characterized ollows: a dependence graph $[V,E,\lambda]$ is a real one iff the past $p{\downarrow} = \{q \in V / (q,p) \in E^*\}$ is e for every vertex p of V. Unfortunately $\mathbb{R}$ is not a submonoid of $\mathbb{G}$. For instance if a and b two dependent letters, the concatenation of the two real graphs a^ω and b^ω is not a real h. More precisely, for a real dependence graph $g = [V,E,\lambda]$, let $alph(g) = \lambda(V)$ be the set of rs which occur in g and let $alphinf(g) = \{a \in X / |\lambda^{-1}(a)| = \omega\}$ be the set of letters which r infinitely often in g. Then the concatenation of two real graphs r and s is a real graph if only if $alphinf(r) \times alph(s) \subseteq I$. In particular the concatenation of a finite trace and a real is always a real trace. Besides, an infinite product of finite traces is always a real trace. that in this paper we **never** use a concatenation or an infinite product of real traces which l a **non** real trace.

η: $\mathbb{M} \to S$ be a morphism into a finite monoid. A real trace language $R \subseteq \mathbb{R}$ is recognized if for any (finite or infinite) sequence $(r_i) \subseteq \mathbb{M}$, we have

$$r_1 \cdot r_2 \cdot r_3 \ldots \in R \Rightarrow \eta^{-1}\eta(r_1) \cdot \eta^{-1}\eta(r_2) \cdot \eta^{-1}\eta(r_3) \ldots \subseteq R$$

denote by Rec $\mathbb{R}$ (resp. Rec $\mathbb{M}$) the family of languages of real traces (resp. finite traces) h are recognized by some morphism. Note that if the independence relation I is empty, definitions reduce to the classical ones for recognizable word languages. In this case, the lies are simply denoted by Rec X^∞ (resp. Rec X^*).

is paper, we use two important results on recognizable real trace languages [Gas91]. The one relates traces and words: a real trace language $R \subseteq \mathbb{R}$ is recognizable if and only if R) is a recognizable word language. The second one gives a rational expression of a real language $R \subseteq \mathbb{R}$ recognized by a morphism η: $\mathbb{M} \to S$. For $s \in S$, let $R_s = n^{-1}(s)$. Then

$$R = \bigcup_{(s,e) \in P} R_s R_e^\omega \text{ where } P = \{(s,e) \in S^2 / se=s, e^2=e \text{ and } R_s R_e^\omega \subseteq R\}.$$

2.3. Automata on finite traces

In a general monoid M, a (non-deterministic) M-automaton [Ber79] is a tuple $\mathcal{A} = \langle Q, \delta,$ where Q is a finite set of states, S, F $\subseteq$ Q are the sets of initial (starting) and final st $\delta: Q \times M \rightarrow \mathcal{P}(Q)$, $\delta(q,m) = qm$ is the transition function which verifies $q \in q1$ and q(m (qm)n for all $q \in Q$ and $m,n \in M$. The recognized set is $L(\mathcal{A}) = \{m \in M / Sm \cap F \neq \emptyset\}$. finite traces, if such an automaton $\mathcal{A}$ recognizes a trace language T, $\mathcal{A}$ can also be consid as a word automaton which recognizes $\varphi^{-1}(T)$. Conversely a word automaton can be viewe a trace automaton provided the I-diamond criterion is verified: qab = qba for any $q \in Q$ $(a,b) \in I$. In fact it suffices in this case to define qt = qx for any trace t and any word x in φ to obtain a trace automaton (see e.g. [Die90]).
Two major subclasses of trace automata, namely asynchronous and asynchronous cell automata, were introduced by Zielonka [Zie87,Zie89]. These automata have a distrib control structure and they can thus perform actions concurrently. Zielonka's theorem states a trace language T $\subseteq$ IM is recognizable if and only if it is recognized by some determin asynchronous (cellular) automaton. In this paper we focus on asynchronous cellular autor but our results can be easily extended to asynchronous automata. The definitio asynchronous cellular automata will be obtained as a special case of the definition of B asynchronous cellular automata given in next section.

3. Büchi asynchronous cellular automata

Although the set of finite and infinite words forms a monoid, the right notion of automa this monoid is not the notion of M-automata presented in Section 2.3. Recall that an autom on infinite words is a classical finite X^*-automaton together with acceptance conditions infinite paths [Eil74,Tho90,PP91]. These conditions are on the set $f(P)$ of states which repeated infinitely often in the path P. More precisely, in a Büchi automaton, a set R of repe states is specified and a path P is Büchi-accepted if $f(P)$ contains some element of R. Bü theorem states that a word language $L \subseteq X^\infty$ is recognizable if and only if it is recognize some Büchi automaton. It is well known that in this result, the Büchi automaton canno chosen deterministic (in other words the class of word languages recognized by determin Büchi automata forms a proper subclass of the family of recognizable word languages).

The principal aim of this paper is to extend Büchi's and Zielonka's theorems to language infinite traces and to (non-deterministic) Büchi asynchronous cellular automata. To this pur we need first a good notion of Büchi asynchronous cellular automata.

Let us first introduce some notations in order to lighten formulas. Let $(Q_z)_{z\in Z}$ be a famil sets. For Y $\subseteq$ Z, we denote by Q_Y the cartesian product $\prod_{y\in Y} Q_y$ and if $q = (q_z)_{z\in Z} \in Q_Z$ denote by q_Y the subfamily $(q_y)_{y\in Y} \in Q_Y$. Finally, $\mathrm{Cell}(Q_X) = \{ \prod_{z\in Z} R_z / R_z \subseteq Q_z$ for all $z \in$ is the set of cells of Q_Z.

üchi asynchronous cellular automaton (BACA for short) is a tuple $\mathcal{A}=\langle(Q_a)_{a\in X},$ $_{a\in X}$, S, F, $\mathcal{R}\rangle$ where, for each $a \in X$, Q_a is the finite set of local states for a, δ_a from $Q_{D(a)}$ (Q_a) is the local transition function for a, $S,F \subseteq Q_X$ are the sets of starting and final states ectively and $\mathcal{R} \subseteq Cell(Q_X)$ is the acceptance table.

best way to describe how an asynchronous automaton works in a distributed way is ably to define paths for dependence graphs. See for instance [Tho89, Tho91] where this oach has been used for asynchronous automata on finite traces. For lack of space, we ose here a more classical (equivalent) formulation using global paths.

global transition function $\Delta : Q_X \times X \to \mathcal{P}(Q_X)$ of $\mathcal{A}$ is defined by:

$$q' \in \Delta(q,a) \text{ if } q'_a \in \delta_a(q_{D(a)}) \text{ and } q'_{X\setminus\{a\}} = q_{X\setminus\{a\}}$$

that, with its global transition function, a BACA can be viewed as a usual word automaton ch turns out to be I-diamond.

nite or infinite path P is a sequence $q_0,a_1,q_1,a_2,q_2,\ldots$ such that $q_i \in \Delta(q_{i-1},a_i)$ for all $i \geq 1$. label of P is the word $\ell(P) = a_1a_2\ldots$, the starting state of P is $s(P) = q_0$. For any $a \in X$, we e the final local information $f_a(P)$ by:

- $f_a(P) = \{q_{i,a} / i = \inf\{j \geq 0 / \forall\ k > j,\ a_k \neq a\}\}$ is the singleton which consists of the final local state for a, if P has finitely many transitions labelled by a

- $f_a(P) = \{p_a \in Q_a / |\{i / a_i = a \text{ and } q_{i,a} = p_a\}| = \omega\}$ is the set of local states for a repeated infinitely often, otherwise.

n the definition of a BACA, it is easy to verify that, in both cases, for each $a \in X$, $f_a(P) =$ $\in Q_a / |\{i / q_{i,a} = p_a\}| = \omega\}$. We denote by $f(P) = \prod_{a\in X} f_a(P)$ the final information of P. that, if $P = q_0,a_1,q_1,\ldots,a_n,q_n$ is a finite path, then $f(P) = \{q_n\}$.

lly, a path P is accepted if $s(P) \in S$ and either P is finite and $f(P) \cap F \neq \emptyset$ or P is infinite $f(P) \supseteq R$ for some $R \in \mathcal{R}$. The word and trace languages accepted by $\mathcal{A}$ are respectively

$L(\mathcal{A}) = \{u \in X^\infty / u = \ell(P) \text{ for some accepted path } P\}$ and

$T(\mathcal{A}) = \{t \in \mathbb{R} / t = \varphi(\ell(P)) \text{ for some accepted path } P\} = \varphi(L(\mathcal{A}))$.

notion of asynchronous cellular automata (ACA for short) introduced by Zielonka is simply otion of BACA with empty acceptance tables ($\mathcal{R} = \emptyset$).

remainder of this section will explain and justify this definition of BACA.

e the global transition function Δ of a BACA is I-diamond, we have for each $q \in Q_X$, $u) = \Delta(q,v)$ if u and v are two equivalent finite words. This fact is strengthened and nded to finite and infinite words in the following proposition.

position 3.1. Let u,v be two equivalent finite or infinite words and let $\mathcal{A}$ be a BACA. n, for each path P such that $\ell(P) = u$, there exists a path P' such that $\ell(P') = v$, $s(P') = s(P)$ $f(P') = f(P)$.

following trivial corollary explains why BACA are trace automata.

ollary 3.2. Let $\mathcal{A}$ be a BACA. Then, $L(\mathcal{A})$ is closed, i.e. $L(\mathcal{A}) = \varphi^{-1}(T(\mathcal{A}))$.

Informally, an infinite path P of a BACA is recognized if there exists an element $R = \prod_{a \in X}$ F

the acceptance table such that, for all $a \in X$ and $r_a \in R_a$, infinitely many states of P have local state. The next remark claims that it would not have been possible to specify only repe local states for some letters or even to specify one repeated local state for each letter.

Remark 3.3. Let $X=\{a,b\}$ with $(a,b) \in I$. If a BACA $\mathcal{A}$ recognizes the trace langu $\{(ab)^{\omega}\}$ then each useful table (R_a,R_b) in the acceptance table of $\mathcal{A}$ verifies $|R_a| \geq 2$ $|R_b| \geq 2$. Indeed, let $\mathcal{A}$ be a BACA recognizing $\{(ab)^{\omega}\}$ and suppose that there exis "useful" table $R = R_a \times R_b$ in the acceptance table $\mathcal{R}$ with, for instance, $|R_a| \leq 1$. Since "useful" there exists an infinite path P such that $R \subseteq f(P)$. If $|R_a| = 0$ then the path P' obta from P by removing transitions with label a is also accepted. If $R_a = \{r_a\}$, we can w $P = P'P''$ with $f_a(P') = r_a$. Then the path P'P''' is also accepted where P''' is obtained fror by removing transitions with label a. In both cases, we obtain an accepted path with a lab $(a+b)^*b^{\omega}$ which is in contradiction with the fact that $\mathcal{A}$ recognizes the trace language $\{(ab)$

The following proposition asserts that the (deterministic) BACA, considered as word auton have the same expressive power as the classical (deterministic) Büchi automata. This expl why we legitimaly called them **Büchi** automata.

Proposition 3.4. Let $\mathcal{A}$ be a (deterministic) BACA, then there exists a class (deterministic) Büchi automaton $\mathcal{A}'$ recognizing the word language $L(\mathcal{A})$.

From Büchi's theorem and Proposition 3.4, we deduce that, if $\mathcal{A}$ is a BACA, the lang $L(\mathcal{A})$ is recognizable. Since $L(\mathcal{A}) = \varphi^{-1}(T(\mathcal{A}))$, the characteristic property of recognizable trace language recalled in Section 2.2 asserts that the language $T(\mathcal{A})$ is recognizable too. H we can claim:

Corollary 3.5. Let $\mathcal{A}$ be a BACA, then the word language $L(\mathcal{A})$ and the trace langu $T(\mathcal{A})$ are both recognizable.

Note that, from Proposition 3.4 and classical results on recognizable word languages, w that there exist recognizable trace languages which are not accepted by **deterministic** BAC

4. Rational constructions on BACA

We have seen in the previous section that BACA accept recognizable real trace languages. aim of this section is, conversely, to construct a BACA for each recognizable real t language. In order to prove this result, we use the rational expression of these langu recalled in Section 2.2. Hence, we need to study closure properties of the family of B under rational operators.

Since we will need this property for the closure under union, we recall that each ACA ca transformed in an equivalent complete ACA [Die90]. An ACA $\mathcal{A} = \langle (Q_a)_{a \in X}, (\delta_a)_{a \in X} \rangle$ is to be complete if $\delta_a(q_{D(a)}) \neq \emptyset$ for all $a \in X$ and $q_{D(a)} \in Q_{D(a)}$.

closure under union and intersection of the BACA family will be treated using a cartesian uct of automata as for word automata.

position 4.1. Let $\mathcal{A}_1$, $\mathcal{A}_2$ be BACA. There exists a BACA $\mathcal{A}$ such that :) = $T(\mathcal{A}_1) \cup T(\mathcal{A}_2)$ (resp. $T(\mathcal{A}) = T(\mathcal{A}_1) \cap T(\mathcal{A}_2)$). Moreover, if $\mathcal{A}_1$, $\mathcal{A}_2$ are rministic, so is $\mathcal{A}$.

of. For i = 1,2, consider the BACA $\mathcal{A}_i = \langle (Q_{i,a})_{a\in X}, (\delta_{i,a})_{a\in X}, S_i, F_i, \mathcal{R}_i \rangle$ and let $= \prod_{a\in X} Q_{i,a}$. Thanks to the canonical bijection between $\prod_{a\in X} (Q_{1,a} \times Q_{2,a})$ and $Q_1 \times Q_2$, we tify these two sets throughout this construction.
ne the cartesian product $\mathcal{A} = \mathcal{A}_1 \times \mathcal{A}_2 = \langle (Q_a)_{a\in X}, (\delta_a)_{a\in X}, S \rangle$ by:

- $Q_a = Q_{1,a} \times Q_{2,a}$,
- $\delta_a((q_1,q_2)_{D(a)}) = \delta_{1,a}((q_1)_{D(a)}) \times \delta_{2,a}((q_2)_{D(a)})$
- $S = S_1 \times S_2$.

i = 1,2, we denote Π_i the projections defined by $\Pi_i(q_1,q_2) = q_i$ for all $(q_1,q_2) \in Q_1 \times Q_2$. n, the BACA $\mathcal{A}$ with final states $F_1 \times F_2$ and acceptance table $\mathcal{R} = \{R \in \text{Cell}(Q_X)$ / $\in \{1,2\}, \Pi_i(R) \supseteq R_i$ for some $R_i \in \mathcal{R}_i\}$ accepts the language $T(\mathcal{A}_1) \cap T(\mathcal{A}_2)$. the union, as we recalled above, we may assume that $\mathcal{A}_1$ and $\mathcal{A}_2$ are complete. Hence, in a lar way, the BACA $\mathcal{A}$ with final states $F = Q_1 \times F_2 \cup F_1 \times Q_2$ and acceptance table $\{R \in \text{Cell}(Q_X)$ / $\exists\, i \in \{1,2\}, \Pi_i(R) \supseteq R_i$ for some $R_i \in \mathcal{R}_i\}$ accepts the language $_1) \cup T(\mathcal{A}_2)$. ◊

v, we turn to the concatenation. To this purpose, we first need an auxiliary lemma which ws to clearly distinguish between finite accepted paths and infinite ones.

nma 4.2. Let $\mathcal{A}$ be a BACA. There exists a BACA $\mathcal{A}'$ such that $T(\mathcal{A}') = T(\mathcal{A})$ and

- for each final state $f=(f_a)_{a\in X}$ of $\mathcal{A}'$, f_a is a dead state for all $a \in X$.
- for each element R of the acceptance table of $\mathcal{A}'$, R_a contains at least two states for some $a \in X$.

of. Let $\mathcal{A} = \langle (Q_a)_{a\in X}, (\delta_a)_{a\in X}, S, F, \mathcal{R} \rangle$ be a BACA and let $Q = \prod_{a\in X} Q_a$. Intuitively, in r to achieve the first condition we use new final states which are copies of the original ones from which no transition are possible. For the second condition, we unfold once each nentary loop $(p_a \in \delta_a(p_{D(a)}))$ which ensure that there is no stationary state in a path and ws to restrict ourselves to tables with at least two states. More formally, the construction ows. Again, we identify the two sets $\prod_{a\in X} (Q_a \times \{-1,0,1\})$ and $Q \times \{-1,0,1\}^X$.
ine the BACA $\mathcal{A}' = \langle (Q'_a)_{a\in X}, (\delta'_a)_{a\in X}, S', F', \mathcal{R}' \rangle$ by:

- $Q'_a = Q_a \times \{-1,0,1\}$,
- $(q_a,k_a) \in \delta'_a((p,j)_{D(a)})$ if $j_a \neq 0$ and $q_a \in \delta_a(p_{D(a)})$ and ($k_a = 0$ or $k_a = -j_a$)
- $S' = S \times \{0,1\}^X$,
- $F' = F \times \{0\}^X$ and

- $\mathcal{R}' = \{R' \in \mathrm{Cell}(Q'_X) \;/\; \Pi_1(R') \supseteq R$ for some $R \in \mathcal{R}$, $\Pi_2(R') \subseteq \{-1,1\}^X$ and $|R'| \geq$

where the projections Π_i are defined by:

- $\Pi_1(q,j) = q$ and $\Pi_2(q,j) = j$ for all $(q,j) \in Q \times \{-1,0,1\}^X$.

The BACA $\mathcal{A}'$ satisfies the required condition and it can be proved that $T(\mathcal{A}') = T(\mathcal{A})$.

Note that in the lemma above, the construction's part which concerns acceptance ta preserves determinism whereas the part for dead final states does not. This is not a prob since we know that the concatenation of two deterministic recognizable word languages ma a non deterministic one. For instance, $(a+b)^*a^\omega$ is non deterministic whereas $(a+b)^*$ and a^ω deterministic (see [PP91] for more details about deterministic word languages).

Proposition 4.3. Let $\mathcal{A}_1$, $\mathcal{A}_2$ be BACA such that $\mathrm{alphinf}(r_1) \times \mathrm{alph}(r_2) \subseteq I$ for $(r_1,r_2) \in T(\mathcal{A}_1) \times T(\mathcal{A}_2)$ (in particular, this condition is satisfied if $T(\mathcal{A}_1) \subseteq IM$ or if $T($ and $T(\mathcal{A}_2)$ are fully independent). Then the concatenation $T(\mathcal{A}_1) \cdot T(\mathcal{A}_2)$ is accepted by sc BACA.

Proof. Consider the BACA $\mathcal{A}_i = \langle (Q_{i,a})_{a\in X}, (\delta_{i,a})_{a\in X}, S_i, F_i, \mathcal{R}_i \rangle$ for i=1,2. Lemma 4.2, we may assume that each final state in F_i is a dead state and that for each elem R of the acceptance tables $\mathcal{R}_i$, R_a contains at least two states for some $a \in X$. For all $a \in$ let $Q_{3,a} = \{1,2\}$. For i=1,2,3, let $Q_i = \prod_{a\in X} Q_{i,a}$. Thanks to the canonical bijection betw $\prod_{a\in X} (Q_{1,a} \times Q_{2,a} \times Q_{3,a})$ and $Q_1 \times Q_2 \times Q_3$, we identify these two sets throughout t construction.

Define the I-shuffle $\mathcal{A} = \mathcal{A}_1 ш_I \mathcal{A}_2 = \langle (Q_a)_{a\in X}, (\delta_a)_{a\in X}, S, F, \mathcal{R} \rangle$ by:

- $Q_a = Q_{1,a} \times Q_{2,a} \times Q_{3,a}$,
- $\delta_a((q_1,q_2,q_3)_{D(a)}) =$

$$\begin{cases} \{q_{1,a}\} \times \delta_{2,a}((q_2)_{D(a)}) \times \{2\} & \text{if } q_{3,b} = 2 \text{ for some } b \in D(a) \\ \{q_{1,a}\} \times \delta_{2,a}((q_2)_{D(a)}) \times \{2\} \cup \delta_{1,a}((q_1)_{D(a)}) \times \{q_{2,a}\} \times \{1\} & \text{otherw} \end{cases}$$

- $S = S_1 \times S_2 \times \{\mathbf{1}\}$ where $\mathbf{1}$ stands for the tuple $(1,\ldots,1)$,
- $F = F_1 \times F_2 \times Q_3$
- $\mathcal{R} = \{R \in \mathrm{Cell}(Q_X) \;/\; \forall i \in \{1,2\}, \Pi_i(R) \cap F_i \neq \emptyset$ or $\Pi_i(R) \supseteq R_i$ for some $R_i \in \mathcal{R}_i\}$

where, for i = 1,2,3, the projections Π_i are defined by:

- $\Pi_i(q_{1,a},q_{2,a},q_{3,a}) = q_{i,a}$ for all $(q_{1,a},q_{2,a},q_{3,a}) \in Q_a$ and
- $\Pi_i(q_1,q_2,q_3) = q_i$ for all $(q_1,q_2,q_3) \in Q_1 \times Q_2 \times Q_3$.

Then $T(\mathcal{A}) = \{r \cdot s \in \mathbb{R} \;/\; r \in T(\mathcal{A}_1), s \in T(\mathcal{A}_2) \;\&\; \mathrm{alphinf}(r) \times \mathrm{alph}(s) \subseteq I\} = T(\mathcal{A}_1) \cdot T(\mathcal{A}_2)$

Note that in order to obtain the main result of this section, we only need the special case of proposition where the first language of the concatenation is a set of finite traces. For restricted case, we can also use a construction which is conceptually a bit simpler. Roug speaking, it is the disjoint union of the automata in which we mix up the final states of the f automata with the initial states of the second one. In fact, these constructions are very cl which explains why we present the more general one.

mains to construct BACA which accepts the infinite iteration of a language accepted by e ACA. Since the infinite iteration of a recognizable language of finite traces is not ssarily recognizable, we need some restriction on the ACA. The following proposition s the major construction which applies when the ACA has a unique initial state, a unique l state and accepts traces with alphabet X only .

position 4.4. Let $\mathcal{A}$ be an ACA with a unique initial state and a unique final state and ch verifies $alph(t) = X$ for all $t \in T(\mathcal{A})$. Then $T(\mathcal{A})^{\omega}$ is accepted by some BACA.

of. Let $\mathcal{A} = \langle (Q_a)_{a\in X}, (\delta_a)_{a\in X}, \{i\}, \{f\}, \emptyset \rangle$ be an ACA and let $Q = \prod_{a\in X} Q_a$. For $N \geq 2$, let be the cyclic group $\mathbb{Z}/_{N\mathbb{Z}}$ and let $\mathbf{G}_N = (G_N)^X = \prod_{a\in X} G_N$. Once again, we identify the two $\prod_{a\in X} (Q_a \times G_N)$ and $Q \times \mathbf{G}_N$ throughout the construction.

$N \geq 2$, define the BACA $\mathcal{A}_N = \langle (Q'_a)_{a\in X}, (\delta'_a)_{a\in X}, \{i'\}, \emptyset, \mathcal{R} \rangle$ by:

- $Q'_a = Q_a \times G_N$,
- $(q'_a, j'_a) \in \delta'_a((q,j)_{D(a)})$ if $\forall\, b \in D(a)$, $(j_b = j'_a)$ or $(j_b = j'_a - 1$ and $q_b = f_b)$

  and $q'_a \in \delta_a(p_{D(a)})$ where $p_b = \begin{cases} q_b & \text{if } j_b = j'_a \\ i_b & \text{otherwise} \end{cases}$
- $i' = i \times \{\mathbf{0}\}$ where $\mathbf{0}$ stands for the tuple $(0,\ldots,0)$ of $\mathbf{G}_N$ and
- $\mathcal{R} = \{R \in Cell(Q'_X) \,/\, \forall\, a \in X, \Pi_2(R_a) = G_N\}$ where $\Pi_2(q_a, j_a) = j_a$ for all (q_a, j_a).

n, for any $N \geq 3$, $T^{\omega} = T(\mathcal{A}_N)$. ◊

ng all the results above, we obtain the following theorem which admits as a trivial corollary major result of [GPZ91]: if T^+ is a recognizable language of finite traces then T^{ω} is ognizable as well.

eorem 4.5. Let $T \in Rec\ \mathbb{M}$ such that $T = T^+$. Then, there exists a BACA which epts T^{ω}.

of. In order to apply Proposition 4.4, we start with some algebraic decompositions of T. t, we use an easy decomposition [GPZ91] : $T^{\omega} = \bigcup_{Y\subseteq X} T \cdot T_Y{}^{\omega}$ where for all $Y \subseteq X$, $= \{t \in T \,/\, alph(t) = Y\}$. Using Zielonka's theorem, Propositions 4.1 and Proposition 4.3, mains to prove that for each $Y \subseteq X$, $T_Y{}^{\omega}$ is accepted by some BACA.

$D' = D \cap (Y \times Y)$ and let $\mathcal{A} = \langle (Q_a)_{a\in Y}, (\delta_a)_{a\in Y}, \{i\}, F, \emptyset \rangle$ be a deterministic ACA over D') which accepts T_Y (Zielonka's theorem). For $t \in T_Y$ we denote by $f(t)$ the final state of path which accepts t. For $f \in F$, let $T_f = \{t \in T_Y \,/\, f(t) = f\}$. Using a Ramsey's type ument, it can be shown that $T_Y{}^{\omega} = \bigcup_{f\in F} T_Y \cdot T_f{}^{\omega}$.

Proposition 4.4, $T_f{}^{\omega}$ is accepted by some BACA over (Y,D'). Using a similar construction in the finite case, it is easy to transform this BACA into a BACA over (X,D) which accepts . We conclude the proof using Proposition 4.1 and Proposition 4.3. ◊

As a consequence of the constructions for union, iteration and infinite iteration, we obtain:

Theorem 4.6. The family of languages accepted by BACA is exactly Rec IR.

Proof. We have seen in Section 3 that if $\mathcal{A}$ is a BACA then T($\mathcal{A}$) is recognizable. Conve let R ∈ Rec IR and let $\eta : \mathbb{M} \rightarrow S$ be a morphism which recognizes R. For s ∈ S $R_s = \eta^{-1}(s)$. We have recalled in Section 2.2 that

$$R = \bigcup_{(s,e)\in P} R_s R_e^{\omega} \text{ where } P = \{(s,e) \in S^2 / se=s, e^2=e \text{ and } R_s R_e^{\omega} \subseteq T\}.$$

For all s ∈ S, R_s ∈ Rec IM and is accepted by some ACA (Zielonka's theorem). Nov e ∈ S is an idempotent ($e^2=e$) then $R_e = R_e^+$. Hence, R_e^{ω} is accepted by some BA (Theorem 4.5). The result follows from Proposition 4.1 and Proposition 4.3.

Note that, in the constructions of Theorems 4.5 and 4.6, we could use for recogniz languages of finite traces non deterministic asynchronous cellular automata which are m more easy to obtain than deterministic ones (see [Pig92]).

5. Conclusion

In this paper we have proved that a real trace language is recognizable if and only if recognized by a Büchi asynchronous cellular automaton. The theory presented in Section 3 easily be extended to define the notion of Büchi asynchronous automata. Then a very sim construction than in the finite case shows that a BACA can be transformed in a Bi asynchronous automaton recognizing the same language. Hence, as BACA, Bi asynchronous automata accept exactly the recognizable real trace languages.
Our main result is a generalization of the deep Zielonka's theorem to infinite real tr languages. This constitutes the main part of the extension of this theorem to complex tra This complete extension is not presented here for lack of space. Note that Büchi asynchron (cellular) automata can recognize real traces only since the label of an infinite path is a real tr. Therefore, to generalize the theory to complex traces it is necessary to transform these auton so that they can take into account the second alphabetic component of a complex trace.
The main drawback of Büchi automata is that non deterministic automata are needed to acc all recognizable languages. On the contrary, in the free monoid X^{∞}, a language is recogniz if and only if it is accepted by some deterministic Muller automaton. All the results presente Section 3 can be easily extended to define the proper classes of Muller asynchronous (cellu automata. Hence, a major open problem is the characterization of recognizable infinite tr languages by such automata.

Acknowledgement: *We thank V. Diekert, H.J. Hoogeboom and W. Zielonka for frui discussions.*

6. References

[AR88] I.J. AALBERSBERG and G. ROZENBERG, "Theory of traces", Theoret Computer Science 60, p. 1-82, 1988.

r79] J. BERSTEL, "Transductions and context free languages", Teubner Studienbucher Informatik, 1979.

IP89] P. BONIZZONI, G. MAURI and G. PIGHIZZINI, "About infinite traces", Proceedings of the ASMICS Workshop on Partially Commutative Monoids, Tech. Rep. TUM-I 9002, Technische Universität München, 1989.

IZ89] R. CORI, Y. METIVIER and W. ZIELONKA, "Asynchronous mappings and asynchronous cellular automata", Tech. Rep. 89-97, LaBRI, Université de Bordeaux, France, 1989.

90] V. DIEKERT, "Combinatorics on traces", Lecture Notes in Computer Science 454, 1990.

:91] V. DIEKERT, "On the concatenation of infinite traces", STACS'91, Lecture Notes in Computer Science 480, p. 105-117, 1991.

P91] V. DIEKERT, P. GASTIN, A. PETIT, "Recognizable complex trace languages", MFCS'91, Lecture Notes in Computer Science 520, p. 131-140, 1991.

74] S. EILENBERG, "Automata, Languages and Machines", Academic Press, New York, 1974.

s90] P. GASTIN, "Infinite traces", Proceedings of the Spring School of Theoretical Computer Science on "Semantics of concurrency", Lecture Notes in Computer Science 469, p. 277-308, 1990.

s91] P. GASTIN, "Recognizable and rational languages of finite and infinite traces", STACS'91, Lecture Notes in Computer Science 480, p. 89-104, 1991.

91] P. GASTIN and A. PETIT, "Asynchronous automata for infinite traces", Tech. rep. 91.68 , LITP, Université Paris 6, France, 1991.

Z91] P. GASTIN, A. PETIT, W. ZIELONKA, "A Kleene theorem for infinite trace languages", ICALP'91, Lecture Notes in Computer Science 510, p. 254-266,1991.

.91] P. GASTIN and B.ROZOY, "The Poset of infinitary traces", to appear in Theoretical Computer Science, Tech. Rep. 91-07, LITP, Université Paris 6, France, 1991.

i90] M.Z. KWIATKOWSKA, "A metric for traces", Information and Processing Letters 35, p. 129-135, 1990.

z77] A. MAZURKIEWICZ, "Concurrent program Schemes and their interpretations", Aarhus University, DAIMI Rep. PB 78, 1977.

z86] A. MAZURKIEWICZ, "Trace theory", Advanced Course on Petri Nets, Lecture Notes in Computer Science 255, p. 279-324, 1986.

h85] E. OCHMANSKI, "Regular behaviour of concurrent systems", Bulletin of EATCS 27, p. 56-67, October 1985.

89] D. PERRIN, "Partial commutations", ICALP'89, Lecture Notes in Computer Science 372, p. 637-651, 1989.

91] D.PERRIN and J.E. PIN, "Mots Infinis", Tech. Rep. 91-06, LITP, Université Paris 6, France, 1991. Book to appear.

;92] G. PIGHIZZINI, "On the iteration of trace languages", Proceed. of the ASMICS Workshop on Infinite Traces, to appear as Tech. Rep., Universität Stuttgart, 1992.

[Tho89] W. THOMAS, "On logical definability of trace languages", Proceedings of ASMICS Workshop on Partially Commutative Monoids, Tech. Rep. TUM-I 9(Technische Universität München, p. 172-182, 1989.

[Tho90] W. THOMAS, "Automata on infinite objects", Handbook of Theoretical Comp Science (J.V. Leeuwen, Ed.), Elsevier, Amsterdam, Vol. B, p. 135-191, 1990.

[Tho91] W. THOMAS, "On logics, tilings, and automata", ICALP'91, Lecture Note Computer Science 510, p. 441-454, 1991.

[Zie87] W. ZIELONKA, "Notes on finite asynchronous automata and trace languag RAIRO Theoretical Informatics and Applications 21, p. 99-135, 1987.

[Zi 89] W. ZIELONKA, "Safe execution of recognizable trace languages by asynchron automata", Lecture Notes in Computer Science 363, p. 278-289, 1989.

A Trace Semantics for Petri Nets

(Extended Abstract)

by

P.W. Hoogers[1], H.C.M. Kleijn[1], P.S. Thiagarajan[2]

[1]Department of Computer Science, Leiden University,
P.O. Box 9512, 2300 RA Leiden, The Netherlands

[2]School of Mathematics, SPIC Science Foundation,
92, G.N. Chetty Road, T. Nagar, Madras - 600 017, India

Abstract. Our aim is to extend the semantic theory of elementary net systems [NRT1,NRT2] to general Petri nets. In doing so, we expect to obtain a theory which would reflect our intuition that general Petri nets are more expressive than elementary net systems (and for that matter, 1-safe Petri nets). As a first step, we provide a semantics for Petri nets using the theory of trace languages due to Mazurkiewicz [Ma]. It turns out that in order to adequately treat the multiplicity of tokens in a Petri net, we must generalize the classical notions underlying trace languages along a number of directions. After doing so, it is easy to associate such a generalized trace language with each Petri net. We then characterize the class of trace languages - called PN-trace languages - that arise in this fashion. It turns out that the new trace languages can also be used to capture the behaviours of other models of concurrency. In particular, we characterize the general event structures of Winskel [W] and their stable subclass in terms of our trace languages. One pleasant consequence of our results is that in this framework, 1-safe Petri nets, event structures, and general Petri nets constitute a strictly ascending chain in terms of expressive power.

ntroduction

his paper is an extended abstract of [HKT] which contains all the technical details d proofs. The aim of this work is to provide a semantics for Petri nets which takes to account both the non-sequential and the branching aspects of their behaviours. or elementary net systems (and in fact for 1-safe Petri nets) the classical notion of nfoldings based on occurrence nets can be used to achieve this purpose. In the case of Petri net it is far from clear what a proper notion of an unfolding is. The proposals at have been made in the literature [GM,GR2,E] more or less view tokens as coloured tities. One consequence is that, in these proposals, general Petri nets are - in terms of the event structures and domains that they yield - no more expressive than 1-sa Petri nets.

Conventional non-sequential processes have also been used to capture the behavio of Petri nets [GR1,BD,BF]. Here again the tokens are viewed as coloured entities sin this notion of non-sequential processes is tied to the intuition concerning the behavio of 1-safe Petri nets.

For elementary net systems, an alternative approach - which is the one follow here - is to use the theory of trace languages initiated by Mazurkiewicz [Ma]. He the concurrency present in the behaviour of a net system is captured by an indepe dence relation over the events of the system. This independence relation induces equivalence relation over the sequences of event occurrences (firing sequences) allow by the system. Each of the equivalence classes, called traces, represents a single no sequential run of the system. The behaviour of the net system can then be describ as a poset of traces ordered by a prefix ordering. This poset of traces also yields, a canonical fashion, a Scott Domain of a certain kind which can be identified wi the prime event structure obtained via the unfolding route. These results follow fro [NPW,NRT1].

This line of attack - which constitutes an attractive alternative to the approach obtaining unfoldings - does not work in the case of a Petri net. The problem is that place in a Petri net may contain many tokens. As a result concurrency (and conflic are no longer global structural relations. Moreover, a *multiset* of transitions may occ concurrently at a marking.

To deal with these phenomena, we generalize the theory of trace languages alo three directions. Firstly we consider *step sequences* instead of ordinary sequences; a step we mean here a finite multiset. Secondly, we consider independence relatio that are context-dependent, where the context is defined by a step sequence. Thirdl we specify in the independence relation, a finite multiset of actions that can occ concurrently at a context (rather than just a pair of symbols that may commute as the classical case). Armed with these generalizations, it is straightforward to lift t standard notions from the theory of trace languages to the - much richer - new settin

Next we show that it is easy to associate a trace language of the new kind wi each Petri net. We then proceed to characterize the class of trace languages, call PN-trace languages, that arise in this fashion.

In Section 3 we introduce the event structures considered by Winskel [W] which a much richer than prime event structures. We show how to associate one of our tra languages with each event structure. We then show that this class of trace language called ES-trace languages, can be characterized by smoothly strengthening the axio capturing PN-trace languages. Finally, we show that by adding one additional axio to the characterization of ES-trace languages we can capture the class of stable eve structures (which "correspond" to the behaviours of 1-safe Petri nets [W]). From the results it then follows that, in the framework of our trace languages, Petri nets a more expressive than event structures which in turn are more expressive than 1-sa Petri nets.

In the concluding section we mention some additional related work and the proble of obtaining unfoldings of Petri nets.

Petri Nets and Generalized Trace Languages

ven an alphabet Σ, we let $\Sigma^\#$ denote the set of (non-empty) *step sequences over* that is $\Sigma^\# = (\Sigma^\otimes)^+$, where $\Sigma^\otimes$ denotes the set of *finite multisets over* Σ, i.e. the of functions $u : \Sigma \to \mathbf{N}$ with $\sum_{a\in\Sigma} u(a) < \infty$. We view $\Sigma^\#$ as a (free) monoid: unit element is $\underline{0} \in \Sigma^\otimes$, where $\underline{0}(a) = 0$ for all $a \in \Sigma$; its product operation is the cordingly modified usual concatenation operation.

efinition 1.1.
(*generalized*) *concurrency alphabet* is a pair (Σ, I), where $I \subseteq \Sigma^\# \times \Sigma^\otimes$ is an *inde-dence relation* (*over* Σ).

- $\doteq_I \subseteq \Sigma^\# \times \Sigma^\#$ is given by: $\rho \doteq_I \rho'$ iff there exist $\rho_1, \rho_2 \in \Sigma^\#$ and there exist $u, v, u', v' \in \Sigma^\otimes$ such that the following conditions are satisfied.
  (1a) $\rho = \rho_1 u v \rho_2$ and $\rho' = \rho_1 u' v' \rho_2$.
  (1b) $u + v = u' + v'$ (where $+$ denotes multiset addition).
  (1c) $(\rho_1, u + v) \in I$.
- $\equiv_I = (\doteq_I)^*$, where $(\doteq_I)^*$ denotes the reflexive transitive closure of $\doteq_I$.
- For $\rho \in \Sigma^\#$, $[\rho]_I = \{\rho' \in \Sigma^\# \mid \rho \equiv_I \rho'\}$ is the *trace* containing ρ.
- $\Sigma^\#/\equiv_I = \{[\rho]_I \mid \rho \in \Sigma^\#\}$ is the *set of traces generated by* (Σ, I).
- A (*generalized*) *trace language* (*over* (Σ, I)) is a subset of $\Sigma^\#/\equiv_I$. It will be convenient to represent a trace language Tr over a concurrency alphabet (Σ, I) as a triple (L, Σ, I) where $L = \bigcup \{[\rho]_I \mid [\rho]_I \in Tr\}$.
-) $\leq_I \subseteq (\Sigma^\#/\equiv_I) \times (\Sigma^\#/\equiv_I)$ is the relation with $[\rho]_I \leq_I [\rho']_I$ iff there exists $\sigma \in \Sigma^\#$ such that $\rho\sigma \equiv_I \rho'$. □

what follows, the index I will be omitted whenever the concurrency alphabet (Σ, I) clear from the context.

let us show now how our generalized trace languages can be used to describe the haviour of Petri nets.

efinition 1.2.
Petri net is a quadruple $N = (S, T, W, M_{in})$ where

) S is a set of *places* and T is a set of *transitions* such that $S \cap T = \emptyset$

) $W : (S \times T) \cup (T \times S) \to \mathbf{N}$ is a *weight function* which satisfies the condition that for all $x \in S \cup T$, there exists $y \in S \cup T$ such that $W(x, y) > 0$ or $W(y, x) > 0$.

) $M_{in} : S \to \mathbf{N}$ is the *initial marking of* N. The set of all markings of N, that is the set of all functions $M : S \to \mathbf{N}$, is denoted by $\mathbf{M}_N$. □

efinition 1.3.
t $N = (S, T, W, M_{in})$ be a Petri net. The set $SFS_N \subseteq T^\#$ of *step firing sequences* N and the *step transition relation* $\Longrightarrow_N \subseteq \{M_{in}\} \times SFS_N \times \mathbf{M}_N$ are the least sets ch that:

) $\underline{0} \in SFS_N$ and $(M_{in}, \underline{0}, M_{in}) \in \Longrightarrow_N$

) if $\rho \in SFS_N$ and $(M_{in}, \rho, M) \in \Longrightarrow_N$ and if $u \in T^\otimes$ is such that $M(s) \geq \sum_{t\in T} u(t) \cdot W(s, t)$ for all $s \in S$, then $\rho u \in SFS_N$ and $(M_{in}, \rho u, M') \in \Longrightarrow_N$ where $M'(s) = M(s) + \sum_{t\in T} u(t) \cdot (W(t, s) - W(s, t))$ for all $s \in$ S. □

If the Petri net N is clear from the context we may omit the subscript N in SFS_N ar $\Longrightarrow_N$.
Intuitively, the steps in the step firing sequences contain "concurrently enabled" occu rences of transitions. Hence we choose these steps to be the independent multisets the independence relation we associate with a Petri net. The step sequence precedi such a step is the context in which the step is independent.

Definition 1.4.
Let $N = (S, T, W, M_{in})$ be a Petri net. Let $\Sigma_N \subseteq T$ be the set of those transitio of N that actually occur in SFS_N. The *independence relation of* N is the set I_N $(\Sigma_N)^\# \times (\Sigma_N)^\otimes$ with $I_N = \{(\rho, u) \mid \rho u \in SFS \text{ with } \rho \in (\Sigma_N)^\# \text{ and } u \in (\Sigma_N)^\otimes\}$. □

Note that SFS_N has the property that $[\rho]_{I_N} \subseteq SFS_N$ for all $\rho \in SFS_N$; we say th SFS_N is *consistent with respect to* (Σ_N, I_N).
Using this observation we can now associate with each Petri net a trace languag Combining this with the prefix ordering of traces we arrive at a poset of traces f each Petri net.

Definition 1.5.
Let $N = (S, T, W, M_{in})$ be a Petri net.
(1) $\mathbf{TR}_N = (SFS, \Sigma_N, I_N)$ is the *trace language generated by N.*
(2) $(\mathbf{TR}_N, \leq_{I_N})$ is the *trace behaviour of* N (where the restriction of $\leq_{I_N}$ to $\mathbf{TR}_N \times \mathbf{TR}_N$ is also denoted by $\leq_{I_N}$). □

2. PN-Trace Languages

Our aim in this section is to characterize those trace languages, called *PN-trace la guages*, which are isomorphic to a trace language associated with a Petri net. O notion of morphism for trace languages is behaviour-preserving in the sense that t underlying language and the independence relation are preserved.

Definition 2.1.
Let (L_i, Σ_i, I_i), $i = 1, 2$, be a pair of trace languages. A (*trace*) *morphism fro* (L_1, Σ_1, I_1) to (L_2, Σ_2, I_2) is a partial function $f : \Sigma_1 \to \Sigma_2$ such that $f(L_1) \subseteq$ and $f(I_1) \subseteq I_2$. □

In this definition the obvious extensions of f to step sequences and independen relations are also denoted by f; note that these extensions are total functions. the standard way we can now define (*trace*) *isomorphisms*. A (trace) isomorphis from a trace language (L_1, Σ_1, I_1) to a trace language (L_2, Σ_2, I_2) is then a bijectic $f : \Sigma_1 \to \Sigma_2$ such that $f(L_1) = L_2$ and $f(I_1) = I_2$. If trace languages Tr_1 and Tr_2 a isomorphic, we denote this fact by $Tr_1 \approx Tr_2$.

Definition 2.2.
A trace language Tr is a *PN-trace language* if there exists a Petri net N such th $Tr \approx \mathbf{TR}_N$. □

order to formulate one of our axioms for characterizing the PN-trace languages, we ed the notion of a (*generalized*) *region*. These regions are defined to model "local" ates. A local state is characterized by an initial value and the change of its value used by each of the actions. The "global" state of the system is then characterized all regions together. The regions discussed here generalize the notion of a region om [ER]. Recently a closely-related idea has been exploited by Mukund [Mu] to aracterize the operational behaviour of Petri nets in terms of transition systems.

efinition 2.3.
t $Tr = (L, \Sigma, I)$ be a trace language. A (*generalized*) *region of* Tr is a function $: L \cup \Sigma \to \mathbf{N} \cup (\mathbf{N} \times \mathbf{N})$ satisfying the following conditions.
) For all $\rho \in L : r(\rho) \in \mathbf{N}$ and for all $a \in \Sigma : r(a) \in \mathbf{N} \times \mathbf{N}$.
) If $\rho, \rho' \in L$ and $u \in \Sigma^{\otimes}$ are such that $\rho u \equiv \rho'$, then $r(\rho) \geq \sum_{a \in \Sigma} u(a) \cdot {}^{r}a$ and $r(\rho') = r(\rho) + \sum_{a \in \Sigma} u(a) \cdot (a^{r} - {}^{r}a)$ where $r(a) = ({}^{r}a, a^{r})$ for all $a \in \Sigma$.
region r of Tr is said to be *non-trivial* iff $r(a) \neq (0,0)$ for some $a \in \Sigma$.
$_{Tr}$ is the set of non-trivial regions of Tr. □

s an example of this notion the following lemma exhibits some (but not all) of the gions of the trace language $\mathbf{TR}_N$ generated by a Petri net N. For each place s of the tri net, there is a region r<s> which gives for each $t \in \Sigma_N$ the number of tokens t kes from s and the number of tokens t puts in s and, for each $\rho \in SFS$, it gives the mber of tokens in s after the step firing sequence ρ.

emma 2.4.
t $N = (S, T, W, M_{in})$ be a Petri net and let $s \in$ S. Define r<s>: $SFS \cup \Sigma_N \to$ $\cup (\mathbf{N} \times \mathbf{N})$ such that
) r<s>$(t) = (W(s,t), W(t,s))$ for all $t \in \Sigma_N$
) r<s>$(\rho) = M(s)$ for all $\rho \in SFS$ where $M \in \mathbf{M}_N$ is such that $(M_{in}, \rho, M) \in \Longrightarrow$.
hen r<s> is a region of $\mathbf{TR}_N$. □

ow we are ready to present the five axioms which characterize PN-trace languages. hese axioms have been stated in terms of a trace language $Tr = (L, \Sigma, I)$. In stating ese axioms, we let $\rho, \rho', \rho_1, \rho_2, \sigma$, etc. range over $\Sigma^{\#}$ and u, v, u', v', etc. range over $^{\otimes}$. We let *alph*: $L \to P_{fin}(\Sigma)$ (with $P_{fin}(\Sigma)$ the set of finite subsets of Σ) be such that *ph*(ρ) is the set of elements from Σ occurring in ρ.

(PN0) $L \neq \emptyset$. (Non-emptiness)

(PN1) $\Sigma = \bigcup alph(L)$. (No redundant elements)

(PN2) $\rho u \in L \Rightarrow \rho \in L$. (Prefix-closedness)

(PN3) $\rho u \in L \Leftrightarrow (\rho, u) \in$ I. (Substep property)

(PN4) $\rho \in L \Rightarrow (\rho u \notin L \Rightarrow (\exists r \in \mathbf{R}_{Tr} : r(\rho) < \sum_{a \in \Sigma} u(a) \cdot {}^{r}a))$.
(Forward closure property)

sing Lemma 2.4, it is easy to see that each trace language generated by a Petri net tisfies (PN0) through (PN4). Since each of these five axioms is preserved under morphisms, this implies that all PN-trace languages satisfy the axioms. In order to prove the converse, we construct, given a trace language $Tr = (L, \Sigma, I)$ satisfying t axioms (PN0) through (PN4), a Petri net N_{Tr} the trace language of which will tu out to be isomorphic with Tr. This Petri net has the non-trivial regions of Tr as places and Σ as its transitions.

Definition 2.5.
Let $Tr = (L, \Sigma, I)$ be a trace language which satisfies (PN0) through (PN4). T *Petri net associated with* Tr is the Petri net $N_{Tr} = (S, T, W, M_{in})$ where $S = \mathbf{R}$ $T = \Sigma$, $W : (\mathbf{R}_{Tr} \times \Sigma) \cup (\Sigma \times \mathbf{R}_{Tr}) \to \mathbf{N}$ is given by: $W(r, a) = {}^{r}a$ and $W(a, r) =$ for all $r \in \mathbf{R}_{Tr}$ and $a \in \Sigma$; $M_{in} : \mathbf{R}_{Tr} \to \mathbf{N}$ is given by: $M_{in}(r) = r(\underline{0})$ for all $r \in \mathbf{R}$
□

It is clear that N_{Tr} is indeed a Petri net. Actually one will have to ensure th $\mathbf{R}_{Tr} \cap \Sigma = \emptyset$ by using a more involved construction, but we shall ignore this here.

Example 2.6.
In Figure 1 a Petri net N, its trace behaviour $\mathbf{TR}_N$, and the Petri net $\hat{N}$ associat with $\mathbf{TR}_N$ are depicted. We have only shown some of the (infinite number of) plac of $\hat{N}$. In depicting $\mathbf{TR}_N$ we have notationally identified steps consisting of one eleme with this element.

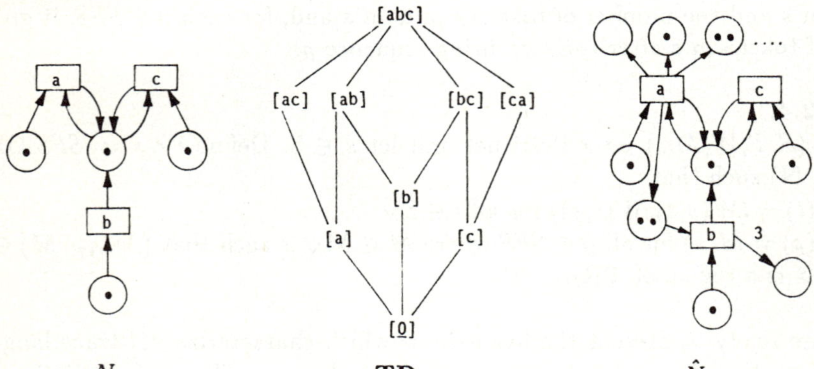

Figure 1. □

Given a trace language Tr which satisfies the axioms (PN0) through (PN4) it can l proved that each region $s \in \mathbf{R}_{Tr}$ agrees with the (non-trivial) region $r\langle s\rangle$ of the tra language generated by the Petri net N_{Tr}. As a consequence the identity function an isomorphism from Tr to $\mathbf{TR}_{N_{Tr}}$. This leads to the following result.

Theorem 2.7.
A trace language is a PN-trace language iff it satisfies the axioms (PN0) throug (PN4). □

Given a Petri net N and its trace language $\mathbf{TR}_N$ the construction in Definition 2. yields a Petri net $\hat{N}$ which is closely related to N: the transitions of $\hat{N}$ form precise the set Σ_N of those transitions of N that actually occur in *SFS*; the places of e all non-trivial regions of $\mathbf{TR}_N$ which include all regions as defined in Lemma 2.4 sociated with places of N which are connected to a transition from Σ_N. Note that plying the construction in Definition 2.5 once more to $\mathbf{TR}_{\hat{N}}$ does not lead to new aces. Thus $\hat{N}$ is a saturated version (w.r.t. places) of N.

Event Structures and Generalized Trace Languages

this section we outline that our generalized trace languages can also be used to pture yet another well-known model of concurrency known as *event structures* [W]. ctually the term event structures is used in various senses and here we shall consider e general event structures specified as a family of configurations by Winskel in [W]. nce we do not deal with infinite step sequences we consider only families of finite nfigurations. This however causes no loss of information due to the axiom of finite uses imposed on families of configurations in [W].

efinition 3.1.
n *event structure* is a pair $W = (E, C)$ where E is a non-empty set of *events* and $\subseteq P_{fin}(E)$ is a family of (*finite*) *configurations* which satisfies the following two quirements.

(W1) For all $c \in C$: if $c \neq \emptyset$, then there exists $e \in c$ such that $c - \{e\} \in C$.

(W2) For all $c, c' \in C$: if $c \uparrow c'$ then $c \cup c' \in C$ (where $c \uparrow c'$ iff there exists $c_1 \in C$ such that $c \subseteq c_1$ and $c' \subseteq c_1$). □

ctually we have slightly tightened up the definition (through (W1)) from [W] to arantee that $\emptyset$ is always a configuration.
ith every event structure a set of step sequences can be associated on the basis of enabling relation for (finite) sets of events.

efinition 3.2.
et $W = (E, C)$ be an event structure.

) Let $c \in C$ and $u \in P_{fin}(E)$. Then u is said to be *enabled* at c, denoted by $c[u{>}_W$, iff $c \cap u = \emptyset$ and $c \cup v \in C$ for all $v \subseteq$ u.

) The set $S_W \subseteq E^{\#}$ of *step sequences of* W and the *step transition relation* $\Longrightarrow_W \subseteq \{\emptyset\} \times S_W \times C$ are the least sets such that:

a) $\underline{0} \in S_W$ and $(\emptyset, \underline{0}, \emptyset) \in \Longrightarrow_W$

b) if $\rho \in S_W$ and $(\emptyset, \rho, c) \in \Longrightarrow_W$ and if $u \in P_{fin}(E)$ is such that $c[u{>}_W$, then $\rho u \in S_W$ and $(\emptyset, \rho u, c \cup u) \in \Longrightarrow_W$. □

this definition we have used the convention that we do not distinguish between a ultiset $u : \Sigma \to \mathbf{N}$ with $u(a) \leq 1$ for all $a \in \Sigma$ and the corresponding set $\{a \in \Sigma \mid (a) = 1\}$.
the event structure W is clear from the context we may omit the subscript W in $>_W$, S_W, and $\Longrightarrow_W$.

Definition 3.3.
Let $W = (E, C)$ be an event structure. Let Σ_W be the set of those events of E th[at] actually occur in S_W.
(1) The *independence relation of* W is the set $I_W \subseteq (\Sigma_W)^\# \times (\Sigma_W)^\otimes$ with $I_W = \{(\rho, u) \mid \rho u \in S \text{ with } \rho \in (\Sigma_W)^\# \text{ and } u \in (\Sigma_W)^\otimes\}$.
(2) $\mathbf{TR}_W = (S, \Sigma_W, I_W)$ is the *trace language generated by* W.
(3) The poset $(\mathbf{TR}_W, \leq_{I_W})$ is the *trace behaviour of* W. □

By the consistency of S the above notions are well-defined.
We now want to give a characterization of the trace languages associated with eve[nt] structures.

Definition 3.4.
A trace language Tr is an *ES-trace language* if there exists an event structure W suc[h] that $Tr \approx \mathbf{TR}_W$. □

ES-trace languages (L, Σ, I) turn out to be characterized by the axioms (PN0) throug[h] (PN3) together with three new axioms (ES1), (ES2), and (ES3). In stating (ES1) w[e] let $\#_a(\rho)$ denote the number of times a occurs in ρ.

(ES1) $\rho \in L \Rightarrow$ for all $a \in \Sigma : \#_a(\rho) \leq 1$. (Unique occurrence property)

(ES2) $\rho, \rho' \in L \Rightarrow (alph(\rho) \subseteq alph(\rho') \Rightarrow [\rho] \leq [\rho'])$. (Prefix-consistency)

(ES3) $\rho u_1 \rho_1, \rho u_2 \rho_2 \in L \Rightarrow (\rho u_1 \rho_1 \equiv \rho u_2 \rho_2 \Rightarrow \rho(alph(u_1 + u_2)) \in L)$.
(Forward diamond propert[y])

The axioms (ES1) through (ES3) are again preserved under isomorphisms. Since f[or] every event structure W its trace language $\mathbf{TR}_W$ satisfies the axioms (PN0) throug[h] (PN3), and (ES1), (ES2), and (ES3), it then follows that every ES-trace languag[e] satisfies the axioms (PN0) through (PN3), (ES1), (ES2), and (ES3). In order to pro[ve] the converse result we define an event structure $W_{Tr} = (E_{Tr}, C_{Tr})$ with $E_{Tr} = \Sigma$ an[d] $C_{Tr} = \{alph(\rho) \mid \rho \in L\}$ for every trace language $Tr = (L, \Sigma, I)$ satisfying the axioms. It turns out that the identity function is an isomorphism from Tr to $\mathbf{TR}_{W_{Tr}}$.

Theorem 3.5.
A trace language is an ES-trace language iff it satisfies the axioms (PN0) throug[h] (PN3), (ES1), (ES2), and (ES3). □

Now we compare the class of ES-trace languages and the class of PN-trace language[s]. It can be proved that for each trace language satisfying the axioms (PN0) throug[h] (PN3), (ES1), (ES2), and (ES3) enough regions can be found so that (PN4) als[o] holds. By Theorem 2.7 and Theorem 3.5 we then get the following result.

Theorem 3.6.
Every ES-trace language is a PN-trace language. □

The converse of this theorem is not true. Let N be the Petri net from Example 2.[6]

ıen $\mathbf{TR}_N$ is a PN-trace language, but it does not satisfy (ES3). Hence $\mathbf{TR}_N$ is not ı ES-trace language. Thus Petri nets are more expressive than event structures under eir representations as generalized trace languages.

s a last point we would like to add that also the *stable event structures* ([W]) can be ıaracterized in terms of their trace languages. An event structure (E, C) is stable if satisfies in addition to (W1) and (W2):

(W3) for all $c, c' \in C$: if $c \uparrow c'$ then $c \cap c' \in C$.

efinition 3.7.
trace language Tr is a *stable* ES-*trace language* if there exists a stable event structure ' such that $Tr \approx \mathbf{TR}_W$. □

ɔ the axioms characterizing event structures we then need to add one more axiom flecting the backward diamond property possessed by their trace languages:

(ES4) $\rho_1 u_1, \rho_2 u_2 \in L \Rightarrow [\rho_1 u_1 \equiv \rho_2 u_2 \Rightarrow$ there exists $\rho \in L$ such that $\rho(alph(u_1 + u_2)) \equiv \rho_1 u_1]$. (Backward diamond property)

heorem 3.8.
trace language Tr is a stable ES-trace language iff Tr satisfies the axioms (PN0) ırough (PN3), and (ES1) through (ES4). □

ːom Winskel's work [W] which shows how to "represent" 1-safe Petri nets as stable ent structures, it follows that in the present framework event structures (and hence etri nets) are more expressive than 1-safe Petri nets.

›iscussion

s demonstrated by our results, the generalization of trace languages proposed here ›pears to be a versatile tool for capturing (and comparing) the behaviours of some ell-known models of concurrency. In fact, we believe that the results reported here ın be extended to a categorical framework and the relationships between the various asses of trace languages and the models stated in terms of reflections, co-reflections ıd categorical equivalences.

As for our main interest, namely the behaviour of Petri nets, the problem of obıining suitable unfoldings still looms large. Due to auto-concurrency, it seems to ɜ a difficult problem to identify a subclass of Petri nets that would correspond to generalization of occurrence nets and to construct an unfold operation that would ıap each Petri net into this subclass. Naturally what one would expect from such an ɔeration is that it becomes the right adjoint to the obvious inclusion functor.

Our work seems to be related to the work of [DMM] and hence to the work of [BD]. seems that the traces that we associate to a Petri net correspond to the equivalence asses of morphisms of [DMM] and the equivalence classes of processes of [BD]. But › far we have not managed to establish any formal connections along these directions. he newly introduced geometric automata due to Pratt [P] might also lend themselves › a characterization by our trace languages.

At present we do not really have a feel for the kind of "event structures" and Sco Domains yielded by the PN-trace languages. This will certainly constitute a futu line of research. We also hope to explore some of the language theoretic aspects of o trace model in our future work.

Acknowledgements

The authors gratefully acknowledge support from the ESPRIT BRA No. 3148 DE MON and the Dutch National Concurrency Project REX sponsored by NFI.

References

[BD] Best, E., and Devillers, R., (1987), Sequential and concurrent behaviour i Petri net theory, Theoretical Computer Science 55, 87-136.

[BF] Best, E., and Fernández C., C., (1988), Nonsequential processes, EATC Monographs on Theoretical Computer Science Vol. 13, Springer Verlag.

[DMM] Degano, P., Meseguer, J., and Montanari, U., (1989), Axiomatizing net com putations and processes, In Proc. Logic in Computer Science, 175-185.

[ER] Ehrenfeucht, A., and Rozenberg, G., (1990), Partial 2-structures; part II: stat spaces of concurrent systems, Acta Informatica, v. 27, 348-368.

[E] Engelfriet, J., (1991), Branching processes of Petri nets, Acta Informatica v.28, 575-591.

[GM] Goltz, U., and Mycroft, A., (1984), On the relationship of CCS and Petri nets Lecture Notes in Computer Science 172, 196-208.

[GR1] Goltz, U., and Reisig, W., (1983), The non-sequential behaviour of Petri nets Information and Control 57, 125-147.

[GR2] Goltz, U., and Reisig, W., (1985), CSP-programs as nets with individual to kens, Lecture Notes in Computer Science 188, 169-196.

[HKT] Hoogers, P.W., Kleijn, H.C.M., and Thiagarajan, P.S., (1992), A trace seman tics for Petri nets, Leiden University Techn. Rep. 92-03.

[Ma] Mazurkiewicz, A., Trace theory, (1987), Lecture Notes in Computer Scienc 255, 279-324.

[Mu] Mukund, M., (1991), A transition system characterization of Petri nets, Repor TCS-91-02, School of Mathematics, SPIC Science Foundation, Madras, India

[NPW] Nielsen, M., Plotkin, G., and Winskel, G., (1981), Petri nets, event structure and domains, part I, Theoretical Computer Science 13, 85-108.

[NRT1] Nielsen, M., Rozenberg, G., and Thiagarajan, P.S., (1990), Behavioural no tions for elementary net systems, Distributed Computing 4, 45-57.

[NRT2] Nielsen, M., Rozenberg, G., and Thiagarajan, P.S., (1990), Elementary tran sition systems, DAIMI-PB 310, Aarhus University, to appear in Theoretica Computer Science.

[P] Pratt, V.R., (1991), Modelling concurrency with geometry, In Proc. 18th Ann ACM Symposium on Principles of Programming Languages, 311-322.

[W] Winskel, G., (1987), Event structures, Lecture Notes in Computer Science 255 325-392.

synchronous Communication of Petri Nets and the Refinement of Transitions

Walter Vogler*

Institut für Informatik, TU München
Arcisstr. 21, D-8000 München 2

Abstract

In this paper we study nets whose boundaries in a context net consist of places only, i.e. nets that communicate asynchronously with their environment. Such nets are called equivalent, if exchanging them in any context preserves deadlock-freeness. We characterize this equivalence internally, i.e. without referring to all possible contexts. Since this equivalence is undecidable in general, we then define a subclass of nets that are to some degree deterministic. For nets from this subclass we can show that the equivalence is decidable and that the exchange of equivalent nets does not only preserve deadlock-freeness, but gives nets that are even bisimilar; these results especially apply to the behaviour preserving refinement of transitions.

Introduction

tri nets offer a rigorous formalism for the specification and design of concurrent systems. the effective use of formal methods in actual engineering practice it is essential to have ucturing methods at hand to reduce and manage system complexity. This is especially e for Petri nets: besides rigorous formalism they offer the additional advantage of a graph- l representation. With growing system complexity the formalism (as any formalism) be- nes more difficult to handle, but even faster the graphical representation looses its clar- . Therefore there is quite an interest in the modular construction of Petri nets, see e.g. nd83,Bau88,Mül85,Sou90,Che91].

This paper contributes some answer to the following question: Which subnets can be changed in an arbitrary context net such that the behaviour of the whole net is preserved? e behaviour we consider here as a starting point is being free of deadlocks, and we study case that the exchanged subnets are combined with their environment by merging places, they communicate asynchronously with their environment. (One can also take other basic haviours as a starting point, e.g. divergence, see [Vog92]).

It is more usual to consider a composition operator $\|_A$ which works by merging transitions n operator motivated by the parallel composition with synchronisation of TCSP. In [Vog89]

*This work was partially supported by the Deutsche Forschungsgemeinschaft, Sonderforschungsbereich 342: thoden und Werkzeuge zur Nutzung paralleler Rechnerarchitekturen, TU München and the ESPRIT Basic search Action No. 3148 DEMON (Design Methods Based on Nets).

the following approach is studied. Let us call nets deadlock-equivalent, if their exchange every $\|_A$-context preserves deadlock-freeness. This is an external equivalence in the sense [Bau88], and its definition involves all possible context nets. To make confirmation of t equivalence feasible one has to look for an internal characterization that does not invo infinitely many context nets. It turns out that nets are deadlock-equivalent if and only if t are failure-equivalent, and that in this case their exchange preserves failure equivalence, it preserves behaviour in a stronger sense than originally required. This justifies to take s a simple property as deadlock-freeness as a starting point. Unfortunately, failure equivale for finite nets is undecidable in general.

In [Vog92] these considerations are continued by restricting attention to the class of deterministic nets, a translation from a TCSP-notion to nets, compare [Bro83,And83,Eng8 For this class failure equivalence is decidable, the class is decidable itself, and it is rich enou to treat especially the refinement of places in a general framework. Exchange of deadlo equivalent nets from this class even transforms a net into a bisimilar net.

While some of these results have their counterparts in action oriented models like CCS TCSP, the contribution of this paper is indigenious to Petri nets. Here we study the co position operator $|||_P$, which combines nets by merging places corresponding to asynchrono communication. Analogously to the above, we call nets P-deadlock-equivalent, if their change in every $|||_P$-context preserves deadlock-freeness. One main result of Section 3 is internal characterization of this external equivalence. For this result, we develop two failu type semantics that take into account not only sets of single actions that can be refused some system state, but also sets of sequences of actions.

Again we meet the necessity to restrict our attention to a subclass of nets. To define t class of so called I,O-nets we split up the interface, i.e. the merged places, into input a output places. Then we isolate three important features which make I,O-nets tractable: $I,$ nets have to be output-deterministic, output-bounded and non-absorbing. In Section 4, show that I,O-nets give behaviour preservation in a strong sense, have favourable decidabili properties, and allow to treat the behaviour preserving refinement of transitions in a gene framework. (See [BGV91] for a survey on refinements of Petri nets.)

The composition by merging places has found little attention in the literature, but see e [Sou90,Che91]; these papers consider unlabelled nets, whereas we are interested in label nets. Due to lack of space no proofs can be given here; they can be found in [Vog92].

2 Basic notions of Petri nets

In this section a very brief introduction to Petri nets is given. For further information t reader is referred to e.g. [Rei85]. We will deal with Petri nets (place/transition-nets) wh transitions are labelled with actions from some infinite alphabet Σ or with the empty sequen λ. These actions are left uninterpreted; the labelling only indicates that a λ-labelled transiti represents an *internal* action, which is not visible for an external observer, and that t transitions with the same label from Σ represent the same *visible* action occurring in differe internal situations.

Thus a *Petri net* $N = (S, T, W, M_N, lab)$ (or just a *net* for short) consists of disjoint (n necessarily finite) sets S of *places* and T of *transitions*, the *weight function* $W : S \times T \cup T \times S$ $\mathbb{N}_0$, the *initial marking* $M_N : S \to \mathbb{N}_0$, and the *labelling* $lab : T \to \Sigma \cup \{\lambda\}$. As usual view a net as a (labelled, arc weighted) bipartite graph whose vertex classes are S and T that has an arc from x to y whenever $W(x,y) \neq 0$. In general, we will not distinguish ween isomorphic nets. If necessary, S, T etc. will be indexed by N, while the components .g. a net N_1 are denoted by S_1, T_1 etc.

If $X \subseteq S \cup T$, then the *subnet induced* by X has place set $X_S = X \cap S$, transition set $= X \cap T$, and the weight function, intial marking and labelling are obtained from those N by the appropriate restriction to $X_S \times X_T \cup X_T \times X_S$, X_S, and X_T respectively.

For $x \in S \cup T$ the sets ${}^\bullet x = \{y \in S \cup T \mid W(y,x) \neq 0\}$ and $x^\bullet = \{y \in S \cup T \mid W(x,y) \neq 0\}$ called the *preset* and the *postset* of x. A *marking* is a function from S into $\mathbb{N}_0$. A transition *enabled* under a marking M, denoted by $M[t\rangle$, if $W(s,t) \leq M(s)$ for all $s \in S$. If $M[t\rangle$ $M'(s) = M(s) + W(t,s) - W(s,t)$, then we denote this by $M[t\rangle M'$ and say that t can *ur* or *fire* under M yielding the *follower marking* M'.

This definition of enabling and occurrence can be extended to sequences of transitions as al: a sequence $w \in T^*$ is *enabled* under a marking M, denoted by $M[w\rangle$, and yields the *ower marking* M' if it *occurs*, denoted by $M[w\rangle M'$, if $w = \lambda$ and $M = M'$ or $w = w't$, $w'\rangle M''$ and $M''[t\rangle M'$ for some marking M''. If w is enabled under the initial marking, then s called a *firing sequence*.

We can extend the labelling of a net to sequences of transitions as usual. Note that internal nsitions do not turn up in the image of a sequence. Next we lift the enabledness and firing initions to the level of actions.

A sequence $v \in \Sigma^*$ is *image-enabled* under a marking M, denoted by $M[v\rangle\rangle$, if there is ne $w \in T^*$ with $M[w\rangle$ and $lab(w) = v$. If additionally $M[w\rangle M'$, we write $M[v\rangle\rangle M'$. If $= M_N$, then v is called an *image firing sequence*; w is one of its *underlying* firing sequences. e *language* $L(N)$ of N is the set of all image firing sequences. We call two nets N_1 and *language-equivalent* if $L(N_1) = L(N_2)$. Similarly, whenever we introduce some semantics $m(N)$ of nets N in the following, nets are called *Sem-equivalent* if they have the same m-semantics.

For a marking M the set $[M\rangle$ of markings *reachable* from M is defined as $\{M' \mid \exists w \in$ $: M[w\rangle M'\}$. The set $[M_N\rangle$ is the set of reachable markings of N. Two nets N_1 and N_2 *marking-equivalent on* a set P of common places, if $\{M|_P \mid M \in [M_{N_1}\rangle\} = \{M|_P \mid M \in$ $_{N_2}\rangle\}$, where | denotes the restriction of functions or relations.

The basic (undesirable) behaviour that we will study is to deadlock, i.e. to reach a state ere no visible action is possible anymore. We say that a net can *deadlock*, if for some chable marking M we have for all $a \in \Sigma$: not $M[a\rangle\rangle$. Also M itself is called a deadlock. the contrary, an action a is *live*, if for all reachable markings M there is some $M' \in [M\rangle$ th $M'[a\rangle\rangle$.

It is well known that for distributed systems a linear-time semantics like the language is t adequate. To take into account some information on the choices that are made we can e.g. (some version of) the failure semantics, which was developed for TCSP [BHR84]. The *lure semantics* of a net N is $\mathcal{F}(N) = \{(w,X) \mid w \in L(N), X \subseteq \Sigma$, and there is some rking M such that $M_N[w\rangle\rangle M$ and $\forall a \in X : \neg M[a\rangle\rangle\}$. In this semantics we add to an age firing sequence w a negative look-ahead of one action, i.e. a set of actions that can be *used* after w. An element (w,X) of $\mathcal{F}(N)$ is called a *failure pair*, and X a *refusal set*.

A more discriminating equivalence is bisimilarity. Two nets N_1 and N_2 are *bisimilar* il83,Par81], if there exists a *bisimulation* between them, i.e. a relation $\mathcal{B} \subseteq [M_{N_1}\rangle \times [M_{N_2}\rangle$

such that

i) $(M_{N_1}, M_{N_2}) \in \mathcal{B}$

ii) If $(M_1, M_2) \in B$ and $M_1[w\rangle\rangle M_1'$, $w \in \Sigma^*$, then there is $M_2' \in [M_{N_2}\rangle$ such that $M_2[w\rangle\rangle$ and $(M_1', M_2') \in \mathcal{B}$.

iii) vice versa, i.e. if $(M_1, M_2) \in \mathcal{B}$ and $M_2[w\rangle\rangle M_2'$, $w \in \Sigma^*$, then there is $M_1' \in [M_{N_1}\rangle$ s that $M_1[w\rangle\rangle M_1'$ and $(M_1', M_2') \in \mathcal{B}$.

3 Merging Places and the Construction of Deadloc Free Nets

The asynchronous communication of nets corresponds to a composition operator that wo by merging places. To keep technicalities simple we will assume that the nets we want compose have just those places that should be merged in common. Since we want to stu the exchange of nets in contexts built with this composition operator, the nets to be merg are not symmetric, i.e. one of them is the context net, the other is a daughter net. require that daughter nets have λ-labelled transitions only; this is necessary in order to stu deadlocks with respect to the actions and transitions of the context net. Also a daughter has effects on the environment only via the common places; in themselves the transitions the daughter net are invisible from the environment.

This way we only capture the refinement of λ-transitions in our approach. But there i natural way to use our results also for the refinement of visible transitions, see [Vog92].

Definition 3.1 Let P be a finite set. A *P-host* is a net N with $P \subseteq S_N$. A *P-daughter-* N_1 is a net with place set $P\dot{\cup}S_1$ such that the marking is zero on P and all transitions λ-labelled.

The *place composition* $N \parallel\!|_P N_1$ of N and N_1 has place set $S_N\dot{\cup}S_1$ and transition $T_N\dot{\cup}T_1$, the arc weight, initial marking and labelling is that of N or N_1 whichever is defin – with the exception that the arc weight is 0 if both arc weights are undefined and the init marking on P equals M_N.

Denoting the place set of N_1 by the disjoint union of P and S_1 is a slight misuse of notati since it implies that S_1 is not the entire place set for a P-daughter-net N_1. It is justified the use of such nets, where P is the interface to a host and 'really' belongs to this host.

Due to lack of space we cannot treat the refinement of transitions in detail. The followi definition just shows in what sense the refinement of λ-transitions is a special case of exchar ing P-daughter-nets in a $\parallel\!|_P$-context. In Section 4 we will restrict our general approach a special class of P-daughter-nets, and then it will be important that we can still treat t special case of an exchange.

Definition 3.2 Let N_h be a net consisting of a λ-labelled transition h and the place $P = {}^\bullet h \cup h^\bullet$ such that ${}^\bullet h \neq \emptyset \neq h^\bullet$, ${}^\bullet h \cap h^\bullet = \emptyset$ and the initial marking is zero. An N_h-λ-h N satisfies $h \in T_N$ and the subnet induced by $\{h\} \cup {}^\bullet h \cup h^\bullet$ equals N_h except for the marki An *N_h-refinement net* R is a P-daughter-net with no arcs leading to ${}^\bullet h$ or from $h^\bullet$.

Let N be an N_h-λ-host, and N^* be obtained from N by deleting h (and the incident arcs);
n $N = N^* \parallel\!\mid_P N_h$ and the *refined net* $N[h \rightarrow R]$ is $N^* \parallel\!\mid_P R$.

Next we define the external equivalence for P-daughter-nets we are interested in.

finition 3.3 Let N_1 and N_2 be P-daughter-nets. They are *P-deadlock-equivalent*, if for P-hosts N we have that $N \parallel\!\mid_P N_1$ can deadlock if and only if $N \parallel\!\mid_P N_2$ can.

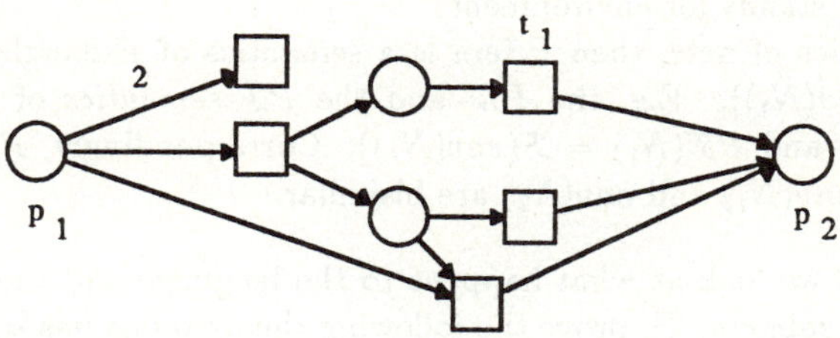

Figure 1

In order to study P-deadlock-equivalence it is important to describe how a P-daughter-net cts to stimuli of the environment, i.e. to tokens put on the places of the interface. For this cription the interplay between P-daughter-net and environment is relevant. Consider the of Figure 1, which has the interface $P = \{p_1, p_2\}$. (Recall that all transitions of P-daughter-s are λ-labelled; these labels are omitted.) If we (taking the part of the environment) put oken onto p_1, then eventually two tokens will be produced on p_2. If after this production put a second token onto p_1, we will get another two tokens on p_2. If instead of waiting for tokens on p_2 we already react to the first token by putting a second token onto p_1, then night happen that we only get two tokens on p_2 all in all instead of four. (This case may pen if the first token on p_2 was produced by t_1.) As a third possibility we could put two ens onto p_1 in the beginning, and then we might get no tokens on p_2; this cannot happen he other two cases.

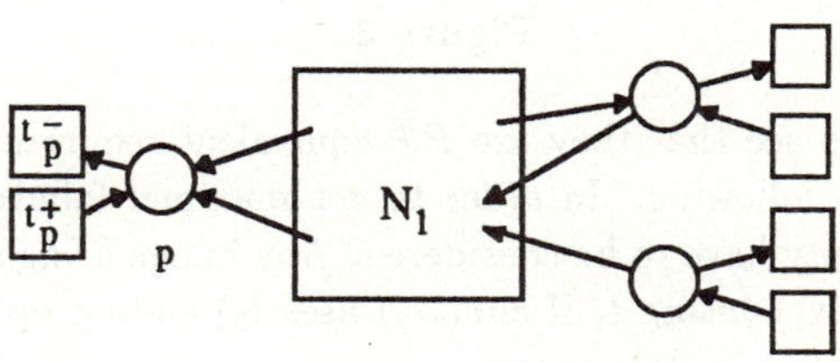

Figure 2

Thus the relevant behaviour of the environment is to put tokens onto interface places, and heck that tokens have been produced, and this check can be done by removing the tokens. order to make this possible behaviour of the environment visible in our action-oriented roach, we add for each interface place p two transitions t_p^+ and t_p^- that respectively add and ove one token from p, see Figure 2. These transitions must be visible, and for simplicity denote the actions they perform again with t_p^+ and t_p^-. In other words, the transitions

t_p^+ and t_p^- represent the environment of N_1, and by adding this environment we can mo
any semantics of nets to a semantics of P-daughter-nets as described in the next definiti
This way we can capture *when* tokens are put onto the interface places by the environm
and *when* the P-daughter-net produces tokens itself, or more precisely when the environm
notices that the P-daughter-net has produced some tokens.

Definition 3.4 Let N_1 be a P-daughter-net. Then $env(N_1)$ is the net obtained by add
for each place $p \in P$ transitions t_p^+, t_p^- with labels t_p^+, t_p^- and arcs of weight 1 from t_p^+ t
and from p to t_p^-. (*env* stands for environment)

If *Sem* is a semantics of nets, then *PSem* is a semantics of P-daughter-nets defined
$PSem(N_1) = Sem(env(N_1))$. E.g. the *PL*- and the *$P\mathcal{F}$-semantics* of N_1 are defined
$PL(N_1) = L(env(N_1))$ and $P\mathcal{F}(N_1) = \mathcal{F}(env(N_1))$. Correspondingly, P-daughter-nets
N_2 are *P-bisimilar*, if $env(N_1)$ and $env(N_2)$ are bisimilar.

To get a first result we look at what happens to the language and the markings of a
when we exchange two subnets. To prove the following theorem one has to compare carefu
the firing sequences of $N \Vert\!\vert_P N_1$, $env(N_1)$, $env(N_2)$, and $N \Vert\!\vert_P N_2$.

Theorem 3.5 *Let N_1, N_2 be PL-equivalent P-daughter-nets, N a P-host. Then $N \Vert\!\vert_P$ and $N \Vert\!\vert_P N_2$ are language-equivalent and marking-equivalent on $S_N - P$.*

Example: Since failure semantics is so useful for the synchronous case, one could hope t
$P\mathcal{F}$-equivalence might give an internal characterization. Figure 3 shows two $P\mathcal{F}$-equival

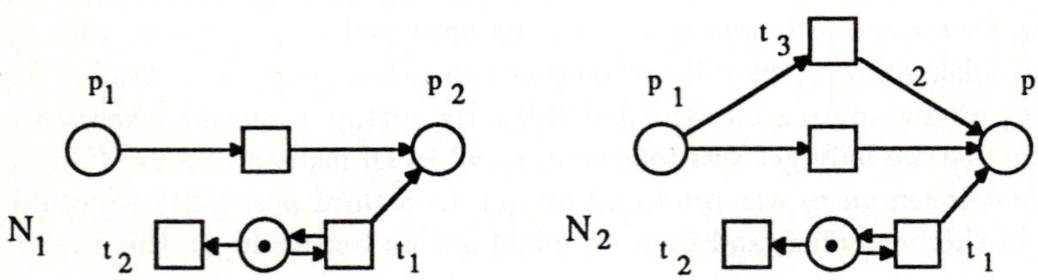

Figure 3

nets for $P = \{p_1, p_2\}$. To see that they are $P\mathcal{F}$-equivalent requires a lengthy case analy
The main points are the following. In order to get maximal failure sets only sequences
$env(N_i)$ that leave p_1 empty have to be considered. Any image firing sequence of $env(N_2)$ c
also be obtained in $env(N_1)$ (using t_1 if $env(N_2)$ uses t_3) ending up with the same marki
and refusal sets; the converse is clear anyway.

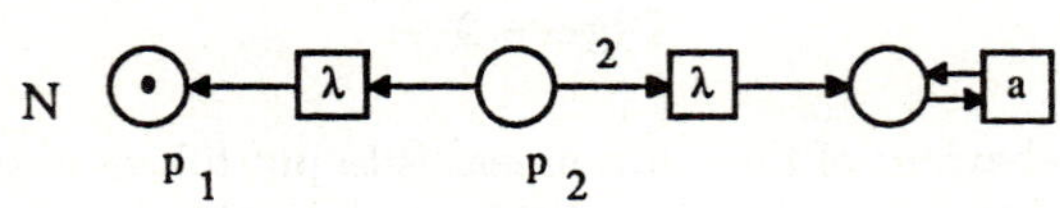

Figure 4

But these nets are not P-deadlock-equivalent. Figure 4 shows a P-host N such th
$N \Vert\!\vert_P N_1$ can deadlock, while $N \Vert\!\vert_P N_2$ cannot. In $N \Vert\!\vert_P N_1$ we can switch off t_1 by firing

l now the only possible behaviour is to let the remaining token cycle between the places p_1 l p_2, which does not involve any visible action. In $N \parallel\!\!|_P N_2$ we can also switch off t_1; but he a-labelled transition is not enabled already, then we can always move a token to p_1 in e p_1 is empty, put two tokens onto p_2 using t_3, and finally activate the action a.

In terms of possible refusals the difference between N_1 and N_2 is the following. While h, $env(N_1)$ and $env(N_2)$, can refuse the single actions $t^-_{p_1}$ and $t^-_{p_2}$ initially (switch off t_1), (N_2) cannot refuse the sequence $t^+_{p_1} t^-_{p_2} t^-_{p_2}$ but $env(N_1)$ can. □

Motivated by this observation we introduce a new semantics that takes into account refusals sequences. One could call it impossible-futures semantics; it corresponds to the possible-ure semantics defined in [RB81] as failure semantics corresponds to readiness semantics 186]. We denote it by $\mathcal{F}^+$, since the refusal sets are subsets of Σ^+ and not necessarily of Σ in the $\mathcal{F}$-semantics.

finition 3.6 The $\mathcal{F}^+$*-semantics* of a net N is

$$\mathcal{F}^+(N) = \{(w,X) \mid w \in L(N),\ X \subseteq \Sigma^+ \text{ and there exists } M \text{ such that } M_N[w\rangle\rangle M \text{ and } \neg M[v\rangle\rangle \text{ for all } v \in X\}.$$

In fact, $P\mathcal{F}^+$-equivalence is strictly stronger than P-deadlock-equivalence. The internal racterization of P-deadlock-equivalence, which we give next, is more elaborate – and less uitive one might feel.

finition 3.7 The $\mathcal{F}^{++}$*-semantics* of a net N is

$$\mathcal{F}^{++}(N) = \{(w,X) \mid w \in L(N),\ X \subseteq \Sigma^+, \text{ and either } X = \emptyset \text{ or there exist some } w' \in X, \text{ a prefix } v \text{ of } w', \text{ and } M \text{ such that } M_N[wv\rangle\rangle M \text{ and } \neg M[v'\rangle\rangle \text{ for all } vv' \in X\}.$$

Note that we do not require anything about sequences $u \in X$ that do not have v as a fix. Thus we have the following relationship between $\mathcal{F}^{++}$-semantics and $\mathcal{F}^+$-semantics. $(wv, Y) \in \mathcal{F}^+(N)$ with $Y \neq \emptyset$ and $X' = \{vv' \mid v' \in Y\}$; now construct X from X' by ling any action sequences that do not have v as a prefix. Then $(w, X) \in \mathcal{F}^{++}(N)$. In s way the element (wv, Y) covers up the element (w, X), which lies in its past. In fact, all $,X) \in \mathcal{F}^{++}(N)$ with nonempty X can be constructed this way, while $(w, \emptyset) \in \mathcal{F}^{++}(N)$ if l only if $(w, \emptyset) \in \mathcal{F}^+(N)$.

Therefore the $\mathcal{F}^{++}$-semantics can be obtained from the $\mathcal{F}^+$-semantics by some closure op-tion or 'covering up'. Especially, $\mathcal{F}^+$-equivalence implies $\mathcal{F}^{++}$-equivalence. (Similarly, the al failure semantics can be obtained from readiness semantics by some closure operation.)

On the other hand, if $(w, X) \in \mathcal{F}^{++}(N)$ and X is a nonempty subset of Σ, then we must e $v = \lambda$ and thus $(w, X) \in \mathcal{F}(N)$. This way it is easy to see that $\mathcal{F}(N) = \{(w, X) \in {}^{+}(N) \mid X \subseteq \Sigma\}$. Especially, $\mathcal{F}^{++}$-equivalence implies $\mathcal{F}$-equivalence.

In the preceeding remarks we have seen the two nonstandard implications stated in the t proposition. That the implications are strict is not easy to see in all cases. But note that nets of Figure 4 are not $P\mathcal{F}^{++}$-equivalent, since $(\lambda, \{(t^+_{p_1} t^-_{p_2})^n t^-_{p_2} \mid n \in \mathbb{N}_0\}) \in P\mathcal{F}^+(N_1)$ $P\mathcal{F}^{++}(N_1)$. This pair is not in $P\mathcal{F}^{++}(N_2)$, since any $(t^+_{p_1} t^-_{p_2})^n$ or $(t^+_{p_1} t^-_{p_2})^n t^+_{p_1}$ leads to a rking of N_2 where $t^+_{p_1} t^-_{p_2} t^-_{p_2}$ or $t^-_{p_2} t^+_{p_1} t^-_{p_2} t^-_{p_2}$ cannot be refused.

Proposition 3.8 *P-bisimilarity implies $P\mathcal{F}^+$-equivalence implies $P\mathcal{F}^{++}$-equivalence imp $P\mathcal{F}$-equivalence implies PL-equivalence. No other implications hold in general.*

Remark: Call nets live equivalent if their exchange in any $\|_A$-context preserves livenes actions. Then it can be shown that $\mathcal{F}^{++}$-equivalence is an internal characterization of equivalence. The details can be found in [Vog92].

Now we will state how the external equivalence is related to the series of semantic eq alences we have just studied. For the following theorem note that the implication given Part i) follows immediately from Part ii) and Proposition 3.8. But my proof that P-deadlo equivalence implies $P\mathcal{F}^{++}$-equivalence uses infinite context nets. Thus it fails, if we o work with finite context nets, in which case we would have a weaker notion of P-deadlo equivalence as we have it here. Therefore Part i) is interesting in its own right.

Theorem 3.9 *i) P-deadlock-equivalence implies $P\mathcal{F}$-equivalence. This also holds if c finite nets are considered in the definition of P-deadlock-equivalence.*

ii) P-daughter-nets are P-deadlock-equivalent if and only if they are $P\mathcal{F}^{++}$-equivalent.

The above results also show that the exchange of P-deadlock-equivalent nets does only preserve deadlock-freeness, but also the language. With the $P\mathcal{F}^{++}$-equivalence we h found for P-deadlock-equivalence an internal characterization, which does not require to c sider all possible context nets. But the $P\mathcal{F}^{++}$-semantics is admittedly somewhat unwiel furthermore, we have the undecidability result below, whose proof is based on the undec ability of language equivalence for labelled nets. For these reasons we will restrict attent to some subclass of P-daughter-nets in the next section, where we will also get a stron result for behaviour preservation. (For the results of the next section it is not necessary use the $P\mathcal{F}^{++}$-semantics; it is sufficient to know that P-deadlock-equivalence lies betw $P\mathcal{F}^+$-equivalence and $P\mathcal{F}$-equivalence.)

Theorem 3.10 *P-deadlock-equivalence is undecidable for finite P-daughter-nets, if $|P| \geq$ This also holds if the equivalence is defined with respect to finite nets only.*

4 I,O-nets and the refinement of transitions

In this section we want to develop a theory of P-daughter-nets that are deterministic in so sense; the aim is to define a notion of determinism that has the following properties.

- It should be decidable whether a P-daughter-net is deterministic.
- P-deadlock-equivalence should be decidable for deterministic nets.
- This notion should allow to treat the refinement of transitions, i.e. nets N_h in the se of Definition 3.2 should be deterministic.
- If the exchange of some given deterministic nets always preserves behaviour in so weak sense (i.e. deadlock-freeness), then it should hold that their exchange also preser behaviour in some strong sense (preferably stronger than failure semantics).

Jnfortunately, the λ-labelled transition h of a net N_h 'decides' nondeterministically whether es or not (provided it is enabled). Suppose, by firing transitions t_p^+ with $p \in {}^\bullet h$ we have just enough tokens on the places in the preset of h to make h enabled. Now, depending on internal action of firing h, transitions t_p^- with $p \in {}^\bullet h$ may be refused or may be fired, i.e. actions t_p^- with $p \in {}^\bullet h$ may occur or may be refused after the same image firing sequence $\imath v(N_h)$. This shows that $env(N_h)$ is not $\mathcal{F}$-deterministic in the sense of [Vog92].

The situation is different for transitions t_p^- with $p \in h^\bullet$. If transition h is enabled, then matter whether h fires first or not – any of these transitions can be fired as the next le transition, i.e. the actions t_p^- with $p \in h^\bullet$ cannot be refused. More generally, $env(N_h)$ terministic with respect to the actions t_p^- with $p \in h^\bullet$, in other words, observing an image g sequence of $env(N_h)$ we can conclude whether t_p^- with $p \in h^\bullet$ is image-enabled after sequence or not. Thus ${}^\bullet h$ and $h^\bullet$ are treated differently with respect to determinism. refore we will study in this section only those P-daughter-nets N_1 where P is split up into -empty sets I of input places and O of output places; this means that no arcs of N_1 lead or from O. We have already required this for N_h-refinement nets in Definition 3.2.

There are three features that we observe for nets N_h and that play an important rôle in proofs of the results below. As already remarked, N_h is deterministic with respect to $h^\bullet$, this leads to the notion of output-determinism. Secondly, N_h produces only bounded out from a given finite input, and this leads to the notion of output-boundedness. Thirdly, is non-absorbing, meaning that it always produces some additional output if it consumes e additional input; hence it does not just absorb the input without any positive effect. course, given some additional input it might happen to be unable to consume this input.) These notions are formalized in our first definition.

inition 4.1 Let P be the disjoint union of nonempty sets I and O. A P-daughter-net N_1 n *I,O-net* if it satisfies the following conditions:

- There are no arcs in N_1 leading to I or from O.
- N_1 is *output-deterministic*, i.e. $\forall p \in O : (w, \{t_p^-\}) \in \mathcal{PF}(N_1) \Rightarrow (wt_p^-, \emptyset) \notin \mathcal{PF}(N_1)$.
- N_1 is *output-bounded*, i.e. for all $p \in O$ and $(w, \emptyset) \in \mathcal{PF}(N_1)$ there is some $n \in \mathbb{N}_0$ with $(w(t_p^-)^n, \emptyset) \notin \mathcal{PF}(N_1)$.
- N_1 is *non-absorbing*, i.e. for $X_I = \{t_p^- \mid p \in I\}$, $w, w' \in \{t_p^+ \mid p \in I\}^*$ with $w' \neq \lambda$ we have: if $(w, X_I), (ww', X_I) \in \mathcal{PF}(N_1)$ then there are $p \in O$ and $n \in \mathbb{N}$ such that $(w(t_p^-)^n, \emptyset) \notin \mathcal{PF}(N_1)$ and $(ww'(t_p^-)^n, \emptyset) \in \mathcal{PF}(N_1)$.

The formal definitions of output-determinism and output-boundedness are quite straightward translations from the verbal explanations given above. In human words the last part Definition 4.1 says the following. The action sequence w describes that some tokens are onto the input places, and $(w, X_I) \in \mathcal{PF}(N_1)$ says that now the transitions of N_1, which internal, can fire in such a way that the input is empty afterwards. The action sequence ′ puts more tokens onto the input places, and these can also be consumed completely by Now the requirements for p and n are that given the input described by w it cannot pen that we find n or more tokens on p, while this can happen if we have the larger input cribed by ww'.

Example: Figure 5 shows the I,O-net N_1. This net is output-deterministic: given s image firing sequence w of $env(N_1)$ there are several cases. Either w visibly (i.e. using puts more tokens onto p_3 than it removes, in which case $t^-_{p_3}$ can fire immediately and action $t^-_{p_3}$ cannot be refused. Or w removes more tokens from p_3 than it puts visibly, in w case the difference must be 1, one of the internal transitions must have fired, and the ac $t^-_{p_3}$ cannot occur next. In the case that w visibly puts the same number of tokens onto p it removes there are two subcases. Either w puts more tokens onto one of p_1 and p_2 tha removes visibly, in which case the corresponding internal transition has already fired or fire next, thus action $t^-_{p_3}$ can occur next and cannot be refused. Or this is not the case, the marking reached after w is the initial marking and $t^-_{p_3}$ cannot occur.

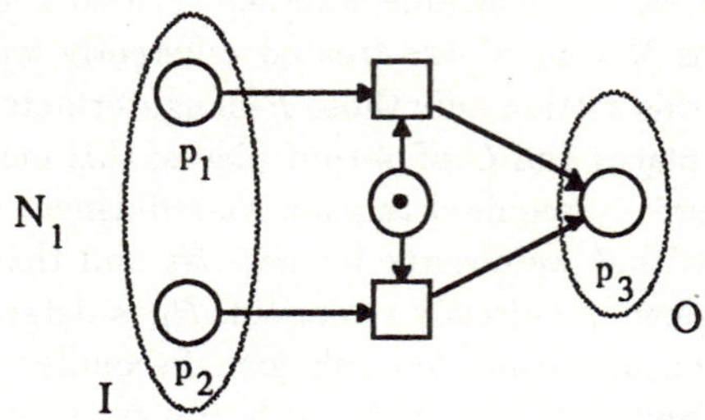

Figure 5

N_1 is output-bounded, since $env(N_1)$ can perform at most one more $t^-_{p_3}$ after some w t there are $t^+_{p_3}$ in w. Finally, N_1 is non-absorbing: if we have $w, ww' \in \{t^+_p \mid p \in I\}^*$ as requ in Definition 4.1, then $(ww', X_I) \in \mathcal{PF}(N_1)$ implies that ww' consists of at most one act i.e. $w = \lambda$, and w' is $t^+_{p_1}$ or $t^+_{p_2}$; we have $(t^-_{p_3}, \emptyset) \notin \mathcal{PF}(N_1)$ but $(t^+_{p_1} t^-_{p_3}, \emptyset), (t^+_{p_2} t^-_{p_3}, \emptyset) \in \mathcal{PF}($

This example also shows that there is some nondeterminism left in I,O-nets. Not only N_1 decide whether it consumes a token from p_1, e.g. $(t^+_{p_1} t^+_{p_2} t^-_{p_1}, \emptyset), (t^+_{p_1} t^+_{p_2}, \{t^-_{p_1}\}) \in \mathcal{PF}(N_1$ also chooses between tokens on p_1 and p_2, i.e. we have $(t^+_{p_1} t^+_{p_2}, \{t^-_{p_1}\}), (t^+_{p_1} t^+_{p_2}, \{t^-_{p_2}\}) \in \mathcal{PF}($ but $(t^+_{p_1} t^+_{p_2}, \{t^-_{p_1}, t^-_{p_2}\}) \notin \mathcal{PF}(N_1)$.

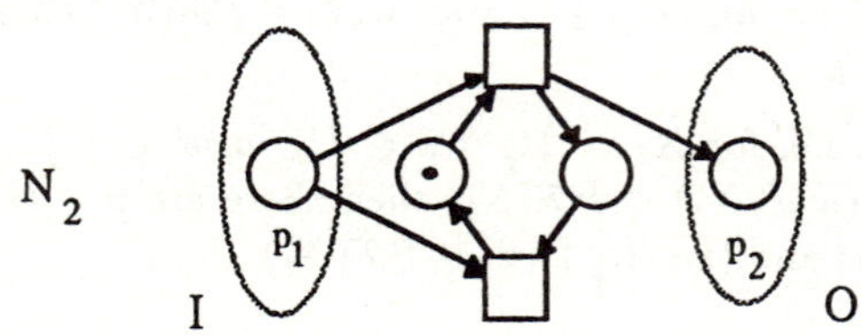

Figure 6

Example: Figure 6 shows another net N_2 that illuminates our definition. Again N_2 choose whether to consume a token on p_1 or not, but otherwise it is very predictable. W consuming tokens from p_1 it strictly alternates between producing a token on p_2 or show no reaction at all. This net is output-deterministic and output-bounded, but it is not r absorbing. For $w = w' = t^+_{p_1}$ we have

- $(t^+_{p_1}, \{t^-_{p_1}\}), (t^+_{p_1} t^+_{p_1}, \{t^-_{p_1}\}) \in \mathcal{PF}(N_2)$

$$(t^+_{p1}(t^-_{p2})^n, \emptyset) \notin \mathcal{PF}(N_2) \Rightarrow n \geq 2 \Rightarrow (t^+_{p1}t^+_{p1}(t^-_{p2})^n, \emptyset) \notin \mathcal{PF}(N_2). \qquad \square$$

)bserve that we have used $\mathcal{PF}(N_1)$ in the definition of I,O-nets, although the $\mathcal{PF}$-valence is in general too weak to imply the external equivalence we are interested in. first justification let us remark that as an immediate consequence of Theorem 3.9 i) the of I,O-nets is closed under the equivalence of interest.

position 4.2 *Let N_1 be an I,O-net, N_2 be a P-daughter-net such that there are no arcs ing to I or from O. If N_1 is P-deadlock-equivalent to N_2, then N_2 is an I,O-net. This holds if P-deadlock-equivalence is defined with respect to finite nets only.*

The next theorem shows that for I,O-nets a large range of equivalences collapses.

orem 4.3 *I,O-nets are P-bisimilar if and only if they are PL-equivalent.*

ollary 4.4 *I,O-nets are P-deadlock-equivalent if and only if they are PL-equivalent. This holds if the external equivalence is defined with respect to finite nets only.*

Now we can show: if the exchange of some I,O-nets guarantees a weak behaviour preser-on, then it automatically preserves behaviour in quite a strong sense.

orem 4.5 *Let N_1, N_2 be I,O-nets, N a P-host. If N_1 and N_2 are P-deadlock-equivalent, $N \parallel\!\mid_P N_1$ and $N \parallel\!\mid_P N_2$ are bisimilar.*

Next we state the decidability results we were aiming for. Their proofs use the decidability veness and different constructions of nets for the three features of I,O-nets and for P-llock-equivalence.

orem 4.6 *It is decidable whether a given finite P-daughter-net is an I,O-net.*

orem 4.7 *It is decidable whether finite I,O-nets are P-deadlock-equivalent.*

Let us finally sketch how these results can be applied to the refinement of transitions. ıme we want to refine some transition h by some N_h-refinement net R as described in nition 3.2, and assume we want to be sure that this refinement always preserves deadlock-ıess. Proposition 4.2 tells us that R must be an I,O-net, and Corollary 4.4 that R and must be PL-equivalent. From Theorems 4.6 and 4.7 we know that these requirements are dable. If these requirements are satisfied, then the refinement with R does not only pre-e the basic feature of deadlock-freeness, but by Theorem 4.5 the refined and the unrefined have the same behaviour in a very strong sense, they are bisimilar.

Further considerations show that nets R that satisfy the above requirements coincide with modules studied in [Vog87]. I,O-nets also allow to cope with the behaviour preserving ement of several independent transitions as studied in [Mül85]. For more details on sition refinement see [Vog92].

References

[And83] C. André. The behaviour of a Petri net on a subset of transitions. *R.A.I.R.O.*, 17:5 1983.

[Bau88] B. Baumgarten. On internal and external characterizations of PT-net building block haviour. In G. Rozenberg, editor, *Advances in Petri Nets 1988*, Lect. Notes Comp. 340, 44–61, 1988.

[BGV91] W. Brauer, R. Gold, and W. Vogler. A survey of behaviour and equivalence preser refinement of petri nets. In G. Rozenberg, editor, *Advances in Petri Nets*, Lect. N Comp. Sci. 483, 1–46, 1991.

[BHR84] S.D. Brookes, C.A.R. Hoare, and A.W. Roscoe. A theory of communicating seque processes. *J. ACM*, 31:560–599, 1984.

[Bro83] S.D. Brookes. A model for communicating sequential processes. Technical Report C CS-83-149, Carnegie-Mellon-Univ., Pittsburgh, 1983.

[Che91] G. Chehaibar. Replacement of open interface subnets and stable state transforma equivalence. In *Proc. 12th Int. Conf. Applications and Theory of Petri Nets, Gjern*, p 390–409, 1991.

[Eng85] J. Engelfriet. Determinacy → (observation equivalence = trace equivalence). *The Comput. Sci.*, 36:21–25, 1985.

[Mil83] R. Milner. Calculi for synchrony and asynchrony. *Theor. Comput. Sci.*, 25:267–310, 1

[Mül85] K. Müller. Constructable Petri nets. *Elektr. Inf. Kybern.*, 21:171–199, 1985.

[OH86] E.R. Olderog and C.A.R. Hoare. Specification-oriented semantics for communicating cesses. *Acta Informatica*, 23:9–66, 1986.

[Par81] D. Park. Concurrency and automata on infinite sequences. In P. Deussen, editor, *P 5th GI Conf. on Theoretical Comp. Sci.*, Lect. Notes Comp. Sci. 104, 167–183, 1981.

[RB81] W.C. Rounds and S.D. Brookes. Possible futures, acceptances, refusals and communica processes. In *Proc. 22nd Ann. Symp. on Foundations of Comp. Sci.*, pages 140–149. IE 1981.

[Rei85] W. Reisig. *Petri Nets*. EATCS Monographs on Theoretical Computer Science. Sprin 1985.

[Sou90] Y. Souissi. On liveness preservation by composition of nets via a set of places. In *P 11th Int. Conf. on Application and Theory of Petri Nets*, pages 104–122, 1990.

[Vog87] W. Vogler. Behaviour preserving refinements of Petri nets. In G. Tinhofer and G. Schm editors, *Graph-Theoretic Concepts in Computer Science, Proc. WG 86, Bernried*, L Notes Comp. Sci. 246, , 82–93, 1987.

[Vog89] W. Vogler. Failures semantics and deadlocking of modular Petri nets. *Acta Informa* 26:333–348, 1989.

[Vog92] W. Vogler. *Modular Construction and Partial Order Semantics of Petri Nets*. Lect. N Comp. Sci. Springer, 1992. To appear.

A Parametric Approach to Localities*

Ugo Montanari
Univ. Pisa, Dip. Informatica

Daniel Yankelevich
Univ. Pisa, Dip. Informatica
HP Labs, Pisa Science Center

Abstract

Location equivalence has been presented in [3] as a bisimulation based equivalence able to take into account the spatial distribution of processes.

In this work, the parametric approach of [9] is applied to location equivalence. An observation domain for localities is identified and the resulting equivalence is shown to coincide with the equivalence introduced in [4, 13]. The observation of a computation is a forest (defined up to isomorphism) whose nodes are the events (labeled by observable actions) and where the arcs describe the sub-location relation.

We show in the paper that our approach is really parametric. By performing minor changes in the definitions, many equivalences are captured: partial and mixed ordering causal semantics, interleaving, and a variation of location equivalence where the generation ordering is not evidenced. It seems difficult to modify the definitions of [4, 13] to obtain the last observation. The equivalence induced by this observation corresponds to the very intuitive assumption that different locations cannot share any common clock, and hence the ordering between events occurring in different places cannot be determined.

Thanks to the general results proved in [9] for the parametric approach, all the observation equivalences described in the paper come equipped with sound and complete axiomatizations.

Introduction

e field of formal models of concurrency has had a recent explosive development. any models and theories supporting them have been defined, which differ on notion, formal basis or assumptions about practical issues. Beside the advantages at this development can offer, the proliferation of models may produce some consion. Even their classification is difficult, because of the fact that each one uses different notation, or introduces ad-hoc definitions, and because of the lack of a mmon framework to compare them. In addition, it is quite likely that there is no st model, since each one emphasizes different aspects of systems, and hence has vantages in proving some kind of properties or in specifying a particular type of plications. One can only choose which model (s)he prefers in a particular situation when specific premises apply.

We restrict ourselves to models equipped with an operational semantics in otkin's SOS style [15], and with some variation of a bisimulation relation [14]. so within this class of models, a variety of approaches have been proposed in the erature. Often a lot of work is redone and several definitions are rewritten in an -hoc manner, changing minor points. This redundancy can be avoided by using general, parametric approach, capable to describe various models. A parametric proach to the definition of concurrent systems has been proposed in [9]. It is the olution of ideas already present in [7]. In its general lines, this approach consists four steps:

*Research partially supported by Hewlett – Packard, Pisa Science Center, within the MDC oject.

1. Define a transition system that captures the desired operational behavicr.

2. Build the possible computations as paths in the transition system, and stru ture them as an *observation tree* (ordering them by prefix).

3. Define what are the observations of computations. These appear as labe.s the nodes of the observation tree.

4. Define an equivalence between observation trees, based on the observatio defined in step (3).

The third step characterizes different semantics according to the observatio they utilize. Once the transition system of step (1) has been fixed, and even choc ing as the equivalence of level (4) the same observational equivalence, different c servations yield very different semantic models. Each observation corresponds distinct premises about what an observer of the system should see. Many theori are captured in this way, just by changing the observation of the computations.

Moreover, this point of view helps in comparing diverse semantics, showing h to obtain one from another and giving a common context to classify them (showi that, for example, they coincide in some of the four steps and differ on others, showing how the different observations are related).

In particular, a general axiomatization of observational congruence for (finit observation trees is given in [9], which is parametric with respect to the observatic Hence, by defining the semantics within this setting one gets an axiomatizati automatically, just as an instance of the general result.

In the present paper, we test this approach with location equivalence. Locati equivalence has been presented in [3] as a bisimulation based equivalence able take into account the spatial distribution of processes. The intuitive idea is th each action occurs at a location and locations may have sublocations (for examp a fork operation at some place may create two sublocations of that place). F CCS, the assumption is that in E|E' the two subprocesses E and E' are at differe locations.

These semantics, which can detect the spatial distribution of events, are ve useful, for example in a context of faulty processors, or when information abo the assignment process-processor has to be taken into account. For an extend motivation of such an approach we refer the reader to [3].

The framework for the equivalence proposed in [3, 4] is given by an observer w can find out where an event takes place, i.e. the location where an action occu More precisely, a dynamic mechanism is given to determine the relative location an event with respect to the locations of previous events. Hence, it is possible discern if a number of events occur in the same place or if they are distributed ov many locations.

This proposal takes spatial information into account, but not causal informatic There are some examples where location equivalence, as defined in [4], is finer th causal equivalence; but there are some examples where the causal approach is mo discriminating.

For example, consider the two following processes in CCS notation: $\alpha.\beta.\gamma.N$ and $(\alpha.\delta.NIL|\bar{\delta}.\beta.\epsilon.NIL|\bar{\epsilon}.\gamma.NIL)\backslash\delta\backslash\epsilon$

Both of them can perform a sequence of three actions, and the causal relati between action occurrences forms a total ordering on both of the processes. B the former executes the actions in one place and the latter executes each action a different location. From a causal point of view both processes are equivalent, b when examining their locations they are different. On the other side, it is easy find processes with the same spatial distribution, but with different causal relatio between the events.

The idea of defining process algebras semantics which are sensitive with respe to the spatial distribution of the bisimulation testing scenario goes back to the wo of [5] on distributed bisimulation. However, the recent developments in [3, 4, 1 have made it easier to embed this approach in the parametric setting outlined abov

In this paper, we fully separate the different conceptual levels of the theory pre-
ıted in [4], and we give an alternative characterization of the proposed semantics.
observation domain for localities is identified and the resulting equivalence is
own to coincide with the equivalence introduced in [4, 13]. It turns out that these
servations constitute a very simple domain: the observation of a computation is a
est (defined up to isomorphism) whose nodes are the events (labeled with observ-
le actions) and where the arcs describe the sub-location relation. The sublocation
ation is statically characterized for CCS by very simple rules. For example, the
ests corresponding to the maximal computations of the terms $\alpha.\beta.\gamma.NIL$ and
$.\delta.NIL|\bar{\delta}.\beta.\epsilon.NIL|\bar{\epsilon}.\gamma.NIL)\backslash\epsilon\backslash\delta$ are shown in figure 1a and 1b respectively.

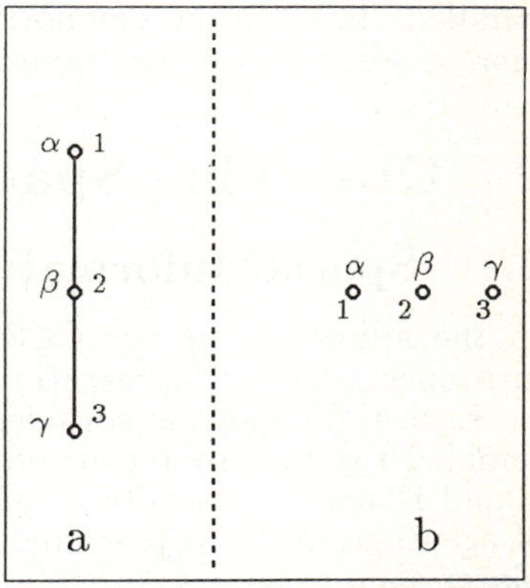

Figure 1: Observations for localities

ıe natural numbers in the events denote the (to-
) generation ordering; i.e. the ordering in which
ents occur.
ducing an ad-hoc definition to a particular in-
ance of a parametric approach is often worth-
ile, because
it is convenient to use a general theory since it
ssibly leads to re-use general theorems, axioma-
ations or parametric tools, and
he resulting characterization possibly clarifies de-
ıls and evidences points, which could be hidden
the ad-hoc definition, helping in the development
new semantics or in testing new ideas.
[5] there is no clear cut separation between
e operational semantics of the language, the description of the observation (i.e.
e assumptions on what an observation of the system is) and the mechanism of
e equivalence (i.e. the assumptions on how computations are organized and the
simulation proceeds).

In [3, 4, 13] for localities and also in [6, 13] for causal mixed orderings, the tran-
ion system that defines the operational semantics is given specifically for the ob-
rvation. In particular, the states are terms of an enriched language which captures
me of the structure of the observations (for example, including causes explicitly
the syntax of the terms). This means that some kind of ad-hoc *unfolding* of the
iginal transition system is performed, where states have informations about the
mputations which can reach them and their observations. For example, in [4], the
nsition system corresponding to $rec\, x.\ \alpha.x$ has no cycle, since in each transition a
efix of the form l:: is introduced. Hence, the right hand side of transitions always
ows. This ad-hoc definition implies also that the transition systems of [3, 4, 13] are
finitely branching, since in each transition a name for a location can be arbitrarily
osen from an infinite set of location names.

We show in the paper that our approach is really parametric; by performing
inor changes in the definitions, many equivalences are obtained: partial and mixed
dering causal semantics, interleaving, and a variation of location equivalence where
e generation ordering is not evidenced. For instance, it seems difficult to modify
e definitions of [4] to obtain the last observation. The equivalence induced by this
servation corresponds to the very intuitive assumption that different locations
ay not share any common clock, and hence the ordering between events occurring
different places cannot be determined.

Thanks to the general results proved in [9] for the parametric approach, all
e observation equivalences described in the paper come equipped with sound and
mplete axiomatizations.

In section two we introduce spatial terms which provide the notation for the spatial distribution of events and we define the observation domain for localities. section three, we recall the notion of observation tree from [9] and we enunciate t main result of the work: the equivalence defined in this paper coincides with t location equivalence of [4, 13]. Section four presents the other observations.

In this volume, another paper [10] is included which also discusses a general a proach to observational equivalences for locality and causality. The main differen is that the first and the second step (and part of the third step) of our approach a merged in [10] in the construction of a tree whose arcs are labeled by proofs. A advantage of their approach is that it is incremental, in the sense that the observ tion of every transition yields a single event. However, their proposal is less gener than ours: for instance abstract localities and partial causal orderings cannot handled. In addition, our notion of observation, being defined on computations, more explicit and makes comparisons easier.

2 Observing Spatial Distribution

2.1 Spatial Information for CCS

In the following, we use CCS [14] as the example language. In this section v introduce a way of representing and collecting spatial information for CCS.

Each CCS agent is considered as representing many sequential components a sembled together by means of the | constructor. That is, in the agent E|E' agen E and E' are supposed to be at different places, and each of them can be compose of agents at different plac, and each of them can be composed of agents at differe places also.

Definition 1 Spatial Terms
Spatial terms are defined by the following (abstract) syntax: $S ::= \bullet, \emptyset, S|S$ *, ar respect the following axiom:* $\emptyset|\emptyset = \emptyset$.

Notice that the | operation is not associative. If we accept that the structu given by the | constructor in a CCS agent determines the distributed nature of number of sequential components, spatial terms reflect this structure. A spatial ter gives the place of a component or of a group of components.

Intuitively, spatial terms are here used to describe information about the physic distribution of events. For instance, the term $\bullet|\emptyset$ describes a place having anoth idle component at the right, and $\bullet|\bullet$ describes the locality of an event which performed simultaneously on both sides of the system. The locality of the occurren of the action α in $(\alpha|\beta)|\gamma)\backslash\gamma$ is described by the term $(\bullet|\emptyset)|\emptyset$.

We will now define a relation between spatial terms which represents the ide of sublocality: a spatial term with exactly one bullet is a sublocality of anoth term with this property if the second is a refinement of the former, i.e., if it can b obtained from the former by substituting the bullet with a spatial term.

Definition 2 Sublocalities
Let $\ll$ *be the least relation defined by the following rules. It is defined only betwee spatial terms with exactly one occurrence of a* $\bullet$.

$$\bullet \ll s \qquad \frac{s_1 \ll s_2}{s_1|\emptyset \ll s_2|\emptyset} \qquad \frac{s_1 \ll s_2}{\emptyset|s_1 \ll \emptyset|s_2}$$

The above relation defines a tree-like partial ordering between spatial terms. Th root is $\bullet$ and a spatial term s is greater than s' if s is obtained from s' by replacin the dot with a spatial term. This partial ordering captures the intuitive idea that

Table 1: Spatial Transition Rules for CCS

Act)	$\mu.E \xrightarrow{\mu@\bullet} E$	*Res*)	$\dfrac{E_1 \xrightarrow{\mu@s} E_2}{E_1\backslash\alpha \xrightarrow{\mu@s} E_2\backslash\alpha}$	$\mu \notin \{\alpha, \bar{\alpha}\}$
Rel)	$\dfrac{E_1 \xrightarrow{\mu@s} E_2}{E_1[\Phi] \xrightarrow{\Phi(\mu)@s} E_2[\Phi]}$	*Sum*)	$\dfrac{E_1 \xrightarrow{\mu@s} E_2}{E_1 + E \xrightarrow{\mu@s} E_2}$	$\dfrac{E_1 \xrightarrow{\mu@s} E_2}{E + E_1 \xrightarrow{\mu@s} E_2}$
Syn)	$\dfrac{E_1 \xrightarrow{\alpha@s} E_2,\ E_1' \xrightarrow{\bar{\alpha}@s'} E_2'}{E_1 \vert E_1' \xrightarrow{\tau@(s\vert s')} E_2 \vert E_2'}$	*Com*)	$\dfrac{E_1 \xrightarrow{\mu@s} E_2}{E_1 \vert E \xrightarrow{\mu@(s\vert\emptyset)} E_2 \vert E}$	$\dfrac{E_1 \xrightarrow{\mu@s} E_2}{E \vert E_1 \xrightarrow{\mu@(\emptyset\vert s)} E \vert E_2}$
Rec)	$\dfrac{E_1[rec\,x.\ E_1/x] \xrightarrow{\mu@s} E_2}{rec\,x.\ E_1 \xrightarrow{\mu@s} E_2}$			

resents the whole system and terms of the form $s|\emptyset$ or $\emptyset|s$ are parts of the system, sublocalities.

Notice that spatial terms having exactly one $\bullet$ are terms associated to events :urring in one place, i.e. which do not correspond to communications.

Let $\mathrm{A} = \Delta \cup \bar{\Delta} \cup \{\tau\}$ be the set of actions. Let A denote the set of CCS agents. w, we will introduce the transition system for CCS. In this transition system, the .tial information is given by decorating the transitions with an extra label. Thus transition relation TR is a subset of $A \times (A \times S) \times A$. The @ is used to construct rs in $A \times S$ and it is read "at", meaning that the action occurs *at* that place.

finition 3 Spatial Transition System
e spatial transition system for CCS is defined by the rules of Table 1. ■

A CCS computation from E to E' is, as usual, a sequence of transitions in the nsition relation, such that sources and targets of transitions coincide: $E_0 \xrightarrow{\mu_1@s_1} \xrightarrow{\mu_2@s_2} \ldots \xrightarrow{\mu_n@s_n} E_n$.

In our approach, the initial step is to define suitable operational rules and then, abstracting some information (i.e. by *observing* the system), to obtain a spectrum different semantics. Hence, initially, a language is conceptually equipped with a lly concrete" operational semantics. For CCS, the most natural such a semantics given by the so-called proved transition system [2], where each arc is labeled h its proof in the SOS deduction system. However, when attention is focused a particular class of semantics (here we consider locality and causality based nantics only) it may be convenient to define an operational semantics which is the most concrete possible. In fact, it is worth choosing the *right* starting point, ving enough information to get the desired observations (i.e. being "fully concrete" y if one wants to reach all possible observations) and abstract enough to avoid ling with irrelevant details.

The transition system we use here is isomorphic neither to Milner's, since some s are duplicated (for example, from the term $\alpha|\mathrm{NIL} + \mathrm{NIL}|\alpha$ to NIL|NIL there are arcs, since the α-actions occur at different places) nor to the proved transition tem (for example, in the proved transition system there are two arcs from $\alpha + \alpha$ NIL while here there is only one).

Our transition system can be seen as the proved transition system modulo certain oms defined on observations. In fact, in the proved transition system, arcs are labeled with proofs, and hence the first Com) rule, for instance, looks like E_1 – E_2 *implies* $E_1|E \stackrel{w|E}{\longrightarrow} E_2|E$, where $\_|E$ is the proof constructor corresponding to t rule. The first rule Com) of table 1 is equivalent to the proved rule given above we throw in the axiom $(\mu@s)|E = \mu@(s|\emptyset)$. In [11] the notion of observation algeb is formally defined, and several algebras dealing also with spatial information a presented.

2.2 Observing Computations

The next step in the parametric approach is to determine what is observed out a single computation. Given a description of a computation of the system, wh is seen by an external observer is established by means of an *observation functic* which maps computations of the system into their corresponding observations.

The only requirement on observation functions is to be monotonic: if one obser a computation and obtains some information, observing a "longer" computati (i.e. a greater computation in the prefix ordering) one obtains more informatic Formally, observation domains are ordered, and the observation function has to monotonic on computations when ordered by prefix. We will call simply *observati* the observation domain together with the function from the computations to t domain. We consider observations modulo isomorphism.

Definition 4 Observation **loc**
Given a computation $E_0 \stackrel{\mu_1@s_1}{\longrightarrow} E_1 \stackrel{\mu_2@s_2}{\longrightarrow} \ldots \stackrel{\mu_n@s_n}{\longrightarrow} E_n$, *we associate to it a struct* $\langle E, <, \prec, l\rangle$, *where* $\langle E, \prec\rangle$ *is a forest of events and* $<$ *is a total* generation *orderi on events, defined in the following way:*
$E = \{e_i | i = 1 \ldots n \wedge \mu_i \neq \tau\}$ *is a set of (at most n) different events*
$e_i \prec e_j$ *iff* $(s_i \ll s_j \wedge i \leq j)$
$e_i < e_j$ *iff* $i \leq j$
$l(e_i) = \mu_i$
This observation is called **loc**.

In the forest $\langle E, <, \prec, l\rangle$ an event e is a successor of e' if e occurs in the sar locality as e', or if e occurs in a sublocality of e'. A branch in the forest correspon intuitively to a "fork" operation, in which two (or more) sublocalities are create Only visible actions are taken into account to construct the structure: in fact tl observation is "weak", in the sense that it does not observe τ's. Moreover, τ's ha no influence in the construction of the observation. This fact corresponds to the id that communications have no locality, and that locality (unlike causality) cannot "transmitted" by synchronizations.

Example 5 *Let c be the computation* $(\alpha.\gamma.\delta|\ \beta.\bar{\gamma}.\epsilon)\backslash\gamma \stackrel{\alpha@\bullet|\emptyset}{\longrightarrow} (\gamma.\delta|\ \beta.\bar{\gamma}.\epsilon)\backslash\gamma \stackrel{\beta@\emptyset|\bullet}{\longrightarrow} (\gamma.\delta|\ \bar{\gamma}.\epsilon)\backslash\gamma \stackrel{\tau@\bullet|\bullet}{\longrightarrow} (\delta|\epsilon)\backslash\gamma \stackrel{\delta@\bullet|\emptyset}{\longrightarrow} (\mathrm{NIL}\ |\epsilon)\backslash\gamma \stackrel{\epsilon@\emptyset|\bullet}{\longrightarrow} (\mathrm{NIL}|\mathrm{NIL})\backslash\gamma$.
The forest $\langle E, <, \prec, l\rangle$ *corresponding to the* **loc** *observation of c is shown in figure 2. The set E has four events, one for each no-tau transition, labeled with* α, β, δ *and* ϵ *respectively. The total ordering* $<$ *is indicated in the figure via the numbers labeling the events and it is obtained directly from the occurrence ordering in the computation c.* ■

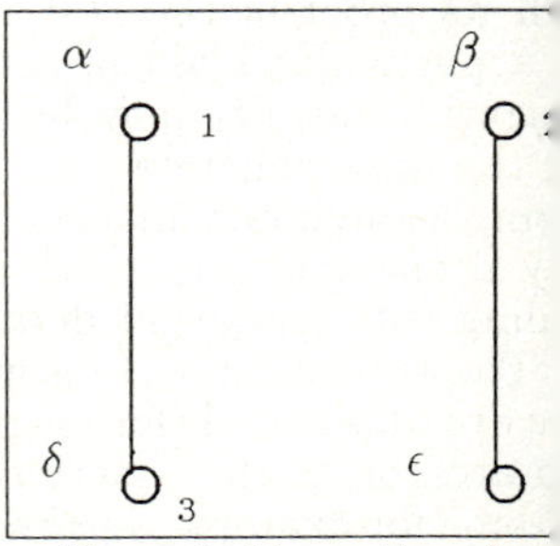

Figure 2: Observation of c.

Observation Trees and Equivalences

to now, the observations of computations have been defined. In this section, we ne observational equivalences with respect to such observations. To this aim, computations an agent can perform are ordered by prefix, generating a tree-structure. Notice that this structure is fixed, independently of the observation. ese structures have been presented in [7] with the name of Nondeterministic Mea-ement Systems (NMS) and have also been studied, with the name of *observation s*, in [9] where a parametric approach has been developed. In particular, there complete axiomatization of observational congruence on observation trees (for finite case), parametric with respect to the observation.

finition 6 Observation Tree
e observation tree corresponding to a CCS agent E with observation **obs** *is the of all the computations from E ordered by prefix, where each node is labeled by observation* **obs** *of the corresponding computation.* ■

We will denote an observation tree as <N, $\leq$, o> , where N is the set of nodes, he ordering relation and o the observation function. A node n' is a *successor* of ode n in <N, $\leq$, o> if n $\leq$ n'; and it is an *immediate* successor if n < n' (i.e. n ı' and n $\neq$ n') and there is no n" such that n < n" < n'.

finition 7 Observational Equivalence and Congruence [1]
t = <N, $\leq$, o> and t' = <N', $\leq$', o'> be two observation trees (over the same ervation domain), and let R be a symmetric binary relation on N$\cup$N'. We will that R is a bisimulation iff n_1 R n_2 implies that the observations of n_1 and n_2 ncide and that for every immediate successor n_1' of n_1 there exists a successor of n_2 such that n_1' R n_2'.
Two trees t and t' are observationally equivalent *if there exists a bisimulation R h that the roots are in the relation R, and they are* observationally congruent *if re exists a bisimulation R* rooted *at the roots of t and t', i.e. such that the roots in the relation R and no other node is in the relation with them.* ■

ample 8 *Figure 3 shows the observation tree responding to the agent $((\alpha+\bar{\beta})|(\alpha.\beta.\gamma))\backslash\beta$ with observations.* ■

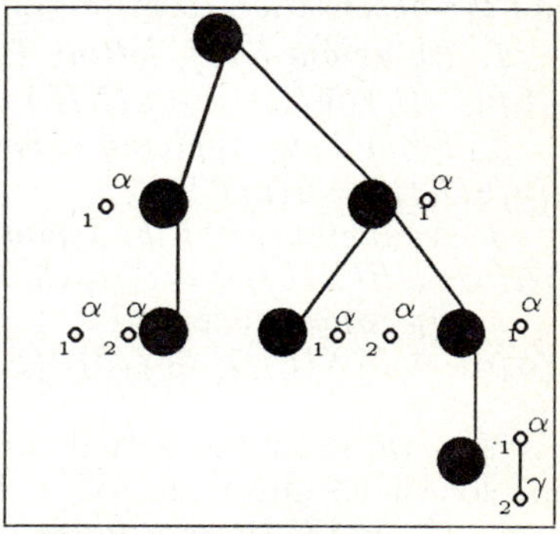

Figure 3: The observation tree of $((\alpha+\bar{\beta})|(\alpha.\beta.\gamma))\backslash\beta$

will say that two CCS agents E and E' are ob-vationally equivalent with respect to an obser-ion **obs** (E $\approx_{obs}$ E') if their corresponding **obs** ervation trees are observationally equivalent.
interesting feature of observation trees is that omplete axiomatization has been given for ob-vational congruence (parametric with respect to observation) in the finite case.
[9] a particular syntax for representing (finite) ervation trees as terms of an algebra is intro-ced. In fact, given a domain of observations D this case a domain of forests), an *observation term* is defined as: E := NIL | $\mathfrak{a}$(E)

[1] Several notions of bisimulation have been defined in [9] on node-labeled structures. In par-lar, a notion of weak bisimulation has been introduced which is the natural extension of the e notion on arc-labeled structures. Instead, the definition and axiomatization given here corre-nds to the *jumping bisimulation* of [9]. However, it is easy to show that for observations without *mic actions*, and this is the case for all the observations presented in this work, jumping and k bisimulation coincide, while the axiomatization for jumping bisimulation is simpler.

Table 2: Axioms for Observational Congruence			
(A1)	P⊕Q	=	Q⊕P
(A2)	(P⊕Q)⊕R	=	P⊕(Q⊕R)
(A3)	$\mathfrak{a}$(NIL)⊕P	=	P
(A4)	P⊕P	=	P
(A5)	$\mathfrak{a}(\mathfrak{b}(\mathfrak{b}(P)\oplus Q)) \oplus \mathfrak{a}(P)$	=	$\mathfrak{a}(\mathfrak{b}(P)\oplus Q)$

| E⊕E, where $\mathfrak{a}$ ∈D. In addition, the operations are partial, and this restriction expressed via a type system. Typed observation terms are denoted by P, Q, R. T interpretation of the operations on trees is as follows. Constant NIL denotes t empty tree. The operation $\mathfrak{a}$(P) prefixes the tree P with a node labeled with $\mathfrak{a}$, a it is defined only if the root of P (if any) is labeled with an observation equal greater than $\mathfrak{a}$ (because the observation function has to be monotonic on comp tations ordered by prefix). The operation ⊕ between observation trees merges tv nonempty trees by collapsing their roots, hence it is defined only if both trees ha the same observation associated to the root.

Observation terms, modulo axioms (A1), (A2) and (A3) of Table 2, are isomc phic with finite observation trees.

In [9] it is proved that the set of axioms (A1)-(A5) shown in Table 2, are consiste and complete with respect to observational congruence over observation trees. Th is, two trees represented by observation terms P and Q are observationally congrue iff (A1)-(A5)⊢P=Q.

Example 9 *Let $p = (\alpha.\gamma|\bar{\gamma}.\beta)\backslash\gamma + (\gamma.\alpha|\beta.\bar{\gamma})\backslash\gamma$ and $q = \alpha|\beta$. Then, $p \approx_{loc} q$.*

Let ϵ denote the empty location observation, α denote the location observati with just one element labeled with α, and let β denote the location observation w one element labeled β.

Let $\alpha\&\beta$ denote the location observation with two elements, one labeled w α and the other with β, which are not related in the partial ordering and in t total ordering the element labeled with α is minor than the element labeled β. T observation term corresponding to p is $\epsilon(\alpha(\alpha(\alpha\&\beta(NIL)))) \oplus \epsilon(\beta(\beta(\beta\&\alpha(NIL)$ and the observation term corresponding to q is $\epsilon(\alpha(\alpha\&\beta(NIL))) \oplus \epsilon(\beta(\beta\&\alpha(NIL)$

1. By axiom (A5), letting P=NIL, Q=$\mathfrak{b}$(R), we deduce:

$\mathfrak{a}(\mathfrak{b}(\mathfrak{b}(NIL)\oplus\mathfrak{b}(R))) \oplus \mathfrak{a}(NIL) = \mathfrak{a}(\mathfrak{b}(NIL)\oplus\mathfrak{b}(R))$

2. From 1.,by applying axiom (A3), we prove the lemma:

$\mathfrak{a}(\mathfrak{b}(\mathfrak{b}(R))) = \mathfrak{a}(\mathfrak{b}(R))$

3. As instances of the lemma in 2., by substitution, we obtain

$\epsilon(\alpha(\alpha(\alpha\&\beta(NIL)))) = \epsilon(\alpha(\alpha\&\beta(NIL)))$ *and* $\epsilon(\beta(\beta(\beta\&\alpha(NIL)))) = \epsilon(\beta(\beta\&\alpha(NIL$

4. By congruence,

$\epsilon(\alpha(\alpha(\alpha\&\beta(NIL)))) \oplus \epsilon(\beta(\beta(\beta\&\alpha(NIL)))) = \epsilon(\alpha(\alpha\&\beta(NIL))) \oplus \epsilon(\beta(\beta\&\alpha(NIL$

Now we show the actual correspondence between the equivalence defined so f and location equivalence of [4, 13].

In [3, 4, 13] the grammar of CCS is extended with an operator of local ca prefixing l::E where l is a location and E a process. Locations form a fixed, counta set. The idea is that each (sequential) process is assigned to a different location whe it executes, and an observer of the system perceives the visible actions of the syste and also the location where they take place. More precisely, location transitio are of the form E $\stackrel{\alpha;l_1 l_2 \ldots l_n}{\longrightarrow}$ E' where α is the action and l_1 l_2 ... l_n is a seque of location names. The prefix ordering in strings of location names represents t concept of sublocality. For example in the process $\alpha.(\beta|\gamma)$, action α may occur y location name, for example l (names of locations are arbitrarily chosen at each p), and β and γ both are performed at different sublocations of l, for example l l' d l l".

The operational semantics of the language is defined by means of two transition ations: one giving *location transitions* discussed above, and another giving *stan- rd transitions*, where no location is assigned to transitions. The second transition ation defines the τ-moves, corresponding to both τ prefixing and communication. bisimulation-like equivalence is defined on these transition systems, and the re- ting equivalence (called location equivalence and denoted with $\approx_l$) is shown to be sistent with the original interleaving semantics for CCS agents.

The following theorem shows that location equivalence and observational equiv- nce with location observations, as defined in this paper, coincide.

eorem 10 *Let E and E' be two CCS agents. Then,* $E \approx_l E'$ *iff* $E \approx_{loc} E'$. ■

The proof of the theorem requires to introduce the definitions of [4] and some ation. Because of lack of space, it is not reported here. It will appear in the ended version of the paper.

Other Observations

this section we consider as fixed the transition system for CCS and the equivalence servational equivalence), and we experiment with different observations. In this y, many different equivalences are defined for the same language, changing the umptions on what an observer should see.

Some of these equivalences can be defined via bisimulation directly on a suitably fined transition system (without explicitly constructing the computations). For ample the equivalence induced by *mixed ordering causal observations*, as presented ow, has been defined on a transition system for a language including CCS in [6]. wever, the equivalence induced by other observations, e.g. *partial ordering causal servations* has not been defined, up to now, directly on a transition system.

The observation corresponding to location equivalence gives information on the est induced by locations but also on the generation ordering. As it is the case the usual approach to causality [8], where one can choose between a mixed and artial ordering (in the former the generation ordering is observed, while in the ter it is not), here it is possible to choose either to observe or not to observe the eration ordering. The model where the generation ordering is not observed is some extent more intuitive: it assumes that it is not possible to determine the ler between two events that occur in different places. It is not obvious how to dify the definitions of [4, 13] to obtain this equivalence. Instead, in our parametric proach, just by performing a simple change on the observations of computations, mely by forgetting the total ordering and by leaving the rest (transition system, imulation) unchanged, this equivalence can be obtained.

For each observation of this chapter, an axiomatization of the observational con- ence induced can be obtained as an instance of the axioms of Table 2.

1 Abstract Localities

ven a computation, the observation **aloc** is defined as a labeled forest (E, $\prec$, l), ere E, $\prec$ and l are defined as in Definition 4 for location observations.

As it has been pointed out above, this observation captures the intuitive idea it there is no global clock, and that it is not possible to establish the order of urrence of events in different, unrelated places. An interesting point is whether s observation induces a different equivalence (and hence different semantics). In fact, the equivalence $\approx_{loc}$ is finer than $\approx_{aloc}$, and the example showing this fact the same example used to show that mixed ordering causal observations induces finer ordering than partial ordering causal observations [8].

Example 11 *Let* $\mathbf{p} = ((\alpha+\delta.\alpha.\beta)|\alpha.\delta.\beta|\ \bar{\delta})\backslash\delta$ *and* $\mathbf{q} = ((\alpha+\alpha.\delta.\beta)|\alpha.\delta.\beta|\ \bar{\delta})\backslash$ *Then, we have that* $\mathbf{p} \approx_{aloc} \mathbf{q}$ *but* $\mathbf{p} \not\approx_{loc} \mathbf{q}$.

The example above, while being finite, is not trivial. It has been checked, togeth with the other examples of this work, by the automatic parametric verification to of [12] (which is based on the theory of observation trees).

The intuitive difference, using **loc** observations, is that, when **p** has perform two α actions in two different locations, it has been decided if the β will take pla in the same location where the first or the second α has been performed. Instea when **q** has executed two α actions in two different locations, may be it still ca choose where to perform the β. This difference is detected by the total orderin but leaves no trace in the tree.

Although the intuitive nature of localities is to assign locations to events up isomorphism, it is sensitive to a global clock, distinguishing two computations whe an action occurs in a location which has been "the first" in executing an α, or in location which has been "the second" in performing an α.

4.2 Observing Causality

A different, well-known notion of observation is suggested by the causal approac In these models, the observer of a system is supposed to be able to determine th causal dependencies between event occurrences.

Here we first define mixed ordering causal observations. This observation h been proposed previously in [8], and the equivalence that it induces coincides wi the *causal bisimulation* of [6] and with the *history preserving bisimulation* of [16].

The main difference between this observation and the one defined for localiti is the treatment of τ's. While localities are not affected by events labeled with τ causes can be inherited by means of communications. Hence, we have to extend th sublocality relation to deal with events labeled with τ's, i.e. with spatial terms wi more than one $\bullet$.

We will now define a relation between spatial terms which includes the idea overlap: two spatial terms are overlapping if they both cover some part of the syste

Definition 12 Overlapping Localities
Let $\leftrightarrow$ *be the least relation defined by the following rules.*

$$\frac{s \neq \emptyset}{\bullet \leftrightarrow s} \qquad \frac{s_1 \leftrightarrow s_2}{s_1|s' \leftrightarrow s_2|s''} \qquad \frac{s_1 \leftrightarrow s_2}{s'|s_1 \leftrightarrow s''|s_2}$$

A dot in a spatial term represents the position where an action takes place. Th idea of overlapping is that some "active" position of one spatial term coincides, is a refinement of, an active position of the other spatial term. Notice that $\leftrightarrow$ neither symmetric nor antisymmetric.

A particular case of overlapping localities is when s_1 and s_2 are restricted have exactly one $\bullet$ and both s' and s'' in the last two rules are $\emptyset$: in that case th sublocality ordering is obtained.

Definition 13 Mixed and Partial Ordering Observations
Given a computation $E_0 \stackrel{\mu_1 @ s_1}{\longrightarrow} E_1 \stackrel{\mu_2 @ s_2}{\longrightarrow} \dots \stackrel{\mu_n @ s_n}{\longrightarrow} E_n$ *we associate to it a structu* $\langle E, <, \sqsubseteq, l \rangle$, *where:*
$E = \{e_i | i = 1 \dots n\}$ *is a set of different events*

$\leftrightarrow s_j$ *and* $i \leq j$ *implies* $e_i \sqsubseteq e_j$
$\sqsupseteq e_k$ *and* $e_k \sqsubseteq e_j$ *implies* $e_i \sqsubseteq e_j$
$< e_j$ *iff* $i \leq j$
$_i) = \mu_i$

The **mo** *observation of the given computation is given by the structure obtained* *m* $\langle E, <, \sqsubseteq, l \rangle$ *by forgetting all elements having a* τ *label, and by restricting the* *lerings and labeling to the remaining events.*

The observation for causal partial orderings is obtained from causal mixed order- *gs as for localities, i.e. by forgetting the total ordering* $\leq$. ■

The construction of the causal partial ordering given above is straightforward, d it is easy to see that the only difference between the approaches based on causal- and on locality is given by the choice of forgetting the τ's after, and respectively fore, constructing the partial ordering. Thus, $\prec \subseteq \sqsubseteq$.

The coincidence of the mixed ordering equivalence of [8] with the observational uivalence with **mo** observations defined above is immediate. In fact, the definition [8] employed NMS, but they are essentially the same as the observation trees. A oof of the coincidence of mixed ordering bisimulation, causal bisimulation and tory preserving bisimulation for event structures can be found in [1].

3 Partial Ordering Localities

cality and mixed ordering observations give different information about the sys- n. Combining them, we have a third kind of observations, namely a structure $, <, \prec, \sqsubseteq, l \rangle$ where $\langle E, <, \prec, l \rangle$ is the **loc** observation and $\langle E, <, \sqsubseteq, l \rangle$ is the **mo** servation. Since $\prec \subseteq \sqsubseteq$, we can give a graphical representation of these observa- ns with colored partial orderings, i.e. with partial ordering diagrams where arcs longing to both relations are painted with a color different from the color used $\sqsubseteq$ (here we use plain lines for arcs belonging to both relations and dotted lines arcs which are only in the causal ordering). We will call this observation **pol**, partial ordering localities. The equivalence induced by this observation coincides th the local/global cause equivalence of [13].

ample 14 *Figure 4 shows the colored partial orderings corresponding to compu-* *ion c of Example 5 and to the computation* $(\alpha.\gamma.\epsilon|\ \beta.\bar{\gamma}.\delta)\backslash\gamma \xrightarrow{\alpha @ \bullet | \emptyset} (\gamma.\epsilon|\ \beta.\bar{\gamma}.\delta)\backslash\gamma \xrightarrow{\beta @ \emptyset | \bullet}$ $\epsilon|\ \bar{\gamma}.\delta)\backslash\gamma \xrightarrow{\tau @ \bullet | \bullet} (\epsilon|.\delta)\backslash\gamma \xrightarrow{\delta @ \emptyset | \bullet} (\epsilon|\mathrm{NIL})\backslash\gamma \xrightarrow{\epsilon @ \bullet | \emptyset} (\mathrm{NIL}|\mathrm{NIL})\backslash\gamma$.

Notice that with location or colored partial (or mixed) orderings they are different. *wever, with partial (or mixed) causal orderings the two computations have the* *me observation.* ■

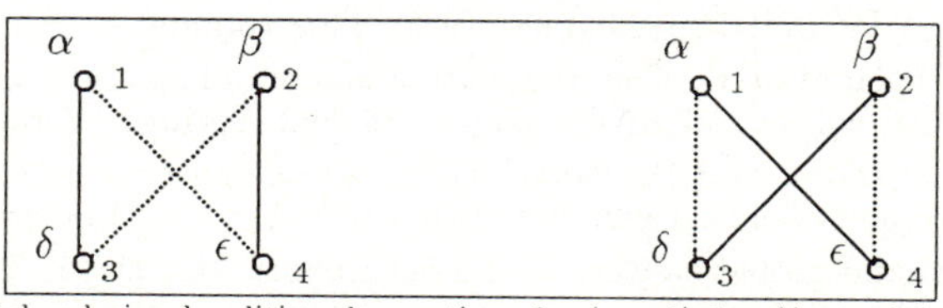

Figure 4: Partial ordering localities observations for $(\alpha.\gamma.\delta|\ \beta.\bar{\gamma}.\epsilon)\backslash\gamma$ and $(\alpha.\gamma.\epsilon|\ \beta.\bar{\gamma}.\delta)\backslash\gamma$.

4 Weak and Strong Interleaving Observations

course, as a particular case, interleaving observations can be defined in this ntext. The interesting point is that the difference between weak and strong equiv- ence is given here at the level of observations, and not at the level of equivalences.

Definition 15 Interleaving Observation
Given a computation $E_0 \stackrel{\mu_1@s_1}{\longrightarrow} E_1 \stackrel{\mu_2@s_2}{\longrightarrow} \ldots \stackrel{\mu_n@s_n}{\longrightarrow} E_n$ *we associate to it the sequence of labels* $\mu_1 ; \mu_2 ; \ldots ; \mu_n$. ■

The weak version of the interleaving observation is obtained by deleting τ labels from the sequence $\mu_1 ; \mu_2 ; \ldots ; \mu_n$.
Figure 5 shows all the weak observations presented in this work. An arrow from x to y means "y is more abstract than x", and hence the equivalence induced by observation x is finer than the equivalence induced by y.

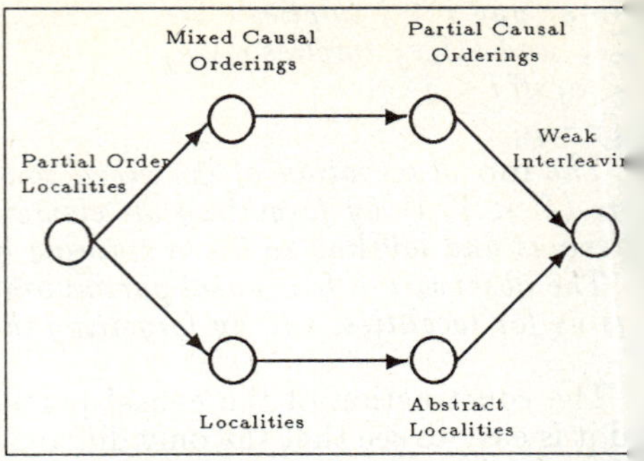

Figure 5: The observations presented in the paper and their relationships.

References

[1] L. Aceto. History preserving, causal and mixed-ordering equivalence for stable event structures. Technical Memo HPL–PSC–91–28, Hewlett-Packard Labs, Pisa, 1991.

[2] G. Boudol and I. Castellani. A non-interleaving semantics for CCS based on proved transitions. *Fundamenta Informaticae*, 11(4):433–452, 1988.

[3] G. Boudol, I. Castellani, M. Hennessy, and A. Kiehn. Observing localities. Technical Report 4/91, University of Sussex, March 1991.

[4] G. Boudol, I. Castellani, M. Hennessy, and A. Kiehn. A theory of processes with localities. Technical Report 13/91, University of Sussex, December 1991.

[5] I. Castellani and M. Hennessy. Distributed bisimulations. *J. ACM*, 36(4):887–91 1989.

[6] P. Darondeau and P. Degano. Causal trees. In Proc. ICALP 89, vol. 372 of *LNCS*, pages 234–248. Springer-Verlag, 1989.

[7] P. Degano, R. De Nicola, and U. Montanari. Observational equivalences for concurrency models. In Proc. of the 3th IFIP WG 2.2 working conference, Ebberup, pages 105–129. North-Holland, 1987.

[8] P. Degano, R. De Nicola, and U. Montanari. Partial orderings descriptions and observations of nondeterministic concurrent processes. In Proc. of the REX School/Workshop, Noordwijkerhout, vol. 354 of *LNCS*, pages 438–466. Springer-Verlag, 1989.

[9] P. Degano, R. De Nicola, and U. Montanari. Universal axioms for bisimulation, 199 Submitted for publication. An extended abstract appeared in Proc. of the Workshop on Concurrency and Compositionality, Goslar.

[10] P. Degano and C. Priami. Proved trees, 1992. This volume.

[11] G. Ferrari and U. Montanari. The observation algebra of spatial pomsets. In *Proceedings CONCUR'91*, vol. 527 of *LNCS*, pages 188–202. Springer-Verlag, 1991.

[12] P. Inverardi, C. Priami, and D. Yankelevich. A parametric verification tool for distributed concurrent systems, 1992. Submitted to CAV'92 - Montreal.

[13] A. Kiehn. Local and global causes. Technical Report 342/23/91, Technische Universitat Munchen, Munich, 1991.

[14] R. Milner. *Communication and Concurrency*. Prentice-Hall International, Englewood Cliffs, 1989.

[15] G. Plotkin. A structural approach to operational semantics. Report DAIMI FN-1 Computer Science Department, Aarhus University, 1981.

[16] A. Rabinovich and B. Trakhtenbrot. Behavior structures and nets. *Fundamenta Informaticae*, 11(4):357–404, 1988.

Proved Trees

Pierpaolo Degano *Corrado Priami*

ipartimento di Matematica, Università di Parma
&
)ipartimento di Informatica, Università di Pisa

Dipartimento di Informatica, Università di Pisa

Abstract. An incremental parametric model is presented that permits to recover different descriptions of concurrent systems. Given a calculus, its semantics is defined through a privileged transition system with transitions labelled by their proofs; then computations, i.e., paths in the transition system, are given a tree structure (*proved trees*). Different representation of the behaviour of a process are obtained by observing, *via* an *observation function*, the proved tree associated to the process. As an example, the paper shows how the interleaving, causal, and locational semantics are recovered, and shows how the corresponding bisimulations are compared, by a simple inspection of the used observation functions. Also, an algebra of proved trees is defined, and a generalized expansion law is exhibited. Through the appropriate observation function, this generalized law originates the analogous expansion law in the selected approach.

Introduction

Many languages have been proposed to specify distributed concurrent systems, among which ; [22], ACP [6], TCSP [5], MEIJE [1]. Commonly, the description of the operational semantics hese languages is given by *transition systems*. A transition system consists of a set of *states* of a set of (labelled) *transitions* between states that specify the kind of activity they repre- t. A *computation* is then a sequence of transitions. In order to mirror nondeterministic as- ts of concurrent systems, such as which choices have been made and when, computations are ally structured as trees. Often, also a denotational semantics is given that directly maps terms he language to their meaning, e.g., to trees.

This operational approach is sometimes too intentional to investigate some properties of a dis- uted concurrent system. In order to catch more extensional aspects, behavioural theories have n defined, mainly based on bisimulations [25]. Intuitively, two systems are considered bisi- ar if, whenever the first may perform a (possibly complex) activity, the other one may as ll, reaching states that are again bisimilar. Actually, only some aspects of the activities are of rest and compared. This information, that we call *observation*, may be simply the label of a nsition or may have some additional structure, e.g., a partial ordering relation representing sality, as done for instance in [2, 8, 12]. The literature reports a great variety of bisimula- ns. Their definitions can roughly be divided in families, according to the visibility of activities l to the granularity of their observations. More precisely, a bisimulation can be either *strong*, en all activities are considered visible, or *weak*, when there are some invisible activities (usu- y denoted by τ) that are skipped in simulations. The second aspect that differentiates bisimula- ns has to do with the length of the computation that is observed in a step of the comparison of tems. Indeed, bisimulations use the observation of either a single transition only, e.g., [22], of a sequence of transitions with more than one visible activity, e.g., [10]. We will say that former bisimulations have a differential shape, while the latter are more integral. Actually, the ferential bisimulations are more natural to us, because computations of transition systems ally grow stepwise, by performing one transition at a time.

Bisimulations provide us with the grounds to define behavioural congruences over distributed ncurrent systems, that in some cases have sound and complete axiomatizations. When the semantic domain is made of trees, axiomatizations and denotational semantics are tigh coupled. The existence of an axiomatization is essentially due to the presence of an *expans law*, that permits to reduce the parallel composition to interleaving and non-determinism. A matter of fact, the expansion law is one of the essential results of the so-called interleaving t ories [21, 6]. An expansion law was considered as a contradiction within the truly concurr theories (e.g. in the Petri net theory [27]), that aim at faithfully representing the causal and spa threads among the activities performed by a system. Instead, recent works [8,17,4] prop versions of this law that take care of causality / locations for fairly general languages. Then, a axiomatizations, and also logical characterizations of bisimulations, can be built for tr concurrent theories similar to those for the interleaving ones, but more expressive.

Above, we sketched some of the many semantic facets that distributed concurrent syst show and that make it hard to compare the various approaches in the field of concurrency. C claim is that many of these difficulties can be overcome if the definition of the semantics of a l guage is driven by the following paradigm, originally introduced by [10, 15].

1) Define in a syntax driven way a "*privileged*" *transition system*; obtain descriptions of syst evolutions from a given initial state, by defining *computations* as paths in the transiti system, and give them a tree structure;
2) Label the arcs of a tree, i.e., the transitions occurring in a computation, with their *observatic*
3) Compare labelled trees *via* bisimulations to determine which terms have an *equivalent haviour* according to the introduced observations.

Indeed, once a particular language is chosen, there exists a privileged transition system defir in the SOS style [26], the transitions of which are labelled by (portions of) their proofs, as dc by [9, 2, 13].[1] We will call them *proved transitions*. Computations are easily built and or nized as trees, named *proved trees*. As a matter of fact, proved trees can be expressed followi the standard syntax, i.e., a proved tree T is $\Sigma_i\ \theta_i \cdot T_i$, where θ_i are the labels (= proofs) of trar tions and T_i are proved trees. A set of operators can also be defined on proved trees, thus obta ing an algebra of them, therefrom a denotational semantics for the language under investigati Note that the usual interpretation of the parallel composition yields an expansion law, possi not within the original language, that is similar to the one presented in [17]. These remarks l to adding the following further item to the above three:

1') Define an algebra of proved trees, and a denotational semantics;

Item (2) requires to introduce *observation functions* that forget some details while keeping information relevant to the aspects under investigation. In the paper, three examples of them reported. The first observation function yields the labelling typical of the interleaving approa (standard synchronization trees are recovered); the second one records the causality relatic between transitions in a computation (causal trees [8] are obtained); the last function retains t locations where activities took place (in the vein of [3]). The sharp separation of the constructi of trees and of the observations of transition occurrences make it easier to compare the differe expressive power that different approaches may have. Actually, it suffices comparing the finitions of the observation functions, as we will do for the causal and locational observatio Note that the definition of an observation function and of a proved transition system fo process description language should help in defining a new transition system in which the lab of transitions carry only the wanted information.

Now, observed behaviours can be compared. Since our trees are labelled on arcs, we can u the classical differential definitions of bisimulations. Indeed, they make no assumption on t

[1] Alternatively, the proofs themselves are the transitions, that make it possible to define a powerful algebra transitions [16, 23].

ture of labels, i.e., they are completely independent of the observation functions. Also, the own sound and complete axiomatizations of congruence relations based on strong and weak imulations hold, without any change at all. Summing up, the behaviour of systems can be mpared with standard, well-known tools, regardless of which aspects of their activity have en observed. As noted above for the operational semantics, also a denotational semantics ould be derivable from that obtained in step (1') equating terms that express only the wanted ormation. In general, these terms will be generated by an extended syntax, taking care of se aspects that are under observation (e.g., causes, localities, etc.) and that are not expressed the original signature of language. Once derived such a denotational semantics $\mathcal{D}$, a chosen ngruence can be checked directly on (extended) terms of the language, through its axioms and laws defining $\mathcal{D}$. Among the latter laws, particularly relevant is the expansion law, the defini-n of which should derive by "interpreting" the analogous expansion law in the proved algebra, cording to the chosen observation. Hence, we are suggesting a general method for instantiating "universal" expansion law of the proved algebra to specific cases. The process of comparing behaviour of systems is exemplified in the paper by considering strong and observational ngruences, with the three observation functions mentioned above (yielding synchronization es, causal trees and locational trees). In the first case, the resulting congruences coincide with original strong and observational congruences of [22]; in the second one, they coincide with eir causal counterparts, as defined in [8]; in the third one, the observational congruence incides with the location congruence of [4]. Remarkably enough, the proof of the three results ove is straighforward. We will discuss also how a version of the "causal" and "locational" pansion laws of [8] and [4], respectively, can be derived by examining the effects that the rresponding observation function has on the "universal" expansion law.

Our claim is that many models for process description languages are clearly compared and u-fied when the definition of their semantics is organized in the four neatly separated phases out-ed above. This programme has been initiated by Degano, DeNicola and Montanari in [10, 14], here so-called observation trees, that have an integral shape, are introduced with the same pur-se. Here, we have worked over their proposal in order to give instead a slightly less general, t differential description of system behaviour, and to define differential notions of bisimula-ns.

Lack of space prevents us from discussing related work, and from including the proofs of the eorems; the reader is referred to the full paper.

Operational Construction of Proved Trees

As anticipated in the Introduction, our study is carried on in a linguistic framework. We assu-e that the operational semantics of the chosen process description language is given by a transi-n system defined by structural operational semantics [26]. The language used here as a test d is the Calculus of Communicating Systems (CCS) [22], the basic notions of which follow.

efinition 2.1 (*CCS*)

t Δ be a set of *names* (ranged over by $\alpha, \beta, \ldots$). Let $^-$ be an involution on Δ, and call *com-ementary names* the elements of the set Δ^-. Then $\Lambda = \Delta \cup \Delta^-$ is the set of *labels* (ranged over λ) and $A = \Lambda \cup \{\tau\}$ is the set of *actions* (ranged over by μ). The set T of closed *CCS terms* expressed by the following BNF-grammar:

$$t ::= nil \mid x \mid \mu.t \mid (t \mid t) \mid (t + t) \mid t\backslash\alpha \mid t[\phi] \mid rec\ x.t$$

ith the restriction that the term t in rec x.t is well-guarded, and where ϕ is a relabelling function eserving τ and $^-$. ◊

Since we want to obtain a very concrete operational semantics, we will define in the SOS st a transition system, the transitions of which are labelled by encodings of their proofs, followi [2, 9, 13]. Actually, we take essentially the definition of Boudol' and Castellani's prov transition system [2]. The only slight difference has to do with the representation of synchro zations, that require in our definition two rules for restriction.

Definition 2.2 (*Proof Terms*)
Let $\vartheta \in \{\|_0, \|_1, +_0, +_1, \backslash_\alpha, \phi\}^*$.
The set of proof terms is $\Theta = \{\vartheta\mu \mid \mu \in A\} \cup \{\vartheta\langle\vartheta_0\lambda, \vartheta_1\lambda^-\rangle \mid \lambda \in \Lambda\}$, ranged over by θ.

Definition 2.3 (*Proved Transition System, PTS [2]*)
The proved transition system PTS = $(T, \Theta, \_\theta\!\!\to)$, where T is the set of CCS terms, Θ is the of proof terms and the transition relation $\_\theta\!\!\to$ is defined by the following axiom and rules:

Act $$\mu.t \xrightarrow{\mu} t$$

Sum $$\frac{t \xrightarrow{\theta} t'}{t + t'' \xrightarrow{+_0\theta} t'} \qquad \frac{t \xrightarrow{\theta} t'}{t'' + t \xrightarrow{+_1\theta} t'}$$

Rel $$\frac{t \xrightarrow{\theta} t'}{\phi(t) \xrightarrow{\phi\theta} \phi(t')}$$

Res $$\frac{t \xrightarrow{\vartheta\mu} t'}{t\backslash\alpha \xrightarrow{\backslash_\alpha\vartheta\mu} t'\backslash\alpha},\ \mu \notin \{\alpha, \alpha^-\} \qquad \frac{t \xrightarrow{\vartheta\langle\theta_0, \theta_1\rangle} t'}{t\backslash\alpha \xrightarrow{\backslash_\alpha\vartheta\langle\theta_0, \theta_1\rangle} t'\backslash\alpha}$$

Asyn $$\frac{t \xrightarrow{\theta} t'}{t \mid t'' \xrightarrow{\|_0\theta} t' \mid t''} \qquad \frac{t \xrightarrow{\theta} t'}{t'' \mid t \xrightarrow{\|_1\theta} t'' \mid t'}$$

Syn $$\frac{t_0 \xrightarrow{\vartheta_0\lambda} t'_0,\ t_1 \xrightarrow{\vartheta_1\lambda^-} t'_1}{t_0 \mid t_1 \xrightarrow{\langle\vartheta_0\lambda, \vartheta_1\lambda^-\rangle} t'_0 \mid t'_1}$$

Rec $$\frac{t[\mathrm{rec}\ x.t/x] \xrightarrow{\theta} t'}{\mathrm{rec}\ x.t \xrightarrow{\theta} t'}$$

As it is already noted by Boudol and Castellani, it is immediate to recover the transition syste of CCS [22] from the above one. Eventually, we define the operational meaning of each term t the unfolding from t of the proved transition system. The obtained tree will be called *proved tr* and will be denoted by $[t]_{pt}$. More formally, we have the following definition, in which w introduce a syntax for proved trees, assuming as usual that + is interpreted as (associative, com mutative, idempotent) sum of trees, $\theta\cdot$ as prefixing of an arc labelled by θ to a tree, and th empty summation NIL as the empty tree, acting as neutral element for +.

Definition 2.4 (*Proved Tree*)
The *proved tree* of a CCS term t is $[t]_{pt} = \Sigma_{i\in I}\, \theta_i\cdot[t_i]_{pt}$, where, $\forall i \in I$, $t \xrightarrow{\theta_i} t_i$ is a transitio of the proved transition system. In the sequel, isomorphic trees will be identified. The set proved trees generated by CCS terms will be denoted by PT.

Although we assume idempotence of sum, x+x = x, proved trees can not be simplified, as ta $+_0$ is different from $+_1$. Therefore, in the formal summation $\Sigma_{i\in I}\, \theta_i\cdot[t_i]_{pt}$, $\forall i,j \in I$, $i \neq j$, we hav $\theta_i \neq \theta_j$. Note also that the set I is always finite, as CCS terms are guarded.

An algebra of Proved Trees

In this section we introduce the algebra A_{PT} of proved trees by generalizing the approach of for causal trees. Then, we prove that the operational semantics induced by the proved tran-on system PTS and the denotational semantics induced by the algebra coincide. The algebra $_T$ will be used as an interpretation for CCS terms, understanding rec x.t as the fixed point at x the functional interpretation [t] of t. The existence and the unicity of fixed points are gua-teed by two facts. The former is the assumption of well-guardedness of recursive definitions, ich induce contracting operators. The latter is the continuity, in the usual metrics of trees, of the operators introduced below.

finition 3.1 (*Algebra of proved trees*)

t $T = \Sigma_i\ \theta_i \cdot T_i$ and $U = \Sigma_j\ \theta_j \cdot T_j$ be two proved trees, with index sets disjoint. The Σ-algebra $_T = \langle PT, \Sigma_{PT} \rangle$ of proved trees is completely determined by the following equations on carrier . where $\Sigma_{PT} = \{nil\} \cup \{\theta \mid \theta \in \Theta\} \cup \{\backslash\alpha \mid \alpha \in \Lambda\} \cup \{\phi \mid \phi$ is a permutation of Λ preserving nplementation and $\tau\} \cup \{+, |\}$, where nil has arity 0; $\theta, \phi, \backslash\alpha$ have arity 1; and+, | have arity 2.

nil	$nil = NIL$	*act*	$\mu(T) = \mu \cdot T$
sum	$T + U = \Sigma_i\ (+_0\theta_i) \cdot T_i + \Sigma_j\ (+_1\theta_j) \cdot U_j$	*rel*	$\phi(T) = \Sigma_i\ (\phi\theta_i) \cdot (\phi\,(T_i))$
nl1	$NIL \mid NIL = NIL$	*nl3*	$\phi(NIL) = NIL$
nl2	$NIL + NIL = NIL$	*nl4*	$NIL\backslash\alpha = NIL$

res $\quad T\backslash\alpha = \Sigma_{\substack{\theta_i = \vartheta_i\mu \\ \mu \notin \{\alpha, \alpha^-\}}} (\backslash_\alpha\theta_i).(T_i\backslash\alpha) + \Sigma_{\theta_i = \vartheta\langle\vartheta_0\lambda,\ \vartheta_1\lambda^-\rangle} (\backslash_\alpha\theta_j) \cdot (T_j\backslash\alpha)$

inter $\quad T \mid U = \Sigma_i\ (||_0\theta_i) \cdot (T_i \mid U) + \Sigma_j\ (||_1\theta_j) \cdot (T \mid U_j) + \Sigma_{\theta_i = \vartheta_0\lambda,\ \theta_j = \vartheta_1\lambda^-} \langle\theta_i, \theta_j\rangle \cdot (T_i \mid U_j)$

e denotational meaning of a CCS term t will be denoted by $[t]_{PT}$ in the sequel. ◊

it was the case for the operational semantics, the equation for restriction deals explicitly with els arising from synchronization. In equation *inter* the parallel composition of T and U is re-ced to interleaving plus nondeterminism. Indeed, either an arc from (the semantics of) T is en, prefixing its label with $||_0$, expressing that a move from the left operand has been taken, d symmetrically for U; or a synchronization may occur, when the labels end with complemen-y actions λ and λ^-. The importance of equation *inter* should not be underestimated, because it roduces an expansion law for our version of CCS, although not *within* the original signature the calculus.

In order to prove that the Σ-algebra PT_A is fully abstract with respect to the operational seman-s $[]_{pt}$, we must show that for each pair of CCS term t and t', $[t]_{pt} = [t']_{pt}$ holds if and only if $_{PT} = [t']_{PT}$. Note that the operational semantics equates two non recursive terms if and only if y are syntactically equal, and so does the denotational interpretation. Moreover, the opera-nal and the denotational semantics equate also recursive terms (possibly not syntactically ual) which define the same fixed point. Then the following claim follows immediately.

oposition 3.1 (*Full abstractness*)

e operational semantics $[]_{pt}$ of proved trees is coincident with the denotational one $[]_{PT}$. ◊

Observing Proved Trees

An arc in a proved tree, representing the occurrence of a transition z, has a label that expresses a very concrete way the effect of z. E.g., the label $||_0\backslash_\beta\phi +_1\alpha$ says that an action α has been rformed, which was in the right alternative of a nondeterministic choice ($+_1$), has been trans-med into a new action *via* ϕ, passed successfully through the filter $\backslash_\beta$, and occurred in the left location of a compound site ($\|_0$). Often, this information is too detailed, and only parts of it a of interest, which we claim to be derivable all from proved trees. Following the approach of [14], the notion of observation of a tree will be introduced, through which some aspects of t behaviour of systems can be disregarded. The reduction of the amount of information encoded proved trees is an abstraction step that is achieved simply by changing the set of possible lab that we call observation set. Then, observing a proved tree will essentially consist of relabelli its arcs, by transforming proof terms into elements of the chosen observation set. The separati of the construction of trees and of their labelling enables us to compare on firm grounds the e pressive power of different observation functions, and more generally of different approaches the semantics of process description languages.

Definition 4.1 (*Observation functions,* $\mathcal{O}$)
Let $\mathcal{A}$ and $\mathcal{B}$ be two sets, and let $T = \langle N, E, f \rangle$ be a labelled tree, where N is the set of nodes the set of arcs and $f : E \to \mathcal{A}$ is the labelling function.
An *observation function* $\mathcal{O}$ re-labels T on $\mathcal{B}$, i.e., $\mathcal{O}(T) = \langle N, E, f' \rangle$, with $f' : E \to \mathcal{B}$.
We will feel free to write $\mathcal{O}(T)$ as $\Sigma_i\ b_i \cdot T_i$ ($b_i \in \mathcal{B}$) assuming summation idempotent, thus ide fying $\mathcal{O}(T)$ with its quotient with respect to the equation x+x=x.

In the rest of the section, three examples of observation functions are examined on the tr obtained by CCS terms. In particular, an interleaving observation function is defined which p mits to obtain from a proved tree the corresponding synchronization tree, recovering the stand interleaving semantics of [22]. In this case, the observation set is simply the set A of actions, defined in Section 2. Then, a causal observation function is introduced which yields the cau tree associated to a given proved tree, by substituting labels of the form $\langle \lambda, K \rangle$ (or τ) for pro terms θ. The semantics for CCS terms obtained in this way coincides with the causal semantics [8]. Finally, we give a location observation function, yielding labels of the form u::λ (or which permits recasting in our framework the semantics based on localities as defined in [4].

4.1. Observing Temporal Relations

The first observation we define is the *interleaving* one. A synchronization tree is obtain from a proved tree simply by retaining only the action labels of the proof terms on the arcs. V need a definition first.

Definition 4.2 (*From proof terms to actions*)
Let $op \in \{\|_0, \|_1, +_0, +_1, \backslash_\alpha, \phi\}$ and let $\ell : \Theta \to A$ be inductively defined as

$$\ell(\mu) = \mu, \qquad\qquad \ell(\langle \vartheta'\lambda, \vartheta''\bar{\lambda} \rangle) = \tau,$$

$$\ell(op\theta) = \begin{cases} \phi(\ell(\theta)) & \text{if } op = \phi \\ \ell(\theta) & \text{otherwise} \end{cases}$$

Definition 4.3 (*Interleaving observation*)
The interleaving observation of a proved tree $T = \Sigma_i\ \theta_i \cdot T_i$ is $Int(T) = \Sigma_i\ \ell(\theta_i) \cdot Int(T_i)$

From the definition of ℓ and the obvious embedding of the original CCS transition syste within the proved transition one, the following theorem holds.

Theorem 4.1 (*Proved Trees and Synchronization Trees*)
Let t be a CCS term. Then, the synchronization tree obtained through the operational semantics [22] and the tree $Int([t]_{pt})$ coincide.

Note that it is immediate recovering the standard expansion law of [22] from the one of t proved algebra (equation *inter*), simply observing the proved terms that occur in it by t observation function *Int*.

2. Observing Causality

Before defining the other observations, some preliminary definitions are needed. We define e function F which erases from a proof term θ the occurrences of constructors μ and $+_i$. Also and ϕ are removed, because they are irrelevant for determining the relations of conflict be- een actions, of causal dependencies and of occurring in the same location.

efinition 4.4 (*F*)

et ε denote the empty string and $\Gamma = \{\|_0, \|_1\}^*$. The function F: $\Theta \to \Gamma \cup (\Gamma \times \Gamma)$ is defined by uctural induction as follows:

$F(\mu) = \varepsilon$ $\qquad F(\backslash_\alpha \theta) = F(\theta)$ $\qquad F(\|_i \theta) = \|_i F(\theta),\ i = 0,1$

$F(+_i \theta) = F(\theta),\ i = 0,1$ $\qquad F(\langle\theta, \theta'\rangle) = \langle F(\theta), F(\theta')\rangle$ $\qquad F(\phi\theta) = F(\theta)$ ◊

e next definition introduces a notion of dependency on occurrences of proved transitions in a e, using the proof terms labelling the arcs in a path. Intuitively, a proof term θ' depends on other one, say θ, if $F(\theta)$ is a prefix of $F(\theta')$. As an example, consider the term $t = \alpha.(\gamma.\mathrm{nil} \mid$ $.\mathrm{nil} + \delta.\mathrm{nil})) \mid \beta.\mathrm{nil} + \alpha.\mathrm{nil}$ where the leftmost β depends on the leftmost α. The labels of the cs corresponding to these actions are $\|_0 +_0 \|_1 \beta$ and $+_0 \|_0 \alpha$, and the causal dependency is correct- recovered, as $\|_0 = F(+_0 \|_0 \alpha)$ is a prefix of $\|_0 \|_1 = F(\|_0 +_0 \|_1 \beta)$. Particular care has to be taken hen θ records a synchronization, i.e., when it has the form $\vartheta\langle\theta_0, \theta_1\rangle$. In this case, the synch- nization is de-coupled in $\vartheta\|_0\theta_0$ and $\vartheta\|_1\theta_1$ (note that $\|_0 / \|_1$ are inserted to record which proof rm was the left / right component of the synchronization). Then, the proof term θ' depends on θ either $F(\vartheta\|_0\theta_0)$ or $F(\vartheta\|_1\theta_1)$ is a prefix of $F(\theta')$. Clearly, this definition is driven by the in- nded meaning of the operator "|" of the language, as reflected by its operational semantics.

efinition 4.5 (*Dependencies of proofs,* $\sqsubset$)

et θ and $\theta' = \vartheta'\theta''$ be the labels of two arcs of the same path of a proved tree. Then, θ' depends 1 θ (written $\theta \sqsubset \theta'$) iff either condition holds:

i. $\theta = \vartheta\mu$ and $F(\vartheta') = F(\theta)$, or

i. $\theta = \vartheta\langle\theta_0, \theta_1\rangle$, and either $F(\vartheta') = F(\vartheta\|_0\theta_0)$ or $F(\vartheta') = F(\vartheta\|_1\theta_1)$. ◊

inally, we need to count the number of synchronizations occurring in a path.

efinition 4.6 (*Number of synchronization actions*)

he function N_τ: $\Theta^* \to \mathbf{N}$ counts the number of the synchronization actions in a path σ of a roved tree, i.e., a sequence of proof terms, and is inductively defined as

$$N_\tau(\varepsilon) = 0 \quad \text{and} \quad N_\tau(\theta\sigma) = \begin{cases} 1 + N_\tau(\sigma) & \text{either } \theta = \vartheta\langle\vartheta_0\lambda, \vartheta_1\bar{\lambda}\rangle \text{ or } \theta = \vartheta\tau \\ N_\tau(\sigma) & \text{otherwise} \end{cases}$$ ◊

We now define the causal observation of a proved tree, that yields a causal tree [8]. Causal ees are a variant of synchronization trees, with enriched action labels which supply indication f the observable causes of observable actions. The added information on labels allows to avoid ny confusion between concurrency and nondeterminism. The two terms $\alpha.\mathrm{nil} \mid \beta.\mathrm{nil}$ and $\alpha.\beta.\mathrm{nil}$ $\beta.\alpha.\mathrm{nil}$, which are identified in the interleaving approach will be distinguished as they originate e causal trees $\langle\alpha,\emptyset\rangle\cdot\langle\beta,\emptyset\rangle + \langle\beta,\emptyset\rangle\cdot\langle\alpha,\emptyset\rangle$ and $\langle\alpha,\emptyset\rangle\cdot\langle\beta,\{1\}\rangle + \langle\beta,\emptyset\rangle\cdot\langle\alpha,\{1\}\rangle$, respectively. ecall that a causal tree is represented as a formal summation where the summand are the labels f the arcs. The labels are pairs in which the first element is the action label and the second is a et of integer interpreted as backwards pointers to the arcs which cause the actual one. Finally, call that τ's are skipped by the pointers.

In order to obtain a causal tree from a proved tree, the label of an arc a_j (at depth j) that de- ends on an arc a_i (at depth i) will be set to $\langle\lambda, K\rangle$, where $\lambda = \ell(\theta_j)$, and the set of backpointers contains j-i-n, if n invisible actions occurred in between a_i and a_j, i.e., $n = N_\tau(a_i \ldots a_j)$. ctually, the set of backpointers K is obtained as the union of two sets, K_1 and K_2. The set K_1

contains the references to the arcs which cause the current one with label of the form $\vartheta\mu$, $\mu\neq$ Since also synchronizations are invisible at all, even if a_i depends on one of them, say on $\vartheta\langle\theta_0, \theta_1\rangle$, its set of causes cannot contain any reference to it. Therefore the other set K_2 will cont only references to those arcs on which $\vartheta\langle\theta_0, \theta_1\rangle$ itself depends. In this way, the new label of t arc a_i will contain backpointers referring to all arcs causing both $\vartheta||_0\theta_0$ and $\vartheta||_1\theta_1$, thus keepi track of the cross inheritance of the causes due to the occurrence of $\vartheta\langle\theta_0, \theta_1\rangle$.

Definition 4.7 (*Causal Observation*)
Given a proved tree T, its associated causal tree Ct(T) is derived by relabelling any pa $\theta_0\theta_1\ldots\theta_n\ldots$ of T as $ct_0\, ct_1\ldots\, ct_n\ldots$, where

$$ct_n = \begin{cases} \tau & \text{if } \theta_n = \vartheta\langle\vartheta'\lambda, \vartheta''\lambda^-\rangle \text{ or } \theta_n = \vartheta\tau \\ \langle \ell(\theta_n), K_1 \cup K_2\rangle & \text{otherwise} \end{cases}$$

$$\text{with}\quad K_1 = \{n - h - N_\tau(\theta_h\ldots\theta_n) \mid \theta_h = \vartheta\lambda \sqsubset \theta_n\} \text{ and}$$

$$K_2 = \{n - h - N_\tau(\theta_h\ldots\theta_n) \mid \vartheta\langle\theta', \theta''\rangle \sqsubset \theta_n \text{ with } \theta_h \sqsubset \vartheta||_0\theta' \text{ or } \theta_h \sqsubset \vartheta||_1\theta''\}.$$

Note that passing from a proved tree T to its associated causal tree Ct(T) may change the tr structure if Ct(T) is written as formal sum, because of axiom x+x = x. The definition above validated by the coincidence (up to x+x = x) of the causal tree obtained from a CCS term t a cording to [8] with the causal tree associated to the proved tree of t. Below, recall that $[t]_{pt}$ (res $[t]_{PT}$) is its proved tree obtained operationally (resp. denotationally) according to Def. 2.4 (3.1,

Theorem 4.2 (*Causal Trees = Causally Observed Proved Trees*)
Let t be a CCS term, and let $[t]_{ct}$ (resp. $[t]_{CT}$) be its causal tree generated by the operational (res denotational) semantics of [8]. Then,

i. $Ct([t]_{pt}) = [t]_{ct}$, up to x+x = x ii. $Ct([t]_{PT}) = [t]_{CT}$.

As announced in the Introduction, we illustrate how to retrieve the causal expansion law sta ing from the "universal" expansion law of the proved algebra and the definition of the observ tion function Ct. By definition of Ct, it is obvious that τ-actions do not affect their sub-tree Hence, in synchronizing two causal trees T and U, the first summand of the proved expansic law must be split into two parts, the first considering τ-actions and the second the visible act ons. Since a τ is simply skipped, the first part has then the shape $\Sigma_i\, \tau\cdot(T_i \mid U)$. Consider now th visible actions in $\Sigma_i\, (||_0\theta_i)\cdot(T_i \mid U)$. From the definition of dependency, all the arcs θ_j in T_i d pending on θ_i will still depend on $||_0\theta_i$, because, inductively, all of them will be prefixed by || Thus, $||_0\theta_i$ will be observed, e.g., as $\langle\lambda_i, K_i\rangle$, and T_i requires no change. Instead, the causes U were computed on a path that now has one arc more. Thus, they must be incremented, as th newly added arc has to be skipped. This is done through the operator **[n]**. (It increments all th causes greater than or equal to n of the sub-tree to which it is applied, thus delaying the corre ponding actions.) Collecting all these arguments, the second part of the first summand will hav the shape $\Sigma_j\, \langle\lambda_j, K_j\rangle\cdot(T_j \mid \mathbf{[1]}\, U)$. Symmetrically, for the symmetric asynchronous summan The synchronization summand must reflect the crossing of causes expressed by the set K_2 def ned in ct_n. It says that in all the arcs in T_i and U_j that contain the pointer n to $\langle\lambda_i, K_i\rangle$ (observa tion of θ_i) synchronized with $\langle\lambda_j^-, K_j\rangle$ (observation of θ_j) must now contain the set $K_i \cup K_j$ i place of n, which is done by the operator $[K_i \cup K_j/\mathbf{n}]$. Then, the last summand can be written a $\Sigma_{\lambda_i = v^-_j}\, \tau\cdot([K_i \cup K_j/\mathbf{1}]\, T_i \mid [K_i \cup K_j/\mathbf{1}]\, U_j)$. All the above leads to the causal expansion law c [8]. In turn, from this law it is immediate deriving the expansion law on causal terms, i.e., o terms of an extended syntax that explicitly express causality [8]. E.g., the term $\{1,5\} \Rightarrow \alpha$.n expresses that the action α is caused by the action which immediately preceded it (by pointer 1 and by an action which occurred five steps before (pointer 5). Assume that terms p and q hav en reduced to $\Sigma_i K_i \Rightarrow \lambda_i \cdot p_i' + \Sigma_j \tau \cdot p_j'$, and $\Sigma_m K_m \Rightarrow \nu_m \cdot q_m' + \Sigma_n \tau \cdot q_n'$, respectively, *via* the vs for the operators different from "|". Then, we get

$$p \mid q = \Sigma_i (K_i \Rightarrow \lambda_i).((\{1\} \cup \delta(K_i)) \Rightarrow p_i' \mid q) + \Sigma_j \tau.(p_j' \mid q) +$$
$$\Sigma_m (K_m \Rightarrow \nu_m).(p \mid (\{1\} \cup \delta(K_m)) \Rightarrow q_m') + \Sigma_n \tau.(p \mid q_n') +$$
$$\Sigma_{\lambda_i = \nu^-_j} \tau.((K_i \cup K_j) \Rightarrow p_i' \mid (K_i \cup K_j) \Rightarrow q_j').$$

4. Observing Localities

As last example, we consider an observation function that retrieves the locational semantics of [4]. Through it, the observer of a system may see the distributed nature of processes, by recording the location, e.g., the address where an action occurred. So the observer can not only test the stem by communicating with it, but can also observe or distinguish which part of the system acted to the experiment. In the sequel, the localities are represented by strings of natural numbers, with the proviso that k encodes the locality of the kth visible action. We define *locational trees* as the unfolding of a variant of the transition system of [4]. Then, proved trees will be observed through the locational observation function defined below. Finally, we will prove that the trees obtained in the two ways are isomorphic.

Definition 4.8 (*Locational Observation*)
Given a proved tree T, its associated locational tree $\mathcal{L}t$(T) is derived by relabelling any path $\theta_1 \ldots \theta_n \ldots$ of T as $\ell t_0\ \ell t_1 \ldots \ell t_n \ldots$, where

$$\ell t_n = \begin{cases} \tau & \text{if } \theta_n = \vartheta\langle\vartheta'\lambda, \vartheta''\lambda^-\rangle \text{ or } \theta_n = \vartheta\tau \\ u\,k::\ell(\theta_n) & \text{otherwise} \end{cases}$$

with $u = \text{MAX}^{2}\,\{w \mid \theta_h \sqsubset \theta_n \text{ and } \ell t_h = w::\lambda\}$ and $k = n - N_\tau(\theta_0,\ldots\theta_n)$. ◊

Note that $k = n - N_\tau(\theta_0 \ldots \theta_n)$ is a fresh sub-locality uniquely linked to the action performed in the arc labelled by θ_n, that indeed encodes the locality of the kth visible action. Differently from what happens in observing causality, here synchronizations are completely ignored. We will come back on this issue at the end of the section. Also in this case, when $\mathcal{L}t$(T) is written as a formal summation, identical sub-trees are collapsed, as assumed in the theorem below.

Theorem 4.3 (*Locational Trees = Locational Observed Proved Trees*)
Let t be a CCS term and let $[t]_{pt}$ and $[t]_{lt}$ be its proved tree and its locational tree, respectively. Then, $\mathcal{L}t([t]_{pt}) = [t]_{lt}$, up to x+x = x. ◊

We now sketch the derivation of a locational expansion law. Also in this case, the first summand of the proved law has to be divided in a part for τ, and one for visible actions. The first part is simply $\Sigma_i\, \tau \cdot (T_i \mid U)$. From the definition of $\mathcal{L}t$, we see that a new number is added to an existing location whenever an action λ_i is performed, as well as to all the locations in which actions depending on λ_i occur. Hence, if $T = u_i::\lambda_i \cdot T_i$, the second part of the first addend is $u_i k::\lambda_i \cdot (k \div T_i \mid U)$, where the auxiliary operator $\div$ postfixes k ($\notin u_i$) to all the locations in T_i. Symmetrically, for the symmetric asynchronous summand. The synchronization summand is derived immediately, as synchronization ignores completely localities. Therefore, we obtain $\Sigma_{\lambda_i=\nu^-_j} \tau \cdot (T_i \mid U_j)$. In turn, from this law it is immediate deriving the following expansion law on location terms, that is the obvious variant of the one given in [4]. Assume that terms p and q have been reduced to $\Sigma_i u_i::\lambda_i \cdot p_i' + \Sigma_j \tau \cdot p_j'$, and $\Sigma_m u_m::\nu_m \cdot q_m' + \Sigma_n \tau \cdot q_n'$, respectively, *via* the obvious laws for the operators different from "|". Then, we get

$$p \mid q = \Sigma_i u_i k_i::\lambda_i \cdot (u_i k_i::p_i' \mid q) + \Sigma_j \tau \cdot (p_j' \mid q) + \qquad (k_i \notin u_i)^{3}$$
$$\Sigma_m u_m k_m::\nu_m \cdot (p \mid u_m k_m::q_m') + \Sigma_n \tau \cdot (p \mid q_n') + \qquad (k_m \notin u_m)$$
$$\Sigma_{\lambda_i=\nu^-_j} \tau \cdot (u_i::p_i' \mid u_j::q_j').$$

[2] Let $I = \{nw \mid n \in \mathbb{N}, w \in \mathbb{N}^*\}$ and $n' = \max\,\{n \mid nw \in I\}$, then we define MAX I = n'w'.
[3] With abuse of notation, we consider $w \in \mathbb{N}^*$ as a set.

Note that the causal and the locational observations have been defined above in an integ way. However, it is possible to define such observations also in a fully inductive way by su ably encoding the information taken from paths.

4.4. Comparing Causal and Locational Observations

As already noted by [20], the causal and locational semantics discussed above are incompa ble, because of the different encoding of dependencies between arcs and of the different tre ment of them when communications occur. The locational approach encodes these dependenci by giving an identity to the locations where actions take place, whereas the causal approach r quires no global name for localities. Thus, there are CCS terms that yield the *same* causal tr but different locational trees, e.g., $t_0 = (\alpha.\delta.nil \mid \delta^-.\beta.nil)\backslash\delta$ and $t'_0 = \alpha.\tau.\beta.nil$. Also, in the ca sal approach *all* the actions on which a τ depends are passed to the actions enabled by it (v K_2), while in the locality one only those actions are passed that lay in the same location (or a pr fix of it). Thus, there are CCS terms that yield the *same* locational tree and different causal tree e.g., $t_1 = \alpha.nil \mid (\delta.nil \mid \delta^-.\beta.nil)\backslash\delta$ and $t'_1 = (\alpha.\delta.nil \mid \delta^-.\beta.nil))\backslash\delta + \tau.(\alpha.nil \mid \beta.nil)$. Also Kie [20] provides two pairs of examples, but bisimulation is used in place of equality. The theore below is an example of how different approaches to the semantics of concurrency can be cor pared formally, and of the easiness of establishing their differences and similarities. Indeed, t two pairs of terms above have been constructed by examining the rôle played by set K_2.

Theorem 4.4 (*Relations between Causality and Locality*)

Let t, t_i be a CCS term and T, T_i be their associated proved trees.

i. If t does not contain "|" then $\mathcal{C}t(T) = \mathcal{L}t(T)$.

ii. There exist t_0, t'_0 and t_1, t'_1 such that
   - $\mathcal{C}t(T_0) = \mathcal{C}t(T'_0)$ and $\mathcal{L}t(T_0) \neq \mathcal{L}t(T'_0)$;
   - $\mathcal{C}t(T_1) \neq \mathcal{C}t(T'_1)$ and $\mathcal{L}t(T_1) = \mathcal{L}t(T'_1)$.

We end this section by noting that the definition of an observation function and of a prove transition system should help in defining a corresponding new, more abstract transition syste in which the labels carry only the wanted information.

5. Equivalences on Proved Trees

In Section 3 we have introduced an algebraic model of proved trees, in which systems a equated if and only if they are syntactically equal. This is so because proved trees are very co crete a model for describing processes. Indeed, we observe also the location of a choice, ar therefore we cannot identify even processes like x + x and x because commutativity (and associ tivity) of the nondeterministic choice do not hold. Moreover, we have introduced observatio functions over proved trees which permit to recover from them less concrete representations o system behaviour. This is done by changing the labelling of the arcs of a tree without changin its structure. Systems can now be compared according to the chosen different observations, e.g *via* bisimulations. The main point of this section is showing that there is no need for definin many different bisimulations, one for each different observation of systems. Indeed, the standar definitions of bisimulations amount essentially to comparing the label, i.e., the observation of computation step of one system with the label of a step of the other system, regardless of the s from which these labels are taken. Thus, a drastic pruning of the variety of bisimulations is pos sible, because of the neat separation between observing and comparing the behaviour of sys tems. It emerges then that the only difference between the various bisimulations is due to ho they deal with activities considered non observable.

Below, we show that some notions of behavioural equivalences presented in the literature are ptured in our framework. In particular, we will study the (strong and) observational congruen- for CCS as originally defined in the interleaving approach [22], their causal variants [8] (re-l that strong causal bisimulation coincides with history preserving bisimulation [28, 11]), and location equivalence of [4]. Of course, any other kind of bisimulation works as well, provi-d that it has a differential definition, e.g., branching bisimulation [18]. We assume in the fol-wing the definitions of (strong $\approx_s$ and) observational $\approx_c$ congruence over observed proved es. Recall that strong congruence on proved trees coincides with isomorphism. However, it is ssible to define a notion of observational congruence on proved trees that ignores certain invi-le actions, e.g., those labelled by τ. In this case, for example, $\alpha.\tau.t$ will be considered obser-tionally congruent to $\alpha.t$. The following theorem is easy to establish.

eorem 5.1 (*Relationships between different behavioural congruences*)
t $\sim_{int}$ and $\cong_{int}$; $\sim_{caus}$ and $\cong_{caus}$; and $\cong_{loc}$ denote the strong and observational congruences of 2]; the strong and weak causal congruences of [8]; and the location congruence of [4], respec-ely. Then, given two terms t and t'

i. $Int([t]_{pt}) \approx_s Int([t']_{pt})$ iff $t \sim_{int} t'$; ii. $Int([t]_{pt}) \approx_c Int([t']_{pt})$ iff $t \cong_{int} t'$;
i. $Ct([t]_{pt}) \approx_s Ct([t']_{pt})$ iff $t \sim_{caus} t'$ iv. $Ct([t]_{pt}) \approx_c Ct([t']_{pt})$ iff $t \cong_{caus} t'$;
. $Lt([t]_{pt}) \approx_c Lt([t']_{pt})$ iff $t \sim_{loc} t'$. ◊

An alternative characterization of strong and observational congruences is given by the well-own axiomatizations of [19], which provide us with a complete system for checking (finite / gular) proved trees congruent, regardless of how they have been observed. As a matter of fact, e algebraic laws of Section 3 permit reasoning directly on terms, although built with an extend- signature taking care of proofs. Moreover, as outlined in Section 4, an algebra of terms can be rived from the proved one, expressing also those aspects of system behaviours that are obser-d (e.g., causes, localities, etc.) and that are not expressed in the original signature of language. nce derived such a denotational semantics $\mathcal{D}$, a chosen congruence can be checked directly on xtended) terms of the language, through its axioms and the laws defining $\mathcal{D}$. These axiomatiza-ons are complete on the extended language, but it is not possible in general to derive from them mplete axiomatizations within the language, showing that the extension to the original signa-re is indeed necessary.

. Conclusions

We have studied the semantics of Process Description Languages, assuming that their opera-nal is defined in three separated phases. The first phase is the definition of transition systems, the SOS style, giving their operational semantics. Then computations, i.e., sequences of ansitions, are given the structure of trees (proved trees), the labels of which are the proofs of e corresponding transition. Also a denotational semantics is defined, which has very concrete expansion law. In the second phase, these trees are "observed" to abstract from unwanted de-ils, and in the third one the observed trees are compared. We claim that the various models for ncurrent languages can be clearly compared and unified when the definition of their semantics organized in this way. Our approach improves over the one of [14], mainly because our repre-ntation is a differential one. This choice enables us to use standard results, such as bisimula-ons and axiomatizations to reason on trees. Milner's CCS has been considered as a case study. he proved transition system of [2] is taken in the first phase, and an algebra of proved trees is efined, one law of which gives a "universal" expansion law. As observations, we considered n interleaving one, yielding synchronization trees [21]; a causal one, recovering causal trees [8]; nd a locational one, producing a model similar to that of [4] (see also [24]). The "universal" expansion law and each observation drive us in obtaining expansion laws specialized to t chosen observation. In this way, it is also possible reasoning on "observed" trees.

Acknowledgements. The authors would like to thank Luca Aceto for his detailed comments, and Ila Castellani, Rocco De Nicola and Ugo Montanari for many discussions.

References

[1] Austry, D. and Boudol, G. *Algébre de Processus et Synchronization*, T.C.S. 30, 1 (1984), 91 - 131.

[2] Boudol, G., Castellani, I., *A non-interleaving semantics for CCS based on proved transitions*, Fundame Informaticae, XI (4) (1988), 433 - 452.

[3] Boudol, G., Castellani, I., Hennessy, M., Kiehn, A., *Observing Localities*, MFCS'91, LNCS 52 Springer-Verlag, 93 - 102.

[4] Boudol, G., Castellani, I., Hennessy, M., Kiehn, A., *A Theory of Processes with Localities*, INRIA Tec Rep., 1992

[5] Brookes, S.D., Hoare, C.A.R., Roscoe, A.D. *A Theory of Communicating Sequential Processes*, Journal A.C.M., 31, 3 (1984), 560 - 599.

[6] Bergstra, J.A. and Klop, J.-W. *Process Algebra for Synchronous Communication*, Info. & Co., 61, (198 109 - 137.

[7] Castellani, I., Private Communication, 1991.

[8] Darondeau, Ph., Degano, P., *Causal Trees*, ICALP '89, LNCS 372, Springer-Verlag, 234 - 248.

[9] Degano, P., De Nicola, R., Montanari, *Partial Ordering Derivations for CCS*, in Proc. FCT, LNCS 19 Springer-Verlag, 1985, 520 - 533.

[10] Degano, P., De Nicola, R. e Montanari, U. *Observational Congruences for Concurrency Models*, In: *Fo mal Description of Programming Concepts* III (M. Wirsing Ed.), North-Holland, 1987, 105 - 132.

[11] Degano, P., De Nicola, R. and Montanari, U. *Partial Ordering Description of Nondeterministic Concurre Systems*, in Linear Time, Branching Time and Partial Orderd in Logic and Models for Concurrency, LNC 354, Springer-Verlag, 1989, 438 - 466.

[12] Degano, P., De Nicola, R., Montanari, U. *A Partial Ordering Semantics for CCS*, T.C.S., 75 (1990), 223 262.

[13] Degano, P., De Nicola, R., Montanari, U. *On the Operational Semantics of Distributed Concurrent Sy tems*, in: Declarative Systems (G. David, R.T. Boute, and B.D. Shriver Eds) Proc. IFIP 10.1 Worksho 1988, North-Holland, Amsterdam, 3-32, 1990.

[14] Degano, P., De Nicola, R., Montanari, U. *Universal Axioms for Bisimulations*, submitted for publication.

[15] Degano, P., Montanari, U. *Concurrent Histories: A Basis for Observing Distributed Systems*, J.C.S.S., 3 1987, 422-461.

[16] Ferrari, G.L., Montanari, U., *Towards the Unification of Models for Concurrency* in CAAP'90, LNCS 43 Springer-Verlag, 162 - 176.

[17] Ferrari, G.L., Gorrieri, R., Montanari, U., *An Extended Expansion Theorem*, in TAPSOFT '91, LNCS 49 Springer-Verlag, 29 - 48.

[18] van Glabbeek,R.J and Weijland,W.P. *Branching Time and Abstraction in Bisimulation Semantics*. Pro IFIP '89 (G.X. Ritter, ed.), 1989, 613 - 618.

[19] Hennessy, M., Milner, R. *Algebraic Laws for Nondeterminism and Concurrency*, Journal of ACM, 3 (1985), 137 - 161.

[20] Kiehn, A., *Local and Global Causes*, Report Technische Universität Munchen 342/23/91 A, 1991.

[21] Milner, R. *A Calculus of Communicating Systems*. LNCS 92, Springer-Verlag, 1980.

[22] Milner, R. *Communicating and Concurrency*. Prentice Hall Int., London 1989.

[23] Montanari, U., Yankelevich, D., *An Algebraic View of Interleaving and Distributed Operational Semantic for CCS*, Category Theory and Computer Science '89, LNCS 389, Springer-Verlag, 5 - 20.

[24] Montanari, U., Yankelevich, D., *An Parametric Approach to Localities*, This Volume.

[25] Park, D. *Concurrency and Automata on Infinite Sequences*, in Proc. GI, LNCS, 104, 1981, 167 - 183.

[26] Plotkin, G. *A Structural Approach to Operational Semantics*. Tech. Rep. DAIMI FN-19, Aarhus Univ., 1981.

[27] Reisig, W. *Petri Nets: An Introduction*, EACTS Monographs on Theoretical Computer Science, 1985.

[28] Rabinovich, A., Trakhtenbrot, B. *Nets of Processes*, Fundamenta Informaticae XI (4), 1988, 357 - 404.

Dipartimento di Informatica, Corso Italia 40, I-56100 PISA, Italy e_mail: {degano,priami}@di.unipi.

Interfaces between Languages for Communicating Systems

Ernst-Rüdiger Olderog
FB Informatik, Universität Oldenburg
Postfach 2503, 2900 Oldenburg, Germany

Abstract

A system design typically involves various languages, each one describing the system at a different level of abstraction. To achieve a trustworthy design, it is essential that the interfaces between these laguages are conceptually well understood and mathematically sound. Recent research in semantics attempts to clarify and structure the development of such interfaces.

This paper considers three languages for communicating, reactive systems – a specification language SL, a programming languages PL and a machine language ML – and outlines two approaches developed in the ESPRIT Basic Research Action "ProCoS" on how to establish the language interfaces. The SL/PL interface is based on a common predicative semantics leading to a mixed term design calculus, and the PL/ML interface is based on an interpreter and algebraic reasoning leading to a prototype compiler written in PROLOG.

Introduction

hen designing complex systems, often various levels of abstraction are considered. Typi- ly these levels are described using different languages, for example requirements' capture guage, specification laguage, programming language and machine language. The question ses how to formally relate these languages, in other words: how to build mathematically nd *interfaces* between these languages. Without such interfaces between the various guages involved in a system design, the correct functioning of the resulting system is ubtful.

Building correct interfaces between the semantics of languages for communicating systems s been one of the primary aims of the ESPRIT Basic Research Action "ProCoS". "ProCoS" nds for "Provably Correct Systems" and is a wide-spectrum verification project where bedded communicating systems are studied at various levels of abstraction ranging from quirements' capture over specification language and programming language down to the chine language [Bjø89].

Right at the start of ProCoS it was decided to use an occam-like programming language d the transputer instraction set as the machine language. During the project the Duration lculus [ZHR92] has been developed as a language for expressing requirements (involving l-time). As the (program) specification language a combination of regular expressions d assertional techniques is used [JROR90] [Old91b]. Summarising, the languages involved the ProCoS project are:

- requirement language RL = Duration Calculus
- specification language SL = regular expressions and assertions
- programming language PL = occam-like programs
- machine language ML = transputer instruction set

While ProCoS has so far not succeeded to close the gap between the languages RL a SL (because real-time is not yet treated in SL, PL and ML), it has built formal interfa between the languages SL, PL and ML. In this paper we describe two different techniq of building such interfaces as developed by the ProCoS partners in Oldenburg and Oxfor

2 Three Languages

2.1 Specification Language SL

The main idea of SL [JROR90] is to split the description of the desired system behaviour two parts: a *trace part* and a *state part*.

The trace part specifies the sequencing constraints on the external system chann whereas the communicated values are ignored. This is done by stating one or more *tr assertions*, each one consisting of an alphabet, i.e. a subset of the external channels, an regular expression over these channels. The regular expression describes the sequencing co straints on the channels mentioned in the alphabet. By stating several such trace assertio we can specify different aspects of the intended system behaviour in a modular fashion. T informal semantics of this part of an SL specification is that the described behaviour m satisfy the sequencing constraints of all trace assertions simultaneously.

The state part of an SL specification describes what the exact values are that can exchanged over the external system channels. To this end, this part starts with the introd tion of certain local state variables. These variables are used in *communication assertio* specifying the link between values and channels but need not appear in an implementati of the specified system. A communication assertion for a given channel has two predicat a *when-predicate* describing when a communication may occur on the channel and a *the predicate* describing the effect of the occurrence of such a communication.

The trace part corresponds to path expressions in the sense of [CH74] or to a regul fragment of trace logic in the sense of [Zwi89] and [Old91a]. The state part describing t effect of a communication in terms of variables is similar to the action notation used in e [Lam91].

An example is the following specification of a simple autopilot for a ship:

```
autopilot = spec  input s of -1..1
                  output a of int
                  trace {s,a} in pref (s.a)*
                  var ps of int init 0
                  var cs of int init 0
                  com s write {ps, cs}
                      when true then ps' = cs /\ cs' = @s
                  com a read {ps, cs}
                      when true then (ps =  0 => @a = 2 * cs) /\
                                     (ps <> 0 => @a = cs)
            end
```

The idea is that the autopilot inputs values in the range $-1..1$ on a channnel s standing "sensor" and – to maintain the correct course of the ship – outputs appropriate corrections he form of integer values on a channel a standing for "actuator". The specification has one trace assertion stating that communications on the channels s and a should occur lternating order starting with s, or equivalently, that at each moment the global channel e of the autopilot should be a prefix of some word in the regular language $(s.a)^*$.

To specify the allowed communication values on the channels s and a, two communication rtions are used referring to local variables *ps* and *cs* standing for "previous state" and rrent state". The communication assertion for channel s states that a communication s may write on both variables *ps* and *cs*, that due to the when-predicate it is always bled, and that due to the then-predicate its effect is as follows: the termination value of s the initial value of *cs* and the termination value of *cs* is the communication value on channel s. We use primed versions of variables to refer to their values at the moment of mination, and we use channel names prefixed by the symbol @ as variables referring to the nmunication values on these channels. The communication assertion for channel a states t a communication on a may only read the variables *ps* and *cs*. Thus their termination ies are identical to their initial values, a fact that need not be stated any more in the n-predicate of this communication assertion.

Although the communication assertions state that communications on both s and a are ays enabled, the meaning of the whole autopilot specification is that these communications occur only if they are simultaneously possible according to the trace assertion. Thus the cified behaviour of the autopilot is that it should be ready for communications on the nnels s and a in alternating order starting with s and that the values communicated on se channels should agree with the values specified in the communication assertions.

Programming Language PL

consider here an occam-like programming language PL [IN88]. Programs are terms structured from a set of n-ary *operators* OP. For PL we consider the 0-ary operators ABORT, P, SKIP, multiple assignments, input and output on channels, the unary operator WHILE describing loops, and the n-ary operators SEQ, IF, ALT for sequential, conditional and ernative composition and PAR for parallel compoisition.

An example is the following program which implements the autopilot specification of the vious section:

```
pilotprog = system  input s of -1..1
                    output a of int
                    var x of int -1..1 init 0
                    var y of int init 0
                    var ps of int init 0
                    var cs of int init 0
                    WHILE( true,
                           SEQ[s?x,
                               ps,cs :=cs,x,
                               IF[ ps =  0 --> y := 2 * cs,
                                   ps <> 0 --> y := cs     ],
                               a!y] )
               end
```

e **system-end** brackets emphasise that programs represent implementations of communi- ting systems. Therefore the external system channels, here the input channel s and the output channel a, are declared explicitly.

2.3 Machine Language ML

We consider here a simplified transputer-like machine [Bow90]. It consists of the follow components:

- An array M from integers to machine instructions representing the *memory* for stor machine programs. The memory cell addressed by the integer n is written as $M[n]$. make here the simplifying assumption that each machine intruction fits into a sin memory cell.
- An array W from integers to integers representing the *workspace* for addresses a data values. The workspace cell addressed by the integer n is written as $W[n]$. neglect here that a real machine can only deal with a finite set of addresses and d values.
- An integer variable p representing the *instruction pointer*. At each moment it poi to the instruction in M that is currently to be executed.
- Three integer variables A, B, C representing the transputer *registers*. These regist are organised as an *evaluation stack* with A being the top register for loading a storing values and for evaluating expressions.

The machine language ML is a subset of the transputer intructions and includes amo others the following instructions:

- `load(n)` – push the value of cell $W[n]$ into the top register A of the evaluation sta
- `store(n)` – pop the contents of register A from the evaluation stack and store in i cell $W[n]$.
- `jump(n)` – increment the instruction pointer by $n + 1$.
- `cond(n)` – if register A contains `false` (represented as 0) then increment the instruct pointer by $n + 1$; otherwise continue with the next instruction.
- `add` – add the values of registers A and B and store the result in register A.
- `stopp` – stop the machine execution.

Machine programs of the language ML are finite sequences of such instructions.

3 The SL/PL Interface

A technique of building the interface between the specification language SL and the progra ming Language PL pursued at Oldenburg and Oxford is to construct a *common seman model* for SL and PL. This model will view specifications and programs as different descr tions of *communicating systems*. Such a system continuously interacts with its envirom through communication channels or global variables:

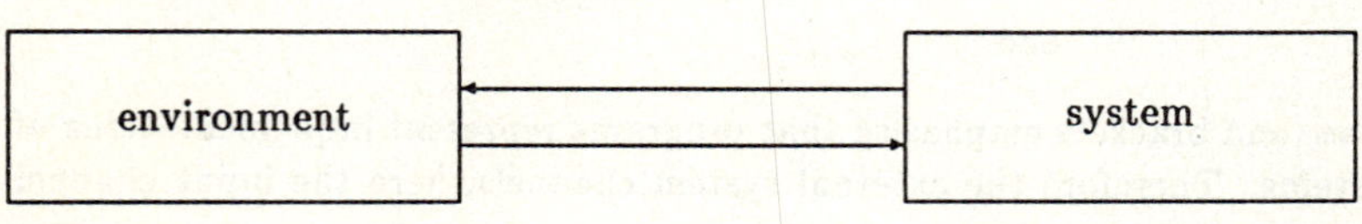

mally, a communicating system is a pair $\Delta : P$ consisting of a system interface Δ and a aviour P. The system interface Δ lists the declarations of the communication channels of system with their direction (input or output) and value type, and of the global variables h their access mode (read or write) and value type. The behaviour P is given by a set *observations* that are considered relevant for understanding the interaction of the system h its environment by communication via the interface channels and by access to the global erface variables.

Observations and Predicates

have chosen a combined *state-trace-readiness model* of communicating systems. This del can represent the typical ingredients of occam-like communicating systems like comnication, synchronisation, concurrrency, nondeterminism and assignable variables. A *nmunication* is a pair $< ch, m >$ consisting of a channel name ch and a value m. By $mm(\Delta)$ we denote the set of all communications $< ch, m >$ where ch is a channel in the tem interface Δ and where m is an element of $\mathcal{I}[\![ty_{ch}]\!]$, the set of values of the declared e ty_{ch} of ch.

The state component of the model refers to the initial values of the global variables of ystem and to their final values when the system has terminated. The trace component cribes the order in which the communications can occur by finite sequences of communiions called *traces*. The readiness component describes the possible continuations of the tem behaviour after a communication trace has occurred with the help of so-called *ready s*. These are the sets of channels that are ready for the next communication when the tem is stable. Since a system may behave nondeterministically, there may be more than ready set associated with the same trace. In particular, the empty ready set represents tem deadlock. If a system is neither stable nor terminated, it pursues some internal acn. If it engages in an infinite sequence of internal actions, it is known as divergent. The te-trace-readiness model considers divergence as an undesirable system behaviour and ats it accordingly.

We pursue here a *predicative approach* as advocated e.g. by Hehner [Heh84] and Hoare a85] whereby the behaviour P is identified with a predicate describing the observable antities just explained. As *predicate language* we take first-order predicates extended by μ-operator $\mu X \bullet P$ for introducing recursive definitions [Par76]. For a given system erface Δ the behaviour is then a predicate in the following free variables:

- a variable st standing for *status* and ranging over the set $\{stable, term, div\}$ where *term* stands for *terminated* and *div* for *divergent*,
- a variable tr standing for *trace* and ranging over $Comm(\Delta)^*$, the set of finite sequences of communications on the interface channnels,
- a variable Rdy standing for *ready set* and ranging over the subsets of $Chans(\Delta)$, the set of interface channels,
- all global read or write variables x of Δ, each one ranging over $\mathcal{I}[\![ty_x]\!]$, the set of values of the declared type ty_x of x,
- a primed version x' for each of the above global variables x ranging over $\mathcal{I}[\![ty_x]\!]$.

uitively, these primed versions represent the values of the global variables at termination, when the variable st holds the value *term*.

3.2 Predicative Semantics

Let SYS be the set of all systems, i.e. pairs $\Delta : P$ where the predicate P fits to Δ as describ above. Then logical implication and equivalence on predicates are lifted to systems as follov

- system implication: $\Delta_1 : P_1 \equiv> \Delta_2 : P_2$ if $\Delta_1 = \Delta_2$ and $\models P_1 \Rightarrow P_2$,
- system equivalence: $\Delta_1 : P_1 \equiv \Delta_2 : P_2$ if $\Delta_1 = \Delta_2$ and $\models P_1 \Leftrightarrow P_2$,

Thus system implication $\Delta_1 : P_1 \equiv> \Delta_2 : P_2$ states that every observation of the behavic P_1 is also an observation of the behaviour P_2.

The SL/PL interface is then established following the principles of predicative semanti

- Identify every specification $\mathcal{S} \in$ SL with a suitable system $\Delta : P$, i.e. define an emb ding
  SL $\subseteq$ SYS.
- Identify every program $\mathcal{P} \in$ PL with a suitable system $\Delta : P$, i.e. define an embeddi PL $\subseteq$ SYS.
- Identify the satisfaction relation **sat** between programs and specifications by

$$\mathcal{P} \text{ sat } \mathcal{S} \text{ iff } \mathcal{P} \equiv> \mathcal{S}.$$

Thus in predicative semantics a program $\mathcal{P}$ *satisfies* or *implements* a specification $\mathcal{S}$ if eve observation of $\mathcal{P}$ is allowed by $\mathcal{S}$. A complete definition of the predicative semantics of and PL will be given in [HO92]. Here we can convey only the style of this semantics.

3.3 Implementability

How can one check that a specification $\mathcal{S}$ is implementable by some program $\mathcal{P}$? It turns c that programs are systems whose behaviour predicates enjoy certain closure or *healthin* properties. We call such systems *processes.*

What are the closure properties of processes $\Delta : P$ in the state-trace-readiness mode Let us mention here a few typical ones:

- P describes a *non-empty* set of observations, i.e. whatever initial values we give to global variables $\overline{y}$, there is some behaviour observable:

$$\forall \overline{y} \in \mathcal{I}[\![ty_{\overline{y}}]\!] \bullet \exists st, tr, \overline{y}', Rdy \bullet P.$$

- P describes a *prefix closed* set of traces:

$$\forall \overline{y} \in \mathcal{I}[\![ty_{\overline{y}}]\!], tr \in Comm(\Delta)^*, t \in Comm(\Delta)^* \bullet \\ (\ \exists st, \overline{y}', Rdy \bullet P[tr.t/tr] \ \Rightarrow \ \exists st, \overline{y}', Rdy \bullet P\ )$$

- P describes *progress* in the sense that if P is stable and a channel ch is ready, then must accept a communication along ch:

$$\forall \overline{y} \in \mathcal{I}[\![ty_{\overline{y}}]\!], tr \in Comm(\Delta)^*, ch \in Chans(\Delta) \bullet \\ (\ P[stable/st] \wedge ch \in Rdy \ \Rightarrow \ \exists st, \overline{y}', Rdy, m \in \mathcal{I}[\![ty_{ch}]\!] \bullet P[tr. < ch, m >$$

- P describes *bounded nondeterminism*, i.e. if P terminates after a trace tr, then the set of all final values of the global variables is finite:

$\forall \overline{y} \in \mathcal{I}[\![ty_{\overline{y}}]\!], tr \in Comm(\Delta)^*, Rdy \in Chans(\Delta) \bullet \{\overline{y}' \mid P[term/st] \wedge \neg P[diver/st]\}$ is finite.

PROC denote the set of all processes. Then the predicative semantics of PL yields

$$\mathrm{PL} \subseteq \mathrm{PROC} \subseteq \mathrm{SYS}$$

ere the inclusions are strict. Thus a necessary condition for implementability of a specifiion $\mathcal{S}$ is that $\mathcal{S} \in \mathrm{PROC}$ holds. It is not sufficient though because the closure properties rocesses do state the computability constraints obeyed by programs.

Operators

us look more closely at the structure of the predicative semantics of PL. Recall that programs are systems that are constructed inductively from a set of n-ary *operators* OP luding STOP, SKIP, ABORT, multiple assignments, input and output on channels, SEQ, , IF, ALT, WHILE.

The predicative semantics of PL is thus given inductively or *compositionally* by stating each n-ary operator OP a definition of the form

$$\mathtt{OP}[\Delta_1 : P_1, ..., \Delta_n : P_n] \equiv \Delta : Q$$

ere Δ is the union of $\Delta_1, ..., \Delta_n$ (usually under some side conditions) and where the dicate Q is obtained from the predicates $P_1, ..., P_n$.

For each n-ary operator OP on systems there is an $n+1$-ary relation op on observations h that the predicate Q is obtained from $P_1, ..., P_n$ as follows:

$$o \models Q \text{ iff } \exists o_1, ..., o_n \bullet\ o_1 \models P_1 \text{ and ... and } o_n \models P_n \text{ and } (o_1, ..., o_n, o) \in op$$

ere $o \models Q$ states that observation o satisfies Q. As a consequence we have the following:

oposition OP is *monotonic* with respect to system implication, i.e. for all $i \in \{1, ..., n\}$

$$\Delta_i : P_i \equiv> \Delta_i : Q_i$$

plies

$$\mathtt{OP}[\Delta_1 : P_1, ..., \Delta_i : P_i, ... \Delta_n : P_n] \equiv> \mathtt{OP}[\Delta_1 : P_1, ..., \Delta_i : Q_i, ..., \Delta_n : P_n].$$

dious is usually the proof of the following:

quirement OP should preserve the closure properties of processes, i.e. if all its argunts $\Delta_1 : P_1, ..., \Delta_n : P_n$ are processes then $\mathtt{OP}[\Delta_1 : P_1, ..., \Delta_n : P_n]$ is a process as well.

As examples we consider the operator definitions for ABORT, STOP, SKIP, input and WHILE a given system interface Δ:

$$\Delta : \mathtt{ABORT} \equiv \Delta : \mathtt{true}\ ,$$

$$\Delta : \mathtt{STOP} \equiv st = stable \wedge tr = \langle\rangle \wedge Rdy = \emptyset\ ,$$

$$\Delta : \mathtt{SKIP} \equiv st = term \wedge tr = <> \wedge \bigwedge_{v \in Vars(\Delta)} v' = v \ ,$$

$$\begin{aligned} \Delta : ch?x \equiv \ & st = stable \wedge tr = <> \ \wedge Rdy = \{ch\} \\ \vee \ & st = term \wedge \exists m \bullet (tr = < ch, m > \wedge x' = m \wedge \bigwedge_{v \in Vars(\Delta) - \{x\}} v' = v) \ , \end{aligned}$$

$$\Delta : \mathtt{WHILE}(b, \Delta : P) \ \equiv \Delta : \mu X \bullet \ \mathtt{IF}[b \ \rightarrow \mathtt{SEQ}[\Delta : P, \Delta : X], \neg \ b \ \rightarrow \mathtt{SKIP}] \ .$$

In the above definitions $Vars(\Delta)$ denotes the set of all global read or write variables of Δ

3.5 Algebraic Laws

Predicative semantics provides a sound basis for the specification and the programmi language but it is tedious to work directly in this semantics. Programs can be transform much easier into other, perhaps more desirable programs when a purely algebraic reasoni about the programming operators OP is possible. This is the motivation behind *proc algebra*, the algebraic study of abstract uninterpreted communicating programs that w initiated by Milner's CCS [Mil80] [Mil89]. In [Hoa et al. 87] it has been vigorously argu that the same algebraic reasoning should also be applied to real programming languages. [RH88] the algebraic laws of the occam operators are described. Most of these algebraic la can be proved valid on the basis of the above predicative semantics for PL. For example, SEQ the following laws are valid:

(seq-1) $\mathtt{SEQ}[\,] \equiv \mathtt{SKIP}$

(seq-2) $\mathtt{SEQ}[\mathcal{P}] \equiv \mathcal{P}$

(seq-3) $\mathtt{SEQ}[\mathcal{P}_1, ..., \mathcal{P}_k, \mathtt{SEQ}[\mathcal{Q}_1, ..., \mathcal{Q}_\ell], \mathcal{R}_1, ..., \mathcal{R}_m] \equiv \mathtt{SEQ}[\mathcal{P}_1, ..., \mathcal{P}_k, \mathcal{Q}_1, ..., \mathcal{Q}_\ell, \mathcal{R}_1, ..., \mathcal{R}$

3.6 Mixed Term Design Calculus

So far we have defined the SL/PL interface "statically", i.e. by saying when a program satisfies a specification $\mathcal{S}$. Now we turn to the "dynamic" aspect of this interface, i.e. h to design a program that satisfies a given specification $\mathcal{S}$?

More precisely this task is as follows: given a system $\Delta_2 : P_2$ described by a specificati $\mathcal{S}$ we wish to find another system $\Delta_1 : P_1$ with $\Delta_1 : P_1 \equiv> \ \Delta_2 : P_2$ such that $\Delta_1 : P_1$ can expressed solely in terms of the operators OP of the programming language PL.

To solve this task we develop a *design calculus* for PL programs. The calculus is collection of *design* or *transformation rules* that (ideally) allow us to transform a giv specification into a program. "Ideally" refers to fact that the specification need not implementable or the calculus need not be complete. In a successful design each desi step should replace a part of the specification by appropriate operators of the programmi language PL.

To express the intermediate design stages we need terms that mix programming a specification notation. We therefore introduce an additional language MIX of *mixed ter* [Old91a] [Zwi89] that contains SL and PL as proper subsets. More generally, in a mixed ter programming operators OP can be applied to specifications or other mixed terms $\mathcal{S}_1, ..., \mathcal{S}$

$$\mathtt{OP}[\mathcal{S}_1, ..., \mathcal{S}_n].$$

The predicative semantics of SL and PL is then extended to MIX. Thus

$$\text{SL, PL} \subseteq \text{MIX} \subseteq \text{SYS}.$$

te that MIX $\subseteq$ PROC does not hold because arbitrary specifications are part of MIX.

A design of a program $\mathcal{P}$ from a specification $\mathcal{S}$ is then presented in a top-down fashion a sequence

$$\begin{array}{c} \mathcal{S} \equiv \ \mathcal{Q}_1 \\ \Uparrow \\ \vdots \\ \Uparrow \\ \mathcal{Q}_n \ \equiv \mathcal{P} \end{array}$$

ystem implications between mixed terms $\mathcal{Q}_1, ..., \mathcal{Q}_n$ where $\mathcal{Q}_1$ is the given specification $\mathcal{S}$ $\mathcal{Q}_n$ is the designed program $\mathcal{P}$. The transitivity of the relation $\equiv>$ ensures the desired rectness result $\mathcal{P} \equiv> \mathcal{S}$.

Each implication $\mathcal{Q}_i \equiv> \mathcal{Q}_{i+1}$ in the sequence is generated by an application of a design e of the form

where ...
$\mathcal{S}_i$ $\Uparrow$ $\text{OP}[\mathcal{S}_{i_1}, ..., \mathcal{S}_{i_n}]$

ting that under the condition "..." the specification $\mathcal{S}_i$ is satisfied by a mixed term sisting of an n-ary occam-like programming operator OP applied to specifications or other ed terms $\mathcal{S}_{i_1}, ..., \mathcal{S}_{i_n}$ that are related to $\mathcal{S}_i$. This rule is applied to the mixed term $\mathcal{Q}_i$ in design sequence by matching $\mathcal{S}_i$ against a subterm of $\mathcal{Q}_i$ and by replacing this subterm $\text{OP}[\mathcal{S}_{i_1}, ..., \mathcal{S}_{i_n}]$. Since all operators appearing in $\mathcal{Q}_i$ are monotonic with respect to the tion $\equiv>$, the implication $\mathcal{Q}_i \equiv> \mathcal{Q}_{i+1}$ follows.

Note that this corresponds to the transformational approach to programming, which is understood in the context of conventional sequential programs due to the pioneering k of Dijkstra, With, Burstall and Darlington. To scale up this approach several major arch projects have been conducted in this area, among them the projects CIP (standing Computer-aided Intuition-guided Programming) [Bau87] and PROSPECTRA (standing PROgram development by SPECification and TRAnsfromation) [Kri89]. However, for nmunicating and concurrent programs the transformational approach is much less devel-d. Mostly the work takes place in the setting of idealised programs [Bro87b] [CM88] c90] [Old91a].

We extend the transformational approach based on mixed terms to the setting of occam-programs. A typical example of a mixed term design rule is the following rule for erating an *output communication* from a given specification $\mathcal{S}$:

$$init(\mathcal{S}) = \{ch\} \subseteq Outputs(\Delta)$$

$$\mathcal{S}$$

$$\Uparrow$$

IF[$when_{ch} \rightarrow$ SEQ[$impl(then_{ch})$, $ch!@ch$, $\mathcal{S}$ after ch]]

Here $init(\mathcal{S})$ denotes the set of channels on which a first communication is ready accord to $\mathcal{S}$, $when_{ch}$ is the when-predicate of channel ch in $\mathcal{S}$ and $then_{ch}$ is the corresponding th predicate. $\mathcal{S}$ `after` ch is a new specification that describes the remainder of the behaviou S *after* the communication on channel ch has occurred. The whole design rule states that specification $\mathcal{S}$ can be expanded into a mixed term consisting of an `IF` operator applied t sequential composition of a program $impl(then_{ch})$ implementing the state transition speci by the effect predicate $then_{ch}$, an output on the channel ch, and the derived specificat $\mathcal{S}$ `after` ch.

For practical application of such design rules a user needs guidance how to apply the In specific cases several rules can be combined to *strategies*, i.e. recipies how to ap them systematically or even mechanically. For example the above output rule is part the so-called *expansion strategy* described in more detail in [Old91b]. Another strategy, *syntax-directed transformation* of SL specifications is described in [RS91]. These strateg yield only sequential programs, i.e. without any `PAR` operator. Current work is concerr with the design of parallel PL programs as indicated in [Old91b] and [RS91].

4 The PL/ML Interface

A technique of building the interface between the programming language PL and the mach language ML pursued at Oxford is to construct an *interpreter* for ML written in PL [HH This way the semantics of PL is used to provide a semantics for ML.

4.1 Interpreter

The interpreter $\Im$ for ML programs takes four parameters M, s, f, D where

- M is the memory for machine instructions introduced in Section 2.3.
- s and f are integer values representing *start address* and *finish address* of a sequence cells in M containing the machine instructions of the machine program which is to interpreted. Thus the memory section $M[s : f-1]$ represents the machine progran
- D is a portion of the workspace W representing the *data space* on which the mach program in $M[s : f-1]$ is to operate.

For given values of these parameters $\Im MsfD$ is then a PL program which is to simul the intended effect of the machine program stored in $M[s : f-1]$ and operating on D. T structure of this program is as follows:

```
ℑMsfD  =  SEQ[ p := s,
               WHILE( s <= p < f,
                      step),
               assert( p = f )]
```

ere p is the instruction pointer and step is a big IF construct that defines the effect for h form of the current instruction:

```
step =   IF[  M[p] = load(n) -->
                A, B, C, p := D[n], A, B, p + 1,
              M[p] = store(n) -->
                A, B, D[n], p := B, C, A, p + 1,
              M[p] = jump(n) -->
                p := p + 1 + n,
              M[p] = cond(n) -->
                IF[ A = false --> p := p + 1 + n, true --> p := p+1],
              M[p] = add -->
                A, B, p := A + B, C, p + 1
              M[p] = stopp -->
                SEQ[p := p+1, STOP],
                  •
                  •
                  •    ]
```

e assertion assert(p = f) checks whether upon termination of the interpreter call $\Im MsfD$ instruction pointer p has properly reached the end of the interpreted machine program without any "wild" jumps outside the range of $[s : f-1]$. This assertion is actually an previation:

```
assert(p = f) = IF[ p = f --> SKIP, p <> f --> ABORT ].
```

us if p = f is false, the assert statement aborts. Thanks to this assertion one can show t the interpretation of two adjacent machine programs correctly implements their sequen- composition:

mma (Composition) If $s \leq \ell \leq f$ then $\Im MsfD$ ≡> SEQ[$\Im Ms\ell D, \Im M\ell fD$].

oof. Using the algebraic laws of WHILE and assert one shows that two WHILE loops in uence can be merged into one. □

Relating ML and PL

rough the interpreter $\Im$ every program $\mathcal{M}$ of the machine language ML can be lifted the level of the programming language PL: just consider the PL program $\Im MsfD$ for ameters M, s, f, D such that the memory section $M[s : f-1]$ stores $\mathcal{M}$. The idea is now use $\Im MsfD$ in order to check whether the machine program $\mathcal{M}$ implements any other en PL program $\mathcal{P}$ using system implication:

$$\Im MsfD \equiv> \mathcal{P}.$$

wever, this implication does not yet take into account that the two PL programs $\Im MsfD$ l $\mathcal{P}$ access rather different data spaces. Whereas $\Im MsfD$ accesses D the program $\mathcal{P}$ esses the program variables that appear in $\mathcal{P}$.

To reconcile this difference, we need the concept of *data refinement* [Hoa72]. Here it comes the form of a mapping ψ which maps every variable x accessed by the PL program $\mathcal{P}$ to address $\psi(x)$ of a corresponding cell in the machine workspace W. The finite mapping represents a *symbol table* as known from compiler construction. We take $D(\psi)$ to denote portion of workspace cells addressed that way.

Based on ψ we consider an *abstraction function* $\hat{\psi}$ that assigns to every program varia x the value of the corresponding cell $W[\psi]$ in the workspace. We represent $\hat{\psi}$ as a PL gram; it contains assignments of the form $x := W[\psi(x)]$ for each program variable x in Then we can formulate a *correctness criterion* defining when a program $\mathcal{P}$ is correctly plemented by a machine program stored in $M[s : f-1]$ under the mapping ψ. The criter is stated as a relationship with the parameters $\mathcal{P}, M, s, f$ and ψ:

Correctness Criterion $\quad \mathcal{C}\,\mathcal{P}Msf\psi$ iff SEQ[$\Im MsfD(\psi), \hat{\psi}$] $\equiv>$ SEQ[$\hat{\psi}, \mathcal{P}$].

It turns out that $\hat{\psi}$ has a *partial inverse* $\hat{\psi}^{-1}$ which can also be represented as a program. Partial inverse means that the following holds.

Proposition $\quad$ SEQ[$\hat{\psi}^{-1}, \hat{\psi}$] $\equiv$ SKIP and SKIP $\equiv>$ SEQ[$\hat{\psi}, \hat{\psi}^{-1}$]

As a consequence we have the following:

Corollary $\quad \mathcal{C}\,\mathcal{P}Msf\psi$ iff $\Im MsfD(\psi)$ $\equiv>$ SEQ[$\hat{\psi}, \mathcal{P}, \hat{\psi}^{-1}$].

From the corollary we see that the correctness criterion $\mathcal{C}$ is just a specific instance the general concept of *simulation-based refinement* (see [BRR90] for an overview).

Proposition $\quad \mathcal{C}\,\mathcal{P}Msf\psi$ if the following diagram commutes:

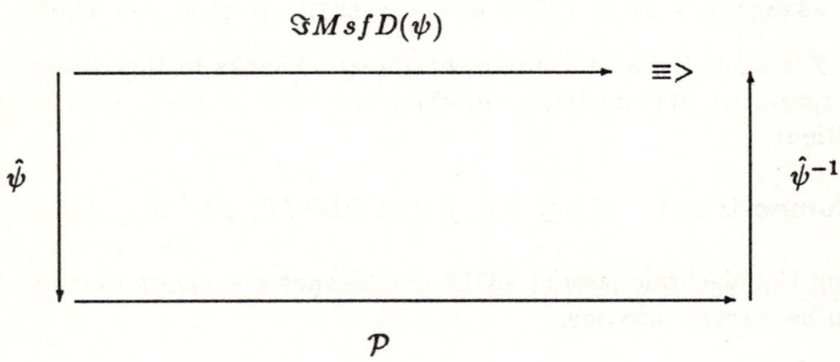

Thus $\Im MsfD(\psi)$ refines $\mathcal{P}$ via the simulation pair $(\hat{\psi}, \hat{\psi}^{-1})$.

4.3 Compiling Theorems

The correctness criterion $\mathcal{C}$ defines the "static" aspect of the PL/ML interface, i.e. whe machine program stored in $M[s : f-1]$ correctly implements a PL program $\mathcal{P}$. We r turn to the "dynamic" aspect of this interface, i.e. how to construct a machine program t implements a given program $\mathcal{P}$? We aim here at a compilation of PL into ML.

To this end, we first prove a number of theorems about the correctness criterion $\mathcal{C}$, for each programming operator OP that may appear in $\mathcal{P}$. These theorems constitut specification of the compiling function from PL to ML and will therefore be referred tc *compiling theorems*. Typical examples of such theorems are the following ones.

Theorem: SKIP $\quad \mathcal{C}$ SKIP $Msf\psi$ if $s = f$

eorem: SEQ $\mathcal{C}\ \mathtt{SEQ}[\mathcal{P}_1,\mathcal{P}_2]\ Msf\psi$ if $\exists\, \ell \bullet s \le \ell \le f \wedge \mathcal{C}\,\mathcal{P}_1 Ms\ell\psi \wedge \mathcal{C}\,\mathcal{P}_2 M\ell f\psi$.

oof. We use the algebraic laws of PL.

$$
\begin{array}{ll}
 & \Im MsfD(\psi) \\
\equiv> & \quad \{ \text{Composition Lemma for } \Im \} \\
 & \mathtt{SEQ}[\ \Im Ms\ell D(\psi), \Im M\ell fD(\psi)\ ] \\
\equiv> & \quad \{ \text{assumption of the theorem, monotonicity of } \mathtt{SEQ} \} \\
 & \mathtt{SEQ[\ SEQ[}\ \hat{\psi},\mathcal{P}_1,\hat{\psi}^{-1}\ \mathtt{],\ SEQ[}\ \hat{\psi},\mathcal{P}_2,\hat{\psi}^{-1}\ \mathtt{]]} \\
\equiv & \quad \{ \text{associativity law (seq-3) of section 3.5} \} \\
 & \mathtt{SEQ[}\ \hat{\psi},\ \mathtt{SEQ[}\ \mathcal{P}_1,\ \mathtt{SEQ[}\ \hat{\psi}^{-1},\hat{\psi}\ \mathtt{]},\ \mathcal{P}_2\ \mathtt{]},\ \hat{\psi}^{-1}\ \mathtt{]} \\
\equiv & \quad \{ \mathtt{SEQ}[\ \hat{\psi}^{-1},\hat{\psi}\ ] \equiv \mathtt{SKIP} \} \\
 & \mathtt{SEQ[}\ \hat{\psi},\ \mathtt{SEQ[}\mathcal{P}_1,\mathcal{P}_2\mathtt{]},\ \hat{\psi}^{-1}\ \mathtt{]}.
\end{array}
$$

□

eorem: WHILE

$$\mathtt{HILE}(b,\mathcal{P})\ Msf\psi \text{ if } s+1 < f-1 \wedge M[s] = \mathtt{load}(\psi(b)) \wedge M[s+1] = \mathtt{cond}(f-s-2) \wedge \mathcal{C}\,\mathcal{P}\ Ms+2\ f-1\ \psi \wedge M[f-1] = \mathtt{jump}(s-f).$$

Implemenentation in PROLOG

as been noticed by J.P. Bowen that all compiling theorems are statements in the from *Iorn clauses.* Therefore these theorems can be systematically transformed into rules of a OLOG program that represents a rapid prototype compiler from PL to ML. For example compiling theorem for SKIP and SEQ is transformed into the following rules:

```
            c(skip, M, S, F, Psi) :-  S = F

    c(seq[ P1 | P2 ], M, S, F, Psi) :-  c(P1, M, S, F, Psi),
                                        c(P2, M, S, F, Psi)
```

details of this implementation technique can be found in [Bow91].

Conclusion

his paper we have explained the main steps in the development of two interfaces be- en different languages for describing communicating systems. For the interface between cification language SL and programming language PL the steps were as follows:

- Choice of observations for a common model for SL and PL,
- Predicative semantics as a concise formalization of this model,
- Algebraic laws proved valid in the predicative semantics,
- Mixed term design calculus based on the predicative semantics.

ed on the SL/PL interface the steps for the interface between programming laguage PL machine laguage ML were as follows:

- Definition of an interpreter for ML written in PL,
- Simulation-based correctness criterion,

- Compiling theorems proved by using the algebraic laws of PL,
- Prototype compiler from PL to ML written in PROLOG.

Full details of these two interfaces will be reported in [HO92]. It should be clearly sta that even in this full version some major gaps will remain. Whereas the SL/PL interf will handle concurrency in the sense of the PAR operator of PL, the PL/ML interface not cover it. Also, as already mentioned in the introduction, there is as yet no form defined interface between of requirements' capture language RL and SL because RL inclu real-time. Another omission is that so far the level of hardware is not addressed at all.

Future work in the follow-up project "ProCoS II" will be concerned with how to c some of these gaps. In particular the inclusion of real-time at the RL/SL and the SL/ interface, and extension to hardware at the bottom level will be studied. Interestingly, w at Oxford indicates that the algebraic techniques employed in development of the PL/ interface can be extended to build also the interface from PL to a *net list description langu* NL for direct hardware implementation of PL. At the PL/NL interface the PAR operator be treated.

We hope to have illustrated that work on language interfaces is important, interest and challenging. We believe that the interface structures explained in this paper are steps towards a more general theory of defining and calculating language interfaces.

References

[Bac90] R.J.R. Back, Refinement Calculus, Part II: Parallel and Reactive Programs, [BRR90] pp. 67-93.

[BRR90] J.W. de Bakker, W.-P. de Roever, G. Rozenberg, Eds., Stepwise Refinemen Distributed Systems: Models, Formalisms, correctness, Lecture Notes in Comput. 430 (Springer-Verlag, 1990).

[Bau87] F.L. Bauer et al., *The Munich Project CIP, Vol. II: The Transformation Sys CIP-S*, LNCS 292 (Springer-Verlag, 1987).

[Bjø89] D. Bjørner, A ProCoS project description, ESPRIT BRA 3104, Bulletin of EATCS 39 (1989) 60-73.

[Bow90] J.P. Bowen, Formal specification of the ProCoS/safemos instruction set, Micro cessors and Microsystems 14 (1990) 631-643.

[Bow91] J.P. Bowen, From programs to object code using logic and logic programming, S. Graham, R. Giegerich, Proc. CODE '91 International Workshop on Code Generat SchloßDagstuhl, Germany, 1991 (to appear).

[Bro87b] M. Broy, Specification and top-down design of distributed systems, J. Com System Sci. 34 (1987) 236-265.

[CH74] R.H. Campbell, A.N. Habermann, The specification of process synchronisation path expressions, Lecture Notes in Comput. Sci. 16 (Springer-Verlag, 1974).

[CM88] K.M. Chandy, J. Misra, *Parallel Program Design: A Foundation*, Addison-Wes 1988.

[HO92] He Jifeng, E.-R. Olderog, Eds., Interfaces for Languages of Concurrent Langua ProCoS Monographs Vol. 2, 1992 (to appear).

h84] E.C.R. Hehner, Predicative programming, Comm. ACM 27 (2), 1984.

a72] C.A.R. Hoare, Proof of correctness of data representation, Acta Inform. 1 (1972) 72-281.

a85] C.A.R. Hoare, Programs are predicates, in: C.A.R. Hoare, J.C. Shepherdson (Eds.), Iathematical Logic and Programming Languages (Prentice-Hall, London, 1985) 141-155.

a et al. 87] C.A.R. Hoare, I.J. Hayes, He Jifeng, C.C. Morgan, A.W. Roscoe, J.W. anders, I.H. Sorensen, J.M. Spivey, B.A. Sufrin, Laws of programming, Comm. ACM 0 (1987) 673-686.

I90] C.A.R. Hoare, He Jifeng, Refinement algebra proves correctness of a compiler, in: I. Broy, Ed., Programming and Mathematical Method, Proc. Marktoberdorf Summer chool 1990 (to appear with Springer-Verlag).

88] INMOS Ltd., occam 2 Reference Manual (Prentice Hall, 1988).

OR90] K.M. Jensen, H. Rischel, E.-R. Olderog, S. Rössig, Syntax and informal semantics or the ProCoS specification language 0, ProCoS Doc. Id. ID/DTH KMJ 4/2, Tech. Univ.)enmark, 1990.

i89] B. Krieg-Brückner, Algebraic specification and functionals for transformational program and meta program development, in: J. Diaz, F. Orejas, Eds., Proc. TAPSOFT '89, ecture Notes in Comput. Sci. 352 (Springer-Verlag, 1989).

m91] L. Lamport, The temporal logic of actions, Manuscript, DEC Systems Research Center, Palo Alto, April 1991.

l80] R. Milner, Calculus of Communicating Processes, Lecture Notes in Comput. Sci. 92 Springer-Verlag, 1980).

l89] R. Milner, Communication and Concurrency (Prentice-Hall, 1989).

d91a] E.-R. Olderog, Nets, Terms and Formulas: Three Views of Concurrent Processes nd Their Relationship (Cambridge University Press, 1991).

d91b] E.-R. Olderog, Towards a design calculus for communicating programs, in: J.C.M. Baeten, J.F. Groote, Eds., CONCUR '91, Lecture Notes in Comput. Sci. 527 (Springer-Verlag, 1991) 61-77.

r76] D. Park, Finiteness is mu-ineffable, Theoret. Comput. Sci. 3 (1976) 173-181.

91] S. Rössig, M. Schenke, Specification and stepwise development of communicating systems, in: S. Prehn, W.J. Toetenel, Eds, Proc. VDM '91 Symposium, Vol. 1, Lecture Notes in Comput. Sci. 551 (Springer-Verlag, 1991) 149-163.

H88] A.W. Roscoe, C.A.R. Hoare, The laws of occam programming, Theoret. Comput. Sci. 60 (1988) 177-229.

IR92] Zhou Chaochen, C.A.R. Hoare, A.P. Ravn, A calculus of durations, Inform. Proc. Letters 40 (1992) 269-276.

vi89] J. Zwiers, Compositionality, Concurrency, and Partial Correctness – Proof Theories for Networks of Processes and Their Relationship, Lecture Notes in Comput. Sci. 321 (Springer-Verlag, 1989).

Toward Formal Development of Programs from Algebraic Specifications: Model–Theoretic Foundations

Donald Sannella
Department of Computer Science
University of Edinburgh
Edinburgh, UK

Andrzej Tarlecki
Institute of Computer Science
Polish Academy of Sciences
Warsaw, Poland

Abstract: This paper presents in an informal way the main ideas underlying our work on the model-theoretic foundations of algebraic specification and program development. We attempt to offer an overall view, rather than new results, and focus on the basic motivation behind the technicalities presented elsewhere and on the conclusions from this work.

Introduction

The long-term goal of work on algebraic specification is to provide a formal basis to sup the systematic development of correct programs from specifications by means of ver refinement steps. There has been a large body of technical work directed towards important goal. Many interesting technical concepts have been introduced and qui number of non-trivial results have been stated and proved (see [BKLOS 91] for a review a comprehensive list of references). Instead of providing yet another piece in the puzzl this paper we try to sketch on a rather informal level our views on how some of the exis pieces fit into an overall picture of what is important in the light of the ultimate goal. focus on the motivations for certain technicalities which we think are of crucial importa and try to draw some conclusions.

Our earlier papers already mentioned many of the points we make here, such as use of "institutions" to ensure sufficient generality of the proposed framework, and use of "constructor implementations" to capture the essence of program development s (including "design" steps which involve a decomposition into independent programr tasks). Some of these ideas were hidden amongst the technical definitions and results, so we think they are worth restating here more prominently, with more careful argum in some cases. For example, we go into a bit more detail to justify our conviction model-classes, not theories, form the appropriate semantic domain for specifications. also present more clearly our current views on the role of behavioural equivalence in development process.

Since the emphasis here is on motivation and intuition, we refer the interested reade the papers we have published on these and related topics over the last few years and forthcoming monograph [ST 93] for the corresponding technical details.

The logical framework

overall aim of work on algebraic specification is to provide semantic foundations for evelopment of programs that are *correct* w.r.t. to their requirements specifications. In words, the program developed must exhibit the required input/output behaviour. We the correctness of a program as its most crucial property. Other desirable properties iency, robustness, reliability etc.) are disregarded in this work. Of course, this does not that we do not care about them, but this approach does not provide any formal means eir analysis.

he assumption that the correctness of the input/output behaviour of a program takes dence over all its other properties allows us to abstract away from concrete details de and algorithms, and to model program functions as mathematical functions. Such ions are never considered in isolation, but always in units (program modules) comprising lection of related functions together with the data domains they operate on. At this of abstraction we are dealing directly with the information essential for the analysis ogram correctness, without the burden of irrelevant details. This leads to the most mental assumption underlying work on algebraic specification: programs are modelled any-sorted *algebras*.

/e refrain from recalling the formal definition of many-sorted algebra (see e.g. [EM 85]). enough to know that an algebra consists of a collection of *carriers* (sets of data) and *tions* on them. Algebras are classified by *signatures*, naming the algebra components s and operations) and thus providing the basic vocabulary for using the program and aking assertions about its properties. The class of all Σ-algebras (algebras over the ture Σ) will be denoted by $Alg(\Sigma)$. For any program P, the algebra it denotes is en as $[\![P]\!] \in Alg(Sig(P))$, where $Sig(P)$ is the underlying signature of P.

or any signature, we need a logical system for describing properties of algebras over signature. Many-sorted equational logic (cf. [GM 85, EM 85]) is the most commonly-system for this purpose, at least in the area of algebraic specification. Properties of ebras (or rather, of their operations) may be described by *equations* over Σ: we know it means for a Σ-algebra A to *satisfy* a Σ-equation φ, written $A \models \varphi$. This also mines a notion of *logical consequence*: a set of axioms Φ entails an equation φ, written φ, if every algebra that satisfies all the axioms in Φ also satisfies φ.

ery rarely in the process of program development does the user work with just a single ture: operations and sorts of data are renamed, added and hidden according to what s most suitable. To take account of this, signatures are equipped with a notion of *ture morphism* (cf. [EM 85]). A signature morphism $\sigma : \Sigma \to \Sigma'$ maps the sort and tion names of Σ to those of Σ'. This determines in a natural way a translation of Σ-equation φ to a Σ'-equation $\sigma(\varphi)$, and on the semantic level, a translation of any gebra $A' \in Alg(\Sigma')$ to its *reduct* $A'|_\sigma \in Alg(\Sigma)$. (Notice the change of direction!)

he above framework is often criticised (quite rightly!) as rather restrictive and cumber- to use in practice. Some important features of programs, for example non-termination igher-order functions, are difficult to model in algebras; equations are often not expres- nough to conveniently capture properties which one may want to state as requirements. nately, this deficiency is relatively easy to overcome using the concept of *institution*. concept was introduced by Goguen and Burstall [GB 84] to capture the informal notion ogical system and was strongly influenced by the understanding of this notion in the y of specifications (see [Bar 74] for an early account of *abstract model theory* covering similar ideas approached from the viewpoint of classical logic and model theory).

An institution defines a notion of a signature together with for any signature Σ, of Σ-sentences, a class of Σ-models and a satisfaction relation between Σ-models an sentences. Moreover, signatures come equipped with a notion of a signature morphism. signature morphism induces a translation of sentences and a translation of models (the l going into the opposite direction as above). The only semantic requirement is that whe change signatures using a signature morphism, the induced translations of sentences a models preserve the satisfaction relation. Many standard logical systems have been pres explicitly as institutions (cf. [GB 90]). It should be easy to see that in fact any usual lo system with a well-defined model theory may be put into this mould.

Everything below, barring concrete examples, works in the framework of an arbi institution, even though for the reader's convenience we avoid "institutional jargon" use more familiar terms (signature, algebra, axiom). Consequently, everything in this p applies to all of the many different concepts of "signature", "algebra" and "axiom" in the theory and practice of software specification. This is a general feature of our li research: most of the technicalities and ideas are parameterised by an arbitrary institu This results in "reusable" methodologies, theorems, and (ultimately) tools. This appr originated in [GB 84]; in [ST 87, ST 88a, ST 88b, ST 91a] various components of a frame for software specification and development have been elaborated following this idea.

Strict followers of the early approaches to algebraic specification might view this a alarming departure, and might protest that what we are doing is not algebraic specific at all. In our view the essential idea of algebraic specification is the stress on "alg like" models and the use of logical axioms to describe such models. The use of ord many-sorted algebras and equations is but a special case of this. Just as it was nece to generalise from classical single-sorted algebras to many-sorted algebras in order to with programs handling several kinds of data, it is necessary to adopt more compli models to deal with other features of programming languages (polymorphism, higher- functions, infinite behaviour, lazy evaluation, etc.). The essence is that we need a noti a semantic structure which is detailed enough to capture the program properties we wa analyse and abstract enough to make this analysis feasible. Moreover, to specify and re about programs, we need a logical system with a model theory based on such structur

2 Specifications

What is a specification? Clearly, since our aim is a formal approach to software develop specifications must be objects as formal as (for example) programs are. That is, we to have a formal language to write specifications down and to provide a vehicle on w formal techniques to manipulate specifications may be based. It is important for s *specification language* to provide a collection of convenient notational conventions whic easy to understand and use. One of the basic constituents of a specification will be a l axioms the specified program is required to satisfy.

Any specification language must be given a precise, formal semantics. Here, the question to ask is what the meanings of specifications are, i.e. what specifications de Whatever the full answer is, a specification at least determines the underlying signatu the specified system. For any specification SP, we write this signature as $Sig(SP)$. one may attempt to give a semantics of specifications on (at least) three different leve

Presentation level: a specification denotes a signature and a set of axioms over this signature (this set may be required to be finite or at least recursive or recursively enumerable). At this level, the meaning of a specification is close to the syntactic form in which specifications are written; the semantics extracts the axioms, resolves references to other specifications, etc.

Theory level: a specification denotes a signature and a set of axioms over this signature which is closed under logical consequence. At this level, the set of axioms is much larger than what has been written explicitly (it is typically infinite, usually not recursive and sometimes not r.e.); thus the meaning of a specification is no longer strictly syntactic. The semantics performs the closure under logical consequence.

Model-class level: a specification denotes a signature and a class of algebras over this signature. At this level, the meaning of a specification is entirely non-syntactic (except for the signature part). The semantics abstracts away from the axioms, taking into account only their possible realizations.

ultimate role of any specification is to describe a class of programs which we want iew as its correct realizations. Since we have already decided to model programs as oras, whichever one of these three levels we choose for the semantic domain, given natural mappings from presentations to theories and from theories to model classes, *ately specifications determine classes of algebras.*

'or any specification SP, the algebras that model programs which are considered to be ct realizations of SP are referred to as the *models* of SP, and the class of all models P is denoted by $[\![SP]\!] \subseteq Alg(Sig(SP))$. This semantics determines a notion of logical equence of a specification: a specification SP entails an axiom φ, written $SP \models \varphi$, if φ s in every model of SP.

)f course, a specification SP may admit a number of different program behaviours, so the class $[\![SP]\!]$ may contain non-isomorphic algebras (hence we cover so-called *loose* ifications). Or it might be empty.

'or any signature Σ, there is a Galois connection between classes of Σ-algebras and sets -axioms, assigning to any set of axioms the class of all algebras that satisfy them, and ny class of algebras the set of all axioms that hold in them (see [GB 84]). The closed ents of this Galois connection are theories, that is, sets of axioms determined by classes of oras; they are in one-to-one correspondence with closed (i.e., definable by sets of axioms) es of algebras. Therefore, it is obvious that as a semantic domain for specifications the ry level is less expressive than either the presentation or the model-class level. The latter are, however, incomparable: there are properties which can be naturally studied at the entation level (for example, finiteness of an axiomatisation) with no natural counterpart e model-class level, and vice versa.

t is not immediately obvious that working at the model-class level brings any essential fits over working with closed classes of algebras only, or equivalently, working at the ry level. It is not clear whether non-closed classes of algebras ever arise as meanings of ifications; even if they do arise, it is not clear whether this makes any difference for the of specifications. The following example exhibits both of these phenomena (we use some *ad hoc* notation for writing specifications with a hopefully clear intuitive interpretation

$$SP\left\{\begin{array}{l}\textbf{enrich}\\ \quad SP_1\left\{\begin{array}{l}\textbf{hide } a \textbf{ in}\\ \quad SP_0\left\{\begin{array}{ll}\textbf{sorts} & s, s'\\ \textbf{opns} & a : s\\ & b, c : s'\end{array}\right.\end{array}\right.\\ \textbf{by}\quad \forall x{:}s.\, b = c\end{array}\right.$$

This example relies on the following well-known phenomenon [GM 85]: $\forall x{:}s.\, b = c$ does imply $b = c$, although it implies $b = c$ for $Sig(SP)$-algebras with non-empty carrier of so

Now, at the model-class level, $[\![SP_0]\!]$ is the class of all algebras (over the indic signature) and $[\![SP_1]\!]$ consists of all algebras that are reducts of $Sig(SP_0)$-algebras, obta by forgetting the constant a. Consequently, $[\![SP_1]\!]$ contains only those algebras havi non-empty carrier of sort s. Then, selecting the algebras that satisfy $\forall x{:}s.\, b = c$ yields class $[\![SP]\!]$ — and all these algebras satisfy $b = c$ (since for the algebras in $[\![SP_1]\!]$, b follows from $\forall x{:}s.\, b = c$). Thus, under the model-class interpretation, the property $b =$ a consequence of the specification SP.

On the other hand, at the theory level, the theory of SP_0 is clearly the trivial equati theory, and so is the theory of SP_1 (there are no equations capable of expressing the fact a carrier is non-empty). Then, the additional axiom $\forall x{:}s.\, b = c$ in the context of the th of SP_1 does not entail the equation $b = c$. Thus, under the theory-level interpretation, b is *not* ensured by the specification SP.

This discrepancy (and similar examples one may construct without relying on the "er carriers" phenomenon) faces us with the necessity to choose between theories and class algebras as the basic semantic domain for specifications. The choice is obvious: the ob of ultimate interest here are programs, which are modelled as algebras, while axioms theories are nothing more than logical means for describing them. The lack of agreer between theories and classes of algebras shows that theories are *not* adequate as denotat of specifications.

The above specification SP is a trivial example of a *structured* specification, in whi complex specification is built from simpler ones. Any specification formalism must offer a possibility if it is to be used in practice: a specification of a real-life system must sta huge number of properties, and building such a specification in an unstructured, monol way would result in a long list of axioms which would be neither understandable nor us Moreover, the structure of a specification may be used to express intangible aspects of specifier's knowledge of the problem. For this purpose, a specification language must pro some *specification-building operations* used to put together small specifications to form r complex ones [BG 77]. Then, an understanding of a large specification is achieved vi understanding of its components. Since we have chosen classes of algebras as mean of specifications, such specification-building operations semantically correspond simpl functions on classes of algebras.

Choosing appropriate specification-building operations to be included in a specifica language is a non-trivial task (even though by now there is a list of typical ones that spe cation languages are based on). It involves a certain trade-off between the expressive p of the specification language and the ease of understanding and dealing with the operati One way to circumvent this problem is to first develop a *kernel* language consisting minimal set of very powerful, but perhaps difficult to use operations, and then build on a higher-level, more user-friendly language, perhaps sacrificing some of the expressive r to achieve ease of use and ease of understanding. Such an approach has been taken the ASL [SW 83, ST 88a, Wir 86] kernel specification language, on top of which such ication languages as PLUSS [Gau 84] and Extended ML [ST 91b] have been built.

'e should stress here that the internal structure of a specification should not constrain nal structure of its implementations. This is one of the consequences of the famous a that a specification should describe only the *what's* of the specified software without raining any of its *how's*. In fact, requiring the structure of the initial specification preserved in its implementation would be highly unrealistic and unreasonable, even ;h this has been explicitly suggested by some (e.g. [GB 80, Mor 90]) and is implicit e approach taken by others. The aims of structuring requirements specifications are contradictory with the aims of structuring software. See for instance [FJ 90] for a nice ssion of a practical example where such a discrepancy occurs.

Specification engineering

oint of constructing a specification is so that it may be used to define a programming by precisely delimiting the range of program behaviours that are to be regarded as issible. This statement of the programming task represents a rather idealized view that equirements specification we start with accurately reflects the real-life requirements the mer expects to be satisfied. Thus formulated, the programming task allows for no ;e to the original specification. Of course, this is not very realistic, and we envisage a programmer working with an appropriate support system will be able to try slightly ent ideas, to modify the original formal specification, to experiment with a whole bunch etter, a tree) of specifications.

he problem of getting the original requirements specification right is the topic of "re-ments engineering" [Par 91]. Even though there has been much work on this, we are onvinced that all the potential of the use of formal specifications has been explored. rally, there always will be a gap between the informal wishes of the customer and their lised version couched as a precise, unambiguous, entirely formal specification. Some ods involved in the process of passing from the former to the latter may and should more precise formulation.

ne major problem within this area is how we can make formal specifications easier to stand and construct. We have already mentioned a part of the answer in the previous n: we need good specification languages with good structuring operations allowing ications to be built and understood in a systematic, modular fashion.

nother aspect here is that once a specification is built, the customer should be able lay with" it, to test whether or not the specification indeed expresses the properties pects. A traditional approach to this is to engage a team of programmers to build *totype*, a quickly assembled but necessarily bad and simplified realization. This can be given to the customer to test. Of course, such an approach is irreplacable for some t of the software to be developed. For example, there can be no better way to test r interface than by playing with some version of it; going through sample sessions the system seems to be the only way for a user to get a feel for what working with ystem will be like. In general, however, the prototyping approach has a number of vantages. First, it involves some extra work to produce a system which is then thrown away. More importantly, if the original specification is loose (and it usually is) any protc will incorporate choices between the alternative behaviours permitted by the specifica and these choices need not necessarily be mirrored in the final implementation. Conseque the user may conclude that the system will have some properties which are not ensure the specification at all, and this undermines the sense of the whole exercise.

The overhead of prototyping may be avoided through the use of a rapid prototy system like RAP [Hus 85]. This demands that requirements specifications be written i executable specification language, not far from high-level programming languages like S dard ML [MTH 90]. In the fundamental trade-off between executability and expressive it is clearly the latter which is of central importance in a language intended for wr requirements specifications, so such a strong restriction seems highly undesirable.

We believe that for many purposes prototyping should be replaced by *theorem pr* (see [GH 80] for a similar observation). To check whether a given specification in embodies a desirable property, it seems most appropriate to state this property explicitly then try to prove that it is a consequence of the specification. This is the most general of specification testing activity; the more usual approaches via rapid prototyping, sym evaluation, term rewriting etc. can easily be seen as some special cases, or rather as special techniques of theorem proving applicable to some particular situations.

This indicates a need for good theorem provers. For use in this area, theorem pr should be able not only to derive consequences of a list of axioms ($\Phi \models \varphi$), but al derive consequences of a specification built in a structured way ($SP \models \varphi$). Theorem pr should be able to exploit the structure of specifications to guide proof search (cf. [S ST 88a, HST 89, Wir 92]). It would also be extremely useful for a theorem prover, i case where it fails to find a proof, to provide the user with readable information on wher proof attempts break down, and perhaps even how the specification may be augment make the proof go through — a desirable feature which few contemporary theorem pr exhibit.

4 Program development

Given a specification SP, the programming task it defines is to construct a program P v is a correct realization of SP, that is such that $[\![P]\!] \in [\![[SP]]\!]$.

There can be no universal recipe which would ensure successful development of a pro implementing a given specification. All we can hope to offer are methodologies, and pe some particular techniques and heuristics oriented towards specific problem areas.

Perhaps the most fundamental point is that it is not possible to leap in a single bound the gap between a high-level user-oriented requirements specification and the very sp realm of programs full of technical decisions and algorithmic details. Program develop should proceed systematically in a stepwise fashion, gradually enriching the original rec ments specification with more and more detail, incorporating more and more design implementation decisions. Such decisions include choosing between the options of beha left open by the specification, between the algorithms which realize this behaviour, bet data representation schemes, etc. Each such decision should be recorded separately, separate step hopefully consisting of a local modification to the specification. The pro development process is then a sequence of such small, easy to understand and easy to

:

$$SP_0 \rightsquigarrow SP_1 \rightsquigarrow \cdots \rightsquigarrow SP_n$$

ch a chain, SP_0 is the original requirements specification and $SP_{i-1} \rightsquigarrow SP_i$ for any $, \ldots, n$ is an individual *refinement* step. The aim is to reach a specification (here, SP_n) ı is an exact description of a program in full detail, with all the technical decisions porated (it may simply *be* a program, if our specification formalism is rich enough). ny formal definition of such refinement steps $SP \rightsquigarrow SP'$ must incorporate the require- that any correct final realization of SP' must be (or, somewhat more generally, must rise to) a correct realization of SP. Recalling that $[\![SP]\!]$ is the class of all admissible ations of SP, this leads to the following straightforward definition [SW 83, ST 88b]:

$$SP \rightsquigarrow SP' \quad \text{iff} \quad [\![SP']\!] \subseteq [\![SP]\!]$$

his definition ensures that the correctness of the final outcome of the stepwise develop- process may be inferred from the correctness of the individual refinement steps:

$$\frac{SP_0 \rightsquigarrow SP_1 \rightsquigarrow \cdots \rightsquigarrow SP_n \qquad A \in [\![SP_n]\!]}{A \in [\![SP_0]\!]}$$

proof is by an easy induction on the length of the refinement sequence.

otice that if the final specification SP_n represents an individual program P, then the usion that $A \in [\![SP_0]\!]$ for all $A \in [\![SP_n]\!]$ is equivalent to our original statement of the am development task: $[\![P]\!] \in [\![SP_0]\!]$.

n indirect way to prove the correctness of the final outcome is to notice a stronger fact, ly that consecutive refinements can be composed (referred to as "vertical composabil- GB 80]):

$$\frac{SP \rightsquigarrow SP' \qquad SP' \rightsquigarrow SP''}{SP \rightsquigarrow SP''}$$

he above gives a formal view of the stepwise development methodology. As mentioned e, there can be no universal recipe for coming up with useful refinements of a given fication — necessarily, this is the place where the developer's invention is required. a refinement step is proposed, though, we should be able to prove it correct, that is, ould have some formalism for proving the inclusion between the corresponding model es. Of course, this must incorporate a theorem prover for the underlying logic. A need which arises here is that such a formalism must also be able to prove entailments een two structured specifications (we write $SP' \models SP$ to state that every model of SP' nodel of SP, yet another formulation of $SP \rightsquigarrow SP'$ which we will use in this context). e structures of SP and SP' match exactly (and the specification-building operations are monotonic w.r.t. inclusion of model classes — this is typically the case) then this lem may be reduced to proving that individual axioms (from SP) are consequences rtain specifications (parts of SP') via the following fact (referred to as "horizontal osability" [GB 80] for the specification-building operation *op*):

$$\frac{SP_1 \rightsquigarrow SP_1' \qquad \cdots \qquad SP_n \rightsquigarrow SP_n'}{op(SP_1, \ldots, SP_n) \rightsquigarrow op(SP_1', \ldots, SP_n')}$$

rtunately, the structures of the two specifications need not coincide, which makes such luction very non-trivial. The only work on this important problem we are aware of is 92].

Another issue which may seem worrying here is that we have not put into our defir of refinement any requirement that the refined specification is consistent (that it has model). Indeed, this can be seen as a problem, since an inconsistent specification cann implemented by any program, and so it opens a blind valley in the program develop process. From this point of view, it would be worthwhile to be able to check consist of a specification as soon as it is formulated. Unfortunately, in general (for any suffici powerful specification framework) this is an undecidable property. Fortunately, inconsist of specifications cannot lead to incorrect programs: if we arrive at a program at some in the development process, then this program is by definition consistent (it has a u model) and consequently, all the specifications leading to it must have been consiste well.

The proposed methodology of stepwise refinement does not and cannot be expect guarantee success. Apart from inconsistencies, there are many sources of "blind val and failures in the development process: there might be no computable realization specification, there might be no "practically computable" realization, we might not be c enough to find a realization, we might run out of money to finish the project, etc. The feature of the methodology we really can ensure is its *safety*: if we arrive at a program, it is a correct realization of the original specification.

5 Constructor implementations

The simple notion of specification refinement is mathematically elegant and powerful en (in the context of a sufficiently rich specification language) to handle all concrete examp interest. However, it is not very convenient to use: in the practice of software developm certain constructions are used so often that it seems tempting to incorporate some treat of them in the notion of refinement. During the process of developing a program successive specifications incorporate more and more details arising from successive d decisions. Thereby, some parts become fully determined, and remain unchanged as a of the specification until the final program is obtained.

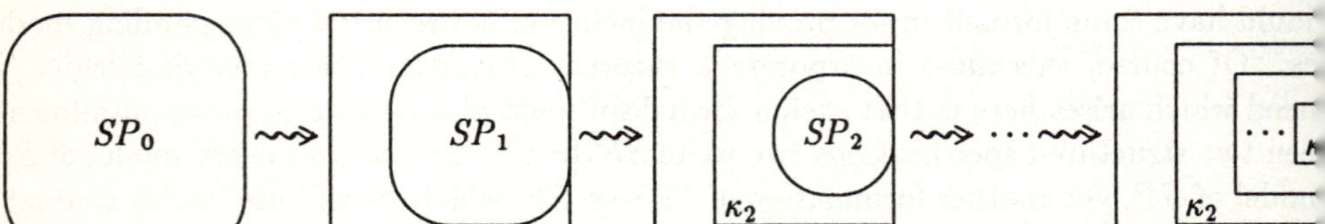

It is more convenient to avoid such clutter by putting the finished parts aside, and proce with the development of the unresolved parts only.

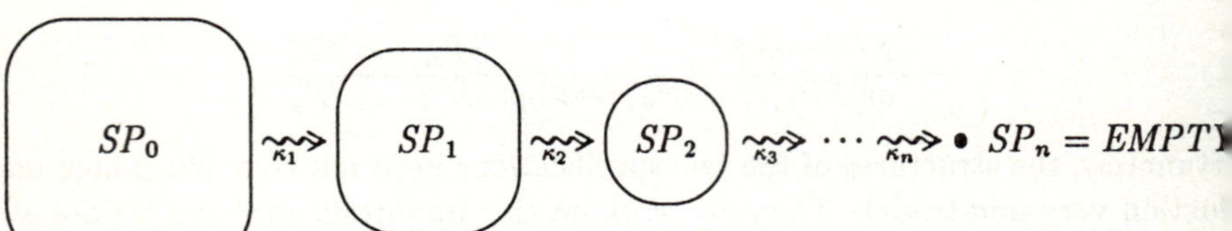

mportant for these finished parts to be independent of the particular choice of realization hat is left: they should act as constructions extending any realization of the unresolved to a realization of what is being refined. Semantically, each κ_i amounts to a function :h we will call a *constructor*) on algebras, $\kappa_i : Alg(Sig(SP_i)) \rightarrow Alg(Sig(SP_{i-1}))$. Once evelopment is finally finished (that is, when nothing is left unresolved) we can put the ructors together to obtain a correct realization of the original specification.

o formally capture the above considerations, we introduce the concept of *constructor* :*mentation* [ST 88b], a more elaborate version of the notion of refinement of the previous on. We write $SP \underset{\kappa}{\leadsto\!\!\!\!>} SP'$ to say that a specification SP' implements a specification SP constructor $\kappa : Alg(Sig(SP')) \rightarrow Alg(Sig(SP))$, and define this as follows:

$$SP \underset{\kappa}{\leadsto\!\!\!\!>} SP' \quad \text{iff} \quad \kappa([\![SP']\!]) \subseteq [\![SP]\!]$$

$\kappa([\![SP']\!])$ is the image of $[\![SP']\!]$ under κ. This suggests how a constructor κ may be :d as a specification-building operation $\overline{\kappa}$, so that constructor implementations may be :d as refinements ($SP \underset{\kappa}{\leadsto\!\!\!\!>} SP'$ is equivalent to $SP \leadsto\!\!\!\!> \overline{\kappa}(SP')$). Then proof techniques :finements may be applied to establish the correctness of constructor implementations. he correctness of the final outcome of the stepwise development process may be inferred the correctness of the individual constructor implementation steps:

$$\frac{SP_0 \underset{\kappa_1}{\leadsto\!\!\!\!>} SP_1 \underset{\kappa_2}{\leadsto\!\!\!\!>} \cdots \underset{\kappa_n}{\leadsto\!\!\!\!>} SP_n = EMPTY}{\kappa_1(\kappa_2(\ldots\kappa_n(\langle\rangle)\ldots)) \in [\![SP_0]\!]}$$

e *EMPTY* is the empty specification over the empty signature and $\langle\rangle$ is its unique ty) model.

is also easy to see that constructor implementations (vertically) compose:

$$\frac{SP \underset{\kappa}{\leadsto\!\!\!\!>} SP' \qquad SP' \underset{\kappa'}{\leadsto\!\!\!\!>} SP''}{SP \underset{\kappa';\kappa}{\leadsto\!\!\!\!>} SP''}$$

colon stands for functions composition, written in the diagrammatic order). As in the of refinement, vertical composability is not really necessary to ensure the correctness of evelopment process. All we need is the condition inherent in the definition of constructor :mentation, namely that implementations reflect realizations:

$$\frac{SP \underset{\kappa}{\leadsto\!\!\!\!>} SP' \qquad A' \in [\![SP']\!]}{\kappa(A') \in [\![SP]\!]}$$

ven though in the above we have viewed constructors as arbitrary functions on alge- in practice more restrictions should be imposed. In particular, we want constructors "effective": given a contructor implementation $SP \underset{\kappa}{\leadsto\!\!\!\!>} SP'$, the ability to compute in gebra $A' \in [\![SP']\!]$ should ensure the ability to compute in the algebra $\kappa(A') \in [\![SP]\!]$. is sense, κ may be viewed as a *parameterised program* [Gog 84] or, equivalently, as a lard ML *functor* [MTH 90] with input interface SP' and output interface SP. A pro- ming language which supports stepwise development in the style suggested here will de syntax and modularisation facilities for coding up such constructors. The modu- tion facilities must ensure that one may successively instantiate constructors in order semble the final program $\kappa_1(\kappa_2(\ldots\kappa_n(\langle\rangle)\ldots))$. This is a much weaker requirement than that constructors be composable (necessary for vertical composability of constr implementations in the specific programming language) — even though any program language with decent modularisation facilities should ensure the latter as well. It seems this stronger requirement becomes important when higher-order parameterised prog and their development from specifications are considered [SST 90].

Much work on vertical composability of implementations (see e.g. [EKMP 82, SV Ore 83]) has aimed at a still stronger requirement. This stems from the choice of a d tion of implementation which is similar to that of constructor implementation but req the constructor to be in a particular fixed form. Then the vertical composition of tw plementations must yield an implementation of the same form. The requirement tha composition of constructors be forced into some given normal form corresponds to req programs to be written in a rather restricted programming language which does not pr sufficiently powerful modularisation facilities for the job.

We have already mentioned that the internal structure of a requirements specific need not be mirrored by programs which realize it. This is why the definitions of refine and constructor implementation above take no account of the structure of specifica However, when developing a large program it is crucial to progressively decompose th into smaller tasks which can be handled separately. Each task is defined by a specifica and solving a task means producing a program component which satisfies this specifica Once all tasks are solved, then producing the final system is a matter of appropri assembling these components.

A development step involving the decomposition of a programming task into sep subtasks is modelled using a constructor implementation with a multi-argument constr

$$SP \underset{\kappa}{\rightsquigarrow} \langle SP_1, \ldots, SP_n \rangle \quad \textbf{iff} \quad \kappa([\![SP_1]\!], \ldots, [\![SP_n]\!]) \subseteq [\![SP]\!]$$

where $\kappa : Alg(Sig(SP_1)) \times \cdots \times Alg(Sig(SP_n)) \rightarrow Alg(Sig(SP))$ is an n-argument constr (an n-argument function on algebras) describing a way to put realizations of $SP_1, \ldots$ together to construct a realization of SP. Now the development process takes on a like shape. This process is finished once a tree is obtained which has empty (sequence specifications as its leaves:

$$SP \underset{\kappa}{\rightsquigarrow} \left\{ \begin{array}{l} SP_1 \underset{\kappa_1}{\rightsquigarrow} \langle\rangle \\ \vdots \\ SP_n \underset{\kappa_n}{\rightsquigarrow} \left\{ \begin{array}{l} SP_{n1} \underset{\kappa_{n1}}{\rightsquigarrow} \left\{ SP_{n11} \underset{\kappa_{n11}}{\rightsquigarrow} \langle\rangle \right. \\ \cdots \\ SP_{nm} \underset{\kappa_{nm}}{\rightsquigarrow} \langle\rangle \end{array} \right. \end{array} \right.$$

Then the composition of the constructors in the tree yields a realization of the ori requirements specification. The above tree yields:

$$\kappa(\kappa_1(), \ldots, \kappa_n(\kappa_{n1}(\kappa_{n11}()), \ldots, \kappa_{nm}())) \in [\![SP]\!].$$

The structure of the final program is determined by the shape of the development which is in turn determined by the decomposition steps. Each such step correspon what software engineers call a *design* specification: it defines the structure of the syste specifying its components and describing how they fit together. This style of develop leads to modular programs, built from fully specified, correct and reusable component orizontal composability for constructor implementations takes the form:

$$\frac{SP_1 \underset{\kappa_1}{\leadsto\!\!>} SP'_1 \quad \cdots \quad SP_n \underset{\kappa_n}{\leadsto\!\!>} SP'_n}{op(SP_1,\ldots,SP_n) \leadsto\!\!> op(\overline{\kappa_1}(SP'_1),\ldots,\overline{\kappa_n}(SP'_n))}$$

ite of this fact, which holds for all monotonic specification-building operations *op*, nposition of a specification $SP = op(SP_1,\ldots,SP_n)$ into separate tasks $SP_1,\ldots,SP_n$ t not be appropriate. It is possible for the design decisions taken in the solutions of these ate tasks to conflict so that even once we have obtained realizations of $SP_1,\ldots,SP_n$, it t not be possible to combine these to form a realization of SP. This problem cannot arise ecification-building operations corresponding to constructors, as in the decomposition via multi-argument constructors above.

Behavioural implementations

ecification should be a precise and complete statement of required properties. We 'd try to avoid including extra requirements, even if they happen to be satisfied by ssible future realization. Such over-specification unnecessarily limits the options left to the implementer. Ideally, the target is to describe exactly the admissible program viours. This suggests that specifications of programming tasks should not distinguish een programs (modelled as algebras) exhibiting the same behaviour.

he intuitive idea of *behaviour* of an algebra has been formalised in a number of ways (see Rei 81, GM 82, SW 83, Sch 87, ST 87]). In most approaches one distinguishes a certain *BS* of sorts as *observable*. Intuitively, these are the sorts of data directly visible to the (integers, booleans, characters, etc.) in contrast to sorts of "internal" data structures, n are observable only via the functions provided by the program. The behaviour of an ra is characterised by the set of *observable computations* taking arguments of sorts in and producing a result of a sort in OBS. Two Σ-algebras A and B are *behaviourally alent* (w.r.t. OBS), written $A \equiv B$, if they exhibit the same behaviour, that is, if all vable computations yield the same results in A and in B. The motivation is related to of so-called *testing* equivalences studied in the context of concurrent systems [DH 84].

hackneyed example is that of stacks of integers, with the usual operations (`empty`, `pty`, `push`, `pop`, `top`). The sorts `int` and `bool` are observable while the sort `stack` is The observable computations are all the terms of the form `isempty(`s`)` and `top(`s`)` e s is a term of sort `stack` with variables (representing the inputs) of sort `int` only. , all intuitively acceptable realizations of stacks are behaviourally equivalent, since for observable computation (like `top(pop(push(`n`,push(4,push(6,empty)))))`) they all er the same result (in this case 4). However not all of these algebras act the same way non-observable computations are considered: for example, the computations `empty` `pop(push(`n`,empty))` yield the same result when stacks are represented as lists, but yield different results when a stack is represented by an array with a pointer to the top ent (the latter computation then leaves n in the position above the pointer).

ur earlier discussion would lead us to expect the class of models of a specification to osed under behavioural equivalence. It is perhaps surprising that this is not easy to ve directly: the class of models of a set of axioms typically does not have this property. tional logic may be modified so as to force this to happen (cf. [NO 88]) but we see raightforward way to achieve this for most logical systems. Instead, we suggest simply

closing the class of models of a specification under behavioural equivalence [SW 83, ST
Any specification SP determines the class $[\![SP]\!] \subseteq Sig(SP)$ of models which "literally" sa
the stated requirements; the ultimate semantics of SP is taken to be the closure of this u
behavioural equivalence:

$$[\![\![SP]\!]\!] = \{A \mid A \equiv B \text{ for some } B \in [\![SP]\!]\}$$

If SP is a list of axioms, then $[\![SP]\!]$ contains exactly the algebras which satisfy all the axi
while $[\![\![SP]\!]\!]$ contains also the algebras which do not satisfy the axioms themselves bu
behaviourally equivalent to algebras which do. For example, if the specification ST
consists of the usual stack axioms, then both the list representation and the array-
pointer representation of stacks are in $[\![\![STACK]\!]\!]$, even though the latter does not lite
satisfy the axiom pop(push(n,s)) $= s$ and so is not in $[\![STACK]\!]$. This approach
extra expressive power to the system: there are classes of algebras which may be fin
characterised in this way, and which cannot be finitely axiomatised directly [Sch 91].
model-oriented specifications [Jon 86] can be handled: if $[\![SP]\!]$ contains just a single alg
$[\![\![SP]\!]\!]$ admits any realisation of the exhibited behaviour.

The basic intuition for the use of behavioural equivalence in the development proce
that it is not necessary to implement a specification SP according to its literal interpret
$[\![SP]\!]$; it is sufficient to implement it up to behavioural equivalence, as captured by its "
mate" semantics $[\![\![SP]\!]\!]$. The crucial novelty, due to [Sch 87], is that when *using* a realiz
of SP, it is convenient (and possible) to pretend that it satisfies the literal interpretati
SP. For example, consider the following program:

```
multipush(n,s) = if n=0 then s else multipush(n-1,push(anything,s))
multipop(n,s) = if n=0 then s else multipop(n-1,pop(s))
id(x,n) = top(multipop(n,multipush(n,push(x,empty))))
```

Given any realization of $STACK$, to verify that id(x,n) $= x$ for all x and $n \geq$
is convenient to assume that the axiom pop(push(n,s)) $= s$ holds literally — th
simple proof by induction goes through — in spite of the fact that this equation is
valid in $[\![\![STACK]\!]\!]$. These considerations lead to the following definition of *behavi
implementation* [ST 88b]:

$$SP \overset{\equiv}{\underset{\kappa}{\rightsquigarrow}} SP' \quad \text{iff} \quad \kappa([\![SP']\!]) \subseteq [\![\![SP]\!]\!]$$

The alert reader will have noticed that there is a problem here: we want to have our cake
eat it. On one hand, we want to allow specifications to be implemented up to behavic
equivalence; on the other hand, we would like to use any realization as if it satisfied its s
ification literally. Behavioural implementations do not compose, and the following cr
property is lost:

$$\frac{SP \overset{\equiv}{\underset{\kappa}{\rightsquigarrow}} SP' \qquad [\![P]\!] \in [\![\![SP']\!]\!]}{[\![P]\!] \in [\![\![SP]\!]\!]}$$

The behavioural implementation $SP \overset{\equiv}{\underset{\kappa}{\rightsquigarrow}} SP'$ ensures only that algebras in $[\![SP']\!]$ give
to correct realizations of SP; this says nothing about the models in $[\![\![SP']\!]\!]$ which are n
$[\![SP']\!]$.

It might seem that all is lost. But there is a way out, originally suggested in [Sch 87].
above crucial property is recovered if we assume that the constructors used are *stable*, th
that any constructor $\kappa : Alg(Sig(SP')) \rightarrow Alg(Sig(SP))$ preserves behavioural equivale

Stability assumption: if $A \equiv B$ then $\kappa(A) \equiv \kappa(B)$

exact definition of stability of constructors in a formal development framework based full-blown programming language is somewhat more complex — see [Sch 87, ST 89]).
nder this assumption, the correctness of the individual implementation steps ensures orrectness of the result:

$$\frac{SP_0 \underset{\kappa_1}{\overset{\equiv}{\leadsto}} SP_1 \underset{\kappa_2}{\overset{\equiv}{\leadsto}} \cdots \underset{\kappa_n}{\overset{\equiv}{\leadsto}} SP_n = EMPTY}{\kappa_1(\kappa_2(\ldots\kappa_n(\langle\rangle)\ldots)) \in [\![SP_0]\!]}$$

ould repeat here the tree-like development picture of Section 5 — developments involving mposition steps based on behavioural implementations with multi-argument (stable) tructors yield correct programs as well. We also recover vertical composability:

$$\frac{SP \underset{\kappa}{\overset{\equiv}{\leadsto}} SP' \qquad SP' \underset{\kappa'}{\overset{\equiv}{\leadsto}} SP''}{SP \underset{\kappa';\kappa}{\overset{\equiv}{\leadsto}} SP''}$$

he correctness of a behavioural implementation $SP \underset{\kappa}{\overset{\equiv}{\leadsto}} SP'$ is easier to verify than the ctness of the corresponding constructor implementation $SP \underset{\kappa}{\leadsto} SP'$ ($\kappa([\![SP']\!]) \subseteq [\![SP]\!]$ aker than $\kappa([\![SP']\!]) \subseteq [\![SP]\!]$). We are still left, though, with the need to establish the lity of constructors, and so one may wonder if it is worthwhile taking advantage of However, the important point is that the constructors which may be used in program lopment are determined by the particular programming language to be used. Thus stacan be checked in advance, for the programming language as a whole (this is simplified what by the fact that the composition of stable constructors is stable) and this frees rogrammer from the need to prove it during the program development process.

here is a close connection between the requirement of stability and the security of psulation mechanisms in programming languages supporting abstract data types. A ramming language ensures stability if the only way to access an encapsulated data type a the operations explicitly provided in its output interface. This suggests that stability nstructors is an appropriate thing to expect; following [Sch 87] we view the stability irement as a methodologically justified design criterion for the modularisation facilities ogramming languages.

Final remarks

have outlined the main ideas of a framework to support the formal development of ect programs from specifications. One of the central themes we did not have space iscuss is the vital role of parameterisation in specification and formal development. wing the technicalities in [SST 90], the framework outlined here may be extended to deal the development of parameterised programs (modelled as functions on algebras) from specifications. Extra flexibility is obtained by allowing higher-order parameterisation. ifications of such parameterised programs (denoting classes of functions on algebras) ld be carefully distinguished from parameterised specifications (denoting functions on es of algebras).

rinciple among the areas in which further work is required is the development of adte proof technology. As mentioned earlier, much more than methods for proving conences of unstructured sets of assumptions is required: we need to prove consequences of structured specifications, entailment between structured specifications, and correct of behavioural implementation steps. Soundness with respect to the underlying m theoretic semantics of specifications and of implementation steps is essential; complete is unfortunately not achievable. In spite of some recent work, it is not clear how to scal the methods and techniques which exist for the unstructured case (e.g. in the area of rewriting) to deal with large structured specifications and with behavioural equivalenc

The main challenge now is to put these ideas into practice in the formal developme non-trivial programs in real programming languages. The above comments concerning p technology indicate just a part of what is required. We are moving in this direction witl work on the Extended ML framework for the formal development of modular Standard programs [ST 91b], though more effort is required. Subjecting foundational work to the of practice is sure to bring fascinating new problems and issues to light.

Acknowledgements: Most of the above ideas have been presented elsewhere, and have discussed with and influenced by many of our colleagues. Thanks especially to Rod Burstall, Farrés, Joseph Goguen, Fernando Orejas, Oliver Schoett and Martin Wirsing. Partial suppor writing this paper was provided by the ESPRIT-funded COMPASS working group and by a g from the UK Science and Engineering Research Council.

References

[Bar 74] J. Barwise. Axioms for abstract model theory. *Ann. Math. Logic* 7, 221–265 (1974).

[BKLOS 91] M. Bidoit *et al* (eds.) *Algebraic System Specification and Development: A Survey Annotated Bibliography.* Springer LNCS 501 (1991).

[BG 77] R. Burstall and J. Goguen. Putting theories together to make specifications. *Proc Intl. Joint Conf. on Artificial Intelligence*, Cambridge, Massachusetts, 1045–1058 (1977

[DH 84] R. De Nicola and M. Hennessy. Testing equivalences for processes. *Theoretical Comp Science* 34, 83–133 (1984).

[EKMP 82] H. Ehrig, H.-J. Kreowski, B. Mahr and P. Padawitz. Algebraic implementatic abstract data types. *Theoretical Computer Science* 20, 209–263 (1982).

[EM 85] H. Ehrig and B. Mahr. *Fundamentals of Algebraic Specification I: Equations and I Semantics.* Springer (1985).

[Far 92] J. Farrés-Casals. Verification in ASL and Related Specification Languages. Ph.D. tl Dept. of Computer Science, Univ. of Edinburgh, to appear (1992).

[FJ 90] J. Fitzgerald and C. Jones. Modularizing the formal description of a database sys *Proc. VDM'90 Conference*, Kiel. Springer LNCS 428 (1990).

[Gau 84] M.-C. Gaudel. A first introduction to PLUSS. Technical report, LRI, Université de P Sud, Orsay (1984).

[Gog 84] J. Goguen. Parameterized programming. *IEEE Trans. on Software Engineering* 10(5), 528–543 (1984).

[GB 80] J. Goguen and R. Burstall. CAT, a system for the structured elaboration of co programs from structured specifications. Technical report CSL-118, SRI International (1

[GB 84] J. Goguen and R. Burstall. Introducing institutions. *Proc. Logics of Programming W shop*, Carnegie-Mellon. Springer LNCS 164, 221–256 (1984).

[GB 90] J. Goguen and R. Burstall. Institutions: abstract model theory for specification programming. Report ECS-LFCS-90-106, Univ. of Edinburgh (1990). *J. ACM*, to appe

[GM 82] J. Goguen and J. Meseguer. Universal realization, persistent interconnection and in mentation of abstract modules. *Proc. Intl. Colloq. on Automata, Languages and Prog ming*, Aarhus. Springer LNCS 140, 265–281 (1982).

[GM 85] J. Goguen and J. Meseguer. Completeness of many-sorted equational logic. *Hou Journal of Mathematics* 11(3), 307–334 (1985).

0] J. Guttag and J. Horning. Formal specification as a design tool. *Proc. ACM Symp. on Principles of Programming Languages*, Las Vegas, 251–261 (1980).

89] R. Harper, D. Sannella and A. Tarlecki. Structure and representation in LF. *Proc. 4th IEEE Symp. on Logic in Computer Science*, Asilomar, 226–237 (1989).

35] H. Hußmann. Rapid prototyping for algebraic specifications: RAP system user's manual. Report MIP-8504, Universität Passau (1985).

36] C. Jones. *Systematic Software Development using VDM.* Prentice Hall (1986).

90] R. Milner, M. Tofte and R. Harper. *The Definition of Standard ML.* MIT Press (1990).

90] J. Morris. A methodology for designing and refining specifications. *Proc. 3rd Refinement Workshop*, Hursley Park. Springer BCS Workshop series (1990).

38] P. Nivela and F. Orejas. Initial behaviour semantics for algebraic specifications. *Selected Papers from the 5th Workshop on Specification of Abstract Data Types*, Gullane. Springer LNCS 332, 184–207 (1988).

33] F. Orejas. Characterizing composability of abstract implementations. *Proc. 1983 Intl. Conf. on Foundations of Computation Theory*, Borgholm. Springer LNCS 158 (1983).

91] H. Partsch. *Requirements Engineering.* Handbuch der Informatik. Oldenbourg (1991).

31] H. Reichel. Behavioural equivalence — a unifying concept for initial and final specification methods. *Proc. 3rd Hungarian Computer Science Conference*, 27–39 (1981).

3] D. Sannella and R. Burstall. Structured theories in LCF. *Proc. Colloq. on Trees in Algebra and Programming*, L'Aquila. Springer LNCS 159, 377–391 (1983).

90] D. Sannella, S. Sokołowski and A. Tarlecki. Toward formal development of programs from algebraic specifications: parameterisation revisited. Report 6/90, FB Informatik, Universität Bremen (1990). *Acta Informatica*, to appear.

7] D. Sannella and A. Tarlecki. On observational equivalence and algebraic specification. *J. of Computer and System Sciences* 34, 150–178 (1987).

8a] D. Sannella and A. Tarlecki. Specifications in an arbitrary institution. *Information and Computation* 76, 165–210 (1988).

8b] D. Sannella and A. Tarlecki. Toward formal development of programs from algebraic specifications: implementations revisited. *Acta Informatica* 25, 233–281 (1988).

9] D. Sannella and A. Tarlecki. Toward formal development of ML programs: foundations and methodology. *Proc. 3rd Joint Conf. on Theory and Practice of Software Development*, Barcelona. Springer LNCS 352, 375–389 (1989).

1a] D. Sannella and A. Tarlecki. A kernel specification formalism with higher-order parameterisation. *Proc. 7th Intl. Workshop on Specification of Abstract Data Types*, Wusterhausen. Springer LNCS 534, 274–296 (1991).

1b] D. Sannella and A. Tarlecki. Extended ML: past, present and future. *Proc. 7th Intl. Workshop on Specification of Abstract Data Types*, Wusterhausen. Springer LNCS 534, 297–322 (1991).

3] D. Sannella and A. Tarlecki. *Foundations of Algebraic Specifications and Formal Program Development.* Cambridge Univ. Press, to appear (1993?).

82] D. Sannella and M. Wirsing. Implementation of parameterised specifications. *Proc. Intl. Colloq. on Automata, Languages and Programming*, Aarhus. Springer LNCS 140, 473–488 (1982).

83] D. Sannella and M. Wirsing. A kernel language for algebraic specification and implementation. *Proc. 1983 Intl. Conf. on Foundations of Computation Theory*, Borgholm. Springer LNCS 158, 413–427 (1983).

87] O. Schoett. Data Abstraction and the Correctness of Modular Programming. Ph.D. thesis, report CST-42-87, Dept. of Computer Science, Univ. of Edinburgh (1987).

91] O. Schoett. An observational subset of first-order logic cannot specify the behaviour of a counter. *Proc. 8th Symp. on Theoretical Aspects of Computer Science*, Hamburg. Springer LNCS 480, 499–510 (1991).

86] M. Wirsing. Structured algebraic specifications: a kernel language. *Theoretical Computer Science* 42, 123–249 (1986).

92] M. Wirsing. Structured specifications: syntax, semantics and proof calculus. Technical report, Universität Passau (1992).

PROGRAM COMPOSITION via UNIFICATION

Limor Fix   Nissim Francez   Orna Grumberg
Computer Science Department, Technion, Haifa 32000, Israel

Abstract

Program composition and compositional proof systems have proven themselves important for simplifying the design and the verification of programs. The paper extends work presented in ICALP'91 and presents a version of the *jigsaw* program composition operator previously defined in [FFG90, FFG91]. Here, the jigsaw operator is defined as the unification of its components by their most general unifier. The jigsaw operator generalizes and unifies the traditional *sequential* and *parallel* program composition operators and the newly proposed *union* and *superposition* operators. We consider a family of frameworks each consisting of a programming language, a specification language and a compositional syntax-directed proof system. We present syntactic rules to augment any given framework in the family with the jigsaw operator. The augmented framework is syntax-directed and compositional. Moreover, it is sound and complete with respect to the given framework.

1 Introduction

Program composition and compositional proof systems have proven themselves import for simplifying the design and verification of programs. In this approach, small parts software system are programmed and verified separately and then composed into a la system. To develop a system such that its syntactic structure reflects the intuitive st ture of the algorithm it implements, program composition operators able to reflect comr relations among components of algorithms are required. Traditional program composit operators are the sequential and the parallel composition operators. Recently, other progr composition operators have been suggested [BF88, CM88, FF90, KFE90].

In this paper a composition operator, called *jigsaw*, is considered. The jigsaw composit operator generalizes and unifies the *sequential* and the *parallel* operators together with newly proposed *union* and *superposition* operators [BF88,CM88].

Two versions of the jigsaw operator have already been presented and studied in [FFG FFG91]. Both versions present a composition operator which composes *gapped programs* gapped program is a program in the language under consideration which includes scatte unspecified statements, to be called *gaps*. The jigsaw operator composes two gapped p grams by "filling" gaps in a component with statements taken from the other compone In [FFG90] the jigsaw operator is added to a CSP-like programming language [Hoa78] a proof rule for partial correctness of a jigsaw-composed program is presented. In [FFG the jigsaw operator is added to a UNITY-like programming language and rules for prov partial correctness and termination of a jigsaw-composed program are presented. In [FFG union and superposition are shown to be special cases of the jigsaw composition.

Here, the jigsaw composition of two gapped programs is defined as their **unification** by r most general unifier. A gap in a program is represented by a (meta) variable ranging programs. A gapped program is a term built over primitive objects, variables and comtion operators. The jigsaw composition of two gapped programs T_1 and T_2 is well-defined d only if those programs are unifiable. The unification procedure simultaneously "fills" of T_1 with sub-programs of T_2 and gaps of T_2 with sub-programs of T_1. In many cases, jigsaw supports development of programs which are syntactically structured according to intuitive structure of the algorithm they implement. Therefore, augmenting an existing ework which does not support such a development with the jigsaw is desirable.

We present a method to extend any given compositional syntax-directed proof system the jigsaw operator. An abstract framework is presented such that any concrete proof em which fits the framework is syntactically extended to include the jigsaw. First, a hod to extend the basic programming language with the jigsaw operator is presented. n, mixed-specifications [Wir71, Bac80, Heh84, Old86] are defined over the basic specifions and constructors. Gapped programs are specified by mixed-specifications. A mixed ification has two syntactic parts, a *promise* and a *conclusion*. The promise introduces mptions on all gaps of the specified gapped program and the conclusion specifies the vior of the gapped program under the assumptions imposed by the promise. Finally, ethod is presented to extend the basic proof system into a system in which satisfaction pecification can be proved for gapped programs. The extended proof system is syntaxcted and compositional, i.e., a specification for a jigsaw-composed program is verified d on the specifications of its components.

The development of a program in the extended system goes through intermediate stages hich gapped programs are built and verified. The program at the final stage must be ogram in which all gaps are bound by the jigsaw operators in the program to gap-free programs. The program at the final stage is therefore always semantically equal to a gram in the basic system. The main theorem of the paper states that the extended system und and complete with respect to the basic system. More formally, given two programs, the basic system and T' in the extended one, such that T' is semantically equal to T, prove that if a specification S can be verified for T' then S is semantically true for T, if S is semantically true for T then it can be verified for T'.

We present an example of a concrete proof system which fits the abstract framework, and actically derive an extended version of it containing the jigsaw operator. The concrete of system is a system for proving partial correctness for CSP-like programs [ZRB85]. The ification language is a two sorted first order predicate language. In the extended system develop a small program consisting of a *Producer* and a *Consumer*. Using the jigsaw ator we separate the development and the verification of the communication management from the development and verification of the other functions of the program.

The rest of the paper is organized as follows. Section 2 presents the abstract basic system. ion 3 introduces the jigsaw-extended system. Finally, Section 4 contains an example of nding a concrete system with the jigsaw operator.

The basic system

follow [KKZ90] and present an abstract framework for the development and verifica- of programs. This framework forms the starting point of our work, to be called the c abstract framework. The framework is defined by introducing programs (in a common braic setup), specifications and correctness formulas over programs and specifications. ddition, a compositional syntax-directed proof system to derive correctness formulas is defined.

Definition: Let $\mathcal{O}$ be a set of primitive objects and let $\mathcal{C}$ be a set of constructors, w $C \in \mathcal{C}$ is of finite and fixed arity. The set of *programs* (terms), $\mathcal{T}$, defined over $\mathcal{O}$ ar is the smallest set s.t. $\mathcal{O} \subseteq \mathcal{T}$ and if $T_1, \ldots, T_n \in \mathcal{T}$, $C \in \mathcal{C}$ and C is of arity n $C(T_1, \ldots, T_n) \in \mathcal{T}$. □

Let $\mathcal{B}$ be a set of semantic objects (behaviors). Let $\widehat{\wp(\mathcal{B})} \subseteq \wp(\mathcal{B})$ be a subset of the p set of $\mathcal{B}$ which has a $\subseteq$-minimal element, denoted by Θ.

Definition: For $\mathcal{T}$ and $\mathcal{B}$ as above, the *semantics* of $\mathcal{T}$ is given by a function $BEH_{\mathcal{T}}$: $\widehat{\wp(\mathcal{B})}$. Such a semantics is *compositional* if for every $C \in \mathcal{C}$ there exists a total func $K_C : (\widehat{\wp(\mathcal{B})})^n \rightarrow \widehat{\wp(\mathcal{B})}$ s.t. $BEH_{\mathcal{T}}[C(T_1, \ldots, T_n)] = K_C(BEH_{\mathcal{T}}[T_1], \ldots, BEH_{\mathcal{T}}[T_n])$. □

The structure of the basic specifications is not explicitly defined and we assume existence of a semantic function which maps the specifications into the set $\widehat{\wp(\mathcal{B})}$. More we assume that a specification does not refer to any program and the association betwe specification and a program is made only in the correctness formulas (defined in the seq

Definition: Let $\mathcal{S}$ be a set of specifications closed under conjunction, $\wedge$. The *semantics* is given by a function $BEH_{\mathcal{S}} : \mathcal{S} \rightarrow \widehat{\wp(\mathcal{B})}$, such that: $BEH_{\mathcal{S}}[S_1 \wedge S_2] = BEH_{\mathcal{S}}[S_1] \bigcap BEH$ A specification S is a *tautology* if $BEH_{\mathcal{S}}[S] = \mathcal{B}$. □

Since both programs and specifications are mapped to the same set of behaviors, ℘ we can relate a program to a specification by relating their respective behaviors.

Definition:

1. The set Φ of *correctness formulas* is the set of formulas of the form "$T\ Sat\ S$", w $T \in \mathcal{T}$ and $S \in \mathcal{S}$.

2. A *structure* is a triple $M = < \mathcal{B}, BEH_{\mathcal{T}}, BEH_{\mathcal{S}} >$, where $\mathcal{B}$ is a set of behaviors $BEH_{\mathcal{T}}$ and $BEH_{\mathcal{S}}$ are semantic functions for $\mathcal{T}$ and $\mathcal{S}$ over the same domain, ℘(

3. A formula "$T\ Sat\ S$"$\in \Phi$ *holds* in a *structure* M, denoted by $M \models T\ Sat\ S$ $BEH_{\mathcal{T}}[T] \subseteq BEH_{\mathcal{S}}[S]$. When there is no confusion we write $\models T\ Sat\ S$.

Definition: A *compositional proof system* $\mathcal{AR} = < \mathcal{A}, \mathcal{R}, \mathcal{G} >$ for Φ consists of:

1. A (recursive) set $\mathcal{A} \subseteq \Phi$ of formulas called *axioms*, s.t. if "$T\ Sat\ S$" $\in \mathcal{A}$ then $T \in$

2. A set $\mathcal{G}$ of predicates s.t. for every $C \in \mathcal{C}$ of arity n, there is $G_C \in \mathcal{G}$ on $\mathcal{S}^{n+1}$.

3. A set $\mathcal{R}$ of *proof-rules*, s.t. $\forall C \in \mathcal{C}$ there exists a unique rule $R_C \in \mathcal{R}$ of the form:

$$\frac{T_i\ Sat\ S_i,\ i = 1, n,\ \ G_C(S_1, \ldots, S_n, S)}{C(T_1, \ldots, T_n)\ Sat\ S}$$

4. A way to determine whether a predicate in $G_C \in \mathcal{G}$ of arity n holds for a se specifications $S_1, \ldots, S_n$.

Definition: A formula $\phi \in \Phi$ is *provable* under no assumption in $\mathcal{AR}$, denoted by $\vdash_{\mathcal{AR}}$ ϕ can be derived from axioms in $\mathcal{A}$ by applications of rules in $\mathcal{R}$.

Definition: A proof system $\mathcal{AR}$ is *sound* w.r.t. a structure M if for each formula $\phi \in$ $\vdash_{\mathcal{AR}} \phi$ implies that $M \models \phi$.

Definition: A proof system $\mathcal{AR}$ is *relatively complete* [1] w.r.t. a structure M if for formula $\phi \in \Phi : M \models \phi$ implies that $\vdash_{\mathcal{AR}} \phi$.

[1] We assume, as usual, that all domain properties that are true can be used as additional axioms [Ap and that all intermediate assertions needed for the proof can be expressed in $\mathcal{S}$.

The *jigsaw*-extended system

The programming language

t, gapped programs and the jigsaw operator, denoted by '#', are presented. Gapped grams are defined by extending the program-algebra with a set of meta variables, $\mathcal{V}$ h $\mathcal{V} \cap \mathcal{O} = \emptyset$). The program $\#(T_1, T_2)$ is well defined only if T_1 and T_2 are *unifiable*, i.e., e exists a substitution θ which binds meta variables to sub-programs, such that applying both T_1 and T_2 results in identical programs, $T_1\theta = T_2\theta$. The meaning of the program $_1, T_2)$ is defined as the meaning of the program obtained by unifying T_1 and T_2 using r *most general unifier.* A substitution θ is the most general unifier of T_1 and T_2 if and if $T_1\theta = T_2\theta$ and moreover for every substitution γ, if $T_1\gamma = T_2\gamma$ then there exists a stitution λ such that $\theta o \lambda = \gamma$, where o stands for composition of substitutions. Intuitively, unification of T_1 and T_2 results in a program T which is the *mutual instantiation* of T_1 T_2, i.e., T can be obtained by a substitution of sub-programs of T_1 for the structurally esponding meta variables of T_2, and vice versa.

inition: Let $\mathcal{O}$ be the basic set of primitive objects, let $\mathcal{C}^{\#} = \mathcal{C} \cup$ '#', and let $\mathcal{V}$ be t of meta variables ranging over programs. The set of *gapped programs*, $\mathcal{T}^{\#}$ (over $\mathcal{O}$ $\mathcal{C}$), is the smallest set s.t. $\mathcal{O} \subset \mathcal{T}^{\#}$, $\mathcal{V} \subset \mathcal{T}^{\#}$ and if $C \in \mathcal{C}^{\#}$, $T_1, \ldots, T_n \in \mathcal{T}^{\#}$ then $_1, \ldots, T_n) \in \mathcal{T}^{\#}$. [2] □

Neither a meaning of a meta variable nor a meaning, in the usual way, of a gapped gram, are defined. Below, a meaning is assigned to a gapped program in a *context.* A text imposes assumptions on the behaviors of the expected "fillings" of the gaps.

The specification language and the correctness formulas

t, a new set of specifications, $\mathcal{S}^{\#}$, and a new set of correctness formulas, are presented. ew correctness formula, of the form "$T\ Sat^{\#}\ S$", binds a gapped program to a specifi- on $S \in \mathcal{S}^{\#}$. In such a formula the specification S defines both a context for T and a racterization of T.

Let T be a gapped program. A meta variable of T is *free* if it is not bound by a jigsaw rator to any sub-program. T can be specified using an *assume-guarantee* [Lam83, Pnu85] e of specification. In such a specification the behavior of T is characterized based on umptions on the behavior of the sub-programs to be substituted for the free variables of hen jigsaw-composed with another program.

We define a set of *assumptions*, $\overline{\mathcal{S}}$, which contains exactly all the specifications of $\mathcal{S}$, er-lined. The upper-line is used to distinguish an assumption from a basic specification. s distinction is needed to define composition of specifications (more in the sequel).

The new set of specifications, $\mathcal{S}^{\#}$, is a set of *mixed-specifications.* A mixed-specification tains two parts separated by a "$\triangleright$" sign. The expression on the right-side of the "$\triangleright$" is a basic specification, considered as the *conclusion* of the specification. The expression the left-side of the "$\triangleright$" sign is syntactically defined as a member of an algebra with a $\mathcal{C}^{\#}$ of operators over the sets of basic specifications and assumptions. This expression is sidered as the *promise* of the specification.

A gapped program T satisfies a mixed-specification S iff sub-programs of T satisfy the *responding* sub-specifications in the promise of S (if any) and T satisfies the conclusion S under the assumption that the free variables of T (if any) satisfy the corresponding umptions in the promise of T. For example, the gapped-program $T_1 :: C_1(C_2(a_1, a_2), X)$

Since a variable denotes a gap in a specific location in the program we assume that each variable appears ost once in a given program.

satisfies the mixed-specification $C_1(S_1, \overline{S_2}) \rhd S_3$ iff $C_2(a_1, a_2)$ satisfies S_1, X is assume satisfy $\overline{S_2}$ and moreover T_1 satisfies S_3 under the assumption that X satisfies $\overline{S_2}$.

Next, we exemplify how mixed-specifications enable a compositional verification jigsaw-composed program. Let T_1 be the above program and $T_2 :: C_1(Y, a_3)$. Assume specification $C_1(S_1, \overline{S_2}) \rhd S_3$ has been verified for T_1 and the specification $C_1(\overline{S_4}, S_5)$ has been verified for T_2. The program $\#(T_1, T_2)$ is semantically equal to the prog $C_1(C_2(a_1, a_2), a_3)$ obtained by substituting $C_2(a_1, a_2)$ for the meta variable Y and subst ing a_3 for the meta variable X. If S_1 and $\overline{S_4}$ are related in such a way that based on the that $C_2(a_1, a_2)$ satisfies S_1 we can conclude that $C_2(a_1, a_2)$ also satisfies $\overline{S_4}$, S_1 *confirm* then it is possible to conclude that $\#(T_1, T_2)$ satisfies S_6. In a similar way, if S_5 confirm then it is possible to conclude that $\#(T_1, T_2)$ satisfies S_3.

Next, we formally define the new specification language.

Definition:

1. The set $\overline{\mathcal{S}} = \{\overline{S} | S \in \mathcal{S}\}$ is called the *assumption* set. The set $\overline{\mathcal{S}}$ is closed u conjunction, and we assume $\overline{S_1} \wedge \overline{S_2} = \overline{S_1 \wedge S_2}$.
2. The set of *mixed-specifications*, $\mathcal{S}^{\#}$, is defined by the following grammar: $\mathcal{S}^{\#} :: \mathcal{S} \mid \overline{\mathcal{S}} \mid C(\mathcal{S}^{\#}, \ldots, \mathcal{S}^{\#}) \rhd \mathcal{S} \mid \#(\mathcal{S}^{\#}, \mathcal{S}^{\#})$, where $C \in \mathcal{C}$.
3. $S \in \mathcal{S}^{\#}$ is *assumption-free* if it does not contain elements of $\overline{\mathcal{S}}$.

Definition: Let $S_1 \in \mathcal{S}$ and suppose that either $S_2 = \overline{S} \in \overline{\mathcal{S}}$ or $S_2 = S \in \mathcal{S}$. We say S_1 *confirms* S_2, to be denoted by $S_1 \Rightarrow S_2$, iff $BEH_{\mathcal{S}}[S_1] \subseteq BEH_{\mathcal{S}}[S]$. □

We assume that the basic system provides a way to determine whether $S_1 \Rightarrow S_2$ h It is possible to redefine $S_1 \Rightarrow S_2$ to hold iff $BEH_{\mathcal{S}}[S_1] = BEH_{\mathcal{S}}[S]$. Using inclusion e the use of the proof system. However, in some concrete systems there might not be a to determine inclusion while equality can be determined.

Definition: For $S \in \mathcal{S}^{\#}$ a mixed-specification, its *conclusion* and *promise*, denote *conc*(S) and *promise*(S), respectively, are defined as follows:

1. if $S \in \mathcal{S}$ then *conc*$(S) = S$ and *promise*(S) is not defined,
2. if $S = \overline{S'} \in \overline{\mathcal{S}}$ then *conc*$(S) = \overline{S'}$ and *promise*(S) is not defined,
3. if $S = C(S_1, \ldots, S_n) \rhd S'$ then *conc*$(S) = S'$ and *promise*$(S) = C(S_1, \ldots, S_n)$,
4. if $S = \#(S_1, S_2)$ then *conc*$(S) =$ *conc*$(S_1) \wedge$ *conc*(S_2) and *promise*$(S) = \#($*promise*$(S_1),$ *promise*$(S_2))$. □

To define the semantics of mixed-specifications, a function $BEH^{\#}_{\mathcal{S}} : \mathcal{S}^{\#} \rightarrow \widehat{\wp(\mathcal{B})}$ is presen Intuitively, if S' is a "consequence" of $C(S_1, \ldots, S_n)$, i.e., $K_C(BEH^{\#}_{\mathcal{S}}[S_1], \ldots, BEH^{\#}_{\mathcal{S}}[S_n]) \subseteq BEH_{\mathcal{S}}[S']$, then $BEH^{\#}_{\mathcal{S}}[C(S_1, \ldots, S_n) \rhd S']$ is def as $BEH_{\mathcal{S}}[S']$. Else, it is defined as Θ. If the structures of S_1 and S_2 are unifiable their assumptions are mutually confirmed then $BEH^{\#}_{\mathcal{S}}[\#(S_1, S_2)]$ is defined as BEH $BEH^{\#}_{\mathcal{S}}[S_2]$. Else, it is defined as Θ. Unification is used to find a match between assumptions and the specifications that should confirm them. However, here we need to specifications to assumptions rather than to variables. We make this precise in the follow definitions which lead to the formal definition of the semantics of mixed-specifications.

Definition: Let $\mathcal{V}$ be a set of variables. The set of *skeleton-terms*, $\mathcal{SK}$, is defined by: $\mathcal{SK} :: \mathcal{S} \mid \mathcal{V} \mid \mathcal{V} \wedge \mathcal{V} \mid C(\mathcal{SK}, \ldots, \mathcal{SK}) \rhd \mathcal{V} \mid \#(\mathcal{SK}, \mathcal{SK})$, where $C \in \mathcal{C}$.

Definition: Let $S \in \mathcal{S}^{\#}$ be a mixed-specification. The *skeleton* of S, denoted by *skel* is a pair in which the first element, denoted by *skel*$_1(S)$, is a skeleton-term and the element, denoted by $skel_2(S)$, is a substitution θ which binds basic specifications or mptions to variables. $skel(S)$ is defined as follows:

1. if $S \in \mathcal{S}$ then $skel(S) = < S, \varepsilon >$,
2. if $S = \overline{S'} \in \overline{\mathcal{S}}$ then $skel(\overline{S'}) = < X, \{X \leftarrow \overline{S'}\} >$, where $X \in \mathcal{V}$,
3. $skel(C(S_1, \ldots, S_n) \rhd S) = < C(skel_1(S_1), \ldots, skel_1(S_n)) \rhd X, skel_2(S_1) \bigcup \ldots \bigcup skel_2(S_n) \bigcup \{X \leftarrow S\} >$ [3],
4. $skel(\#(S_1, S_2)) = < \#(skel_1(S_1), skel_1(S_2)), skel_2(S_1) \bigcup skel_2(S_2) >$.

finition: A substitution $\theta_1 = \{X_1 \leftarrow S_1, \ldots, X_n \leftarrow S_n\}$ is *valid* w.r.t. another substiion $\theta_2 = \{Y_1 \leftarrow S'_1, \ldots, Y_m \leftarrow S'_m\}$ if for all $1 \leq i \leq n$, $1 \leq j \leq m$ if $X_i = Y_j$ then $c(S_i) \Rightarrow conc(S'_j)$ holds. □

Next, a transformation on mixed-specifications, $elim : \mathcal{S}^{\#} \rightarrow \mathcal{S}^{\#}$, is defined, mapping ixed-specification containing jigsaw occurrences to a mixed-specification free of jigsaw, h the same meaning (to be defined). The function *elim* is defined for $\#(S_1, S_2)$ iff S_1 and are unifiable and their assumption are mutually confirmed. Intuitively, the function *elim* nes a composition of specifications which is to be used to obtain a specification of the aw-composed program from the specifications of its components.

finition: Let $S \in \mathcal{S}^{\#}$ be a mixed-specification. Let $elim(S)$ be:

1. if $S \in \mathcal{S} \cup \overline{\mathcal{S}}$ then $elim(S) = S$,
2. $elim(C(S_1, \ldots, S_n) \rhd S')$ is defined to be $C(elim(S_1), \ldots, elim(S_n)) \rhd S'$, if for every $1 \leq i \leq n$, $elim(S_i)$ is defined, and otherwise it is not defined.
3. $elim(\#(S_1, S_2))$ is defined to be $\#(elim(S_1), elim(S_2)) = skel_1(elim(S_1))\theta$, if $skel_1(elim(S_i))$, $i = 1, 2$ are defined and are unifiable [4] and moreover θ is a valid substitution both w.r.t. $skel_2(S_1)$ and w.r.t. $skel_2(S_2)$, where $\theta = (mgu^*(skel_1(S_1), skel_1(S_2)) \, o \, skel_2(S_1) \, o \, skel_2(S_2))$. [5] Otherwise, it is not defined.

lim$(\#(S_1, S_2))$ is defined, $\#(S_1, S_2)$ is called *well-defined.*

finition: Let $\mathcal{S}^{\#}$ be the set of mixed-specifications, let $\mathcal{B}$ be the basic set of behaviors let $BEH_{\mathcal{S}}$ be the basic semantic function for $\mathcal{S}$. The *semantics* of $\mathcal{S}^{\#}$ is defined by the ction $BEH^{\#}_{\mathcal{S}} : \mathcal{S}^{\#} \rightarrow \widehat{\wp(\mathcal{B})}$:

1. if $S \in \mathcal{S}$ then $BEH^{\#}_{\mathcal{S}}[S] = BEH_{\mathcal{S}}[S]$,
2. if $\overline{S} \in \overline{\mathcal{S}}$ then $BEH^{\#}_{\mathcal{S}}[\overline{S}] = BEH_{\mathcal{S}}[S]$,
3. if $S = C(S_1, \ldots, S_n) \rhd S'$ and $C \neq$ '#' then $BEH^{\#}_{\mathcal{S}}[C(S_1, \ldots, S_n) \rhd S'] = BEH^{\#}_{\mathcal{S}}[S']$, if $K_C(BEH^{\#}_{\mathcal{S}}[S_1], \ldots, BEH^{\#}_{\mathcal{S}}[S_n]) \subseteq BEH^{\#}_{\mathcal{S}}[S']$ and the set Θ, otherwise.
4. if $S = \#(S_1, S_2)$ then $BEH^{\#}_{\mathcal{S}}[\#(S_1, S_2)] = BEH^{\#}_{\mathcal{S}}[elim(\#(S_1, S_2))]$, if $\#(S_1, S_2)$ is well-defined and the set Θ, otherwise. □

---

[3] For simplicity assume that $skel_2(S_i)$, $i = 1, n$ have disjoint sets of variables and X is a fresh variable.

[4] mgu^* is the most general unifier except that if it binds of a variable to another variable, for example $X \leftarrow Y$, then this assignment is replaced by $X \leftarrow X \wedge Y$, $Y \leftarrow X \wedge Y$. Intuitively, the assumptions on the esponding gaps are 'collected'.

[5] The substitution composition in this case is commutative since the substitutions have disjoint sets of ables, therefore, $elim(\#(S_1, S_2)) = elim(\#(S_2, S_1))$.

The meaning of a gapped program T in *context* S is defined by a semantic func $BEH^{\#}_{T-S} : \mathcal{T}^{\#} \times \mathcal{S}^{\#} \rightarrow \widehat{\wp(\mathcal{B})}$, to be denoted by $BEH^{\#}_{T-S}[T \downarrow_S]$. The meaning of T context S is defined iff T and S are *identically structured*, i.e., the structural elements i correspond to the structure of T. Intuitively, if the program T is a basic program then meaning is defined to be the meaning given to it in the basic system. If T contains a variable then the meaning of this variable is taken to be the meaning of the correspond assumption in the promise of S. If T is of the form $\#(T_1, T_2)$ then a meaning is defi for T iff T_1 ad T_2 are unifiable. The meaning of $\#(T_1, T_2)$ is defined to be the meanin the program obtained by unifying T_1 and T_2 using their most general unifier. In orde formally define the meaning of T in context S we need the following definitions.

Definition: Let $T \in \mathcal{T}^{\#}$ and $S \in \mathcal{S}^{\#}$. T and S are *identically structured* iff

1. $T \in \mathcal{T}$ and $S \in \mathcal{S}$, or
2. $T \in \mathcal{V}$ and $S \in \overline{\mathcal{S}}$, or
3. $T = C(T_1, \ldots, T_n)$, $S = C(S_1, \ldots, S_n) \triangleright S$ and T_i and S_i are identically structu $1 \leq i \leq n$, or
4. $T = \#(T_1, T_2)$, $S = \#(S_1, S_2)$ and T_i and S_i are identically structured, $i = 1, 2$.

Definition: Let $T \in \mathcal{T}^{\#}$ be a gapped program. *elim*(T) is defined by:

1. if $T \in \mathcal{O} \cup \mathcal{V}$ then *elim*$(T) = T$,
2. *elim*$(C(T_1, \ldots, T_n)) = C($*elim*$(T_1), \ldots,$ *elim*$(T_n))$, if $C \neq$'#' and for every $1 \leq i \leq$ *elim*(T_i) is defined, and it is not defined, otherwise,
3. *elim*$(\#(T_1, T_2)) = \#($*elim*$(T_1),$ *elim*$(T_2)) =$ *elim*(T_1)*mgu*$($*elim*$(T_1),$ *elim*$(T_2))$, if *eli* $i = 1, 2$ are defined and $\#(T_1, T_2)$ is well-defined, and it is not defined, otherwise.

Next the meaning of a term $T \in \mathcal{T}^{\#}$ in a context $S \in \mathcal{S}^{\#}$ is defined by a semantic fu tion $BEH^{\#}_{T-S}$. This function is defined only for well-defined terms and specifications a moreover only for terms and specifications with identical structures.

Definition: Let $T \in \mathcal{T}^{\#}$ and $S \in \mathcal{S}^{\#}$ be such that both *elim*(T) and *elim*(S) are defi and moreover *elim*(T) and *elim*(S) are identically structured. The *meaning of the tern in the context of* S is next given by the function $BEH^{\#}_{T-S} : \mathcal{T} \times \mathcal{S} \rightarrow \widehat{\wp(\mathcal{B})}$:

1. if *elim*$(T) \in \mathcal{T}$ then $BEH^{\#}_{T-S}[T \downarrow_S] = BEH^{\#}_{T}[$*elim*$(T)]$,
2. if *elim*$(T) \in \mathcal{V}$ then $BEH^{\#}_{T-S}[T \downarrow_S] = BEH^{\#}_{S}[S]$,
3. if *elim*$(T) = C(T_1, \ldots, T_n)$ and *elim*$(S) = C(S_1, \ldots, S_n) \triangleright S'$ then $BEH^{\#}_{T-S}[T \downarrow_S] = K_C(BEH^{\#}_{T-S}[T_1 \downarrow_{S_1}], \ldots, BEH^{\#}_{T-S}[T_n \downarrow_{S_n}])$.

Definition: Let $T \in \mathcal{T}^{\#}$ and $S \in \mathcal{S}^{\#}$. A subterm $T' \in T$ and a subterm $S' \in S$ *correspo* w.r.t. T and S if one on the following conditions holds:

1. $T' = T$ and $S' = S$,
2. There exist subterms T'' and S'' of T and S which correspond w.r.t. T and S and o of the following conditions holds:

(a) $T'' = C(T_1,\ldots,T_n)$, $S'' = C(S_1,\ldots S_n) \triangleright S^*$, $T' = T_i$ and $S' = S_i$, $1 \le i \le n$,
(b) $T'' = \#(T_1,T_2)$, $S'' = \#(S_1,S_2)$, $T' = T_i$ and $S' = S_i$, $1 \le i \le 2$.

inition: Let $T \in \mathcal{T}^\#$ and $S \in \mathcal{S}^\#$. The formula "$T\ Sat^\#\ S$" *holds* in the structure $, BEH^\#_{T-S}, BEH^\#_S >$ iff $BEH^\#_{T-S}[T \downarrow_S]$ is defined and for each pair of subterms $T' \in T$ $S' \in S$, if T' and S' correspond w.r.t. T and S then $BEH^\#_{T-S}[T' \downarrow_{S'}] \subseteq BEH^\#_S[S']$.

The proof system

t, a proof system for the new correctness formulas is presented.

inition: Let $\mathcal{AR} =< \mathcal{A}, \mathcal{R}, \mathcal{G} >$ be a basic compositional proof system. The composi-al proof system $\mathcal{AR}^\# =< \mathcal{A}^\#, \mathcal{R}^\#, \mathcal{G} >$ for formulas of the form "$T\ Sat^\#\ S$" is defined

. The set $\mathcal{G}$ is the basic set of criteria,

:. The set $\mathcal{A}^\#$ is an extension of $\mathcal{A}$ with the following formulas:

The variable-axioms

$$X\ Sat^\#\ \overline{S}, \quad where\ X \in \mathcal{V}\ and\ \overline{S} \in \overline{\mathcal{S}}.$$

. The set $\mathcal{R}^\#$ is defined as the basic set of rules, $\mathcal{R}$, except that:

(a) **The composition rules:** each rule R_C in $\mathcal{R}$ of the form

$$\frac{T_i\ Sat\ S_i,\ i = 1,n,\ \ G_C(S_1,\ldots,S_n,S)}{C(T_1,\ldots,T_n)\ Sat\ S}$$

is transformed into

$$\frac{T_i\ Sat^\#\ S_i,\ i = 1,n,\ \ G_C(conc(S_1),\ldots,conc(S_n),S)}{C(T_1,\ldots,T_n)\ Sat^\#\ C(S_1,\ldots,S_n) \triangleright S}$$

(b) Three additional rules:

i. **The jigsaw rule**

$$\frac{T_i\ Sat^\#\ S_i,\ i = 1,2}{\#(T_1,T_2)\ Sat^\#\ \#(S_1,S_2)}$$

Provided that $\#(S_1,S_2)$ *is well – defined*

ii. **The substitution rule**

$$\frac{T\ Sat^\#\ S}{T\ Sat^\#\ S[skel_1(S_1)\theta\ /\ \#(S_1,S_2)]}$$

For every subterm $\#(S_1,S_2)$ in S, where
$\theta = (mgu^*(skel_1(S_1), skel_1(S_2))\ o\ skel_2(S_1)\ o\ skel_2(S_2))$

iii. **The assumption-freeness rule**

$$\frac{T\ Sat^\#\ M \triangleright S}{T\ Sat^\#\ S[S_1/M \triangleright S_1]}$$

For every subterm $M \triangleright S_1$ in S, where M *is assumption – free.*

mments:

1. The variable axioms allow to assume any basic specification for a variable. If the $\mathcal{S}$ is not recursive neither is the set of additional axioms presented here.
2. A predicate $G \in \mathcal{G}$ is redefined over $(\mathcal{S} \cup \overline{\mathcal{S}})^n$ in the following way: $G(S_1, \ldots, \overline{S_i}, \ldots$ holds iff $G(S_1, \ldots, S_i, \ldots S_n)$ holds.
3. The composition rules allow to preserve the component specifications which the c clusion S is based upon.
4. The jigsaw rule allows to derive a mixed specification for a jigsaw-composed prog based on the specifications of its constituents. Thus, the compositionality of the b proof system is preserved.
5. The substitution rule allows to eliminate a jigsaw operator and to reduce the num of assumptions in a specification.
6. The assumption-freeness rule allows to replace a mixed-specification which is fre assumptions by a semantically equivalent basic specification. □

Next the main theorem of the paper is presented. The theorem claims the soundness completeness of the extended system with respect to the basic one, i.e., given two progra T in the basic system and T' in the extended one, such that T' is semantically equal t we prove that any basic specification that can be verified for T' is semantically true fo and every basic semantically true specification of T can be verified for T'.

Definition: A predicate $G_C \in \mathcal{G}$ is *monotone* iff

1. If $S_i' \Rightarrow S_i$ and $G_C(S_1, \ldots, S_i, \ldots, S_k, S_{k+1})$ holds then $G_C(S_1, \ldots, S_i', \ldots, S_k, S_k$ holds, for every $1 \leq i \leq k$.
2. If $S_{k+1} \Rightarrow S_{k+1}'$ and $G_C(S_1, \ldots, S_i, \ldots, S_k, S_{k+1})$ holds then $G_C(S_1, \ldots, S_i, \ldots, S_k, S$ holds.

Definition: A predicate $G_C \in \mathcal{G}$ is *compositional* iff, if $G_C(S_1, \ldots, S_i, \ldots, S_k, S')$ holds $G_C(S_1, \ldots, S_i, \ldots, S_k, S'')$ holds then $G_C(S_1, \ldots, S_i, \ldots, S_k, S' \wedge S'')$ holds.

Theorem 3.1: (Soundness and Completeness) Assume all predicates in $\mathcal{G}$ are monot and compositional. For every $T' \in \mathcal{T}^{\#}$ and $T \in \mathcal{T}$ such that $T = elim(T')$ and for ev basic specification S: $\models T\ Sat\ S$ *iff* $\vdash_{\mathcal{AR}^{\#}} T'\ Sat^{\#}\ S$. □

In [FFG92] a proof of Theorem 3.1 is presented. The extended proof system is also pro in [FFG92] to be sound and complete, using the common definitions for these notions. this case, additional properties of the basic framework are assumed.

4 Example

In this section we present an example of a concrete proof system and show that it fits abstract framework presented in Section 2. Consequently, an extended proof system wh allows for jigsaw compositions can be syntactically derived from the basic proof system. T soundness and the completeness of the concrete extended system with respect to the concr basic system is derived from the soundness and completeness of the abstract framew (Theorem 3.1). The concrete system is a simple version of the system presented in [ZRB8

The programming language

owing [ZRB85], a compositional syntax-directed proof system for proving partial cor-ness for a CSP-like language is presented. The set $\mathcal{T}$ of programs is defined over the owing sets of primitive objects, $\mathcal{O}$, and constructors, $\mathcal{C}$:

$$\mathcal{O} = \{skip\} \cup \{\overline{x} := \overline{e}\} \cup \{b\} \cup \{D!e\} \cup \{D?x\}$$
$$\mathcal{C} = \{\|,\ ;,\ \Box,\ While\}$$

skip statement and the multiple assignment statement $\overline{x} := \overline{e}$ have their usual meanings. statement b (a boolean condition over $\overline{x}$) functions as a guard. Whenever b is evaluated *rue* it can be passed. When b evaluates to *false* the guard cannot be passed and the aputation either terminates, if b guards a '*While*' statement or fails, if b does not guard *Vhile*' statement. Processes in the network are not allowed to have shared program ables. Processes can communicate with each other only along named, directed channels. nmunication along a channel, say D, occurs when an output command $D!e$ and an input amand $D?x$ are executed simultaneously by the processes attached to D. The value e is n assigned to the program variable x and both processes continue their execution. The ements $(P_1 \| P_2)$[6] and $(P_1; P_2)$ stand for parallel and sequential compositions of P_1 and respectively. The statement $(P_1 \Box P_2)$ stands for nondeterministic choice between P_1 and Finally, the statement $(While\ b, P)$ stands for guarded iteration, where the guard is b the body is P.

The set of behaviors, $\mathcal{B}$, and a proof that the above programming language and its antics fit the basic abstract framework, are presented in [FFG92].

The specification

set $\mathcal{S}$ of specifications is a set of formula-triples with a typical element $[I, p, q]$, where I, p q are formulas in a first order predicate language with two sorts of *auxiliary* variables, *ger* variables v and *trace* variables t. A trace is a sequence of communications each hich contains a channel name and a value. The meaning of the specifications in $\mathcal{S}$ is nally presented in [FFG92]. Intuitively, a correctness formula "$T\ Sat\ [I, p, q]$" holds if never T is started in an initial state and with an initial communication history for which lds, then the invariant I holds for the communication history of T at any moment during execution of T. Also, whenever T terminates, q holds for the final state and the final amunication history of T. The invariant I does not contain program variables.

The syntax of the formulas I, p, q is defined as:

expression	$e :: 0 \mid 1 \mid \omega \mid v \mid x \mid val(texp[e]) \mid \lvert texp \rvert \mid e_1 + e_2$
channel – name	$c :: D \mid chan(texp[e])$
trace	$texp :: t \mid \pi_{cset} \mid (texp \lvert cset)$
formula	$p :: e_1 = e_2 \mid c_1 = c_2 \mid p_1 \wedge p_2 \mid \neg p \mid \exists v.[p] \mid \exists t.[p]$

mments: ω is a special value of program variables, standing for "undefined". $val(texp[e])$ he value of the e-th communication in trace $texp$. $|texp|$ is the length of the trace $texp$. $n(texp[e])$ is the channel-name of the e-th communication in the trace $texp$. π is a special iliary trace variable which must occur free in a formula. π denotes the communication ory of the execution up to the current state. Moreover, only a projection of π on a set hannels $cset$, to be denoted π_{cset}, can occur in a formula. $(texp|cset)$ is the projection

For clarity we use infix notation.

of trace *texp* on a set of channels, *cset*. Auxiliary variables, except for π, that occur fre formulas are interpreted as if they were universally quantified.

We will freely use abbreviations such as $texp_1 = texp_2$ or $p \rightarrow q$, defined as usual.

The basic proof system contains a set of axioms and for each constructor in C ther exactly one rule of the form

$$\frac{T_i\ Sat\ S_i,\ \ G_C(S_1,\ldots,S_n,S)}{C(T_1,\ldots,T_n)\ Sat\ S}.$$

For example, the **parallel composition** rule is:

$$\frac{T_1\ Sat\ S_1,\ \ T_2\ Sat\ S_2,\ \ G_{||}(S_1,S_2,S)}{(T_1 \,||\, T_2)\ Sat\ S}$$

Where, $S_i = [I_i, p_i, q_i]$, $i = 1,2$, $S = [I,p,q]$ and $G_{||}(S_1,S_2,S)$ holds iff $((I_1 \wedge I_2) \rightarrow I$ $(p \rightarrow (p_1 \wedge p_2)) \wedge ((q_1 \wedge q_2) \rightarrow q)$. With the restrictions: $var(p_i,q_i) \cap var(s_j) = \emptyset$ $(i,j) = (1,2)$, $(2,1)$, $chan(I_i,p_i,q_i) \cap chan(S_j) \subset chan(S_i)$ for $(i,j) = (1,2)$, $(2,1)$.

The proof system is fully presented in [ZRB85, FFG92].

4.3 Application

Showing that the version of the system [ZRB85] presented above fits the basic abstract s tem, we can develop and verify programs containing jigsaw operators based on the exten system. The jigsaw operator allows the development and verification of a program i compositional manner in which the units of modularity are not necessarily derived by top-level constructors of the programming language. To make this point clear we apply extended concrete system to an example.

Assume we are given the following informal description of a system. The network c sists of two nodes. The Producer node is required to produce a sequence of consecu prime numbers and send them to the other node, called the Consumer. The Consumer n receives the numbers and signals the Producer to halt after receiving a prime number wl meets some requirements. In addition, assume that due to the nature of the communicat system the values to be communicated should be representable by a binary number of more than six digits (should not be greater than 63). Thus, some communication man ment subtask should be added, that handles numbers greater than 63. It is clear that program should be of the general form (*Producer* || *Consumer*). Therefore, to develop program in a compositional syntax-directed manner would traditionally mean to deve and verify the Producer subtask and the Consumer subtask separately and than para compose them. However, the program consists of two distinct subtasks that one might w to handle separately in a syntax-directed and compositional manner. One subtask inclu the production and the acceptance of the prime numbers. The other subtask includes communication managment. The jigsaw operator makes it possible to handle separa those subtasks (Figure 1). Mixed specifications which can be proved for $Producer^c$ $Consumer^c$, are:

$S_{producer^c}$ ::
$(While\ [true, more = 1, more = 1],$
$(\overline{[I, Ass_1, Ass_1]};$
$[I, Ass_1, true]) \rhd [I, Ass_1, true]$
$) \rhd [I, Initial, true]$

$S_{Consumer^c}$ ::
$(While\ [true, more = 1, more = 1],$
$((\overline{[I, more = 1, prime(new) \wedge more = 1]};$
$[I, prime(new) \wedge more = 1, Ass_2]) \rhd [I, more = 1,$
$[I, Ass_2, Ass_2]) \rhd [I, more = 1, true]) \rhd [I, Initia$

system is defined as $\#(Communication, Primes)$, where $Communication$ :: $\!oducer^c \,\|\, Consumer^c)$ and

```
Producer^c ::                          Consumer^c ::
(While more = 1,                       (While more = 1,
   Y;                                     D12?temp; new := temp;
   val := prod;                           (While temp = 63,
   (While val ≥ 63,                          D12?temp; new := new + temp);
     D12!63; val := val − 63);            M;
   D12!val; D21?more)                     D21!more)
```

mes :: $(Producer^p \,\|\, Consumer^p)$, where

```
Producer^p ::                                             Consumer^p ::
(While more = 1,                                          (While more = 1,
  (While found = 0,                                          V;
   prod, j, found := prod + 1, 1, 1;                         (new ≥ 100; more := 0
   (While j ≤ max ∧ found ≠ 0,                               □
     found, j := mod(prod, array[j]), j + 1));                 new < 100; skip);
   max, array[max + 1], found := max + 1, prod, 0;            N)
  Z)
```

Figure 1: The Communication and the Primes subtasks

:ed specifications which can be proved for $Producer^p$ and $Consumer^p$, are:

$S_{producer^p}$ ::
$While\ [true, more = 1, more = 1],$
$[I, Ass_1, Ass_1];$
$\overline{true, Ass_1, true]}) \triangleright [true, Ass_1, true]$
$\triangleright [true, Initial, true]$

$S_{Consumer^p}$ ::
$While[true, more = 1, more = 1],$
$((\overline{[true, more = 1, prime(new) \wedge more = 1]};$
$[I, prime(new) \wedge more = 1, Ass_2]) \triangleright$
$[true, more = 1, Ass_2];$
$\overline{[true, Ass_2, Ass_2]} \triangleright [true, more = 1, Ass_2]) \triangleright$
$[true, Initial, prime(new) \wedge new \geq 100]$

ere $I \stackrel{df.}{=} (\forall v.val(\pi_{D12,D21}[v]) \leq 63) \wedge (\forall v.val(\pi_{D12}[v]) \neq 63 \rightarrow prime(\Sigma_{j \in A}\pi_{D12}[j])$ and $\{v', v'+1, \ldots, v\}$ where $(\forall j \in A.j \neq v \rightarrow val(\pi_{D12}[j]) = 63) \wedge val(\pi_{D12}[v'-1]) \neq 63$.

$Initial \stackrel{df.}{=} found = 0 \wedge prod = 2 \wedge array[1] = 2 \wedge max = 1$

$Ass_1 \stackrel{df.}{=} prod \in array[1..max] \wedge (\forall 1 \leq l \leq max.prime(array[l])) \wedge (\forall k.prime(k) \wedge k < prod \rightarrow k \in array[1..max])$

$Ass_2 \stackrel{df.}{=} prime(new) \wedge (more \neq 1 \rightarrow new \geq 100)$

For the Communication subtask we can conclude using the rule for parallel composition t it satisfies: $S_{communication} :: (S_{producer^c} \,\|\, S_{consumer^c}) \triangleright [I, Initial, true]$.

For the Primes subtask we can conclude using the rule for parallel composition that it sfies: $S_{primes} :: (S_{producer^p} \,\|\, S_{consumer^p}) \triangleright [true, Initial, prime(new) \wedge new \geq 100]$.

For the jigsaw composition $\#(Communication, Primes)$ we can conclude by using the aw rule, the substitution rule and the assumption freeness rule that it satisfies $nitial, prime(new) \wedge new \geq 100]$.

5 Conclusions

In [FFG90, FFG91] the jigsaw operator has been introduced into two different framewo In both frameworks the jigsaw composition rules are syntax-directed but not compositio This raises the question whether the non compositionality of those rules is due to the na of the jigsaw composition or due to the non compositionality of the basic frameworks. H we address this question and show that adding the jigsaw to an algebraic and compositi framework results in a compositional rule for jigsaw composition.

6 Acknowledgments

The part of the first and second authors was partially supported by the Israeli Nati Academy of Sciences (basic research). The second author was also supported by the f for promotion of research in the Technion. The third author was partially supported by United States-Israel Binational Science Foundation, Jerusalem, Israel.

7 References

[Apt81] K.R. Apt: "Ten years of Hoare's logic: a survey Part I, ACM TOPLAS 3, 1981.

[Bac80] R.J.R. Back: "Correctness preserving program refinements: proof theory and applicati Mathematical Centre Tracts 131, Mathematical Centre, Amsterdam, 1980.

[BF88] L. Bouge, N. Francez: "A Compositional Approach to Superimposition," 15th ACM Sy on Principles of Programming Languages, San Diego, CA, January 1988.

[CM88] M. Chandy, J. Misra: **"Parallel programs design,"** Addison-Wesly, 1988.

[FF90] N. Francez, I.R. Forman: "Superimposition for interacting processes," CONCUR'90, Am dam, August 1990. LNCS 458 J.C.M. Baeten, J.W. Klop (Eds.), Springer-Verlag, 1990.

[FFG90] L. Fix, N. Francez, O. Grumberg: "Semantics-driven decompositions for the verificati distributed programs," Proc. of the IFIP working group 2.2/2.3 working conference on Program concepts and Methods, Sea of Galilee, Israel, April 1990, North-Holland, pp. 101-123.

[FFG91] L. Fix, N. Francez, O. Grumberg: "Program composition and modular verification," I 18th ICALP, Madrid, July 1991. LNCS 510 J. Leach Albert, B. Monien, M. Rodriguez Art (Eds.), Springer-Verlag, 1991.

[FFG92] L. Fix, N. Francez, O. Grumberg: "Program composition via unification," TR 722, Tech - Israel Institute of Technology, 1992.

[Heh84] E.C.R. Hehner: "Predictive programming," part I and II, Communication ACM 27, 19

[Hoa78] C.A.R. Hoare: "Communicating sequential processes," CACM 21, 8, August 1978.

[KFE90] S. Katz, I. Forman, M. Evangelist: "Language constructs for distributed systems," Pro the IFIP working group 2.2/2.3 working conference on Programming concepts and Methods, Se Galilee, Israel, April 1990, North-Holland, pp. 73-100.

[KKZ90] R. Koymans, R. Kuiper, E. Zijlstra: "Specification specified," in R. Kuiper Ph.D. T University of Eindhoven, 1989.

[Old86] E.R. Olderog: "Process theory: semantics, specification and verification," LNCS 224 J.W Bakker, W.P. de Roever, G. Rozenberg (Eds.), Springer-Verlag, 1986.

[Pnu85] A. Pnueli: "In transition from global to modular temporal reasoning about programs," P Logics and Models of Concurrent Systems, Ed. K.R. Apt, Springer-Verlag, 1985.

[Wir71] N. Wirth: "Program development by stepwise refinement," Communication ACM 14, 19

[ZRB85] J. Zwiers, W.P. de Roever, P. van Emde Boas: "Compositionality and concurrent netw soundness and completeness of a proof system," Proc. 12th ICALP, Nafplion, Greece, July 1 Springer LNCS 194, pp. 509-519.

Barbed Bisimulation

Robin Milner Davide Sangiorgi[1]

LFCS - Department of Computer Science
Edinburgh University, Edinburgh EH9 3JZ, U.K.

Abstract. We propose in this paper barbed bisimulation as a tool to describe bisimulation-based equivalence *uniformly* for any calculi possessing *(a)* a reduction relation and *(b)* a convergency predicate which simply detects the possibility of performing some observable action. This opens interesting perspectives for the adoption of a *reduction semantics* in process algebras. As a test-case we prove that strong bisimulation of CCS coincides with the congruence induced by barbed bisimulation.

Introduction

ocess algebras have emerged as a key model to reason about nondeterministic d concurrent computations. The *transition semantics* on which they are sed is given by defining the appropriate congruence on top of the *labelled* ansition system describing the possible evolutions of a process.

This contrasts with what happens in term-rewriting systems, as based on *reduction semantics* where the equivalence is defined on the top of an *un-elled* transition system (the *reduction system*). It is indeed the simplicity their semantics which makes term-rewriting systems attractive, and in turn cilitates a satisfactory mathematical analysis.

In the lambda calculus, probably the most well-known term-rewriting sys-m, what makes a reduction system possible is that two terms having to teract are naturally in contiguous positions. This is not the case in process lculi, where interaction is not dependent upon physical contiguity. To put is another way, a *redex* of a lambda term is a subterm, while a "redex" in a ocess calculus is distributed over the term.

A guideline for the definition of reduction systems in process algebras is fered by Milner [8], inspired by Berry and Boudol's Chemical Abstract Ma-ine [2]. In this technique, axioms for a *structural congruence relation* are troduced prior to the reduction system, in order to break a rigid, geometri-l vision of concurrency; then reduction rules can easily be presented in which dexes are indeed subterms again. It can then be shown that modulo struc-ral congruence the reduction relation exactly represents the silent action $\xrightarrow{\tau}$ the labelled transition semantics.

[1]Supported by a grant from Consiglio Nazionale delle Ricerche, Italy

It is left as an open problem in [8] how to recover from such a formulati the familiar congruences which are based upon a labelled transition syste It turns out that this is not a trivial problem. We tackle it in this paper the simple case of CCS and strong observational equivalence ($\sim$). Because t reduction relation coincides with the silent action of the labelled transiti system, we can remain within the latter framework. But we wish to retain t spirit of the reduction semantics as far as possible, in the sense that we wi to equip a global observer with a minimal ability to observe actions and/ process states. We then obtain an equivalence, namely indistinguishabili under global observations. This in turn induces a congruence over agen namely equivalence in all contexts. The question is: what minimal power observation is needed so that the congruence induced in this way coincid with the familiar bisimilarity congruence based upon the labelled transiti semantics?

If we look for a characterization of $\sim$ in this way, then it is reasonal to examine first the case in which there are *no* observables. Unfortunate the *reduction congruence* which is so obtained is in general less discriminato than $\sim$ (it is the same as $\sim$ only for the divergence-free processes). Therefc it is necessary to add the capability of observing some properties of the stat It seems natural in concurrency that the extra power provided is in terms action observability; in this sense the minimal capability is given by the ado tion of a predicate detecting the possibility of performing *some* unidentifi observable action. It is therefore a "*may*" *convergency* predicate, in this sen the same used by Boudol in [3]. We call the resulting bisimulation relati and its induced congruence respectively *barbed bisimulation* and *barbed co gruence*. This paper is mainly devoted to proving that barbed congruence a $\sim$ coincide.

Barbed bisimulation strongly recalls Abramsky's applicative bisimulati for lambda calculus [1], as both are defined in terms of reduction and co vergency. In fact the generality of such predicates, which are independent the underlying transition system, allows one to use barbed bisimulation as tool to describe *uniformly* bisimulation-based equivalences in calculi whi syntactically may be quite far from each other. This is useful in different a pects. First of all it helps a lot in understanding the relationship between tv different calculi. It is by using barbed bisimulation, for instance, that we we able to conduct the study in [11] on the expressive power of the π-calculus f modeling lambda calculus and higher-order π-calculus.

Barbed equivalence also provides for free any "new" calculus with an ir portant equivalence; it is then an excellent test for the calculus itself and i operators to find out whether barbed congruence can be characterized in direct way, not involving quantification over all contexts.

An important point emerging in [8] is that in some cases a reduction se-antics may be much more enlighting and simple than a transition semantics. ır study in [11] suggests that in higher-order process calculi such benefits n become decisive.

Structure of the paper: We review the basic concepts of pure CCS in ction 2; we study the reduction bisimulation/congruence and the barbed simulation/congruence respectively in sections 3 and 4; in section 5 we look other issues and uses of barbed bisimulations.

Background

'e review here the syntax and the operational semantics of 'pure' CCS [7] gether with the concept of strong bisimulation.

Let $N = \{a, b, \ldots\}$ be a set of names, $\Lambda = \{a, \overline{a} \mid a \in \Lambda\}$ be the set of labels ilt on N and $Act = \Lambda \cup \{\tau\}$ where τ is a distinguished action (the silent tion) not in Λ . We use the convention that if $a \in N$, $\overline{\overline{a}} = a$ and we let ℓ to nge over Λ and α over Act. We will also use a set Ψ of variables ranged over X and a set Φ of agent constant ranged over by A.

The class of the CCS expressions with names in N is the smallest class which includes Ψ and Φ and such that contains the following expressions, here E, E_i are already in $\Im$:

$$\alpha.E, \quad E \setminus L\,(L \subseteq N), \quad E_1|E_2, \quad \textstyle\sum_{i\in I} E_i \quad \text{(I is an indexing set)}$$

is supposed that for each constant A there is a defining equation of the form $\stackrel{def}{=} E$, for $E \in \Im$. A CCS expression with no free variable in it is a *process*. the following P, Q, R, T represent processes. A CCS expression with a ngle occurrence of a variable in it is a *context*. Given a context C and a ocess P, we write $C(P)$ to denote the closed term obtained by substituting for the variable occurring in C.

SSUMPTION. We suppose that processes do not use all names in N, i.e. given process P, we can always find a name which does not appear in P. The proof Theorem 1 depends upon this assumption. We leave as an open problem hether it can be removed.

The operational semantics of a CCS process is modelled by a labelled tran-tion system. It is shown in table 1. The transition $P \stackrel{\alpha}{\longrightarrow} Q$, for $\alpha \in Act$ has be interpreted as 'P may perform the action α and become Q in doing so'.

We will write $\mathbf{0}$ to denote the inactive agent, defined as $\mathbf{0} \stackrel{def}{=} \sum_{i\in\emptyset} E_i$. /e will also write $P \stackrel{\alpha}{\longrightarrow}$ to mean that P' exists such that $P \stackrel{\alpha}{\longrightarrow} P'$; we will bbreviate $P \stackrel{\tau}{\longrightarrow} P'$ as $P \longrightarrow P'$ and $\alpha.\,\mathbf{0}$ as α; we will use $\equiv$ to denote

$$\frac{}{\alpha.E \xrightarrow{\alpha} E} \qquad \frac{E_j \xrightarrow{\alpha} E'_j}{\sum_{i \in I} E_i \xrightarrow{\alpha} E'_j} (j \in I)$$

$$\frac{E \xrightarrow{\alpha} E'}{E \setminus L \xrightarrow{\alpha} E' \setminus L} (\alpha, \overline{\alpha} \notin L) \qquad \frac{P \xrightarrow{\alpha} P'}{A \xrightarrow{\alpha} P'} (A \stackrel{def}{=} P)$$

$$\frac{E \xrightarrow{\alpha} E'}{E|F \xrightarrow{\alpha} E'|F} \qquad \frac{E \xrightarrow{\alpha} E'}{F|E \xrightarrow{\alpha} F|E'} \qquad \frac{E \xrightarrow{\ell} E' \quad F \xrightarrow{\overline{\ell}} F}{E|F \longrightarrow E'|F'}$$

Table 1: The operational semantics of CCS

syntactical identity.

Definition 1 *A process Q is a* descendant *of a process P if a sequence actions α_1, α_2 ..., α_n $\alpha_i \in Act$, $1 \leq i \leq n$, $n \geq 0$ exist such that $P \xrightarrow{\alpha_1}$ ... $\xrightarrow{\alpha_n} Q$ (where for $\alpha, \alpha' \in Act$, $\xrightarrow{\alpha}\xrightarrow{\alpha'}$ is defined in the obvious way composition of relations). The set $L \subseteq N$ is a* sort *for the process P if whene Q is a descendant of P and $Q \xrightarrow{\ell}$, then ℓ or $\overline{\ell}$ belong to L. Then we wr $P : L$.*

Strong Bisimulation is commonly accepted as the finest extensional or b havioral equivalence on processes one would like to impose.

Definition 2 *A relation $S \subseteq Pr^2$ is a* strong simulation *if for each $(P, Q) \in$ whenever $P \xrightarrow{\alpha} P'$ for some action α, then $Q \xrightarrow{\alpha} Q'$ and $(P', Q') \in S$; S* a strong bisimulation *if S and S^{-1} are strong simulations. Two processes and Q are* strongly bisimilar, *written $P \sim Q$ if $(P, Q) \in S$, for some stro bisimulation S.*

3 Reduction Bisimulation

In the definition of strong bisimulation, all actions in *Act* are treated uniform Can we get a characterization for $\sim$ if instead we focus only on the silent actio the simplest form of action? Consider the following definition:

Definition 3 *A relation $S \subseteq Pr^2$ is a* reduction simulation *if for each (P, Q) S, whenever $P \longrightarrow P'$ then $Q \longrightarrow Q'$ and $(P', Q') \in S$. S is a* reducti bisimulation *if S and S^{-1} are reduction simulation. Two processes P a Q are* reduction bisimilar, *written $P \dot{\sim}^{\tau} Q$ if $(P, Q) \in S$, for some reducti bisimulation S.*

By itself, $\dot{\sim}^\tau$ is not very interesting; it is rather weak and for instance it not preserved by parallel composition.

xample 1 *Consider* $P \stackrel{def}{=} a.\mathbf{0}$ *and* $Q \stackrel{def}{=} \mathbf{0}$. *Then we have* $P \dot{\sim}^\tau Q$, *but* $|\,\overline{a} \not\dot{\sim}^\tau Q\,|\,\overline{a}$.

It is natural then to consider the congruence induced by $\dot{\sim}^\tau$.

efinition 4 *Two processes P and Q are* reduction congruent, *written* $P \sim^\tau$ *if for each context C, it holds that* $C(P) \dot{\sim}^\tau C(Q)$.

EMARK: We are trying to use a dot to denote an equivalence which is not cessarily a congruence; so in this case, and for future equivalences, the ab- nce of a dot means a congruence.

oposition 1 $\sim \subseteq \sim^\tau$

oof. Clearly, $\sim \subseteq \dot{\sim}^\tau$; since $\sim$ is a congruence, $P \sim Q$ implies $C(P) \sim C(Q)$, ich in turns implies $C(P) \dot{\sim}^\tau C(Q)$, for every context C. □

EMARK: The proof of the above proposition would still be correct if we lowed non standard CCS operators in the context C, as long as such operators e "well-formed"; intuitively an operator is well-formed if its behavior depends ly on the semantics - not on the syntax - of its operands. In fact, $\sim$ is a ngruence over well-formed operators. The meaning of well-formed operators d well-formed transition systems is formally studied in [5] and [4].

Unfortunately the converse of Proposition 1 not true.

efinition 5 *P is* always divergent *if for every descendant P′ of P, we have* $' \longrightarrow$.

Processes which are always divergent are not necessarily strong bisimilar ee Proposition 3); however they are reduction congruent; intuitively since y state that they can reach is divergent, no CCS context can make the stinction between them as long as only τ actions are taken into account.

roposition 2 *If P and Q are always divergent, then* $P \sim^\tau Q$.

roof (outline) We shall prove that

$S = \{(C(P), C(Q)) \mid C$ is a context, and P, Q are always divergent$\}$

a reduction bisimulation.

Let $(C(P), C(Q)) \in S$. Suppose that $C(P) \longrightarrow C'(P)$ and C itself has oduced the τ-action (i.e. P has not been "used"): then also $C(Q) \longrightarrow C'(Q)$ d $(C'(P), C'(Q) \in S$.

Suppose now that $C(P) \longrightarrow C'(P')$ and P has contributed to such tion (either with an interaction with C or by simply performing itself τ-action): then P' cannot be appear as a subterm in a summation and he it cannot be discharged by C'. Since P' is always divergent, this means t also $C'(P')$ is always divergent. It remains to find a process R always diverg s.t. $C'(Q) \longrightarrow R$, as this would yield $(C'(P'), R) \in S$ (in the definition o take the empty context X).

Now, since P has contributed to the action of $C(P)$, P is not guar in $C(P)$, i.e. it is *active* and can perform reductions. Then also Q is act in $C(Q)$. Therefore if $Q \longrightarrow Q'$, then for some C'', also $C(Q) \longrightarrow C''(Q$ Moreover, for the same reason as $C'(P')$, also $C''(Q)$ is always divergent. $C''(Q') = R$, this concludes the proof.

Proposition 3 $\sim \neq \sim^{\tau}$.

Proof. Take $Q \stackrel{def}{=} \tau.Q$ and $P \stackrel{def}{=} \tau.P + a.Q$. By Proposition 2, $P \sim^{\tau}$ however it is $P \not\sim Q$.

We will prove however in section 3 that under some restrictions, nam divergence-freeness, we do have $\sim^{\tau} = \sim$.

Notice that we could distinguish the processes P and Q in the proof Proposition 3 if we added a disabling operator $\triangleright$, very similar in the intent Lotos's desrupting operator [6], described by the following rules:

i) if $Q \stackrel{\alpha}{\longrightarrow} Q'$ then $P \triangleright Q \stackrel{\alpha}{\longrightarrow} P \triangleright Q'$;

ii) if $P \stackrel{\ell}{\longrightarrow} P'$ and $Q \stackrel{\bar{\ell}}{\longrightarrow} Q'$, then $P \triangleright Q \longrightarrow P'$.

Then for P and Q as defined in Proposition 3 it would indeed hold th $\bar{a}.\,\mathbf{0} \triangleright P \dot{\not\sim}^{\tau} \bar{a}.\,\mathbf{0} \triangleright Q$.

We believe that, at least when image-finiteness is guaranteed, the additi of $\triangleright$ would make $\sim^{\tau}$ and $\sim$ coincide. We have not pursued further this dir tion. Instead we wanted to stick to CCS-contexts; what we have tried to is to increase as little as possible the power of the global observer, but enou to induce a congruence identical with $\sim$.

4 Barbed Bisimulation

Let us write $P\downarrow$ to mean that ℓ exists such that $P \stackrel{\ell}{\rightarrow}$ (that is, P can perfo some observable action) and $P \not\downarrow$ if it is not the case that $P\downarrow$.

Definition 6 *A process is called* inactive *if $P : \emptyset$ (i.e. P can only perform actions).*

finition 7 *A relation $S \subseteq Pr^2$ is a* barbed simulation *if for each $(P,Q) \in$*

whenever $P \longrightarrow P'$ then $Q \longrightarrow Q'$ and $(P',Q') \in S$;
$P\downarrow$ implies $Q\downarrow$.
s a barbed bisimulation *if S and S^{-1} are barbed simulation. Two processes and Q are* barbed-bisimilar, *written $P \dot{\simeq} Q$ if $(P,Q) \in S$, for some barbed imulation S.*

Notice that the global observer has visibility of the silent actions produced l of the state reached; he can also recognize the production of an observable ion, but in this case he cannot see neither the identity of the action produced the state reached. (In some sense, we are adding "barbs" to the reduction nantics: hence the name chosen.)

finition 8 *Two processes P and Q are* barbed-congruent, *written $P \simeq Q$ or each context C, it holds that $C(P) \dot{\simeq} C(Q)$.*

Again, clearly, it holds that $\sim \subseteq \simeq$. The remaining of this section is devoted the proof that $\simeq \subseteq \sim$.

Let b be a name. Consider the sequence $\{N^b_\gamma\}_{\gamma\ ordinal}$ of processes so deed:
$N^b_0 \stackrel{def}{=} \tau.b$,
$N^b_\gamma \stackrel{def}{=} \tau.(b + \tau.N^b_{\gamma-1})$, γ ordinal successor,
$N^b_\gamma \stackrel{def}{=} \sum_{\delta<\gamma} N^b_\delta$, γ ordinal limit.

mma 1 *Let P, Q, be two inactive processes; then for all ordinals γ and δ, $\neq \delta$ implies $P|N^b_\gamma \not\simeq Q|N^b_\delta$.*

oof. Use transfinite induction. □

Now we can prove the main theorem:

eorem 1 $\simeq = \sim$

Proof. We have only to prove $\simeq \subseteq \sim$. Let $P \simeq Q$. We will define a strong imulation S containing the pair (P,Q). Let $P, Q : L$, where $L = \{\ell_i | i \in I\}$ d $b \notin L$.

Because I is a set and $\{N^b_\gamma\}$ is built over the ordinals, we are sure that it possible to associate a distinguished N^b_i for each $i \in I$.
Define the context $C(X) \stackrel{def}{=} (X|V) \setminus L$ where
$V \stackrel{def}{=} \sum_{i\in I} \overline{\ell}_i.\tau.V_i, \qquad V_i \stackrel{def}{=} (b + \tau.\tau.(\tau.N^b_i + b) + \tau.V)$.
ow take $S = \{(P,Q) | P,Q : L \text{ and } C(P) \dot{\simeq} C(Q)\}$.

We claim that S is a strong bisimulation. Let $(P,Q) \in S$ and let $P \xrightarrow{\ell_i}$ We have to find Q' such that $Q \xrightarrow{\ell_i} Q'$ and $(P',Q') \in S$.

First note that $C(P) \longrightarrow\longrightarrow R$, for $R \equiv (P'|V_i) \setminus L$. Since $C(P) \dot{\simeq} C(Q$ we must have $C(Q) \longrightarrow\longrightarrow T$, for some T such that $R \dot{\simeq} T$. But $R\downarrow$; hen for some j, $Q \xrightarrow{\ell_j} Q'$ and $T \equiv (Q'|V_j) \setminus L$.

Also, it must be $i = j$. For if not, then $R \longrightarrow\longrightarrow R' \equiv (P'|(\tau.N_i^b + b)) \setminus$ where after the first move b is not observable, while it is after the second. Th forces $T \longrightarrow\longrightarrow T' \equiv (Q'|(\tau.N_j^b + b)) \setminus L$; but then $R' \longrightarrow R'' \equiv (P'|N_i^b) \setminus L$ a $R'' \not\downarrow$, which could only be matched with $T' \longrightarrow T'' \equiv (Q'|N_j^b) \setminus L$. Howev we can prove that $R'' \not\dot{\simeq} T''$. In fact, since $b \notin L$ and for any γ it is $N_\gamma^b : \{$ we have that $R'' \sim (P' \setminus L)|N_i^b$ and using Proposition 1, also $R'' \dot{\simeq} (P' \setminus L)|N$ similarly, $T'' \dot{\simeq} (Q' \setminus L)|N_j^b$. Both $(P' \setminus L)|N_i^b$ and $(Q' \setminus L)|N_j^b$ are inactive; th the conclusion follows from Lemma 1.

It therefore remains to prove $(P',Q') \in S$, i.e. $C(P') \dot{\simeq} C(Q')$. Usi analogous reasoning as above, one can prove that the move $R \longrightarrow (P'|V) \setminus L$ can only be matched by $T \longrightarrow (Q'\,|\,V) \setminus L$ obtaining $(P'\,|\,V) \setminus L \dot{\simeq} (Q'\,|\,V) \setminus$ which is exactly $C(P') \dot{\simeq} C(Q')$.

The case when $P \longrightarrow P'$ is simpler (V does not need to move).

Remark: Notice that the summations used in the proof of Theorem 1 becor finite when P and Q have finite sortings (which in practice is usually the cas

Proposition 3 asserted that in general $\sim \neq \sim^\tau$. This was shown by usi divergent processes. We can prove now that divergence is just what makes t two relations different.

Definition 9 *Let us call* divergent *a process which can perform an infin number of τ transitions and* divergent free *a process with no divergent desce dant.*

Definition 10 *Let $TotDiv$ be the largest class of processes such that whenev $P \in TotDiv$ and Q is a descendant of P, then $Q \longrightarrow$. We will call* total divergent *the processes in $TotDiv$.*

The process $\Omega \stackrel{def}{=} \tau.\Omega$ is the classical totally divergent process. Notice th a process may be divergent but not totally divergent. As an example, take t process $\tau.b + \Omega$.

Theorem 2 *If P and Q are divergent free, then $\sim = \sim^\tau$*

Proof. (outline) Because of Proposition 1 we have only to prove that $\sim^\tau$ $\sim$. Let $\{N_\gamma^\Omega\}_{\gamma\ ordinal}$, be defined as $\{N_\gamma^b\}_{\gamma\ ordinal}$ with $\Omega + \tau$ replacing b. No en P and Q are divergent free, for N_i^{Ω} instead of N_i^b, $\Omega + \tau$ instead of b, the dicate "$R \longrightarrow R'$, for some R' totally divergent" instead of "$R \downarrow$", and $\dot{\sim}^{\tau}$, instead of $\dot{\simeq}$, $\simeq$, the proofs of Lemma 1 and Theorem 1 remain correct, refore yielding $\sim^{\tau} \subset \sim$. □

Other issues on Barbed Bisimulations

mention here some other issues on barbed bisimulations that we have amined; in particular the use of barbed bisimulation in the weak case and higher-order process calculi.

. The Weak case

he weak case the relation $\stackrel{\alpha}{\Longrightarrow}$ replaces $\stackrel{\alpha}{\longrightarrow}$, where $\Longrightarrow$ is defined as reflexive l transitive closure of $\longrightarrow$, and $\stackrel{\alpha}{\Longrightarrow}$ is $\Longrightarrow\stackrel{\alpha}{\longrightarrow}\Longrightarrow$. Then $P \Downarrow$, which replaces , has to be interpreted as "for some ℓ, $P \stackrel{\ell}{\Longrightarrow}$".

Notice that since in the weak case τ-actions have no weight, weak reduction gruence corresponds to the universal relation (i.e. for every P and Q, $\approx^{\tau} Q$).

Let $\dot{\approx}$ be the obvious generalization of $\dot{\simeq}$ in the weak case. We have tried understand if it is possible to recover the CCS observational equivalence), probably the most important bisimulation-based equivalence for CCS, ng $\dot{\approx}$. First of all, since $\approx$ is not a congruence over the sum operator, we mot hope to get $\approx$ simply by parametrizing $\dot{\approx}$ over *all* the possible contexts is would lead more towards the CCS observational congruence).

Instead we need to parametrize $\dot{\approx}$ over a *subclass* of all the possible conts. The obvious representative is the subclass $\mathfrak{R}$ of *structural* contexts, uitively contexts which are built by composing variables and processes by ans of operators which cannot be discharged, i.e. their rules have the consion of the form $\otimes P_1 ... P_n \stackrel{\alpha}{\longrightarrow} \otimes P_1' ... P_n'$ (this excludes prefix and sum). We ve called barbed congruence over $\mathfrak{R}$ (written $\approx_{\mathfrak{R}}$) the resulting relation (i.e. $\approx_{\mathfrak{R}} Q$ if for every $C \in \mathfrak{R}$, $C(P) \dot{\approx} C(Q)$).

So far we have been able to prove that $\approx_{\mathfrak{R}} = \approx$ only for processes which image-finite w.r.t. the relation $\stackrel{\ell}{\Longrightarrow}$. It is still an open problem whether equality still holds when the image-finiteness requirement is dropped.

Because $\dot{\approx}$ seems not easy to handle, we have also considered the possiity of increasing the power of the observer. One way to obtain this is by rtitioning the observable actions into subsets and then allowing the observer be able to discriminate between observable actions from different subsets it still preventing him/her from discriminating actions in the same sub). Then, one gets $\approx$ just by using a partition into two subsets and by the parametrization of the resulting relation over the contexts in $\Re$.

As an aside let us just point out that the possibility of parametrizing t barbed-based bisimulations over a specific class of contexts seems to have otl interesting applications. In [11] it has been used to obtain some fully abstr model for lazy lambda calculus from its encoding into π-calculus. We plan considering also the use of parametrization in the framework of action refi ment. Intuitively, since the refinement of an action modifies the communicati protocol of a process, only contexts which "respect" such protocol should considered when verifying the equivalence between two refined processes.

5.2 Higher Order Process Calculi

We have considered π-calculus, Plain CHOCS and HOπ as case study.

In π-calculus. The π-calculus [9] is a generalization of CCS where nan can be exchanged as a result of a communication. Although it is perha questionable whether π-calculus should be considered an higher order proc calculus, it has been proved ([11]) that name passing is enough to give a n encoding of higher-order communications like process passing.

There are two major philosophies for defining in π-calculus the correspo dent of the CCS $\sim$ and $\approx$, respectively referred as "early version" and "la version" (see [10]); it is not clear yet which one should be preferred. We w interested in finding out which of the two – if any – could be captured usi the barbed bisimulation machinery. The answer has been the early versi a good point in favor of this one; in fact, the same results described here CCS can be recovered for the corresponding early equivalences in π-calculu

In Plain CHOCS and HOπ. Plain CHOCS ([12]) is a CCS-based la guage where process passing is adopted; HOπ ([11]) is an enrichment of calculus where - besides names - also processes and parametrized processes arbitrarily high order can be transmitted.

The adoption of the barbed bisimulations/congruences seems to solve t non trivial problem of obtaining a natural bisimulation-based equivalence such calculi. In particular, in Plain CHOCS the adoption of the barbed bisi ulations avoids some counterintuitive equalities induced by the higher-ord bisimulation proposed by Thomsen for his Plain CHOCS. Precise results this topic will appear in the second author's PhD thesis [11].

6 Acknowledgement

Thanks to David Walker for his comments on an earlier draft of this paper

eferences

Abramsky, S., *The Lazy Lambda Calculus*, in Declarative Programming, ed. Addison Wesley, 1988.

Berry, G. and Boudol, G., *The Chemical Abstract Machine*, Proc 17th Annual Symposium on Principle of Programming Languages, 1990.

Boudol, G., *A Lambda Calculus for Parallel Functions*, Rapport de Recherche 1231, INRIA-Sophia Antipolis, 1990.

Groote, J.F., *Transition System Specifications with Negative Premises*, Report CS-R8950, Centrum voor Wiskunde in Informatica, Amsterdam.

Groote, J.F. and Vaandrager, F.W., *Structured Operational Semantics and Bisimulation as a Congruence*, in Proc. ICALP '89, LNCS 372, 423-438, Springer Verlag, 1989.

ISO DIS 8807, Information Processing Systems - Open System Interconnection, LOTOS, A Formal Description Technique Based on the Temporal Ordering of Observational Behaviour, 1987.

] Milner, R., *Communication and Concurrency*, Prentice Hall, 1989.

] Milner, R., *Functions as Processes*, Report N.1154, INRIA-Roquencourt, 1990, to appear in Journal of Mathematical Structures in Computer Science.

] Milner, R., Parrow, J.G. and Walker, D.J., *A Calculus of Mobile Processes, Parts I and II*, Report ECS-LFCS-89-85 and -86, Laboratory for Foundations of Computer Science, Computer Science Department, Edinburgh University, to appear in Journal of Information and Computation.

0] Milner, R., Parrow, J.G. and Walker, D.J., *Modal Logics for Mobile Processes*, in Proc. CONCUR '91, LNCS 527, pp. 45-60, Springer Verlag, 1991.

1] Sangiorgi, D., forthcoming PhD thesis, Computer Science Department, Edinburgh University, 1992.

2] Thomsen, B., *Plain CHOCS*, Report DOC 89/4, Department of Computing, Imperial College, London, 1989.

Checking Equivalences Between Concurrent Systems of Finite Agents (Extended Abstract)

Alexander Rabinovich
IBM Research Division
T.J. Watson Research Center
P.O. Box 218
Yorktown Heights, NY 10598
e.mail: alik@watson.ibm.com

Abstract

Consider two synchronously communicating systems p and q over finite agents. Assume that one wants to check whether $p \sim q$ for one of the commonly used equivalences. We show that this question is PSPACE hard for all equivalences which lie between strong bisimulation and trace equivalences. For some equivalences exponential lower and upper bounds are proven. We also show that this problem is NP hard and co-NP hard even for a class of very simple finite agents.

Dataflow nets are the simplest systems of asynchronously communicating agents. Let $\sim$ be an equivalence which lies between strong bisimulation and trace equivalences. We show that there is no algorithm for checking $\sim$-equivalence between dataflow nets over finite state agents.

1. Introduction

1.1. Equivalences

There is a variety of semantics for concurrency which reflect the alternatives: li time vs branching time, interleaving vs causality. Here we consider only interlea semantics. In such a setting concurrent systems are described by labeled transi systems [Plo] or in a more classical terminology by automata (may be with infi number of states). Usually a behavior equivalence $\sim$ is introduced on labeled transi systems and the semantical objects are the $\sim$ equivalence classes. The question whe system p behaves like (or implements) system q is mathematically reformulated question whether $p \sim q$. In the literature on concurrency many such equivalences proposed. For example, in classical automata theory, two automata are equivalen they accept the same language. This equivalence is sometimes called *language* eq alence. *Weak trace* equivalence is an equivalence on labeled transition systems w ven coarser than language equivalence. Two labeled transition systems are (weak) e equivalent if the automata obtained from them by marking all states as accepting es are language equivalent. Weak trace equivalence is considered as the coarsest avior equivalence of interest.

guage and weak trace equivalences completely ignore branching. For example, they tify the automata *described* by expressions $a(b+c)$ and $ab+ac$, which are considered ifferent in most theories of concurrency. Informally, their difference is justified as ows: the first automaton after performing a can choose between b and c; on the er hand, the second automaton after performing a is unable to choose; only one of ons b, c is available in the state it has reached.

ther extreme equivalence is *(strong) bisimulation* equivalence. It catches very subtle erences between labeled transition systems on the basis of their branching structure. as a very appealing mathematical theory and is accepted as the finest behavior ivalence of interest for concurrency (it is often argued that strongly bisimilar labeled sition systems are indistinguishable for all reasonable notions of observations).

ny equivalences on labeled transition systems were studied in the literature. Failure ivalence [BHR], acceptance equivalence [He], weak bisimulation equivalence [Mi], ervational congruence [Mi] are only few among many well investigated equivalences. re is no consensus what is the best equivalence or what criteria it should satisfy. it seems that there exists a consensus that a good equivalence should lie between e and bisimulation equivalences. This consensus is supported by

pirical Fact: All equivalences studied in the literature lie between bisimulation trace equivalences.[1]

. Complexity of Checking Equivalences between Finite Agents

Kanellakis and S. Smolka examined in [KS] the computational complexity of checking erent equivalences between finite state automata. The classical result of L. Stock-er and A. R. Meyer [SM] states that the complexity of checking language equivalence SPACE complete. It is shown in [KS] that trace equivalence, failure equivalence a number of other equivalences are PSPACE complete.

vever, polynomial algorithms for checking bisimulation equivalence, weak bisimula- equivalence and observational congruence were given in [KS]. Paige and Tarjan [PT] e a very efficient algorithm for checking bisimulation equivalence; its complexity is $n \log m$, where n is a number of transitions and m is number of states. In [GV] and polynomial algorithms are given for checking branching bisimulation equivalence readiness equivalence. It is proved in [ABGS] that the bisimulation equivalence blem is PTIME complete.

As every empirical fact this one also has an exception. Some equivalences which are congruences action refinement distinguish between strong bisimilar automata

1.3. State Explosion

Concurrent CCS-like languages, in addition to sequential combinators like prefix choice and iteration use two new fundamental combinators: parallel composition and hiding. Roughly speaking, parallel composition describes communication betw components of a systems and hiding describes which communication events will bec invisible and which will remain observable.

Complex systems are assembled from simple agents by applying parallel composi and hiding. The role of these operations is prominent for nets of communicating age For example, *hide* $a_1, \ldots a_l$ *in* $(p_1 \| p_2 \ldots \| p_n)$ describes the network of synchronous c municating agents $p_1, \ldots p_n$ in which the communication events $a_1, \ldots a_n$ are hidde

From the complexity point of view, parallel composition is different from other op tions: the size of automata $p \| q$ is of the order $|p| \times |q|$. Therefore, if n_i is the size c then the size of system $p_1 \| p_2 \ldots \| p_k$ is $n_1 \times n_2 \times \cdots \times n_k$ and it is exponential in size $n_1 + n_2 \cdots + n_k + k$ of its description. This fact is known as state explosion.

Given descriptions *hide* $a_1, \ldots a_l$ *in* $(p_1 \| p_2 \ldots \| p_n)$ and *hide* $b_1, \ldots b_m$ *in* $(q_1 \| q_2 \ldots \|$ of two synchronously communicating systems p, q. Assume that one wants to ch whether $p \sim q$ for one of the commonly used equivalences. A straightforward algori will construct automata p and q and then will check their $\sim$-equivalence. Since sizes of p and q are exponential in the sizes of their description, the complexity of algorithm is at least EXPTIME. Can descriptions of p, q in terms of their compone p_i, q_j be used in order to obtain an efficient algorithm? J. F. Groote and F. Mo [GM] considered the problem of checking bisimulation equivalence between two syst $p = p_1 \| p_2 \ldots \| p_n$ and $q = q_1 \| q_2 \ldots \| q_n$ of finite automata. They developed a met that avoids the 'state explosion'-problem. It works only in the case when there is communication between the components of the systems. Their algorithm is polynon and works not only for strong bisimulation equivalence, but also for other equivaler which satisfy a certain set of axioms.

1.4. Our Contribution

We investigate the complexity of checking equivalences between networks of comm cating finite agents.

1.4.1. Synchronous Communication First we consider the problem of check equivalences between networks of synchronously communicating finite agents. Suc network p can be described as *hide* $a_1, \ldots a_l$ *in* $(p_1 \| p_2 \ldots \| p_n)$ where p_i are finite st automata and a_i are communication events. Here not only communications betw components of a system are allowed, but also some communication events can be hid (to become invisible τ-moves).

Given two networks p and q and an equivalence $\sim$ on automata. We show that problem whether $p \sim q$ is: (1) PSPACE-hard for all equivalences which lie betw strong bisimulation and trace equivalence. (2) it is NP hard and co-NP hard for equivalences which lie between strong bisimulation and trace equivalence even in

gents	equivalences	lower bound	upper bound
inite	language and trace	DSPACE($\Omega(c^{n/\log n})$)	DSPACE(d^n)
utomata	bisimulation	PSPACE	EXPTIME
	any equivalence between trace and bisimulation	PSPACE	
inite Acyclic utomata	any equivalence between trace and bisimulation	co-NP hard NP hard	

Figure 1: Complexity results for synchronous communication

e when p_i, q_j are acyclic automata. (3) for language and trace equivalences the blem is EXPSPACE-complete and we provide DSPACE($\Omega(c^{n/\log n})$) lower bound DSPACE($O(d^n)$) upper bound.

e that the algorithm which first constructs p and q and then checks whether they are mulation equivalent using Paige and Tarjan algorithm, has EXPTIME complexity gives us an upper bound for checking bisimulation. These results are summarized igure 1.

use parallel composition and hiding à la Hoare. However, the results stated above valid for Milner parallel composition and restriction. They also remain valid in any guage which is able to describe net operations.

.2. Asynchronous Communication Dataflow is the simplest model of asynonously communicating agents. A semantics of a dataflow net N over agents $p_1 \dots p_k$ be expressed as *hide* $c_1, \dots c_m$ *in* $((p_1 || p_2 \dots || p_n) || (ch_1 || ch_2 \dots || ch_r))$, where p_i are components of the nets, ch_j are the communication channels which behave like ounded buffers, and c_l are the internal ports of the net. Let $\sim$ be an equivalence ch lies between strong bisimulation and trace equivalences. We show that there is algorithm for checking $\sim$-equivalence between dataflow nets over finite state agents.

tion 2 provides basic definitions. Section 3 states main results. Ideas of proofs are y outlined. Section 4 - conclusion.

Basic Definitions

tion 2.1 introduces labeled transition systems. In section 2.2 trace and strong bisimtion equivalences are defined. The former is considered as the coarsest equivalence the later as the most discriminating equivalence for interest of interleaving seman- for concurrency. In section 2.3 synchronization and hiding operations on labeled nsition systems are defined. Section 2.4 briefly describes dataflow networks - the plest asynchronous model.

. Labeled Transition Systems

finition 1: *A Labeled Transition System (LTS) consists of*

- *Set of states Q.*
- *Initial state $q_0 \in Q$.*
- *Communication alphabet A and a special invisible action τ ($\tau \notin A$).*
- *Transition Relation: $\longrightarrow$ a subset of $Q \times \{A \cup \{\tau\}\} \times Q$.*

An action is either a communication or τ. We use $q \xrightarrow{a} q'$ as a notation for a transit from state q via action a to state q'; we say that a labeled transition system T is *fi* if it has a finite set of states, a finite alphabet and a finite transition relation. We that T is *acyclic* if it does not contain a cyclic path (a path of the from $q_1 \xrightarrow{a_1} q_2$ $\cdots \xrightarrow{a_{k-1}} q_k \xrightarrow{a_k} q_1$).

Remarks: *(Contrasting the above definition with classical automata theory).* (1) T alphabet is implicitly presented in automata theory. Usually it is extracted from transition diagram which defines an automaton. The role of the alphabet in concurre is much more important. The appearance of a in the alphabet of T, but not as a la of a transition of T implies that a will be blocked in any system which runs in para with T. (2) In our definition, unlike the definition of automaton, the set of accept states was not mentioned. Implicitly, all states are accepting states. It is a techni decision which simplifies our presentation.

2.2. Trace and Strong Bisimulation Equivalences

A finite alternating sequence $q_0,\ a_0,\ q_1,\ a_1, \cdots a_{n-1},\ q_n \cdots$ of states of LTS T a actions of T is an *execution sequence* of T if q_0 is the initial state of T and $q_{i-1} \xrightarrow{a_i}$ are transitions of T.

A *trace* of T is the sequence of communications which is obtained from an executi sequence by deleting the states of T and τ actions. We use the notation *trace*(T) the sets of traces of labeled transition system T.

Definition 2: *Labeled transition systems T and T_1 over the same alphabet are tr equivalent (notation $T \sim_{trace} T_1$) if trace$(T) =$ trace(T_1).*

The notion of strong bisimulation was introduced by D. Park [Pa] and plays a ve important role in concurrency [Mi].

Definition 3: *(Strong Bisimulation [Pa].) Let $T^1 = < Q^1,\ q_0^1,\ A,\ \longrightarrow_1>$ a $T^2 = < Q^2,\ q_0^2,\ A,\ \longrightarrow_2>$ be labeled transition systems over the same alphabet. relation R between the states of T^1 and T^2 is called a (strong) bisimulation if*

- $q_0^1 R q_0^2$
- *Whenever qRp and $q \xrightarrow{a}_1 q'$, then $p \xrightarrow{a}_2 p'$, for some p' for which $q'Rp'$.*
- *Whenever qRp and $p \xrightarrow{a}_2 p'$, then $q \xrightarrow{a}_1 q'$, for some q' for which $q'Rp'$.*

inition 4: *Labeled transition systems T and T_1 are (strong) bisimulation equivalent ation $T \sim_{bis} T_1$) if there exists a (strong) bisimulation relation between their states.*

say that an equivalence $\sim_1$ refines an equivalence $\sim_2$ (notations $\sim_1 \leq \sim_2$) if $T^1 \sim_1 T^2$ lies $T^1 \sim_2 T^2$. We say that $\sim$ lies between $\sim_1$ and $\sim_2$ if $\sim_1 \leq \sim \leq \sim_2$.

Operations on Labeled Transition Systems

define here synchronization (parallel composition) and hiding operations à la Hoare ch are more convenient for our purposes. But all results given in this paper are valid ilner's parallel composition and restriction are used or if one uses combinators from er CCS-like languages which are able to express net operations.

chronization. (infix notations - $||$) The synchronization T of two labeled transition ems $T^1 =< Q^1, q_0^1, A^1, \longrightarrow_1>$ and $T^2 =< Q^2, q_0^2, A^2, \longrightarrow_2>$ is defined as follows:

- States: $Q = Q^1 \times Q^2$
- The initial state is the pair consisting of the initial state of T^1 and the initial state of T^2.
- Communication alphabet $A = A^1 \cup A^2$.
- Transitions are given by the following inference rules:
  1. if a is not in $A^1 \cap A^2$ then
     $\dfrac{q_1 \xrightarrow{a}_1 q_2}{(q_1, q) \xrightarrow{a} (q_2, q)} \qquad \dfrac{q_1 \xrightarrow{a}_2 q_2}{(q, q_1) \xrightarrow{a} (q, q_2)}$
  2. τ-transitions
     $\dfrac{q_1 \xrightarrow{\tau}_1 q_2}{(q_1, q) \xrightarrow{\tau} (q_2, q)} \qquad \dfrac{q_1 \xrightarrow{\tau}_2 q_2}{(q, q_1) \xrightarrow{\tau} (q, q_2)}$
  3. if a is in $A^1 \cap A^2$
     $\dfrac{q_1 \xrightarrow{a}_1 q_1' \quad q_2 \xrightarrow{a}_2 q_2'}{(q_1, q_2) \xrightarrow{a} (q_1', q_2')}$

ling . *hide a in T^1* is the labeled transition system defined as follows:

- States: the same as the states of T^1.
- Communication alphabet: $A = A^1 - \{a\}$.
- The initial state is the initial state of T^1
- Transitions are defined by the following inference rules:
  1. if $a' \neq a$ then
     $\dfrac{q_1 \xrightarrow{a'}_1 q_2}{q_1 \xrightarrow{a'} q_2}$
  2. τ-transitions:
     $\dfrac{q_1 \xrightarrow{a}_1 q_2}{q_1 \xrightarrow{\tau} q_2} \qquad \dfrac{q_1 \xrightarrow{\tau} q_2}{q_1 \xrightarrow{\tau} q_2}$

t $A = \{a_1, a_2, \cdots a_n\}$ be a set of actions. We use notation *hide A in T* for le a_1 *in* (*hide* a_2 *in* $\cdots$ (*hide* a_n *in* T)...); Since *hide a in* (*hide b in T*)) = le b *in* (*hide a in T*)) this is a well defined notation.

2.4. Dataflow Networks

A dataflow is the simplest model of asynchronously communicating agents. A dataf network consists of agents communicating through unbounded buffer channels (see 2). Agents communicate with each other and with an environment by passing data c the channels. As for the synchronous case, an agent is described by a labeled transit system; the communication alphabet of a dataflow agent is structured as Cartes product $PORTS \times DATA$. An action $< c,\ d >$ is said to be a communication thro port c which passes the data value d. The ports of the net are partitioned into in output and internal. Data are coming from the environment through input ports are going to the environment through output ports. For example, I_1, I_2 are the in ports, O_1, O_2 are the output ports and all others are the internal ports for the ne Fig. 2. The channels of a network behave like unbounded buffers (FIFO queues).

The buffer from port I to port O over data set Δ is described by the following labe transition system:

Definition 5: *(Buffer) $BUF(I \rightarrow O)$ over Δ has the following labeled transition s tem:*

- *States: Δ^* - finite strings over Δ.*
- *Initial state - the empty string ϵ.*
- *Communication alphabet - $\{I,\ O\} \times \Delta$.*
- *Transition Relation: for every $d \in \Delta$ and every state $s \in \Delta^*$ there are transitic $s \xrightarrow{<I,d>} sd$ and $ds \xrightarrow{<O,d>} s$*

Therefore, the buffer is a simple infinite state labeled transition system.

A dataflow network has a very clear operational semantics. This semantics equivalent to the result obtained first by taking synchronization of all its com nent agents and the buffer channels which connect them and then hiding the tions on the internal ports. For example, the semantics of the net in Fig. 2 *hide* Int *in* $(A_1\|A_2\|A_3\|BUF_1\|\cdots\|BUF_7)$, where Int is the set of the communi tions through internal ports $c_1 \ldots c_{10}$.

3. Main Results

In all theorems stated in this section $p_1 \ldots p_n$, q are finite labeled transition syste and $\sim$ is an equivalence which lies between strong bisimulation and trace equivalen

Theorem 3.1: *The problem whether hide* $a_1, \ldots a_k$ in $(p_1\|p_2 \ldots p_n) \sim$ *hide* $b_1, \ldots$ in $(q_1\|q_2 \ldots q_r)$ *is NP hard and co-NP hard, even in the case when p_i and q_j are acy transition systems.*

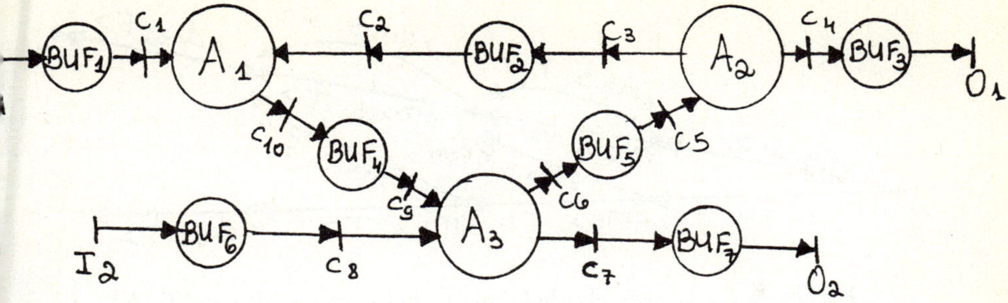

Figure 2: A dataflow net

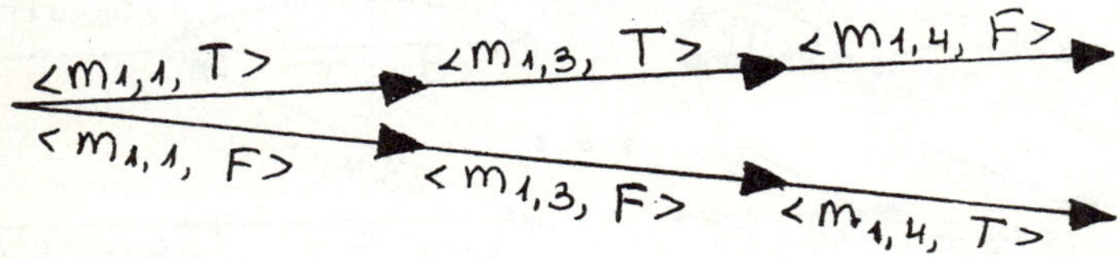

Figure 3: X_1 is over alphabet $\{m_{1,1},\ m_{1,3},\ m_{1,4}\} \times \{T,\ F\}$

of: We will show here only co-NP hardness. The proof for NP-hardness is similar. provide a reduction from the tautology problem. The tautology problem is defined ollows. There is a finite set $\{x_1,\ x_2, \ldots x_n\}$ of propositional variables. A literal is er a variable or its negation. A formula A is in 3-Disjunctive Normal Form if it is he form $\bigvee_i\ c_i$, where each c_i is a conjunction of three literals $c_i\ =\ \bigwedge_{j=1}^{3} l_{i,j}$. The blem whether $\forall x_1 \ldots x_n\ A$ is true is known as 3-DNF tautology problem, and it is NP complete.

are going to construct an expression which simulates this formula. It will have the n: $X_1 || \cdots || X_n || C_1 || C_2 \cdots || C_k || D$

labeled transition system X_i will 'simulate' variable x_i; C_i will 'simulate' conjunct nd D will 'simulate' disjunction of conjuncts.

will use structured alphabet - labels are of the form $< ch,\ v >$.

her than giving formal definitions we provide a generic example of the construction.

ume that x_1 occurs positively in c_1 and c_3 and negatively in c_4. Then X_1 is the 3 in Fig. 3. Assume that c_3 is $x_1 \wedge x_2 \wedge \neg x_4$; C_3 will be as in Fig. 4. Disjunction imulated by D in Fig. 5. Let us sketch some ideas underlying constructions of X_1, and D. X_1 is a tree with two branches. The upper (lower) branch corresponds to choice of value *True* (*False*) for x_1. For every clause c_i in which x_1 occurs, X_1 ds the value contributed by x_1 to this clause along port $m_{1,i}$; if x_1 occurs positively gatively) in C_i then X_1 contains action $< m_{1,i},\ T >$ ($< m_{1,i},\ F >$) on its upper nch and action $< m_{1,i},\ F >$ ($< m_{1,i},\ T >$) on its lower branch. C_3 reads the ies contributed by the variables to c_3. If all variables contribute value *True* then C_3 puts $< d_3,\ T >$, otherwise it outputs $< d_3,\ F >$. D reads the values contributed by c_i. If at least one clause contributes *True*, then D outputs $< Result,\ T >$, otherwise utputs $< Result,\ F >$.

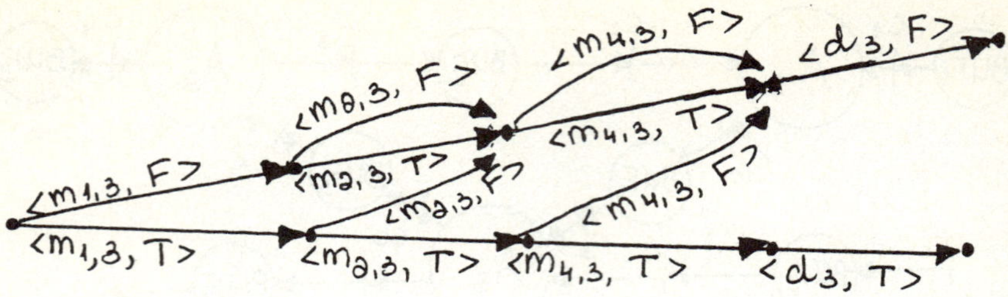

Figure 4: C_3 is over alpabet $\{m_{1,3},\ m_{2,3},\ m_{4,3},\ d_3\} \times \{T,\ F\}$

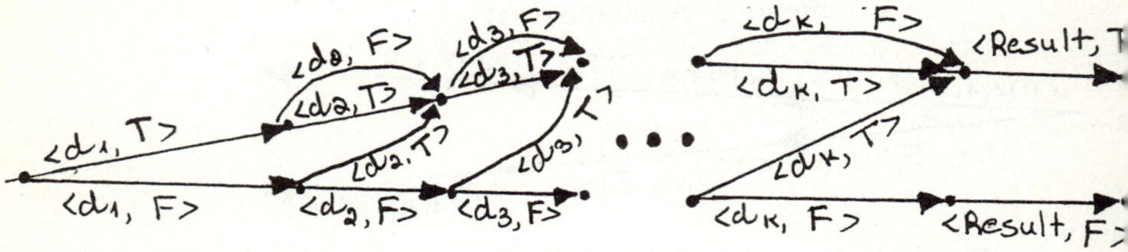

Figure 5: D is over alphabet $\{d_1,\ d_2, \ldots d_k,\ Result\} \times \{T,\ F\}$

Let M be the alphabet of $X_1 \| \cdots \| X_n \| C_1 \| C_2 \cdots \| C_k \| D$ and B is defi as $M - \{< Result,\ T >,\ < Result,\ F >\}$. Consider labeled transition sys $Sys = hide\ B\ in\ X_1 \| \cdots \| X_n \| C_1 \| C_2 \cdots \| C_k \| D$. Its description is polynomial in size of 'simulated' formmula $\forall x_1 \ldots x_n.(c_1 \vee c_2 \vee \cdots \vee c_k)$. It is easy to observe tha

1. Sys is an acyclic labeled transition system.
2. All its maximal paths have the same length l which is equal to the numbe occurrences of variables +k+1=4k+1.
3. The last action in any maximal path of Sys is either $< Result,\ T >$ $< Result,\ F >$.
4. All last actions in the maximal paths are $< Result,\ T >$ iff the formula is a tau ogy; in this case the Sys is bisimulation equivalent to the system $\tau^l < Result,$ (see Fig. 6).
5. Moreover, for any equivalence $\sim$ between trace and bisimulation equivalences $Sys \sim \tau^l < Result,\ T >$ iff the original formula is tautology.

Remarks: Hiding is not essential for co-NP hardness; one can compare syst $X_1 \| \cdots \| X_n \| C_1 \| C_2 \cdots \| C_k \| D$ with $X_1 \| \cdots \| X_n \| C_1 \| C_2 \cdots \| C_k \| D \| TRUE$, were TR

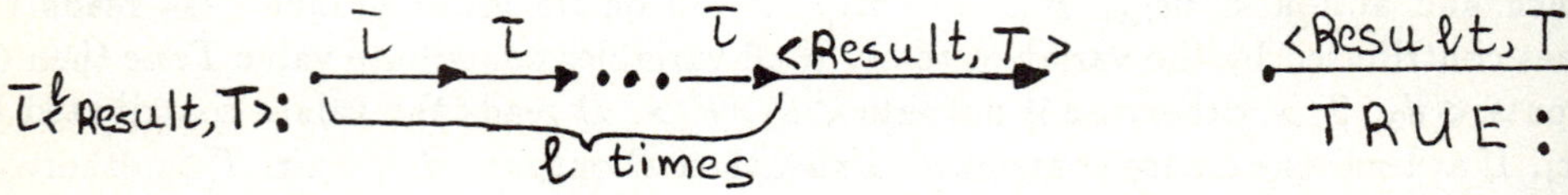

Figure 6: $\tau^l < Result,\ T >$ and $TRUE$ are over alphabet $\{Result\} \times \{T,\ F\}$

e LTS over alphabet $\{< Result,\ T >, < Result,\ F >\}$ given in Fig. 6. They are valent iff the simulated formula is a tautology.

eorem 3.2: *The problem whether hide* $a_1, \ldots a_n$ in $(p_1||p_2 \ldots ||p_k)$ *is* $\sim$*-equivalent ide* $b_1, \ldots b_m$ in $(q_1||q_2 \ldots ||q_r)$ *is PSPACE-hard.*

of: (Outline) For each deterministic Turing machine M that is space bounded by a nomial $p(n)$, we give a polynomial algorithm that takes a string x as an input and uces two expressions in the form *hide* $a_1, \ldots a_n$ *in* $(p_1||p_2 \ldots p_n)$. The expressions equivalent iff M accepts x. The central idea is to construct a short description of stem SIM which 'simulates' the behavior of M on x. In our proof, for x of length e SIM will have a form: $CONTROL_n||CELL_1||CELL_2||\cdots||CELL_{p(n)}$. $CELL_i$ esents the contents of the tape cell i in the computation of M on x. $CONTROL_n$ lates the control of M. Cells are communicating with control; there is no di- communication among the cells. The size of $CELL_i$ is independent of n and $NTROL_n$ has a size $O(n \log n)$. Space limitations prevent us from giving more ils about the construction. □

eorem 3.3: *The problem whether hide* $a_1, \ldots a_k$ in $(p_1||p_2 \ldots p_n)$ *is trace equivalent ide* $b_1, \ldots a_m$ in $(q_1||q_2 \ldots q_r)$ *is EXPSPACE-complete.*

of: (Outline of ideas) H. B. Hunt [Hu] showed that the problem of inequiva- e of regular expressions over operations *union, concatanetion, Kleene star*, and *rsection* is EXPSPACE-complete. We emulate[2] the simplified proof of this fact n by Furer [Fu]. The major difficulties in the emulation are: (1) How inter- ion operation can be 'simulated' by synchronization; (2) To show that the prob- is still EXPSPACE-hard even when restricted to the expressions in the *canonical* *hide A in* $(Prefix(e_1)||Prefix(e_2)\cdots||Prefix(e_n))$, where e_i are expressions over *on, concatanetion* and *Kleene star*, and *Prefix(L)* is a notation for the operation on uages which returns all the prefixes of the words in the language L. The proof gives $PACE(c^{n/\log n})$ lower bound. The upper bound of $DSPACE(O(d^n))$ is obtained larly to the proof of theorem 13.14 in [HU]. □

eorem 3.4: *The problem of* $\sim$*-equivalence of two dataflow nets over finite labeled sition systems is not recursively enumerable.*

of: (Hints) In the full paper we show that any Turing machine can be 'simulated' a finite transition system and two buffers (buffers simulate the tape and a finite sition system simulates the control). The divergence problem for Turing machines educible to the equivalence problem for dataflow nets. □

Conclusion

demonstrated that the problem of equivalence of synchronously communicating ems of finite agents is PSPACE-hard for any equivalence between bisimulation and

[2] Our emulation was inspired by A. Mayer and L. Stockmeyer paper [MS].

trace equivalences. As it was mentioned in the introduction, in order to check bisi lation equivalence between $p =$ *hide* $a_1, \ldots a_l$ *in* $(p_1 || p_2 \ldots || p_n)$ and $q =$ *hide* $b_1, \ldots$ *in* $(q_1 || q_2 \ldots || q_k)$, one can first construct p and q and then apply the algorithm g by Paige and Tarjan [PT]. This procedure requires exponential time. Our conjuc is: EXPTIME-lower bound can be proved for all equivalences between bisimulation trace. Recently, Larry Stockmeyer [Sto] proved $EXPTIME$ lower bound for chec bisimulation and weak bisimulation equivalences.

We proved EXPSPACE-completness for the problem of checking trace equivalence tween synchronous systems of finite agents. Let us mention other equivalences for w we can prove EXPSPACE lower bound. Bisimulation equivalence was originally defi as the limit of the sequence $\sim_0, \sim_1, \cdots \sim_k \cdots$ of successively finer equivalences (see page 224). We can show that for every fixed k checking $\sim_k$ equivalence between syst of synchronously communicating agents is EXPSPACE-hard. Language equivalenc defined on automata like in classical automata theory.[3] Synchronization and hiding be defined in a natural way on the automata. The proof of EXPSPACE-complete for language equivalence is very similar to the proof for trace equivalence.

Theorems 3.1, 3.2 provide lower bounds on complexity of checking an equivalence tween communicating systems of the form *hide* $a_1, \ldots a_k$ *in* $(p_1 || \ldots || p_n)$. Actually, proofs give stronger results. Namely, the lower bounds of these theorems hold even the systems of the form $p_1 || \ldots || p_n$ with *binary communications* (i.e., every action in the alphabets of at most two among $p_1, \ldots p_n$).

The results and proofs of the paper work not only for parallel composition and hidin la Hoare that were considered here, but also for CCS parallel composition and rest tion [Mi] and for other languages in which *net operations* can be described. Moreo the results hold not only for all equivalences between trace and bisimulation equi lences, but for all equivalences which satisfy a cumbersome semantical requirement. preferred to state the theorems in this way, a bit sacrificing generality, but with introducing these quite unnatural semantical conditions.

Acknowledgments

The author is grateful to Jan Friso Groote, Albert R. Meyer, Alain Mayer and La Stockmeyer for fruitful discussions and comments.

References

[ABGS] C. Alvarez, J. Balcazar, J. Gabarro and M. Santa. Parallel Complexity in design and analysis of concurrent systems. In *PARL 91*, volume 505 of *L Notes in Computer Science*. Springer Verlag, 1991.

[Bl] B. Bloom. Ready Simulation, Bisimulation, and the Semantics of CCS-l languages. *Ph.D. Thesis*, Technical Report MIT/LCS/TR-491, 1990.

[3] An automaton is a labeled transition system with a set of accepting states

R] S. Brookes, C. Hoare, and A. Roscoe. A theory of communicating sequential processes. *J. ACM*, 31(3):560–599, 1984.

M. Furer. The complexity of inequivalence problem for regular expressions with intersection. In *Proc. 7th ICALP*, volume 85 of *Lect. Notes in Computer Science*. Springer Verlag, 1980.

J. F. Groote and F. Moller. Verification of Parallel System via Decomposition. *Preliminary Report, CWI 1991.*

J. F. Groote and F. Vaandrager. An Efficient Algorithm for Branching Bisimulation and Stuttering Equivalence. In *International Conference on Automata, Languages and Programming*, volume 443 of *Lect. Notes in Computer Science*. Springer Verlag, 90.

M. Hennessy. *Algebraic Theory of Processes.* MIT Press, Cambridge, Massachusetts, 1988.

C. Hoare. *Communicating Sequential Processes.* Prentice-Hall International, Englewood Cliffs, 1985.

J. Hopcroft and J. Ullman. *Introduction to Automata Theory, Lamguages and Computations*, Addison-Wesley, Reading, MA, 1979.

H. Hunt. The equivalence problem for regular expressions with intersection is not polynomial in tape, *Tech. Rep. TR 73-161*, Cornell University, 1973.

A. J. Mayer and L. J. Stockmeyer. World Problems - This Time with Interleaving. *Technical Report, IBM RJ8180, July, 1991.*

R. Milner. *Communication and Concurrency.* Prentice-Hall International, Englewood Cliffs, 1989.

P. C. Kanellakis and S. A. Smolka. CCS Expressions, Finite State Processes, and Three Problems of Equivalence. In *Information and Computation*, volume 86, 1990.

R. Paige and R. Tarjan. Three partition refinement algorithms. *SIAM J. Comput.*, 16(6):973–989, 1987.

D. Park. Concurrency and Automata on infinite sequences. volume 104 of *Lect. Notes in Computer Science*. Springer Verlag, 1981.

G. Plotkin. A structured approach to operational semantics. FN 19 DAIMI, Aarhus Univ., 1981.

L. J. Stockmeyer and A. R. Meyer. Word Problems Requiring Exponential Time. In *Proc. 5th ACM Symp. on Theory of Computing, 1973.*

L. J. Stockmeyer. Private communication, Jan 1992.

Testing Preorders for Probabilistic Processes

Rance Cleaveland
Computer Science Dept.
N.C. State Univ.
Raleigh, NC 27695
USA
rance@csc.ncsu.edu

Scott A. Smolka
Computer Science Dept.
SUNY at Stony Brook
Stony Brook, NY 11794
USA
sas@cs.sunysb.edu

Amy Zwarico
Computer Science Dept.
Johns Hopkins Univ.
Baltimore, MD 21218
USA
amy@cs.jhu.edu

Abstract

We present a testing preorder for probabilistic processes based on the natural notion of a process passin a test with a certain probability. The theory enjoys close connections with the classical testing theory o Hennessy and DeNicola in that whenever a process passes a test with probability 1 (respectively some non- probability) in our setting, then the process *must* (respectively *may*) pass the test in the classical theory In addition, we develop an alternative characterization of the probabilistic testing preorder that is base on the "must sets" characterization of De Nicola. Finally, we extend our theory of testing to *substochasti processes*, in which the sum of the probabilities of a process's outgoing transitions may be strictly less tha 1, with the deficit representing the process' capacity for *undefined behavior.* A simple example involvin the construction of pipelines from faulty buffer cells is given to illustrate how substochastic processes, an the resulting preorder, can be used to model fault-tolerant systems and to reason about system reliability.

1 Introduction

Real-life processes often exhibit behavior of a statistical or probabilistic nature. For example, through re observations, a failure rate of 1.6×10^{-3} *failures/sec* can be attributed to an off-the-shelf memory chi communication medium can be seen to be dropping two messages over a one-minute interval with prob 0.0945.

In this paper, we present a semantic framework for *probabilistic reactive processes* based on the *preorder* of Hennessy and De Nicola [DNH83, Hen88]. We begin by formalizing the notion of a process p a test with a certain probability. This leads naturally to the definition of a *probabilistic testing preorder* process $\mathcal{P}$ is less than process $\mathcal{P}'$ in the preorder when, for all tests $\mathcal{T}$, the probability that $\mathcal{P}$ passes $\mathcal{T}$ greater than the probability that $\mathcal{P}'$ passes $\mathcal{T}$. This testing preorder has an alternative characterizatio extends the characterization of [DN87] for the nonprobabilistic preorder by taking into account inform about the actions a process may *refuse* (i.e. execute with probability 0).

We establish close connections with the testing theory of Hennessy and De Nicola as follows: let $\mathcal{P}$ probabilistic process, $\mathcal{T}$ a probabilistic test, and $h(\mathcal{P})$, $h(\mathcal{T})$ the "deprobabilized" process and test ob by erasing the probabilities in the underlying transition systems. Then we have that $h(\mathcal{P})$ *may* pass h $\mathcal{P}$ passes $\mathcal{T}$ with some non-zero probability, and $h(\mathcal{P})$ *must* pass $h(\mathcal{T})$ if $\mathcal{P}$ passes $\mathcal{T}$ with probability 1. these results, we show that if two probabilistic processes are related by our probabilistic testing preorder their deprobabilized images are related by the testing preorder of De Nicola and Hennessy. Thus, the t theory of probabilistic processes is a refinement of the testing theory of nonprobabilistic processes.

Finally, we extend the theory to handle *substochastic processes*. A substochastic process has the pr that the sum of the probabilities of its outgoing transitions may be strictly less than 1. The "missing" p the probability distribution can be interpreted as a measure of the *undefined behavior* of the process, ar be seen as a probabilistic analogue of the phenomenon of divergence in the testing theory of nonprobab processes. Substochastic processes enable the modeling of faulty and fault-tolerant processes, while the t preorder is useful in the analysis of system reliability. This is illustrated by a simple example involvi construction of pipelines of faulty buffer cells.

The structure of the rest of this paper is as follows. Section 2 defines probabilistic processes, test interaction systems, as well as our testing preorder for probabilistic processes. Section 3 connects our the testing for probabilistic processes with the theory of Hennessy and De Nicola for nondeterministic pro n 4 proposes another characterization of probabilistic testing, and Section 5 extends the theory of ilistic testing to substochastic processes. Section 6 contains our conclusions and directions for future

ed Work

pproaches to testing probabilistic processes have previously been proposed. Larsen and Skou [LS89] n *probabilistic bisimulation*, a natural and elegant extension of Milner and Park's strong bisimulation to ilistic processes. Based on a simple, yet powerful, language for tests (in which copying is permitted) they a testing characterization of probabilistic bisimulation. They also present a probabilistic extension of ssy-Milner logic that induces an equivalence on probabilistic processes that coincides with probabilistic lation.

ristoff [Chr90] introduces a testing framework based on the probability distributions over finite sequences ervable events induced by probabilistic processes in their interactions with deterministic tests. An ative characterization of the resulting preorders is given. His approach differs from our own in three ays: 1) his notion of testing is based on the probability of event sequences in interaction systems, where s based on the notion of a process passing a test with a certain probability; 2) he uses deterministic where in our setup, processes and tests are essentially indistinguishable; 3) as a result of 1) and 2), our ach is more closely grounded in the testing theory of De Nicola and Hennessy.

her work has focused on the development of models of probabilistic processes. In [vGSST90], van eek et al. introduce three such models—namely, *reactive*, *generative*, and *stratified*— and show that orm a hierarchy: the reactive model is an abstraction of the generative model which in turn is an ction of the stratified model. We use generative processes for the processes being tested, for tests and *interaction systems* resulting from the application of tests to processes. Christoff also uses generative sses but deterministic tests. Larsen and Skou use reactive processes and nonprobabilistic tests with g.

[JS90], probabilistic extensions of a number of other well-known process equivalences are examined, ing probabilistic trace, completed trace, failure, and ready.

Probabilistic Processes, Tests and Interaction Systems

s section we define the basic elements for our theory: probabilistic processes, tests, and interaction ns. This leads us to a natural notion of probabilistic testing and a probabilistic testing preorder.

Probabilistic Processes and Tests

bilistic processes are modeled using generative transition probabilities [vGSST90], meaning that from iven state, the sum of the probabilities on outgoing transitions is either 0 or 1; i.e. the process is either nated or not. Even though our interest is in reactive systems, we have chosen this model rather than active one of [vGSST90, LS89]. Reactive systems often make more than one possible action available environment, which may in turn attempt to execute any one of them. The generative model permits sulting nondeterminism as to which action actually takes place to be resolved probabilistically. This ssiveness is not available in the probabilistic reactive model, in which the sum of the probabilities of all ing actions *labeled by the same action* must be either 0 or 1.

ition 1 *A* probabilistic process *is a 4-tuple* (P, Act, μ, p_0) *where*

- P *is a countable set of* states;
- *Act is a countable set of observable* actions, *and* τ *is a distinguished "silent action" that is assumed not to be in Act;*
- $\mu : P \times (Act \cup \{\tau\}) \times P \to [0,1]$, *the* transition probability distribution function, *satisfies: for all* $p \in P$, $\sum_{p' \in P, a \in Act \cup \{\tau\}} \mu(p, a, p') \in \{0, 1\}$;
- $p_0 \in P$ *is the* start state.

lass of all probabilistic processes is denoted **PP**.

In the sequel we sometimes write $p \xrightarrow{a,\pi} p'$ when $\mu(p,a,p') = \pi > 0$ and $p \xrightarrow{a} p'$ if $\mu(p,a,p') > 0$. W use $\mu(p,a)$ as shorthand for $\sum_{p' \in P} \mu(p,a,p')$, and if $A \subseteq Act \cup \{\tau\}$ then we use $\mu(p,A)$ for $\sum_{a \in A}$ In the remainder of the paper we restrict our attention to *image-finite, divergence-free* processes. A $\mathcal{P} = (P, Act, \mu, p_0)$ is image-finite if for any $p \in P$ and $a \in Act \cup \{\tau\}$, the set $\{p' \mid p \xrightarrow{a} p'\}$ is finit divergence-free if no $p \in P$ is divergent. Here a state p is divergent if there is an infinite seqence $\langle p_0,$ with $p = p_0$ and $p_i \xrightarrow{\tau} p_{i+1}$.

We now turn our attention to defining the *tests* that will be applied to processes. Probabilistic te simply probabilistic processes with a set of states designated as *successful*. The reader should note the departure from the standard formulation [Hen88], in which a state of a test is considered successful if perform a w-transition.

Definition 2 *A* probabilistic test *is a 5-tuple* (P, Act, μ, p_0, S) *where* (P, Act, μ, p_0) *is a probabilistic and* $S \subseteq P$ *is a set of* success states.

In contrast to [LS89], we require that transitions in tests be assigned probabilities. This is beca permit τ-transitions to occur in tests, and since τ actions are different from other actions (becaus cesses/tests can execute them autonomously), the model must account for the probability of their occu In the remainder of the paper, we only consider tests that are finite in the usual sense (i.e. finite-bra and incapable of an infinite execution sequence). For both probabilistic processes and tests, we ofte the adjective "probabilistic" when this is clear from the context.

For notational conciseness, we make use of the following simple algebra of probabilistic processes/t the remainder of the paper.

Definition 3 *The* algebra of finite serial probabilistic processes and tests *is defined inductively as foll*

- nil *is the finite serial process given by* $(\{nil\}, \emptyset, \mu_{nil}, nil)$ *with* $\mu_{nil}(nil, \tau) = 0$.
- good *is the finite serial test obtained by designating the start state of* nil *to be successful, and the finite serial test obtained by designating the start state of* nil *to be unsuccessful. Formally, is the finite serial test given by* $(\{nil\}, \emptyset, \mu_{nil}, nil, \{nil\})$, *while* bad *is the finite serial test gi* $(\{nil\}, \emptyset, \mu_{nil}, nil, \emptyset)$.
- *Let* $\mathcal{P}_i = (P_i, Act_i, \mu_i, p_i)$ *be a finite set of finite serial processes with the property that for any* $p \in$ $\mu_i(p,a,p') = \mu_j(p,a,p')$ *for all* $a \in Act_i \cup Act_j$ *and* $p' \in P_i \cup P_j$. *Then* $a_1[\pi_1].\mathcal{P}_1 + \cdots + a_n[\pi,$ *with* $\sum \pi_i = 1$, *is a finite serial process. It is given by* (P, Act, μ, p_0), *where* $P = \bigcup_{i=1}^{n} P_i \cup \{p_0\}$ $p_0 \notin \bigcup_{i=1}^{n} P_i$), $Act = (\bigcup_{i=1}^{n} A_i \cup \{a_1, \ldots, a_n\}) - \{\tau\}$, *and* μ *given as follows.*

$$\mu(p,a,p') = \begin{cases} \pi_i & \text{if } p = p_0, a = a_i \text{ and } p' = p_i \\ \mu_i(p,a,p') & \text{if } p, p' \in P_i \\ 0 & \text{otherwise} \end{cases}$$

Similarly, if $\mathcal{T}_i = (T_i, Act_i, \mu_i, t_i, S_i)$ *is a finite set of finite serial tests with the appropriate pro then* $a_1[\pi_1].\mathcal{T}_1 + \cdots + a_n[\pi_n].\mathcal{T}_n$, *with* $\sum \pi_i = 1$, *is a finite serial test given by* (T, Act, μ, t_0, S), (T, Act, μ, t_0) *is as specified above and* $S = \bigcup_{i=1}^{n} S_i$.

So $a[1].nil$, $a[\frac{1}{3}].nil + b[\frac{2}{3}].c[1].nil$, and $a[1].(b[\frac{1}{2}].nil + \tau[\frac{1}{2}].nil)$ all represent finite serial pro In what follows we sometimes leave off trailing $.nil$'s and probabilities of 1; thus, e.g., $b.c$ is shortha $b[1].c[1].nil$. Readers should note the (intended) similarity with Milner's CCS [Mil80, Mil89].

2.2 Interaction Systems

We use *interaction systems* to formalize the notion of applying a test $\mathcal{T}$ to a process $\mathcal{P}$. Intuitive definition of an interaction system can be interpreted in terms of a *coin-flipping protocol*. Consid interaction between a process $\mathcal{P}$ and test $\mathcal{T}$ where initially $\mathcal{P}$ has n transitions of non-zero probabilit $\mathcal{T}$ has m transitions of non-zero probability. To commence interacting, $\mathcal{P}$ flips an n-sided coin (we in accordance with the probability distribution on its outgoing transitions) and, simultaneously, $\mathcal{T}$ fl m-sided coin (similarly weighted). Should $\mathcal{T}$'s coin flip indicate a transition a (with a possibly being $\mathcal{P}$'s coin flip indicate an a-transition as well, then the interaction system will perform an a-transitio ansition between the process and the test. On the other hand, if either $\mathcal{P}$'s or $\mathcal{T}$'s coin, but not ndicates a τ-transition, then it will do its τ independently of the other. Otherwise, the coin flips are d. The protocol continues until the process and test reach a configuration where no further transitions sible. Due to the finite nature of tests, the protocol is guaranteed to terminate in the sense that the lity of not terminating is 0.

tion 4 *Let $\mathcal{P} = (P, Act, \mu_P, p_0)$ be a probabilistic process and $\mathcal{T} = (T, Act, \mu_T, t_0, S)$ be a test. Then eraction system for $\mathcal{P}$ and $\mathcal{T}$, $\mathcal{P}||\mathcal{T}$, has the form $(P \times T, Act, \mu_I, \langle p_0, t_0 \rangle)$ where μ_I is defined with to a normalization function $\nu : P \times T \to [0,1]$, with*

$$\nu(p,t) = (\sum_{a \in Act} \mu_P(p,a) * \mu_T(t,a)) + \mu_T(t,\tau) + \mu_P(p,\tau) - \mu_T(t,\tau) * \mu_P(p,\tau)$$

ws:

$$\mu_I(\langle p,t \rangle, a, \langle p',t' \rangle) = \begin{cases} 0 & \text{if } \nu(p,t) = 0 \\ \frac{\mu_T(t,\tau,t')*(1-\mu_P(p,\tau))}{\nu(p,t)} & \text{if } a = \tau \ \wedge \ p = p' \ \wedge \ \nu(p,t) \neq 0 \\ \frac{\mu_P(p,\tau,p')*(1-\mu_T(t,\tau))}{\nu(p,t)} & \text{if } a = \tau \ \wedge \ t = t' \ \wedge \ \nu(p,t) \neq 0 \\ \frac{\mu_P(p,a,p')*\mu_T(t,a,t')}{\nu(p,t)} & \text{otherwise} \end{cases}$$

normalization factor $\nu(p,t)$ records the sum of the probabilities of the transitions possible from state and is used to compute the conditional probabilities of transitions from this interaction system state. t three terms can be seen to represent the probability that the test performs a τ *or* the process performs the definition of μ_I, the first clause pertains to a terminated interaction system state; the second clause es the conditional probability that the test performs an autonomous τ action; the third clause is similar second with the roles of p and t reversed; and the fourth clause computes the conditional probability e process and test synchronize (possibly on a τ).

ce tests are finite and processes are image-finite and divergence-free, interaction systems turn out to e. The normalization that is performed on the transition probabilities reflects our desire to model e processes. In a reactive system in which several actions (including, possibly, a τ) are enabled, if a test ble of activating at least one of these actions (or a τ) with a certain probability then some interaction n the system and test is guaranteed to take place. Viewed in this light, the normalized probabilities nt the conditional probability of a particular interaction taking place, *given that some interaction takes* A by-product of the normalization is that interaction systems themselves are stochastic.

clarify, consider the process $a[1]$ and the test $a[\frac{1}{3}].\,good + b[\frac{2}{3}].\,bad$. By our definition we end up with eraction system that is isomorphic to $a[1].\,nil$; in other words, the process and test synchronize on a obability 1. Without normalization, we would obtain a substochastic interaction system whose start erforms an a-transition with probability $\frac{1}{3}$, and no other transitions are possible.

borne out by the definitions, processes, tests, and interaction systems are all objects of essentially ne type. This homogeneity, which is not present in the approaches to probabilistic testing found 9, Chr90], mirrors that found in the original testing theory of De Nicola and Hennessy. One desirable ence of this is that there is a strong connection between applying a test to a process and running a in a network with other processes.

Testing Probabilistic Processes

w formalize the notion of a process passing a test with a certain probability, and we use this idea to a probabilistic testing preorder on processes.

tion 5 *Let $\mathcal{P} = (P, Act, \mu_P, p_0)$ be a process and $\mathcal{T} = (T, Act, \mu_T, t_0, S)$ a test, with $\mathcal{P}||\mathcal{T}$ the associated tion system.*

ach pair $\langle p,t \rangle \in P \times T$ is called a configuration*; A configuration $\langle p,t \rangle$ is* successful *if $t \in S$;*

computation *from a configuration $\langle p,t \rangle$ is a (maximal) sequence of the form*

$$\langle p,t \rangle = \langle p_0,t_0 \rangle \xrightarrow{a_1,\pi_1} \langle p_1,t_1 \rangle \xrightarrow{a_2,\pi_2} \dots \xrightarrow{a_n,\pi_n} \langle p_n,t_n \rangle$$

here

- *for all* $1 \le i \le n$, $\mu_I(\langle p_{i-1}, t_{i-1}\rangle, a_i, \langle p_i, t_i\rangle) = \pi_i$;
- *for all* $0 \le i \le n-1$, $\langle p_i, t_i\rangle$ *is unsuccessful,* i.e., $t_i \notin S$; *and*
- *either* $\langle p_n, t_n\rangle$ *is successful or* $\sum_{a\in Act\cup\{\tau\}} \mu_I(\langle p_n, t_n\rangle, a) = 0$.

- $\mathcal{C}_{p,t}$ *is the set of all computations starting from* $\langle p,t\rangle$, $\mathcal{C} = \bigcup_{p,t}\mathcal{C}_{p,t}$ *is the set of all computation* $\mathcal{S}_{p,t} \subseteq \mathcal{C}_{p,t}$ *is the set of all successful computations from* $\langle p,t\rangle$. $\mathcal{C}, \mathcal{C}_{p,t}, \mathcal{S}_{p,t}$ *are ranged over by* C.
- *The* **computation probability distribution function** $Pr : \mathcal{C} \to [0,1]$ *is defined inductively as follow*

$$Pr(\langle p,t\rangle) = 1$$
$$Pr(\langle p,t\rangle \xrightarrow{a,\pi} C) = \pi * Pr(C)$$

We now define the notion of passing a test with a particular probability and from this derive a preorder for probabilistic processes.

Definition 6 *Let* $\mathcal{P} = (P, Act, \mu_P, p_0)$ *and* $\mathcal{P}' = (P', Act, \mu_{P'}, p_0')$ *be processes and* $\mathcal{T} = (T, Act, \mu_T, t_0$ *a test.*

- $\mathcal{P}$ *pass*$_\pi$ $\mathcal{T}$ *if* $\sum_{C\in\mathcal{S}_{p_0,t_0}} Pr(C) \ge \pi$.
- $\mathcal{P} \sqsubseteq \mathcal{P}'$ *if for all* $\pi \in [0,1]$ *if* $\mathcal{P}$ *pass*$_\pi$ $\mathcal{T}$ *then* $\mathcal{P}'$ *pass*$_\pi$ $\mathcal{T}$.

In the following section, we will show that there are close connections between the *pass*$_\pi$ relatic the *may* and *must* relations of the nonprobabilistic setting. We end this section with some examples probabilistic preorder.

Example 1

1. Let $\mathcal{P}_1 = a[\frac{1}{3}].b + a[\frac{2}{3}].c$ and $\mathcal{P}_1' = a[\frac{1}{2}].b + a[\frac{1}{2}].c$. Then $\mathcal{P}_1 \not\sqsubseteq \mathcal{P}_1'$ and $\mathcal{P}_1' \not\sqsubseteq \mathcal{P}_1$ as can be s considering the tests $\mathcal{T}_1 = a.c.good$, $\mathcal{T}_1' = a.b.good$. $\mathcal{P}_1$ *pass*$_{\frac{2}{3}}$ $\mathcal{T}_1$ but only $\mathcal{P}_1'$ *pass*$_{\frac{1}{2}}$ $\mathcal{T}_1$, and $\mathcal{P}_1'$ *pa* but only $\mathcal{P}_1$ *pass*$_{\frac{1}{3}}$ $\mathcal{T}_1'$.
2. Let $\mathcal{P}_2 = a[\frac{1}{2}].b.c + a[\frac{1}{2}].b.d$ and $\mathcal{P}_2' = a.(b[\frac{1}{2}].c + b[\frac{1}{2}].d)$. Then $\mathcal{P}_2 = \mathcal{P}_2'$. This example a probabilization of the well-known pair of processes that illustrates that bisimulation is a refin of testing equivalence. We have, as desired, that these processes are distinguished by proba bisimulation [LS89].
3. Let $\mathcal{P}_3 = a[\frac{1}{2}].b + a[\frac{1}{2}].c$ and $\mathcal{P}_3' = a.(b[\frac{1}{2}] + c[\frac{1}{2}])$. Then $\mathcal{P}_3 \not\sqsubseteq \mathcal{P}_3'$ and $\mathcal{P}_3' \not\sqsubseteq \mathcal{P}_3$. Consider th $\mathcal{T}_3 = a.(b[\frac{1}{4}].good + c[\frac{3}{4}].bad)$. $\mathcal{P}_3$ *pass*$_{\frac{1}{2}}$ $\mathcal{T}_3$ but only $\mathcal{P}_3'$ *pass*$_{\frac{1}{4}}$ $\mathcal{T}_3$, and therefore $\mathcal{P}_3 \not\sqsubseteq \mathcal{P}_3'$. Th $\mathcal{T}_3' = a.(b[\frac{3}{4}].good + c[\frac{1}{4}].bad)$ can be similarly used to establish that $\mathcal{P}_3' \not\sqsubseteq \mathcal{P}_3$.

In the last example we have a slight departure from the nonprobabilistic case, where the deproba version of $\mathcal{P}_3$ is strictly less than the deprobabilized version of $\mathcal{P}_3'$ in the testing preorder (due to the nondeterministic nature of $\mathcal{P}_3$). Intuitively, the reason why these probabilistic processes are not rel that after executing an a, the test $\mathcal{T}_3$ is sufficiently "biased" in favor of $c.bad$ to ensure that $\mathcal{P}_3$ passes a greater probability (since half of the time its c is disabled after performing an a) than $\mathcal{P}_3'$ (which nev its c disabled after engaging in an a).

3 Recovering Nonprobabilistic from Probabilistic Testing

We now establish that the probabilistic testing preorder $\sqsubseteq$ is a strict refinement of the classical testing pr $\sqsubseteq_{test}$ of [DNH83, Hen88]. The proof relies on a correspondence between a probabilistic process passing with probability 1 (respectively a non-zero probability) and the associated nonprobabilistic process *passing* (respectively *may-passing*) the associated nonprobabilistic test.

We begin by introducing our renderings of the definitions of classical (nonprobabilistic) testing [D Hen88]. Although our framework differs slightly from theirs, the differences are shown to be inconsequ

tion 7 *A (nonprobabilistic) process is a quadruple* $(P, Act, \rightarrow, p_0)$, *where* P *is a set of states,* Act *a set* *ns,* $\rightarrow \subseteq P \times Act \times P$ *the* transition relation, *and* $p_0 \in P$ *the* start state. *We use* $\mathbf{P}$ *to denote the class of* *babilistic processes. A (nonprobabilistic) test is a quintuple* $(T, Act, \rightarrow, t_0, S)$, *where* $(T, Act, \rightarrow, t_0) \in \mathbf{P}$ $\subseteq T$ *is the set of* sucessful states.

ur definition of probabilistic tests (Definition 2), our definition of nonprobabilistic tests differs slightly hat of [DNH83, Hen88]; there, a test is a process, and the successful states are defined to be those e of performing a special "success action," w. This difference, however, is minor. In what follows, we $\stackrel{a}{\rightarrow}$ (resp. $p \stackrel{a}{\not\rightarrow}$) if there exists (resp. does not exist) a p' with $p \stackrel{a}{\rightarrow} p'$. We also write $p \not\rightarrow$ if $p \stackrel{a}{\not\rightarrow}$ holds $a \in Act \cup \{\tau\}$.

e result of DeNicola and Hennessy is that it suffices to consider only finite tests when defining the l testing preorders. Accordingly, we only consider such tests in what follows. We now define what it for a nonprobabilistic process to pass a test. We do so by introducing (nonprobabilistic) interaction s, which record the computations in which a process may engage in response to a test.

tion 8 *Let* $\mathcal{P} = (P, Act, \rightarrow_P, p_0)$ *be a nonprobabilistic process and* $\mathcal{T} = (T, Act, \rightarrow_T, t_0, S)$ *be a non-* *listic test. Then the nonprobabilistic interaction system* $\mathcal{P}||'\mathcal{T}$ *is a quadruple* $(P \times T, Act, \rightarrow, \langle p_0, t_0 \rangle)$, $\rightarrow$ *is defined as follows.*

$\stackrel{\tau}{\rightarrow}_T t', p \not\rightarrow_P \ \vee\ \exists a \neq \tau . p \stackrel{a}{\rightarrow}_P \ \Rightarrow\ \langle p,t \rangle \stackrel{\tau}{\rightarrow} \langle p,t' \rangle.$

$\stackrel{\tau}{\rightarrow}_P p', t \not\rightarrow_T \ \vee\ \exists a \neq \tau . t \stackrel{a}{\rightarrow}_T \ \Rightarrow\ \langle p,t \rangle \stackrel{\tau}{\rightarrow} \langle p',t \rangle.$

$\stackrel{a}{\rightarrow}_P p', t \stackrel{a}{\rightarrow}_T t' \ \Rightarrow\ \langle p,t \rangle \stackrel{a}{\rightarrow} \langle p',t' \rangle.$

definition of $\rightarrow$ above differs considerably from that of [DNH83, Hen88]. In the first clause, autonomous s by t are disallowed if the only action available to p is τ; in this case, t must synchronize with p to τ-move. This restriction on the behavior of t is needed to reflect the factor of $(1 - \mu_P(p,t))$ in the clause of the definition of μ_I (Definition 4). The second clause is similar to the first with the roles of t eversed. Finally, the third clause permits synchronization on τ. As the motivation for the differences two definitions is technical, it is difficult to justify it in isolation. However, it turns out that the babilistic testing preorders induced by this definition coincide exactly with those of De Nicola and sy. We return to this point after defining the notions of a process $\mathcal{P}$ *may-passing* and *must-passing* a

tion 9 *Let* $\mathcal{P} = (P, Act, \rightarrow_P, p_0)$ *be a nonprobabilistic process and* $\mathcal{T} = (T, Act, \rightarrow_T, t_0)$ *be a nonprob-* *c test, with* $\mathcal{P}||'\mathcal{T}$ *the associated interaction system.*

onfigurations *and* successful configurations *are defined as in Definition 5.*

computation *from a configuration* $\langle p,t \rangle$ *is a (maximal) sequence of the form*

$$\langle p,t \rangle = \langle p_0,t_0 \rangle \stackrel{a_1}{\rightarrow} \langle p_1,t_1 \rangle \stackrel{a_2}{\rightarrow} \ldots \stackrel{a_n}{\rightarrow} \langle p_n,t_n \rangle$$

here

- *for all* $0 \leq i \leq n-1$, $\langle p_i,t_i \rangle$ *is unsuccessful; and*
- *either* $\langle p_n,t_n \rangle$ *is successful or* $\langle p_n,t_n \rangle \not\rightarrow$.

$_{p,t}$, $\mathcal{C}$, *and* $\mathcal{S}_{p,t}$ *are defined as in Definition 5.*

may $\mathcal{T}$ *if* $\mathcal{S}_{p_0,t_0} \neq \emptyset$.

must $\mathcal{T}$ *if* $\mathcal{S}_{p_0,t_0} = \mathcal{C}_{p_0,t_0}$.

noted above, our definition of $\rightarrow$ between configurations differs markedly from the one of De Nicola nnessy, and one might therefore worry that our notions of *may* and *must* are also different. This ut not to be the case; $\mathcal{P}$ *may* (resp. *must*)$\mathcal{T}$ holds in our framework exactly when $\mathcal{P}$ *may* (resp. *must*)$\mathcal{T}$ n the framework of [DNH83, Hen88]. Thus, the preorders defined next coincide with theirs.

Definition 10 *Let $\mathcal{P}, \mathcal{P}' \in \mathbf{P}$.*

- *$\mathcal{P} \sqsubseteq_{may} \mathcal{P}'$ if for all $\mathcal{T}, \mathcal{P}$ may $\mathcal{T}$ implies $\mathcal{P}'$ may $\mathcal{T}$.*
- *$\mathcal{P} \sqsubseteq_{must} \mathcal{P}'$ if for all $\mathcal{T}, \mathcal{P}$ must $\mathcal{T}$ implies $\mathcal{P}'$ must $\mathcal{T}$.*
- *$\mathcal{P} \sqsubseteq_{test} \mathcal{P}'$ if $\mathcal{P} \sqsubseteq_{may} \mathcal{P}'$ and $\mathcal{P} \sqsubseteq_{must} \mathcal{P}'$.*

In the remainder of this section, we show that $\sqsubseteq$ refines $\sqsubseteq_{test}$. We begin by establishing that Pr (I tion 5) is indeed a probability distribution function for all maximal computations in a probabilistic inter system.

Lemma 1 *Let $\mathcal{P} = (P, Act, \mu_P, p_0)$ and $\mathcal{T} = (T, Act, \mu_T, t_0)$. Then $\sum_{C \in \mathcal{C}_{p_0,t_0}} Pr(C) = 1$.*

Proof: We prove something stronger: for all $\langle p, t \rangle \in P \times T$, $\sum_{C \in \mathcal{C}_{p,t}} Pr(C) = 1$. The proof is by ind on the length of the longest computation in $\mathcal{C}_{p_0,t_0}$.

We next define the function h that maps probabilistic processes to nonprobabilistic ones essenti erasing the probabilities on transitions.

Definition 11 *The mapping $h : \mathbf{PP} \to \mathbf{P}$ is defined by $h((P, Act, \mu_P, p_0)) = (P, Act, \to_P, p_0)$, where $(p, a, p') \in \to_P$ if $\mu_P(p, a, p') > 0$.*

Note that we abuse notation slightly by applying h to tests in the obvious way: $h((T, Act, \mu_T, t_0,$ $(T, Act, \to_T, t_0, S)$. The next lemma establishes that h is a homomorphism with respect to the inte system constructor $||$.

Lemma 2 *Let $\mathcal{P}$ be a probabilistic process, $\mathcal{T}$ a test, and $\mathcal{I} = \mathcal{P}||\mathcal{T}$ an interaction system. Then I $h(\mathcal{P}) ||' h(\mathcal{T})$.*

Proof: Let $\langle p, t \rangle$ be a state in $h(\mathcal{I})$ and $a \neq \tau$. $\langle p, t \rangle \xrightarrow{a} \langle p', t' \rangle$ iff $\mu_I(\langle p, t \rangle, a, \langle p', t' \rangle) > 0$ iff $\mu_P(p, a,$ and $\mu_T(t, a, t') > 0$ iff $p \xrightarrow{a} p'$ and $t \xrightarrow{a} t'$ iff $\langle p, t \rangle \xrightarrow{a} \langle p', t' \rangle$ in $h(\mathcal{P})||'h(\mathcal{T})$. The proof is similar for τ actic

The proof that $\sqsubseteq$ implies $\sqsubseteq_{test}$ is broken into several steps. We first establish a relationship between and *may* and *must*. We then show that $\sqsubseteq$ implies both $\sqsubseteq_{may}$ and $\sqsubseteq_{must}$. From this we obtain the that the probabilistic testing preorder is a refinement of the nonprobabilistic testing preorder. In the fol lemmas, let $\mathcal{P}, \mathcal{P}'$ be probabilistic processes and $\mathcal{T}$ a probabilistic test.

Lemma 3 *Let $\mathcal{P}$ and $\mathcal{T}$ be a probabilistic process and test, respectively.*

- *If $\mathcal{P}$ $pass_\pi$ $\mathcal{T}$ for some $\pi > 0$, then $h(\mathcal{P})$ may $h(\mathcal{T})$.*
- *If $\mathcal{P}$ $pass_1$ $\mathcal{T}$, then $h(\mathcal{P})$ must $h(\mathcal{T})$.*

This lemma follows directly from Lemmas 1 and 2. Using it, we may now prove the following.

Lemma 4 *Let $\mathcal{P}$ and $\mathcal{P}'$ be probabilistic processes.*

- *$\mathcal{P} \sqsubseteq \mathcal{P}'$ implies $h(\mathcal{P}) \sqsubseteq_{may} h(\mathcal{P}')$.*
- *$\mathcal{P} \sqsubseteq \mathcal{P}'$ implies $h(\mathcal{P}) \sqsubseteq_{must} h(\mathcal{P}')$.*

The next theorem is a trivial consequence of this lemma.

Theorem 1 *$\mathcal{P} \sqsubseteq \mathcal{P}'$ implies $h(\mathcal{P}) \sqsubseteq_{test} h(\mathcal{P}')$.*

Note that the converse of Theorem 1 does not hold. For example, consider once again processe $a[\frac{1}{3}] . b + a[\frac{2}{3}] . c$ and $\mathcal{P}_1' = a[\frac{1}{2}] . b + a[\frac{1}{2}] . c$ of Example 1: $h(\mathcal{P}_1) =_{test} h(\mathcal{P}_1')$ but $\mathcal{P}_1 \not\sqsubseteq \mathcal{P}_1'$ and $\mathcal{P}_1' \not\sqsubseteq \mathcal{P}_1$. the testing theory of nonprobabilistic processes is strictly more abstract than the testing theory of proba processes.

'oward an Alternative Characterization

ntuitively appealing, the operational definition of our preorder can be difficult to reason about. In lar, establishing that one process is related to another requires a consideration of the behavior of rocesses in the context of all possible tests. Accordingly, in this section we propose an alternative erization of the preorder that is designed to ease the task of establishing relationships between systems. aracterization also has a denotational flavor that may be useful in future developments of denotational .

characterization is reminiscent of De Nicola's account of the testing preorder [DN87] in that it relates es on the basis of the sets of actions they can execute after engaging in a sequence of visible actions arbitrary amounts of internal computation may be performed in between). In addition, we also take of the actions that a process *refuses* after a sequence.

rder to define our characterization we first introduce the probabilistic counterpart of the $\stackrel{a}{\Rightarrow}$ found in l process theory.

tion 12 *Let $\mathcal{P} = (P, Act, \mu_P, p_0)$ be a probabilistic process.*

$M_{\mathcal{P},\tau} : P \times Int \times P \to [0,1]$ is defined as follows.

- $M_{\mathcal{P},\tau}(p, 0, p') = \begin{cases} 1 & \text{if } p = p' \\ 0 & \text{otherwise} \end{cases}$
- $M_{\mathcal{P},\tau}(p, i+1, p') = \sum_{q \in P} \mu_P(p, \tau, q) * M_{\mathcal{P},\tau}(q, i, p')$

'he function $M_{\mathcal{P},\epsilon} : P \times P \to [0,1]$ is defined as follows.

$$M_{\mathcal{P},\epsilon}(p, p') = \sum_{i=0}^{\infty} M_{\mathcal{P},\tau}(p, i, p')$$

'he function $M_{\mathcal{P}} : P \times Act \times P \to [0,1]$ is defined as follows.

$$M_{\mathcal{P}}(p, a, p') = \sum_{q,q' \in P} M_{\mathcal{P},\epsilon}(p, q) * \mu(q, a, q') * M_{\mathcal{P},\epsilon}(q', p')$$

ely, $M_{\mathcal{P},\tau}(p, i, p')$ records the probability that after i steps of internal computation, process $\mathcal{P}$ moves ate p to p'. $M_{\mathcal{P},\epsilon}(p, p')$ then represents the probability that $\mathcal{P}$ may move from p to p' after some number nal computation steps, while $M_{\mathcal{P}}(p, a, p')$ computes the probability that, from p, $\mathcal{P}$ may perform some computation, followed by an a action, followed by more internal computation, and enter state p'. ll-formedness of these definitions (in particular, of $M_{\mathcal{P},\epsilon}$) depends crucially on our assumption that esses are divergence-free; thus, there can be no τ-cycles, and hence, no possibility of $M_{\mathcal{P},\epsilon}(p, p)$ being han 1. We sometimes write $M_{\mathcal{P}}(p, a)$ to represent $\sum_{p' \in P} M_{\mathcal{P}}(p, a, p')$.

also introduce the following function, $C_{\mathcal{P}} : P \times 2^{Act} \times 2^{Act}$.

$$C_{\mathcal{P}}(p, S_1, S_2) = \begin{cases} 1 & \text{if } \exists a \in S_1 . M_{\mathcal{P}}(p, a) > 0 \wedge \forall a \in S_2 . M_{\mathcal{P}}(p, a) = 0 \\ 0 & \text{otherwise} \end{cases}$$

ely, $C_{\mathcal{P}}(p, S_1, S_2) = 1$ if p is capable of executing (with non-zero probability) at least one action in S_1 action in S_2.

alternative characterization now relies on the function $\mathrm{AC}_{\mathcal{P}}$.

tion 13 *Let $\mathcal{P} = (P, Act, \mu_P, p_0)$ be a probabilistic process. Then $\mathrm{AC}_{\mathcal{P}} : P \times Act^* \times 2^{Act} \times 2^{Act} \to [0,1]$ given inductively as follows.*

$$\mathrm{AC}_{\mathcal{P}}(p, \epsilon, S_1, S_2) = (1 - \mu_P(p, \tau)) * C_{\mathcal{P}}(p, S_1, S_2) + \sum_{p' \in P} \mu_P(p, \tau, p') * \mathrm{AC}_{\mathcal{P}}(p', \epsilon)$$

$$\mathrm{AC}_{\mathcal{P}}(p, as, S_1, S_2) = \begin{cases} 1 & \text{if } M_{\mathcal{P}}(p, a) = 0 \\ \frac{1}{M_{\mathcal{P}}(p,a)} * \sum_{p' \in P} M_{\mathcal{P}}(p, a, p') * \mathrm{AC}_{\mathcal{P}}(p', s, S_1, S_2) & \text{otherwise} \end{cases}$$

itively, $\mathrm{AC}_{\mathcal{P}}(p, s, S_1, S_2)$ records the probability that, after engaging in sequence s from state p, the $\mathcal{P}$ can be in states in which some action in S_1 is possible and every action in S_2 is refused. We now to define the alternative characterization.

Definition 14 *Let $\mathcal{P}, \mathcal{Q} \in$ **PP**, with start states p_0 and q_0, respectively. Then $\mathcal{P} \ll \mathcal{Q}$ if for all $s \in Ac$ finite $S_1, S_2 \subseteq Act$, $AC_{\mathcal{P}}(p_0, s, S_1, S_2) \leq AC_{\mathcal{Q}}(q_0, s, S_1, S_2)$.*

We now conjecture the following.

Conjecture 1 *Let $\mathcal{P}, \mathcal{Q} \in$ **PP**. Then $\mathcal{P} \ll \mathcal{Q}$ if and only if $\mathcal{P} \sqsubseteq \mathcal{Q}$.*

We have proven the $\Leftarrow$ direction of the conjecture by giving a technique for building a test from s, S_2 in such a way that for some monotonic function f and state p, AC$(p, s, S_1, S_2) = \pi$ if and only if *p pas* Such a test can be constructed using a technique similar to De Nicola's for nonprobabilistic processes [We have not yet been able to prove (or disprove) the $\Rightarrow$ direction. Our proposed proof technique i analyzing the computations of $\mathcal{P}||\mathcal{T}$ for a test $\mathcal{T}$ to show that the associated probabilities correspond to values of AC$_{\mathcal{P}}$. The negative information in S_2 is important for showing that normalization factors at stages in the computations of $\mathcal{P}||\mathcal{T}$ and $\mathcal{Q}||\mathcal{T}$ are related appropriately.

5 Substochastic Processes

In this section we investigate the family of *substochastic* probabilistic processes. A substochastic proc the property that the sum of the probabilities of its outgoing transitions may be less than 1. The "m part of the probability distribution can be viewed as a measure of the undefined behavior of the proce example, the behavior of process $\mathcal{P} = a[\frac{1}{4}].b + a[\frac{1}{2}].c$ is undefined with probability $\frac{1}{4}$ in its initial sta

Substochastic processes prove useful in the modeling of fault-tolerant systems and in the analysis of reliability. System faults and unreliability can be captured by a process exhibiting undefined behavi the probabilistic testing preorder we define in this section will relate substochastic processes $\mathcal{P}$ and $\mathcal{Q}$ $\mathcal{P}$ is less reliable than $\mathcal{Q}$.

In order to introduce substochasticness into our framework, we augment the definition of a proba process with a function δ that represents the probability that a process will fail in a particular sta $\delta(p) = \pi$ means that, in state p, the behavior of the process will be undefined with probability π. The definition of substochastic processes is as follows.

Definition 15 *A* substochastic process *is a 5-tuple $(P, Act, \delta, \mu, p_0)$ where*

- *P, Act and p_0 are as in Definition 1;*
- *$\delta : P \rightarrow [0,1]$ is the* undefinedness distribution*;*
- *$\mu : P \times (Act \cup \{\tau\}) \times P \rightarrow [0,1]$ is the* transition probability distribution function *such that, $p \in P$, $(\sum_{a \in Act} \mu(p, a)) + \delta(p) \in \{0, 1\}$*

Substochastic processes will be subjected to *stochastic* tests; that is, tests are considered relia general, this restriction seems reasonable, since if a process fails an unreliable test, it would not in gen possible to identify the source of unreliability in the interaction. Nonetheless, our theory could be m in a straightforward way to accommodate substochastic tests.

The definition of an interaction system for substochastic processes is the same as Definition 4, exc two points: (1) The normalization factor $\nu(p,t)$, used to compute the probabilities of transitions em from configuration $\langle p, t \rangle$, contains the additional term $\delta(p)$; i.e.,

$$\nu(p,t) = (\sum_{a \in Act} \mu_P(p,a) * \mu_T(t,a)) + \mu_T(t,\tau) + \mu_P(p,\tau) - \mu_T(t,\tau) * \mu_P(p,\tau) + \delta_P(p)$$

As such, the normalized transition probabilities will reflect the measure of substochasticness present original process, and thus the interaction system itself will be substochastic. (2) Interaction syst substochastic processes are equipped with an undefinedness distribution function δ_I, which is the δ f of the process normalized by $\nu(p,t)$ defined above.

Given these notions, the definitions of $pass_\pi$ and the probabilistic testing preorder do not chang those given earlier for stochastic processes (Definition 6). The following examples illustrate the relati induced by the probabilistic testing preorder on substochastic processes.

ple 2

et $\mathcal{P}_1 = a[\frac{3}{4}]$ and $\mathcal{P}_1' = a$. Then $\mathcal{P}_1 \sqsubset \mathcal{P}_1'$, i.e., $\mathcal{P}_1 \sqsubseteq \mathcal{P}_1'$ and $\mathcal{P}_1' \not\sqsubseteq \mathcal{P}_1$, as $\mathcal{P}_1$ is more substochastic han $\mathcal{P}_1'$. To see this one need only consider the test $a\,.\,good$, which $\mathcal{P}_1$ passes with probability $\frac{3}{4}$ and $\mathcal{P}_1'$ passes with probability 1.

or a more complex example, let $\mathcal{P}_2 = a[\frac{1}{4}]\,.\,b[\frac{1}{2}] + a[\frac{1}{4}]\,.\,c[\frac{1}{2}]$ and $\mathcal{P}_2' = a[\frac{1}{4}]\,.\,b + a[\frac{1}{4}]\,.\,c$. Then $\mathcal{P}_2 \sqsubset \mathcal{P}_2'$

et $\mathcal{P}_3 = a[\frac{1}{4}]\,.\,b[\frac{1}{2}] + a[\frac{1}{4}]\,.\,c[\frac{1}{2}]$ and $\mathcal{P}_3' = a\,.\,(b[\frac{1}{2}] + c[\frac{1}{2}])$. Then $\mathcal{P}_3$ and $\mathcal{P}_3'$ are unrelated by probabilistic esting. The test which prevents $\mathcal{P}_3 \sqsubseteq \mathcal{P}_3'$ is $\mathcal{T} = a\,.\,(b[\frac{1}{16}]\,.\,good + c[\frac{15}{16}]\,.\,bad)$ because $\mathcal{P}_3\, pass_{\frac{1}{8}}\, \mathcal{T}$ ut $\mathcal{P}_3'\, pass_{\frac{1}{32}}\, \mathcal{T}$. Thus, even for substochastic processes, a sufficiently biased test can be found which revents the more deterministic process from being greater than the less deterministic one.

ochasticness and Reliability

trate the utility of our testing theory for substochastic processes, we present a simple example of how el unreliable system components. The example also illustrates what it means for one system to be eliable (or unreliable) than another in our testing preorder.

first define a probabilistic *chaining* operator that permits probabilistic processes to be linked together tworks. This requires some notation. If $S \subseteq Act_P \times Act_Q$ is a set of pairs, then the first and second ion operators, $(\cdot)_1$ and $(\cdot)_2$, respectively, are defined in the usual way:

$$(S)_1 = \{a \in Act_P \mid \exists b \in Act_Q.\langle a, b\rangle \in S\}$$
$$(S)_2 = \{b \in Act_Q \mid \exists a \in Act_P.\langle a, b\rangle \in S\}.$$

dered pairs in S will be used to specify which pairs of ports are to be linked in the chaining operation next.

tion 16 *Let $\mathcal{P} = (P, Act_P, \delta_P, \mu_P, p_0)$ and $\mathcal{Q} = (Q, Act_Q, \delta_Q, \mu_Q, q_0)$ be substochastic processes. Also $Act_P \times Act_Q$ and $0 < \pi < 1$. Then $\mathcal{P}|S|_\pi\mathcal{Q}$ is the substochastic process $(P \times Q, Act, \mu_{||}, \langle p_0, q_0\rangle, \delta_{||})$, $Act = (Act_P - (S)_1) \cup (Act_Q - (S)_2)$ and where $\delta_{||}$ and $\mu_{||}$ are defined with respect to a normalization*

$$\nu_{||}(p,q) = \sum_{\langle p',q'\rangle \in P\times Q} Syn(\langle p,q\rangle, \langle p',q'\rangle) + \sum_{a\in Act\cup\{\tau\},\langle p',q'\rangle\in P\times Q} (ASyn_1(\langle p,q\rangle, a, \langle p',q'\rangle) + ASyn_2(\langle p,q\rangle, a, \langle p',q'\rangle)) + \delta_P(p) + \delta_Q(q) - \delta_P(p) * \delta_Q(q)$$

ws:

$$\delta_{||}(\langle p,q\rangle) = \frac{\delta_P(p) + \delta_Q(q) - \delta_P(p) * \delta_Q(q)}{\nu_{||}(p,q)}$$

$$\mu_{||}(\langle p,q\rangle, c, \langle p',q'\rangle) = \begin{cases} 0 & \text{if } \nu_{||}(p,q) = 0 \\ \frac{Syn(\langle p,q\rangle,\langle p',q'\rangle) + ASyn_1(\langle p,q\rangle,\tau,\langle p',q'\rangle + ASyn_2(\langle p,q\rangle,\tau,\langle p',q'\rangle)}{\nu_{||}(p,q)} & \text{if } c = \tau \\ \frac{Asyn_1(\langle p,q\rangle,c,\langle p',q'\rangle) + Asyn_2(\langle p,q\rangle,c,\langle p',q'\rangle)}{\nu_{||}(p,q)} & \text{otherwise} \end{cases}$$

$$Syn(\langle p,q\rangle, \langle p',q'\rangle) = \sum_{\langle a,b\rangle\in S} \mu_P(p,a,p') * \mu_Q(q,b,q')$$

$$ASyn_1(\langle p,q\rangle, a, \langle p',q'\rangle) = \begin{cases} \mu_P(p,a,p') * \mu_Q(q, (Act_Q - (S)_2) \cup \{\tau\}) * \pi + \\ \quad \mu_P(p,a,p') * \mu_Q(q, (S)_2) & \text{if } q = q' \text{ and } a \notin (S)_1 \\ 0 & \text{otherwise} \end{cases}$$

$$ASyn_2(\langle p,q\rangle, a, \langle p',q'\rangle) = \begin{cases} \mu_Q(q,a,q') * \mu_P(p, (Act_P - (S)_1) \cup \{\tau\}) * (1-\pi) + \\ \quad \mu_Q(q,a,q') * \mu_P(p, (S)_1) & \text{if } p = p' \text{ and } a \notin (S)_2 \\ 0 & \text{otherwise} \end{cases}$$

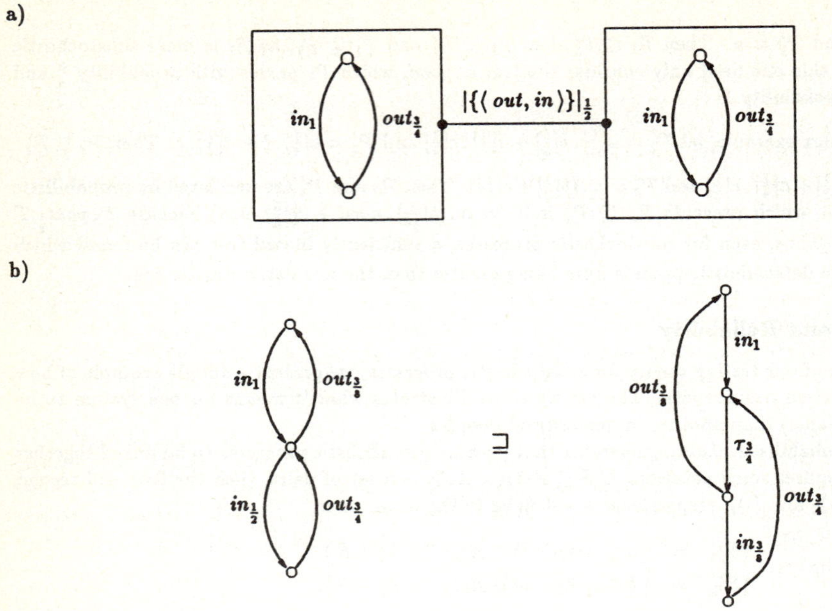

Figure 1: Example of substochastic processes

Intuitively, in $\mathcal{P}|S|_\pi\mathcal{Q}$, $\mathcal{P}$ and $\mathcal{Q}$ execute "in parallel" in an interleaved fashion, with synchron required on the pairs of ports mentioned in S. A synchronization results in a τ. Note that Syn ca the unnormalized probabilities of $\mathcal{P}$ and $\mathcal{Q}$ synchronizing on a pair in S, while $ASyn_1$ and $ASyn_2$ c the unnormalized probabilities of the indicated asynchronous transitions for the first and second com respectively. The normalization factor follows from the fact that we want the operator to have the p that, if any interaction is possible, then some interaction will take place (modulo undefinedness).

The formal complexity of the definition belies the relatively simple "coin flipping" intuition un the semantics of this chaining operator, which is given as follows. Assume that $\mathcal{P}$ initially has m tra and let $l_P = 1$ if $\delta(p_0) > 0$ and 0 otherwise. Likewise, assume that $\mathcal{Q}$ initially has n transitions an be defined similarly to l_P. To determine the transitions for $\mathcal{P}|S|_\pi\mathcal{Q}$, we flip an $(m + l_P)$-sided coin (w according to the relevant probabilities) for $\mathcal{P}$ and a similarly weighted $(n + l_Q)$-sided coin for $\mathcal{Q}$. If eit indicates that $\mathcal{P}$ or $\mathcal{Q}$ is undefined, then so is $\mathcal{P}|S|_\pi\mathcal{Q}$. If the labels of the indicated transitions (with t for $\mathcal{P}$ being ordered before the label for $\mathcal{Q}$) constitute a pair in S, then $\mathcal{P}|S|_\pi\mathcal{Q}$ executes a τ, and both $\mathcal{Q}$ change state. Otherwise, if one of $\mathcal{P}$'s and $\mathcal{Q}$'s coins (but not both) indicates a transition whose not mentioned in S, then the associated component ($\mathcal{P}$ or $\mathcal{Q}$) engages asynchronously in this transitio the other component remains idle. Finally, if both $\mathcal{P}$'s and $\mathcal{Q}$'s coins indicate that each process w engage in an asynchronous transition, then a two-sided coin, with faces weighted by π (for $\mathcal{P}$) and 1 - $\mathcal{Q}$) is flipped to determine which process executes its transition. If none of these cases holds, the coi and $\mathcal{Q}$ are flipped again. If $\mathcal{P}$ and $\mathcal{Q}$ are capable of any interaction (synchronous or asynchronous), t procedure is guaranteed, with probability 1, to terminate with some action being performed.

We note here that $|S|_\pi$ could be defined in terms of more primitive operators (such as prob counterparts of hiding, restriction, renaming, and CSP parallel composition). To illustrate the utilit testing theory for substochastic processes, we consider, in Figure 1a, two unreliable 1-place buffers co by the chaining operator $|\{\langle out, in\rangle\}|_{\frac{1}{2}}$. Each process may fail to output the contents of its buf probability $\frac{1}{4}$. In the left-hand side of Figure 1b we have an unreliable 2-place buffer which, like the buffers, one quarter of the time fails to produce output when an *out*-transition is attempted. (Tec $\delta(p) = \frac{1}{8}$ in the middle state as, ideally, the *out*-transition should occur with probability equal to the *in*-transition.) Finally, the right-hand side of Figure 1b depicts the composite process resulti aining of the two 1-place buffers in Figure 1a. The τ-transition is a result of the synchronization ng when the left-hand process passes along its buffer contents to the right-hand process. Note that the ilities associated with the actions *in* and *out* in third state from the top are both $\frac{3}{8}$. This results from definedness associated with the state from which *out* can occur.

cording to our probabilistic testing preorder we have that the 2-place buffer is more reliable than the d 1-place buffers. For example, the 2-place buffer passes the test $\mathcal{T} = in\,.\,out\,.\,good$ with probability $\frac{3}{4}$ the chained buffer passes $\mathcal{T}$ with probability only $\frac{9}{20}$. This difference in behavior can be attributed to tra communication (and concomitant possibility for failure) inherent to the chained architecture.

Conclusions

paper we have developed a testing theory for probabilistic processes based on the notion of a process g a test with a certain probability. Close connections with the testing theory of De Nicola and Hennessy een demonstrated, providing a well-established foundation for our approach. By extending our theory stochastic processes, we have obtained techniques for modeling faulty systems and for distinguishing ses based on their degree of reliability.

eral topics for future study can be identified. Modifying the theory to accommodate processes with ry τ transitions would likely necessitate the use of measure theory to handle infinite computations raction systems appropriately. The same can also be said of allowing infinite (as opposed to finite) Other interesting topics to pursue are the integration of real-time into our probabilistic testing preorder ork reported in [CZ91] is a natural candidate); and axiomatizations of our preorders with respect to a ilistic process algebra.

rences

B. Bloom and A. R. Meyer. A remark on bisimulation between probabilistic processes. In Meyer and Tsailin, editors, *Logik at Botik*, Springer-Verlag, 1989.

I. Christoff. *Testing Equivalences for Probabilistic Processes.* Technical Report DoCS 90/22, Ph.D. Thesis, Department of Computer Science, Uppsala University, Uppsala, Sweden, 1990.

R. Cleaveland and A. Zwarico. A theory of testing for real-time. In *Proceedings of the Sixth IEEE Symposium on Logic in Computer Science*, July 1991.

R. De Nicola. Extensional equivalences for transition systems. *Acta Informatica*, 24:211–237, 1987.

3] R. De Nicola and M. C. B. Hennessy. Testing equivalences for processes. *Theoretical Computer Science*, 34:83–133, 1983.

M. C. B. Hennessy. *Algebraic Theory of Processes.* MIT Press, Boston, Mass., 1988.

C.-C. Jou and S. A. Smolka. Equivalences, congruences, and complete axiomatizations for probabilistic processes. In J. C. M. Baeten and J. W. Klop, editors, *Proceedings of CONCUR '90*, pages 367–383, Springer-Verlag, Berlin, 1990.

K. G. Larsen and A. Skou. Bisimulation through probabilistic testing. In *Proceedings of 16th Annual ACM Symposium on Principles of Programming Languages*, 1989.

R. Milner. *A Calculus of Communicating Systems.* Volume 92 of *Lecture Notes in Computer Science*, Springer-Verlag, Berlin, 1980.

R. Milner. *Communication and Concurrency. International Series in Computer Science*, Prentice Hall, 1989.

90] R. J. van Glabbeek, S. A. Smolka, B. Steffen, and C. M. N. Tofts. Reactive, generative, and stratified models of probabilistic processes. In *Proceedings of the 5th IEEE Symposium on Logic in Computer Science*, pages 130–141, Philadelphia, PA, 1990.

AUTHOR INDEX

ture Notes in Computer Science

formation about Vols. 1–535
contact your bookseller or Springer-Verlag

: J. E. Tomayko, Software Engineering Education. Pro-
, 1991. VIII, 296 pages. 1991.

7: A. J. Menezes, S. A. Vanstone (Eds.), Advances in
gy – CRYPTO '90. Proceedings. XIII, 644 pages. 1991.

: M. Kojima, N. Megiddo, T. Noma, A. Yoshise, A
Approach to Interior Point Algorithms for Linear
nentarity Problems. VIII, 108 pages. 1991.

: H. F. Mattson, T. Mora, T. R. N. Rao (Eds.), Applied
, Algebraic Algorithms and Error-Correcting Codes.
ings, 1991. XI, 489 pages. 1991.

: A. Prieto (Ed.), Artificial Neural Networks. Proceed-
91. XIII, 476 pages. 1991.

: P. Barahona, L. Moniz Pereira, A. Porto (Eds.), EPIA
ceedings, 1991. VIII, 292 pages. 1991. (Subseries

: Z. W. Ras, M. Zemankova (Eds.), Methodologies for
nt Systems. Proceedings, 1991. X, 644 pages. 1991.
s LNAI).

: J. Dix, K. P. Jantke, P. H. Schmitt (Eds.), Non-
ic and Inductive Logic. Proceedings, 1990. X, 243
91. (Subseries LNAI).

: M. Broy, M. Wirsing (Eds.), Methods of Program-
1, 268 pages. 1991.

: H. Alblas, B. Melichar (Eds.), Attribute Grammars ,
ons and Systems. Proceedings, 1991. IX, 513 pages.

O. Herzog, C.-R. Rollinger (Eds.), Text Understand-
LOG. XI, 738 pages. 1991. (Subseries LNAI).

: D. W. Davies (Ed.), Advances in Cryptology –
RYPT '91. Proceedings, 1991. XII, 556 pages. 1991.

R. Kruse, P. Siegel (Eds.), Symbolic and Quantitative
es to Uncertainty. Proceedings, 1991. XI, 362 pages.

E. Ardizzone, S. Gaglio, F. Sorbello (Eds.), Trends in
Intelligence. Proceedings, 1991. XIV, 479 pages. 1991.
s LNAI).

A. van Lamsweerde, A. Fugetta (Eds.), ESEC '91.
ngs, 1991. XII, 515 pages. 1991.

S. Prehn, W. J. Toetenel (Eds.), VDM '91. Formal
Development Methods. Volume 1. Proceedings, 1991.
pages. 1991.

S. Prehn, W. J. Toetenel (Eds.), VDM '91. Formal
Development Methods. Volume 2. Proceedings, 1991.
pages. 1991.

H. Bieri, H. Noltemeier (Eds.), Computational Ge-
Methods, Algorithms and Applications '91. Proceed-
. VIII, 320 pages. 1991.

G. Grahne, The Problem of Incomplete Information
nal Databases. VIII, 156 pages. 1991.

H. Maurer (Ed.), New Results and New Trends in
Science. Proceedings, 1991. VIII, 403 pages. 1991.

Vol. 556: J.-M. Jacquet, Conclog: A Methodological Approach to Concurrent Logic Programming. XII, 781 pages. 1991.

Vol. 557: W. L. Hsu, R. C. T. Lee (Eds.), ISA '91 Algorithms. Proceedings, 1991. X, 396 pages. 1991.

Vol. 558: J. Hooman, Specification and Compositional Verification of Real-Time Systems. VIII, 235 pages. 1991.

Vol. 559: G. Butler, Fundamental Algorithms for Permutation Groups. XII, 238 pages. 1991.

Vol. 560: S. Biswas, K. V. Nori (Eds.), Foundations of Software Technology and Theoretical Computer Science. Proceedings, 1991. X, 420 pages. 1991.

Vol. 561: C. Ding, G. Xiao, W. Shan, The Stability Theory of Stream Ciphers. IX, 187 pages. 1991.

Vol. 562: R. Breu, Algebraic Specification Techniques in Object Oriented Programming Environments. XI, 228 pages. 1991.

Vol. 563: A. Karshmer, J. Nehmer (Eds.), Operating Systems of the 90s and Beyond. Proceedings, 1991. X, 285 pages. 1991.

Vol. 564: I. Herman, The Use of Projective Geometry in Computer Graphics. VIII, 146 pages. 1992.

Vol. 565: J. D. Becker, I. Eisele, F. W. Mündemann (Eds.), Parallelism, Learning, Evolution. Proceedings, 1989. VIII, 525 pages. 1991. (Subseries LNAI).

Vol. 566: C. Delobel, M. Kifer, Y. Masunaga (Eds.), Deductive and Object-Oriented Databases. Proceedings, 1991. XV, 581 pages. 1991.

Vol. 567: H. Boley, M. M. Richter (Eds.), Processing Declarative Kowledge. Proceedings, 1991. XII, 427 pages. 1991. (Subseries LNAI).

Vol. 568: H.-J. Bürckert, A Resolution Principle for a Logic with Restricted Quantifiers. X, 116 pages. 1991. (Subseries LNAI).

Vol. 569: A. Beaumont, G. Gupta (Eds.), Parallel Execution of Logic Programs. Proceedings, 1991. VII, 195 pages. 1991.

Vol. 570: R. Berghammer, G. Schmidt (Eds.), Graph-Theoretic Concepts in Computer Science. Proceedings, 1991. VIII, 253 pages. 1992.

Vol. 571: J. Vytopil (Ed.), Formal Techniques in Real-Time and Fault-Tolerant Systems. Proceedings, 1992. IX, 620 pages. 1991.

Vol. 572: K. U. Schulz (Ed.), Word Equations and Related Topics. Proceedings, 1990. VII, 256 pages. 1992.

Vol. 573: G. Cohen, S. N. Litsyn, A. Lobstein, G. Zémor (Eds.), Algebraic Coding. Proceedings, 1991. X, 158 pages. 1992.

Vol. 574: J. P. Banâtre, D. Le Métayer (Eds.), Research Directions in High-Level Parallel Programming Languages. Proceedings, 1991. VIII, 387 pages. 1992.

Vol. 575: K. G. Larsen, A. Skou (Eds.), Computer Aided Verification. Proceedings, 1991. X, 487 pages. 1992.

Vol. 576: J. Feigenbaum (Ed.), Advances in Cryptology - CRYPTO '91. Proceedings. X, 485 pages. 1992.

Vol. 577: A. Finkel, M. Jantzen (Eds.), STACS 92. Proceedings, 1992. XIV, 621 pages. 1992.

Vol. 578: Th. Beth, M. Frisch, G. J. Simmons (Eds.), Public-Key Cryptography: State of the Art and Future Directions. XI, 97 pages. 1992.

Vol. 579: S. Toueg, P. G. Spirakis, L. Kirousis (Eds.), Distributed Algorithms. Proceedings, 1991. X, 319 pages. 1992.

Vol. 580: A. Pirotte, C. Delobel, G. Gottlob (Eds.), Advances in Database Technology – EDBT '92. Proceedings. XII, 551 pages. 1992.

Vol. 581: J.-C. Raoult (Ed.), CAAP '92. Proceedings. VIII, 361 pages. 1992.

Vol. 582: B. Krieg-Brückner (Ed.), ESOP '92. Proceedings. VIII, 491 pages. 1992.

Vol. 583: I. Simon (Ed.), LATIN '92. Proceedings. IX, 545 pages. 1992.

Vol. 584: R. E. Zippel (Ed.), Computer Algebra and Parallelism. Proceedings, 1990. IX, 114 pages. 1992.

Vol. 585: F. Pichler, R. Moreno Díaz (Eds.), Computer Aided System Theory – EUROCAST '91. Proceedings. X, 761 pages. 1992.

Vol. 586: A. Cheese, Parallel Execution of Parlog. IX, 184 pages. 1992.

Vol. 587: R. Dale, E. Hovy, D. Rösner, O. Stock (Eds.), Aspects of Automated Natural Language Generation. Proceedings, 1992. VIII, 311 pages. 1992. (Subseries LNAI).

Vol. 588: G. Sandini (Ed.), Computer Vision – ECCV '92. Proceedings. XV, 909 pages. 1992.

Vol. 589: U. Banerjee, D. Gelernter, A. Nicolau, D. Padua (Eds.), Languages and Compilers for Parallel Computing. Proceedings, 1991. IX, 419 pages. 1992.

Vol. 590: B. Fronhöfer, G. Wrightson (Eds.), Parallelization in Inference Systems. Proceedings, 1990. VIII, 372 pages. 1992. (Subseries LNAI).

Vol. 591: H. P. Zima (Ed.), Parallel Computation. Proceedings, 1991. IX, 451 pages. 1992.

Vol. 592: A. Voronkov (Ed.), Logic Programming. Proceedings, 1991. IX, 514 pages. 1992. (Subseries LNAI).

Vol. 593: P. Loucopoulos (Ed.), Advanced Information Systems Engineering. Proceedings. XI, 650 pages. 1992.

Vol. 594: B. Monien, Th. Ottmann (Eds.), Data Structures and Efficient Algorithms. VIII, 389 pages. 1992.

Vol. 595: M. Levene, The Nested Universal Relation Database Model. X, 177 pages. 1992.

Vol. 596: L.-H. Eriksson, L. Hallnäs, P. Schroeder-Heister (Eds.), Extensions of Logic Programming. Proceedings, 1991. VII, 369 pages. 1992. (Subseries LNAI).

Vol. 597: H. W. Guesgen, J. Hertzberg, A Perspective of Constraint-Based Reasoning. VIII, 123 pages. 1992. (Subseries LNAI).

Vol. 598: S. Brookes, M. Main, A. Melton, M. Mislove, D. Schmidt (Eds.), Mathematical Foundations of Programming Semantics. Proceedings, 1991. VIII, 506 pages. 1992.

Vol. 599: Th. Wetter, K.-D. Althoff, J. Boose, B. R. Gaines, M. Linster, F. Schmalhofer (Eds.), Current Developments in Knowledge Acquisition - EKAW '92. Proceedings. XIII, 444 pages. 1992. (Subseries LNAI).

Vol. 600: J. W. de Bakker, C. Huizing, W. P. de Roever, G. Rozenberg (Eds.), Real-Time: Theory in Practice. Proceedings, 1991. VIII, 723 pages. 1992.

Vol. 601: D. Dolev, Z. Galil, M. Rodeh (Eds.), Theory of Computing and Systems. Proceedings, 1992. VIII, 220 pages. 1992.

Vol. 602: I. Tomek (Ed.), Computer Assisted Lea
Proceedigs, 1992. X, 615 pages. 1992.

Vol. 603: J. van Katwijk (Ed.), Ada: Moving Towards
Proceedings, 1992. VIII, 324 pages. 1992.

Vol. 604: F. Belli, F.-J. Radermacher (Eds.), Industr
Engineering Applications of Artificial Intelligence and
Systems. Proceedings, 1992. XV, 702 pages. 1992. (Su
LNAI).

Vol. 605: D. Etiemble, J.-C. Syre (Eds.), PARLE '92. F
Architectures and Languages Europe. Proceedings, 1992
984 pages. 1992.

Vol. 606: D. E. Knuth, Axioms and Hulls. IX, 109 pages

Vol. 607: D. Kapur (Ed.), Automated Deduction – CA
Proceedings, 1992. XV, 793 pages. 1992. (Subseries L

Vol. 608: C. Frasson, G. Gauthier, G. I. McCalla (Eds.),
gent Tutoring Systems. Proceedings, 1992. XIV, 686
1992.

Vol. 609: G. Rozenberg (Ed.), Advances in Petri Net
VIII, 472 pages. 1992.

Vol. 610: F. von Martial, Coordinating Plans of Auto
Agents. XII, 246 pages. 1992. (Subseries LNAI).

Vol. 612: M. Tokoro, O. Nierstrasz, P. Wegner (Eds.),
Based Concurrent Computing. Proceedings, 1991. X, 26
1992.

Vol. 613: J. P. Myers, Jr., M. J. O'Donnell (Eds.), Constr
in Computer Science. Proceedings, 1991. X, 247 pages

Vol. 614: R. G. Herrtwich (Ed.), Network and Operati
tem Support for Digital Audio and Video. Proceeding
XII, 403 pages. 1992.

Vol. 615: O. Lehrmann Madsen (Ed.), ECOOP '92. E
Conference on Object Oriented Programming. Proceed
426 pages. 1992.

Vol. 616: K. Jensen (Ed.), Application and Theory of P
1992. Proceedings, 1992. VIII, 398 pages. 1992.

Vol. 617: V. Mařík, O. Štěpánková, R. Trappl (Eds.), A
Topics in Artificial Intelligence. Proceedings, 1992.
pages. 1992. (Subseries LNAI).

Vol. 618: P. M. D. Gray, R. J. Lucas (Eds.), Advanced
Systems. Proceedings, 1992. X, 260 pages. 1992.

Vol. 620: A. Nerode, M. Taitslin (Eds.), Logical Found
Computer Science – Tver '92. Proceedings. IX, 514 pag

Vol. 621: O. Nurmi, E. Ukkonen (Eds.), Algorithm
SWAT '92. Proceedings. VIII, 434 pages. 1992.

Vol. 623: W. Kuich (Ed.), Automata, Languages and
ming. Proceedings, 1992. XII, 721 pages. 1992.

Vol. 624: A. Voronkow (Ed.), Logic Programming a
mated Reasoning. Proceedings, 1992. XIV, 509 page
(Subseries LNAI).